Integer and Combinatorial Optimization

WILEY-INTERSCIENCE
SERIES IN DISCRETE MATHEMATICS AND OPTIMIZATION

A complete list of titles in this series appears at the end of this volume.

Integer and Combinatorial Optimization

GEORGE NEMHAUSER
School of Industrial
and Systems Engineering
Georgia Institute of Technology
Atlanta, Georgia

LAURENCE WOLSEY
Center for Operations Research and Econometrics
Université Catholique de Louvain
Louvain-la-Neuve, Belgium

A Wiley-Interscience Publication
JOHN WILEY & SONS, INC.
New York • Chichester • Weinheim • Brisbane • Singapore • Toronto

Published simultaneously in Canada.

Library of Congress Cataloging in Publication Data:

Nemhauser, George L.
Integer and combinatorial optimization.
(Wiley-Interscience series in discrete mathematics and optimization)
"A Wiley-Interscience publication."
Includes bibliographies and index.
1. Mathematical optimization. 2. Integer programming.
3. Combinatorial optimization. I. Wolsey, Laurence A.
II. Title. III. Series.
QA402.5.N453 1988 519.7'7 87-34067
ISBN 0-471-82819-X
ISBN 0-471-35943-2 (paperback)

Printed in the United States of America

19 18 17 16 15 14 13

To our parents

Preface

The explosion of new results in integer and combinatorial optimization that began about fifteen years ago inspired us to write a book that would unify theory and algorithms and could serve as a graduate text and reference for researchers and practitioners. We have been very excited about many of the new developments that have made it possible to solve large-scale integer programming problems and that have opened up new areas of research which surely will yield more robust and efficient algorithms. Little did we realize the enormity of the task. Both of us worked steadily on this project for more than four years. The end result was a manuscript of nearly 1400 typewritten pages which, although it does not come close to covering all of the literature, covers those topics that we believe constitute the most significant theoretical and algorithmic developments.

Optimization means to maximize (or minimize) a function of many variables subject to constraints. The distinguishing feature of *discrete*, *combinatorial*, or *integer* optimization is that some of the variables are required to belong to a discrete set, typically a subset of integers. These discrete restrictions allow the mathematical representation of phenomena or alternatives where indivisibility is required or where there is not a continuum of alternatives.

Discrete optimization problems abound in everyday life. An important and widespread area of applications concerns the management and efficient use of scarce resources to increase productivity. These applications include operational problems such as the distribution of goods, production scheduling, and machine sequencing. They also include planning problems such as capital budgeting, facility location and portfolio selection, and design problems such as telecommunication and transportation network design, VLSI circuit design and the design of automated production systems. Discrete optimization problems also arise in statistics (data analysis), physics (determination of minimum energy states), cryptography (designing unbreakable codes), politics (selecting fair election districts), and mathematics (as a powerful technique for proving combinatorial theorems). Moreover, applications of discrete optimization are in a period of rapid development because of the widespread use of microcomputers and the data provided by information systems. This is particularly relevant in the manufacturing sector of the economy where increased competition and flexibility provided by new technology make it imperative to seek better solutions from larger and more complex sets of alternatives.

This book is about the mathematics of discrete optimization, which includes the representation of problems by mathematical models and, especially, the solution of the models. The focus is on understanding the mathematical underpinnings of the algorithms that make it possible to solve (exactly or approximately) the large and complex models that arise in practical applications.

Chapter I.1 discusses problem formulation, which is important not only to demonstrate the scope of applications, but also because the structure of the formulation is of crucial importance to solving the model. Chapter I.1 gives a comprehensive treatment of this subject.

The remainder of Part I presents mathematics and algorithms that are the foundations for the discrete optimization theory and techniques of Parts II and III. There are chapters on well-established subjects including linear programming (Chapter I.2), graphs and networks (Chapter I.3), and computational complexity (Chapter I.5). The presentation of polyhedral theory (Chapter I.4) begins with basic results from linear algebra and then emphasizes precisely those results that are essential to a fundamental understanding of the algebra and geometry of the convex hull of a discrete set. Chapter I.6 gives new algorithms and results on linear programming and, in particular, establishes the fundamental connection between separation and optimization. Chapter I.7 presents a modern treatment of the classical problem of solving linear equations in integers and also includes an introduction to the recent work on reduced bases for integer lattices.

Parts II and III present basic approaches and algorithms for solving discrete optimization problems. Part II deals with general problems and those that contain some structure. These are the problems that are hard to solve but, for the most part, they are the ones that arise in practical applications.

Chapters II.1 and II.2 treat the problem of describing the set of feasible solutions to an integer program by a set of linear inequalities. It begins with elementary ideas, but also includes a thorough development of advanced topics such as superadditive valid inequalities and the use of structure to obtain facet-defining inequalities. Objective functions for integer programs are introduced in Chapter II.3 where the fundamental approaches of relaxation and duality are developed for the purpose of obtaining upper bounds on the optimal value. Most of the advanced material in these chapters has appeared only in research articles and monographs, but is essential for the development of future generation algorithms for solving integer programs.

Algorithms are presented in Chapters II.4, II.5 and II.6. Chapter II.4 presents classical branch-and-bound and cutting plane algorithms. Specialized algorithms that use varying degrees of structure to obtain exact or approximate solutions are presented in Chapters II.5 and II.6. Here we study and illustrate a number of techniques that, for the most part, have been developed over the last decade and are not covered in the currently available textbooks. These include strong cutting plane algorithms, primal and dual heuristic analysis, decomposition and reduced bases, and their applications to 0-1 integer programs, the traveling salesman problem and fixed-charge network flow problems.

Part III treats highly structured combinatorial optimization problems for which elegant results are known. Chapter III.1 studies polyhedra with integral extreme points. It includes classical results on total unimodularity and recent results on totally balanced, balanced, and perfect matrices and on the blocking and antiblocking theory of polyhedra. Chapters III.2 and III.3 are on the classical combinatorial problems of matching and matroids, respectively. In both of these chapters the emphasis is on optimization algorithms, polyhedral combinatorics and duality. Chapter III.3 also introduces the significant role of submodular and supermodular functions in combinatorial optimization.

Notes appear at the end of each chapter. Their purpose is to reference our source materials, and to comment briefly on extensions and related topics that are not discussed in the body of the text. The citations and references are selective. With the exception of Chapter I.1, in Part I our objective is to provide foundation material, and thus the notes are limited to a small number of references that cover the corresponding topics in much greater detail than is done here. However, in Parts II and III we have attempted to cite the original papers in which the material appears as well as some other influential works.

The book can be used as a graduate text or for self-guided reading in several ways. Since we cannot imagine a reader who would want to undertake a straight cover-to-cover reading and since our experience has shown that it is not possible to cover the whole book in even a two-semester, graduate level course, it is necessary to be selective in a first reading.

For graduate students in mathematical programming, especially those planning to undertake research in discrete optimization, we suggest a full academic year course (course AY). Three one-semester options are: a course emphasizing practical algorithms (course PA), a course emphasizing general theory (course GT), and a course in polyhedral combinatorics (course PC).

Each course should begin with some exposure to Chapter I.1 on model formulation, which is important not only to demonstrate the scope of applications, but also because the structure of the formulation is of crucial importance in solving the model.

Chapters I.2 and I.3 are only for review, since it is wise for any reader of the book to have studied linear programming as a prerequisite. But a typical linear programming course, unfortunately, does not cover polyhedral theory. Therefore, all courses should cover Chapter I.4. In course PA, just enough of the first four sections should be covered (without proofs) for the student to understand the concept of facets of polyhedra and the idea of Theorem 6.1 on the convex hull of a discrete set of points.

The coverage of Chapter I.5 on computational complexity will depend on the students' backgrounds and the instructor's taste; but at the very least, students in all courses should be introduced to the concepts of polynomial computation and NP-completeness. Similarly, students in all courses should be introduced to the concept of separation and the polynomial equivalence of separation and optimization (Section I.6.3). This should be done very informally in course PA. Sections I.6.2, I.6.4 and I.6.5 are independent reading and should be omitted in a first reading of the book. Chapter I.7, and then Section II.6.5, might be covered only in courses AY and GT if time permits at the end of the course. They can also be omitted in a first reading.

Courses PA and GT focus on different parts of Part II. Course PC can omit Part II altogether, but would be more interesting if Sections II.1.1, II.1.2 (first-half), II.2.1, II.2.3, and II.6.3 also were included.

The following sections from Part II are common to courses AY, PA, and GT: II.1.1, II.2.1, II.2.2, II.3.1, II.3.6, II.3.7, II.4.1, II.5.1, II.5.2 and II.5.3.

Course PA should also cover Sections II.4.2, II.5.4, II.5.5, II.6.1 (knapsack problem) and II.6.2, and, if time permits, Sections II.2.4 and II.6.4. The instructor may find some time for the important class of problems and algorithms discussed in the later two sections by omitting or only sketching some proofs from the earlier sections.

Course GT should also cover Sections II.1.2 (leaving out the subsection on bounded integer variables), II.1.3, II.1.4, II.1.5, II.2.3, II.3.2, and II.3.3, and the first two sections of Chapter III.1. If time permits, additional theoretical material could be selected from Sections II.1.6, II.1.7, II.3.4, II.3.5, or some algorithms could be studied from Sections II.4.3, II.5.4 and the first three sections of Chapter II.6.

With respect to Part II, course AY is the union of the material covered in courses PC and GT. From Part III, course AY should also cover sections III.1.1, III.1.2, III.1.4 and the first three sections of Chapter III.2. Any remaining time could be spent on either sections III.1.5, III.1.6, or the first few sections of Chapter III.3.

The material to be selected from Part III for course PC can vary according to taste. We suggest all of Chapter III.1 except for Section III.1.3, the first 3 sections of Chapter III.2 and the first 5 sections of Chapter III.3.

This book could not have been written without the tremendous support that we received from the Center for Operations Research and Econometrics (CORE) of the Université Catholique de Louvain, and thus we are extremely grateful to Jacques Dreze, the founder and intellectual leader of CORE.

We met at CORE in the winter of 1970. Nemhauser spent the academic year 1969–1970 at CORE and Wolsey presented a seminar on his early work on the group-theoretic approach to integer programming. Subsequently, Wolsey became a permanent member of CORE and Nemhauser returned to CORE for the period 1975–1977 as Research Director. During this period, the authors collaborated extensively on research in the analysis of heuristics and other topics in integer and combinatorial optimization stimulated by an active research group that included Jack Edmonds, Bob Bland, Guy de Ghellinck, Rick Giles, Bob Jeroslow, Tom Magnanti, Bill Pulleyblank, Mike Ball and Gerard Cornuejols. All of these people, as well as our Dutch neighbors, Jan-Karel Lenstra and Alexander Rinnooy Kan, contributed to our understanding of the subject and motivation to write a book.

A NATO research grant made it possible for us to continue our research collaboration through the late seventies and early eighties, and we began to draft the manuscript earnestly during Nemhauser's fourth year at CORE in 1983–1984. During the writing of the book we benefitted from numerous discussions with our friends and professional colleagues including Egon Balas, Vasek Chvátal, Marshall Fisher, Martin Grötschel, Ellis Johnson, Manfred Padberg, Lex Schrijver, Jorgen Tind and Les Trotter. We are particularly grateful to Gerard Cornuejols who read Parts I and II and provided extensive comments and suggestions and to Bill Pulleyblank who did the same for Part III. We are also thankful for the comments we received on various drafts of the text from Jorgen Tind, Bob Jeroslow, Alan Goldman, Anton Kolen, Jan-Karel Lenstra, Lex Schrijver, Donna Crystal Llewellyn, Martin Dyer, Mike Todd, Jean-Philippe Vial and John Vande Vate. Our students in courses given at CORE, Cornell and Georgia Tech found typos and other mistakes that otherwise would have been missed; special thanks are due to Ronny Aboudi, Yves Pochet and Gabriele Sigismondi.

The chores of deciphering our untidy handwritten drafts and of retyping endless revisions were done graciously and with utmost care and patience by the late Elizabeth Pecquereau, formerly a secretary at CORE. We are very sad that we will not be able to share the joy of seeing the final product with our dear friend Elizabeth. Fabienne Henry of CORE and Yvonne Kissi of Georgia Tech also did excellent jobs in typing parts of the manuscript. Sheila Verkaren of CORE always managed to spare some of Elizabeth's or Fabienne's time for our book, even though we were using far more than our fair share of CORE's secretarial resources.

Over a period of four years, Ellen Nemhauser and Marguerite Wolsey were frequently ignored while their husbands spent evenings and weekends writing, and occasionally were imposed upon by a boarder who ate and slept at their house, but otherwise was too involved in mathematics to engage in civil conversation or to wash the dishes. We thank them for their love and patience and hope to make amends.

GEORGE L. NEMHAUSER
LAURENCE A. WOLSEY

Atlanta, Georgia, USA
Louvain-la-Neuve, Belgium
February, 1988.

Contents

Integer and Combinatorial Optimization

Part I
FOUNDATIONS

I.1

The Scope of Integer and Combinatorial Optimization

1. INTRODUCTION

Integer and combinatorial optimization deals with problems of maximizing or minimizing a function of many variables subject to (a) inequality and equality constraints and (b) integrality restrictions on some or all of the variables. Because of the robustness of the general model, a remarkably rich variety of problems can be represented by discrete optimization models.

An important and widespread area of application concerns the management and efficient use of scarce resources to increase productivity. These applications include operational problems such as the distribution of goods, production scheduling, and machine sequencing. They also include (a) planning problems such as capital budgeting, facility location, and portfolio analysis and (b) design problems such as communication and transportation network design, VLSI circuit design, and the design of automated production systems.

In mathematics there are applications to the subjects of combinatorics, graph theory, and logic. Statistical applications include problems of data analysis and reliability. Recent scientific applications involve problems in molecular biology, high-energy physics, and x-ray crystallography. A political application concerns the division of a region into election districts.

Some of these discrete optimization models will be developed later in this chapter. But their number and variety are so great that we only can provide references for some of them. The main purpose of this book is to present the mathematical foundations of integer and combinatorial optimization models along with the algorithms that can be used to solve the problems.

Throughout most of this book, we assume that the function to be maximized and the inequality restrictions are linear. Note that minimizing a function is equivalent to maximizing the negative of the same function and that an equality constraint can be represented by two inequalities. It is also common to require the variables to be nonnegative. Hence we write the *linear mixed-integer programming problem* as

(MIP) $$\max\{cx + hy: Ax + Gy \leq b, x \in Z^n_+, y \in R^p_+\},$$

where Z^n_+ is the set of nonnegative integral n-dimensional vectors, R^p_+ is the set of nonnegative real p-dimensional vectors, and $x = (x_1, \ldots, x_n)$ and $y = (y_1, \ldots, y_p)$ are the *variables* or *unknowns*. An *instance* of the problem is specified by the *data* (c, h, A, G, b), with c an n-vector, h a p-vector, A an $m \times n$ matrix, G an $m \times p$ matrix and b an m-vector. We do not distinguish between row and column vectors unless the clarity of the presentation makes it necessary to do so. This problem is called mixed because of the presence of both integer and continuous (real) variables.

We assume throughout the text that all of the data sets are rational, that is, that each of the individual numbers is rational. Although in making this assumption we sacrifice some theoretical generality, it is a natural assumption for solving problems on a digital computer.

The set $S = \{x \in Z_+^n, y \in R_+^p, Ax + Gy \leqslant b\}$ is called the *feasible region*, and an $(x, y) \in S$ is called a *feasible solution*. An instance is said to be *feasible* if $S \neq \emptyset$. The function

$$z = cx + hy$$

is called the *objective function*. A feasible point (x^0, y^0) for which the objective function is as large as possible, that is,

$$cx^0 + hy^0 \geqslant cx + hy \quad \text{for all } (x, y) \in S,$$

is called an *optimal solution*. If (x^0, y^0) is an optimal solution, $cx^0 + hy^0$ is called the *optimal value* or *weight* of the solution.

A feasible instance of MIP may not have an optimal solution. We say that an instance is *unbounded* if for any $\omega \in R^1$ there is an $(x, y) \in S$ such that $cx + hy > \omega$. We use the notation $z = \infty$ for an unbounded instance.

In Section I.4.6, we will show that every feasible instance of MIP either has an optimal solution or is unbounded. This result requires the assumption of rational data. With irrational data, it is possible that no feasible solution attains the least upper bound on the objective function.

Thus to solve an instance of MIP means to produce an optimal solution or to show that it is either unbounded or infeasible.

The *linear (pure) integer programming problem*

$$\text{(IP)} \qquad \max\{cx\colon Ax \leqslant b, x \in Z_+^n\}$$

is the special case of MIP in which there are no continuous variables. The *linear programming problem*

$$\text{(LP)} \qquad \max\{hy\colon Gy \leqslant b, y \in R_+^p\}$$

is the special case of MIP in which there are no integer variables.

In many models, the integer variables are used to represent logical relationships and therefore are constrained to equal 0 or 1. Thus we obtain the 0-1 MIP (respectively 0-1 IP) in which $x \in Z_+^n$ is replaced by $x \in B^n$, where B^n is the set of n-dimensional binary vectors.

While there is no generally agreed-upon definition of a combinatorial optimization problem, most problems so named are 0-1 IPs that deal with finite sets and collections of subsets. The following is a generic combinatorial optimization problem. Let $N = \{1, \ldots, n\}$ be a finite set and let $c = (c_1, \ldots, c_n)$ be an n-vector. For $F \subseteq N$, define $c(F) = \Sigma_{j \in F}\, c_j$. Suppose we are given a collection of subsets $\mathcal{F}$ of N. The *combinatorial optimization* problem is

$$\text{(CP)} \qquad \max\{c(F): F \in \mathscr{F}\}.$$

Some examples of combinatorial optimization problems will be given later in this chapter.

This book is divided into three parts. This chapter is concerned with the formulation of integer optimization problems, which means how to translate a verbal description of a problem into a mathematical statement of the form MIP, IP, or CP. The rest of Part I contains prerequisites, including linear programming, graphs and networks, polyhedral theory, and computational complexity, which are necessary for Parts II and III.

Part II is concerned with the theory and algorithms for problems IP and MIP. Part III is devoted to some combinatorial optimization problems whose structure makes them relatively easy to solve.

2. MODELING WITH BINARY VARIABLES I: KNAPSACK, ASSIGNMENT AND MATCHING, COVERING, PACKING AND PARTITIONING

An important and very common use of 0-1 variables is to represent binary choice. Consider an event that may or may not occur, and suppose that it is part of the problem to decide between these two possibilities. To model such a dichotomy, we use a binary variable x and let

$$x = \begin{cases} 1 & \text{if the event occurs} \\ 0 & \text{if the event does not occur.} \end{cases}$$

The event itself may be almost anything, depending on the specific situation being considered. Several examples follow.

The 0-1 Knapsack Problem

Suppose there are n projects. The jth project, $j = 1, \ldots, n$, has a cost of a_j and a value of c_j. Each project is either done or not, that is, it is not possible to do a fraction of any of the projects. Also there is a budget of b available to fund the projects. The problem of choosing a subset of the projects to maximize the sum of the values while not exceeding the budget constraint is the 0-1 *knapsack problem*

$$\max\left\{\sum_{j=1}^{n} c_j x_j: \sum_{j=1}^{n} a_j x_j \leq b, x \in B^n\right\}.$$

Here the jth event is the jth project. This problem is called the knapsack problem because of the analogy to the hiker's problem of deciding what should be put in a knapsack, given a weight limitation on how much can be carried. In general, problems of this sort may have several constraints. We then refer to the problem as the *multidimensional knapsack problem*.

The Assignment and Matching Problems

Another classical problem involves the assignment of people to jobs. Suppose there are n people and m jobs, where $n \geq m$. Each job must be done by exactly one person; also, each person can do, at most, one job. The cost of person j doing job i is c_{ij}. The problem is to assign the people to the jobs so as to minimize the total cost of completing all of the jobs. To formulate this problem, which is known as the *assignment problem*, we introduce 0-1 variables x_{ij}, $i = 1, \ldots, m, j = 1, \ldots, n$ corresponding to the ijth event of assigning person j to job i. Since exactly one person must do job i, we have the constraints

$$\text{(2.1)} \qquad \sum_{j=1}^{n} x_{ij} = 1 \quad \text{for } i = 1, \ldots, m.$$

Since each person can do no more than one job, we also have the constraints

$$\text{(2.2)} \qquad \sum_{i=1}^{m} x_{ij} \leqslant 1 \quad \text{for } j = 1, \ldots, n.$$

It is now easy to check that if $x \in B^{mn}$ satisfies (2.1) and (2.2), we obtain a feasible solution to the assignment problem. The objective function is min $\Sigma_{i=1}^{m} \Sigma_{j=1}^{n} c_{ij}x_{ij}$.

In the assignment problem the $m + n$ elements are partitioned into disjoint sets of jobs and people. But in other models of this type, we cannot assume such a partition. Suppose $2n$ students are to be assigned to n double rooms. Here each student must be assigned exactly one roommate. Let the ijth event, $i < j$, correspond to assigning students i and j to the same room; also suppose that there is a value of c_{ij} when students i and j are roommates. The problem

$$\text{(2.3)} \qquad \left\{\max \sum_{i=1}^{2n-1} \sum_{j=i+1}^{2n} c_{ij}x_{ij}\colon \sum_{k<i} x_{ki} + \sum_{j>i} x_{ij} = 1,\ i = 1, \ldots, 2n,\ x \in B^{n(2n-1)}\right\}$$

is known as the *perfect matching problem*. We will see later that it is a generalization of the assignment problem. If the equality constraints in (2.3) are replaced by equal-to-or-less-than inequalities, then the problem is called the *matching problem*.

Each of the above problems fits into the context of CP. In the knapsack problem, $N = \{1, \ldots, n\}$ and $F \in \mathcal{F}$ if and only if $\Sigma_{j\in F}\, a_j \leqslant b$. In the assignment problem, $N = \{ij\colon i = 1, \ldots, m, j = 1, \ldots, n\}$ and $F \in \mathcal{F}$ if and only if $|F \cap \{i1, \ldots, in\}| = 1$ for all i and $|F \cap \{1j, \ldots, mj\}| \leqslant 1$ for all j.

Set-Covering, Set-Packing, and Set-Partitioning Problems

A common way of defining $\mathcal{F}$ leads to important classes of combinatorial optimization problems known as set-covering, set-packing, and set-partitioning problems. Let $M = \{1, \ldots, m\}$ be a finite set and let $\{M_j\}$ for $j \in N = \{1, \ldots, n\}$ be a given collection of subsets of M. For example, the collection might consist of all subsets of size k, for some $k \leqslant m$. We say that $F \subseteq N$ *covers* M if $\bigcup_{j\in F} M_j = M$. In the CP known as the *set-covering problem*, $\mathcal{F} = \{F\colon F \text{ covers } M\}$. We say that $F \subseteq N$ is a *packing* with respect to M if $M_j \cap M_k = \emptyset$ for all $j, k \in F, j \neq k$. In the CP known as the *set-packing problem*, $\mathcal{F} = \{F\colon F \text{ is a packing with respect to } M\}$. If $F \subseteq N$ is both a covering and a packing, then F is said to be a *partition* of M. In the set-covering problem, c_j is the cost of M_j and we seek a minimum-cost cover; in the set-packing problem, however, c_j is the weight or value of M_j and we seek a maximum-weight packing.

These problems are readily formulated as 0-1 IPs. Let A be the $m \times n$ incidence matrix of the family $\{M_j\}$ for $j \in N$; that is, for $i \in M$,

$$a_{ij} = \begin{cases} 1 & \text{if } i \in M_j \\ 0 & \text{if } i \notin M_j \end{cases} \qquad x_j = \begin{cases} 1 & \text{if } j \in F \\ 0 & \text{if } j \notin F. \end{cases}$$

Then F is a cover (respectively packing, partition) if and only if $x \in B^n$ satisfies $Ax \geqslant 1$ (respectively $Ax \leqslant 1$, $Ax = 1$), where 1 is an m-vector all of whose components equal 1. We see, for example, that the set-packing problem is the special case of the 0-1 IP with A a 0-1 matrix (i.e., a matrix all of whose elements equal 0 or 1) and $b = 1$. Note that an assignment problem with m jobs and m people is a set-partitioning problem in which $M = \{1, \ldots, m, m+1, \ldots, 2m\}$ and M_j for $j = 1, \ldots, m^2$ is a subset of M consisting of one job and one person.

Many practical problems can be formulated as set-covering problems. A typical application concerns facility location. Suppose we are given a set of potential sites $N = \{1, \ldots, n\}$ for the location of fire stations. A station placed at j costs c_j. We are also given a set of communities $M = \{1, \ldots, m\}$ that have to be protected. The subset of communities that can be protected from a station located at j is M_j. For example, M_j might be the set of communities that can be reached from j in 10 minutes. Then the problem of choosing a minimum-cost set of locations for the fire stations such that each community can be reached from some fire station in 10 minutes is a set-covering problem. There are many other applications of this type, including assigning customers to delivery routes, airline crews to flights, and workers to shifts.

3. MODELING WITH BINARY VARIABLES II: FACILITY LOCATION, FIXED-CHARGE NETWORK FLOW, AND TRAVELING SALESMAN

The set-packing, set-partitioning, and set-covering models of the previous section illustrated how we can use linear constraints on binary variables to represent relationships among the variables or the events that they represent. A packing constraint, $\Sigma_j x_j \leqslant 1$, states that at most one of a set of events is allowed to occur. Similarly, covering and partitioning constraints state, respectively, that at least one and exactly one of the events can occur. Here we show how more complex relationships can be modeled with binary variables, and we also formulate some models that use these relationships.

The relation that neither or both events 1 and 2 must occur is represented by the linear equality $x_2 - x_1 = 0$ in the binary variables x_1 and x_2. Similarly, the relation that event 2 can occur only if event 1 occurs is represented by the linear inequality $x_2 - x_1 \leqslant 0$. More generally, consider an activity that can be operated at any level y, $0 \leqslant y \leqslant u$. Now suppose that the activity can be undertaken only if some event represented by the binary variable x occurs. This relation is represented by the linear inequality $y - ux \leqslant 0$ since $x = 0$ implies $y = 0$ and $x = 1$ yields the original constraint $y \leqslant u$. We now consider two models that use this relationship.

Facility Location Problems

These problems, as does our illustration of the set-covering model, concern the location of facilities to serve clients economically. We are given a set $N = \{1, \ldots, n\}$ of potential facility locations and a set of clients $I = \{1, \ldots, m\}$. A facility placed at j costs c_j for $j \in N$. This problem is more complicated than the set-covering application because each client has a demand for a certain good, and the total cost of satisfying the demand of client i from a facility at j is h_{ij}. The optimization problem is to choose a subset of the locations at which to place facilities and then to assign the clients to these facilities so as to minimize total cost. In the uncapacitated facility location problem, there is no restriction on the number of clients that a facility can serve.

In addition to the binary variable $x_j = 1$, if a facility is placed at j and $x_j = 0$ otherwise, we introduce the continuous variable y_{ij}, which is the fraction of the demand of client i

that is satisfied from a facility at j. The condition that each client's demand must be satisfied is given by

$$(3.1) \qquad \sum_{j \in N} y_{ij} = 1 \quad \text{for } i \in I.$$

Moreover, since client i cannot be served from j unless a facility is placed at j, we have the constraints

$$(3.2) \qquad y_{ij} - x_j \leqslant 0 \quad \text{for } i \in I \text{ and } j \in N.$$

Hence the *uncapacitated facility location problem* is the MIP

$$\min \sum_{j \in N} c_j x_j + \sum_{i \in I} \sum_{j \in N} h_{ij} y_{ij}$$

subject to the constraints (3.1), (3.2) and $x \in B^n$, $y \in R_+^{mn}$.

It may be unrealistic to assume that a facility can serve any number of clients. Suppose a facility located at j has a capacity of u_j and the ith client has a demand of b_i. Now we let y_{ij} be the quantity of goods sent from facility j to client i and let h_{ij} be the shipping cost per unit. To formulate the *capacitated facility location problem* as an MIP, we replace (3.1) by

$$(3.3) \qquad \sum_{j \in N} y_{ij} = b_i \quad \text{for } i \in I,$$

and (3.2) by

$$(3.4) \qquad \sum_{i \in I} y_{ij} - u_j x_j \leqslant 0 \quad \text{for } j \in N.$$

The Fixed-Charge Network Flow Problem

We are given a network (see Figure 3.1) with a set of nodes V (facilities) and a set of arcs $\mathscr{A}$. An arc $e = (i, j)$ that points from node i to node j means that there is a direct shipping route from node i to node j. Associated with each node i, there is a demand b_i. Node i is a demand, supply, or transit point depending on whether b_i is, respectively, positive, negative, or zero. We assume that the net demand is zero, that is, $\Sigma_{i \in V} b_i = 0$. Each arc (i, j) has a flow capacity u_{ij} and a unit flow cost h_{ij}.

Let y_{ij} be the flow on arc (i, j). A flow is feasible if and only if it satisfies

$$(3.5) \qquad y \in R_+^{|\mathscr{A}|}$$

$$(3.6) \qquad y_{ij} \leqslant u_{ij} \quad \text{for } (i, j) \in \mathscr{A}$$

$$(3.7) \qquad \sum_{j \in V} y_{ji} - \sum_{j \in V} y_{ij} = b_i \quad \text{for } i \in V.$$

The constraints (3.7) are the *flow conservation* constraints. The problem

$$(3.8) \qquad \min\left\{ \sum_{(i,j) \in \mathscr{A}} h_{ij} y_{ij} : y \text{ satisfies (3.5), (3.6) and (3.7)} \right\}$$

is known as the *network flow problem*. It will be discussed in Chapter I.3.

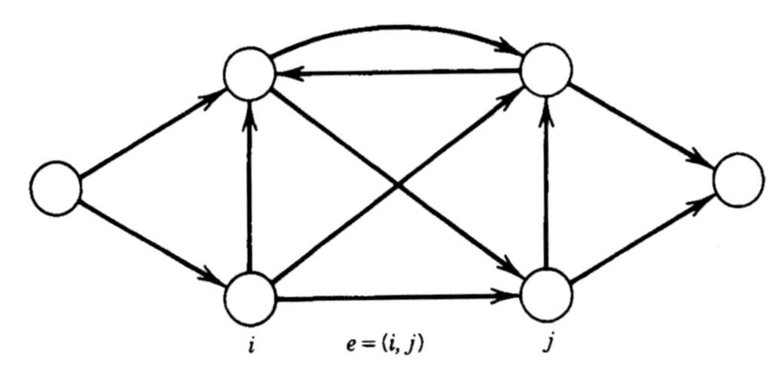

Figure 3.1

The *fixed-charge network flow problem* is obtained by imposing a fixed cost of c_{ij} if there is positive flow on arc (i, j). Now we introduce a binary variable x_{ij} to indicate whether arc (i, j) is used. The constraint $y_{ij} = 0$ if $x_{ij} = 0$ is represented by

$$y_{ij} - u_{ij}x_{ij} \leqslant 0 \quad \text{for } (i, j) \in \mathscr{A}. \tag{3.9}$$

Hence we obtain the formulation

$$\min\left\{ \sum_{(i,j)\in\mathscr{A}} (c_{ij}x_{ij} + h_{ij}y_{ij}) \colon x \in B^{|\mathscr{A}|},\ y \in R_+^{|\mathscr{A}|} \text{ satisfies (3.7), (3.9)}\right\}. \tag{3.10}$$

The fixed-charge flow model is useful for a variety of design problems that involve material flows in networks. These include water supply systems, heating systems, and road networks.

The formulations of the traveling salesman problem given below provide another example of the use of binary variables in the modeling of logical relations. They also exhibit another important property of integer programming formulations, namely, that it may be appropriate to use an extraordinarily large number of constraints in order to obtain a good formulation.

The Traveling Salesman Problem

We are again given a set of nodes $V = \{1, \ldots, n\}$ and a set of arcs $\mathscr{A}$. The nodes represent cities, and the arcs represent ordered pairs of cities between which direct travel is possible. For $(i, j) \in \mathscr{A}$, c_{ij} is the direct travel time from city i to city j. The problem is to find a tour, starting at city 1, that (a) visits each other city exactly once and then returns to city 1 and (b) takes the least total travel time.

To formulate this problem, we introduce variables $x_{ij} = 1$ if j immediately follows i on the tour, $x_{ij} = 0$ otherwise. Hence

$$x \in B^{|\mathscr{A}|}. \tag{3.11}$$

The requirements that each city is entered and left exactly once are stated as

$$\sum_{\{i:\,(i,j)\in\mathscr{A}\}} x_{ij} = 1 \quad \text{for } j \in V \tag{3.12}$$

and

$$\sum_{\{j:\,(i,j)\in\mathscr{A}\}} x_{ij} = 1 \quad \text{for } i \in V. \tag{3.13}$$

The constraints (3.11)–(3.13) are not sufficient to define the tours since they are also satisfied by subtours; for example for $n = 6$, $x_{12} = x_{23} = x_{31} = x_{45} = x_{56} = x_{64} = 1$ satisfies (3.11)–(3.13) but does not correspond to a tour (see Figure 3.2).

One way to eliminate subtours is to observe that in any tour there must be an arc that goes from {1, 2, 3} to {4, 5, 6} and an arc that goes from {4, 5, 6} to {1, 2, 3}. In general, for any $U \subset V$ with $2 \leqslant |U| \leqslant |V| - 2$, the constraints

$$\sum_{\{(i,j)\in\mathscr{A}:\, i\in U, j\in V\setminus U\}} x_{ij} \geqslant 1 \tag{3.14}$$

are satisfied by all tours, but every subtour violates at least one of them. Hence the traveling salesman problem can be formulated as

$$\min\left\{ \sum_{(i,j)\in\mathscr{A}} c_{ij}x_{ij}\colon x \text{ satisfies (3.11)–(3.14)} \right\}. \tag{3.15}$$

An alternative to the set of constraints (3.14) is

$$\sum_{\{(i,j)\in\mathscr{A}:\, i\in U, j\in U\}} x_{ij} \leqslant |U| - 1 \quad \text{for } 2 \leqslant |U| \leqslant |V| - 2, \tag{3.16}$$

which also excludes all subtours but no tours.

However, regardless of whether we use (3.14) or (3.16), the number of these constraints is nearly $2^{|V|}$. This huge number of constraints might motivate us to seek a more compact formulation. In fact, we will give such a formulation in Section I.1.5. But we will argue that the compact formulation is inferior and we will show, in Parts II and III, that a very large number of constraints can frequently be handled successfully.

4. MODELING WITH BINARY VARIABLES III: NONLINEAR FUNCTIONS AND DISJUNCTIVE CONSTRAINTS

In this section, we present two important uses of binary variables in the modeling of optimization problems. The first concerns the representation of nonlinear objective functions of the form $\Sigma_j f_j(y_j)$ using linear functions and binary variables. The second concerns the modeling of disjunctive constraints. In the usual statement of an optimization problem, it is assumed that all of the constraints must be satisfied. But in some applications, only one of a pair (or, more generally, k of m) constraints must hold. In this case, we say that the constraints are *disjunctive*.

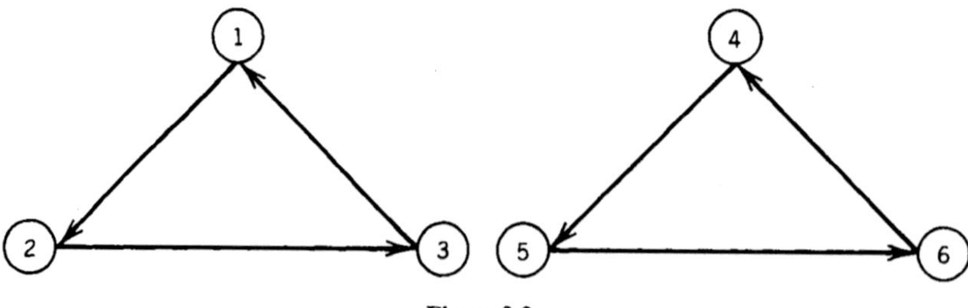

Figure 3.2

Piecewise Linear Functions

A function of the form $f(y_1, \ldots, y_p) = \Sigma_{j=1}^p f_j(y_j)$ is said to be a *separable* function. Here we consider separable objective functions and suppose that $f_j(y_j)$ is piecewise linear for each j (see Figure 4.1). Note that an arbitrary continuous function of one variable can be approximated by a piecewise linear function, with the quality of the approximation being controlled by the size of the linear segments.

Suppose we have a piecewise linear function $f(y)$ specified by the points $\{a_i, f(a_i)\}$ for $i = 1, \ldots, r$. Then, any $a_1 \leqslant y \leqslant a_r$ can be written as

$$y = \sum_{i=1}^{r} \lambda_i a_i, \quad \sum_{i=1}^{r} \lambda_i = 1, \quad \lambda = (\lambda_1, \ldots, \lambda_r) \in R_+^r.$$

The λ_i are not unique, but if $a_i \leqslant y \leqslant a_{i+1}$ and λ is chosen so that $y = \lambda_i a_i + \lambda_{i+1} a_{i+1}$ and $\lambda_i + \lambda_{i+1} = 1$, then we obtain $f(y) = \lambda_i f(a_i) + \lambda_{i+1} f(a_{i+1})$. In other words,

$$f(y) = \sum_{i=1}^{r} \lambda_i f(a_i), \quad \sum_{i=1}^{r} \lambda_i = 1, \quad \lambda \in R_+^r \tag{4.1}$$

if at most two of the λ_i's are positive and if λ_j and λ_k are positive, then $k = j - 1$ or $j + 1$. This condition can be modeled using binary variables x_i for $i = 1, \ldots, r - 1$ (where $x_i = 1$ if $a_i \leqslant y \leqslant a_{i+1}$ and $x_i = 0$ otherwise) and the constraints

$$\begin{aligned} &\lambda_1 \leqslant x_1 \\ &\lambda_i \leqslant x_{i-1} + x_i \quad \text{for } i = 2, \ldots, r-1 \\ &\lambda_r \leqslant x_{r-1} \\ &\sum_{i=1}^{r-1} x_i = 1 \\ &\quad x \in B^{r-1}. \end{aligned} \tag{4.2}$$

Note that if $x_j = 1$, then $\lambda_i = 0$ for $i \neq \{j, j+1\}$.

Piecewise linear functions that are convex (concave) can be minimized (maximized) by linear programming because the slope of the segments are increasing (decreasing) (see Figure 4.2). But general piecewise linear functions are neither convex nor concave, so binary variables are needed to select the correct segment for a given value of y.

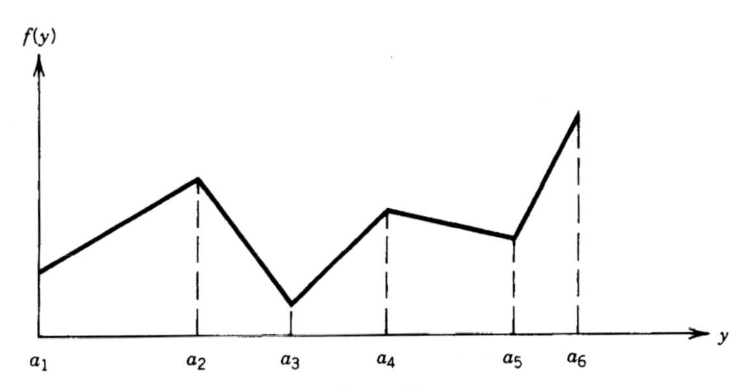

Figure 4.1

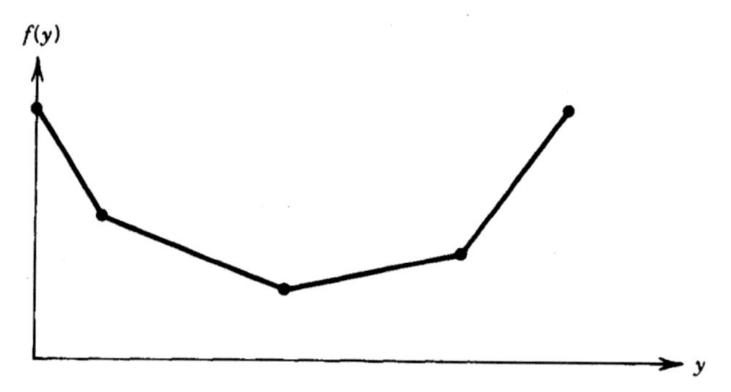

Figure 4.2. A convex piecewise linear function.

Disjunctive Constraints

Disjunctive constraints arise naturally in many models. A simple illustration is when we need to define a variable equal to the minimum of two other variables, that is, $y = \min(u_1, u_2)$. This can be done with the two inequalities

$$y \leqslant u_1 \quad \text{and} \quad y \leqslant u_2$$

together with one of two inequalities

$$y \geqslant u_1 \quad \text{or} \quad y \geqslant u_2.$$

A typical disjunctive set of constraints states that a point must satisfy at least k of m sets of linear constraints. The case of $k = 1$, $m = 2$ is shown in Figure 4.3, where the feasible region is shaded.

Suppose $P^i = \{y \in R^p_+ : A^i y \leqslant b^i, y \leqslant d\}$ for $i = 1, \ldots, m$. Note that there is a vector ω such that, for all i, $A^i y \leqslant b^i + \omega$ is satisfied for any y, $0 \leqslant y \leqslant d$. Hence there is a y contained in at least k of the sets P^i if and only if the set

$$\begin{aligned} A^i y &\leqslant b^i + \omega(1 - x_i) \quad \text{for } i = 1, \ldots, m \\ \sum_{i=1}^{m} x_i &\geqslant k \\ y &\leqslant d \\ x \in B^m, &\quad y \in R^p_+ \end{aligned} \tag{4.3}$$

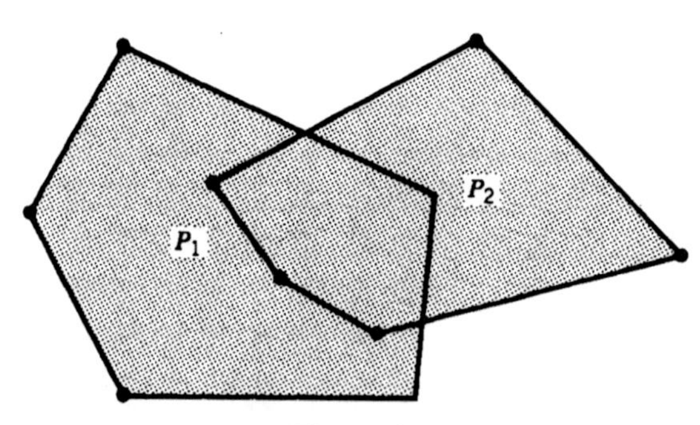

Figure 4.3

is nonempty. This follows since $x_i = 1$ yields the constraint $A^i y \leqslant b^i$ while $x_i = 0$ yields the redundant constraints $A^i y \leqslant b^i + \omega$.

When $k = 1$, an alternative formulation is

$$
\begin{aligned}
&A^i y^i \leqslant x_i b^i \quad \text{for } i = 1, \ldots, m \\
&y^i \leqslant x_i d \quad \text{for } i = 1, \ldots, m \\
&\sum_{i=1}^{m} x_i = 1 \\
&\sum_{i=1}^{m} y^i = y \\
&x \in B^m, \quad y \in R^p_+, \quad y^i \in R^p_+ \quad \text{for } i = 1, \ldots, m.
\end{aligned} \tag{4.4}
$$

Now we claim that $\cup_{i=1}^m P^i \neq \emptyset$ if and only if (4.4) is nonempty. First, given that $y \in \cup_{i=1}^m P^i$, suppose without loss of generality that $y \in P^1$. Then a solution to (4.4) is $x_1 = 1$, $x_i = 0$ otherwise, $y^1 = y$, and $y^i = 0$ otherwise. On the other hand, suppose (4.4) has a solution and, without loss of generality, suppose $x_1 = 1$ and $x_i = 0$ otherwise. Then we obtain $y^i = 0$ for $i = 2, \ldots, m$ and $y = y^1$. Thus $y \in P^1$ and $\cup_{i=1}^m P^i \neq \emptyset$.

The models (4.3) and (4.4) are quite different formulations of the same problem. This choice of formulation is typical. A significant issue to be discussed in the next section is what constitutes a good formulation?

A Scheduling Problem

Disjunctive constraints arise naturally in scheduling problems where several jobs have to be processed on a machine and where the order in which they are to be processed is not specified. Thus we obtain disjunctive constraints of the type either "job k precedes job j on machine i" or vice versa.

Suppose there are n jobs and m machines and each job must be processed on each machine. For each job, the machine order is fixed, that is, job j must first be processed on machine $j(1)$ and then on machine $j(2)$, and so on. A machine can only process one job at a time, and once a job is started on any machine it must be processed to completion. The objective is to minimize the sum of the completion times of all the jobs. The data that specify an instance of the problem are (a) m, n, and p_{ij} for $j = 1, \ldots, n$ and $i = 1, \ldots, m$, which is the processing time of job j on machine i, and (b) the machine order, $j(1), \ldots, j(m)$, for each job j.

Let t_{ij} be the start time of job j on machine i. Since the $(r + 1)$st operation on job j cannot start until the rth operation has been completed, we have the constraints

$$t_{j(r+1),j} \geqslant t_{j(r),j} + p_{j(r),j} \quad \text{for } r = 1, \ldots, m-1 \text{ and all } j. \tag{4.5}$$

To represent the disjunctive constraints for jobs j and k on machine i, let $x_{ijk} = 1$ if job j precedes job k on machine i and $x_{ijk} = 0$ otherwise where $j < k$. Thus

$$t_{ik} \geqslant t_{ij} + p_{ij} \quad \text{if } x_{ijk} = 1$$

and

$$t_{ij} \geqslant t_{ik} + p_{ik} \quad \text{if } x_{ijk} = 0.$$

Given an upper-bound ω on $t_{ij} - t_{ik} + p_{ij}$ for all i, j, and k, we obtain the disjunctive constraints

$$(4.6) \qquad \begin{aligned} t_{ij} - t_{ik} &\leqslant -p_{ij} + \omega(1-x_{ijk}) \\ t_{ik} - t_{ij} &\leqslant -p_{ik} + \omega x_{ijk} \quad \text{for all } i, j \text{ and } k. \end{aligned}$$

Hence the problem is to minimize $\Sigma_{j=1}^{n} t_{j(m),j}$ subject to (4.5), (4.6), $t_{ij} \geqslant 0$ for all i and j and $x_{ijk} \in \{0, 1\}$ for all i, j, and k.

This model requires $m\binom{n}{2}$ binary variables. In contrast to the integer programming models introduced previously, this mixed-integer programming model has not been successfully solved for values of m and n that are of practical interest. This formulation, which is based on (4.3), is cumbersome partly because of the large number of binary variables needed to represent the large number of disjunctions. Note that a formulation based on (4.4) would also have a large number of binary variables. In fact, a large number of binary variables may be unavoidable for this scheduling problem.

Good formulations are essential to solving integer programming problems efficiently. In the next section, we will give some reasons why some formulations may be better than others; we will also suggest how formulations can be improved.

5. CHOICES IN MODEL FORMULATION

We have formulated several integer optimization problems in this chapter to motivate the richness and variety of applications. Although a formulation may give insight into the structure of the problem, our goal is to solve the problem for an optimal or nearly optimal solution. As we have already indicated, most integer programming problems can be formulated in several ways. Moreover, in contrast to linear programming:

> *In integer programming, formulating a "good" model is of crucial importance to solving the model.*

Indirectly, the subject of "good" model formulation is a major topic of this book and is closely related to the algorithms themselves (see Chapters II.2 and II.5).

A model is specified by the variables, objective function, and constraints. Typically, defining the variables is the first question addressed in formulating a model. Often the variables are chosen simply from the definition of a solution. That is a solution specifies the values of certain unknowns, and we define a variable for each unknown. Once the variables and an objective function have been defined, say in an IP, we can speak of an implicit representation of the problem

$$\max\{cx: x \in S \subset Z_{+}^{n}\},$$

where S represents the set of feasible points in Z_{+}^{n}. Now we say that

$$\max\{cx: Ax \leqslant b, x \in Z^n_+\}$$

is a *valid* IP *formulation* if $S = \{x \in Z^n_+: Ax \leqslant b\}$.

In general, when there is a valid formulation, there are many choices of (A, b), and it is usually easy to find some (A, b) that yields one. But an obvious choice may not be a good one when it comes to solving the problem. We believe that the most important aspect of model formulation is the choice of (A, b).

The following example illustrates different representations of an $S \subseteq Z^n_+$ by linear inequality and integrality restrictions.

Example 5.1

$$S = \{(0000), (1000), (0100), (0010), (0001), (0110), (0101), (0011)\} \subseteq B^4.$$

The reader can easily check that

(a) $$S = \{x \in B^4: 93x_1 + 49x_2 + 37x_3 + 29x_4 \leqslant 111\}$$

gives a valid formulation. Two other formulations that are easily established to be valid are:

(b) $$S = \{x \in B^4: 2x_1 + x_2 + x_3 + x_4 \leqslant 2\}$$

(c) $$\begin{aligned} S = \{x \in B^4: 2x_1 + x_2 + x_3 + x_4 &\leqslant 2 \\ x_1 + x_2 &\leqslant 1 \\ x_1 + x_3 &\leqslant 1 \\ x_1 + x_4 &\leqslant 1\}. \end{aligned}$$

We will see that, in a certain sense, formulation (b) is better than (a), and (c) is better than (b).

How should we compare different formulations? Later we will see that most integer programming algorithms require an upper bound on the value of the objective function, and the efficiency of the algorithm is very dependent on the sharpness of the bound. An upper bound is determined by solving the linear program

$$z_{\text{LP}} = \{\max cx: Ax \leqslant b, x \in R^n_+\}$$

since $P = \{x \in R^n_+: Ax \leqslant b\} \supseteq S$. Now given two valid formulations, defined by (A^i, b^i) for $i = 1, 2$, let $P^i = \{x \in R^n_+: A^i x \leqslant b^i\}$ and $z^i_{\text{LP}} = \max\{cx: x \in P^i\}$. Note that if $P^1 \subseteq P^2$, then $z^1_{\text{LP}} \leqslant z^2_{\text{LP}}$. Hence we get the better bound from the formulation based on (A^1, b^1) and we say that it is the better formulation. We leave it to the reader to check that in Example 5.1, formulation (c) gives a better bound than (b), which, in turn, gives a better bound than (a).

A striking example of one formulation being better than another, in the sense just described, is provided by the uncapacitated facility location problem. We obtain a formulation with fewer constraints than the one given in Section 3 by replacing (3.2) with

(5.1) $$\sum_{i \in I} y_{ij} - mx_j \leqslant 0 \quad \text{for all } j \in N.$$

When $x_j = 0$, (5.1) says that no clients can be served from facility j; and when $x_j = 1$, there is no restriction on the number of clients that can be served from facility j. In fact, by summing (3.2) over $i \in I$ for each j, we obtain (5.1). Although with $x \in B^n$, (3.2) and (5.1) give the same set of feasible solutions, with $x \in R^n_+$, (3.2) gives a much smaller feasible set than (5.1). Our ability to solve the formulation with (3.2) is remarkably better than with the more compact formulation that uses (5.1).

We belabor this point because it is instinctive to believe that computation time increases and computational feasibility decreases as the number of constraints increases. But, trying to find a formulation with a small number of constraints is often a very bad strategy. In fact, one of the main algorithmic approaches involves the systematic addition of constraints, known as cutting planes (see Part II).

A nice illustration of the suitability of choosing (A, b) with a very large number of rows concerns the traveling salesman problem. In Section 3, we gave two different sets of constraints, (3.14) and (3.16), for eliminating subtours. Both formulations contain a huge number of constraints, far too many to write down explicitly. Nevertheless, algorithms for the traveling salesman problem that solve these formulations have been successful on problems with more than 2000 cities. On the other hand, there is a more subtle way of eliminating subtours that only requires a small number of constraints.

Let $u \in R^{n-1}$ and consider the constraints

$$(5.2) \qquad u_i - u_j + nx_{ij} \leq n - 1 \quad \text{for } (i, j) \in \mathcal{A},\ i \neq 1, j \neq 1.$$

If $x \in B^{|\mathcal{A}|}$ satisfies (3.12) and (3.13) and does not represent a tour, then x represents at least two subtours, one of which does not contain node 1. By summing (5.2) over the arc set $\mathcal{A}'$ of some subtour that does not contain node 1, we obtain

$$(5.3) \qquad \sum_{(i,j) \in \mathcal{A}'} x_{ij} \leq |\mathcal{A}'| \cdot (1 - 1/n).$$

Thus (5.2) excludes all subtours that do not contain node 1 and hence excludes all solutions that contain subtours.

Now we prove that no tours are excluded by (5.2) by showing that for any tour there exists a corresponding u satisfying (5.2). In particular, we set $u_i = k$, where k is the position $(2 \leq k \leq n)$ of node i in the tour. Now if $x_{ij} = 0$, $u_i - u_j + nx_{ij} \leq n - 2$, while if $x_{ij} = 1$, $u_i = k$ and $u_j = k + 1$ for some k, and so $u_i - u_j + nx_{ij} = n - 1$. Hence $\{x \in B^{|\mathcal{A}|}$: x satisfies (3.12), (3.13), and (5.2)$\}$ is the set of incidence vectors of tours.

Now let $P^1 = \{x \in R^{|\mathcal{A}|}_+$: x satisfies (3.12), (3.13), (3.16)$\}$ and $P^2 = \{x \in R^{|\mathcal{A}|}_+$: x satisfies (3.12), (3.13), and (5.2) for some $u\}$. It is easy to see that $P^2 \not\subseteq P^1$. For example, if $n \geq 4$, then $u_2 = u_3 = u_4 = 0$ and $x_{23} = x_{34} = x_{42} = (n - 1)/n > \frac{2}{3}$ satisfies (5.2) but not (3.16). In fact, it can be shown that $P^1 \subseteq P^2$.

We have emphasized the choice of constraints in obtaining a good formulation, given that the variables have already been defined, because for most problems this is the part of the formulation where there is the greatest freedom of choice. There are, however, problems in which the quality of the formulation depends on the choice of variables.

In our formulation of network flow problems, we defined the variables to be the arc flows. However, in certain situations it is more advantageous to define variables that represent the flow on each path between two given nodes. Such a formulation involves many more variables but eliminates the need for some flow conservation constraints and can be preferable for finding integral solutions.

We now give two radically different formulations of a production lot-sizing problem that depend on the choice of variables. The object is to minimize the sum of the costs of production, storage, and set-up, given that known demands in each of T periods must be satisfied. For $t = 1, \ldots, T$, let d_t be the demand in period t, and let c_t, p_t, and h_t be the set-up, unit production, and unit storage costs, respectively, in period t.

One formulation is obtained by defining y_t, s_t as the production and end storage in period t and by defining a binary variable x_t, indicating whether $y_t > 0$ or not. This leads to the model

$$\min \sum_{t=1}^{T} (p_t y_t + h_t s_t + c_t x_t)$$

$$(5.4)\qquad \begin{aligned} y_1 &= d_1 + s_1 \\ s_{t-1} + y_t &= d_t + s_t \quad \text{for } t = 2, \ldots, T \\ y_t &\leqslant \omega x_t \qquad \text{for } t = 1, \ldots, T \\ s_T &= 0 \\ s, y \in R_+^T, &\qquad x \in B^T, \end{aligned}$$

where $\omega = \sum_{t=1}^{T} d_t$ is an upper bound on y_t for all t.

A second possibility is to define q_{it} as the quantity produced in period i to satisfy the demand in period $t \geqslant i$, and x_t as above. Now we obtain the model

$$\min \sum_{t=1}^{T} \sum_{i=1}^{t} (p_i + h_i + h_{i+1} + \ldots + h_{t-1}) q_{it} + \sum_{t=1}^{T} c_t x_t$$

$$(5.5)\qquad \begin{aligned} \sum_{i=1}^{t} q_{it} &= d_t \qquad \text{for } t = 1, \ldots, T \\ q_{it} &\leqslant d_t x_i \quad \text{for } i = 1, \ldots, T \text{ and } t = i, \ldots, T \\ q \in R_+^{T(T+1)/2}, &\qquad x \in B^T. \end{aligned}$$

In (5.5) if we replace $x \in B^T$ by $0 \leqslant x_t \leqslant 1$ for all t, then the resulting linear programming problem has an optimal solution with $x \in B^T$. But this is not necessarily the case for (5.4), which is the inferior formulation for soliving the problem by certain integer programming techniques. It is interesting to observe that (5.5) is a special case of the uncapacitated facility location problem. This can be seen by substituting $y_{it} = q_{it}/d_t$ for all i and $t \geqslant i$.

There is a similar result for the formulations (4.3) and (4.4) for finding a point that satisfies one of m sets of linear constraints. In (4.4), one can replace the condition $x \in B^m$ with $0 \leqslant x \leqslant 1$ and use linear programming to find a point in one of the P^i. But this is not true for (4.3), which is therefore considered to be the inferior formulation.

6. PREPROCESSING

Given a formulation, preprocessing refers to elementary operations that can be performed to improve or simplify the formulation by tightening bounds on variables, fixing values, and so on. Preprocessing can be thought of as a phase between formulation and solution. It can greatly enhance the speed of a sophisticated algorithm that might, for example, be unable to recognize the fact that some variable can be fixed and then eliminated from the model. Occasionally a small problem can be solved in the preprocessing phase or by combining preprocessing with some enumeration. Although this approach had been advocated as a solution technique in the early development of integer programming, under the name of implicit enumeration, this is not the important role of these simple techniques. Their main purpose is to prepare a formulation quickly and automatically for a more sophisticated algorithm. Unfortunately, it has taken a long time for researchers to recognize the fact that there is generally a need for both phases in the solution of practical problems.

Tightening Bounds

We have seen that a common constraint in MIPs is $y_j \leqslant u_j x_j$, where u_j is an upper bound on y_j and x_j is a binary variable. Provided that $x_j \in \{0, 1\}$, the tightness of the upper bound doesn't matter. But if we replace $x_j \in \{0, 1\}$ by $0 \leqslant x_j \leqslant 1$, it becomes important to have a tight bound. Suppose, for example, that the largest feasible value of y_j is $u_j' < u_j$ and that there is a fixed cost $f_j > 0$ associated with x_j. If $y_j = u_j'$ in an optimal solution, and we use the constraint $y_j \leqslant u_j x_j$, we will obtain $x_j = u_j'/u_j < 1$. On the other hand, if we use the constraint $y_j \leqslant u_j' x_j$, we obtain $x_j = 1$.

In some cases, good bounds can be determined analytically. For example, in the lot-sizing problem, rather than using a common bound for each y_t, it is more efficient to use the bounds $y_t \leqslant (\Sigma_{i=t}^{T} d_i)x_t$. In general, tight bounds can be determined by solving a linear program with the objective of maximizing y_j. Doing this for each variable with an upper bound constraint may be prohibitively time consuming, so a good compromise is to approximate the upper bounds heuristically.

Example 6.1. We show a fixed-charge model in Figure 6.1 with the accompanying formulation:

$$
\begin{aligned}
y_1 + y_2 \qquad\qquad\qquad\qquad &= 1.46 \\
y_3 + y_4 \qquad\qquad &= 0.72 \\
-y_2 - y_3 \qquad + y_5 \qquad\qquad &= 0 \\
y_6 \qquad &= 0.32 \\
-y_5 - y_6 + y_7 &= 0 \\
0 \leqslant y_i \leqslant \omega x_i, \quad x_i \in B^1 \quad &\text{for all } i,
\end{aligned}
$$

where ω is a large positive number because the arcs do not have capacity constraints.

It is easy to tighten the bounds, giving

$$
\begin{aligned}
&y_1 \leqslant 1.46x_1, \quad y_2 \leqslant 1.46x_2 \\
&y_3 \leqslant 0.72x_3, \quad y_4 \leqslant 0.72x_4 \\
&y_5 \leqslant (1.46 + 0.72)x_5 \\
&y_6 = 0.32 \\
&y_7 \leqslant (1.46 + 0.72 + 0.32)x_7.
\end{aligned}
$$

In addition, we can set $x_6 = x_7 = 1$ because the flow into node 7 must use these arcs.

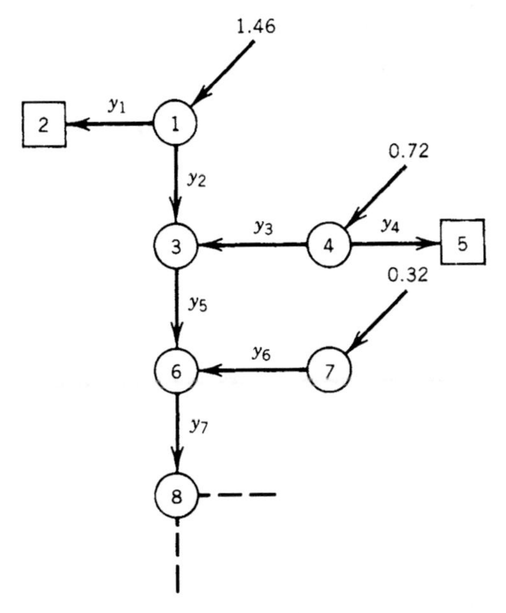

Figure 6.1

Adding Logical Inequalities, Fixing Variables, and Removing Redundant Constraints

Preprocessing of this sort is most useful for binary IPs. Consider a single inequality in binary variables, that is, $S = \{x \in B^n: \Sigma_{j\in N} a_j x_j \leqslant b\}$. If $a_j < 0$, we can replace x_j by $1 - x_j'$ and obtain the constraint $\Sigma_{j\in N: a_j>0} a_j x_j + \Sigma_{j\in N: a_j<0} |a_j| x_j' \leqslant b - \Sigma_{j\in N: a_j<0} a_j$. Thus without loss of generality, we can assume that $a_j > 0$ for $j \in N$. Now if $\Sigma_{j\in C} a_j > b$ for $C \subseteq N$, we obtain the inequality

$$\sum_{j\in C} x_j \leqslant |C| - 1. \tag{6.1}$$

Obviously, the best inequalities of this type are obtained when $\Sigma_{j\in C\setminus\{k\}} a_j \leqslant b$ for all $k \in C$.

Once some inequalities of this type have been obtained, it may be possible to combine some of them to fix variables. For example, $x_1 + x_2 \leqslant 1$ and $x_1 + (1 - x_2) \leqslant 1$ yield $x_1 = 0$.

The application of these simple ideas is easy to see by considering an example.

Example 6.2

$$\begin{aligned} -3x_2 - 2x_3 &\leqslant -2 \quad (\qquad 3x_2' + 2x_3' \leqslant 3) \\ -4x_1 - 3x_2 - 3x_3 &\leqslant -6 \quad (4x_1' + 3x_2' + 3x_3' \leqslant 4) \\ 2x_1 - 2x_2 + 6x_3 &\leqslant 5 \quad (2x_1 + 2x_2' + 6x_3 \leqslant 7) \\ x &\in B^3. \end{aligned}$$

The first constraint yields $x_2' + x_3' \leqslant 1$ or $x_2 + x_3 \geqslant 1$. The third constraint yields $x_2' + x_3 \leqslant 1$ or $x_3 \leqslant x_2$. Combining these two yields $x_2 = 1$. Now the first constraint is redundant and the second and third reduce to $4x_1' + 3x_3' \leqslant 4$ and $2x_1 + 6x_3 \leqslant 7$. From these two, we obtain $x_1' + x_3' \leqslant 1$ and $x_1 + x_3 \leqslant 1$, or $x_1 + x_3 = 1$. Thus, by substitution, we can eliminate either x_1 or x_3.

Other simplifications of this type are considered as exercises.

A second stage of preprocessing can be carried out after an upper bound has been obtained by linear programming. In particular, variables can be fixed by using the reduced prices that are obtained from a linear programming solution (see Section II.5.2).

7. NOTES

Section I.1.1

Here we list bibliographies, other books, proceedings, and some of the main journals that contain a great deal of material on integer programming and/or combinatorial optimization. Four volumes of comprehensive bibliographies on integer programming have been prepared at Bonn University [see Kastning (1976), Hausmann (1978) and von Randow (1982, 1985)]. Each volume contains an alphabetical listing by authors, a subject classification, and a third part that enables one to find items by an author who is not listed first. The first volume contains items published through 1975 and includes 4704 entries classified under 41 subject headings. The last volume covers items published in the period 1981–1984 and contains 4751 entries classified under 50 subject headings. A much briefer, but annotated, bibliography is the subject of O'hEigertaigh et al. (1985).

Several books on integer programming and combinatorial optimization have appeared in the 1980s. In chronological order, these are Papadimitriou and Steiglitz (1982), Gondran and Minoux (1984), Lawler, Lenstra et al. (1985), Schrijver (1986a), and Grötschel, Lovasz, and Schrijver (1988). Papadimitriou and Steiglitz emphasize algorithms and computational complexity from the point of view of computer scientists. Gondran and Minoux also stress algorithms and focus on problems associated with graphs. Lawler et al. is restricted to the traveling salesman problem, but we mention it here because of the prominent role played by the traveling salesman problem as a generic difficult combinatorial optimization problem. Schrijver gives an encyclopedic treatment of the theory of linear and integer programming from the polyhedral point of view. Grotschel et al. is a monograph whose subject matter is motivated by the consequences of ellipsoid algorithms in combinatorial optimization. It also contains information on algorithmic approaches to problems in geometric number theory. The applications of this branch of mathematics in discrete optimization have just begun to be investigated.

Earlier general textbooks on integer programming are Hu (1969), Greenberg (1971), Garfinkel and Nemhauser (1972a), Salkin (1975), and Taha (1975). Lawler (1976) emphasizes the roles of network flows and matroids in combinatorial optimization. Christofides (1975a) studies a variety of combinatorial optimization problems associated with graphs. Johnson (1980a) is a monograph on integer programming theory that emphasizes subadditivity and group theory.

Beale (1968) and Williams (1978a) are general texts on mathematical programming that are of some interest here because they emphasize modeling and problem formulation.

General survey articles appeared early in the development of the field [see Beale (1965), Balinski (1965, 1967, 1970a), Balinski and Spielberg (1969), Garfinkel and Nemhauser (1973), Geoffrion and Marsten (1972) and Geoffrion (1976)]. Some recent surveys on combinatorial optimization are by Klee (1980), Pulleyblank (1983), Schrijver (1983a), and Grötschel (1984); Grötschel (1985) gives an annotated bibliography. More specialized surveys will be cited in the appropriate chapters.

Numerous proceedings and study volumes have been devoted to integer and combinatorial optimization. These include Balinski (1974), Hammer, Johnson, Korte, and Nemhauser (1977), Balinski and Hoffman (1978), Hammer, Johnson, and Korte

(1979a,b), Christofides, Mingozzi et al. (1979), Padberg (1980a), Hansen (1981), Pulleyblank (1984), and Monma (1986). The Hammer, Johnson, and Korte volumes and the book by Christofides et al. are collections of surveys. For the most part, the others are collections of research articles that complement the journals that contain a substantial number of papers on integer programming and combinatorial optimization.

Some of the more prominent journals published in English are *Mathematical Programming*, *Mathematical Programming Studies*, *Operations Research*, *Operations Research Letters*, *Annals of Operations Research*, *Networks*, *SIAM Journal on Algebraic and Discrete Methods*, *Discrete Mathematics*, *Discrete Applied Mathematics*, *Annals of Discrete Mathematics*, *Combinatorica*, *Journal of the Association for Computing Machinery*, *Management Science*, *Operational Research Quarterly*, *The European Journal of Operations Research*, *Naval Research Logistics Quarterly*, *IIE Transactions*, and *Transportation Science*.

The scope of each of these journals relative to their coverage of integer and combinatorial optimization is difficult to specify. A rough guideline is the following. The first five purport to cover the subject broadly, although there is unfortunately a dearth of papers on applications. The same can be said for *Networks* within its more narrowly defined scope of problems. The next five emphasize theory. The remainder contain some methodology oriented toward specific models and a few applications.

The periodical *Interfaces* publishes an annual issue on successful case studies in operations research and management science. Some of these studies involve the use of integer programming techniques. Applications of integer programming are also discussed in journals of finance, marketing, production, economics, and the various branches of engineering.

Sections I.1.2–I.1.4

Dantzig (1957, 1960) formulated several integer programming models and showed how a variety of nonlinear and nonconvex optimization problems could be formulated as mixed-integer programs. References on the models presented in these sections will be given in the notes for the chapters in which the models are discussed in detail. In particular, knapsack problems are considered in Sections II.2.2 and II.6.1, matching problems are discussed in Chapter III.2, set covering is presented in Section II.6.2 and Chapter III.1, fixed-charge network problems are considered in Sections II.2.4 and II.6.4, and the traveling salesman problem is discussed in Sections II.2.3 and II.6.3.

Section I.1.5

Strong formulations is one of the major themes of this book. See Williams (1974, 1978b) and Jeroslow and Lowe (1984) for a comparison of alternative formulations for some general integer programs.

Systematic reformulation of knapsack problems was treated by Bradley et al. (1974). Formulation (5.2) appears in Miller et al. (1960). The strength of reformulation (5.5) was shown by Krarup and Bilde (1977), and that of the disjunctive formulation (4.4) was shown by Balas (1979). Many other citations will be made in the notes for Chapters II.2, II.5, and II.6.

Section I.1.6

Preprocessing techniques are frequently attributed to folklore because the references are difficult to pin down. Bound tightening, variable fixing, and row elimination schemes used in mathematical programming systems are discussed in Brearley et al. (1975).

Preprocessing techniques that use boolean inequalities have been studied by Guignard and Spielberg (1977, 1981). Also see Guignard (1982), Johnson and Suhl (1980), Crowder, Johnson, and Padberg (1983), Johnson and Padberg (1983), and Johnson, Kostreva, and Suhl (1985).

8. EXERCISES

1. Show that the integer program with irrational data $\max\{x_1 - (2)^{1/2}x_2: x_1 \leqslant (2)^{1/2}x_2, x_1 \geqslant 1, x \in Z_+^2\}$ has no optimal solution, even though there exist feasible solutions with value arbitrarily close to zero.

2. The BST Delivery Company must make deliveries to 10 customers whose respective demands are d_j for $j = 1, \ldots, 10$. The company has four trucks available with capacities L_k and daily operating costs c_k for $k = 1, \ldots, 4$. A single truck cannot deliver to more than five customers, and customer pairs $\{1, 7\}$, $\{2, 6\}$, and $\{2, 9\}$ cannot be visited by the same truck. Formulate a model to determine which trucks to use so as to minimize the cost of delivering to all the customers.

3. An airline has fixed its daily timetable for flights between five cities. It now has the problem of scheduling the crews. There are certain legal limits on how much time each crew can work within any 24-hour period. The problem is to propose a crew schedule using the minimum number of crews in which each flight leg is covered. Formulate a generic problem of this type as a set covering problem.

4. The DuFour Bottling Company has two machines for its bottle production. The problem each year is to devise a maintenance schedule. Maintenance of each machine lasts 2 months. In addition, only half the workforce is available in July and August, so that only one machine can be used during that period. Monthly demands for bottles are d_t, $t = 1, \ldots, 12$. Machine k, $k = 1, 2$, produces bottles at the rate of a_k bottles per month but can produce less. There is also a labor constraint. Machine k requires l_k labor days to produce a_k, and the total available days per month are L_t for $t = 1, \ldots, 12$. Formulate the problem of finding a feasible maintenance schedule in which all demands are satisfied. Modify your formulation to handle the following objectives.

   i) Minimize the sum of the monthly fluctuations in labor utilization.

   ii) Minimize the largest monthly fluctuation.

5. Integer and mixed-integer programming models are used on Wall Street to select bond portfolios. The idea is to pick a mix of bonds to maximize average yield subject to constraints on quality, length of maturity, industrial and government percentages, and total budget. Integrality arises because certain bonds only come in 100-unit lots. Formulate a model for this generic problem.

6. A company has two products $k = 1, 2$, one factory, two distribution centers $i = 1, 2$, and five major clients $j = 1, \ldots, 5$ whose product demands d_{jk} are known. The company must decide which products should be handled by each center and how each client should be serviced. The problem is to minimize total costs, where the costs include:

   i) a fixed cost f_{ik} if product k is handled by distribution center i;

   ii) fixed costs f_{ijk} if the demand of client j for product k is satisfied by center i; and

iii) unit shipping costs c_{ijk} per unit of product k shipped to client j via center i.

How does your model change if demands can be split between distribution centers?

7. Formulate the traveling salesman problem using the variables x_{ijk}, where $x_{ijk} = 1$ if (i, j) is the kth arc of the tour and $x_{ijk} = 0$ otherwise.

8. **a)** Given a graph $G = (V, E)$ with weights w_e for $e \in E$, formulate the following problems (see Chapter I.3 for some of the definitions) as integer programs.

i) Find a maximum-weight tree.

ii) Find a maximum-weight s-t cut.

iii) Find a minimum-weight covering of nodes by edges.

iv) Find a maximum-weight cycle with an odd number of edges.

v) Find a maximum-weight bipartite subgraph.

vi) Find a maximum-weight eulerian subgraph.

b) Given a graph $G = (V, E)$ with weights c_j for $j \in V$, formulate the following problems.

i) Find a maximum-weight clique.

ii) Find a minimum-weight dominating set (a set of nodes $U \subseteq V$ such that every node of V is adjacent to some node in U).

9. Suppose k trucks can be used to serve n clients from a single depot. Each client must be visited once. The time for truck k to travel from i to j is c_{ijk}. The tour of each truck cannot take longer than L_k. Formulate the problem of finding a feasible schedule.

10. Consider the quadratic 0-1 knapsack problem

$$\min\left\{\sum_{j=1}^{n} c_j x_j + \sum_{i=1}^{n-1} \sum_{j=i+1}^{n} c_{ij} x_j x_j \colon \sum_{j=1}^{n} a_j x_j \geqslant b,\, x \in B^n\right\}.$$

By introducing a variable y_{ij} to represent $x_i x_j$, reformulate the problem as a linear mixed-integer programming problem.

11. Show that the BIP $\max\{cx\colon Ax \leqslant b,\, x \in B^n\}$ may be solved by solving the quadratic program

$$\max\{cx - Mx^T(1 - x)\colon Ax \leqslant b,\, 0 \leqslant x_j \leqslant 1 \quad \text{for all } j\},$$

where M is a large positive number. Given A, b, c, how large should M be?

12. Let $H \in R_+^{m \times n}$ and $c \in R_+^n$. Let $\mathcal{F}$ be the collection of all the nonempty subsets of $\{1, 2, \ldots, n\}$. For $F \in \mathcal{F}$ define

$$z(F) = \sum_{i=1}^{m} \max_{j \in F} h_{ij} - \sum_{j \subset F} c_j.$$

i) Show that the problem $\max\{z(F)\colon F \in \mathcal{F}\}$ can be formulated as the following integer program:

$$\max \sum_{i=1}^{m}\sum_{j=1}^{n} h_{ij}y_{ij} - \sum_{j=1}^{n} c_j x_j$$

$$\sum_{j=1}^{m} y_{ij} = 1 \quad \text{for } i = 1, \ldots, m$$

$$y_{ij} \leqslant x_j \quad \text{for } i = 1, \ldots, m \text{ and } j = 1, \ldots, n$$

$$y \in B^{mxn}, \qquad x \in B^n.$$

ii) Show that the problem $\max\{z(F): F \in \mathcal{F}\}$ can also be formulated as the integer program:

$$\max \sum_{i=1}^{m} u_i - \sum_{j=1}^{n} c_j x_j$$

$$u_i \leqslant h_{ik} + \sum_{j=1}^{n} (h_{ij} - h_{ik})^+ x_j \quad \text{for } k = 0, \ldots, n \text{ and } i = 1, \ldots, m$$

$$\sum_{j=1}^{n} x_j \geqslant 1$$

$$u \in R_+^m, \qquad x \in B^n,$$

where a^+ denotes $\max(0, a)$ and $h_{i0} = 0$ for $i = 1, \ldots, m$.

13. Consider the scheduling problem of Section 4 with only one machine. Each job has processing time p_j, a deadline d_j, and a weight $w_j > 0$.

i) Formulate the problem of finding a feasible schedule in which the weighted sum of completion times is minimized. Avoid using ω as in (4.6) by writing an exact expression for the finish time of job j.

ii) Give an alternative formulation using the variables x_{jt}, where $x_{jt} = 1$ if job j is completed at time t. (Assume p_j, d_j are integers).

14. Suppose the departure times of trucks A and B have to be scheduled. Each truck can leave at 1, 2, 3, or 4 p.m. Truck B cannot leave until at least 1 hour after truck A. Let x_i (y_i) = 1 if truck A (B) leaves at time i. Give two formulations of the feasible region and compare them.

15. Show that

$$\begin{aligned} S &= \{x \in B^4: 97x_1 + 32x_2 + 25x_3 + 20x_4 \leqslant 139\} \\ &= \{x \in B^4: 2x_1 + x_2 + x_3 + x_4 \leqslant 3\} \\ &= \left\{x \in B^4: \begin{array}{l} x_1 + x_2 + x_3 \leqslant 2 \\ x_1 + x_3 + x_4 \leqslant 2 \\ x_1 + x_2 + x_4 \leqslant 2 \end{array}\right\} \end{aligned}$$

Which formulation do you think is most effective for solving $\max\{cx: x \in S\}$?

16. Consider the 0-1 feasible region

$$S = \left\{ x \in B^n: \sum_{j \in N} a_j x_j \leqslant b \right\} \quad \text{with } a_j, b \in Z_+^1 \text{ for } j \in N.$$

Formulate as an integer program the problem of finding weights c_j, $d \in Z_+^1$ such that

$$S = \left\{ x \in B^n: \sum_{j \in N} c_j x_j \leqslant d \right\}$$

and d is minimized. Formulate and solve the example with

$$S = \{x \in B^4: 93x_1 + 62x_2 + 37x_3 + 12x_4 \leqslant 140\}.$$

17. Consider the two formulations of the traveling salesman problem in Section 5. Show that $P_1 \subset P_2$.

18. To show that (4.4) gives a tight formulation of $\cup_{i=1}^m P_i$ when

$$P_i = \{z \in R_+^p: A^i z \leqslant b_i,\ z \leqslant d\},$$

let

$$T^* = \left\{ (y^i, y, x): A^i y^i \leqslant b_i x_i,\ y^i \leqslant d x_i \text{ for } i = 1, \ldots, m, \quad \sum_{i=1}^m x_i = 1,\ \sum_{i=1}^m y^i = y,\ y^i \in R_+^p,\ y \in R_+^p,\ x \in R_+^m \right\}$$

and

$$T^{**} = T^* \cap \{(y^i, y, x): x \in B^m\}.$$

i) Show that if $y^* \in \cup_{i=1}^m P_i$, there exists (y^i, x) such that $(y^i, y^*, x) \in T^{**}$.

ii) Show that if $(y^i, y^*, x) \in T^{**}$, then $y^* \in \cup_{i=1}^m P_i$.

iii) Show that if $(y^i, y^*, x) \in T^*$, then $y^* \in \text{conv}(\cup_{i=1}^m P_i)$.

iv) What difficulties can arise if the polyhedra P_i are unbounded, that is, the constraints $z \leqslant d$ are not present?

19. Given a linear inequality in 0-1 variables and the region

$$S = \left\{ x \in B^{|N_1|+|N_2|}: \sum_{j \in N_1} a_j x_j - \sum_{j \in N_2} a_j x_j \leqslant b \right\}$$

where $a_j > 0$ for $j \in N_1 \cup N_2$, write necessary and sufficient conditions for

i) $S = \emptyset$,

ii) $S = B^n$,

iii) $x_j = 0$ for all $x \in S$,

iv) $x_j = 1$ for all $x \in S$,

v) $x_i + x_j \leqslant 1$ for all $x \in S$,

vi) $x_i \leqslant x_j$ for all $x \in S$, and

vii) $x_i + x_j \geqslant 1$ for all $x \in S$.

20. If $x \in B^n$, what is implied by

i) $x_i + x_j \leqslant 1$ and $x_i \leqslant x_j$,

ii) $x_i + x_j \leqslant 1$ and $x_i + x_j \geqslant 1$, and

iii) $x_i \leqslant x_j$ and $x_j + x_k \leqslant 1$?

21. Use the results of Exercises 19 and 20 to solve the following problem without having recourse to enumeration:

$$\max 2x_1 - 2x_2 + 3x_3 + 1x_4 + 2x_5$$
$$7x_1 + 3x_2 + 9x_3 - 2x_4 + 2x_5 \leqslant 7$$
$$-6x_1 + 2x_2 - 3x_3 + 4x_4 + 9x_5 \leqslant -2$$
$$x \in B^5.$$

I.2

Linear Programming

1. INTRODUCTION

The general linear programming problem is

$$\text{(LP)} \qquad z_{LP} = \max\{cx: Ax \leqslant b, x \in R^n_+\},$$

where the data are rational and are given by the $m \times n$ matrix A, the $1 \times n$ matrix c, and the $m \times 1$ matrix b. This notation is different from that of Section I.1.1 but is preferable here because of its widespread use in linear programming. Recall that, as we observed in Section I.1.1, equality constraints can be represented by two inequality constraints.

Problem LP is well-defined in the sense that if it is feasible and does not have unbounded optimal value, then it has an optimal solution.

A good understanding of the theory and algorithms of linear programming is essential for understanding integer programming for several reasons that can be summed up by the statement that "one has to learn to walk before one can run". Integer programming is a much harder problem than linear programming, and neither the theory nor the computational aspects of integer programming are as developed as they are for linear programming. So, first of all, the theory of linear programming serves as a guide and motivating force for developing results for integer programming.

Computationally, linear programming algorithms are very often used as a subroutine in integer programming algorithms to obtain upper bounds on the value of the integer program. Let

$$\text{(IP)} \qquad z_{IP} = \max\{cx: Ax \leqslant b, x \in Z^n_+\}$$

and observe that $z_{LP} \geqslant z_{IP}$ since $Z^n_+ \subset R^n_+$. The upper bound z_{LP} sometimes can be used to prove optimality for IP; that is, if x^0 is a feasible solution to IP and $cx^0 = z_{LP}$, then x^0 is an optimal solution to IP.

A deeper connection between linear and integer programming is that corresponding to any integer programming problem there is a linear programming problem $\max\{cx: Ax \leqslant b, A^1x \leqslant b^1, x \in R^n_+\}$ that has the same answer as IP.

Our presentation of linear programming is by necessity very terse and is not intended as a substitute for a full treatment. The reader who has already studied linear programming is advised to scan this section to become familiar with our notation or, perhaps, to review an unfamiliar topic.

In the next section, we consider the duality theory of linear programming which, among other things, provides necessary and sufficient optimality conditions. In the following two sections, we present algorithms for solving linear programs.

The simplex algorithms are used to prove the main duality theorem and also to show that every feasible instance of LP that is not unbounded has an optimal solution. But, more importantly, they are the practical algorithms that are part of linear programming software systems and many integer programming software systems as well. The performance of simplex algorithms, observed over years of practical experience, shows that they are very robust and efficient. Typically the number of iterations required is a small multiple of m. Although there exist simplex algorithms that converge finitely, these are inefficient; and the ones used in practice can fail to converge. Moreover, there are examples which show that finitely convergent simplex algorithms may require an exponential number of iterations. But this bad behavior does not seem to occur in the solution of practical problems.

Section 4 deals with subgradient optimization. There are convergent subgradient algorithms, but, as described, they are not finite. However, on certain classes of linear programs that arise in solving integer programs, they tend to produce good solutions very quickly.

In Chapter I.6, we consider two other linear programming algorithms. These have been deferred to a later chapter because some of the motivation for considering them concerns the theoretical complexity of computations, which is studied in Chapter I.5.

2. DUALITY

Duality deals with pairs of linear programs and the relationships between their solutions. One problem is called the primal and the other the dual.

We state the *primal* problem as

$$\text{(P)} \qquad z_{\mathrm{LP}} = \max\{cx\colon Ax \leqslant b,\, x \in R^n_+\}.$$

Its *dual* is defined as the linear program

$$\text{(D)} \qquad w_{\mathrm{LP}} = \min\{ub\colon uA \geqslant c,\, u \in R^m_+\}.$$

It does not matter which problem is called the primal because:

Proposition 2.1. *The dual of the dual is the primal.*

Proof. To take the dual of the dual, we need to restate it as a maximization problem with equal-to-or-less-than constraints. Once this is done, the result follows easily. We leave the details to the reader. ■

Feasible solutions to the dual provide upper bounds on z_{LP} and feasible solutions to the primal yield lower bounds on w_{LP}. In particular:

Proposition 2.2 *(Weak Duality). If x^* is primal feasible and u^* is dual feasible, then $cx^* \leqslant z_{\mathrm{LP}} \leqslant w_{\mathrm{LP}} \leqslant u^*b$.*

Proof. $cx^* \leqslant u^*Ax^* \leqslant u^*b$, where the first inequality uses $u^*A \geqslant c$ and $x^* \geqslant 0$, and the second uses $Ax^* \leqslant b$ and $u^* \geqslant 0$. Hence $w_{\mathrm{LP}} \geqslant cx$ for all feasible solutions x to P, and $z_{\mathrm{LP}} \leqslant ub$ for all feasible solutions u to D, so that $w_{\mathrm{LP}} \geqslant z_{\mathrm{LP}}$. ■

Corollary 2.3. *If problem* P *has unbounded optimal value, then* D *is infeasible.*

Proof. By weak duality, $w_{LP} \geqslant \lambda$ for all $\lambda \in R^1$. Hence D has no feasible solution. ■

We now come to the fundamental result of linear programming duality, which says that if both problems are feasible their optimal values are equal. A constructive proof will be given in the next section.

Theorem 2.4 *(Strong Duality). If z_{LP} or w_{LP} is finite, then both P and D have finite optimal value and $z_{LP} = w_{LP}$.*

Corollary 2.5. *There are only four possibilities for a dual pair of problems P and D.*

i. *z_{LP} and w_{LP} are finite and equal.*
ii. *$z_{LP} = \infty$ and D is infeasible.*
iii. *$w_{LP} = -\infty$ and P is infeasible.*
iv. *Both P and D are infeasible.*

A problem pair with property iv is $\max\{x_1 + x_2 : x_1 - x_2 \leqslant -1, -x_1 + x_2 \leqslant -1, x \in R_+^2\}$ and its dual.

Another important property of primal–dual pairs is *complementary slackness*. Let $s = b - Ax \geqslant 0$ be the vector of slack variables of the primal and let $t = uA - c \geqslant 0$ be the vector of surplus variables of the dual.

Proposition 2.6. *If x^* is an optimal solution of P and u^* is an optimal solution of D, then $x_j^* t_j^* = 0$ for all j, and $u_i^* s_i^* = 0$ for all i.*

Proof. Using the definitions of s^* and t^*, we have

$$\begin{aligned} cx^* &= (u^*A - t^*)\,x^* = u^*Ax^* - t^*x^* \\ &= u^*(b - s^*) - t^*x^* = u^*b - u^*s^* - t^*x^*. \end{aligned}$$

By Theorem 2.4, $cx^* = u^*b$. Hence $u^*s^* + t^*x^* = 0$ with $u^*, s^*, t^*, x^* \geqslant 0$ so that the result follows. ■

Example 2.1. The dual of the linear program

$$\text{(P)} \qquad \begin{aligned} z_{LP} = \max\ & 7x_1 + 2x_2 \\ & -x_1 + 2x_2 \leqslant 4 \\ & 5x_1 + \ \ x_2 \leqslant 20 \\ & -2x_1 - 2x_2 \leqslant -7 \\ & x \in R_+^2 \end{aligned}$$

is

$$\text{(D)} \qquad \begin{aligned} w_{LP} = \min\ & 4u_1 + 20u_2 - 7u_3 \\ & -u_1 + \ 5u_2 - 2u_3 \geqslant 7 \\ & 2u_1 + \ \ \ u_2 - 2u_3 \geqslant 2 \\ & u \in R_+^3. \end{aligned}$$

It is easily checked that $x^* = (\frac{36}{11} \;\; \frac{40}{11})$ is feasible in P, and hence $z_{LP} \geq cx^* = 30\frac{2}{11}$. Similarly, $u^* = (\frac{3}{11} \;\; \frac{16}{11} \;\; 0)$ is feasible in D, and hence, by weak duality, $z_{LP} \leq u^*b = 30\frac{2}{11}$. The two points together yield a proof of optimality, namely, x^* is optimal for P and u^* is optimal for D.

Note also that the complementary slackness condition holds. The slack variables in P are $(s_1^*, s_2^*, s_3^*) = (0 \;\; 0 \;\; 6\frac{9}{11})$, and the surplus variables in D are $(t_1^*, t_2^*) = (0 \;\; 0)$. Hence $x_j^* t_j^* = 0$ for $j = 1, 2$ and $u_i^* s_i^* = 0$ for $i = 1, 2, 3$. ∎

It is important to be able to verify whether a system of linear inequalities is feasible or not. Duality provides a very useful characterization of infeasibility.

Theorem 2.7 *(Farkas' Lemma). Either $\{x \in R_+^n: Ax \leq b\} \neq \emptyset$ or (exclusively) there exists $v \in R_+^m$ such that $vA \geq 0$ and $vb < 0$.*

Proof. Consider the linear program $z_{LP} = \max\{0x: Ax \leq b, x \in R_+^n\}$ and its dual $w_{LP} = \min\{vb: vA \geq 0, v \in R_+^m\}$. As $v = 0$ is a feasible solution to the dual problem, only possibilities i and iii of Corollary 2.5 can occur.

i. $z_{LP} = w_{LP} = 0$. Hence $\{x \in R_+^n: Ax \leq b\} \neq \emptyset$ and $vb \geq 0$ for all $v \in R_+^m$ with $vA \geq 0$;
iii. $z_{LP} = w_{LP} = -\infty$. Hence $\{x \in R_+^n: Ax \leq b\} = \emptyset$ and there exists $v \in R_+^m$ with $vA \geq 0$ and $vb < 0$. ∎

There are many other versions of Farkas' Lemma. Some are presented in the following proposition.

Proposition 2.8. *(Variants of Farkas' Lemma)*

a. Either $\{x \in R_+^n: Ax = b\} \neq \emptyset$, or $\{v \in R^m: vA \geq 0, vb < 0\} \neq \emptyset$.
b. Either $\{x \in R^n: Ax \leq b\} \neq \emptyset$, or $\{v \in R_+^m: vA = 0, vb < 0\} \neq \emptyset$.
c. If $P = \{r \in R_+^n: Ar = 0\}$, either $P \setminus \{0\} \neq \emptyset$, or $\{u \in R^m: uA > 0\} \neq \emptyset$.

3. THE PRIMAL AND DUAL SIMPLEX ALGORITHMS

Here it is convenient to consider the primal linear program with equality constraints:

(LP) $$z_{LP} = \max\{cx: Ax = b, x \in R_+^n\}.$$

Its dual is

(DLP) $$w_{LP} = \min\{ub: uA \geq c, u \in R^m\}.$$

We suppose that $\text{rank}(A) = m \leq n$, so that all redundant equations have been removed from LP.

Bases and Basic Solutions

Let $A = (a_1, a_2, \ldots, a_n)$ where a_j is the jth column of A. Since $\text{rank}(A) = m$, there exists an $m \times m$ nonsingular submatrix $A_B = (a_{B_1}, \ldots, a_{B_m})$. Let $B = \{B_1, \ldots, B_m\}$ and let $N =$

$\{1, \ldots, n\} \setminus B$. Now permute the columns of A so that $A = (A_B, A_N)$. We can write $Ax = b$ as $A_B x_B + A_N x_N = b$, where $x = (x_B, x_N)$. Then a solution to $Ax = b$ is given by $x_B = A_B^{-1}b$ and $x_N = 0$.

Definition 3.1

a. The $m \times m$ nonsingular matrix A_B is called a *basis*.
b. The solution $x_B = A_B^{-1}b$, $x_N = 0$ is called a *basic solution* of $Ax = b$.
c. x_B is the vector of *basic variables* and x_N is the vector of *nonbasic variables*.
d. If $A_B^{-1}b \geqslant 0$, then (x_B, x_N) is called a *basic primal feasible solution* and A_B is called a *primal feasible basis*.

Now let $c = (c_B, c_N)$ be the corresponding partition of c, that is, $cx = c_B x_B + c_N x_N$, and let $u = c_B A_B^{-1} \in R^m$. This solution is complementary to $x = (x_B, x_N)$, since

$$uA - c = c_B A_B^{-1}(A_B, A_N) - (c_B, c_N) = (0, c_B A_B^{-1} A_N - c_N)$$

and $x_N = 0$. Observe that u is a feasible solution to the dual if and only if $c_B A_B^{-1} A_N - c_N \geqslant 0$. This motivates the next definition.

Definition 3.2. If $c_B A_B^{-1} A_N \geqslant c_N$, then A_B is called a *dual feasible basis*.

Note that a basis A_B defines the point $x = (x_B, x_N) = (A_B^{-1}b, 0) \in R^n$ and the point $u = c_B A_B^{-1} \in R^m$. A_B may be only primal feasible, only dual feasible, neither, or both. Bases that are both primal and dual feasible are of particular importance.

Proposition 3.1. *If A_B is primal and dual feasible, then $x = (x_B, x_N) = (A_B^{-1}b, 0)$ is an optimal solution to* LP *and $u = c_B A_B^{-1}$ is an optimal solution to* DLP.

Proof. $x = (A_B^{-1}b, 0)$ is feasible to LP with value $cx = c_B A_B^{-1} b$. $u = c_B A_B^{-1}$ is feasible in DLP and $ub = c_B A_B^{-1} b$. Hence the result follows from weak duality. ∎

Changing the Basis

We say that *two bases A_B and $A_{B'}$ are adjacent* if they differ in only one column, that is $|B \setminus B'| = |B' \setminus B| = 1$. If A_B and $A_{B'}$ are adjacent, the basic solutions they define are also said to be adjacent. The simplex algorithms to be presented in this section work by moving from one basis to another adjacent one.

Given the basis A_B, it is useful to rewrite LP in the form

$$\text{LP}(B) \qquad \begin{aligned} z_{\text{LP}} = c_B A_B^{-1} b + &\max(c_N - c_B A_B^{-1} A_N) x_N \\ x_B + A_B^{-1} A_N x_N &= A_B^{-1} b \\ x_B, x_N &\geqslant 0. \end{aligned}$$

It is simple to show that problems LP(B) and LP have the same set of feasible solutions and objective values.

We now define some additional notation that allows us to state things more concisely. Let $\bar{A}_N = A_B^{-1} A_N$, $\bar{b} = A_B^{-1} b$, and $\bar{c}_N = c_N - c_B A_B^{-1} A_N$ so that

$$\text{LP}(B) \qquad \begin{aligned} z_{\text{LP}} &= c_B\bar{b} + \max \bar{c}_N x_N \\ x_B &+ \bar{A}_N x_N = \bar{b} \\ x_B&, x_N \geqslant 0. \end{aligned}$$

Also, for $j \in N$, we let $\bar{a}_j = A_B^{-1}a_j$ and $\bar{c}_j = c_j - c_B\bar{a}_j$ so that

$$\text{LP}(B) \qquad \begin{aligned} z_{\text{LP}} &= c_B\bar{b} + \max \sum_{j\in N} \bar{c}_j x_j \\ x_B &+ \sum_{j\in N} \bar{a}_j x_j = \bar{b} \\ x_B &\geqslant 0, \quad x_j \geqslant 0 \quad \text{for } j \in N. \end{aligned}$$

Finally, we sometimes write the equations of LP(B) as

$$x_{B_i} + \sum_{j\in N} \bar{a}_{ij} x_j = \bar{b}_i \quad \text{for } i = 1, \ldots, m,$$

that is, $\bar{a}_j = (\bar{a}_{1j}, \ldots, \bar{a}_{mj})$ and $\bar{b} = (\bar{b}_1, \ldots, \bar{b}_m)$.

Let $\bar{c}_N = c_N - c_B\bar{A}_N$ be the *reduced price* vector for the nonbasic variables. Then, by Definition 3.2, dual feasibility of basis A_B is equivalent to $\bar{c}_N \leqslant 0$.

Now given the representation LP(B), we show how to move from one basic primal feasible solution to another in a systematic way.

Definition 3.3. A primal basic feasible solution $x_B = \bar{b}$, $x_N = 0$ is *degenerate* if $\bar{b}_i = 0$ for some i.

Proposition 3.2. *Suppose all primal basic feasible solutions are nondegenerate. If A_B is a primal feasible basis and a_r is any column of A_N, then matrix (A_B, a_r) contains, at most, one primal feasible basis other than A_B.*

Proof. We consider the system

$$\text{(3.1)} \qquad \begin{aligned} x_B + \bar{a}_r x_r &= \bar{b} \\ x_B \geqslant 0, \quad x_r &\geqslant 0, \end{aligned}$$

that is, all components of x_N except x_r equal zero.

Case 1. $\bar{a}_r \leqslant 0$. Suppose $x_r = \lambda > 0$. Then for all $\lambda > 0$ we obtain

$$x_B = \bar{b} - \bar{a}_r \lambda \geqslant \bar{b} > 0.$$

Thus for every feasible solution to (3.1) with $x_r > 0$, we have $x_B > 0$ so that A_B is the only primal feasible basis contained in (A_B, a_r).

Case 2. At least one component of $\bar{a}_r$ is positive. Let

$$\text{(3.2)} \qquad \lambda_r = \min\left\{\frac{\bar{b}_i}{\bar{a}_{ir}} : \bar{a}_{ir} > 0\right\} = \frac{\bar{b}_s}{\bar{a}_{sr}}.$$

Hence $\bar{b} - \bar{a}_r\lambda_r \geqslant 0$ and $\bar{b}_s - \bar{a}_{sr}\lambda_r = 0$. So we obtain an adjacent primal feasible basis $A_{B^{(r)}}$ by deleting B_s from B and replacing it with r, that is, $B^{(r)} = B \cup \{r\} \setminus \{B_s\}$. Note that the nondegeneracy assumption implies that $\bar{b}_i - \bar{a}_{ir}\lambda_r > 0$ for $i \neq s$ so that the minimum in (3.2) is unique. Consequently, any basis $A_{\hat{B}}$ with $\hat{B} = B \cup \{r\} \setminus \{k\}$ for $k \in B \setminus \{B_s\}$ is not primal feasible. ■

The new solution is calculated by:

1. Dividing

$$x_{B_s} + \bar{a}_{sr}x_r + \sum_{j\in N\setminus\{r\}} \bar{a}_{sj}x_j = \bar{b}_s$$

by $\bar{a}_{sr}$, which yields

$$\frac{1}{\bar{a}_{sr}}x_{B_s} + x_r + \sum_{j\in N\setminus\{r\}} \left(\frac{\bar{a}_{sj}}{\bar{a}_{sr}}\right)x_j = \frac{\bar{b}_s}{\bar{a}_{sr}}. \tag{3.3}$$

2. Eliminating x_r from the remaining equations by adding $-\bar{a}_{ir}$ multiplied by (3.3) to

$$x_{B_i} + \bar{a}_{ir}x_r + \sum_{j\in N\setminus\{r\}} \bar{a}_{ij}x_j = \bar{b}_i \quad \text{for } i \neq s$$

and eliminating x_r from the objective function.

This transformation is called a *pivot*. It corresponds precisely to a step in the well-known Gaussian elimination technique for solving linear equations. The coefficient $\bar{a}_{sr}$ is called the *pivot element*.

Corollary 3.3. *Suppose A_B is a primal feasible nondegenerate basis that is not dual feasible and $\bar{c}_r > 0$.*

a. *If $\bar{a}_r \leqslant 0$, then $z_{LP} = \infty$.*
b. *If at least one component of $\bar{a}_r$ is positive, then $A_{B^{(r)}}$, the unique primal feasible basis adjacent to A_B that contains a_r, is such that $c_{B^{(r)}}x_{B^{(r)}} > c_Bx_B$.*

Proof

a. $x_B = \bar{b} - \bar{a}_r\lambda$, $x_r = \lambda$, $x_j = 0$ otherwise is feasible for all $\lambda > 0$ and

$$c_Bx_B + c_rx_r = c_B\bar{b} + \bar{c}_r\lambda \to \infty \quad \text{as} \quad \lambda \to \infty.$$

b.

$$c_{B^{(r)}}x_{B^{(r)}} = c_B\bar{b} + \bar{c}_r\lambda_r > c_B\bar{b} = c_Bx_B,$$

where the inequality holds since λ_r defined by (3.2) is positive and $\bar{c}_r > 0$ by hypothesis. ■

Primal Simplex Algorithm

We are now ready to describe the main routine of the primal simplex method called *Phase 2*. It begins with a primal feasible basis and then checks for dual feasibility. If the basis is not dual feasible, either an adjacent primal feasible basis is found with (in the absence of degeneracy) a higher objective value or $z_{LP} = \infty$ is established.

Phase 2

Step 1 (Initialization): Start with a primal feasible basis A_B.

Step 2 (Optimality Test): If A_B is dual feasible (i.e., $\bar{c}_N < 0$), stop. $x_B = \bar{b}$, $x_N = 0$ is an optimal solution. Otherwise go to Step 3.

Step 3 (Pricing Routine): Choose an $r \in N$ with $\bar{c}_r > 0$.

a. *Unboundedness test.* If $\bar{a}_r \leqslant 0$, $z_{LP} = \infty$.
b. *Basis change.* Otherwise, find the unique adjacent primal feasible basis $A_{B^{(r)}}$ that contains a_r. Let $B \leftarrow B^{(r)}$ and return to Step 2.

Note that in Step 3, we can choose any $j \in N$ with $\bar{c}_j > 0$. A pricing rule commonly used is to choose $r = \arg(\max_{j \in N} \bar{c}_j)$, since it gives the largest increase in the objective function per unit increase of the variable that becomes basic. But this computation can be time consuming when n is large, so that various modifications of it are used in practice.

Theorem 3.4. *Under the assumption that all basic feasible solutions are nondegenerate, Phase 2 terminates in a finite number of steps either with an unbounded solution or with a basis that is primal and dual feasible.*

Proof. At each step the value of the basic feasible solution increases. Thus no basis can be repeated. Because there is only a finite number of bases, this procedure must terminate finitely. ■

When basic solutions are degenerate, and this happens often in practice, Proposition 3.2 and Corollary 3.3 are not true. Consequently, the finiteness argument given in the proof of Theorem 3.4 does not apply.

Note that when the basic feasible solution is degenerate, the arg(min) of (3.2) may not be unique. In this case, (A_B, a_r) contains more than one primal feasible basis adjacent to A_B, and in Step 3b of the algorithm an arbitrary choice is made. A complication arises when $\lambda_r = 0$ in (3.2) since each primal feasible basis in (A_B, a_r) defines the same solution, namely, $x_B = \bar{b}$ and $x_N = 0$. A sequence of such degenerate changes of basis can, although it rarely happens in practice, lead back to the original basis. This phenomenon is called *cycling*.

Two methods for eliminating the possibility of cycling are known. One involves a lexicographic rule for breaking ties in (3.2), and the other involves both the choice of r in Step 3 and a tie-breaking rule for (3.2). By eliminating cycling, these algorithms establish the finiteness of Phase 2 for any linear programming problem. Hence there are primal simplex methods for which Theorem 3.4 holds without a nondegeneracy assumption.

Example 3.1

$$\begin{aligned} z_{LP} = \max\ & 7x_1 + 2x_2 \\ & -x_1 + 2x_2 + x_3 \qquad\quad = 4 \\ & 5x_1 + \ \ x_2 \qquad + x_4 = 20 \end{aligned}$$

$$-2x_1 - 2x_2 \qquad\qquad + x_5 = -7$$

$$x \geqslant 0.$$

Step 1 (Initialization): The basis $A_B = (a_3, a_4, a_1)$ with

$$A_B^{-1} = \begin{pmatrix} 1 & 0 & -\frac{1}{2} \\ 0 & 1 & \frac{5}{2} \\ 0 & 0 & -\frac{1}{2} \end{pmatrix}$$

yields the primal feasible solution

$$x_B = (x_3, x_4, x_1) = A_B^{-1}b = \bar{b} = \begin{pmatrix} 7\frac{1}{2} & 2\frac{1}{2} & 3\frac{1}{2} \end{pmatrix}$$

and $x_N = (x_2, x_5) = (0 \quad 0)$,

Iteration 1

Step 2:

$$\bar{A}_N = (\bar{a}_2, \bar{a}_5) = A_B^{-1}A_N = \begin{pmatrix} 3 & -\frac{1}{2} \\ -4 & \frac{5}{2} \\ 1 & -\frac{1}{2} \end{pmatrix},$$

$$\bar{c}_N = c_N - c_B\bar{A}_N = (2 \quad 0) - (0 \quad 0 \quad 7)\bar{A}_N = \begin{pmatrix} -5 & \frac{7}{2} \end{pmatrix}.$$

Thus LP(B) can be stated as

$$z_{LP} = 24\frac{1}{2} + \max - 5x_2 + \frac{7}{2}x_5$$

$$3x_2 - \frac{1}{2}x_5 + x_3 \qquad = 7\frac{1}{2}$$

$$-4x_2 + 2\frac{1}{2}x_5 \qquad + x_4 \qquad = 2\frac{1}{2}$$

$$x_2 - \frac{1}{2}x_5 \qquad\qquad + x_1 = 3\frac{1}{2}$$

$$x \geqslant 0.$$

Step 3: The only choice for a new basic variable is x_5. By (3.2),

$$\lambda_5 = \min\left\{ -, \frac{2\frac{1}{2}}{2\frac{1}{2}}, - \right\} = 1.$$

Hence x_4 is the leaving variable.

$$A_B \leftarrow A_{B^{(r)}} = (a_3, a_5, a_1).$$

Iteration 2

Step 2: $A_B^{-1} = \begin{pmatrix} 1 & \frac{1}{5} & 0 \\ 0 & \frac{2}{5} & 1 \\ 0 & \frac{1}{5} & 0 \end{pmatrix}$, $\quad x_B = (x_3, x_5, x_1) = \bar{b} = (8 \quad 1 \quad 4),$

$$\bar{c}_N = (\bar{c}_2, \bar{c}_4) = \begin{pmatrix} \frac{3}{5} & -\frac{7}{5} \end{pmatrix}.$$

x_2 is the entering variable.

Step 3: $\bar{a}_2 = (\frac{11}{5} \quad -\frac{8}{5} \quad \frac{1}{5})$. By (3.2), $\lambda_2 = \min(\frac{8}{11/5}, -, \frac{4}{1/5}) = \frac{40}{11}$. Hence x_3 is the leaving variable. $A_B \leftarrow (a_2, a_5, a_1)$.

Iteration 3

Step 2: $A_B^{-1} = \begin{pmatrix} \frac{5}{11} & \frac{1}{11} & 0 \\ \frac{8}{11} & \frac{6}{11} & 1 \\ -\frac{1}{11} & \frac{2}{11} & 0 \end{pmatrix}$, $\quad x_B = (x_2, x_5, x_1) = \begin{pmatrix} \frac{40}{11} & \frac{75}{11} & \frac{36}{11} \end{pmatrix},$

$$\bar{c}_N = (\bar{c}_3, \bar{c}_4) = \begin{pmatrix} -\frac{3}{11} & -\frac{16}{11} \end{pmatrix} \leqslant 0.$$

Hence $x = (x_1, x_2, x_3, x_4, x_5) = (\frac{36}{11} \quad \frac{40}{11} \quad 0 \quad 0 \quad \frac{75}{11})$ is an optimal solution to LP, and $u = c_B A_B^{-1} = (\frac{3}{11} \quad \frac{16}{11} \quad 0)$ is an optimal solution to DLP.

We have shown that if LP has a basic primal feasible solution, it either has unbounded optimal value or it has an optimal basic solution. It remains to show that if it has a feasible solution, then it has a basic feasible solution. This is accomplished by Phase 1 of the simplex algorithm.

Phase 1. By changing signs in each row if necessary, write LP as $\max\{cx: Ax = b, x \in R_+^n\}$ with $b \geqslant 0$. Now introduce *artificial variables* x_i^a for $i = 1, \ldots, m$, and consider the linear program

$$(\text{LP}^a) \qquad z_a = \max\left\{-\sum_{i=1}^{m} x_i^a: Ax + Ix^a = b, (x, x^a) \in R_+^{n+m}\right\}.$$

1. LP^a is a feasible linear program for which a basic feasible solution $x^a = b, x = 0$ is available. Hence LP^a can be solved by the Phase 2 simplex method. Moreover $z_a \leqslant 0$ so that LP^a has an optimal solution.
2. i) A feasible solution (x, x^a) to LP^a yields a feasible solution x to LP if and only if $x^a = 0$. Thus if $z_a < 0$, LP^a has no feasible solution with $x^a = 0$ and hence LP is infeasible.
   ii) If $z_a = 0$, then any optimal solution to LP^a has $x^a = 0$ and hence yields a feasible solution to LP. In particular, if all the artificial variables are nonbasic in some basic optimal solution to LP^a, a basic feasible solution for LP has been found.

On the other hand, if one or more artificial variables are basic, it may be possible to remove them from the basis by degenerate basis changes. When this is not possible it can be shown that certain constraints in the original problem are redundant, and the equations with basic artificial variables can be dropped. Again this leads to a basic feasible solution to LP.

By combining Phases 1 and 2, we obtain a finite algorithm for solving any linear program. This establishes Theorem 2.4 and also Theorem 3.5:

Theorem 3.5

a. *If* LP *is feasible*, it *has a basic primal feasible solution*.
b. *If* LP *has a finite optimal value*, it *has an optimal basic feasible solution.*

Example 3.1 (continued). We will use Phase 1 to construct the initial basis (a_3, a_4, a_1) that we used previously. The Phase 1 problem is

$$
\begin{aligned}
z_a = \max \quad & -x_1^a - x_2^a - x_3^a \\
& -x_1 + 2x_2 + x_3 + x_1^a = 4 \\
& 5x_1 + x_2 + x_4 + x_2^a = 20 \\
& 2x_1 + 2x_2 - x_5 + x_3^a = 7 \\
& x, x^a \geqslant 0.
\end{aligned}
$$

Observe, however, that because x_3, x_4 are slack variables and b_1 and b_2 are nonnegative, the artificial variables x_1^a and x_2^a are unnecessary. Hence we can start with (x_3, x_4, x_3^a) as basic variables. Since $-x_3^a = -7 + 2x_1 + 2x_2 - x_5$, the Phase 1 problem is

$$
\begin{aligned}
z_a = \max \quad & -7 + 2x_1 + 2x_2 - x_5 \\
& -x_1 + 2x_2 + x_3 = 4 \\
& 5x_1 + x_2 + x_4 = 20 \\
& 2x_1 + 2x_2 - x_5 + x_3^a = 7 \\
& x \geqslant 0, x_3^a \geqslant 0.
\end{aligned}
$$

Using the simplex algorithm (Phase 2) we introduce x_1 into the basis, and x_3^a leaves. The resulting basis (a_3, a_4, a_1) is a feasible basis for the original problem.

Dual Simplex Algorithm

The primal simplex algorithm works by moving from one primal feasible basis to another. In contrast, the dual simplex algorithm works by moving from one dual feasible basis to another. This latter approach is useful when we know a basic dual feasible solution but not a primal one. This occurs, for example, when we have an optimal solution to a linear programming problem that becomes infeasible because additional constraints have been added.

Proposition 3.6. *Let A_B be a dual feasible basis with $\bar{b}_s < 0$.*

a. *If $\bar{a}_{sj} \geqslant 0$ for all $j \in N$, then* LP *is infeasible.*
b. *Otherwise there is an adjacent dual feasible basis $A_{B^{(r)}}$, where $B^{(r)} = B \cup \{r\} \setminus \{B_s\}$ and $r \in N$ satisfies $\bar{a}_{sr} < 0$ and*

$$r = \arg\min_{j \in N} \left\{ \frac{\overline{c}_j}{\overline{a}_{sj}} : \overline{a}_{sj} < 0 \right\}.$$

Proof.

a. $x_{B_s} + \Sigma_{j \in N} \overline{a}_{sj} x_j = \overline{b}_s < 0$. Hence if $\overline{a}_{sj} \geqslant 0$ for all $j \in N$, every solution to $Ax = b$ with $x_j \geqslant 0$ for all $j \in N$ has $x_{B_s} < 0$.

b. If x_r enters the basis and x_{B_s} leaves we have

$$z = c_B \overline{b} + \sum_{j \in N} \overline{c}_j x_j - \lambda(x_{B_s} + \sum_{j \in N} \overline{a}_{sj} x_j) + \lambda \overline{b}_s$$

$$= c_B \overline{b} + \lambda \overline{b}_s + \sum_{j \in N} (\overline{c}_j - \lambda \overline{a}_{sj}) x_j - \lambda x_{B_s},$$

where $\lambda = \dfrac{\overline{c}_r}{\overline{a}_{sr}} \geqslant 0$. The basis $A_{B^{(r)}}$ is dual feasible since $\lambda \geqslant 0$, $\overline{c}_j - \lambda \, \overline{a}_{sj} \leqslant \overline{c}_j$ for all j with $\overline{a}_{sj} \geqslant 0$, and $\overline{c}_j - \lambda \, \overline{a}_{sj} \leqslant 0$ for all j with $\overline{a}_{sj} < 0$ by the choice of r. ∎

Dual Simplex Algorithm (Phase 2)

Step 1 (Initialization): A dual feasible basis A_B.

Step 2 (Optimality Test): If A_B is primal feasible, that is, $\overline{b} = A_B^{-1} b \geqslant 0$, then $x_B = \overline{b}$ and $x_N = 0$ is an optimal solution. Otherwise go to Step 3.

Step 3 (Pricing Routine): Choose an s with $\overline{b}_s < 0$.

a. *Feasibility Test.* If $\overline{a}_{sj} \geqslant 0$ for all $j \in N$, LP is infeasible.

b. *Basis change.* Otherwise let

$$r = \arg\min_{j \in N} \left\{ \frac{\overline{c}_j}{\overline{a}_{sj}} : \overline{a}_{sj} < 0 \right\}$$

and $B^{(r)} = B \cup (r) \setminus (B_5)$. Return to Step 2 with $B \leftarrow B^{(r)}$.

In contrast to the primal algorithm, in the dual simplex algorithm the objective function is nonincreasing. The magnitude of the decrease at each step is $|\overline{c}_r \overline{b}_s / \overline{a}_{rs}|$. In the absence of dual degeneracy, $\overline{c}_r < 0$ and the decrease is strict. As with the primal algorithm, it is possible to give more specific rules that guarantee finiteness. Such an algorithm is presented in Section II.4.3. A Phase 1 may be required to find a starting dual feasible basic solution.

Example 3.1 (continued). We apply the dual simplex algorithm.

Step 1 (Initialization): Consider the basis $A_B = (a_3, a_2, a_5)$, which is dual feasible since $\overline{c}_N = (\overline{c}_1, \overline{c}_4) = (-3 \quad -2)$.

Iteration 1

Step 2: The basis is not primal feasible since $x_B = (x_3, x_2, x_5) = (-36 \quad 20 \quad 33)$.

Step 3: The only possible choice is $s = 1$. We have $\overline{a}_{11} = -11$, $\overline{a}_{14} = -2$, and $\min(\frac{3}{11}, \frac{2}{2}) = \frac{3}{11}$. Hence $x_{B_1} = x_3$ leaves the basis, x_1 enters the basis, and $A_B \leftarrow (a_1, a_2, a_5)$.

Iteration 2. We have seen earlier that A_B is primal and dual feasible and hence optimal.

The Simplex Algorithm with Simple Upper Bounds

It is desirable for computational purposes to distinguish between upper-bound constraints of the form $x_j \leqslant h_j$ and other more general constraints. Hence we consider the problem

$$\text{(ULP)} \qquad z_{\text{LP}} = \max\{cx: Ax = b, 0 \leqslant x_j \leqslant h_j \text{ for } j \in \{1, \ldots, n\}\}.$$

Whereas the primal simplex algorithm described earlier would treat ULP as a problem with $m + n$ constraints, the *simplex algorithm with upper bounds* treats it as a problem with m constraints.

Now the columns of A are permuted so that $A = (A_B, A_{N_1}, A_{N_2})$, where A_B is a basis matrix as before, but the index set of the nonbasic variables N is partitioned into two sets N_1 and N_2. N_1 is the index set of variables at their lower bound ($x_j = 0$), and N_2 is the index set of variables at their upper bound ($x_j = h_j$).

Now we need to modify Definition 3.1.

Definition 3.4

a. The $m \times m$ nonsingular matrix A_B is called a *basis*.
b. For each partition N_1, N_2 of N, we associate the *basic solution* $x_B = A_B^{-1}(b - A_{N_2}h_{N_2}) = \bar{b} - \bar{A}_{N_2}h_{N_2}$, $x_{N_1} = 0$, $x_{N_2} = h_{N_2}$.
c. If $0 \leqslant \bar{b} - \bar{A}_{N_2}h_{N_2} \leqslant h_B$, then (x_B, x_{N_1}, x_{N_2}) is a *basic primal feasible solution*, and (B, N_1, N_2) indexes a primal feasible basis.

Now consider the dual of ULP,

$$\begin{aligned} \min\ & ub + vh \\ & uA + v \geqslant c \\ & v \geqslant 0, \end{aligned}$$

and let $v = (v_B, v_{N_1}, v_{N_2})$ and $c = (c_B, c_{N_1}, c_{N_2})$. The dual basic solution complementary to (x_B, x_{N_1}, x_{N_2}) is $(u, v_B, v_{N_1}, v_{N_2}) = (c_B A_B^{-1}, 0, 0, c_{N_2} - c_B\bar{A}_{N_2})$. Observe that (u, v) is a feasible solution to the dual if and only if $\bar{c}_{N_1} = c_{N_1} - c_B\bar{A}_{N_1} \leqslant 0$ and $\bar{c}_{N_2} = c_{N_2} - c_B\bar{A}_{N_2} \geqslant 0$.

Proposition 3.7. *If (A_B, A_{N_1}, A_{N_2}) is primal and dual feasible, then $x = (x_B, x_{N_1}, x_{N_2}) = (\bar{b} - \bar{A}_{N_2}h_{N_2}, 0, h_{N_2})$ is an optimal solution to ULP and $(u, v_{B_1}, v_{N_1}, v_{N_2}) = (c_B A_B^{-1}, 0, 0, c_{N_2} - c_B\bar{A}_{N_2})$ is an optimal solution to its dual.*

The modifications to the simplex algorithms are straightforward. Bases A_B and $A_{B'}$, are adjacent if (i) $|B \setminus B'| = |B' \setminus B| = 1$ or (ii) $B = B'$, and in both cases $|N_1' \setminus N_1| + |N_2' \setminus N_2| = 1$. In the latter case, one nonbasic variable changes from its lower to its upper bound, or vice versa. It is then easy to write out the rules for the choice of entering and leaving variable, leading to primal and dual simplex algorithms for ULP. Note that these algorithms choose the same pivots as the standard simplex algorithms, so the advantage lies in handling a basis that is $m \times m$ rather than $(m + n) \times (m + n)$.

Addition of Constraints or Variables

After solving LP to optimality, it is common that one or more new constraints or columns have to be added. In Part II, we will discuss cutting-place algorithms that add a constraint cutting off the optimal solution of LP; we will also discuss problems having such a large number of variables that we do not wish to introduce them all a priori.

If LP has been solved by a simplex algorithm, there is a straightforward way to use the current optimal basis A_B to solve the new problem. Suppose an inequality $\Sigma_{j=1}^n d_j x_j \leq d_0$ is added that is violated by the optimal solution $(x_B, x_N) = (A_B^{-1}b, 0)$. Now if x_{n+1} is the slack variable of the new constraint, then $B' = B \cup \{n+1\}$ indexes a new basis, and we obtain LP(B'):

$$\begin{aligned} z'_{LP} = \max\ & c_B\bar{b} + \bar{c}_N x_N \\ & x_B + \bar{A}_N x_N = \bar{b} \\ & x_{n+1} + (d_N - d_B\bar{A}_N)x_N = d_0 - d_B\bar{b} \\ & x_B, x_N, x_{n+1} \geq 0 \end{aligned}$$

since

$$dx + x_{n+1} = d_B x_B + d_N x_N + x_{n+1} = d_B\bar{b} + (d_N - d_B\bar{A}_N)x_N + x_{n+1}.$$

We see immediately that this basis is dual feasible and that it is primal feasible in all but the last row, that is, $d_B\bar{b} > d_0$. It is therefore desirable to reoptimize using the dual simplex algorithm. Since the current solution is "nearly" primal feasible, it is likely that only a few iterations will be required.

The procedure to be followed in adding new columns is dual to that described above. Given a new variable x_{n+1} with column $\binom{c_{n+1}}{a_{n+1}}$, we calculate its reduced price $\bar{c}_{n+1} = c_{n+1} - c_B A_B^{-1} a_{n+1}$ to check if the basis A_B remains optimal. If $\bar{c}_{n+1} \leq 0$, A_B is still optimal and the solution is unchanged. If $\bar{c}_{n+1} > 0$, we can use the primal simplex algorithm as A_B remains primal feasible.

Example 3.1 (continued). We add the upper-bound constraint $x_1 \leq 3$, cutting off the optimal solution $x = (\frac{36}{11}\ \ \frac{40}{11}\ \ 0\ \ 0\ \ \frac{75}{11})$. Let $x_1 + x_6 = 3$, so that x_6 is the new basic variable. Starting from the optimal basis $A_B = (a_2, a_5, a_1)$, we have $d_B = (0\ \ 0\ \ 1)$, $d_N = (0\ \ 0)$, $d_0 = 3$, and $A_{B'} = (a_2, a_5, a_1, a_6)$.

Iteration 1

Step 2: $x_{B'} = (\frac{40}{11}\ \ \frac{75}{11}\ \ \frac{36}{11}\ \ -\frac{3}{11})$.
Step 3: x_6 leaves the basis

$$\bar{c}_N = \left(-\frac{3}{11}\ \ -\frac{16}{11}\right),\quad d_N - d_B\bar{A}_N = \left(\frac{1}{11}\ \ -\frac{2}{11}\right)$$

$$\min\left\{-, \frac{16}{11}\Big/\frac{2}{11}\right\} = 8.$$

Hence x_4 enters the basis. $A_{B'} \leftarrow (a_2, a_5, a_1, a_4)$.

Iteration 2

Step 2: $x_{B'} = (\frac{7}{2}\ \ 6\ \ 3\ \ \frac{3}{2}) \geq 0$. Hence $x = (3\ \ \frac{7}{2}\ \ 0\ \ \frac{3}{2}\ \ 6\ \ 0)$ is an optimal solution to the revised problem.

Noting that the added constraint is an upper-bound constraint means that we can also reoptimize without increasing the size of the basis by using the dual simplex algorithm with upper bounds. In this case we have:

Iteration 1

$$A_B = (a_2, a_5, a_1), \quad A_{N_1} = (a_3, a_4), \quad N_2 = \emptyset$$

$$c_{N_1} = \left(-\frac{3}{11} \quad -\frac{16}{11}\right) \leqslant 0,$$

so the basis is dual feasible.

$$x_B = \left(\frac{40}{11} \quad \frac{75}{11} \quad \frac{36}{11}\right).$$

Because $x_{B_1} = x_1 > h_1$, the basis is not primal feasible.

The dual simplex algorithm then removes x_1 from the basis at its upper bound and calculates (as above) that x_4 enters the basis.

Iteration 2

$$A_B = (a_2, a_5, a_4), \quad A_{N_1} = (a_3), \quad A_{N_2} = (a_1).$$

$$\bar{c}_{N_1} = (-1) \leqslant 0, \quad \bar{c}_{N_2} = (8) \geqslant 0,$$

so the basis remains dual feasible.

$x_B = (\frac{7}{2} \quad 6 \quad \frac{3}{2})$. Because $0 \leqslant x \leqslant h_B$, the basis is primal feasible and hence optimal.

4. SUBGRADIENT OPTIMIZATION

Here we consider an algorithm for solving linear programs whose roots are in nonlinear, nondifferentiable optimization. Consider the linear program

$$\zeta = \min \sum_{i=1}^{l} \lambda_i d_i$$

$$\sum_{i=1}^{l} \lambda_i g_{ij} = c_j \quad \text{for } j = 1, \ldots, n$$

$$0 \leqslant \lambda_i \leqslant h_i \quad \text{for } i = 1, \ldots, l.$$

By duality it can be shown (see Section II.3.6) that this problem can be restated as

$$\zeta = \max_{x \in R^n} \min_{0 \leqslant \lambda \leqslant h} \left[\sum_{j=1}^{n} c_j x_j + \sum_{i=1}^{l} \lambda_i \left(d_i - \sum_{j=1}^{n} g_{ij} x_j\right)\right].$$

Now to solve the inner optimization problem for fixed x, we can set $\lambda_i = 0$ if $d_i - \sum_{j=1}^{n} g_{ij} x_j > 0$, and $\lambda_i = h_i$ otherwise. Thus there are a finite number of candidate solutions $\lambda^k \in R_+^l$, $k \in K$, where $\lambda_i^k \in \{0, h_i\}$. So we can rewrite the problem as

$$\zeta = \max_{x \in R^n} \min_{k \in K} [(c - \lambda^k G)x + \lambda^k d],$$

or more generally as

$$\zeta = \max_{x \in R^n} f(x), \tag{4.1}$$

where

$$f(x) = \min_{i \in I} (a^i x - b_i) \quad \text{and} \quad I = \{1, \ldots, m\} \quad \text{is a finite set.} \tag{4.2}$$

In other words a general linear program can be transformed to the nonlinear optimization problem (4.1), where typically m is much larger than n. In this section, we present an algorithm for problem (4.1).

Figure 4.1 illustrates f given by (4.2) for $n = 1$. The heavy lines give $f(x)$, and point B is the optimum solution x^* with value $\zeta = f(x^*)$.

We now develop an important property of the function f.

Definition 4.1. A function $g: R^n \to R^1$ is *concave* if

$$g(\alpha x^1 + (1 - \alpha)x^2) \geq \alpha g(x^1) + (1 - \alpha)\, g(x^2) \quad \text{for all } x^1, x^2 \in R^n \text{ and all } 0 \leq \alpha \leq 1.$$

Note that the definition simply states that the function is underestimated by linear interpolation (see Figure 4.2).

This suggests the following proposition.

Proposition 4.1. *Let* $f(x) = \min_{i=1,\ldots,m} (a^i x - b_i)$. *Then* $f(x)$ *is concave.*

Proof. Let $x^3 = \alpha x^1 + (1 - \alpha)x^2$ and $f(x^j) = a^{i(j)}x^j - b_{i(j)}$ for $j = 1, 2, 3$. Then

$$\begin{aligned} f(x^3) = a^{i(3)}x^3 - b_{i(3)} &= a^{i(3)}(\alpha x^1 + (1 - \alpha)x^2) - b_{i(3)} \\ &= \alpha(a^{i(3)}x^1 - b_{i(3)}) + (1 - \alpha)(a^{i(3)}x^2 - b_{i(3)}) \\ &\geq \alpha(a^{i(1)}x^1 - b_{i(1)}) + (1 - \alpha)(a^{i(2)}x^2 - b_{i(2)}) \\ &= \alpha f(x^1) + (1 - \alpha) f(x^2). \end{aligned}$$

■

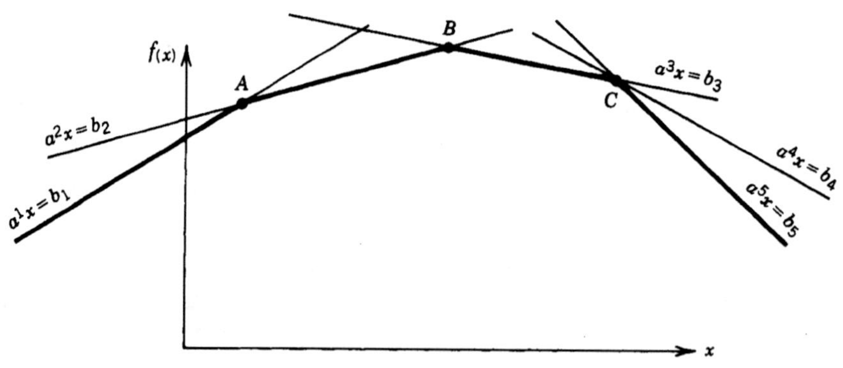

Figure 4.1

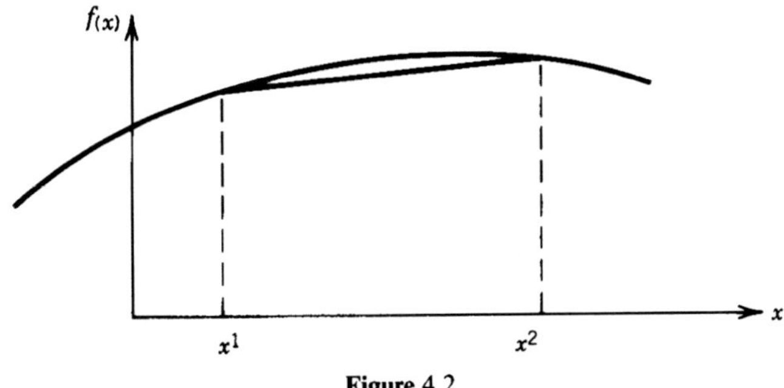

Figure 4.2

An alternative characterization of concave functions is given by the following proposition.

Proposition 4.2. *A function g: $R^n \to R^1$ is concave if and only if for any $x^* \in R^n$ there exists an $s \in R^n$ such that $g(x^*) + s(x - x^*) \geqslant g(x)$ for all $x \in R^n$.*

The characterization is illustrated in Figure 4.3. Note that s is the slope of the hyperplane that supports the set $\{(x, z) \in R^{n+1}: z \leqslant g(x)\}$ at $(x, z) = (x^*, g(x^*))$.

Comparing Figures 4.1 and 4.3, we see that in Figure 4.1 there is not a unique supporting hyperplane at the points A, B, and C, while for the smooth function g in Figure 4.3, the supporting hyperplane is unique at each point.

Figure 4.4 illustrates Proposition 4.2 for $x \in R^2$. Contours of $\{x: g(x) = c\}$ are shown for different values of c along with the supporting hyperplane given by $s(x - x^*) = 0$. By Proposition 4.2, if x satisfies $s(x - x^*) \leqslant 0$, then $g(x) \leqslant g(x^*)$. In other words, if $g(x) > g(x^*)$, then $s(x - x^*) > 0$. Thus if we are at the point x^* and want to increase $g(x)$, we should move to a point x' with $s(x' - x^*) > 0$. One possibility is to move in a direction normal to the hyperplane $s(x - x^*) = 0$. This direction is given by the vector s, which is, when g is differentiable at x^*, the *gradient vector* $\nabla g(x^*) = (\partial g(x^*)/\partial x_1, \ldots, \partial g(x^*)/\partial x_n)$ at $x = x^*$. It is well known that the gradient vector is the local direction of maximum increase of $g(x)$, and $\nabla g(x^*) = 0$ implies that x^* solves $\max\{g(x): x \in R^n\}$.

The classical steepest ascent method for maximizing $g(x)$ is given by the sequence of iterations

$$x^{t+1} = x^t + \theta^t \nabla g(x^t), \quad t = 1, 2, \ldots.$$

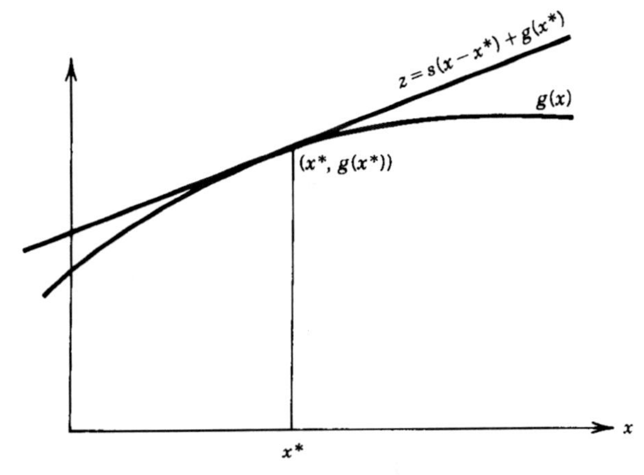

Figure 4.3

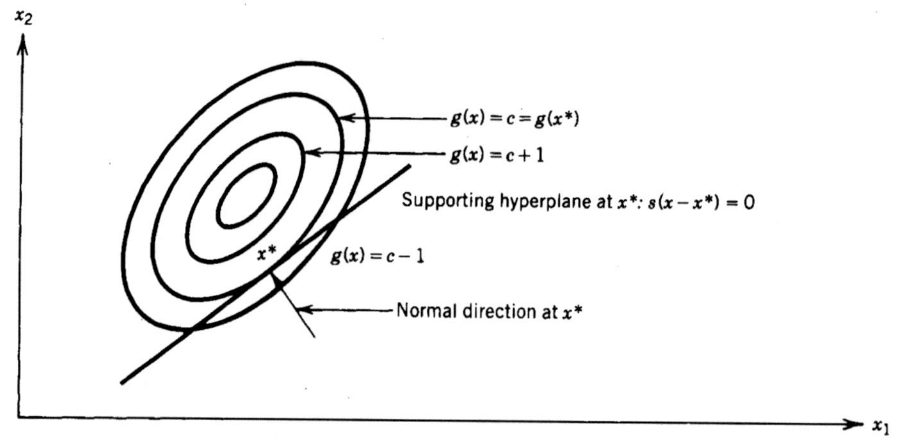

Figure 4.4

With appropriate assumptions on the sequence of step sizes $\{\theta^t\}$, the iterates $\{x^t\}$ converge to a maximizing point.

The potential problems that arise in applying this idea to a nondifferentiable concave function are illustrated in Example 4.1.

Example 4.1

$$f(x_1, x_2) = \min\{-x_1, x_1 - 2x_2, x_1 + 2x_2\}.$$

The contours $f(x) = c$ for $c = 0, -1$, and -2 are shown in Figure 4.5.

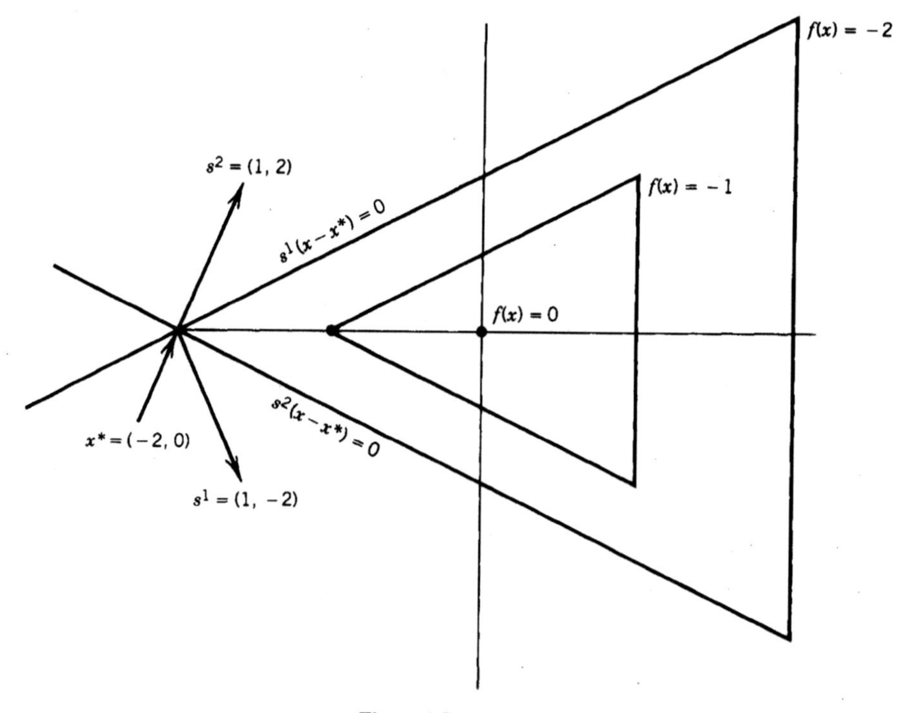

Figure 4.5

In addition, at the point $x^* = (-2 \quad 0)$ we show the supporting hyperplanes $s^i(x - x^*) = 0$, for $i = 1, 2$ where $s^1 = (1 \quad -2)$ and $s^2 = (1 \quad 2)$.

Now consider what happens when we move from x^* in the direction s^1. We have

$$\begin{aligned} f(x^* + \theta s^1) = f(-2 + \theta, 0 - 2\theta) &= \min\{2 - \theta, -2 + 5\theta, -2 - 3\theta\} \\ &= -2 - 3\theta \quad \text{for all } \theta \geqslant 0. \end{aligned}$$

Hence $f(x^* + \theta s^1) < f(x^*)$ for all $\theta > 0$. Similar behavior is observed for s^2.

The example illustrates the nonuniqueness of the supporting hyperplanes and also shows that a direction normal to a supporting hyperplane may not be a direction of increase.

There is, however, an alternative point of view, which provides the intuitive justification for moving in a direction normal to any supporting hyperplane at x^*. As we have already noted, if $s(x - x^*) = 0$ is *any* supporting hyperplane at x^*, then any point with a larger objective value than x^* is contained in the half-space $s(x - x^*) > 0$. Now it is a simple geometric exercise to show that if $\hat{x}$ is an optimal solution, a small move in the direction s gives a point that is closer to $\hat{x}$. In particular, there exists $\hat{\theta}$ such that for any $0 < \theta < \hat{\theta}$,

$$\|\hat{x} - (x^* + \theta s)\| < \|\hat{x} - x^*\|$$

(see Figure 4.6.). The notation $\|u\|$, $u \in R^n$, represents the euclidean distance from 0 to u, that is, $\sqrt{u^T u}$.

We now formalize the discussion given above.

Definition 4.2. If $g\colon R^n \to R^1$ is concave, $s \in R^n$ is a *subgradient* of g at x^* if $s(x - x^*) \geqslant g(x) - g(x^*)$ for all $x \in R^n$.

Definition 4.3. The set $\partial g(x) = \{s \in R^n\colon s \text{ is a subgradient of } g \text{ at } x\}$ is called the *subdifferential* of g at x.

Note that by Proposition 4.2, $\partial g(x) \neq \varnothing$.

Proposition 4.3. *If g is concave on R^n, x^* is an optimal solution of $\max\{g(x)\colon x \in R^n\}$ if and only if $0 \in \partial g(x^*)$.*

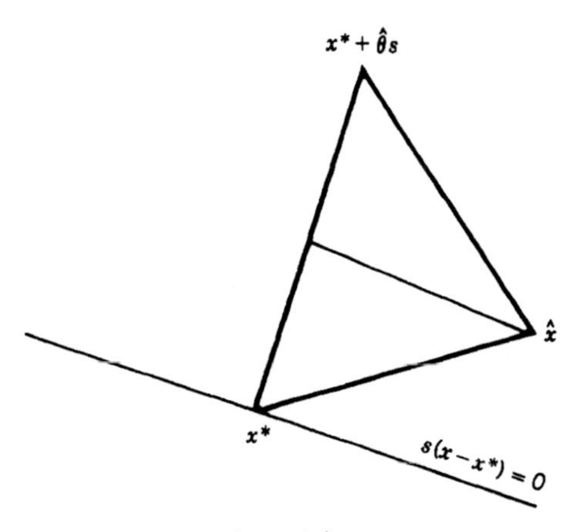

Figure 4.6

Proof. $0 \in \partial g(x^*)$ if and only if $0(x - x^*) \geqslant g(x) - g(x^*)$ for all $x \in R^n$ if and only if $g(x) \leqslant g(x^*)$ for all $x \in R^n$. ■

Now we characterize the subdifferential of $f(x)$ given by (4.2).

Proposition 4.4. *Let* $f(x) = \min_{i=1,\ldots,m}(a^i x - b_i)$ *and let* $I(x^*) = \{i: f(x^*) = a^i x^* - b_i\}$.

1. a^i *is a subgradient of* f *at* x^* *for all* $i \in I(x^*)$.
2. $\partial f(x^*) = \{s \in R^n: s = \Sigma_{i \in I(x^*)} \lambda_i a^i, \Sigma_{i \in I(x^*)} \lambda_i = 1, \lambda_i \geqslant 0$ *for* $i \in I(x^*)\}$.

Proof.

1. If $i \in I(x^*)$, then $a^i(x - x^*) = (a^i x - b_i) - (a^i x^* - b^i) \geqslant f(x) - f(x^*)$ for all $x \in R^n$, so that $a^i \in \partial f(x^*)$.
2. A proof is obtained by using statement 1 of Proposition 4.4 along with the Farkas lemma. ■

The following algorithm can use any subgradient at each step, but for computational purposes one of the extreme directions a^i will be chosen.

The Subgradient Algorithm for (4.1)

Step 1 (Initialization): Choose a starting point x^1 and let $t = 1$.

Step 2: Given x^t, choose any subgradient $s^t \in \partial f(x^t)$. If $s^t = 0$, then x^t is an optimal solution. Otherwise go to Step 3.

Step 3: Let $x^{t+1} = x^t + \theta_t s^t$ for some $\theta_t > 0$. (Procedures for selecting θ_t are given below.) Let $t \leftarrow t + 1$ and return to Step 2.

Two schemes for selecting $\{\theta_t\}$ are the following:

i. A divergent series: $\Sigma_{t=1}^{\infty} \theta_t \to \infty$, $\theta_t \to 0$ as $t \to \infty$.
ii. A geometric series: $\theta_t = \theta_0 \rho^t$, or $\theta_t = [\bar{f} - f(x^t)]\rho^t/\|s^t\|^2$ where $0 < \rho < 1$ and $\bar{f}$ is a target, or upper bound on the optimal value ζ of (4.1).

Series i is satisfactory theoretically, since it converges to an optimal point. But in practice the convergence is much too slow. Series ii, which is recommended in practice, is less satisfactory theoretically. The convergence is "geometric", but the limit point is only an optimal point if the initial choices of (θ_0, ρ) or $(\bar{f}, \rho)$ are sufficiently large. In practice, appropriate values can typically be found after a little testing, and step sizes closely related to a geometric series of type ii will be used in our applications of the subgradient algorithm in Part II.

Ideally the subgradient algorithm can be stopped when, on some iteration t, we find $s^t = 0 \in \partial f(x^t)$. However, in practice this rarely happens, since the algorithm just chooses one subgradient s^t and has no way of showing $0 \in \partial f(x^t)$ as a convex combination of subgradients. Hence the typical stopping rule is either to stop after a fixed number of iterations or to stop if the function has not increased by at least a certain amount within a given number of iterations.

Example 4.2. Consider $\max\{f(x): x \in R^2\}$, where

$$f(x) = \min\{f_i(x): i = 1, \ldots, 5\}$$

and

$$\begin{aligned} f_1(x) &= \quad x_1 - 2x_2 + 4 \\ f_2(x) &= -5x_1 - x_2 + 20 \\ f_3(x) &= \quad 2x_1 + 2x_2 - 7 \\ f_4(x) &= \quad x_1 \\ f_5(x) &= \qquad\qquad x_2. \end{aligned}$$

We apply the subgradient algorithm with $\theta_t = (0.9)^t$ and initial point $x^1 = (0 \quad 0)$. The results of 25 iterations are shown in Table 4.1, in which the last column, $i(t)$, gives the index of the function that defines the subgradient. The best solution of value 2.30 is found at iteration 13. The optimal solution is $(x_1 \quad x_2) = (\frac{52}{17} \quad \frac{40}{17})$ of value $\frac{40}{17} = 2.353$.

Table 4.1.

t	x_1^t	x_2^t	$f(x^t)$	ρ^t	$i(t)$
1	0.000	0.000	−7.000	0.900	3
2	1.800	1.800	0.200	0.810	3
3	3.420	3.420	−0.520	0.729	2
4	−0.225	2.691	−2.068	0.656	3
5	1.087	4.003	−2.919	0.590	1
6	1.678	2.822	0.033	0.531	1
7	2.209	1.759	0.937	0.478	3
8	3.166	2.716	1.455	0.430	2
9	1.013	2.285	−0.402	0.387	3
10	1.788	3.060	−0.332	0.349	1
11	2.137	2.363	1.411	0.314	1
12	2.451	1.735	1.372	0.282	3
13	3.016	2.300	2.300	0.254	5
14	3.016	2.554	1.907	0.229	1
15	3.244	2.097	1.681	0.206	2
16	2.215	1.891	1.212	0.185	3
17	2.585	2.262	2.062	0.167	1
18	2.752	1.928	1.928	0.150	5
19	2.752	2.078	2.078	0.135	5
20	2.752	2.213	2.213	0.122	5
21	2.752	2.335	2.083	0.109	1
22	2.862	2.116	2.116	0.098	5
23	2.862	2.214	2.214	0.089	5
24	2.862	2.303	2.256	0.080	1
25	2.941	2.144	2.144	0.072	5

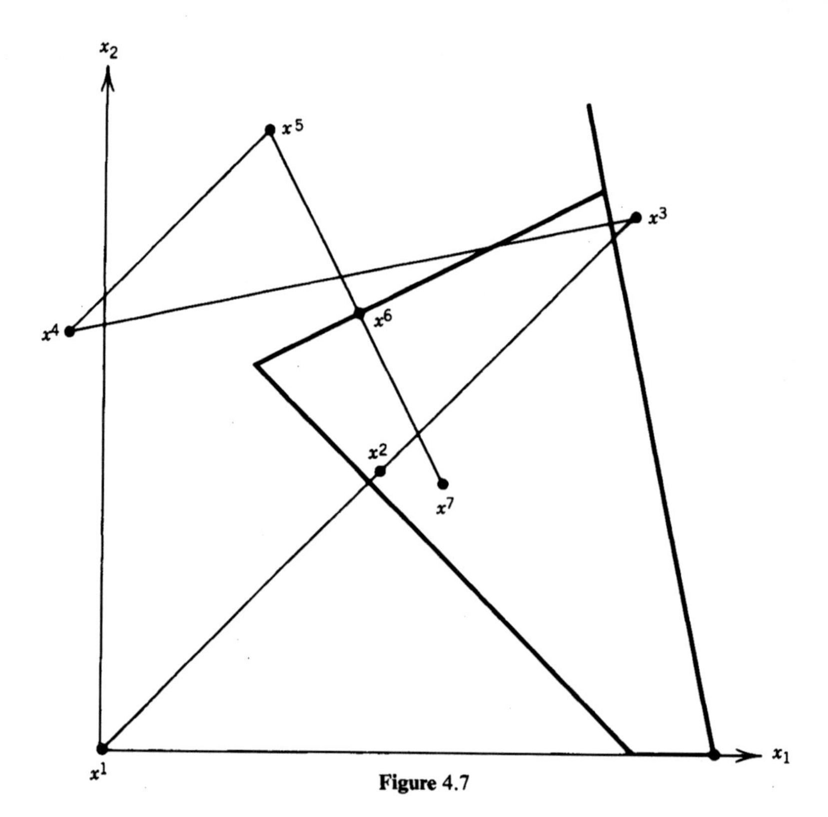

Figure 4.7

We can also view the problem as one of finding (x_1, x_2) such that the smallest slack variable y_i of the constraints

$$\begin{array}{rrrrrrrl}
-x_1 & +2x_2 & +y_1 & & & & & = 4 \\
5x_1 & +x_2 & & +y_2 & & & & = 20 \\
-2x_1 & -2x_2 & & & +y_3 & & & = -7 \\
-x_1 & & & & & +y_4 & & = 0 \\
& -x_2 & & & & & +y_5 & = 0
\end{array}$$

is as large as possible (see Figure 4.7). With this geometry, each subgradient step is in the direction of the normal to the constraint whose slack variable is smallest.

Because the magnitudes of the constraint coefficients are different, the five subgradients have different magnitudes which can substantially bias the progress of the algorithm. This suggests the use of normalized subgradients $s/\|s\|$ in the subgradient algorithm. For Example 4.2, this gives the iterations shown in Table 4.2. Note that more rapid convergence is achieved using normalized subgradients.

Finally suppose that $x \in R^n$ must satisfy some linear constraints, say $x \in C$. Thus we have the problem

$$\text{(4.3)} \qquad \eta = \max\{f(x): x \in C\}, \quad \text{where } f(x) = \min_{i=1,\ldots,m} (a^i x - b_i).$$

The subgradient algorithm for (4.3) is as before, except that Step 3 is modified to maintain feasibility.

Table 4.2.

t	x_1^t	x_2^t	$f(x^t)$	ρ^t	$i(t)$
1	0.000	0.000	−7.000	0.900	3
2	0.636	0.636	−4.454	0.810	3
3	1.209	1.209	−2.163	0.729	3
4	1.725	1.725	−0.101	0.656	3
5	2.189	2.189	1.754	0.590	3
6	2.606	2.606	1.394	0.531	1
7	2.844	2.131	2.131	0.478	5
8	2.844	2.609	1.626	0.430	1
9	3.036	2.224	2.224	0.387	5
10	3.036	2.611	1.813	0.349	1
11	3.192	2.300	1.739	0.314	2
12	2.885	2.238	2.238	0.282	5
13	2.885	2.520	1.844	0.254	1
14	2.998	2.293	2.293	0.229	5
15	2.998	2.522	1.954	0.206	1
16	3.090	2.338	2.211	0.185	2
17	2.909	2.301	2.301	0.167	5
18	2.909	2.468	1.972	0.150	1
19	2.976	2.334	2.308	0.135	1
20	3.036	2.213	2.213	0.122	5
21	3.036	2.335	2.335	0.109	5
22	3.036	2.444	2.148	0.098	1
23	3.080	2.356	2.243	0.089	2
24	2.993	2.339	2.316	0.080	1
25	3.029	2.267	2.267	0.072	5

Step 3′: Let $y^{t+1} = x^t + \theta_t s^t$ for some $\theta_t > 0$ and let $x^{t+1} = \arg\min_{x \in C} \|x - y^{t+1}\|$.

In other words, x^{t+1} is the projection of y^{t+1} onto the feasible region C. A typical application is to have $C = R_+^n$, in which case $x_j^{t+1} = \max(x_j^t + \theta_t s_j^t, 0)$ for $j = 1, \ldots, n$

5. NOTES

Sections I.2.1–I.2.3.

Chvátal (1983) gave a modern and comprehensive treatment of linear programming, with the exception of the significant post-1983 developments covered in Sections I.6.2–I.6.4. Some earlier books are Charnes and Cooper (1961), Dantzig (1963), Gass (1975), Hadley (1962), and Murty (1976).

Section I.2.4

The use of subgradient directions in the solution of large-scale linear programs that arise from combinatorial optimization problems was instigated by Held and Karp (1970, 1971) in a study of the traveling salesman problem. Held et al. (1974) investigated the behavior of a subgradient algorithm in a variety of combinatorial problems. A theoretical analysis of the convergence of subgradient algorithms is given by Goffin (1977). Subgradients and subgradient algorithms are also discussed by Grinold (1970, 1972), Camerini et al. (1975), Shapiro (1979a, b), and Sandi (1979).

I.3

Graphs and Networks

1. INTRODUCTION

In this section we give the terminology and some elementary results of graph theory. For our purposes the language of graphs is nearly as important as the results, which are elementary and given without proof.

In the remaining sections, we define some classical optimization problems on graphs and present algorithms to solve them. All of these problems are linear programming problems and, excluding the minimum-weight spanning tree problem, are in the class of linear programming problems known as *network flow problems*. Their structure makes it possible to solve them by special-purpose algorithms that are more efficient than the simplex method.

These problems are of interest to us because they frequently arise as subproblems in the solution of integer programs. The algorithms presented in the following sections are examples of classes of algorithms that are used to solve some of the problems considered in Parts II and III. We will introduce the ideas of recursive, greedy, augmenting, primal–dual, and specialized simplex algorithms. So this chapter also has the pedagogical objective of introducing different algorithmic approaches in a simple setting. To explain the basic ideas succinctly, we have deliberately chosen to present simple, rather than efficient, versions of the algorithms. Thus, in this chapter, the reader should not necessarily expect the algorithmic details that yield efficient implementations.

A *graph* $G = (V, E)$ consists of a finite, nonempty set $V = \{1, 2, \ldots, m\}$ and a set $E = \{e_1, e_2, \ldots, e_n\}$ whose elements are subsets of V of size 2, that is, $e_k = (i, j)$, where $i, j \in V$. The elements of V are called *nodes*, and the elements of E are called *edges*. Thus graphs are a mechanism for specifying certain pairs of a set.

Graphs can be represented pictorially in R^2 by points and lines. The points or nodes are placed arbitrarily in the plane, and a line connects points i and j if $e = (i, j) \in E$. A graph with five nodes and seven edges is shown in Figure 1.1.

Graphs are useful models for many of the problems considered in combinatorial optimization. We have used graphs informally in Chapter I.1 to model network flow problems, the traveling salesman problem, and so on. Generic examples of graph models are derived from transportation and communication networks. Here V is a set of cities, and E consists of those pairs of cities that are connected by a direct transportation or communication link. Another set of generic examples concerns relationships between objects. For example, V is a set of people; and E are those pairs that are married, or of the same sex, religion, and so on. The list of examples could go on and on. We are just going to give one more that relates directly to some examples discussed in Chapter I.1.

A graph $G = (V, E)$ is called *bipartite* if there is a partition of V into disjoint sets V_1 and V_2 such that each edge joins a node in V_1 to a node in V_2 (see Figure 1.2). Bipartite

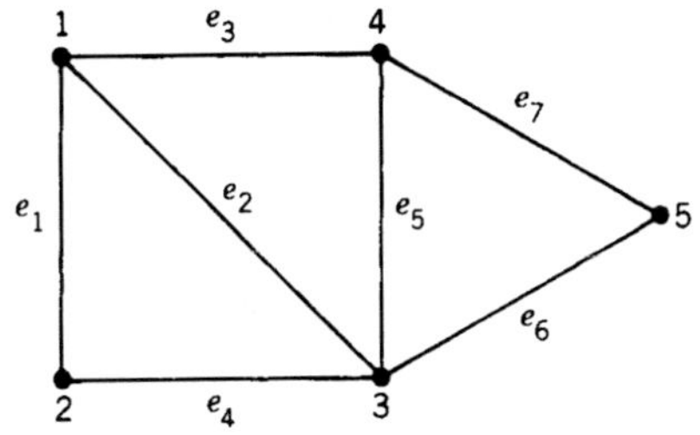

Figure 1.1. $V = \{1, 2, 3, 4, 5\}$ and $E = \{e_1 = (1, 2), e_2 = (1, 3), e_3 = (1, 4), e_4 = (2, 3), e_5 = (3, 4), e_6 = (3, 5), e_7 = (4, 5)\}$.

graphs arise in many applications. For example, in the assignment problem, V_1 is the set of workers, V_2 is the set of jobs, and $(i, j) \in E$ if and only if worker i can do job j. In facility location problems, V_1 is the set of customers, V_2 is the set of facilities, and $(i, j) \in E$ if and only if facility j can serve customer i.

Unless otherwise specified, we assume that the edges are distinct and if $e = (i, j)$, then $i \neq j$. Such graphs are called *simple*.

We say that $e_i \in E$ *meets* or is *incident to* $v \in V$ or that v is an *endpoint* of e_i if $v \in e_i$. One way to represent a graph is by its $m \times n$ *node-edge incidence matrix* $A = (a_{ij})$, where

$$a_{ij} = \begin{cases} 1 & \text{if } e_j \text{ is incident to node } i \\ 0 & \text{otherwise.} \end{cases}$$

The incidence matrix of the graph of Figure 1.1 is

$$A = \begin{array}{c} \begin{matrix} e_1 & e_2 & e_3 & e_4 & e_5 & e_6 & e_7 \end{matrix} \\ \begin{pmatrix} 1 & 1 & 1 & 0 & 0 & 0 & 0 \\ 1 & 0 & 0 & 1 & 0 & 0 & 0 \\ 0 & 1 & 0 & 1 & 1 & 1 & 0 \\ 0 & 0 & 1 & 0 & 1 & 0 & 1 \\ 0 & 0 & 0 & 0 & 0 & 1 & 1 \end{pmatrix} \end{array} \begin{matrix} \\ 1 \\ 2 \\ 3 \\ 4 \\ 5 \end{matrix}$$

Note that each column of A contains exactly two 1's. The number of 1's in row i equals the number of edges incident to node i and is called the *degree* of node i. The set of edges incident to node i is denoted by $\delta(i)$. We have $0 \leqslant |\delta(i)| \leqslant m - 1$ for all $i \in V$. A graph is called *complete* if it contains all possible edges, that is, $|\delta(i)| = m - 1$ for all $i \in V$.

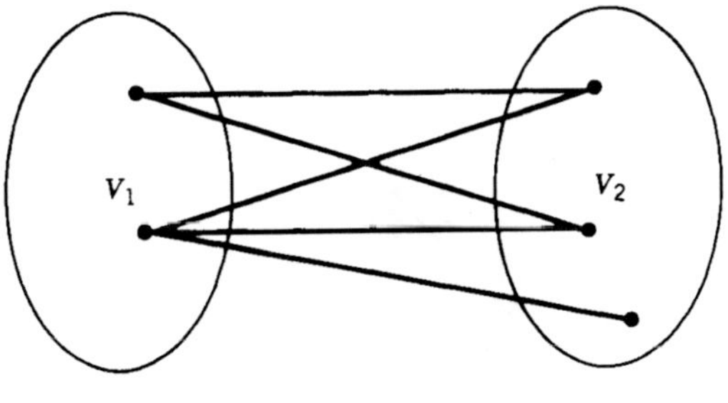

Figure 1.2

Another way to represent a graph is by its $m \times m$ adjacency matrix $A' = (a'_{ij})$, where

$$a'_{ij} = \begin{cases} 1 & \text{if } (i, j) \in E \\ 0 & \text{otherwise.} \end{cases}$$

The adjacency matrix for the graph of Figure 1.1 is

$$A' = \begin{pmatrix} 0 & 1 & 1 & 1 & 0 \\ 1 & 0 & 1 & 0 & 0 \\ 1 & 1 & 0 & 1 & 1 \\ 1 & 0 & 1 & 0 & 1 \\ 0 & 0 & 1 & 1 & 0 \end{pmatrix}.$$

The *complement* of $G = (V, E)$ is $\overline{G} = (V, \overline{E})$, where $\overline{E} = \{e\colon e \notin E\}$. The complement of the graph of Figure 1.1 is shown in Figure 1.3.

For $U \subseteq V$, let $E(U) = \{(i, j)\colon (i, j) \in E, i \in U, j \in U\}$. $E(U)$ is the set of edges with both endpoints in U. If $V' \subseteq V$ and $E' \subseteq E(V')$, then $G' = (V', E')$ is said to be a *subgraph* of $G = (V, E)$. G' is a *spanning subgraph* if $V' = V$. G' is the *subgraph induced by* V' if $E' = E(V')$. Figure 1.4 gives the subgraph induced by $V' = \{1, 2, 3, 4\}$ of the graph of Figure 1.1.

Two of the most important definitions that we need are paths and cycles. To define these terms, we need another definition. A node sequence $v_0, v_1, \ldots, v_k$, $k \geq 1$, is called a v_0-v_k *walk* if $(v_{i-1}, v_i) \in E$ for $i = 1, \ldots, k$. Node v_0 is called the *origin*, node v_k is called the *destination*, and nodes $\{v_1, \ldots, v_{k-1}\}$ are *intermediate nodes*. We can also represent a walk by its edge sequence $e_1, e_2, \ldots, e_k$, where $e_i = (v_{i-1}, v_i)$ for $i = 1, \ldots, k$. The *length* of the walk $v_0, v_1, \ldots, v_k$ or $e_1, \ldots, e_k$ is k, the number of edges in it. A walk is called a *path* if there are no node repetitions. In the graph of Figure 1.1, 1, 3, 4, 5 is a 1-5 path. Its edge sequence is e_2, e_5, e_7. A v_0-v_k walk is said to be *closed* if $v_0 = v_k$. A closed walk is said to be a *cycle* if $k \geq 3$ and $v_0, v_1, \ldots, v_{k-1}$ is a path. In the graph of Figure 1.1, 1, 3, 5, 4, 1 is a cycle of length 4. A graph is said to be *acyclic* if it does not contain any cycles.

Let w be a v_0-v_k walk with node repetitions. Consider a subsequence of nodes $v_i, v_{i+1}, \ldots, v_j = v_i$ that contains no node repetitions other than the beginning and end nodes. (The subsequence is a cycle unless it contains three nodes). By deleting $v_{i+1}, \ldots, v_j$ from w we obtain a v_0-v_k walk of smaller length. And by deleting all such subsequences, we obtain a v_0-v_k path. Referring to Figure 1.1, by deleting the indicated subsequences from

$$1, \underbrace{3, 4, 1}, 2, 3, \underbrace{4, 3}, 5$$

we obtain the 1-5 path 1, 2, 3, 5.

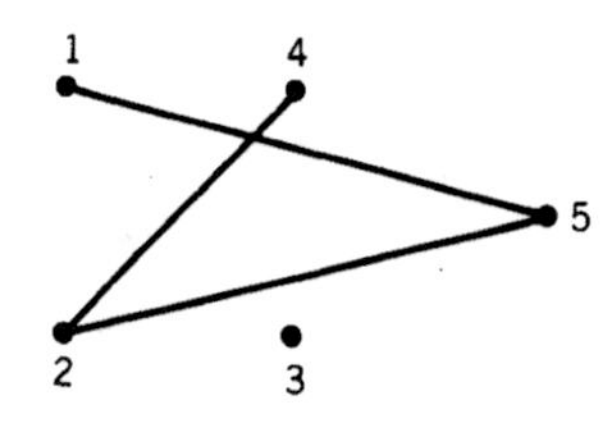

Figure 1.3

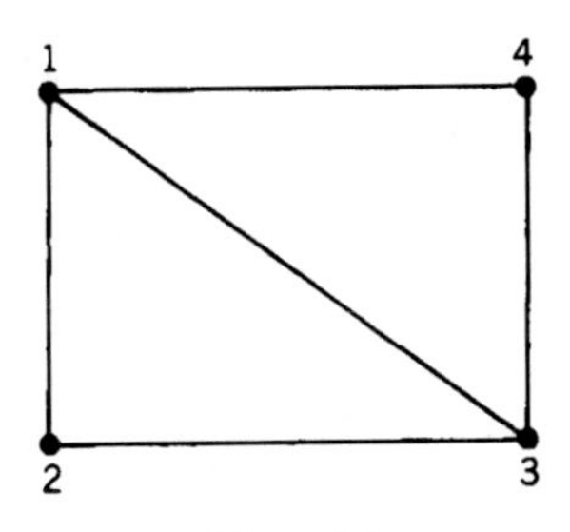

Figure 1.4

Proposition 1.1. *There is a unique partition of the nodes of a graph G into subsets $V_1, \ldots, V_p$ with the property that nodes i and j are in the same subset if and only if G contains an i-j path.*

Let V_k be a subset of the partition. The subgraph $G_k = (V_k, E(V_k))$ is called a *component* of G. G is said to be *connected* if it has one component. This means that there is a path between each pair of nodes. The graph of Figure 1.1 is connected. The graph of Figure 1.3 has two components defined by $V_1 = \{1, 2, 4, 5\}$ and $V_2 = \{3\}$. When a component contains only one node, that node is said to be *isolated*.

An acyclic graph is called a *forest*. A connected forest is called a *tree*. The subgraph obtained by deleting edges $\{e_4, e_5, e_6\}$ from the graph of Figure 1.1 is a spanning tree (see Figure 1.5).

The following proposition gives four useful characterizations of trees.

Proposition 1.2. *Let $G = (V, E)$ be a graph on m nodes. The following statements are equivalent.*

1. *G is a tree.*
2. *There is a unique path between each pair of nodes in G.*
3. *G contains $m - 1$ edges and is connected.*
4. *G contains $m - 1$ edges and is acyclic.*
5. *G is connected and acyclic.*

Trees are minimal (with respect to the number of edges) connected graphs. A *leaf* of a graph is a node of degree 1. It is easy to show that every component of a forest with at least two nodes contains at least two leaves.

Corollary 1.3

a. *If $G = (V, E)$ is a tree and $e' \notin E$, then $G' = (V, E \cup \{e'\})$ contains exactly one cycle.*
b. *If C is the edge set of the cycle of G' and $e^* \in C \setminus \{e'\}$, $G^* = (V, E \cup \{e'\} \setminus \{e^*\})$ also is a tree.*

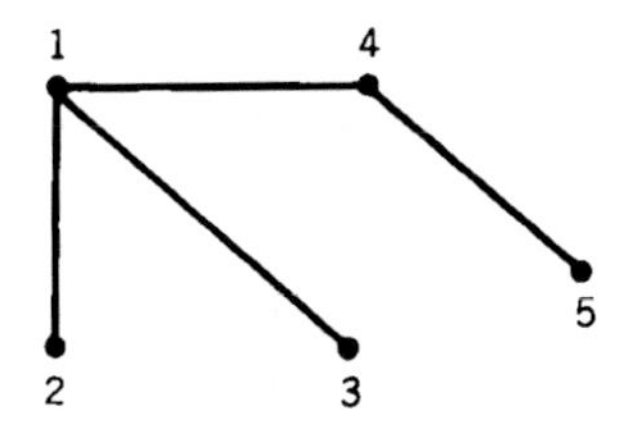

Figure 1.5

A walk is called *odd* or *even* according to whether its length is odd or even. The following proposition characterizes bipartite graphs.

Proposition 1.4. *A graph is bipartite if and only if it has no odd cycles.*

An important generalization of graphs is directed graphs. A *directed graph* or *digraph* $\mathscr{D} = (V, \mathscr{A})$ consists of a finite, nonempty set $V = \{1, \ldots, m\}$ and a set $\mathscr{A} = \{e_1, e_2, \ldots, e_n\}$ whose elements are *ordered* subsets of V of size 2 called *arcs*. (Note that we use e for both an edge of a graph and an arc of a digraph.) In a digraph, (i, j) and (j, i) are different elements and we may have neither, one, or both of these elements. In the pictorial representation of a digraph, arrows are used to indicate order. Figure 1.6 gives a digraph.

Digraphs are useful for modeling one-way relationships. For example, it is possible to go directly from intersection i to intersection j directly (by a one-way street) but not conversely, i is the father of j but not conversely, and so on.

By removing the directions from the arcs of a digraph $\mathscr{D}$, that is, replacing the arcs by edges and removing any edge duplications, we obtain a graph G that is said to *underlie* $\mathscr{D}$.

The *node-arc incidence matrix* of a digraph $\mathscr{D}$ with m nodes and n edges is the $m \times n$ matrix A with

$$a_{ij} = \begin{cases} 1 & \text{if } e_j = (k, i) \text{ for some } k \in V \setminus \{i\} \\ -1 & \text{if } e_j = (i, k) \text{ for some } k \in V \setminus \{i\} \\ 0 & \text{otherwise.} \end{cases}$$

The node-arc incidence matrix of the graph of Figure 1.6 is

$$A = \begin{array}{c} \begin{matrix} e_1 & e_2 & e_3 & e_4 & e_5 & e_6 & e_7 & e_8 & e_9 \end{matrix} \\ \begin{pmatrix} -1 & -1 & -1 & 0 & 0 & 0 & 0 & 1 & 0 \\ 1 & 0 & 0 & -1 & 1 & 0 & 0 & 0 & 0 \\ 0 & 1 & 0 & 1 & -1 & -1 & -1 & 0 & 0 \\ 0 & 0 & 1 & 0 & 0 & 1 & 0 & -1 & -1 \\ 0 & 0 & 0 & 0 & 0 & 0 & 1 & 0 & 1 \end{pmatrix} \end{array} \begin{matrix} 1 \\ 2 \\ 3 \\ 4 \\ 5 \end{matrix}$$

The node sequence $v_0, v_1, \ldots, v_k$, $k \geqslant 1$, is a v_0-v_k *directed walk* in $\mathscr{D} = (V, \mathscr{A})$ if $(v_{i-1}, v_i) \in \mathscr{A}$ for $i = 1, \ldots, k$. The walk is called a v_0-v_k *directed path* if there are no node repetitions and is called a *directed cycle* if $k \geqslant 2$, and the only node repetition is $v_0 = v_k$. In Figure 1.6, 1, 3, 4, 1 is a directed cycle, but 1, 4, 3, 5 is not a directed path since $(4, 3) \notin \mathscr{A}$.

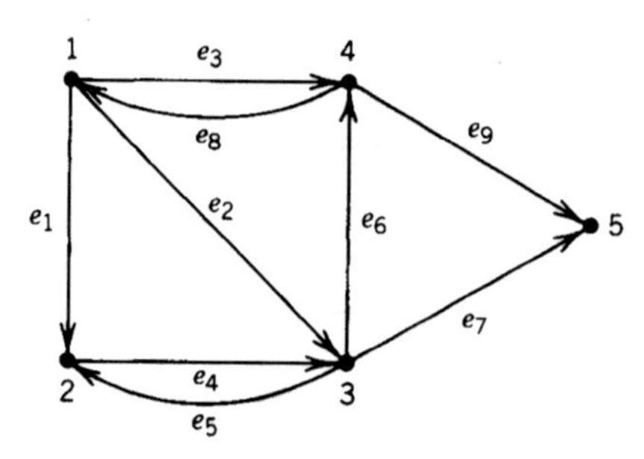

Figure 1.6. $V = \{1, 2, 3, 4, 5\}$ and $\mathscr{A} = \{e_1 = (1, 2), e_2 = (1, 3), e_3 = (1, 4)$
$e_4 = (2, 3), e_5 = (3, 2), e_6 = (3, 4),$
$e_7 = (3, 5), e_8 = (4, 1), e_9 = (4, 5)\}$.

By deleting the cycles from a v_0-v_k directed walk with $v_k \neq v_0$, we obtain a v_0-v_k directed path (see Figure 1.7). Note that the figure does not unambiguously specify the walk.

Generally when we deal with digraphs we use the term *path* to mean a directed path. However, there are times when we need to distinguish between a directed path in $\mathscr{D}$ and a path in its underlying graph. Then we use the term *dipath* to refer to the directed path in $\mathscr{D}$. The same terminology applies to cycles.

A directed graph $\mathscr{D}$ is called *strongly connected* if there is a directed path between each pair of nodes. When we say that $\mathscr{D}$ is connected, we mean that the underlying graph is connected.

A digraph is called a *tree* if the underlying graph is a tree. A subgraph of $\mathscr{D}$ that is a tree and spans $\mathscr{D}$ is called a *spanning tree*. A tree is called a *branching* if there is a node called the *root* such that there is a directed path from the root to every other node. If the root r is specified a priori, we will refer to an r-branching or branching with specified root r.

2. THE MINIMUM-WEIGHT OR SHORTEST-PATH PROBLEM

One of the simplest and most widely applicable combinatorial optimization problems is the minimum-weight or shortest-path problem. An instance of the shortest-path problem is given by a digraph $\mathscr{D} = (V, A)$, a function $w : \mathscr{A} \to R^1$ (where w_e is the weight of arc e), and designated origin and destination nodes 1 and m, respectively. The weight of a $1-m$ path is the sum of the arc weights over all arcs in the path. (All paths considered here are directed.) The problem is to find a $1-m$ path of minimum weight. Such a path is generally called a *shortest path*, but it may not be a minimum-length path unless all arcs have equal weight. Clearly if $\mathscr{D}$ is strongly connected, there is a shortest path since no path can contain more than $m - 1$ arcs where $|V| = m$.

A generic example of the shortest-path problem is to find a minimum cost route between two cities where, if $e = (i, j)$, then w_e is the cost of a direct route between nodes i and j. We will encounter many other examples throughout the text, including the finding of shortest paths as a subroutine in the solution of more complex problems.

We first consider the special case in which all arc weights are nonnegative, that is, $w: \mathscr{A} \to R^1_+$. Thus, if p is a $1-m$ path contained in a $1-m$ walk p', then the weight of p is not greater than the weight of p'.

The algorithm we present for solving this problem actually solves the slightly more general problem of finding minimum-weight paths from node 1 to all other nodes. It is based on the following fundamental property of minimum-weight paths.

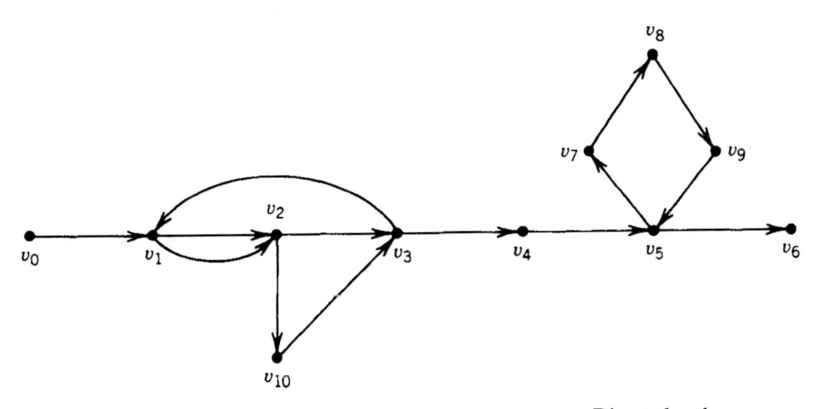

Figure 1.7. Directed walk: $v_0, v_1, v_2, v_{10}, v_3, v_1, v_2, v_3, v_4, v_5, v_7, v_8, v_9, v_5, v_6$. Directed path: $v_0, v_1, v_2, v_{10}, v_3, v_4, v_5, v_6$.

Proposition 2.1. *Suppose k is an intermediate node on a minimum-weight $1-i$ path p_i. Then the $1-k$ subpath p_k of p_i is a minimum-weight $1-k$ path.*

Proof. Let $w(p)$ be the weight of path p. The proof is by contradiction. So we suppose that $\hat{p}_k$ is a $1-k$ path and $w(\hat{p}_k) < w(p_k)$ (see Figure 2.1).

Let $p_i = (p_k, p_{ki})$. Then $\hat{p}_i = (\hat{p}_k, p_{ki})$ is a $1-i$ walk and

$$w(\hat{p}_i) = w(\hat{p}_k) + w(p_{ki}) < w(p_k) + w(p_{ki}) = w(p_i).$$

This is a contradiction because $\hat{p}_i$ contains a $1-i$ path $\tilde{p}_i$ and $w(\tilde{p}_i) \leq w(\hat{p}_i) < w(p_i)$. ■

Now let $g(i)$ be the weight of a minimum-weight 1–i path and define $g(1) = 0$.

Dijkstra's Minimum-Weight Path Algorithm

Step 1 (Initialization): $g(1) = 0$, $U = \{1\}$, $h(j) = w_{1j}$ if $(1, j) \in \mathscr{A}$, $h(j) = \infty$ otherwise.

Step 2: Let $i = \arg(\min_{j \notin U} h(j))$. If the minimum is not unique, select any i that achieves the minimum. Set $U \leftarrow U \cup \{i\}$ and $g(i) = h(i)$. If $U = V$, stop.

Step 3: For all $j \notin U$ with $(i, j) \in \mathscr{A}$, $h(j) \leftarrow \min(g(i) + w_{ij}, h(j))$. Return to Step 2.

As stated, the algorithm determines only the weights of paths. To determine the path, we simply keep a record of the node before j on the path that has weight $h(j)$. Thus, in the initialization, we let $p(j) = 1$ if $(1, j) \in \mathscr{A}$ and j otherwise, and in Step 3 we set $p(j)$ to i if $h(j) = g(i) + w_{ij}$. Thus when the algorithm terminates, $p(j)$ is the node before j on some minimum-weight $1-j$ path.

Theorem 2.2. *Dijkstra's algorithm is correct.*

Proof. The proof is inductive. The induction hypothesis is that after t passes through Step 3, $g(j)$ is correct for all $j \in U$, and $h(j)$ is the weight of a minimum-weight 1–j path restricted to having intermediate nodes in the set U. This is true initially with $U = \{1\}$ and $g(1) = 0$.

From the induction hypothesis, $h(j) \geq g(j)$ for all $j \notin U$. Suppose now that $h(i) > g(i)$, where i is as defined in Step 2. Then the minimum-weight 1–i path must contain some intermediate node not in U. Let $k \notin U$ be the first such node. Then by Proposition 2.1 the subpath from 1 to k must be a minimum-weight $1-k$ path so that its weight is $g(k)$. But this $1-k$ path contains only intermediate nodes in U. Thus $h(k) = g(k) \leq g(i) < h(i)$, contradicting the choice of i. Hence $h(i) = g(i)$.

To see that for $j \notin U \cup \{i\}$, $h(j)$ now represents the weight of a minimum weight $1-j$ path with intermediate nodes in $U \cup \{i\}$, it suffices to observe that any such path either remains as before or contains i as its last node, in which case $h(j) = g(i) + w_{ij}$. ■

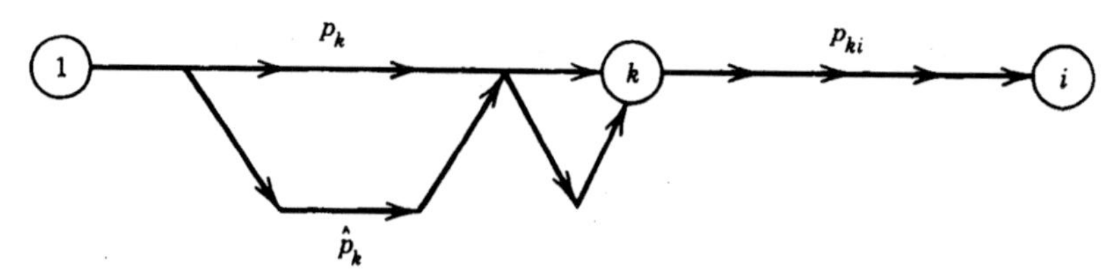

Figure 2.1

In order to consider the number of computations required in Dijkstra's algorithm and other algorithms to be given later, we need to introduce some new notation. Given functions $f(n)$ and $g(n)$ from Z^1_+ to Z^1_+, we say that $f(n)$ is $O(g(n))$ if there is a constant $c > 0$ and $n' \in Z^1_+$ such that $f(n) \leqslant cg(n)$ for all $n \geqslant n'$. Thus, for example if

$$f(n) = 7.2n^3 + 4n^2 + 9n + 4,$$

then $f(n)$ is $O(n^3)$. In other words, the "big O" notation allows us to approximate f from above by a simpler function cg with c unspecified.

Let $f(m)$ be the maximum number of basic operations (additions and comparisons) required by Dijkstra's algorithm on a graph with m nodes. At each step of the algorithm, $|U|$ is increased by 1. When $|V \setminus U| = k$, $1 \leqslant k \leqslant m - 1$, Step 3 requires no more than k additions and comparisons, and Step 2 requires finding the minimum of k numbers. Hence $f(m)$ is bounded by $c \sum_{k=1}^{m-1} k$ for some constant c. Thus Dijkstra's algorithm is $O(m^2)$.

The efficiency of the algorithm can be seen by observing that each arc is examined only once. Note that a slight improvement can be obtained by including in U at Step 2 all nodes for which the minimum is achieved.

Example 2.1. We determine minimum-distance paths from Chicago to nine other midwestern cities. The distances shown in Table 2.1 are miles/10, and $w_{ij} = w_{ji}$ for all $i \neq j$.

Table 2.2 gives the $h(j)$ and $p(j)$ at each iteration if they have changed from the previous iteration. An asterisk indicates that $h(j) = g(j)$.

Table 2.1.

		2	3	4	5	6	7	8	9	10
1.	Chicago	96	105	50	41	86	46	29	56	70
2.	Dallas		78	49	94	21	64	63	41	37
3.	Denver			60	84	61	54	86	76	51
4.	Kansas City (MO)				45	35	20	26	17	18
5.	Minneapolis					80	36	55	59	64
6.	Oklahoma City						46	50	28	8
7.	Omaha							45	37	30
8.	St. Louis								21	45
9.	Springfield (MO)									25
10.	Wichita									

Table 2.2.

Iteration	2	3	4	5	6	7	8	9	10
0	$(\infty, 2)$	$(\infty, 3)$	$(\infty, 4)$	$(\infty, 5)$	$(\infty, 6)$	$(\infty, 7)$	$(\infty, 8)$	$(\infty, 9)$	$(\infty, 10)$
1	96, 1	105, 1	50, 1	41, 1	86, 1	46, 1	29, 1*	56, 1	70, 1
2	92, 8			*	79, 8			50, 8	
3						*			
4		100, 7	*					*	
5	91, 9				78, 9				68, 4*
6					76, 10*				
7	*								
8		*							

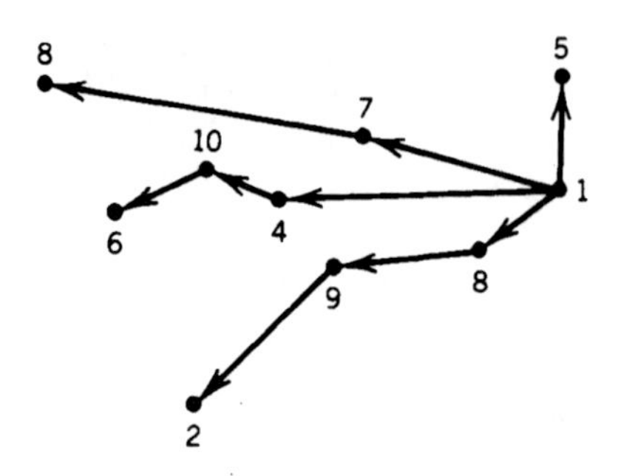

Figure 2.2

Figure 2.2 gives the solution.

It is easy to see that the algorithm can fail when there are negative arc weights. An example is shown in Figure 2.3. The algorithm would set $g(3) = 3$ at iteration 1, but $g(3) = w_{2,3} + g(2) = -2 + 4 = 2$. In particular, it is not valid to set $g(i) = h(i)$ just because $h(i)$ is the smallest value of $h(j)$ for $j \notin U$.

However, if the graph does not contain any cycles of negative weight, the algorithm can be modified to treat negative arc weights. The essential modification is that none of the $h(j)$ are set equal to $g(j)$ until m iterations have taken place.

Bellman–Ford Minimum-Weight Path Algorithm

Step 1 (Initialization): $h^0(1) = 0$, $h^0(j) = \infty$ for $j \in V \setminus \{1\}$, $k = 1$.

Step 2: For all $j \in V$,

$$h^k(j) = \min\left\{\min_{i:(i,j)\in\mathscr{A}} (w_{ij} + h^{k-1}(i)),\ h^{k-1}(j)\right\}.$$

Step 3: If $h^k(j) = h^{k-1}(j)$ for all $j \in V$, then $g(j) = h^k(j)$ for all $j \in V$. Otherwise if $k < m$, $k \leftarrow k + 1$ and return to Step 2. If $k = m$, the graph contains a cycle of negative weight.

Theorem 2.3. *The Bellman–Ford algorithm is correct.*

Proof. We claim that $h^k(j)$ is the weight of a minimum-weight 1–j walk containing no more than k arcs. This is trivially true for $k = 0$. Suppose it is true for $k - 1$. At iteration k, we consider all possible ways of adding an arc (i, j) to the end of a minimum-weight 1–i walk containing no more than $k - 1$ arcs, and then we compare the weights of these walks to the weight of a minimum-weight 1–j walk containing $k - 1$ or fewer arcs. Thus, by

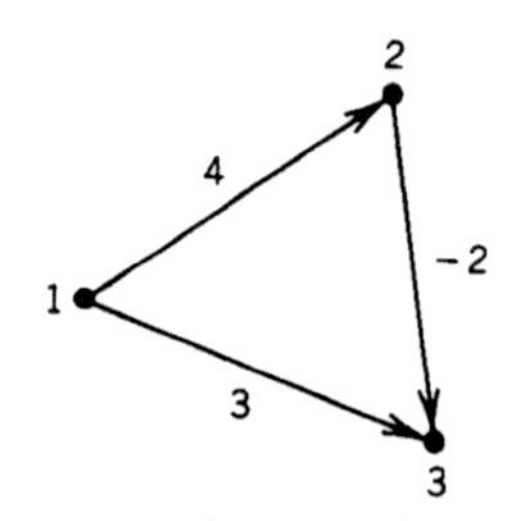

Figure 2.3

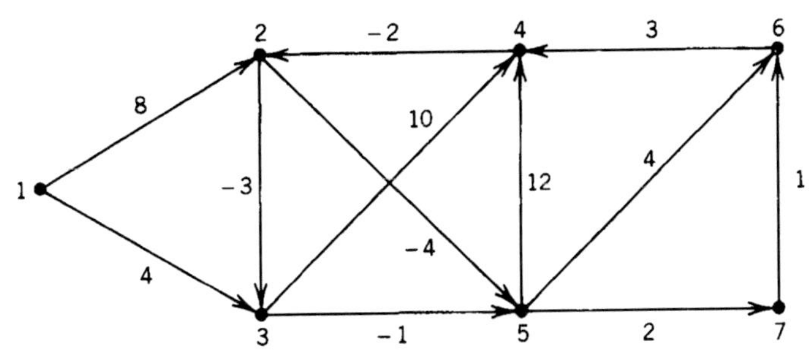

Figure 2.4

enumeration, $h^k(j)$ is the weight of a minimum-weight $1-j$ walk containing no more than k arcs. Now if $h^m(j) = h^{m-1}(j)$ for all $j \in V$, then $h^k(j) = h^m(j)$ for all $k > m$. Hence the minimum-weight $1-j$ walk is of bounded weight for all $j \in V$. This implies that $\mathcal{D}$ contains no cycles of negative weight so that $h^{m-1}(j)$ is the weight of a minimum-weight $1-j$ path containing $m-1$ or fewer arcs. But since any $1-j$ path contains no more than $m-1$ arcs, $h^{m-1}(j) = g(j)$. On the other hand, if there exists a j^* such that $h^m(j^*) < h^{m-1}(j^*)$, there is a $1-j^*$ walk containing m arcs that has lower weight than any $1-j^*$ walk containing $m-1$ arcs. Hence this walk contains a cycle of negative weight. ■

To find a minimum-weight path or a negative-weight cycle, we use the bookkeeping scheme proposed above for Dijkstra's algorithm. In other words if $h^k(j) = w_{ij} + h^{k-1}(i)$, then we set $p^k(j) = i$. To avoid having cycles of zero weight, set $p^k(j) = p^{k-1}(j)$ whenever $h^k(j) = h^{k-1}(j)$.

At each of the m steps of the algorithm, we do an addition for each of the n arcs and then for each node take the minimum over m numbers. Hence the number of computations is $cm(n + m)$, where c is a constant. In the case of a complete digraph, the number of computations is $O(m^3)$.

Thus the price we pay for being able to deal with negative arc weights in the absence of negative-weight cycles is an increase in computation time by a factor of m. Although the algorithm is able to detect a negative-weight cycle, it is unable to find a minimum-weight path in this case. The general minimum-weight path problem is much more difficult.

Example 2.2. The numbers on the edges of the digraph of Figure 2.4 are the weights. The problem is to find minimum-weight paths from node 1 to all other nodes or to detect a negative-weight cycle. Table 2.3 gives $h^k(j)$ and $p^k(j)$ for $k = 1, \ldots, 7$.

The solution is shown in Figure 2.5.

Table 2.3.

Iteration	1	2	3	4	5	6	7
0	(0, 1)	$(\infty, 2)$	$(\infty, 3)$	$(\infty, 4)$	$(\infty, 5)$	$(\infty, 6)$	$(\infty, 7)$
1		(8, 1)	(4, 1)				
2				(14, 3)	(3, 3)		
3						(7, 5)	(5, 5)
4				(10, 6)		(6, 7)	
5				(9, 6)			
6		(7, 4)					
7		No change					

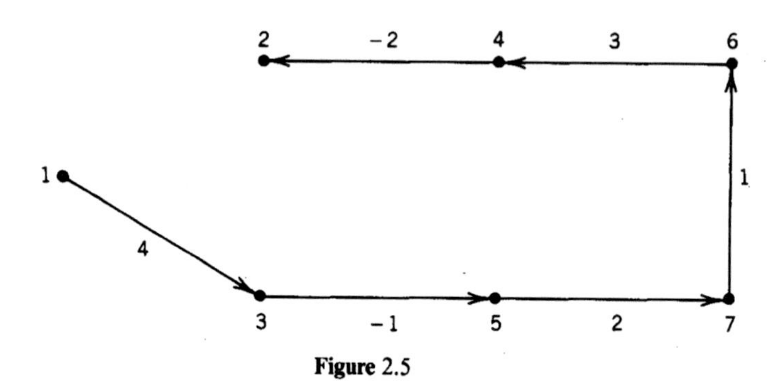

Figure 2.5

3. THE MINIMUM-WEIGHT SPANNING TREE PROBLEM

Spanning trees are used in the design of communication networks in which each node must be able to communicate with every other node. If the communication links are expensive, then it is desirable to have just one path between each pair of nodes so that the resulting network is a spanning tree.

Given a connected graph $G = (V, E)$, let E be those pairs of nodes that can be joined directly by a communication link. The weight of an edge $e \in E$ is $w_e \geqslant 0$. The problem is to build a spanning tree of G of minimum weight, where the weight of a tree $T = (V, E(T))$, $E(T) \subseteq E$, is $\Sigma_{e \in E(T)} w_e$.

It is easy to build a spanning tree from a connected graph. We scan the edges in any order, say $e_1, e_2, \ldots, e_n$, and include e_i in the tree if and only if it does not create a cycle with those edges already chosen from $\{e_1, \ldots, e_{i-1}\}$. More precisely, we have

Algorithm for Constructing a Spanning Tree

Step 1 (Initialization): Edge ordering $e_1, e_2, \ldots, e_n$, $E^0 = \emptyset$, $k = 1$.

Step 2: If $H = (V, E^{k-1} \cup \{e_k\})$ is acyclic, then $E^k = E^{k-1} \cup \{e_k\}$. Otherwise $E^k = E^{k-1}$.

Step 3: If $|E^k| = m - 1$, stop, (V, E^k) is a spanning tree. Otherwise $k \leftarrow k + 1$, and return to Step 2.

To execute Step 2, we keep track of the components of (V, E^{k-1}). Then e_k is included if and only if it joins two nodes that are in different components of (V, E^{k-1}). Thus each time we add an edge, the number of components is decreased by 1.

Now to find a minimum-weight spanning tree we simply order the edges according to increasing weight. Thus Step 1 is replaced by:

Step 1′ (Initialization): Edge ordering $e_1, e_2, \ldots, e_n$ such that $w(e_1) \leqslant w(e_2) \leqslant \ldots \leqslant w(e_n)$. $E^0 = \emptyset$, $k = 1$.

The algorithm consisting of Steps 1', 2, and 3 is called a *greedy algorithm* because at each iteration the edge of least weight is considered and included in the tree if it does not create a cycle. The greedy algorithm does what is locally best without regard to future consequences.

We now show that the greedy algorithm produces a minimum-weight spanning tree. However, for most combinatorial optimization problems, greedy algorithms are merely heuristics for finding a good feasible solution (see Section II.5.3).

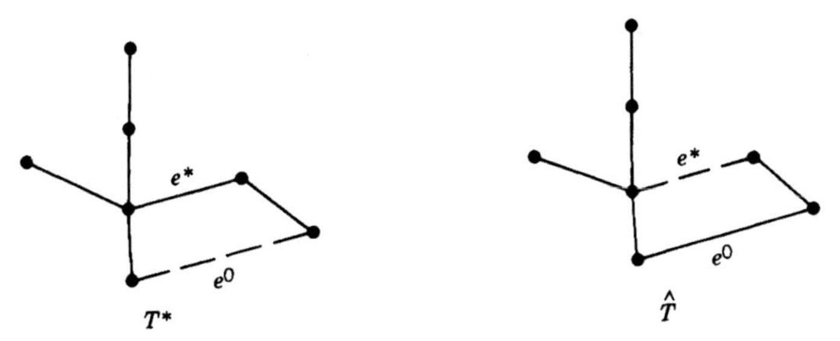

Figure 3.1

Theorem 3.1. *The greedy algorithm produces a minimum-weight spanning tree.*

Proof. Suppose the greedy algorithm produces the tree $T^0 = (V, E^0)$ and T^0 is not optimal. Let $T^* = (V, E^*)$ be an optimal tree with the property that $|E^* \setminus E^0|$ is minimum over all optimal trees. Note that $E^* \setminus E^0 \neq \emptyset$ and $E^0 \setminus E^* \neq \emptyset$. Let e^0 be a smallest-weight edge in $E^0 \setminus E^*$. Consider the set of edges $E^* \cup \{e^0\}$, which, by Corollary 1.3, contains a unique cycle. Let C be the edge set of the cycle. Now by Corollary 1.3, there is an edge $e^* \in C \setminus E^0$ such that the graph $(V, E^* \cup \{e^0\} \setminus \{e^*\})$ is a tree, say $\hat{T}$ (see Figure 3.1). Moreover, $\hat{T}$ is an optimal tree, since $w_{e^0} \leq w_{e^*}$, where the inequality holds because the greedy algorithm selected e^0. Finally $|\hat{E} \setminus E^0| = |E^* \setminus E^0| - 1$, which contradicts the choice of T^*. So T^0 is optimal. ■

Unless G is a sparse graph, that is, contains a very small number of edges, the dominant step of the greedy algorithm with respect to the number of computations is Step 1′. Since it takes $n \log n$ computations to order the edges by increasing weight, the total number of computations is $O(n \log n)$. There are, in fact, more efficient greedy-like algorithms as well as others designed specifically for sparse graphs.

The greedy algorithm is still applicable if the graph contains edges with negative weight. It also applies to the problem of finding a maximum-weight spanning tree. Here we order the edges by decreasing weight. Note that if there are some edges of negative weight, we might prefer to solve the problem of finding a maximum-weight acyclic subgraph. To solve this problem, we simply terminate the greedy algorithm as soon as the last edge of positive weight has been considered.

Example 3.1. A minimum-weight spanning tree for the graph of Example 2.1 is shown in Figure 3.2. After including the two edges of weight 21, edges (9, 10) and (4, 8) are skipped because they would create cycles. Several other edges are skipped before the final edge (3, 10) is included. The example suggests why a full sort is not needed. Note that after a tree has been found on $V' = \{2, 4, 6, 7, 8, 9, 10\}$, only edges that are incident to $\{1, 3, 5\}$ need to be considered.

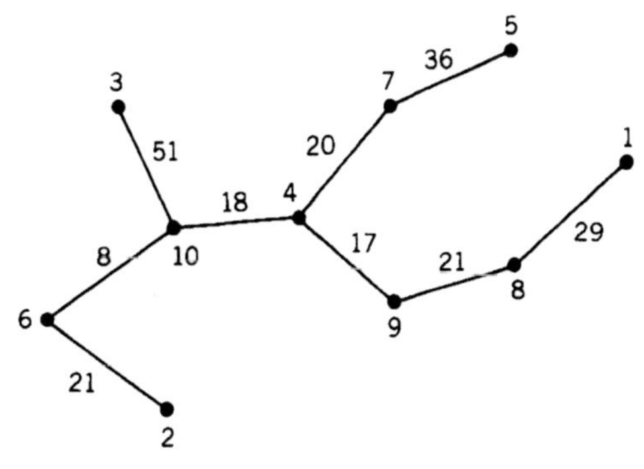

Figure 3.2

4. THE MAXIMUM-FLOW AND MINIMUM-CUT PROBLEMS

Network flow problems were introduced in Section I.1.3. In the general linear minimum cost network flow problem, we are given a digraph $\mathscr{D} = (V, \mathscr{A})$, a function $d: \mathscr{A} \to R^n_+$ where d_{ij} is called the *capacity* of arc (i, j), a function $w: \mathscr{A} \to R^n$ where w_{ij} is the *unit cost* of flow on arc (i, j), a function $b: V \to R^m$ where b_i is called the *supply* at node i ($b_i < 0$ is called a *demand*), and $\Sigma_{i \in V} b_i = 0$. A *feasible flow* in $\mathscr{D}$ is an $x: \mathscr{A} \to R^n_+$ that satisfies

$$\text{(4.1)} \qquad \sum_{j \in \delta^+(i)} x_{ij} - \sum_{j \in \delta^-(i)} x_{ji} = b_i \quad \text{all } i \in V$$

$$\text{(4.2)} \qquad x_{ij} \leqslant d_{ij} \quad \text{all } (i, j) \in \mathscr{A},$$

where $\delta^+(i) = \{j: (i, j) \in \mathscr{A}\}$ and $\delta^-(i) = \{j: (j, i) \in \mathscr{A}\}$.

The equations (4.1) express the *node conservation* relations indicating that flow out − flow in = supply, and (4.2) indicates that the flow in each arc has a specified upper bound. When there is no upper bound on x_{ij}, we take $d_{ij} = \infty$.

The *general minimum-cost network flow problem* is to find a feasible flow that minimizes the objective function

$$\text{(4.3)} \qquad \sum_{(i,j) \in \mathscr{A}} w_{ij} x_{ij}.$$

We will consider this problem in Section 6.

An important special case is the *transportation problem*. Here $\mathscr{D} = (V_1 \cup V_2, \mathscr{A})$ is bipartite and $b_i > 0$ for all $i \in V_1$ and $b_i < 0$ for all $i \in V_2$. We will study the transportation problem in Section 5.

In this section we consider the *maximum-flow problem*. Two nodes s and t, called the *source* and *sink*, respectively, are specified, $b_i = 0$ for all $i \in V$, $w_{ts} = -1$, $w_{ij} = 0$ otherwise, and $d_{ts} = \infty$. In other words, the problem is to find a feasible flow that maximizes the flow on arc (t, s) with no exogenous supplies or demands. Observe that any feasible flow that maximizes x_{ts} will have $x_{is} = 0$ for $i \neq t$ and $x_{tj} = 0$ for $j \neq s$ (see Figure 4.1). Hence $x_{ts} = \Sigma_{j \in \delta^+(s)} x_{sj} = \Sigma_{i \in \delta^-(t)} x_{it}$.

So, stated in its customary form, the maximum-flow problem is to maximize the flow out of the source or, equivalently, the flow into the sink, subject to the constraints of flow out equal to flow in for all the other nodes. Thus the maximum-flow problem asks the

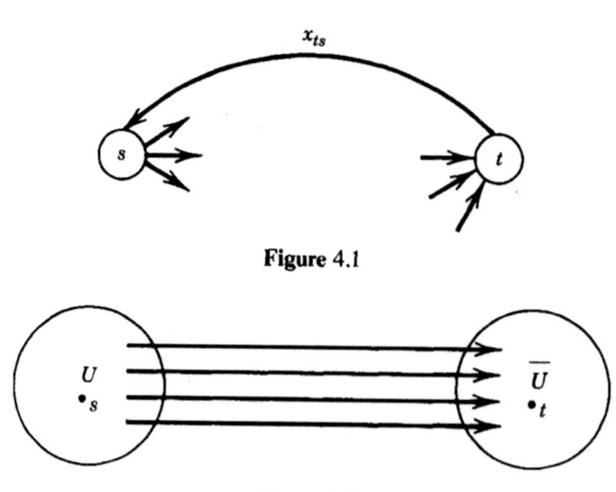

Figure 4.1

Figure 4.2

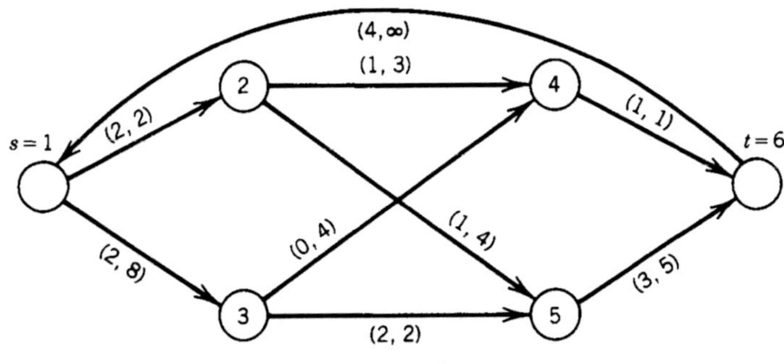

Figure 4.3

question of how much flow can be sent from the source to the sink subject to conservation at the nodes and capacities on the arcs.

We now introduce the minimum-cut problem. Let $(U, \bar{U})$ be a partition of V such that $s \in U$ and $t \in \bar{U}$. The set of arcs $\delta^+(U) = \{(i,j) \in \mathcal{A}: i \in U, j \in \bar{U}\}$ is called an *s–t cut* (see Figure 4.2). The *capacity* of the cut $\delta^+(U)$ is $\Sigma_{(i,j)\in\delta^+(U)} d_{ij}$. The *minimum-cut problem* is to find a cut of minimum capacity.

It is apparent from Figure 4.2 that all flow from s to t must pass through the arcs of $\delta^+(U)$. Hence for any feasible flow, we have

$$x_{ts} \leqslant \sum_{(i,j)\in\delta^+(U)} d_{ij} \text{ for all } s\text{–}t \text{ cuts } U,$$

and, in particular,

$$\max_{x \text{ feasible}} x_{ts} \leqslant \min_{\{U: s\in U, t\notin U\}} \sum_{(i,j)\in\delta^+(U)} d_{ij}. \tag{4.4}$$

The algorithm we present in this section finds a maximum flow and minimum cut for any maximum-flow problem and also proves the following two theorems.

Theorem 4.1. *The value of a maximum flow equals the capacity of a minimum cut.*

Theorem 4.2. *If all of the arc capacities are integer-valued, then there is a maximum flow* $x \in Z^n_+$.

An important concept in finding a maximum flow is that of an augmenting path. Given a flow x, we say that arc (i,j) is *saturated* if $x_{ij} = d_{ij}$. Let x be any feasible flow and let p be the arcs of an s–t path with no saturated arcs. Then $\min_{(i,j)\in p} (d_{ij} - x_{ij}) = \Delta > 0$, and x is not a maximum flow because we can increase x_{ts} by Δ by increasing x_{ij} by Δ for all $(i,j) \in p$. If no such path exists, x is said to be a *blocking flow*. A blocking flow may not be maximum.

Example 4.1. In Example 4.1 (see Figure 4.3), the numbers on arc (i,j) are the pair (x_{ij}, d_{ij}). It is easy to check that each of the four s–t paths contains a saturated arc. But we can increase x_{ts} by 1 as shown in Figure 4.4, to obtain the flow given in Figure 4.5. The arc from 2 to 4 in Figure 4.4 indicates that we have returned to node 2 the unit of flow previously shipped from 2 to 4.

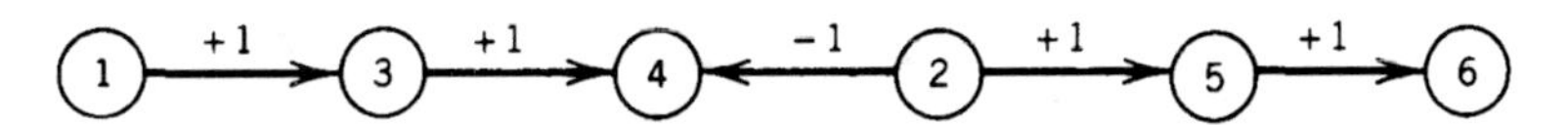

Figure 4.4

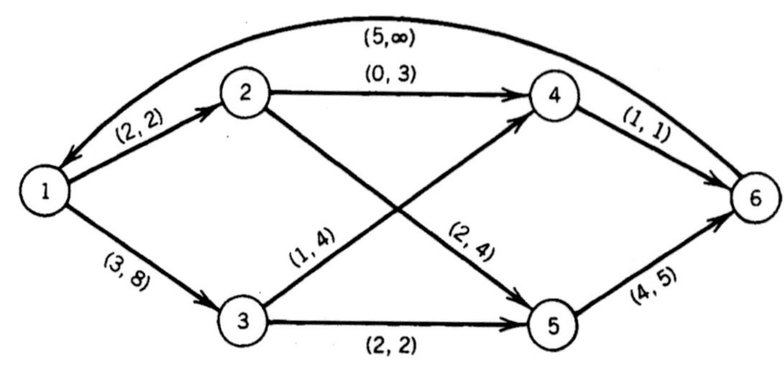

Figure 4.5

Note that the flow in Figure 4.5 is maximum because $U = \{1, 3, 4\}$ generates the cut $\delta^+(U) = \{(1, 2), (3, 5), (4, 6)\}$ of capacity 5 (see Figure 4.6).

Given a flow x, define the digraph $\mathscr{D}(x) = (V, \mathscr{A}(x))$ by

$$\begin{aligned}\mathscr{A}(x) &= \{(i, j): (i, j) \in \mathscr{A}, x_{ij} < d_{ij}\} \cup \{(i, j): (j, i) \in \mathscr{A}, x_{ji} > 0\} \\ &= \mathscr{A}_f(x) \cup \mathscr{A}_r(x).\end{aligned}$$

We say that $\mathscr{A}_f(x)$ is the set of *forward arcs* and $\mathscr{A}_r(x)$ is the set of *reverse arcs*. Corresponding to Figure 4.3, we obtain the graph shown in Figure 4.7.

An $s-t$ path in $\mathscr{D}(x)$ is called an *augmenting path* with respect to x.

Proposition 4.3. *A feasible flow x is not maximum, if there is an augmenting path with respect to x.*

Proof. Let p be the set of arcs in an augmenting path and let $p_f = p \cap \mathscr{A}_f(x)$ and $p_r = p \cap \mathscr{A}_r(x)$. Let

$$\Delta = \min\left\{\min_{(i,j)\in p_f} (d_{ij} - x_{ij}),\ \min_{(i,j)\in p_r} x_{ji}\right\}.$$

By the definition of $\mathscr{A}_f(x)$ and $\mathscr{A}_r(x)$, $\Delta > 0$. We claim that by increasing x_{ij} by Δ for all $(i, j) \in p_f$ and decreasing x_{ji} by Δ for all $(i, j) \in p_r$, x_{ts} increases by Δ. By choice of Δ, the capacity constraints are still satisfied. Also the flow out of s increases by Δ, and the flow into t increases by Δ, so x_{ts} increases by Δ. Now consider a node j on the path. If the arcs in and out of j are both forward (reverse) arcs, then the flow in and the flow out of i goes up (down) by Δ. On the other hand, if one of the arcs is a forward arc and the other is a reverse arc, there is no change of flow in or out. Hence conservation of flow is maintained. ■

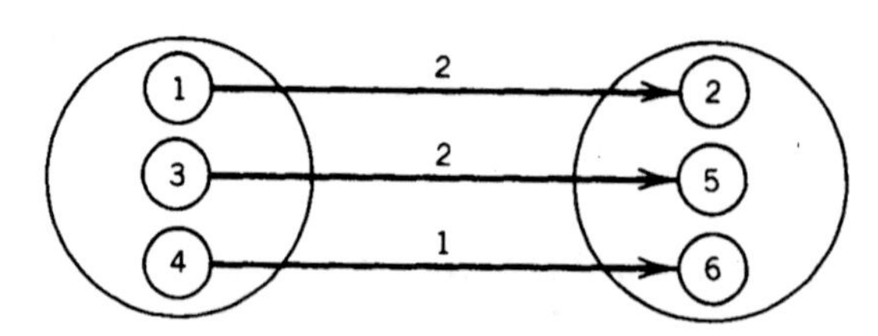

Figure 4.6

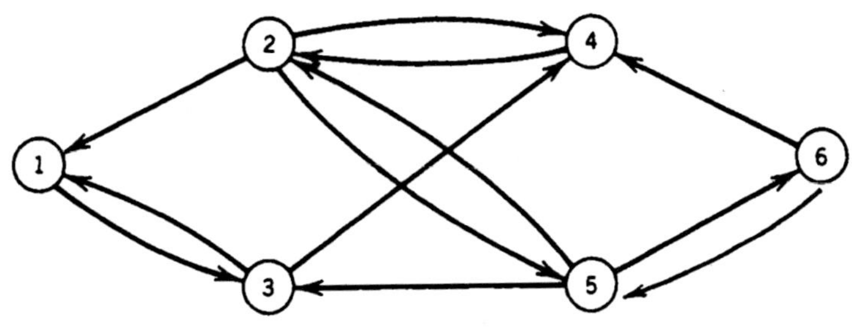

Figure 4.7

We will prove the converse of Proposition 4.3 by showing that when no augmenting path exists, there is a cut of capacity x_{ts}. Before doing so, we present a simple algorithm for finding an augmenting path if one exists.

In the algorithm, nodes $j \neq s$ get a label of the form $(p(j), \Delta)$, where $p(j)$ is the node from which j receives flow and Δ is the amount of flow sent from $p(j)$ to j. The source s is initialized with the label (s, ∞) which means that s can receive any amount of flow exogenously. Figure 4.8 shows the labeling for forward and reverse arcs on $\mathscr{D}(x)$.

The labeling can also be done directly on the original graph, which is what we do in the algorithm given below, as shown in Figure 4.9.

Augmenting Path Algorithm

Step 1 (Initialization): $x = 0$ (or any feasible flow). Source is labeled (s,∞). All nodes are unscanned, and all nodes except s are unlabeled. Let $i = s$.

Step 2 (Scan node i): For all j such that $(i, j) \in \mathscr{A}(x)$, $x_{ij} < d_{ij}$, and j is unlabeled, label j $(i, \min(\Delta, d_{ij} - x_{ij}))$. For all j such that $(j, i) \in \mathscr{A}(x)$, j is unlabeled, and $x_{ji} > 0$, label j $(i, \min(\Delta, x_{ji}))$. Node i is scanned.

Step 3: If the sink is labeled, go to Step 4. If not, choose a labeled and unscanned node i and go to Step 2. If none exists, the current flow is maximum.

Step 4: Suppose t has the label $(p(t), \Delta)$. An augmenting path has been found. Use the first element of each label to trace the path back to s. Increase the flow by Δ on all forward arcs of the path, and decrease the flow by Δ on all reverse arcs. Erase all labels and return to Step 1.

Note that to find one augmenting path, the number of computations is proportional to n, since each arc is considered no more than once.

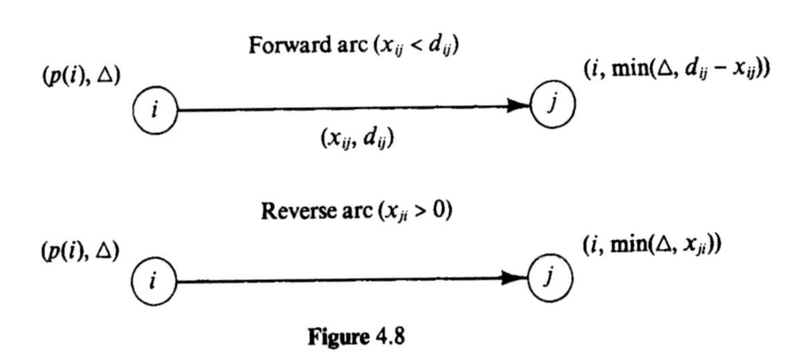

Figure 4.8

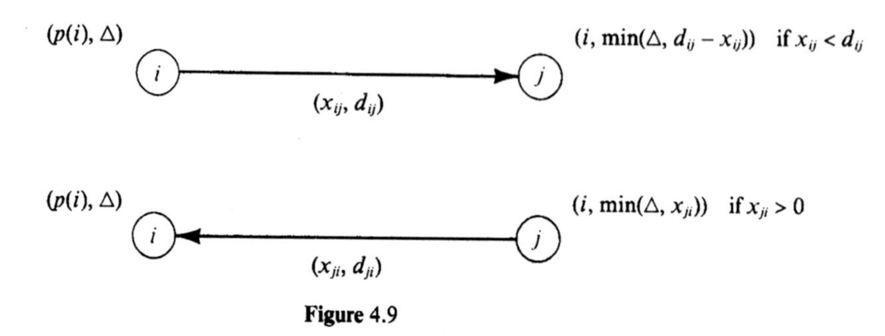

Figure 4.9

Example 4.1 (continued). The algorithm is applied to the example shown in Figure 4.3. The labels are shown in Figure 4.10. Nodes are scanned in the order (1, 3, 4, 2, 5, 6). The augmenting path has been shown in Figure 4.4. We now label again as shown in Figure 4.11.

Nodes 1, 3, and 4 are scanned and no further labeling is possible. Now observe that the cut generated by the set of labeled nodes $U = \{1, 3, 4\}$, that is, $\delta^+(U) = \{(1, 2), (3, 5), (4, 6)\}$, has capacity equal to 5 so that this flow is maximum.

Theorem 4.4. *A feasible flow x is maximum if and only if there is no augmenting path with respect to x.*

Proof. We have already shown (Proposition 4.3) that the existence of an augmenting path implies that the flow is not maximum. Now suppose there is no augmenting path and let $U = \{i \in V: i$ is scanned in the augmenting path algorithm$\}$. Then $s \in U$, $t \notin U$, $x_{ij} = d_{ij}$ for all $(i, j) \in \delta^+(U)$, and $x_{ij} = 0$ if $i \in \bar{U}$ and $j \in U$. Hence the flow into node t equals $\Sigma_{(i,j)\in\delta^+(U)} x_{ij} = \Sigma_{(i,j)\in\delta^+(U)} d_{ij}$. In other words, we have shown that if the algorithm does not find an augmenting path, it defines a cut of capacity equal to the flow into node t. ■

Note that we have also proved Theorem 4.1, which also can be proved by linear programming duality. Theorem 4.2 also is a consequence of the augmenting path algorithm. The flow change in Step 4 either equals x_{ij} for some $(i, j) \in \mathscr{A}$ with positive flow or $d_{ij} - x_{ij}$ for some $(i, j) \in \mathscr{A}$ with $x_{ij} < d_{ij}$. Hence if we begin the algorithm with any integral flow, we terminate with an integral maximum flow when all of the arc capacities are integral.

Therefore, when the capacities are integral, the number of augmentations is bounded above by the value of the maximum flow. In fact the bound can be achieved with a poor

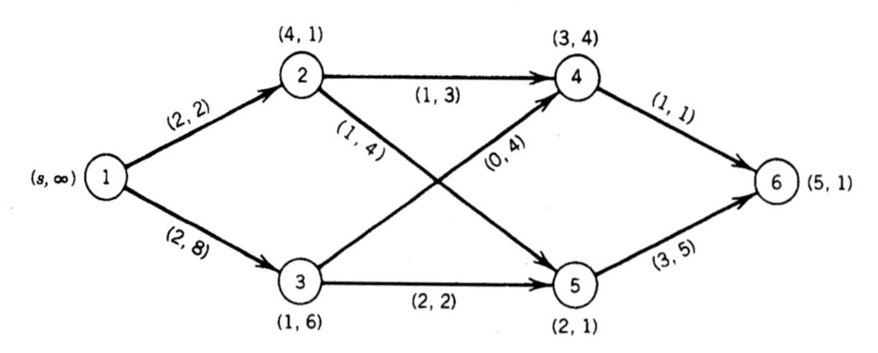

Figure 4.10

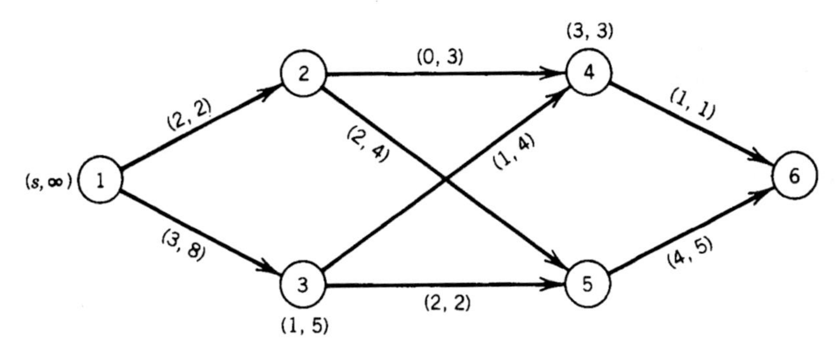

Figure 4.11

choice of augmenting paths. In the example of Figure 4.12, each of the arcs except (2, 3) has capacity K, where K is a large positive integer. The maximum flow of $2K$ can be found with augmentations of K along the paths s, 2, t and s, 3, t. On the other hand, it is possible to send one unit of flow along the augmenting paths s, 2, 3, t, then one unit along the augmenting path s, 3, 2, t, and so on. To achieve the maximum flow in this way requires $2K$ augmentations.

Fortunately, a very natural way of selecting a next node to be scanned in the augmenting path algorithm yields a bound on the number of augmentations that is independent of the capacities.

Proposition 4.5. *If at each step of the augmenting path algorithm a shortest-length augmenting path is found, then the number of augmentations is bounded by mn.*

Although we omit the details, the essential idea of the proof is to show that after, at most, n augmentations, the length of an augmenting path increases.

Note that we don't need a general shortest-path algorithm to find an augmenting path with the fewest number of arcs. We simply use *breadth-first search* to choose the next node to be scanned. That is, after s is scanned, all labeled nodes j with $(s, j) \in \mathscr{A}$ are scanned. These are the labeled nodes of distance 1 from s. In general, all labeled nodes of distance k from s are scanned before any of distancc $k + 1$ from s.

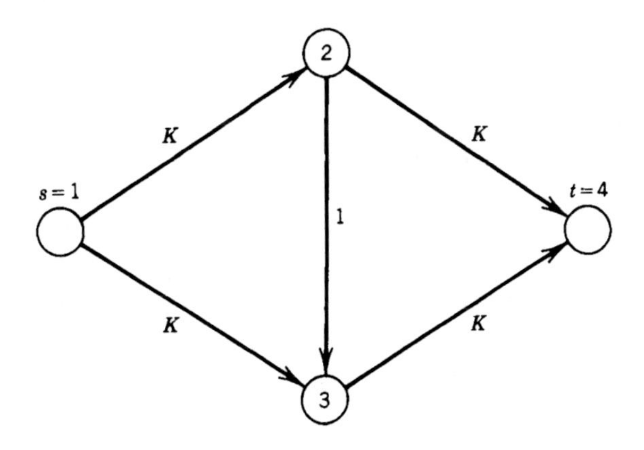

Figure 4.12

There is another class of algorithms for the maximum-flow problem that are not based on augmenting paths, and some of these have a smaller bound on the number of computations than the breadth-first algorithm for finding augmenting paths. We will not give the details here. The basic idea is that a set of, at most, $n - 1$ blocking flows are found and then combined into a maximum flow.

5. THE TRANSPORTATION PROBLEM: A PRIMAL-DUAL ALGORITHM

The transportation problem introduced in the previous section can be formulated as a minimum-cost flow problem on a bipartite digraph $\mathscr{D} = (V_1 \cup V_2, \mathscr{A})$, where $V_1 = \{1, \ldots, m_1\}$ is the set of sources, $V_2 = \{m_1 + 1, \ldots, m\}$ is the set of sinks, and $\mathscr{A} = \{(i, j): i \in V_1, j \in V_2\}$. Thus we make the assumption, without loss of generality, that there is an arc from each supply node to each demand node. The unit shipping cost from $i \in V_1$ to $j \in V_2$ is w_{ij}. Thus if there is really no arc from i to j, we take w_{ij} to be very large. Node $i \in V_1$ has a positive integral supply a_i, and node $j \in V_2$ has a positive integral demand of b_j. The flow out of a source is required to equal its supply, and the flow into a sink must equal its demand. Thus a necessary condition for feasibility is $\Sigma_{i\in V_1} a_i = \Sigma_{j\in V_2} b_j$.

The *transportation problem* is to find a flow $x \in R^n_+$, $n = |\mathscr{A}|$, that satisfies the supply-and-demand conservation equations at minimum cost. It can be formulated as the linear program

$$\begin{aligned} &\min \sum_{i\in V_1} \sum_{j\in V_2} w_{ij}x_{ij} \\ &\sum_{j\in V_2} x_{ij} = a_i \quad \text{for } i \in V_1 \\ (5.1) \qquad &\sum_{i\in V_1} x_{ij} = b_j \quad \text{for } j \in V_2 \\ &x \in R^n_+. \end{aligned}$$

Note that the problem remains unchanged by adding a constant to all of the w_{ij}, so there is no loss of generality in assuming $w_{ij} \geqslant 0$ for all i and j.

When $a_i = b_j = 1$ for all i and j and $m = 2m_1$, (5.1) is the *assignment problem* (see Section I.1.2).

It is easy to accommodate some variations of the transportation problem in the formulation (5.1). For example, if $\Sigma_{i\in V_1} a_i > \Sigma_{j\in V_2} b_j$ and the source node constraints are $\Sigma_{j\in V_2} x_{ij} \leqslant a_i$, then we add a "dummy" sink with demand $\Sigma_{i\in V_1} a_i - \Sigma_{j\in V_2} b_j$ and set the unit shipping costs to zero for arcs from V_1 to the dummy node.

The dual of (5.1) is

$$\begin{aligned} (5.2) \qquad &\max \sum_{i\in V_1} a_i u_i + \sum_{j\in V_2} b_j v_j \\ &u_i + v_j \leqslant w_{ij} \quad \text{for } i \in V_1, j \in V_2. \end{aligned}$$

The complementary slackness conditions for this pair of linear programs are

$$x_{ij}\,(w_{ij} - u_i - v_j) = 0 \quad \text{for } i \in V_1, j \in V_2$$

or

$$(5.3) \qquad x_{ij}\overline{w}_{ij} = 0 \quad \text{for } i \in V_1, j \in V_2$$

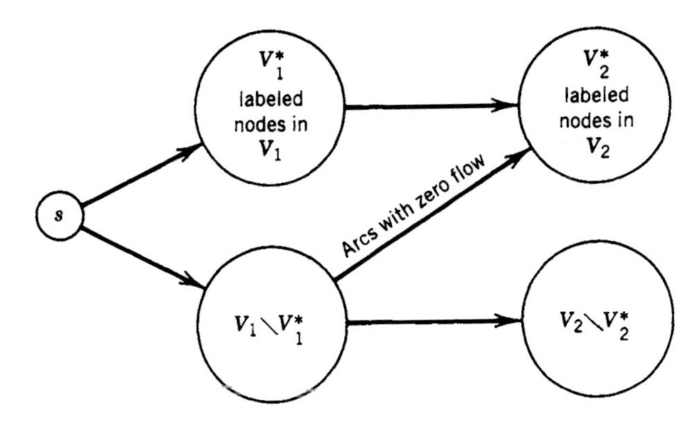

Figure 5.1

where $\overline{w}_{ij} = w_{ij} - u_i - v_j$. Thus $(u, v) \in R^m$ is dual feasible if

$$\overline{w}_{ij} \geqslant 0 \quad \text{for } i \in V_1, j \in V_2. \tag{5.4}$$

(We implement w_{ij} "very large" simply by assuming $\overline{w}_{ij} > 0$; then, by complementarity, $x_{ij} = 0$.) Thus $x \in R^n_+$ and $(u, v) \in R^m$ are optimal solutions to the primal and dual if they satisfy (5.3), (5.4), and

$$\sum_{j \in V_2} x_{ij} = a_i \quad \text{for } i \in V_1, \quad \sum_{i \in V_1} x_{ij} = b_j \quad \text{for } j \in V_2. \tag{5.5}$$

The *primal–dual* algorithm for the transportation problem maintains (5.3), (5.4), and

$$\sum_{j \in V_2} x_{ij} \leqslant a_i \quad \text{for } i \in V_1, \quad \sum_{i \in V_1} x_{ij} \leqslant b_j \quad \text{for } j \in V_2. \tag{5.6}$$

It is easy to find an initial solution that satisfies these conditions. For example, take $u_i^0 = \min_{j \in V_2} w_{ij}$ for $i \in V_1$ and $v_j^0 = \min_{i \in V_1} (w_{ij} - u_i^0)$ for $j \in V_2$, and $x^0 = 0$. At each major iteration the algorithm increases $\sum_{i \in V_1} \sum_{j \in V_2} x_{ij}$ by an integer and stops when (5.5) is satisfied.

Given $\overline{w}$, to see whether (5.5) can be satisfied, we consider the problem of maximizing $\sum_{i \in V_1} \sum_{j \in V_2} x_{ij}$ subject to (5.3), (5.6), and $x \in R^n_+$. This is an s–t maximum-flow problem on the digraph $\mathcal{D}(\overline{w}) = (V_1 \cup V_2 \cup \{s, t\}, \mathcal{A}(\overline{w}))$, where

$$\mathcal{A}(\overline{w}) = \{(i, j) \in \mathcal{A} \colon \overline{w}_{ij} = 0\} \cup \{(s, i) : i \in V_1\} \cup \{(j, t) : j \in V_2\}.$$

The capacity of arc (s, i) is a_i for $i \in V_1$, and the capacity of arc (j, t) is b_j for $j \in V_2$. All other arcs have "very large" capacities. If the maximum flow equals $\sum_{i \in V_1} a_i$, we have found an optimal solution. If not, we change the dual variables.

Consider the status of the node labels when the maximum-flow algorithm terminates (see Figure 5.1). Note that if $i \in V_1^*$ and $j \in V_2 \setminus V_2^*$, then $\overline{w}_{ij} > 0$ otherwise, we could label j from i. Also if $i \in V_1 \setminus V_1^*$ and $j \in V_2^*$, then $x_{ij} = 0$ otherwise, we could label i from j.

We now change the dual solution as follows. Let $h = \min_{i \in V_1^*, j \in V_2 \setminus V_2^*} \overline{w}_{ij} > 0$ and define new values for the dual variables by

$$\hat{u}_i = \begin{cases} u_i + h & i \in V_1^* \\ u_i & i \in V_1 \setminus V_1^* \end{cases}, \quad \hat{v}_j = \begin{cases} v_j - h & j \in V_2^* \\ v_j & j \in V_2 \setminus V_2^* \end{cases}. \tag{5.7}$$

Hence the new reduced costs are

$$\hat{\overline{w}} = \begin{cases} \overline{w}_{ij} - h & i \in V_1^*, \ j \in V_2 \setminus V_2^* \\ \overline{w}_{ij} + h & i \in V_1 \setminus V_1^*, \ j \in V_2^* \\ \overline{w}_{ij} & \text{otherwise.} \end{cases} \tag{5.8}$$

By the choice of h, dual feasibility is maintained. Complementary slackness is maintained, since for $i \in V_1^*$ and $j \in V_2 \setminus V_2^*$ we have $x_{ij} = 0$ by definition of $\mathcal{D}(\overline{w})$, and for $i \in V_1^* \setminus V_1$ and $j \in V_2^*$ we have $x_{ij} = 0$ by the labeling rules.

The important outcome is that $\hat{\overline{w}}_{i^*j^*} = 0$ for some $i^* \in V_1^*$ and $j^* \in V_2 \setminus V_2^*$ so that at least one new arc from $i \in V_1^*$ to $j \in V_2 \setminus V_2^*$ is added to $\mathcal{D}(\overline{w})$ to obtain $\mathcal{D}(\hat{\overline{w}})$. Some arcs may also be deleted from $i \in V_1 \setminus V_1^*$ to $j \in V_2^*$. Now we can transfer the final labels from $\mathcal{D}(\overline{w})$ to $\mathcal{D}(\hat{\overline{w}})$ and continue with the maximum-flow algorithm, with the assurance that at least one node in $V_2 \setminus V_2^*$ will be labeled. This proves that after, at most $m - m_1$ such dual changes, the maximum flow will be increased by at least one unit. Thus the whole process is applied, at most, $\Sigma_{i \in V_1} a_i$ times.

Primal–Dual Algorithm for the Transportation Problem

Step 1 (Initialization): $t = 0$, $x^0 = 0$, $u_i^0 = \min_{j \in V_2} w_{ij}$ for $i \in V_1$, $v_j^0 = \min_{i \in V_1} (w_{ij} - u_i^0)$ for $j \in V_2$, and $\overline{w}_{ij} = w_{ij}^0 - u_i^0 - v_j^0$ for all i and j.

Iteration t

Step 2: Solve the maximum-flow problem over $\mathcal{D}(\overline{w}^t)$. Let x_{ij}^t be the flow from $i \in V_1$ to $j \in V_2$ for all i and j.

Step 3: If the maximum flow equals $\Sigma_{i \in V_1} a_i$, then $x^t = (x_{ij}^t)$ is an optimal solution. Otherwise, adjust the dual variables and reduced costs using (5.7) and (5.8). Keep the labels from the solution of the maximum-flow problem and return to Step 2 with $t \leftarrow t + 1$.

We have already proved the following theorem.

Theorem 5.1. *The primal–dual algorithm solves the transportation problem with, at most, $\Sigma_{i \in V_1} a_i$ applications of the maximum-flow routine.*

Corollary 5.2. *There is an integral optimal solution to the transportation problem.*

Proof. At each iteration, the solution x^t is obtained as the solution to a maximum-flow problem with integral capacities and hence is integral. ■

Example 5.1

$$w = \begin{pmatrix} 5 & 3 & 7 & 3 & 8 & 5 \\ 5 & 6 & 12 & 5 & 7 & 11 \\ 2 & 8 & 3 & 4 & 8 & 2 \\ 11 & 6 & 10 & 5 & 10 & 9 \end{pmatrix}, \quad \begin{array}{l} a = (4 \quad 5 \quad 3 \quad 5) \\ b = (3 \quad 3 \quad 6 \quad 2 \quad 1 \quad 2) \end{array}$$

$u^0 = (3 \quad 5 \quad 2 \quad 5)$, $v^0 = (0 \quad 0 \quad 1 \quad 0 \quad 2 \quad 0)$, and

$$\overline{w}^0 = \begin{pmatrix} 2 & 0 & 3 & 0 & 3 & 2 \\ 0 & 1 & 6 & 0 & 0 & 6 \\ 0 & 6 & 0 & 2 & 4 & 0 \\ 6 & 1 & 4 & 0 & 3 & 4 \end{pmatrix}.$$

Solving the maximum-flow problem on $\mathscr{D}(\overline{w}^0)$ yields

$$x^0 = \begin{array}{c} \\ \checkmark \\ \checkmark \\ \\ \checkmark \end{array} \begin{array}{cccccc} \checkmark & \checkmark & & \checkmark & \checkmark & \\ \end{array}$$

$$x^0 = \begin{array}{c|cccccc} & \checkmark & \checkmark & & \checkmark & \checkmark & \\ \hline \checkmark & 0 & 3 & 0 & 1 & 0 & 0 \\ \checkmark & 3 & 0 & 0 & 1 & 1 & 0 \\ & 0 & 0 & 3 & 0 & 0 & 0 \\ \checkmark & 0 & 0 & 0 & 0 & 0 & 0 \end{array}.$$

Rows and columns corresponding to labeled nodes are noted with a check mark. Hence $h = \overline{w}^0_{16} = 2$, $u^1 = (5 \quad 7 \quad 2 \quad 7)$, $v^1 = (-2 \quad -2 \quad 1 \quad -2 \quad 0 \quad 0)$,

$$\overline{w}^1 = \begin{array}{c|cccccc} & \checkmark & & & \checkmark & \checkmark & \\ \hline & 2 & 0 & 1 & 0 & 3 & 0 \\ \checkmark & 0 & 1 & 4 & 0 & 0 & 4 \\ & 2 & 8 & 0 & 4 & 6 & 0 \\ \checkmark & 6 & 1 & 2 & 0 & 3 & 2 \end{array},$$

and

$$x^1 = \begin{array}{c|cccccc} & \checkmark & & & \checkmark & \checkmark & \\ \hline & 0 & 3 & 0 & 0 & 0 & 1 \\ \checkmark & 3 & 0 & 0 & 1 & 1 & 0 \\ & 0 & 0 & 3 & 0 & 0 & 0 \\ \checkmark & 0 & 0 & 0 & 1 & 0 & 0 \end{array}.$$

Now $h = \overline{w}^1_{22} = \overline{w}^1_{42} = 1$, $u^2 = (5 \quad 8 \quad 2 \quad 8)$, $v^2 = (-3 \quad -2 \quad 1 \quad -3 \quad -1 \quad 0)$,

$$\overline{w}^2 = \begin{pmatrix} 3 & 0 & 1 & 1 & 4 & 0 \\ 0 & 0 & 3 & 0 & 0 & 3 \\ 3 & 8 & 0 & 5 & 7 & 0 \\ 6 & 0 & 1 & 0 & 3 & 1 \end{pmatrix},$$

and

$$\begin{array}{cc} & \begin{array}{cccccc} \checkmark & \checkmark & & \checkmark & \checkmark & \checkmark \end{array} \\ x^2 = \begin{array}{c} \checkmark \\ \checkmark \\ \\ \checkmark \end{array} & \begin{pmatrix} 0 & 2 & 0 & 0 & 0 & 2 \\ 3 & 0 & 0 & 1 & 1 & 0 \\ 0 & 0 & 3 & 0 & 0 & 0 \\ 0 & 1 & 0 & 1 & 0 & 0 \end{pmatrix}. \end{array}$$

Now $h = \overline{w}^2_{13} = \overline{w}^2_{43} = 1$, $u^3 = (6 \quad 9 \quad 2 \quad 9)$, $v^3 = (-4 \quad -3 \quad 1 \quad -4 \quad -2 \quad -1)$,

$$\overline{w}^3 = \begin{pmatrix} 3 & 0 & 0 & 1 & 4 & 0 \\ 0 & 0 & 2 & 0 & 0 & 3 \\ 4 & 9 & 0 & 6 & 8 & 1 \\ 6 & 0 & 0 & 0 & 3 & 1 \end{pmatrix},$$

and we obtain an optimal flow given by

$$x^3 = \begin{pmatrix} 0 & 2 & 0 & 0 & 0 & 2 \\ 3 & 0 & 0 & 1 & 1 & 0 \\ 0 & 0 & 3 & 0 & 0 & 0 \\ 0 & 1 & 3 & 1 & 0 & 0 \end{pmatrix}.$$

When the total supply is large, there is a simple way to reduce the maximum number of possible augmentations from $\Sigma_{i\in V_1} a_i$ to $m \lceil \log_2 (\max_{i,j} (a_i, b_j)) \rceil$. The technique is called *scaling*. An integer $a < 2^k$ can be written as $a = \Sigma_{i=0}^{k-1} \delta_i 2^i$, where $\delta_i \in \{0, 1\}$ for $i = 1, \ldots, k-1$. The *binary representation* of a is the string $(\delta_{k-1}\ \delta_{k-2} \ldots \delta_0)$. The scaling technique represents each supply and demand in binary. If $2^{k-1} \leq \max_{i,j} (a_i, b_j) < 2^k$, then the length of each string is k. Hence in Example 5.1, $a_1 = 100$, $a_2 = 101$, $a_3 = 011$, and so on.

We now consider an approximate problem with supply-and-demand vectors (a^0, b^0) in which only the leading digit is considered, that is, $a_i^0 = 1$ if $a_i \geq 2^{k-1}$, $a_i^0 = 0$ otherwise, or $a_i^0 = \lfloor a_i/2^{k-1} \rfloor$ for $i \in V_1$ and $b_j^0 = \lfloor b_j/2^{k-1} \rfloor$ for $j \in V_2$. In example 5.1, $k = 3$, $a^0 = (1 \quad 1 \quad 0 \quad 1)$ and $b^0 = (0 \quad 0 \quad 1 \quad 0 \quad 0 \quad 0)$. As the example shows, supply and demand may now be unequal. Without loss of generality, assume $\Sigma_{i\in V_1} a_i^0 \geq \Sigma_{j\in V_2} b_j^0$, so there may be a need for a dummy sink node. Since $\Sigma_{i\in V_1} a_i^0 + \Sigma_{j\in V_2} b_j^0 \leq m$, the first approximation can be solved with m or fewer augmentations. Suppose the solution is $(x^0, \overline{w}^0)$, where $x^0 \in Z_+^m$ does not include shipments to the dummy sink.

We now begin the next approximation with the optimal reduced costs $\overline{w}^0$, the flow $2x^0$, and the supplies and demands $a_i^1 = \lfloor a_i/2^{k-2} \rfloor$ for $i \in V_1$ and $b_j^1 = \lfloor b_j/2^{k-2} \rfloor$ for $j \in V_2$. Since

$\Sigma_{i \in V_1} a_i^0 \geqslant \Sigma_{j \in V_2} b_j^0$, we have $\Sigma_{i \in V_1} x_{ij}^0 = b_j$ for all $j \in V_2$. Thus the *unsatisfied* supplies and demands are

$$\tilde{a}_i^1 = a_i^1 - 2 \sum_{j \in V_2} x_{ij}^0 \leqslant 3 \quad \text{for } i \in V_1$$

and

$$\tilde{b}_j^1 = b_j^1 - 2b_j^0 \leqslant 1 \quad \text{for } j \in V_2.$$

Hence $\Sigma_{j \in V_2} \tilde{b}_j^1 \leqslant m$, so no more than m augmentations are required, other than the trivial ones to the dummy source or sink.

The procedure continues in this way. In the pth approximation, $a_i^{p-1} = \lfloor a_i/2^{k-p} \rfloor$ for $i \in V_1$ and $b_j^{p-1} = \lfloor b_j/2^{k-p} \rfloor$ for $j \in V_2$. The primal solution from the previous approximation is doubled to get the unsatisfied supply and demand, at least one of which does not exceed m. The dual variables are kept from one iteration to the next. The procedure is applied $k = \lceil \log_2 (\max_{i,j} (a_i, b_j)) \rceil$ times to find an optimal solution.

Example 5.1 (continued). We apply the scaling technique to solve this problem. The initial supplies and demands are $a^0 = (1 \quad 1 \quad 0 \quad 1)$ and $b^0 = (0 \quad 0 \quad 1 \quad 0 \quad 0 \quad 0)$, so to accommodate the imbalance, we add a dummy sink with a demand of 2 and costs of $w_{i7} = 0$ for all i. An optimal solution is given by $u^0 = (0 \quad 0 \quad -4 \quad 0)$, $v^0 = (0 \quad 0 \quad 7 \quad 0 \quad 0 \quad 0 \quad 0)$,

$$\overline{w}^0 = \begin{pmatrix} 5 & 3 & 0 & 3 & 8 & 5 & 0 \\ 5 & 6 & 5 & 5 & 7 & 11 & 0 \\ 6 & 12 & 0 & 8 & 12 & 6 & 4 \\ 11 & 6 & 3 & 5 & 10 & 9 & 0 \end{pmatrix},$$

and

$$x^0 = \begin{pmatrix} 0 & 0 & 1 & 0 & 0 & 0 & 0 \\ 0 & 0 & 0 & 0 & 0 & 0 & 1 \\ 0 & 0 & 0 & 0 & 0 & 0 & 0 \\ 0 & 0 & 0 & 0 & 0 & 0 & 1 \end{pmatrix}.$$

The initial flow for the second approximations is $2x^0$, and the supplies and demands are given by $a^1 = (2 \quad 2 \quad 1 \quad 2)$ and $b^1 = (1 \quad 1 \quad 3 \quad 1 \quad 0 \quad 1)$. Thus $\tilde{a}^1 = (0 \quad 2 \quad 1 \quad 2)$ and $\tilde{b}^1 = (1 \quad 1 \quad 1 \quad 1 \quad 0 \quad 1)$. Now there are five units of unsatisfied supply. Since $\Sigma_{i \in V_1} a_i^1 = \Sigma_{j \in V_2} b_j^1$, no extra source or sink is needed.

An optimal solution to the second approximation is $u^1 = (0 \quad 7 \quad -4 \quad 3)$, $v^1 = (0 \quad 1 \quad 5 \quad 0 \quad 2 \quad -2)$,

$$\overline{w}^1 = \begin{pmatrix} 3 & 0 & 0 & 1 & 4 & 0 \\ 0 & 0 & 2 & 0 & 0 & 3 \\ 4 & 9 & 0 & 6 & 8 & 1 \\ 6 & 0 & 0 & 0 & 3 & 1 \end{pmatrix},$$

and

$$x^1 = \begin{pmatrix} 0 & 0 & 1 & 0 & 0 & 1 \\ 1 & 1 & 0 & 0 & 0 & 0 \\ 0 & 0 & 1 & 0 & 0 & 0 \\ 0 & 0 & 1 & 1 & 0 & 0 \end{pmatrix}.$$

The initial flow for the third approximation is $x = 2x^1$, and $a^2 = a = (4 \quad 5 \quad 3 \quad 5)$, $b^2 = b = (3 \quad 3 \quad 6 \quad 2 \quad 1 \quad 2)$, $\tilde{a}^2 = (0 \quad 1 \quad 1 \quad 1)$, and $\tilde{b}^2 = (1 \quad 1 \quad 0 \quad 0 \quad 1 \quad 0)$. Hence there are three units of unsatisfied supply. No dual variable change is required, and we immediately obtain an optimal solution given by

$$x^2 = \begin{pmatrix} 0 & 0 & 2 & 0 & 0 & 2 \\ 3 & 1 & 0 & 0 & 1 & 0 \\ 0 & 0 & 3 & 0 & 0 & 0 \\ 0 & 2 & 1 & 2 & 0 & 0 \end{pmatrix}.$$

There is another interpretation and implementation of the primal–dual algorithm that is also of interest. Note that for any u and v, the instances of the transportation problem with cost matrix $w = (w_{ij})$ and $\overline{w} = (w_{ij} - u_i - v_j)$ have the same optimal solutions, since for any feasible x we have

$$\begin{aligned} \sum_{i \in V_1} \sum_{j \in V_2} w_{ij} x_{ij} - \sum_{i \in V_1} \sum_{j \in V_2} \overline{w}_{ij} x_{ij} &= \sum_{i \in V_1} \sum_{j \in V_2} (u_i + v_j)\, x_{ij} \\ &= \sum_{i \in V_1} u_i a_i + \sum_{j \in V_2} v_j b_j. \end{aligned}$$

The dual part of the primal–dual algorithm eventually finds a matrix $\overline{w} \geq 0$ such that there is a feasible solution x of cost $\Sigma_{i \in V_1} \Sigma_{j \in V_2} \overline{w}_{ij} x_{ij} = 0$. Since zero is a lower bound on the cost of any solution with $\overline{w} \geq 0$, such a solution must be optimal. The primal part of the algorithm uses maximum flow to find the solution of cost zero when one exists. In other words, with respect to the matrix $\overline{w}$, all flow is sent over paths of zero cost.

We want to point out that this can be achieved by a different implementation that uses a minimum-cost path algorithm to calculate the dual variables. Consider the digraph $\mathcal{D} = (V_1 \cup V_2 \cup \{s, t\}, \mathcal{A})$, where there is a directed arc from the source s to each node in V_1, a directed arc from each node in V_2 to the sink t, and arcs (i, j), $i \in V_1, j \in V_2$ if it is possible to ship directly from i to j. Arcs going out of the source or into the sink have zero cost and a capacity equal to the corresponding supply or demand. Arc (i, j), $i \in V_1, j \in V_2$, has cost w_{ij} and infinite capacity.

Minimum-Cost Path Augmentation Algorithm

Step 1 (Initialization): $k = 0$, $x^0 = 0$, $\mathscr{D}^0 = \mathscr{D}$.

Step 2 (Iteration k): Let the current flow be x^k. Find a minimum-cost path from s to t of the form $(s, i^0, \ldots, j^0, t)$. Let

$$\Delta_1 = \min(a_{i^0} - x^k_{si^0}, b_{j^0} - x^k_{j^0t}).$$

Let

$$\Delta_2 = \min\{x^k_{ij}: (j, i) \text{ is on the path with } i \in V_1, \text{ and } j \in V_2\}.$$

If no arcs from $j \in V_2$ to $i \in V_1$ are on the path, let $\Delta_2 = \infty$. Let $\Delta = \min(\Delta_1, \Delta_2)$.

Step 3 (Flow augmentation): Increase the flow in (s, i^0) and (j^0, t) by Δ. For all arcs (i, j) on the path with $i \in V_1$ and $j \in V_2$, increase the flow by Δ. For all arcs (j, i) on the path with $i \in V_1$ and $j \in V_2$, decrease the flow in (i, j) by Δ. If the new flow x^{k+1} satisfies $x^{k+1}_{si} = a_i$ for all $i \in V_1$, stop. x^{k+1} is an optimal solution.

Step 4 (Arc and cost change): Add (j, i), $j \in V_2$, $i \in V_1$, to the graph if $x^k_{ij} = 0$ and $x^{k+1}_{ij} > 0$, and assign it the cost $(-w_{ij})$. Delete (s, i^0) if $x^{k+1}_{si^0} = a_{i^0}$, delete (j^0, t) if $x^{k+1}_{j^0t} = b_{j^0}$, and delete (j, i) $j \in V_2$, $i \in V_1$ if $x^{k+1}_{ij} = 0$. $k \leftarrow k + 1$.

Each time a minimum-cost path is found, we can interpret the costs on the nodes of that path as the incremental values of dual variables such that when the dual change $w_{ij} - u_j - v_j$ is made, the cost of the path is reduced to zero. By augmenting over minimum-cost paths, the flow at iteration k is the minimum-cost solution to the transportation problem with supplies $\Sigma_{j\in V_2} x^k_{ij}$ for $i \in V_1$ and demands $\Sigma_{i\in V_1} x^k_{ij}$ for $j \in V_2$. In fact it is possible to implement the primal–dual approach given above to produce the same augmentations as those determined by minimum-cost paths.

Example 5.1 (continued). We find an optimal solution by finding minimum-cost path augmentations. Table 5.1 shows the paths, quantity of flow, and cost per unit of flow for each augmentation. Arcs from V_2 to V_1 are noted by overbars.

Table 5.1.

Augmentation	Path	Flow	Cost
1	$(s, 3), (3, 1), (1, t)$	3	2
2	$(s, 1), (1, 2), (2, t)$	3	3
3	$(s, 1), (1, 4), (4, t)$	1	3
4	$(s, 4), (4, 4), (4, t)$	1	5
5	$(s, 2), (2, 1), \overline{(1, 3)}, (3, 6), (6, t)$	2	5
6	$(s, 2), (2, 1), \overline{(1, 3)}, (3, 3), (3, t)$	1	6
7	$(s, 2), (2, 5), (5, t)$	1	7
8	$(s, 4), (4, 4), \overline{(4, 1)}, (1, 6), \overline{(6, 3)}, (3, 3), (3, t)$	1	8
9	$(s, 2), (2, 3), \overline{(2, 1)}, (1, 6), \overline{(6, 3)}, (3, 3), (3, t)$	1	9
10	$(s, 3), (3, 3), (3, t)$	3	20

The optimal solution found is

$$x^{10} = \begin{pmatrix} 0 & 2 & 0 & 0 & 0 & 2 \\ 2 & 1 & 0 & 0 & 1 & 0 \\ 0 & 0 & 3 & 0 & 0 & 0 \\ 0 & 0 & 3 & 2 & 0 & 0 \end{pmatrix}.$$

The primal–dual method is readily extended to handle the general minimum-cost flow problem. However, in the next section we give a primal simplex algorithm that seems to be more practical for solving large-scale minimum-cost network flow problems.

6. A PRIMAL SIMPLEX ALGORITHM FOR NETWORK FLOW PROBLEMS

The simplex method works very efficiently on network flow problems because the basis matrices have a very simple structure that greatly simplifies the calculations required in the pivot operations. Graphically, the arcs corresponding to basic variables induce subgraphs that are spanning trees. The trees provide a very simple way of calculating primal and dual solutions and the other quantities needed to do simplex pivots.

Let $\mathscr{D} = (V, \mathscr{A})$, where $V = \{1, \ldots, m\}$ and $\mathscr{A} = \{e_1, \ldots, e_n\}$ be the connected digraph of an instance of a network flow problem, and let A be the coefficient matrix of the conservation equations (4.1). Note that $A = (a_{ij})$ is the node-arc incidence matrix of $\mathscr{D}$, that is, if $e_j = (k, l)$ then $a_{kj} = -1$, $a_{lj} = 1$, and $a_{ij} = 0$ otherwise.

Proposition 6.1. *If A is the node-arc incidence matrix of a connected digraph $\mathscr{D}$ with m nodes, then* $\text{rank}(A) = m - 1$ (see Definition 1.3 of Section I.4.1).

Proof. $\Sigma_{i=1}^{m} a_{ij} = 0$ for all j, hence $\text{rank}(A) < m$.

To show that $\text{rank}(A) = m - 1$, let $T = (V, \mathscr{A}')$ be a spanning tree of $\mathscr{D}$ and let A_T be the $m \times (m-1)$ incidence matrix of T. The idea of the proof is to permute the rows and columns of A_T so that the $(m-1) \times (m-1)$ submatrix consisting of the first $m-1$ rows is lower triangular with the magnitude of each diagonal element equal to 1.

Let i_1 be a leaf of T so that the row of A_T corresponding to i_1 is a unit vector or its negative. Put row i_1 and the column corresponding to the arc e_{i_1} incident to node i_1 as the first row and column, respectively. Then delete node i_1 from T. The resulting graph is again a tree and thus contains a leaf, say i_2. Now the row corresponding to i_2 contains, at most, two nonzero elements, one corresponding to an arc $e_{i_2} \neq e_{i_1}$, and if there is another it corresponds to e_{i_1}. Hence by putting row i_2 and the column corresponding to e_{i_2} second, the first two rows are in lower triangular form. Now a straightforward induction yields the hypothesized lower triangular matrix with 1's (or −1's) on the diagonal. ■

We have seen, in the proof of Proposition 6.1, how a spanning tree on $\mathscr{D}$ yields an $(m-1) \times (m-1)$ nonsingular incidence matrix. But if $(V, \mathscr{A}')$, $\mathscr{A}' \subseteq \mathscr{A}$, $|\mathscr{A}'| = m - 1$ is not a spanning tree, then the underlying graph contains a cycle. Thus the incidence matrix of $(V, \mathcal{A}')$ contains a submatrix, which, after appropriate permutation of columns and multiplication of some columns by -1, is of the form

$$\begin{pmatrix} 1 & 0 & 0 & -1 \\ -1 & 1 & 0 & 0 \\ 0 & -1 & 1 & 0 \\ 0 & 0 & -1 & 1 \end{pmatrix}.$$

Hence the incidence matrix of $(V, \mathcal{A}')$ is singular, and we have shown the following:

Proposition 6.2. *There is a one-to-one correspondence between spanning trees on $\mathcal{D}$ and $(m-1) \times (m-1)$ nonsingular submatrices of A.*

Thus each spanning tree on $\mathcal{D}$ yields a basis matrix for the conservation equations (4.1), and if there are no upper-bound constraints (4.2), the tree corresponds to a primal feasible basis if the corresponding solution to (4.1) is nonnegative. Moreover, it is simple to compute the unique solution of (4.1) given that $x_{ij} = 0$ for all $(i, j) \in \mathcal{A}$ that are not tree arcs. We arbitrarily designate some node to be the root of the tree, say node r. Then we compute the solution of (4.1) recursively along each path from a leaf to the root, beginning with the arcs adjacent to the leaves.

An example of this computation is shown in Figure 6.1. Suppose we are given $T = (V, \mathcal{A}')$, a spanning tree of $\mathcal{D}$. We first compute the flows for the arcs incident to the leaves, that is, $x_{2r} = b_2$, $x_{61} = -b_1$, $x_{36} = b_3$, $x_{47} = b_4$, and $x_{75} = -b_5$. Then $x_{67} = -(b_7 + x_{47} - x_{75})$ is determined, and finally $x_{r6} = -(b_6 - x_{61} + x_{36} - x_{67})$. Note that flows balance at node r since $b_2 + b_r = x_{r6} = -(b_1 + b_3 + b_4 + b_5 + b_6 + b_7)$, and we have assumed that $\Sigma_{i=1}^{7} b_i + b_r = 0$.

Our computational scheme is nothing more than the obvious way of solving the lower triangular system beginning with the first variable, and so on. It illustrates that if the b_i are integral, then the solution will be integral, which is, of course, a consequence of the diagonal elements of the lower triangular basis matrix having a magnitude of 1.

The primal feasibility of a basis depends only on the vector b. For example, if $b = (b_1, b_2, \ldots, b_7) = (-3\ 2\ 3\ 4\ -5\ -1\ 0)$, the induced spanning tree of Figure 6.1 yields the basic feasible primal solution $x_{2r} = 2$, $x_{61} = 3$, $x_{36} = 3$, $x_{47} = 4$, $x_{75} = 5$, $x_{67} = 1$, $x_{r6} = 2$.

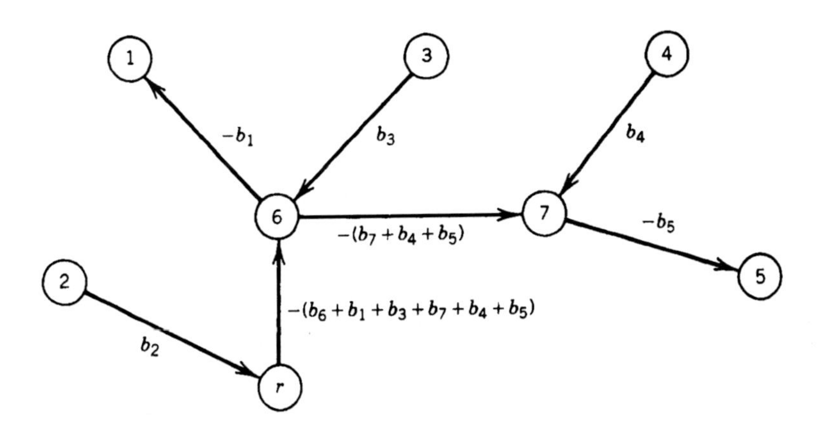

Figure 6.1

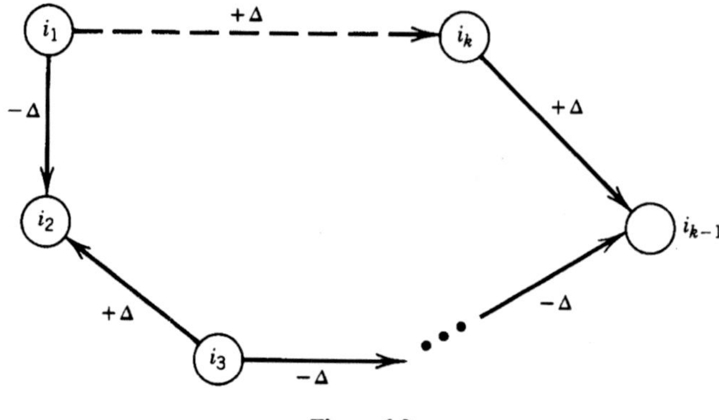

Figure 6.2

A Phase I procedure that uses artificial arcs may be necessary to determine an initial primal feasible basic solution.

Now suppose we have a basic feasible primal solution that is not optimal. The criterion for optimality (i.e., dual feasibility) will be discussed subsequently. A primal simplex pivot corresponds to adding an arc to the tree and then deleting an arc from the cycle (in the underlying undirected graph). The arc to be deleted is chosen to maintain primal feasibility.

The cycle of Figure 6.2 has been created by adding the arc $(i_1, i_k) \in \mathcal{A}$. Now observe that if we set $x_{i_1 i_k} = \Delta > 0$ the conservation equations will be satisfied by increasing the flow by Δ on all arcs of the cycle that have the same orientation as (i_1, i_k) and by decreasing the flow by Δ on all arcs of the cycle that have the opposite orientation. Thus if all arcs of the cycle have the same orientation as (i_1, i_k), the flow can be increased without bound. Otherwise, there is a unique largest value of $\Delta \geqslant 0$ (> 0 in the absence of degeneracy) given by

$$\Delta = \min\{x_{ij}: (i, j) \text{ is an arc of the cycle whose orientation is opposite from } (i_1, i_k)\}.$$

Suppose $\Delta = x_{i_p i_{p+1}}$. Then we obtain a new basis by deleting arc (i_p, i_{p+1}) from the cycle. The new solution $\hat{x}$ is given by

$$\hat{x}_{i_1 i_k} = \Delta, \ \hat{x}_{i_p i_{p+1}} = 0$$

and

$$\hat{x}_{ij} = \begin{cases} x_{ij} + \Delta & \text{if } (i, j) \text{ has the same orientation as } (i_1, i_k) \text{ in the cycle} \\ x_{ij} - \Delta & \text{if } (i, j) \text{ has the opposite orientation from } (i_1, i_k) \text{ in the cycle} \\ x_{ij} & \text{otherwise.} \end{cases}$$

In the absence of degeneracy, $\Delta > 0$ and $\hat{x} \neq x$.

Suppose in the example of Figure 6.1 with $b = (-3\ 2\ 3\ 4\ -5\ -1\ 0)$ we add the arc (3, 4) (see Figure 6.3). Then $\Delta = \min(x_{36}, x_{67}) = \min(3, 1) = 1$ and arc (6, 7) is deleted. The resulting spanning tree is shown in Figure 6.4.

We now consider the complementary dual solution and the computation of the reduced costs to establish optimality conditions and to find the arc to enter the tree when the optimality conditions do not hold.

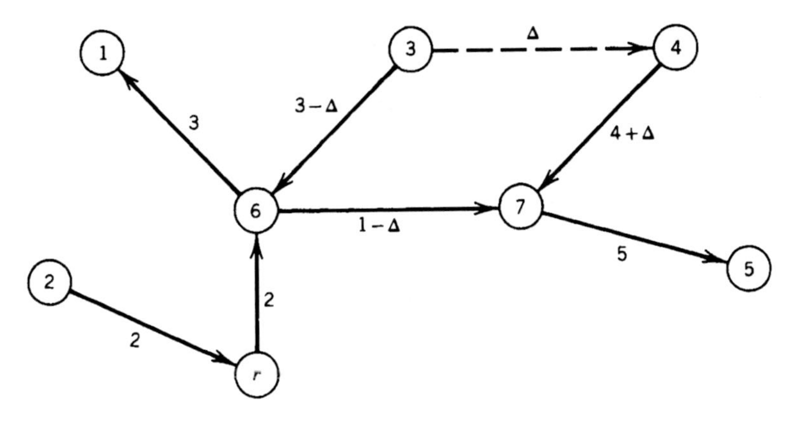

Figure 6.3

Corresponding to the equations (4.1) and $x \in R^n_+$ we obtain the dual constraints $y_i - y_j \leqslant c_{ij}$ for all $(i, j) \in \mathscr{A}$. By complementary slackness, $y_i - y_j = c_{ij}$ for all tree arcs. We can arbitrarily set $y_r = 0$ and then use these $m - 1$ equations to compute the remainder of the dual variables. Then if $y_i - y_j \leqslant c_{ij}$ is satisfied for all nontree arcs, the present solution is optimal. Otherwise, following the standard simplex criterion, we introduce an arc (i, j) for which the reduced price $\bar{c}_{ij} = c_{ij} - y_i + y_j$ is minimum.

The dual variables are computed by starting at the root of the tree with $y_r = 0$ and progressing toward the leaves (see Figure 6.5).

As with the primal variables, after changing the basis it is not necessary to recalculate all of the dual variables. For example, if we add the arc (3, 4) and delete the arc (6, 7), then the dual variables change only at nodes 4, 7, and 5, and we obtain the new solution $\hat{y}_i = y_i$ for $i = 1, 2, 3, 6$, $\hat{y}_4 = y_4 - \bar{c}_{34}$, $\hat{y}_7 = y_7 - \bar{c}_{34}$, and $\hat{y}_5 = y_5 - \bar{c}_{34}$.

As in the general primal simplex algorithm, the dual variables are needed only to calculate the reduced prices. But since the number of nontree arcs is generally much greater than the number of nodes, it makes sense to calculate and store the dual variables.

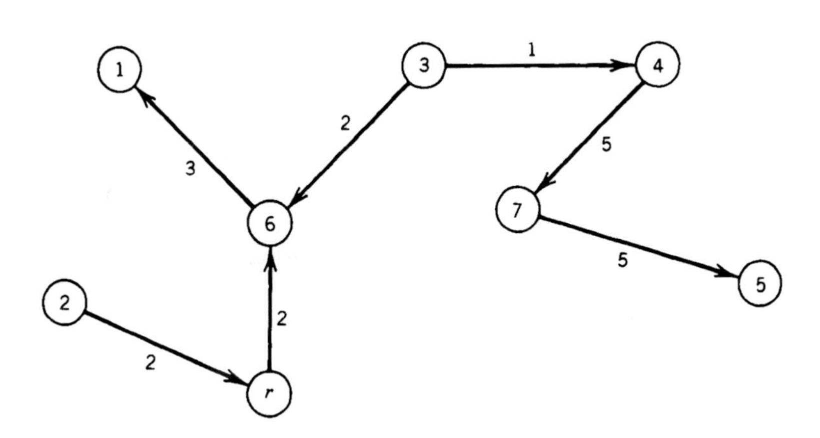

Figure 6.4

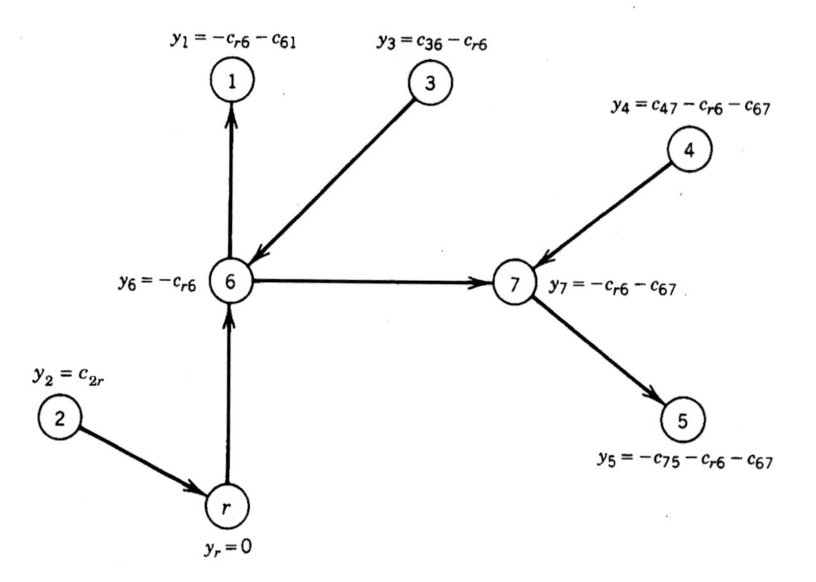

Figure 6.5

Example 6.1. We continue with the example that has been used to demonstrate the calculations. The data are

$$
C = \begin{pmatrix}
- & 8 & 5 & - & - & 9 & 2 & 4 \\
7 & - & 3 & 1 & - & - & 4 & - \\
2 & 7 & - & - & - & - & 5 & - \\
- & 0 & - & - & 1 & - & 7 & 7 \\
- & - & - & 6 & - & 4 & - & 2 \\
9 & - & - & - & 1 & - & 3 & 3 \\
5 & 2 & 2 & 4 & - & 8 & - & 2 \\
6 & - & - & 2 & 2 & 3 & 4 & -
\end{pmatrix}
\begin{matrix} r \\ 1 \\ 2 \\ 3 \\ 4 \\ 5 \\ 6 \\ 7 \end{matrix}
\qquad
b = \begin{pmatrix} 0 \\ -3 \\ 2 \\ 3 \\ 4 \\ -5 \\ -1 \\ 0 \end{pmatrix}.
$$

$$
\begin{matrix} r & 1 & 2 & 3 & 4 & 5 & 6 & 7 \end{matrix}
$$

In Figure 6.6, the number adjacent to the nodes are the dual variables for the first primal basic feasible solution, solid lines are tree arcs, and the numbers adjacent to them

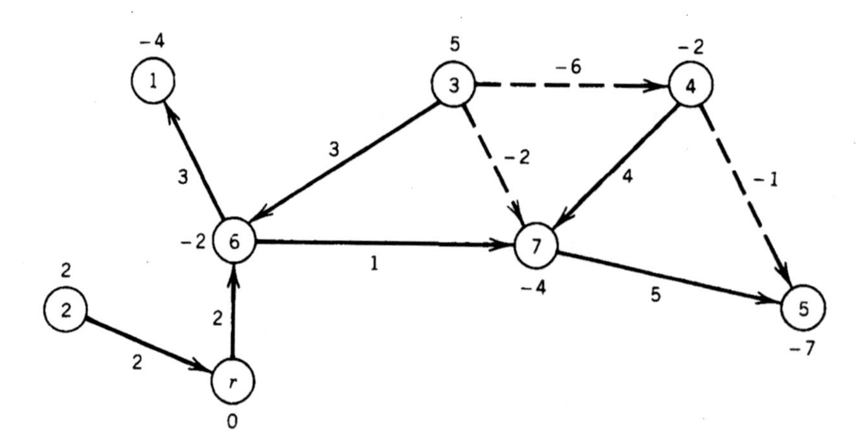

Figure 6.6

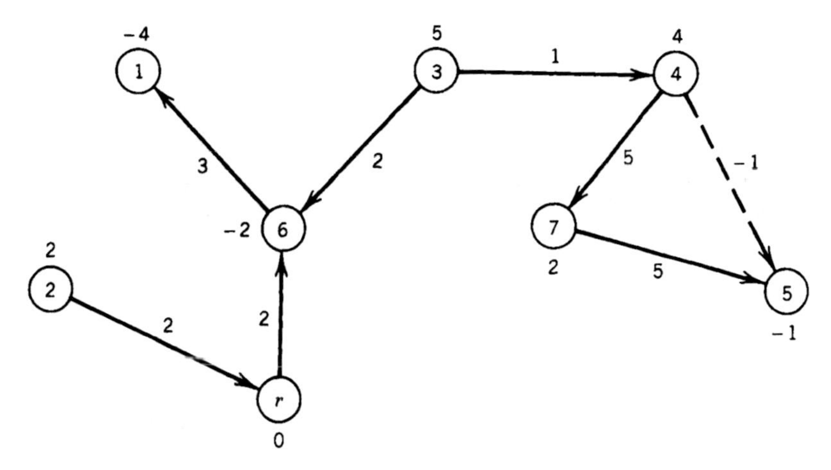

Figure 6.7

are the flows. With dotted lines, we show the arcs with negative reduced price, that is, the ones that want to enter the basis; the adjacent numbers are the reduced costs.

Arc (3, 4) enters the solution and arc (6, 7) leaves (see Figure 6.7). Now arc (4, 5) enters the solution. There is a tie for the leaving arc between (4, 7) and (7, 5). We choose (7, 5) and obtain the degenerate optimal solution shown in Figure 6.8. Now it can be checked that all reduced prices are nonnegative, so the solution shown in Figure 6.8 is optimal.

It is a simple matter to include arc capacities in the network simplex algorithm by treating upper-bound constraints implicitly. Thus, in the absence of degeneracy, if a variable is at its upper bound, the corresponding arc is not in the tree. The optimality conditions and pivot rules need to be modified accordingly.

Finally, it is important to observe that the effectiveness of the network simplex algorithm depends very substantially on the use of appropriate data structures for representing trees so that the calculations can be done efficiently.

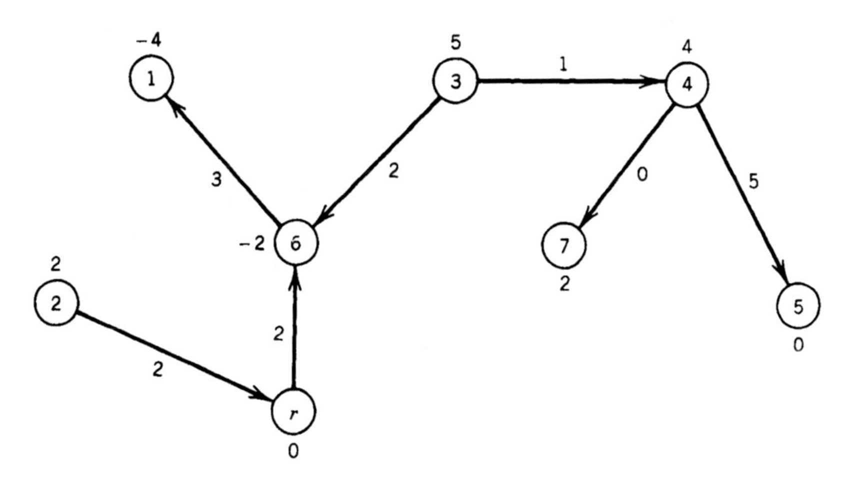

Figure 6.8

7. NOTES

Section I.3.1

Berge (1973), and Bondy and Murty (1976) are general books on graph theory.

Data structures are extremely important to the implementation of efficient graph algorithms (see Tarjan, 1983). Although we have not dealt with this important aspect of graph and network algorithms, the notes for each of the following sections contains a reference to an article that gives some of the recent results on efficient algorithms.

Section I.3.2

The shortest-path algorithm for nonnegative arc weights is due to Dijkstra (1959). The other algorithm appears in Ford and Fulkerson (1962). An earlier variant was given by Bellman (1958).

Gallo and Pallotino (1986) presented a survey of shortest-path algorithms.

Section I.3.3

The minimum-weight spanning tree algorithm is due to Kruskal (1956). Another classical algorithm is that of Prim (1957).

Gabow et al. (1986) have presented results on efficient spanning tree algorithms.

Section I.3.4

The classical reference for network flows is Ford and Fulkerson (1962). More recent texts are Bazarra and Jarvis (1977), Christofides (1975a), Jensen and Barnes (1980), Kennington and Helgason (1980), and Lawler (1976).

The maximum-flow algorithm presented is that of Ford and Fulkerson (1956).

Edmonds and Karp (1972) showed the efficiency of augmenting along shortest-length paths.

Tarjan (1986) gave a survey of efficient maximum-flow algorithms.

Section I.3.5

The primal–dual algorithm is due to Ford and Fulkerson (1962). Scaling was introduced by Edmonds and Karp (1972).

Bertsekas (1985) gave a unified framework of primal–dual network flow algorithms.

Section I.3.6

Kennington and Helgason (1980) gave a detailed presentation of primal simplex network flow algorithms, including a computer code for solving large-scale problems. Also see Glover, Karney, and Klingman (1974), Bradley et al. (1977) and Bland and Jensen (1987).

Ikura and Nemhauser (1986) gave a polynomial time dual simplex algorithm for the transportation problem and also investigated the use of scaling. A strongly polynomial network flow algorithm was described by Tardos (1985).

I.4

Polyhedral Theory

1. INTRODUCTION AND ELEMENTARY LINEAR ALGEBRA

A considerable portion of this book involves the description of a set of points in R^n by a set of linear inequalities. In linear programming, we are given a description of the feasible set of points by a set of linear inequalities $P = \{x \in R^n_+ : Ax \leq b\}$. When we solve a linear program by the simplex method, issues such as the dimension of P and which inequalities are necessary for the description of P do not need to be addressed.

Integer programming is different. Typically, we are given a set $S \subseteq Z^n_+$ of feasible points described implicitly, for example, the set of integer solutions to a linear inequality system $S = \{x \in Z^n_+ : Ax \leq b\}$, the set of binary vectors corresponding to tours in a graph, and so on. One of our objectives is to find a linear inequality description of the set.

Definition 1.1. Given a set $S \subseteq R^n$, a point $x \in R^n$ is a *convex combination* of points of S if there exists a finite set of points $\{x^i\}_{i=1}^t$ in S and a $\lambda \in R^t_+$ with $\Sigma_{i=1}^t \lambda_i = 1$ and $x = \Sigma_{i=1}^t \lambda_i x^i$. The *convex hull* of S, denoted by conv(S), is the set of all points that are convex combinations of points in S.

Figure 1.1 shows the convex hull of a set of integral points in R^2. We see that conv(S) can be described by a finite set of linear inequalities and that $\max\{cx: x \in S\} = \max\{cx: x \in \text{conv}(S)\}$. Moreover, the latter problem is a linear program. The validity of these observations for general integer programs is shown in Section 6.

Finding an inequality description of conv(S) is not easy, and questions such as the dimension of conv(S), the necessity of a certain inequality for the description of conv(S), and so on, are very important. Most of Chapter II.1 is devoted to finding such a description. To facilitate the later developments, we collect together in this chapter some basic results on polyhedra.

In this section we give, without proof, some standard results from linear algebra.

Definition 1.2. A set of points $x^1, \ldots, x^k \in R^n$ is *linearly independent* if the unique solution of $\Sigma_{i=1}^k \lambda_i x^i = 0$ is $\lambda_i = 0$, $i = 1, \ldots, k$.

Note that the maximum number of linearly independent points in R^n is n.

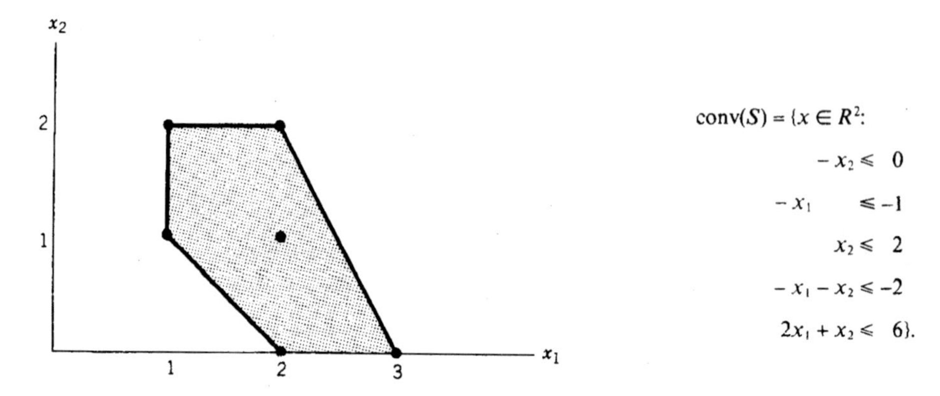

Figure 1.1. The black dots represent points in S; conv(S) is shaded.

Proposition 1.1. *If A is an $m \times n$ matrix, the maximum number of linearly independent rows of A, viewed as vectors $a^i \in R^n$, equals the maximum number of linearly independent columns of A, viewed as vectors $a_j \in R^m$.*

Definition 1.3. The maximum number of linearly independent rows (columns) of A is the *rank* of A and is denoted by rank(A).

Now we give a basic result for systems of linear equalities.

Proposition 1.2. *The following statements are equivalent:*

a. $\{x \in R^n: Ax = b\} \neq \emptyset$.
b. $\text{rank}(A) = \text{rank}(A, b)$.

When dealing with linear equalities and inequalities it is often more appropriate to use the concept of affine independence.

Definition 1.4. A set of points $x^1, \ldots, x^k \in R^n$ is *affinely independent* if the unique solution of $\Sigma_{i=1}^k \alpha_i x^i = 0$, $\Sigma_{i=1}^k \alpha_i = 0$ is $\alpha_i = 0$ for $i = 1, \ldots, k$.

Linear independence implies affine independence, but the converse is not true.

Proposition 1.3. *The following statements are equivalent:*

a. *$x^1, \ldots, x^k \in R^n$ are affinely independent.*
b. *$x^2 - x^1, \ldots, x^k - x^1$ are linearly independent.*
c. *$(x^1, -1), \ldots, (x^k, -1) \in R^{n+1}$ are linearly independent.*

Note that the maximum number of affinely independent points in R^n is $n + 1$ (e.g., n linearly independent points and the zero vector).

The following proposition will be used frequently in proving results concerning polyhedra.

Proposition 1.4. *If $\{x \in R^n: Ax = b\} \neq \emptyset$, the maximum number of affinely independent solutions of $Ax = b$ is $n + 1 - \text{rank}(A)$.*

Example 1.1. Suppose

$$(A, b) = \begin{pmatrix} 1 & -4 & -3 \\ -2 & 8 & 6 \end{pmatrix}.$$

Then $\text{rank}(A) = \text{rank}(A, b) = 1$. By Proposition 1.4, the maximum number of affinely independent solutions of $Ax = b$ is $3 - 1 = 2$. Two such solutions are $x^1 = (5 \quad 2)$ and $x^2 = (1 \quad 1)$.

Definition 1.5. $H \subseteq R^n$ is a *subspace* if $x \in H$ implies $\lambda x \in H$ for all $\lambda \in R^1$ and if $x, y \in H$ implies $x + y \in H$.

Proposition 1.5. *The following are equivalent:*

a. *$H \subseteq R^n$ is a subspace.*
b. *There is an $m \times n$ matrix A such that $H = \{x \in R^n: Ax = 0\}$.*
c. *There is a $k \times n$ matrix B such that $H = \{x \in R^n: x = uB, u \in R^k\}$.*

Proposition 1.6. *If $H \subseteq R^n$ is a subspace, then $\{x \in R^n: xy = 0 \text{ for } y \in H\}$ is a subspace.*

This subspace is called the *orthogonal subspace* of H and is denoted by $H^\perp$.

Proposition 1.7. *If $H = \{x \in R^n: Ax = 0\}$, with A being an $m \times n$ matrix, then $H^\perp = \{x \in R^n: x = A^T u, u \in R^m\}$.*

Example 1.2. $H = \{x \in R^2: x_1 = 2x_2\}$ is a subspace. Here $A = (1 \quad -2)$ and $B = (2 \quad 1)$.

$$H^\perp = \left\{x \in R^2: x = \begin{pmatrix} 1 \\ -2 \end{pmatrix} u\right\}$$

$$= \{x \in R^2: 2x_1 + x_2 = 0\} \quad \text{(see Figure 1.2).}$$

Definition 1.6. If $p \in R^n$ and H is a subspace, the *projection of p on H* is the vector $q \in H$ such that $p - q \in H^\perp$. The *projection of S on H* is denoted by $\text{proj}_H(S) = \{q: q$ is the projection of p on H for some $p \in S\}$.

2. DEFINITIONS OF POLYHEDRA AND DIMENSION

Definition 2.1. A *polyhedron* $P \subseteq R^n$ is the set of points that satisfy a finite number of linear inequalities; that is, $P = \{x \in R^n: Ax \leqslant b\}$, where (A, b) is an $m \times (n + 1)$ matrix. A polyhedron is said to be *rational* if there exists an $m' \times (n + 1)$ matrix (A', b') with rational coefficients such that $P = \{x \in R^n: A'x \leqslant b'\}$.

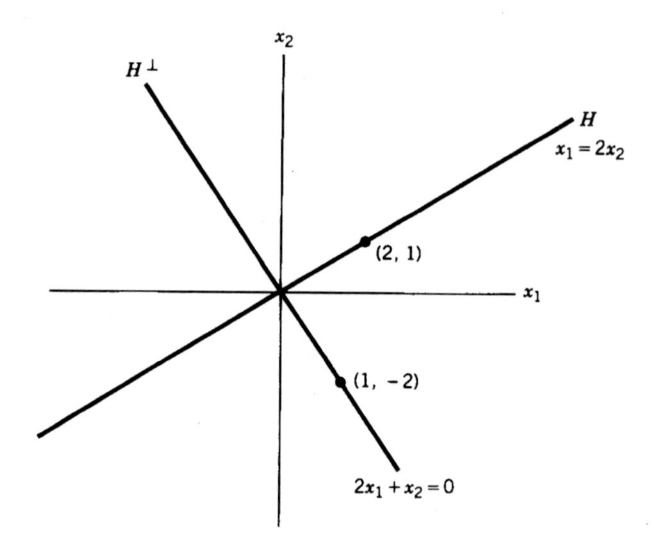

Figure 1.2

Throughout the text we consider only rational polyhedra and assume that if P is stated as $\{x \in R^n: Ax \leqslant b\}$, then (A, b) has rational coefficients.

Definition 2.2. A polyhedron $P \subseteq R^n$ is *bounded* if there exists an $\omega \in R^1_+$ such that $P \subseteq \{x \in R^n: -\omega \leqslant x_j \leqslant \omega$ for $j = 1, \dots, n\}$. A bounded polyhedron is called a *polytope*.

Definition 2.3. $T \subseteq R^n$ is a *convex* set if $x^1, x^2 \in T$ implies that $\lambda x^1 + (1 - \lambda)x^2 \in T$ for all $0 \leqslant \lambda \leqslant 1$.

Proposition 2.1. *A polyhedron is a convex set.*

Definition 2.4. $C \subseteq R^n$ is a *cone* if $x \in C$ implies $\lambda x \in C$ for all $\lambda \in R^1_+$.

Proposition 2.2. *The polyhedron* $\{x \in R^n: Ax \leqslant 0\}$ *is a cone.*

Definition 2.5. A polyhedron P is of *dimension* k, denoted by $\dim(P) = k$, if the maximum number of affinely independent points in P is $k + 1$.

Definition 2.6. A polyhedron $P \subseteq R^n$ is *full-dimensional* if $\dim(P) = n$.

Below we will show that if P is not full-dimensional, then at least one of the inequalities $a^i x \leqslant b_i$ is satisfied at equality by all points of P.

Let $M = \{1, 2, \dots, m\}$, $M^= = \{i \in M: a^i x = b_i$ for all $x \in P\}$ and let $M^\leqslant = \{i \in M: a^i x < b_i$ for some $x \in P\} = M \setminus M^=$. Let $(A^=, b^=)$, $(A^\leqslant, b^\leqslant)$ be the corresponding rows of (A, b). We refer to the *equality* and *inequality* sets of the representation (A, b) of P, that is, $P = \{x \in R^n: A^= x = b^=, A^\leqslant x \leqslant b^\leqslant\}$. Note that if $i \in M^\leqslant$, then (a^i, b_i) cannot be written as a linear combination of the rows of $(A^=, b^=)$.

Definition 2.7. $x \in P$ is called an *inner point* of P if $a^i x < b_i$ for all $i \in M^\leqslant$.

Definition 2.8. $x \in P$ is called an *interior point* of P if $a^i x < b_i$ for all $i \in M$.

Proposition 2.3. *Every nonempty polyhedron P has an inner point.*

Proof. If $M^{\leqslant} = \emptyset$, every point of P is inner. Otherwise, for each $i \in M^{\leqslant}$ there exists a point $x^i \in P$ with $a^i x^i < b_i$. Now $\hat{x} = (1/|M^{\leqslant}|)\ \Sigma_{i \in M^{\leqslant}}\, x^i \in P$ since P is convex. Since $a^i \hat{x} < b_i$ for all $i \in M^{\leqslant}$, $\hat{x}$ is an inner point. ■

Now we relate the dimension of P to the rank of its equality matrix $(A^=, b^=)$. Below we will always assume that $P \neq \emptyset$. However, the next result is still valid with the convention that if $P = \emptyset$, then $\dim(P) = -1$.

Proposition 2.4. *If* $P \subseteq R^n$, *then* $\dim(P) + \text{rank}(A^=, b^=) = n$.

Proof. Suppose $\text{rank}(A^=) = \text{rank}(A^=, b^=) = n - k$, where $0 \leqslant k \leqslant n$. Then by Proposition 1.4 there are $k + 1$ affinely independent solutions of $A^= x = 0$. Let $y^1, \ldots, y^{k+1}$ denote any such solutions, and let $\hat{x}$ be an inner point of P. Now for ε sufficiently small, $\hat{x} + \varepsilon y^i$ for $i = 1, \ldots, k + 1$ are affinely independent points in P. Thus $\dim(P) \geqslant k$ and we have that $\dim(P) + \text{rank}(A^=, b^=) \geqslant n$.

Now suppose that $\dim(P) = k$ and that $x^1, \ldots, x^{k+1}$ are affinely independent points of P. Since $A^= x^j = b^=$ for $j = 1, \ldots, k + 1$, by Proposition 1.4 we have $\text{rank}(A^=, b^=) \leqslant (n + 1) - (k + 1) = n - k$. Hence $\dim(P) + \text{rank}(A^=, b^=) \leqslant n$. ■

Corollary 2.5. *A polyhedron P is full-dimensional if and only if it has an interior point.*

Note that we have shown that $\text{rank}(A^=, b^=)$ is independent of the particular inequality description of P.

Example 2.1. Suppose $P \subset R^3$ is given by

$$
\begin{aligned}
x_1 + x_2 + x_3 &\leqslant 1\\
-x_1 - x_2 - x_3 &\leqslant -1\\
x_1 \phantom{{}+x_2} + x_3 &\leqslant 1\\
-x_1 \phantom{{}-x_2-2x_3} &\leqslant 0\\
-x_2 \phantom{{}-2x_3} &\leqslant 0\\
x_3 &\leqslant 2\\
x_1 + x_2 + 2x_3 &\leqslant 2
\end{aligned}
$$

(see Figure 2.1).

The three points $(1\ \ 0\ \ 0)$, $(0\ \ 1\ \ 0)$, $(0\ \ 0\ \ 1)$ lie in P and are affinely independent. Hence $\dim(P) \geqslant 2$. Because all points of P satisfy the equality $x_1 + x_2 + x_3 = 1$, we have $\text{rank}(A^=, b^=) \geqslant 1$; hence, by Proposition 2.4, $\dim(P) \leqslant 2$. Therefore $\dim(P) = 2$.

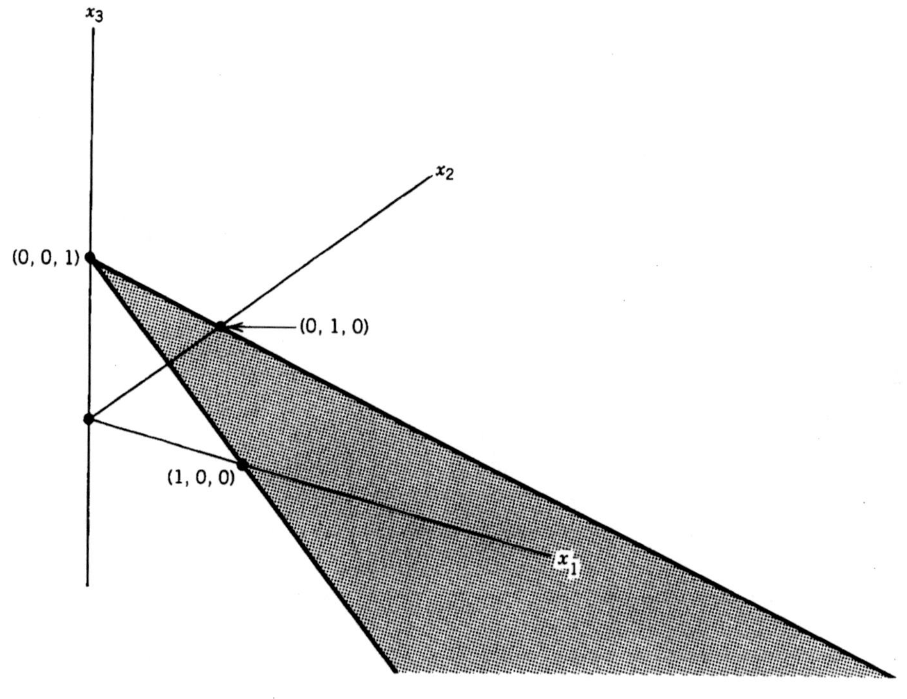

Figure 2.1

3. DESCRIBING POLYHEDRA BY FACETS

Given a polyhedron $P = \{x \in R^n: Ax \leqslant b\}$, the question we address below is to find out which of the inequalities $a^i x \leqslant b_i$ are necessary in the description of P and which can be dropped. In fact we will show that those necessary to describe P are the same, whatever the initial inequality description of P.

Definition 3.1. The inequality $\pi x \leqslant \pi_0$ [or (π, π_0)] is called a *valid inequality* for P if it is satisfied by all points in P.

Note that (π, π_0) is a valid inequality if and only if P lies in the half-space $\{x \in R^n: \pi x \leqslant \pi_0\}$, or equivalently if and only if $\max\{\pi x: x \in P\} \leqslant \pi_0$ (see Figure 3.1).

Definition 3.2. If (π, π_0) is a valid inequality for P, and $F = \{x \in P: \pi x = \pi_0\}$, F is called a *face* of P, and we say that (π, π_0) *represents* F. A face F is said to be *proper* if $F \neq \emptyset$ and $F \neq P$.

The face F represented by (π, π_0) is nonempty if and only if $\max\{\pi x: x \in P\} = \pi_0$. When F is nonempty, we say that (π, π_0) *supports* P.

As a first step in discarding superfluous inequalities, note that we can discard inequalities $a^i x \leqslant b_i$ that are not supports of P. Hence from now on we suppose that all the inequalities $a^i x \leqslant b_i$ for $i \in M$ support P and therefore represent nonempty faces.

Proposition 3.1. *If $P = \{x \in R^n: Ax \leqslant b\}$ with equality set $M^= \subseteq M$, and F is a nonempty face of P, then F is a polyhedron and $F = \{x \in R^n: a^i x = b_i$ for $i \in M_F^=$, $a^i x \leqslant b_i$ for $i \in M_F^\leqslant\}$ where $M_F^= \supseteq M^=$ and $M_F^\leqslant = M \setminus M_F^=$. The number of distinct faces of P is finite.*

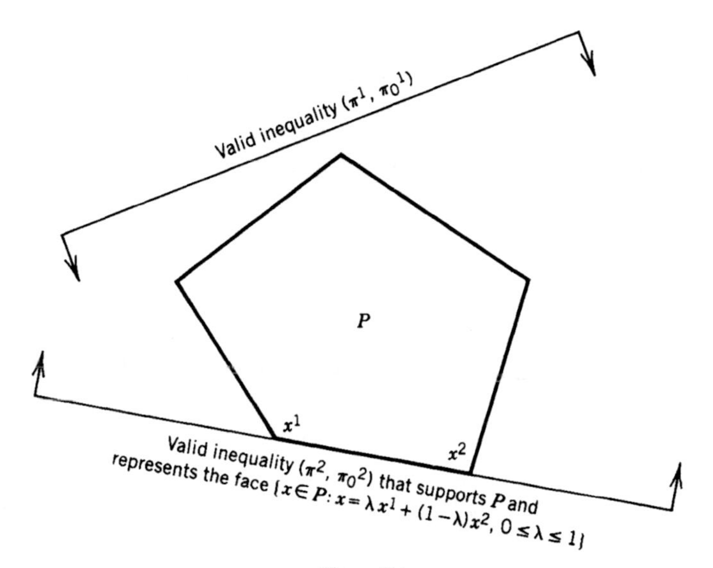

Figure 3.1

Proof. Suppose F is the set of optimal solutions to the linear program $\pi_0 = \max\{\pi x: Ax \leqslant b\}$. Let u^* be an optimal solution to the dual linear program $\min\{ub: uA = \pi, u \in R_+^m\}$, and let $I^* = \{i \in M: u_i^* > 0\}$. Now consider the polyhedron $F^* = \{x \in R^n: a^i x = b_i \text{ for } i \in I^*, a^i x \leqslant b_i \text{ for } i \in M \setminus I^*\}$. We claim that $F = F^*$.

Note first that if $x \in F^*$, then

$$\pi x = u^* A x = \sum_{i \in I^*} u_i^* a^i x = \sum_{i \in I^*} u_i^* b_i = \pi_0.$$

But if $x \in P \setminus F^*$, then $a^k x < b_k$ for some $k \in I^*$, so $u_k^* > 0$ and $\pi x = \Sigma_{i \in I^*} u_i^* a^i x < \Sigma_{i \in I^*} u_i^* b_i = \pi_0$. Hence $F = F^*$ and F is a polyhedron. Since $F \subseteq P$, the equality set $(A_F^=, b_F^=)$ of F must have the required property.

Finally, since M is finite, the possible equality subsets $M_F^=$ [corresponding to the rows of $(A_F^=, b_F^=)$] are finite in number, so the number of distinct faces is finite. ∎

Note that by Proposition 2.4, if F is a proper face of P, then $\dim(F) < \dim(P)$. In particular, the dimension of F is k if the maximum number of affinely independent points that lie in F is $k + 1$.

Definition 3.3. A face F of P is a *facet* of P if $\dim(F) = \dim(P) - 1$.

Proposition 3.2. *If F is a facet of P, there exists some inequality $a^k x \leqslant b_k$ for $k \in M^{\leqslant}$ representing F.*

Proof. Since $\dim(F) = \dim(P) - 1$, it follows from Proposition 2.4 that $\text{rank}(A_F^=, b_F^=) = \text{rank}(A^=, b^=) + 1$. The result follows. ∎

Example 2.1 (continued). $(\pi, \pi_0) = (-1 \quad -1 \quad 1, \quad 1)$ is a valid inequality for P because $\max\{-x_1 - x_2 + x_3: x \in P\} = 1 = \pi_0$. Also $F_1 = \{x \in P: -x_1 - x_2 + x_3 = 1\} = \{(0 \quad 0 \quad 1)\}$ is a face of P. Note that the face F_1 is not generated by any of the inequalities $a^i x \leqslant b_i$ in the description of P.

Now consider the face F_2 generated by the valid inequality $2x_1 - 7x_2 + 2x_3 \leq 2$, that is, $F_2 = \{x \in P: 2x_1 - 7x_2 + 2x_3 = 2\}$. The two points $(1 \quad 0 \quad 0)$ and $(0 \quad 0 \quad 1)$ lie in F_2 and are affinely independent. In addition, the point $(0 \quad 1 \quad 0) \in P$ does not lie on F_2, so $F_2 \subset P$. Since $\dim(P) = 2$ and $\dim(F_2) \geq 1$, we have $\dim(F_2) = 1$. Thus F_2 is a facet of P.

Now from Proposition 3.2, one of the initial inequalities must represent F_2. In fact, both $x_1 + x_3 \leq 1$ and $-x_2 \leq 0$ represent the facet F_2.

Finally consider $(\pi, \pi_0) = (0 \quad 0 \quad 1, \quad 2)$. Now $\max\{x_3: x \in P\} = 1 < \pi_0$, so $x_3 \leq 2$ is a valid inequality but not a support of P. Hence $x_3 \leq 2$ can be discarded from the description of P.

Proposition 3.3. *For each facet F of P, one of the inequalities representing F is necessary in the description of P.*

Proof. Let P_F be the polyhedron obtained from P by removing all the inequalities representing F. We will show that $P_F \setminus P \neq \emptyset$ so that at least one of the inequalities is necessary. Let $\hat{x}$ be an inner point of the facet F and let $a^r x \leq b_r$ be an inequality representing F. Since a^r is linearly independent of the rows of $A^=$, it follows from the Farkas lemma that there exists a $y \in R^n$ such that $A^= y = 0$ and $a^r y > 0$. Because $\hat{x}$ is an inner point of F, $a^i \hat{x} < b_i$ for all inequalities $i \in M^{\leq}$ that do not represent F. But now $\hat{x} + \varepsilon y \in P_F \setminus P$ for sufficiently small $\varepsilon > 0$. ■

Besides being necessary, the facets are sufficient for the description of P.

Proposition 3.4. *Every inequality $a^r x \leq b_r$ for $r \in M^{\leq}$ that represents a face of P of dimension less than $\dim(P) - 1$ is irrelevant to the description of P.*

Proof. Suppose $a^r x \leq b_r$ represents a face F of P of dimension $\dim(P) - k$ with $k > 1$, and the inequality is not irrelevant. In other words, there exists $x^* \in R^n$ such that $A^= x^* = b^=$, $a^i x^* \leq b_i$ for $i \in M^{\leq} \setminus \{r\}$, and $a^r x^* > b_r$. Let $\hat{x}$ be an inner point of P. Then on the line between x^* and $\hat{x}$ there exists a point z in F satisfying $A^= z = b^=$, $a^i z < b_i$ for $i \in M^{\leq} \setminus \{r\}$, and $a^r z = b_r$. Hence the equality set of F is $(A^=, b^=)$ and (a^r, b_r), which is of rank $n - \dim(P) + 1$. Therefore the dimension of F is $\dim(P) - 1$, which is a contradiction. ■

Example 2.1 (continued). We verify that the face $F_2 = \{x \in R^3: x_1 + x_3 = 1, x \in P\}$ is a facet of P. The equality set of F_2 is

$$(A^=_{F_2}, b^=_{F_2}) = \begin{pmatrix} -1 & -1 & -1 & -1 \\ 1 & 1 & 1 & 1 \\ 1 & 0 & 1 & 1 \\ 0 & -1 & 0 & 0 \end{pmatrix},$$

which is a matrix of rank 2. Hence, by Proposition 2.4, F_2 is of dimension 1. Thus F_2 is a facet represented either by $x_1 + x_3 \leq 1$ or $-x_2 \leq 0$. In fact, since

$$2x_1 - 7x_2 + 2x_3 - 2 = 2(x_1 + x_3 - 1) - 7x_2,$$

F_2 is also represented by $2x_1 - 7x_2 + 2x_3 \leq 2$, which is the representation we gave earlier.

Similarly it can be shown that $-x_1 \leq 0$ defines a facet. Now consider $x_1 + x_2 + 2x_3 \leq 2$, which is a support of P. Let

$$\begin{aligned} F_3 &= \{x \in P: x_1 + x_2 + 2x_3 = 2\} \\ &= \{x \in P: x_1 + x_2 + x_3 = 1, x_1 + x_2 + 2x_3 = 2, x_1 + x_3 = 1, -x_1 = 0, -x_2 = 0\}. \end{aligned}$$

Hence $(A^=_{F_3}, b^=_{F_3})$ is of rank 3. Thus the face F_3 is of dimension 0. In fact $F_3 = F_1 = \{(0 \quad 0 \quad 1)\}$, and hence $x_1 + x_2 + 2x_3 \leq 2$ is redundant. Therefore a minimal description of P is given by

$$\begin{array}{rcl} x_1 + x_2 + x_3 &=& 1 \\ -x_1 \qquad\qquad &\leq& 0 \\ x_1 \qquad + x_3 &\leq& 1 \\ x \in R^3 && \end{array} \quad \text{or} \quad \begin{array}{rcl} x_1 + x_2 + x_3 &=& 1 \\ -x_1 \qquad\qquad &\leq& 0 \\ - x_2 \qquad &\leq& 0 \\ x \in R^3. && \end{array}$$

The example raises the question as to when two inequalities (e.g., $x_1 + x_3 \leq 1, -x_2 \leq 0$) are "equivalent". The answer is straightforward. The set $\{x: A^=x = b^=, \pi x \leq \pi_0\} = \{x: A^=x = b^=, (\lambda\pi + uA^=)x \leq \lambda\pi_0 + ub^=\}$ for all $\lambda > 0$ and all $u \in R^{|M^=|}$. Hence we say that (π^1, π_0^1) and (π^2, π_0^2) are *equivalent, or identical inequalities* with respect to P when $(\pi^2, \pi_0^2) = \lambda(\pi^1, \pi_0^1) + u(A^=, b^=)$ for some $\lambda > 0$ and $u \in R^{|M^=|}$. Now we can summarize the main result given so far.

Theorem 3.5

a. *A full-dimensional polyhedron P has a unique (to within scalar multiplication) minimal representation by a finite set of linear inequalities. In particular, for each facet F_i of P there is an inequality $a^ix \leq b_i$ (unique to within scalar multiplication) representing F_i and $P = \{x \in R^n: a^ix \leq b_i$ for $i = 1, \ldots, t\}$.*

b. *If $\dim(P) = n - k$ with $k > 0$, then $P = \{x \in R^n: a^ix = b_i$ for $i = 1, \ldots, k, a^ix \leq b_i$ for $i = k + 1, \ldots, k + t\}$. For $i = 1, \ldots, k$, (a^i, b_i) are a maximal set of linearly independent rows of $(A^=, b^=)$, and for $i = k + 1, \ldots, k + t$, (a^i, b_i) is any inequality from the equivalence class of inequalities representing the facet F_i.*

We now give a theorem that characterizes facets and that is useful in establishing when a valid inequality is a facet.

Theorem 3.6. *Let $(A^=, b^=)$ be the equality set of $P \subseteq R^n$ and let $F = \{x \in P: \pi x = \pi_0\}$ be a proper face of P. The following two statements are equivalent:*

i. *F is a facet of P.*
ii. *If $\lambda x = \lambda_0$ for all $x \in F$ then*

(3.1) $\quad (\lambda, \lambda_0) = (\alpha\pi + uA^=, \alpha\pi_0 + ub^=)$ *for some $\alpha \in R^1$ and some $u \in R^{|M^=|}$.*

Proof. ii $\Rightarrow$ i. Let $L = \{(\lambda, \lambda_0) \in R^{n+1}: (\lambda, \lambda_0)$ is of the form (3.1)$\}$ and $L' = \{(\lambda, \lambda_0) \in R^{n+1}: \lambda x = \lambda_0$ for all $x \in F\}$. $L \subseteq L'$, since

$$\alpha\pi x + uA^{=}x = \alpha\pi_0 + ub^{=} \quad \text{for all } x \in F.$$

By the hypothesis, $L' \subseteq L$. Hence $L = L'$.

Suppose that $\dim(P) = n - k$ so that $\text{rank}(A^{=}, b^{=}) = k$. Since F is a proper face, (π, π_0) is not a linear combination of the rows of $(A^{=}, b^{=})$. Thus L is a $(k + 1)$-dimensional subspace.

Now let $x^1, \ldots, x^r$ be a maximal set of affinely independent points in F and let

$$D = \begin{pmatrix} x^1 & -1 \\ \vdots & \vdots \\ x^r & -1 \end{pmatrix}$$

be an $r \times (n + 1)$ matrix. Clearly $r \leqslant n - k$.

By Proposition 1.4, the maximum number of affinely independent solutions of $(\lambda, \lambda_0)D^T = 0$ is $(n + 1) + 1 - \text{rank}(D) = n + 2 - r$. Thus L' is an $(n + 1 - r)$-dimensional subspace. Since $L = L'$, $r = n - k$. Hence F is a facet of P.

i $\Rightarrow$ ii. As above, $L \subseteq L'$. Here we need to show that $L = L'$. Suppose $\dim(P) = n - k$. Since F is a facet of P, F contains $n - k$ affinely independent points. Thus, as in the proof of ii $\Rightarrow$ i, $\dim(L') = k + 1$. Since $\dim(L) = k + 1$ and $L \subseteq L'$, $L = L'$. ■

4. DESCRIBING POLYHEDRA BY EXTREME POINTS AND EXTREME RAYS

Here we consider a representation of polyhedra in terms of lowest-dimensional faces.

Proposition 4.1. *If* $P = \{x \in R^n: Ax \leqslant b\} \neq \emptyset$ *and* $\text{rank}(A) = n - k$, *P has a face of dimension k and has no proper face of lower dimension.*

Proof. For any face $F \neq \emptyset$ of P, $\text{rank}(A_F^{=}, b_F^{=}) \leqslant n - k$. Hence, by Proposition 2.4, the dimension of F is greater than or equal to k. Now let F be a face of P of minimum dimension. If $\dim(F) = k = 0$, there is nothing to prove. So suppose $\dim(F) > 0$.

Let $\hat{x}$ be an inner point of F. Since $\dim(F) > 0$, there exists some other point y of F. Consider the line joining $\hat{x}$ and y, that is, $z(\lambda) = \hat{x} + \lambda(y - \hat{x})$ where $\lambda \in R^1$. Suppose that the line intersects $a^i x = b_i$ for some $i \in M_F^{\leqslant}$. Let $\lambda^* = \min\{|\lambda^i|: i \in M_F^{\leqslant}, z(\lambda^i)$ lies in $a^i x = b_i\}$, and $\lambda^* = |\lambda^{i^*}|$. Then $\lambda^* \neq 0$ because $\hat{x}$ is an inner point. Thus $F_{i^*} = \{x \in P: A_F^{=}x = b_F^{=}, a^{i^*}x = b_{i^*}\} \neq \emptyset$ is a face of P of smaller dimension than F, which is a contradiction.

Therefore the line does not intersect $a^i x = b_i$ for any $i \in M_F^{\leqslant}$. But this means that $A\hat{x} + A\lambda(y - \hat{x}) \leqslant b$ for all $\lambda \in R^1$. Since $A\hat{x} \leqslant b$, this implies that $A(y - \hat{x}) = 0$ for all $y \in F$. Thus $F = \{y: Ay = A\hat{x}\}$. Since $\text{rank}(A) = n - k$, Proposition 2.4 implies that $\dim(F) = k$. ■

Example 4.1. $P = \{x \in R^2: x_1 + x_2 \leqslant 1\}$. See Figure 4.1. We have $\text{rank}(A) = 1$. A face of minimum dimension is the one-dimensional face $F = \{x \in R^2: x_1 + x_2 = 1\}$.

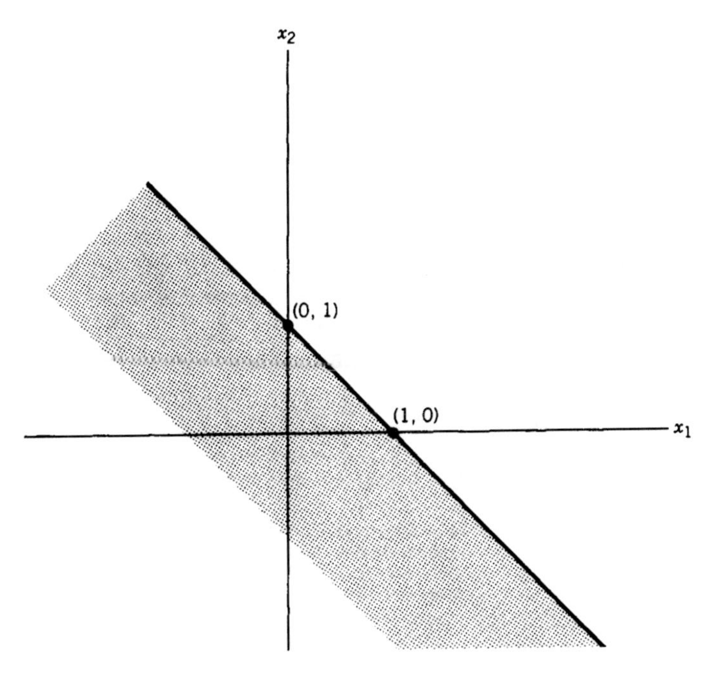

Figure 4.1

In practice, we frequently deal with polyhedra lying within the nonnegative orthant R^n_+. For such polyhedra, $\text{rank}(A) = n$; and by Proposition 4.1, these polyhedra have zero-dimensional faces. For this reason and for simplicity, we assume for the next two sections that $\text{rank}(A) = n$. Note also that if $P = \{x \in R^n: Ax \leqslant b\}$ is a polytope, then $\text{rank}(A) = n$.

Definition 4.1. $x \in P$ is an *extreme point* of P if there do not exist $x^1, x^2 \in P$, $x^1 \neq x^2$, such that $x = \frac{1}{2}x^1 + \frac{1}{2}x^2$.

For $x \in P$, let $(A^=_x, b^=_x)$ be the equality set of x, i.e. $(A^=_x, b^=_x) = (A^=_F, b^=_F)$ where F is the face of minimum dimension containing x, and x is an inner point of F.

Proposition 4.2. *x is an extreme point of P if and only if x is a zero-dimensional face of P.*

Proof. Suppose x is a zero-dimensional face of P. By Proposition 2.4, $\text{rank}(A^=_x) = n$. Let $(\tilde{A}, \tilde{b})$ be a submatrix of $(A^=_x, b^=_x)$ with $\tilde{A}$ $n \times n$ and nonsingular, so $x = \tilde{A}^{-1}\tilde{b}$. If $x = \frac{1}{2}x^1 + \frac{1}{2}x^2$, $x^1, x^2 \in P$, then since $\tilde{A}x^i \leqslant \tilde{b}$ for $i = 1, 2$, $\tilde{A}x^1 = \tilde{A}x^2 = \tilde{b}$. Hence $x^1 = x^2 = x$, so x is an extreme point.

If $x \in P$ is not a zero-dimensional face of P, then by Proposition 2.4 we have $\text{rank}(A^=_x) < n$. But now there exists $y \neq 0$ satisfying $A^=_x y = 0$, and for sufficiently small ε, $x^1 = x + \varepsilon y \in P$ and $x^2 = x - \varepsilon y \in P$. Now $x = \frac{1}{2}x^1 + \frac{1}{2}x^2$, so x is not an extreme point. ■

Definition 4.2. Let $P^0 = \{r \in R^n: Ar \leqslant 0\}$. If $P = \{x \in R^n: Ax \leqslant b\} \neq \emptyset$, then $r \in P^0 \setminus \{0\}$ is called a *ray* of P.

A point $r \in R^n$ is a ray of P if and only if for any point $x \in P$, the set $\{y \in R^n: y = x + \lambda r, \lambda \in R^1_+\} \subseteq P$.

Definition 4.3. A ray r of P is an *extreme ray* if there do not exist rays $r^1, r^2 \in P^0$, $r^1 \neq \lambda r^2$ for any $\lambda \in R^1_+$, such that $r = \frac{1}{2}r^1 + \frac{1}{2}r^2$.

Proposition 4.3. *If $P \neq \varnothing$, r is an extreme ray of P if and only if $\{\lambda r: \lambda \in R^1_+\}$ is a one-dimensional face of P^0.*

Proof. Let $A_r^= = \{a^i: i \in M, a^i r = 0\}$. If $\{\lambda r: \lambda \in R^1_+\}$ is a one-dimensional face of P^0, $\text{rank}(A_r^=) = n - 1$. Hence all solutions of $A_r^= y = 0$ are of the form $y = \lambda r, \lambda \in R^1$. If $r = \frac{1}{2}r^1 + \frac{1}{2}r^2$, we obtain a contradiction as in the previous proposition.

If $r \in P^0$ and rank $(A_r^=) < n - 1$, there exists $\hat{r} \neq \lambda r$, $\lambda \in R^1$, such that $A_r^= \hat{r} = 0$. The rays $r^1 = r + \varepsilon\hat{r}$, $r^2 = r - \varepsilon\hat{r}$ show that r is not an extreme ray. ■

Corollary 4.4. *A polyhedron has a finite number of extreme points and extreme rays.*

Example 2.1 (continued). Since the inequalities describing P include

$$\begin{aligned} -x_1 \quad\quad\quad\quad &\leq 0 \\ -x_2 \quad\quad &\leq 0 \\ x_3 &\leq 2, \end{aligned}$$

it is clear that $\text{rank}(A) = 3$.

The face $F_1 = \{(0\ 0\ 1)\}$ has the equality set

$$\begin{aligned} x_1 + x_2 + x_3 &= 1 \quad \text{(also the negative of this row which is omitted)} \\ x_1 \quad\quad\quad\quad &= 0 \\ x_2 \quad\quad &= 0 \end{aligned}$$

and since $\begin{pmatrix} 1 & 1 & 1 \\ 1 & 0 & 0 \\ 0 & 1 & 0 \end{pmatrix}$ is of rank 3, $(0\ \ 0\ \ 1)$ is a zero-dimensional face, or extreme point.

Note that $r^1 = (1\ \ 0\ \ -1)$ satisfies

$$\begin{aligned} x_1 + x_2 + x_3 &= 0 \\ x_1 \quad\quad + x_3 &= 0 \\ x_2 \quad\quad &= 0 \end{aligned}$$

and $a^i r^i < 0$ for all other constraints. Since $\begin{pmatrix} 1 & 1 & 1 \\ 1 & 0 & 1 \\ 0 & 1 & 0 \end{pmatrix}$ is of rank 2, r^1 is an extreme ray. A similar argument shows that $r^2 = (0\ \ 1\ \ -1)$ is another extreme ray. The polyhedron P^0 for Example 2.1 is shown in Figure 4.2.

A polyhedron can be represented in terms of its extreme points and extreme rays. Some preliminaries are needed to obtain this fundamental result.

Theorem 4.5. *If $P \neq \varnothing$, $\text{rank}(A) = n$, and $\max\{cx: x \in P\}$ is finite, then there is an optimal solution that is an extreme point.*

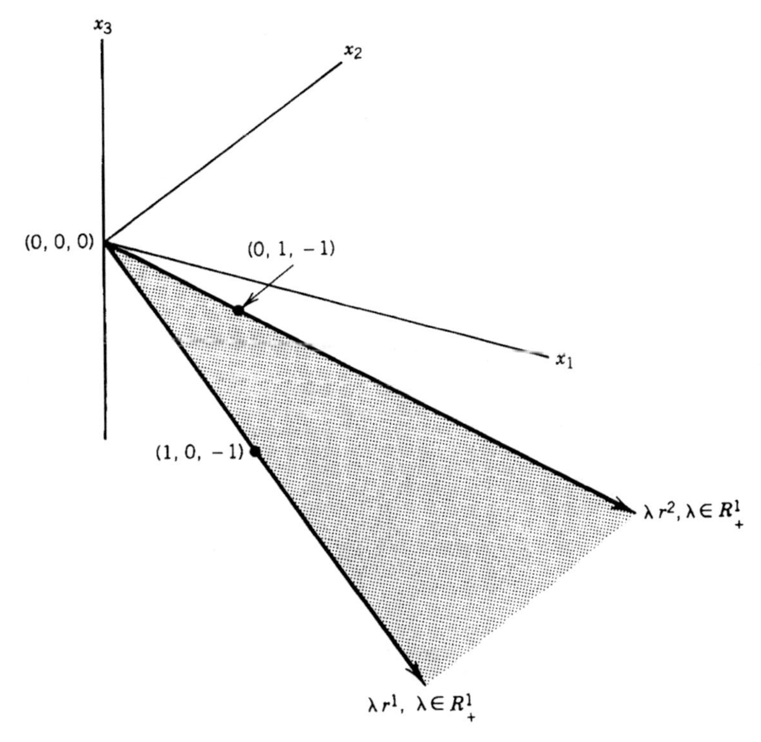

Figure 4.2

Proof. The set of optimal solutions is a nonempty face $F = \{x \in P: cx = c_0\}$. By Proposition 4.1, F contains an $(n - \text{rank}(A))$-dimensional face. Since $n - \text{rank}(A) = 0$, by Proposition 4.2, F contains an extreme point. ■

Theorem 4.6. *For every extreme point* $\{x^k\}_{k \in K}$ *of* P, *there exists a* $c \in Z^n$ *such that* x^k *is the unique optimal solution of* $\max\{cx: x \in P\}$.

Proof. Let $M^=_{x^k}$ be the equality set of x^k. Let $c^* = \sum_{i \in M^=_{x^k}} a^i$. Since the a^i are rational vectors, there exists a $\lambda > 0$ such that $c = \lambda c^* \in Z^n$. Since x^k is a zero-dimensional face of P, for all $x \in P \setminus \{x^k\}$ there exists an $i \in M^=_{x^k}$ such that $a^i x < b_i$. Hence for $x \in P \setminus \{x^k\}$,

$$cx = \sum_{i \in M^=_{x^k}} \lambda a^i x < \sum_{i \in M^=_{x^k}} \lambda b_i = \sum_{i \in M^=_{x^k}} \lambda a^i x^k = cx^k.$$ ■

Theorem 4.7. *If* $P \neq \emptyset$, $\text{rank}(A) = n$, *and* $\max\{cx: x \in P\}$ *is unbounded,* P *has an extreme ray* r^* *with* $cr^* > 0$.

Proof. Since $\max\{cx: Ax \leqslant b\}$ is unbounded, by linear programming duality, the set $\{u \in R^m_+: uA = c\} = \emptyset$. By Farkas' lemma, this implies there exists an $r \in R^n$ such that $Ar \leqslant 0$ and $cr > 0$. Now consider the linear program $\max\{cr: Ar \leqslant 0, cr \leqslant 1\} = 1$. By Theorem 4.5, this linear program has an optimal extreme point solution. An optimal extreme point is a point $r^* \in P^0$ such that the equality set $A^=_{r^*}$ is of rank $n - 1$, and $cr^* > 0$. Now by Proposition 4.3, r^* is an extreme ray of P. ■

We now prove one of the fundamental results on the representation of polyhedra.

Theorem 4.8 (Minkowski's Theorem). *If $P \neq \emptyset$ and* rank$(A) = n$, *then*

$$P = \left\{x \in R^n\colon x = \sum_{k \in K} \lambda_k x^k + \sum_{j \in J} \mu_j r^j,\ \sum_{k \in K} \lambda_k = 1,\ \lambda_k \geqslant 0 \text{ for } k \in K,\ \mu_j \geqslant 0 \text{ for } j \in J\right\},$$

where $\{x^k\}_{k \in K}$ is the set of extreme points of P and $\{r^j\}_{j \in J}$ is the set of extreme rays of P.

Proof. Let

$$Q = \left\{x\colon x = \sum_{k \in K} \lambda_k x^k + \sum_{j \in J} \mu_j r^j,\ \sum_{k \in K} \lambda_k = 1,\ \lambda_k \geqslant 0 \text{ for } k \in K,\ \mu_j \geqslant 0 \text{ for } j \in J\right\}.$$

Since $x^k \in P$ for $k \in K$, and P is convex, $x' = \Sigma_{k \in K} \lambda_k x^k \in P$ for any λ satisfying $\Sigma_{k \in K} \lambda_k = 1$, $\lambda_k \geqslant 0$ for $k \in K$. Also since r^j for $j \in J$ are rays, $x' + \Sigma_{j \in J} \mu_j r^j \in P$ for any $\mu_j \geqslant 0$ for $j \in J$. Hence $Q \subseteq P$.

Now suppose that $Q \subsetneq P$, so there exists $y \in P \setminus Q$. In other words there do not exist λ, μ satisfying

$$\begin{aligned}
\sum_{k \in K} \lambda_k x^k + \sum_{j \in J} \mu_j r^j &= y \\
-\sum_{k \in K} \lambda_k \qquad\qquad &= -1 \\
\lambda_k \geqslant 0 \quad \text{for } k &\in K, \\
\mu_j \geqslant 0 \quad \text{for } j &\in J.
\end{aligned}$$

Then by Farkas' lemma, there exists $(\pi, \pi_0) \in R^{n+1}$ such that $\pi x^k - \pi_0 \leqslant 0$ for $k \in K$, $\pi r^j \leqslant 0$ for $j \in J$ and $\pi y - \pi_0 > 0$. Now consider the linear program $\max\{\pi x\colon x \in P\}$. If it has a finite optimal value, by Theorem 4.5 the optimum value is attained at an extreme point. However, $y \in P$ and $\pi y > \pi x^k$ for all extreme points $\{x^k\}_{k \in K}$, which is a contradiction. On the other hand, if the linear program has an unbounded optimum, by Theorem 4.7 there exists an extreme ray r^j with $\pi r^j > 0$. Again there is a contradiction. Hence $Q = P$. ∎

Example 2.1 (continued). Since P has one extreme point $x^1 = (0 \quad 0 \quad 1)$ and two extreme rays $r^1 = (1 \quad 0 \quad -1)$ and $r^2 = (0 \quad 1 \quad -1)$, we have an alternative description of P given by

$$P = \left\{x \in R^3\colon x = \begin{pmatrix} 0 \\ 0 \\ 1 \end{pmatrix} \lambda_1 + \begin{pmatrix} 1 \\ 0 \\ -1 \end{pmatrix} \mu_1 + \begin{pmatrix} 0 \\ 1 \\ -1 \end{pmatrix} \mu_2,\ \lambda_1 = 1,\ \lambda_1, \mu_1, \mu_2 \geqslant 0\right\}.$$

Combining Minkowski's theorem and linear programming duality leads to a characterization of certain projections of polyhedra; it also leads to an important converse to Minkowski's theorem, which says that every set obtained as a convex combination of a finite set of vectors in R^n plus a nonnegative combination of some other finite set of vectors in R^n is a polyhedron.

First we restate the basic results for the dual pair of linear programs

$$z = \max\{cx: x \in P\} \quad \text{with} \quad P = \{x \in R^n_+: Ax \leqslant b\}$$

and

$$w = \min\{ub: u \in Q\} \quad \text{with} \quad Q = \{u \in R^m_+: uA \geqslant c\}$$

in terms of extreme points and extreme rays. Note that this is partially a repeat of Theorems 4.5 and 4.7. Let $\{x^k\}_{k\in K}$ and $\{u^i\}_{i\in I}$ be the sets of extreme points of P and Q, respectively, and let $\{r^j\}_{j\in J}$ and $\{v^t\}_{t\in T}$ be the sets of extreme rays of P^0 and Q^0, respectively.

Theorem 4.9

i. *The following are equivalent*:
   a) *The primal problem is feasible, that is, $P \neq \varnothing$;*
   b) *$v^t b \geqslant 0$ for all $t \in T$.*
ii. *The following are equivalent when the primal problem is feasible*:
   a) *z is unbounded from above;*
   b) *there exists an extreme ray r^j of P with $cr^j > 0$;*
   c) *the dual problem is infeasible, that is, $Q = \varnothing$.*
iii. *If the primal problem is feasible and z is bounded, then*

$$z = \max_{k\in K} cx^k = w = \min_{i\in I} u^i b.$$

Proof

i. By the Farkas lemma, $P \neq \varnothing$ if and only if $vb \geqslant 0$ for all $v \in R^m_+$ with $vA \geqslant 0$. By Minkowski's theorem,

$$Q^0 = \{v \in R^m_+: vA \geqslant 0\} = \left\{v \in R^m_+: v = \sum_{t\in T} \mu_t v^t, \mu_t \geqslant 0 \text{ for } t \in T\right\}.$$

Hence $vb \geqslant 0$ for all $v \in Q^0$ if and only if $v^t b \geqslant 0$ for all $t \in T$.

ii. Again by Minkowski's theorem,

$$P = \left\{x \in R^n: x = \sum_{k\in K} \lambda_k x^k + \sum_{j\in J} \mu_j r^j, \sum_{k\in K} \lambda_k = 1, \right.$$
$$\left. \lambda_k \geqslant 0 \text{ for } k \in K, \mu_j \geqslant 0 \text{ for } j \in J \right\} \neq \varnothing.$$

Thus z is bounded if and only if $cr^j \leqslant 0$ for all $j \in J$. The equivalence of statements ii.b and ii.c is obtained by applying statement i to the dual problem.

iii. This also follows from strong duality and Minkowski's theorem applied to P and Q. ■

Now we consider the projection of a polyhedron. Note first that the projection of a point $(x, y) \in R^n \times R^p$ onto the subspace $H = \{(x, y): y = 0\}$ is the point $(x, 0)$. Therefore it is natural to consider a projection of a polyhedron $P \subseteq R^n \times R^p$ onto $y = 0$ as a projection from the (x, y)-space to the x-space, denoted by $\text{proj}_x(P)$.

Theorem 4.10. *Let $P = \{(x, y) \in R^n \times R^p: Ax + Gy \leqslant b\}$, then*

$$\text{proj}_x(P) = \{x \in R^n: v^t(b - Ax) \geqslant 0 \text{ for all } t \in T\},$$

where $\{v^t\}_{t \in T}$ are the extreme rays of $Q = \{v \in R^m_+: vG = 0\}$.

Proof. If $H = \{(x, y) \in R^n \times R^p: y = 0\}$, then $\text{proj}_H(P) = \{(x, 0) \in R^n \times R^p: (x, y) \in P\}$. Applying statement i.b of Theorem 4.9 to $\{y \in R^p: Gy \leqslant b - Ax\}$ gives

$$\text{proj}_H(P) = \{(x, 0) \in R^n \times R^p: v^t(b - Ax) \geqslant 0 \text{ for } t \in T\}.$$ ■

Corollary 4.11. *The projection of a polyhedron is a polyhedron.*

Given two polyhedra $P \subset R^n \times R^p$ and $Q \subset R^n$, the question will arise of showing whether $Q = \text{proj}_x(P)$ or not.

Corollary 4.12. *If $P = \{(x, y) \in R^n \times R^p: Ax + Gy \leqslant b\}$ and $Q = \{x \in R^n: Dx \leqslant d\}$, where D is $q \times n$, then $Q = \text{proj}_x(P)$ if and only if:*

i. *For $i = 1, \ldots, q$, $d^i x \leqslant d_0^i$ is a valid inequality for P.*
ii. *For each $x^* \in Q$, there exists a y^* such that $(x^*, y^*) \in P$.*

Proof

i. Equivalent to $Q \supseteq \text{proj}_x(P)$.
ii. Equivalent to $Q \subseteq \text{proj}_x(P)$. ■

Another immediate consequence of Theorem 4.10 is the converse of Minkowski's theorem.

Theorem 4.13 (Weyl's theorem). *If A is a rational $m_1 \times n$ matrix, B is a rational $m_2 \times n$ matrix, and*

$$Q = \left\{x \in R^n: x = yA + zB, \sum_{k=1}^{m_1} y_k = 1, y \in R^{m_1}_+, z \in R^{m_2}_+\right\},$$

then Q is a rational polyhedron.

Proof. $Q = \text{proj}_x(P)$, where

$$P = \left\{(x, y, z) \in R^n \times R^{m_1}_+ \times R^{m_2}_+: x - yA - zB = 0, \sum_{k=1}^{m_1} y_k = 1\right\}.$$ ■

5. POLARITY

Here we consider a polyhedron $\Pi \subseteq R^{n+1}$ whose feasible points are the valid inequalities of P. We will characterize the facets of P in terms of the extreme rays of Π and establish a duality between P and Π.

Definition 5.1. $\Pi = \{(\pi, \pi_0) \in R^{n+1}: \pi x - \pi_0 \leqslant 0 \text{ for all } x \in P\}$ is called the *polar* of the polyhedron $P = \{x \in R^n: Ax \leqslant b\}$.

Note that $(\pi, \pi_0) \in \Pi$ if and only if (π, π_0) is a valid inequality for P. For simplicity, assume that $\text{rank}(A) = n$.

Proposition 5.1. *Given a nonempty polyhedron $P \subseteq R^n$ with* $\text{rank}(A) = n$, *Π is a polyhedral cone described by*

$$\begin{aligned} \pi x^k - \pi_0 &\leqslant 0 \quad \text{for } k \in K \\ \pi r^j \quad &\leqslant 0 \quad \text{for } j \in J \end{aligned}$$

where $\{x^k\}_{k\in K}$, $\{r^j\}_{j\in J}$ are the extreme points and extreme rays of P.

Proof. Let $\Pi' = \{(\pi, \pi_0) \in R^{n+1}: \pi x^k - \pi_0 \leqslant 0 \text{ for } k \in K, \pi r^j \leqslant 0 \text{ for } j \in J\}$. Suppose $(\pi, \pi_0) \in \Pi$. Since $x^k + \mu r^j \in P$ for any x^k, any r^j, and all $\mu \geqslant 0$, we have $\pi(x^k + \mu r^j) \leqslant \pi_0$ for all $\mu \geqslant 0$. But this implies $\pi x^k \leqslant \pi_0$ and $\pi r^j \leqslant 0$. Hence $(\pi, \pi_0) \in \Pi'$, so $\Pi \subseteq \Pi'$.

Conversely if $(\pi, \pi_0) \in \Pi'$ and $x \in P$, then, by Theorem 4.8, $x = \Sigma_{k\in K} \lambda_k x^k + \Sigma_{j\in J} \mu_j r^j$ for some λ, μ satisfying $\Sigma_{k\in K} \lambda_k = 1$, $\lambda_k \geqslant 0$ for $k \in K$, and $\mu_j \geqslant 0$ for $j \in J$. Hence

$$\pi x = \sum_{k\in K} \lambda_k (\pi x^k) + \sum_{j\in J} \mu_j (\pi r^j) \leqslant \left(\sum_{k\in K} \lambda_k\right)\pi_0 + 0 = \pi_0.$$

Therefore $(\pi, \pi_0) \in \Pi$, so $\Pi' \subseteq \Pi$. ■

Example 2.1 (continued). A polyhedral description of $\Pi \subseteq R^4$ is as follows:

$$\begin{aligned} \pi_3 - \pi_0 &\leqslant 0 \\ \pi_1 \quad - \pi_3 \quad &\leqslant 0 \\ \pi_2 - \pi_3 \quad &\leqslant 0. \end{aligned}$$

Now we are ready to prove the main result on polarity.

Theorem 5.2. *If* $\dim(P) = n$, $\text{rank}(A) = n$, *and $\pi^* \neq 0$, then (π^*, π_0^*) is an extreme ray of Π if and only if (π^*, π_0^*) defines a facet of P.*

Proof. By Proposition 4.3, $(\pi^*, \pi_0^*) \neq 0$ is an extreme ray of Π if and only if its equality set is of rank $(n + 1) - 1 = n$. Using the description of Π from Proposition 5.1, this means there exist $(x^1, \ldots, x^t, r^{t+1}, \ldots, r^n)$ such that $\pi^* x^i - \pi_0^* = 0$ for i $= 1, \ldots, t$ and $\pi^* r^j = 0$ for $j = t + 1, \ldots, n$ and

$$\begin{pmatrix} x^1 & -1 \\ \vdots & \\ x^t & -1 \\ r^{t+1} & 0 \\ \vdots & \\ r^n & 0 \end{pmatrix}$$

is of rank n. (Note that $t \geq 1$, since $\pi^* r^j = 0$ for $j = 1, \ldots, n$ would imply $\pi^* = 0$.) But this implies that the vectors $(x^1, -1), \ldots, (x^t, -1), (x^1 + r^{t+1}, -1), \ldots, (x^1 + r^n, -1)$ are linearly independent. Hence by Proposition 1.3, $x^1, \ldots, x^t, x^1 + r^{t+1}, \ldots, x^1 + r^n$ are affinely independent. Therefore (π^*, π_0^*) defines a facet of P.

Conversely if (π^*, π_0^*) defines a facet of P, there exist n affinely independent points $\{x^i\}_{i=1}^n$ of P, with $\pi^* x^i = \pi_0^*$ for $i = 1, \ldots, n$. But now considering the polyhedral cone Π, the equality set of (π^*, π_0^*) includes $(x^1, -1), \ldots, (x^n, -1)$ and hence is of rank at least n. If the equality set is of rank $n + 1$, then $(\pi^*, \pi_0^*) = (0, 0)$. Hence its rank is n, so $\{(\pi, \pi_0) \in R^{n+1}: (\pi, \pi_0) = \lambda(\pi^*, \pi_0^*), \lambda \in R_+^1\}$ is a one-dimensional face of Π. It follows from Proposition 4.3 that (π^*, π_0^*) is an extreme ray of Π. ∎

We have also implicitly proved a dual result to Theorem 5.2.

Theorem 5.3. *If* $\dim(P) = n$ *and* $\text{rank}(A) = n$, $\pi x^* - \pi_0 \leq 0$ *defines a facet of* Π *if and only if* x^* *is an extreme point of* P, *and* $\pi r^* \leq 0$ *defines a facet of* Π *if and only if* r^* *is an extreme ray of* P.

Proof. By Proposition 5.1, every facet of Π is either of the required form $\pi x^k - \pi_0 \leq 0$ for $k \in K$ or $\pi r^j \leq 0$ for $j \in J$. To show that each of these inequalities defines a facet, remember that x^* is an extreme point only if its equality set $(A^=_{x^*}, b^=_{x^*})$ is of rank n. Hence there exist $(\pi^1, -\pi_0^1), \ldots, (\pi^n, -\pi_0^n)$ such that $\pi^1, \ldots, \pi^n$ are linearly independent, and $\pi^t x^* - \pi_0^t = 0$ for $t = 1, \ldots, n$. Now these n vectors plus $(0, 0)$ are affinely independent, and hence $\pi x^* - \pi_0 \leq 0$ defines a facet. A similar argument shows that $\pi r^* \leq 0$ defines a facet. ∎

Now we specialize further and assume that P is a full-dimensional polytope. By translation we can take the origin 0 to be an interior point, so if $a^i x \leq b_i$ is an equality describing P, then $b_i > 0$. Hence we can rewrite P as $P = \{x \in R^n: Ax \leq 1\}$, where $1 = (1 \ldots 1)$. Now every valid inequality (π, π_0) must also have $\pi_0 > 0$, so we can normalize the polar and consider the so-called *1-polar* of P: $\Pi^1 = \{\pi \in R^n: (\pi, 1) \in \Pi\}$. Furthermore, since P is a polytope, by Theorem 4.7 we have

$$P = \left\{x \in R^n: x = \sum_{k \in K} \lambda_k x^k, \sum_{k \in K} \lambda_k = 1, \lambda_k \geq 0 \text{ for } k \in K\right\},$$

and by Proposition 5.1 we have $\Pi^1 = \{\pi \in R^n: \pi x^k \leq 1 \text{ for } k \in K\}$.

Proposition 5.4 *If* $P = \{x \in R^n: Ax \leq 1\}$ *is a full-dimensional polytope, then* Π^1 *is a full-dimensional polytope and* P *is the 1-polar of* Π^1.

Proof. Since 0 is an interior point of Π^1, by Corollary 2.5, Π^1 is full-dimensional. Suppose Π^1 has a ray γ, so that $\gamma x^k \leq 0$ for $k \in K$. This implies that $(\gamma, 0)$ is a valid inequality for P, which is a contradiction. Hence Π^1 is bounded.

Now consider $\overline{P} = \{y: \pi y \leq 1 \text{ for all } \pi \in \Pi^1\}$. If $x \in P$, then $\pi x \leq 1$ for all $\pi \in \Pi^1$ and hence $P \subseteq \overline{P}$. Suppose $y \in \overline{P} \setminus P$. Then there exists no solution to

$$y = \sum_{k \in K} \lambda_k x^k$$

$$-1 = -\sum_{k \in K} \lambda_k$$

$$\lambda_k \geq 0 \quad \text{for } k \in K.$$

So there exists a (π, π_0) such that $\pi x^k - \pi_0 \leq 0$ for all $k \in K$ and $\pi y - \pi_0 > 0$. Since $0 \in P$, we can again normalize so that $\pi_0 = 1$. Then $\pi \in \Pi^1$ but $\pi y > 1$, which is a contradiction. Hence $\bar{P} = P$. ■

Now we observe that there is complete symmetry between P and Π^1.

Theorem 5.5 *If P is full-dimensional and bounded, and 0 is an interior point of P, then*

a. $P = \{x\colon \pi^t x \leq 1$ *for* $t \in T$, *where* $\{\pi^t\}_{t\in T}$ *are the extreme points of* $\Pi^1\}$, *and*
b. $\Pi^1 = \{\pi\colon \pi x^k \leq 1$ *for* $k \in K$, *where* $\{x^k\}_{k\in K}$ *are the extreme points of* $P\}$.

Moreover, each of the inequalities in descriptions a *and* b *define facets.*

Proof. We have already proven b, and a follows from Proposition 5.4. ■

Corollary 5.6. *If P is as described in Theorem 5.5, then*

$$x^* \in P \text{ if and only if } \max\{\pi x^*\colon \pi \in \Pi^1\} \leq 1$$

and

$$\pi^* \in \Pi^1 \text{ if and only if } \max\{\pi^* x\colon x \in P\} \leq 1.$$

Proof. $x^* \in P$ if and only if $\pi x^* \leq 1$ for all $\pi \in \Pi^1$, which holds if and only if $\max\{\pi x^*\colon \pi \in \Pi^1\} \leq 1$. The second equivalence is merely a dual statement. ■

Corollary 5.6 is important in establishing the equivalence between separation and optimization (see Section I.6.3).

Example 5.1 (See Figure 5.1.)

$$(P)\quad \begin{aligned} -x_1 \quad\quad &\leq 1 \\ x_2 &\leq 1 \\ -x_1 - x_2 &\leq 1 \\ 3x_1 + x_2 &\leq 1 \end{aligned} \qquad\qquad (\Pi^1)\quad \begin{aligned} \pi_2 &\leq 1 \\ -\pi_1 + \pi_2 &\leq 1 \\ -\pi_1 \quad\quad &\leq 1 \\ \pi_1 - 2\pi_2 &\leq 1. \end{aligned}$$

There are two other polars that are of special interest in combinatorial optimization. For the remainder of this section we assume that A is a nonnegative $m \times n$ matrix.

Suppose that $P = \{x \in R^n_+\colon Ax \geq 1\}$, where A has no zero rows. The *blocker* P^B of P is the polyhedron:

$$P^B = \{\pi \in R^n_+\colon \pi x \geq 1 \text{ for all } x \in P\}.$$

Let B be a $|K| \times n$ matrix whose rows are the extreme points $\{x^k\}_{k\in K}$ of P.

Proposition 5.7. *Let $P = \{x \in R^n_+\colon Ax \geq 1\}$, where A is a nonnegative matrix with no zero rows. Then*

i. $P^B = \{\pi \in R^n_+\colon B\pi \geq 1\}$ *and*
ii. $(P^B)^B = P$.

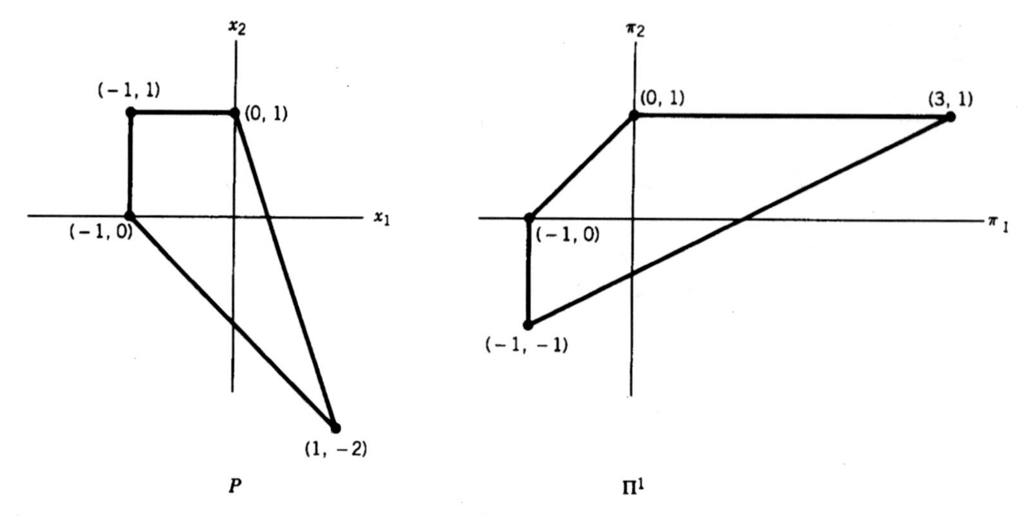

Figure 5.1

Proof. Note that since A has no zero rows, $P \neq \emptyset$. Also, since $P^0 = R^n_+$, the extreme rays of P are the unit vectors e_j for $j = 1, \ldots, n$. This means that $\pi_j \leq 0$ for all $(\pi, \pi_0) \in \Pi$, so $\Pi = \{(\pi, \pi_0) \in R^n_- \times R^1: \pi x \leq \pi_0 \text{ for } x \in P\}$ and so $\pi \in P^B$ if and only if $(-\pi, -1) \in \Pi$. Now by Proposition 5.1, $P^B = \{\pi \in R^n_+: \pi x^k \geq 1 \text{ for } k \in K\}$, where $\{x^k\}$ are the extreme points of P, and statement i is verified.

Since P is full-dimensional and rank $\binom{A}{I} = n$, we obtain from Theorem 5.2 that $(-\pi^*, -1)$ is an extreme ray of Π if and only if $\pi^* x \geq 1$ defines a facet of P. But since $P^B = \{(\pi, \pi_0): (-\pi, -\pi_0) \in \Pi, \pi_0 = -1\}$, it follows that $(-\pi^*, -1)$ is an extreme ray of Π if and only if π^* is an extreme point of P^B. Now we consider $(P^B)^B$. By statement i, $(P^B)^B = \{x \in R^n_+: Qx \geq 1\}$, where the rows of Q are the extreme points of P^B, or, as we have just shown, the facets of P. Hence statement ii holds. ■

Example 5.2. Let

$$A = \begin{pmatrix} 1 & 1 & 0 \\ 1 & 0 & 1 \\ 0 & 1 & 1 \end{pmatrix}.$$

The reader can check that the extreme points of $P = \{x \in R^3_+: Ax \geq 1\}$ are $(1 \quad 1 \quad 0)$, $(1 \quad 0 \quad 1)$, $(0 \quad 1 \quad 1)$, and $(\frac{1}{2} \quad \frac{1}{2} \quad \frac{1}{2})$. Hence

$$B = \begin{pmatrix} 1 & 1 & 0 \\ 1 & 0 & 1 \\ 0 & 1 & 1 \\ \frac{1}{2} & \frac{1}{2} & \frac{1}{2} \end{pmatrix}.$$

We note that all the extreme points of P are minimal points because $P^0 = R^n_+$ and they are all necessary in the description of P^B.

Finally, we consider polytopes of the form $P = \{x \in R^n_+: Ax \leq 1\}$, where A is a nonnegative matrix with no zero columns. The *antiblocker* P^C of P is the polytope

$$P^C = \{\pi \in R^n_+: \pi x \leq 1 \text{ for all } x \in P\}.$$

Let C be an $r \times n$ matrix whose rows are the extreme points of P.

Proposition 5.8. *If $P = \{x \in R^n_+: Ax \leqslant 1\}$, where A is a nonnegative matrix with no zero columns, then*

i. $P^C = \{\pi \in R^n_+: C\pi \leqslant 1\}$ *and*
ii. $(P^C)^C = P$.

Proof. Since A has only nonzero columns, P is a polytope. Statement i follows by replicating the proof of Proposition 5.1 with $\pi \geqslant 0$, $\pi_0 = 1$, and $J = \emptyset$. To establish statement ii, observe that if $x \in P$, it follows that $\pi x \leqslant 1$ for all $\pi \in P^C$ and hence $x \in (P^C)^C$. Now suppose that $x \in (P^C)^C$. Since $a^i \in P^C$, it follows that $a^i x \leqslant 1$. Hence $P = (P^C)^C$. ■

Example 5.3. Let

$$A = \begin{pmatrix} 1 & 1 & 0 \\ 1 & 0 & 1 \\ 0 & 1 & 1 \end{pmatrix}.$$

The reader can check that the extreme points of $P = \{x \in R^3_+: Ax \leqslant 1\}$ are $(1 \quad 0 \quad 0)$, $(0 \quad 1 \quad 0)$, $(0 \quad 0 \quad 1)$, and $(\frac{1}{2} \quad \frac{1}{2} \quad \frac{1}{2})$. Hence

$$P^C = \{\pi \in R^3_+: \pi_j \leqslant 1 \text{ for } j = 1, 2, 3 \text{ and } \tfrac{1}{2}\pi_1 + \tfrac{1}{2}\pi_2 + \tfrac{1}{2}\pi_3 \leqslant 1\}.$$

The extreme points of P^C are the rows of A and the points $(1 \quad 0 \quad 0)$, $(0 \quad 1 \quad 0)$, $(0 \quad 0 \quad 1)$, and $(0 \quad 0 \quad 0)$.

This example shows the difference between the blocking and antiblocking cases. We see that not all the extreme points of P are needed to describe its antiblocker. In fact, it is not difficult to show:

Proposition 5.9. *If $P = \{x \in R^n_+: Ax \leqslant 1\}$ where $A \geqslant 0$ and has no zero columns, then*

i. *The facet defining inequalities of P^C are the inequalities $x^r\pi \leqslant 1$, $r = 1, \ldots, R$, where $\{x^r\}_{r=1}^R$ are the extreme points of P that are maximal in P.*
ii. *If x^0 is an extreme point of P that is not maximal in P, there exists a maximal extreme point x^r for which $x_j^0 = x_j^r$ for all j such that $x_j^0 > 0$.*

The main results of blocking and antiblocking can also be interpreted as problems involving the (fractional) packing and (fractional) covering by rows of A.

Definition 5.2. If A and B are nonnegative matrices with the property that $\{\pi \in R^n_+: B\pi \geqslant 1\}$ is the blocker of $\{x \in R^n_+: Ax \geqslant 1\}$, then A, B is called a *blocking pair*. *Antiblocking pairs* are defined similarly.

Definition 5.3

i. The *max–min inequality* holds for a pair of $m \times n$ and $r \times n$ nonnegative matrices A, B if for all $w \in R^n_+$

$$\max\{1y: yA \leqslant w, y \in R^m_+\} = \min_{1 \leq j \leq r} b^j w.$$

ii. The *min–max inequality* holds for a pair of $m \times n$ and $r \times n$ nonnegative matrices A, C if for all $w \in R^n_+$

$$\min\{1y\colon yA \geqslant w,\ y \in R^m_+\} = \max_{1 \leq j \leq r} c^j w.$$

Theorem 5.10. *The max–min (min–max) inequality holds for a pair A, B if and only if A and B form a blocking (antiblocking) pair of matrices.*

Proof. We consider only the blocking case. If the max–min inequality holds, then

$$\begin{aligned}\max\{1y\colon yA \leqslant w,\ y \in R^m_+\} &= \min\{wx\colon Ax \geqslant 1,\ y \in R^n_+\}\\ &= \min\{wx^k\colon x^k \text{ is an extreme point of } P\}.\end{aligned}$$

It follows that the rows of B are precisely the extreme points of P and that any other row of B is equal to or greater than a convex combination of these extreme points. Hence B is a blocking matrix associated with A. The converse is an immediate consequence of linear programming duality. ∎

6. POLYHEDRAL TIES BETWEEN LINEAR AND INTEGER PROGRAMS

Now, as promised in the introduction to this chapter, we will show that an integer program can, in theory, be reduced to a linear program.

Given $P = \{x \in R^n_+\colon Ax \leqslant b\}$, where (A, b) is an integer $m \times (n+1)$ matrix, and $S = P \cap Z^n$, we are going to show that conv(S) is a rational polyhedron. Whenever P is bounded, S is either empty or a finite set of points, so the result is a consequence of Theorem 4.13.

To obtain the result when S contains an infinite number of points, we will show that conv(S) can be generated from a finite number of points in S and a finite number of integral-valued rays. The idea of the proof is shown in Figure 6.1. Geometrically, we see that conv(S) is the polyhedron generated by convex combinations of the points {(1, 2), (2, 1), (4, 0)} plus nonnegative linear combinations of the rays r^1 and r^2, which are the extreme rays of P.

The important step in the proof is to show that the set of integer points in a polyhedron can be finitely generated. We will give a finite set $Q \subset S$ (in Figure 6.1, the integral points in the shaded region of P) and then show that S can be generated by taking a point in Q plus a nonnegative integer linear combination of the extreme rays of P.

Theorem 6.1. *If $P = \{x \in R^n_+\colon Ax \leqslant b\} \neq \emptyset$ and $S = P \cap Z^n$, where (A, b) is an integer $m \times (n+1)$ matrix, then the following statements are true:*

i. *There exist a finite set of points $\{q^l\}_{l \in L}$ of S and a finite set of rays $\{r^j\}_{j \in J}$ of P such that*

$$S = \left\{x \in R^n_+\colon x = \sum_{l \in L} \alpha_l q^l + \sum_{j \in J} \beta_j r^j,\ \sum_{l \in L} \alpha_l = 1,\ \alpha \in Z^{|L|}_+,\ \beta \in Z^{|J|}_+\right\}.$$

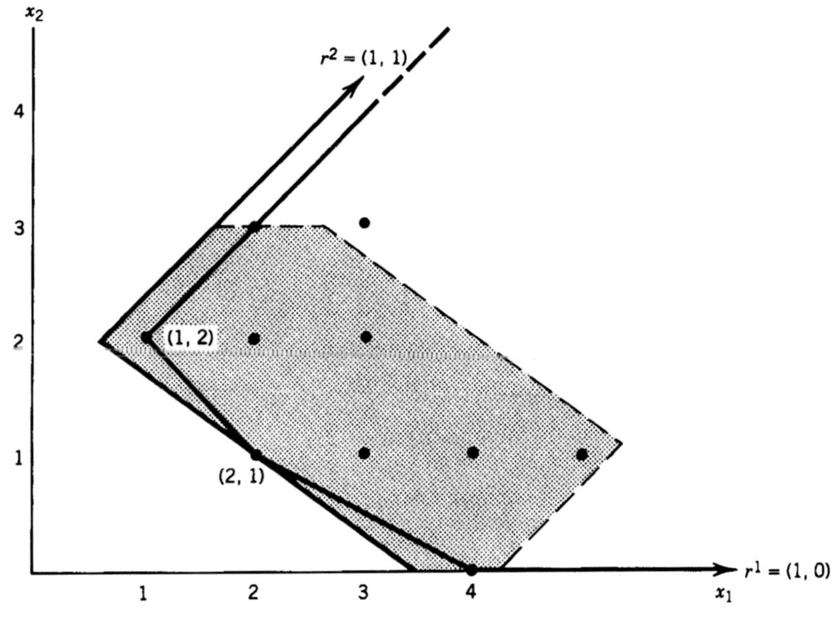

Figure 6.1

ii. *If P is a cone (b = 0), there exists a finite set of rays* $\{v^h\}_{h \in H}$ *of P such that*

$$S = \left\{x \in R^n_+ : x = \sum_{h \in H} \gamma_h v^h, \ \gamma \in Z^{|H|}_+\right\}.$$

Proof

i. Let $\{x^k \in R^n_+ : k \in K\}$ be the finite set of extreme points of P and let $\{r^j \in R^n_+ : j \in J\}$ be the finite set of extreme rays of P. Since P is a rational polyhedron, all of these extremal vectors have rational coordinates. We have

$$P = \left\{x \in R^n_+ : x = \sum_{k \in K} \lambda_k x^k + \sum_{j \in J} \mu_j r^j, \ \sum_{k \in K} \lambda_k = 1, \ \lambda_k, \mu_j \geqslant 0 \text{ for } k \in K \text{ and } j \in J\right\}.$$

Without loss of generality, we can assume that $\{r^j\}$ for $j \in J$ are integer vectors.

Let

$$Q = \left\{x \in Z^n_+ : x = \sum_{k \in K} \lambda_k x^k + \sum_{j \in J} \mu_j r^j, \ \sum_{k \in K} \lambda_k = 1, \ \lambda_k \geqslant 0 \text{ for } k \in K, \ 0 \leqslant \mu_j < 1 \text{ for } j \in J\right\}.$$

Q is a finite set, say $Q = \{q^l \in Z^n_+ : l \in L\}$, and $Q \subseteq S$. Now observe that $x^i \in S$ if and only if $x^i \in Z^n_+$ and

$$x^i = \left\{\sum_{k \in K} \lambda^i_k x^k + \sum_{j \in J} (\mu^i_j - \lfloor \mu^i_j \rfloor) r^j\right\} + \left\{\sum_{j \in J} \lfloor \mu^i_j \rfloor r^j\right\}, \ \sum_{k \in K} \lambda_k = 1, \ \lambda_k, \mu_j \geqslant 0$$

(6.1) for $k \in K$ and $j \in J$.

The first term of (6.1) is a point of Q, so there exists $l(i) \in L$ such that

$$x^i = q^{l(i)} + \sum_{j\in J} \beta^i_j r^j, \quad \beta^i_j = \lfloor \mu^i_j \rfloor \quad \text{for all } j \in J. \tag{6.2}$$

The result follows.

ii. Observe that if P is a cone, then $q^l \in S$ implies $\gamma q^l \in S$ for all $\gamma \in Z^1_+$. Therefore it suffices to take

$$\{v^h: h \in H\} = \{q^l: l \in L\} \cup \{r^j: j \in J\}$$

from part i. ■

Now we easily obtain the following theorem.

Theorem 6.2. *If $P = \{x \in R^n_+: Ax \leqslant b\}$, where (A, b) is an integer $m \times (n+1)$ matrix, and $S = P \cap Z^n$, then* conv(S) *is a rational polyhedron.*

Proof. Since any point $x^i \in S$ can be written in the form (6.2), any convex combination of points $\{x^i \in S, i \in I\}$ can be written as

$$\begin{aligned} x &= \sum_{i\in I} \gamma_i x^i = \sum_{i\in I} \gamma_i \left(q^{l(i)} + \sum_{j\in J} \beta^i_j r^j \right) \\ &= \sum_{l\in L} \left(\sum_{\{i\in I:\, l(i)=l\}} \gamma_i \right) q^l + \sum_{j\in J} \left(\sum_{i\in I} \gamma_i \beta^i_j \right) r^j \\ &= \sum_{l\in L} \alpha_l q^l + \sum_{j\in J} \beta_j r^j, \end{aligned}$$

where $\alpha_l = \Sigma_{\{i\in I:\, l(i)=l\}} \gamma_i \geqslant 0$ for $l \in L$, $\Sigma_{l\in L} \alpha_l = \Sigma_{i\in I} \gamma_i = 1$, and $\beta_j = \Sigma_{i\in I} \gamma_i \beta^i_j \geqslant 0$ for $j \in J$.

Now it follows that

$$\text{conv}(S) = \left\{ x \in R^n_+: x = \sum_{l\in L} \alpha_l q^l + \sum_{j\in J} \beta_j r^j, \sum_{l\in L} \alpha_l = 1, \alpha_l, \beta_j \geqslant 0 \text{ for } l \in L \text{ and } j \in J \right\},$$

with $q^l, r^j \in Z^n_+$ for $l \in L$ and $j \in J$. Hence by Theorem 4.13, conv(S) is a rational polyhedron. ■

The above proof extends straightforwardly to mixed-integer sets with rational data. As a consequence, all of the following results given in this section apply to mixed-integer sets and mixed-integer programs. The above proof also shows that if $P \cap Z^n \neq \emptyset$, then the extreme rays of $P = \{x \in R^n_+: Ax \leqslant b\}$ and conv$(P \cap Z^n)$ coincide.

Theorem 6.2 suggests that we can solve the integer program

$$\text{(IP)} \qquad \max\{cx: x \in S\} \quad \text{where } S = P \cap Z^n$$

by solving the linear program

$$\text{(CIP)} \qquad \max\{cx: x \in \text{conv}(S)\}.$$

This important, but elementary, result is formalized in the following theorem.

Theorem 6.3. *Given $S = P \cap Z^n \neq \emptyset$, $P = \{x \in R^n_+ : Ax \leqslant b\}$, and any $c \in R^n$, it follows that:*

a. *The objective value of* IP *is unbounded from above if and only if the objective value of* CIP *is unbounded from above.*
b. *If* CIP *has a bounded optimal value, then it has an optimal solution (namely, an extreme point of* conv(S)) *that is an optimal solution to* IP.
c. *If x^0 is an optimal solution to* IP, *then x^0 is an optimal solution to* CIP.

Proof. Let z^0 and z^* be the optimal values of IP and CIP, respectively, with the convention that z^0 or $z^* = \infty$ if the objective value is unbounded from above. Note that conv(S) $\supseteq S$ implies that

$$z^* \geqslant z^0. \tag{6.3}$$

a. Inequality (6.3) implies that if $z^0 = \infty$, then $z^* = \infty$. On the other hand, if $z^* = \infty$, then there is an integral extreme point $x^0 \in \text{conv}(S)$ and a ray $r \in Z^n_+$ such that $cr > 0$ and $x^0 + \theta r \in \text{conv}(S)$ for all $\theta \geqslant 0$. But then $x^0 + \theta r \in S$ for all $\theta \in Z^1_+$, which implies that $z^0 = \infty$.
b. Since conv(S) is a polyhedron, if CIP has an optimal solution, then it has an extreme point optimal solution, say x^0. Thus $x^0 \in S$, so $z^0 \geqslant cx^0 = z^*$. By (6.3), $z^0 = z^*$.
c. This follows from parts a and b along with $x^0 \in \text{conv}(S)$. ■

Corollary 6.4. IP *is either infeasible or unbounded or has an optimal solution.*

Theorem 6.3 states that we can solve the integer program IP by solving the linear program CIP. In fact, if we knew a polyhedral representation of conv(S) in terms of linear inequalities, this would be a nice way to describe our integer program. But generally we do not know a set of linear inequalities that define conv(S). Thus we formulate our integer program using some polyhedron $P = \{x \in R^n_+ : Ax \leqslant b\}$ such that $S = P \cap Z^n$. Viewed in this framework, reducing an integer program to a linear program amounts to deducing a linear inequality representation of conv(S), or at least the relevant inequalities with respect to an objective function c, from the linear inequality representation of P and the integrality requirement. This is the principal topic of Chapter II.1.

Until now we have only considered valid inequalities for polyhedra. We say that (π, π_0) is a *valid inequality for a set S* if $\pi x \leqslant \pi_0$ for all $x \in S$.

Proposition 6.5 *If $\pi x \leqslant \pi_0$ is valid for S, it is also valid for* conv(S).

Proof. Consider an $x \in \text{conv}(S)$. Then $x = \sum_{j \in J} \lambda_j x^j$, where $x^j \in S$ for $j \in J$, and $\sum_{j \in J} \lambda^j = 1$ and $\lambda_j \geqslant 0$ for $j \in J$. Hence

$$\pi x = \sum_{j \in J} \lambda_j(\pi x^j) \leqslant \sum_{j \in J} \lambda_j \pi_0 = \pi_0.$$

■

To establish the dimensionality of a face of conv(S), it suffices to consider points of S.

Proposition 6.6. *If $\pi x \leqslant \pi_0$ defines a face of dimension $k-1$ of* conv(S)*, there are k affinely independent points $x^1, \ldots, x^k \in S$ such that $\pi x^i = \pi_0$ for $i = 1, \ldots, k$.*

Proof. By definition, there are k affinely independent points $\bar{x}_1, \ldots, \bar{x}_k \in \text{conv}(S)$ such that $\pi\bar{x}^i = \pi_0$ for $i = 1, \ldots, k$. If $\bar{x}^i \in S$ for $i = 1, \ldots, k$, there is nothing more to prove, that is, take $x^i = \bar{x}^i$ for $i = 1, \ldots, k$. So suppose $\bar{x}^1 \notin S$. Then $\bar{x}^1 = \Sigma_{j\in J} \lambda_j \hat{x}^j$, where $\hat{x}^j \in S$ and $\lambda_j > 0$ for all $j \in J$, and $\Sigma_{j\in J} \lambda_j = 1$. Now $\pi\bar{x}^2 = \pi_0$ and $\pi\hat{x}^j \leqslant \pi_0$ for $j \in J$ imply that $\pi\hat{x}^j = \pi_0$ for all $j \in J$. Since $\bar{x}^1, \ldots, \bar{x}^k$ are affinely independent, there exists $j^* \in J$ such that $\hat{x}^{j^*}, \bar{x}^2, \ldots, \bar{x}^k$ are affinely independent. The proof is completed by repeating this process until the resulting set contains only elements of S. ■

Consider the problem

(LP) $$\max\{cx\colon Ax \leqslant b, x \in R^n_+\}.$$

Previously we have related IP to CIP. Now we relate IP to LP.

Let

$$z(d) = \max\{cx\colon Ax \leqslant d, x \in Z^n_+\}$$

and

$$z_{\text{LP}}(d) = \max\{cx\colon Ax \leqslant d, x \in R^n_+\},$$

so that $z(b) = \max\{cx\colon x \in P \cap Z^n\}$ and $z_{\text{LP}}(b) = \max\{cx\colon x \in P\}$ with $P = \{x \in R^n_+\colon Ax \leqslant b\}$.

Proposition 6.7

a. $z_{\text{LP}}(0) = z(0)$.
b. $z(0) = 0$ *if and only if* $Q = \{u \in R^m_+\colon uA \geqslant c\} \neq \emptyset$.
c. $z(0) = \infty$ *if and only if* $Q = \emptyset$.
d. *If* $Q \neq \emptyset$, *then* $S = P \cap Z^n = \emptyset$ *or* $z(b)$ *is finite.*
e. *If* $Q = \emptyset$, *then* $S = \emptyset$ *or* $z(b) = \infty$.

Proof. Clearly $0 \leqslant z(0) \leqslant z_{\text{LP}}(0)$. If $Q \neq \emptyset$, then $z_{\text{LP}}(0) = 0$ by duality and hence $z(0) = z_{\text{LP}}(0) = 0$. If $Q = \emptyset$, then from Theorem 4.9 there exists an extreme ray r^j of P with $cr^j > 0$. Since r^j can be taken to be integer, $z(0) = \infty = z_{\text{LP}}(0)$. This proves statements a, b, and c.

If $Q \neq \emptyset$, then it follows, by duality, that $P = \emptyset$ or $z_{\text{LP}}(b)$ is finite. Hence statement d follows. Similarly if $Q = \emptyset$, then it follows, by duality, that $P = \emptyset$ or $z_{\text{LP}}(b) = z_{\text{LP}}(0) = \infty$. If $S \neq \emptyset$, then from statement c it follows that $z(0) = z(b) = \infty$. Hence statement e follows. ■

Corollary 6.8

a. *If* $P = \emptyset$, *then* $S = \emptyset$.
b. *If* $z_{\text{LP}}(b)$ *is finite, then* $S = \emptyset$ *or* $z(b)$ *is finite.*
c. *If* $z_{\text{LP}}(b) = \infty$, *then* $S = \emptyset$ *or* $z(b) = \infty$.

Note that by solving the linear program $\max\{cx: x \in P\}$, we establish which of the cases a, b, or c occurs. Corollary 6.8 says that, except for the fact that S may be empty when P is not empty, IP and LP have the same status.

7. NOTES

Sections I.4.1–I.4.4

Halmos (1959) and Strang (1976) are basic reference books on linear algebra. The fundamental works on general polyhedral theory and convexity are Grunbaum (1967), Rockafellar (1970), and Stoer and Witzgall (1970).

Chapter I of Pulleyblank's Ph.D. dissertation (1973) focuses on the aspects of polyhedral theory used in combinatorial optimization. Also see Bachem and Grötschel (1982) and Pulleyblank (1983).

Section I.4.5

The basic reference on polarity is Rockafellar (1970). The study of blocking and antiblocking polyhedra is due to Fulkerson (1968, 1970a, 1971, 1972). Also see Tind (1974, 1977, 1979).

Section I.4.6

The proof of Theorem 6.1 is taken from Giles and Pulleyblank (1979). Also see Meyer (1974, 1975) and Meyer and Wage (1978).

8. EXERCISES

1. Consider the polyhedron P described by

$$
\begin{aligned}
x_1 - x_2 &\le 0 \\
-x_1 + x_2 &\le 1 \\
2x_2 &\ge 5 \\
8x_1 - x_2 &\le 16 \\
x_1 + x_2 &\ge 4 \\
x &\in R^2.
\end{aligned}
$$

   i) Find the dimension of P.
   ii) Find an interior point (if one exists).
   iii) Describe all the faces of P.
   iv) Consider each of the faces $F^i = P \cap \{x \in R^2\colon a^i x = b_i\}$ for $i = 1, \ldots, 5$. What is the dimension of F^i? Which inequalities define facets of P?
   v) Give a "minimal" representation of P.

2. Consider the assignment polytope

$$P = \left\{ x \in R_+^{n^2}: \sum_{j=1}^{n} x_{ij} = 1, i = 1, \ldots, n, \sum_{i=1}^{n} x_{ij} = 1, j = 1, \ldots, n \right\}.$$

   i) Determine its dimension and its facets.

   ii) What happens if we replace the equality constraints by

$$\sum_{j=1}^{n} x_{ij} \leq 1, \; i = 1, \ldots, n \quad \text{and} \quad \sum_{i=1}^{n} x_{ij} \geq 1, \; j = 1, \ldots, n.$$

3. A wheel $W_n = (V, E)$ is a graph defined by $V = \{v_0, v_1, \ldots, v_n\}$ and $E = \{(v_0, v_i): i = 1, \ldots, n\} \cup \{(v_i, v_{i+1}): i = 1, \ldots, n-1\} \cup \{(v_n, v_1)\}$.
Let $P = \{x \in R^{|E|}: \Sigma(x_e: e$ contains node $v) = 2$ for all $v \in V$, $0 \leq x_e \leq 1$ for all $e \in E\}$.

   i) Find the dimension of P.

   ii) Show that the inequalities $x_e \geq 0$ are redundant.

   iii) Show that the inequalities $x_e \leq 1$ are redundant for $e = (v_0, v_i)$ for $i = 1, \ldots, n$.

   iv) Give a minimal representation of P by a system of linear inequalities and equalities.

   v) Give a representation of P by means of its extreme points.

4. Let F be the face of optimal solutions of the linear program $\max\{cx: x \in P\}$, where $P = \{x \in R^n: Ax \leq b\}$. Let $M_F^=$ be the equality set of F and let u^* be any optimal solution of the dual linear program. Show that $u_i^* = 0$ if $i \notin M_F^=$.

5. Show that if H and G are two faces of a polyhedron P of dimension r and $r + s$, respectively, and H is a face of G, there exists a sequence of faces $\{F_i\}_{i=0}^{s}$ with:

   i) $F_0 = H$, $F_s = G$;

   ii) F_i is a facet of F_{i+1} for $i = 0, \ldots, s-1$.

6. Find all extreme points and extreme rays of:

   i) the polyhedron in Exercise 1;

   ii) the polyhedron

$$\begin{aligned} x_1 + x_2 \qquad\quad &\geq 1 \\ x_1 \qquad + 2x_3 &\geq 2 \\ -x_2 + \; x_3 &\geq -4 \\ x \in R^3&. \end{aligned}$$

7. For each face F of P in Exercise 1, find the values of c such that $\max\{cx: x \in P\}$ has F as the set of optimal solutions.

8. (Fourier–Motzkin Elimination). Given a polyhedron $P \subseteq R^{n+1}$ described by the inequalities

$$\begin{aligned} a^l x + y &\le a_0^l \quad \text{for } l = 1, \ldots, L \\ b^k x - y &\le b_0^k \quad \text{for } k = 1, \ldots, K \\ c^i x \quad &\le c_0^i \quad \text{for } i = 1, \ldots, I, \end{aligned}$$

where $x \in R^n$, $y \in R^1$:

i) Show that

$$\text{proj}_x(P) = \{x \in R^n: b^k x - b_0^k \le a_0^l - a^l x \text{ for all } l \text{ and } k,\ c^i x \le c_0^i \text{ for all } i\}.$$

ii) Find $\text{proj}_x(P)$, where

$$P = \{(x, y) \in R^2: x - y \ge -2, x + y \le 3, x - y \le -1, y \ge 0\}.$$

9. **i)** Given a polyhedron P, let $\{F_i\}_{i \in I}$ and $\{G_j\}_{j \in J}$ be polyhedra with $F_i, G_j \subseteq P$. Show that

$$\text{conv}\Big(\bigcup_{ij} (F_i \cap G_j)\Big) \subseteq \text{conv}\bigcup_{j}\Big[G_j \cap \text{conv}\Big(\bigcup_{i} F_i\Big)\Big].$$

ii) Show that the inequality is strict in the following example:

Let $F_i = P \cap \{x \in R_+^2: x_1 = i\}$ for $i = 0, 1, 2$

and $G_j = P \cap \{x \in R_+^2: x_2 = i\}$ for $j = 0, 1, 2$

with P as in Figure 8.1. Note that $\bigcup_{i,j} (F_i \cap G_j) = P \cap Z_+^2$.

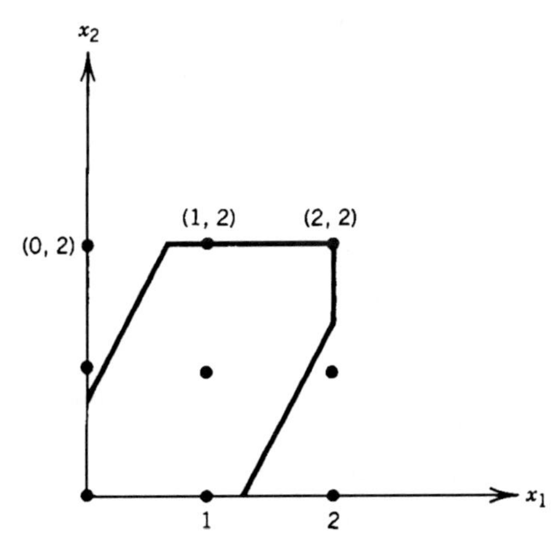

Figure 8.1

iii) Show that equality holds in part i when the $\{F_i\}$, $\{G_j\}$ are faces of P.

iv) Suppose P is contained in the unit hypercube in R^2. Take $F_\delta = P \cap \{x \in R^2: x_1 = \delta\}$ and $G_\delta = P \cap \{x \in R^2: x_i \in \delta\}$ for $\delta \in \{0, 1\}$. Interpret the equality in this case.

10. Find the polar and its extreme rays for the polyhedron in Exercise 1.

11. Let $S = \{x \in Z^2_+: x_1 - x_2 \leq 1, 4x_1 + x_2 \leq 28, x_1 + 4x_2 \leq 27\}$.

i) Find an inequality description of conv(S).

ii) Find the extreme points of conv(S).

iii) Find the polar of conv(S).

iv) Find the extreme rays of the polar of conv(S).

12. Find the 1-polar of

$$P = \{x \in R^2: x_1 + 2x_2 \leq 1, -x_1 \leq 1, -x_1 - x_2 \leq 2, x_1 - x_2 \leq 1\}.$$

13. Find the blocker of

$$P = \{x \in R^2_+: \tfrac{3}{4}x_1 + \tfrac{1}{4}x_2 \geq 1, \tfrac{1}{3}x_1 + \tfrac{1}{3}x_2 \geq 1, \tfrac{1}{4}x_1 + \tfrac{1}{2}x_2 \geq 1\}.$$

14. Prove Proposition 5.9.

15. Prove the min–max version of Theorem 5.10.

16. Let $P = \{x \in R^n: Ax \leq b\}$, where rank$(A) = k < n$. Let $L = \{x \in R^n: Ax = 0\}$, $L^\perp = \{x \in R^n: Bx = 0\}$ and $P^* = P \cap L^\perp$.

i) Show that $P = P^* + L$.

ii) Derive Minkowski's theorem when rank$(A) < n$.

iii) Demonstrate it for

$$P = \{x \in R^3: x_1 + x_2 + x_3 \leq 1, -x_1 - x_2 - x_3 \leq 2, x_3 = 1\}.$$

17. Find a finite set of generators for the set $S = P \cap Z^2$, where

$$P = \{x \in R^2_+: 5x_1 + 3x_2 \geq 10, 5x_1 - 5x_2 \geq -1, -x_1 + 2x_2 \geq -2\}.$$

18. Give examples of pairs,

(LP) $\max\{cx: Ax \leq b, x \in R^n_+\}$

(IP) $\max\{cx: Ax \leq b, x \in Z^n_+\}$,

where

i) LP and IP are unbounded,

ii) LP and IP have finite optimum value,

iii) LP is unbounded, and IP is infeasible,

iv) LP is bounded, and IP is infeasible.

19. Using the polyhedron of Exercise 1 of Section I.1.8, show that Theorem 6.2 does not hold for irrational polyhedra.

20. Consider the graph $G = (V, E)$, where $V = \{1, 2, \ldots, 2k + 1\}$, $k \geq 2$, and $E = \{(1, 2), (2, 3), \ldots, (2k + 1, 1)\}$. Let $S = P \cap Z^{2k+1}$ be the set of node packings on G where $P = \{x \in R_{+}^{2k+1} : x_i + x_j \leq 1 \text{ for } (i, j) \in E\}$. Show that $\text{conv}(S) \subsetneqq P$. Find another facet of conv(S). Now do you have conv(S)? Why?

I.5

Computational Complexity

1. INTRODUCTION

The purpose of this chapter is to describe a theory of computational complexity that yields insights into how difficult a problem may be to solve.

At the easy end of our spectrum, there are problems like the minimum-weight spanning tree problem. Recall that in Section I.3.3 we gave an algorithm for the minimum-weight spanning tree problem with running time $O(n \log n)$ for a graph with n edges. One fundamental issue to be discussed here is when a problem can be solved in time $O(l^k)$, where k is a constant and l is an appropriate measure of the length of the input needed to describe the data.

For most integer programming problems, no such algorithm is known. We will show that there are integer programming problems much more specific than the general pure-integer programming problem (e.g., maximum cardinality node packing) with the following property. If maximum cardinality node packing for a graph with m nodes can be solved in time $O(m^k)$ for some fixed k, then there exists a $\hat{k}$ such that the pure-integer programming problem can be solved in $O(l^{\hat{k}})$, where l measures the input needed to describe the data A, b, c.

A very important concept introduced in this chapter is a "certificate of optimality." Given a certificate of optimality, one can prove in $O(l^k)$ time, for some fixed k, that a given solution is indeed optimal.

After introducing some basic concepts in this section and Section 2, we will show in Section 3 that primal and dual basic optimal solutions provide a certificate of optimality for linear programming.

Although no certificate of optimality is known for the general pure-integer programming problem, in Section 4 we will develop some results for pure-integer programs that will enable us to establish a weaker result, namely, a "certificate of feasibility." This means that given an appropriate feasible solution, we will be able to check feasibility quickly. The result is not trivial, since one can imagine feasible integer programs for which the only solutions have a large number of digits, so checking feasibility by substitution is a formidable task.

In Section 5, we formalize the concept of a feasibility problem and the class of feasibility problems with a certificate of feasibility. In Section 6, we show that there are hardest feasibility problems in the above class and relate these results to optimization problems. In Section 7, we consider the complexity of problems associated with polyhedra such as whether $\pi x \leqslant \pi_0$ is satisfied by all points in a given polyhedron.

The presentation here represents a compromise between the rigor found in computer science texts, which would require many new definitions and concepts, and a very informal presentation that can lead to fundamental misconceptions. Thus it is necessary for us to define rather precisely the meanings of terms such as problem, instance of a problem, polynomial solvability, and so on. But we will avoid using terms such as Turing machine, language, and so forth.

Mixed-integer programming is the problem written generically as

$$\max\{cx + hy: Ax + Gy \leqslant b, x \in Z_+^n, y \in R_+^p\},$$

where m is any positive integer, p and n are any nonnegative integers with $p + n \geqslant 1$, and c, h, A, G, b are matrices with integral coefficients; the dimensions of these matrices are as follows: c is $1 \times n$, h is $1 \times p$, A is $m \times n$, G is $m \times p$, and b is $m \times 1$. We could just as well have assumed that the matrices have rational coefficients, but the assumption of integer coefficients is no less general and is more convenient.

A problem consists of an infinite number of instances. An instance is specified by assigning numerical values called *data* to the problem parameters. In the case of mixed-integer programming, the data that specify an instance are integers m, n, and p as well as integral matrices c, h, A, G, and b of appropriate dimension.

It is desirable to delineate special cases of the mixed-integer programming problem. This is done by restricting the parameters in natural ways. *Pure-integer programming* is the special case of mixed-integer programming in which $p = 0$, and hence the matrices h and G do not appear. *Linear programming* is the special case in which $n = 0$, and hence the matrices c and A do not appear.

Every instance of a linear or pure-integer program is also an instance of a mixed-integer program. Thus an algorithm that can solve all instances of mixed-integer programming can, by definition, solve all instances of the special cases of pure-integer and linear programming. An obvious conclusion is that mixed-integer programming is at least as hard as pure-integer and linear programming.

Figure 1.1 is a directed graph that shows relationships among some of the problems that have been formulated in Chapter I.1. The problem at the head of an arc is a special case of the problem at the tail. The relationship extends transitively to directed paths. Thus, if there is a directed path from problem X_1 to problem X_2, then every instance of X_2 is also an instance of X_1.

Most of the arcs in Figure 1.1 are easily justified. For example, 0-1 *integer programming* contains those instances of pure-integer programming in which

$$A = \begin{pmatrix} A' \\ I \end{pmatrix} \quad \text{and} \quad b = \begin{pmatrix} b' \\ 1 \end{pmatrix},$$

where I is an $n \times n$ identity matrix and 1 is an $n \times 1$ matrix of 1's. *Set packing* contains those instances of 0-1 integer programming in which each coefficient of matrix A is 0 or 1 and b is a column of 1's. *Node packing* contains those instances of set packing in which each row of A has exactly two 1's.

In attempting to classify these problems, an extreme view is to ignore the special cases. All of our problems are just mixed-integer programs to be solved by the same algorithm. While there are good reasons for having a robust algorithm, by carrying it to this extreme we would fail to take advantage of the structure and simplicity of important special cases. On the other hand, there are so many interesting special cases of mixed-integer programming that it would be foolish, if not hopeless, to consider each separately. The fundamental issue is to find natural divisions. One possible way of achieving this is to attempt to add arcs to the graph of Figure 1.1 to create directed cycles. Then if problems X_1 and X_2 are contained in a directed cycle, they are equivalent in the sense that each is a special case of the other.

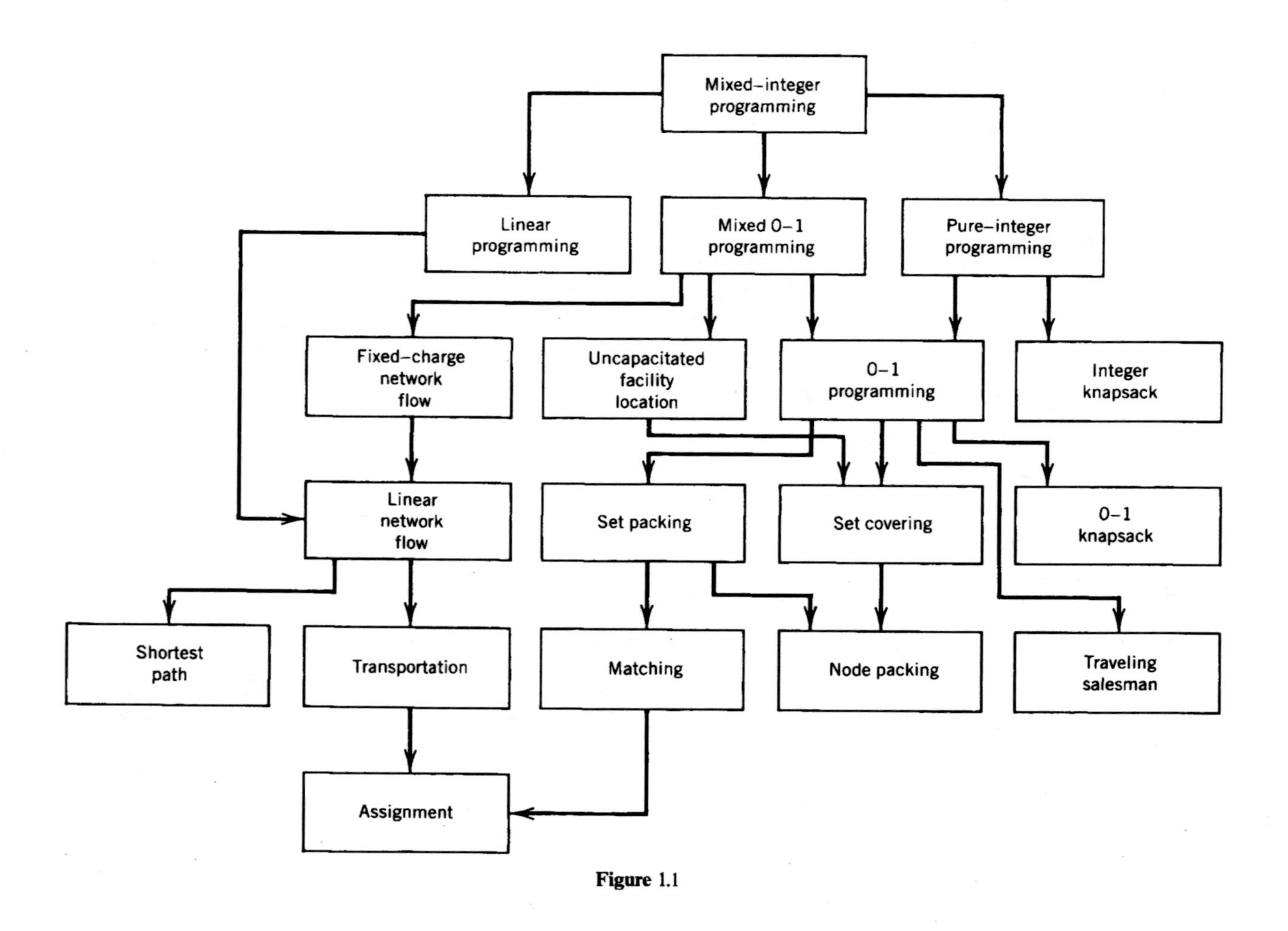

Mixed–integer programming
Linear programming
Mixed 0–1 programming
Pure–integer programming
Fixed–charge network flow
Uncapacitated facility location
0–1 programming
Integer knapsack
Linear network flow
Set packing
Set covering
0–1 knapsack
Shortest path
Transportation
Matching
Node packing
Traveling salesman
Assignment

Figure 1.1

By the end of this chapter we will have shown that apart from the problems that are well-solved, all the other problems in Figure 1.1 lie on common cycles and are theoretically "equally difficult". However, in parts II and III we will see that there remain many reasons to distinguish between problem classes by using structure in developing algorithms.

Example 1.1 *(Set Packing is a special case of Node Packing).* Let A be an $m \times n$ matrix, all of whose coefficients are 0 or 1, and let 1 be a column vector of 1's. We will construct a $p \times n$ 0-1 matrix A' with $p \leqslant n(n-1)/2$ and $\Sigma_{j=1}^{n} a'_{ij} = 2$ for $i = 1, \ldots, p$ such that

$$\{x \in B^n: Ax \leqslant 1\} = \{x \in B^n: A'x \leqslant 1\}.$$

Since a 0-1 matrix with exactly two 1's in each row and no duplicate rows is the edge-node incidence matrix of a graph, it suffices to specify matrix A' as the edge-node incidence matrix of some graph. Let $G_A = (V, E)$ be the graph with $V = \{1, 2, \ldots, n\}$ and $E = \{(j, k): a_{ij} = a_{ik} = 1 \text{ for some } i\}$. G_A is called the *intersection graph* of matrix A. The construction is illustrated in Figure 1.2.

Now it is easy to see that $x^0 \in B^n$ satisfies $Ax^0 \leqslant 1$ if and only if the node set $V^0 = \{j \in V: x_j^0 = 1\}$ is such that $j, k \in V^0$ implies $(j, k) \notin E$. Hence there is one-to-one correspondence between feasible solutions to the set packing problem with matrix A and feasible solutions to the node packing problem on graph G_A. Finally, we note that V^0 is a feasible solution to the node packing problem if and only if $A'x^0 \leqslant 1$, where A' is the edge-node incidence matrix of G_A.

Later in this chapter we will be more precise in what we mean by a special case. We want to have a definition that insofar as possible conveys the idea that if X_2 is a special case of X_1, then an algorithm that can solve instances of X_1 efficiently can also efficiently solve those instances of X_1 that belong to X_2. For example, we could say that integer programming is a special case of linear programming by replacing the constraint set of an integer program by a linear inequality description of its convex hull. While this is true, it is misleading because the efficiency of the algorithm for linear programming may be a function of the linear inequality description of the convex hull, and in addition it may be extremely difficult to find these linear inequalities.

2. MEASURING ALGORITHM EFFICIENCY AND PROBLEM COMPLEXITY

It is common practice to relate computation time to problem size. Traditionally, the size of an instance of an optimization problem has been described by its number of variables and number of constraints. These two parameters, however, may not be adequate. There are algorithms whose number of steps depends explicitly on the magnitude of the numerical data. For example, there is an algorithm for the integer knapsack problem

$$A = \begin{pmatrix} 1 & 0 & 0 & 0 \\ 1 & 1 & 0 & 0 \\ 1 & 0 & 1 & 0 \\ 0 & 1 & 1 & 1 \end{pmatrix} \qquad A' = \begin{pmatrix} 1 & 1 & 0 & 0 \\ 1 & 0 & 1 & 0 \\ 0 & 1 & 1 & 0 \\ 0 & 1 & 0 & 1 \\ 0 & 0 & 1 & 1 \end{pmatrix}$$

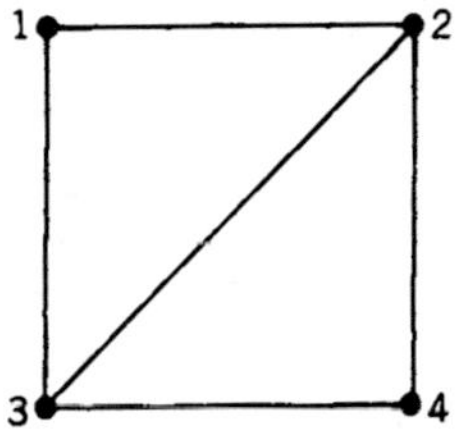

1 2 3 4

G_A

Figure 1.2

whose number of computations is proportional to the number of variables times the right-hand side of the constraint. In the ellipsoid method for linear programming, the number of computations depends on the volume of the initial ellipsoid, which in turn depends on the magnitude of the numerical data. The size of numbers involved in elementary calculations, such as additions and multiplications, may also be of concern. It is frequently reasonable to assume that these operations are done in constant or unit time. For example, if a and b are integers that are part of the data, then a reasonable assumption is that a is read in unit time, b is read in unit time and $a + b$ is calculated in unit time. However, it may not be reasonable to assume that huge integers such as factorials can be added in unit time.

We say that the *size of a problem instance* is the amount of information required to represent the instance. The data needed are generally obvious; for example, an instance of integer programming is specified by integers m and n and matrices A, b, and c. How should we represent the information? A model that is robust with respect to representing the essence of real computation is to use a two-symbol or *binary* (0, 1) alphabet for the representation of numerical and logical data. In this model, a positive integer x, $2^n \leqslant x < 2^{n+1}$, is represented by the vector $(\delta_0, \delta_1, \ldots, \delta_n)$, where

$$x = \sum_{i=0}^{n} \delta_i 2^i \quad \text{and} \quad \delta_i \in \{0, 1\} \quad \text{for } i = 1, \ldots, n.$$

Note that $n \leqslant \log_2 x < n + 1$. An additional digit is necessary to represent the sign of x, and rational numbers are represented by pairs of integers. We always assume that the initial numerical data are integral or rational. Only rarely do we have to be concerned with irrational numbers in intermediate calculations. In such situations (e.g., in the ellipsoid algorithm, which requires square roots), we have to take care to specify the precision of the arithmetic calculations. However, for the most part, integer arithmetic suffices.

Subsets of a set can be represented by *incidence* or *characteristic* vectors. Thus if $Q = \{1, 2, \ldots, n\}$, the subset Q_j is given by the vector $(q_1, q_2, \ldots, q_n)$, where $q_i = 1$ if $i \in Q_j$ and $q_i = 0$ otherwise. This, of course, is a way of representing graphs, since an edge of a graph is just a subset of nodes of cardinality 2. Thus a graph $G = (V, E)$ with m nodes and n edges can be represented by an $m \times n$ node-edge incidence matrix. Alternatively, it can be represented by the $m \times m$ symmetric *adjacency matrix* A, where $a_{ij} = 1$ if nodes i and j are joined by an edge and $a_{ij} = 0$ otherwise. Another data structure for subsets and graphs is to represent a subset by a list of its elements.

While choosing a good data structure can be very important in implementing an algorithm efficiently, it is fortunate that our primary classification scheme of algorithm efficiency is very insensitive to the choice of data representation. There are, however, some restrictions.

The alphabet used to represent data must contain at least two symbols. In particular, for reasons to be explained later, a one-symbol representation of integers is not permitted. The second restriction deals with the amount of information that we agree to call data. We will explain this point by considering the symmetric traveling salesman problem on a complete graph $G = (V, E)$, where c_e for $e \in E$ is the cost of edge e. The natural representation of the data is a list of edges, named by their endpoints, and their costs. A representation of the data that is not permitted is a list of all $(n - 1)!/2$ tours and their costs. The number of tours grows exponentially with the size of the graph, and if an algorithm required this information, we would regard its generation to be part of the algorithm, not part of the data description. Similarly, the integer programming formulation of the symmetric traveling salesman problem given in Chapter I.1, which requires an inequality for each $U \subseteq V$ with $2 \leqslant |U| \leqslant |V| - 2$, is not permitted as a description of the data. Here the number of rows in the constraint coefficient matrix A grows exponentially with the size of the graph. If an algorithm required this formulation, we would regard its generation to be part of the algorithm.

Having set up a model for describing a problem and the data of its instances, we now consider computation time. We want our measure of time to be independent of the characteristics of particular computers, so we basically count the number of elementary operations such as additions, multiplications, comparisons, and so on; that is, we assume that each elementary operation is done in unit time. This is a reasonable assumption as long as the size of the numbers does not grow too rapidly as the calculations progress. We will see later that one may need to be very careful in checking that this is the case.

Consider an optimization problem X consisting of an infinite number of instances $(d_1, d_2, \ldots)$, where the data for the instance d_i is given by a binary string of length $l_i = l(d_i)$. Let A be an algorithm that can solve every instance of X in finite time. We assume that the running time of A is specified by a function $g_A: X \to R^1_+$. We would like to express running time as a function of l. Since it is not necessarily the case that if two instances have the same length, they have the same running time, we must use some statistic to aggregate the running times for all instances of the same length. Our approach is to use a *worst-case* analysis. In other words, the running time that we associate with all instances of size k is

$$f_A(k) = \max\{g_A(d_i): l(d_i) = k\}.$$

This highly conservative measure of running time, which only considers the worst-possible outcome for each size, has three advantages:

1. It gives an absolute guarantee on running time.
2. It is independent of a probability distribution of the instances.
3. It appears to be the easiest measure to analyze.

However, it also has disadvantages. Foremost among these is its failure to give a true picture when a large percentage of instances of a given size can be solved rapidly and only a very small percentage require considerably more time. In these situations, measures such as expected running time may be preferable. But measures that require a probability distribution of the instances appear to be more difficult to analyze and require assumptions about an underlying probability distribution.

Rather than attempting to get a precise expression for the function $f_A(k)$, it will suffice here to approximate it from above. Recall that we say $f(k)$ is $O(g(k))$ whenever there exists a positive constant c and a positive integer k' such that $f(k) \leqslant cg(k)$ for all integers $k \geqslant k'$. With this convention, a polynomial $\Sigma_{i=0}^{p} c_i k^i$ is $O(k^p)$. In other words, we ignore all of the terms of degree less than p and all of the constants. This means that only the asymptotic behavior of the function as $k \to \infty$ is being considered, since for "small" values of k, depending on the constants, the lower-degree terms may dominate.

Algorithm A is said to be a *polynomial time algorithm* for problem X if $f_A(k)$ is $O(k^p)$ for some fixed p. Let $\mathcal{P}$ be *the class of problems that can be solved in polynomial time.* Problem X is in $\mathcal{P}$ if and only if there is a polynomial time algorithm for solving X. A main theme of computational complexity is the inherent difference between problems known to be in $\mathcal{P}$ and others for which no polynomial time algorithm is known.

The function f is said to be *exponential* if for some constants $c_1, c_2 > 0$ and $d_1, d_2 > 1$ and a positive integer k' we have

$$c_1 d_1^k \leq f(k) \leq c_2 d_2^k \quad \text{for all integers } k \geq k'.$$

A typical example of exponential time calculation is the enumeration of the 2^k 0-1 k-dimensional vectors. The function $f(k) = 2^k$ is not bounded by any polynomial in k, and it does not require very large values of k for 2^k to exceed polynomial functions of reasonably small degree. For example, with $k = 60$, an algorithm that required 2^k calculations, each of which took a microsecond, could not be completed in 300 centuries, whereas one that required k^5 calculations would be done in less than 15 minutes.

Although most of the algorithms that we consider can be shown to either run in polynomial or exponential time, there are other possibilities. There are functions whose rate of growth is faster than polynomial but slower than exponential—for example, $f(k) = k^{\log k}$. There are also functions whose rate of growth is faster than exponentially—for example, $f(k) = k^{k^k}$.

Exponential time can also occur when the number of computations is a function of the size of the numbers in the input. Let θ be the largest integer in a given instance. Since the binary encoding of θ only requires a string of length $O(\log \theta)$, an algorithm that requires θ steps is at least exponential. This is one reason for our having stressed the encoding of numbers earlier in this section. We ruled out a one-symbol alphabet because it would permit θ steps to be carried out in polynomial time.

Also, if an algorithm computed very large numbers, such as 2^θ, that are not bounded by a polynomial function in θ, their encoding would require strings of length not polynomially bounded in $\log \theta$. However, as long as the numbers remain polynomially bounded in θ, the assumption of unit time calculations has no bearing on whether the algorithm runs in polynomial time. Besides being convenient, this assumption is made because computers work with "words" in unit time, and quite large integers are represented by a single word. Thus when we say that an algorithm runs in $O(k^p)$ time, we generally have ignored a factor in $\log \theta$ that would be required if we had assumed that the time for an elementary calculation was proportional to the logarithm of the numbers involved. However, we will not ignore the possibility of exponential growth in the size of numbers.

In this regard, we must consider the representation of rationals that are not integers. We assume that a rational a/b is encoded by the pair of integers a and b. Thus $2^{-\theta}$ represented as $(1, 2\theta)$ is a very large number. A more subtle point is that 2 represented as $2^{\theta+1}/2^\theta$ is a very large number. We avoid the latter problem by assuming that a rational a/b is represented by two relatively prime integers p and q (i.e., $a/b = p/q$ and the greatest common divisor of p and q equals 1). In fact, in Section I.7.2 we will give a version of the euclidean algorithm, which, given a and b, finds p and q in polynomial time. So the assumption of representing rationals by two relatively prime integers is theoretically justified.

While the distinction between polynomial time algorithms and the rest is important theoretically, it is not a satisfactory practical division. We will begin to see in the next section, and then in Part II, that some polynomial time algorithms are inefficient and that some algorithms known to be exponential in the worst case are very reliable algorithms for solving practical problems. Of course, polynomial time algorithms that run in, say, linear time are fast. The problems with the division occur with polynomial time algorithms in which the degree of the polynomial is not small and with exponential time algorithms that are fast in most cases.

We have chosen here to emphasize computation time, but the space or memory needed to solve a problem is also important. Observe that if $X \in \mathcal{P}$ there must be an algorithm for X whose space requirements are a polynomial function of the length of the input. However, the converse is false. In other words, there are exponential time algorithms whose space requirements are polynomially bounded.

So far, we have ignored the question of whether integer programming problems can be solved finitely. Obviously they can be when the variables are bounded, since the enumeration of all points within the hypercube defined by the bounds is a finite process. In Section 4, we will show that upper bounds on the variables can be found as a function of (A, b, c, m, n) for pure-integer programming problems with the property that if $\max\{cx: Ax \leqslant b, x \in Z_+^n\}$ has an optimal solution, then it has an optimal solution within the specified bounds. This result, along with schemes for resolving infeasibility and unboundedness, shows that every pure-integer programming problem can be solved finitely. It also can be proved that mixed-integer programs can be solved finitely. Thus it is interesting to observe that some nonlinear problems with integer variables are impossible to solve. For example, it is impossible to describe an algorithm that decides whether $\{x \in Z^m: f(x) = 0\}$ is nonempty or not, where f can be any polynomial function.

3. SOME PROBLEMS SOLVABLE IN POLYNOMIAL TIME

In this section, we briefly discuss the complexity of some of the problems in Figure 1.1 that are known to be in $\mathcal{P}$.

To point out some distinctions between the complexity of problems in $\mathcal{P}$, we consider five problems.

1. *The minimum-weight path problem with nonnegative data* (see Section I.3.2). An instance is specified by any m node graph and integral nonnegative edge weights.

   Dijkstra's algorithm requires $O(m^2)$ elementary calculations. Note that the number of calculations is independent of the numerical values of the edge weights. Moreover, each of the numerical operations is an addition or comparison, so the numbers involved only grow slowly. The performance of this algorithm is very satisfactory, since a complete graph on m nodes contains $m(m - 1)/2$ edges, and thus $O(m^2)$ integers are needed to describe some of the m node instances.
2. *The minimum-weight path problem* (see Section I.3.2). An instance is specified by any m node graph and integral edge weights.

   The Bellman-Ford algorithm either finds a minimum-weight path or detects a negative weight cycle with $O(m^3)$ elementary calculations. It is not known whether more theoretically efficient algorithms [e.g., an algorithm with running time $O(m^2)$ or $O(m^2 \log m)$] are possible. In general, establishing lower bounds on the complexity of a problem is extremely difficult.
3. *Solving linear equations*. Given an $n \times n$ system of equations $Ax = b$, where A is nonsingular, $x = A^{-1}b$ can be found by Gaussian elimination. The basic elimination method requires n pivots, each of which requires $O(n^2)$ calculations.

   The size of the numbers that occur is bounded from above by the largest magnitude of the determinant of any square submatrix of (A, b). Now since $\det A$ involves $n! < n^n$ terms, the largest number is less than $(n\theta)^n$, where $\theta_A = \max_{i,j} |a_{ij}|$, $\theta_b = \max_i |b_i|$, and $\theta = \max(\theta_A, \theta_b)$. Hence Gaussian elimination is polynomial in n. By considering (A, I), where A is $m \times n$, Gaussian elimination also yields polynomial time algorithms for calculating $\operatorname{rank}(A)$ and $\det(A)$ and for solving $m \times n$ linear systems.

4. *The transportation problem* (see Section I.3.5). An instance is specified by an $m_1 \times m_2$ $(m_1 + m_2 = m)$ integral matrix C, where c_{ij} is the unit shipping cost from supply point i to demand point j, an m_1-vector of integral supplies $(a_1, \ldots, a_{m_1})$, and an m_2-vector of integral demands $(b_1, \ldots, b_{m_2})$, where $\Sigma_{i=1}^{m_1} a_i = \Sigma_{j=1}^{m_2} b_j = \alpha$.

   The primal–dual algorithm (without scaling) requires no more than α steps and $O(m^2)$ computations in each step. This is not a polynomial time algorithm, since the number of steps is exponential in log α. However, when scaling is included, the number of steps is reduced to $m\lceil \log \theta \rceil$, where $\theta = \max(\max_i a_i, \max_j b_j)$. Thus we obtain a polynomial time algorithm whose running time is $O(m^3 \log \theta)$.

   Very recently, polynomial time algorithms with running time bounds that are independent of the numerical data have been found. The practical efficiency of these algorithms is not yet known. Furthermore, the practical significance of scaling is unresolved. Presently, it is generally believed that the most practical algorithm is a primal simplex algorithm, which is known to be exponential. So here we have an indication that the polynomial/exponential dichotomy is a dubious way to measure the practical efficiency of algorithms.

5. *The linear programming problem* (see Chapter I.2). An instance is given by $\max\{cx: Ax \leqslant b, x \in R^n_+\}$, where (A, b) is an integral $m \times (n+1)$ matrix and c is an integral n-vector.

   The simplex method, which is used in all commercial linear programming codes, is *not* a polynomial time algorithm. There are classes of linear programs for which the simplex method takes exponential time. This fact is the outstanding evidence for the argument against worst-case analysis of algorithms, since the simplex method has been enormously successful in the solution of real-world instances. Recently, the efficiency of the simplex method has been supported even further by probabilistic analysis, which shows that under rather general assumptions on the underlying distribution of instances, the *expected* running time of the simplex method is bounded by a polynomial in m and n.

In Chapter I.6 we will give two polynomial time algorithms for linear programming. The older of these two methods is the ellipsoid algorithm. It certainly seems to be inferior to the simplex algorithm as a computational tool. However, it provides an important proof technique for showing that some combinatorial optimization problems are in $\mathcal{P}$. The more recent method, Karmarkar's algorithm, appears to be a promising technique for practical computation. But as of this writing, not enough empirical evidence is available.

Prior to the development and analysis of the ellipsoid algorithm, many researchers believed that linear programming was in $\mathcal{P}$, but no proof was known. The reason for this conjecture assumes a central role in the development and analysis of algorithms for integer optimization problems. Here we give an informal explanation of the reason. In Section 5, we reexamine it in the language of computational complexity.

Suppose we owned a supercomputer that ran as fast as an exponential number of standard computers running in parallel. We could then solve a bounded instance of linear programming by using Gaussian elimination to enumerate all basic solutions. Each basic solution could be checked for nonnegativity, and from the feasible ones we could pick one that maximizes the objective function.

Having determined an optimal solution in this way, how could we, in polynomial time, convince someone else, who did not have access to the supercomputer, that we really had found an optimal solution? One answer, of course, is to apply a polynomial-time ellipsoid algorithm. But there is a much simpler answer that was known long before the ellipsoid algorithm.

Suppose we ask our computer to produce an optimal dual solution u^0 as well as an optimal primal solution x^0. Then given (x^0, u^0), with $O(mn)$ calculations, we could convince anyone that x^0 and u^0 were optimal. We would show the feasibility of x^0 and u^0 (i.e., $\{Ax^0 \leqslant b, x^0 \in R^n_+\}$ and $\{u^0A \geqslant c, u^0 \in R^m_+\}$) and then show that $cx^0 = u^0b$. Thus the duality theorem of linear programming gives the proof. One subtle point remains. We must show that the coefficients of x^0 and u^0 are polynomial in the length of the encoding of the data. Fortunately this is true for basic solutions and extreme rays. The argument is essentially a repeat of that used above to observe that the intermediate numbers in Gaussian elimination are polynomial in the input length.

The following notation will be used throughout the text.

$$\theta_A = \max_{i,j} |a_{ij}|, \; \theta_b = \max_i |b_i|,$$

$$\theta = \theta_{A,b} = \max(\theta_A, \theta_b).$$

Proposition 3.1. *Let x^0, r^0 be an extreme point and extreme ray of $P = \{x \in R^n_+: Ax \leqslant b\}$, where (A, b) is an integral $m \times (n+1)$ matrix. Then for $j = 1, \ldots, n$:*

i. *$x^0_j = p_j/q$, where p_j, q are integers such that $0 \leqslant p_j < n\theta_b(n\theta_A)^{n-1}$ and $1 \leqslant q < (n\theta_A)^n$.*
ii. *$r^0_j = p_j/q$, where p_j and q are integers such that $0 \leqslant p_j < ((n-1)\theta_A)^{n-1}$ and $1 \leqslant q < ((n-1)\theta_A)^{n-1}$.*

Proof. i. By the characterization of extreme points of P, x^0 is a solution to $A'x = b'$, where A' is $n \times n$ and nonsingular and each row of A' is either of the form $a^i x = b_i$ or $x_j = 0$. Then, by Cramer's rule, $x^0_j = p_j/q$, where $q \geqslant 1$ is the magnitude of the determinant of A' and p_j is the magnitude of the determinant of the matrix obtained by replacing the jth column of A' by b'. Each of these determinants contains $n! < n^n$ terms. Hence $1 \leqslant q < (n\theta_A)^n$ and $0 \leqslant p_j < n\theta_b(n\theta_A)^{n-1}$.

ii. Similarly r^0 is determined by $n - 1$ equations, either of the form $a^i x = 0$ or $x_j = 0$. ■

The bound of Proposition 3.1 states that the number of binary digits needed to represent x^0 is less than $2n \log(n\theta)^n = 2n^2 \log(n\theta)$, which is polynomial in n and $\log \theta$. Intuitively, Proposition 3.1 states that if a polyhedron has extreme points with both large and small integral coordinates, then it has very sharp angles (see Figure 3.1). But in order to obtain very sharp angles, the defining hyperplanes must have some very large coefficients.

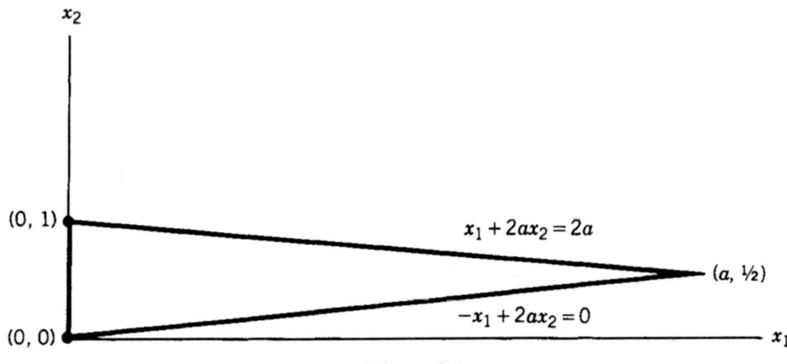

Figure 3.1

A theoretical consequence of this bound is that if $P = \{x \in R^n_+ : Ax \leqslant b\}$ is an unbounded polyhedron and $\max\{cx: x \in P\}$ is finite, it suffices to solve $\max\{cx: x \in P'\}$, where $P' = \{x \in R^n_+ : Ax \leqslant b, x \leqslant (n\theta)^n\}$ is a polytope whose length of description $l' = l + O(n^2(\log n\theta))$ is not significantly longer than the description length l of P. This will be used in the ellipsoid and projection algorithms in the next chapter.

Information that can be used to check optimality in polynomial time is called a *certificate of optimality* or a *good characterization*. A binary string is said to be *short* if its length is a polynomial function of the length of the input.

For linear programming, a certificate consists of basic optimal primal and dual solutions. To use it we simply verify primal and dual feasibility and the equality of the objective function values. Of course, if a problem is in $\mathcal{P}$ it has a good characterization. Although it is not known whether a good characterization implies that a problem is in $\mathcal{P}$, for nearly all optimization problems for which a good characterization is known, a polynomial-time algorithm is also known. Until 1979, linear programming was one of the rare exceptions. Some other exceptions in combinatorial optimization were also resolved through the use of the ellipsoid algorithm.

We now consider a problem that may properly be designated an integer optimization problem and is a generalization of the assignment problem.

6. *The weighted matching problem*. An instance is specified by a graph $G = (V, E)$ with m nodes, n edges, and integral weights c_e for $e \in E$.

   We have previously given the integer programming formulation

$$
\begin{aligned}
&\max \sum_{e \in E} c_e x_e \\
(3.1)\qquad &\sum_{e \in \delta(i)} x_e \leqslant 1 \quad \text{for } i \in V \\
&x \in Z^n_+,
\end{aligned}
$$

   where $\delta(i)$ is the set of edges incident to node i. Here the linear program obtained by replacing $x \in Z^n_+$ by $x \in R^n_+$ does not necessarily have an integral solution. However, there is an algorithm for weighted matching that only requires $O(m^3)$ calculations.

All of the known polynomial-time algorithms for weighted matching implicitly use a linear inequality description of the convex hull of matchings. We will show later that $x \in R^n_+$ is a matching if and only if it is an extreme point of the polytope given by $x \in R^n_+$, (3.1), and the *odd set constraints*

$$\sum_{e \in E(U)} x_e \leqslant \frac{|U| - 1}{2} \quad \text{for all } U \subseteq V \text{ such that } |U| = 2k + 1,\ k = 1, 2, \ldots,$$

where $E(U)$ is the set of edges with both ends in U. An odd set constraint states the obvious fact that when $|U|$ is odd, no matching can have more than $(|U| - 1)/2$ edges internal to U.

One should note that this formulation, together with the fact that linear programming is in $\mathcal{P}$, does not immediately yield a polynomial-time algorithm for weighted matching. The reason is that the linear programming formulation has a number of constraints that are exponential in the size of the data needed to describe the weighted matching problem.

Nevertheless, this formulation readily produces a good characterization. Again, duality provides the certificate of optimality. Although there is a very large number of dual variables, a basic dual solution contains no more than n positive variables. Moreover, it can be shown that in a basic dual solution, the values of the dual variables are not "too large". A certificate then consists of an optimal matching, an optimal dual solution, and those odd sets with positive dual variables. Note that it is not necessary to check the odd set constraints to verify the feasibility of a matching.

4. REMARKS ON 0-1 AND PURE-INTEGER PROGRAMMING

In the previous section we mentioned that linear programming and matching, and hence all special cases of them in Figure 1.1, are in $\mathcal{P}$. The status of all of the other problems shown in Figure 1.1 is much less settled. It is not known whether any are in $\mathcal{P}$, but there is a theory that leads us to believe that none are in $\mathcal{P}$. This theory is the subject of Sections 5 and 6. In this section we will make a few remarks on worst-case running times for some of the problems and, in particular, give bounds on the magnitude of the values of variables in an optimal solution to the pure-integer programming problem.

1. *The 0-1 integer programming problem*. An instance of the general problem $\max\{cx: Ax \leqslant b, x \in B^n\}$ is specified by an integral $m \times (n+1)$ matrix (A, b) and an integral n-vector c.

   It can be solved by a brute-force enumerative algorithm in $O(2^n mn)$ time. Even for such special cases as the node packing problem, no significantly better worst-case result is known. However, there are many special cases (e.g., matching, node packing on appropriately restricted classes of graphs, and some matroid optimization problems) that are in $\mathcal{P}$. These problems are the subject of Part III of this book.
2. *The integer knapsack problem*. An instance of the general problem $\max\{cx: ax \leqslant b, x \in Z_+^n\}$ is specified by integral n-vectors c and a, and an integer b.

   There is an $O(nb)$ algorithm, but it is exponential unless we restrict b to be a polynomial function of n. Although in some applications the magnitude of b can be restricted, large values cannot be dismissed. One reason is that rather general integer programs can be easily transformed into an equality-constrained version of the knapsack problem with constraints $ax = b$ and upper bounds $x_j \leqslant d_j$ for $j = 1, \ldots, n$. What makes this transformation uninteresting is that the magnitudes of the resulting constraint coefficients are generally exponential in the length of the encoding of the data.
3. *The pure-integer programming problem*. An instance of the general problem $\max\{cx: Ax \leqslant b, x \in Z_+^n\}$ is specified by an integral $m \times (n+1)$ matrix (A, b) and an integral n-vector c.

   Let $P = \{x \in R_+^n: Ax \leqslant b\}$. If P is bounded, by Proposition 3.1, we know that $x_j \leqslant (n\theta)^n$ for $j = 1, \ldots, n$. Hence it is possible to solve the problem $\max\{cx: x \in P \cap Z_+^n\}$ by enumerating the finite number of points in Z_+^n satisfying $x_j \leqslant (n\theta)^n$ for $j = 1, \ldots, n$.

   We now show that even if P is unbounded, the integer programming problem can be solved by enumeration. By Theorem 6.3 of Section I.4.6 we know that if the problem has a finite optimum value, there is an optimal solution at an extreme point of conv(S), where $S = P \cap Z^n$. We will obtain a bound on the magnitude of any such extreme point.

Theorem 4.1. *Let $P = \{x \in R_+^n: Ax \leqslant b\}$ and $S = P \cap Z^n$. If x^0 is an extreme point of* conv(S), *then*

$$x_j^0 \leqslant ((m+n)n\theta)^n \quad \textit{for } j = 1, \ldots, n.$$

Proof. In the proofs of Theorem 6.1 and 6.2 of Section I.4.6 we have shown that

$$\text{conv}(S) = \left\{x \in R_+^n: x = \sum_{l \in L} \alpha_l q^l + \sum_{j \in J} \mu_j r^j, \sum_{l \in L} \alpha_l = 1, \alpha \in R_+^{|L|}, \mu \in R_+^{|J|}\right\},$$

where $q^l, r^j \in Z_+^n$ for $l \in L$ and $j \in J$. Any extreme point of conv(S) must be one of the points $\{q^l\}_{l \in L}$, that is, $x^0 \in Q$, where

$$Q = \left\{x \in Z_+^n: x = \sum_{k \in K} \lambda_k x^k + \sum_{j \in J} \mu_j r^j, \sum_{k \in K} \lambda_k = 1, \mu_j < 1 \text{ for } j \in J, \lambda \in R_+^{|K|}, \mu \in R_+^{|J|}\right\},$$

where $\{x^k\}_{k \in K}$ are the extreme points of P and $\{r^j\}_{j \in J}$ are the extreme rays.

Since $|J| \leqslant \binom{m+n}{n-1}$, $|x_i^k| \leqslant (n\theta)^n$, and $|r_i^j| \leqslant (n\theta)^n$, it follows that

$$|x_i^0| \leqslant (n\theta)^n (1 + |J|) < ((m+n)n\theta)^n. \qquad \blacksquare$$

Note that $((m+n)n\theta)^n \leqslant (2\tilde{n}^2\theta)^{\tilde{n}}$, where $\tilde{n} = \max(m, n)$. We will use $\omega_{A,b} = (2\tilde{n}^2\theta)^{\tilde{n}}$ as notation for this value from now on.

Theorem 4.1 combined with Theorem 6.3 of Section I.4.6 implies that we can add the constraints $|x_j| \leqslant \omega_{A,b}$ to any integer program, and because no extreme points are removed we can test for feasibility (unboundedness) and optimality by enumerating the integer points in $S \cap \{x \in Z_+^n: x \leqslant \omega_{A,b}\}$. We can now show that any instance of a pure-integer program can be transformed in polynomial time to an instance of a 0-1 integer program.

For $j = 1, \ldots, n$ let $x_j = \sum_{k=0}^d 2^k x_{jk}$, where $(x_{j0}, \ldots, x_{jd}) \in B^{d+1}$ and $d = \lceil \tilde{n} \log (2\tilde{n}^2\theta) \rceil$. With this substitution, we obtain the 0-1 integer program $\max\{c'x': A'x' \leqslant b, x' \in B^{n(d+1)}\}$, where c' is $1 \times n(d+1)$ and A' is $m \times n(d+1)$. Note that the largest coefficient of A' has magnitude less than $2^d\theta = \theta(2\tilde{n}^2\theta)^{\tilde{n}}$ and that the largest coefficient of c' has magnitude less than $(2\tilde{n}^2\theta)^{\tilde{n}} \times (\max_{j=1,\ldots,n} c_j)$. Thus the length of the data needed to describe the transformed 0-1 integer program is a polynomially bounded function of the length of the data needed to describe the original integer program. Hence we have the following proposition.

Proposition 4.2. *An instance of a pure-integer programming problem can be transformed in polynomial time to an instance of a 0-1 integer programming problem.*

We have observed that Theorem 4.1 gives a finite algorithm—namely, enumeration—for integer programming. Now consider the class of integer programs with n fixed. For 0-1 integer programming, enumeration is polynomial. However, for pure-integer programming, enumeration is not polynomial, since the upper bound $\omega_{A,b}$ depends polynomially on θ. Furthermore, the transformation of pure-integer programming to 0-1 integer programming given above yields $d + 1$ variables and 2^d is polynomial in θ, so enumeration on the transformed problem is not polynomial for n fixed. In fact, it is a theorem that integer programming with a fixed number of variables is in $\mathcal{P}$, but the proof requires much deeper results than Proposition 4.2 (see Section II.6.5).

Analogous to the results we have given on the size of numbers in feasible and optimal solutions to integer programs, there is a result on the size of numbers that can arise in a description of the convex hull of feasible solutions by linear inequalities. The following theorem can be obtained from Theorem 4.1 and polarity.

Theorem 4.3. *Suppose $S = \{x \in Z_+^n\colon Ax \leq b\}$, where (A, b) is an integral $(m + 1) \times n$ matrix. If (π, π_0) defines a facet of* conv(S), *then the length of the description of the coefficients of (π, π_0) is bounded by a polynomial function of m, n, and* $\log \theta$.

5. NONDETERMINISTIC POLYNOMIAL-TIME ALGORITHMS AND $\mathcal{NP}$ PROBLEMS

The theoretical model that we study in both this section and the next one addresses the question of whether integer programming and many special cases are solvable in polynomial time. The model does not provide a definite answer, but one of the main conclusions is that it is just as unlikely that there are polynomial-time algorithms for most special cases of integer programming (e.g., integer knapsack, node packing) as there are for the general integer programming problem. We will prove that if, for example, integer knapsack or node packing is solvable in polynomial time, then general integer programming is solvable in polynomial time.

Although we can use the model to draw conclusions about optimization problems, it has been developed for so-called *decision*, *recognition*, or *feasibility problems*. We will use the term *feasibility problem* because of the close connection with feasibility testing in an optimization problem.

A *feasibility problem* X is a pair (D, F) with $F \subseteq D$, where the elements of D are finite binary strings. D is called the *set of instances* of X, and F is called the *set of feasible instances*. Given an instance $d \in D$, we want to determine whether $d \in F$. Given $d \in D$, the answer is either yes or no.

In the remainder of this chapter we will follow the notation commonly used in complexity theory and we will define $\mathcal{P}$ to be the class of feasibility problems that are solvable in polynomial time.

Associated with an optimization problem we define a feasibility problem in which an instance corresponds to a description of a constraint set. F is the set of instances for which the constraint set is nonempty.

Example 5.1 *(0-1 integer programming feasibility).* D is the set of all integral matrices (A, b), where b contains one column and the same number of rows as A. An instance is specified by integers m and n, the dimensions of A, and numerical values for the coefficients of A and b.

This is the feasibility problem for $S = \{x \in B^n\colon\ Ax \leq b\}$. Hence $F = \{(A, b)\colon \{x \in B^n\colon Ax \leq b\} \neq \emptyset\}$. Here a yes answer is commonly established by exhibiting a feasible x.

A second feasibility problem concerns a lower bound on the objective function. Here we augment each instance by an objective function c and an integer z. The *lower-bound feasibility problem* is the feasibility problem with the additional constraint $cx \geq z$.

Example 5.2 *(0-1 integer programming lower-bound feasibility).* $D = \{(A, b, c, z)\}$ is the set of all integral matrices A, b, c and integers z, where b (respectively, c) contains one column (row) and the same number of rows (columns) as A. $F = \{(A, b, c, z)\colon \{x \in B^n\colon Ax \leq b, cx \geq z\} \neq \emptyset\}$.

Note that if $b \in Z_+^m$, the feasibility problem for 0-1 integer programming is trivial, but the lower-bound feasibility problem is not. This is frequently the situation as, for example, in node packing.

The lower-bound feasibility problem is closely connected to the optimization problem. If $(A, b, c, z^0) \in F$ and $(A, b, c, z^0 + 1) \notin F$, then $\max\{cx: Ax \leq b, x \in B^n\} = z^0$. Thus if it is known that $z_L \leq z^0 \leq z_U$, we can find z^0 by solving, at most, $z_U - z_L + 1$ lower-bound feasibility problems. Note that $z_U - z_L$ is not polynomial in the input length.

There is, however, a more efficient method for finding z^0, called *binary search*. Suppose we are given a function $h: Z^1 \to B^1$ of the form

$$h(x) = \begin{cases} 0 & \text{for } x \leq x_0 \\ 1 & \text{for } x > x_0, \end{cases}$$

where x_0 is unknown. We are also given integers x_L and x_U with $h(x_L) = 0$ and $h(x_U) = 1$. The problem is to find x_0. By putting $h(z) = 0$ if $(A, b, c, z) \in F$ and $h(z) = 1$ otherwise, we see that the problem of finding z^0 is of this form.

The following binary search algorithm finds x_0 with, at most, $\lceil \log(x_U - x_L) \rceil$ evaluations of the function h.

Step 1: If $x_U - x_L \leq 1$, stop. $x_0 = x_L$. Otherwise go to Step 2.
Step 2: Let $x = \lfloor (x_L + x_U)/2 \rfloor$. If $h(x) = 0$, set $x_L = x$; otherwise set $x_U = x$. Go to Step 1.

Each function evaluation halves the length of the interval that contains x_0. Hence the number of evaluations is bounded by $\lceil \log(x_U - x_L) \rceil$. An example is shown in Figure 5.1.

Thus with binary search, we can find z^0 by solving $\lceil \log(z_U - z_L + 1) \rceil$ lower-bound feasibility problems. Since $\lceil \log(z_U - z_L + 1) \rceil$ is polynomial in the length of the input of the 0-1 lower-bound feasibility problem, we obtain the following proposition.

Proposition 5.1. *If the 0-1 integer programming lower-bound feasibility problem can be solved in polynomial time, the 0-1 integer programming problem can be solved in polynomial time.*

This proposition has an obvious generalization to other optimization problems. In particular, it applies to the integer programming problem, where Theorem 4.1 is used to give bounds z_U and z_L such that $\lceil \log(z_U - z_L + 1) \rceil$ is polynomial in the length of the input.

Certificates of Feasibility, the Class $\mathcal{NP}$, and Nondeterministic Algorithms

Analogous to certificates of optimality, information that can be used to check feasibility in polynomial time is called a *certificate of feasibility*. Given $X = (D, F)$, for each instance $d \in F$ we let Q_d denote such a certificate. We know that if Q_d exists it must be short. Here we are interested in the class of feasibility problems having a certificate of feasibility.

One might imagine an algorithm that makes a large number of guesses in the hope of eventually guessing Q_d. This leads to the concept of a nondeterministic algorithm for a feasibility problem $X = (D, F)$. The reader should take note that such algorithms cannot be realized in practical computation.

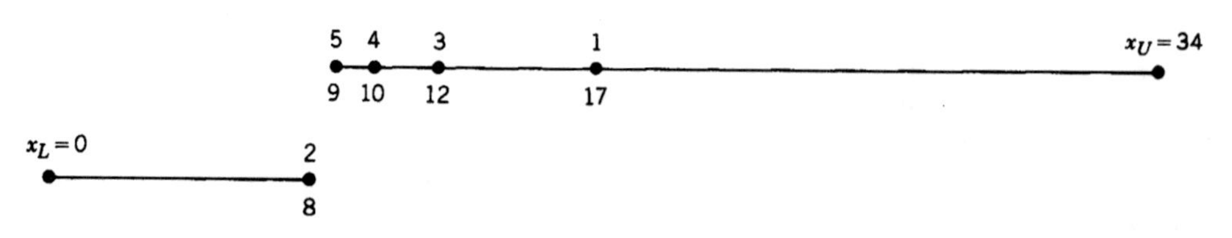

Figure 5.1

A nondeterministic algorithm consists of two stages. The input to the algorithm is a $d \in D$. The first stage is a guessing stage. Here we guess a binary string Q which is then passed on to the second stage. The second stage, called the *checking stage*, is an algorithm that works with the pair (d, Q) and may provide the output that $d \in F$. For example, the checking stage may verify that $x \in S$ and thus output that $d \in F$. Two properties are required:

1. If $d \in F$, there is a certificate Q_d such that when the pair (d, Q_d) is given to the checking stage, the algorithm gives the answer that $d \in F$.
2. If $d \notin F$, there is no output. Hence whenever there is output, $d \in F$.

We measure the work done by a nondeterministic algorithm only in the checking stage and only when the checking stage is given a $d \in F$ and a certificate of feasibility. We say that the *nondeterministic algorithm is polynomial* if, for each $d \in F$, its running time in the checking stage is a polynomial function of the length of the encoding of d for some Q_d for which it replies that $d \in F$. This means that when $d \in F$, there is a *short (polynomial-time) proof of feasibility*.

We define $\mathcal{NP}$ to be *the class of feasibility problems such that for each instance with $d \in F$, the answer $d \in F$ is obtained in polynomial time by some nondeterministic algorithm*. Nothing is said when $d \notin F$.

We will also encounter feasibility problems that are not in $\mathcal{NP}$. Many of these are in a related set called $\mathcal{C}o\mathcal{NP}$, which will be defined and discussed later in this section.

Example 5.3 *(Nondeterministic polynomial-time algorithm for 0-1 integer feasibility).*

Guessing stage: Guess an $x \in B^n$.
Checking stage: If $Ax \leqslant b$, output $(A, b) \in F$; otherwise return.

The algorithm for 0-1 integer programming lower-bound feasibility is similar. Now consider general integer programming feasibility. The same algorithm works with the guesses being $x \in Z_+^n$ because Theorem 4.1 stipulates that if $\{x \in Z_+^n: Ax \leqslant b\} \neq \emptyset$, then there is a feasible x such that the logarithm of its largest coefficient is bounded by a polynomial in the length of the encoding of (A, b). This is one of the few nontrivial $\mathcal{NP}$ results that we need.

Proposition 5.2. *General integer programming feasibility is in $\mathcal{NP}$.*

The Hamiltonian cycle problem is to determine whether a graph $G = (V, E)$ contains a Hamiltonian cycle. A *Hamiltonian cycle* is a cycle that contains all of the nodes of G.

Proposition 5.3. *Hamiltonian cycle is in $\mathcal{NP}$.*

Proof. We give a nondeterministic polynomial-time algorithm for Hamiltonian cycle.

Input: A graph $G = (V, E)$.
Guessing stage: Guess an $E' \subseteq E$.
Checking stage: *Step a*. If the degree of each node of $G' = (V, E')$ is two, go to Step b; otherwise return.
Step b. If $G' = (V, E')$ is connected, output $G \in F$; otherwise return. ■

We have simply used the facts that (a) a graph G' is a Hamiltonian cycle if each node is of degree 2 and the graph is connected and (b) each of these properties is easily checked in polynomial time. The upper-bound feasibility problem associated with the minimum-cost traveling salesman problem is also in $\mathcal{NP}$. This is shown by slightly generalizing the algorithm given in Proposition 5.3.

A nondeterministic algorithm does not completely solve the feasibility problem, since it ignores $d \notin F$. However, by being just a little bit intelligent about our guesses, we can simulate a nondeterministic polynomial-time algorithm by a deterministic exponential-time algorithm. For each $d \in F$ there is a structure Q_d whose length $l(Q_d)$ is polynomial in the length of d, say $l(Q_d) = c(l(d))^p$. Therefore, for a given $d \in D$ we can limit our guesses to binary strings of length equal to or less than $L = c(l(d))^p$. Hence we need to consider, at most, 2^{L+1} structures. But there is a polynomial function $f(l(d))$ that gives an upper bound on the running time of the checking stage for $d \in F$ when Q_d is guessed. Hence for each of the 2^{L+1} structures, we run the checking stage for $f(l(d))$ time and then go on to the next structure if the checking stage has not verified $d \in F$. Thus if a feasibility problem is in $\mathcal{NP}$, it can be completely solved in exponential time.

The Class $\mathcal{C}o\mathcal{NP}$

Each feasibility problem $X = (D, F)$ has a related feasibility problem $\overline{X} = (D, \overline{F})$, where $\overline{F} = D \setminus F$, called the *complement* of X. In the complement of 0-1 integer programming feasibility we have $\overline{F} = \{(A, b): \{x \in B^n: Ax \leqslant b\} = \emptyset\}$. It is not known whether the complement of 0-1 integer programming feasibility is in $\mathcal{NP}$. In fact, it is not known whether the complements of any of the feasibility problems mentioned so far in this section are in $\mathcal{NP}$.

For the complement of the 0-1 integer programming lower-bound feasibility problem $\overline{F} = \{(A, b, c, z): \{x \in B^n: Ax \leqslant b, cx \leqslant z\} = \emptyset\}$, which is equivalent to showing that $cx < z$ is a valid inequality for $\{x \in B^n: Ax \leqslant b\}$. Thus if the lower-bound feasibility problem and its complement are in $\mathcal{NP}$ we would have a good characterization for the optimization problem.

To establish terminology for complements of $\mathcal{NP}$ problems, let $\mathcal{C}o\mathcal{NP} = \{X: X$ is a feasibility problem, $\overline{X} \in \mathcal{NP}\}$.

Proposition 5.4. *If X is a feasibility problem and $X \in \mathcal{P}$, then $X \in \mathcal{NP} \cap \mathcal{C}o\mathcal{NP}$.*

Proof. Every polynomial-time algorithm is also a nondeterministic polynomial-time algorithm. We simply ignore the guessing stage and apply the polynomial-time algorithm in the checking stage. Hence $X \in \mathcal{P} \Rightarrow X \in \mathcal{NP}$. But if $X \in \mathcal{P}$ so is $\overline{X} \in \mathcal{P}$, since if $d \notin F$, it follows that our polynomial-time algorithm, which needs no guesses, will also tell us this in polynomial time. Hence $\overline{X} \in \mathcal{P}$ implies $\overline{X} \in \mathcal{NP}$ or, equivalently, $X \in \mathcal{C}o\mathcal{NP}$. ∎

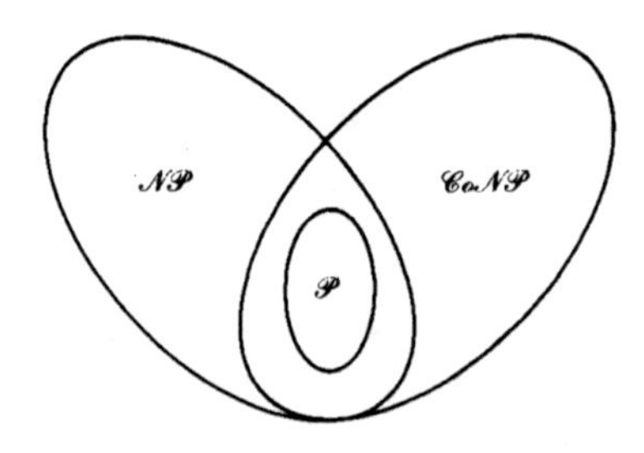

Figure 5.2

The linear programming feasibility problem for sets of the form $\{x \in R^n_+: Ax \leqslant b\}$ is in $\mathcal{P}$ by virtue of the ellipsoid method. Hence we can use Proposition 5.4 to establish its membership in $\mathcal{NP} \cap \mathcal{C}o\mathcal{NP}$. But this fact was known long before the ellipsoid method, since membership in $\mathcal{C}o\mathcal{NP}$ is a consequence of linear programming duality. We leave it to the reader to show membership in $\mathcal{NP}$. Here we show membership in $\mathcal{C}o\mathcal{NP}$. The reader should refer back to the good characterization of linear programming given in Section 3 which essentially does the same thing.

Example 5.4 *(Nondeterministic algorithm for linear programming infeasibility).*

Input:	An integral $m \times n$ matrix (A, b).
Guessing stage:	Guess a $u \in R^m_+$.
Checking stage:	If $uA \geqslant 0$ and $ub < 0$, output $(A, b) \in \bar{F}$; otherwise return.

We have used the Farkas lemma—that is, if there exists a $u \in R^m_+$ such that $uA \geqslant 0$ and $ub < 0$, then $\{x \in R^n_+: Ax \leqslant b\} = \emptyset$—and Proposition 3.1, which guarantees the existence of suitably small rational u so that the checking can be done in polynomial time.

The sets $\mathcal{P}$, $\mathcal{NP}$, and $\mathcal{C}o\mathcal{NP}$ for feasibility problems are shown in Figure 5.2.

The answers to the following questions are unknown.

1. Does $\mathcal{P} = \mathcal{C}o\mathcal{NP} \cap \mathcal{NP}$?
2. Does $\mathcal{C}o\mathcal{NP} = \mathcal{NP}$?
3. Does $\mathcal{P} = \mathcal{NP}$?

An affirmative answer to question 3 implies affirmative answers to question 1 and 2, since, by Proposition 5.4, we have $\mathcal{P} \subseteq \mathcal{NP} \cap \mathcal{C}o\mathcal{NP}$. Similarly, affirmative answers to questions 1 and 2 imply an affirmative answer to question 3.

In the next section, we will study the class $\mathcal{NP}$ further; at the end of that section, we will make some remarks about the impact on integer programming of answers to the above questions.

6. THE MOST DIFFICULT $\mathcal{NP}$ PROBLEMS: THE CLASS $\mathcal{NPC}$

The main result of this section is that $\mathcal{NP}$ contains hardest problems. By this we mean that there is a subset of $\mathcal{NP}$, called $\mathcal{NPC}$, such that if there exists $X \in \mathcal{NPC} \cap \mathcal{P}$, then every problem in $\mathcal{NP}$ is in $\mathcal{P}$, that is, $\mathcal{P} = \mathcal{NP}$. Problems in $\mathcal{NPC}$ are called *$\mathcal{NP}$-complete.* Moreover, we will show that amongst these hardest problems are feasibility problems associated with integer optimization and many special cases.

The technique used here is that of polynomially transforming one problem into another. Suppose $X_i = (D_i, F_i)$, $i = 1, 2$, are two feasibility problems and there exists a function $g: D_1 \to D_2$ such that for every $d \in D_1$ we have $g(d) \in F_2$ if and only if $d \in F_1$. If the function g is computable in time that is polynomial in the length of the encoding of d, then X_1 is said to be *polynomially transformable* to X_2. The consequence of this definition is clear.

Proposition 6.1. *If X_1 is polynomially transformable to X_2 and $X_2 \in P$, then $X_1 \in P$.*

Proof. The polynomial-time algorithm for X_1 is to compute the function g and then apply the polynomial-time algorithm for X_2. ■

The transformation idea is surely familiar. When confronted with a new problem, a traditional approach to solving it is to restate it as a problem we already know how to solve. The only thing we have added is the requirement that the transformation be done in polynomial time.

We say that X_1 is a "special case" of X_2 if $D_1 \subset D_2$ and $F_1 = D_1 \cap F_2$. Here $g(d) = d$ is the *identity transformation*. Many of the arcs in the graph of Figure 1.1 were determined by identity transformations. But we have also done nontrivial transformations. In particular, we have shown that integer programming feasibility is polynomially transformable to 0-1 integer programming feasibility (see Proposition 4.2). Also, in Example 1.1 we have shown that set-packing lower-bound feasibility is polynomially transformable to node-packing lower-bound feasibility. This means that the problem "Given a 0-1 $m \times n$ matrix A, an integral n-vector c, and an integer z, determine whether $\{x \in B^n: Ax \leqslant 1, cx \geqslant z\} \neq \varnothing$" is polynomially transformable to the problem "Given a 0-1 $m' \times n$ matrix A' with, at most, two 1's per row, an integral n-vector c, and an integer z, determine whether $\{x \in B^n: A'x \leqslant 1, cx \geqslant z\} \neq \varnothing$".

There is a technique, called *polynomial reduction*, that appears to be a more general approach than polynomial transformation for establishing that one problem can be solved in polynomial time given that another can. We say that X_1 is *polynomially reducible* to X_2 if there is an algorithm for X_1 that uses an algorithm for X_2 as a subroutine and runs in polynomial time under the assumption that each call of the subroutine takes unit time. Note that transformation is the special case of reduction in which the subroutine is used only once; that is, it is applied directly to the transformed data $g(d)$.

A generalization of Proposition 6.1 is the following.

Proposition 6.2. *If X_1 is polynomially reducible to X_2* and $X_2 \in \mathcal{P}$, then $X_1 \in \mathcal{P}$.

Although polynomial reducibility appears to be more general than polynomial transformability, it is not known whether it really is. In any case, all of the polynomial reductions needed in this section can be accomplished through the simpler technique of polynomial transformation.

We now address the question of the existence of hardest problems in $\mathcal{NP}$. $X \in \mathcal{NP}$ is said to be *$\mathcal{NP}$-complete* if all problems in $\mathcal{NP}$ can be polynomially reduced to X. The set of $\mathcal{NP}$-complete problems, which we will soon claim to be nonempty, is denoted by $\mathcal{NPC}$. The implication of the existence of an $\mathcal{NP}$-complete problem is given by the following proposition.

Proposition 6.3. *If X is $\mathcal{NP}$-complete, then $\mathcal{P} = \mathcal{NP}$ if and only if $X \in \mathcal{P}$.*

Proof. $X \in \mathcal{NP}$ and $\mathcal{P} = \mathcal{NP}$ implies $X \in \mathcal{P}$. On the other hand, if X is $\mathcal{NP}$-complete and in $\mathcal{P}$, then by Proposition 6.2 there is a polynomial algorithm for any problem in $\mathcal{NP}$. ■

Once we have an $\mathcal{NP}$-complete problem, we may be able to find others by polynomial reduction.

Proposition 6.4. *If X_1 is $\mathcal{NP}$-complete and X_1 is polynomially reducible to $X_2 \in \mathcal{NP}$, then X_2 is $\mathcal{NP}$-complete.*

The proof is obvious, but it is important to note the direction of the statement to avoid making the mistake of concluding that X_2 is $\mathcal{NP}$-complete by reducing X_2 to X_1.

The *satisfiability problem*, which is a classical problem in logic, is of historical interest because it was the first problem in $\mathcal{NP}$ shown to be $\mathcal{NP}$-complete. It is described by a finite set $N = \{1, \ldots, n\}$ and m pairs of subsets of N, denoted by $C_i = (C_i^+, C_i^-)$ for $i = 1, \ldots, m$. An instance is feasible if the set

$$\left\{x \in B^n \colon \sum_{j \in C_i^+} x_j + \sum_{j \in C_i^-} (1 - x_j) \geq 1 \text{ for } i = 1, \ldots, m\right\} \tag{6.1}$$

is nonempty.

The satisfiability problem is in $\mathcal{NP}$. We use subsets N^0 of N as guesses, set $x_j = 1$ if $j \in N^0$ and $x_j = 0$ otherwise, and then simply check for feasibility in (6.1).

Theorem 6.5 *(Cook). The satisfiability problem is $\mathcal{NP}$-complete.*

We will not prove this famous theorem. The proof is technical but is not very difficult mathematically. To comprehend it, one must understand the formal model of a nondeterministic Turing machine that can solve any problem in $\mathcal{NP}$ in polynomial time. The proof then amounts to a polynomial transformation of the nondeterministic Turing machine into the satisfiability problem.

Since we have described the satisfiability problem as a 0-1 integer feasibility problem, we obtain the following proposition.

Proposition 6.6. *The 0-1 integer programming feasibility problem is $\mathcal{NP}$-complete.*

Soon after the appearance of Cook's theorem, the list of $\mathcal{NP}$-complete problems was substantially enriched. This list includes lower-bound feasibility versions of all of the problems in Figure 1.1 that we have not already stated are in $\mathcal{P}$. It is important to understand that showing that a problem is in $\mathcal{NPC}$ is a negative result about the likelihood of finding a polynomial time algorithm for it.

To illustrate the use of polynomial transformations, we now show that some problems are $\mathcal{NP}$-complete. In choosing candidates, it is important to try to get as close to the boundary (if it exists) between $\mathcal{P}$ and $\mathcal{NPC}$. By this we mean the following: Given a problem in $\mathcal{P}$, what are the most simple generalizations of it that make it $\mathcal{NP}$-complete? For example, lower-bound feasibility testing for matching can be solved in polynomial time. In terms of linear inequalities, this problem is to determine if $\{x \in B^n \colon Ax \leq 1, cx \geq z\} \neq \emptyset$, where A is an $m \times n$ 0-1 matrix with two 1's in each column (the node-edge incidence matrix of a graph), c is an integral n-vector, and z is an integer. However, if we allow matrix A to contain three 1's in each column, the problem becomes $\mathcal{NP}$-complete, even if we restrict c to be a vector of 1's.

A similar situation occurs when we limit the number of 1's in each row of A. When the 0-1 matrix A contains one 1 in each row, the feasibility problem for the set $\{x \in B^n \colon Ax \leq 1, cx \geq z\}$ is trivial. However, if we allow matrix A to contain two 1's in each row, the problem becomes $\mathcal{NP}$-complete, even if we restrict c to be a vector of 1's. We now prove this result by a polynomial transformation from the satisfiability problem.

An instance of the *unweighted node-packing problem* is: Given a graph $G = (V, E)$ and an integer k, is there a $U \subseteq V$ such that $|U| \geq k$ and U is a node packing? Alternatively, is $\{x \in B^{|V|} \colon Ax \leq 1, \Sigma_{j \in V} x_j \geq k\} \neq \emptyset$, where A is the edge-node incidence matrix of G (i.e., where A is a 0-1 matrix with two 1's in each row)?

Proposition 6.7. *The lower-bound feasibility problem for unweighted node packing is $\mathcal{NP}$-complete.*

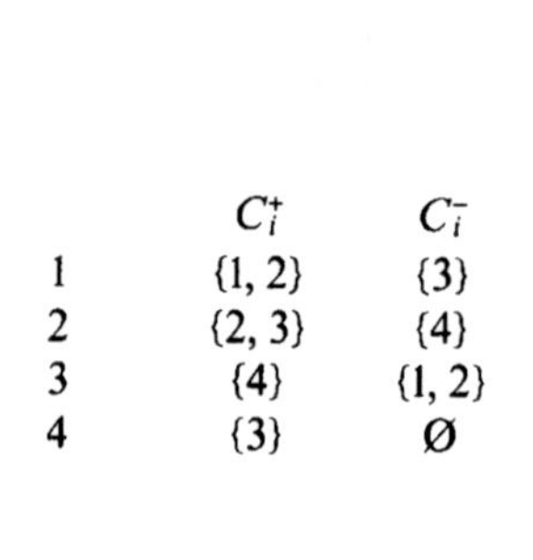

	C_i^+	C_i^-
1	{1, 2}	{3}
2	{2, 3}	{4}
3	{4}	{1, 2}
4	{3}	∅

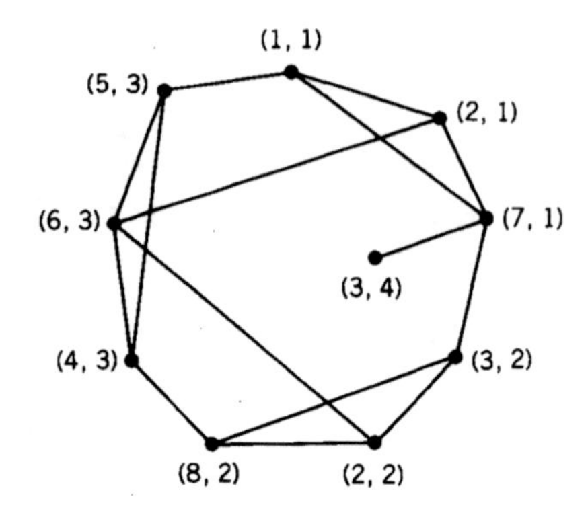

Figure 6.1.

Proof. The problem is a special case of 0-1 integer programming feasibility, so it is in $\mathcal{NP}$. Membership in $\mathcal{NPC}$ is established by polynomial transformation from the satisfiability problem.

Given an instance of the satisfiability problem specified by $N = \{1, \ldots, n\}$ and $C_i = (C_i^+, C_i^-)$ for $i = 1, \ldots, m$, we set $k = m$ and construct $G = (V, E)$ as follows. Let

$$V_i^+ = \{(j, i): j \in C_i^+\}, \; V_i^- = \{(n + j, i): j \in C_i^-\},$$

$$V_i = V_i^+ \cup V_i^- \quad \text{for } i = 1, \ldots, m \text{ and } V = \bigcup_{i=1}^{m} V_i.$$

Each pair of nodes in V_i is joined by an edge; and for $j = 1, \ldots, n$ and $l \neq i$, nodes (j, i) and $(n + j, l)$ are joined by an edge.

An example of the construction of G is shown in Figure 6.1. A feasible solution to the satisfiability problem is $N^0 = \{1, 3\}$ or $x_1 = x_3 = 1$ and $x_2 = x_4 = 0$. A node packing of size 4 is $\{(1, 1), (8, 2), (6, 3), (3, 4)\}$.

In general, any node packing of size m is of the form $\{(\alpha_1, 1), (\alpha_2, 2), \ldots, (\alpha_m, m)\}$ and such a packing exists if and only if $N^0 = \{\alpha_i: \alpha_i \leq n\}$ is a solution to the satisfiability problem. ■

An instance of the *set partitioning feasibility problem* is: Given an $m \times n$ 0-1 matrix A, is $\{x \in B^n: Ax = 1\} \neq \emptyset$?

Proposition 6.8. *The set partitioning feasibility problem is $\mathcal{NP}$-complete.*

Proof. The problem is a special case of 0-1 integer programming feasibility, so it is in $\mathcal{NP}$. We prove membership in $\mathcal{NPC}$ by transformation from the unweighted node-packing lower-bound feasibility problem. Given a graph $G = (V, E)$ and an integer k, let I_E be an $|E| \times |E|$ identity matrix, let A_G be the edge-node incidence matrix of G, and let 1 be a row vector of $|V|$ 1's. Construct the $(|E| + k) \times (|E| + k|V|)$ matrix

$$A = \left(\begin{array}{c|c|c|c|c} I_E & A_G & A_G & \cdots & A_G \\ \hline & 1 & 0 & & 0 \\ & 0 & 1 & \cdots & 0 \\ 0 & \vdots & \vdots & & \vdots \\ & 0 & 0 & & 1 \end{array}\right) = (B_0, B_1, \ldots, B_k).$$

Suppose $Ax = 1$, where $x \in B^{|E|+k|V|}$. This can be the case if and only if the columns of A for which $x_j = 1$ have the following structure. There is exactly one column from B_i for $i = 1, \ldots, k$. If b_{ip} is the column chosen from B_i, and b_{lq} is the column chosen from B_l, then the nodes corresponding to these columns are not joined by an edge. Hence the k columns chosen from $(B_1, \ldots, B_k)$ define a node packing of size k. The partition is completed by choosing appropriate columns from B_0; these correspond to edges of G that are not met by any nodes in the packing. An example is shown in Figure 6.2. Nodes 1 and 3 yield a packing of size 2, and a partition is indicated by the checked columns. ∎

Proposition 6.9. *The set partitioning feasibility problem in which matrix A has, at most, three 1's per column is $\mathcal{NP}$-complete.*

Proof. We polynomially transform the general set partitioning feasibility problem with an arbitrary 0-1 $m \times n$ matrix A into a 0-1 $m' \times n'$ matrix A' such that matrix A' has no more than three 1's per column and there is a one-to-one correspondence between solutions of $\{x \in B^n: Ax = 1\}$ and $\{y \in B^{n'}: A'y = 1\}$, where $n' \leqslant n(2m - 1)$ and $m' \leqslant m + 2n(m - 1)$. We assume that A has at least one column, say a_1, with $t \geqslant 4$ 1's; otherwise there is nothing to prove. Let $A = (a_1, A_{n-1})$ and

$$A_1' = \begin{pmatrix} \Delta_1 & A_{n-1} & 0_1 \\ H_1 & 0 & K_1 \end{pmatrix},$$

where Δ_1 is an $m \times t$ matrix of unit columns such that e_i is a column of Δ_1 if and only if $a_{i1} = 1$, and 0_1 is an $m \times (t - 1)$ null matrix. H_1 and K_1 are 0-1 matrices that will be described subsequently. Consider the equations

$$\begin{aligned} \Delta_1 y^1 + A_{n-1} y^2 \qquad\qquad &= 1 \\ H_1 y^1 \qquad\qquad + K_1 y^3 &= 1 \\ y^1 \in B^t, \; y^2 \in B^{n-1}, \; y^3 &\in B^{t-1}. \end{aligned}$$

Suppose H_1 and K_1 are such that the only two solutions to

$$H_1 y^1 + K_1 y^3 = 1, \qquad y^1 \in B^t, \; y^3 \in B^{t-1}$$

are $y^1 = (1, 1, \ldots, 1)$, $y^3 = (0, 0, \ldots, 0)$, and $y^1 = (0, 0, \ldots, 0)$, $y^3 = (1, 1, \ldots, 1)$. This condition can be achieved if H_1 and K_1 each have $2t - 2$ rows and the following structure:

$$H_1 = \begin{pmatrix} 1 & & & & & \\ & 1 & & & & \\ & 1 & & & & \\ & & 1 & & & \\ & & 1 & & & \\ & & & \ddots & & \\ & & & & 1 & \\ & & & & 1 & \\ & & & & & 1 \end{pmatrix}, \qquad K_1 = \begin{pmatrix} 1 & & & \\ 1 & & & \\ & 1 & & \\ & 1 & & \\ & & 1 & \\ & & 1 & \\ & & & \ddots \\ & & & & 1 \\ & & & & 1 \end{pmatrix}.$$

Note that if $y_1^1 = 1$, then $y_1^3 = 0$ and $y_2^1 = 1$. Similarly, if $y_1^3 = 1$, then $y_1^1 = y_2^1 = 0$ and $y_2^3 = 1$. We then proceed inductively to obtain the result.

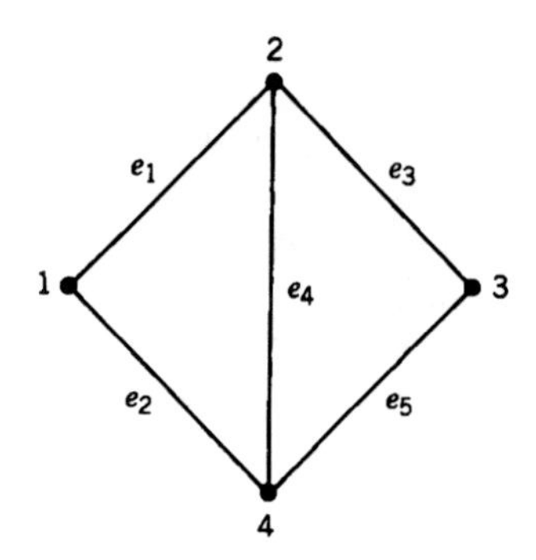

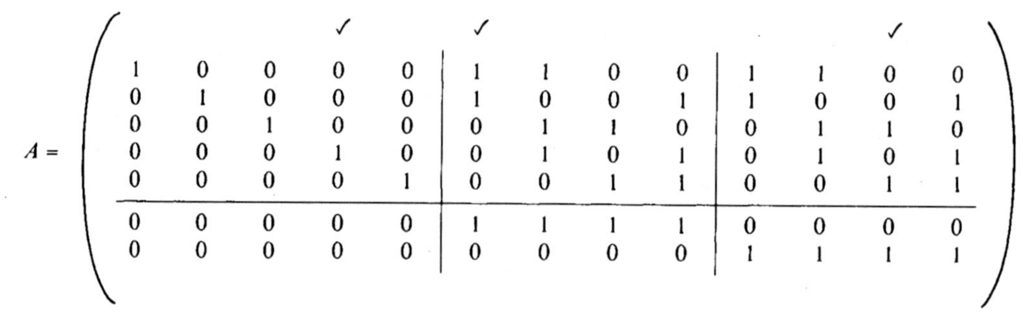

$$A = \left(\begin{array}{ccccc|cccc|cccc} & & & \checkmark & & \checkmark & & & & & & \checkmark & \\ 1 & 0 & 0 & 0 & 0 & 1 & 1 & 0 & 0 & 1 & 1 & 0 & 0 \\ 0 & 1 & 0 & 0 & 0 & 1 & 0 & 0 & 1 & 1 & 0 & 0 & 1 \\ 0 & 0 & 1 & 0 & 0 & 0 & 1 & 1 & 0 & 0 & 1 & 1 & 0 \\ 0 & 0 & 0 & 1 & 0 & 0 & 1 & 0 & 1 & 0 & 1 & 0 & 1 \\ 0 & 0 & 0 & 0 & 1 & 0 & 0 & 1 & 1 & 0 & 0 & 1 & 1 \\ \hline 0 & 0 & 0 & 0 & 0 & 1 & 1 & 1 & 1 & 0 & 0 & 0 & 0 \\ 0 & 0 & 0 & 0 & 0 & 0 & 0 & 0 & 0 & 1 & 1 & 1 & 1 \end{array}\right)$$

Figure 6.2.

The solution with $y^3 = (1, 1, \ldots, 1)$ yields $A_{n-1}y^2 = 1$, and we note that $\{x \in B^n: Ax = 1, x_1 = 0\} = \{(0, y^2) \in B^n: A_{n-1}y^2 = 1\}$. Also the solution with $y^1 = (1, 1, \ldots, 1)$ yields $\Delta_1 y^1 = a_1$, so

$$\{x \in B^n: Ax = 1, x_1 = 1\} = \{(1, y^2) \in B^n: a_1 + A_{n-1}y^2 = 1\}.$$

It is important to observe that H_1 and K_1 have been chosen so that each column of

$$\begin{pmatrix} \Delta_1 & O_1 \\ H_1 & K_1 \end{pmatrix}$$

has no more than three 1's.

Now suppose that $A = (A_k, A_{n-k})$, where each column of A_k has more than three 1's and each column of A_{n-k} has three or fewer 1's. By applying the above procedure recursively, we eventually obtain the desired $m' \times n'$ matrix

$$A' \quad \begin{pmatrix} \Delta_1 & \Delta_2 & \ldots & \Delta_k & A_{n-k} & 0_1 & 0_2 & \ldots & 0_k \\ H_1 & 0 & \ldots & 0 & 0 & K_1 & 0 & \ldots & 0 \\ 0 & H_2 & \ldots & 0 & 0 & 0 & K_2 & \ldots & 0 \\ \vdots & & & & & & & & \\ 0 & 0 & \ldots & H_k & 0 & 0 & 0 & \ldots & K_k \end{pmatrix}.$$

Since $k \leqslant n$ and $t \leqslant m$, it follows that $m' \leqslant m + 2n(m-1)$ and $n' \leqslant n(2m-1)$. ■

Next we consider the 0-1 integer programming feasibility problem with only one linear equation. Let $N = \{1, \ldots, n\}$. An instance of the *subset sum problem* is: Given an integer n, an integral n-vector $(a_1, \ldots, a_n)$, and an integer b, is $\{x \in B^n: \Sigma_{j \in N} a_j x_j = b\} \neq \emptyset$?

Proposition 6.10. *The subset sum problem is $\mathcal{NP}$-complete.*

Proof. Membership in $\mathcal{NP}$ is shown by guessing subsets of N. We show membership in $\mathcal{NPC}$ by polynomially transforming the set-partitioning feasibility problem to the subset sum problem. Given a 0-1 $m \times n$ matrix A, define

$$a_j = \sum_{i=1}^{m} (n+1)^{i-1} a_{ij} \quad \text{for } j = 1, \ldots, n$$

and

$$b = \sum_{i=1}^{m} (n+1)^{i-1} = \frac{(n+1)^m - 1}{n}.$$

Since $\Sigma_{j \in N} a_j x_j = b$ is a linear combination of $Ax = 1$ obtained by weighting the ith row by $(n+1)^{i-1}$, we have $\{x \in B^n: Ax = 1\} \subseteq \{x \in B^n: \Sigma_{j \in N} a_j x_j = b\}$. Now to show that the two sets are identical, we note that the unique solution to $\Sigma_{i=1}^{m} (n+1)^{i-1} u_i = b$, $u_i \in B^1$ is $u_i = 1$ for $i = 1, \ldots, m$. Thus if $\Sigma_{j \in S} a_j = b$ so that $\Sigma_{j \in S} \Sigma_{i=1}^{m} (n+1)^{i-1} a_{ij} = \Sigma_{i=1}^{m} (n+1)^{i-1}$, then $\Sigma_{j \in S} a_{ij} = 1$ for $i = 1, \ldots, m$. ■

An instance of the 0-1 *knapsack lower-bound feasibility problem* is: Given an integer n, integral n-vectors $(a_1, \ldots, a_n)$ and $(c_1, \ldots, c_n)$, and integers b and z, is

$$\left\{x \in B^n: \sum_{j \in N} a_j x_j \leq b,\ \sum_{j \in N} c_j x_j \geq z\right\} \neq \emptyset?$$

Corollary 6.11. *The 0-1 knapsack lower-bound feasibility problem is $\mathcal{NP}$-complete.*

Proof. The problem is in $\mathcal{NP}$, since it is a special case of the 0-1 feasibility problem. The subset sum problem can be reformulated as the feasibility problem for the set $\{x \in B^n: \Sigma_{j \in N} a_j x_j \leq b, \Sigma_{j \in N} a_j x_j \geq b\}$. Hence it is a special case of the 0-1 knapsack lower-bound feasibility problem. ■

Membership in $\mathcal{NPC}$ for the 0-1 knapsack lower-bound feasibility problem does not immediately imply that the integer knapsack lower-bound feasibility problem is in $\mathcal{NPC}$ because upper-bound constraints are needed in the obvious transformation of a 0-1 knapsack problem to an integer problem. Nevertheless, there is a polynomial transformation of the 0-1 knapsack problem to the integer knapsack problem. This is left as an exercise.

Figure 6.3 shows the class $\mathcal{NP}$ and the two subsets $\mathcal{P}$ and $\mathcal{NPC}$, which are disjoint unless $\mathcal{P} = \mathcal{NP}$. If $\mathcal{P} \neq \mathcal{NP}$, it can be shown that $\mathcal{P} \cup \mathcal{NPC} \neq \mathcal{NP}$.

Within the class $\mathcal{NPC}$, it is useful to make some distinctions. The subset sum problem can be solved in $O(nb)$ time. Although this is not polynomial, it is less formidable than $O(2^n)$. We say that an algorithm runs in *pseudopolynomial* time, if its running time is a polynomial function of the length of the data encoded in *unary* (a one-symbol alphabet). The principal significance of unary encoding is that an integer k is represented by a string of k symbols. The $O(nb)$-time algorithm for the subset sum problem is pseudopolynomial because the unary encoding of the integer b requires a string of length b. It should be noted that a polynomial transformation of $X_1 \in \mathcal{NP}$ to X_2, which is solvable in pseudopolynomial time (e.g., set partition feasibility to subset sum), does not imply a pseudopolynomial algorithm for X_1. (Why?)

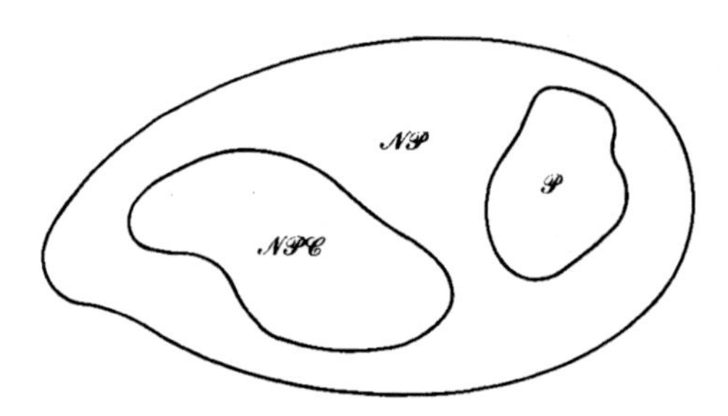

Figure 6.3

At the other extreme, there are $\mathcal{NP}$-complete problems for which the existence of a pseudopolynomial algorithm would imply $\mathcal{P} = \mathcal{NP}$. These problems are called *strongly* $\mathcal{NP}$-complete. Existence is obvious because there are $\mathcal{NP}$-complete problems for which the length of a unary encoding of the data is a polynomial function of the length of a binary encoding. An example is 0-1 integer feasibility in which all of the constraint coefficients are either 0 or ± 1. Figure 6.4 shows the relationships among the subsets of $\mathcal{NP}$ that we have discussed.

Note that $\mathcal{NPC} \cap \mathcal{C}o\mathcal{NP}$ appears in Figure 6.4. The implication of $\mathcal{NPC} \cap \mathcal{C}o\mathcal{NP} \neq \emptyset$ is given in the following proposition.

Proposition 6.12. *If* $\mathcal{NPC} \cap \mathcal{C}o\mathcal{NP} \neq \emptyset$, *then* $\mathcal{NP} = \mathcal{C}o\mathcal{NP}$.

A polynomial-time algorithm for determining $z_0 = \max\{cx: x \in S\}$ obviously implies a polynomial-time algorithm for the lower-bound feasibility problem $\{x \in S: cx \geqslant z\} \neq \emptyset$. Hence if the feasibility problem is $\mathcal{NP}$-complete, a polynomial-time algorithm for the optimization problem would imply $\mathcal{P} = \mathcal{NP}$. But since the optimization problem is not in $\mathcal{NP}$, it is not $\mathcal{NP}$-complete.

In speaking about these problems, we need to extend the notion of polynomial reducibility to problems other than feasibility problems. We call a problem $\mathcal{NP}$-*hard* if there is an $\mathcal{NP}$-complete problem that can be polynomially reduced to it. Thus if a problem is $\mathcal{NP}$-hard it is at least as difficult as any $\mathcal{NP}$-complete problem. It also follows that a polynomial algorithm for an $\mathcal{NP}$-hard problem implies $\mathcal{P} = \mathcal{NP}$.

There is also a converse for optimization problems such as the integer programming problem. We have already observed that a polynomial-time algorithm for the lower-bound feasibility problem and binary search yields a polynomial-time algorithm for the optimi-

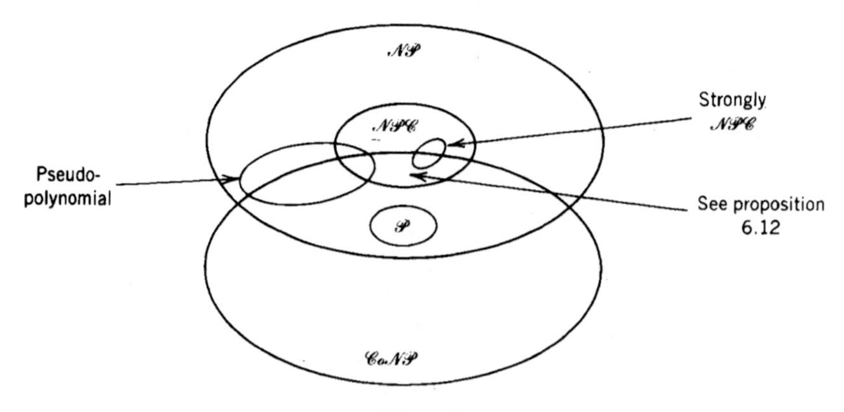

Figure 6.4

zation problem (see Proposition 5.1). Hence if $\mathcal{P} = \mathcal{NP}$ the integer programming problem is solvable in polynomial time. The point is that when an optimization problem has this property, it serves the same role as an $\mathcal{NP}$-complete problem with regard to the question of whether $\mathcal{P} = \mathcal{NP}$.

The classification scheme presented in this chapter can be very useful when we begin to study an integer optimization problem. Our knowledge of $\mathcal{P}$ and $\mathcal{NP}$-hard problems makes it likely that most problems that we encounter will be classifiable. If $\mathcal{P} \neq \mathcal{NP}$, our expectations of what can be accomplished algorithmically should be guided by the classification. If we know that the problem is $\mathcal{NP}$-hard, we can expect that some large instances will be difficult for any algorithm. However, this definitely does not mean that we will be unable to solve many large instances in a reasonable amount of time. And even when we cannot find an optimal solution or prove that a known solution is optimal, it may very well be possible to obtain a good feasible solution and to show this feasible solution is within a specified tolerance of being optimal. If this were not the case, we would not have written this book.

Part II of this book develops theory and algorithms largely for dealing with $\mathcal{NP}$-hard integer optimization problems. Once one departs from the worst-case point of view, we will see that much can be accomplished.

Above we observed that $\mathcal{P} = \mathcal{NP}$ would imply the existence of a polynomial-time algorithm for integer programming. However, it is difficult for us to imagine the impact of the existence of polynomial-time algorithms on the computational aspects of integer programming, since it is not clear what kind of algorithms would result. In the next chapter we will give two polynomial-time algorithms for linear programming. One of these certainly appears to be computationally inferior to exponential-time simplex algorithms. The other is more promising, but its practical implementations ignore some of the details required to prove polynomiality.

7. COMPLEXITY AND POLYHEDRA

We begin this section by considering the relationships among three feasibility problems associated with polyhedra.

1. The *membership problem* for a family of polyhedra. An instance is given by an integer n, a polyhedron in the family $P \subset R^n$, and an $x \in R^n$. The instance is feasible if $x \in P$.
2. The *validity problem* for a family of polyhedra. An instance is given by an integer n, a polyhedron in the family $P \subset R^n$, and a $(\pi, \pi_0) \in R^{n+1}$. The instance is feasible if (π, π_0) is a valid inequality for P.
3. The *lower-bound feasibility problem* for a family of polyhedra. An instance is given by an integer n, a polyhedron in the family $P \subset R^n$, and a $(\pi, \pi_0) \in R^{n+1}$. The instance is feasible if $P \cap \{x \in R^n: \pi x \geq \pi_0\} \neq \emptyset$.

Proposition 7.1. *The following problems are equivalent:*

a. *the validity problem;*
b. *the membership problem for the family of polars; and*
c. *the complement of the lower-bound feasibility problem.*

Proof. The equivalence of problems a and b follows immediately from the definition of polarity, that is, (π, π_0) is a valid inequality for P if and only if (π, π_0) belongs to the polar of P. Recall that the polar of P is a polyhedron in R^{n+1}.

By definition of validity, (π, π_0) is valid for P if and only if $\{x \in P: \pi x > \pi_0\} = \emptyset$. But this is just the complement of the lower-bound feasibility problem with input P and $(\pi, \pi_0 + \varepsilon)$ for some suitably small $\varepsilon > 0$. ■

The complexity of these problems depends on the description of the polyhedra and the points x or (π, π_0). We will assume throughout this section that the allowable inputs for the points x and (π, π_0) are polynomial in the description of the polyhedra. Such vectors are called *short*. By restricting the input in this way, it is sufficient to consider the description of P alone. The results of Section 4 on the size of numbers that can arise in optimal solutions and coefficients of facet defining inequalities (Theorems 4.1 and 4.3) justify the assumption of short vectors in integer programming.

Suppose P is described by a set of linear inequalities, that is, $P = \{x \in R^n: Ax \leqslant b\}$. Then the membership problem for P is solved by substitution. The validity and lower-bound feasibility problems for P can be answered by solving the linear program $z = \max\{\pi x: Ax \leqslant b\}$, since (π, π_0) is valid for P if and only if $\pi_0 \geqslant z$, and $P \cap \{x \in R^n: \pi x \geqslant \pi_0\} \neq \emptyset$ if and only if $\pi_0 \leqslant z$. Hence, given a linear inequality description of P, there are polynomial-time algorithms for all three problems. We also obtain polynomial algorithms when P is described by a list of its extreme points and rays.

However, in many integer and combinatorial optimization problems, P is the convex hull of a set of integral points. We may have an implicit description of the extreme points of P (e.g, the node packings of a graph), or we may have a set of linear inequalities such that P is the convex hull of integral points that satisfy these inequalities, or we may even have a linear inequality description of P, but with the number of inequalities exponential in the natural description of P. With these descriptions, there is not an obvious polynomial-time algorithm for any of the three problems.

***Example** 7.1 (The family of polytopes for 0-1 integer programming).* An instance is specified by integers m and n, an $m \times (n + 1)$ matrix (A, b), and a short vector $x \in R^n$ or $(\pi, \pi_0) \in R^{n+1}$. The polytope for an instance is $P = \text{conv}\{x \in B^n: Ax \leqslant b\}$.

a. *Lower-bound feasibility.* $P \cap \{x \in R^n: \pi x \geqslant \pi_0\} \neq \emptyset$ if and only if $\{x \in B^n: Ax \leqslant b, \pi x \geqslant \pi_0\} \neq \emptyset$. We have already established that the latter lower-bound feasibility problem is $\mathcal{NP}$-complete. Hence lower-bound feasibility for the family of 0-1 integer programming polytopes is also $\mathcal{NP}$-complete.

b. *Validity.* By Proposition 7.1, the validity problem for 0-1 integer programming polytopes is in $\mathcal{C}o\mathcal{NP}$. However, if it was in $\mathcal{NP}$, the lower-bound feasibility problem for 0-1 integer programming polytopes would be in $\mathcal{NP} \cap \mathcal{C}o\mathcal{NP}$.

c. *Membership.* We claim that the membership is in $\mathcal{NP}$. First, an instance is trivial if either $x \notin \{x \in R^n_+: Ax \leqslant b, x_j \leqslant 1 \text{ for all } j\}$ or if $x \in B^n$. So suppose $x \in \{x \in R^n_+ \setminus B^n: Ax \leqslant b, x_j \leqslant 1 \text{ for all } j\}$. Observe that if $\dim(P) = n$, any $x \in P$ can be written as a convex combination of a set of $n + 1$ binary vectors in P. Now the nondeterministic algorithm for membership is to guess vectors $\hat{x}^i \in B^n$ for $i = 1, \ldots, n + 1$. If each of these vectors are in P, we continue. Otherwise we guess a new set. Next we consider the linear system

$$\sum_{i=1}^{n+1} \lambda_i \hat{x}^i = x \quad \text{and} \quad \sum_{i=1}^{n+1} \lambda_i = 1.$$

If this system has a solution $\lambda^0 \in R_+^{n+1}$, we conclude that $x \in P$; otherwise we return to the guessing stage.

Proposition 7.2. *If lower-bound feasibility (validity) for a family of polyhedra is $\mathcal{NP}$-complete and validity (lower bound feasibility) is in $\mathcal{NP}$, then $\mathcal{NP} = \mathcal{C}o\mathcal{NP}$.*

Proof. Suppose validity is in $\mathcal{NP}$. Then by Proposition 7.1, lower-bound feasibility is in $\mathcal{C}o\mathcal{NP}$. Hence lower-bound feasibility is in $\mathcal{NPC} \cap \mathcal{C}o\mathcal{NP}$. Now by Proposition 6.12 we obtain $\mathcal{NP} = \mathcal{C}o\mathcal{NP}$. ∎

In other words, if one of the problems is $\mathcal{NP}$-complete, it is very unlikely that the other is in $\mathcal{NP}$. We frequently encounter the case (as in Example 7.1) where the lower-bound feasibility problem is $\mathcal{NP}$-complete, so it is unlikely that the validity problem is in $\mathcal{NP}$. The following example, however, illustrates an $\mathcal{NP}$ validity problem.

Example 7.2 *(Fractional node-packing polytopes).* An instance is specified by a graph $G = (V, E)$ and a $(\pi, \pi_0) \in R_+^{|V|+1}$. Let A be the incidence matrix of maximal cliques by nodes of G and $P = \{x \in R_+^{|V|}: Ax \leqslant 1\}$. Note that the number of rows of A is generally exponential in the size of G. Here (π, π_0) is valid if and only if, for some $k \leqslant n$,

$$\pi \leqslant \sum_{i=1}^{k} u_i a^i \quad \text{and} \quad \pi_0 \geqslant \sum_{i=1}^{k} u_i,$$

where $\{a^i\}_{i=1}^k$ are rows of A and $u_i \geqslant 0$ for $i = 1, \ldots, k$. Since there is a polynomial-time algorithm for determining whether a 0-1 vector a^i is the incidence vector of a maximal clique of G, there is an $\mathcal{NP}$ algorithm for the validity problem.

In Examples 7.1 and 7.2 we have implicitly considered the *extreme point membership problem* for a family of polytopes. The input is the same as in the membership problem, but it is feasible only if x is an extreme point of P. Note that in Example 7.2, extreme point membership was with respect to the polar; that is, a 0-1 vector a^i is the incidence vector of a maximal clique of G only if it is an extreme point of the polar of the fractional node-packing polytope. In both examples, we have sketched proofs of the following proposition.

Proposition 7.3. *If the extreme point membership problem for a family of polytopes is in $\mathcal{NP}$, then the membership problem for the family is also in $\mathcal{NP}$.*

We now put together Propositions 7.2 and 7.3 by considering the *facet validity problem* for a family of polyhedra. The input is the same as in the validity problem, but it is feasible only if (π, π_0) defines a facet of P.

Proposition 7.4. *If lower-bound feasibility is $\mathcal{NP}$-complete for a family of polyhedra and facet validity is in $\mathcal{NP}$, then $\mathcal{NP} = \mathcal{C}o\mathcal{NP}$.*

Proof. Suppose facet validity is in $\mathcal{NP}$. Then by Proposition 7.3, applied to the family of polars, validity is in $\mathcal{NP}$. Now Proposition 7.2 implies that $\mathcal{NP} = \mathcal{C}o\mathcal{NP}$. ∎

Proposition 7.4 says that for an $\mathcal{NP}$-complete lower-bound feasibility problem, a good characterization of all of the facets of the family of polyhedra is not possible unless $\mathcal{NP} = \mathcal{C}o\mathcal{NP}$. In other words, there is some class of facets for the family of polyhedra for which there is no short proof that they are facets unless $\mathcal{NP} = \mathcal{C}o\mathcal{NP}$.

Example 7.3 *(Node-packing polytopes).* An instance is specified by a graph $G = (V, E)$ and a $(\pi, \pi_0) \in R_+^{|V|+1}$. Here P is the convex hull of node packings. If facet validity is in $\mathcal{NP}$, then $\mathcal{NP} = \mathcal{C}o\mathcal{NP}$, since lower-bound feasibility is $\mathcal{NP}$-complete. The reader should note the subtle difference between Examples 7.2 and 7.3.

8. NOTES

Sections I.5.1 and I.5.2

Basic reference books on computational complexity are Aho et al. (1974), Garey and Johnson (1979), Knuth (1979, 1981), and Lewis and Papadimitriou (1981). Two surveys and an annotated bibliography prepared for the combinatorial optimization community are, respectively, Lenstra and Rinnooy Kan (1979), Johnson and Papadimitriou (1985a), and Papadimitriou (1985).

Jeroslow (1972) discusses the unsolvability of quadratic integer programs.

Section I.5.3

Polynomial-time algorithms for the minimum-weight path problem were presented in Section I.3.2.

Edmonds (1967a) pointed out that very large numbers could arise in Gaussian elimination if rationals were not necessarily represented by a pair of relatively prime numbers. He also gave a modified elimination scheme and proved that with this scheme the size of integer numbers used to represent rationals was polynomially bounded.

Tardos (1985) gave a polynomial-time algorithm for the transportation problem with the bound being independent of the numerical data. Her approach will be presented in Section I.6.5 in the more general setting of linear programming.

Klee and Minty (1972) have shown that the simplex algorithm with a standard pivoting rule does not have a polynomially bounded number of pivots. The expected behavior of the simplex algorithm has been analyzed by Borgwardt (1982a, b), Smale (1983a, b), and others. Shamir (1987) gives a survey of these results.

Edmonds (1965a, c) proposed the concept of a good characterization. This was done in the context of the maximum-weight matching problem (see Chapter III.2).

Section I.5.4

Bell (1977) proved that the formulation of a feasible n-variable integer program with linear inequalities and integrality restrictions requires no more than $2^n - 1$ inequalities.

The results on the size of numbers that arise in general integer programming problems have been obtained independently by several people, including Borosh and Treybig (1976), Von zur Gathen and Sieveking (1978), Kannan and Monma (1978), and Papadimitriou (1981a). The simple proof given in the text was suggested to us by Gerard Cornuejols.

Sections I.5.5 and I.5.6

The basic references for these sections are Garey and Johnson (1979) and the more recent survey by Johnson and Papadimitriou (1985a).

The class $\mathcal{NP}$ was formally introduced by Cook (1971). A slightly different definition of $\mathcal{NP}$ was used by Karp (1972, 1975). Cook used polynomial reducibility in the definition of $\mathcal{NP}$ and proved the fundamental result of the existence of complete problems in $\mathcal{NP}$. Karp defined $\mathcal{NP}$ by polynomial transformability and showed that numerous combinatorial optimization problems are $\mathcal{NP}$-complete. The proofs of Propositions 6.9 and 6.10 are taken from Lenstra and Rinnooy Kan (1979).

Section I.5.7

Facet complexity problems have been studied by Karp and Papadimitriou (1982) and by Papadimitriou and Yannakakis (1984). A survey of these results is contained in Papadimitriou (1984).

Also see the notes for Section I.6.3.

9. EXERCISES

1. Verify the relations implied in Figure 1.1.
2. Give a tight bound for the magnitude of coefficients in the extreme points of $P = \{x \in R^n_+: \Sigma^n_{j=1} a_j x_j \leqslant b\}$, where $a_j, b \in Z^1_+$. Compare this bound with the bound of Proposition 3.1.
3. Can you find an example for which the bounds of Proposition 3.1 are tight?
4. Give a certificate of optimality that $x = [\frac{36}{11} \;\; \frac{40}{11}]$ is optimal in Example 3.1 of Chapter I.2.
5. Give a short proof that $M = \{(1, 2), (3, 5)\}$ is a maximum-weight matching in the graph of Figure 9.1.
6. Show that if there is a polynomial algorithm to test feasibility of $P = \{x \in R^n: Ax \leqslant b\}$, there is a polynomial algorithm to find a minimal face of P.
7. Give a tight bound for the magnitude of coefficients in extreme points of conv(S), where $S = P \cap Z^n$ and $P = \{x \in R^n_+: \Sigma^n_{j=1} a_j x_j \leqslant b\}$. Compare it to the bound of Theorem 4.1.
8. Can you find an example for which the bound of Theorem 4.1 is tight?
9. Prove Proposition 5.7 of Section I.6.5.
10. Given A and b, prove that if $x \in R^n_+$ satisfies

$$\left| \sum_{j=1}^{n} a_{ij} x_j - b_i \right| \leqslant \varepsilon \text{ for } i = 1, \ldots, m$$

with $\varepsilon = (2^{mn \log \theta} mn)^{-1}$, $\{x \in R^n_+: Ax = b\} \neq \emptyset$. (See Proposition 4.6 of Section I.6.4).

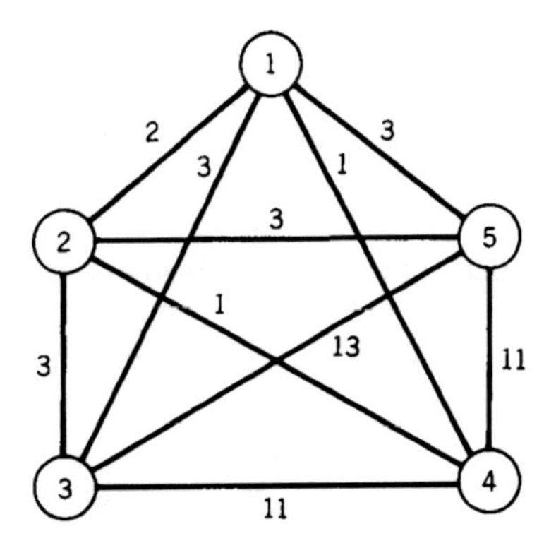

Figure 9.1

11. Give algorithms to show that the following problems on graphs are in $\mathcal{P}$.

i) Does G contain a cycle?

ii) Is G bipartite?

In i and ii, how would you give a short proof when the answer is no?

12. A graph $G = (V, E)$ is a *hole* if it contains a single cycle through all of the nodes and no other edges. Show that the problem "Does G or its complement contain a node-induced subgraph that is a hole of odd length?" is in $\mathcal{C}o\mathcal{NP}$.

13. Is the problem "Is $b \in Z^1_+ \setminus \{0\}$ a prime number?" in $\mathcal{NP}$, $\mathcal{C}o\mathcal{NP}$, neither, or both?

14. Given that the node-packing problem is $\mathcal{NP}$-complete, show that the following problems are $\mathcal{NP}$-complete:

i) *Node cover*. Given a graph $G = (V, E)$ and an integer K, is there a subset $S \subseteq V$ with $|S| \leqslant K$ such that every edge of E is incident to a node of S?

ii) *Uncapacitated facility location*. Given sets M and N and integers c_{ij} for $i \in M, j \in N$, f_j for $j \in N$ and K, is there a set $S \subseteq N$ such that $\Sigma_{i \in M} \min_{j \in S} c_{ij} + \Sigma_{j \in S} f_j \leqslant K$?

15. Show that the following problems are $\mathcal{NP}$-complete:

i) *Set covering*: Given an $m \times n$ 0-1 matrix A and an integer K, does there exist $x \in B^n$ such that $Ax \geqslant 1$ and $\Sigma_{i=1}^n x_i \leqslant K$?

ii) *Directed Hamiltonian circuit*. Given a directed graph $\mathcal{D} = (V, \mathcal{A})$, is there a directed cycle passing through each vertex exactly once?

iii) *Matching with bonds*. Given a graph $G = (V, E)$, pairwise disjoint subsets B_i for $i = 1, \ldots, p$ of E, and an integer K, does there exist a matching M in G such that $|M| \geqslant K$ and, for $i = 1, \ldots, p$, either $B_i \cap M = B_i$ or $B_i \cap M = \emptyset$ (i.e., either all the edges in B_i are in the matching or none are)? The subsets B_i are called *bonds*.

16. A set function $f(S) = \Sigma_{T \subseteq S} c_T$ for $S \subseteq N$ is described by the data $\{T, c_T\}$, where $c_T \neq 0$. A set function is *submodular* on N if

$$f(S) + f(T) \geqslant f(S \cup T) + f(S \cap T) \quad \text{for all } S, T \subseteq N.$$

Show that the problem: "Is f not submodular?" is $\mathcal{NP}$-complete.

17. Show that the traveling salesman problem is $\mathcal{NP}$-hard.

18. Show that the minimum-weight path problem (with positive and negative edge weights) is $\mathcal{NP}$-hard.

19. Give a polynomial transformation of 0-1 knapsack to integer knapsack.

20. Show that the fixed-charge network flow problem is $\mathcal{NP}$-hard.

21. Show that the maximum-cut problem "Given $\mathcal{D} = (V, \mathcal{A})$ and $c \in R^n_+$, find $\max_{(i,j) \in \delta^+(U)} c_{ij}$" is $\mathcal{NP}$-hard, where $\delta^+(U) = \{(i, j) \in \mathcal{A}: i \in U, j \in V \setminus U\}$.

22. Show that the problem

$$\max \sum_{i=1}^{n} \sum_{j=1}^{n} c_{ij} x_{ij}$$

$$\sum_{j=1}^{n} x_{ij} = 1 \text{ for all } i$$

$$\sum_{i=1}^{n} x_{ij} = 1 \text{ for all } j$$

$$\sum_{i=1}^{n} \sum_{j=1}^{n} t_{ij} x_{ij} \leq T$$

$$x \in B^{n^2}$$

is $\mathcal{NP}$-hard.

23. Show that the single-machine scheduling problem with due dates is $\mathcal{NP}$-hard.

24. Let $S = \{x \in Z_+^n : Ax \leq b\}$. Which of the following problems (if any) are known to be in $\mathcal{P}$, $\mathcal{NP}$, $\mathcal{C}o\mathcal{NP}$? Which are unlikely to be in $\mathcal{NP}$? Justify your answers.

- **i)** Membership for conv(S).
- **ii)** Extreme point membership for conv(S).
- **iii)** Validity for conv(S).
- **iv)** Facet validity for conv(S).

I.6

Polynomial-Time Algorithms for Linear Programming

1. INTRODUCTION

Simplex methods (see Chapter I.2) are practical techniques for solving linear programs. But, according to the model of computational complexity presented in the previous chapter, they are unsatisfactory because their running time can grow exponentially with the size of the input. Here we give some polynomial-time algorithms for linear programming and discuss their consequences in combinatorial optimization.

The ellipsoid algorithm, which will be presented in Section 2, was acclaimed on the front pages of newspapers throughout the world when it appeared in 1979. Although the algorithm turned out to be computationally impractical, it yielded important theoretical results. It was the first polynomial-time algorithm for linear programming. Also, as will be discussed in Section 3, it is a tool for proving that certain combinatorial optimization problems can be solved in polynomial time.

In Section 4, we will present a version of a polynomial-time projective algorithm for linear programming. Remarkably good computational results have been claimed for projective algorithms, but only time will tell whether they are superior to, or a serious rival of, simplex methods.

The running times of these polynomial-time algorithms typically depend on m, n, and $\log \theta_{A,b,c}$ where

$$\theta_{A,b,c} = \max\{\max |a_{ij}|, \max |b_i|, \max |c_j|\}.$$

In Section 5, it will be shown how the dependence on b and c can be eliminated. Thus, for example, when A is a (0, 1) matrix, there are linear programming algorithms that are polynomial in m and n.

To present polynomial-time versions of the ellipsoid and projective algorithms, some basic questions about linear programming must be addressed.

1. Unlike the simplex methods, the ellipsoid and projective algorithms are naturally described as algorithms to find a feasible point in a polyhedron. Hence, we must convert a feasibility algorithm into an optimization algorithm. The standard approach is to formulate a linear program as the feasibility problem: Find $x \in R^n_+$, $u \in R^m_+$ satisfying

$$Ax \leqslant b, \quad uA \geqslant c, \quad cx \geqslant ub.$$

But this approach is computationally unsatisfactory, so we will need to consider other methods.

2. Neither the ellipsoid nor the projective algorithm search extreme points. Whereas extreme points and extreme rays can be described in polynomial time, arbitrary points cannot. So care has to be taken that the points obtained have polynomial descriptions.

3. The last step of the ellipsoid and projective algorithms requires the conversion of an "almost feasible/almost optimal" point to a basic feasible/optimal solution. We need to show that this operation can be executed in polynomial time. An intermediate step in this process is the perturbation of a constraint or of the objective function so that the resulting linear program has a unique primal or dual feasible solution.

2. THE ELLIPSOID ALGORITHM

To describe the ellipsoid algorithm, we need a few basic properties of ellipsoids.

Definition 2.1. An $n \times n$ symmetric matrix D is *positive definite* if $x^TDx > 0$ for all $x \in R^n$ except $x = 0$.

Definition 2.2. An *ellipsoid* with center y is a set $E = \{x \in R^n: (x - y)^T D^{-1}(x - y) \leq 1\}$, written as $E(D, y)$, where D is an $n \times n$ positive definite matrix and $y \in R^n$.

Definition 2.3. A *sphere* with center y and radius r is a set of the form $S = \{x \in R^n: (x - y)^T(x - y) \leq r^2\}$, written as $S(y, r)$.

Evidently a sphere is a special case of an ellipsoid with $D = r^2 I_n$, where I_n is the $n \times n$ identity matrix (see Figure 2.1). We let S^n denote the unit sphere in R^n with center 0, that is, $S^n = S(0, 1)$.

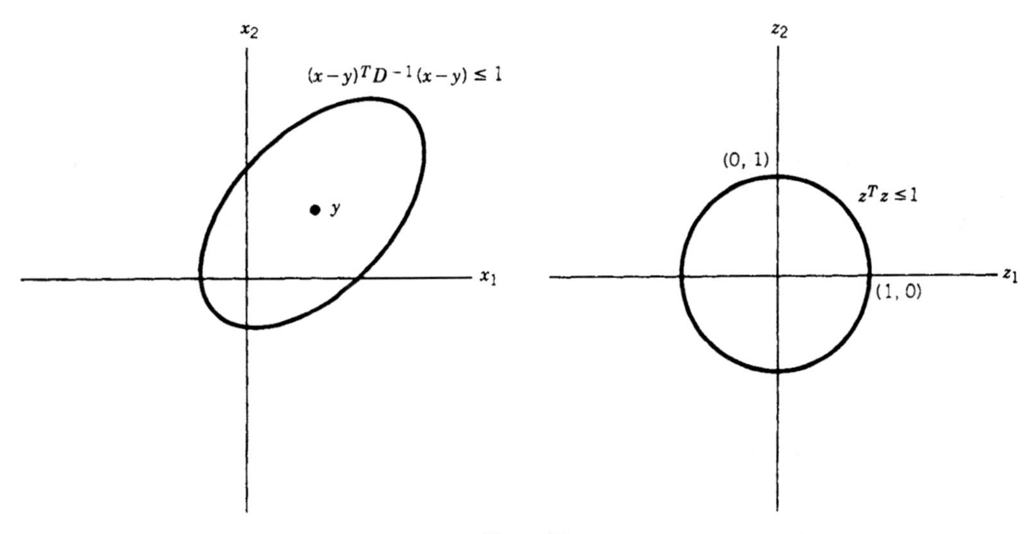

Figure 2.1

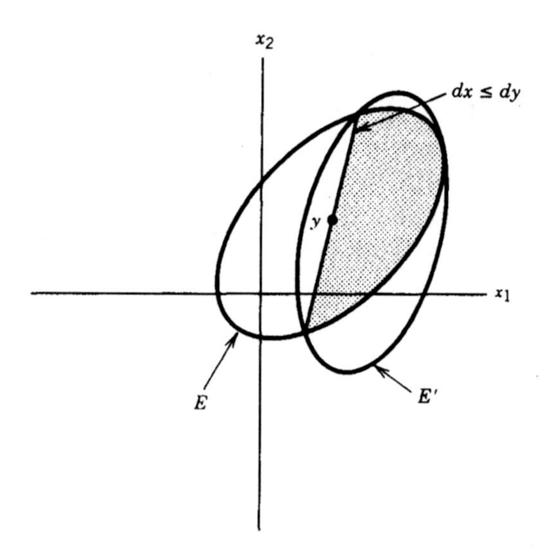

Figure 2.2

The following property is crucial to the ellipsoid algorithm.

The Ellipsoid Property. Given an ellipsoid $E = E(D, y)$, the half-ellipsoid $H = E(D, y) \cap \{x \in R^n\colon dx \leqslant dy\}$ obtained by intersecting E with any inequality $dx \leqslant dy$ through its center is contained in an ellipsoid E' with the property that $\text{vol}(E')/\text{vol}(E) \leqslant e^{-1/2(n+1)}$, where vol denotes volume (see Figure 2.2). A constructive proof of this property will be given below.

We begin by describing the ellipsoid algorithm for a membership problem.

Strict Membership Problem. Given integers m and n, an integer $m \times n$ matrix A, and an integer m-vector b, find a point in $P_< = \{x \in R^n\colon Ax < b\}$ or show that $P_< = \emptyset$. [The notation $Ax < b$ means that $a^i x < b_i$ for $i = 1, \ldots, m$ where (a^i, b_i) is the ith row of (A, b).]

Throughout this section we will assume that given a point $y \in R^n$ we check whether $y \in P_<$ by testing whether $a^i y < b_i$ for $i = 1, \ldots, m$.

We will also assume that $P_<$ is bounded, so that there exists an ω such that if $x \in P_<$, then $|x_j| < \omega$ for all $j \in N$. This means that $P_< \subseteq S(0, s)$, where $s = \omega\sqrt{n}$, and hence $\text{vol}(P_<) \leqslant \text{vol}(S(0, s)) = s^n \text{vol}(S^n)$.

A second important observation concerns the volume of polyhedra.

The Strict Feasibility Property. If $P_< \neq \emptyset$, then $\text{vol}(P_<) > 0$. More precisely, given a point $y \in P_<$, there is an $r > 0$ such that $S(y, r) \subseteq P_<$. This implies that $\text{vol}(P_<) \geqslant \text{vol}(S(y, r)) = r^n \text{vol}(S^n)$.

Now suppose we are given (a) a number v such that $\text{vol}(P_<) > v$ if $P_< \neq \emptyset$ and (b) a number V such that $\text{vol}(E_0) = V$, where $E_0 = E(D_0, x_0)$ is an ellipsoid containing $P_<$. Let $t^* = \lceil 2(n + 1)(\log_e V - \log_e v) \rceil$.

The Ellipsoid Algorithm for $P_<$

Initialization: $E_0 = E(D_0, x_0)$.
Set $t = 0$.

Iteration t: If $x_t \in P_<$, stop. A feasible solution has been found.
If $t \geqslant t^*$, stop. $P_< = \emptyset$.
If $x_t \notin P_<$, suppose $a^{i(t)}x_t \geqslant b_{i(t)}$.

(Note that we have departed from our usual notation here in that $x_t \in R^n$; that is, x_t is not the tth component of x.)

Find an ellipsoid E_{t+1} containing the half-ellipsoid $H_t = E_t \cap \{x \in R^n: a^{i(t)}x \leqslant a^{i(t)}x_t\}$ as specified by the Ellipsoid Property.

Let $E_{t+1} = E(D_{t+1}, x_{t+1})$ and $t \leftarrow t + 1$.

Theorem 2.1. *Given v, V, and t^* as defined above, the ellipsoid algorithm for $P_<$ terminates correctly after no more than t^* iterations.*

Proof. Since $x_t \in P_<$ is readily verifiable, we only have to show that if $x_t \notin P_<$ for $t = 0, \ldots, t^*$, then $P_< = \emptyset$.

First we use induction to show that $P_< \subseteq E_t$ for all $t \leqslant t^*$. We have constructed E_0 so that $P_< \subseteq E_0$. Now suppose that $P_< \subseteq E_k$. Then as $a^{i(k)}x_k \geqslant b_{i(k)}$, we have

$$P_< \subseteq \{x \in R^n: a^{i(k)}x \leqslant b_{i(k)}\} \subseteq \{x \in R^n: a^{i(k)}x \leqslant a^{i(k)}x_k\}.$$

Hence

$$P_< \subseteq E_k \cap \{x \in R^n: a^{i(k)}x \leqslant a^{i(k)}x_k\} = H_k \subset E_{k+1}.$$

Now consider the volume of E_{t^*}. Since $\text{vol}(E_{t+1})/\text{vol}(E_t) \leqslant e^{-1/2(n+1)}$, it follows that $\text{vol}(E_{t^*})/\text{vol}(E_0) \leqslant e^{-t^*/2(n+1)}$. Hence

$$\begin{aligned}\text{vol}(E_{t^*}) &\leqslant Ve^{-t^*/2(n+1)} = Ve^{-[2(n+1)(\log V - \log v)]/2(n+1)} \\ &\leqslant Ve^{-\log(V/v)} = v.\end{aligned}$$

But now if $P_< \neq \emptyset$, it would follow that $\text{vol}(P_<) > v$, $\text{vol}(E_{t^*}) \leqslant v$, and $P_< \subseteq E_{t^*}$, which is impossible. Hence $P_< = \emptyset$. ∎

The actual details of how $E_{t+1} = E(D_{t+1}, x_{t+1})$ is constructed from $E_t = E(D_t, x_t)$ are given by the following expressions. We assume $n > 1$.

Let $d = a^{i(t)}$ and $D = D_t$.

$$x_{t+1} = x_t - \frac{1}{n+1}\frac{Dd}{\sqrt{d^TDd}} \tag{2.1}$$

$$D_{t+1} = \frac{n^2}{n^2-1}\left(D - \frac{2}{n+1}\frac{(Dd)(Dd)^T}{d^TDd}\right). \tag{2.2}$$

We will now show that these transformations lead to a new ellipsoid satisfying the ellipsoid property. Without loss of continuity the reader can go directly to Example 2.1.

Proposition 2.2. *Every symmetric positive-definite $n \times n$ matrix D has a decomposition $D = Q_1^T Q_1$, where Q_1 is an $n \times n$ nonsingular matrix.*

Definition 2.4. If A is an $n \times n$ nonsingular matrix, $b \in R^n$, and T_A: $R^n \to R^n$ is defined by $T_A(x) = Ax + b$, then T_A is called an *affine transformation*.

Affine transformations have several important properties. We let $T_A(L) = \{\xi \in R^n: \xi = Ax + b, x \in L\}$. Affine transformations preserve set inclusion.

Proposition 2.3. *If $L \subseteq L' \subseteq R^n$, then $T_A(L) \subseteq T_A(L') \subseteq R^n$.*

Volumes are changed by a constant factor, so relative volumes are preserved.

Proposition 2.4. *If $L \subseteq R^n$ is full-dimensional and convex, then $vol(T_A(L)) = |\det A| \operatorname{vol}(L)$.*

Given an ellipsoid, there exists an affine transformation mapping it into a sphere centered at the origin.

Proposition 2.5. *Let $E = E(D, y)$ be an ellipsoid with $D = Q_1^T Q_1$ and let T be the affine transformation given by $T(x) = (Q_1^T)^{-1}x - (Q_1^T)^{-1}y$. Then $T(E) = S^n$.*

Proof.

$$\begin{aligned} T(E) &= \{\xi: \xi = (Q_1^T)^{-1}(x - y): (x - y)^T D^{-1}(x - y) \leqslant 1\} \\ &= \{\xi: ((Q_1^T)\xi)^T D^{-1}(Q_1^T)\xi \leqslant 1\} \\ &= \{\xi: \xi^T Q_1 D^{-1} Q_1^T \xi \leqslant 1\} \\ &= \{\xi: \xi^T \xi \leqslant 1\}. \end{aligned}$$

■

In Figure 2.3 we see what happens to E, E', and the half-ellipsoid H when the above transformation is applied to $E = E(D, y)$.

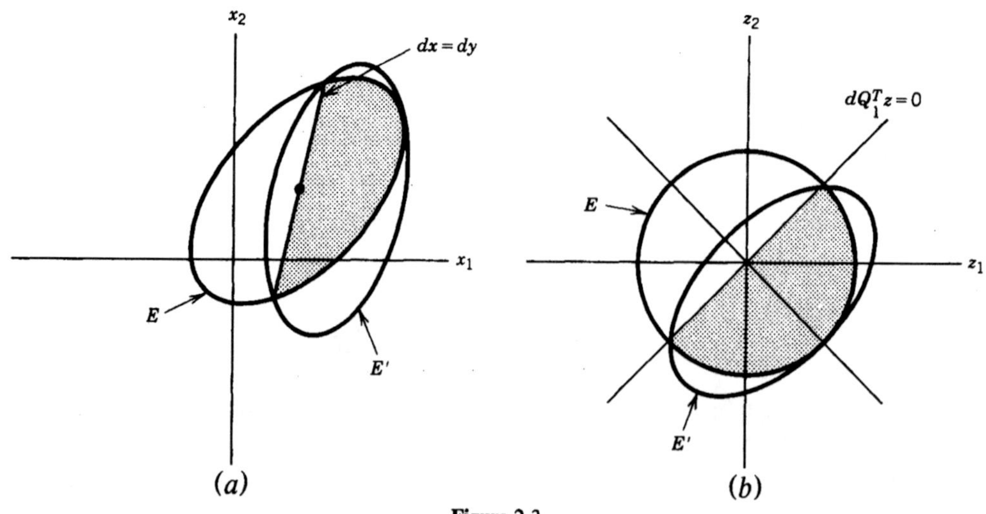

Figure 2.3

Affine transformations corresponding to rotations can be represented by transformation matrices Q_2 with the property that $Q_2^T Q_2 = I$. Such matrices are called *orthonormal.*

Proposition 2.6. *Given an arbitrary nonzero vector $d \in R^n$, there exists an $n \times n$ orthonormal matrix Q_2 such that $Q_2 d = -\|d\| e_1$, where $\|d\|$ is the length of d and where $e_1 = (1, 0, \ldots, 0)^T$*

Applying this proposition to the vector $Q_1 d$, we can rotate the sphere in Figure 2.3(*b*) so that the shaded area is just the half-sphere with $\xi_1 \geq 0$. Setting $Q = Q_1 Q_2$, the effect of the transformation $\xi = (Q^T)^{-1}(x - y)$ is to map E, E', and H as shown in Figure 2.4.

Now we use the above transformations to show that the ellipsoid property holds for (2.1) and (2.2) with $E = E_t$, $E' = E_{t+1}$, and $y = x_t$.

We define

1. Q_1 to be any matrix such that $Q_1^T Q_1 = D$,
2. Q_2 to be the rotation matrix such that $Q_2 Q_1 d = -\|Q_1 d\| e_1$, and
3. $Q = Q_2 Q_1$,

and we use the transformation $x = Q^T \xi + x_t$, or $T(x) = (Q^T)^{-1}(x - x_t)$. Simple calculations give the following proposition.

Proposition 2.7

i. $\|Qd\| = \sqrt{d^T D d}$.
ii. $QD^{-1}Q^T = I$.
iii. $(Q^T)^{-1} D = Q$.
iv. $T(E_t) = \{\xi : \xi^T \xi \leq 1\}$.
v. $T(\{x : dx \leq dx_t\}) = \{\xi : \xi_1 \geq 0\}$.
vi. (a) $T(x_{t+1}) = \frac{1}{n+1} e_1$.
(b) $T(E_{t+1}) = \{\xi : (\xi - \frac{e_1}{n+1})^T (\frac{n^2}{n^2-1}(I - \frac{2}{n+1} e_1 e_1^T))^{-1} (\xi - \frac{e_1}{n+1}) \leq 1\}$.

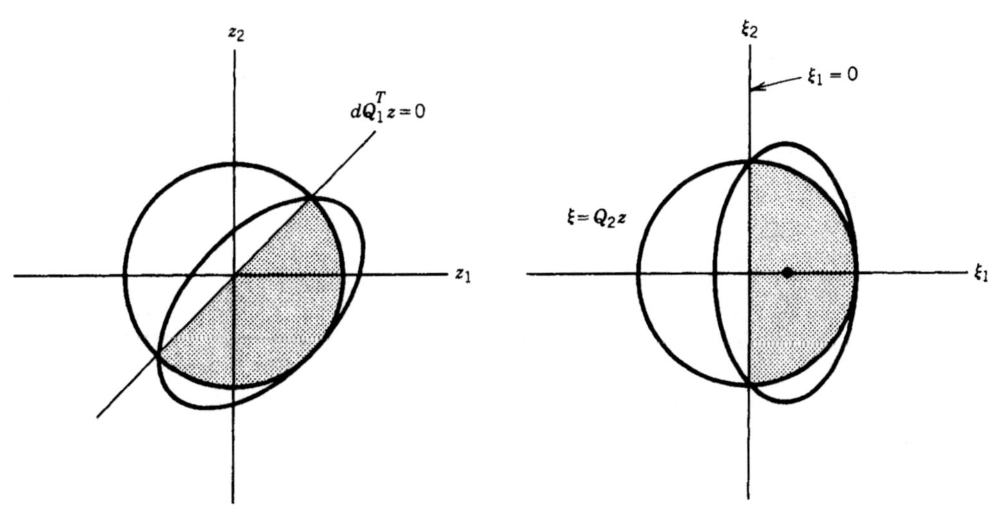

Figure 2.4

Proof

i. $\|Q_2Q_1 d\| = \sqrt{(Q_2Q_1d)^T(Q_2Q_1d)} = \sqrt{d^TQ_1^TQ_2^TQ_2Q_1d} = \sqrt{d^TDd}$ because $Q_2^T = Q_2^{-1}$ since Q_2 is orthonormal.

ii. $QD^{-1}Q^T = Q_2Q_1(Q_1^TQ_1)^{-1}(Q_2Q_1)^T = Q_2Q_1Q_1^{-1}(Q_1^T)^{-1}Q_1^TQ_2^T = Q_2Q_2^T = I.$

iii. By ii, $QD^{-1}Q^T = I$, and hence $(Q^T)^{-1}D = Q$.

iv. $T(E_t) = \{\xi: (Q^T\xi)^TD^{-1}(Q^T\xi) \leqslant 1\} = \{\xi: \xi^T(QD^{-1}Q^T)\xi \leqslant 1\} = \{\xi: \xi^T\xi \leqslant 1\}.$

v. $T(\{x: dx \leqslant dx_t\}) = \{\xi: d^TQ^T\xi \leqslant 0\} = \{\xi: (Q_2Q_1d)^T\xi \leqslant 0\} = \{\xi: \xi_1 \geqslant 0\}.$

vi. $T(x_t) = (Q^T)^{-1}(x_t - x_t) = 0.$

$$T(x_{t+1}) = T\left(x_t - \frac{1}{n+1}\frac{Dd}{\sqrt{d^TDd}}\right) = -\frac{1}{n+1}\frac{(Q^T)^{-1}Dd}{\sqrt{d^TDd}}$$

$$= -\frac{1}{n+1}\frac{Qd}{\|Qd\|} = \frac{1}{n+1}e_1 \text{ (by i and iii).}$$

Letting $\xi_{t+1} = T(x_{t+1})$, that is, $x_{t+1} - x_t = Q^t\xi_{t+1}$, we have

$$\begin{aligned} T(E_{t+1}) &= \{\xi: Q^T\xi = x - x_t, (x - x_{t+1})^TD_{t+1}^{-1}(x - x_{t+1}) \leqslant 1\} \\ &= \{\xi: (Q^T(\xi - \xi_{t+1}))^TD_{t+1}^{-1}Q^T(\xi - \xi_{t+1}) \leqslant 1\} \\ &= \{\xi: (\xi - \xi_{t+1})^TQD_{t+1}^{-1}Q^T(\xi - \xi_{t+1}) \leqslant 1\} \\ &= \{\xi: (\xi - \xi_{t+1})^T((Q^T)^{-1}D_{t+1}Q^{-1})^{-1}(\xi - \xi_{t+1}) \leqslant 1\}. \end{aligned}$$

We note that (2.2) yields

$$\begin{aligned} (Q^t)^{-1}D_{t+1}Q^{-1} &= \frac{n^2}{n^2-1}\left[(Q^T)^{-1}DQ^{-1} - \frac{2}{n+1}\frac{(Q^T)^{-1}(Dd)(Dd)^TQ^{-1}}{d^TDd}\right] \\ &= \frac{n^2}{n^2-1}\left[I - \frac{2}{n+1}\frac{(Qd)(Qd)^T}{d^TDd}\right] \quad \text{using iii} \\ &= \frac{n^2}{n^2-1}\left[I - \frac{2}{n+1}e_1e_1^T\right] \quad \text{using i and the definition of } Q. \end{aligned}$$

■

Now we have what is needed to show that E_{t+1} satisfies the ellipsoid property.

Proposition 2.8

i. *D_{t+1} is positive definite.*

ii. $\mathrm{vol}(E_{t+1})/\mathrm{vol}(E_t) \leqslant e^{-1/2(n+1)}$.

iii. $H_t = E_t \cap \{x: dx \leqslant dx_t\} \subseteq E_{t+1}$.

Proof. i. From statement vi of Proposition 2.7, we see that $D_{t+1} = Q^T\Delta Q$, where Δ is a diagonal matrix with positive diagonal entries δ_i for $i = 1, \ldots, n$. Let $\Delta^{1/2}$ denote the diagonal matrix with diagonal entries $\delta_i^{1/2}$ for $i = 1, \ldots, n$. It follows that

$$x^TD_{t+1}x = x^TQ^T(\Delta^{1/2})^T\Delta^{1/2}Qx = \|\Delta^{1/2}Qx\| > 0 \quad \text{for all } x \in R^n \setminus \{0\}.$$

ii. Using Propositions 2.5 and 2.6, we have

$$\begin{aligned}
\mathrm{vol}(E_{t+1}) / \mathrm{vol}(E_t) &= \mathrm{vol}(T(E_{t+1})) / \mathrm{vol}(T(E_t)) \\
&= \mathrm{vol}\left(E\left(\Delta, \frac{1}{n+1}e_1\right)\right) \Big/ \mathrm{vol}(S^n) = \sqrt{\det\Delta} \\
&= \sqrt{\det\left(\frac{n^2}{n^2-1}\left(I - \frac{2}{n+1}\, e_1 e_1^T\right)\right)} \\
&= \frac{n}{n+1}\left(\frac{n^2}{n^2-1}\right)^{(n-1)/2} \leqslant e^{-1/(n+1)}\left(\frac{1}{e^{n^2-1}}\right)^{(n-1)/2} \\
&= e^{-1/2(n+1)},
\end{aligned}$$

where the inequality is derived by two applications of the standard inequality $(1+\alpha) \leqslant e^{\alpha}$ for all $|\alpha|$.

iii. Under the transformation, we have

$$T(H_t) = \{\xi\colon \xi^T\xi \leqslant 1 \text{ and } \xi_1 \geqslant 0\}$$

and

$$\begin{aligned}
T(E_{t+1}) &= \left\{\xi\colon \left(\xi - \frac{1}{n+1}e_1\right)^T \Delta^{-1}\left(\xi - \frac{1}{n+1}\, e_1\right) \leqslant 1\right\} \\
&= \left\{\xi\colon \left(\frac{n+1}{n}\right)^2\left(\xi_1 - \frac{1}{n+1}\right)^2 + \frac{n^2-1}{n^2}\sum_{i=2}^{n} \xi_i^2 \leqslant 1\right\} \\
&= \left\{\xi\colon \frac{n^2-1}{n^2}\sum_{i=1}^{n}\xi_i^2 + \frac{2(n+1)}{n^2}\,\xi_1^2 + \left(\frac{n+1}{n}\right)^2\left(\frac{-2\xi_1}{n+1} + \frac{1}{(n+1)^2}\right) \leqslant 1\right\} \\
&= \left\{\xi\colon \frac{n^2-1}{n^2}\sum_{i=1}^{n}\xi_i^2 + \frac{1}{n^2} + \frac{2(n+1)\xi_1}{n^2}\,(\xi_1 - 1) \leqslant 1\right\}.
\end{aligned}$$

It follows that $T(H_t) \subseteq T(E_{t+1})$ since $0 \leqslant \xi_1 \leqslant 1$ for $\xi \in T(H_t)$. Applying Proposition 2.3 to the inverse transformation of T, it follows that $H_t \subseteq E_{t+1}$. ■

Example 2.1. $P_< = \{x \in R^2\colon x_1 + x_2 < 2,\ -2x_1 + 2x_2 < 1,\ -x_2 < 0\}$.

We suppose it is known that $|x_j| \leqslant 3$ for $j = 1, 2$ if $x \in P_<$ and that $\mathrm{vol}(P_<) > \frac{1}{100}$ if $P_< \neq \varnothing$.

We take $E_0 = \{x \in R^2\colon \frac{1}{18}x_1^2 + \frac{1}{18}x_2^2 \leqslant 1\}$ with $x_0 = (0\ 0)$ and $D_0 = \begin{pmatrix}18 & 0\\ 0 & 18\end{pmatrix}$ so that $\mathrm{vol}(E_0) \leqslant \mathrm{vol}\{x \in R^2\colon |x_j| \leqslant 3 \text{ for } j = 1, 2\} = 81$.

We then calculate $t^* = \lceil 2(n+1)(\log 81 - \log \frac{1}{100})\rceil = \lceil 6 \log_e 8100\rceil = 54$.

The numerical calculations of the iterations are given below. The shrinking of the ellipses is shown in Figure 2.5 and the solutions are shown in Figure 2.6.

Iteration 0. $x_0 = (0\quad 0) \notin P_<$ because $-2x_1 + 2x_2 < -1$ is violated.

Using the updating formulas (2.1) and (2.2) with $d = (-2\quad 2)$ and $D = \begin{pmatrix}18 & 0\\ 0 & 18\end{pmatrix}$ gives

$$x_1 = (1\quad -1) \quad \text{and} \quad D_1 = \begin{pmatrix}16 & 8\\ 8 & 16\end{pmatrix}.$$

Iteration 1. $x_1 = (1 \quad -1) \notin P_<$ because $-x_2 < 0$ is violated.

$$x_2 = (\tfrac{5}{3} \quad \tfrac{1}{3}) \quad \text{and} \quad D_2 = \begin{pmatrix} \frac{160}{9} & \frac{32}{9} \\ \frac{32}{9} & \frac{64}{9} \end{pmatrix}$$

Iteration 2. $x_2 = (\tfrac{5}{3} \quad \tfrac{1}{3}) \notin P_<$ because $x_1 + x_2 < 2$ is violated.

$$x_3 = (0.41 \quad -0.30) \quad \text{and} \quad D_3 = \begin{pmatrix} 11.06 & -1.58 \\ -1.58 & 6.32 \end{pmatrix}$$

Iteration 3. $x_3 = (0.41 \quad -0.30) \notin P_<$ because $-x_2 < 0$ is violated.

$$x_4 = (0.20 \quad 0.54) \quad \text{and} \quad D_4 = \begin{pmatrix} 14.40 & -0.70 \\ -0.70 & 2.81 \end{pmatrix}.$$

Iteration 4. $x_4 = (0.20 \quad 0.54) \notin P_<$ because $-2x_1 + 2x_2 < -1$ is violated.

$$x_5 = (1.37 \quad 0.27) \quad \text{and} \quad D_5 = \begin{pmatrix} 8.31 & 1.60 \\ 1.60 & 3.16 \end{pmatrix}.$$

Iteration 5. Since $x_5 \in P_<$, the algorithm terminates.

Note that in contrast to the subgradient algorithm of Section I.2.4, the steps $x_t - x_{t+1}$ taken in the ellipsoid algorithm are not normal to the violated inequality $a^{i(t)}x < b_{i(t)}$ except in special cases such as when the ellipsoid E_t is a sphere.

Now we describe how the ellipsoid algorithm can be modified to find nearly optimal solutions to a linear program. For convenience we will distinguish between the problem of maximizing cx over an arbitrary polytope

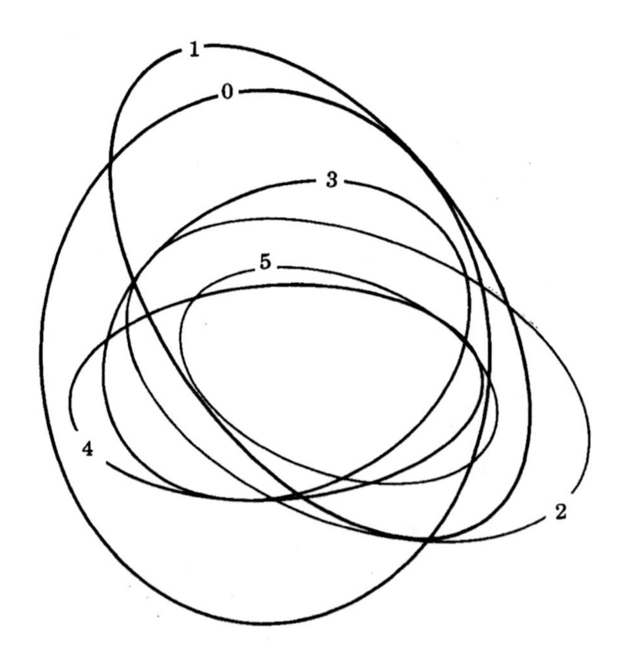

Figure 2.5

(2.3) $$z_{LP} = \max\{cx: x \in P\}$$

and the linear program where P is explicitly described by a set of linear constraints

(2.4) $$z_{LP} = \max\{cx: x \in P\}, \quad \text{where } P = \{x \in R^n: Ax \leqslant b\}.$$

In both cases we assume that P is nonempty and bounded and that $P_< \neq \emptyset$. Then by first running the ellipsoid algorithm for $P_<$ we can determine an initial point $a_0 \in P_<$. We then set $x_0 = a_0$ and consider the strict inequality system $P_< \cap \{x \in R^n: -cx < -cx_0\}$.

The idea behind the modification is simple. Every time a better feasible point $x_t \in P_<$ is found, we take $P_< \cap \{x \in R^n: -cx < -cx_t\}$ as our new strict inequality system and reapply the ellipsoid algorithm. This approach, called the *sliding objective function method*, has the nice feature that the algorithm is always being applied to a feasible system unless x_t is an optimal point.

The Sliding Objective Function Approximate Ellipsoid Algorithm for (2.1)

Initial assumptions: A feasible point $a_0 \in P_<$ is given. There exists a sphere $S(a_0, r) \subset P$. There exists a sphere $S(a_0, s) \supset P$. A value for ε is chosen.

$$N = 2n(n+1)\left\lceil \log \frac{2s^2\|c\|}{r\varepsilon} \right\rceil.$$

Initialization: $x_0 = a_0$, $D_0 = s^2 I$, $\zeta_0 = ca_0$, $t = 0$.

Iteration t: If $x_t \notin P_<$, set $d = a^{i(t)}$, where $a^{i(t)}x_t \geqslant b_{i(t)}$ and $\zeta_{t+1} = \zeta_t$. If $x_t \in P_<$, set $d = -c$, $\zeta_{t+1} = \max\{\zeta_t, cx_t\}$, and $\hat{x} \leftarrow x_t$ if $cx_t > \zeta_t$. Use formulas (2.1) and (2.2) to obtain x_{t+1} and D_{t+1}. If $t > N$, stop $\hat{x} \in P_<$ and $c\hat{x} \geqslant z_{LP} - \varepsilon$. Otherwise set $t \leftarrow t + 1$.

To analyze this algorithm we need the following results on volumes.

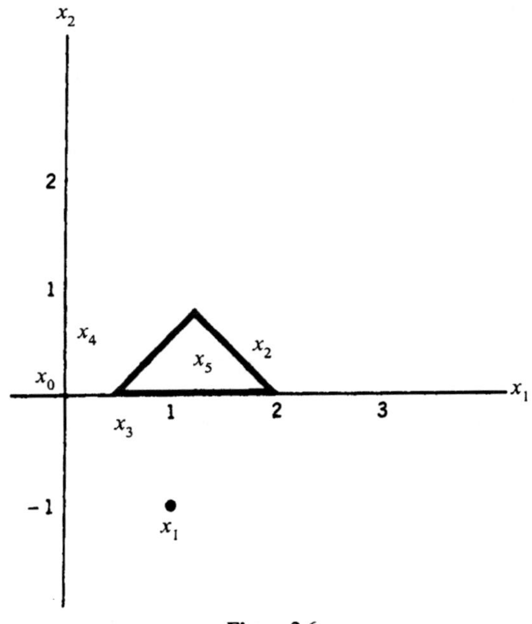

Figure 2.6

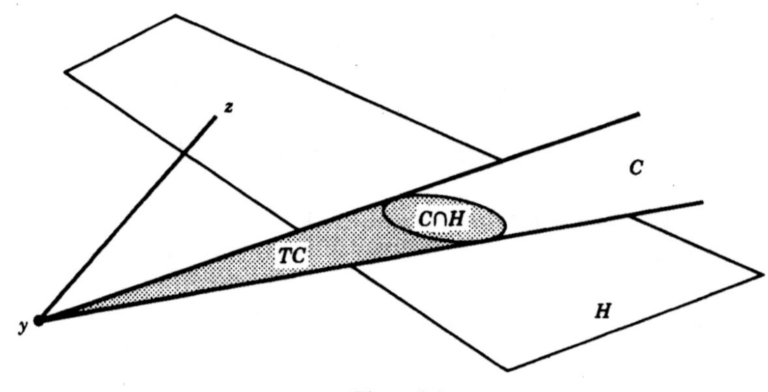

Figure 2.7

Proposition 2.9

a. $\text{vol}(S^{n+1})/\text{vol}(S^n) < 2^n/n$.
b. $\text{vol}(E(D, y)) = \sqrt{|\det D|}\,\text{vol}(S^n)$.
c. *If C is a cone with vertex y, and H is a hyperplane intersecting C, then the truncated cone $TC = \text{conv}(\{y\}, C \cap H)$ with base $C \cap H$ and vertex y has volume given by*

$$\text{vol}(TC) = \frac{1}{n}\,\text{vol}(C \cap H)d(y, H), \quad \textit{where } d(y, H) = \min\{\|y - z\|: z \in H\}$$

(see Figure 2.7).

Theorem 2.10. *When the approximate ellipsoid algorithm terminates, we have $\hat{x} \in P$ and $\zeta_N = c\hat{x} \geqslant z_{\text{LP}} - \varepsilon$.*

Proof. We use the volume argument given in the proof of Theorem 2.1. Suppose x^* is an optimal solution to LP. Because $P_< \supset S(a_0, r)$, the initial feasible region $P_< \cap \{x \in R^n: cx > ca_0\}$ contains the truncated cone TC with base $S(a_0, r) \cap \{x: cx = ca_0\}$ and vertex x^*, and after N iterations the final ellipsoid E_N contains the truncated cone $TC' = TC \cap \{x \in R^n: cx > \zeta_N\}$. By Proposition 2.9, we have

$$\text{vol}(TC) = \frac{1}{n}\,\text{vol}(S(a_0, r) \cap \{x: cx = ca_0\})\,\frac{c(x^* - a_0)}{\|c\|},$$

since the distance from x^* to the hyperplane $cx = ca_0$ is $c(x^* - a_0)/\|c\|$. Also

$$\text{vol}(S(a_0, r) \cap \{x: cx = ca_0\}) = r^{n-1}\,\text{vol}(S^{n-1}),$$

since the intersection of a sphere with a hyperplane through its center is again a sphere with the same radius but of one dimension less. Finally

$$\text{vol}(TC') = \left(\frac{z_{\text{LP}} - \zeta_N}{z_{\text{LP}} - ca_0}\right)^n \text{vol}(TC),$$

since the height of TC' is $(z_{\text{LP}} - \zeta_N)/(z_{\text{LP}} - ca_0)$ times the height of TC. Hence

$$\text{vol}(TC') = \frac{1}{n}\left(\frac{z_{\text{LP}} - \zeta_N}{z_{\text{LP}} - ca_0}\right)^n r^{n-1} \text{vol}(S^{n-1}) \frac{cx^* - ca_0}{\|c\|}.$$

Since $TC' \subseteq E_N$, it follows that

$$\text{vol}(TC') \leqslant e^{-N/2(n+1)} \text{vol}(E_0) = e^{-N/2(n+1)} s^n \text{vol}(S^n).$$

Therefore

$$(z_{\text{LP}} - \zeta_N) \leqslant e^{-N/2n(n+1)} s \left(\frac{n \text{ vol}(S^n)}{\text{vol}(S^{n-1})}\right)^{1/n} \|c\|^{1/n} \left(\frac{z_{\text{LP}} - ca_0}{r}\right)^{(n-1)/n}.$$

Also

$$z_{\text{LP}} - ca_0 = c(x^* - a_0) \leqslant \|c\| \, \|x^* - a_0\| \leqslant s \, \|c\|,$$

and hence

$$z_{\text{LP}} - \zeta_N \leqslant e^{-N/2n(n+1)} \frac{s^2}{r} \|c\| \left(\frac{n \text{ vol}(S^n)}{\text{vol}(S^{n-1})}\right)^{1/n} < 2e^{-N/2n(n+1)} \frac{s^2}{r} \|c\|.$$

Since $N = 2n(n+1)\lceil \log(2s^2 \|c\| / r\varepsilon) \rceil$, it follows that $z_{\text{LP}} - \zeta_N < \varepsilon$. ■

Several results are needed to show that an ellipsoid algorithm solves the linear programming problem in polynomial time.

We must deal with the precision of the arithmetic calculations. Square roots occur in the ellipsoid updating formula (2.2), and the assumption we have made so far is that these irrational numbers are found exactly. But, of course, this precludes the possibility of digital calculation, which requires finite representation of numbers. Thus we must specify a maximum number of digits permitted in the calculations. In particular, we now assume that the initial data and all intermediate numbers produced during the calculations are rational numbers represented by the ratio of two integers, each of which is specified with ρ binary digits of precision.

With finite precision calculation it is still possible to obtain an approximate solution to linear programming problems. We need, however, to ensure that the feasible region, if any, remains inside the half-ellipsoid, given the numerical errors produced by the finite precision. This is done by using slightly larger ellipsoids. Then we compensate by using a larger number of iterations.

The algorithm so modified is called the *finite precision approximate ellipsoid algorithm*. The following theorem, which we will not prove, gives the precision and number of iterations required provided there exists a_0 such that $S(a_0, r) \subset P \subset S(a_0, s)$.

Theorem 2.11. *When* $N = 4n^2\lceil \log(2s^2\|c\| / r\varepsilon) \rceil$, $\rho = 5N$, *formula (2.2) is replaced by*

$$D_{t+1} = \frac{2n^2+3}{2n^2}\left(D - \frac{2}{n+1} \frac{(Dd)\, Dd)^T}{d^T Dd}\right), \tag{2.5}$$

and D_{t+1} *and* x_{t+1} *are calculated to* ρ *binary digits of precision, the finite precision ellipsoid algorithm applied to (2.3) terminates with a solution* $\hat{x} \in P$ *such that* $c\hat{x} \geq z_{\text{LP}} - \varepsilon$.

The precise values of N and ρ are not important in this theorem. What is important is that N and ρ are polynomial functions of n, $\log s$, $\log r$, $\log \|c\|$, and $\log(1/\varepsilon)$. In addition, the amount of calculation needed to update x_t and D_t at each iteration is polynomial in n and ρ. Furthermore, the theorem applies to (2.4) provided a violated inequality $a^{i(t)}x \le b_{i(t)}$ can be expressed with ρ digits of precision (independently of the question of how it is found).

Now we consider how to relate the values of r and s to the initial description of the polytope P.

Definition 2.5. T is the largest numerator or denominator of any component of an extreme point of P. T' is the largest numerator or denominator of any component of a facet-defining inequality. (The components of these vectors are rationals expressed as the ratio of integers.)

Proposition 2.12. *For any full-dimensional polytope P, the following statements are true:*

i. $T' \le (nT)^{n^2+n}$.
ii. $T \le (nT')^n$,
iii. *There exists $a_0 \in P$ such that $S(a_0, r) \subset P \subset S(a_0, s)$ with $r = (nT)^{-2n^2-2n}$ and $s = 2nT$.*

Proof. i. Suppose $ax \le b$ is a facet-defining inequality of P. Without loss of generality, we assume that $b = \pm 1$ or 0. Then there are extreme points x^i for $i = 1, \ldots, n$ such that a is the unique solution to $ax^i = b$ for $i = 1, \ldots, n$.

Now we can write $x^i_j = p_{ij}/q_{ij}$ with $|p_{ij}|$, $|q_{ij}|$ integers not exceeding T. Taking $\theta^i = \Pi^n_{j=1} q_{ij}$, the system $(a\theta^i)x^i = \theta^i b$, $i = 1, \ldots, n$, has integer coefficients bounded in magnitude by T^{n+1}. Then by Cramer's rule we have $a_j = p_j/q$, where p_j and q are integers with $|p_j|$, $|q| \le n!\,(T^{n+1})^n < (nT)^{n^2+n}$.

ii. The proof is similar to i using polarity.

iii. For all $x \in P$, we have $|x_i| \le T$. Hence $(x - a_0)^T(x - a_0) \le n(2T)^2 < (2nT)^2$, and $P \subset S(a_0, s)$ with $s = 2nT$.

Finally we show that there is an inscribed sphere of the given radius. Take $n + 1$ affinely independent extreme points $\{x^i\}^{n+1}_{i=1}$, and let $a_0 = [1/(n + 1)]\, \Sigma^{n+1}_{i=1} x^i$. Clearly, $a_0 \in P_<$. The distance from a_0 to any facet $ax = b$ is $(b - a^Ta_0)/\ \|a\|$. Our goal is to find a lower-bound r on this distance, since this will provide us with the imbedded sphere $S(a_0, r)$. Since $(a, b) \in Z^{n+1}$, it follows that $b - a^Ta_0 \ge 1/\psi$, where ψ is the common denominator of the components a_{0j} of a_0. Since $x^i_j = p/q$ with $p, q \in Z^1$, it follows that $|p|$, $|q| \le T$, $a_{0j} = p'_j/q'$ with $p'_j, q' \in Z^1$, and $|q'| \le (n + 1)T^{n+1}$. Thus $a_0 = (p''_1, \ldots, p''_n)/q''$ with $|q''| \le (n + 1)^n T^{n^2+n} \le (nT)^{n^2+n}$. Hence $b - a^Ta_0 \ge 1/\psi = 1/q'' \ge (nT)^{-n^2-n}$.

Now $\|a\| \le n(nT)^{n^2+n}$, and for each a_i we have $|a_i| < (nT)^{n^2+n}$. Hence $(b - a^Ta_0)/\|a\| > (nT)^{-2n^2-2n} = r$. ■

We have established that $\log s$ and $\log r$ are polynomial in n and $\log T$, or equivalently in n and $\log T'$. Hence N and ρ in Theorem 2.11 are polynomial in n, $\log T$, $\log \|c\|$, and $\log (1/\varepsilon)$

To convert the ε-approximate solution obtained in Theorem 2.11 to an optimal solution, we perturb the objective function of (2.3) so that the resulting linear program has a unique solution.

Proposition 2.13. *Given P, let $Q = 2T^{2n}$ and $c' = Q^n c + (1, Q, \ldots, Q^{n-1})$. Then the linear program*

$$(2.6) \qquad z_{\mathrm{LP'}} = \max\{c'x: x \in P\}$$

has a unique optimal solution x^*, and x^* is an optimal solution of (2.3).

Proof. Let x^1 be some other vertex of P. Letting $x^* = (p_1^*/q_1^*, \ldots, p_n^*/q_n^*)$ and $x^1 = (p_1^1/q_1^1, \ldots, p_n^1/q_n^1)$, we have

$$x_j^* - x_j^1 = \left(\prod_{i=1}^{n} q_i^* q_i^1\right)^{-1} \left(\prod_{i \neq j} q_i^* q_i^1\right)(p_j^* q_j^1 - p_j^1 q_j^*).$$

Hence we can write $x^* - x^1 = z/\alpha$, where $\alpha < T^{2n}$, and z is an integer vector with $|z_j| < 2T^{2n}$ for all j.

Since x^* is optimal,

$$c'(x^* - x^1) = \frac{1}{\alpha} c'z = \frac{1}{\alpha}\left(Q^n cz + \sum_{j=1}^{n} Q^{j-1} z_j\right) \geqslant 0.$$

But because $|z_j| < Q$ for all j, it follows that

$$\left|\sum_{j=1}^{n} Q^{j-1} z_j\right| \leqslant \left|(Q-1)\sum_{j=1}^{n} Q^{j-1}\right| = |Q^n - 1|.$$

Since $Q^n cz \geqslant -\Sigma_{j=1}^n Q^{j-1} z$, and cz is integer, we have $cz \geqslant 0$. Hence we have shown that $cx^* \geqslant cx^1$ for any extreme point x^1, and x^* is optimal to (2.3).

Finally we observe that because $z \neq 0$ and $|z_j| < Q$ for all j, $\Sigma_{j=1}^n Q^{j-1} z_j \neq 0$. This implies that $c'(x^* - x^1) \neq 0$, and hence x^* is the unique optimum of (2.6). ■

The next step is to show that if we choose ε appropriately in the finite precision approximate ellipsoid algorithm we can get very close to x^*.

Proposition 2.14. *If the finite precision approximate ellipsoid algorithm is applied to (2.6) and $\varepsilon = (1/4n)T^{-4n-2}$, the algorithm terminates with $\hat{x} \in P$ satisfying $c'\hat{x} \geqslant z_{\mathrm{LP'}} - \varepsilon$ and $\|x^* - \hat{x}\| \leqslant 1/2T^2$, where x^* is an extreme point optimal solution of (2.3).*

Proof. Since $\hat{x} \in P$, $\hat{x}$ is a convex combination of extreme points, say $\{x^i\}_{i=1}^t$. We claim that one of these extreme points is x^*. If not, let $cx^1 = \min_i cx^i$. Then

$$c'x^* - c'\hat{x} \geqslant c'x^* - c'x^1 = \frac{1}{\alpha} c'z \geqslant \frac{1}{\alpha} > \frac{1}{T^{2n}},$$

contradicting $c'x^* - c'\hat{x} \leqslant \varepsilon$.

Therefore we can write $\hat{x} = \sum_{i=1}^{n} \lambda_i x^i + \lambda_* x^*$ with $\sum_{i=1}^{n} \lambda_i + \lambda_* = 1$, $\lambda_i \geqslant 0$ for $i = 1, \ldots, n$, and $\lambda_* > 0$. Now observe that since $c'(x^* - x^i) > 1/T^{2n}$ for all extreme points $x^i \neq x^*$, it follows that

$$\begin{aligned} c'\hat{x} &= \sum_{i=1}^{n} c'\lambda_i x^i + c'\lambda_* x^* \\ &< \sum_{i=1}^{n} \lambda_i \left(c'x^* - \frac{1}{T^{2n}} \right) + c'\lambda_* x^* \\ &= c'x^* - \frac{1}{T^{2n}} (1 - \lambda_*). \end{aligned}$$

But since $c'\hat{x} \geqslant c'x^* - \varepsilon$, it follows that $\varepsilon > (1/T^{2n})(1 - \lambda_*)$.

Now

$$\|\hat{x} - x^*\| = (1 - \lambda_*) \left\| \sum_{i=1}^{n} \frac{\lambda_i}{1 - \lambda_*} x^i - x^* \right\| = (1 - \lambda_*)\|y - x^*\|$$

for some $y \in P$, since $\sum_{i=1}^{n} \lambda_i = 1 - \lambda^*$, and y is thus a convex combination of x^i for $i = 1, \ldots, n$. However, $\|y - x^*\| \leqslant 2nT^{2n}$, and therefore

$$\|\hat{x} - x^*\| \leqslant \varepsilon T^{2n}\, 2nT^{2n} < \frac{1}{2T^2}.$$

■

Proposition 2.15. *Let $\hat{x}$ and x^* be as described in Proposition 2.14. The vector obtained by rounding each coefficient $\hat{x}_j$ of $\hat{x}$ to the nearest rational multiple of $\{1, 1/2, \ldots, 1/T\}$ is x^*.*

Proof. Since x^* is an extreme point of P, it follows that x_j^* is some rational multiple of $\{1, 1/2, \ldots, 1/T\}$ for all $j = 1, \ldots, n$. Suppose x^* is not obtained as described. Then $x_j^* - \hat{x}_j > 1/2T^2$ for some j, and hence $\|\hat{x} - x^*\| > 1/2T^2$, contradicting Proposition 2.14. ■

One approach to finding x^* from $\hat{x}$ is by the method of continued fractions. In Section I.7.3, we will describe this method and show that it can be executed in time polynomial in n and $\log T$.

We have established how the finite precision approximate ellipsoid algorithm produces an optimal solution to (2.3) or (2.4) when P is a full-dimensional polytope. We have also shown that each step (except that of finding a violated inequality) requires a number of calculations that are polynomial in n, $\log T$, and $\log \|c\|$ (see Theorem 2.11 and Propositions 2.13–2.15). For the linear program (2.3), this number is polynomial in the length of the input description, since to find a violated inequality we simply check whether the p-digit number x^t satisfies each constraint. This requires $O(mnp)$ calculations. Hence we obtain the following theorem.

Theorem 2.16. *There is a polynomial algorithm to solve the linear programming problem over full-dimensional polytopes with description (m, n, A, b, c).*

3. THE POLYNOMIAL EQUIVALENCE OF SEPARATION AND OPTIMIZATION

The importance of the ellipsoid algorithm in combinatorial optimization is as a tool to prove that certain problems can be solved in polynomial time.

To illustrate this idea, consider the minimum-weight s–t cut problem (see Section I.3.4). Given a graph $G = (V, E)$ with $|E| = n$, $s, t \in V$, and a nonnegative weight vector $c \in R^n_+$ on the edges of G, the problem is to find a minimum-weight set of edges that intersects every s–t path in G. Let $\mathcal{K}$ be the family of s–t paths in G.

One way to formulate this problem is as the linear program

$$\begin{aligned} &\min \sum_{e \in E} c_e y_e \\ (3.1) \qquad &\sum_{e \in K_j} y_e \geq 1 \quad \text{for } K_j \in \mathcal{K} \\ &0 \leq y_e \leq 1 \quad \text{for } e \in E. \end{aligned}$$

This is true because $y \in B^n$ is a feasible solution if and only if it is the incidence vector of an s–t cut and, as will be shown in Section III.1.6, all of the extreme points of the polytope defined by the constraint set of (3.1) are in B^n.

There is one difficulty in solving (3.1) by the ellipsoid algorithm given in Section 2. This is that $|\mathcal{K}|$ is an exponential function of n, which means that the feasibility of a point y^* determined from an iteration of the ellipsoid algorithm cannot be decided efficiently by the usual method of substitution. However, in this case there is an alternate way of checking feasibility.

Suppose $0 \leq y^* \leq 1$, since this can be checked by substitution. Consider the problem of finding a minimum-weight s–t path with weight vector y^*. This can be done efficiently by the algorithm given in Section I.3.2. Let ξ be the weight of a minimum-weight path. Then $\Sigma_{e \in K_j} y_j^* \geq 1$ for all $K_j \in \mathcal{K}$ if and only if $\xi \geq 1$. Moreover, if $\xi < 1$, any minimum-weight path yields an inequality that is most violated by y^*.

So for this problem we have overcome the apparent difficulty of a large number of constraints by providing an efficient subroutine that implicitly checks feasibility and provides a violated inequality when the point in question is not feasible.

This example motivates the separation problem, which combines both the membership and validity problems.

The Separation Problem for a Family of Polyhedra. An instance is given by an integer n, a description of a polyhedron $P \subseteq R^n$ in the family, and an $x^* \in R^n$.

A solution is an answer to the membership problem and, if $x^* \notin P$, a valid inequality (π, π_0) for P such that $\pi x^* > \pi_0$. (Note that the separation problem is not a feasibility problem because it requires that we exhibit a valid inequality.)

Our objective here is to relate the complexity of the separation problem to the complexity of the linear programming problem over the family of polyhedra.

We now formally describe the linear programming problem.

The Linear Programming Problem for a Family of Polyhedra. An instance is given by an integer n, a description of a polyhedron $P \subset R^n$ in the family, and a $c \in R^n$. Assuming that $P \neq \emptyset$ and cx is bounded for all $x \in P$, a solution is an $x^0 \in P$ such that $cx^0 = \max\{cx: x \in P\}$.

The principal result is that, under certain technical assumptions, the linear programming problem for a family of polyhedra is solvable in polynomial time if and only if the separation problem is solvable in polynomial time.

As in the previous section, we confine the analysis to full-dimensional polytopes. But here we do not assume that the number of constraints is part of the description of P. Instead, we assume that for $P \subset R^n$ the length of the input l needed to encode P is bounded from below by a polynomial in n and $\log T$, where T is given in Definition 2.5. This assumption enables us to work with the parameters n and T as well as to establish the polynomiality of an algorithm by showing that its running time is polynomial in n and $\log T$. Note that by Proposition 2.12 this assumption is equivalent to assuming that the length of the input needed to encode P is bounded from below by a polynomial in n and $\log T'$ (T' is also given in Definition 2.5).

This assumption is reasonable for the families of polytopes of interest to us, because if $\log T$ was superpolynomial in the true input length, we would have no hope of describing an optimal solution to the linear programming problem over the family in polynomial time.

More specifically, consider the case where we are dealing with a family of full-dimensional polytopes where either $P = \{x \in R^n: Ax \leq b\}$ or $P = \text{conv}(S)$ with $S = \{x \in Z^n_+: Ax \leq b\}$ and where the standard input is the $m \times (n + 1)$ integer matrix (A, b). The input length needed to describe these problems is $l = O(mn \log \theta)$, where $\theta = \max(\max_{ij} |a_{ij}|, \max_i |b_i|)$.

When $P = \{x \in R^n: Ax \leq b\}$, by Proposition 3.1 of Section I.5.3, the largest value that can be taken by the numerator or denominator of any extreme point is $T = (n\theta)^n$. Hence, since $\log T = n \log \theta + n \log n$, it follows that $\log T$ is certainly, at most, a polynomial function of l. A similar result holds when $P = \text{conv}(S)$ (see Theorem 4.1 of Section I.5.4).

As illustrated by problem (3.1), it may not be efficient to solve the separation problem by substitution. Moreover, if $x \notin P$, we need to establish that we can find a violated inequality whose encoding length is polynomial in l. But this follows, since if $x \notin P$, then x does not satisfy some facet-defining inequality whose encoding length is $O(\log(nT'))$. Moreover, a facet-defining inequality can be described exactly using the precision specified in Theorem 2.11.

In the previous section we gave an ellipsoid algorithm in which each step, except that of finding a violated inequality, requires a number of calculations that are polynomial in n, $\log T$, and $\log \|c\|$. We have just seen that a violated inequality can be described by a polynomial in n and $\log T$, and it is easily checked that this meets the requirement of the precision required in the algorithm.

The only step that remains is to solve the separation problem at each iteration. Immediately we can conclude that Theorem 2.16 generalizes to:

Theorem 3.1. *Given a family of full-dimensional polytopes $P(n, T)$ whose description length is at least a polynomial in n and* $\log T$*, if the separation problem over the family is solvable in polynomial time, then the linear programming problem over the family is solvable in polynomial time.*

We can also make use of polarity to give a polynomial algorithm for the separation problem based on a polynomial algorithm for the linear programming problem.

Theorem 3.2. *The following statements are equivalent for a family of full-dimensional polytopes having the origin in their interior and whose input length is at least a polynomial in n and* $\log T$.

a. *There is a polynomial-time algorithm for the separation problem.*
b. *There is a polynomial-time algorithm for the linear programming problem.*
c. *There is a polynomial-time algorithm for the separation problem over the family of 1-polars.*
d. *There is a polynomial-time algorithm for the linear programming problem over the family of 1-polars.*

Proof. $a \Rightarrow b$. This is Theorem 3.1.

$b \Rightarrow c$. We apply the results of Section I.4.5 to relate the optimization problem for P to the separation problem for its 1-polar Π^1. By Proposition 5.4, Π^1 is also full-dimensional and bounded and contains 0 in its interior.

Given $\pi^* \in R^n$, let x^* be an optimal solution to $\max\{\pi^* x: x \in P\}$. By Corollary 5.6, $\pi^* \in \Pi^1$ if and only if $\pi^* x^* \leq 1$. If $\pi^* x^* > 1$, then by Theorem 5.5 it follows that $(x^*, 1)$ is a valid inequality for Π^1 that cuts off π^*. Finally, by Proposition 2.12, T' is a polynomial function of n and $\log T$, so the length of the input needed to described Π^1 is polynomial in n and $\log T$.

$c \Rightarrow d$. Because Π^1 is a full-dimensional polytope, we can apply Theorem 3.1 to the family of 1-polars.

$d \Rightarrow a$. The proof is the same as $b \Rightarrow c$ with the roles of P and Π^1 interchanged. ■

Now observe that any family of full-dimensional polytopes can be translated so that the origin becomes an interior point. The above argument then shows the equivalence of statements a and b for full-dimensional polytopes. This result extends to arbitrary rational polyhedra, though certain steps of the algorithms need modification, and for unbounded polyhedra we need to redefine T to include extreme rays.

The following theorem justifies what we call the *equivalence of separation and optimization* throughout this book.

Theorem 3.3. *For a family of rational polyhedra $P(n, T)$ whose input length is at least polynomial in n and* $\log T$, *there is a polynomial-time reduction of the linear programming problem over the family to the separation problem over the family, and conversely there is a polynomial-time reduction of the separation problem to the linear programming problem.*

Theorem 3.3 implies that the linear programming problem is solvable in polynomial time if and only if the separation problem is solvable in polynomial time. In Part III we will use this result to develop polynomial-time algorithms for some combinatorial optimization problems by giving polynomial-time algorithms for the separation problem. In contrast to the minimum-cut example, we will see an example where this provides the only known polynomial-time algorithm.

Theorem 3.3 also implies that the linear programming problem is $\mathcal{NP}$-hard if and only if the separation problem is $\mathcal{NP}$-hard. However, as we shall see in Chapters II.5 and II.6, separation is extremely important in the solution of $\mathcal{NP}$-hard optimization problems where we know some classes of strong valid inequalities and are able to solve or approximate the solution of the separation problem efficiently.

We close this section by using Theorem 3.3 to show that the linear optimization problem over the fractional node packing polytope is $\mathcal{NP}$-hard.

Example 3.1 *(Example 7.2 of Section I.5.7 continued).* The 1-polar of the fractional node-packing polytope is a polytope whose extreme points are the incidence vectors of the maximal cliques of G. Hence the linear programming problem over the 1-polar is $\mathcal{NP}$-hard because it is equivalent to the maximum-weight clique problem on G or the maximum-weight node-packing problem on the complement of G. Now, by polarity, the separation problem over the fractional node-packing polytope is $\mathcal{NP}$-hard. Finally, by Theorem 3.3, we can reduce the linear program problem to the separation problem, so the linear programming problem over the fractional node-packing polytope is also $\mathcal{NP}$-hard.

4. A PROJECTIVE ALGORITHM FOR LINEAR PROGRAMMING

Recently, remarkable claims have been made concerning the computational efficiency of a new algorithm based on projections. Here we describe a conceptually simple variant of that algorithm whose geometric rate of convergence is easily established. Our objective is just to show the significance of projections in solving linear programs. We neither claim that the version of the algorithm given here is efficient, nor do we prove polynomiality.

We will need the formula for the projection of a vector onto a subspace.

Proposition 4.1. *Let A be an $m \times n$ matrix, with* $\text{rank}(A) = m$ *and* $H = \{x \in R^n: Ax = 0\}$. *The projection of p onto H is given by* $q = [I - A^T(AA^T)^{-1}A]p$.

Proof. $q \in H$ so $Aq = 0$. Also, since $p - q \in H^{\perp}$, by Proposition 1.6 of Section I.4.1, there exists $u \in R^m$ such that $A^Tu = p - q$. Therefore

$$AA^Tu = Ap - Aq = Ap \quad \text{and} \quad u = (AA^T)^{-1}Ap$$

because AA^T is nonsingular when $\text{rank}(A) = m$.

Hence $q = p - A^Tu = [I - A^T(AA^T)^{-1}A]p$. ■

We first apply the projective algorithm to the homogeneous feasibility problem.

Homogeneous Feasibility Problem. Given an integer $m \times n$ matrix A with $\text{rank}(A) = m$, find a ray $r \in P \setminus \{0\}$, where $P = \{r \in R^n_+: Ar = 0\}$, or show that $P = \{0\}$.

The algorithm works with candidate rays $r^k > 0$ for $k = 1, 2, \ldots$ and attempts to satisfy $Ar = 0$. Suppose $r^1 = 1 \notin P$ because $A1 \neq 0$. To obtain a point closer to being a solution than r^1, we attempt to preserve nonnegativity by finding the closest point to 1 that satisfies $Aq = 0$. This is the problem $\min\{\|1 - q\|: Aq = 0\}$. Its solution is $\tilde{q}$, the projection of 1 onto $Aq = 0$.

Now if $\tilde{q} \geq 0$, then $\tilde{q} \in P$ and we are done. Otherwise (unless we can deduce that $P = \emptyset$), we modify $\tilde{q}$ to obtain a vector $q' > 0$ that is "closer" to being a solution in P than the initial vector 1. We take a positive linear combination of $\tilde{q}$ and 1, giving $q' = 1 + \alpha\tilde{q}$ with $\alpha > 0$. α must be chosen so that $q' > 0$, so α cannot be chosen arbitrarily large. Alternatively, the larger α is, the "closer" q' is to $\tilde{q}$ and hence to satisfying $Aq = 0$. We have described one iteration of the algorithm from the point $r^k = 1$ to the new point $r^{k+1} = 1 + \alpha\tilde{q}$.

We observe that even if $r^k \neq 1$, the same iterative step can be applied when $r^k > 0$ but $r^k \notin P$. Let D^k be a diagonal matrix with $d^k_{jj} = r^k_j > 0$ for $j = 1, \ldots, n$, let $A^k = AD^k$, and let $P^k = \{r \in R^n_+: A^kr = 0\}$ be a cone. Clearly $P \neq \{0\}$ if and only if $P^k \neq \{0\}$. Also, $Ar^k = A^k1$. Now we can describe iteration k in terms of the feasibility problem for P^k. A candidate vector 1 is given. If $A^k1 \neq 0$, so $1 \notin P^k$, we derive a new vector $q' > 0$ that is "closer" to being a solution of P^k. Let $q' = 1 + \alpha_kq^k$, where $\alpha_k > 0$ and q^k is the projection of 1 onto

$A^k q = 0$. Restated with respect to P, the new point is $r^{k+1} = D^k(1 + \alpha_k q^k)$ because $Ar^{k+1} = A^k(1 + \alpha_k q^k) = A^k q'$.

We need to make precise what it means for $r^{k+1} > 0$ to be "closer" to the cone P than $r^k > 0$. Comparing the values of Ar^{k+1} and Ar^k is relatively meaningless. To obtain a useful measure of comparison we need to work with a homogeneous version of r, such as

$$\frac{r}{(1/n)\sum_{j=1}^n r_j} \quad \text{or} \quad \frac{r}{(\prod_{j=1}^n r_j)^{1/n}},$$

which is invariant under the transformation $r \leftarrow \lambda r$ with $\lambda > 0$. Throughout this section we use the homogeneous version

$$\tilde{r} = \frac{r}{(\prod_{j=1}^n r_j)^{1/n}}.$$

Thus when comparing violations, if $|A\tilde{r}^{k+1}| < |A\tilde{r}^k|$, it makes sense to say than r^{k+1} is closer than r^k to being a ray of P.

Without specifying α_k, we can already analyze the basic behavior of an iteration of the algorithm.

Proposition 4.2

i. $\sum_{j=1}^n a_{ij} r_j^{k+1} = \sum_{j=1}^n a_{ij} r_j^k$ *for* $i = 1, \ldots, m$.

ii. $\sum_{j=1}^n a_{ij} \tilde{r}_j^{k+1} = \beta_k \sum_{j=1}^n a_{ij} \tilde{r}_j^k$,

where

$$\beta_k = \left[\prod_{j=1}^n \left(\frac{1}{1 + \alpha_k q_j^k}\right)\right]^{1/n}$$

Proof.

i. Substituting $r_j^{k+1} = r_j^k(1 + \alpha_k q_j^k)$ yields

$$\sum_{j=1}^n a_{ij} r_j^{k+1} = \sum_{j=1}^n a_{ij} r_j^k + \alpha_k \sum_{j=1}^n a_{ij} r_j^k q_j^k$$

$$= \sum_{j=1}^n a_{ij} r_j^k \quad \text{since } A^k q^k = 0.$$

ii. $r_j^k / r_j^{k+1} = 1/(1 + \alpha_k q_j^k)$. Therefore

$$\left[\prod_{j=1}^n \left(\frac{r_j^k}{r_j^{k+1}}\right)\right]^{1/n} = \left[\prod_{j=1}^n \left(\frac{1}{1 + \alpha_k q_j^k)}\right)\right]^{1/n} = \beta_k$$

and by i it follows that

$$\begin{aligned}\sum_{j=1}^{n} a_{ij}\tilde{r}_j^{k+1} &= \left(\sum_{j=1}^{n} a_{ij} r_j^{k+1}\right)\left[\prod_{j=1}^{n}\left(\frac{1}{r_j^{k+1}}\right)\right]^{1/n} \\ &= \left(\sum_{j=1}^{n} a_{ij} r_j^{k}\right)\left[\prod_{j=1}^{n}\left(\frac{1}{r_j^{k+1}}\right)\right]^{1/n} \\ &= \left(\sum_{j=1}^{n} a_{ij}\tilde{r}_j^{k}\right)\left[\prod_{j=1}^{n}\left(\frac{r_j^k}{r_j^{k+1}}\right)\right]^{1/n} \\ &= \beta_k\left(\sum_{j=1}^{n} a_{ij}\tilde{r}_j^{k}\right).\end{aligned}$$

■

Hence we see that from one iteration to the next, the values of all the terms of $A\tilde{r}^k$ change by a constant factor β_k. By specifying a choice of α_k for which $\beta_k < \beta < 1$, we will obtain an algorithm with the property that $A\tilde{r}^k \to 0$ geometrically.

The Projective Algorithm to Find an ε-Approximate Ray

Initialization. $r^1 = 1, D^1 = I, A^1 = A, k = 1, \beta < 1, \varepsilon > 0$.

Iteration k

Step 1: If $|\Sigma_{j=1}^{n} a_{ij}\tilde{r}_j^k| \leq \varepsilon$ for all i, stop. $\tilde{r}^k$ is an ε-approximate ray.
Step 2: Find the projection q^k of 1 onto the subspace $A^k q = 0$.
Step 3: If $q^k \geq 0$, stop. $r = D^k q^k \in P \setminus \{0\}$.
Step 4: If $\max_j q_j^k < 1$, stop. $P = \{0\}$.
Step 5: Find a point α_k in the set

$$\{\alpha \in R_+^1\colon 1 + \alpha q_j^k > 0 \quad \text{for } j = 1, \ldots, n, \beta_k = \left[\prod_{j=1}^{n}\left(\frac{1}{1+\alpha q_j^k}\right)\right]^{1/n} \leq \beta < 1\}.$$

Step 6: $r_j^{k+1} = r_j^k(1 + \alpha_k q_j^k)$ for $j = 1, \ldots, n$,

$$D^{k+1} = \begin{bmatrix} r_1^{k+1} & & \\ & \ddots & \\ & & r_n^{k+1}\end{bmatrix}, \qquad A^{k+1} = AD^{k+1}.$$

Set $k \leftarrow k + 1$ and return to Step 1.

The validity of Step 4 and the feasibility of Step 5 require verification. First we consider Step 4.

Proposition 4.3. *If* $\max_j q_j^k < 1$, *there is no ray in* $P/\{0\}$.

Proof. Since q^k is the projection of 1 onto $H = \{q\colon A^k q = 0\}$, it follows that $1 - q^k \in H^\perp$, and hence there exists $u \in R^m$ such that $uA^k = 1 - q^k > 0$ or $uA = (1 - q^k)(D^k)^{-1} > 0$.

Therefore by Farkas' lemma applied to the cone $P = \{r \in R_+^n\colon Ar = 0\}$, we obtain $P = \{0\}$. ■

The following result, which is not difficult to prove, shows that for certain values of $\beta < 1$, the set in Step 5 is nonempty and so the algorithm converges geometrically.

Proposition 4.4. *Taking* $\alpha_k = 1/(1 + \|q^k\|)$ *as the step size in Step 5 of the projective algorithm, it follows that* $r^{k+1} > 0$ *and*

$$\beta_k = \left(\prod_{j=1}^{n} (1 + \alpha_k q_j^k)\right)^{-1/n} \le \left(\frac{2}{e}\right)^{1/2}.$$

Theorem 4.5. *The Projective Algorithm to find an ε-approximate ray terminates after no more than*

$$\left\lceil \log\left(\max_i \frac{\left|\sum_{j=1}^{n} a_{ij}\right|}{\varepsilon}\right) \Big/ \log(1/\beta) \right\rceil$$

iterations.

Proof. After k iterations, we have

$$\left|\sum_{j=1}^{n} a_{ij} \tilde{r}_j^{k+1}\right| = \prod_{t=1}^{k} \beta_t \left|\sum_{j=1}^{n} a_{ij}\right| \le \beta^k \left|\sum_{j=1}^{n} a_{ij}\right| \quad \text{for all } i.$$

With k as claimed, $\beta^k \,|\sum_{j=1}^{n} a_{ij}| \le \varepsilon$ for all i.

Now we consider the nonhomogeneous feasibility problem:

(4.1) Find $x \in P$, where $P = \{x \in R_+^n : Ax = b\}$, or show that $P = \emptyset$

Observe that if we take $r = (x, r_{n+1}) \in R^{n+1}$ and $\tilde{A} = (A, -b)$, we can apply the projective algorithm to the cone $\tilde{P} = \{r \in R_+^{n+1} : \tilde{A}r = 0\}$.

We suppose that P is bounded, so there does not exist $(r_1, \ldots, r_n, 0) \in R_+^{n+1} \setminus \{0\}$ such that $\tilde{A}r = 0$. Hence if the cone $\tilde{P}$ has a nonzero ray r, then necessarily $r_{n+1} > 0$, and then $x = (r_1/r_{n+1}, \ldots, r_n/r_{n+1})$ is a solution of (4.1). This suggests that at each iteration of the projective algorithm, we should normalize candidate rays so that $r_{n+1} = 1$. In other words, we will choose a normalization factor ρ_k so that $r_{n+1}^k = 1$ at each iteration.

The Projective Algorithm for Problem (4.1)

Initialization. $r^1 = 1, D^1 = I_{n+1}, \tilde{A}^1 = \tilde{A}, k = 1$.

Iteration k

Step 1: $(x^k, 1) = r^k$. If $|\tilde{A}r^k| \le \varepsilon$, then $x^k = (r_1^k, \ldots, r_n^k)$ is an ε-approximate solution to (4.1) with $|Ax^k - b| \le \varepsilon$ and $x^k > 0$.

Step 2: This is unchanged.

Step 3: If $q^k > 0$, stop. $x^k \in P$ where, $x_j^k = (1/q_{n+1}^k) r_j^k q_j^k$ for $j = 1, \ldots, n$.

Step 4: If $\max_{j=1,\ldots,n+1} q_j^k < 1$, stop. $P = \emptyset$.
Step 5: This is unchanged except that $n \leftarrow n + 1$.
Step 6: $r_j^{k+1} = \rho_k r_j^k(1 + \alpha_k q_j^k)$ for $j = 1, \ldots, n + 1$, where $\rho_k = 1/(1 + \alpha_k q_{n+1}^k)$. The rest of the step is as before.

Now using Proposition 4.2, we can analyze the behavior of the algorithm on the feasibility problem (4.1). We also use the following result, which can be proved using estimates of the size of solutions as in Section I.4.5.

Proposition 4.6 *(The Perturbation Lemma). Given A and b, there exists* $\varepsilon(A, b) > 0$ *such that if* $x \in R_+^n$ *satisfies*

$$\left|\sum_{j=1}^{n} a_{ij}x_j - b_i\right| \le \varepsilon(A, b) \quad \text{for } i = 1, \ldots, m,$$

then there exists $x^* \in R_+^n$ *satisfying* $Ax = b$.

Proposition 4.7. *If the projective algorithm is applied to Problem (4.1) and* $r^{k+1} = (x^{k+1}, 1)$, *then:*

i. $\left|\sum_{j=1}^{n} a_{ij}x_j^{k+1} - b_i\right| = \rho_k\left|\sum_{j=1}^{n} a_{ij}x_j^k - b_i\right|$ *for all i;*
ii. *There exists* $\omega(A, b) \in R_+^1$ *such that*
$\left|\sum_{j=1}^{n} a_{ij}x_j^{k+1} - b_i\right| \le \beta^k\omega(A, b)\left|\sum_{j=1}^{n} a_{ij} - b_i\right|$ *for all i.*

Proof. We observe first that because

$$\beta_t = \left[\prod_{j=1}^{n+1} \frac{1}{1 + \alpha_t q_j^t}\right]^{1/(n+1)} \quad \text{and} \quad \frac{x_j^{t+1}}{x_j^t} = \rho_t(1 + a_t q_j^t),$$

we have

$$\beta_t\left[\prod_{j=1}^{n}\left(\frac{x_j^{t+1}}{x_j^t}\right)\right]^{1/(n+1)} = \rho_t.$$

Statement i follows immediately from statement i of Proposition 4.2. Now substituting for ρ_t in statement i, we obtain

$$\left|\sum_{j=1}^{n} a_{ij}x_j^{k+1} - b_i\right| = \left(\prod_{t=1}^{k}\beta_t\right)\left[\prod_{j=1}^{n}\left(\frac{x_j^{k+1}}{x_j^1}\right)\right]^{1/(n+1)}\left|\sum_{j=1}^{n} a_{ij}x_j^1 - b_i\right|$$

$$= \left(\prod_{t=1}^{k}\beta_t\right)\left(\prod_{j=1}^{n} x_j^{k+1}\right)^{1/(n+1)}\left|\sum_{j=1}^{n} a_{ij} - b_i\right|$$

$$\le \beta^k\left(\frac{\sum_{j=1}^{n} x_j^{k+1} + 1}{n + 1}\right)\left|\sum_{j=1}^{n} a_{ij} - b_i\right|$$

since $\beta_t \leq \beta$, and because the geometric mean does not exceed the arithmetic mean. Now dividing both sides by $\Sigma_{j=1}^n x_j^{k+1}$ and setting $y_j = x_j^{k+1}/\Sigma_{j=1}^n x_j^{k+1} > 0$, we obtain

$$\left|\sum_{j=1}^n a_{ij}y_j - \frac{b_i}{\Sigma_{j=1}^n x_j^{k+1}}\right| \leq \beta^k\left(\frac{1 + 1/\Sigma_{j=1}^n x_j^{k+1}}{n+1}\right)\left|\sum_{j=1}^n a_{ij} - b_i\right|.$$

If $\Sigma_{j=1}^n x_j^{k+1}$ is unbounded as $k \to \infty$, then as $\beta^k \to 0$, there exists a k for which $|\Sigma_{j=1}^n a_{ij}y_j| \leq \varepsilon(A, 0)$, where $\varepsilon(A, 0)$ is as in Proposition 4.6. Hence $Ay = 0$, $y \in R_+^n$, has a solution with $y \neq 0$, contradicting the boundedness assumption. Hence $\Sigma_{j=1}^n x_j^{k+1}$ remains bounded, and the claim follows. ■

Note that this algorithm has the property that the violation in each constraint decreases at exactly the same rate.

Example 4.1. We apply the projective algorithm to the feasibility problem

$$\begin{aligned} -x_1 + 2x_2 &\leq 4 \\ 5x_1 + x_2 &\leq 20 \\ -2x_1 - 2x_2 &\leq -7 \\ cx = 7x_1 + 2x_2 &\geq \lambda \\ x &\in R_+^2 \end{aligned}$$

with $\lambda = 30$.

Converting into equality form, we have $\tilde{A} = (A \quad -b)$ with

$$\tilde{A} = \begin{pmatrix} -1 & 2 & 1 & 0 & 0 & 0 & -4 \\ 5 & 1 & 0 & 1 & 0 & 0 & -20 \\ -2 & -2 & 0 & 0 & 1 & 0 & 7 \\ 7 & 2 & 0 & 0 & 0 & -1 & -30 \end{pmatrix}.$$

We use the starting point $x^1 = (1 \ 1 \ 1 \ 1 \ 1 \ 1)$ and $r^1 = (x^1, 1)$.

Initialization. $D^1 = I_7$, $\tilde{A}^1 = \tilde{A}$.

Iteration 1

Step 1:

$$\tilde{A}r^1 = \begin{pmatrix} -2 \\ -13 \\ 4 \\ -22 \end{pmatrix}.$$

Step 2: The projection of 1 on $\tilde{A}q = 0$ is the vector

$$q^1 = (0.9907 \ 0.6622 \ 0.8254 \ -0.1806 \ 1.337 \ 0.08659 \ 0.2698)$$

Step 5: $\alpha_1 = 4.433$, $\beta_2 = 0.425$, $\rho_1 = 0.455$.

Step 6: $r^2 = (x^2, 1) = (2.455\ \ 1.711\ \ 2.122\ \ 0.091\ \ 3.155\ \ 0.630\ \ 1)$

$$\tilde{A}^2 = \begin{pmatrix} -2.455 & 3.423 & 2.122 & 0 & 0 & 0 & -4 \\ 12.277 & 1.711 & 0 & 0.0909 & 0 & 0 & -20 \\ -4.911 & -3.423 & 0 & 0 & 3.155 & 0 & 7 \\ 17.188 & 3.423 & 0 & 0 & 0 & -0.6302 & -30 \end{pmatrix}.$$

Iteration 2

$$Ax^2 - b = \tilde{A}^2 1 = \begin{pmatrix} -0.911 \\ -5.921 \\ 1.821 \\ -10.02 \end{pmatrix} = \rho_1(Ax^1 - b).$$

Step 2: $q^2 = (0.8328\ \ 1.205\ \ 0.1850\ \ 0.7620\ \ 1.232\ \ -0.1514\ \ 0.6178)$

Step 5: $\alpha_2 = 5.431, \beta_2 = 0.326, \rho_2 = 0.230.$

Step 6: $r^3 = (x^3, 1) = (3.114\ \ 2.964\ \ 0.9766\ \ 0.1072\ \ 5.574\ 0.02569\ \ 1)$

Iteration 3

$$Ax^3 = \begin{pmatrix} -0.209 \\ -1.36 \\ 0.418 \\ -2.30 \end{pmatrix}.$$

Step 2: $q^3 = (1.060\ \ 1.189\ \ 0.2750\ \ 0.5488\ \ 1.1878\ \ 0.9252\ \ 1.0044).$

Step 3: Because $q^3 > 0$, it follows that

$$x^3 = (3.286\ \ 3.509\ \ 0.2674\ \ 0.05859\ \ 6.592\ \ 0.02366).$$

Stop. $x^* = (3.286\ \ 3.509)$ is feasible and $cx^* > 30$.

Now we consider the linear programming problem:

$$z_{LP} = \max\{cx\colon Ax = b, x \in R^n_+\}. \tag{4.2}$$

Viewed in terms of feasibility, we solve

$$\max\{\zeta\colon P(\zeta) \neq \varnothing\},$$

where $P(\zeta) = \{x \in R_+^n: Ax - b = 0, cx - \zeta = 0\}$. We let $\tilde{A} = (A \quad -b)$, $\tilde{c}_\zeta = (c \quad -\zeta)$, and

$$\tilde{A}_\zeta = \begin{pmatrix} A & -b \\ c & -\zeta \end{pmatrix}.$$

We only describe an algorithm for Phase 2, that is, we assume a feasible point $x^1 \in R_+^n$ satisfying $Ax^1 = b$ and $x^1 > 0$ is known.

We apply the feasibility algorithm to $P(\zeta)$. The algorithm is motivated by the following observations.

i. Let $r^k = (x^k, 1)$. By the choice of x^1, we have $\tilde{A}r^1 = 0$. Because $\tilde{A}r^{k+1} = \rho_k \tilde{A}r^k$ for each k, we have $\tilde{A}r^k = 0$ or $Ax^k = b$ for all k.

ii. Since the only violated constraint is $cx - \zeta = 0$, the projective algorithm works to decrease $\zeta - cx^k$ geometrically at each iteration. To maintain geometric convergence, we need projection vectors with $\max_j q_j^k \geq 1$. However, if ζ is a strict upper bound on z_{LP}, then $P(\zeta) = \emptyset$ and hence the algorithm will stop with $\max_j q_j^k < 1$.

   We overcome this difficulty by viewing (and calculating) the projection vector q^k as a function of ζ. q^k is of the form $\alpha + \beta\zeta$ with $\alpha, \beta \in R^{n+1}$. Therefore if $\max_j q_j^k(\zeta) < 1$, it is easy to calculate $\zeta' < \zeta$ such that $\max_j q_j^k(\zeta') = 1$. ζ' is a new upper bound for z_{LP}, and the algorithm can now proceed to find a new iterate r^{k+1}.

iii. If ζ_k is the value of ζ at iteration k we can associate a dual feasible solution u^k to (4.2) with $u^k b \leq \zeta_k$, so the algorithm simultaneously produces primal and dual feasible solutions.

   As before we assume that $\{x \in R_+^n: Ax = b\}$ is a polytope.

The Projective Algorithm for the Linear Program (4.2)

Initialization. Given x^1 feasible in (4.2) with $x^1 > 0$, set $r^1 = (x^1, 1)$. If a specific upper bound on z_{LP} is known, set ζ_1 to be this bound. If none is known, take

$$\zeta_1 = \max_j |c_j| n\omega(A, b), \qquad \tilde{A}^1 = (A \quad -b)\begin{pmatrix} r_1^1 & & \\ & \ddots & \\ & & r_{n+1}^1 \end{pmatrix},$$

$$\tilde{c}_\zeta^1 = (c \quad -\zeta_1)\begin{pmatrix} r_1^1 & & \\ & \ddots & \\ & & r_{n+1}^1 \end{pmatrix}.$$

Iteration k

Step 1: Let $v_k = \zeta_k - cx^k$. If $v_k \leq \varepsilon$, stop. x^k is an ε-optimal solution of (4.2) with $Ax^k = b$, $x^k \in R_+^n$, and $z_{LP} - cx^k \leq \varepsilon$.

Step 2: Find the projection $q^k(\zeta)$ of 1 onto the subspace $\tilde{A}^k r = 0$, $\tilde{c}_\zeta^k r = 0$.

Step 3: If $q^k(\zeta_k) \geq 0$, stop. Let $r = D^k q^k(\zeta_k)$. Then $x = [r_1/r_{n+1}, \ldots, r_n/r_{n+1}]$ is optimal in (4.2).

Step 4: If $q^k(\zeta_k) < 1$, find $\zeta^* < \zeta_k$ such that $\max_{j=1,\ldots,n+1} q_j^k(\zeta^*) = 1$, and set $\zeta_{k+1} = \zeta^*$. Otherwise set $\zeta_{k+1} = \zeta_k$.

Step 5: Take $q^k = q^k(\zeta_{k+1})$, and find α_k as before.
Step 6: This is unchanged.

Proposition 4.8. *Suppose the projective algorithm is applied to problem (4.2) starting from a feasible point x^1. At iteration $k+1$, x^{k+1} is a feasible point and ζ_{k+1} is an upper bound on z_{LP} satisfying*

i. $\zeta_{k+1} - cx^{k+1} \leq \rho_k(\zeta_k - cx^k)$ *and*
ii. $\zeta_{k+1} - cx^{k+1} \leq \beta^k \omega(A, b)(\zeta_1 - cx^1)$.

Proof. i. We just need to verify what happens during one iteration of the algorithm. Initially $v_k = \zeta_k - cx^k = |\tilde{c}_{\zeta_k} r^k|$. After Step 5,

$$
\begin{aligned}
v_{k+1} &= \zeta_{k+1} - cx^{k+1} &&= |\tilde{c}_{\zeta_{k+1}} r^{k+1}| \\
&= \rho_k |\tilde{c}_{\zeta_{k+1}} r^k| &&= \rho_k(\zeta_{k+1} - cx^k) \quad \text{by statement i of Proposition 4.2} \\
&\leq \rho_k(\zeta_k - cx^k) &&= \rho_k v_k \quad \text{since } \zeta_{k+1} \leq \zeta_k \text{ from Step 4.}
\end{aligned}
$$

ii. This follows from statement ii of Proposition 4.7 because x^1 is feasible and hence bounded. ■

Example 4.2. The Projection Algorithm is applied to the linear program

$$
\begin{aligned}
z_{\text{LP}} = \max \quad & 7x_1 + 2x_2 \\
& -x_1 + 2x_2 + x_3 = 4 \\
& 5x_1 + x_2 + x_4 = 20 \\
& -2x_1 - 2x_2 + x_5 = -7 \\
& x \in R^5_+
\end{aligned}
$$

with $\varepsilon = 10^{-5}$.

We initialize with $x^1 = (1.01 \ \ 2.5 \ \ 0.01 \ \ 12.45 \ \ 0.02)$, $\zeta_1 = 200$. The calculations are shown in Table 4.1.

It can be shown that at each iteration the algorithm yields a dual feasible solution. Also, by combining the algorithms for problems (4.1) and (4.2), one can describe a single-phase primal–dual algorithm for linear programming. The remarkable convergence rate of the projective algorithm means that, in practice, never more than about 30 iterations seem to be necessary. The key practical question is how to carry out the projection step efficiently.

5. A STRONGLY POLYNOMIAL ALGORITHM FOR COMBINATORIAL LINEAR PROGRAMS

We generally bound the running time of an optimization or feasibility algorithm by the number of variables n, the number of constraints m, and the size of the largest coefficient θ in the data. Polynomial-time algorithms require a bound that is a polynomial function of m, n, and $\log \theta$. An algorithm is *strongly polynomial* if the bound only depends on m and n. Thus for the family of instances in which $\log \theta$ is a polynomial function of m and n, we can trivially eliminate the dependence on θ and say that the algorithm is strongly polynomial.

Table 4.1.

k	x^k					ζ_k	$\zeta_k - cx^k$	$\zeta_{k+1} - cx^k$	$u^k b$	α_k	β_k	ρ_k
1	1.01000	2.50000	0.01000	12.45000	0.02000	200.00000	187.93000	187.93000	152.86228	201.47566	0.172	0.930
2	2.89112	2.50302	1.88509	3.04137	3.78828	200.00000	174.75611	8.55008	32.48775	1.29514	0.703	0.577
3	3.37029	2.63617	2.09795	0.51237	5.01292	33.79397	4.92958	1.63481	30.34192	2.29413	0.601	0.376
4	3.33667	3.26400	0.80867	0.05263	6.20135	30.49920	0.61447	0.34921	30.21156	1.96888	0.565	0.371
5	3.27225	3.59924	0.07377	0.03951	6.74298	30.23393	0.12970	0.07829	30.18217	3.71527	0.383	0.213
6	3.27690	3.61377	0.04936	0.00175	6.78133	30.18251	0.01670	0.01620	30.18193	4.97273	0.307	0.168
7	3.27261	3.63549	0.00162	0.00144	6.81622	30.18201	0.00273	0.00253	30.18182	5.02502	0.305	0.166
8	3.27284	3.63577	0.00129	0.00005	6.81722	30.18182	0.00042	0.00042	30.18182	5.37459	0.291	0.157
9	3.27272	3.63634	0.00004	0.00004	6.81813	30.18182	0.00007	0.00007	30.18182	5.35710	0.292	0.157
10	3.27273	3.63635	0.00003	0.00000	6.81816	30.18182	0.00001	0.00001	30.18182	5.36843	0.291	0.157

Here we show that for the linear programming problem

$$\max\{cx: Ax \leq b, x \geq 0\}$$

it is possible to eliminate the dependence of the running time on θ_b and θ_c for any algorithm that finds primal and dual feasible and complementary solutions. This is done by replacing the original problem by a sequence of problems in which the coefficients b and c are a polynomial function of m, n, and $\log \theta_A$. In particular, the bounds in the projective and ellipsoid algorithm will depend only on m, n, and $\log \theta_A$.

Also, if A is a (0, 1)-matrix, as in a network flow problem or the linear programming relaxation of a set covering problem, a polynomial-time linear programming algorithm can be refined to a strongly polynomial algorithm. For example, the primal–dual algorithm for the transportation problem given in Section I.3.5 enjoys this property.

First we will show how the dependence on c is eliminated. The problem is reduced to solving a sequence of no more than n linear programs in each of which the coefficients of the cost vector are bounded by $n^2\Delta$, where Δ is an upper bound on $|\Delta(A)|$, the maximum absolute value of any subdeterminant of A. We will assume throughout this section that either $\Delta(A)$ is known and we set $\Delta = \Delta(A)$ or that $\Delta = (n\theta_A)^n$ which, given $m \leq n$, is known to be an upper bound on $\Delta(A)$. Now $\log(n^2\Delta)$ is a polynomial function of m, n, and $\log \theta_A$, and so the dependence on c disappears.

We consider linear programs of the form

$$\max\{cx: x \in P\}, \quad \text{where } P = \{x \in R^n_+: Ax = b\}. \tag{5.1}$$

The dual problem is

$$\min\{ub: u \in U\}, \quad \text{where } U = \{u \in R^m: uA \geq c\}. \tag{5.2}$$

The first result we need concerns primal and dual solutions (x, u) that are close to satisfying the optimality conditions. We let $U(\varepsilon) = \{u \in R^m: uA \geq c - \varepsilon 1\}$.

Definition 5.1. (x, u) are a pair of *ε-approximate solutions* for problem (5.1) if

i. $x \in P$
ii. $u \in U(\varepsilon)$;
iii. if $ua_j > c_j$, then $x_j = 0$.

Proposition 5.1. *Given an ε-approximate pair (x, u) for problem (5.1), let $J = \{j \in N: ua_j \geq c_j + \varepsilon n\Delta\}$. Then if x^* is any optimal solution to (5.1), $x^*_j = 0$ for all $j \in J$.*

Proof. Suppose the contrary, so there exists a k such that $ua_k \geq c_k + \varepsilon n\Delta$ and an optimal solution x^* to (5.1) with $x^*_k > 0$.

The vector $z = x^* - x$ satisfies $Az = 0$ because $Ax^* = Ax = b$, $z_k > 0$ since $ua_k > c_k$ implies $x_k = 0$; also, $z_j \geq 0$ whenever $x_j = 0$ because $x^*_j \geq 0$. Hence z is a feasible solution to

$$\begin{aligned} &Az = 0 \\ &z_k > 0 \\ &z_j \geq 0 \quad \text{for all } j \text{ with } x_j = 0. \end{aligned} \tag{5.3}$$

Moreover, $cz \geq 0$ because x^* is optimal to (5.1).

Now there exists an integer basic feasible solution $\bar{z}$ to (5.3) with $c\bar{z} \geq 0$. Furthermore, we know from Cramer's rule that $\max_j |\bar{z}_j| \leq \Delta(A) \leq \Delta$. In addition, if $\bar{z}_j < 0$ we know that $x_j > 0$ and hence, by condition iii of Definition 5.1, that $ua_j \leq c_j$.

Now

$$\begin{aligned}
c\bar{z} &= (c - uA)\bar{z} && \text{since } A\bar{z} = 0 \\
&\leq (c_k - ua_k)\bar{z}_k + \varepsilon \sum_{j \neq k} |\bar{z}_j| && \text{since } u \in U(\varepsilon) \text{ implies } c_j - ua_j \leq \varepsilon \\
&\leq -\varepsilon n\Delta + \varepsilon(n-1)\Delta && \text{since } \bar{z}_k \geq 1,\ |z_j| \leq \Delta \\
&= -\varepsilon\Delta
\end{aligned}$$

However, $c\bar{z} \geq 0$, so there is a contradiction. ■

A simple way to obtain 1-approximate solutions for (5.1) is to let $c' = \lfloor c \rfloor$ and to solve the linear program

$$(5.4) \qquad \max\{c'x\colon x \in P\}.$$

Proposition 5.2. *Let (x, u) be an optimal solution pair for the linear program (5.4), then (x, u) is a 1-approximate pair for the linear program (5.1).*

Proof. Since x is optimal in (5.4), $x \in P$. Since u is dual feasible,

$$uA \geq \lfloor c \rfloor = c - (c - \lfloor c \rfloor) \geq c - 1.$$

Finally, $ua_j > c_j$ implies $ua_j > \lfloor c_j \rfloor$; and hence, by complementary slackness for (5.4), $x_j = 0$. ■

Combining Propositions 5.2 and 5.1, we obtain

Corollary 5.3. *Let u be a dual optimal solution of (5.4) and let $J = \{j \in N\colon ua_j \geqslant c_j + n\Delta\}$. If x^* is any optimal solution to (5.1), then $x_j^* = 0$ for all $j \in J$.*

Proof. Let (x, u) be an optimal solution pair for (5.4). Then by Proposition 5.2, (x, u) is a 1-approximate pair for (5.1). Hence by Proposition 5.1, if $J = \{j \in N\colon ua_j \geq c_j + n\Delta\}$ and x^* is any optimal solution to (5.1), then $x_j^* = 0$ for all $j \in J$. ■

Example 5.1. We consider the transportation problem of Example 5.1 of Section I.3.5 with weights $w'_{ij} = 4w_{ij} + \phi_{ij}$ for all i and j, where $0 \leq \phi_{ij} < 1$. Since the constraint matrix A of a transportation problem is totally unimodular (see Section III.1.2), $\Delta = 1$.

To find a 1-approximate pair (x^*, u^*), we solve problem (5.4), which has weights $c_{ij} = \lfloor w'_{ij} \rfloor = 4w_{ij}$. From Section I.3.5 we know a dual optimal solution of (5.4):

$$u^* = (24 \quad 36 \quad 8 \quad 36)$$

$$v^* = (-16 \quad -12 \quad 4 \quad -16 \quad -8 \quad -4)$$

with

$$\bar{c} = \begin{pmatrix} 12 & 0 & 0 & 4 & 16 & 0 \\ 0 & 0 & 8 & 0 & 0 & 12 \\ 16 & 36 & 0 & 24 & 32 & 4 \\ 24 & 0 & 0 & 0 & 12 & 4 \end{pmatrix}.$$

Now applying Corollary 5.3, we see that $n\Delta = 24$. Hence $J = \{(i, j): \bar{c}_{ij} \geq 24\}$, and $x_{32} = x_{34} = x_{35} = x_{41} = 0$ in any optimal solution.

In Example 5.1, 4 of the 24 variables were eliminated on the basis of Corollary 5.3. The next proposition shows that if the weight vector c is appropriately normalized and scaled, it is always possible to set at least one variable to zero. Hence after applying the procedure no more than n times, the original linear program is solved.

Definition 5.2. For any Δ satisfying $\Delta(A) \leq \Delta \leq (n\theta)^n$, we say that c is *polynomially normalized* for (5.1) if $c = 0$ or $Ac = 0$ and $\max_j |c_j| = n^2\Delta$.

Proposition 5.4. *Given the linear program (5.1) with $n > 1$ and a polynomially normalized weight vector $c \neq 0$, let u be a dual optimal solution to the linear program (5.4). Then $J = \{j: ua_j > c_j + n\Delta\} \neq \emptyset$.*

Proof. Let $\tilde{c}_j = ua_j - c_j$. We need to show that $\max_j \tilde{c}_j \geq n\Delta$.
Since u is dual feasible in (5.4), $ua_j \geq c'_j = \lfloor c_j \rfloor \geq c_j - 1$, and hence $\tilde{c}_j \geq -1$ for all j.
The projection of $\tilde{c}$ onto $H = \{x: Ax = 0\}$ is the vector $-c$ since $uA \in H^\perp$ and $-c \in H$. Therefore, $\|\tilde{c}\| \geq \|c\|$.
Now observe that

$$\|\tilde{c}\|^2 = \sum_{j=1}^{n} \tilde{c}_j^2 \leq n \left(\max_j |\tilde{c}_j|\right)^2 < \left(n \max_j |\tilde{c}_j|\right)^2.$$

Therefore

$$\max_j |\tilde{c}_j| \geq \frac{1}{n} \|\tilde{c}\| \geq \frac{1}{n} \|c\| \geq \frac{1}{n} \left(\max_j |c_j|\right) = \frac{1}{n} n^2\Delta = n\Delta.$$

Since $\tilde{c}_j \geq -1$ for all j, we must have $\max_j \tilde{c}_j \geq n\Delta$ as required. ■

The final step is to show that the objective function vector of any linear program (5.1) can be put in polynomially normalized form without affecting the set of optimal solutions.

An Algorithm to Polynomially Normalize c

Step 1: Find the projection $\bar{c}$ of c onto $H = \{x: Ax = 0\}$. If $\bar{c} = 0$, set $d = 0$ and stop.

Step 2: Set $\alpha = n^2\Delta/\max_j |\bar{c}_j|$ and $d = \alpha\bar{c}$, where $\Delta(A) \leq \Delta \leq (n\theta_A)^n$. d is the required objective function vector.

Proposition 5.5. *Let $F(c)$ be the face of P of optimal solutions to (5.1) with weight vector c, and let d be as derived above:*

i. $F(c) = F(d)$
ii. *If* $d = 0$, *then* $F(c) = P$.

Proof. i. The scaling in Step 2 does not affect the set of optimal solutions, so $F(\bar{c}) = F(d)$. Since $\bar{c}$ is the projection of c onto H, and $H = \{z: z = yA, y \in R^m\}$, it follows that $\bar{c} = c - yA$ for some $y \in R^m$. But for all $x \in P$, we have $\bar{c}x = cx - yAx = cx - yb$. Since yb is a constant, x is optimal for $\bar{c}$ if and only if it is optimal for c. Hence $F(c) = F(\bar{c}) = F(d)$.
ii. $F(0) = P$, and hence $F(c) = P$. ■

Now we can describe the algorithm.

An "Objective Rounding" Algorithm for the Linear Program (5.1) with $P \neq \emptyset$

Step 1 (Initialization): $N^1 = N, t = 1, n_1 = |N^1|$.
Step 2 (Iteration t): Consider the linear program

$$(5.1^t) \qquad \max\left\{cx: \sum_{j \in N^t} a_j x_j = b, x_j \in R_+^1 \text{ for } j \in N^t\right\}.$$

a. Put c into polynomially normalized form with $\Delta(A) \leq \Delta \leq (n\theta_A)^n$.
b. If $c = 0$, stop. The feasible solutions of (5.1^t) are the set of optimal solutions to (5.1).
c. Otherwise solve the linear program

$$(5.4^t) \qquad \max\left\{\sum_{j \in N^t} \lfloor c_j \rfloor x_j: \sum_{j \in N^t} a_j x_j = b, x_j \in R_+^1 \text{ for } j \in N^t\right\}.$$

d. If (5.4^t) is unbounded, stop. (5.1) is unbounded (since if $ua_j \geq \lfloor c_j \rfloor$ for $j \in N^t$ is infeasible, then $uA \geq c$ is infeasible).
e. Otherwise, let u^t be an optimal dual solution and let

$$J^t = \{j \in N^t: u^t a_j > c_j + n_t \Delta\}$$
$$N^{t+1} = N^t \setminus J^t, \qquad n_{t+1} = |N^{t+1}|.$$

Set $t \leftarrow t + 1$

Theorem 5.6. *If the linear program (5.1) is feasible, the objective rounding algorithm either shows that (5.1) is unbounded or it terminates with the set of optimal feasible solutions to (5.1) after no more than n iterations.*

Example 5.2. We apply the objective rounding algorithm to the feasible linear programming problem (5.1)

$$\max 931x_1 + 724x_2 + 296x_3$$
$$8x_1 + 5x_2 + 3x_3 = 527$$
$$x \in R_+^3$$

We initialize with $\Delta = \Delta(A) = 8$, $N^1 = \{1, 2, 3\}$, $n_1 = 3$, $t = 1$.

Iteration 1. Consider $(5.1^1) = (5.1)$.

a. Projecting c onto $8x_1 + 5x_2 + 3x_3 = 0$, we obtain $\bar{c} = (-45 \quad 114 \quad -70)$. Also, $\max_j |c_j| = 114$. Therefore $\alpha = n^2\Delta/\max|c_j| = 72/114$. Now $d = \alpha\bar{c} = (-28.42 \quad 72 \quad -44.21)$.
b. $d \neq 0$.
c. We solve the linear program (5.4^1) with objective function $\lfloor d \rfloor$,

$$\begin{aligned} \max \; -29x_1 + 72x_2 - 45x_3 & \\ 8x_1 + \;\; 5x_2 + \;\; 3x_3 &= 527 \\ x &\geq 0, \end{aligned}$$

giving an optimal dual solution $\bar{u} = 72/5$.
d. Now

$$\begin{aligned} \bar{u}a_1 - d_1 &= \frac{576}{5} + 28.42 > n\Delta = 24 \\ \bar{u}a_2 - d_2 &= 0 \\ \bar{u}a_3 - d_3 &= \frac{216}{5} + 44.21 > 24. \end{aligned}$$

Hence $J_1 = \{1, 3\}$, $N^2 = \{2\}$, $n_2 = 1$, $t \leftarrow 2$.

Iteration 2. Solve (5.1^2)

$$\max\{724x_2\colon 5x_2 = 527,\; x_2 \in R^1_+\}.$$

a. Projecting $c = (724)$ onto $5x_2 = 0$, we obtain $\bar{c} = (0)$.
b. The set of optimal solutions to (5.1) is given by

$$\{x \in R^3_+\colon 8x_1 + 5x_2 + 3x_3 = 527,\; x_1 = x_3 = 0\}.$$

Given a feasible linear program (5.1), we have seen how the dependence of the running time on c can be eliminated. Now we consider the dependence on b. Note that the running time of the linear program (5.4^t) still depends on b. Furthermore, although the objective rounding algorithm terminates with the face $P^* = \{x \in R^n_+\colon Ax = b,\; x_j = 0, j \in N \setminus N^*\}$ of optimal solutions, finding an optimal solution still involves finding a feasible solution to a linear program when the face P^* is not just a point.

So now we consider the elimination of the dependence on b. For convenience we consider inequalities here, that is $P = \{x \in R^n\colon Ax \leq b\}$. Also, we first consider the *feasibility problem*:

$$\text{Is } P = \emptyset\text{? If not, find } x^* \in P.$$

We will need the following result, which makes precise how to perturb the objective function so that all dual feasible basic solutions are nondegenerate. (Applied to b, this is precisely the perturbation approach to avoiding cycling in the simplex algorithm; see Section I.2.3).

Proposition 5.7. *Consider the linear program* $\max\{cx: x \in P\}$ *with* $P = \{x \in R^n: Ax \leq b\}$ *and* $c = \sum_{i=1}^m (\Delta + 1)^i a^i$. *Suppose its dual* $\min\{ub: uA = c,\ u \in R_+^m\}$ *has a finite optimal solution* u^*. *Let* $I^* = \{i \in M = \{1, \ldots, m\}: u_i^* > 0\}$ *and let* x^* *be any solution of* $a^i x = b_i$ *for* $i \in I^*$. *Then* x^* *is an optimal solution to the linear program.*

Feasibility Algorithm for $P = \{x \in R^n: Ax \leq b\}$.

1. Take c as in Proposition 5.7. Note that c is polynomial in the size of A.
2. The dual problem

   $$\min\{ub: uA = c,\ u \in R_+^m\} \tag{5.5}$$

   has a feasible solution $u = (\Delta + 1, (\Delta + 1)^2, \ldots, (\Delta + 1)^m)$. Hence we can apply the objective rounding algorithm to (5.5).
3. If the dual problem is unbounded, then $P = \emptyset$.
4. Otherwise the algorithm terminates with the face of optimal solutions $\{u \in R_+^m: uA = c,\ u_i = 0 \text{ for } i \in M \setminus I'\}$.
5. Find a point $\bar{u}$ in this face using a linear programming algorithm. Let $I^* = \{i: \bar{u}_i > 0\}$.
6. Use Gaussian elimination to find a solution x^* to the system of linear equations $a^i x = b_i$ for $i \in I^*$.
7. By Proposition 5.7, $x^* \in P$.

Now observe that in Step 2 when we apply the objective rounding algorithm, the subproblems (5.4′) have a modified objective $\tilde{b}$ and the right-hand side c that are polynomial in m, n, and $\log \theta_A$. Finally, in Step 6 the running time of Gaussian elimination does not depend on b. Hence, if a polynomial-time algorithm is used to solve each of the linear programs, Steps 1–7 can be executed in polynomial time.

Proposition 5.8. *The feasibility problem can be solved in time polynomial in m, n, and* $\log \theta_A$.

Putting together the objective rounding and feasibility algorithms it is now possible to solve any linear program in time polynomial in m, n, and $\log \theta_A$.

The only question that needs to be dealt with is the resolution of the feasible linear programs (5.4′) in Step c of the objective rounding algorithm, which are of the form

$$\max\{\tilde{c}^t x: Ax = b,\ x \in R_+^n\},$$

where $\tilde{c}^t$ is polynomial in the size of A. The steps are:

1. Check whether $\{u \in R^m: uA \geq \tilde{c}^t\}$ is feasible with a polynomial linear programming algorithm. If not, the primal is unbounded.
2. Use the objective rounding algorithm to solve the dual of (5.4′): $\min\{ub: uA \geq \tilde{c}^t\}$. Here the subproblems are linear programs with both the modified objective $\tilde{b}$ and the right-hand side $\tilde{c}^t$ polynomial in the size of A. The basic algorithm terminates with the face of optimal solutions.

3. Use a polynomial linear programming algorithm to find an optimal solution $\overline{u}^t$ lying on the face.
4. An optimal dual solution $\overline{u}^t$ for problem (5.4′) is precisely what is required in Step e of the objective rounding algorithm.

Again it is easy to check that the data for each problem solved are polynomial in the size of A. Hence we have shown the following:

Theorem 5.9. *The linear programming problem can be solved in time polynomial in m, n, and* log θ_A.

6. NOTES

Section I.6.1

Until the summer of 1979, it was not known whether linear programming was solvable in polynomial time. However, there were theoretical and empirical reasons for believing that there existed a polynomial-time algorithm for linear programming. It had been known for several years that linear programming belonged to the complexity class $\mathcal{NP} \cap \mathcal{C}o\mathcal{NP}$, and although the simplex method does not have a polynomial time bound (see Klee and Minty, 1972), it is empirically very efficient.

Section I.6.2

An ellipsoid algorithm for linear programming, as well as a proof of its polynomial-time bound, first appeared in a brief Russian article by Khachian (1979). It was brought to the attention of Western researchers at a meeting in Oberwolfach, West Germany in June 1979. An English version of Khachian's results, including many missing details, was produced by Gacs and Lovasz (1981). Their article appeared as a technical report toward the end of the summer of 1979, and the results were announced to the research community at the Xth Mathematical Programming Symposium in Montreal, Canada in August 1979. A flood of articles on ellipsoid algorithms subsequently appeared [see Bland, Goldfarb, and Todd (1981) for a survey]. However, the flood subsided nearly as quickly as it had appeared when it was realized that ellipsoid algorithms were not empirically efficient. Lawler (1980) discussed the reaction of the popular press.

Section I.6.3

"The use of ellipsoid algorithms in combinatorial optimization—in particular, the polynomial equivalence of optimization and separation—is due to Grötschel, Lovasz, and Schrijver (1981) and was also studied by Karp and Papadimitriou (1982). Also see Grötschel, Lovasz, and Schrijver (1984a,b,c); and their monograph (1987), which gives all of the technical details.

Section I.6.4

Projective algorithms for linear programming and their polynomiality were introduced by Karmarkar (1984). The variant of Karmarkar's approach given here is taken from de Ghellinck and Vial (1986, 1987).

The choice of homogenization $\tilde{r} = r/\Pi_{j=1}^{n} r_j^{1/n}$ is one of the key ideas in Karmarkar's algorithm. It encourages the successive iterates to stay away from the boundaries and thereby avoid the combinatorial problems associated with extreme points. This idea had been used earlier by Huard (1967) in the method of centers and in Frisch's (1955) barrier method for nonlinear programming.

Also, de Ghellinck and Vial (1987) show that the projective algorithm to find an ε-approximate ray produces the same sequence of points as Karmarker's algorithm applied to the phase I problem:

$$\min\{y: Ax - (A\,1)y = 0,\ 1x + y = n + 1,\ x \in R^n_+,\ y \in R^1_+\}.$$

The idea of a primal–dual approach to Karmarkar's algorithm is due to Todd and Burell (1986).

At this time there is some indication that projective algorithms may compete with, or be superior to, simplex algorithms for some, or maybe even all, classes of linear programs. This issue is unresolved and is the subject of some controversy. Todd (1987) gives a survey of results on this topic.

Section I.6.5

The strongly polynomial-time algorithm for linear programming is due to Tardos (1986). This article was a sequel to her article on a strongly polynomial-time algorithm for network flow problems [Tardos (1985); also see Orlin (1984)]. Fujishige (1986) and Orlin (1986) discuss dual versions of Tardos' algorithm. Frank and Tardos (1987) have extended these results to linear programs in which the number of constraints is not polynomially bounded in the number of variables.

I.7
Integer Lattices

1. INTRODUCTION

In this chapter the basic problem is:

The Linear Equation Integer Feasibility Problem. Given m, n, and an integral $m \times (n + 1)$ matrix (A, b), find a point $x \in Z^n$ satisfying $Ax = b$ or show that no such point exists.

Definition 1.1 The set $L(A) = \{y \in R^m: y = Ax, x \in Z^n\}$, where A is an $m \times n$ matrix, is called the *lattice* generated by the columns of A.

In lattice terms, the linear equation integer feasibility problem becomes

(1.1) Determine if $b \in L(A)$ and if so give a representation of b as an integral linear combination of the columns of A.

A natural generalization of problem (1.1) is:

The Closest Vector Problem. Given m, n, and A as above, along with $b \in R^m$, find

$$\min_y\{\|b - y\|: y \in L(A)\}. \tag{1.2}$$

Taking $b = 0$ and excluding $y = 0$ yields:

The Shortest Vector Problem: Given m, n, and A, find

$$\min_y\{\|y\|: y \in L(A), y \neq 0\}. \tag{1.3}$$

Example 1.1. In Figure 1.1 we see the lattice generated by $\begin{pmatrix} 7 & 6 \\ 6 & 4 \end{pmatrix}$. The closest lattice point to $b = \begin{pmatrix} 6\frac{1}{2} \\ 1\frac{1}{4} \end{pmatrix}$ is $y \binom{5}{2}$, and the shortest vector is $v = \binom{1}{2}$.

An important difference between problem (1.1) and problems (1.2) and (1.3) is their respective complexities. There is a polynomial-time algorithm for solving (1.1), whereas (1.2) is $\mathcal{NP}$-hard, and (1.3) is suspected to be $\mathcal{NP}$-hard. Also, the problems obtained by replacing the euclidean norm by the maximum norm in (1.2) and (1.3) are $\mathcal{NP}$-hard. However, there are algorithms for problems (1.2) and (1.3) that run in polynomial time for a fixed value of n. Since these algorithms and the algorithm for (1.1) depend upon finding an appropriate representation of the lattice $L(A)$, this is the main theme of the chapter.

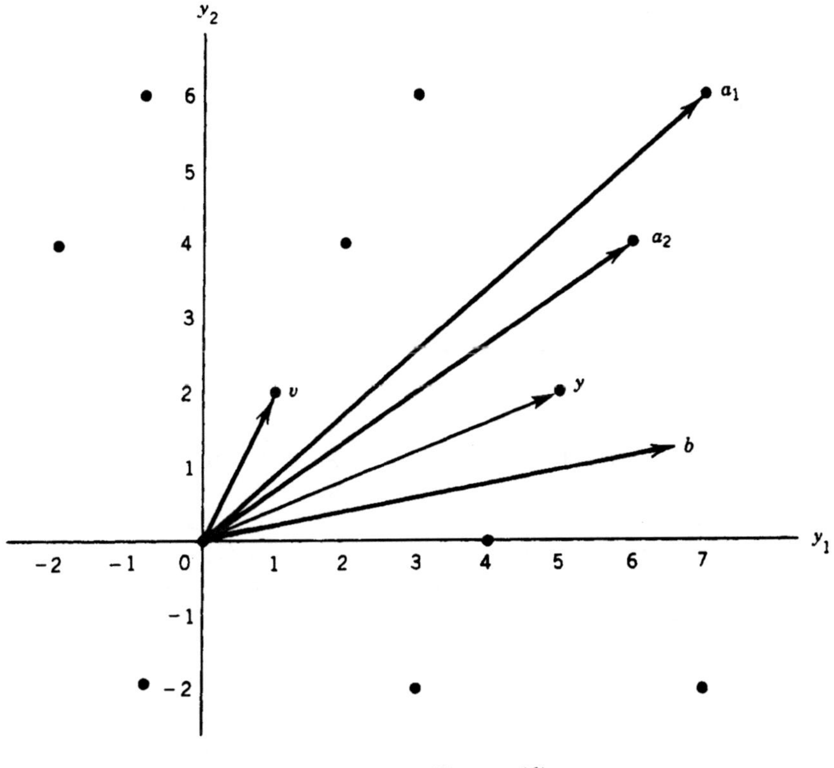

Figure 1.1. $a_1 = \begin{pmatrix} 7 \\ 6 \end{pmatrix}$, $a_2 = \begin{pmatrix} 6 \\ 4 \end{pmatrix}$

Definition 1.2. The *greatest common divisor* of the integers a_1 and a_2, not both zero, denoted by $\gcd(a_1, a_2)$, is the largest positive integer r such that r divides a_1 and a_2 exactly; that is, there exist integers z_i such that $rz_i = a_i$ for $i = 1, 2$. If $\gcd(a_1, a_2) = 1$, then a_1 and a_2 are *relatively prime*.

When $n = 1$, problems (1.1)–(1.3) essentially reduce to the problem of finding $\gcd(a_1, a_2)$. In the next section, we will describe the euclidean algorithm to find $\gcd(a_1, a_2)$. This will be interpreted as an algorithm that either solves the system

$$a_1x_1 + a_2x_2 = a_0, \quad x \in Z^2 \tag{1.4}$$

or shows it to be infeasible. It also provides the basic step in finding alternative descriptions of the lattice $L(A)$.

In Section 3, we will establish the connection between the euclidean algorithm and the continued fraction expansion of a rational number a_1/a_2. This allows us to solve, in polynomial time, the following problem.

The Rational Approximation Problem. Given positive integers a_1, a_2, and K, determine integers p and q that solve

$$\min\left\{\left|\frac{a_1}{a_2} - \frac{q}{p}\right| : (p, q) \in Z_+^2, p \leq K\right\}.$$

A solution to this problem is required in the ellipsoid algorithm (see Section I.6.2).
An important generalization of this problem is:

The Simultaneous Diophantine Approximation Problem. Given n positive rationals $\{\alpha_i\}_{i=1}^n$ and an integer K, determine positive integers $q_1, \ldots, q_n$ and p that solve

$$\min\left\{\max_i\left|\alpha_i - \frac{q_i}{p}\right| : p \in Z_+^1, q \in Z_+^n, p \leqslant K\right\}. \tag{1.5}$$

This problem is $\mathcal{NP}$-hard.

In Section 4, we will introduce some basic properties of the lattice $L(A)$. We will then develop a canonical representation of $L(A)$, called the *Hermite normal form*, and sketch a polynomial-time algorithm for finding the Hermite normal form. This also provides a polynomial-time algorithm for problem (1.1).

In Section 5, we will introduce an alternative representation of $L(A)$, called a *reduced basis*, which can also be obtained in polynomial time. Such bases have various interesting properties. We use them to give an algorithm for problem (1.3) that is polynomial for fixed n and a polynomial algorithm for an approximate version of (1.5). Other applications—in particular, an outline of a polynomial-time algorithm for the linear inequality integer feasibility problem with a fixed number of variables—will be given in Section II.6.5.

2. THE EUCLIDEAN ALGORITHM

In this section we present a polynomial algorithm to find $\gcd(a, b)$, where a and b are integers satisfying $a \geqslant b > 0$. We use the notation $u \mid v$ to mean that u divides v. The algorithm will terminate with integers p and q such that p and q are relatively prime and $pa - qb = \gcd(a, b)$.

Note that for any positive integers a and b, with $b \leqslant a$, we have $a = \lfloor a/b \rfloor b + c$, where $0 \leqslant c < b$. The basic idea of the euclidean algorithm is embodied in the following proposition.

Proposition 2.1. *Suppose a and b, $a \geqslant b$, are positive integers and $c = a - \lfloor a/b \rfloor b$.*

i. *If $c \neq 0$, then* $\gcd(a, b) = \gcd(b, c)$.
ii. *If $c = 0$, then* $\gcd(a, b) = b$.

Proof. i. Let $r = \gcd(a, b)$ and $s = \gcd(b, c)$. Since $c = a - db$ and d is an integer, $r \mid a$ and $r \mid b$ imply that $r \mid c$. Hence $r \mid s$. Similarly $s \mid r$.
ii. This is obvious. ∎

Note that because $c < b \leqslant a$, we can apply the proposition first to the pair (a, b), then to the pair (b, c), and so on.

In the description of the euclidean algorithm given below, c_t is the remainder at iteration t. We also carry along integers (p_t, q_t), which will be used later.

The Euclidean Algorithm To Find gcd(a, b)

Initialization: Order so that $a \geqslant b$.
$(c_{-1}, c_0) = (a, b)$, $(p_{-1}, p_0) = (1, 0)$, $(q_{-1}, q_0) = (0, 1)$.
Set $t = 1$.

Iteration t: $d_t = \lfloor \frac{c_{t-2}}{c_{t-1}} \rfloor$
$c_t = c_{t-2} - d_t c_{t-1}$
$p_t = p_{t-2} + d_t p_{t-1}$
$q_t = q_{t-2} + d_t q_{t-1}$
If $c_t = 0$, stop. Set $T = t$. $\gcd(a, b) = c_{T-1}$.
Otherwise set $t \leftarrow t + 1$.

Proposition 2.2. *The euclidean algorithm is correct and*

$$c_t = (-1)^{t+1}[p_t a - q_t b] \quad \text{for } t = -1, 0, \ldots, T.$$

Proof. We use Proposition 2.1. By using statement i repeatedly, we have $\gcd(a, b) = \gcd(c_{-1}, c_0) = \gcd(c_0, c_1) = \cdots = \gcd(c_{T-2}, c_{T-1})$. Since $c_{T-2} - d_T c_{T-1} = 0$, we note that $\gcd(c_{T-2}, c_{T-1}) = c_{T-1}$ follows from statement ii.

Note also that

$$c_{-1} = (-1)^0 (p_{-1}a - q_{-1}b) = a$$

and

$$c_0 = (-1)^1 (p_0 a - q_0 b) = b.$$

Thus by induction,

$$\begin{aligned} c_t &= c_{t-2} - d_t c_{t-1} \\ &= (-1)^{t-1}(p_{t-2}a - q_{t-2}b) - (-1)^t d_t(p_{t-1}a - q_{t-1}b) \\ &= (-1)^{t+1}[(p_{t-2} + d_t p_{t-1})a - (q_{t-2} + d_t q_{t-1})b] \\ &= (-1)^{t+1}[p_t a - q_t b]. \end{aligned}$$

■

Example 2.1. Find gcd(51, 36). Using the euclidean algorithm we get

t	d_t	c_t	p_t	q_t
−1		51	1	0
0		36	0	1
1	1	15	1	1
2	2	6	2	3
3	2	3	5	7
4	2	0	12	17

Hence

$$T = 4, \gcd(51, 36) = 3,$$

$$(-1)^4(5 \cdot 51 - 7 \cdot 36) = 3,$$

and

$$(-1)^5(12 \cdot 51 - 17 \cdot 36) = 0, \text{ etc.}$$

To show that the euclidean algorithm runs in polynomial time, we need to show that the number of iterations and the size of the numbers produced are polynomial in $\log(a)$. For later use, we also consider the values taken by p_t, q_t as t increases.

Proposition 2.3

i. $(p_t q_{t+1} - p_{t+1} q_t) = (-1)^{t+1}$ *for* $t = -1, 0, \ldots, T$;
ii. $\gcd(p_t, q_t) = 1$ *for* $t = 1, \ldots, T$;
iii. $a/b = q_T/p_T$;
iv. $F_T \leqslant q_T \leqslant a$, *where* F_t *is the Fibonacci number given by* $F_{-1} = 0$, $F_0 = 1$, *and* $F_t = F_{t-2} + F_{t-1}$ *for* $t = 1, \ldots, T$.

Proof. i. For $t = -1$, we obtain $(p_t q_{t+1} - p_{t+1} q_t) = 1 = (-1)^{t+1}$. Using induction,

$$\begin{aligned}(p_t q_{t+1} - p_{t+1} q_t) &= p_t(q_{t-1} + d_t q_t) - (p_{t-1} + d_t p_t) q_t \\ &= (-1)(p_{t-1} q_t - p_t q_{t-1}) = (-1)(-1)^t \\ &= (-1)^{t+1}.\end{aligned}$$

ii. Since $p_{t-1} q_t - p_t q_{t-1} = (-1)^t$ and $p_t, q_t > 0$ for $t \geq 1$, it follows that $\gcd(p_t, q_t) = 1$.
iii. $c_T = 0 = p_T a - q_T b$. Hence $q_T/p_T = a/b$.
iv. Since d_t is a positive integer for all $t > 0$, it follows that

$$q_t = q_{t-2} + d_t q_{t-1} \geqslant q_{t-2} + q_{t-1}.$$

Since $F_{-1} = q_{-1} = 0$, $F_0 = q_0 = 1$, and $F_t = F_{t-2} + F_{t-1}$, we have $q_t \geqslant F_t$ for all t. Finally we observe from statements ii and iii that $q_T \leqslant a$. ∎

Proposition 2.4. *The euclidean algorithm runs in polynomial time.*

Proof. Since the Fibonnaci series grows exponentially fast and $F_T \leqslant q_T \leqslant a$, it follows that T is, at most, $O(\log a)$. Furthermore, the size of the numbers p_t and q_t never exceeds the size of a. ∎

Now consider the equation

$$a_1 x_1 + a_2 x_2 = a_0, \quad x \in Z^2. \tag{2.1}$$

Proposition 2.5. *Let* $r = \gcd(a_1, a_2)$ *with* $r = pa_1 - qa_2$, *where* p *and* q *are relatively prime. Equation (2.1) has a solution if and only if* $r \mid a_0$. *If* $r \mid a_0$ *the set of solutions of (2.1) is described by*

$$x = \begin{pmatrix} x_1 \\ x_2 \end{pmatrix} = \frac{a_0}{r}\begin{pmatrix} p \\ -q \end{pmatrix} + \frac{z}{r}\begin{pmatrix} a_2 \\ -a_1 \end{pmatrix}, \quad z \in Z^1.$$

Proof. For any x satisfying (2.1), we obtain $(a_1/r)x_1 + (a_2/r)x_2 = a_0/r$. Since $(a_1/r)x_1 + (a_2/r)x_2 \in Z^1$, (2.1) is infeasible if $(a_0/r) \notin Z^1$. On the other hand, if $a_0/r \in Z^1$, then $x^* = (a_0/r)\binom{p}{-q}$ is a solution, since $pa_1 - qa_2 = r$. But any solution can be written as

$x = x^* + y$, where $a_1y_1 + a_2y_2 = 0$, $y \in Z^2$. Hence $y = y_1\binom{1}{-a_1/a_2} \in Z^2$. Since $\gcd(a_1, a_2) = r$, it follows that y_1 must be a multiple of a_2/r, and x is as claimed. ■

Example 2.1. (continued). We determine the set of solutions to

$$51x_1 + 36x_2 = 27, \quad x \in Z^2.$$

By Proposition 2.5 with $p = 5$, $q = 7$, and $r = 3$, the solution set is nonempty because $3 \mid 27$. The complete set of solutions is

$$\begin{pmatrix} x_1 \\ x_2 \end{pmatrix} = 9\begin{pmatrix} 5 \\ -7 \end{pmatrix} + z\begin{pmatrix} 12 \\ -17 \end{pmatrix}, \quad z \in Z^1.$$

3. CONTINUED FRACTIONS

The problem of finding the gcd of two positive integers a and b is equivalent to the problem

$$\min\left\{p: \frac{a}{b} - \frac{q}{p} = 0,\ p, q > 0 \text{ and integer}\right\}.$$

By adding the constraint $p \leqslant K$, we obtain the *diophantine approximation problem*

$$\min\left\{\left|\frac{a}{b} - \frac{q}{p}\right|: p, q > 0 \text{ and integer},\ p \leqslant K\right\}. \tag{3.1}$$

This problem arose in Section I.6.2, where we needed to find the rational point $\hat{x} = (q_1/p_1, \ldots, q_n/p_n)$ nearest to x^* with p_i and q_i being integers for $i = 1, \ldots, n$ and $1 \leqslant p_i \leqslant K$.

To solve this problem we need to represent rationals as continued fractions.

Definition 3.1. Given a rational β, its *continued fraction expansion*, denoted by $\langle d_1, \ldots, d_j \rangle$, is an expression of the form

$$\beta = d_1 + \cfrac{1}{d_2 + \cfrac{1}{d_3 + \cfrac{1}{\ddots \, d_{j-1} + \cfrac{1}{d_j}}}}$$

where $d_1, \ldots, d_j$ are integers, all positive except possibly d_1.

We also use partial expansions of the form

$$\beta = d_1 + \cfrac{1}{d_2 + \cfrac{1}{d_3 + \cdots + \cfrac{1}{d_{i-1} + \cfrac{1}{x}}}}$$

with $x \geqslant 1$, denoted by $<d_1, \ldots, d_{i-1}; x>$.

Example 3.1. Let $\beta = a/b = \frac{51}{36}$.

$$\begin{aligned}
\beta &= 1 + \frac{1}{36/15} = <1; \tfrac{36}{15}> \\
&= 1 + \cfrac{1}{2 + \cfrac{1}{15/6}} = <1, 2; \tfrac{15}{6}> \\
&= \quad \cdots = <1, 2, 2; \tfrac{6}{3}> \\
&= 1 + \cfrac{1}{2 + \cfrac{1}{2 + \frac{1}{2}}} = <1, 2, 2, 2>.
\end{aligned}$$

The example indicates that the continued fraction expansion is unique, and a comparison with Example 2.1 suggests that there is a close relationship between the continued fraction expansion of a rational $\beta = a/b$ and the euclidean algorithm applied to (a, b). We suppose, without loss of generality, that $\beta = a/b \geqslant 1$.

Proposition 3.1. *Let d_t, c_t, p_t, q_t and T be as in the euclidean algorithm:*

i. $\beta = <d_1, d_2, \ldots, d_t; c_{t-1}/c_t>$ *for* $t = 0, \ldots, T$;
ii. $\beta = <d_1, d_2, \ldots, d_T> = q_T/p_T$;
iii. $<d_1, d_2, \ldots, d_t> = q_t/p_t$ *for* $t = 1, \ldots, T$.

Proof. i. We use induction. For $r = 0$, we have $\beta = a/b = c_{-1}/c_0$. Now assume $\beta = <d_1, d_2, \ldots, d_{r-1}; c_{r-2}/c_{r-1}>$. Expanding $1/x$ with $x = c_{r-2}/c_{r-1}$ we obtain

$$\begin{aligned}
\frac{1}{x} &= \left(\frac{c_{r-2}}{c_{r-1}}\right)^{-1} = \left(\left\lfloor\frac{c_{r-2}}{c_{r-1}}\right\rfloor + \frac{c_{r-2} - \lfloor c_{r-2}/c_{r-1}\rfloor c_{r-1}}{c_{r-1}}\right)^{-1} \\
&= \left(d_r + \frac{c_r}{c_{r-1}}\right)^{-1} = \left(d_r + \frac{1}{c_{r-1}/c_r}\right)^{-1}.
\end{aligned}$$

Hence $\beta = <d_1, \ldots, d_r; c_{r-1}/c_r>$.

ii. We know from the euclidean algorithm that $0 = c_T = p_T a - q_T b$ with p_T and q_T relatively prime.

iii. This follows from applying the euclidean algorithm to the rational $\beta' = <d_1, \ldots, d_t>$. ■

To solve the diophantine approximation problem (3.1), note that the successive values q_t/p_t approach β. Therefore suppose we truncate the continued fraction expansion just before the value of p_t exceeds K.

Letting $j = \max\{t: p_t \leqslant K\}$, a candidate for the best approximation is $q_j/p_j = <d_1, d_2, \ldots, d_j>$. However, this does not always solve the problem.

Proposition 3.2. *Let* $j = \max\{t: p_t \leqslant K\}$ *and* $k = \lfloor (K - p_{j-1})/p_j \rfloor$. *Then either* q_j/p_j *or* $(q_{j-1} + kq_j)/(p_{j-1} + kp_j)$ *solves problem (3.1).*

The idea of the proof is to show that any rational between these two values necessarily has a denominator exceeding K.

Example 3.1 (continued). Suppose we wish to find the best approximation q/p to $\frac{51}{36}$ with $p \leqslant 10$.

For $t = 1, 2, 3, 4$, we have $q_t/p_t = 1, \frac{3}{2}, \frac{7}{5}, \frac{17}{12}$, respectively. Since $p_3 \leqslant 10 < p_4$, we take $j = 3$, and one estimate is $q_3/p_3 = \frac{7}{5}$. Since $k = \lfloor (10 - 2)/5 \rfloor = 1$, the other estimate is $(3 + 1 \cdot 7)/2 + 1 \cdot 5) = \frac{10}{7}$. Since $|\frac{17}{12} - \frac{7}{5}| = \frac{1}{60}$, and $|\frac{10}{7} - \frac{17}{12}| = \frac{1}{84}$, the best approximation is $\frac{10}{7}$.

4. LATTICES AND HERMITE NORMAL FORM

Here we consider the lattice

$$L(A) = \{y \in Z^m: y = Ax, x \in Z^n\},$$

where A is an $m \times n$ integer matrix. In this and the next sections we will consider different ways to represent $L(A)$. The basic operations that can be carried out on the matrix A are column operations that do not change the lattice.

Definition 4.1. An $n \times n$ matrix C is *unimodular* if it is integer and $|\det C| = 1$.

Proposition 4.1. *If* A *is an integer* $m \times n$ *matrix, and* C *is a unimodular* $n \times n$ *matrix, then* $L(AC) = L(A)$.

Proof. By substituting $x = Cw$, we have that

$$L(A) = \{y \in Z^m: y = Ax, x \in Z^n\} = \{y \in Z^m: y = ACw, Cw \in Z^n\}.$$

The result follows by showing that $\{w: Cw \in Z^n\} = \{w: w \in Z^n\}$. Since C is an integer matrix, $w \in Z^n$ implies $Cw \in Z^n$. Conversely, since C is unimodular, C^{-1} is an integer matrix and hence $Cw \in Z^n$ implies $C^{-1}Cw = w \in Z^n$. ■

Definition 4.2. An $m \times m$ nonsingular integer matrix H is said to be in *Hermite normal form* if:

a. H is lower triangular and $h_{ij} = 0$ for $i < j$;
b. $h_{ii} > 0$ for $i = 1, \ldots, m$; and
c. $h_{ij} \leqslant 0$ and $|h_{ij}| < h_{ii}$ for $i > j$.

The main result of this section is:

Theorem 4.2. *If A is an $m \times n$ integer matrix with* $\text{rank}(A) = m$*, then there exists an $n \times n$ unimodular matrix C such that:*

a. $AC = (H, 0)$ *and H is in Hermite normal form;*
b. $H^{-1}A$ *is an integer matrix.*

$(H, 0)$ is called the *Hermite normal form of A*. We will outline a polynomial-time algorithm for finding C and H which will serve as a constructive proof of Theorem 4.2. It also can be shown that H is unique.

Example 4.1

$$A = \begin{pmatrix} 2 & 6 & 1 \\ 4 & 7 & 7 \end{pmatrix}.$$

It is readily verified that the matrices

$$H = \begin{pmatrix} 1 & 0 \\ -3 & 5 \end{pmatrix} \quad \text{and} \quad C = \begin{pmatrix} 1 & 3 & -7 \\ 0 & -1 & 2 \\ -1 & 0 & 2 \end{pmatrix}$$

satisfy the conditions of Theorem 4.2.

There are several immediate consequences of Theorem 4.2. The first, a canonical description of $L(A)$, uses Proposition 4.1.

Proposition 4.3. $L(H) = L(A)$.

Definition 4.3. If $L(A) = L(B)$ and B is nonsingular, then B is a *basis* for the lattice $L(A)$.

Corollary 4.4. *Every lattice $L(A)$ with* $\text{rank}(A) = m$ *has a basis.*

Given the above characterization of $L(A)$, we can solve the Linear Equation Integer Feasibility Problem. Let $S = \{x \in Z^n: Ax = b\}$ and let H and $C = (C_1, C_2)$ be as in Theorem 4.2, with C_1 an $n \times m$ matrix and C_2 an $n \times (n - m)$ matrix. Observe that $S \neq \emptyset$ if and only if $b \in L(A)$.

Theorem 4.5

i. *$S \neq \emptyset$ if and only if $H^{-1}b \in Z^m$.*
ii. *If $S \neq \emptyset$, every solution of S is of the form*

$$x = C_1 H^{-1} \cdot b + C_2 z, \quad z \in Z^{n-m}.$$

Proof

$$\begin{aligned} S &= \{x \in Z^n\colon Ax = b\} \\ &= \{x\colon x = Cw, ACw = b, w \in Z^n\} \text{ (since } C \text{ is unimodular)} \\ &= \{x\colon x = Cw, (H, 0)\, w = b, w \in Z^n\} \\ &= \{x\colon x = C_1 w_1 + C_2 w_2, H w_1 = b, w_1 \in Z^m, w_2 \in Z^{n-m}\} \\ &= \{x\colon x = C_1 H^{-1} b + C_2 w_2, H^{-1} b \in Z^m, w_2 \in Z^{n-m}\}. \end{aligned}$$

■

Example 4.2. Find the set of integer solutions, if any, to

$$2x_1 + 6x_2 + 1x_3 = 7$$

$$4x_1 + 7x_2 + 7x_3 = 4.$$

H and C were given in Example 4.1. Now, by Theorem 4.5, the solution set is nonempty since

$$H^{-1}\begin{pmatrix} 7 \\ 4 \end{pmatrix} = \frac{1}{5}\begin{pmatrix} 5 & 0 \\ 3 & 1 \end{pmatrix}\begin{pmatrix} 7 \\ 4 \end{pmatrix} = \begin{pmatrix} 7 \\ 5 \end{pmatrix} \in Z^2.$$

The general solution is

$$\begin{aligned} \begin{pmatrix} x_1 \\ x_2 \\ x_3 \end{pmatrix} &= C_1 \begin{pmatrix} 7 \\ 5 \end{pmatrix} + C_2 w_2 \\ &= \begin{pmatrix} 1 & 3 \\ 0 & -1 \\ -1 & 0 \end{pmatrix}\begin{pmatrix} 7 \\ 5 \end{pmatrix} + \begin{pmatrix} -7 \\ 2 \\ 2 \end{pmatrix} w_2 \\ &= \begin{pmatrix} 22 \\ -5 \\ -7 \end{pmatrix} + \begin{pmatrix} 7 \\ -2 \\ -2 \end{pmatrix} w_2, \; w_2 \in Z^1. \end{aligned}$$

We also obtain an integer version of Farkas' lemma.

Corollary 4.6. *Either $S = \{x\colon Ax = b, x \in Z^n\} \neq \emptyset$ or (exclusively) there exists $u \in R^m$ such that $uA \in Z^m$, $ub \notin Z^1$.*

Proof. Both cannot hold, since this would imply $uAx = ub$ with $uAx \in Z^1$ and $ub \notin Z^1$. If $S = \emptyset$, then by Theorem 4.5 we have $H^{-1}b \notin Z^m$. Suppose the ith coefficient of $H^{-1}b \notin Z^1$, and then take u to be the ith row of H^{-1}. ■

Example 4.2 (continued). If $b = \binom{8}{4}$, then

$$H^{-1}b = \frac{1}{5}\begin{pmatrix} 5 & 0 \\ 3 & 1 \end{pmatrix}\begin{pmatrix} 8 \\ 4 \end{pmatrix} = \begin{pmatrix} 8 \\ \frac{36}{5} \end{pmatrix}$$

and hence $S = \emptyset$.

Taking $u = (\frac{3}{5} \;\; \frac{1}{5})$, we we obtain $uA = (2 \;\; 5 \;\; 2)$ and $ub = \frac{36}{5}$; in other words, any $x \in S$ must satisfy

$$2x_1 + 5x_2 + 2x_3 = \frac{36}{5}, \quad x \in Z^3,$$

which is impossible.

Now let $\gcd(a_1, \ldots, a_n)$ denote the greatest common divisor of $a_1, \ldots, a_n$. An observation that is used later is:

Corollary 4.7. *Let $S = \{x \in Z^n \colon \sum_{j=1}^n a_j x_j = b\}$ with $a_j, b \in Z^1$. If $\gcd(a_1, \ldots, a_n) \mid b$, there exist n affinely independent points in S.*

Proof. Let $A = (a_1, \ldots, a_n)$ and let H and $C = (C_1, C_2)$ be as described in Theorem 4.2. C_1 is an $n \times 1$ matrix, and C_2 is $n \times (n-1)$ with $AC_1 = h_{11} = \gcd(a_1, a_2, \ldots, a_n)$, and $AC_2 = (0, \ldots, 0)$. Hence using Theorem 4.5, we obtain

$$S = \left\{x \colon x = \frac{b}{h_{11}} C_1 + C_2 w,\ w \in Z^{n-1}\right\}.$$

Since $\text{rank}(C) = n$, it follows that $\text{rank}(C_2) = n - 1$, and the claim is true. ■

We now turn to the proof of Theorem 4.2 and define the elementary column operations that correspond to right multiplication by a unimodular matrix C.

The *elementary column operations* of interest are:

a. Interchange columns j and k.
b. Multiply column j by -1.
c. Add $\lambda \in Z^1$ times column k to column j.

The corresponding unimodular matrices C are easily constructed. Figure 4.1 shows an example with $m = n = 6$.

$$C = \begin{pmatrix} 1 & & & & & \\ & 0 & & & 1 & \\ & & 1 & & & \\ & & & 1 & & \\ & 1 & & & 0 & \\ & & & & & 1 \end{pmatrix} \quad C = \begin{pmatrix} 1 & & & & & \\ & -1 & & & & \\ & & 1 & & & \\ & & & 1 & & \\ & & & & 1 & \\ & & & & & 1 \end{pmatrix} \quad C = \begin{pmatrix} 1 & & & & & \\ & 1 & & & & \\ & & 1 & & & \\ & & & 1 & & \\ & 3 & & & 1 & \\ & & & & & 1 \end{pmatrix}$$

$j = 2, k = 5$ (a) $\quad$ $j = 2$ (b) $\quad$ $j = 2, k = 5, \lambda = 3$ (c)

Figure 4.1

Now we give an algorithm that constructs the Hermite normal form of an $m \times n$ integer matrix A with $\text{rank}(A) = m$. The basic operation of the algorithm involves a row i, as well as columns s and t of the matrix A with $s < t$. A sequence of elementary column operations are performed so that $a_{is} \leftarrow \gcd(a_{is}, a_{it})$, $a_{it} \leftarrow 0$.

Proposition 4.8. *Let $A = (a_1, \ldots, a_n)$, $\gcd(a_{is}, a_{it}) = r$, and $pa_{is} + qa_{it} = r$, where p and q are relatively prime. There exists an $n \times n$ unimodular integer matrix C such that $AC = A'$, where*

$$a'_l = a_l \quad \text{for} \quad l \neq s, t$$

$$a'_s = pa_s + qa_t$$

$$a'_t = -\frac{a_{it}}{r}a_s + \frac{a_{is}}{r}a_t.$$

In particular, $a'_{is} = r$ and $a'_{it} = 0$.

Proof. Take C to be an identity matrix in all but columns s and t. In column s, we have $c_{ss} = p$, $c_{ts} = q$, and $c_{is} = 0$ otherwise. In column t, we have $c_{st} = -a_{it}/r$, $c_{tt} = a_{is}/r$, and $c_{it} = 0$ otherwise. It is readily verified that $AC = A'$, and $\det C = pa_{is}/r + qa_{it}/r = 1$. ∎

The Hermite Normal Form Algorithm

Initialization: $i = 1$.

Step 1: Work on row i. Set $j \leftarrow i + 1$.

Step 2: Work on row i and columns i and $j > i$. If $a_{ij} = 0$, do nothing. Otherwise use the euclidean algorithm to find $r = \gcd(a_{ii}, a_{ij})$ and p, q relatively prime such that $pa_{ii} + qa_{ij} = r$. Set $A \leftarrow AC$, where C is the unimodular matrix described in Proposition 4.8, with $s = i$, $t = j$. If $j < n$, set $j \leftarrow j + 1$ and return to Step 2. If $j = n$, go to Step 3.

Step 3: Work on row i and column i. If $a_{ii} < 0$, set $A \leftarrow AC$, where C multiplies column a_i by -1.

Step 4: Work on row i and column $j < i$. Set $j \leftarrow 1$. Set $A \leftarrow AC$, where C replaces column a_j by $a_j - \lceil a_{ij}/a_{ii} \rceil a_i$. If $j = i - 1$, set $i \leftarrow i + 1$. If $i > m$, stop. Otherwise return to Step 1. If $j < i - 1$, set $j \leftarrow j + 1$ and return to Step 4.

Proposition 4.9. *The Hermite normal form algorithm terminates with matrices H and C as described in Theorem 4.2.*

Proof. All the operations performed are column operations corresponding to right multiplication by a unimodular matrix. Hence the product C of these matrices is unimodular. Let $H' = AC$ be the final matrix. Note that after Step 2, $h'_{ij} = 0$ for all $j > i$; after Step 3, $h'_{ii} \geqslant 0$; and after Step 4, $h'_{ij} < 0$ and $|h'_{ij}| < h'_{ii}$ for $j < i$ unless $h'_{ii} = 0$. Furthermore, these values are never changed in later steps. Hence we only need to show that after completing Step 2 for row i, we obtain $|h'_{ii}| > 0$.

Suppose $h'_{ij} > 0$ for $j < i$ but $h'_{ii} = 0$. Let A_1 be the matrix consisting of the first i rows of A. Then the algorithm has produced a unimodular matrix C^* such that $A_1 C^* = H^*$, where $h^*_{kj} = 0$ for all $k \leqslant i$ and $j \geqslant i$. Hence $\text{rank}(H^*) = i - 1$. However, since C^* is unimodular, $\text{rank}(A_1) = \text{rank}(A_1 C^*) = i - 1$, which contradicts $\text{rank}(A) = m$. ∎

Example 4.3. We find the Hermite Normal Form of matrix A given below.

$$A = \begin{pmatrix} 2 & 6 & 1 \\ 4 & 7 & 7 \\ 0 & 0 & 1 \end{pmatrix} \quad \begin{array}{l} i = 1, j = 2. \\ (a_{11}, a_{12}) = (2 \ \ 6). \\ (p, q) = (1 \ \ 0), r = 2. \end{array} \quad C^1 = \begin{pmatrix} 1 & -3 & 0 \\ 0 & 1 & 0 \\ 0 & 0 & 1 \end{pmatrix}$$

$$A = \begin{pmatrix} 2 & 0 & 1 \\ 4 & -5 & 7 \\ 0 & 0 & 1 \end{pmatrix} \quad \begin{array}{l} i = 1, j = 3. \\ (a_{11}, a_{13}) = (2 \ \ 1). \\ (p, q) = (1 \ \ 0), r = 1. \end{array} \quad C^2 = \begin{pmatrix} 0 & 0 & -1 \\ 0 & 1 & 0 \\ 1 & 0 & 2 \end{pmatrix}$$

$$A = \begin{pmatrix} 1 & 0 & 0 \\ 7 & -5 & 10 \\ 1 & 0 & 2 \end{pmatrix} \quad \begin{array}{l} i = 1, j = 1. \text{ No change.} \\ i = 2, j = 3. \\ (a_{22}, a_{23}) = (-5 \ \ 10). \\ (p, q) = (-1 \ \ 0), r = 5. \end{array} \quad C^3 = \begin{pmatrix} 1 & 0 & 0 \\ 0 & -1 & -2 \\ 0 & 0 & -1 \end{pmatrix}$$

$$A = \begin{pmatrix} 1 & 0 & 0 \\ 7 & 5 & 0 \\ 1 & 0 & -2 \end{pmatrix} \quad \begin{array}{l} i = 2, j = 2. \text{ No change.} \\ i = 2, j = 1. \end{array} \quad C^4 = \begin{pmatrix} 1 & 0 & 0 \\ -2 & 1 & 0 \\ 0 & 0 & 1 \end{pmatrix}$$

$$A = \begin{pmatrix} 1 & 0 & 0 \\ -3 & 5 & 0 \\ 1 & 0 & -2 \end{pmatrix} \quad i = 3, j = 3. \quad C^5 = \begin{pmatrix} 1 & & \\ & 1 & \\ & & -1 \end{pmatrix}$$

$$A = \begin{pmatrix} 1 & 0 & 0 \\ -3 & 5 & 0 \\ 1 & 0 & 2 \end{pmatrix} \quad i = 3, j = 1. \quad C^6 = \begin{pmatrix} 1 & & \\ & 1 & \\ -1 & & 1 \end{pmatrix}$$

$$A = \begin{pmatrix} 1 & 0 & 0 \\ -3 & 5 & 0 \\ -1 & 0 & 2 \end{pmatrix} \quad i = 3, j = 2. \text{ No change.} \quad H = \begin{pmatrix} 1 & 0 & 0 \\ -3 & 5 & 0 \\ -1 & 0 & 2 \end{pmatrix}.$$

Finally,

$$C = \prod_{i=1}^{6} C^i = \begin{pmatrix} 1 & 3 & -7 \\ 0 & -1 & 2 \\ -1 & 0 & 2 \end{pmatrix}.$$

Although the number of iterations of the HNF algorithm is polynomially bounded, it is not known whether the size of the numbers is polynomially bounded. In practice, they get so large that the algorithm is difficult to execute on a computer. We now modify the algorithm to guarantee that the numbers remain sufficiently small. We assume first that A is $m \times m$.

Observe that if $b \in L(A)$, then $L(A, b) = L(A)$. Hence the Hermite normal form of (A, b) is the same as that of $(A, 0)$.

The following proposition gives multiples of the unit vectors in $L(A)$. Note that $D = \det H = \prod_{i=1}^{m} h_{ii}$.

Proposition 4.10. *Let $d_i = \Pi_{k=i}^{m} h_{kk}$ for $i = 1, \ldots, m$. Then $d_i e_k \in L(A)$ for $k = i, \ldots, m$.*

Proof. The vector $x^k = (DA^{-1})e_k$ is integer because DA^{-1} is integer and $Ax^k = ADA^{-1}e_k = De_k$. Hence $De_k \in L(A)$. Now observe that $h_i, \ldots, h_m \in L(A)$ and apply the above result to the lattice $L(h_i, \ldots, h_m)$ with D replaced by $d_i = \Pi_{k=i}^{m} h_{kk}$. ■

The way to use these results is to calculate D first and then add the vectors $d_i e_k$ for $k = i, \ldots, m$ and $i = 1, \ldots, m$ to the lattice $L(A)$ as soon as we have calculated $h_{11}, \ldots, h_{i-1,i-1}$. This allows us to reduce all components in rows $i, \ldots, m$ modulo (d_i).

Based on this observation, a modified Hermite normal form algorithm is easily described in which no intermediate numbers in the computation ever exceed $2D^2$ in absolute value. The resulting algorithm runs in polynomial time.

Simple modifications of the algorithm give the Hermite normal form of a general $m \times n$ integer matrix A when $\text{rank}(A) = k < \min(m, n)$. Specifically we obtain

$$AC = \begin{pmatrix} H & 0 \\ 0 & 0 \end{pmatrix},$$

where C is an $n \times n$ unimodular matrix and H is a $k \times k$ Hermite matrix.

The construction of the Hermite normal form of a matrix A depends on column operations. By also doing similar transformations on the rows, we obtain another canonical representation.

Theorem 4.11. *If A is an $m \times m$ nonsingular integer matrix, there exist unimodular matrices R and C such that*

i. $RAC = \Delta$,
ii. *Δ is a diagonal matrix with diagonal entries $\delta_i \in Z_+^1 \setminus \{0\}$,*
iii. $\delta_1 | \delta_2 \ldots | \delta_m$,
iv. *Δ is unique.*

The matrix Δ is called the *Smith normal form of A*. Using an algorithm based on the Hermite normal form algorithm, Δ can be constructed in polynomial time.

5. A REDUCED BASIS OF A LATTICE

Suppose that $L(A) \subseteq R^n$ is a full-dimensional lattice. In the previous section the Hermite basis for $L(A)$ was used to solve the Linear Equation Integer Feasibility Problem. Here we construct another basis B for $L(A)$, called a *reduced basis*, and indicate how it can be used to solve the shortest vector problem. In addition, two applications of reduced bases in integer programming will be given in Section II.6.5.

Before we can define and explain the use of a reduced basis, we need to introduce several facts about bases and how they relate to determinants. First we introduce the Gram–Schmidt orthogonalization of a basis.

Definition 5.1. A basis $B = (b_1, \ldots, b_n)$ is said to be *orthogonal* if $b_i b_j = 0$ for $i, j = 1, \ldots, n, i \neq j$.

The Gram–Schmidt Orthogonalization of a Basis *B*

$$
\begin{aligned}
b_1^* &= b_1 \\
b_2^* &= b_2 - \alpha_{12} b_1^* \\
&\vdots \\
b_k^* &= b_k - \sum_{j=1}^{k-1} \alpha_{jk} b_j^* \quad \text{for } k = 3, \ldots, n,
\end{aligned}
\tag{5.1}
$$

where $\alpha_{ij} = b_i^* b_j / \|b_i^*\|^2$ for $i < j$.

Proposition 5.1

i. *The Gram–Schmidt procedure constructs an orthogonal basis* $B^* = (b_1^*, \ldots, b_n^*)$ *for* R^n,
ii. b_k^* *is the component of* b_k *orthogonal to the subspace generated by* $b_1^*, \ldots, b_{k-1}^*$,
iii. $|\det B| = |\det B^*| = \prod_{j=1}^n \|b_j^*\|$.

To examine how the lengths of the vectors of a basis are related to det B, we need to work with a more general definition of a determinant.

Definition 5.2. $|\det(b_1, \ldots, b_k)| = \prod_{j=1}^k \|b_j^*\|$ for all $k \leq n$.

We see from the definition that $\det(b_1, \ldots, b_k)$ is the k-dimensional volume of the parallelepiped with vertices given by $\sum_{j=1}^k b_j x_j$, where $x_j \in \{0, 1\}$ for $j = 1, \ldots, k$ (see Figure 5.1). Furthermore, since b_j^* is the component of b_j orthogonal to the subspace generated by $b_1^*, \ldots, b_{j-1}^*$, it follows that $\|b_j^*\| \leq \|b_j\|$.

Given a full-dimensional lattice L, we know by statement iii of Proposition 5.1 that $|\det B|$ has the same value for all bases B of the lattice. Let $d(L)$ be this common value and let $\alpha(B) = \prod_{j=1}^n \|b_j\|$. From the above observations we obtain the following proposition.

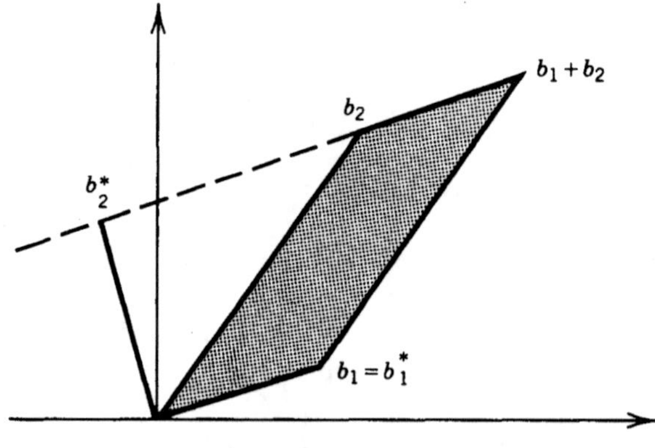

Figure 5.1

Proposition 5.2 *(The Hadamard Inequality). For all bases B of L, we have $\alpha(B) \geq d(L)$.*

Example 5.1. We consider a full-dimensional lattice L with basis

$$B = \begin{pmatrix} 2 & 6 & 1 \\ 4 & 7 & 7 \\ 0 & 0 & 1 \end{pmatrix}.$$

We have $d(L) = |\det B| = 10$. Note that $\alpha(B) = (20 \cdot 85 \cdot 51)^{1/2} = (86{,}700)^{1/2}$ is much larger than $|\det B|$. The Gram–Schmidt Orthogonalization Procedure applied to B gives

$$b_1^* = b_1 = \begin{pmatrix} 2 \\ 4 \\ 0 \end{pmatrix},$$

$$b_2^* = b_2 - \alpha_{12} b_1^* = \begin{pmatrix} 6 \\ 7 \\ 0 \end{pmatrix} - 2\begin{pmatrix} 2 \\ 4 \\ 0 \end{pmatrix} = \begin{pmatrix} 2 \\ -1 \\ 0 \end{pmatrix},$$

$$b_3^* = b_3 - \alpha_{13} b_1^* - \alpha_{23} b_2^* = \begin{pmatrix} 1 \\ 7 \\ 1 \end{pmatrix} - \frac{3}{2}\begin{pmatrix} 2 \\ 4 \\ 0 \end{pmatrix} - \frac{-5}{5}\begin{pmatrix} 2 \\ -1 \\ 0 \end{pmatrix} = \begin{pmatrix} 0 \\ 0 \\ 1 \end{pmatrix}.$$

Now we show how $\alpha(B)$ relates to the shortest vector problem (1.3).

Proposition 5.3. *Given a full-dimensional lattice $L \subseteq R^n$ and a basis B of L, let x^0 be a solution of the problem*

$$\min\left\{ \|Bx\| \colon |x_j| \leq \frac{\alpha(B)}{|\det B|} \quad \text{for } j = 1, \ldots, n, x \in Z^n \setminus \{0\} \right\}.$$

Then $v = Bx^0$ is a shortest vector in the lattice L.

Proof. Let $v = Bx$ be a shortest nonzero vector with $x \in Z^n \setminus \{0\}$. Using Cramer's rule, we have $|x_j| = |\det B_j| / |\det B|$, where $B_j = (b_1, \ldots, b_{j-1}, v, b_{j+1}, \ldots, b_n)$. By Hadamard's inequality (Proposition 5.2), we have

$$|\det B_j| \leq \|b_1\| \cdots \|b_{j-1}\| \, \|v\| \cdots \|b_n\|.$$

However, since $b_j \in L$, it follows that $\|v\| \leq \|b_j\|$. Hence $|\det B_j| \leq |\alpha(B)|$, and so $|x_j| \leq \alpha(B)/|\det B|$. ∎

As a consequence of Proposition 5.3, we are motivated to find a basis such that $\alpha(B)/|\det B|$ is small; in particular, we would like to find one with $\lceil \log(\alpha(B)/|\det B|) \rceil$ polynomial in n.

We also have the following lower bound on the shortest vector in the lattice.

Proposition 5.4. *If $b \in L$, $b \neq 0$, and B is a basis of L, with B^* being its Gram–Schmidt orthogonalization, then $\|b\| \geq \min_j \|b_j^*\|$.*

Proof. Since $b \in L$, there exists a $k \leqslant n$ such that $b = \sum_{j=1}^{k} b_j z_j$ with $z \in Z^k$ and $z_k \neq 0$. By substituting for b_j, using (5.1), we obtain $b = \sum_{j=1}^{k} b_j^* z_j^*$ with $z_k^* = z_k \in Z^1$. Since the $\{b_j^*\}$ are orthogonal, $\|b\| \geqslant |z_k|\ \|b_k^*\| \geqslant \min_j \|b_j^*\|$. ■

Unfortunately, B^* is typically not a basis for L because $\{\alpha_{ij}\}_{i<j}$ are not all integer. This does suggest, however, the need to find a basis that is "nearly orthogonal".

Definition 5.3. Let L be a full-dimensional lattice, let B be a basis of L, and let B^* be the basis obtained from the Gram–Schmidt orthogonalization procedure. B is a *reduced basis* if:

a. $\alpha_{ij} \leqslant \frac{1}{2}$ for all $i < j$ and
b. $\|b_{j+1}^* + \alpha_{j,j+1} b_j^*\|^2 \geqslant \frac{3}{4} \|b_j^*\|^2$ for $j = 1, \ldots, n-1$.

The interest in a reduced basis lies in the following results, giving an upper bound on $\|b_1\|$ (and hence on the shortest vector) and an upper bound on $\alpha(B)$.

Theorem 5.5. *Let B be a reduced basis for the full-dimensional lattice L. Then*

i. $\|b_j^*\|^2 \leqslant 2\|b_{j+1}^*\|^2$.
ii. $\|b_1\| \leqslant 2^{(n-1)/4} (d(L))^{1/n}$.
iii. $\|b_1\| \leqslant 2^{(n-1)/2} \min\{\|b\|: b \in L, b \neq 0\}$.
iv. $\alpha(B) \leqslant 2^{n(n-1)/4}\, d(L)$.

Proof. i. Since b_j^* and b_{j+1}^* are orthogonal, by statement b in the definition of a reduced basis we have

$$\|b_{j+1}^* + \alpha_{j,j+1} b_j^*\|^2 = \|b_{j+1}^*\|^2 + \alpha_{j,j+1}^2 \|b_j^*\|^2 \geqslant \tfrac{3}{4}\|b_j^*\|^2.$$

By statement a of Definition 5.3, we have $\alpha_{j,j+1}^2 \leqslant \frac{1}{4}$ and hence $\|b_{j+1}^*\|^2 \geqslant \frac{1}{2}\|b_j^*\|^2$.
ii. By i, it follows that $\|b_j^*\|^2 \geqslant 2^{-(j-1)} \|b_1^*\|^2$ and since $b_1 = b_1^*$, we obtain

$$(d(L))^2 = \prod_{j=1}^{n} \|b_j^*\|^2 \geqslant 2^{-\sum_{j=1}^{n}(j-1)} \|b_1\|^{2n} = 2^{-n(n-1)/2} \|b_1\|^{2n}.$$

iii. From the proof of ii, we have

$$\|b_j^*\|^2 \geqslant 2^{-(j-1)} \|b_1\|^2 \geqslant 2^{-(n-1)} \|b_1\|^2.$$

Hence, using Proposition 5.4, we obtain

$$\|b\| \geqslant \min_j \|b_j^*\| \geqslant 2^{-(n-1)/2} \|b_1\|.$$

iv. Since $b_j = \sum_{i=1}^{j} \alpha_{ij} b_i^*$ and the vectors b_i^* are orthogonal, it follows that

$$\|b_j\|^2 = \sum_{i=1}^{j} \alpha_{ij}^2 \|b_i^*\|^2 \text{ (with } \alpha_{jj} = 1)$$

$$\leqslant \|b_j^*\|^2 + \tfrac{1}{4} \sum_{i=1}^{j-1} \|b_i^*\|^2 \quad \text{by statement a of Definition 5.3.}$$

Now, using i, we obtain $\|b_i^*\|^2 \leqslant 2^{j-i}\|b_j^*\|^2$ and hence

$$\|b_j\|^2 \leqslant \|b_j^*\|^2\left(1 + \tfrac{1}{4}\sum_{i=1}^{j-1} 2^{j-i}\right) \leqslant 2^{j-1}\|b_j^*\|^2.$$

Finally,

$$(\alpha(B))^2 = \prod_{j=1}^{n} \|b_j\|^2 \leqslant 2^{n(n-1)/2} \prod_{j=1}^{n} \|b_j^*\|^2 = 2^{n(n-1)/2}(d(L))^2$$

■

Example 5.1 (continued). The following computations show that

$$B = \begin{pmatrix} 0 & 2 & 1 \\ 0 & -1 & 2 \\ 2 & 0 & 1 \end{pmatrix}$$

is a reduced basis for the lattice L:

$$b_1^* = b_1 = \begin{pmatrix} 0 \\ 0 \\ 2 \end{pmatrix} \quad \text{so} \quad \alpha_{12} = 0 \quad \text{and} \quad \alpha_{13} = \tfrac{1}{2}.$$

$$b_2^* = b_2 - \alpha_{12}b_1^* = \begin{pmatrix} 2 \\ -1 \\ 0 \end{pmatrix} \quad \text{so} \quad \alpha_{23} = 0.$$

$$b_3^* = b_3 - \alpha_{13}b_1^* - \alpha_{23}b_2^* = \begin{pmatrix} 1 \\ 2 \\ 0 \end{pmatrix}.$$

Hence part a in the definition of a reduced basis holds. We will now check b of Definition 5.3 as follows:

$$\|b_2^* + \alpha_{12}b_1^*\|^2 = 5 \geqslant \tfrac{3}{4}\|b_1^*\|^2 = \tfrac{3}{4}\cdot 4 = 3$$
$$\|b_3^* + \alpha_{23}b_2^*\|^2 = 5 \geqslant \tfrac{3}{4}\|b_2^*\|^2 = \tfrac{3}{4}\cdot 5 = \tfrac{15}{4}.$$

Hence B is a reduced basis.

Checking the bounds in Theorem 5.5, we observe

ii. $\|b_1\| \leq 2^{1/2}\,(d(L))^{1/3} = 2^{1/2}\,10^{1/3}$ and
iv. $\alpha(B) = \Pi_{j=1}^{3}\,\|b_j\| = (120)^{1/2} \leqslant 2^{3/2}d(L) = 20\cdot 2^{1/2}$.

A Reduced Basis Algorithm for a Full-Dimensional Lattice L

Step 0: Let B be a basis of the lattice L.

Step 1: Let $(b_1^*, \ldots, b_n^*)$ be the Gram–Schmidt orthogonalization of $(b_1, \ldots, b_n)$ with $\alpha_{ij} = b_i^* b_j / \|b_i^*\|^2$.

Step 2: For $j = 2, \ldots, n$ and for $i = j-1, \ldots, 1$ replace b_j by $b_j - \hat{\alpha}_{ij}b_i$, where $\hat{\alpha}_{ij}$ is the integer closest to α_{ij}.

Step 3: If $\|b_{j+1}^* + \alpha_{j,j+1}b_j^*\|^2 < \tfrac{3}{4}\|b_j^*\|^2$ for some j, interchange b_j and b_{j+1} and return to Step 1 with the new basis B.

Theorem 5.6. *The above algorithm finds a reduced basis for L in polynomial time.*

We will not prove this result. However, the reader may observe that even though the basis B changes in Step 2, the corresponding B^* does not change. This implies that on termination of Step 2, $|\alpha_{ij}| \leq \frac{1}{2}$ for all $i < j$, so condition a for a reduced basis is satisfied.

Now we return to the shortest vector problem. We have seen that a reduced basis can be found in polynomial time and that for such a basis we obtain $\alpha(B)/|\det B| \leq 2^{n(n-1)/4}$. By Proposition 5.3 it suffices to enumerate over $x \in Z^n \setminus \{0\}$ with $|x_j| \leq 2^{n(n-1)/4}$. Hence we have shown the following:

Theorem 5.7. *For fixed n, the shortest vector problem can be solved in polynomial time.*

Example 5.1 (continued). Find a shortest vector in L. We already have a reduced basis

$$B = \begin{pmatrix} 0 & 2 & 1 \\ 0 & -1 & 2 \\ 2 & 0 & 1 \end{pmatrix}.$$

For $n = 3$, it suffices to enumerate the $5^3 - 1$ vectors $v = Bx$ with $|x_j| \leq \lfloor 2^{3/2} \rfloor = 2$, and $x \in Z^3 \setminus \{0\}$, giving

$$v = \begin{pmatrix} 0 \\ 0 \\ 2 \end{pmatrix}$$

as a shortest vector.

The final item in this section is to show how a reduced basis can be used to solve "approximately" a feasibility version of (1.5), namely:

The Simultaneous Diophantine Approximation Feasibility Problem (SDAF). Given rationals $\alpha_1, \ldots, \alpha_n$, ε, and an integer $K > 0$, decide if there exist integers $q_1, \ldots, q_n$ and $p > 0$ such that $p \leq K$ and $|p\alpha_i - q_i| \leq \varepsilon$ for $i = 1, \ldots, n$.

The approach is to reformulate a "weak" version of SDAF as a problem of finding a short vector in a lattice.

Theorem 5.8. *There is a polynomial-time algorithm which either*

i. *determines that* SDAF *is infeasible or*
ii. *finds integers $q_1, \ldots, q_n$ and $p > 0$ such that $|p\alpha_i - q_i| < 2^{n/2} \varepsilon (n + 1)^{1/2}$ for $i = 1, \ldots, n$ and $p < 2^{n/2} K (n + 1)^{1/2}$.*

Proof. Let $a = (\alpha_1, \ldots, \alpha_n, \varepsilon/K) \in R^{n+1}$ and let e_i be the ith unit vector in R^{n+1}. Consider the lattice L generated by $(e_1, \ldots, e_n, -a)$. For any $(q_1, \ldots, q_n, p) \in Z^{n+1}$, we have $w = \Sigma_{i=1}^n q_i e_i - pa \in L$. Furthermore, if $(q_1, \ldots, q_n, p)$ is a solution to SDAF, then $|w_i| = |q_i - \alpha_i p| \leq \varepsilon$ for $i = 1, \ldots, n$ and $|w_{n+1}| = |p\varepsilon/K| \leq \varepsilon$. Hence $\|w\| \leq \varepsilon(n + 1)^{1/2}$.

Now let B be a reduced basis for L with b_1 being its first column. By Theorem 5.6, B can be found in polynomial time.

By iii of Theorem 5.5, if $\|b_1\| > 2^{(n-1)/2}\,\varepsilon\,(n+1)^{1/2}$, then $\|b\| \geq 2^{-(n-1)/2}\|b_1\| > \varepsilon(n+1)^{1/2}$ for all $b \in L \setminus \{0\}$. Hence SDAF has no solution. If $\|b_1\| \leq 2^{(n-1)/2}\,\varepsilon\,(n+1)^{1/2}$, choose $(q', p') \in Z^{n+1}$ such that $b_1 = \Sigma_{i=1}^{n} q_i' e_i - p'a$. If $p' \neq 0$, then (q', p') satisfies the conditions of ii because

$$|q_i' - p'\alpha_i| \leq 2^{(n-1)/2}\,\varepsilon(n+1)^{1/2} \quad \text{for } i = 1, \ldots, n$$

and

$$\left|p'\,\frac{\varepsilon}{K}\right| \leq 2^{(n-1)/2}\,\varepsilon(n+1)^{1/2}.$$

If $p' = 0$, then $b_1 = (q_1', \ldots, q_n', 0)$ and $\|b_1\| \geq 1$. Hence $2^{(n-1)/2}\,\varepsilon(n+1)^{1/2} \geq 1$, and $p = 1$, $q_i = \lfloor \alpha_i \rfloor$ satisfies ii. ■

6. NOTES

Section I.7.1

The problems considered in this chapter are related to classical topics in number theory (see, e.g., Cassels, 1971). Our study of these topics arises from the interest in polynomial-time algorithms and applications in integer programming. Also see Section II.6.5.

Section I.7.2

The euclidean algorithm is surely the oldest algorithm in this book. Its complexity is analyzed in detail by Knuth (1981).

Section I.7.3

In the preparation of this section and Section 5, we have made liberal use of both the article and the monograph by Grötschel, Lovasz, and Schrijver (1984b, 1988). Proposition 3.2 is due to Khintchine (1930).

Section I.7.4

Kannan and Bachem (1979) give a provably polynomial-time algorithm for computing Hermite normal form. The algorithm sketched here is due to Domich et al. (1987).

The integral Farkas lemma, Corollary 4.6, appears in Edmonds and Giles (1977).

Section I.7.5

Lagarias (1985) has shown that the simultaneous diophantine approximation problem is $\mathcal{NP}$-complete. It is also the case that the nearest vector problem is $\mathcal{NP}$-hard (see Van Emde Boas, 1981).

The polynomial-time algorithm for finding a reduced basis and a short vector in a lattice is due to Lenstra, Lenstra and Lovasz (1982). These and related problems are also discussed by Bachem and Kannan (1984) and Kannan (1987b).

7. EXERCISES

1. **i)** Find the gcd of 27,692 and 100,000.

ii) Find the continued fraction expansion of 100,000/27,692.

iii) Find the best rational approximation of 100,000/27,692 with denominator of 100 or less.

2. Show that the continued fraction expansion of a rational number is unique.

3. Prove Proposition 3.2.

4. **i)** Find the Hermite normal form of $\begin{pmatrix} 12 & 6 & 7 \\ 2 & 9 & 4 \end{pmatrix}$.

ii) Find all solutions (if any) to the system of equations

$$12x_1 + 6x_2 + 7x_3 = 8$$
$$2x_1 + 9x_2 + 4x_3 = 7$$
$$x \in Z^3.$$

5. Show that the Hermite normal form is unique.

6. Describe, in detail, a polynomial Hermite normal form algorithm based on Proposition 4.10. Apply the algorithm to the matrix

$$A = \begin{pmatrix} 2 & -6 & 4 \\ -1 & -1 & -22 \\ 0 & -2 & 23 \end{pmatrix}.$$

7. Describe a polynomial algorithm to find the Smith normal form of a matrix.

8. Use the reduced basis algorithm to find a reduced basis of $L(B)$, where

$$B = \begin{pmatrix} 2 & 6 & 1 \\ 4 & 7 & 7 \\ 0 & 0 & 1 \end{pmatrix}$$

is the basis of Example 5.1.

9. Solve the simultaneous diophantine approximation feasibility problem with $(\alpha_1, \alpha_2, \alpha_3) = (\frac{1}{4}\ \frac{2}{7}\ \frac{4}{5})$, $\varepsilon = \frac{1}{3}$ and $K = 18$.

Part II
GENERAL INTEGER PROGRAMMING

II.1

The Theory of Valid Inequalities

1. INTRODUCTION

We consider the discrete optimization problem $\max\{cx: x \in S\}$, where $S \subseteq Z^n_+$, and we formulate it as an integer program by specifying a rational polyhedron $P = \{x \in R^n_+: Ax \leqslant b\}$ such that $S = Z^n \cap P$. Hence $S = \{x \in Z^n_+: Ax \leqslant b\}$, and the integer program can be written as

$$\max\{cx: Ax \leqslant b, x \in Z^n_+\}.$$

Throughout this chapter, unless otherwise specified, A and b are $m \times n$ and $m \times 1$ matrices, respectively, with rational coefficients.

The topics to be studied in this chapter and the next one concern the representation of an integer program by a linear program that has the same optimal solution. We have already established the existence of such a representation, namely,

$$\max\{cx: x \in S\} = \max\{cx: x \in \text{conv}(S)\}$$

(see Theorem 6.3 of Section I.4.6).

Here we are interested in the constructive aspects of the representation. This will be done by using integrality and valid inequalities for P to construct suitable valid inequalities for the set S.

Example 1.1. $S = \{x \in Z^n_+: Ax \leqslant b\}$.

$$A = \begin{pmatrix} -1 & 2 \\ 5 & 1 \\ -2 & -2 \end{pmatrix}, \qquad b = \begin{pmatrix} 4 \\ 20 \\ -7 \end{pmatrix}.$$

Figure 1.1 shows the polytope defined by the constraints $Ax \leqslant b$ and $x \geqslant 0$ (the outer polytope), the feasible integral points (black dots), and conv(S) (the inner polytope).

We have

$$S = \left\{\begin{pmatrix}2\\2\end{pmatrix}, \begin{pmatrix}2\\3\end{pmatrix}, \begin{pmatrix}3\\1\end{pmatrix}, \begin{pmatrix}3\\2\end{pmatrix}, \begin{pmatrix}3\\3\end{pmatrix}, \begin{pmatrix}4\\0\end{pmatrix}\right\} = \{x^1, x^2, \dots, x^6\}.$$

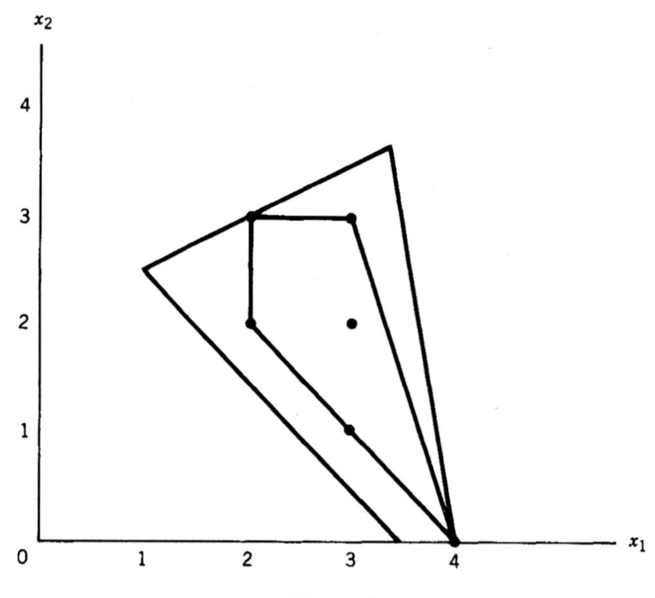

Figure 1.1

Thus, $\text{conv}(S) = \{x\colon x = \Sigma_{i=1}^{6} \lambda_i x^i,\ \lambda_i \geqslant 0 \text{ for } i = 1, \ldots, 6 \text{ and } \Sigma_{i=1}^{6} \lambda_i = 1\}$. Since

$$\begin{pmatrix}3\\2\end{pmatrix} = \tfrac{1}{2}\begin{pmatrix}3\\1\end{pmatrix} + \tfrac{1}{2}\begin{pmatrix}3\\3\end{pmatrix} \quad \text{and} \quad \begin{pmatrix}3\\1\end{pmatrix} = \tfrac{1}{2}\begin{pmatrix}2\\2\end{pmatrix} + \tfrac{1}{2}\begin{pmatrix}4\\0\end{pmatrix},$$

conv(S) is a polytope defined by the four extreme points

$$\begin{pmatrix}2\\2\end{pmatrix}, \begin{pmatrix}2\\3\end{pmatrix}, \begin{pmatrix}3\\3\end{pmatrix}, \begin{pmatrix}4\\0\end{pmatrix}.$$

In this small example, it is a simple matter to obtain a linear inequality representation of conv(S) from the four lines defined by the adjacent pairs of extreme points. In particular, conv(S) is defined by the constraints $A'x \leqslant b'$, where

$$A' = \begin{pmatrix} -1 & 0 \\ 0 & 1 \\ -1 & -1 \\ 3 & 1 \end{pmatrix}, \qquad b' = \begin{pmatrix} -2 \\ 3 \\ -4 \\ 12 \end{pmatrix}.$$

From the extreme points of conv(S) we can obtain its polar set, which is the set of valid inequalities for conv(S) (see Proposition 5.1 of Section I.4.5). This is also the set of valid inequalities for S, since $S \subseteq \text{conv}(S)$ and any valid inequality for S is also valid for conv(S) (see Proposition 6.5 of Section I.4.6).

Thus the set of valid inequalities for S is given by

$$\begin{aligned} 2\pi_1 + 2\pi_2 - \pi_0 &\leqslant 0 \\ 2\pi_1 + 3\pi_2 - \pi_0 &\leqslant 0 \\ 3\pi_1 + 3\pi_2 - \pi_0 &\leqslant 0 \\ 4\pi_1 \qquad - \pi_0 &\leqslant 0. \end{aligned}$$

The polyhedral cone defined by these four half-spaces is shown in Figure 1.2.

The valid inequalities $\pi x \leqslant \pi_0$ and $\gamma x \leqslant \gamma_0$ are said to be *equivalent* if $(\gamma, \gamma_0) = \lambda(\pi, \pi_0)$ for some $\lambda > 0$. If they are not equivalent and there exists $\mu > 0$ such that $\gamma \geqslant \mu\pi$ and $\gamma_0 \leqslant \mu\pi_0$, then $\{x \in R^n_+ : \gamma x \leqslant \gamma_0\} \subset \{x \in R^n_+ : \pi x \leqslant \pi_0\}$. In this case we say that $\gamma x \leqslant \gamma_0$ *dominates* or *is stronger than* $\pi x \leqslant \pi_0$ or that $\pi x \leqslant \pi_0$ *is dominated by* or *is weaker than* $\gamma x \leqslant \gamma_0$. A *maximal* valid inequality is one that is not dominated by any other valid inequality. Any maximal valid inequality for S defines a nonempty face of conv(S), and the set of maximal valid inequalities contains all of the facet-defining inequalities for conv(S).

Example 1.1 (continued). The valid inequality $3x_1 + 4x_2 \leqslant 24$ is not maximal since it is dominated by the maximal valid inequality $x_1 + x_2 \leqslant 6$. The valid inequality $x_1 \leqslant 4$ defines the zero-dimensional face $\{(4\ 0)\}$, but it is not maximal since it is dominated by the facet-defining inequality $3x_1 + x_2 \leqslant 12$. (see Figure 1.3).

Given $P = \{x \in R^n_+ : Ax \leqslant b\}$ and $S = P \cap Z^n$, facets of conv(S) can be constructed iteratively using integrality and the linear inequality description of P. This means that we start with the valid inequalities $Ax \leqslant b$ and, if they are not enough to define conv(S), we progressively construct stronger valid inequalities.

We obtain valid inequalities for P by taking nonnegative linear combinations of rows of $Ax \leqslant b$. (These can be weakened by adding in nonnegative linear combinations of $-x \leqslant 0$). This gives the infinite family of valid inequalities

(1.1) $$(uA - v)x \leqslant ub + \alpha \quad \text{for all } u \in R^m_+,\ v \in R^n_+, \text{ and } \alpha \geqslant 0.$$

Moreover, under some technical assumptions stated below, all valid inequalities for P can be obtained in this way, and the linear combinations can be restricted to using, at most, min(m, n) rows of A.

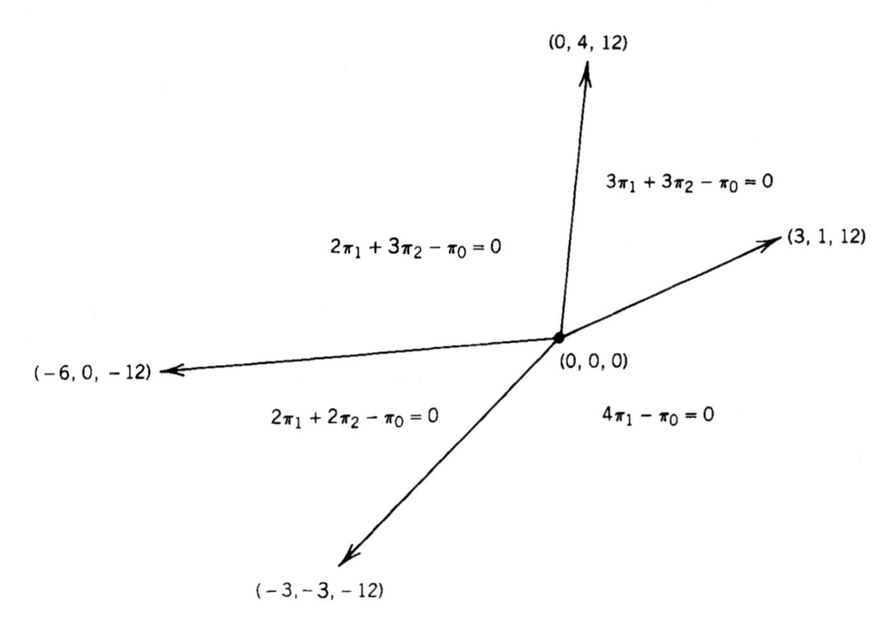

Figure 1.2

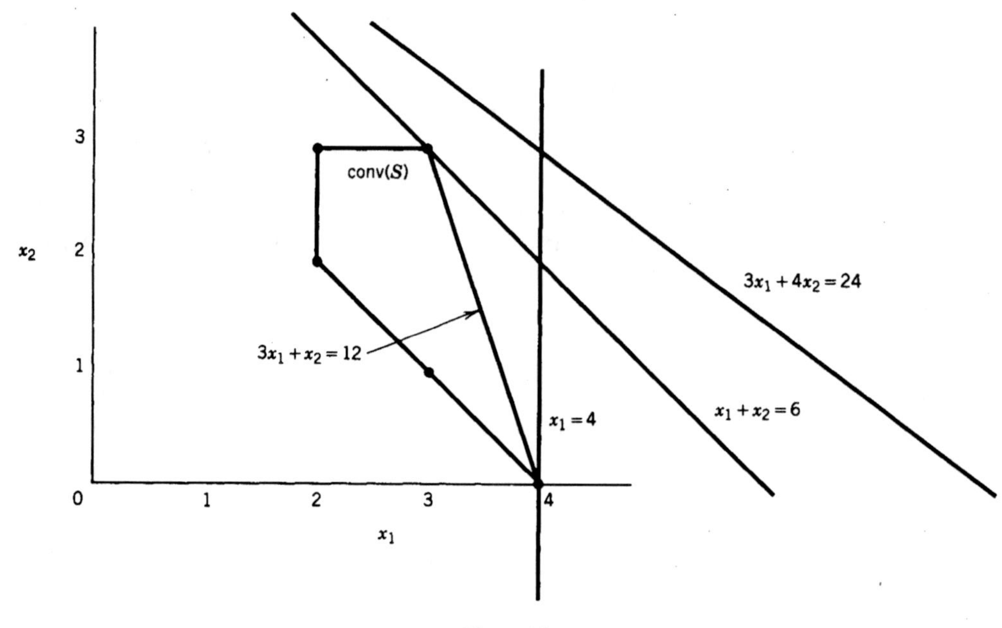

Figure 1.3

Proposition 1.1. *Let $\pi x \leqslant \pi_0$ be any valid inequality for $P = \{x \in R_+^n: Ax \leqslant b\}$. Then $\pi x \leqslant \pi_0$ is either equivalent to or dominated by an inequality of the form $uAx \leqslant ub$, $u \in R_+^m$ if any of the following conditions hold:*

a. $P \neq \emptyset$ *(in this case no more than* $\min(m, n)$ *components of u need be positive).*
b. $\{u \in R_+^m: uA \geqslant \pi\} \neq \emptyset$.
c. $A = \binom{A'}{I}$, *where I is an $n \times n$ identity matrix.*

Proof. a. Since (π, π_0) is valid and $P \neq \emptyset$, the linear program $\max\{\pi x: Ax \leqslant b, x \in R_+^n\}$ has a feasible solution and its value is bounded by π_0. Hence the dual linear program has a basic feasible solution $u^0 \in R_+^m$ with $u^0 A \geqslant \pi$ and $u^0 b \leqslant \pi_0$. The vector u^0 has no more than $\min(m, n)$ positive components and $\pi x \leqslant (u^0 A)x \leqslant u^0 b \leqslant \pi_0$.

b. If $P \neq \emptyset$, see part a. So assume $P = \emptyset$ and $\hat{u} \in R_+^m$ satisfies $\hat{u}A \geqslant \pi$. If $\hat{u}b \leqslant \pi_0$, we are done. Otherwise, since $P = \emptyset$ there exists $\bar{u} \in R_+^m$ such that $\bar{u}A \geqslant 0$ and $\bar{u}b < 0$. Hence for some $\beta > 0$ we obtain $(\hat{u} + \beta\bar{u})A \geqslant \pi$ and $(\hat{u} + \beta\bar{u})b \leqslant \pi_0$.

c. It is a simple exercise in linear programming duality to show that for any π there exists a $u \in R_+^m$ with $uA \geqslant \pi$. ■

When $P = \emptyset$, every inequality is valid for P. However, if conditions a and b fail, which is equivalent to the primal and dual linear programs being infeasible, then we cannot generate the valid inequality by nonnegative linear combinations.

Example 1.2

$$
\begin{aligned}
0x_1 + x_2 &\leqslant 1 \\
0x_1 - x_2 &\leqslant -2 \\
x &\in R_+^2.
\end{aligned}
$$

Here $P = \emptyset$. Consider the valid inequality $x_1 \leqslant 1$. The dual feasibility region given by

$$\begin{aligned} 0u_1 + 0u_2 &\geqslant 1 \\ u_1 - u_2 &\geqslant 0 \\ u &\in R_+^2 \end{aligned}$$

is also empty. Thus conditions a and b fail. Moreover, it is obvious that $x_1 \leqslant 1$ is not equivalent to or dominated by an inequality of the form $(0u_1 + 0u_2)x_1 + (u_1 - u_2)x_2 \leqslant u_1 - 2u_2$.

To avoid the trouble that can arise when $P = \emptyset$ we frequently assume that the linear inequality description of P contains explicit bounding constraints, that is, $A = \binom{A'}{I}$.

Since $S \subseteq P$, the inequalities (1.1) are also valid for S. However, unless $\text{conv}(S) = P$, there are valid inequalities for S that are not valid for P and hence cannot be obtained just from nonnegative linear combinations.

Integrality must be used to obtain the inequalities for S that are not valid for P. We now consider techniques that use integrality to obtain valid inequalities.

Integer Rounding

This approach is based on the simple principle that if a is an integer and $a \leqslant b$, then $a \leqslant \lfloor b \rfloor$, where $\lfloor b \rfloor$ is the largest integer less than or equal to b.

Consider the matching problem on the graph $G = (V, E)$. For $U \subseteq V$, let $E(U)$ be the set of edges with both ends in U and let $\delta(U)$ be the set of edges with one end in U. A subset of edges is a matching if

$$\sum_{e \in \delta(\{i\})} x_e \leqslant 1 \quad \text{for all } i \in V \tag{1.2}$$

$$x \in Z_+^{|E|}, \tag{1.3}$$

where $x_e = 1$ if e is in the matching and $x_e = 0$ otherwise.

For any set $U \subseteq V$, the number of edges in a matching with both ends in U is at most $\lfloor \frac{1}{2} |U| \rfloor$. Thus if $|U| = 2k + 1$, then

$$\sum_{e \in E(U)} x_e \leqslant k \tag{1.4}$$

is a valid inequality for all $k \geqslant 1$.

The inequalities (1.4) cannot be obtained just by taking nonnegative linear combinations of the constraints (1.2). However, they can be justified algebraically by the following three-step argument.

i. Take a linear combination of the constraints (1.2) with weights $u_i = \frac{1}{2}$ for all $i \in U$ and $u_i = 0$ for all $i \in V \setminus U$. This yields the valid inequality

$$\sum_{e \in E(U)} x_e + \frac{1}{2} \sum_{e \in \delta(U)} x_e \leqslant \frac{1}{2} |U|. \tag{1.5}$$

ii. Since $x_e \geqslant 0$ for all $e \in E$, it follows that

$$-\frac{1}{2}\sum_{e\in\delta(U)} x_e \leqslant 0 \tag{1.6}$$

is a valid inequality. Adding (1.5) and (1.6) yields

$$\sum_{e\in E(U)} x_e \leqslant \frac{1}{2}|U|. \tag{1.7}$$

iii. From (1.3), the left-hand side of (1.7) is an integer. Therefore, the right-hand side can be replaced by the largest integer equal to or less than it; that is, if $|U| = 2k + 1$, then $\lfloor\frac{1}{2}|U|\rfloor = k$ is a valid right-hand side. Thus (1.4) is a valid inequality for all $k \geqslant 1$.

In Section III.2.3 we will prove that the convex hull of matchings is given by (1.2) and (1.4) for all odd sets U with $|U| \geqslant 3$, and $x_e \geqslant 0$ for all $e \in E$. But, in general, the three-step procedure must be applied recursively.

For $S = \{x \in Z^n_+ : Ax \leqslant b\}$, where $A = (a_1, a_2, \ldots, a_n)$ and $N = \{1, \ldots, n\}$, the three-step procedure yields the following:

i. $\Sigma_{j\in N} (ua_j)x_j \leqslant ub$ for all $u \geqslant 0$;
ii. $\Sigma_{j\in N} (\lfloor ua_j\rfloor)x_j \leqslant ub$, since $x \geqslant 0$ implies $-\Sigma_{j\in N} (ua_j - \lfloor ua_j\rfloor)x_j \leqslant 0$; and
iii. $\Sigma_{j\in N} (\lfloor ua_j\rfloor)x_j \leqslant \lfloor ub\rfloor$, since $x \in Z^n$ implies $\Sigma_{j\in N} (\lfloor ua_j\rfloor)x_j$ is an integer.

The crucial step is iii, where we invoke integrality to round down the right-hand side.

The valid inequality

$$\sum_{j\in N} (\lfloor ua_j\rfloor)x_j \leqslant \lfloor ub\rfloor \tag{1.8}$$

can be added to $Ax \leqslant b$, and then the procedure can be repeated by combining generated inequalities and/or original ones. As noted in Proposition 1.1, in one application of the procedure it is sufficient to combine, at most, n inequalities. The procedure is called the *Chvatal–Gomory (C–G) rounding method*, and the inequalities it produces are called *C–G inequalities*. In Section 2, we will prove that by repeating the C–G procedure a finite number of times, all of the valid inequalities for S can be generated.

In Example 1.1, $u = (\frac{5}{11} \;\; \frac{3}{22} \;\; 0)$ in step i yields

$$\frac{5}{22}x_1 + \frac{23}{22}x_2 \leqslant \frac{100}{22}.$$

Then

$$x_2 \leqslant \frac{100}{22} \qquad \text{(step ii)}$$

and

$$x_2 \leqslant 4 \qquad \text{(step iii)}.$$

A geometric explanation of the procedure is shown in Figure 1.4.

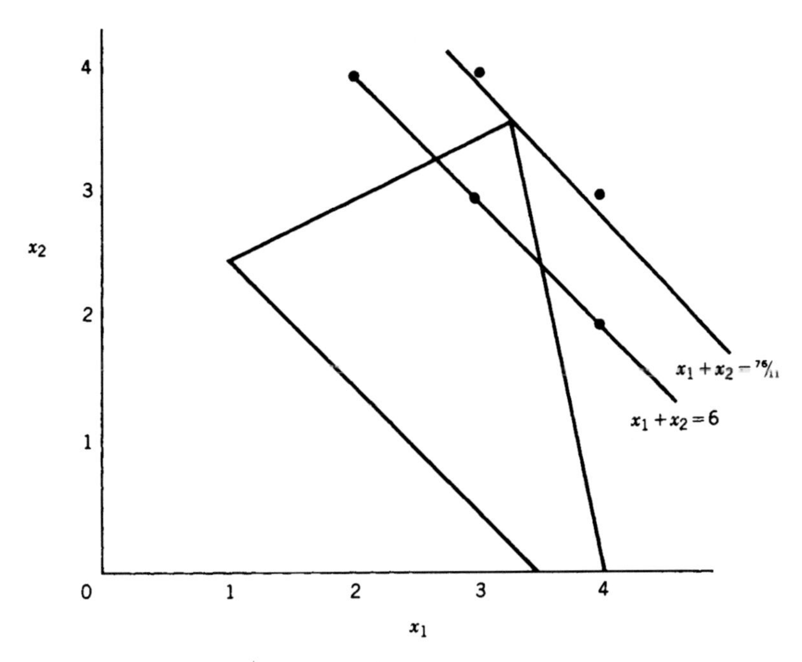

Figure 1.4

In Example 1.1, $u = (\frac{4}{11}\ \frac{3}{11}\ 0)$ yields $x_1 + x_2 \leqslant \frac{76}{11}$. But since there are no points of S such that $x_1 + x_2 = \frac{76}{11}$, it is permissible to push the hyperplane inward until it meets a point in S. However, step iii does not, in general, allow the hyperplane to be pushed in this far. It can be pushed in no further than an intersection with some point of Z^2, and perhaps not even this far if the integral coefficients of x_1 and x_2 are not relatively prime. It is fortuitous in the example that the line $x_1 + x_2 = 6$ happens to contain a point in S. Note, however, that if we started with the equivalent inequality $2x_1 + 2x_2 = \frac{152}{11}$, then step iii would have yielded $2x_1 + 2x_2 \leqslant 13$. The line $2x_1 + 2x_2 = 13$ contains no points of Z^2.

The above discussion is summarized by the following result. Let the greatest common divisor of the integers a and b be denoted by gcd$\{a, b\}$.

Proposition 1.2. *Let* $S = \{x \in Z^n\colon \Sigma_{j\in N} a_jx_j \leqslant b\}$, *where* $a_j \in Z^1$ *for* $j \in N$, *and let* $k = \text{gcd}\{a_1, \ldots, a_n\}$. *Then* $\text{conv}(S) = \{x \in R^n\colon \Sigma_{j\in N} (a_j/k)x_j \leqslant \lfloor b/k \rfloor\}$.

Proof. Using steps i and iii of the C–G procedure with $u = 1/k$, it follows that $\Sigma_{j\in N} (a_j/k)a_j \leqslant \lfloor \frac{b}{k} \rfloor$ is a valid inequality for S. Since gcd$\{a_1/k, \ldots, a_n/k\} = 1$, it follows from Corollary 4.7 of Section I.7.4 that $\Sigma_{j\in N} (a_j/k)x_j = \lfloor b/k \rfloor$ contains an infinite number of points of S—in particular, n affinely independent ones. Therefore the inequality represents a facet of conv(S). But conv(S) contains no other facets, since if $\pi x \leqslant \pi_0$ is a different facet, then $\pi x \leqslant \pi_0$ for all $x \in R^n$ such that $\Sigma_{j\in N} (a_j/k)x_j = \lfloor b/k \rfloor$, which is impossible. ■

Proposition 1.2 shows the limitations of one application of the C–G rounding method. The step i inequality $\Sigma_{j\in N} (ua_j)x_j \leqslant ub$ must be weakened in step ii if there exists a j such that ua_j is not integral. Then the best we can hope for from step iii is to intersect $\{x \in R^n_+\colon Ax \leqslant b\}$ with $\text{conv}\{x \in Z^n\colon \Sigma_{j\in N} \lfloor ua_j \rfloor x_j \leqslant \lfloor ub \rfloor\}$.

Many other arguments can be used to generate valid inequalities. Some are explained and illustrated below.

Modular Arithmetic

Here we derive a valid inequality for the set of solutions to one linear equation in nonnegative integers, that is, $S = \{x \in Z_+^n: \Sigma_{j\in N} a_j x_j = a_0\}$, where $a_j \in R^1$ for all j.

Let d be a positive integer and

$$S_d = \left\{x \in Z_+^n: \sum_{j\in N} a_j x_j = a_0 + kd \text{ for some integer } k\right\}.$$

We are going to derive a valid inequality for S_d. Then, since $S \subseteq S_d$, the inequality is valid for S.

Let $a_j = b_j + \alpha_j d$ for $j = 0, 1, \ldots, n$, where $0 \leqslant b_j < d$ and α_j is an integer, that is, b_j is the remainder when a_j is divided by d. Then

$$S_d = \left\{x \in Z_+^n: \sum_{j\in N} b_j x_j = b_0 + kd \text{ for some integer } k\right\}.$$

Now $\Sigma_{j\in N} b_j x_j \geqslant 0$ and $b_0 < d$ imply $k \geqslant 0$; hence we obtain the valid inequality

$$(1.9) \qquad \sum_{j\in N} b_j x_j \geqslant b_0.$$

Inequality (1.9) is nontrivial only if d does not divide a_0, that is, only if $b_0 > 0$.

For the set S given by

$$37x_1 - 68x_2 + 78x_3 + x_4 = 141, \qquad x \in Z_+^4,$$

inequality (1.9) with $d = 12$ yields $x_1 + 4x_2 + 6x_3 + x_4 \geqslant 9$.

An important set of inequalities of this type arises when $d = 1$ and a_0 is not an integer. Then (1.9) yields the valid inequality

$$(1.10) \qquad \sum_{j\in N} (a_j - \lfloor a_j \rfloor) x_j \geqslant a_0 - \lfloor a_0 \rfloor,$$

which is called a *Gomory cutting plane* (see Section 3).

For example, suppose $x_0 = 3\frac{3}{4} - \frac{1}{2}x_1 + \frac{7}{4}x_2 - \frac{11}{4}x_3$ is an equation obtained in the solution of a linear program and we also require the variables to be nonnegative integers. Then (1.10) yields $\frac{1}{2}x_1 + \frac{1}{4}x_2 + \frac{3}{4}x_3 \geqslant \frac{3}{4}$.

Disjunctive Constraints

Proposition 1.3. *If $\Sigma_{j\in N} \pi_j^1 x_j \leqslant \pi_0^1$ is valid for $S_1 \subset R_+^n$ and $\Sigma_{j\in N} \pi_j^2 x_j \leqslant \pi_0^2$ is valid for $S_2 \subset R_+^n$, then*

$$(1.11) \qquad \sum_{j\in N} \min(\pi_j^1, \pi_j^2) x_j \leqslant \max(\pi_0^1, \pi_0^2)$$

is valid for $S_1 \cup S_2$.

In other words, if we must satisfy one set of constraints or another set, but not necessarily both, and we know valid inequalities for each set, then (1.11) is a valid inequality for the disjunction of the two sets.

This yields another systematic way, called the *disjunctive procedure*, of generating valid inequalities for the region $S = \{x \in Z_+^n: Ax \leq b\}$. The two steps of the procedure are:

i. $\Sigma_{j\in N} (ua_j)x_j \leq ub$ for all $u \geq 0$.
ii. Given $\delta \in Z_+^1$, if

(a) $$\sum_{j\in N} \pi_j x_j - \alpha(x_k - \delta) \leq \pi_0$$

is valid for S for some $\alpha \geq 0$ and

(b) $$\sum_{j\in N} \pi_j x_j + \beta(x_k - \delta - 1) \leq \pi_0$$

is valid for S for some $\beta \geq 0$, then

(c) $$\sum_{j\in N} \pi_j x_j \leq \pi_0$$

is valid for S.

Note that (a) shows that (c) is valid for $S_1 = S \cap \{x \in Z_+^n: x_k \leq \delta\}$ and (b) shows that (c) is valid for $S_2 = S \cap \{x \in Z_+^n: x_k \geq \delta + 1\}$. Since $S = S_1 \cup S_2$, Proposition 1.3 establishes that (c) is valid for S.

Inequalities generated by repeated application of the disjunctive procedure are called *D-inequalities*.

Example 1.3. An example of the D-inequalities is shown in Figure 1.5, where

$$P = \left\{x \in R_+^2: -x_1 + x_2 \leq \frac{1}{2}, \frac{1}{2}x_1 + x_2 \leq \frac{5}{4}, x_1 \leq 2\right\}.$$

The first two inequalities can be rewritten as

$$-\frac{1}{4}x_1 + x_2 - \frac{3}{4}x_1 \qquad \leq \frac{1}{2}$$

and

$$-\frac{1}{4}x_1 + x_2 - \frac{3}{4}(1 - x_1) \leq \frac{1}{2}.$$

Using the disjunction $x_1 \leq 0$ or $x_1 \geq 1$ leads to the valid inequality $-\frac{1}{4}x_1 + x_2 \leq \frac{1}{2}$ for $S = P \cap Z^2$.

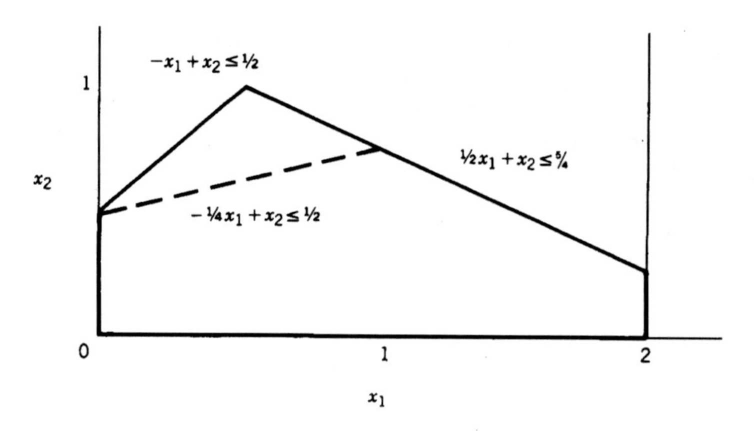

Figure 1.5

An interesting property of the disjunctive procedure is that if we just consider a disjunction on a single variable, then every valid inequality for the disjunction is a D-inequality.

Proposition 1.4. *Let* $P = \{x \in R_+^n : Ax \leqslant b\}$ *and suppose* $A = \binom{A'}{I}$. *Then* $\pi x \leqslant \pi_0$ *is a valid inequality for*

$$P' = (P \cap \{x : x_k \leqslant \delta\}) \cup (P \cap \{x : x_k \geqslant \delta + 1\})$$

only if there exist $\alpha,\ \beta \geqslant 0$ *such that* $\sum_{j\in N} \pi_j x_j - \alpha(x_k - \delta) \leqslant \pi_0$ *and* $\sum_{j\in N} \pi_j x_j + \beta(x_k - \delta - 1) \leqslant \pi_0$ *are valid for P.*

Proof. Suppose that $\pi x \leqslant \pi_0$ is valid for P'. Then $\pi x \leqslant \pi_0$ is valid for $P \cap \{x : x_k \leqslant \delta\}$ and $P \cap \{x : x_k \geqslant \delta + 1\}$. Hence by Proposition 1.1, there exists $(u^1, \alpha) \geqslant 0$ and $(u^2, \beta) \geqslant 0$ such that

$$u^1 A + \alpha e_k \geqslant \pi, \qquad u^1 b + \alpha\delta \leqslant \pi_0$$

and

$$u^2 A - \beta e_k \geqslant \pi, \qquad u^2 b - \beta(\delta + 1) \leqslant \pi_0,$$

where e_k is the kth unit vector. Since $u^1 \geqslant 0$ and $u^2 \geqslant 0$, the inequalities $u^i Ax \leqslant u^i b$ for $i = 1, 2$, are valid for P and are equal to or dominate

$$\pi x - \alpha(x_k - \delta) \leqslant \pi_0 \quad \text{and} \quad \pi x - \beta(\delta + 1 - x_k) \leqslant \pi_0. \qquad \blacksquare$$

In the next section we will show that for 0-1 problems it is possible to use this procedure variable by variable to produce all valid inequalities for $S = P \cap B^n$.

Another application of Proposition 1.3 is to derive a valid inequality for the system

$$(1.12) \qquad S = \left\{ x \in R_+^n \colon x_0 = a_0 - \sum_{j \in N} a_j x_j \in Z^1, \right\}$$

where $a_j \in R^1$ for all j. The feasible set S of (1.12) is contained in $S_1 \cup S_2$, where

$$S_1 = \left\{ x \in R_+^n \colon \sum_{j \in N} a_j x_j \geqslant a_0 - \lfloor a_0 \rfloor \right\}$$

and

$$S_2 = \left\{ x \in R_+^n \colon \sum_{j \in N} a_j x_j \leqslant a_0 - \lfloor a_0 \rfloor - 1 \right\}.$$

Note that when a_0 is not integral, $S_1 \cup S_2$ is the standard disjunction used in enumeration, that is $x_0 \leqslant \lfloor a_0 \rfloor$ or $x_0 \geqslant \lfloor a_0 \rfloor + 1$. Since

$$S_1 = \left\{ x \in R_+^n \colon \frac{1}{\lfloor a_0 \rfloor - a_0} \sum_{j \in N} a_j x_j \leqslant -1 \right\}$$

and

$$S_2 = \left\{ x \in R_+^n \colon \frac{1}{\lfloor a_0 \rfloor + 1 - a_0} \sum_{j \in N} a_j x_j \leqslant -1 \right\}$$

we obtain from (1.11) the valid inequality $\sum_{j \in N} \pi_j x_j \leqslant -1$, where

$$(1.13) \qquad \pi_j = \min\left(\frac{a_j}{\lfloor a_0 \rfloor - a_0}, \frac{a_j}{\lfloor a_0 \rfloor + 1 - a_0} \right) \quad \text{for } j \in N.$$

If $x_0 = 3\frac{3}{4} - \frac{1}{2}x_1 + \frac{7}{4}x_2 - \frac{11}{4}x_3$, where $x_0 \in Z^1$ and $x \in R_+^3$, then (1.13) yields the valid inequality $\frac{2}{3}x_1 + 7x_2 + \frac{11}{3}x_3 \geqslant 1$. Note that this inequality is weaker than the one obtained previously where we required $x \in Z_+^3$.

Boolean Implications

Here we derive some valid inequalities that require the assumption $S \subset B^n = \{x \in Z_+^n \colon x_j \leqslant 1 \text{ for } j \in N\}$.

Suppose

$$(1.14) \qquad S = \left\{ x \in B^n \colon \sum_{j \in N} a_j x_j \leqslant b \right\},$$

where the a_j's and b are positive integers, that is, S is the feasible set for a 0-1 knapsack problem. Let $C \subset N$ be such that $\sum_{j \in C} a_j > b$. Then a valid inequality for (1.14) is

$$(1.15) \qquad \sum_{j \in C} x_j \leqslant |C| - 1.$$

Valid inequalities of this form are used in the solution of general 0-1 integer programs (see Sections II.2.2 and II.6.2).

The system

$$T = \left\{ x \in B^1, y \in R^n_+ : \sum_{j \in N} y_j \leq nx, y_j \leq 1 \text{ for } j \in N \right\} \tag{1.16}$$

arises in many models. Here we see that $y_j = 0$ if $x = 0$ and $y_j \leq 1$ if $x = 1$. Hence

$$y_j \leq x \quad \text{for } j \in N \tag{1.17}$$

are valid inequalities for (1.16).

These are just two examples of a variety of ad hoc tricks for obtaining valid inequalities from Boolean implications.

Geometric or Combinatorial Implications

The valid inequalities that we illustrate here are related to the logical implications considered above but are associated with combinatorial systems such as graphs and matroids. One example is the set of constraints (1.4) for the matching problem.

To give another, consider the node-packing problem on the graph $G = (V, E)$. A subset of nodes is a packing if

$$x_i + x_j \leq 1 \quad \text{for all } (i, j) \in E \tag{1.18}$$

and

$$x \in B^{|V|}, \tag{1.19}$$

where $x_i = 1$ if node i is in the packing and $x_i = 0$ otherwise. In the graph of Figure 1.6, no more than one of the pairwise adjacent nodes {4, 5, 6, 7} can be in a packing and no more than two of the nodes from the cycle {1, 2, 3, 4, 5} can be in a packing. Thus we obtain the valid inequalities

$$x_4 + x_5 + x_6 + x_7 \leq 1 \tag{1.20}$$

and

$$x_1 + x_2 + x_3 + x_4 + x_5 \leq 2, \tag{1.21}$$

neither of which can be obtained from nonnegative linear combinations of (1.18) and nonnegativity.

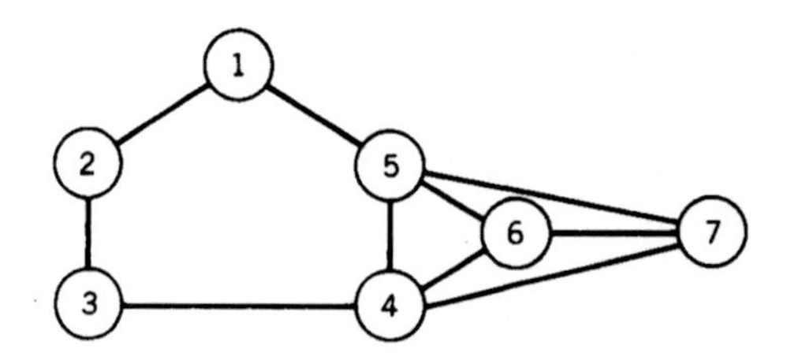

Figure 1.6

Valid Inequalities and Model Formulation

At this point, we suggest that the reader go back to Chapter I.1 to observe that valid inequalities, particularly those derived from Boolean and combinatorial implications, have been used to formulate models. So where should we draw the line between using valid inequalities in formulating models and using them in the solution of the model? There is no right answer to this question. It is sufficient to formulate the model so that $\{x \in Z_+^n: Ax \leqslant b\}$ is precisely the set of feasible points. But the formulation is not unique and, in particular, any valid inequalities can be added.

In some cases, such valid inequalities imply the ones from which they are derived, so the original ones can be deleted from the formulation. For example, by summing the constraints (1.17) we obtain the linear constraint of (1.16), which can then be deleted from the formulation. By using the constraint (1.20), we can delete the constraints (1.18) for the six edges joining nodes 4, 5, 6, and 7. However, (1.21) does not imply any of the edge constraints (1.18).

With hindsight, the valid inequalities that we would like to have in our formulation are precisely those that are active in an optimal solution. Thus, given a formulation, a guiding principle is to include any additional valid inequalities that we know, which we believe might be active in an optimal solution.

The methods given above for generating valid inequalities do not exhaust all possibilities, and others will be introduced later. Usually, a valid inequality can be obtained by a variety of different arguments, although one method may produce it directly while another requires repeated applications of the procedure. For example, two applications of the C–G procedure are needed to derive (1.20) from (1.18) and (1.19).

2. GENERATING ALL VALID INEQUALITIES

Given $S = \{x \in Z_+^n: Ax \leqslant b\}$, where $A = (a_1, \ldots, a_n)$, in Section 1 we used the C–G procedure to develop the valid inequalities

$$\sum_{j \in N} \lfloor ua_j \rfloor x_j \leqslant \lfloor ub \rfloor$$

for any $u \geqslant 0$ and showed how integrality can be used to develop D-inequalities for S. We will show now, by using these procedures a finite number of times, that it is possible to generate all valid inequalities for S. To simplify exposition, we say that any valid inequality dominated by a C–G inequality (D-inequality) is also a C–G inequality (D-inequality).

0-1 Problems

We consider 0-1 problems with $P = \{x \in R_+^n: Ax \leqslant b, x \leqslant 1\}$ and $S = P \cap Z^n$. First we will show that all valid inequalities for S are D-inequalities.

Define $P^n = P$, and for $t = n - 1, \ldots, 0$ define

$$P^t = \text{conv}[(P^{t+1} \cap \{x: x_{t+1} = 0\}) \cup (P^{t+1} \cap \{x: x_{t+1} = 1\})].$$

For $n = 2$, $P^2 = P$, P^1 and P^0 are shown in Figure 2.1. Note that in this example, $\text{conv}(S) = P^0$.

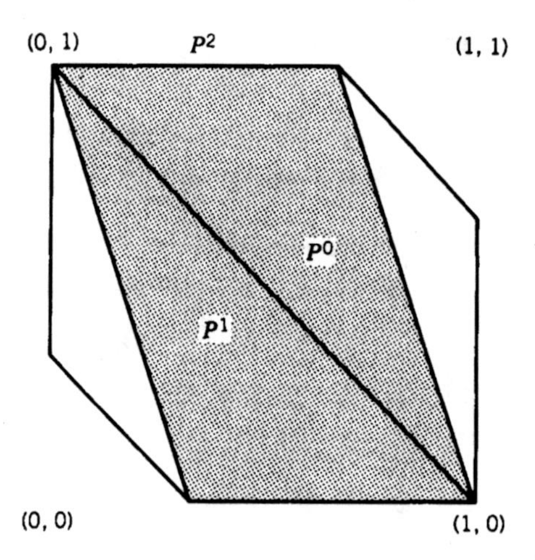

Figure 2.1. P^2 is the outer polytope. P^1 is the shaded polytope. P^0 is the line joining (0, 1) and (1, 0).

Since P^0 contains all of the integral points in P, in general we have $\text{conv}(S) \subseteq P^0$. We prove that all valid inequalities for S are D-inequalities by showing that all valid inequalities for S are D-inequalities for P^0. This also yields $P^0 \subseteq \text{conv}(S)$, so $P^0 = \text{conv}(S)$.

Suppose $\pi x \leqslant \pi_0$ is valid for S and $\omega > 0$, then so are the inequalities

$$I_t(N^0, N^1) \quad \sum_{j\in N} \pi_j x_j - \omega \sum_{j\in N^0} x_j - \omega \sum_{j\in N^1} (1 - x_j) \leqslant \pi_0,$$

where $N^0 \cap N^1 = \emptyset$, $N^0 \cup N^1 = \{1, \ldots, t\}$, and $t = 0, \ldots, n$ ($t = 0$ means that $N^0 = N^1 = \emptyset$). We will show that if $\pi x \leqslant \pi_0$ is valid for S, then $I_t(N^0, N^1)$ is a D-inequality for P^t. In particular, when $t = 0$, we obtain that $\pi x \leqslant \pi_0$ is a D-inequality for P^0.

The proof uses a backward induction that is convenient to represent by an enumeration tree in which the nodes at level t represent all partitions of $\{1, \ldots, t\}$ (see Figure 2.2). The intention of the figure is to show that $I_t(N^0, N^1)$ is derived from $I_{t+1}(N^0 \cup \{t + 1\}, N^1)$ and $I_{t+1}(N^0, N^1 \cup \{t + 1\})$.

Proposition 2.1. *If $I_{t+1}(N^0 \cup \{t+1\}, N^1)$ and $I_{t+1}(N^0, N^1 \cup \{t + 1\})$ are D-inequalities for P^{t+1}, then $I_t(N^0, N^1)$ is a D-inequality for P^t.*

Proof. By hypothesis, we have

$$\pi x - \omega \sum_{j\in N^0\cup(t+1)} x_j - \omega \sum_{j\in N^1} (1 - x_j) \leqslant \pi_0$$

and

$$\pi x - \omega \sum_{j\in N^0} x_j - \omega \sum_{j\in N^1\cup(t+1)} (1 - x_j) \leqslant \pi_0$$

are D-inequalities for P^{t+1}. Since $x_{t+1} \in \{0, 1\}$, step ii of the disjunctive procedure establishes that $I_t(N^0, N^1)$ is a D-inequality for P^t. ■

Therefore if we can establish that the inequalities $I_n(N^0, N^1)$ are D-inequalities for $P = P^n$, then Proposition 2.1 yields that $I_t(N^0, N^1)$ is a D-inequality for P^t for all partitions (N^0, N^1) of $\{1, \ldots, t\}$ and all $t = 0, 1, \ldots, n-1$. In fact, it suffices to show that all of the inequalities $I_n(N^0, N^1)$ are valid for P, since any valid inequality for P can, by Proposition 1.1, be obtained by step 1 of the disjunctive procedure.

Proposition 2.2. *If $\sum_{j \in N} \pi_j x_j \leqslant \pi_0$ is a valid inequality for S, there exists an $\omega \geqslant 0$ such that all of the inequalities $I_n(N^0, N^1)$ are valid for P.*

Proof. If $P = \emptyset$, the result follows immediately from Proposition 1.1. If $P \neq \emptyset$, consider the extreme points $\{x^k\}_{k \in K}$ of P. If $x^k \in S$, then x^k satisfies the inequality $I_n(N^0, N^1)$ because $\pi x^k \leqslant \pi_0$ by hypothesis and $\omega \geqslant 0$.

If $x^k \notin S$, let

$$\alpha^k = \min_{\{N^0 : N^0 \cup N^1 = N\}} \left(\sum_{j \in N^0} x_j^k + \sum_{j \in N^1} (1 - x_j^k) \right) > 0$$

and $\alpha = \min_{\{k \in K : x^k \notin S\}} \alpha^k > 0$. Thus for $x^k \notin S$ we obtain

$$-\omega \sum_{j \in N^0} x_j^k - \omega \sum_{j \in N^1} (1 - x_j^k) \leqslant -\omega\alpha.$$

Now let $\gamma = \max_{x \in P}(\pi x - \pi_0)$. Hence $\pi x^k \leqslant \gamma + \pi_0$ for all $k \in K$. Thus for $x^k \notin S$ we have

$$\pi x^k - \omega \sum_{j \in N^0} x_j^k - \omega \sum_{j \in N^1} (1 - x_j^k) \leqslant \gamma + \pi_0 - \omega\alpha.$$

The result follows by taking $\omega \geqslant \gamma/\alpha$ and observing that an inequality valid for the extreme points of P is valid for P. ∎

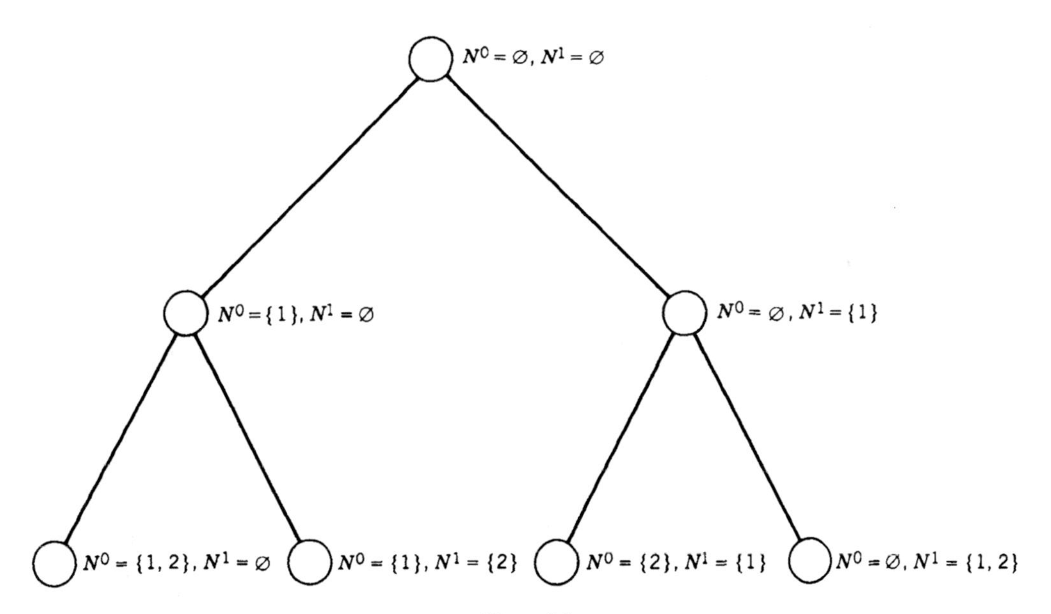

Figure 2.2

Theorem 2.3. *Every valid inequality for $S = P \cap Z^n$ with $P = \{x \in R^n_+: Ax \leqslant b, x \leqslant 1\}$ is a D-inequality.*

Theorem 2.4. $P^0 = \text{conv}(S)$.

Theorem 2.4 indicates a surprising property of 0-1 integer programs—namely, that it suffices to integralize one variable at a time to obtain the convex hull.

We have actually shown a somewhat stronger result that will be used in Section 6. Let $Q^t = \text{conv}\{x \in P: (x_{t+1}, \ldots, x_n) \in Z^{n-t}\}$, so $Q^t \subseteq P^t$ for $t = 0, 1, \ldots, n$ and $Q^0 = \text{conv}(S)$. By assuming that $\pi x \leqslant \pi_0$ is a valid inequality for Q^t, the proof we have given shows:

Theorem 2.5. *Every valid inequality for Q^t, $t = 0, \ldots, n$, is a D-inequality and $P^t = Q^t$.*

Now we show that all valid inequalities for the 0-1 problem are also C–G inequalities. The structure of the proof is the same as for D-inequalities, with two differences. The induction step requires the inequalities

$$I'_t(N^0, N^1) \qquad \sum_{j \in N} \pi_j x_j - \sum_{j \in N^0} x_j - \sum_{j \in N^1} (1 - x_j) \leqslant \pi_0.$$

Note that $I'_t(N^0, N^1)$ is the inequality $I_t(N^0, N^1)$ with $\omega = 1$. Also, we need to assume that $\pi x \leqslant \pi_0 + 1$ is a C–G inequality to prove that $\pi x \leqslant \pi_0$ is a C–G inequality.

Proposition 2.6. *If $(\pi, \pi_0) \in Z^{n+1}$, $\Sigma_{j \in N} \pi_j x_j \leqslant \pi_0 + 1$ is a C–G inequality, and $\Sigma_{j \in N} \pi_j x_j \leqslant \pi_0$ is valid for S, then*

$$I'_n(N^0, N^1) \qquad \sum_{j \in N} \pi_j x_j - \sum_{j \in N^0} x_j - \sum_{j \in N^1} (1 - x_j) \leqslant \pi_0$$

is a C–G inequality for S for all partitions N^0, N^1 of N.

Proof. By Proposition 2.2, inequality $I_n(N^0, N^1)$ is a C–G inequality. If $\omega < 1$, we add $(1 - \omega)$ times the inequality $x_j \leqslant 1$ for $j \in N^1$. The resulting inequality is then dominated by $I'_n(N^0, N^1)$. If $\omega > 1$, then combining $I_n(N^0, N^1)$ with weight $1/\omega$ and $\Sigma_{j \in N} \pi_j x_j \leqslant \pi_0 + 1$ with weight $(\omega - 1)/\omega$ and rounding gives $I'_n(N^0, N^1)$. ■

Theorem 2.7. *If $(\pi, \pi_0) \in Z^{n+1}$, $\Sigma_{j \in N} \pi_j x_j \leqslant \pi_0 + 1$ is a C–G inequality, and $\Sigma_{j \in N} \pi_j x_j \leqslant \pi_0$ is valid for S, then $I'_t(N^0, N^1)$ is a C–G inequality for all disjoint subsets N^0, N^1 of N.*

Proof. Here again we argue by working up through the nodes of an enumeration tree. Suppose $N^0 \cup N^1 = \{1, \ldots, t\}$. Then by the induction hypothesis, it follows that

$$\sum_{j \in N} \pi_j x_j - \sum_{j \in N^0 \cup \{t+1\}} x_j - \sum_{j \in N^1} (1 - x_j) \leqslant \pi_0$$

and

$$\sum_{j \in N} \pi_j x_j - \sum_{j \in N^0} x_j - \sum_{j \in N^1 \cup \{t+1\}} (1 - x_j) \leqslant \pi_0$$

are C–G inequalities. Combining these two inequalities with weights of $\frac{1}{2}$ and rounding establishes that $I'_t(N^0, N^1)$ is a C–G inequality. ■

We now show that every valid inequality for S with integral coefficients is a C–G inequality.

Theorem 2.8. *Let $\pi x \leqslant \pi_0$ with $(\pi, \pi_0) \in Z^{n+1}$ be a valid inequality for $S = P \cap Z^n$ with $P = \{x \in R^n_+: Ax \leqslant b, x \leqslant 1\}$. Then $\pi x \leqslant \pi_0$ is a C–G inequality for S.*

Proof. Let

$$\pi_0^{\text{LP}} = \max\{\pi x: x \in P\}. \tag{2.1}$$

If $P = \emptyset$, the result is immediate from Case b of Proposition 1.1. Otherwise π_0^{LP} is finite because P is bounded. Let (v^0, w^0) be an optimal solution to the dual of (2.1). Consider the C–G inequality

$$\sum_{j \in N} \lfloor (v^0 a_j + w_j^0) \rfloor x_j \leqslant \left\lfloor v^0 b + \sum_{j \in N} w_j^0 \right\rfloor = \lfloor \pi_0^{\text{LP}} \rfloor. \tag{2.2}$$

By dual feasibility, it follows that $\lfloor v^0 a_j + w_j^0 \rfloor \geqslant \pi_j$ for $j \in N$, so if $\lfloor \pi_0^{\text{LP}} \rfloor \leqslant \pi_0$ we are done. Otherwise we apply Theorem 2.7 to (2.2) $\lfloor \pi_0^{\text{LP}} \rfloor - \pi_0$ times. ∎

Example 2.1. Given a set $S = P \cap Z^2$, where P is given by

$$\begin{aligned} 5x_1 + 6x_2 &\leqslant 7 \\ x_1 \qquad &\leqslant 1 \\ x_2 &\leqslant 1 \\ x \in R^2_+, \end{aligned}$$

we show that the valid inequality $9x_1 + 7x_2 \leqslant 10$ is a C–G inequality. To prove this it suffices to show that $9x_1 + 7x_2 - x_1 \leqslant 10$ and $9x_1 + 7x_2 - (1 - x_1) \leqslant 10$ are C–G inequalities.

Solving $\max\{9x_1 + 7x_2: x \in P\}$, we obtain an optimal dual solution $u = (\frac{7}{6}\ \frac{19}{6}\ 0)$ and the inequality $9x_1 + 7x_2 \leqslant \frac{68}{6}$. Rounding gives us the C–G inequality

$$9x_1 + 7x_2 \leqslant 11. \tag{2.3}$$

Now we construct a tree as in Figure 2.2. For the leaves of the tree we determine inequalities that are equivalent to or dominate

$$9x_1 + 7x_2 - \omega \sum_{j \in N^0} x_j - \omega \sum_{j \in N^1} (1 - x_j) \leqslant 10.$$

By examining all the leaves of the tree we establish a priori that it suffices to take $\omega = 9$. We then consider the leaves with $N^0 \cup N^1 = \{1, 2\}$ and $1 \in N^1$.

For $N^0 = \{2\}$ and $N^1 = \{1\}$, weights of $(0\ 18\ 0)$ on the original inequalities give the valid inequality $18x_1 \leqslant 18$, or

$$9x_1 + 7x_2 - 9(1 - x_1) - 7x_2 \leqslant 9, \tag{2.4}$$

which dominates $9x_1 + 7x_2 - 9(1 - x_1) - 9x_2 \leqslant 10$.

Now as explained in the proof of Proposition 2.6, we weight (2.4) by $\frac{1}{9}$ and (2.3) by $\frac{8}{9}$ and round to obtain

(2.5) $$9x_1 + 7x_2 - (1 - x_1) - x_2 \leq 10.$$

For $N^0 = \emptyset$ and $N^1 = \{1, 2\}$, weights of (2 8 4) on the original inequalities give the valid inequality $18x_1 + 16x_2 \leq 26$ for P, which is the same as

(2.6) $$9x_1 + 7x_2 - 9(1 - x_1) - 9(1 - x_2) \leq 8.$$

Using the rounding procedure to combine (2.3) and (2.6) with respective weights of $\frac{8}{9}$ and $\frac{1}{9}$ yields

(2.7) $$9x_1 + 7x_2 - (1 - x_1) - (1 - x_2) \leq 10.$$

Now we move up the tree as explained in Theorem 2.7. For $N^0 = \emptyset$ and $N^1 = \{1\}$, (2.5) and (2.7) combined with weights $(\frac{1}{2}\,\frac{1}{2})$ and rounded yield

(2.8) $$9x_1 + 7x_2 - (1 - x_1) \leq 10.$$

Similarly, for $N^0 = \{1\}$ and $N^1 = \emptyset$, we obtain the C–G inequality

(2.9) $$9x_1 + 7x_2 - x_1 \leq 10.$$

The integer rounding procedure applied to (2.8) and (2.9) with weights $(\frac{1}{2}\,\frac{1}{2})$ gives the desired result.

Bounded Integer Variables

We now consider the case where $S = P \cap Z^n$ and where $P = \{x \in R^n_+: Ax \leq b, x \leq d\}$ is a polytope but is not contained in the unit cube. Again we will establish constructively that every valid inequality is a C–G inequality by showing that if $\pi x \leq \pi_0 + 1$ is a C–G inequality and $\pi x \leq \pi_0$ is valid for S, then $\pi x \leq \pi_0$ is also a C–G inequality.

The approach, however, is different and more complicated than our approach to the 0-1 case because for bounded integer variables we cannot obtain conv(S) by imposing integrality one variable at a time.

We are given that $\pi x \leq \pi_0$ is valid for S and that $x \leq \bar{d} = (d, d, \ldots, d)$ is valid for P—the bounded variable assumption. Now let s_i be any integer between 0 and d and let k be any integer between 0 and n. Then the following inequality is valid for S:

$$\prod_{i=1}^{k} (d + 1 - s_i)(\pi x - \pi_0) + \sum_{i=1}^{k} \prod_{j=i+1}^{k} (d + 1 - s_j)(x_i - d) \leq 0.$$

We denote this inequality by $L(s_1, \ldots, s_k)$. If $k = 0$, the inequality is simply $\pi x - \pi_0 \leq 0$ and is denoted by $L(\emptyset)$. $N(s_1, \ldots, s_k)$ denotes the same inequality with right-hand side of 1, and M_k denotes the inequality $x_k - d \leq 0$. Thus we are given that $N(\emptyset)$ is a C–G inequality, and we wish to establish that $L(\emptyset)$ is a C–G inequality.

Example 2.2. With $d = 2$ and $n = 2$, the inequalities $L(s_1, \ldots, s_k)$ are:

$$
\begin{array}{lll}
L(2,2) & (\pi x - \pi_0) + (x_1 - 2) + (x_2 - 2) & \leqslant 0 \\
L(2,1) & 2(\pi x - \pi_0) + 2(x_1 - 2) + (x_2 - 2) & \leqslant 0 \\
L(2,0) & 3(\pi x - \pi_0) + 3(x_1 - 2) + (x_2 - 2) & \leqslant 0 \\
L(2) & (\pi x - \pi_0) + (x_1 - 2) & \leqslant 0 \\
L(1,2) & 2(\pi x - \pi_0) + (x_1 - 2) + (x_2 - 2) & \leqslant 0 \\
L(1,1) & 4(\pi x - \pi_0) + 2(x_1 - 2) + (x_2 - 2) & \leqslant 0 \\
L(1,0) & 6(\pi x - \pi_0) + 3(x_1 - 2) + (x_2 - 2) & \leqslant 0 \\
L(1) & 2(\pi x - \pi_0) + (x_1 - 2) & \leqslant 0 \\
L(0,2) & 3(\pi x - \pi_0) + (x_1 - 2) + (x_2 - 2) & \leqslant 0 \\
L(0,1) & 6(\pi x - \pi_0) + 2(x_1 - 2) + (x_1 - 2) & \leqslant 0 \\
L(0,0) & 9(\pi x - \pi_0) + 3(x_1 - 2) + (x_1 - 2) & \leqslant 0 \\
L(0) & 3(\pi x - \pi_0) + (x_1 - 2) & \leqslant 0 \\
L(\emptyset) & (\pi x - \pi_0) & \leqslant 0.
\end{array}
$$

An important component of the proof is the order in which we show that the inequalities $L(s_1, \ldots, s_k)$ are C–G inequalities. We say that $t = (t_1, \ldots, t_n)$ is *lexicographically larger* than $s = (s_1, \ldots, s_n)$, $t \overset{L}{>} s$, if, for some i, $1 \leqslant i \leqslant n$, $t_j = s_j$ for $j < i$ and $t_i > s_i$. We then show that the inequalities are C–G in lexicographically decreasing order using the convention that if $k < n$, $(s_1, \ldots, s_k)$ is regarded as the n-vector $(s_1, \ldots, s_k, -1, \ldots, -1)$. Thus $(t_1, \ldots, t_l) \overset{L}{>} (s_1, \ldots, s_k)$ either if, for some $i \leqslant \min(l, k)$, $t_j = s_j$ for $j < i$ and $t_i > s_i$ or if $l > k$ and $t_j = s_j$ for $j = 1, \ldots, k$. This order is shown in Example 2.2.

The order can also be interpreted as a right-to-left search through an enumeration tree (see Figure 2.3).

We will be repeatedly adding together valid inequalities.

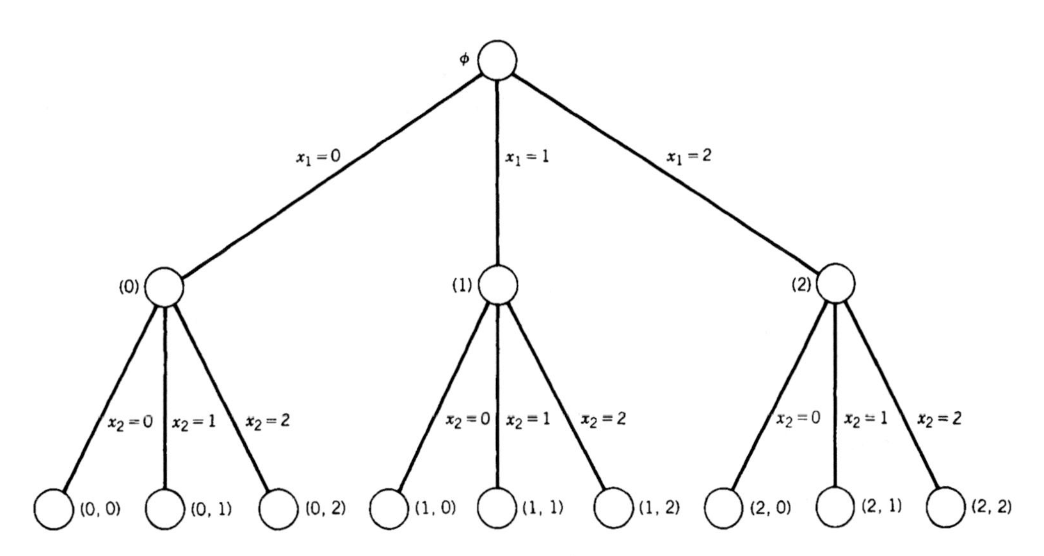

Figure 2.3

Proposition 2.9

i. $L(s_1, \dots, s_k - 1) = L(s_1, \dots, s_k) + L(s_1, \dots, s_{k-1})$ for $1 \leqslant s_k \leqslant d$.
ii. $L(s_1, \dots, s_{k-1}, d) = L(s_1, \dots, s_{k-1}) + M_k$
iii. $L(s_1, \dots, s_{k-1}, s_k) = L(s_1, \dots, s_{k-1})\,(d + 1 - s_k) + M_k$

By repeated application of equalities i and ii, we obtain the following proposition.

Proposition 2.10

i. $L(s_1, \dots, s_k) = \sum_{\{i:i\leq k, s_i<d\}} L(s_1, \dots, s_{i-1}, s_i + 1) + L(\emptyset) + \sum_{\{i:i\leq k, s_i=d\}} M_i;$
ii. $N(s_1, \dots, s_k) = \sum_{\{i:i\leq k, s_i<d\}} L(s_1, \dots, s_{i-1}, s_i + 1) + N(\emptyset) + \sum_{\{i:i\leq k, s_i=d\}} M_i.$

The proofs of these two propositions are elementary exercises.

Observe from statement ii of Proposition 2.10 that if $L(t_1, \dots, t_l)$ is a C–G inequality for all $(t_1, \dots, t_l) \overset{L}{>} (s_1, \dots, s_k)$, it follows that $N(s_1, \dots, s_k)$ is a sum of C–G inequalities and hence is a C–G inequality. Therefore the critical step is to deduce from this that $L(s_1, \dots, s_k)$ is also a C–G inequality.

Proposition 2.11. *For any* $s \in Z^k$ *with* $(s_1, \dots, s_k) \overset{L}{\geqslant} 0$, *if* $x \in R^n$ *satisfies at equality the inequalities* $L(s_1, \dots, s_i + 1)$ *for* $i \leqslant k$, $s_i < d$, M_i *for* $i \leqslant k$, $s_i = d$, *and* $N(\emptyset)$, *then* $x_i = s_i$ *for* $i = 1, \dots, k$.

Proof. We argue by induction. If $k = 1$ and $s_{k-1} < d$, we obtain

$$(d + 1 - (s_1 + 1))(\pi x - \pi_0) + (x_1 - d) = 0,$$

or $(d - s_1) + x_1 - d = 0$, or $x_1 = s_1$, since $\pi x - \pi_0 = 1$. If $s_1 = d$, then $x_1 = s_1$ is immediate, since M_1 holds at equality. Now suppose by induction that the claim holds for $k - 1$, that is, $x_i = s_i$ for $i = 1, \dots, k - 1$. If $s_k = d$, the result is immediate, since M_k is satisfied at equality. If $s_k < d$, we observe that since $N(s_1, \dots, s_{k-1})$ is a sum of inequalities satisfied at equality, it also is satisfied at equality; it follows that $L(s_1, \dots, s_{k-1})$ has a slack of 1. But by statement iii of Proposition 2.9,

$$L(s_1, \dots, s_k + 1) = L(s_1, \dots, s_{k-1})\,(d - s_k) + M_k.$$

Since $L(s_1, \dots, s_k + 1)$ is satisfied at equality, it follows that $d - s_k + x_k - d = 0$. ∎

We now associate with $(s_1, \dots, s_k)$ a polytope $P(s_1, \dots, s_k)$ given by

$$\{x \in R^n_+: Ax \leqslant b, x \leqslant d, \pi x \leqslant \pi_0 + 1, x \text{ satisfies } L(t_1, \dots, t_l) \text{ for } (t_1, \dots, t_l) \overset{L}{>} (s_1, \dots, s_k)\}.$$

Proposition 2.12. *$L(s_1, \ldots, s_k)$ is a C–G inequality for $P(s_1, \ldots, s_k) \cap Z^n$.*

Proof. If $k < n$, then $L(s_1, \ldots, s_k, 0)$ is an inequality defining $P(s_1, \ldots, s_k)$. By statement iii of Proposition 2.9,

$$L(s_1, \ldots, s_k, 0) = (d + 1)\, L(s_1, \ldots, s_k) + M_{k+1}.$$

Multiplying the inequality $L(s_1, \ldots, s_k, 0)$ by $1/(d + 1)$, followed by rounding, establishes that $L(s_1, \ldots, s_k)$ is a C–G inequality for $P(s_1, \ldots, s_k) \cap Z^n$.

Now suppose $k = n$ and $s \neq \bar{d}$. By statement ii of Proposition 2.10, the inequality $N(s_1, \ldots, s_n)$ is valid for $P(s_1, \ldots, s_n)$. Suppose it is satisfied at equality. Then each of the inequalities appearing in the statement of Proposition 2.11 is satisfied at equality, and it follows that $x_i = s_i$ for $i = 1, \ldots, n$. Moreover, since $N(\emptyset)$ is satisfied at equality, we obtain $\pi x = \pi_0 + 1$. However, since $S = P(s_1, \ldots, s_n) \cap Z^n$ and $\pi x \leq \pi_0$ is valid, there is no feasible integer point with $\pi x = \pi_0 + 1$. Hence $N(s_1, \ldots, s_n)$ cannot be satisfied at equality for any point in $P(s_1, \ldots, s_n)$. This means that $N(s_1, \ldots, s_n)$ with its right-hand side reduced by $\varepsilon > 0$ is a valid inequality for $P(s_1, \ldots, s_n)$. Hence by rounding this inequality we obtain $L(s_1, \ldots, s_n)$. Thus $L(s_1, \ldots, s_n)$ is a C–G inequality for $P(s_1, \ldots, s_n) \cap Z^n$.

Finally, the proof that $L(\bar{d})$ is a C–G inequality for $P(\bar{d}) \cap Z^n$ is similar to the proof of Proposition 2.2 and is not repeated here. ■

In particular, Proposition 2.12 states that $L(\emptyset)$ is a C–G inequality for $P(\emptyset) \cap Z^n = S$. Thus

Theorem 2.13. *Let $P = \{x \in R^n_+ : Ax \leq b, x \leq \bar{d}\}$ and let $S = P \cap Z^n$. If $\pi x \leq \pi_0 + 1$ is a C–G inequality for S and $\pi x \leq \pi_0$ is valid for S, then $\pi x \leq \pi_0$ is a C–G inequality for S.*

Corollary 2.14. *If $S = \emptyset$, then $0x \leq -1$ is a C–G inequality for S.*

Finally, using the argument given in the proof of Theorem 2.8, we establish the generality of C-G inequalities.

Theorem 2.15. *Let $\pi x \leq \pi_0$ with $(\pi, \pi_0) \in Z^{n+1}$ be a valid inequality for $S = P \cap Z^n$ with $P = \{x \in R^n_+ : Ax \leq b, x \leq \bar{d}\}$. Then $\pi x \leq \pi_0$ is a C–G inequality for S.*

Theorem 2.15 also holds for unbounded sets in Z^n, but the only known proof of the result uses a very different technique.

Theorem 2.16 *Let $\pi x \leq \pi_0$ with $(\pi, \pi_0) \in Z^{n+1}$ be a valid inequality for $S = \{x \in Z^n_+ : Ax \leq b\} \neq \emptyset$. Then $\pi x \leq \pi_0$ is a C–G inequality for S.*

We now consider how many applications of step iii of the C–G procedure are necessary to define the convex hull of $S = P \cap Z^n$, when P is a nonempty rational polytope. We saw earlier that if $\max\{\pi x : x \in P\} = \pi_0^{\text{LP}}$, then the inequality $\pi x \leq \lfloor \pi_0^{\text{LP}} \rfloor$ can be obtained by one application of the C–G procedure.

Define the *elementary closure of P* to be

$$e(P) = \{(\pi, \pi_0) : \pi_j = \lfloor u a_j \rfloor \text{ for } j \in N,\ \pi_0 = \lfloor ub \rfloor \text{ for some } u \in R^m_+\}.$$

Then, by definition of the C–G procedure, $e(P)$ contains all of the nondominated C–G inequalities that can be obtained by one application of the procedure.

Proposition 2.17. *If* $(\pi, \pi_0) \in e(P)$, *then* $\pi_0 \geq \lfloor \pi_0^{\text{LP}} \rfloor$.

Proof. Since $(\pi, \pi_0) \in e(P)$, there exists $u \in R_+^m$ such that $\lfloor ua_j \rfloor = \pi_j$ for $j \in N$ and $\lfloor ub \rfloor = \pi_0$. Consider any such u. Since $ua_j \geq \lfloor ua_j \rfloor$ for $j \in N$, it follows that u is a feasible solution to the dual of $\max\{\pi x: x \in P\}$. Thus $ub \geq \pi_0^{\text{LP}}$ and $\pi_0 = \lfloor ub \rfloor \geq \lfloor \pi_0^{\text{LP}} \rfloor$. ■

For Example 2.1, the inequality $9x_1 + 7x_2 \leq 11$ is of the form $\Sigma_{j \in N} \lfloor ua_j \rfloor x_j \leq \lfloor ub \rfloor$ since it has been obtained from an optimal dual solution. Hence $(\pi = (9, 7), \pi_0 = 11) \in e(P)$. Proposition 2.17 implies that if $\pi = (9, 7)$ and $\pi_0 \leq 10$, then $(\pi, \pi_0) \notin e(P)$. Thus, it would be interesting to know, for example, the minimum number of repetitions of the C–G procedure needed to derive $9x_1 + 7x_2 \leq 9$ and, more generally, any valid inequality.

We say that a valid inequality $\pi x \leq \pi_0$ for $S = Z^n \cap P \neq \emptyset$ is of *rank k* with respect to $P \subseteq R_+^n$ if $\pi x \leq \pi_0$ is not equivalent to or dominated by any nonnegative linear combination of C–G inequalities, each of which can be determined by no more than k−1 applications of the C–G procedure, but is equivalent to or dominated by a nonnegative linear combination of some C–G inequalities that require no more than k applications of the procedure. Thus the rank 0 inequalities are those that are equivalent to or dominated by a nonnegative linear combination of the defining inequalities of P, and the rank 1 inequalities are those that are not of rank 0 but are equivalent to or dominated by a nonnegative linear combination of the defining inequalities of P and those in $e(P)$.

Theorem 2.16 shows that every valid inequality for $S = Z^n \cap P \neq \emptyset$ is of finite rank for any rational polyhedron P.

In Example 2.1, $9x_1 + 7x_2 \leq 11$ is of rank 1. The construction of $9x_1 + 7x_2 \leq 10$ shows that its rank is, at most, 4. By constructing an inequality by the C–G procedure, we determine an upper bound on its rank, but determining the actual rank appears to be very difficult.

We use the notation $r(\pi, \pi_0) = k$ to represent the rank of a valid inequality $\pi x \leq \pi_0$ for $S = Z^n \cap P$. The rank of P is defined to be

$$\rho(P) = \max\{r(\pi, \pi_0): (\pi, \pi_0) \text{ is valid for } S = P \cap Z^n\}.$$

Thus $\rho(P)$ is the number of applications of the C–G procedure needed to determine some facet of conv(S) if we begin with $S = P \cap Z^n$. Note that $\rho = 0$ if and only if conv(S) $= P$. If S is the set of matchings of a graph and $P = \{x \in R_+^n: x \text{ satisfies (1.2)}\}$, then conv($S$) $= P \cap \{x: x \text{ satisfies (1.4)}\}$. Since inequalities (1.4) are rank 1, it follows that $\rho(P) = 1$. Matching is a rare example of a family of polyhedra of positive and bounded rank.

For most integer programming problems, the rank of the polyhedron increases without bound as a function of the dimension of the polyhedron. For example, suppose

$$P^n = \{x \in R_+^n: x_i + x_j \leq 1 \text{ for } i, j \in N, i \neq j\} \text{ and } S^n = P^n \cap Z^n.$$

We note that $\Sigma_{j \in N} x_j \leq 1$ is a valid inequality for S^n, and it is not hard to show that conv(S^n) $= \{x \in R_+^n: \Sigma_{j \in N} x_j \leq 1\}$. But the rank of $\Sigma_{j \in N} x_j \leq 1$ is $O(\log(n))$.

Even when the dimension of P is fixed, there are families of polyhedra such that the rank increases without bound as a function of the magnitude of the coefficients in the linear inequality description of P. For example, suppose P^t is defined by the inequalities

$$\begin{aligned} tx_1 + x_2 &\leqslant 1 + t \\ -tx_1 + x_2 &\leqslant 1 \\ x_1 \quad &\leqslant 1 \\ x_1, x_2 &\geqslant 0 \end{aligned}$$

and $S^t = P^t \cap Z^2$. Here it can be shown that $\rho(P^t) = t - 1$ for $t = 1, 2, \ldots$.

An infinite family of polyhedra $\mathcal{F}$ is said to have *bounded rank* if there is an integer k such that $\rho(P) \leqslant k$ for all $P \in \mathcal{F}$. Thus if $P \in \mathcal{F}$ and $\pi x \leqslant \pi_0$ is valid for $S = P \cap Z^n$, we have $r(\pi, \pi_0) \leqslant k$. Hence to verify the validity of $\pi x \leqslant \pi_0$, we need to produce no more than n inequalities that are obtained by rounding inequalities of rank no higher than $k - 1$ and weights $(u_1, \ldots, u_n)$ to combine them. Each of these lower-rank inequalities can be produced from n inequalities of still lower rank. Thus, altogether we need the original inequalities and $1 + n + \ldots + n^{k-1} \leqslant n^k$ weight vectors to prove the validity of $\pi x \leqslant \pi_0$ for any $P \in \mathcal{F}$ of dimension n.

This observation leads to an important implication concerning the computational complexity of integer programs. Let $\mathcal{F}$ be an infinite family of polyhedra and consider the integer programming problem whose instances are given by $\max\{cx: x \in S\}$, where $S = Z^n_+ \cap P$ for each $P \in \mathcal{F}$ of dimension n. The optimality of $x^0 \in S$ can be established by showing that $cx \leqslant z^0$ is a valid inequality, where $cx^0 = z^0$. Hence if $\mathcal{F}$ is of bounded rank, the optimality of a proposed solution can be checked by displaying no more than n of the original inequalities of P and n^k weight vectors for some fixed integer k. Thus, provided that the weight vectors are polynomial in the description of P, we have an optimality proof whose length is a polynomial function of n, which suggests that it is highly unlikely that the problem is $\mathcal{NP}$-hard. In other words, it may well be the case that if an integer programming problem is $\mathcal{NP}$-hard, then the family of polyhedra over which it is defined does not have bounded rank.

3. GOMORY'S FRACTIONAL CUTS AND ROUNDING

Although the results on rounding in the previous section were developed by V. Chvatal, we have attributed the procedure to him and R. Gomory. The reason is that, from a rather different viewpoint, these results appear in Gomory's much earlier work on finite cutting-plane algorithms. In this section, we will show the relationship between the valid inequalities used by Gomory and the rounding procedure.

Here we write the constraints $S = \{x \in Z^n_+: Ax \leqslant b\}$ in equality form as $S^e = \{x \in Z^{n+m}_+: (A, I)x = b\}$, where the original variables are $(x_1, \ldots, x_n)$ and the slack variables are $(x_{n+1}, \ldots, x_{n+m})$. We assume that (A, b) is an integral $m \times (n + 1)$ matrix.

Let $\lambda = (\lambda_1, \ldots, \lambda_m)$ be a weight vector and consider the linear combination of equations given by $\lambda(A, I)x = \lambda b$. Suppose $N = \{1, \ldots, n\}$, $M = \{1, \ldots, m\}$, $A = (a_1, \ldots, a_n)$, and $I = (e_1, \ldots, e_m)$ and define $\overline{a}_j = \lambda a_j$ for $j \in N$ and $\overline{b} = \lambda b$. Then $\lambda(A, I)x = \lambda b$ can be written as

$$\sum_{j \in N} \overline{a}_j x_j + \sum_{i \in M} \lambda_i x_{n+i} = \overline{b}. \tag{3.1}$$

In Section 1 [see (1.10)], we used modular arithmetic and $x \in Z_+^{n+m}$ to derive the valid inequality

$$\sum_{j \in N} f_j x_j + \sum_{i \in M} g_i x_{n+i} \geqslant f_0, \tag{3.2}$$

where $f_j = \bar{a}_j - \lfloor \bar{a}_j \rfloor$ for $j \in N$, $g_i = \lambda_i - \lfloor \lambda_i \rfloor$ for $i \in M$, and $f_0 = \bar{b} - \lfloor \bar{b} \rfloor$.

Inequality (3.2) is the *Gomory fractional cut*.

Example 3.1. We return to Example 1.1, where $S^e = \{x \in Z_+^5: (A, I)x = b\}$ and

$$(A, I) = \begin{pmatrix} -1 & 2 & 1 & 0 & 0 \\ 5 & 1 & 0 & 1 & 0 \\ -2 & -2 & 0 & 0 & 1 \end{pmatrix}, \quad b = \begin{pmatrix} 4 \\ 20 \\ -7 \end{pmatrix}.$$

Let $\lambda = (\frac{-1}{11}\ \frac{2}{11}\ 0)$, which yields the equation

$$x_1 - \frac{1}{11}x_3 + \frac{2}{11}x_4 = \frac{36}{11}. \tag{3.3}$$

Applying (3.2) to (3.3) yields the valid inequality

$$\frac{10}{11}x_3 + \frac{2}{11}x_4 \geqslant \frac{3}{11}. \tag{3.4}$$

Now we eliminate the slack variables from (3.4) to obtain

$$\frac{10}{11}(4 + x_1 - 2x_2) + \frac{2}{11}(20 - 5x_1 - x_2) \geqslant \frac{3}{11}$$

or

$$2x_2 \leqslant 7. \tag{3.5}$$

To obtain (3.5) by rounding, use the weight vector $u = (\frac{10}{11}, \frac{2}{11}, 0)$ on the original inequalities and round.

Observe that in the example $u_i = \lambda_i - \lfloor \lambda_i \rfloor$ for $i \in M$. This, in fact, is the general relationship.

Theorem 3.1. *Let $S = \{x \in Z_+^n: Ax \leqslant b\}$, where (A, b) is an $m \times (n + 1)$ matrix with integral coefficients. The fractional cut (3.2) derived from (3.1) is a C–G inequality for S obtained with weights $u_i = \lambda_i - \lfloor \lambda_i \rfloor$ for $i \in M$.*

Proof. Let $\lfloor \lambda \rfloor = (\lfloor \lambda_1 \rfloor, \ldots, \lfloor \lambda_m \rfloor)$ and $u = \lambda - \lfloor \lambda \rfloor \geqslant 0$. Then

$$uAx = \lambda Ax - \lfloor \lambda \rfloor Ax \leqslant \lambda b - \lfloor \lambda \rfloor b = ub \tag{3.6}$$

or

$$\sum_{j \in N} \bar{a}_j x_j - \sum_{j \in N} \sum_{i \in M} \lfloor \lambda_i \rfloor a_{ij} x_j \leqslant \bar{b} - \sum_{i \in M} \lfloor \lambda_i \rfloor b_i. \tag{3.7}$$

Since the a_{ij}'s and b_i's are integers, rounding (3.7) yields

$$\sum_{j\in N} \lfloor \bar{a}_j \rfloor x_j - \sum_{j\in N} \sum_{i\in M} \lfloor \lambda_i \rfloor a_{ij} x_j \leq \lfloor \bar{b} \rfloor - \sum_{i\in M} \lfloor \lambda_i \rfloor b_i,$$

or since $x_{n+i} = b_i - \Sigma_{j\in N} a_{ij} x_j$ we obtain

$$\sum_{j\in N} \lfloor \bar{a}_j \rfloor x_j + \sum_{i\in M} \lfloor \lambda_i \rfloor x_{n+i} \leq \lfloor \bar{b} \rfloor. \tag{3.8}$$

Subtracting (3.8) from (3.1) yields (3.2). ■

There is an obvious converse to Theorem 3.1, which states that every rank 1 C–G inequality can be obtained as a fractional cut. Thus, analogous to Theorem 2.16, we have that every valid inequality for $S = \{x \in Z_+^{n+m}: (A, I)x = b\}$ is equivalent to or dominated by an inequality obtained from the recursive generation of fractional cuts of the form (3.2). Gomory's proof of this result was algorithmic. He showed that the integer program $\max\{cx: x \in S\}$ could be solved by solving a finite sequence of linear programs, each of which was obtained from its predecessor by the addition of an inequality (3.2). Thus if $z^0 = cx^0 = \max\{cx: x \in S\}$, the algorithm derived the valid inequality $cx \leq z^0$. We will study this algorithm in Section II.4.3.

4. SUPERADDITIVE FUNCTIONS AND VALID INEQUALITIES

Suppose $S = Z^n \cap P$, where $P = \{x \in R_+^n: Ax \leq b\}$. Our first objective in this section is to give a functional description of valid inequalities for S. For example, the C–G rank 1 inequality $\Sigma_{j\in N} \lfloor ua_j \rfloor x_j \leq \lfloor ub \rfloor$, where a_j is the jth column of A and $u \in R_+^m$, can be described functionally by

$$\sum_{j\in N} F(a_j) x_j \leq F(b), \tag{4.1}$$

where $F(d) = \lfloor ud \rfloor$ for all $d \in R^m$.

A function $F: D \subseteq R^m \to R^1$ is called *superadditive* over D if

$$F(d_1) + F(d_2) \leq F(d_1 + d_2) \quad \text{for all } d_1, d_2, d_1 + d_2 \in D. \tag{4.2}$$

Note that $d_1 = 0$ yields $F(0) + F(d_2) \leq F(d_2)$ or $F(0) \leq 0$. Throughout the book, when F is superadditive, we assume $F(0) = 0$ and $0 \in D$.

A function $F: D \to R^1$ is called *nondecreasing over D* if $d_1, d_2 \in D$ and $d_1 \leq d_2$ implies $F(d_1) \leq F(d_2)$.

Functions with these two properties yield valid inequalities.

Proposition 4.1. *If $F: R^m \to R^1$ is superadditive and nondecreasing, then (4.1) is a valid inequality for $S = Z^n \cap \{x \in R_+^n: Ax \leq b\}$ for any (A, b).*

Proof. There are three steps in showing that (4.1) holds for all $x \in S$:

i. $\Sigma_{j\in N} F(a_j) x_j \leq \Sigma_{j\in N} F(a_j x_j)$.
ii. $\Sigma_{j\in N} F(a_j x_j) \leq F(Ax)$.
iii. $F(Ax) \leq F(b)$.

Since $Ax \leqslant b$ for all $x \in S$ and F is nondecreasing, inequality iii holds. The first two steps use superadditivity, and the first also uses $F(0) = 0$.

i. It suffices to show that $F(a_j)x_j \leqslant F(a_jx_j)$ for all j. If $x_j = 0$, then $F(a_j)x_j = 0 = F(0) = F(a_jx_j)$. If $x_j = 1$, then $F(a_j)x_j = F(a_j) = F(a_jx_j)$. Suppose it is true for $x_j = k - 1$. Then

$$\begin{aligned} kF(a_j) &= F(a_j) + (k-1)F(a_j) \\ &\leqslant F(a_j) + F((k-1)a_j) \\ &\leqslant F(a_j + (k-1)a_j) \leqslant F(ka_j). \end{aligned}$$

ii.

$$\begin{aligned} \sum_{j=1}^{n} F(a_jx_j) &= (F(a_1x_1) + F(a_2x_2)) + \sum_{j=3}^{n} F(a_jx_j) \\ &\leqslant F(a_1x_1 + a_2x_2) + \sum_{j=3}^{n} F(a_jx_j) \leqslant \cdots \leqslant F(Ax). \end{aligned}$$

■

When the linear constraints are equalities (i.e., $Ax = b$), then step iii of the proof is irrelevant. Thus we obtain the following corollary.

Corollary 4.2. *If $F: R^m \to R^1$ is superadditive and $F(0) = 0$, then (4.1) is a valid inequality for $S^e = Z^n \cap \{x \in R^n_+: Ax = b\}$.*

Corollary 4.2 (and Proposition 4.1) suggest the following terminology. If F is superadditive (and nondecreasing) with $F(0) = 0$, we call (4.1) a *superadditive valid inequality for* $S^e(S)$.

Linear functions are obviously superadditive. Starting with this simple fact and applying some elementary operations that preserve superadditivity allows us to construct some useful superadditive functions.

Proposition 4.3. *Let $H: R^k \to R^1$ be superadditive and nondecreasing and let $F_i: R^m \to R^1$ be superadditive for $i = 1, \ldots, k$.*

a. *The composite function $H(F_1, \ldots, F_k)$ is superadditive.*
b. *If, in addition, F_i, for $i = 1, \ldots, k$, is nondecreasing, then $H(F_1, \ldots, F_k)$ is nondecreasing.*

Proof. a. We have

$$\begin{aligned} H(F_1(d_1 + d_2), \ldots, F_k(d_1 + d_2)) &\geqslant H(F_1(d_1) + F_1(d_2), \ldots, F_k(d_1) + F_k(d_2)) \\ &\geqslant H(F_1(d_1), \ldots, F_k(d_1)) + H(F_1(d_2), \ldots, F_k(d_2)), \end{aligned}$$

where the first inequality holds since the F_i's are superadditive and H is nondecreasing, and the second inequality holds since H is superadditive.

b. Suppose $d_2 \geqslant 0$ in the proof of a. Since the F_i's are nondecreasing, $F_i(d_2) \geqslant 0$ for $i = 1, \ldots, k$. Since H is nondecreasing, $H(F_1(d_2), \ldots, F_k(d_2)) \geqslant 0$. Hence

$$H(F_1(d_1 + d_2), \ldots, F_k(d_1 + d_2)) \geqslant H(F_1(d_1), \ldots, F_k(d_1)),$$

so $H(F_1, \ldots, F_k)$ is nondecreasing. ■

Corollary 4.4. *Let $F, G: R^m \to R^1$ be superadditive. Then the following functions are superadditive.*

1. $K = \lambda F$ *for all* $\lambda \geq 0$.
2. $K = \lfloor F \rfloor$.
3. $K = F + G$.
4. $K = \min(F, G)$.

Proof. We apply Proposition 4.3.

1. $F_1 = F$ and $H: R^1 \to R^1$ is given by $H(d) = \lambda d$.
2. $F_1 = F$ and $H: R^1 \to R^1$ is given by $H(d) = \lfloor d \rfloor$. Clearly, H is nondecreasing. H is superadditive since

$$H(a+b) = \begin{cases} \lfloor a \rfloor + \lfloor b \rfloor + 1 & \text{if } a + b - (\lfloor a \rfloor + \lfloor b \rfloor) \geq 1 \\ \lfloor a \rfloor + \lfloor b \rfloor & \text{if } a + b - (\lfloor a \rfloor + \lfloor b \rfloor) < 1 \end{cases}$$
$$\geq \lfloor a \rfloor + \lfloor b \rfloor.$$

3. $F_1 = F$, $F_2 = G$, and $H: R^2 \to R^1$ is given by $H(a, b) = a + b$, which is linear and nondecreasing.
4. $H: R^2 \to R^1$ is given by $H(a, b) = \min(a, b)$. Clearly, H is nondecreasing. Also, H is superadditive since

$$\begin{aligned} H(a_1, b_1) + H(a_2, b_2) &= \min(a_1, b_1) + \min(a_2, b_2) \\ &\leq \min\{(a_1 + a_2), (b_1 + b_2)\} \\ &= H(a_1 + a_2, b_1 + b_2). \end{aligned}$$

■

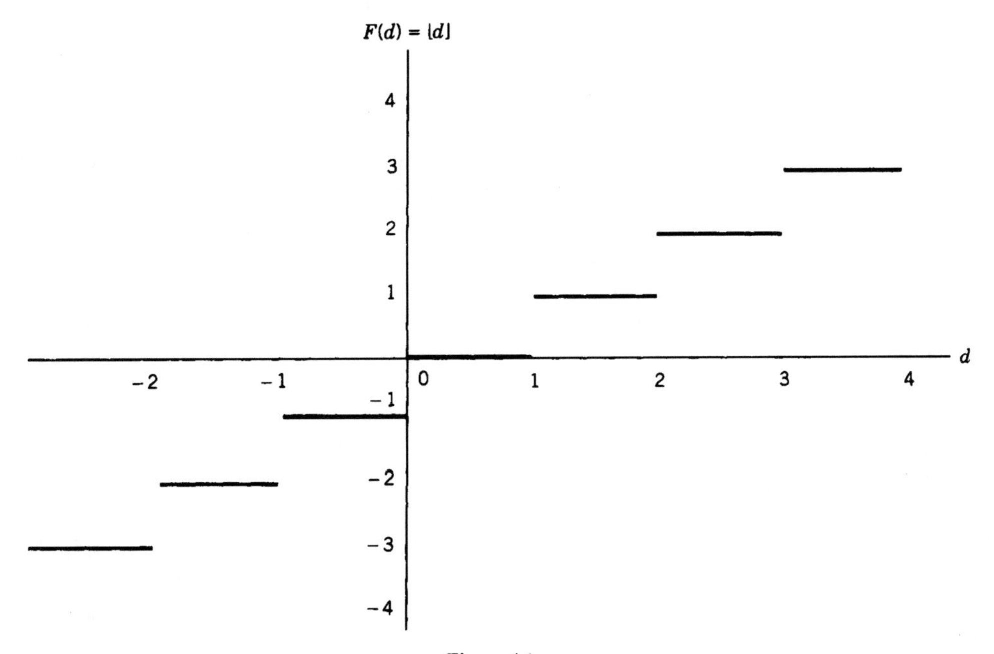

Figure 4.1

We now give several illustrations of superadditive valid inequalities. Some of them have been developed previously in the text.

1. *Integer Rounding: C–G Rank 1 Inequalities*. Let $F: R^m \to R^1$ be defined by $F(d) = \lfloor ud \rfloor$, $u \in R^m_+$. Here we apply statement 2 of Corollary 4.4 to a linear function to conclude that F is superadditive. Moreover, F is also nondecreasing since if $d_1 < d_2$, then $u \in R^m_+$ implies $\lfloor ud_1 \rfloor \leqslant \lfloor ud_2 \rfloor$. This function is illustrated in Figure 4.1 with $m = 1$ and $u = 1$.

2. *Integer Rounding: General Inequalities Constructed by the C–G Rounding Procedure*. This is illustrated by the example presented below.

Example 4.1. To construct the function that yields the inequality $9x_1 + 7x_2 \leqslant 10$ for Example 2.1, we use the earlier calculations. Note that $t = |N^0 \cup N^1|$.

$t = 0$: 1. $9x_1 + 7x_2 \leqslant 11$ is given by $F_1(d) = \lfloor \frac{7}{6}d_1 + \frac{19}{6}d_2 \rfloor$

$t = 2$: 2. $9x_1 + 7x_2 - 9x_1 - 7x_2 \leqslant 0$ is given by $F_2(d) = 0$

3. $9x_1 + 7x_2 - 9x_1 - 9(1 - x_2) \leqslant 7$ is given by $F_3(d) = 16d_3$

4. $9x_1 + 7x_2 - 9(1 - x_1) - 7x_2 \leqslant 9$ is given by $F_4(d) = 18d_2$

5. $9x_1 + 7x_2 - 9(1 - x_1) - 9(1 - x_2) \leqslant 9$ is given by $F_5(d) = \frac{7}{4}d_1 + \frac{37}{4}d_2 + \frac{11}{4}d_3$

⋮

$t = 1$: 11. $9x_1 + 7x_2 - (1 - x_1) \leqslant 10$ is given by $F_{11}(d) = \lfloor \frac{1}{2}\lfloor \frac{8}{9}F_1 + \frac{1}{9}F_4 \rfloor + \frac{1}{2}\lfloor \frac{8}{9}F_1 + \frac{1}{9}F_5 \rfloor \rfloor$

$t = 0$: 12. $9x_1 + 7x_2 \leqslant 10$ is given by

$$F_{12} = \lfloor \tfrac{1}{2}\lfloor \tfrac{1}{2}\lfloor \tfrac{8}{9}F_1 + \tfrac{1}{9}F_2 \rfloor + \tfrac{1}{2}\lfloor \tfrac{8}{9}F_1 + \tfrac{1}{9}F_3 \rfloor \rfloor + \tfrac{1}{2}\lfloor \tfrac{1}{2}\lfloor \tfrac{8}{9}F_1 + \tfrac{1}{9}F_4 \rfloor + \tfrac{1}{2}\lfloor \tfrac{8}{9}F_1 + \tfrac{1}{9}F_5 \rfloor \rfloor \rfloor.$$

We see that each of these functions is superadditive because it is obtained by taking nonnegative linear combinations of superadditive functions and then rounding. In general, we have the following result.

Proposition 4.5. *If $\pi x \leqslant \pi_0$ is a valid inequality for $S = Z^n \cap \{x \in R^n_+ : Ax \leqslant b\}$ constructed by the C–G rounding procedure, then there is a superadditive and nondecreasing $F: R^m \to R^1$ such that $\pi_j = F(a_j)$ for $j \in N$ and $\pi_0 = F(b)$.*

Proof. Suppose $\pi x \leqslant \pi_0$ is of rank k. Then the C–G construction procedure yields $\pi_j = F(a_j)$ for $j \in N$ and $\pi_0 = F(b)$, where F is obtained by recursive application of nonnegative linear combinations and rounding and hence is nondecreasing and superadditive. ■

Theorem 4.6. *Every valid inequality for a nonempty $S = Z^n \cap \{x \in R^n_+ : Ax \leqslant b\}$ is equivalent to or dominated by a superadditive valid inequality.*

Proof. By Theorem 2.16, every valid inequality for S is equivalent to or dominated by an inequality constructed by the C–G procedure. By Proposition 4.5, every inequality constructed by the C–G procedure can be obtained from a superadditive nondecreasing function. ■

We have shown that all maximal valid inequalities for S are superadditive. Moreover, they can be obtained from the family of superadditive functions generated by the recursive application of linear combinations and rounding. But as we have seen in Example 4.1, to determine a particular valid inequality from one of these functions can require a very long expression. In other words, although the basic formula is simple, it must be applied recursively to obtain particular inequalities. Perhaps with more complex basic functions, the number of recursive applications can be decreased.

3. *Strengthened Integer Rounding*. Consider the set

$$S = \left\{x \in Z_+^4 \colon 3\frac{1}{4}x_1 - 6\frac{1}{2}x_2 + 2\frac{2}{3}x_3 + x_4 \leq 4\frac{1}{3}\right\}.$$

Applying the function $F(d) = \lfloor d \rfloor$ gives the valid inequality

$$3x_1 - 7x_2 + 2x_3 + x_4 \leq 4.$$

We will show that this inequality is not maximal by producing a superadditive nondecreasing function that yields an inequality that dominates it.

Consider the family of functions $F_\alpha \colon R^1 \to R^1$ with $0 \leq \alpha \leq 1$ defined by

$$(4.3) \qquad F_\alpha(d) = \begin{cases} \lfloor d \rfloor & \text{for } f_d \leq \alpha \\ \lfloor d \rfloor + \dfrac{f_d - \alpha}{1 - \alpha} & \text{for } f_d > \alpha, \end{cases}$$

where $f_d = d - \lfloor d \rfloor$.

Let $(a)^+ = \max(0, a)$ for any $a \in R^1$. Then

$$(4.4) \qquad F_\alpha(d) = \lfloor d \rfloor + \frac{(f_d - \alpha)^+}{1 - \alpha} \quad \text{for } \alpha < 1.$$

The function $F_1(d) = \lfloor d \rfloor$, but here we are interested in $\alpha < 1$. The function $F_{1/3}$ is drawn in Figure 4.2.

Proposition 4.7. *F_α is continuous, nondecreasing, and superadditive for* $0 \leq \alpha < 1$.

Proof. F_α is continuous and nondecreasing because it is piecewise linear with slope of either 0 or $1/(1 - \alpha)$ and has no jumps. To prove superadditivity, let $f_i = d_i - \lfloor d_i \rfloor$ for $i = 1, 2$.

Case 1. $f_1 + f_2 < 1$.

$$\begin{aligned} F_\alpha(d_1) + F_\alpha(d_2) &= \lfloor d_1 \rfloor + \frac{(f_1 - \alpha)^+}{1 - \alpha} + \lfloor d_2 \rfloor + \frac{(f_2 - \alpha)^+}{1 - \alpha} \\ &\leq \lfloor d_1 + d_2 \rfloor + \frac{(f_1 + f_2 - \alpha)^+}{1 - \alpha} = F_\alpha(d_1 + d_2). \end{aligned}$$

Case 2. $f_1 + f_2 \geqslant 1, f_2 \leqslant \alpha$.

$$F_\alpha(d_1) + F_\alpha(d_2) = \lfloor d_1 \rfloor + \frac{(f_1 - \alpha)^+}{1 - \alpha} + \lfloor d_2 \rfloor$$
$$< \lfloor d_1 \rfloor + \lfloor d_2 \rfloor + 1 = \lfloor d_1 + d_2 \rfloor \leqslant F_\alpha(d_1 + d_2).$$

(The same argument applies if $f_1 \leqslant \alpha$.)

Case 3. $f_1 + f_2 \geqslant 1,\ f_1,\ f_2 > \alpha$.

$$F_\alpha(d_1) + F_\alpha(d_2) = \lfloor d_1 \rfloor + \frac{f_1 - \alpha}{1 - \alpha} + \lfloor d_2 \rfloor + \frac{f_2 - \alpha}{1 - \alpha}$$
$$= \lfloor d_1 \rfloor + \lfloor d_2 \rfloor + 1 + \frac{f_1 + f_2 - 1 - \alpha}{1 - \alpha} \leqslant F_\alpha(d_1 + d_2).$$ ■

Consider $S = \{x \in Z_+^n : \Sigma_{j \in N} a_j x_j \leqslant b\}$ with $b \in R^1$. If $f_0 = b - \lfloor b \rfloor > 0$, then when $f_0 \leqslant \alpha < 1$ it follows that $\Sigma_{j \in N} F_\alpha(a_j) x_j \leqslant F_\alpha(b)$ dominates $\Sigma_{j \in N} F_1(a_j) x_j \leqslant F_1(b)$ since $F_\alpha(b) = F_1(b)$ and $F_\alpha(a_j) \geqslant F_1(a_j)$ for all j. Moreover, the strongest of these cuts is obtained with $\alpha = f_0$ since for $f_0 \leqslant \alpha < 1$ we have $F_{f_0}(a_j) \geqslant F_\alpha(a_j)$ for all j.

In the above example with $\alpha = \frac{1}{3}$ we obtain the valid inequality

$$3x_1 - 6\frac{3}{4}x_2 + 2\frac{1}{2}x_3 + x_4 \leqslant 4.$$

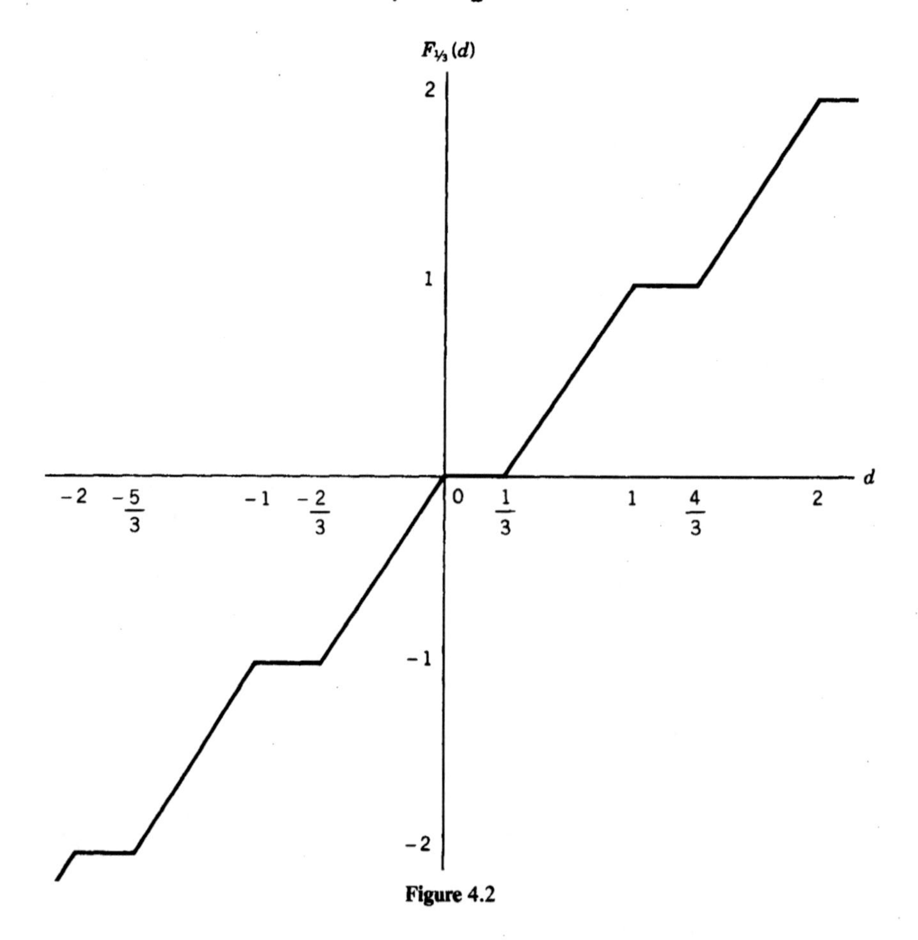

Figure 4.2

The practical disadvantage of this inequality in comparison with the C–G inequality is that when $x \in Z^n_+$, the slack variable $F_\alpha(b) - \Sigma_{j\in N} F_\alpha(a_j)x_j$ is not necessarily integer.

4. *A Two-Dimensional Function.* The only nonlinear superadditive functions considered so far have been one-dimensional. Here we introduce a two-dimensional function, based on F_α. Let

(4.5) $$\tilde{F}_\alpha(d_1, d_2) = \frac{1}{1-\alpha} d_1 + F_\alpha(d_2 - d_1) \quad \text{for } 0 \leqslant \alpha < 1.$$

The contours of this function are exhibited in Figure 4.3.

Proposition 4.8. *The function $\tilde{F}_\alpha$ given by (4.5) is nondecreasing and superadditive.*

Proof. Since F_α is nondecreasing, $\tilde{F}_\alpha$ is nondecreasing in d_2. With respect to d_1, the first term of (4.5) has slope $1/(1-\alpha)$ and the second term is piecewise linear with slope of $-1/(1-\alpha)$ or 0. Hence $\tilde{F}_\alpha$ is nondecreasing in d_1.

Since the first term in (4.5) is linear, to prove that $\tilde{F}_\alpha$ is superadditive it suffices to show that the second term is superadditive. The second term is $F_\alpha(G(d))$, where $G(d) = d_2 - d_1$ is linear. Hence by Propositions 4.3 and 4.7, the second term is superadditive. ∎

Combining Propositions 4.3 and 4.8 yields the following corollary.

Corollary 4.9. *If F_1 and F_2 are superadditive on R^m, the function $F(F_1, F_2) = 1/(1-\alpha)F_1 + F_\alpha(F_2 - F_1)$ is superadditive on R^m. If F_1 and F_2 are also nondecreasing, then $F(F_1, F_2)$ is also nondecreasing.*

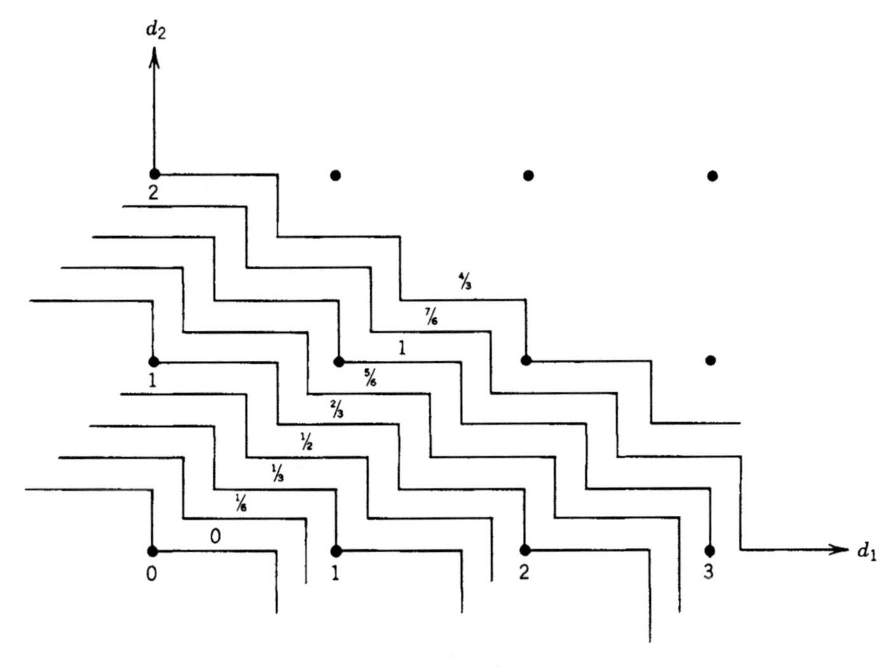

Figure 4.3. Contours of $(1-\alpha)\tilde{F}_\alpha(d_1, d_2)$ for $\alpha = \frac{1}{3}$.

As an example of the use of the function $\tilde{F}_\alpha$ to construct a valid inequality, consider the set

$$cx - \beta(\pi x - \pi_0) \leqslant c_0$$
$$cx + \beta(\pi x - \pi_0 - 1) \leqslant c_0$$
$$x \in Z_+^n$$

with $(\pi, \pi_0) \in Z^{n+1}$, $(c, c_0) \in R^{n+1}$ and $\beta > 0$.

Let $F(d_1, d_2) = \beta\tilde{F}_{1/2}(d_1/2\beta, d_2/2\beta)$. Then

$$\beta\tilde{F}_{1/2}\left(\frac{c_j - \beta\pi_j}{2\beta}, \frac{c_j + \beta\pi_j}{2\beta}\right) = \beta\left[2\left(\frac{c_j - \beta\pi_j}{2\beta}\right) + F_{1/2}(\pi_j)\right] = \beta\left[\frac{c_j}{\beta} - \pi_j + \pi_j\right] = c_j,$$

$$\beta\tilde{F}_{1/2}\left(\frac{c_0 - \beta\pi_0}{2\beta}, \frac{c_0 + \beta\pi_0 + \beta}{2\beta}\right) = \beta\left[2\left(\frac{c_0 - \beta\pi_0}{2\beta}\right) + F_{1/2}\left(\pi_0 + \frac{1}{2}\right)\right] = \beta\left[\frac{c_0}{\beta} - \pi_0 + \pi_0\right] = c_0,$$

and we obtain the valid inequality $cx \leqslant c_0$.

This shows that F_α permits us to generate the disjunctive inequality in one step, while with $F(d) = \lfloor \frac{1}{2}d_1 + \frac{1}{2}d_2 \rfloor$ and c, c_0, and β restricted to be integral, we only obtain $cx \leqslant c_0 + \lfloor \beta/2 \rfloor$. Thus, except for $\beta = 1$, this example suggests that it can be advantageous to use functions other than the rounding function $F(d) = \lfloor ud \rfloor$.

5. *Modular Arithmetic and Gomory Fractional Cuts.* Let $F: R^1 \to R^1$ be defined by $F(d) = -d (\text{mod } \delta)$, where δ is a positive integer, that is, $-F(d)$ is the remainder when d is divided by δ (see Figure 4.4).

Since $d/\delta = \lfloor d/\delta \rfloor - F(d)/\delta$, we have $F(d) = \delta(\lfloor \frac{d}{\delta} \rfloor - \frac{d}{\delta})$. We claim that $F(d)$ is superadditive. Note that $F(d) = \delta(\lfloor G(d) \rfloor - G(d))$, where $G(d)$ is the linear function d/δ. Since $\lfloor G \rfloor$ is superadditive (by statement 2 of Corollary 4.4) and $-G$ is linear, $\lfloor G \rfloor - G$ is superadditive by statement 3 of Corollary 4.4. Finally $\delta > 0$ and statement 1 of Corollary 4.4 yield that F is superadditive.

When $\delta = 1$, we obtain $F(d) = \lfloor d \rfloor - d = -f_d$. The m-dimensional version of this function, $F: R^m \to R^1$, given by $F(d) = \lfloor ud \rfloor - ud$ for $u \in R_+^m$ generates the Gomory fractional cut, $\Sigma_{j \in N} f_j x_j \geqslant f_0$, where $f_j = ua_j - \lfloor ua_j \rfloor$ for $j \in N$ and $f_0 = ub - \lfloor ub \rfloor$.

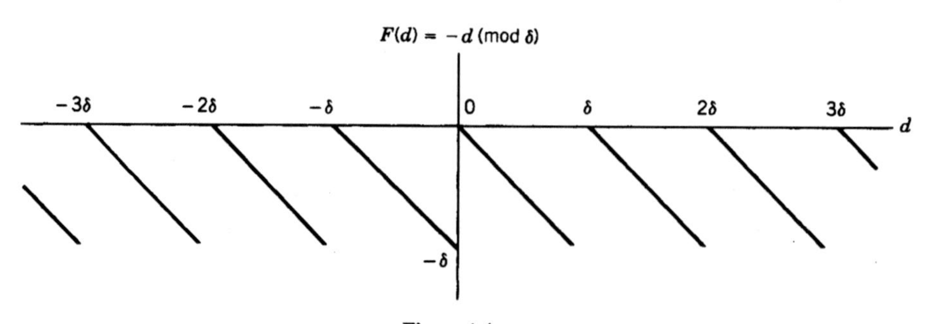

Figure 4.4

6. *A Stronger Fractional Cut.* Let $\phi_\alpha: R^1 \to R^1$ be defined by

$$\phi_\alpha(d) = \begin{cases} -f_d & \text{for } 0 \leqslant f_d \leqslant \alpha \\ \dfrac{(f_d - 1)\alpha}{1 - \alpha} & \text{for } \alpha < f_d < 1, \end{cases}$$

where $0 < \alpha \leqslant 1$. Note that $\phi_1(d) = -f_d$ for all d. The function is shown in Figure 4.5 for $\alpha = \frac{1}{3}$.

Since $\phi_\alpha(d) = F_\alpha(d) - d$, the superadditivity of ϕ_α is a corollary to Proposition 4.7. The fractional cut obtained from ϕ_{f_0} is

$$\sum_{\{j \in N: f_j \leqslant f_0\}} f_j x_j + \frac{f_0}{1 - f_0} \sum_{\{j \in N: f_j > f_0\}} (1 - f_j)x_j \geqslant f_0, \tag{4.6}$$

which dominates $\Sigma_{j \in N} f_j x_j \geqslant f_0$. Valid inequalities of the form (4.6) are important for mixed-integer regions (see Section 7).

5. A POLYHEDRAL DESCRIPTION OF SUPERADDITIVE VALID INEQUALITIES FOR INDEPENDENCE SYSTEMS

An $S \subset Z_+^n$ is called an *independence system* if

i. $0 \in S$ and
ii. $x^1 \in S, x^2 \in Z_+^n$ and $x^2 \leqslant x^1 \Rightarrow x^2 \in S$.

It is easy to see that $\{x \in Z_+^n: Ax \leqslant b\}$ is an independence system if all of the coefficients of (A, b) are nonnegative integers. Here we consider independence systems generated in this way. We also assume that $b_i \geqslant \max\{a_{ij}: \text{for all } j\}$ so that the n vectors e_j for $j \in N$ are in S.

All valid inequalities have $\pi_0 \geqslant 0$ since $\pi x \leqslant -1$ is not satisfied by $x = 0$. The n constraints $x \geqslant 0$ are valid and define facets of conv(S) since $x_j = 0$ is satisfied by the n affinely independent points $(0, e_1, \ldots, e_{j-1}, e_{j+1}, \ldots, e_n)$ for $j = 1, \ldots, n$. Since $e_j \in S$ for all j, any other valid inequality of the form $\pi x \leqslant 0$ has $\pi \leqslant 0$ and therefore is not maximal. Thus, except for $x \geqslant 0$, all facets of conv(S) are of the form $\pi x \leqslant 1$. Moreover, all of these facets have $\pi \geqslant 0$ because if $\pi x \leqslant 1$ is valid and $\pi_j < 0$, then π^1 with $\pi_k^1 = \pi_k$, $k \neq j$, and $\pi_j^1 = 0$ is valid.

Let $A = (a_1, \ldots, a_n)$, where $a_j \in D(b) = \{d \in Z_+^m: d \leqslant b\}$ for all j. Here we give a polyhedral description of the valid inequalities of the form $\Sigma_{j \in N} F(a_j)x_j \leqslant F(b)$, where F is superadditive and nondecreasing, that contains all of the maximal valid inequalities other than $x \geqslant 0$.

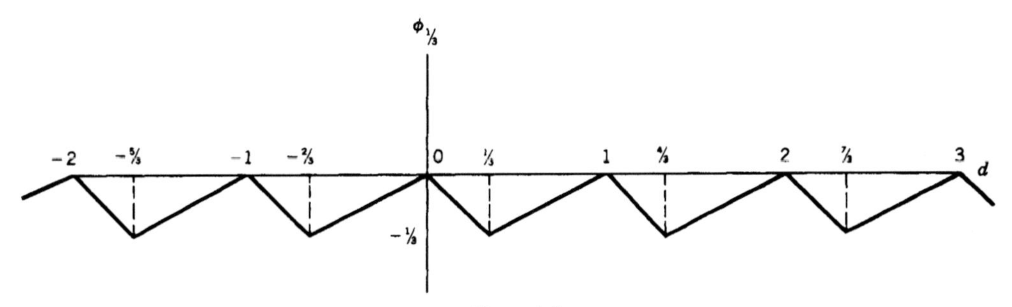

Figure 4.5

Since we are dealing with functions on the finite domain $D(b)$, any such function F can be represented by a vector $(F(0), F(e_1), F(e_2), \ldots, F(b))$ with $\Pi_{i=1}^{m} (b_i + 1)$ components. In addition, we assume that $F(0) = 0$ for all F, and we normalize the functions so that $F(b) = 1$ for all F.

Proposition 5.1. *$F: D(b) \to [0, 1]$ is nondecreasing and superadditive if and only if its corresponding vector is in the polytope given by*

$$
\begin{aligned}
&F(d_1) + F(d_2) - F(d_1 + d_2) \leq 0 \quad && \text{for } d_1, d_2 \in D(b),\ d_1 + d_2 \leq b \\
&F(d) \geq 0 && \text{for } d \in D(b) \\
&F(b) = 1. &&
\end{aligned}
\tag{5.1}
$$

Since F is defined for all $d \in D(b)$, it is natural to consider the constraint set where the matrix A has a column for each $d \in D(b)$. In other words, $S(b) = Z^{|D(b)|} \cap P(b)$, where $P(b) = \{x \in R_+^{|D(b)|}: \Sigma_{d \in D(b)}\, dx(d) \leq b\}$.

We call conv$(S(b))$ the *master polytope* for the independence system with right-hand side b. In this section we will first derive results for $S(b)$ and then show how they carry over to our given constraint set S involving only a subset $\{a_j\}_{j \in N}$ of the vectors in $D(b)$.

Example 5.1. Consider $S(3) = \{x \in Z_+^3: x(1) + 2x(2) + 3x(3) \leq 3\}$. The system (5.1) yields

$$
\begin{aligned}
2F(1) - F(2) &\leq 0 \\
F(1) + F(2) - F(3) &\leq 0 \\
F(1), F(2) &\geq 0 \\
F(3) &= 1.
\end{aligned}
$$

The feasible solutions are shown in $(F(1), F(2))$ space in Figure 5.1. The feasible region contains all of the maximal superadditive inequalities. We will see that the maximal extreme points $(0 \quad 1)$, $(\frac{1}{3} \quad \frac{2}{3})$ define the facets of conv$(S(3))$ (other than $x \geq 0$). Thus conv$(S(3))$ is defined by the inequalities

$$
\begin{aligned}
\frac{1}{3}x(1) + \frac{2}{3}x(2) + x(3) &\leq 1 \\
x(2) + x(3) &\leq 1 \\
x(1), x(2), x(3) &\geq 0.
\end{aligned}
$$

Since the maximal points of $S(3)$ are $\{(3 \quad 0 \quad 0), (1 \quad 1 \quad 0), (0 \quad 0 \quad 1)\}$, the 1-polar restricted to $\pi \geq 0$ and $\pi_0 = 1$ yields

$$
\begin{aligned}
3\pi_1 &\leq 1 \\
\pi_1 + \pi_2 &\leq 1 \\
\pi_1, \pi_2 &\geq 0
\end{aligned}
$$

(see Figure 5.2).

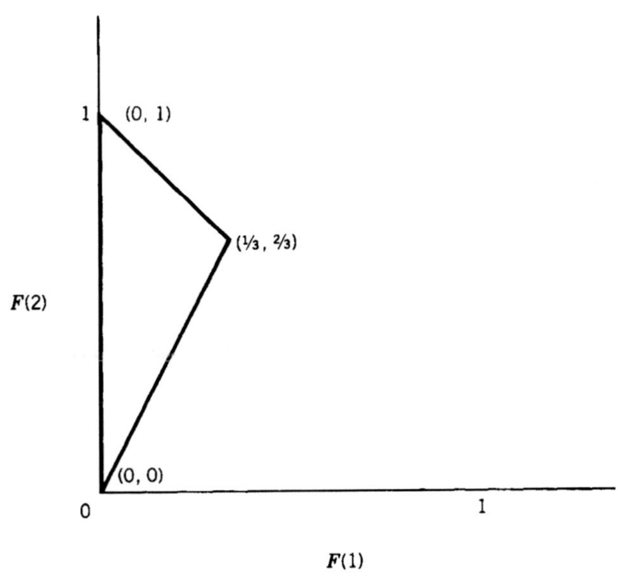

Figure 5.1

Thus we see that the superadditive inequalities are properly contained in the 1-polar.

Note that in the example, all of the maximal inequalities lie on the line $F(1) + F(2) = F(3) = F(b) = 1$. This is a necessary and sufficient condition for maximality, which is made precise in the following proposition.

From Theorem 4.6, we have that all maximal valid inequalities for $S(b)$ are superadditive. Hence the statement that F is a maximal feasible solution to (5.1) is identical to the statement that the superadditive inequality $\Sigma_{d \in D(b)} F(d)x(d) \leq 1$ is a maximal valid inequality for $S(b)$.

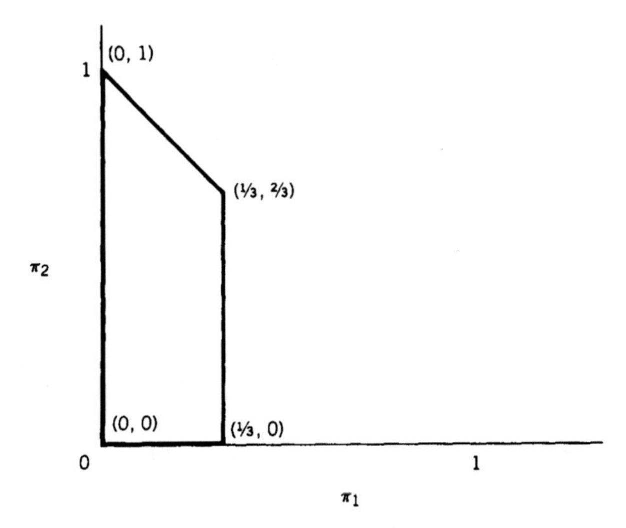

Figure 5.2

Proposition 5.2. *F is a maximal feasible solution to (5.1) if and only if F is feasible and $F(d) + F(b - d) = 1$ for all $d \in D(b)$.*

Proof. If F satisfies (5.1) and $F(d) + F(b - d) = 1$ for all d, then no component of F can be increased while maintaining feasibility. Hence F is maximal.

We now prove that if there exists a d^0 such that $F(d^0) + F(b - d^0) < 1$, then $\Sigma_{d\in D(b)} F(d)x(d) \leqslant F(b)$ is not a maximal inequality for $S(b)$. There are two cases, namely $d^0 = b/2$ and $d^0 \neq b/2$.

Case 1. $d^0 = b/2$ and $F(d^0) < \frac{1}{2}$. We will show that $\Sigma_{d\in D(b)} \pi(d)x(d) \leqslant 1$ is valid for $S(b)$, where $\pi(d) = F(d)$ for $d \neq d^0$ and $\pi(d^0) = \frac{1}{2}$. We have $x(d^0) \in \{0, 1, 2\}$ for all $x \in S(b)$. If $x(d^0) = 0$, then $\Sigma_{d\in D(b)} \pi(d)x(d) = \Sigma_{d\in D(b)} F(d)x(d) \leqslant 1$ is valid for $S(b)$. If $x(d^0) = 2$, then $x(d) = 0$ for $d \neq d^0$ and $\Sigma_{d\in D(b)} \pi(d)x(d) = 2\pi(d^0) = 1$. If $x(d^0) = 1$, then $\Sigma_{d\neq d^0} dx(d) \leqslant d^0$. Hence

$$\sum_{d\neq d^0} F(d)x(d) \leqslant F\left(\sum_{d\neq d^0} dx(d)\right) \leqslant F(d^0) < \frac{1}{2},$$

where the first inequality follows from superadditivity and the second one follows from monotonicity. Thus

$$\sum_{d\in D(b)} \pi(d)x(d) = \sum_{d\neq d^0} F(d)x(d) + \pi(d^0) < \frac{1}{2} + \frac{1}{2} = 1.$$

Case 2. $d^0 \neq b/2$ and $F(d^0) + F(b - d^0) < 1$. Without loss of generality, we can assume that for some i we have $d_i^0 > b_i/2$. Hence $x(d^0) \in \{0, 1\}$ for all $x \in S(b)$. We will show that $\Sigma_{d\in D(b)} \pi(d)x(d) \leqslant 1$ is valid for $S(b)$, where $\pi(d) = F(d)$ for $d \neq d^0$ and $\pi(d^0) = 1 - F(b - d^0)$. If $x(d^0) = 0$, then $\Sigma_{d\in D(b)} \pi(d)x(d) = \Sigma_{d\in D(b)} F(d)x(d) \leqslant 1$ is valid for $S(b)$. If $x(d^0) = 1$, then $\Sigma_{d\neq d^0} dx(d) \leqslant b - d^0$. Hence

$$\sum_{d\neq d^0} F(d)x(d) \leqslant F\left(\sum_{d\neq d^0} dx(d)\right) \leqslant F(b - d^0).$$

Thus

$$\begin{aligned}\sum_{d\in D(b)} \pi(d)x(d) &= \sum_{d\neq d^0} F(d)x(d) + \pi(d^0)\\ &\leqslant F(b - d^0) + (1 - F(b - d^0)) = 1.\end{aligned}$$

■

Proposition 5.2 allows us to tighten the system (5.1) by ruling out functions that fail to satisfy $F(d_1) + F(d_2) = 1$ when $d_1 + d_2 = b$. Thus we obtain the polytope defined by

$$\begin{aligned}
&F(d_1) + F(d_2) - F(d_1 + d_2) \leqslant 0 && \text{for all } d_1, d_2 \in D(b),\ d_1 + d_2 < b\\
&F(d) + F(b - d) = 1 && \text{for all } d \in D(b)\\
&F(d) \geqslant 0 && \text{for all } d \in D(b)\\
&F(b) = 1. &&
\end{aligned} \tag{5.2}$$

In Example 5.1, the system (5.2) in $(F(1), F(2))$ space is given by

$$2F(1) - F(2) \leqslant 0$$
$$F(1) + F(2) = 1$$
$$F(1), F(2) \geqslant 0.$$

The feasible region is the line joining the points $(\frac{1}{3} \quad \frac{2}{3})$ and $(0 \quad 1)$ in Figure 5.1, that is, precisely the set of maximal points.

Besides characterizing the maximal valid inequalities for $\text{conv}(S(b))$ (other than nonnegativity), the system (5.2) gives a useful description of the facets of $\text{conv}(S(b))$. We thus obtain the main result of this section.

Theorem 5.3. $\Sigma_{d\in D(b)} F(d)x(d) \leqslant 1$ *is a facet of* $\text{conv}(S(b))$ *(other than nonnegativity) if and only if F is an extreme point solution of (5.2).*

Proof. All feasible solutions to (5.2) generate valid inequalities for $S(b)$. Suppose that F generates a facet of $\text{conv}(S(b))$ but that $F = \frac{1}{2}F^1 + \frac{1}{2}F^2$, where F^1 and F^2 satisfy (5.2). Then F^1 and F^2 generate valid inequalities, and $\Sigma_{d\in D(b)} F(d)x(d) \leqslant 1$ is a convex combination of $\Sigma_{d\in D(b)} F^k(d)x(d) \leqslant 1$ for $k = 1, 2$, which is a contradiction.

On the other hand, suppose F is an extreme point solution of (5.2) but does not generate a facet of $\text{conv}(S(b))$. By Proposition 5.2, $\Sigma_{d\in D(b)} F(d)x(d) \leqslant 1$ is maximal and, by the hypothesis, is a convex combination of valid inequalities. In other words, it is dominated by a convex combination of maximal valid inequalities. Thus there exist $F^i \neq F$ for $i = 1, \ldots, p$ such that for all $d \in D(b)$ we have $F(d) \leqslant \Sigma_{i=1}^{p} \lambda_i F^i(d)$, where $\Sigma_{i=1}^{p} \lambda_i = 1$ and $\lambda_i \geqslant 0$ for $i = 1, \ldots, p$. However, maximality of the inequality generated by F implies that $F(d) = \Sigma_{i=1}^{p} \lambda_i F^i(d)$ for all $d \in D(b)$, which is a contradiction. ■

Now we use projection and Theorem 5.3 to go from the master polytope to the general polytope.

Theorem 5.4. *Let* $S = Z^n \cap P$, *where* $P = \{x \in R_+^n\colon Ax \leqslant b\}$ *and all coefficients of* (A, b) *are nonnegative and integral. Then*

$$\text{conv}(S) = \text{conv}(S(b)) \cap \{x\colon x(d) = 0 \text{ if } d \neq a_j \text{ for some } j \in N\}.$$

Proof. Since $x(d) \geqslant 0$ is a facet of $\text{conv}(S(b))$ for all $d \in D(b)$, it follows that $\text{conv}(S)$ is the face of $\text{conv}(S(b))$ obtained by setting $x(d) = 0$ if $d \neq a_j$ for some $j \in N$. ■

Example 5.2. Consider $S(5)$. The extreme points of (5.2) with $F(5) = 1$ are

$$(F(1), F(2), F(3), F(4)) = \left(\frac{1}{5} \quad \frac{2}{5} \quad \frac{3}{5} \quad \frac{4}{5}\right)$$
$$(F(1), F(2), F(3), F(4)) = (0 \quad 0 \quad 1 \quad 1)$$
$$(F(1), F(2), F(3), F(4)) = \left(0 \quad \frac{1}{2} \quad \frac{1}{2} \quad 1\right).$$

Hence conv($S(5)$) is given by the inequalities

$$\frac{1}{5}x(1) + \frac{2}{5}x(2) + \frac{3}{5}x(3) + \frac{4}{5}x(4) + x(5) \leq 1$$

$$x(3) + x(4) + x(5) \leq 1$$

$$\frac{1}{2}x(2) + \frac{1}{2}x(3) + x(4) + x(5) \leq 1$$

$$x(d) \geq 0 \text{ for } d = 1, \ldots, 5.$$

Now given $S = \{x \in Z^3_+: x(1) + 2x(2) + 4x(4) \leq 5\}$ it follows from Theorem 5.4 that conv(S) is given by the inequalities

$$\frac{1}{5}x(1) + \frac{2}{5}x(2) + \frac{4}{5}x(4) \leq 1$$

$$x(4) \leq 1$$

$$\frac{1}{2}x(2) + x(4) \leq 1$$

$$x(1), x(2), x(4) \geq 0.$$

Note here that the first and third inequalities are facets of conv(S) but the second one is redundant.

6. VALID INEQUALITIES FOR MIXED-INTEGER SETS

Suppose we are given the mixed-integer region

$$T = \{x \in Z^n_+, y \in R^p_+: Ax + Gy \leq b\},$$

where (A, G, b) is an $m \times (n + p + 1)$ rational matrix. Our objective in this section is to develop a procedure for generating valid inequalities for T. Note that the C–G procedure does not work when there are continuous variables. In particular, we cannot round down the right-hand side of an inequality to its integer part when all of the coefficients on the left-hand side are integers. However, we will be able to obtain a procedure, related to the disjunctive procedure, that generalizes the C–G procedure.

To motivate the approach, consider the example with T defined by

$$3x_1 - 7x_2 + 2x_3 + y_1 - y_2 \leq 4\frac{1}{3}$$

$$x \in Z^3_+, y \in R^2_+.$$

In the absence of the y variables, we obtained the valid inequality $3x_1 - 7x_2 + 2x_3 \leq 4$. Can one find a valid inequality for T of the form

$$(6.1) \qquad 3x_1 - 7x_2 + 2x_3 + \mu^+ y_1 - \mu^- y_2 \leq 4?$$

i. *A Bound on* μ^+. Suppose there is a feasible solution with $3x_1 - 7x_2 + 2x_3 = 4$, $y_2 = 0$, and $y_1 > 0$. The inequality (6.1) can only be valid if $4 + \mu^+ y_1 - 0 \leqslant 4$ or $\mu^+ \leqslant 0$.

ii. *A Bound on* μ^-. Suppose there is a feasible solution with $3x_1 - 7x_2 + 2x_3 = 5$, $y_1 = 0$, and $y_2 = \frac{2}{3}$. Validity of (6.1) implies that $5 - 2\mu^-/3 \leqslant 4$ or $\mu^- \geqslant 3/2$.

Letting $b = \lfloor b \rfloor + f_0$, the example indicates that $\mu^+ \leqslant 0, \mu^- \geqslant 1/(1 - f_0)$ and motivates the following proposition.

Proposition 6.1. *Let* $T = \{x \in Z^n_+, y \in R^p_+ : \Sigma_{j\in N} a_j x_j + \Sigma_{j \in J} g_j y_j \leqslant b\}$, *where* $N = \{1, \ldots, n\}$, $J = \{1, \ldots, p\}$, *and* a_j, g_j, $b \in R^1$ *for all* j. *The inequality*

$$\sum_{j\in N} \lfloor a_j \rfloor x_j + \frac{1}{1 - f_0} \sum_{j \in J^-} g_j y_j \leqslant \lfloor b \rfloor, \tag{6.2}$$

where $J^- = \{j \in J: g_j < 0\}$ *and* $f_0 = b - \lfloor b \rfloor$, *is valid for* T.

Proof. Suppose $\Sigma_{j \in J} g_j y_j > f_0 - 1$. Then

$$\sum_{j\in N} \lfloor a_j \rfloor x_j \leqslant \sum_{j \in N} a_j x_j \leqslant b - \sum_{j\in J} g_j y_j < b - (f_0 - 1) = \lfloor b \rfloor + 1.$$

Since $\Sigma_{j\in N} \lfloor a_j \rfloor x_j$ is an integer, we have $\Sigma_{j \in N} \lfloor a_j \rfloor x_j \leqslant \lfloor b \rfloor$. Adding this inequality to $\frac{1}{1-f_0} \Sigma_{j \in J^-} g_j y_j \leqslant 0$ yields (6.2).

Now suppose that $\Sigma_{j\in J} g_j y_j \leqslant f_0 - 1$ so that $\Sigma_{j \in J^-} g_j y_j \leqslant f_0 - 1$. Hence

$$\begin{aligned}
\sum_{j\in N} \lfloor a_j \rfloor x_j + \frac{1}{1-f_0} \sum_{j\in J^-} g_j y_j &\leqslant \sum_{j \in N} a_j x_j + \frac{1}{1-f_0} \sum_{j \in J^-} g_j y_j \\
&\leqslant b - \sum_{j\in J} g_j y_j + \frac{1}{1-f_0} \sum_{j\in J^-} g_j y_j \leqslant b + \left(\sum_{j\in J^-} g_j y_j\right)\left(\frac{1}{1-f_0} - 1\right) \\
&= b + \frac{f_0}{1-f_0}\left(\sum_{j \in J^-} g_j y_j\right) \leqslant b - f_0 = \lfloor b \rfloor. \qquad \blacksquare
\end{aligned}$$

Example 6.1. $T = \{x_1 \in Z^1_+, y_1 \in R^1_+ : x_1 + y_1 \leqslant \frac{5}{2}\}$. From (6.2), we obtain the valid inequality $x_1 \leqslant 2$. The geometry is shown in Figure 6.1. Note that

$$\left\{(x_1, y_1) \in R^2 : x_1 + y_1 \leqslant \frac{5}{2}, x_1 \leqslant 2, y_1 \geqslant 0\right\} = \text{conv}\left\{x_1 \in Z^1, y_1 \in R^1_+ : x_1 + y_1 \leqslant \frac{5}{2}\right\}.$$

Now let $T = \{x_1 \in Z^1_+, y_1 \in R^1_+ : x_1 - y_1 \leqslant \frac{5}{2}\}$. From (6.2), we obtain the valid inequality $x_1 - 2y_1 \leqslant 2$ (see Figure 6.2). Note that

$$\{(x_1, y_1) \in R^2 : x_1 - y_1 \leqslant \tfrac{5}{2}, x_1 - 2y_1 \leqslant 2, y_1 \geqslant 0\} = \text{conv}\{x_1 \in Z^1, y_1 \in R^1_+ : x_1 - y_1 \leqslant \tfrac{5}{2}\}.$$

Example 6.1 illustrates the following proposition.

Proposition 6.2. *Let* $T = \{x \in Z^n, y \in R^p_+ : \Sigma_{j \in N} a_j x_j + \Sigma_{j\in J} g_j y_j \leqslant b\}$, *where* $a_j \in Z^1$ *for* $j \in N$, $\gcd\{a_1, \ldots, a_n\} = 1$, *and* $b \notin Z^1$. *Then (6.2) is a facet of* conv(T).

Proof. We have already shown that (6.2) is valid for T. To show that it is a facet of conv(T), we first take n affinely independent points, $\bar{x}^1, \ldots, \bar{x}^n \in Z^n$, that satisfy

$\Sigma_{j\in N} a_j x_j = \lfloor b \rfloor$. Let $p_1 = |J \setminus J^-|$. We represent points in T as triples (u, v, w), where $u \in Z^n$, $v \in R_+^{p_1}$, and $w \in R_+^{p-p_1}$. We get n affinely independent points in T that satisfy (6.2) at equality by taking $u^i = \bar{x}^i$, $v^i = 0$, and $w^i = 0$ for $i = 1, \ldots, n$, and we get another p_1 points by taking $u^i = \bar{x}^1$, $v^i = \varepsilon e_i$, and $w^i = 0$, where e_i is the ith unit vector in $R_+^{p_1}$ and $\varepsilon > 0$ is suitably small. Now let $\hat{x} \in Z^n$ be a solution to $\Sigma_{j\in N} a_j x_j = \lfloor b \rfloor + 1$. The final set of $p - p_1$ points are obtained by taking $u^i = \hat{x}$, $v^i = 0$ and $w^i = \gamma_i e_i$, where e_i is the ith unit vector in $R_+^{p-p_1}$ and $\gamma_i = (f_0 - 1) / g_i$. These last points satisfy (6.2) since

$$\sum_{j\in N} a_j \hat{x}_j + \frac{1}{1-f_0}\left(\frac{f_0 - 1}{g_i}\right) g_i = \lfloor b \rfloor + 1 - 1 = \lfloor b \rfloor$$

and are in T since

$$\sum_{j\in N} a_j \hat{x}_j + \left(\frac{f_0 - 1}{g_i}\right) g_i = \lfloor b \rfloor + 1 + f_0 - 1 = b. \qquad \blacksquare$$

As we saw in the derivation of (6.2), it was necessary to use the nonnegativity of the continuous variables. In particular, we must use the nonnegativity of slack variables to generate other valid inequalities. We now give a procedure based on (6.2) for generating valid inequalities for the set $T = \{x \in Z_+^n, y \in R_+^p : Ax + Gy \leq b\}$.

Mixed-Integer Rounding (MIR) Procedure

Step 1: The inequalities

$$\sum_{j\in N} (ua_j)x_j + \sum_{j\in J} (ug_j)y_j \leq ub \quad \text{are valid for all } u \in R_+^m.$$

Step 2: Given two valid inequalities

$$\sum_{j\in N} \pi_j^i x_j + \sum_{j\in J} \mu_j^i y_j \leq \pi_0^i \quad \text{for } i = 1, 2, \tag{6.3}$$

construct the third valid inequality

$$\sum_{j\in N} \lfloor \pi_j^2 - \pi_j^1 \rfloor x_j + \frac{1}{1-f_0}\left(\sum_{j\in N} \pi_j^1 x_j + \sum_{j\in J} \min(\mu_j^1, \mu_j^2) y_j - \pi_0^1\right) \leq \lfloor \pi_0^2 - \pi_0^1 \rfloor, \tag{6.4}$$

where $\pi_0^2 - \pi_0^1 = \lfloor \pi_0^2 - \pi_0^1 \rfloor + f_0$.

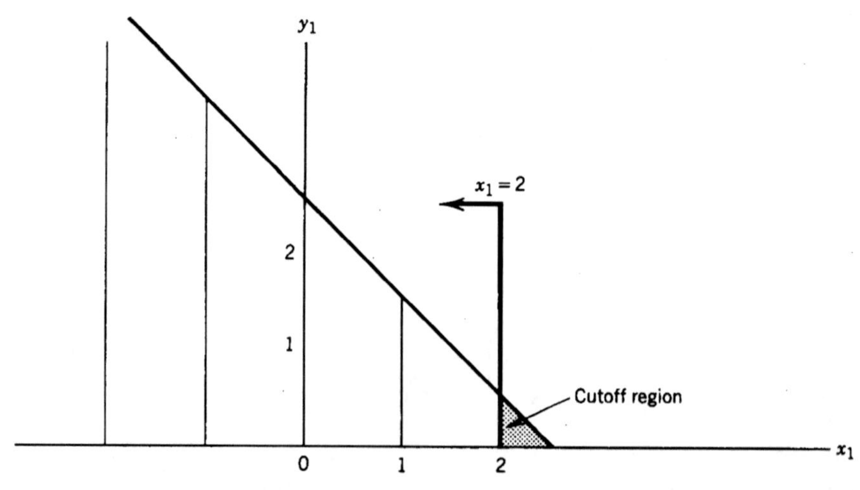

Figure 6.1

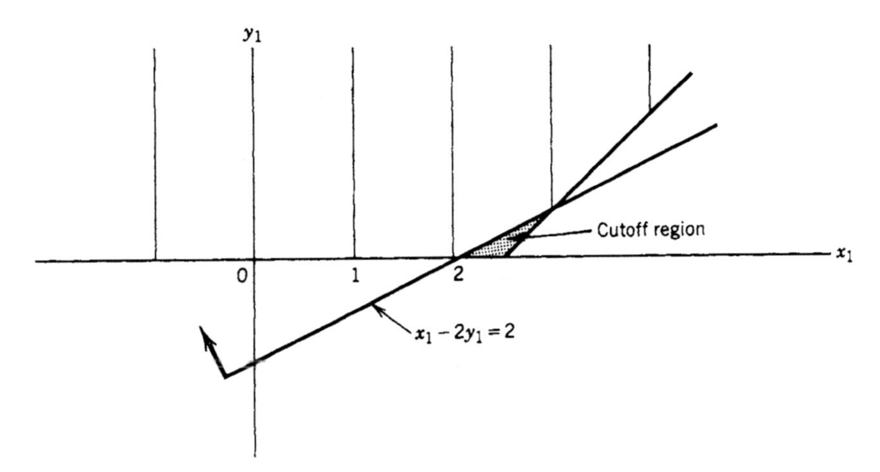

Figure 6.2

Proposition 6.3. *Given the two valid inequalities (6.3) for T, it follows that (6.4) is also valid for T.*

Proof. Since (6.3) is valid for T and $y \geqslant 0$, it follows that

$$\sum_{j \in N} \pi_j^i x_j + \sum_{j \in J} \min(\mu_j^1, \mu_j^2) y_j \leqslant \pi_0^i \quad \text{for } i = 1, 2 \tag{6.5}$$

is valid for T. Rewrite (6.5) for $i = 2$ as

$$\sum_{j \in N} (\pi_j^2 - \pi_j^1) x_j - \left(\pi_0^1 - \sum_{j \in N} \pi_j^1 x_j - \sum_{j \in J} \min(\mu_j^1, \mu_j^2)\, y_j \right) \leqslant \pi_0^2 - \pi_0^1.$$

Now (6.5) with $i = 1$ implies

$$s = \pi_0^1 - \sum_{j \in N} \pi_j^1 x_j - \sum_{j \in J} \min(\mu_j^1, \mu_j^2)\, y_j \geqslant 0.$$

Thus we can apply Proposition 6.1 to

$$\sum_{j \in N} (\pi_j^2 - \pi_j^1) x_j - s \leqslant \pi_0^2 - \pi_0^1$$

to obtain (6.4). ■

We say that any valid inequality equivalent to or dominated by an inequality constructed by the MIR procedure is an *MIR inequality*.

Example 6.2. $T = \{x \in B^2, y \in R_+^2 \colon y_1 + y_2 \leqslant 7, y_i \leqslant 5x_i \text{ for } i = 1, 2\}$. Using Step 1, we obtain the two valid inequalities

$$\frac{1}{3}(y_1 + y_2) \leqslant \frac{7}{3}$$

$$-\frac{5}{3}(x_1 + x_2) + \frac{1}{3}(y_1 + y_2) \leqslant 0.$$

Taking the first of these as the $i = 1$ inequality, and the second as the $i = 2$ inequality, we obtain from (6.4) the valid inequality given by:

$$\left\lfloor -\frac{5}{3} \right\rfloor (x_1 + x_2) + \frac{1}{1 - \frac{2}{3}} \left(\frac{1}{3} (y_1 + y_2) - \frac{7}{3} \right) \leq \left\lfloor -\frac{7}{3} \right\rfloor$$

or

$$-2(x_1 + x_2) + (y_1 + y_2) \leq 4.$$

Proposition 6.4. *Let $T = \{x \in Z^n_+, y \in R^p_+ : -\alpha x_k + cx + hy \leq c_0, \beta x_k + cx + hy \leq c_0 + \beta\}$ with $\alpha, \beta > 0$. Then $cx + hy \leq c_0$ is an MIR inequality for T.*

Proof. Scale each of the inequalities in the definition of T by $1/(\alpha + \beta)$ to obtain

$$\frac{-\alpha}{\alpha + \beta} x_k + \frac{1}{\alpha + \beta}(cx + hy) \leq \frac{c_0}{\alpha + \beta} \quad [i = 1 \text{ in } (6.3)]$$

$$\frac{\beta}{\alpha + \beta} x_k + \frac{1}{\alpha + \beta}(cx + hy) \leq \frac{c_0 + \beta}{\alpha + \beta} \quad [i = 2 \text{ in } (6.3)].$$

We obtain from (6.4) the valid inequality:

$$x_k + \frac{1/(\alpha + \beta)}{\alpha/(\alpha + \beta)}(-\alpha x_k + cx + hy - c_0) \leq \left\lfloor \frac{\beta}{\alpha + \beta} \right\rfloor = 0$$

or

$$x_k - x_k + \frac{1}{\alpha}(cx + hy) \leq \frac{c_0}{\alpha}$$

or

$$cx + hy \leq c_0. \qquad \blacksquare$$

Proposition 6.4 shows that the MIR procedure accomplishes what is done in the disjunctive procedure. Now if $x \in B^n$, we can invoke Theorem 2.5 to conclude that every valid inequality for conv(T) is a D-inequality. Thus we obtain the following theorem.

Theorem 6.5. *Suppose $T = \{x \in B^n, y \in R^p_+ : Ax + Gy \leq b, x \leq 1\} \neq \emptyset$ where (A, G, b) is an $m \times (n + p + 1)$ matrix with rational coefficients. Any valid inequality $\pi x + \mu y \leq \pi_0$ for T is an MIR inequality.*

Theorem 6.5 is false for bounded integer variables. A counterexample is discussed in Exercise 22.

7. SUPERADDITIVITY FOR MIXED-INTEGER SETS

In this section, we extend the development of superadditive valid inequalities to mixed-integer constraint sets. Suppose $T = \{x \in Z^n_+, y \in R^p_+ : Ax + Gy \leq b\}$ and F and H are functions from R^m to R^1. We first consider conditions for which

$$\sum_{j\in N} F(a_j)x_j + \sum_{j\in J} H(g_j)y_j \leqslant F(b) \tag{7.1}$$

is valid for T for all A, G, and b. Since we want to generalize the results for the pure-integer constraint set, we assume throughout this section that F is nondecreasing and superadditive and that $F(0) = 0$. The problem is to determine the appropriate conditions to be imposed on H.

We first develop two necessary conditions.

Positive Homogeneity. Since the substitution $y_j' = \lambda y_j$ for $\lambda > 0$ is permissible and makes no essential change to T, we must have

$$\text{(a)} \qquad H(\lambda d) = \lambda H(d) \quad \text{for all } \lambda \geqslant 0 \text{ and } d \in R^m.$$

Dominance. If some valid inequality is satisfied at equality by a solution (x, y) with $Ax = d$, and some continuous activity is a multiple of d (i.e., $g_j = \lambda d$), then we must have $H(\lambda d)/\lambda \leqslant F(d)$. Positive homogeneity then implies

$$\text{(b)} \qquad H(d) \leqslant F(d) \quad \text{for all } d \in R^m.$$

Conditions (a) and (b) also are sufficient for the generation of valid inequalities.

Theorem 7.1. *If F is superadditive and nondecreasing and $F(0) = 0$, and H satisfies conditions (a) and (b), then (7.1) is a valid inequality for T for all A, G, and b.*

Proof. We have $\Sigma_{j\in N} F_j(a_j)x_j \leqslant F(Ax)$ by superadditivity and

$$\Sigma_{j\in J} H(g_j)y_j = \Sigma_{j\in J} H(g_j y_j) \leqslant \Sigma_{j\in J} F(g_j y_j) \leqslant F(Gy)$$

where we use, respectively, property (a), property (b), and the superadditivity of F. Finally, since F is superadditive and nondecreasing and $Ax + Gy \leqslant b$, it follows that $F(Ax) + F(Gy) \leqslant F(Ax + Gy) \leqslant F(b)$. ■

Note that superadditivity of H is not required here. However, we shall see below why it is natural also to impose the conditions that H be superadditive and nondecreasing.

From conditions (a) and (b), it follows that

$$H(d) = \frac{H(\lambda d)}{\lambda} \leqslant \frac{F(\lambda d)}{\lambda} \quad \text{for all } d \in R^m \text{ and } \lambda > 0. \tag{7.2}$$

The condition (7.2) restricts the class of superadditive nondecreasing functions that can be used in (7.1) for F. For example, suppose F is the C–G function $F(d) = \lfloor d \rfloor$. Then (7.2) implies, with $d = -1$,

$$H(-1) \leqslant \frac{F(-\lambda)}{\lambda} = \frac{-1}{\lambda} \quad \text{for all } 0 < \lambda \leqslant 1$$

or $H(-1) = -\infty$.

We define $\lim_{\lambda\searrow 0_+} g(x, \lambda) = \bar{g}(x)$ to mean that for each x and any $\varepsilon > 0$ there exists a $\delta(x, \varepsilon) > 0$ such that $|g(x, \lambda) - \bar{g}(x)| < \varepsilon$ whenever $0 < \lambda \leqslant \delta(x, \varepsilon)$. The C–G function is not useful for T since $\lim_{\lambda\searrow 0_+} F(\lambda d)/\lambda \to -\infty$ for $d = -1$.

However, suppose F is superadditive and nondecreasing and

$$\overline{F}(d) = \lim_{\lambda \searrow 0_+} \frac{F(\lambda d)}{\lambda} \tag{7.3}$$

exists and is finite for all d. Then from (7.2) we see that

$$H(d) \leqslant \overline{F}(d) \tag{7.4}$$

is necessary in (7.1).

Since for a given F we would like to have the function H that gives the strongest possible valid inequality, we would like to choose H as large as possible, subject to the conditions (a), (b), and (7.4). Thus if $\overline{F}$ satisfies conditions (a) and (b), the desired function is $H = \overline{F}$. Fortunately, whenever $\overline{F}$ exists and is finite for all $d \in R^m$, it satisfies conditions (a) and (b).

Proposition 7.2. *Given a nondecreasing superadditive function F, for which $\overline{F}$ given by (7.3) is defined and finite for all d, it follows that* (i) *$\overline{F}$ is positively homogeneous and* (ii) *$\overline{F}$ is dominated by F.*

Proof. i. For any given $\mu > 0$ and any $d \in R^m$, we have

$$\mu\overline{F}(d) = \mu \lim_{\lambda \searrow 0_+} \frac{F(\lambda d)}{\lambda} = \lim_{\lambda \searrow 0_+} \frac{F(\lambda d)}{\lambda/\mu} = \lim_{\lambda/\mu \searrow 0_+} \frac{F((\lambda/\mu)\mu d)}{\lambda/\mu} = \overline{F}(\mu d).$$

ii. For any $k > 0$, let $t = \lfloor k \rfloor$ and $r = k - \lfloor k \rfloor$. Then

$$\begin{aligned} F(d) = F\left(k\left(\frac{d}{k}\right)\right) &= F\left(t\left(\frac{d}{k}\right) + r\left(\frac{d}{k}\right)\right) \\ &\geqslant F\left(t\left(\frac{d}{k}\right)\right) + F\left(r\left(\frac{d}{k}\right)\right) \quad \text{by superadditivity} \\ &\geqslant tF\left(\frac{d}{k}\right) + F\left(r\left(\frac{d}{k}\right)\right) \quad \text{by superadditivity} \\ &= kF\left(\frac{d}{k}\right) + F\left(r\left(\frac{d}{k}\right)\right) - rF\left(\frac{d}{k}\right). \end{aligned}$$

Now let $\lambda = 1/k$ so that

$$F(d) \geqslant \frac{F(\lambda d)}{\lambda} + r\lambda\left(\frac{F(r\lambda d)}{r\lambda} - \frac{F(\lambda d)}{\lambda}\right).$$

Taking the limit as $\lambda \searrow 0_+$ gives $F(d) \geqslant \overline{F}(d)$. ∎

We get a bonus by taking $H = \overline{F}$ since $\overline{F}$ shares the properties of F.

Proposition 7.3. *If the $\overline{F}$ given by (7.3) is defined and finite for all d, then it is superadditive and nondecreasing.*

Proof. Since $\overline{F}$ exists, given any $\varepsilon > 0$ there exists λ_i for $i = 1, 2$ such that $|\overline{F}(d_i) - F(\lambda d_i) / \lambda| \leqslant \varepsilon$ for all $0 < \lambda \leqslant \lambda_i$, and there exists a λ_3 such that $|\overline{F}(d_1 + d_2) - F(\lambda(d_1 + d_2)) / \lambda| \leqslant \varepsilon$ for all $0 < \lambda \leqslant \lambda_3$. Taking $\lambda \leqslant \min\{\lambda_1, \lambda_2, \lambda_3\}$, we have that

$$\begin{aligned}\overline{F}(d_1) + \overline{F}(d_2) &\leqslant \frac{F(\lambda d_1)}{\lambda} + \varepsilon + \frac{F(\lambda d_2)}{\lambda} + \varepsilon \\ &\leqslant \frac{F(\lambda(d_1 + d_2))}{\lambda} + 2\varepsilon \\ &\leqslant \overline{F}(d_1 + d_2) + 3\varepsilon.\end{aligned}$$

Hence $\overline{F}(d_1) + \overline{F}(d_2) \leqslant \overline{F}(d_1 + d_2)$, and $\overline{F}$ is superadditive. Since F is nondecreasing, $F(\lambda d_2) / \lambda - F(\lambda d_1) / \lambda \geqslant 0$ for all $\lambda > 0$ and $d_2 \geqslant d_1$. Hence, taking the limit as $\lambda \searrow 0_+$ yields $\overline{F}(d_2) - \overline{F}(d_1) \geqslant 0$. Thus $\overline{F}$ is nondecreasing. ■

Thus we have justified the use of $\overline{F}$ in place of H in (7.1).

Theorem 7.4. *If F is superadditive and nondecreasing, $F(0) = 0$, and $\overline{F}$ exists, then*

$$\sum_{j \in N} F_j(a_j) x_j + \sum_{j \in J} \overline{F}(g_j)\, y_j \leqslant F(b) \tag{7.5}$$

is a superadditive valid inequality for T for all A, G, and b.

The Function F_α and the Gomory Mixed-Integer Cuts. Consider the function

$$F_\alpha(d) = \lfloor d \rfloor + \frac{(f_d - \alpha)^+}{1 - \alpha} \quad \text{for } 0 \leqslant \alpha < 1$$

given by (4.4). We obtain

$$\overline{F}_\alpha(d) = \begin{cases} 0 & \text{for } d \geqslant 0 \\ \dfrac{1}{1 - \alpha} d & \text{for } d < 0 \end{cases}$$

(see Figure 7.1). Thus, with $H = \overline{F}_\alpha$, we obtain a generalization of Proposition 6.1.

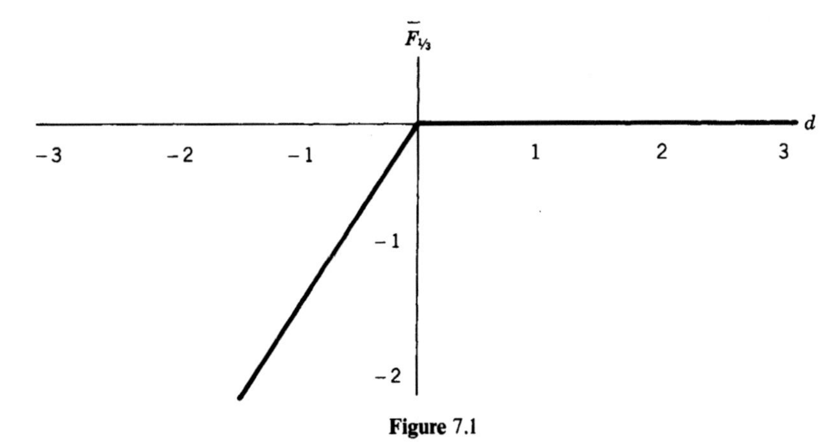

Figure 7.1

Proposition 7.5. *Let* $T = \{x \in Z^n_+, y \in R^p_+: \sum_{j \in N} a_j x_j + \sum_{j \in J} g_j y_j \leqslant b\}$, *where* $a_j, g_j, b \in R^1$ *for all* j. *The inequality*

$$\sum_{j \in N} F_\alpha(a_j)x_j + \frac{1}{1-\alpha}\sum_{j \in J^-} g_j y_j \leqslant F_\alpha(b)$$

is valid for T, *where* $J^- = \{j \in J: g_j < 0\}$.

Now consider the system

$$(7.6) \qquad \begin{aligned} &\sum_{j \in N} a_j x_j + \sum_{j \in J} g_j y_j = b \\ &x \in Z^n_+, \ y \in R^p_+ \end{aligned}$$

with $b = \lfloor b \rfloor + f_0$, $0 < f_0 < 1$. Replacing the equality in (7.6) by an inequality and then applying Proposition 7.5 with $\alpha = f_0$, we obtain the valid inequality

$$\sum_{j \in N} F_{f_0}(a_j)x_j + \frac{1}{1-f_0}\sum_{j \in J^-} g_j y_j \leqslant \lfloor b \rfloor$$

for the system (7.6).

Combining the equality of (7.6) with the above inequality yields

$$\sum_{j \in N} (a_j - F_{f_0}(a_j))x_j + \sum_{j \in J^+} g_j y_j - \frac{f_0}{1-f_0}\sum_{j \in J^-} g_j y_j \geqslant f_0$$

or

$$\sum_{\{j \in N: f_j \leq f_0\}} f_j x_j + \frac{f_0}{1-f_0}\sum_{\{j \in N: f_j > f_0\}} (1-f_j)x_j + \sum_{j \in J^+} g_j y_j - \frac{f_0}{1-f_0}\sum_{j \in J^-} g_j y_j \geqslant f_0,$$

where $J^+ = \{j \in J: g_j > 0\}$ and $f_j = a_j - \lfloor a_j \rfloor$ for $j \in N$. This is the *Gomory mixed-integer cut*.

We can derive this cut for the original mixed-integer set $T = \{x \in Z^n_+, y \in R^p_+: Ax + Gy \leqslant b\}$ and can also express it in terms of the original variables. The procedure to do this involves the introduction of slack variables $s \in R^m_+$, the use of row multipliers $u \in R^m$ to produce the system

$$\{x \in Z^n_+, y \in R^p_+, s \in R^m_+: uAx + uGy + uIs = ub\}$$

of the form (7.6), generation of the cut, and then elimination of the slack variables by substitution. We leave as an exercise the task of showing that the resulting inequality is

$$\sum_{j \in N} F(a_j)x_j + \sum_{j \in J} \overline{F}(g_j)y_j \leqslant F(b),$$

where

$$F(d) = F_\alpha(ud) - \frac{1}{1-\alpha}\sum_{\{i \in M: u_i < 0\}} u_i d_i = \tilde{F}_\alpha\Big(-\sum_{\{i: u_i < 0\}} u_i d_i, \sum_{\{i: u_i > 0\}} u_i d_i\Big)$$

and $\tilde{F}_\alpha(d_1, d_2) = [1/(1-\alpha)]d_1 + F_\alpha(d_2 - d_1)$ is the two-dimensional function given by (4.5).

We now derive a property of $\tilde{F}_\alpha$.

Proposition 7.6. *If $F = [1/(1-\alpha)]F_1 + \underline{F_\alpha}(F_2 - F_1)$, where F_1 and F_2 are superadditive and nondecreasing and where $\overline{F}_1$ and $\overline{F}_2$ exist and are finite with $\alpha > 0$, then $\overline{F} = [1/(1-\alpha)]\min(\overline{F}_1, \overline{F}_2)$.*

Proof. First we show that $F(\lambda d)/\lambda \geqslant \min(\overline{F}_1(d), \overline{F}_2(d))$. For $\lambda \geqslant 0$, we obtain

$$\begin{aligned}
\frac{F(\lambda d)}{\lambda} &= \frac{1}{1-\alpha}\frac{F_1(\lambda d)}{\lambda} + \frac{1}{\lambda}F_\alpha(F_2(\lambda d) - F_1(\lambda d)) \\
&\geqslant \frac{1}{1-\alpha}\frac{F_1(\lambda d)}{\lambda} + \frac{1}{\lambda}\overline{F}_\alpha(F_2(\lambda d) - F_1(\lambda d)) \\
&= \frac{1}{1-\alpha}\frac{F_1(\lambda d)}{\lambda} + \frac{1}{1-\alpha}\frac{1}{\lambda}\min(F_2(\lambda d) - F_1(\lambda d), 0) \\
&= \frac{1}{1-\alpha}\min\left(\frac{F_1(\lambda d)}{\lambda}, \frac{F_2(\lambda d)}{\lambda}\right) \\
&\geqslant \frac{1}{1-\alpha}\min(\overline{F}_1(d), \overline{F}_2(d)).
\end{aligned}$$

Now we must show the inequality in the opposite direction for sufficiently small positive λ. Since $\overline{F}_1$ and $\overline{F}_2$ exist, given d and $\varepsilon > 0$, there exists λ^* such that $F_i(\lambda d)/\lambda \leqslant \overline{F}_i(d) + \varepsilon$ for $i = 1, 2$ and for all $0 < \lambda < \lambda^*$. Hence

$$\begin{aligned}
\frac{F(\lambda d)}{\lambda} &= \frac{1}{1-\alpha}\frac{F_1(\lambda d)}{\lambda} + \frac{1}{\lambda}F_\alpha(F_2(\lambda d) - F_1(\lambda d)) \\
&\leqslant \frac{1}{1-\alpha}\frac{F_1(\lambda d)}{\lambda} + \frac{1}{\lambda}F_\alpha(\lambda(\overline{F}_2(d) - \overline{F}_1(d) + \varepsilon))
\end{aligned}$$

since F_α is nondecreasing, and $F_1(\lambda d) \geqslant \overline{F}_1(\lambda d) = \lambda\overline{F}_1(d)$. Now for λ sufficiently small, we see from Figure 4.2 that $F_\alpha(\lambda x) = \dfrac{1}{1-\alpha}\min(0, \lambda x)$. So

$$\begin{aligned}
\frac{F(\lambda d)}{\lambda} &\leqslant \frac{1}{1-\alpha}\frac{F_1(\lambda d)}{\lambda} + \frac{1}{1-\alpha}\min(0, \overline{F}_2(d) - \overline{F}_1(d) + \varepsilon) \\
&= \frac{1}{1-\alpha}\min\left(\frac{F_1(\lambda d)}{\lambda}, \overline{F}_2(d) + \frac{F_1(\lambda d)}{\lambda} - \overline{F}_1(d) + \varepsilon\right)
\end{aligned}$$

and

$$\frac{F(\lambda d)}{\lambda} \leqslant \frac{1}{1-\alpha}[\min(\overline{F}_1(d), \overline{F}_2(d)) + 2\varepsilon].$$

Hence $\overline{F} = [1/(1-\alpha)]\min(\overline{F}_1, \overline{F}_2)$. ■

Thus $\overline{\tilde{F}}_\alpha(d_1, d_2) = [1/(1-\alpha)]\min(d_1, d_2)$. Using $F = \tilde{F}_\alpha$ and $H = \overline{\tilde{F}}_\alpha$ in (7.1) we obtain the following proposition.

Proposition 7.7. *If* $T = \{x \in Z^n_+, y \in R^p_+: (c - \beta\pi)x + hy \leqslant c_0 - \beta\pi_0, \ (c + \beta\pi)x + hy \leqslant c_0 + \beta\pi_0 + \beta\}$, *where* $(c, h, c_0) \in R^{n+p+1}$ *and* $(\pi, \pi_0) \in Z^{n+1}$, *then* $cx + hy \leqslant c_0$ *is a superadditive valid inequality for T.*

Proof. Taking the function $F(d_1, d_2) = \beta\tilde{F}_{1/2}(d_1/2\beta, d_2/2\beta)$, we obtain

$$F(c_j - \beta\pi_j, c_j + \beta\pi_j) = \beta\left[\frac{2(c_j - \beta\pi_j)}{2\beta} + F_{1/2}\left(\frac{c_j + \beta\pi_j}{2\beta} - \frac{c_j - \beta\pi_j}{2\beta}\right)\right]$$

$$= c_j - \beta\pi_j + \beta F_{1/2}(\pi_j) = c_j - \beta(\pi_j - \pi_j) = c_j,$$

$$\overline{F}(h_j, h_j) = \beta\left[2\min\left\{\frac{h_j}{2\beta}, \frac{h_j}{2\beta}\right\}\right] = h_j,$$

$$F(c_0 - \beta\pi_0, c_0 + \beta\pi_0 + \beta) = \beta\left[\frac{2\,(c_0 - \beta\pi_0)}{2\beta} + F_{1/2}\left(\frac{c_0 + \beta\pi_0 + \beta}{2\beta} - \frac{c_0 - \beta\pi_0}{2\beta}\right)\right]$$

$$= c_0 - \beta\pi_0 + \beta F_{1/2}\left(\pi_0 + \frac{1}{2}\right) = c_0.$$

This yields the superadditive valid inequality $cx + hy \leqslant c_0$. ■

As a result of Proposition 7.7 and Theorem 6.5 we can establish the generality of superadditive inequalities.

Theorem 7.8. *Given* $T = \{x \in Z^n, y \in R^p_+: Ax + Gy \leqslant b, x \leqslant 1\} \neq \emptyset$, *every valid inequality* $\pi x + \mu y \leqslant \pi_0$ *is equal to or dominated by some superadditive valid inequality*

$$\sum_{j \in N} F(a_j)x_j + \sum_{j \in J} \overline{F}(g_j)y_j \leqslant F(b).$$

Theorem 7.8 holds for mixed-integer regions of the form $T = \{x \in Z^n, y \in R^p\} \cap P$, where P is any rational polyhedron. However, the proofs are not constructive.

Note that the function F in Theorem 7.8 can be constructed iteratively using nonnegative linear functions and $\tilde{F}_{1/2}$ a finite number of times. Furthermore, since the procedure starts with linear functions and $\overline{\tilde{F}}_{1/2}$ is the minimum of linear functions, the corresponding function $\overline{F}$ is the minimum of a finite number of linear functions and is therefore piecewise linear and concave.

Example 7.1. $T = \{x \in B^2, y \in R^2_+: y_1 + y_2 \leqslant 7, y_i \leqslant 5x_i, i = 1, 2\}$. We construct the functions representing the valid inequality $y_1 + y_2 - 2x_1 - 2x_2 \leqslant 3$. Consider the enumeration tree shown in Figure 7.2. Let the linear constraints be given in matrix form by

$$\begin{pmatrix} 0 & 0 & 1 & 1 \\ -5 & 0 & 1 & 0 \\ 0 & -5 & 0 & 1 \\ 1 & 0 & 0 & 0 \\ 0 & 1 & 0 & 0 \end{pmatrix} \begin{pmatrix} x_1 \\ x_2 \\ y_1 \\ y_2 \end{pmatrix} \leqslant \begin{pmatrix} 7 \\ 0 \\ 0 \\ 1 \\ 1 \end{pmatrix}$$

$$x, y \geqslant 0.$$

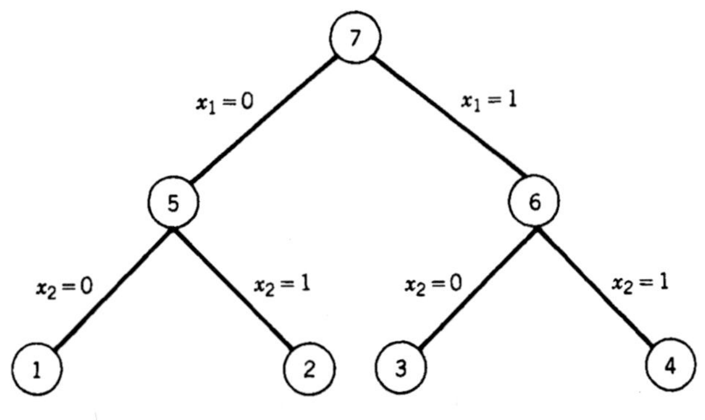

Figure 7.2

At each node (N^0, N^1) with $N^0 \cup N^1 = N$, we use a linear function to construct an inequality dominating the inequality

$$(7.7) \qquad -3\sum_{j\in N^0} x_j - 3\sum_{j\in N^1}(1-x_j) - 2x_1 - 2x_2 + y_1 + y_2 \leq 3.$$

1. $N^0 = \{1, 2\}, N^1 = \emptyset$. $F_1(d) = (0 \quad 1 \quad 1 \quad 0 \quad 0)d$ gives

$$-3x_1 - 3x_2 + (-2x_1 - 2x_2 + y_1 + y_2) \leq 0,$$

   which is stronger than (7.7).
2. $N^0 = \{1\}, N^1 = \{2\}$. $F_2(d) = (0 \quad 1 \quad 1 \quad 0 \quad 6)d$ gives

$$-3x_1 - 3(1-x_2) + (-2x_1 - 2x_2 + y_1 + y_2) \leq 3.$$

3. $N^0 = \{2\}, N^1 = \{1\}$. $F_3(d) = (0 \quad 1 \quad 1 \quad 6 \quad 0)d$ gives

$$-3(1-x_1) - 3x_2 + (-2x_1 - 2x_2 + y_1 + y_2) \leq 3.$$

4. $N^0 = \emptyset, N^1 = \{1, 2\}$. $F_4(d) = (1 \quad 0 \quad 0 \quad 1 \quad 1)d$ gives

$$-3(1-x_1) - 3(1-x_2) + (-2x_1 - 2x_2 + y_1 + y_2) \leq 3.$$

Now to obtain the inequalities that dominate (7.7) for the sets (N^0, N^1) with $N^0 \cup N^1 = \{1\}$, we combine the superadditive functions generating the above inequalities as in the proof of Proposition 7.7.

Combining the function F_1 generating the $N^0 = \{1, 2\}$, $N^1 = \emptyset$ inequality and the function F_2 generating the $N^0 = \{1\}$, $N^1 = \{2\}$ inequality yields the following:

5. $N^0 = \{1\}, N^1 = \emptyset$. $F_5 = 3\tilde{F}_{1/2}(F_1/6, F_2/6)$ gives

$$-3x_1 + (-2x_1 - 2x_2 + y_1 + y_2) \leq 3.$$

   Combining F_3 and F_4 yields the following:
6. $N^0 = \emptyset, N^1 = \{1\}$. $F_6 = 3\tilde{F}_{1/2}(F_3/6, F_4/6)$ gives

$$-3(1-x_1) + (-2x_1 - 2x_2 + y_1 + y_2) \leq 3.$$

To obtain the inequality at the root, we combine F_5 and F_6:

7. $N^0 = \emptyset$, $N^1 = \emptyset$. $F_7 = 3\tilde{F}_{1/2}(F_5/6, F_6/6)$ gives

$$-2x_1 - 2x_2 + y_1 + y_2 \leq 3.$$

8. NOTES

Section II.1.1

Valid inequalities that are implied by integrality constraints were introduced by Dantzig, Fulkerson, and Johnson (1954, 1959) in a study of the traveling salesman problem. Their pioneering work demonstrates the derivation of logical or combinatorial inequalities that can be obtained from problem structure and 0-1 variables. Another early study of this type is Markowitz and Manne (1957).

The initial study of valid inequalities for general integer programs was carried out almost single-handedly by Gomory in the late 1950s and early 1960s. His work emphasized the generation of a finite number of valid inequalities to solve the general integer programming problem. The integer rounding procedure appears implicitly in Gomory (1958, 1960a, 1963a,b) and explicitly in Chvátal (1973a). Its derivation uses a modular argument which was exploited to a greater extent by Gomory (1965) in his derivation of all valid inequalities for the group relaxation of an integer program (see Section II.3.5).

The valid inequalities of exercise 11 of Section II.4.5 were introduced by Dantzig (1959) and refined by Charnes and Cooper (1961) and Bowman and Nemhauser (1970).

Surveys on algebraic methods for obtaining valid inequalities were given by Garfinkel and Nemhauser (1972a, Chapter 4), and Jeroslow (1978, 1979a,c).

Gomory (1960b) used a disjunctive argument to develop valid inequalities for mixed-integer regions. A general disjunctive approach for obtaining valid inequalities appears in Balas (1975b). The D-inequalities were studied by Blair (1976). Jeroslow (1977, 1979a,c) and Balas (1979) gave surveys of disjunctive methods.

Valid inequalities that can be deduced from combinatorial structures and 0-1 variables appear throughout the text and, in particular, in Chapter II.2. References will be given in the notes for the corresponding sections.

Section II.1.2

Chvátal (1973a) contains all of the results of this section with the exception of Theorem 2.16, although a few of the results are given only implicitly.

Blair (1976) also showed that the D-inequalities suffice for 0-1 problems (Theorem 2.3). The close connection between this theorem when P is empty with the inequality $0x \leq -1$ and the resolution method of propositional logic of Davis and Putnam (1960) is discussed in Blair, Jeroslow, and Lowe (1986). General conditions under which the convex hull can be obtained sequentially by imposing disjunctions one-by-one, as in the proof of Theorem 2.5, were studied by Balas (1979); see Section 6 of that article and Exercise 9 of Section I.4.8.

Theorem 2.16 is due to Schrivjer (1980). He showed that for any integer k, the linear inequality system consisting of all of the inequalities of rank equal to or less than k defines a rational polyhedron, and then he used total dual integrality (see Section III.1.1) to show that there existed some k for which the system defines the convex hull of integer solutions.

The example of the two-dimensional family of polyhedra of unbounded rank is from Chvátal (1973a). Related results are given in Jeroslow (1971) and Jeroslow and Kortanek (1971).

Section II.1.3

The fractional cuts are due to Gomory (1958). The connection between them and integer rounding in the space of the original variables is implicit in that article and Chvátal (1973a).

Section II.1.4

Connections between valid inequalities for integer programs and superadditive functions originated with the work of Gomory (1965, 1967, 1969, 1970) on the group problem relaxation of a general integer program (see Section II.3.5). The explicit use of superadditive functions in the generation of valid inequalities for general (pure) integer programs was developed in a series of articles by Gomory and Johnson (1972, 1973) and Burdet and Johnson (1974), again in the context of the group problem.

Araoz (1973) investigated superadditive valid inequalities for packing and covering problems and showed that the modular arithmetic requirement of the group relaxation was not essential to the superadditive theory. For general (pure) integer programs, the superadditive representation of all facet-defining inequalities (Proposition 4.5 and Theorem 4.6) appears in the articles by Burdet and Johnson (1977) and Jeroslow (1978). The function of Figure 4.5 was used by E. L. Johnson (1974), and the two-dimensional function given by (4.5) and exhibited in Figure 4.3 appears in Nemhauser and Wolsey (1984). Some other classes of superadditive functions that have been proposed for the purpose of generating valid inequalities are given by Burdet and Johnson (1974, 1977).

Surveys by Jeroslow (1978, 1979a,c) and Johnson (1979), and a monograph by Johnson (1980a) provide comprehensive treatments of the use of superadditivity in integer and mixed-integer programming. These references are also relevant to the following three sections.

Section II.1.5

A polyhedral description of superadditive valid inequalities for the group problem was given by Gomory (1967, 1969, 1970). He also introduced the concept of master polytopes in these articles and showed how facets for lower-dimensional polytopes could be obtained from the master polytope by projection.

Gomory's approach was extended to independence systems or packing problems and to dependence systems or covering problems by Araoz (1973) as well as to general pure-integer programs by Burdet and Johnson (1977). See also Johnson (1979, 1980a, 1981a) and Araoz and Johnson (1981).

Superadditive inequalities for 0-1 problems were studied by Wolsey (1977), and those for multiple right-hand side problems were studied by Johnson (1981b).

Section II.1.6

This section is based on Nemhauser and Wolsey (1984). The motivation for the MIR inequalities came from the mixed-integer cuts of Gomory (1960b).

Schrijver gave us the example in Exercise 22, which shows that Theorem 6.5 is false unless each integer variable belongs to the set $\{0, 1\}$. This is related to the absence of finite convergence of Gomory's mixed-integer cutting-plane algorithm, as shown by White (1961). White's counterexample appears in Salkin (1975).

Section II.1.7

The extension of the superadditive theory to mixed-integer programs began with the work of E. L. Johnson (1974) on a mixed-integer group problem.

Theorem 7.8, for bounded mixed-integer constraint sets, appears in Jeroslow (1979b). Its generalization to unbounded sets is given by Bachem and Schrader (1980) and by Bachem, Johnson, and Schrader (1982). Also see Blair (1978) and Jeroslow (1985).

9. EXERCISES

1. Let $S = \{x \in Z^2_+: 4x_1 + x_2 \leq 28, x_1 + 4x_2 \leq 27, x_1 - x_2 \leq 1\}$. Determine the facets of conv(S) graphically (see Exercise 10 of Section I.4.8). Then derive each of the facets of conv(S) as a C–G inequality.
2. Let $S = \{x \in Z^3_+: 19x_1 + 28x_2 - 184x_3 = 8\}$. Derive the valid inequality $x_1 + x_2 + 5x_3 \geq 8$ using modular arithmetic.
3. For $S = \{x \in B^4: 9x_1 + 7x_2 - 2x_3 - 3x_4 \leq 12, 2x_1 + 5x_2 + 1x_3 - 4x_4 \leq 10\}$ show that $4x_1 + 5x_2 - 2x_3 - 4x_4 \leq 12$ is a valid inequality by disjunctive arguments.
4. Consider the node-packing problem on the graph of Figure 9.1. Show that $\Sigma_{i=1}^7 x_i \leq 2$ is a valid inequality, both combinatorially and algebraically.
5. Prove the following:
   i) Let $P = \{x \in R^n: Ax \leq b\} \neq \emptyset$. $\pi x \leq \pi_0$ is a valid inequality for P if and only if there exists $u \in R^m_+$ such that $uA = \pi$ and $ub \leq \pi_0$.
   ii) Let $P = \{x \in R^m_+: Ax \leq b, x \leq d\}$. $\pi x \leq \pi_0$ is a valid inequality for P if and only if there exist $u \in R^m_+$ and $w \in R^n_+$ such that $uA + w \geq \pi$ and $ub + wd \leq \pi_0$.
6. Let $P_i = \{x \in R^n_+: A^i x \leq b_i\}$ for $i = 1, 2$. Show that $\pi x \leq \pi_0$ is a valid inequality for $P_1 \cup P_2$ if there exists $u^i \in R^m_+$ such that $u^i A^i \geq \pi$ and $u^i b_i \leq \pi_0$ for $i = 1, 2$. Under what restrictions on P_1 and P_2 does the converse hold?
7. (The Davis–Putnam Procedure). Consider the satisfiability problem for $S \subseteq B^n$ defined by

$$\sum_{j \in C_k} x_j + \sum_{j \in \tilde{C}_k} (1 - x_j) \geq 1 \quad \text{for } k = 1, \ldots, K, x \in B^n$$

   where $C_k \cap \tilde{C}_k = \emptyset$ and $C_k, \tilde{C}_k \subseteq N$ for $k = 1, \ldots, K$.

   i) Given $q \in N$ and a pair of constraints k, l such that $q \in C_k \cap \tilde{C}_l$, show that

$$\sum_{j \in (C_k \cup C_l) \setminus \{q\}} x_j + \sum_{j \in (\tilde{C}_k \cup \tilde{C}_l) \setminus \{q\}} (1 - x_j) \geq 1$$

   is a valid inequality for S.

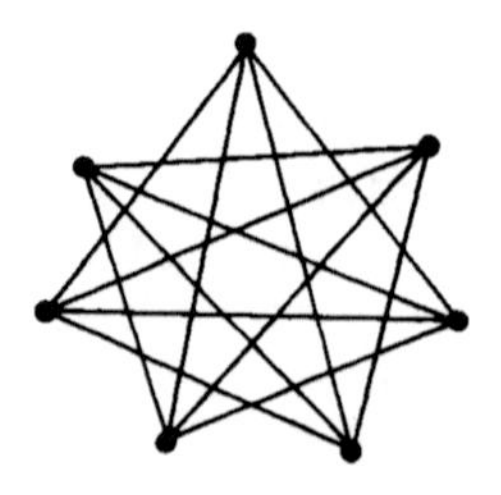

Figure 9.1

ii) Show that the inequality is a D-inequality.

iii) Show that if $S = \emptyset$, it is possible to generate the valid inequality $0x \leq -1$ by a finite number of replications of the procedure i.

iv) Show that the resulting algorithm is polynomial if $|C_k \cup \tilde{C}_k| \leq 2$ for all k.

8. What is the rank of conv(S) in Exercise 1?

9. Prove Propositions 2.9 and 2.10.

10. Show that the rank of conv(S^t) is $t - 1$, where $S^t = P^t \cap Z^2$ and

$$P^t = \{x \in R_+^2 \colon tx_1 + x_2 \leq 1 + t,\ -tx_1 + x_2 \leq 1,\ x_1 \leq 1\}.$$

11. Show that if $P = \{x \in R_+^n \colon x_i + x_j \leq 1 \text{ for } 1 \leq i < j \leq n\}$ and $S = P \cap B^n$, the rank of $\Sigma_{j=1}^n x_j \leq 1$ is $O(\log n)$.

12. Use Theorem 2.5 to show that every valid inequality is a D-inequality for mixed 0-1 programs.

13. Consider the integer program $\max\{2x_1 + 5x_2 \colon x \in S\}$, where S is given in Exercise 1. Using the optimal basis of the corresponding linear program, the problem can be rewritten as

$$\begin{aligned} &\max z \\ &z + \frac{1}{5}x_3 + \frac{6}{5}x_4 = 38 \\ &x_1 + \frac{4}{15}x_3 - \frac{1}{15}x_4 = \frac{17}{3} \\ &x_2 - \frac{1}{15}x_3 + \frac{4}{15}x_4 = \frac{16}{3} \\ &-\frac{1}{3}x_3 + \frac{1}{3}x_4 + x_5 = \frac{2}{3} \\ &x \in Z_+^5. \end{aligned}$$

Derive a Gomory fractional cut from each equation. Express each cut in terms of the original variables (x_1, x_2). Derive each cut as a rank 1 C–G inequality.

14. For $S = P \cap Z^2$ as given in Exercise 1 show that

i) $x_1 \leq 5$,

ii) $x_1 + 2x_2 \leq 15$, and

iii) $2x_1 + 5x_2 \leq 36$
are superadditive valid inequalities.

15. What conditions must be imposed on F so that $\Sigma_{j=1}^n F(a_j)x_j \leq F(b)$, is a valid inequality for $S = \{x \in Z^n \colon Ax \leq b\}$?

16. Show that the following functions are superadditive:

i) $G(d) = \max\{\alpha, F(d)\}$, where $\alpha < 0$ and F is superadditive.

ii) $G(d) = \max_{h \in Z^m} \{F_1(h) + F_2(d - h)\}$, where F_1 and F_2 are superadditive on Z^m.

iii) $G_\alpha(d) = \max\{\alpha, \min(0, d)\}$ for $d \in R^1$ and $\alpha < 0$.

17. **i)** Draw G_α, which was defined in iii of Exercise 16.

ii) Use G_α to show that $-3x_1 - 2x_2 - 2x_2 \le -3$ is a valid inequality for $S = \{x \in Z_+^3: -7x_1 - 4x_2 - 4x_3 \le -6\}$.

iii) Can you show by repeated use of G_α that $-2x_1 - x_2 - x_3 \le -2$ is valid for S?

iv) For $S = \{x \in B^3: -7x_1 - 4x_2 - 4x_3 \le -6\}$ use G_α to show that $-x_1 - x_2 \le -1$ is valid for S.

18. Suppose that we define the disjunctive rank of an inequality via the function $\tilde{F}_\alpha$. What is the maximum disjunctive rank of conv(S), where

i) $S = \{x \in R^n: Ax \le b, x \le 1, x \ge 0\} \cap Z^n$,

ii) $S = \{x \in R^n: Ax \le b, x \le d, x \ge 0\} \cap Z^n$,

iii) $S = \{x \in R_+^n: Ax \le b\} \cap Z^n$?

19. Let $\Phi(d) = \max\{cx: Ax \le d, x \in Z_+^n\}$ and suppose the problem is feasible for all $d \in R^m$. Show that Φ is superadditive on R^m.

20. Find the convex hull of $S = (x \in Z_+^4; x_1 + 2x_2 + 3x_3 + 4x_4 \le 4)$.

21. Write an implicit polyhedral description of the set of valid inequalities for

i) $x_1 + x_2 + x_3 + x_4 = 4, \quad x \in Z_+^4$.

ii) $x_1 + x_2 + x_3 + x_4 \ge 4, \quad x \in Z_+^4$.

iii) $\binom{1}{0}x_1 + \binom{0}{1}x_2 + \binom{2}{0}x_3 + \binom{1}{1}x_4 + \binom{0}{2}x_5 + \binom{2}{1}x_6 + \binom{1}{2}x_7 + \binom{2}{2}x_8 \le \binom{2}{2}, \quad x \in Z_+^8$.

22. Show that the set

$$T = \{(x, y) \in Z_+^2 \times R_+^1: x_1 + x_2 + y \le 2, -x_1 + y \le 0, -x_2 + y \le 0\}$$

and the valid inequality $y \le 0$ give a counterexample to Theorem 6.5 when the constraints $x \le 1$ are not present.

23. Given $T = \{(x_0, x, y) \in Z^1 \times Z_+^n \times R_+^p: x_0 + \Sigma_{j \in N} a_j x_j + \Sigma_{j \in J} g_j y_j = b\}$ with $b = \lfloor b \rfloor + f_0$ and $0 < f_0 < 1$, derive (by a disjunctive and modular argument) the Gomory mixed-integer cut

$$\sum_{\{j \in N: f_j \le f_0\}} f_j x_j + \frac{f_0}{1 - f_0} \sum_{\{j \in N: f_j > f_0\}} (1 - f_j) x_j + \sum_{j \in J^+} g_j y_j - \frac{f_0}{1 - f_0} \sum_{j \in J^-} g_j y_j \ge f_0,$$

where $J^+ = \{j \in J: g_j > 0\}$, $J^- = J \setminus J^+$, and $f_j = a_j - \lfloor a_j \rfloor$ for $j \in N$.

24. Verify that the Gomory mixed-integer cut for

$$T' = \{(x, y, s) \in Z_+^n \times R_+^p \times R_+^1: uAx + uGy + us = ub\}$$

is equivalent to the superadditive valid inequality

$$\sum_{j \in N} F(a_j) x_j + \sum_{j \in J} \overline{F}(g_j) y_j \le F(b)$$

for $T = \{(x, y) \in Z_+^n \times R_+^p: Ax + Gy \le b\}$, where $F(d) = \tilde{F}_\alpha(-\Sigma_{u_i<0} u_i d_i, \Sigma_{u_i>0} u_i d_i)$ and $\alpha = ub - \lfloor ub \rfloor$.

II.2

Strong Valid Inequalities and Facets for Structured Integer Programs

1. INTRODUCTION

In the preceding chapter we presented a general theory of valid inequalities for integer and mixed-integer programs and techniques for generating all valid inequalities. However, these general techniques can be quite inefficient in deriving facets or even lower-dimensional faces of the convex hull of a set of integral points.

The theme of this chapter is to use structure to determine strong valid inequalities for the constraint sets of some $\mathcal{NP}$-hard integer programming problems. The determination of families of strong valid inequalities is more of an art than a formal methodology. Thus our presentation will largely be a series of examples that convey the basic ideas. The mathematics enters in proving that classes of inequalities, which are often easily shown to be valid, are indeed strong in the sense that they define facets or faces of reasonable dimension. A related mathematical problem, which is considered in Part III, is to prove that a given family of inequalities represents all of the facets of the convex hull. We defer this topic because the results are limited almost exclusively to those combinatorial optimization problems for which polynomial-time algorithms are known.

There are many interesting problems for which strong valid inequalities have been obtained. Only a small selection of these results can be given here, so we have picked a few prototype problems. To motivate some basic ideas, in this section we consider the node-packing polytope. In the following sections, we study the 0-1 knapsack polytope, the symmetric traveling salesman polytope, and a class of generic mixed-integer sets that we call 0-1 variable upper-bound flow models. The attention given to polyhedra for which the integer variables are binary reflects the fact that most of the known results are in this domain.

In the preceding chapter, we derived some valid inequalities for the node-packing problem. Here we will establish the strength of the inequalities. Recall that a node packing in a graph $G = (V, E)$ is a set of nodes such that no pair in the set is joined by an edge. Thus the set of node packings S is given by

$$S = \{x \in B^n: x_i + x_j \leqslant 1 \text{ for all } (i, j) \in E\},$$

where $n = |V|$. The vector $x \in S$ is the characteristic vector of a packing; that is, $x_i = 1$ if node i is in the packing and $x_i = 0$ otherwise. Since S contains the zero vector and the n unit vectors, $\dim(\text{conv}(S)) = n$.

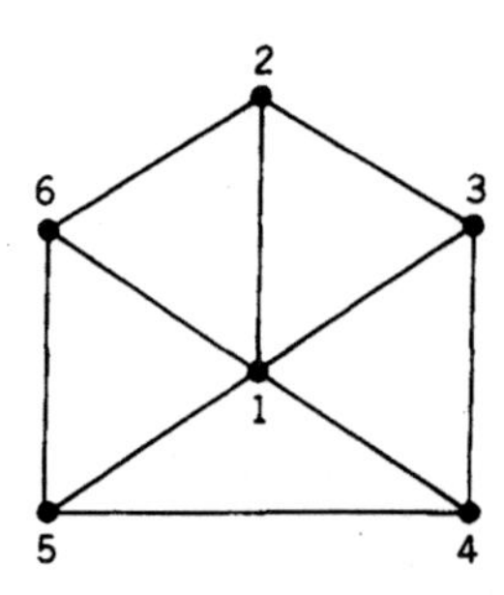

Figure 1.1

A set $C \subseteq V$ is called a *clique* if each pair of nodes in C is joined by an edge. Thus a node packing can contain no more than one node from each clique. For the graph of Figure 1.1, the maximal cliques yield the inequalities

$$
(1.1) \qquad \begin{pmatrix} 1 & 1 & 1 & & & \\ 1 & & 1 & 1 & & \\ 1 & & & 1 & 1 & \\ 1 & & & & 1 & 1 \\ 1 & 1 & & & & 1 \end{pmatrix} \begin{pmatrix} x_1 \\ x_2 \\ x_3 \\ x_4 \\ x_5 \\ x_6 \end{pmatrix} \leqslant \begin{pmatrix} 1 \\ 1 \\ 1 \\ 1 \\ 1 \end{pmatrix}
$$

corresponding to the cliques {1, 2, 3}, {1, 3, 4}, {1, 4, 5}, {1, 5, 6}, and {1, 2, 6}.

When C is a maximal clique, the *clique constraint*

$$
(1.2) \qquad \sum_{j \in C} x_j \leqslant 1
$$

defines a facet of conv(S). This is an easy result to prove directly from the definition of a facet. A facet of conv(S) is of dimension $n - 1$ and thus contains n affinely independent points of conv(S). Moreover, as noted in Proposition 6.6 of Section I.4.6, a facet contains n affinely independent points of S. Since the hyperplane $\Sigma_{j \in C} x_j = 1$ does not contain the origin, any set of affinely independent points on it are also linearly independent. Thus we will exhibit n linearly independent points of S that satisfy (1.2) at equality.

Suppose, for simplicity of notation, that $C = \{1, \ldots, k\}$. Since C is maximal, for each $j \notin C$ there is a node $l(j)$ such that $l(j) \leqslant k$ and $\{j, l(j)\}$ is a node packing. The characteristic vectors of the packings $\{1\}, \ldots, \{k\}, \{k + 1, l(k + 1)\}, \ldots, \{n, l(n)\}$ are easily shown to be linearly independent.

The rows of the matrix given below are six linearly independent vectors which establish that $x_1 + x_2 + x_3 \leqslant 1$ is facet for the graph of Figure 1.1.

$$
\begin{pmatrix} 1 & & & & & \\ & 1 & & & & \\ & & 1 & & & \\ & 1 & & 1 & & \\ & 1 & & & 1 & \\ & & 1 & & & 1 \end{pmatrix}.
$$

Although there is an important class of node-packing problems for which the maximal clique constraints and nonnegativity give all the facets of conv(S), this is not true in our example. In particular, $x^1 = \frac{1}{2}(0\ 1\ 1\ 1\ 1\ 1)$ is an extreme point of the polytope given by (1.1) and $x \geqslant 0$. This can be seen by solving the linear program max $\Sigma_{j=1}^{6} x_j$ subject to (1.1) and $x \geqslant 0$. The unique optimal solution is x^1.

To cut off x^1, we consider another family of valid inequalities. Suppose there is an $H \subseteq V$ that induces a *chordless cycle*, that is, the nodes of H can be ordered as $(i_1, i_2, \ldots, i_p)$ such that $(i_r, i_s) \in E$ if and only if $s = r + 1$ or $s = 1$ and $r = p$. If p is odd and at least 5, then H is called an *odd hole*. If H is an odd hole, then

$$\sum_{j \in H} x_j \leqslant \frac{|H| - 1}{2} \tag{1.3}$$

is satisfied by all node packings. Moreover, the clique constraints $x_i + x_j \leqslant 1$ for $i, j \in H$ do not imply (1.3).

In our example, $H = \{2, 3, 4, 5, 6\}$ is an odd hole and we obtain the constraint

$$x_2 + x_3 + x_4 + x_5 + x_6 \leqslant 2, \tag{1.4}$$

which cuts off the solution x^1.

Since (1.4) is satisfied at equality by the five linearly independent characteristic vectors corresponding to the packings {2, 4}, {2, 5}, {3, 5}, {3, 6}, and {4, 6}, inequality (1.4) gives a facet of the convex hull of node packings for the subgraph with node set H. But it does not give a facet of conv(S) for the graph G, since there are no other packings that satisfy (1.4) at equality. If we added (1.4) to the clique constraints, we would obtain the new extreme point $\frac{1}{5}(1\ 2\ 2\ 2\ 2\ 2)$.

Since (1.4) is a four-dimensional face of conv(S) but not a facet, it can perhaps be strengthened by tilting it to produce a facet. In other words, is there a valid inequality of the form

$$\alpha x_1 + (x_2 + x_3 + x_4 + x_5 + x_6) \leqslant 2 \tag{1.5}$$

with $\alpha > 0$? And if so, what is the largest value of α that preserves validity? To answer these questions, we must consider $x_1 = 0$ and $x_1 = 1$. When $x_1 = 0$, (1.5) is valid for any $\alpha > 0$. When $x_1 = 1$, we have $\alpha \leqslant 2 - (x_2 + x_3 + x_4 + x_5 + x_6)$. But $x_1 = 1$ implies $x_2 = x_3 = x_4 = x_5 = x_6 = 0$, so $\alpha \leqslant 2$. Thus

$$2x_1 + x_2 + x_3 + x_4 + x_5 + x_6 \leqslant 2 \tag{1.6}$$

is a valid inequality. Moreover, it gives a facet of conv(S) since it is satisfied at equality by the characteristic vector of {1} and the characteristic vectors of the five packings given above that satisfy (1.4) at equality.

We have just illustrated a general principle called *lifting* whereby a valid inequality for $S \cap \{x \in B^n: x_1 = 0\}$ is extended to a valid inequality for S.

Proposition 1.1. *Suppose* $S \subseteq B^n$, $S^\delta = S \cap \{x \in B^n: x_1 = \delta\}$ *for* $\delta \in \{0, 1\}$, *and*

$$\sum_{j=2}^{n} \pi_j x_j \leqslant \pi_0 \tag{1.7}$$

is valid for S^0. *If* $S^1 = \varnothing$, *then* $x_1 \leqslant 0$ *is valid for* S. *If* $S^1 \neq \varnothing$, *then*

(1.8) $$\alpha_1 x_1 + \sum_{j=2}^{n} \pi_j x_j \leqslant \pi_0$$

is valid for S *for any* $\alpha_1 \leqslant \pi_0 - \zeta$, *where* $\zeta = \max\{\sum_{j=2}^{n} \pi_j x_j : x \in S^1\}$. *Moreover, if* $\alpha_1 = \pi_0 - \zeta$ *and (1.7) gives a face of dimension* k *of* conv(S^0), *then (1.8) gives a face of dimension at least* $k + 1$ *of* conv(S). [*If (1.7) gives a facet of* conv(S^0), *then (1.8) gives a facet of* conv(S).]

Proof. If $\bar{x} \in S^0$, then

$$\alpha_1 \bar{x}_1 + \sum_{j=2}^{n} \pi_j \bar{x}_j = \sum_{j=2}^{n} \pi_j \bar{x}_j \leqslant \pi_0$$

since (1.7) is valid for S^0.

If $\bar{x} \in S^1$, then

$$\alpha_1 \bar{x}_1 + \sum_{j=2}^{n} \pi_j \bar{x}_j = \alpha_1 + \sum_{j=2}^{n} \pi_j \bar{x}_j \leqslant \alpha_1 + \zeta \leqslant \pi_0$$

by definition of the quantities α_1 and ζ.

Since (1.7) gives a k-dimensional face of conv(S^0), there exist $x^i \in S^0$ for $i = 1, \ldots, k + 1$ that are affinely independent and satisfy (1.7) at equality. Since $x_1^i = 0$, it follows that x^i satisfies (1.8) at equality for $i = 1, \ldots, k + 1$. Let $\zeta = \sum_{j=2}^{n} \pi_j x_j^*$, where $x^* \in S^1$. With $\alpha_1 = \pi_0 - \zeta$, x^* satisfies (1.8) at equality. Finally, since $x_1^* = 1$, it follows that x^* cannot be written as an affine combination of $\{x^1, \ldots, x^{k+1}\}$, so the $k + 2$ vectors $\{x^*, x^1, \ldots, x^{k+1}\}$ are affinely independent. ■

The lifting principle is also applicable to extending a valid inequality from S^1 to S. Using the same notation as in Proposition 1.1, we have the analogous result:

Proposition 1.2. *Suppose (1.7) is valid for* S^1. *If* $S^0 = \emptyset$, *then* $x_1 \geqslant 1$ *is valid for* S. *If* $S^0 \neq \emptyset$, *then*

(1.9) $$\gamma_1 x_1 + \sum_{j=2}^{n} \pi_j x_j \leqslant \pi_0 + \gamma_1$$

is valid for S *for any* $\gamma_1 \geqslant \zeta - \pi_0$, *where* $\zeta = \max\{\sum_{j=2}^{n} \pi_j x_j : x \in S^0\}$. *Moreover, if* $\gamma_1 = \zeta - \pi_0$ *and (1.7) gives a face of dimension* k *of* conv(S^1), *then (1.9) gives a face of dimension at least* $k + 1$ *of* conv(S).

When $\alpha_1 = \pi_0 - \zeta$ in Proposition 1.1 or when $\gamma_1 = \zeta - \pi_0$ in Proposition 1.2, we say that the *lifting is maximum*.

Propositions 1.1 and 1.2 are meant to be used sequentially. Given an $N_1 \subset N = \{1, \ldots, n\}$ and an inequality $\sum_{j \in N_1} \pi_j x_j \leqslant \pi_0$ that is valid for $S \cap \{x \in B^n : x_j = 0 \text{ for } j \in N \setminus N_1\}$, we lift one variable at a time to obtain a valid inequality

(1.10) $$\sum_{j \in N \setminus N_1} \alpha_j x_j + \sum_{j \in N_1} \pi_j x_j \leqslant \pi_0$$

for S.

The coefficients $\{\alpha_j\}$ in (1.10) are dependent on the order in which the variables are lifted. So by considering different orderings of the elements of $N \setminus N_1$, we can get a family of valid inequalities for S.

It is insightful to examine the lifting process in the polar space $\Pi^1 = \{\pi \in R^n: \pi x \leq 1$ for all $x \in S \subseteq B^n\}$. If $\sum_{j=2}^n \pi_j x_j \leq 1$ is valid for S^0, maximum lifting can be described by the one-dimensional optimization problem in Π^1-space:

$$\max(\alpha: (0, \pi_2, \ldots, \pi_n) + \alpha(1, 0, \ldots, 0) \in \Pi^1\}.$$

The geometry is illustrated in Figure 1.2 for the case $n = 2$. We suppose that Π^1 has the three extreme points $\{\pi^0, \pi^1, \pi^2\}$. Since $\pi_2^0 > \max(\pi_2^1, \pi_2^2)$, we have that $\pi_2^0 x_2 \leq 1$ gives a facet of conv(S^0), where $S^0 = S \cap \{x: x_1 = 0\}$. Maximum lifting is equivalent to moving from $(0, \pi_2^0)$ in the direction $(1 \quad 0)$ to obtain the extreme point π^0 of Π^1 or, equivalently, the facet of conv(S) defined by $\pi_1^0 x_1 + \pi_2^0 x_2 \leq 1$. Similarly, by a maximum lifting from $\pi_1^2 x_1 \leq 1$, we obtain the facet of conv(S) defined by $\pi_1^2 x_1 + \pi_2^2 x_2 \leq 1$. We also see that there is no way to generate the facet of conv(S) defined by $\pi_1^1 x_1 + \pi_2^1 x_2 \leq 1$ by sequential lifting.

To interpret sequential lifting geometrically, suppose we begin with the trivial inequality $0 \leq 1$. Maximum lifting in the order (1, 2) yields the facet corresponding to the extreme point π^2, and maximum lifting in the order (2, 1) yields the facet corresponding to the extreme point π^0. Neither order gives π^1.

In principle, lifting is not restricted to choosing one coefficient at a time. If we observe that maximum sequential lifting is equivalent to finding an extreme point in a one-dimensional polyhedron, it is not surprising that in the simultaneous lifting of k coefficients, the "best" liftings are obtained by finding the extreme points of a k-dimensional polyhedron. Hence if we start from the inequality $0 \leq 1$ and allow the simultaneous lifting of $(\pi_1 \ \pi_2)$, we can indeed obtain π^0, π^1, and π^2.

As we have already seen, the values of the coefficients in (1.10) depend on the ordering of the variables in the sequential lifting. The following proposition, which will be useful in the next section, indicates how the coefficient of one variable depends on the ordering.

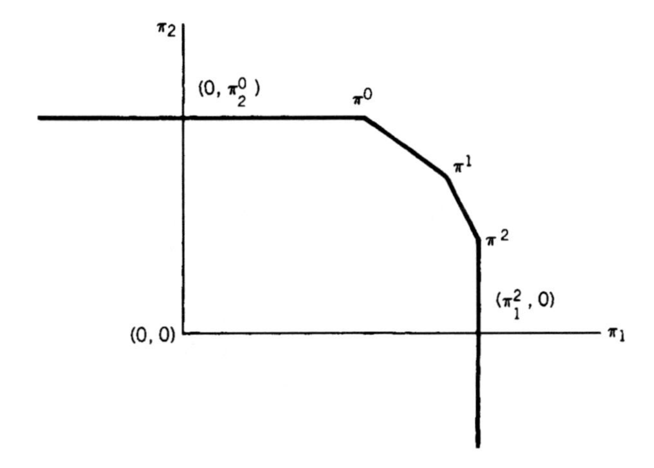

Figure 1.2

Proposition 1.3. *Let $N \setminus N_1 = \{1, 2, \ldots, t\}$ and suppose when Proposition 1.1 is applied sequentially using maximum lifting in the order $(i_1, i_2, \ldots, i_{k-1}, i_k, \ldots, i_t)$, the inequality*

$$\sum_{j \in N \setminus N_1} \alpha_j x_j + \sum_{j \in N_1} \pi_j x_j \leqslant \pi_0$$

is obtained. Then for any order $(i'_1, i'_2, \ldots, i'_{k-1}, i'_k, \ldots, i'_t)$ with $i'_j = i_j$ for $j = 1, \ldots, k-1$, the resulting inequality

$$\sum_{j \in N \setminus N_1} \alpha'_j x_j + \sum_{j \in N_1} \pi_j x_j \leqslant \pi_0$$

obtained by maximum lifting has $\alpha'_{i_k} \leqslant \alpha_{i_k}$.

Proof. $i_k = i'_s$ for some $s > k$. Then

$$\alpha_{i_k} = \pi_0 - \max\left\{\sum_{j=1}^{k-1} \alpha_{i_j} x_{i_j} + \sum_{j \in N_1} \pi_j x_j : x \in S \cap \{x \in B^n : x_{i_k} = 1 \text{ and } x_{i_j} = 0 \text{ for } j > k\}\right\}$$

$$= \pi_0 - \max\left\{\sum_{j=1}^{k-1} \alpha_{i_j} x_{i_j} + \sum_{j \in N_1} \pi_j x_j : x \in S \cap \{x \in B^n : x_{i'_s} = 1 \text{ and } x_{i'_j} = 0 \text{ for } k \leqslant j \leqslant t, j \neq s\}\right\}$$

$$\geqslant \pi_0 - \max\left\{\sum_{j=1}^{k-1} \alpha_{i_j} x_{i_j} + \sum_{j \in N_1} \pi_j x_j + \sum_{j=k}^{s-1} \alpha'_{i'_j} x_{i'_j} : x \in S \cap \{x \in B^n : x_{i'_s} = 1 \text{ and } x_{i'_j} = 0 \text{ for } j > s\}\right\}$$

$$= \alpha'_{i_k}. \qquad \blacksquare$$

Corollary 1.4. *In any sequential maximum lifting of the variables in $N \setminus N_1$, the minimum value of α_{i_k} is obtained by lifting x_{i_k} last and the maximum value is obtained by lifting x_{i_k} first.*

Although this discussion has focused on maximum lifting, ζ can be hard to compute. Thus, in practice, α and γ are generally determined from easily computable upper bounds on ζ. We will illustrate these computations in Section II.6.2.

We have given two ways of showing that a valid inequality gives a facet of conv(S). The first approach was to apply the definition, the second approach was by maximum lifting of a lower-dimensional facet. We now consider a third approach, which is to apply Proposition 3.6 of Section I.4.3.

We illustrate this approach by showing that (1.6) gives a facet of conv(S) in the node-packing example of Figure 1.1. Consider a valid inequality $\Sigma_{j=1}^{6} \pi_j x_j \leqslant \pi_0$ and suppose that it is satisfied at equality by the packings {2, 4}, {2, 5}, {3, 5}, {3, 6}, and {4, 6}. From {2, 4} and {2, 5} we obtain $\pi_2 + \pi_4 = \pi_2 + \pi_5 = \pi_0$ or $\pi_4 = \pi_5$. Similarly from {2, 5} and {3, 5} we obtain $\pi_2 = \pi_3$, from {3, 5} and {3, 6} we obtain $\pi_5 = \pi_6$, and from {2, 4} and {4, 6} we obtain $\pi_2 = \pi_6$. Hence any equality that is satisfied by these five packings must be of the form

$$\pi_1 x_1 + \alpha(x_2 + \cdots + x_6) = 2\alpha.$$

Now if the packing {1} also lies on the above hyperplane, we must have $\pi_1 = 2\alpha$ and the equality must be of the form

$$\alpha(2x_1 + x_2 + \cdots + x_6) = 2\alpha.$$

Finally, the inequality $\alpha(2x_1 + x_2 + \cdots + x_6) \leqslant 2\alpha$ must hold for $x = 0$. Thus $\alpha > 0$, and it suffices to take $\alpha = 1$.

Here we have applied Proposition 3.6 of Section I.4.3 with $n = 6$, $k = 0$, and $x^1, \ldots, x^6$ being the characteristic vectors of the six packings given above. The argument shows that all solutions to the linear system $\lambda x^i = \lambda_0$ for $i = 1, \ldots, 6$ are of the form $\lambda = \alpha\pi$ and $\lambda_0 = \alpha\pi_0$ with $\alpha \in R^1$. Finally, we used $x = 0$ to establish that $\alpha > 0$. Other applications of this technique will be given in Sections 3 and 4.

We close this section with a pessimistic reminder regarding the possibility of obtaining all facets of the convex hull of a feasible set of points for an $\mathcal{NP}$-hard optimization problem, but we add a note of optimism with respect to using the strong inequalities that can be obtained.

In Proposition 7.4 of Section I.5.7, it was established that for an $\mathcal{NP}$-complete lower-bound feasibility problem, a good characterization of all of the facets of the convex hull of feasible solutions is not possible unless $\mathcal{NP} = \mathcal{C}o\mathcal{NP}$. Thus our use of structure to obtain a polyhedral representation of the constraint set is limited by the inherent complexity of the problem. For this reason the results of this chapter are only partial descriptions of the convex hull of the constraint set of the problem being studied. However, there are some experimental results which indicate that simple classes of strong valid inequalities that can be identified efficiently are extremely useful in solving a variety of integer programming problems by cutting-plane algorithms. In Chapter II.5, we will show how the results of this chapter can be incorporated in such cutting-plane algorithms.

2. VALID INEQUALITIES FOR THE 0-1 KNAPSACK POLYTOPE

We consider the constraint set of a 0-1 knapsack problem

(2.1) $$S = \left\{x \in B^n\colon \sum_{j \in N} a_j x_j \leqslant b\right\},$$

where $N = \{1, \ldots, n\}$, $a_j \in Z^1_+$ for $j \in N$, and $b \in Z^1_+$. Note that S is an independence system (see Section II.1.5). Since $a_j > b$ implies $x_j = 0$ for all $x \in S$, we assume $a_j \leqslant b$ for all $j \in N$. Thus $\dim(\text{conv}(S)) = n$. It is convenient to order the coefficients monotonically so that $a_1 \geqslant a_2 \geqslant \cdots \geqslant a_n$. We represent elements of B^n by characteristic vectors so that for $R \subseteq N$ the vector x^R has components $x^R_j = 1$ if $j \in R$ and $x^R_j = 0$ otherwise. If $x^C \in S$, we say that C is an *independent set*; otherwise C is a *dependent set*.

As we observed in Section II.1.5, the n constraints $x \geqslant 0$ give facets of conv(S). In addition, $x_j \leqslant 1$ gives a facet if $\{j, k\}$ is an independent set for all $k \in N \setminus \{j\}$. We leave these results as exercises and go on to more interesting inequalities.

Proposition 2.1. *If C is a dependent set, then*

(2.2) $$\sum_{j \in C} x_j \leqslant |C| - 1$$

is a valid inequality for S.

Proof. Suppose $x^R \in S$ and $\Sigma_{j\in C} x_j^R \geqslant |C|$. This means that $R \supseteq C$ so that R is dependent, which contradicts $x^R \in S$. ■

A dependent set is *minimal* if all of its subsets are independent. Note that if a dependent set C is not minimal, then $\Sigma_{j\in C} x_j \leqslant |C| - 1$ is the sum of $\Sigma_{j\in C'} x_j \leqslant |C'| - 1$ and $x_j \leqslant 1$ for $j \in C \setminus C'$, where C' is a minimal dependent set.

Example 2.1. $S = \{x \in B^5: 79x_1 + 53x_2 + 53x_3 + 45x_4 + 45x_5 \leqslant 178\}$. The minimal dependent sets and corresponding valid inequalities are:

$$\begin{aligned}
C_1 &= \{1, 2, 3\} & x_1 + x_2 + x_3 &\leqslant 2\\
C_2 &= \{1, 2, 4, 5\} & x_1 + x_2 + x_4 + x_5 &\leqslant 3\\
C_3 &= \{1, 3, 4, 5\} & x_1 + x_3 + x_4 + x_5 &\leqslant 3\\
C_4 &= \{2, 3, 4, 5\} & x_2 + x_3 + x_4 + x_5 &\leqslant 3.
\end{aligned}$$

While the constraints (2.2) are quite simple, they are nontrivial with respect to the polytope $P \supseteq S$ obtained by replacing $x \in B^n$ by $x \in R^n_+$ and $x_j \leqslant 1$ for all $j \in N$, that is, the linear programming relaxation with $P = \{x \in R^n_+: \Sigma_{j\in N} a_j x_j \leqslant b,\ x_j \leqslant 1 \text{ for } j \in N\}$. If $\Sigma_{j\in N} a_j > b$, then every nonintegral extreme point $\hat{x}$ of P is of the form

$$\begin{aligned}
\hat{x}_j &= 1 \text{ for } j \in C \setminus \{k\}\\
\hat{x}_j &= 0 \text{ for } j \in N \setminus C\\
\hat{x}_k &= \left(b - \sum_{j\in C\setminus\{k\}} a_j\right) / a_k > 0,
\end{aligned}$$

where C is a dependent set, $k \in C$, and $C \setminus \{k\}$ is independent. However, $\hat{x}$ does not satisfy the inequality (2.2).

Proposition 2.1 applies to any independence system. We now begin to use some particular properties of the knapsack problem.

The *extension* $E(C)$ of a minimal dependent set C is the set $C \cup \{k \in N \setminus C: a_k \geqslant a_j$ for all $j \in C\}$. In Example 2.1, $E(C_i) = C_i$ for $i = 1, 2, 3$ and $E(C_4) = C_4 \cup \{1\}$.

Proposition 2.2. *If C is a minimal dependent set, then*

$$\sum_{j\in E(C)} x_j \leqslant |C| - 1 \tag{2.3}$$

is a valid inequality for S.

Proof. Suppose $x^R \in S$ and $\Sigma_{j\in E(C)} x_j^R \geqslant |C|$ so that $|R \cap E(C)| \geqslant |C|$. Now $\Sigma_{j\in R} a_j \geqslant \Sigma_{j\in R\cap E(C)} a_j$ and by definition of $E(C)$ we obtain $\Sigma_{j\in R\cap E(C)} a_j \geqslant \Sigma_{j\in C} a_j > b$, which contradicts $x^R \in S$. ■

In Example 2.1, $\Sigma_{j=1}^5 x_j \leqslant 3$ is a valid inequality obtained from Proposition 2.2 with $E(C_4)$. It dominates the inequalities (2.2) generated by C_2, C_3, and C_4.

In some instances the inequalities (2.3) give facets of conv(S).

Proposition 2.3. *Let $C = \{j_1, \ldots, j_r\}$ be a minimal dependent set with $j_1 < j_2 < \cdots < j_r$. If any of the following conditions holds, then (2.3) gives a facet of* conv(S).

a. $C = N$.
b. $E(C) = N$ *and* (i) $(C \setminus \{j_1, j_2\}) \cup \{1\}$ *is independent.*
c. $C = E(C)$ *and* (ii) $(C \setminus \{j_1\}) \cup \{p\}$ *is independent, where* $p = \min\{j: j \in N \setminus E(C)\}$.
d. $C \subset E(C) \subset N$ *and* (i) *and* (ii).

Proof. The following n independent sets satisfy (2.3) at equality.

1. $I_i = C \setminus \{j_i\}$ for $j_i \in C$. There are $|C|$ of these.
2. $I'_k = (C \setminus \{j_1, j_2\}) \cup \{k\}$ for $k \in E(C) \setminus C$. $|I'_k \cap E(C)| = |C| - 1$ and I'_k is independent by (i) and $a_k \leqslant a_1$. There are $|E(C) \setminus C|$ of these.
3. $\tilde{I}_j = (C \setminus \{j_1\}) \cup \{j\}$ for $j \in N \setminus E(C)$. $|\tilde{I}_j \cap E(C)| = |C| - 1$ and $\tilde{I}_j$ is independent by (ii) and $a_j \leqslant a_p$.

We leave it to the reader to show that the corresponding characteristic vectors are linearly independent. ■

In Example 2.1, Proposition 2.3 establishes that (2.3) with $C = C_1$ gives a facet of conv(S) since $C_1 = E(C_1)$ and $(C_1 \setminus \{j_1\}) \cup \{p\} = \{2, 3, 4\}$ is independent. Also, since $E(C_4) = N$ and $(C_4 \setminus \{2, 3\}) \cup \{1\} = \{1, 4, 5\}$ is independent, (2.3) with $C = C_4$ gives a facet of conv(S).

A simple consequence of Proposition 2.3 is:

Corollary 2.4. *If C is a minimal dependent set for S and (C_1, C_2) is any partition of C with $C_1 \neq \emptyset$, then $\sum_{j \in C_1} x_j \leqslant |C_1| - 1$ gives a facet of* conv($S(C_1, C_2)$), *where*

$$S(C_1, C_2) = S \cap \{x \in B^n: x_j = 0 \text{ for } j \in N \setminus C, x_j = 1 \text{ for } j \in C_2\}.$$

Proof. For any C_2, $\emptyset \subseteq C_2 \subset C$, it follows that $C_1 = C \setminus C_2$ is a minimal dependent set for $S(C_1, C_2)$ since

$$S(C_1, C_2) = \left\{x \in B^{|C_1|}: \sum_{j \in C_1} a_j x_j \leqslant b - \sum_{j \in C_2} a_j\right\},$$

$\sum_{j \in C_1} a_j > b - \sum_{j \in C_2} a_j$, and $\sum_{j \in C_1 \setminus \{k\}} a_j \leqslant b - \sum_{j \in C_2} a_j$ for all $k \in C_1$. Now Proposition 2.3 applies with $S = S(C_1, C_2)$ and $N = E(C_1) = C_1$. ■

We can use Corollary 2.4 and the lifting results of Section 1 to generate facets of conv(S).

Proposition 2.5. *If C is a minimal dependent set for S and (C_1, C_2) is any partition of C with $C_1 \neq \emptyset$, then* conv(S) *has a facet represented by*

$$\sum_{j \in N \setminus C} \alpha_j x_j + \sum_{j \in C_2} \gamma_j x_j + \sum_{j \in C_1} x_j \leqslant |C_1| - 1 + \sum_{j \in C_2} \gamma_j,$$

where $\alpha_j \geqslant 0$ for all $j \in N \setminus C$ and where $\gamma_j \geqslant 0$ for all $j \in C_2$.

Proof. We start with the inequality $\sum_{j \in C_1} x_j \leqslant |C_1| - 1$, which gives a facet of conv($S(C_1, C_2)$), and do lifting by applying Proposition 1.1 for each $j \in N \setminus C$ and Proposition 1.2 for each $j \in C_2$. The nonnegativity of the coefficients is implied by their definitions in Propositions 1.1 and 1.2. ■

As we observed previously, the order of the variables in the lifting affects the coefficients. However, we should begin with a $j \in N \setminus C$, because beginning with $k \in C_2$ is equivalent to starting with $\Sigma_{j \in C_1 \cup \{k\}} x_j \leq |C_1|$.

Example 2.2. $S = \{x \in B^5: 3x_1 + x_2 + x_3 + x_4 + x_5 \leq 4\}$.

a. $C = \{1, 4, 5\}$ is a minimal dependent set and $E(C) = \{1, 4, 5\}$. By Proposition 2.3, $x_1 + x_4 + x_5 \leq 2$ gives a facet of conv(S).
b. $C = \{1, 4, 5\}$, $C_1 = \{4, 5\}$, $C_2 = \{1\}$. By Corollary 2.4, $x_4 + x_5 \leq 1$ gives a facet of

$$\text{conv}\{(x_4, x_5) \in B^2: x_4 + x_5 \leq 4 - 3 = 1\} = \text{conv}\{S \cap \{x \in B^5: x_1 = 1, x_2 = x_3 = 0\}\}.$$

First we lift with respect to the variable x_3 by applying Proposition 1.1. This yields

$$\alpha_3 = 1 - \max[x_4 + x_5: \{(x_4, x_5) \in B^2: x_4 + x_5 \leq 4 - 3x_1 - x_3, x_1 = x_3 = 1\}].$$

Hence $\alpha_3 = 1$ and $x_3 + x_4 + x_5 \leq 1$ gives a facet of conv$\{x \in B^3: x_3 + x_4 + x_5 \leq 1\}$. Now we lift with respect to x_1 by applying Proposition 1.2. Hence

$$\gamma_1 = \max[x_3 + x_4 + x_5: \{x \in B^3: x_3 + x_4 + x_5 \leq 4\}] - 1 = 2.$$

Thus $2x_1 + x_3 + x_4 + x_5 \leq 3$ gives a facet of conv$\{x \in B^4: 3x_1 + x_3 + x_4 + x_5 \leq 4\}$. Finally, we lift with respect to x_2 by applying Proposition 1.1. Hence

$$\alpha_2 = 3 - \max[2x_1 + x_3 + x_4 + x_5: \{x \in B^4: 3x_1 + x_3 + x_4 + x_5 \leq 3\}].$$

Thus $\alpha_2 = 0$ and $2x_1 + x_3 + x_4 + x_5 \leq 3$ gives a facet of conv(S).

By symmetry, lifting in the order (x_2, x_1, x_3) yields the facet represented by $2x_1 + x_2 + x_4 + x_5 \leq 3$. The orders (x_2, x_3, x_1) and (x_3, x_2, x_1) show that the original inequality $3x_1 + x_2 + x_3 + x_4 + x_5 \leq 4$ also gives a facet of conv(S). We have not considered lifting x_1 first because, as explained before the example, this yields $x_1 + x_4 + x_5 \leq 2$, which we already know gives a facet.

To apply Proposition 2.5, we must solve $|N \setminus C_1|$ 0-1 knapsack problems. However, unlike the general 0-1 knapsack problem, these knapsack problems can be solved in polynomial-time by dynamic programming (see Section II.5.5) because the objective coefficients are polynomial in n. Nevertheless, for computational purposes, it may suffice to get lower bounds on the α_j and upper bounds on the γ_j. We will return to these computational issues in Section II.6.2, where we will give an algorithm for solving general 0-1 integer programs that uses strong valid inequalities derived from 0-1 knapsack problems.

When $C_2 = \emptyset$ in Proposition 2.5, there is a formula that nearly determines all of the lifting coefficients.

Proposition 2.6. *Let $C = \{j_1, \ldots, j_r\}$ be a minimal dependent set with $j_1 < j_2 < \cdots < j_r$. Let $\mu_h = \Sigma_{k=1}^{h} a_{j_k}$ for $h = 1, \ldots, r$; also let $\mu_0 = 0$ and $\lambda = \mu_r - b \geq 1$. Every valid inequality of the form*

$$\sum_{j \in N \setminus C} \alpha_j x_j + \sum_{j \in C} x_j \leq |C| - 1 \tag{2.4}$$

that represents a facet of conv(S) *satisfies the following conditions:*

i. *If* $\mu_h \leqslant a_j \leqslant \mu_{h+1} - \lambda$, *then* $\alpha_j = h$.
ii. *If* $\mu_{h+1} - \lambda + 1 \leqslant a_j \leqslant \mu_{h+1} - 1$, *then* (a) $\alpha_j \in \{h, h+1\}$ *and* (b) *there is at least one facet of the form* (2.4) *with* $\alpha_j = h + 1$.

Proof. The proof is based on lifting $\Sigma_{j\in C} x_j \leqslant |C| - 1$. Suppose, for $j^* \in N \setminus C$, that $a_{j^*} \geqslant \mu_h$. We will prove that $\alpha_{j^*} \geqslant h$ in any lifting in which x_{j^*} is lifted last. Then from Corollary 1.4, it follows that $\alpha_{j^*} \geqslant h$ in all liftings.

Suppose we have obtained the inequality

$$\sum_{j\in N\setminus(C\cup\{j^*\})} \alpha_j x_j + \sum_{j\in C} x_j \leqslant |C| - 1 \tag{2.5}$$

after determining all of the lifting coefficients except α_j^*. Let

$$\begin{aligned} G(d) = \max & \sum_{j\in N\setminus(C\cup\{j^*\})} \alpha_j x_j + \sum_{j\in C} x_j \\ & \sum_{j\in N\setminus\{j^*\}} a_j x_j \leqslant d \\ & x \in B^{n-1}. \end{aligned} \tag{2.6}$$

Then $\alpha_{j^*} = |C| - 1 - G(b - a_{j^*})$. Since (2.5) is valid when $x_{j^*} = 0$, we have $G(b) \leqslant |C| - 1$ so that $\alpha_{j^*} \geqslant G(b) - G(b - a_{j^*})$.

Now we show that $G(b) - G(b - a_{j^*}) \geqslant h$. Consider (2.6) with $d = b - a_{j^*}$. We have

$$\sum_{k=h+1}^{r} a_{j_k} = b + \lambda - \sum_{k=1}^{h} a_{j_k} \geqslant b + \lambda - a_{j^*} > b - a_{j^*}$$

since $a_{j^*} \geqslant \Sigma_{k=1}^{h} a_{j_k}$ and $\lambda > 0$. Hence there is no feasible solution with $x_{j_k} = 1$ for $k = h+1, \ldots, r$, and since $\min_{k=1,\ldots,h} a_{j_k} \geqslant \max_{k=h+1,\ldots,r} a_{j_k}$ there exists an optimal solution $\hat{x}$ with $\hat{x}_{j_k} = 0$ for $k = 1, \ldots, h$. Define $\tilde{x}$ by $\tilde{x}_{j_k} = 1$ for $k = 1, \ldots, h$ and $\tilde{x}_j = \hat{x}_j$ otherwise. Since $\Sigma_{k=1}^{h} a_{j_k} \leqslant a_{j^*}$, it follows that $\tilde{x}$ is a feasible solution to (2.6) with $d = b$. Hence

$$G(b) \geqslant G(b - a_{j^*}) + \sum_{k=1}^{h} \tilde{x}_{j_k} = G(b - a_{j^*}) + h.$$

Thus we have shown that $\alpha_{j^*} \geqslant h$ in all liftings when $a_{j^*} \geqslant \mu_h$.

Now suppose that $\mu_h \leqslant a_k \leqslant \mu_{h+1} - \lambda$ and x_k is lifted first. We will show that $\alpha_k = h$, so by Corollary 1.4 we obtain $\alpha_k \leqslant h$ in all liftings. From Proposition 1.1, $\alpha_k = (r-1) - \zeta$, where

$$\begin{aligned} \zeta &= \max\left\{ \sum_{j\in C} x_j : \sum_{j\in C} a_j x_j \leqslant b - a_k, x \in B^r \right\} \\ &= \max\left\{ r + 1 - i : \sum_{l=i}^{r} a_{j_l} \leqslant b - a_k \right\} \end{aligned}$$

since $a_{j_r} \leqslant a_{j_{r-1}} \leqslant \cdots \leqslant a_{j_1}$. Now

$$\sum_{l=h+2}^{r} a_{j_l} = b - (\mu_{h+1} - \lambda) \leq b - a_k$$

since $a_k \leq \mu_{h+1} - \lambda$. Also,

$$\sum_{l=h+1}^{r} a_{j_l} = b + \lambda - \mu_h > b - a_k$$

since $\lambda > 0$ and $a_k \geq \mu_h$. Hence $\zeta = r + 1 - (h + 2) = r - \mathrm{h} - 1$ and $\alpha_k = (r - 1) - (r - h - 1) = h$.

Putting these two results together establishes i and ii(a). To obtain ii(b), note that if $a_k > \mu_{h+1} - \lambda$, it follows that $\Sigma_{l=h+2}^{r} a_{j_l} > b - a_k$, which implies $\alpha_k = h + 1$ if x_k is lifted first. ■

Example 2.3. $S = \{x \in B^{10}: 35x_1 + 27x_2 + 23x_3 + 19x_4 + 15x_5 + 15x_6 + 12x_7 + 8x_8 + 6x_9 + 3x_{10} \leq 39\}$.

Let $C = \{6, 7, 8, 9\}$. Then $\mu_0 = 0$, $\mu_1 = 15$, $\mu_2 = 27$, $\mu_3 = 35$, $\mu_4 = 41$, and $\lambda = 2$. Proposition 2.6 yields

$$\alpha_j = \begin{cases} 0 & \text{if } 0 \leq a_j \leq 13 \\ 0 \text{ or } 1 & \text{if } a_j = 14 \\ 1 & \text{if } 15 \leq a_j \leq 25 \\ 1 \text{ or } 2 & \text{if } a_j = 26 \\ 2 & \text{if } 27 \leq a_j \leq 33 \\ 2 \text{ or } 3 & \text{if } a_j = 34 \\ 3 & \text{if } 35 \leq a_j \leq 39. \end{cases}$$

Hence the only facet that can be obtained from lifting $x_6 + x_7 + x_8 + x_9 \leq 3$ is represented by

$$3x_1 + 2x_2 + x_3 + x_4 + x_5 + x_6 + x_7 + x_8 + x_9 \leq 3.$$

3. VALID INEQUALITIES FOR THE SYMMETRIC TRAVELING SALESMAN POLYTOPE

A *Hamiltonian cycle* or *tour* of a graph is a cycle that contains all of the nodes. Thus, given a graph $G = (V, E)$, the edge set $E' \subseteq E$ induces a *tour* if and only if the subgraph $G' = (V, E')$ is connected and each node is met by exactly two edges. Our reason for studying tours is that they are the feasible solutions to the symmetric traveling salesman problem.

The results of this section are of two types. We develop inequalities that are valid for all graphs and prove that some of these inequalities are facets for complete graphs. Thus it is convenient to assume throughout the section that G is a *complete graph* on m nodes, that is, there is an edge between each pair of nodes so that $|E| = n = m(m - 1)/2$. The reader should observe, however, that all of the classes of valid inequalities given subsequently are derived without assumptions about which edges are in the graph.

We represent subsets of edges by their characteristic vectors $x \in B^n$ so that E' is represented by the vector $x^{E'}$, where $x_e^{E'} = 1$ if $e \in E'$ and $x_e^{E'} = 0$ otherwise. Thus the set of feasible solutions S is the set of characteristic vectors whose edge sets induce tours. We will study conv(S) and another closely related polytope.

Let $T = \{x \in B^n: x \leqslant x' \text{ for some } x' \in S\}$. Note that T is the independence system whose maximal members define S. Because $T \supset S$, any valid inequality for T is also valid for S. Since $0 \in T$ and the n unit vectors are in T, dim(conv(T)) $= n$. Our reason for considering T is that conv(T) is full-dimensional and thus easier to analyze than conv(S), which is not. Later in this section, we will show that dim(conv(S)) $= n - m$.

T is also of practical interest since we can construct an objective function such that x^0 is optimal over S if and only if x^0 is optimal over T.

Proposition 3.1. *For any $c \in R^n$ and $\omega >$ max$\{|c_e|: e \in E\}$, the following statements are equivalent.*

1. *x^0 is an optimal solution to the symmetric traveling salesman problem* $\min\{cx: x \in S\}$.
2. *x^0 is an optimal solution to* $\max\{\bar{c}x: x \in S\}$, *where $\bar{c}_e = \omega - c_e$ for all $e \in E$.*
3. *x^0 is an optimal solution to* $\max\{\bar{c}x: x \in T\}$.

Proof. $1 \Leftrightarrow 2$. x^0 is an optimal solution to $\min\{cx: x \in S\}$ if and only if x^0 is an optimal solution to $\max\{-cx: x \in S\}$. But for any $x \in S$ we have $\omega \sum_{e \in E} x_e = m\omega$, so 1 and 2 are equivalent.

$2 \Leftrightarrow 3$. Since $\bar{c}_e > 0$ for all $e \in E$, it follows that if x^0 is an optimal solution to $\max\{\bar{c}x: x \in T\}$, then x^0 is a maximal element of T. But $x^0 \in S$ if and only if x^0 is a maximal element of T. ■

We begin our study of valid inequalities by first considering the lower- and upper-bound constraints

$$x_e \geqslant 0 \quad \text{for all } e \in E \tag{3.1}$$

$$x_e \leqslant 1 \quad \text{for all } e \in E, \tag{3.2}$$

which are obviously valid for T and S.

Proposition 3.2. *For all $e \in E$, (3.1) and (3.2) give facets of* conv(T).

Proof. All of the inequalities (3.1) are facets since T is a full-dimensional independence system.

For any $e, e' \in E$, we have $x^{\{e,e'\}} \in T$. The n vectors $x^{\{e\}}$ and $x^{\{e,e'\}}$ for all $e' \neq e$ are linearly independent and satisfy $x_e = 1$. Hence, all of the inequalities (3.2) are facets. ■

The relative complexity of conv(S) in comparison with conv(T) is already seen by observing that for $m = 3$, conv(S) contains the single point $x = (1\ 1\ 1)$, so, for example, (3.1) is not even a supporting hyperplane for any $e \in E$. It can be shown, however, that (3.1) yields facets of conv(S) for all $e \in E$ when $m \geqslant 5$, and all of the inequalities (3.2) yield facets for $m \geqslant 4$.

We now introduce the two sets of constraints that are usually used in the integer programming formulation of the symmetric traveling salesman problem. For $U \subseteq V$ let $\delta(U) = \{e \in E: e \text{ has exactly one end in } U\}$. If $x \in S$, then

$$\sum_{e \in \delta(\{v\})} x_e = 2 \quad \text{for all } v \in V; \tag{3.3}$$

and if $x \in T$, then

$$\sum_{e \in \delta(\{v\})} x_e \leqslant 2 \quad \text{for all } v \in V. \tag{3.4}$$

The constraints (3.3) and (3.4) are the *degree constraints* for S and T, respectively.

Proposition 3.3. *For all $v \in V$, the inequality (3.4) gives a facet of* conv(T).

Proof. Suppose that $\delta(\{v\}) = \{e_1, e_2, \ldots, e_{m-1}\}$ and that $\{e_1, e_2, e_n\}$ forms a cycle. Consider the n vectors: $x^{\{e_1,e_j\}}$ for $j = 2, \ldots, m-1$; $x^{\{e_2,e_3\}}$, $x^{\{e_1,e_2,e_j\}}$ for $j = m, \ldots, n-1$; and $x^{\{e_1,e_3,e_n\}}$. Each of these vectors is in T and satisfies (3.4) at equality, and it is easy to check that they are linearly independent. ∎

We now consider the dimension of conv(S).

Proposition 3.4. $\dim(\text{conv}(S)) = n - m = m(m-1)/2 - m$.

Proof. Let $Q = \{x \in B^n: x \text{ satisfies (3.3)}\}$. The equation system (3.3) defines a constraint matrix of rank m. Hence, by Proposition 2.4 of Chapter I.4, we have $\dim(\text{conv}(Q)) = n - m$. Since $\text{conv}(S) \subseteq \text{conv}(Q)$, it follows that $\dim(\text{conv}(S)) \leqslant n - m$.

To prove that $\dim(\text{conv}(S)) = \dim(\text{conv}(Q)) = n - m$, it suffices to show that if the hyperplane $\pi x = \pi_0$, $\pi \neq 0$, contains the incidence vector of every tour, then $\pi x = \pi_0$ is a linear combination of the constraints (3.3).

The edge set of the graph G is $E = \{(i, j): i = 1, \ldots, m-1, j = i+1, \ldots, m\}$. The variable x_e for $e = (i, j)$ is written as x_{ij}.

Let $j \in \{4, \ldots, m\}$ and P_{j3} be a path from j to 3 through all of the points $\{4, \ldots, m\}$. Now consider the pairs of tours $T_j^1 = P_{j3} \cup \{(1, j), (1, 2), (2, 3)\}$ and $T_j^2 = P_{j3} \cup \{(2, j), (1, 2), (1, 3)\}$, shown in Figure 3.1. Since T_j^1 and T_j^2 lie on the hyperplane $\pi x = \pi_0$, it follows that $\pi_{1j} + \pi_{23} = \pi_{2j} + \pi_{13}$ or $\pi_{2j} - \pi_{1j} = \pi_{23} - \pi_{13}$ for $j = 3, \ldots, m$. Let $\lambda_1 = \pi_{2j} - \pi_{1j}$ for $j = 3, \ldots, m$. By an identical argument, we obtain the following for $i = 1, \ldots, m$: $\lambda_i = \pi_{i+1,j} - \pi_{ij}$ for $j > i + 1$ and $\lambda_i = \pi_{j,i+1} - \pi_{ji}$ for $j < i$.

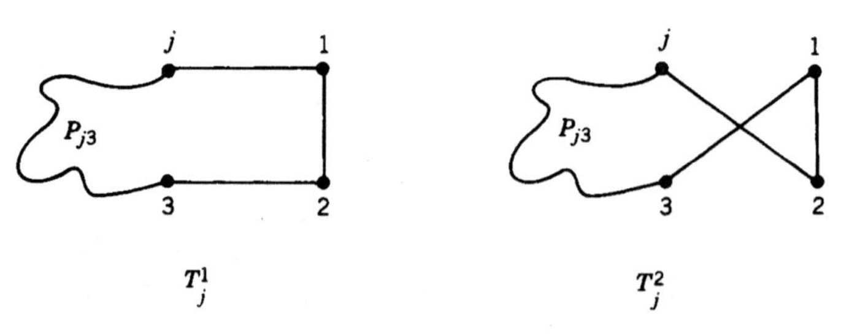

Figure 3.1

Thus for any coefficient π_{ij}, we have

$$\begin{aligned}
\pi_{ij} &= (\pi_{ij} - \pi_{i-1,j}) + \pi_{i-1,j} \\
&= (\pi_{ij} - \pi_{i-1,j}) + \cdots + (\pi_{2j} - \pi_{1j}) + \pi_{1j} \\
&= \sum_{t=1}^{i-1} \lambda_t + \pi_{1j} \\
&= \sum_{t=1}^{i-1} \lambda_t + (\pi_{1j} - \pi_{1,j-1}) + \pi_{1,j-1} \\
&= \sum_{t=1}^{i-1} \lambda_t + (\pi_{1j} - \pi_{1,j-1}) + \cdots + (\pi_{13} - \pi_{12}) + \pi_{12} \\
&= \sum_{t=1}^{i-1} \lambda_t + \sum_{t=2}^{j-1} \lambda_t + \pi_{12} \\
&= u_i + u_j - u_2 + \pi_{12},
\end{aligned}$$

where $u_i = \Sigma_{t=1}^{i-1} \lambda_t$ for $i > 1$ and $u_1 = 0$. Let $\alpha = \pi_{12} - u_2$. Hence

$$\begin{aligned}
\sum_{e\in E} \pi_e x_e = \sum_{i=1}^{m-1} \sum_{j=i+1}^{m} \pi_{ij} x_{ij} &= \sum_{i=1}^{m-1} \sum_{j=i+1}^{m} (u_i + u_j + \alpha)\, x_{ij} \\
&= \sum_{i=1}^{m} \left[\left(u_i + \frac{\alpha}{2}\right)\left(\sum_{j<i} x_{ji} + \sum_{j>i} x_{ij}\right)\right] \\
&= \sum_{v\in V} \left[\left(u_v + \frac{\alpha}{2}\right)\left(\sum_{e\in\delta(\{v\})} x_e\right)\right],
\end{aligned}$$

which establishes that the constraint is a linear combination of degree constraints with $\pi_0 = 2\Sigma_{v\in V}\, u_v + m\alpha$. ■

A cycle of G that does not contain all of the nodes is called a *subtour*. In a cycle, each vertex is of degree 2. Hence if $x^{E'} \in B^n$ satisfies (3.3) for all $v \in V$, then the subgraph $G' = (V, E')$ is either a tour or a set of disjoint subtours (see Figure 3.2). Such subgraphs are called *2-matchings*.

We now introduce a set of constraints that are valid for T and are not satisfied by any subtours. For $W \subseteq V$, let $E(W) = \{e \in E\text{: both ends of } e \text{ are in } W\}$. If $E' \subseteq E$ and $|E' \cap E(W)| \geq |W|$, the subgraph $G' = (V, E')$ contains at least one subtour. This yields the *subtour elimination* constraints

$$(3.5) \qquad \sum_{e\in E(W)} x_e \leq |W| - 1 \quad \text{for all } W \subset V,\ 2 \leq |W| \leq m - 1.$$

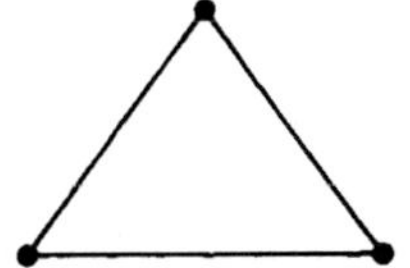

Figure 3.2

We have included the case $|W| = 2$ in (3.5), although it is not a subtour elimination constraint, since these are simply the upper-bound constraints $x_e \leqslant 1$ for $e = (u, v)$ and $W = \{u, v\}$. Thus we no longer need to consider (3.2).

In addition, if the degree constraints (3.3) are satisfied for all $v \in V$, then (3.5) is superfluous for all W with $|W| \geqslant \lfloor m/2 \rfloor + 1$. This is obvious when $|W| \geqslant m - 2$. In general, when each node is of degree 2 and $\Sigma_{e \in E(W)} x_e \geqslant |W|$, then every $v \in W$ is in a subtour and there can be no edges between W and $V \setminus W$. Hence $\Sigma_{e \in E(W)} x_e = |W|$ and $\Sigma_{e \in E(V \setminus W)} x_e = |V \setminus W|$. Thus it suffices to use (3.5) for either W or $V \setminus W$.

Proposition 3.5. *The subtour elimination inequalities (3.5) give facets of* conv(S) *for* $m \geqslant 4$ *for all* W *with* $2 \leqslant |W| \leqslant \lfloor m/2 \rfloor$.

Proof. We show the result for $m \geqslant 6$ and $3 \leqslant |W|$, where $W = \{1, \ldots, k\}$ and $k \leqslant \lfloor m/2 \rfloor$. The remaining cases are left as exercises. Note that the inequalities (3.5) represent proper faces since each of them is satisfied at equality by some tour and is a strict inequality for some other tour.

We prove the result by showing that the conditions of Theorem 3.6 of Section I.4.3 hold. Here $\pi x \leqslant \pi_0$ represents a subtour elimination inequality (3.5), $A^{=}x = b^{=}$ represents the degree constraints (3.3), and we are concerned with solutions to the linear system $\lambda x^{T_i} = \lambda_0$, where $\{T_i\}_{i=1}^{n-m}$ is a set of tours that satisfy $\pi x^{T_i} = \pi_0$. Hence it suffices to demonstrate that all solutions (λ, λ_0) to $\lambda x^{T_i} = \lambda_0$ for $i = 1, \ldots, n - m$ are of the form $\lambda = \alpha\pi + uA^{=}$, $\lambda_0 = \alpha\pi_0 + ub^{=}$ for some $\alpha \in R^1$ and $u \in R^m$.

First observe that if (λ, λ_0) is a solution, there is a solution (λ', λ_0') with $\lambda' = \lambda + u^1 A^{=}$, where $\lambda'_{1j} = 1$ for $j = 2, \ldots, k$, $\lambda'_{23} = 1$, and $\lambda'_{1j} = 0$ for $j = k + 1, \ldots, m$. To see this, we observe that the $m \times m$ node-edge incidence matrix

$$
B = \begin{array}{c} \\ 1 \\ 2 \\ 3 \\ 4 \\ \cdot \\ \cdot \\ m \end{array}
\begin{array}{c} \begin{array}{ccccccc} e_{12} & e_{13} & e_{23} & e_{14} & \ldots & e_{1m} \end{array} \\
\left(\begin{array}{ccc|ccc}
1 & 1 & 0 & 1 & \ldots & 1 \\
1 & 0 & 1 & 0 & \ldots & 0 \\
0 & 1 & 1 & 0 & \ldots & 0 \\ \hline
 & & & & & \\
 & 0 & & & I & \\
 & & & & &
\end{array}\right) \end{array}
$$

is nonsingular. Hence the appropriate m components of λ' can be fixed by solving the $m \times m$ system

$$u^1 B = (1 - \lambda_{12}, 1 - \lambda_{13}, 1 - \lambda_{23}, \ldots, 1 - \lambda_{1k}, -\lambda_{1,k+1}, \ldots, -\lambda_{1m}).$$

We now show in the following series of steps that: $\lambda'_{ij} = 1$ if $i, j \in W$; $\lambda'_{ij} = 0$ if $i \in W$, $j \notin W$; and $\lambda'_{ij} = \beta$ for $i, j \notin W$.

Consider the two tours $T_1 = P_1 \cup \{1, 3\} \cup P_2 \cup \{2, i\}$ and $T_2 = P_1 \cup \{1, i\} \cup P_2 \cup \{2, 3\}$ shown in Figure 3.3 that are assumed to satisfy (3.5) at equality. We leave it to the reader to establish the existence of such tours. Since T_1 and T_2 contain $k - 1$ edges in $E(W)$, we require $\lambda' x^{T_1} = \lambda' x^{T_2} = \lambda_0'$. Thus $\lambda'_{13} + \lambda'_{2i} = \lambda'_{1i} + \lambda'_{23}$, so $\lambda'_{2i} = 1$ if $4 \leqslant i \leqslant k$ and $\lambda'_{2i} = 0$ if $k < i \leqslant m$.

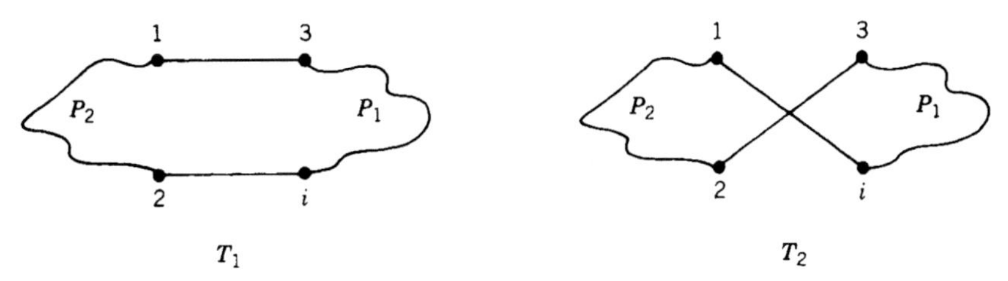

Figure 3.3

The remaining cases are similar. We stipulate tours T_1 and T_2 with $k - 1$ edges in $E(W)$ and $n - k - 1$ edges in $E(V \setminus W)$ containing paths P_1 and P_2 with specified endpoints whose intermediate nodes are the remaining nodes of W and $V \setminus W$, respectively.

Suppose $3 \leqslant i \leqslant k$ and consider two tours $T_1 = P_1 \cup \{i, j\} \cup P_2 \cup \{1, 2\}$ and $T_2 = P_1 \cup \{1, j\} \cup P_2 \cup \{2, i\}$ with P_1, P_2 having endnodes 2, j and 1, i, respectively. This gives $\lambda'_{12} + \lambda'_{ij} = \lambda'_{1j} + \lambda'_{2i}$, or $\lambda'_{ij} = 1$ for all j with $3 \leqslant i < j \leqslant k$ and $\lambda'_{ij} = 0$ for all j with $k < j \leqslant m$.

The final case involves $p, q, r \notin W$, tours $T_1 = P_1 \cup \{1, p\} \cup P_2 \cup \{q, r\}$ and $T_2 = P_1 \cup \{1, r\} \cup P_2 \cup \{p, q\}$ and paths P_1, P_2 with endpoints 1, q and p, r, respectively. Then we have $\lambda'_{1p} + \lambda'_{qr} = \lambda'_{1r} + \lambda'_{pq}$ so that λ'_{pq} is a constant β for all $p, q \notin W$.

Hence we have shown that $\lambda' x^{T_i} = \lambda'_0$ is of the form

$$\sum_{e \in E(W)} x_e^{T_i} + \beta \sum_{e \in E(V \setminus W)} x_e^{T_i} = |W| - 1 + \beta(|V \setminus W| - 1)$$

for any x^{T_i} that satisfies (3.5) at equality.

Now defining $u^2 \in R^m$ by $u_i^2 = \beta/2$, $i \in W$, $u_i^2 = -\beta/2$, $i \in V \setminus W$, we have that $(\lambda' + u^2 A^=)x^{T_i} = (\lambda' + u^2 b^=)$ is of the form

$$\begin{aligned}(1 + \beta) \sum_{e \in E(W)} x_e^{T_i} &= |W| - 1 + \beta(|V \setminus W| - 1) + \beta|W| - \beta|V \setminus W| \\ &= (1 + \beta)(|W| - 1),\end{aligned}$$

so that $(1 + \beta)\pi = \lambda' + u^2 A^=$ and $(1 + \beta)\pi_0 = \lambda'_0 + u^2 b^=$. Hence Theorem 3.6 of Section 1.4.3 applies with

$$(\lambda, \lambda_0) = ((1 + \beta)\pi - (u^1 + u^2)A^=, (1 + \beta)\pi_0 - (u^1 + u^2)b^=). \quad \blacksquare$$

Let $P^{\text{LP}} = \{x \in R^n: x \text{ satisfies (3.1), (3.3), and (3.5)}\}$. For $m \leqslant 5$, it can be shown that $\text{conv}(S) = P^{\text{LP}}$. A subgraph on six nodes is shown in Figure 3.4. The reader can check that $x^0_{e_i} = \frac{1}{2}$ for $i = 1, \ldots, 6$, $x^0_{e_i} = 1$ for $i = 7, 8, 9$, and $x^0_{e_i} = 0$ otherwise is an extreme point of P^{LP} since it is the unique optimal solution to $\min\{cx: x \in P^{\text{LP}}\}$, where $c_{e_i} = 1$ for $i = 1, \ldots, 6$, $c_{e_i} = 0$ for $i = 7, 8, 9$, and c_{e_i} is suitably large otherwise. To define a polytope that contains $\text{conv}(S)$ but not x^0, we use a rank 1 C–G inequality. Use weights of $\frac{1}{2}$ on the degree constraints for nodes 1, 2, and 3, weights of $\frac{1}{2}$ for the constraints $x_{e_i} \leqslant 1$, for $i = 7, 8, 9$, weights of $\frac{1}{2}$ on $-x_{e_i} \leqslant 0$ for all other edges with one end in $\{1, 2, 3\}$, and round down the right-hand side. This yields

$$x_{e_1} + x_{e_2} + x_{e_3} + x_{e_7} + x_{e_8} + x_{e_9} \leqslant \left\lfloor 4\frac{1}{2} \right\rfloor = 4.$$

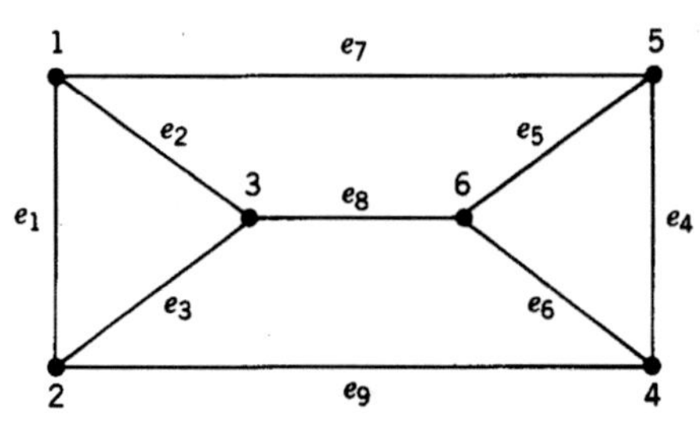

Figure 3.4

In general, let H be any subset of nodes with $3 \leqslant |H| \leqslant |V| - 1$ and let $\hat{E} \subset E$ be an odd set of disjoint edges, each of which has one end in H. Then using weights of $\frac{1}{2}$ on the degree constraints for all $v \in H$, weights of $\frac{1}{2}$ on $-x_e \leqslant 0$ for all $e \in \delta(H) \setminus \hat{E}$, weights of $\frac{1}{2}$ on $x_e \leqslant 1$ for all $e \in \hat{E}$, and rounding yields that

$$\sum_{e \in E(H)} x_e + \sum_{e \in \hat{E}} x_e \leqslant |H| + \left\lfloor \frac{|\hat{E}|}{2} \right\rfloor \tag{3.6}$$

is a valid inequality for T. Note that if $|\hat{E}| = 1$, (3.6) is dominated by subtour elimination constraints, so we only consider (3.6) for $|\hat{E}| \geqslant 3$.

The inequalities (3.6) are called *2-matching inequalities* since they are needed to define the convex hull of 2-matchings. Now we have that $\text{conv}(S) \subseteq P^{\text{LP}_1} = \{x \in R^n : x \text{ satisfies } (3.6)\} \cap P^{\text{LP}}$. In fact it can be shown that $P^{\text{LP}_1} = \text{conv}(S)$ on all graphs with six or fewer nodes. But, for $m \geqslant 7$, more general inequalities are needed.

A subgraph for generating a 2-matching inequality is shown in Figure 3.5. It resembles a comb with handle H and teeth $W_i = \{u_i, v_i\}$ for $i = 1, \ldots, k$, where $k \geqslant 3$ is odd. We can restate (3.6) as

$$\sum_{e \in E(H)} x_e + \sum_{i=1}^{k} \sum_{e \in E(W_i)} x_e \leqslant |H| + k - \frac{k+1}{2} = |H| + \sum_{i=1}^{k} (|W_i| - 1) - \frac{k+1}{2}.$$

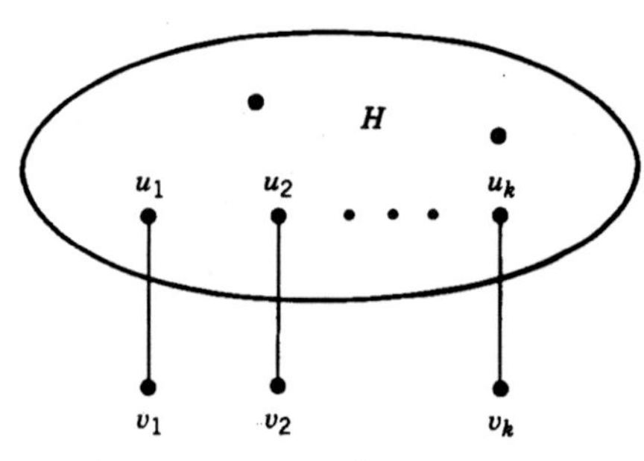

Figure 3.5

A general comb is shown in Figure 3.6. Here the teeth W_i for $i = 1, \ldots, k$, can contain more than two nodes and can have more than one node in common with the handle. Specifically a *comb* is a subgraph generated by a node set $\{H, W_1, \ldots, W_k\}$ with the following properties:

1. $|H \cap W_i| \geq 1$ for $i = 1, \ldots, k$.
2. $|W_i \setminus H| \geq 1$ for $i = 1, \ldots, k$.
3. $2 \leq |W_i| \leq m - 2$ for $i = 1, \ldots, k$.
4. $W_i \cap W_j = \emptyset$ for $i \neq j$.
5. k is odd and at least 3.

Proposition 3.6. *For any subgraph of G that is a comb, the comb inequality*

$$\sum_{e \in E(H)} x_e + \sum_{i=1}^{k} \sum_{e \in E(W_i)} x_e \leq |H| + \sum_{i=1}^{k} (|W_i| - 1) - \frac{k+1}{2} \tag{3.7}$$

is valid for T.

Proof. First weight the degree constraints for $v \in H$ by $\frac{1}{2}$ and sum them. This yields

$$\sum_{e \in E(H)} x_e + \frac{1}{2} \sum_{e \in \delta(H)} x_e \leq |H|. \tag{3.8}$$

Now add $-\frac{1}{2}x_e \leq 0$ for all $e \in \delta(H) \setminus \cup_{i=1}^{k} E(W_i)$ to (3.8) to obtain

$$\sum_{e \in E(H)} x_e + \frac{1}{2} \sum_{i=1}^{k} \sum_{e \in \delta(H) \cap E(W_i)} x_e \leq |H|. \tag{3.9}$$

Consider the subtour elimination constraints for W_i, $H \cap W_i$, and $W_i \setminus H$, respectively:

$$\sum_{e \in E(W_i)} x_e \leq |W_i| - 1 \quad \text{for } i = 1, \ldots, k \tag{3.10}$$

$$\sum_{e \in E(H \cap W_i)} x_e \leq |H \cap W_i| - 1 \quad \text{for } i = 1, \ldots, k \tag{3.11}$$

$$\sum_{e \in E(W_i \setminus H)} x_e \leq |W_i \setminus H| - 1 \quad \text{for } i = 1, \ldots, k. \tag{3.12}$$

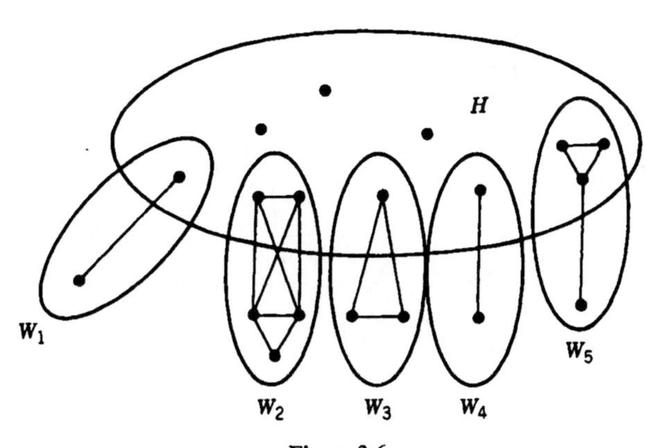

Figure 3.6

The edge e appears in (a) (3.10) and (3.11) if $e \in E(H \cap W_i)$, (b) (3.10) and (3.12) if $e \in E(W_i \setminus H)$, and (c) (3.10) and the left-hand side of (3.9) with a coefficient of $\frac{1}{2}$ if $e \in \delta(H) \cap E(W_i)$. Also note that (3.11) [respectively, (3.12)] is trivial when $|H \cap W_i| = 1$ [respectively, $|W_i \setminus H| = 1$]. Hence by multiplying each of the inequalities (3.10)–(3.12) by $\frac{1}{2}$ and adding them to (3.9), the result is

$$\sum_{e \in E(H)} x_e + \sum_{i=1}^{k} \sum_{e \in E(W_i)} x_e$$

$$\leqslant |H| + \frac{1}{2} \sum_{i=1}^{k} [(|W_i| - 1) + (|H \cap W_i| - 1) + (|W_i \setminus H| - 1)]$$

$$= |H| + \frac{1}{2} \sum_{i=1}^{k} [(|W_i| - 1) + (|H \cap W_i| - 1) + (|W_i| - 1 - |H \cap W_i|)]$$

$$= |H| + \sum_{i=1}^{k} (|W_i| - 1) - \frac{1}{2}k.$$

Then, since k is odd, by rounding we obtain

$$\sum_{e \in E(H)} x_e + \sum_{i=1}^{k} \sum_{e \in E(W_i)} x_e \leqslant |H| + \sum_{i=1}^{k} (|W_i| - 1) - \frac{k+1}{2}.$$

■

Consider the comb C shown in Figure 3.7. The comb inequality (3.7) is

$$\sum_{i=1}^{12} x_{e_i} + x_{e_2} \leqslant |H| + |W_1| - 1 + |W_2| - 1 + |W_3| - 1 - \frac{3+1}{2}$$

$$= 4 + 2 + 2 + 1 - 2 = 7.$$

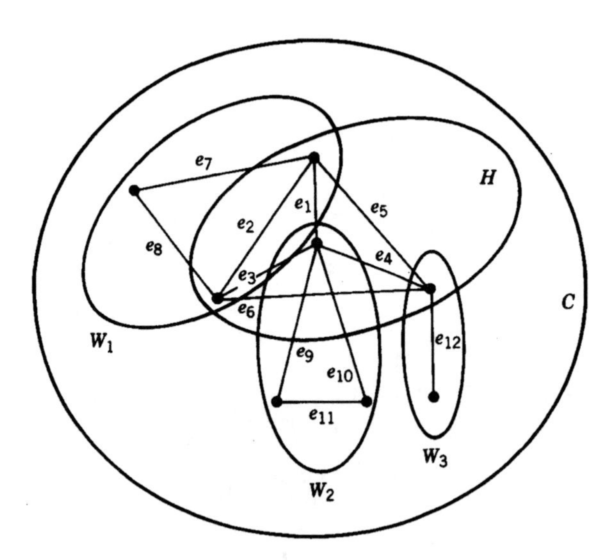

Figure 3.7

The comb inequalities have coefficients in {0, 1, 2} and the 2's appear on x_e if $e \in E(W_i \cap H)$ for some i. These inequalities are rank 1 C–G inequalities with respect to the inequalities (3.1), (3.3), and (3.5).

The comb inequalities can be generalized to obtain higher-rank C–G inequalities by considering generalized combs that have teeth which themselves are combs. Consider the graph of Figure 3.8. The handle H_1 has three teeth, namely, W_1, W_2, and C, where C is comb. We require that $C \cap H_1$ contain no vertices of H_2 and that each original tooth W_i has at least one node that is not contained in any handle. To derive a valid inequality for the graph of Figure 3.8, we proceed as we did in deriving the comb inequalities. Hence the following inequalities are weighted by $\frac{1}{2}$ and summed, and then the resulting right-hand side is rounded down:

1. degree constraints for H_1;
2. nonnegativity constraints for $e \in \delta(H_1) \setminus (E(W_1) \cup E(W_2) \cup E(W_5))$;
3. subtour elimination constraints for W_i, $W_i \cap H_1$ and $W_i \setminus H_1$ for $i = 1, 2$ and for $H_1 \cap W_5$;
4. comb inequalities (3.7) for C and $C \setminus H_1$.

The result for the graph of Figure 3.8 is

$$(3.13)\quad \sum_{i=1}^{2} \sum_{e \in E(H_i)} x_e + \sum_{i=1}^{5} \sum_{e \in E(W_i)} x_e \leqslant \sum_{i=1}^{2} |H_i| + \sum_{i=1}^{4} (|W_i| - 1) + (|W_5| - 2) - \frac{k+1}{2},$$

where $k = 5$. The left-hand side of (3.13) is clear. The contributions to the right-hand side are, respectively, from

1. $|H_1|$,
3. $\frac{1}{2}[(2|W_1| - 3) + (2|W_2| - 3) + |H_1 \cap W_5| - 1] = |W_1| - 1 + |W_2| - 1 - \frac{3}{2} + \frac{1}{2}|H_1 \cap W_5|$, and
4. $\frac{1}{2}[(2|W_3| - 2) + (2|W_4| - 2) + (|W_5| - 1) + (|W_5 \setminus H_1| - 1)] - 2 + |H_2|$.

Hence rounding yields

$$\begin{aligned} &|H_1| + |H_2| + \sum_{i=1}^{5} (|W_i| - 1) + \left\lfloor -\frac{7}{2} \right\rfloor \\ &= |H_1| + |H_2| + \sum_{i=1}^{4} (|W_i| - 1) + (|W_5| - 2) - \frac{k+1}{2}. \end{aligned}$$

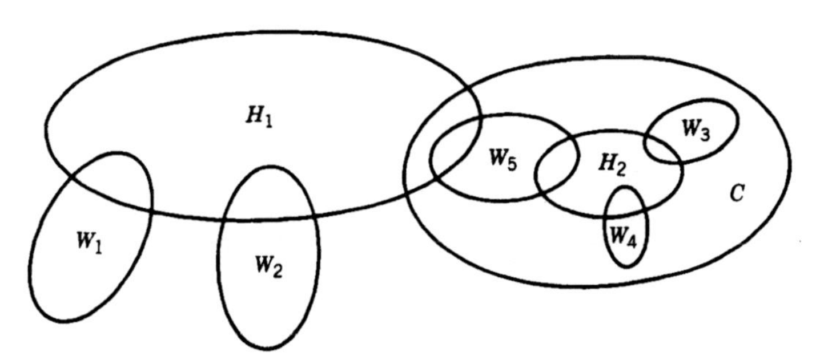

Figure 3.8

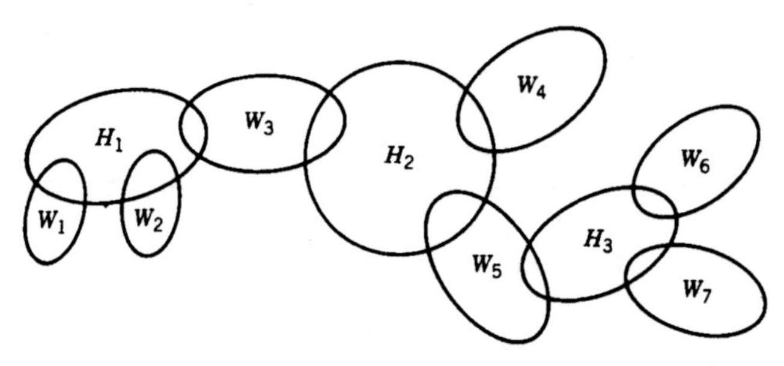

Figure 3.9

Inductively, we can build still more complicated generalized combs (see Figure 3.9). Besides properties 1–5 given in the definition of a comb, it is required for a graph with k teeth and r handles that:

6. $H_i \cap H_j = \emptyset$ for $i \neq j$;
7. $W_j \setminus \bigcup_{i=1}^{r} H_i \neq \emptyset$ for $j = 1, \ldots, k$; and
8. if $H_i \cap W_j \neq \emptyset$ and $H_i \cap W_j$ is deleted from the generalized comb graph, then the resulting graph is disconnected.

When these conditions are satisfied it can be shown inductively that *the generalized comb* inequality

$$(3.14) \qquad \sum_{i=1}^{r} \sum_{e \in E(H_i)} x_e + \sum_{i=1}^{k} \sum_{e \in E(W_i)} x_e \leqslant \sum_{i=1}^{r} |H_i| + \sum_{i=1}^{k} (|W_i| - w_i) - \frac{k+1}{2},$$

where w_i is the number of handles met by W_i, is valid for T. Moreover, for a complete graph we have the following theorem.

Theorem 3.7. *The generalized comb inequalities (3.14) give facets of* conv(S).

The proof of this theorem is much too long to give here.

Theorem 3.7 generates a very large class of facets, but there are yet other classes. For example, the famous Petersen graph $G = (V, E')$ on 10 nodes (see Figure 3.10) does not contain a tour, which means $\sum_{e \in E'} x_e \leqslant 9$ is valid for the complete graph on 10 nodes. In fact, it can be proved to represent a facet of conv(S). But it does not belong to any of the families of valid inequalities introduced in this section.

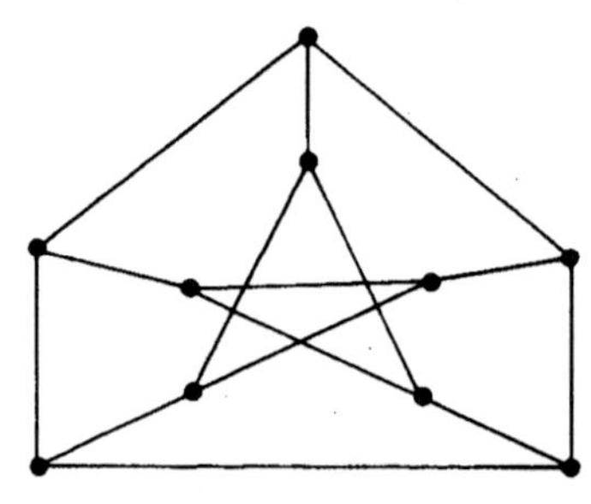

Figure 3.10

The Petersen graph belongs to a certain infinite family of graphs $G = (V, E')$ that do not contain any tours. From some graphs in this class, we obtain facets of conv(S) represented by valid inequalities of the form $\Sigma_{e \in E'} x_e \leqslant |V| - 1$. Yet these graphs are not likely to have a good characterization. So one cannot expect to have a good characterization of the corresponding facets.

Fortunately, such facets have not been necessary in the solution of many symmetric traveling salesman problems in the literature by algorithms that use cutting planes and branch-and-bound.

4. VALID INEQUALITIES FOR VARIABLE UPPER-BOUND FLOW MODELS

We consider a single-node flow model with an exogenous supply of b and n outflow arcs (see Figure 4.1). For each $j \in N = \{1, \ldots, n\}$ the flow $y_j \in R_+^1$ on the jth arc is bounded by the capacity a_j if arc j is open ($x_j = 1$) and 0 otherwise. We call this relationship a *variable upper bound on the flow* y_j. Since the total outflow cannot exceed b, this model can be represented by the mixed-integer region

$$T = \left\{ x \in B^n, y \in R_+^n : \sum_{j \in N} y_j \leqslant b, \; y_j \leqslant a_j x_j \text{ for } j \in N \right\}. \tag{4.1}$$

Our initial objective is to find strong valid inequalities for T. Consider the polytope

$$P = \left\{ x \in R_+^n, y \in R_+^n : \sum_{j \in N} y_j \leqslant b, \; y_j \leqslant a_j x_j, \; x_j \leqslant 1 \text{ for } j \in N \right\}$$

used in the formulation of T, that is, $T = P \cap \{x \in Z^n, y \in R^n\}$. The fractional extreme points of P are characterized in the following proposition.

Proposition 4.1. *All fractional extreme points of P are of the form*

$$\hat{y}_j = a_j, \; \hat{x}_j = 1 \quad \text{for } j \in C \setminus \{k\}$$

$$\hat{y}_k = b - \sum_{j \in C \setminus \{k\}} a_j, \qquad \hat{x}_k = \frac{1}{a_k}\left(b - \sum_{j \in C \setminus \{k\}} a_j\right) > 0$$

$$\hat{y}_j = 0, \; \hat{x}_j \in \{0, 1\} \quad \text{for } j \notin C.$$

where $C \subseteq N$ *is a dependent set of* $S = \{x \in B^n : \Sigma_{j \in N} a_j x_j \leqslant b\}$, $k \in C$ *and* $C \setminus \{k\}$ *is independent.*

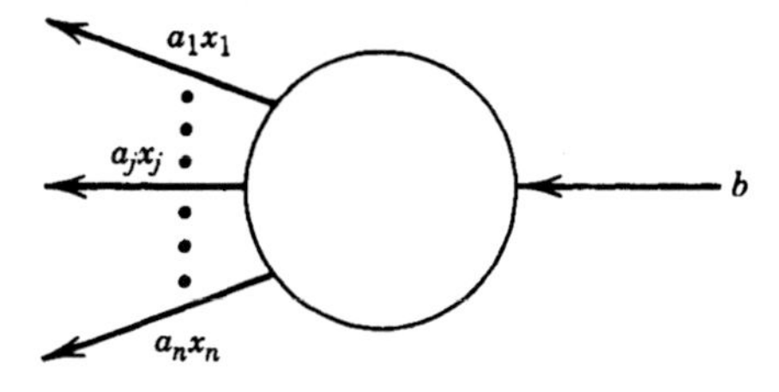

Figure 4.1

There are simple valid inequalities for T that cut off these fractional extreme points of P. Let $\lambda = \Sigma_{j\in C}\, a_j - b$ be the excess capacity of the arcs in a dependent set C. For $k \in C$, the capacity of the set $C \setminus \{k\}$ is

$$\min\left\{\sum_{j\in C\setminus\{k\}} a_j,\, b\right\} = b - (a_k - \lambda)^+;$$

and for any $C' \subset C$, the capacity of the set $C \setminus C'$ is

$$b - \left(\sum_{j\in C'} a_j - \lambda\right)^+ \leqslant b - \sum_{j\in C'} (a_j - \lambda)^+.$$

Thus we have proved

Proposition 4.2. *If $C \subseteq N$ is a dependent set of S and $\lambda = \Sigma_{j\in C}\, a_j - b$, then*

$$\sum_{j\in C} y_j \leqslant b - \sum_{j\in C} (a_j - \lambda)^+ (1 - x_j) \tag{4.2}$$

is a valid inequality for T given by (4.1).

Since the point $(\hat{x}, \hat{y})$ given in Proposition 4.1 is such that $\Sigma_{j\in C}\, \hat{y}_j = b$, $a_k - \lambda > 0$, and $\hat{x}_k < 1$, it follows that $(\hat{x}, \hat{y})$ does not satisfy (4.2).

Example 4.1. Consider the set T given by:

$$T = \left\{x \in B^4,\, y \in R^4_+\colon \sum_{j\in N} y_j \leqslant 9,\, y_1 \leqslant 5x_1,\, y_2 \leqslant 5x_2,\, y_3 \leqslant x_3,\, y_4 \leqslant 3x_4\right\}.$$

$C = \{1, 2, 3, 4\}$, $\lambda = 5$: (4.2) yields $y_1 + y_2 + y_3 + y_4 \leqslant 9$ (the original inequality)

$C = \{1, 2, 4\}$, $\lambda = 4$: (4.2) yields $y_1 + y_2 + y_4 \leqslant 9 - (1 - x_1) - (1 - x_2)$ or $y_1 + y_2 + y_4 - x_1 - x_2 \leqslant 7$.

$C = \{1, 2, 3\}$, $\lambda = 2$: (4.2) yields $y_1 + y_2 + y_3 \leqslant 9 - 3(1 - x_1) - 3(1 - x_2)$ or $y_1 + y_2 + y_3 - 3x_1 - 3x_2 \leqslant 3$

$C = \{1, 2\}$, $\lambda = 1$: (4.2) yields $y_1 + y_2 \leqslant 9 - 4(1 - x_1) - 4(1 - x_2)$ or $y_1 + y_2 - 4x_1 - 4x_2 \leqslant 1$.

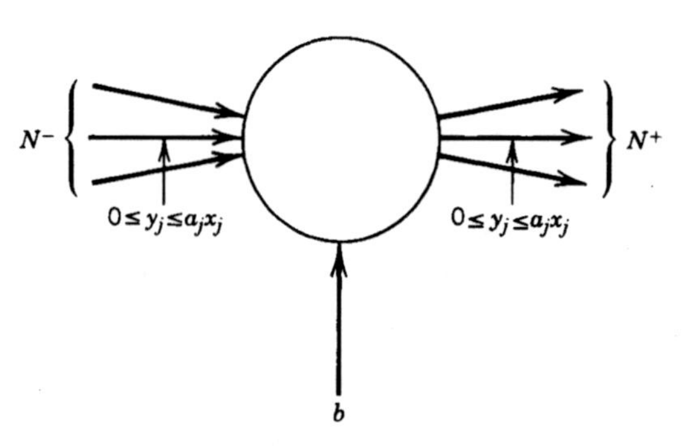

Figure 4.2

Each of the inequalities of Example 4.1 can be shown to give a facet of conv(T). Moreover if $\max_{j\in C} a_j > \lambda$, then the inequality (4.2) gives a facet of conv(T). We postpone the proof of this result to consider a more general model that also includes inflow arcs.

Let

$$(4.3) \qquad T = \left\{x \in B^n, y \in R_+^n\colon \sum_{j\in N^+} y_j - \sum_{j\in N^-} y_j \leqslant b,\ y_j \leqslant a_j x_j \text{ for } j \in N\right\},$$

where $N^+ \cup N^- = N$ (see Figure 4.2). Here $a_j \in R_+^1$ for $j \in N$ and $b \in R^1$, that is, b can be negative. We say that $C \subseteq N^+$ is a *dependent set* if $\Sigma_{j\in C} a_j > b$. Note, for example, that if $b < 0$, every subset of N^+ is dependent.

We can now generalize Proposition 4.2.

Proposition 4.3. *If $C \subseteq N^+$ is a dependent set, $\lambda = \Sigma_{j\in C} a_j - b$, and $L \subseteq N^-$, then*

$$(4.4) \qquad \sum_{j\in C} [y_j + (a_j - \lambda)^+ (1 - x_j)] \leqslant b + \sum_{j\in L} \lambda x_j + \sum_{j\in N^-\setminus L} y_j$$

is a valid inequality for T given by (4.3).

Proof. Let $C^+ = \{j \in C\colon a_j > \lambda\}$. Suppose a feasible point $(x, y) \in T$ is given and $N^1 = \{j \in N\colon x_j = 1\}$. Note that if $j \notin N^1$, then $y_j = x_j = 0$.

Case 1. $C^+ \setminus N^1 = \emptyset$ and $L \cap N^1 = \emptyset$.

$$\begin{aligned}
\sum_{j\in C} [y_j + (a_j - \lambda)^+ (1 - x_j)] &= \sum_{j\in C\cap N^1} y_j + \sum_{j\in C^+\setminus N^1} (a_j - \lambda) \\
&= \sum_{j\in C\cap N^1} y_j && (\text{since } C^+ \setminus N^1 = \emptyset) \\
&\leqslant \sum_{j\in N^+\cap N^1} y_j && (\text{since } C \subseteq N^+) \\
&\leqslant b + \sum_{j\in N^-\cap N^1} y_j && [\text{by (4.3) and } y_j = 0 \text{ if } j \notin N^1]. \\
&\leqslant b + \sum_{j\in N^-\setminus L} y_j && (\text{since } L \cap N^1 = \emptyset) \\
&= b + \sum_{j\in N^-\setminus L} y_j + \sum_{j\in L} \lambda x_j && (\text{since } L \cap N^1 = \emptyset \text{ and } x_j = 0 \text{ for } j \notin N^1).
\end{aligned}$$

Case 2. $(C^+ \setminus N^1) \cup (L \cap N^1) \neq \emptyset$.

$$\begin{aligned}
\sum_{j\in C} [y_j + (a_j - \lambda)^+ (1 - x_j)] &= \sum_{j\in C\cap N^1} y_j + \sum_{j\in C^+\setminus N^1} (a_j - \lambda) \\
&\leqslant \sum_{j\in C\cap N^1} a_j + \sum_{j\in C^+\setminus N^1} a_j - \lambda\, |C^+ \setminus N^1| && (\text{since } y_j \leqslant a_j \text{ for all } j) \\
&\leqslant \sum_{j\in C} a_j - \lambda + \lambda\, |L \cap N^1| && (\text{since } C^+ \subseteq C \text{ and } |C^+ \setminus N^1| \geqslant 1 - |L \cap N^1|) \\
&= b + \sum_{j\in L} \lambda x_j && \left(\text{since } x_j = 1 \text{ for } j \in N^1 \text{ and } \lambda = \sum_{j\in C} a_j - b\right) \\
&\leqslant b + \sum_{j\in L} \lambda x_j + \sum_{j\in N^-\setminus L} y_j && (\text{since } y_j \geqslant 0 \text{ for } j \in N).
\end{aligned}$$

■

Example 4.2. The feasible set T is given by

$$y_1 + y_2 + y_3 + y_4 \leqslant 9 + y_5 + y_6$$

$$y_1 \leqslant 5x_1,\ y_2 \leqslant 5x_2,\ y_3 \leqslant x_3,\ y_4 \leqslant 3x_4,\ y_5 \leqslant 3x_5,\ y_6 \leqslant x_6,$$

$$y \in R^6_+,\ x \in B^6.$$

Taking $C = \{1, 2, 3\}$ and $L = \{5\}$, we have $\lambda = 2$ and (4.4) yields

$$[y_1 + 3(1 - x_1)] + [y_2 + 3(1 - x_2)] + y_3 \leqslant 9 + 2x_5 + y_6, \text{ or}$$
$$y_1 + y_2 + y_3 - y_6 - 3x_1 - 3x_2 - 2x_5 \leqslant 3. \tag{4.5}$$

We can establish the dimension of the face formed by (4.5) by specifying a set of linearly independent points that satisfy it at equality:

C						$N^+ \setminus C$		L		$N^- \setminus L$		
y_1	x_1	y_2	x_2	y_3	x_3	y_4	x_4	y_5	x_5	y_6	x_6	
3	1	5	1	1	1	0	0	0	0	0	0	z^1
5	1	3	1	1	1							z^2
$3+\varepsilon$	1	5	1	$1-\varepsilon$	1							z^3
0	0	5	1	1	1							$\tilde{z}^1$
5	1	0	0	1	1							$\tilde{z}^2$
4	1	5	1	0	0							$\tilde{z}^3$
0	0	5	1	1	1	0	1					$\hat{z}^4$
0	0	5	1	1	1	ε	1					$\bar{z}^4$
5	1	5	1	1	1	0	0	2	1			w^5
5	1	5	1	1	1	0	0	$2+\varepsilon$	1			$\tilde{w}^5$
3	1	5	1	1	1	0	0	0	0	0	1	$\hat{w}^6$
$3+\varepsilon$	1	5	1	1	1	0	0	0	0	ε	1	$\overline{w}^6$

where ε is a small positive number.

An ad hoc argument shows that these 12 points are linearly independent so that (4.5) gives a facet of conv(T).

More generally, we have the following theorem.

Theorem 4.4. *If* $\max_{j \in C} a_j > \lambda$, *and* $a_j > \lambda$ *for* $j \in L$, *then (4.4) gives a facet of* conv(T), *where T is given by (4.3).*

Proof. We prove the theorem by giving $2n$ points of T that define the coefficients in (4.4) up to a scalar multiple; that is, the unique solution (π, μ) to $\pi x^i + \mu y^i = \pi_0$ for $i = 1, \ldots, 2n$ is a scalar multiple of the coefficients in (4.4).

Let $z^i = (y^i, x^i) \in T$ for $i = 1, \ldots, 2n$. For clarity, we write

$$z^i = ({}^1y^i, {}^1x^i, {}^2y^i, {}^2x^i, {}^3y^i, {}^3x^i, {}^4y^i, {}^4x^i),$$

where $({}^1y^i, {}^1x^i)$ are the (y, x) values for the arcs in C, $({}^2y^i, {}^2x^i)$ are those for the arcs in $N^+ \setminus C$, $({}^3y^i, {}^3x^i)$ are those for the arcs in L, and $({}^4y^i, {}^4x^i)$ are those for the arcs in $N^- \setminus L$.

Suppose that $a_1 = \max_{j \in C} a_j > \lambda$. Let e_i be the ith unit vector, $\bar{1} = (1, 1, \ldots, 1)$ and let ε be a small positive number. By ${}^1y = a$, we mean ${}^1y_j = a_j$ for all $j \in C$. The points given below are the general versions of the points given in Example 4.2. We leave it to the reader to check that they are in T and satisfy (4.4) at equality. We first describe a set of $2|N^+|$ points:

i. $z^k = (a - \lambda e_k, \bar{1}, 0, 0, 0, 0, 0, 0)$ for $k \in C$ with $a_k \geqslant \lambda$
$\quad = (a - (\lambda - \varepsilon)e_1 - \varepsilon e_k, \bar{1}, 0, 0, 0, 0, 0, 0)$ for $k \in C$ with $a_k < \lambda$.

ii. $\tilde{z}^k = (a - a_k e_k, \bar{1} - e_k, 0, 0, 0, 0, 0, 0)$ for $k \in C$ with $a_k \geqslant \lambda$
$\quad = (a - (\lambda - a_k)e_1 - a_k e_k, \bar{1} - e_k, 0, 0, 0, 0, 0\ 0)$ for $k \in C$ with $a_k < \lambda$.

iii. $\hat{z}^j = (a - a_1 e_1, \bar{1} - e_1, 0, e_j, 0, 0, 0, 0)$ for $j \in N^+ \setminus C$.

iv. $\bar{z}^j = (a - a_1 e_1, \bar{1} - e_1, \varepsilon e_j, e_j, 0, 0, 0, 0)$ for $j \in N^+ \setminus C$.

Suppose these points satisfy $\Sigma_{j\in N} (\pi_j x_j + \mu_j y_j) = \pi_0$. By comparing $\tilde{z}^1$, $\hat{z}^j$, and $\bar{z}^j$, we see that $\pi_j = \mu_j = 0$ for $j \in N^+ \setminus C$. For each of the points z^k, we have $\Sigma_{j\in C} \mu_j y_j^k = \pi_0 - \Sigma_{j\in C} \pi_j$. It can then be seen that $\mu_j = \mu_0$ for all $j \in C$. Moreover, since $\Sigma_{j\in C} y_j^k = b$, we also have $\mu_0 b + \Sigma_{j\in C} \pi_j = \pi_0$.

From the points $\tilde{z}^k$, we see that when $k \in C$ and $a_k \geqslant \lambda$ we obtain

$$\mu_0\left(\sum_{j\in C\setminus\{k\}} a_j\right) + \sum_{j\in C\setminus\{k\}} \pi_j = \pi_0.$$

Thus $\mu_0(b - \Sigma_{j\in C\setminus\{k\}} a_j) + \pi_k = 0$. Since $\Sigma_{j\in C} a_j = b + \lambda$, we obtain $\mu_0(\lambda - a_k) = \pi_k$ when $a_k \geqslant \lambda$. On the other hand, when $a_k < \lambda$, we have

$$\sum_{j\in C} \tilde{y}_j^k = \sum_{j\in C\setminus\{1,k\}} a_j + a_1 + a_k - \lambda = b.$$

Hence

$$\mu_0 b + \sum_{j\in C\setminus\{k\}} \pi_j = \pi_0$$

or $\pi_k = 0$.

In summary, our inequality must be of the form

$$\mu_0 \sum_{j\in C} y_j - \mu_0 \sum_{j\in C} (a_j - \lambda)^+ x_j + \sum_{j\in N^-} (\pi_j x_j + \mu_j y_j) \leqslant \mu_0\left(b - \sum_{j\in C} (a_j - \lambda)^+\right).$$

Now we describe another set of $2|L|$ points:

v. $w^k = (a, \bar{1}, 0, 0, \lambda e_k, e_k, 0, 0)$ for $k \in L$

$\quad \tilde{w}^k = (a, \bar{1}, 0, 0, (\lambda + \varepsilon)e_k, e_k, 0, 0)$ for $k \in L$.

From w^k and $\tilde{w}^k$, we obtain $\mu_k = 0$ for $k \in L$ and that

$$\mu_0\left(\sum_{j\in C} a_j\right) - \mu_0\left(\sum_{j\in C} (a_j - \lambda)^+\right) + \pi_k = \mu_0 b - \mu_0\left(\sum_{j\in C} (a_j - \lambda)^+\right)$$

or

$$\pi_k = \mu_0\left(b - \sum_{j\in C} a_j\right) = -\mu_0\lambda \quad \text{for } k \in L.$$

The final $2|N^- \setminus L|$ points are:

vi. $$\hat{w}^k = (a - \lambda e_1, \bar{1}, 0, 0, 0, 0, 0, e_k) \qquad \text{for } k \in N^- \setminus L$$

$$\bar{w}^k = (a - (\lambda - \varepsilon)e_1, \bar{1}, 0, 0, 0, 0, \varepsilon e_k, e_k) \qquad \text{for } k \in N^- \setminus L.$$

Comparing z^1 and $\hat{w}^k$, we see that $\pi_k = 0$ for $k \in N^- \setminus L$. Comparing $\hat{w}^k$ and $\bar{w}^k$, we see that $\varepsilon(\mu_0 + \mu_k) = 0$ or $\mu_k = -\mu_0$ for $k \in N^- \setminus L$.

Using the results of v and vi, the inequality must be of the form

$$\mu_0\left[\sum_{j\in C} (y_j + (a_j - \lambda)^+ (1 - x_j)) - \sum_{j\in L} \lambda x_j - \sum_{j\in N^-\setminus L} y_j\right] \leqslant \mu_0 b. \tag{4.6}$$

Now with $\mu_0 = 1$, we obtain (4.4).

It remains to show that not all points $(x, y) \in T$ satisfy (4.6) at equality. Since $b > 0$ is implied, the point given by $y_j = 0$ for $j \in N$, $x_j = 1$ for $j \in N^+$, and $x_j = 0$ for $j \in N^-$ is in T, and, when substituted in (4.6), one obtains zero on the left-hand side and $b > 0$ on the right.

Additional results along these lines are known. For example, if we require $\Sigma_{j\in N^+} y_j - \Sigma_{j\in N^-} y_j = b$ in (4.3), then (4.4) is, of course, still a valid inequality for T. Moreover, under some mild additional assumptions, (4.4) still gives a facet of conv(T). Also, some other valid inequalities for T given by (4.3) are known (see Section II.6.4).

The flow model with constraint set T given by (4.3) is much more general than it appears. With some additional simple constraints, it can be used to represent any linear inequality involving both continuous and 0-1 variables in which some of the continuous variables have simple upper bounds while the others have variable upper bounds.

Suppose T' is the set of feasible solutions to

$$\sum_{j\in J_1} (\alpha_j z_j + \alpha_j' x_j) + \sum_{j\in J_2} \alpha_j z_j + \sum_{j\in J_3} \alpha_j' x_j \leqslant b$$

$$0 \leqslant z_j \leqslant k_j x_j \quad \text{for } j \in J_1$$

$$0 \leqslant z_j \leqslant k_j \quad \text{for } j \in J_2$$

$$x_j \in \{0, 1\} \quad \text{for } j \in J_1 \cup J_3.$$

In addition we assume for simplicity that $\alpha_j' > 0$ for $j \in J_3$ and $\alpha_j'\alpha_j \geqslant 0$ for $j \in J_1$.

Now let $J_1^+ = \{j \in J_1: \alpha_j > 0\}$, $J_1^- = J_1 \setminus J_1^+$, $J_2^+ = \{j \in J_2: \alpha_j > 0\}$ and $J_2^- = J_2 \setminus J_2^+$. Define $x_j \in \{0, 1\}$ for $j \in J_2$,

$$y_j = \begin{cases} \alpha_j z_j + \alpha_j' x_j & \text{for } j \in J_1^+ \\ -(\alpha_j z_j + \alpha_j' x_j) & \text{for } j \in J_1^- \\ \alpha_j z_j & \text{for } j \in J_2^+ \\ -\alpha_j z_j & \text{for } j \in J_2^- \\ \alpha_j' x_j & \text{for } j \in J_3, \end{cases} \tag{4.7}$$

and

$$a_j = \begin{cases} \alpha_j k_j + \alpha_j' & \text{for } j \in J_1^+ \\ -(\alpha_j k_j + \alpha_j') & \text{for } j \in J_1^- \\ \alpha_j k_j & \text{for } j \in J_2^+ \\ -\alpha_j k_j & \text{for } j \in J_2^- \\ \alpha_j' & \text{for } j \in J_3. \end{cases}$$

Now T' is given by the flow model constraints

$$\sum_{j \in N^+} y_j \leq b + \sum_{j \in N^-} y_j$$

$$0 \leq y_j \leq a_j x_j \quad \text{for } j \in N^+ \cup N^-,$$

where $N^+ = J_1^+ \cup J_2^+ \cup J_3$ and $N^- = J_1^- \cup J_2^-$, together with the additional constraints $x_j = 1$ for $j \in J_2$, $y_j = a_j x_j$ for $j \in J_3$, and (4.7). Thus (4.4) is a valid inequality for T'.

We now give some examples of the use of (4.4) in different models.

Example 4.3. *(The 0-1 Knapsack Problem:* $S = \{x \in B^n: \Sigma_{j \in N} a_j x_j \leq b\}$ with $a_j \in R_+^1$ for $j \in N$ and $b \in R_+^1$). Here $N^+ = N$ and $y_j = a_j x_j$ for $j \in N$. Let C be a minimal dependent set so that $\lambda = \Sigma_{j \in C} a_j - b > 0$ and $a_j > \lambda$ for $j \in C$. Then (4.4) yields

$$\sum_{j \in C} (a_j x_j + (a_j - \lambda)(1 - x_j)) \leq b$$

or

$$\lambda \sum_{j \in C} x_j \leq b - \sum_{j \in C} a_j + \lambda\, |C| = \lambda(|C| - 1),$$

which is precisely the constraint (2.2).

Example 4.4 *(Facility Location).* Suppose

$$T' = \left\{ x_0 \in B^1, y \in R_+^n: \sum_{j \in N^+} y_j \leq a_0 x_0, y_j \leq a_j \text{ for } j \in N^+ \right\},$$

where $0 < a_j < a_0$ for all $j \in N^+$. Here a_0 is the capacity of a facility and $x_0 = 1$ if and only if the facility is open. The flow from the facility to client j is y_j, and a_j is the maximum requirement of client j. Here $b = 0$, $N^- = \{0\}$, and $y_0 = a_0 x_0$. Take $C = \{j\}$ so that $\lambda = a_j$ and take $L = N^-$. Then (4.4) yields $y_j \leq a_j x_0$ for $j \in N^+$, that is, $y_j = 0$ if $x_0 = 0$ and $y_j \leq a_j$ if $x_0 = 1$.

Example 4.5 *(Machine Scheduling).* Suppose that two jobs must be executed on the same machine. The ith job for $i = 1, 2$ has an earliest start time of l_i and a processing time of $p_i > 0$. The machine can only process one job at a time, and our objective is to model this restriction.

Let $\delta = 1$ if job 1 is processed before job 2 and let $\delta = 0$ otherwise; for $i = 1, 2$, let t_i be the time at which the machine begins to process job i. Then we have the model

$$\begin{aligned} t_1 - t_2 &\geqslant p_2 - \omega\delta \\ -t_1 + t_2 &\geqslant p_1 - \omega(1 - \delta) \\ t_i &\geqslant l_i \quad \text{for } i = 1,2 \text{ and } \delta \in B^1, \end{aligned}$$

where ω is a suitably large number so that the first constraint is valid when $\delta = 1$ and the second is valid when $\delta = 0$.

Suppose $l_2 + p_2 > l_1$. By substituting $y_i = t_i - l_i$ and $x_3 = \delta$, the first constraint becomes

$$y_2 \leqslant l_1 - l_2 - p_2 + y_1 + \omega x_3,$$

where $y_1, y_2 \geqslant 0$ and $x_3 \in B^1$. Here $N^+ = \{2\}$, $N^- = \{1, 3\}$, $y_3 = \omega x_3$, and $b = l_1 - l_2 - p_2 < 0$. Take $C = \emptyset$ so that $\lambda = -b > 0$ and take $L = \{3\}$. Then (4.4) yields $0 \leqslant -\lambda + \lambda x_3 + y_1$. Translating back into the original variables, we obtain

$$t_1 \geqslant l_1 + (l_2 + p_2 - l_1)(1 - \delta),$$

that is, $t_1 \geqslant l_1$ and if $\delta = 0$, then $t_1 \geqslant l_2 + p_2$.

While the general inequalities (4.4) can be quite useful, still more valid inequalities may be obtained by using the structure of a problem. We illustrate this by considering the constraint set of an *uncapacitated lot-size problem* that involves production planning over a horizon of T periods [see (5.4) of Section I.1.5].

In period t, $t = 1, \ldots, T$, there is a given demand of $d_t \in R^1_+$ that must be satisfied by production in period t and by inventory carried over from previous periods. The production in period t is y_t, $0 \leqslant y_t \leqslant \omega x_t$, where ω is a large positive number, and $x_t \in B^1$ equals 1 if the plant operates during period t and equals 0 otherwise. Let s_t be the inventory at the end of period t. Thus we obtain the constraints

$$(4.8) \qquad \begin{aligned} y_1 &= d_1 + s_1 \\ s_{t-1} + y_t &= d_t + s_t \quad \text{for } t = 2, \ldots, T \\ y_t &\leqslant \omega x_t \quad \text{for } t = 1, \ldots, T \\ s_T = 0,\ s &\in R^T_+,\ y \in R^T_+,\ x \in B^T. \end{aligned}$$

The constraints for a single period, namely,

$$\begin{aligned} & s_{t-1} + y_t = d_t + s_t \\ & 0 \leqslant y_t \leqslant \omega x_t, \qquad s_{t-1}, s_t \geqslant 0, \qquad x_t \in B^1, \end{aligned}$$

are an equality-constrained version of the flow model (4.3). Thus from (4.4), we obtain the valid inequalities

$$(4.9) \qquad y_t \leqslant d_t x_t + s_t \quad \text{for } t = 1, \ldots, T,$$

which simply state the obvious facts that $s_t \geqslant 0$ when $x_t = 0$ and $s_t \geqslant y_t - d_t$ when $x_t = 1$.

We now develop a more general set of inequalities for the system given by (4.8).

Proposition 4.5. *For any* $1 \leqslant l \leqslant T$, $L = \{1, \ldots, l\}$, *and* $C \subseteq L$,

$$\sum_{i \in C} y_i \leqslant \sum_{i \in C} \left(\sum_{t=i}^{l} d_t \right) x_i + s_l \tag{4.10}$$

is a valid inequality for (4.8).

Proof. Take any feasible solution (y, s, x) to (4.8). If $x_i = 0$ for all $i \in C$, then $y_i = 0$ for all $i \in C$ and (4.10) reduces to $s_l \geqslant 0$.

Now suppose that $x_i = 1$ for some $i \in C$ and let $k = \min\{i \in C: x_i = 1\}$. Hence $y_i = 0$ for all $i \in C$ with $i < k$ and thus

$$\begin{aligned} \sum_{i \in C} y_i &\leqslant \sum_{t=k}^{l} y_t = \sum_{t=k}^{l} d_t + s_l - s_{k-1} \leqslant \sum_{t=k}^{l} d_t + s_l \\ &\leqslant \sum_{i \in C} \left(\sum_{t=i}^{l} d_t \right) x_i + s_l \qquad (\text{since } x_k = 1). \end{aligned}$$ ■

Note that when $C = \{l\}$, (4.10) yields (4.9). There is, in fact, a much stronger result here whose proof will not be given.

Theorem 4.6. *The convex hull of solutions to (4.8) is given by the constraints* $s \in R^T_+$, $y \in R^T_+$, $x \in R^T_+$, $x_t \leqslant 1$ *for all* t, $s_T = 0$, $y_1 = d_1 + s_1$, *and* $s_{t-1} + y_t = d_t + s_t$ *for* $t = 2, \ldots, T$ *and by the inequalities (4.10) for all* l *and* $C \neq \emptyset$.

Example 4.6. Suppose $(d_1, d_2, d_3, d_4) = (4\ 2\ 7\ 3)$. The convex hull of solutions to (4.8) is given by the inequalities

$$\begin{array}{llcl}
l = 1, C = \{1\} & y_1 & \leqslant & 4x_1 + s_1 \\
l = 2, C = \{2\} & y_2 & \leqslant & 2x_2 + s_2 \\
l = 3, C = \{2\} & y_2 & \leqslant & 9x_2 + s_3 \\
l = 3, C = \{3\} & y_3 & \leqslant & 7x_3 + s_3 \\
l = 3, C = \{2, 3\} & y_2 + y_3 & \leqslant & 9x_2 + 7x_3 + s_3 \\
l = 4, C = \{2\} & y_2 & \leqslant & 12x_2 + s_4 \\
l = 4, C = \{3\} & y_3 & \leqslant & 10x_3 + s_4 \\
l = 4, C = \{4\} & y_4 & \leqslant & 3x_4 + s_4 \\
l = 4, C = \{2, 3\} & y_2 + y_3 & \leqslant & 12x_2 + 10x_3 + s_4 \\
l = 4, C = \{2, 4\} & y_2 + y_4 & \leqslant & 12x_2 + 3x_4 + s_4 \\
l = 4, C = \{3, 4\} & y_3 + y_4 & \leqslant & 10x_3 + 3x_4 + s_4 \\
l = 4, C = (2, 3, 4) & y_2 + y_3 + y_4 & \leqslant & 12x_2 + 10x_3 + 3x_4 + s_4
\end{array}$$

$$y_1 = d_1 + s_1$$

$$s_{t-1} + y_t = d_t + s_t \quad \text{for } t = 2, 3, 4$$

$$s_4 = 0, \quad s \in R^4_+, \ y \in R^4_+, \ x \in R^4_+, \ x_t \leqslant 1 \quad \text{for } t = 1, \ldots, 4.$$

The reader is asked to check that the $l = 1$, $C = \{1\}$ inequality is equivalent to $x_1 = 1$, since $y_1 = 4 + s_1$ and $x_1 \leqslant 1$. It then follows that all the other inequalities that could be generated with $1 \in C$ are superfluous. All the inequalities given above with $1 \notin C$ give facets except the last four. The last four inequalities are superfluous because $s_4 = 0$.

Although most production planning problems are much more complicated than our simple model in that they involve plant capacities, multiple items, and multistage production, they frequently have the system (4.8) as part of their formulation. Hence the theoretical results for the system (4.8) can be used in improving the formulation of more realistic production planning problems (see Section II.6.4).

5. NOTES

Section II.2.1

The idea of using structure to obtain strong valid inequalities for $\mathcal{NP}$-hard integer programs has its roots in the work of Dantzig, Fulkerson and Johnson (1954, 1959) on the traveling salesman problem and in the work of Gomory (1965, 1967, 1969, 1970) on the group problem.

Facet-defining inequalities for the node-packing problem were given by Padberg (1973), Nemhauser and Trotter (1974), Chvátal (1975), Trotter (1975), and Giles and Trotter (1981).

Gomory (1969) introduced the idea of lifting in the context of the group problem. Its computational possibilities were emphasized by Padberg (1973), and the approach was generalized by Wolsey (1976), Zemel (1978), and Balas and Zemel (1984).

The significance of having a partial description of the convex hull of integer solutions is strongly emphasized in the survey by Padberg (1979).

Section II.2.2

Facet-defining inequalities for the knapsack polytope were studied simultaneously by Balas (1975a), Hammer, Johnson and Peled (1975), and Wolsey (1975). Proposition 2.6 is due to Balas (1975a). Also see Balas and Zemel (1978), Padberg (1980b), and Zemel (1986).

The problem of extending these results to two or more general constraints remains an important open question.

Section II.2.3

The study of the convex hull of tours for the symmetric traveling salesman problem is largely due to Grötschel and Padberg (1979a,b, 1985). The proof of Proposition 3.4 is due to Maurras (1975). A different proof is given by Grötschel and Padberg (1979a).

Subtour elimination constraints were introduced by Dantzig, Fulkerson and Johnson (1954, 1959) and were shown to define facets of the convex hull of tours by Grötschel and Padberg (1979b).

Comb inequalities in which each tooth contains only one node of the handle are due to Chvátal (1973b). Chvátal's combs were generalized and were shown to define facets by Grötschel and Padberg (1979b). The inequalities (3.14) are due to Grötschel and Pulleyblank (1986). They called them *clique-tree inequalities* and proved that they define facets of the convex hull of tours.

The facet-defining inequality obtained from the Petersen graph is due to Chvátal (1973b). The Petersen graph is the smallest of a large class of graphs known as *hypohamiltonian graphs* that give facets for which no good characterization is known (see

Grötschel, 1980b). Another such class of graphs has been studied by Papadimitriou and Yannakakis (1984).

Other polyhedral results for the symmetric traveling salesman problem have been obtained by Cornuejols and Pulleyblank (1982), Cornuejols, Naddef and Pulleyblank (1983), and Cornuejols, Fonlupt and Naddef (1985).

Facets for the convex hull of tours on a directed graph have been studied by Grötschel and Padberg (1975) and Grötschel and Wakabayashi (1981a,b). Grötschel and Padberg (1985) surveyed these results.

Section II.2.4

The basic results for the variable upper-bound flow model are from Padberg, Van Roy and Wolsey (1985). Martin and Schrage (1985) obtained similar inequalities using different arguments. Van Roy and Wolsey (1986) have generalized these results to handle variable lower bounds.

The facet-defining inequalities for the lot-size model (4.8) were developed in Barany et al. (1984). Extensions to handle capacities are given in Leung and Magnanti (1986) and Pochet (1988), and those to treat backlogging are given in Pochet and Wolsey (1988). Valid inequalities for more general fixed-cost network problems are given in Van Roy and Wolsey (1985).

6. EXERCISES

1. Use clique inequalities, odd hole inequalities, and lifting to derive facets for the convex hull of node packings for the graph in Figure 6.1.
2. Prove Proposition 1.2.
3. Consider the uncapacitated facility location problem (UFL) introduced in Section I.1.3, with

$$T = \left\{ x \in B^n, y \in R_+^{mn}: \sum_{j \in N} y_{ij} = 1 \text{ for } i \in M, y_{ij} \leqslant x_j \text{ for all } i \in M, j \in N \right\}.$$

   i) Show that $\dim(\text{conv}(T)) = mn - m + n$.
   ii) Show that $y_{ij} \leqslant x_j$ define facets of conv(T).

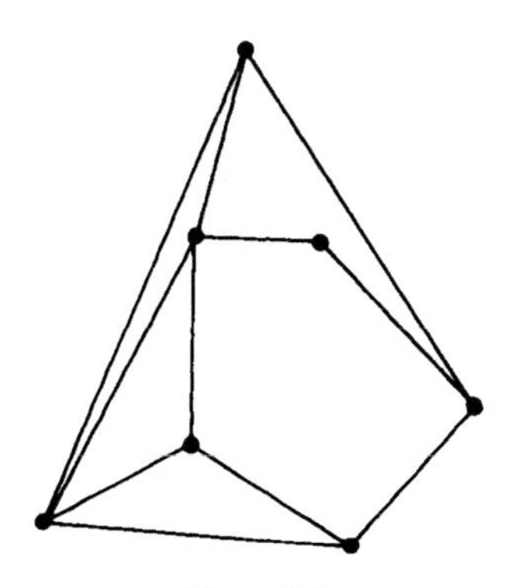

Figure 6.1

4. Let $G = (V, E)$ be a graph where each node has degree at least 3. Consider the set

$$S = \left\{x \in B^{|E|}: x_j - \sum_{e \in \delta(v)\setminus\{j\}} x_e \leqslant 0 \text{ for all } j \in \delta(v) \text{ and } v \in V\right\}$$

where $\delta(v)$ denotes the set of edges incident to node v.

i) Show that the inequalities $x_e \leqslant 1$ define facets of conv(S).

ii) Show that the inequalities $x_j - \sum_{e\in\delta(v)\setminus\{j\}} x_e \leqslant 0$ define facets of conv(S).

5. Consider the linear ordering problem of determining a permutation π: $\{1, \ldots, n\} \to \{1, \ldots, n\}$ formulated as

$$\max \sum_{ij} c_{ij}\delta_{ij}$$

$$\delta_{ij} + \delta_{ji} = 1 \quad \text{for all } i < j$$

$$\delta_{j_1 j_2} + \cdots + \delta_{j_r j_1} \leqslant |C| - 1 \quad \text{for all cycles } C = \{j_1, \ldots, j_r\}$$

$$\delta \in B^{n(n-1)},$$

where $\delta_{ij} = 1$ if i precedes j.

i) Show that the inequalities with $|C| \geqslant 4$ are unnecessary in the description of the problem.

ii) Show that for $|C| = 3$, the inequalities define facets.

6. For $S = \{x \in B^n: \sum_{j\in N} a_j x_j \leqslant b\}$, show that $x_j \geqslant 0$ and $x_j \leqslant 1$ define facets of conv(S) when $a \in Z_+^n$ and $a_j + a_k \leqslant b$ for all $j, k \in N$ with $j \neq k$.

7. For Example 2.3 use Propositions 2.3 and 2.6 to find as many facets as you can. Use these results to solve

$$\max 12x_1 + 5x_2 + 8x_3 + 7x_4 + 5x_5 + 5x_6 + 4x_7 + 3x_8 + 2x_9 + x_{10}$$

$$35x_1 + 27x_2 + 23x_3 + 19x_4 + 15x_5 + 15x_6 + 12x_7 + 8x_8 + 6x_9 + 3x_{10} \leqslant 39$$

$$x \in B^{10}$$

as a linear programming problem.

8. Let $S = \{x \in B^6: 27x_1 + 23x_2 + 17x_3 + 12x_4 + 8x_5 + 2x_6 \leqslant 40\}$.

i) Describe as many facet-defining inequalities as possible for S based on Proposition 2.3 and Corollary 2.4.

ii) What inequalities are obtained for S from Proposition 2.6?

9. Let $S = \{x \in B^n: \sum_{i\in I}\sum_{j\in Q_i} a_j x_j \leqslant b, \sum_{j\in Q_i} x_j \leqslant 1 \text{ for } i \in I\}$ with $N = \cup_{i\in I} Q_i$.

i) Show that if C is a minimal dependent set with $|C \cap Q_i| \leqslant 1$, $C \cap Q_i = \{j(i)\}$ when $C \cap Q_i \neq \emptyset$, and

$$\tilde{E}(C) = E(C) \cup \bigcup_{\{i:\, C\cap Q_i \neq \emptyset\}} \{j \in Q_i: a_j \geqslant a_{j(i)}\},$$

then $\sum_{j\in\tilde{E}(C)} x_j \leqslant |C| - 1$ is a valid inequality for S.

ii) Specify conditions under which this valid inequality defines a facet of conv(S).

10. Let

$$S = \{x \in B^7\colon\ 5x_1 + 7x_2 + 11x_3 - 8x_4 - 10x_5 - 15x_6 \leqslant -2$$
$$x_1 + x_2 + x_3 \leqslant 1$$
$$x_4 + x_5 + x_6 \leqslant 1\}$$

(see Exercise 14 of Section I.1.8).

i) Derive facets for conv(S).

ii) Can you show that these facet-defining inequalities give conv(S)?

11. Consider the symmetric traveling salesman polytope for the complete graphs on 5 and 7 nodes, respectively. Try to write down all of the facet-defining inequalities and see if you can give a proof that you have them all.

12. Give a nontrivial lower bound on the number of facets of the symmetric traveling salesman polytope for complete graphs with $n = 5, 7, 10, 100$, and 1000 nodes.

13. Prove the validity of the generalized comb inequalities (3.14).

14. Prove that $\Sigma_{e\in E'} x_e \leqslant 9$ is valid for the complete graph on 10 nodes, where $G = (V, E')$ is the Petersen graph, by showing it to be a C–G inequality.

15. Prove Proposition 4.1.

16. i) Use Proposition 4.3 to derive valid inequalities for

$$T = \{x \in B^6,\ y \in R^6_+\colon\ y_1 + y_2 + y_3 + y_4 - y_5 - y_6 \leqslant 12,$$
$$y_1 \leqslant 8x_1,\ y_2 \leqslant 7x_2,\ y_3 \leqslant 4x_3,\ y_4 \leqslant 2x_4,\ y_5 \leqslant 3x_5,\ y_6 \leqslant x_6\}.$$

ii) Which of these inequalities define facets?

17. Under what conditions does (4.4) define a facet of

$$T' = \left\{x \in B^n,\ y \in R^n_+\colon\ \sum_{j\in N^+} y_j - \sum_{j\in N^-} y_j = b,\ y_j \leqslant a_j x_j \text{ for } j \in N = N^+ \cup N^-\right\}?$$

18. Consider the capacitated facility problem with feasible region

$$T = \left\{x \in B^n,\ y \in R^{mn}_+\colon\ \sum_j y_{ij} = a_i \text{ for } i \in M,\ \sum_i y_{ij} \leqslant b_j x_j \text{ for } j \in N\right\}.$$

Let $I \subseteq M$ and $z_j = \Sigma_{i\in I}\, y_{ij}$ so that the z_j satisfy

$$\sum_{j\in N} z_j = \sum_{i\in I} a_i \text{ and } \qquad z_j \leqslant b_j x_j.$$

i) Derive valid inequalities for T.

ii) Can you show that the inequalities define facets?

19. Consider the mixed 0-1 region with lower and upper bounds

$$T = \left\{ x \in B^n, y \in R_+^n \colon \sum_{j \in N} y_j \leqslant b,\ l_j x_j \leqslant y_j \leqslant a_j x_j \text{ for } j \in N \right\}$$

with $l_j, a_j \geqslant 0$ and the region

$$T' = \left\{ x \in B^n, y \in R_+^n \colon \sum_{j \in N} (y_j + p_j x_j) \leqslant B,\ y_j \leqslant m_j x_j \text{ for } j \in N \right\},$$

where y_j is a variable representing production time, and p_j is the associated set-up time.

i) Show the equivalence between T and T'.

ii) Derive valid inequalities for T (or T').

20. For the fixed-cost networks shown below, show that the proposed inequalities are valid.

a)

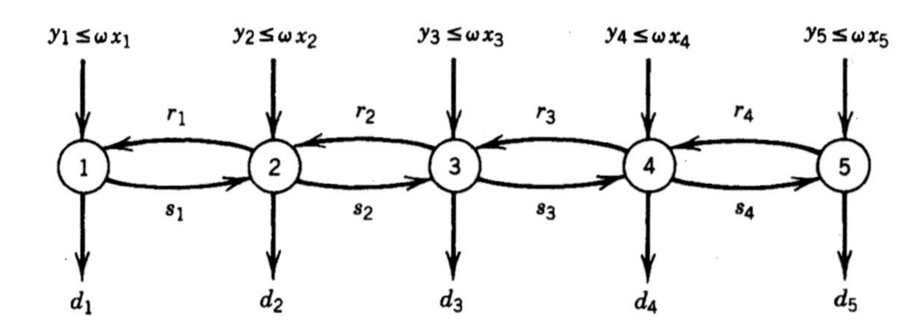

$$y_2 + y_3 + y_4 \leqslant r_1 + r_2 + s_3 + s_4 + (d_2 + d_3)x_2 + d_3 x_3 + (d_3 + d_4)x_4.$$

b)

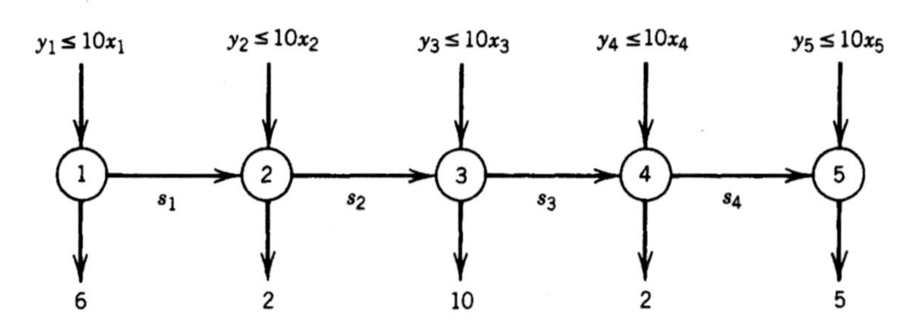

$$y_2 + y_3 + y_4 \leqslant 6 + 4x_2 + 4x_3 + 4x_4 + s_4.$$

c)

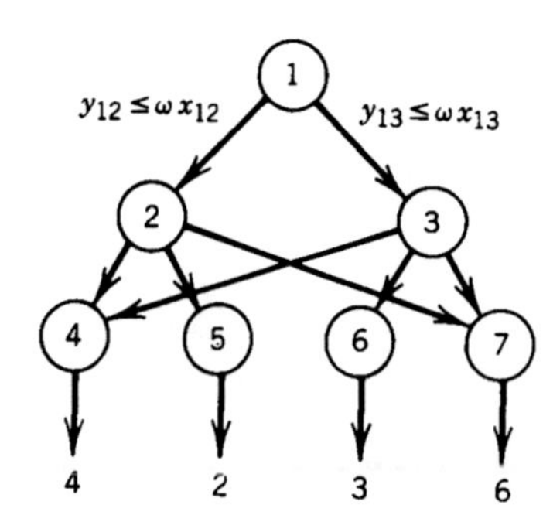

$$y_{12} + y_{13} \leqslant 6x_{12} + 4x_{13} + y_{27} + y_{36} + y_{37}.$$

II.3

Duality and Relaxation

1. INTRODUCTION

In the preceding two chapters we studied polyhedral descriptions of the set of feasible solutions to linear inequalities in nonnegative integer variables. Now we introduce an objective function and consider the integer optimization problem

$$\text{(IP)} \qquad z_{\text{IP}} = \max\{cx: x \in S\}, \qquad S = \{x \in Z^n_+ : Ax \leq b\},$$

where c is an n-vector with integral coefficients and (A, b) is an $m \times (n + 1)$ matrix with integral coefficients.

The theme of this chapter is to develop a theory for determining z_{IP}, or at least a good upper bound on z_{IP}, without explicitly solving IP. This can be considered to be a theory of optimality, since a tight bound on z_{IP} provides the fundamental way of proving optimality of a feasible solution to IP. Suppose we are given an $x^0 \in S$ that is claimed to be an optimal solution to IP. How can we decide whether this claim is true?

Our previous results provide one answer. Consider the linear program

$$z^* = \max\{cx: x \in \text{conv}(S)\}.$$

Then x^0 is an optimal solution to IP if and only if $cx^0 = z^*$ (see Theorem 6.4 of Section I.4.6). Although this answer is correct, it depends on knowing conv(S), which is an assumption we do not make here.

Observe that the answer just given tells us if x^0 is optimal or not; that is, $cx^0 = z^*$ is a necessary and sufficient condition for the optimality of x^0. Suppose we just ask for a sufficient condition for the optimality of x^0. We prefer to focus on sufficiency rather than necessity because if a sufficient condition is satisfied, the optimality claim is proved.

Here is a simple, but rather naive, sufficient condition. Consider the linear program

$$\text{(1.1)} \qquad z_{\text{LP}} = \max\{cx: x \in P\},$$

where $P = \{x \in R^n_+: Ax \leq b\}$. Then $x^0 \in S$ is an optimal solution to IP if $cx^0 = z_{\text{LP}}$. We said this condition is naive because without further assumptions, it is unlikely to hold.

An equivalent sufficient condition arises from considering the linear programming dual of IP:

$$z_{\mathrm{LP}} = \min\{ub : u \in P_D\}, \tag{1.2}$$

where $P_D = \{u \in R_+^m : uA \geqslant c\}$. But now we can phrase the sufficient condition in a subtly different way; that is, x^0 is optimal to IP if there is a $u^0 \in P_D$ such that $cx^0 = u^0 b$.

Problem (1.1) is called a *relaxation* of IP, and problem (1.2) is called a *(weak) dual* of IP. Relaxation and duality are the two fundamental ways of determining z_{IP} and upper bounds on z_{IP}. These notions will be made precise after we give an example.

Example 1.1. Consider the maximum cardinality node-packing problem on the graph shown in Figure 1.1. We use the clique constraints, that is no more than one node can be chosen from each clique, to obtain the integer programming formulation

$$\begin{aligned}
z_{\mathrm{IP}} = \max x_1 + x_2 + x_3 + x_4 + x_5 + x_6 & \\
x_1 + x_2 \qquad\qquad\qquad\qquad + x_6 &\leqslant 1 \\
x_2 + x_3 \qquad\qquad\qquad &\leqslant 1 \\
x_3 + x_4 \qquad\qquad &\leqslant 1 \\
x_4 + x_5 \qquad &\leqslant 1 \\
x_5 + x_6 &\leqslant 1 \\
x \in B^6. &
\end{aligned}$$

The solution $x_1^0 = x_3^0 = x_5^0 = 1$, $x_j^0 = 0$ otherwise, is feasible. We want to prove that it is optimal. The relaxation (1.1) is obtained by replacing $x \in B^6$ by $x \in R_+^6$, and the dual of this linear program is

$$\begin{aligned}
z_{\mathrm{LP}} = \min u_1 + u_2 + u_3 + u_4 + u_5 & \\
u_1 \qquad\qquad\qquad\qquad &\geqslant 1 \\
u_1 + u_2 \qquad\qquad\qquad &\geqslant 1 \\
u_2 + u_3 \qquad\qquad &\geqslant 1 \\
u_3 + u_4 \qquad &\geqslant 1 \\
u_4 + u_5 &\geqslant 1 \\
u_1 \qquad\qquad\qquad + u_5 &\geqslant 1 \\
u \geqslant 0. &
\end{aligned}$$

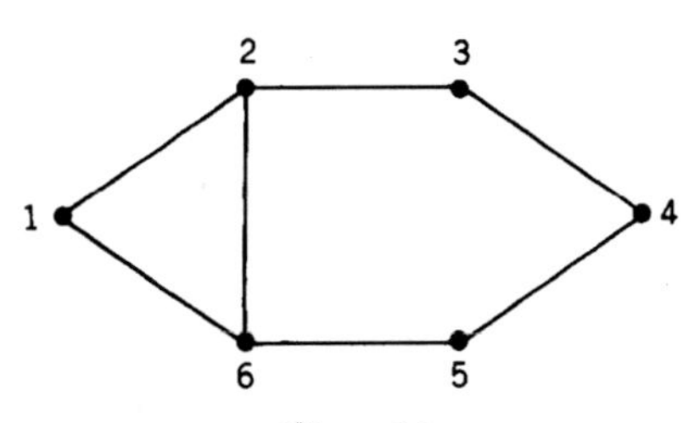

Figure 1.1

A feasible solution to the dual is $u_1^0 = u_3^0 = u_5^0 = 1$, $u_i^0 = 0$ otherwise. Since $\Sigma_{j=1}^6 x_j^0 = \Sigma_{i=1}^5 u_i^0 = 3$, it follows that x^0 is an optimal solution to IP and that $z_{\text{IP}} = z_{\text{LP}} = 3$. The reader can check that in this example the clique constraints and nonnegativity are not sufficient to give the convex hull of node packings. However, for the given objective function, we have the good fortune that the linear programming relaxation has an integral optimal solution. This example shows that it is not necessary to have the convex hull of feasible solutions to obtain or prove the optimality of an integral solution.

Another simple argument that does not use linear programming also establishes the optimality of x^0. Consider any set of cliques such that each node is contained in at least one of them, for example, $C_1 = \{1, 2, 6\}$, $C_2 = \{3, 4\}$, and $C_3 = \{4, 5\}$. Such a set of cliques is called a *clique cover*. Any node packing contains no more than one node from each of the cliques in a clique cover. Hence we obtain the max–min relationship that the maximum number of nodes in any node packing is equal to or less than the minimum number of cliques in any clique cover. Thus from the cover $\{C_1, C_2, C_3\}$, we obtain $z_{\text{IP}} \leqslant 3$. This is an example of a combinatorial duality, which is a principle that is fundamental to the solution of combinatorial optimization problems.

A *relaxation* of IP is any maximization problem

$$\text{(RP)} \qquad z_R = \max\{z_R(x): x \in S_R\}$$

with the following two properties:

$$\text{(R1)} \qquad S \subseteq S_R$$

$$\text{(R2)} \qquad cx \leqslant z_R(x) \quad \text{for } x \in S.$$

Proposition 1.1. *If* RP *is infeasible, so is* IP. *If* IP *is feasible, then* $z_{\text{IP}} \leqslant z_R$.

Proof. From (R1), if $S_R = \emptyset$, then $S = \emptyset$, so the first statement holds.

Now suppose that z_{IP} is finite and let x^0 be an optimal solution to IP. Then $z_{\text{IP}} = cx^0 \leqslant z_R(x^0) \leqslant z_R$, where the first inequality follows from (R2) and the second one follows from (R1). Finally, if $z_{\text{IP}} = \infty$, (R1) and (R2) imply that $z_R = \infty$. ■

If $x^* \in S$ satisfies $cx^* \geqslant z_{\text{IP}} - \varepsilon$ for some fixed $\varepsilon > 0$, then we say that x^* is an *ε-optimal solution* to IP. Since it is sometimes too costly to find a provably optimal solution, we may have to be satisfied with a provably ε-optimal solution for a given tolerance ε. Although the upper-bound z_R may fail to prove optimality, RP may allow us to establish ε-optimality. In particular, if $x^* \in S$ satisfies $cx^* \geqslant z_R - \varepsilon$, then x^* is an ε-optimal solution.

The most common way to obtain a relaxation is to satisfy (R1) by dropping one or more of the constraints that define S and to satisfy (R2) by setting $z_R(x) = cx$.

The *linear programming relaxation* of IP is (1.1). The so-called *group relaxation* is obtained by dropping certain nonnegativity conditions. In many problems, the constraints can be partitioned into a set of simple ones that can be handled easily and complicated ones. A relaxation is obtained by removing the complicated constraints and including them in the objective function in such a way that (R2) is satisfied. This technique is called *Lagrangian relaxation*. The latter two approaches will be considered in Sections 5 and 6 of this chapter.

Dropping constraints is not the only way to satisfy (R1). We can combine equalities by taking linear combinations and inequalities by taking nonnegative linear combinations. A right-hand side b of a constraint can be replaced by a set of right-hand sides that contains b. In particular, if $S = \{x \in R_+^n: Ax \leqslant b\}$ and $\tilde{S} = \bigcup_{d \in B}\{x \in R_+^n: Ax \leqslant d\}$, where $B \subseteq R^m$, then $S \subseteq \tilde{S}$.

Adding and/or changing variables can also be used to obtain relaxations. For example, we obtain a relaxation if $S = \{x \in Z_+^n: Ax \leqslant b\}$ is replaced by $S' = \{(x, x') \in Z_+^{n+p}: Ax + A'x' \leqslant b\}$ since $S = \{x \in Z_+^n: (x, 0) \in S'\}$. Such a relaxation can be useful if matrix (A, A') is easier to work with than A. These ideas for relaxation will be used in the algorithms to be developed subsequently.

A distinct disadvantage of using relaxation to obtain bounds is that only an optimal solution to the relaxed problem guarantees an upper bound on z_{IP}. Duality eliminates this difficulty since the dual problem is defined so that any dual feasible solution yields an upper bound on z_{IP}.

A *weak dual* of IP is any minimization problem

(DP) $$z_D = \min\{z_D(u): u \in S_D\}$$

that satisfies

(D1) $$z_D(u) \geqslant cx \quad \text{for all } x \in S \text{ and } u \in S_D.$$

Analogous to Proposition 1.1, we have the following proposition.

Proposition 1.2. *If* DP *is feasible, then* $z_{\text{IP}} \leqslant z_D$. *If* DP *has an unbounded objective value, then* IP *is infeasible.*

A *strong dual* of IP is a weak dual that also satisfies

(D2) If $S \neq \emptyset$ and z_{IP} is bounded from above, then there exists $u^0 \in S_D$ and $x^0 \in S$ such that $z_D(u^0) = cx^0$.

By solving a strong dual we find z_{IP}, since $z_{\text{IP}} = z_D$ when both problems have finite optimum values. By solving a weak dual we can approximate z_{IP} from above. We call $\Delta_D = z_D - z_{\text{IP}}$ the *absolute value of the duality gap*.

Weak duals are easy to construct. For example, by taking the dual of a linear programming relaxation of IP we obtain a weak dual to IP.

Combinatorial structures are used to construct dual problems. A typical combinatorial optimization problem exemplified by the node-packing problem is the following. Let $V = \{1, 2, \ldots, n\}$ be a finite set and let $\mathscr{C} = \{C_1, C_2, \ldots, C_m\}$ be a finite collection of subsets of V. A subset $V^0 \subseteq V$ is called a *packing* if $|V^0 \cap C_i| \leqslant 1$ for $i = 1, \ldots, m$. A subset $\mathscr{C}^0 \subseteq \mathscr{C}$ is called a *cover* if $\bigcup_{C_i \in \mathscr{C}^0} C_i = V$. Suppose V^0 is any packing and $\mathscr{C}^0$ is any cover. Then

$$|V^0| \leqslant \sum_{\{i: C_i \in \mathscr{C}^0\}} |V^0 \cap C_i| \leqslant |\mathscr{C}^0|,$$

where the first inequality follows from $\bigcup_{C_i \in \mathscr{C}^0} C_i \supseteq V^0$ and the second one follows from $|V^0 \cap C_i| \leqslant 1$ for all i. In other words, the cardinality of any packing is equal to or less than the cardinality of any covering, so the minimum covering problem is a weak dual of the maximum packing problem. A fundamental problem of combinatorial optimization is to characterize packing and covering problems for which strong duality holds.

The general relationship between duality and relaxation is given in the following proposition.

Proposition 1.3. *If a problem is dual to a relaxation of* IP, *then it is also dual to* IP.

Proof. Suppose $z_{DR} = \min\{z_{DR}(u): u \in S_{DR}\}$ is dual to RP. Then $z_R(x) \leqslant z_{DR}(u)$ for all $x \in S_R$ and all $u \in S_{DR}$. By relaxation, $cx \leqslant z_R(x)$ for all $x \in S \subseteq S_R$. Hence $cx \leqslant z_{DR}(u)$ for all $x \in S$ and $u \in S_{DR}$. ■

As with relaxations, algorithms generally use a weak dual to obtain bounds and iteratively refine the dual to strengthen the bounds.

2. DUALITY AND THE VALUE FUNCTION

Here we consider a family of integer programs

$$\text{(2.1)} \qquad z(d) = \max\{cx: x \in S(d)\}, \quad S(d) = \{x \in Z_+^n: Ax \leqslant d\} \quad \text{for } d \in D,$$

where A and c are fixed and d is a parameter in $D \subseteq R^m$. Depending on our need we may take $D = R^m$ or $D = Z^m$ or $D = \{d \in R^m: S(d) \neq \emptyset\}$. The function $z(d)$ for $d \in D$ is called the *value function* of IP. We say that $z(d) = -\infty$ if $S(d) = \emptyset$ and that $z(d) = +\infty$ if the objective value is unbounded from above.

The following propositions give some elementary properties of the value function.

Proposition 2.1. *The value function of* IP *is nondecreasing over* R^m.

Proof. If $d^1 \leqslant d^2$, then $S(d^2) \supseteq S(d^1)$, which implies $z(d^2) \geqslant z(d^1)$. ■

Proposition 2.2. $z(0) \in \{0, \infty\}$. *If* $z(0) = \infty$, *then* $z(d) = \pm\infty$ *for all* $d \in R^m$. *If* $z(0) = 0$, *then* $z(d) < \infty$ *for all* $d \in R^m$.

Proof. See Proposition 6.7 of Section I.4.6. ■

Problems with $z(d) = \pm\infty$ for all $d \in R^m$ (e.g., $\max\{x_1: 2x_1 - x_2 \leqslant d, x \in Z_+^2\}$) reduce to feasibility problems. Thus, for simplicity of exposition, it is convenient to ignore them here. Hence, unless otherwise specified, we assume $z(0) = 0$, so $z(d) < \infty$ for all $d \in R^m$.

Proposition 2.3. *The value function of* IP *is superadditive over* $D = \{d \in R^m: S(d) \neq \emptyset\}$.

Proof. Suppose $x^i \in Z_+^n$ and $Ax^i \leqslant d^i$ for $i = 1, 2$. Then $(x^1 + x^2) \in Z_+^n$ and $A(x^1 + x^2) \leqslant d^1 + d^2$. Thus if $c\hat{x}^i = z(d^i)$ for $i = 1, 2$, then

$$z(d^1) + z(d^2) = c(\hat{x}^1 + \hat{x}^2) \leqslant z(d^1 + d^2). \qquad ■$$

The problem of finding an upper bound on the optimal value of IP can be generalized to the problem of finding a function $g(d)$: $R^m \rightarrow R^1$ such that $g(d) \geqslant z(d)$ for all $d \in R^m$ (see Figure 2.1). Thus a dual problem to IP can be formulated as

$$\text{(2.2)} \qquad \min\{g(b): g(d) \geqslant z(d) \text{ for } d \in R^m, g: R^m \rightarrow R^1\}$$

or, equivalently, as

$$\text{(2.3)} \qquad \min\{g(b): g(d) \geqslant cx \text{ for } x \in S(d) \text{ and } d \in R^m\}.$$

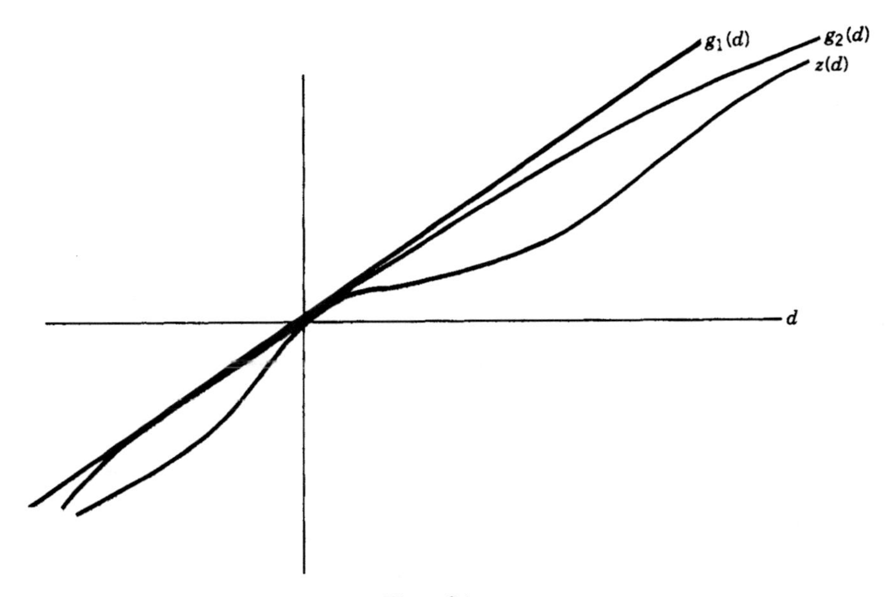

Figure 2.1

This dual is strong since there are feasible solutions with $g(b) = z(b)$; for example, $g(d) = z(d)$ when $z(d) > -\infty$ and $g(d) = 0$ otherwise.

Some restrictions on g are needed to obtain a useful dual problem. Since $z(d)$ is nondecreasing, it is natural to assume that $g(d)$ is nondecreasing. Then g satisfies $g(d) \geqslant cx$ for $x \in S(d)$ if and only if $g(Ax) \geqslant cx$ for $x \in Z^n_+$. Thus when g is nondecreasing, (2.3) can be stated as

$$\begin{aligned} &\min g(b) \\ (2.4) \qquad &g(Ax) \geqslant cx \quad \text{for } x \in Z^n_+ \\ &g \text{ nondecreasing.} \end{aligned}$$

Now suppose that g is linear; that is, $g(d) = ud$ with $u \in R^m_+$. Thus we require $uAx \geqslant cx$ for all $x \in Z^n_+$. This last condition is equivalent to $uA \geqslant c$. Thus we obtain the weak dual

$$(2.5) \qquad \min\{ub\colon uA \geqslant c,\ u \in R^m_+\},$$

which is the dual of the linear programming relaxation of IP.

Linear functions are generally too restrictive to obtain strong duality. In the following example, we first consider the value function and the linear dual and then we give two illustrations of strong dual functions.

Example 2.1

$$\begin{aligned} z(d) = \max\ & 3x_1 + 6x_2 + 11x_3 + 12x_4 \\ & x_1 + 2x_2 + 3x_3 + 4x_4 \leqslant d \\ & x \in Z^4_+ \end{aligned}$$

with $d \in R^1$.

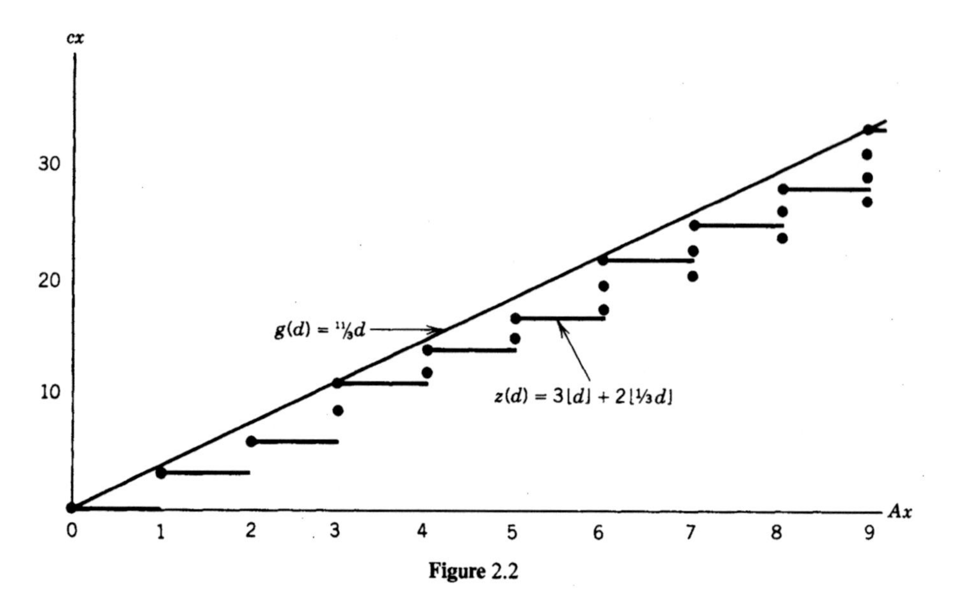

Figure 2.2

Figure 2.2 gives a plot of (Ax, cx) for the feasible points. The upper envelope of these points gives the value function

$$z(d) = \begin{cases} \dfrac{11}{3}d, & d = 0, 3, 6, \ldots \\ 3 + \dfrac{11}{3}(d-1), & d = 1, 4, 7, \ldots \\ 6 + \dfrac{11}{3}(d-2), & d = 2, 5, 8, \ldots . \end{cases}$$

$z(d) = z(\lfloor d \rfloor)$ for d positive and not integral, and $z(d) = -\infty$ if $d < 0$. We can also express the value function over R^1_+ by $z(d) = 3\lfloor d \rfloor + 2\lfloor \frac{1}{3}d \rfloor$, which shows that z is superadditive over R^1_+.

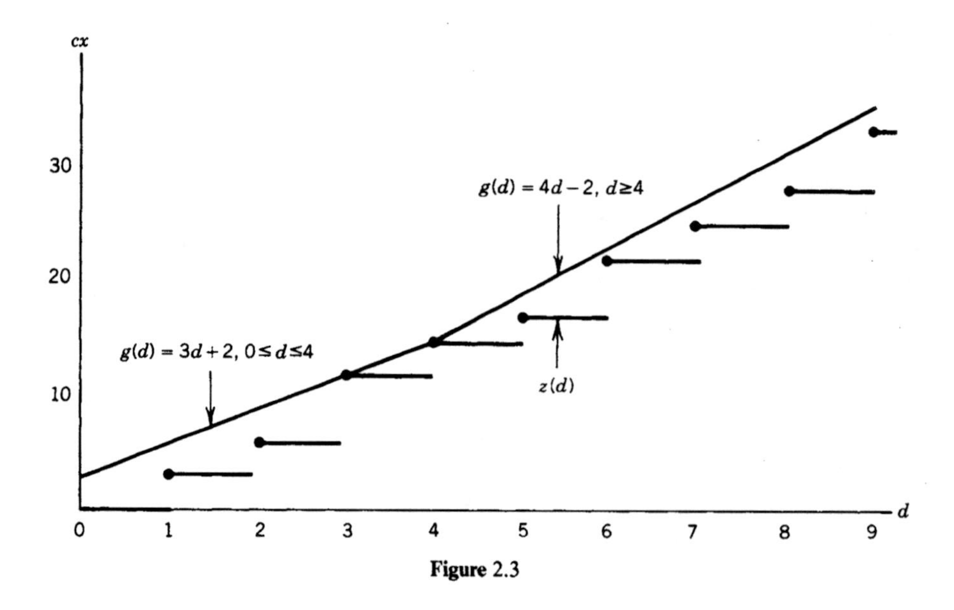

Figure 2.3

Figure 2.2 also shows the function $g(d) = \frac{11}{3}d$, which is the optimal dual solution when g is restricted to be linear; that is, $u = \frac{11}{3}$ is the optimal solution to (2.5). We see this graphically by observing that any line through the origin with slope $< \frac{11}{3}$ is not dual feasible and that if the slope is greater than $\frac{11}{3}$, $ud > \frac{11}{3}d$ for all $d \in R^1_+$. Note that the optimal linear function only provides a strong dual when d is an integer multiple of 3.

Figure 2.3 shows $z(d)$ and the function

$$g_0(d) = \max(g_1(d), g_2(d)),$$

where $g_1(d) = 3d + 2$ and $g_2(d) = 4d - 2$. We can see from the picture that g_0 is dual feasible.

We now give an algebraic justification of its dual feasibility. We have

$$\begin{aligned} g_1(Ax) &= 3x_1 + 6x_2 + 9x_3 + 12x_4 + 2 \\ &= cx + 2(1 - x_3) \\ &\geq cx \quad \text{for } x \in \{Z^4_+: x_3 \leq 1\} \end{aligned}$$

and

$$\begin{aligned} g_2(Ax) &= 4x_1 + 8x_2 + 12x_3 + 16x_4 - 2 \\ &\geq cx + (x_3 - 2) \\ &\geq cx \quad \text{for } x \in \{Z^4_+: x_3 \geq 2\}. \end{aligned}$$

Hence $g_0(Ax) \geq cx$ for all $x \in Z^4_+$. Note that $g_0(4) = z(4)$ so that strong duality is obtained for $d = 4$.

Figure 2.4 shows $z(d)$ and the superadditive function $F(d) = 3d + \lfloor \frac{2}{3}d \rfloor$. Note that $F(1) = 3 = c_1$, $F(2) = 7 > c_2$, $F(3) = 11 = c_3$, and $F(4) = 14 > c_4$. Thus $F(a_j) \geq c_j$ for $j = 1, \ldots, 4$ and hence superadditivity implies

$$F(Ax) \geq \sum_{j=1}^{4} F(a_j)x_j \geq \sum_{j=1}^{4} c_j x_j \quad \text{for } x \in Z^4_+.$$

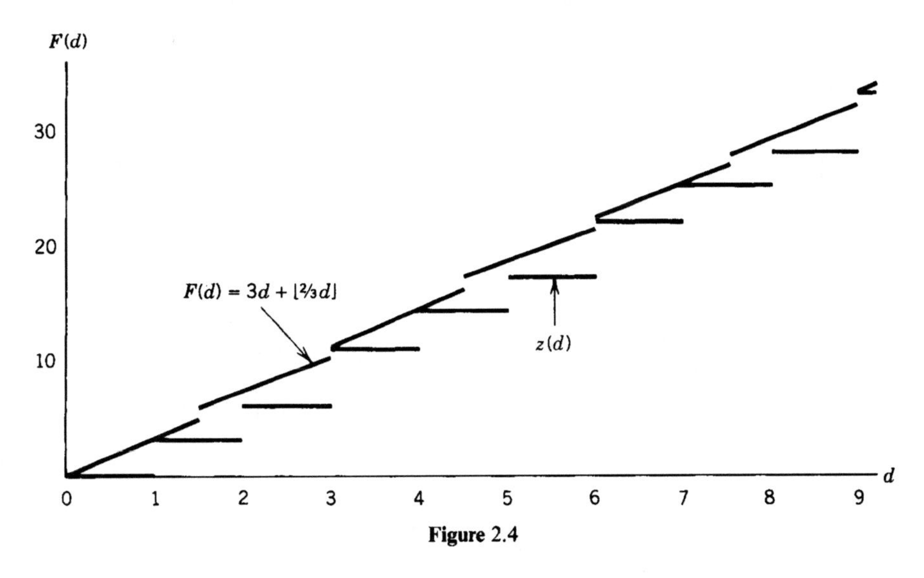

Figure 2.4

Thus F is dual feasible. Strong duality is obtained for $d = 4$ since $F(4) = 14$.

The three functions used in the example illustrate important classes of dual functions that are used in integer programming algorithms. Linear functions are the simplest, but they do not generally yield strong duality. The function $g_0(d)$ exemplifies the type of dual function used to prove optimality in branch-and-bound algorithms with linear programming relaxations. Superadditive functions are used to prove optimality in cutting-plane algorithms.

3. SUPERADDITIVE DUALITY

There are two important reasons for restricting the function g to be superadditive in the dual problem (2.4):

a. The purpose of the dual problem is to estimate the value function from above, and the value function is superadditive over the domain for which it is finite.
b. If g is superadditive, the condition $g(Ax) \geqslant cx$ for $x \in Z_+^n$ is equivalent to $g(a_j) \geqslant c_j$ for $j \in N$. This is true since $g(Ae_j) \geqslant ce_j$ is the same as $g(a_j) \geqslant c_j$ for $j \in N$; and if g is superadditive, then $g(a_j) \geqslant c_j$ for $j \in N$ implies

$$g(Ax) \geqslant \sum_{j \in N} g(a_j)x_j \geqslant \sum_{j \in N} c_j x_j = cx$$

for $x \in Z_+^n$.

Condition b enables us to state a *superadditive dual problem* independent of x.

$$\text{(SDP)} \quad \begin{aligned} w = \min\ & F(b) \\ & F(a_j) \geqslant c_j \quad \text{for } j \in N \\ & F(0) = 0 \\ & F: R^m \to R^1, \quad \text{nondecreasing and superadditive.} \end{aligned}$$

We now establish results analogous to linear programming duality for the primal problem IP and the dual problem SDP.

Proposition 3.1. *(Weak Duality). If F is feasible to* SDP *and x is feasible to* IP, *then $cx \leqslant F(b)$.*

Proof

$$\begin{aligned} \sum_{j \in N} c_j x_j &\leqslant \sum_{j \in N} F(a_j)x_j && \text{since } c_j \leqslant F(a_j) \quad \text{for } j \in N \text{ and } x \in R_+^n \\ &\leqslant F(Ax) && \text{since } F \text{ is superadditive, } F(0) = 0, \text{ and } x \in Z_+^n \\ &\leqslant F(b) && \text{since } F \text{ is nondecreasing.} \end{aligned}$$

■

Weak duality allows us to take care of the case of an unbounded primal objective function that we dismissed earlier.

Corollary 3.2 *(Unbounded Primal Objective Function). If* IP *is feasible and* $z(b) = \infty$, *then the superadditive dual is infeasible.*

Proof. If $z(b) = \infty$, then $z(0) = \infty$, so no dual solution can satisfy $F(0) = 0$. ■

Weak duality establishes that if F is a feasible solution to the superadditive dual, then F provides an upper bound on the value function for all $d \in R^m$. It remains to be shown that the objective min $F(b)$, which we did not use in the proof of weak duality, yields strong duality.

For the remainder of this section, we assume that $P = \{x \in R^n_+: Ax \leqslant b\}$ contains explicit bound constraints so that we can use Corollary 2.14 of Section II.1.2.

Theorem 3.3 *(Strong Duality).*

1. *If* IP *is feasible, then* SDP *is feasible and* $w = z(b)$.
2. *If* IP *is infeasible, then the dual objective function is unbounded from below* ($w = -\infty$).

Proof. 1. $\Sigma_{j \in N} c_j x_j \leqslant z(b)$ is a valid inequality for S. Hence Theorem 4.6 of Section II.1.4 implies that there exists a superadditive and nondecreasing function F (with $F(0) = 0$) such that $\Sigma_{j \in N} F(a_j) x_j \leqslant F(b)$ is valid for S and dominates $\Sigma_{j \in N} c_j x_j \leqslant z(b)$. This means that $F(a_j) \geqslant c_j$ for $j \in N$ and $F(b) \leqslant z(b)$. Hence F is dual feasible. But then $F(b) \geqslant z(b)$, so F is an optimal dual solution with $w = F(b) = z(b)$.
2. Since P contains explicit bound constraints, there exists $u \in R^m_+$ such that $uA \geqslant c$ (see Proposition 1.1 of Section II.1.1). Let $F_1(a) = ua$ for all $a \in R^m$. By Corollary 2.14 of Section II.1.2, we have $0x \leqslant -1$ is a C–G inequality for S. Hence there is a superadditive and nondecreasing function F_2 with $F_2(a_j) \geqslant 0$ for $j \in N$ and $F_2(b) \leqslant -1$. Thus for any $\lambda \in R^1_+$, it follows that $F_1 + \lambda F_2$ is a feasible dual solution; also, $F_1(b) + \lambda F_2(b) \to -\infty$ as $\lambda \to \infty$. ■

The familiar *complementary slackness* property of linear programming duality carries over to superadditive integer programming duality. In particular, if x^0 is an optimal solution to IP and F^0 is an optimal superadditive dual solution, then

$$(F^0(a_j) - c_j)x_j^0 = 0 \quad \text{for } j \in N. \tag{3.1}$$

We prove (3.1) as a corollary to a slightly more general result.

Theorem 3.4. *If* x^0 *is an optimal solution to* IP *and* F^0 *is an optimal solution to the superadditive dual, then*

$$F^0(Ax) = cx \quad \text{and} \quad F^0(Ax) + F^0(b - Ax) = F^0(b)$$

for all $x \in Z^n_+$ *such that* $x \leqslant x^0$.

Proof

$$\begin{aligned} cx^0 = cx + c(x^0 - x) &\leqslant \sum_{j \in N} F^0(a_j)x_j + \sum_{j \in N} F^0(a_j)(x_j^0 - x_j) \\ &\leqslant F^0(Ax) + F^0(A(x^0 - x)) \leqslant F^0(Ax) + F^0(b - Ax) \\ &\leqslant F^0(b) = cx^0. \end{aligned}$$

Hence the second equality holds. The first equality holds since we also have $F^0(Ax) \geqslant cx$ and $F^0(A(x^0 - x)) \geqslant c(x^0 - x)$. ∎

Note that (3.1) is trivial when $x_j^0 = 0$. When $x_j^0 \geqslant 1$, we obtain (3.1) from $F^0(Ae_j) = ce_j$.

The next two results describe optimal solutions to the superadditive dual. The first result comes from a superadditive description of conv(S) and linear programming duality.

Theorem 3.5. *If $S \neq \emptyset$ and* $\max\{cx: x \in S\} < \infty$, *then there exist a $u \in R_+^t$ and finite rank C–G functions F^k for $k = 1, \ldots, t$ with $t \leqslant n$ such that $F^c = \sum_{k=1}^{t} u_k F^k$ is an optimal solution to the superadditive dual.*

Proof. By Proposition 4.5 of Section II.1.4 there exist C–G functions F^k for $k = 1, \ldots, t$ such that

$$\operatorname{conv}(S) = \left\{x \in R_+^n: \sum_{j \in N} F^k(a_j)x_j \leqslant F^k(b) \text{ for } k = 1, \ldots, t\right\}.$$

Now apply linear programming duality. ∎

The value function $z(d)$ of IP would be a feasible (and hence optimal) solution to SDP except for the fact that it is superadditive only on the domain D where IP is feasible. The following theorem tells us that z can always be extended to a (finite-valued) superadditive function over R^m.

Theorem 3.6. *There are C–G functions F^i for $i = 1, \ldots, q$ such that $z(d) = \min_{i=1,\ldots,q} F^i(d)$ for all d with $z(d) > -\infty$.*

We will not prove this theorem. We observe, however, that $F(d) = \min_{i=1,\ldots,q} F^i(d)$ is superadditive over R^m since it is the minimum of a finite number of superadditive functions. The functions F^i are optimal solutions to the superadditive dual for certain values of d. Hence, implicit in the result is that it is possible to calculate $z(d)$ for all $d \in D$ by solving IP for only a finite number of $d \in D$.

Example 3.1 (Example 2.1 continued). We showed in the previous section that $F(d) = 3d + \lfloor \frac{2}{3}d \rfloor$ is an optimal dual solution for $d = 4$. This solution can be obtained by applying Theorem 3.5 as explained below. It can be shown that conv(S) is given by the inequalities

$$\begin{aligned} x_1 + 2x_2 + 3x_3 + 4x_4 &\leqslant 4 \\ x_2 + 2x_3 + 2x_4 &\leqslant 2 \\ x &\in R_+^4 \end{aligned}$$

and thus is generated from P by the functions $F^1(d) = d$ and $F^2(d) = \lfloor \frac{2}{3}d \rfloor$. For $c = (3 \quad 6 \quad 11 \quad 12)$, an optimal solution to the dual of the linear program $\max\{cx: x \in \operatorname{conv}(S)\}$ is $u = (3 \quad 1)$. Hence $F = 3F^1 + F^2$ is an optimal solution to the superadditive dual. Note that the optimal solution to IP is $x = (1 \quad 0 \quad 1 \quad 0)$. Since $F(1) = c_1$ and $F(3) = c_3$, the complementary slackness conditions are satisfied.

Theorem 3.6 is trivial for this example. We take $F(d) = 3\lfloor d \rfloor + 2\lfloor \frac{1}{3}d \rfloor$ for $d \in R^1$ and note that $F(d) = z(d)$ whenever $z(d)$ is finite.

Example 3.2. This integer program has the constraint set of the example presented in Section II.1.1.

$$\max 7x_1 + 2x_2$$
$$-x_1 + 2x_2 \leqslant 4$$
$$5x_1 + x_2 \leqslant 20$$
$$-2x_1 - 2x_2 \leqslant -7$$
$$x \in Z_+^2.$$

The superadditive dual is

$$\min F\begin{pmatrix} 4 \\ 20 \\ -7 \end{pmatrix}$$
$$F\begin{pmatrix} -1 \\ 5 \\ -2 \end{pmatrix} \geqslant 7, \quad F\begin{pmatrix} 2 \\ 1 \\ -2 \end{pmatrix} \geqslant 2$$
$$F(0) = 0, \quad F \text{ superadditive and nondecreasing.}$$

1. A dual feasible solution is $F(d) = \frac{3}{11}d_1 + \frac{16}{11}d_2 + 0d_3$. This is the linear solution obtained from an optimal dual solution to the linear programming relaxation. It yields the bound $z_{\text{IP}} \leqslant 30\frac{2}{11}$.
2. Rounding yields the better dual solution $F(d) = \lfloor \frac{3}{11}d_1 + \frac{16}{11}d_2 \rfloor$.
3. An optimal dual solution (see Section 4 of Chapter II.1) is given by the complicated function F_{12}.

Example 3.3. We reconsider the node-packing example of Section 1 (see Figure 1.1) with the objective function $c = (1 \quad 3 \quad 3 \quad 3 \quad 3 \quad 3)$. Solutions to the dual of the linear programming relaxation correspond to assigning nonnegative weights u_i to the cliques C_i so that for all $j \in V$ the sum of the weights over all cliques containing node j is at least c_j. Given the cliques $C_1 = \{1, 2, 6\}$, $C_2 = \{2, 3\}$, $C_3 = \{3, 4\}$, $C_4 = \{4, 5\}$, and $C_5 = \{5, 6\}$, we see that a feasible solution is $u = (1 \quad 2 \quad 1 \quad 2 \quad 2)$, which yields the superadditive dual feasible solution $F_1(d) = d_1 + 2d_2 + d_3 + 2d_4 + 2d_5$. Thus we obtain $z_{\text{IP}} \leqslant F_1(b) = 1 + 2 + 1 + 2 + 2 = 8$. Now the odd hole induced by the nodes $\{2, 3, \ldots, 6\}$ yields the valid inequality $x_2 + x_3 + \cdots + x_6 \leqslant 2$, which is generated by the superadditive function $F_2(d) = \lfloor \frac{1}{2}d_1 + \cdots + \frac{1}{2}d_5 \rfloor$. Note that $F_2(a_1) = 0$ and $F_2(a_j) = 1$ for $j > 1$. Hence a feasible dual solution is given by $F(d) = 3F_2(d) + d_1$. Since $F(b) = 3 \times 2 + 1 = 7$, we have $z_{\text{IP}} \leqslant 7$. To show that F is an optimal dual solution, we observe that $x^0 = (1 \quad 0 \quad 1 \quad 0 \quad 1 \quad 0)$ is a feasible node packing and $cx^0 = 7$.

Neither the extended value function of Theorem 3.6 nor the C–G function of Theorem 3.5 are useful for computing bounds. The value function is not available, even after the problem is solved, and the C–G function depends on having a linear inequality description of conv(S). Both functions, in a sense, provide more information than we need. The extended value function is optimal for all $d \in R^m$ for which IP is feasible, and the C–G function is a nonnegative linear combination of the same C–G functions for all c. Unfortunately, we do not know how to characterize a locally optimal function (e.g., one that is optimal only in a neighborhood of a particular b and c of interest). Thus for algorithmic purposes, we must restrict the class of dual feasible functions to computable ones that do not necessarily yield strong duality. In the following sections we will consider some classes of dual feasible functions that are useful algorithmically.

To complete this section, we state without proof the analogous result on superadditive duality for mixed-integer programs.

Theorem 3.7. *Let $T = \{x \in Z^n_+, y \in R^p_+: Ax + Gy \leqslant b\}$ and $z(b) = \max\{cx + hy: (x, y) \in T\}$. A strong dual to the mixed-integer programming problem is*

$$w = \min F(b)$$

$$F(a_j) \geqslant c_j \qquad \text{for } j \in N$$

$$\overline{F}(g_j) \geqslant h_j \qquad \text{for } j \in J$$

$$F \text{ nondecreasing and superadditive,} \qquad F(0) = 0$$

$$\overline{F}(d) = \lim_{\lambda \searrow 0_+} \frac{F(\lambda d)}{\lambda}.$$

If $z(0) = 0$ and $T \neq \emptyset$, then $z(b) = w$. If $z(0) = 0$ and $T = \emptyset$, the dual is infeasible or its objective value is unbounded. If $z(0) = \infty$, the dual is infeasible.

4. THE MAXIMUM-WEIGHT PATH FORMULATION AND SUPERADDITIVE DUALITY

Consider the integer programming problem $z_{\text{IP}} = \max\{cx: x \in S\}$, where $S = P \cap Z^n_+$, $P = \{x \in R^n_+: Ax \leqslant b\}$, and $(A, b) \geqslant 0$ with integral coefficients. By using the polyhedral characterization of superadditive functions developed in Section II.1.5, the superadditive dual problem can be written explicitly as a linear program:

$$w = \min F(b)$$

$$\text{(SDLP)} \qquad \begin{aligned} F(a_j) &\geqslant c_j && \text{for } j \in N \\ F(d_1) + F(d_2) - F(d_1 + d_2) &\leqslant 0 && \text{for } d_1, d_2, d_1 + d_2 \in D(b) \\ F(0) = 0,\ F(d) &\geqslant 0 && \text{for } d \in D(b), \end{aligned}$$

where $D(b) = \{d \in Z^m_+: d \leqslant b\}$ and F is a vector with $|D(b)|$ coordinates.

Example 4.1 (Example 2.1 continued). For the knapsack problem

$$\max 3x_1 + 6x_2 + 11x_3 + 12x_4$$

$$x_1 + 2x_2 + 3x_3 + 4x_4 \leqslant 4$$

$$x \in Z^4_+,$$

SDLP is

$$
\begin{aligned}
\min\ & F(4) \\
& F(1) \geqslant 3 \\
& F(2) \geqslant 6 \\
& F(3) \geqslant 11 \\
& F(4) \geqslant 12 \\
& 2F(1) - F(2) \leqslant 0 \\
& F(1) + F(2) - F(3) \leqslant 0 \\
& F(1) + F(3) - F(4) \leqslant 0 \\
& 2F(2) - F(4) \leqslant 0 \\
& F(0) = 0,\ F(\mathrm{d}) \geqslant 0 \quad \text{for } d \in D(4).
\end{aligned}
$$

Since SDLP is a linear program that is strongly dual to IP, we can use linear programming duality to express IP as a linear program. The purpose of this section is to study the structure of this dual pair of linear programs. We will discover that IP can be formulated as the linear program of finding a maximum-weight path joining two specified nodes in a directed graph and that SDLP can be interpreted as the dual of this maximum-weight path problem. We will establish the duality after formulating IP as a maximum-weight path problem.

To formulate IP as a maximum-weight path problem when S is an independence system, consider the digraph $\mathscr{D} = (V, \mathscr{A})$, where $V = D(b) = \{d \in Z_+^m: d \leqslant b\}$, and $\mathscr{A} = \mathscr{A}_1 \cup \mathscr{A}_2$, where

$$\mathscr{A}_1 = \{(d, d + a_j): d, d + a_j \in D(b), j \in N\} \quad \text{and} \quad \mathscr{A}_2 = \{(d, b): d \in D(b)\}.$$

Since $a_j > 0$ for all $j \in N$ and $b \geqslant d$ for all $d \in V$, $\mathscr{D}$ has no cycles.

The arc $e = (d, d + a_j)$ for $j \in N$ is called a *variable j arc* and is assigned weight $w_e = c_j$. For $d \neq b$, node d represents the subset of feasible solutions $S^*(d) = \{x \in Z_+^n: Ax = d\}$ since every path from node 0 to node d has the property that $\Sigma_{j \in N} a_j x_j = d$, where x_j is the number of variable j arcs in the path. The weight of any such path is $\Sigma_{j \in N} c_j x_j$. Arcs $(d, b) \in \mathscr{A}_2$ are called *slack arcs* and are assigned a weight of 0. If there is a j such that $d + a_j = b$, then a variable j arc and a slack arc join the same pair of nodes. Node b represents the set of all feasible solutions since every path from node 0 to node b has the property that $\Sigma_{j \in N} a_j x_j \leqslant b$, where x_j is the number of variable j arcs in the path. Hence any maximum-weight path from node 0 to node b corresponds to an optimal solution to IP.

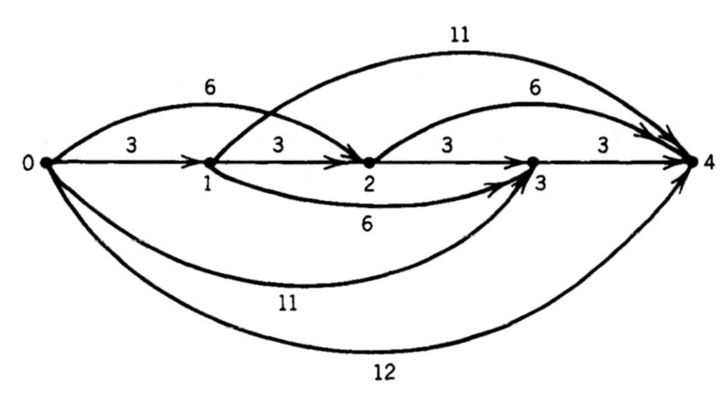

Figure 4.1

Table 4.1.

Path	x	Weight
[(0, 1), (1, 2), (2, 3), (3, 4)]	(4 0 0 0)	12
[(0, 1), (1, 2), (2, 4)]	(2 1 0 0)	12
[(0, 1), (1, 3), (3, 4)]	(2 1 0 0)	12
[(0, 2), (2, 3), (3, 4)]	(2 1 0 0)	12
[(0, 1), (1, 4)]	(1 0 1 0)	14
[(0, 3), (3, 4)]	(1 0 1 0)	14
[(0, 4)]	(0 0 0 1)	12

The digraph for Example 4.1 is shown in Figure 4.1. The slack arcs have been omitted. Actually they are unnecessary in the example since each slack arc is "parallel" to a non-slack arc of positive weight. The paths from node 0 to node 4, along with the corresponding feasible solutions, are given in Table 4.1. We see that there is at least one path corresponding to each feasible solution and that the weight of the path equals the value of the corresponding solution. In general, many paths correspond to the same feasible solution because each path is an ordering of the set of arcs that represent the solution.

We now give the standard flow formulation of the maximum-weight path representation of IP. Each arc is represented by a binary variable with the interpretation that an arc is in the solution if and only if the corresponding variable equals 1. The variable for the arc $(d, d + a_j)$ is $y_j(d)$, and the variable for a slack arc (d, b) is $y_0(d)$.

The constraints that are satisfied only by paths from node 0 to node b are:

i. Exactly one arc leaves node 0, that is,

$$-\sum_{j\in N} y_j(0) - y_0(0) = -1. \tag{4.1}$$

ii. Exactly one arc enters node b, that is,

$$\sum_{j\in N} y_j(b - a_j) + \sum_{\substack{d\in D(b)\\ d\neq b}} y_0(d) = 1. \tag{4.2}$$

iii. For $d \neq 0, b$, the number of arcs that enter node d equals the number that leave, that is,

$$\sum_{\{j\in N:\, d-a_j\geq 0\}} y_j(d - a_j) - \sum_{\{j\in N:\, d+a_j\leq b\}} y_j(d) - y_0(d) = 0 \quad \text{for } d \neq 0, b. \tag{4.3}$$

Note that (4.1) implies that the number of arcs entering each node $d \neq 0$ is already constrained to be 0 or 1. Finally note that (4.1) also implies that it suffices to allow the variables to be nonnegative integers. Thus we obtain an integer program representation of the maximum-weight path formulation of IP given by

$$z_{MP} = \max \sum_{\{j\in N, d\in D(b):\, d+a_j\leq b\}} c_j y_j(d) \tag{MP}$$

subject to (4.1)–(4.3) and

$$y_j(d) \in Z_+^1 \quad \text{for all } j \text{ and } d. \tag{4.4}$$

We have $z_{MP} = z_{IP}$ as explained above; also, $z_{IP} = w$ by strong duality (Theorem 3.3).

We now consider the dual of the linear programming relaxation of MP, which is given by

$$
\text{(4.5)} \quad
\begin{aligned}
z_{LP} = \min\ & (u(b) - u(0)) \\
& u(d + a_j) - u(d) \geqslant c_j \quad \text{for } j \in N,\ d \in D(b),\ d + a_j \leqslant b \\
& u(b) - u(d) \geqslant 0 \qquad\qquad \text{for } d \in D(b),\ d \neq b,
\end{aligned}
$$

where $u(d)$ is the dual variable for the node d constraint. Note that if $\overline{u}$ is a feasible solution to (4.5) with $\overline{u}(0) \neq 0$, then so is u^* where $u^*(\mathrm{d}) = \overline{u}(d) - \overline{u}(0)$ for all $d \in D(b)$. Since $u^*(b) = u^*(b) - u^*(0) = \overline{u}(b) - \overline{u}(0)$, we can set $u(0) = 0$.

In Example 4.1, (4.5) is

$$
\begin{aligned}
\min\ & u(4) \\
& u(1) && && && \geqslant 3 \\
& && u(2) && && \geqslant 6 \\
& && && u(3) && \geqslant 11 \\
& && && && u(4) \geqslant 12 \\
& -u(1) + u(2) && && && \geqslant 3 \\
& -u(1) && + u(3) && && \geqslant 6 \\
& -u(1) && && + u(4) && \geqslant 11 \\
& && - u(2) + u(3) && && \geqslant 3 \\
& && - u(2) && + u(4) && \geqslant 6 \\
& && && - u(3) + u(4) && \geqslant 3.
\end{aligned}
$$

We have omitted $u(0) = 0$ and the constraints $u(b) - u(d) \geqslant 0$, which are superfluous here.

Now we return to SDLP, introduced at the beginning of this section. Consider the constraints

$$
\text{(4.6)} \qquad F(d_1) + F(d_2) - F(d_1 + d_2) \leqslant 0.
$$

We are going to relax (4.6) in three different ways, depending on d_1 and d_2.

i. If $d_1 = a_j$ for some $j \in N$, then we replace (4.6) by $F(a_j + d_2) - F(d_2) \geqslant c_j$. This is a relaxation since $F(a_j) \geqslant c_j$.
ii. If $d_1 + d_2 = b$, then we replace (4.6) by $F(d_2) - F(b) \leqslant 0$. This is a relaxation since $F(d_1) \geqslant 0$.
iii. Otherwise we drop (4.6).

Finally, we omit the nonnegativity constraints.

This yields the relaxation of SDLP:

$$
\begin{array}{llll}
& w_R = \min F(b) & & \\
& & F(a_j) \geqslant c_j & \text{for } j \in N \\
(4.7) & & F(d + a_j) - F(d) \geqslant c_j & \text{for } j \in N, d \in D(b), d + a_j \leqslant b \\
& & F(b) - F(d) \geqslant 0 & \text{for } d \in D(b), d \neq b \\
& & F(0) = 0. &
\end{array}
$$

By relaxation, $w_R \leqslant w$.

Now observe that (4.5) and (4.7) are identical. Hence $w_R = z_{\text{LP}}$. Thus

$$z_{\text{LP}} \geqslant z_{\text{MP}} = z_{\text{IP}} = w \geqslant w_R = z_{\text{LP}},$$

and all of these objective values are equal.

In conclusion, we have interpreted the superadditive dual of an integer program whose constraints generate an independence system as the linear programming dual of a maximum-weight path formulation of the integer program.

The maximum-weight path formulation is of limited use in computation because of the size of the digraph. The number of nodes $|D(b)|$ grows exponentially with the number of constraints m and, even for fixed m, grows linearly with the size of the coefficients of $b \in Z_+^m$. Generally, its use is restricted to knapsack problems that have constraint coefficients of modest size.

Despite these practical limitations, the fundamental idea is applicable to any integer program. In particular, any integer programming constraint set $S = \{x \in Z_+^n: Ax \leqslant b\}$ can be represented by a digraph with the property that directed walks from node 0 to node $d \in Z^m$ correspond to solutions with $Ax = d$. As before, an arc $e = (d, d + a_j)$ is a variable j arc and is assigned weight $w_e = c_j$. However, when matrix A has negative coefficients, we must determine a finite subset $\hat{D}(b) \subset Z^m$ to which we can restrict d. Note that $\hat{D}(b) = D(b)$ does not suffice since $d + a_j \geqslant d$ is no longer true.

To determine $\hat{D}(b)$, we use the result that if $S \neq \emptyset$ there is an ω which depends on A, b, and c, such that $x_j \leqslant \omega$ for all $j \in N$ in some optimal solution (see Theorem 4.1 of Section I.5.4). Then $x_j \leqslant \omega$ for $j \in N$ implies $d^- \leqslant Ax \leqslant d^+$, where $d_i^- = \omega \sum_{j \in N} \min(0, a_{ij})$ and $d_i^+ = \omega \sum_{j \in N} \max(0, a_{ij})$ for $i = 1, \ldots, m$. Thus $\hat{D}(b) = \{d \in Z^m: d^- \leqslant d \leqslant d^+\}$.

5. MODULAR ARITHMETIC AND THE GROUP PROBLEM

In this section, we consider relaxations of the maximum-weight path formulation that reduce the size of the digraph. In fact, we will be able to choose the number of nodes, although the quality of the bounds produced by the relaxation will generally deteriorate as the digraph gets smaller.

Consider the set $S = \{x \in Z_+^n: \sum_{j \in N} a_j x_j = b\}$, where $a_j \in Z^1$ for $j \in N$ and $b \in Z^1$. Suppose $k \in Z_+^1$ and we relax S to

$$S_k = \left\{x \in Z_+^n: \sum_{j \in N} a_j x_j = b + kw,\ w \in Z^1\right\}.$$

This means that multiples of k can be subtracted from each coefficient and the right-hand side of the original constraint. Thus $x \in S_k$ if and only if $x \in Z_+^n$ and for each $(\lambda_0, \lambda_1, \ldots, \lambda_n) \in Z^{n+1}$ there exists a $w' \in Z^1$ such that

$$\sum_{j \in N} (a_j - k\lambda_j)x_j = (b - k\lambda_0) + kw'.$$

By choosing $\lambda_j = \lfloor a_j/k \rfloor$ for $j \in N$ and $\lambda_0 = \lfloor b/k \rfloor$, we have that $x \in S_k$ if and only if $x \in Z_+^n$ satisfies

$$\sum_{j \in N} \phi_k(a_j)x_j = \phi_k(b) \qquad \text{(modulo } k),$$

where $\phi_k(d) = d - k \lfloor d/k \rfloor$; that is, $\phi_k(d) = d(\text{mod } k)$ is the remainder when d is divided by k.

We can now represent S_k by directed walks in a digraph that contain only k nodes. The graph is $\mathcal{D}_k = (V_k, \mathcal{A}_k)$, where

$$V_k = \{\phi_k(d) \colon d \in Z^1\} = \{0, 1, \ldots, k-1\} \quad \text{and} \quad \mathcal{A}_k = \{(d, \phi_k(d + a_j)) \colon d \in V_k, j \in N\}.$$

The arc $e_j(d) = (d, \phi_k(d + a_j))$ is called a *variable j arc*. A directed walk in $\mathcal{D}_k$ from node 0 to node $\phi_k(b)$ with $\bar{x}_j$ variable j arcs generates a feasible solution $\bar{x} \in S_k$; conversely, any $x \in S_k$ generates directed walks from node 0 to node $\phi_k(b)$.

Any directed walk from node 0 to node $\phi_k(b)$ can be decomposed into a dipath from node 0 to node $\phi_k(b)$ and (possibly) directed cycles. Correspondingly, any $x \in S_k$ can be decomposed into $x^* + \Sigma_{i=1}^t x^i$, where $\phi_k(\Sigma_{j \in N} a_j x_j^*) = \phi_k(b)$ and x^* generates dipaths from node 0 to node $\phi_k(b)$ and where $\phi_k(\Sigma_{j \in N} a_j x_j^i) = 0$ and x^i generates directed cycles, for $i = 1, \ldots, t$.

Later in this section, we will consider the problem of finding a maximum-weight dipath from node 0 to node $\phi_k(b)$ in $\mathcal{D}_k$, where the arcs have nonpositive weights. Hence the solution will be a dipath corresponding to an $x^* \in S_k$. However, there is no guarantee that dipaths correspond to elements of S; that is, it could be the case that some, or even all, elements of S correspond to directed walks from node 0 to node $\phi_k(b)$ that contain directed cycles.

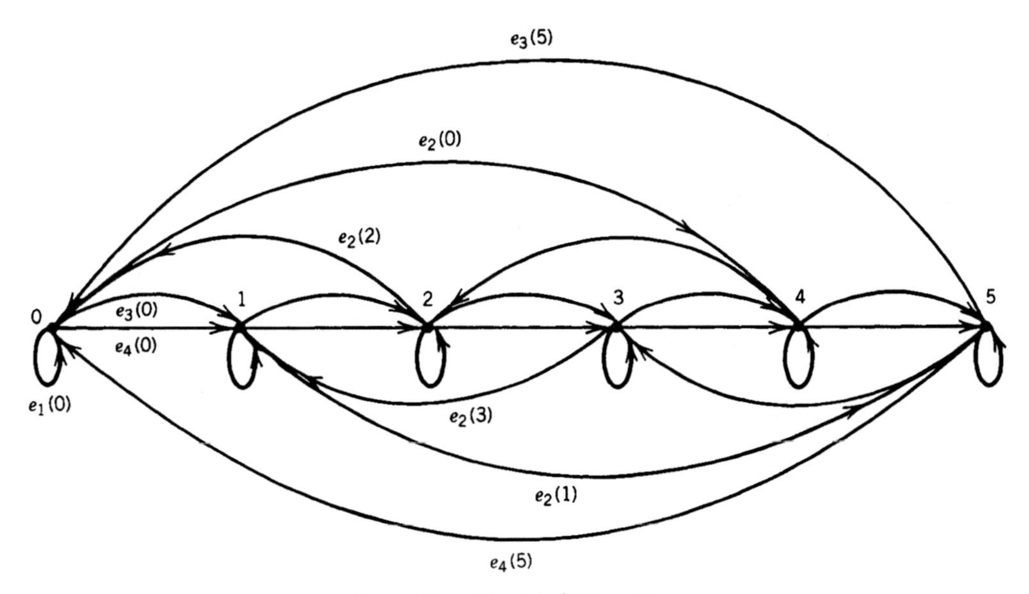

Figure 5.1. $e_j(d) = (d, \phi_6(d + a_j))$.

Example 5.1. $S = \{x \in Z_+^4: 78x_1 - 68x_2 + 37x_3 + x_4 = 141\}$ and $k = 6$. We have $\phi_6(a_1) = 0$, $\phi_6(a_2) = 4$, $\phi_6(a_3) = \phi_6(a_4) = 1$, and $\phi_6(b) = 3$. Hence $S_6 = \{x \in Z_+^4: 0x_1 + 4x_2 + x_3 + x_4 = 3 \pmod 6\}$.

The digraph $\mathscr{D}_6$ is shown in Figure 5.1. Table 5.1 gives all of the solutions corresponding to dipaths, and Table 5.2 gives some directed cycles. More cycles can be generated by replacing any of the variable 3 arcs by variable 4 arcs.

Any x in Table 5.1 plus a nonnegative integer multiple of an x in Table 5.2 is in S_6; for example,

$$(4\ \ 3\ \ 1\ \ 2) = (0\ \ 0\ \ 1\ \ 2) + 4\,(1\ \ 0\ \ 0\ \ 0) + (0\ 3\ 0\ 0).$$

Any such x is in S if and only if $\Sigma_{j\in N}\, a_j x_j = b$; for example,

$$(9\ \ 11\ \ 5\ \ 2) = (0\ \ 0\ \ 1\ \ 2) + 9\,(1\ \ 0\ \ 0\ \ 0) + 3\,(0\ \ 3\ \ 0\ \ 0) + 2\,(0\ \ 1\ \ 2\ \ 0).$$

Now we introduce weights on the arcs of $\mathscr{D}_k$. For any positive integer k and $\beta \in R^1$, the problem

$$\begin{aligned} z_k(b) = \max \sum_{j\in N} c_j x_j - \beta w \\ \sum_{j\in N} a_j x_j - kw = b \\ x \in Z_+^n, \qquad w \in Z^1 \end{aligned} \tag{5.1}$$

is a relaxation of our original problem

$$\text{(IP)} \qquad z(b) = \max\left\{ \sum_{j\in N} c_j x_j: \sum_{j\in N} a_j x_j = b,\, x \in Z_+^n \right\}.$$

This is true because any feasible solution to IP can be extended to a feasible solution to (5.1) of the same value by putting $w = 0$. We now show how (5.1) can be formulated as a maximum-weight path problem from node 0 to node $\phi_k(b)$ in the digraph $\mathscr{D}_k$.

We can eliminate w from the objective function by substituting

$$w = \Big(\sum_{j\in N} a_j x_j - b\Big)/k$$

and we have already shown how to describe the relaxed solution set S_k. Hence (5.1) can be reformulated as the *group problem*

Table 5.1.

Acyclic Paths from 0 to 3	x
$e_3(0), e_3(1), e_3(2)$	(0 0 3 0)
$e_3(0), e_3(1), e_4(2)$	(0 0 2 1)
$e_3(0), e_4(1), e_4(2)$	(0 0 1 2)
$e_4(0), e_4(1), e_4(2)$	(0 0 0 3)
$e_2(0), e_2(4), e_3(2)$	(0 2 1 0)
$e_2(0), e_2(4), e_4(2)$	(0 2 0 1)

Table 5.2.

Simple Cycles	x
$e_1(0)$	(1 0 0 0)
$e_2(0)$, $e_2(4)$, $e_2(2)$	(0 3 0 0)
$e_2(0)$, $e_3(4)$, $e_3(5)$	(0 1 2 0)
$e_3(0)$, $e_3(1)$, $e_3(2)$, $e_3(3)$, $e_3(4)$, $e_3(5)$	(0 0 6 0)

$$z_k(b) = \frac{\beta b}{k} + \max \sum_{j \in N} \left(c_j - \frac{\beta a_j}{k}\right) x_j$$

$$\text{(GP)} \qquad \sum_{j \in N} \phi_k(a_j) x_j = \phi_k(b) \qquad (\text{mod } k)$$

$$x \in Z_+^n.$$

We use the term *group* because of the modulo addition in the constraint, which is equivalent to addition in the cyclic abelian group of order k.

If $c_j - \beta a_j/k > 0$, we can choose x_j so that $\phi_k(a_j)x_j = 0$ and $(c_j - \beta a_j/k)x_j$ is arbitrarily large. Thus if $S_k \neq \emptyset$, we can choose x so that GP has an unbounded optimal value. Hence we impose the condition $c_j - \beta a_j/k \leqslant 0$ for $j \in N$ or

$$\beta_1 = \max\left\{\frac{c_j}{a_j}: j \in N, a_j > 0\right\} \leqslant \frac{\beta}{k} \leqslant \min\left\{\frac{c_j}{a_j}: j \in N, a_j < 0\right\} = \beta_2,$$

which implies that $z_k(b) \leqslant \beta b/k$. Thus we can restate GP as

$$z_k(b) = ub + \max \sum_{j \in N} (c_j - ua_j) x_j$$

$$\sum_{j \in N} \phi_k(a_j) x_j = \phi_k(b) \qquad (\text{mod } k)$$

$$x \in Z_+^n,$$

where $u = \beta/k$ and $\beta_1 \leqslant u \leqslant \beta_2$. Note that $\beta_1 \leqslant u \leqslant \beta_2$ if and only if u is a dual feasible solution to the linear programming relaxation of (5.1). Moreover, the term ub, which is independent of x, is minimized by an optimal dual solution to the linear programming relaxation of (5.1); that is, $u = \beta_1$ if $b > 0$. With $u = \beta_1$, we obtain

$$z_k(b) = z_{\text{LP}}(b) - \min \sum_{j \in N} -\bar{c}_j x_j$$

$$\sum_{j \in N} \phi_k(a_j) x_j = \phi_k(b) \qquad (\text{mod } k)$$

$$x \in Z_+^n,$$

where $\bar{c}_j = c_j - \beta_1 a_j \leqslant 0$ for $j \in N$. Thus $z_k(b) \leqslant z_{\text{LP}}(b)$, and we also observe that GP yields a minimum-weight correction to the linear programming relaxation subject to the constraint $x \in S_k$.

The correction term

$$\psi_k(b) = -\min\left\{\sum_{j\in N} -\bar{c}_j x_j : x \in S_k\right\}$$

is the same for all $d \in \phi_k^{-1}(b)$. Hence $z_k(b)$ is the sum of a linear term in b and a correction term that is cyclic with period k.

An interesting observation is that $\psi_k(d)$ is superadditive; that is, for any $d^1, d^2 \in Z^1$ we obtain

$$\psi_k(d^1) + \psi_k(d^2) = \psi_k(d^1) + \psi_k(d^1 + d^2 - d^1) \leqslant \psi_k(d^1 + d^2).$$

The inequality holds because the left-hand side can be interpreted as the weight of a maximum-weight path from node 0 to node $d^1 + d^2$ that is constrained to contain node d^1.

Example 5.1 (continued). Suppose $c = (4 \quad -4 \quad 1 \quad 0)$. Then $\frac{2}{39} = \beta_1 = c_1/a_1 \leqslant \beta/6 \leqslant c_2/a_2 = \beta_2 = \frac{1}{17}$. Since $b > 0$, we choose $u = \beta/6 = \beta_1$ and obtain the group problem

$$z_6(141) = z_{\text{LP}}(141) + \frac{1}{39}\max(0x_1 - 20x_2 - 35x_3 - 2x_4)$$

$$0x_1 + 4x_2 + x_3 + x_4 = 3 \qquad (\text{mod } 6)$$

$$x \in Z_+^4.$$

To find a maximum-weight path from node 0 to node 3 in $\mathcal{D}_6$, we can eliminate the loop arcs $e_1(d)$ since $\bar{c}_1 = 0$, and we can also eliminate the $e_3(d)$ arcs since they are parallel to the $e_4(d)$ arcs and $\bar{c}_3 < \bar{c}_4$. This yields the digraph shown in Figure 5.2, where the number on each arc is its weight times 39.

A path of maximum weight is $(e_4(0), e_4(1), e_4(2))$, corresponding to the solution $x = (0 \quad 0 \quad 0 \quad 3)$. Since $\Sigma_{j\in N} a_j x_j = 3$, it follows that $x \notin S$. The weight of this path is $\psi_6(b) = -\frac{6}{39}$, so we obtain the upper bound of $z_6(b) = z_{\text{LP}}(b) + \psi_6(b) = [2(141) - 6]/39 = 7\frac{1}{13}$.

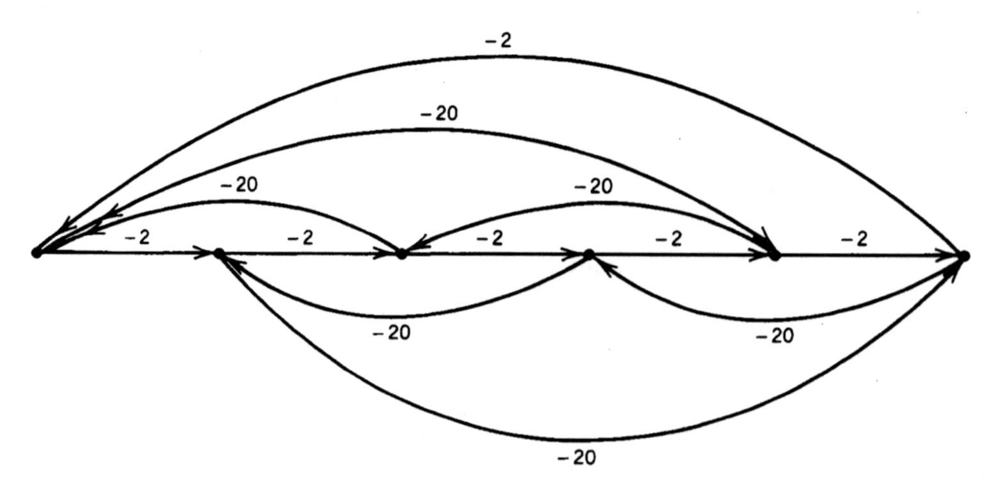

Figure 5.2

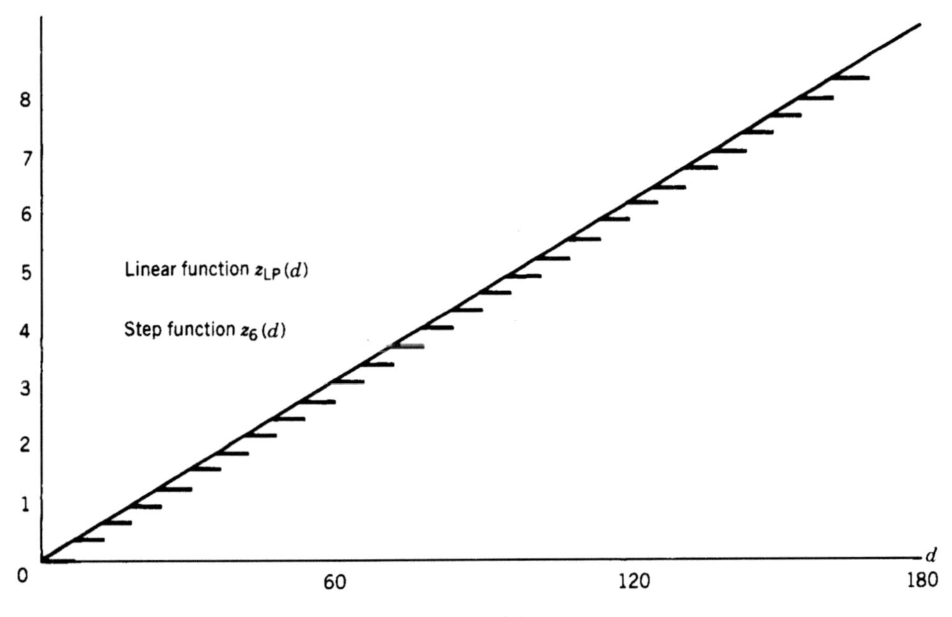

Figure 5.3

It is clear from the digraph of Figure 5.2 that, for any $d \in Z^1$, only the variable 4 arcs are used in a maximum-weight path from node 0 to node $\phi_6(d)$. Thus $\psi_6(d) = -\frac{2}{39}\phi_6(d)$ for all $d \in Z^1$. The functions $z_{LP}(d)$ and $z_6(d)$ are shown in Figure 5.3 for $d \in Z^1_+$.

So far, the development of the group problem has been done for an arbitrary positive integer k. Now we consider a meaningful choice of k that is motivated by trying to enhance the possibility of an optimal solution to GP being feasible to IP.

Suppose $b \in Z^1_+$ and $c_1/a_1 = \max\{c_j/a_j\colon a_j > 0, j \in N\}$. With $k = a_1$, GP can be restated as

$$
(5.2) \qquad
\begin{aligned}
z_{a_1}(b) = z_{LP}(b) + \max \sum_{j \in N \setminus \{1\}} \left(c_j - \frac{c_1 a_j}{a_1}\right) x_j \\
\phi_{a_1}\left(\sum_{j \in N \setminus \{1\}} a_j x_j\right) = \phi_{a_1}(b) \\
x_j \in Z^1_+, \qquad j = 2, \ldots, n.
\end{aligned}
$$

Note that any feasible solution to (5.2) yields an integer value for x_1. This is true since $\sum_{j \in N \setminus \{1\}} a_j x_j = t a_1 + \phi_{a_1}(b)$ for some $t \in Z^1$. Thus

$$
x_1 = \frac{1}{a_1}\left(b - \sum_{j \in N \setminus \{1\}} a_j x_j\right) = \frac{b - \phi_{a_1}(b)}{a_1} - t,
$$

which is integer for all $b \in Z^1$ since $\phi_{a_1}(b - \phi_{a_1}(b) = 0$. However, x_1 may be negative. This should not be surprising since, in this case, (5.1) is

$$z_{a_1}(b) = \max \sum_{j\in N} c_j x_j - c_1 w$$
$$\sum_{j\in N} a_j x_j - a_1 w = b$$
$$x \in Z^n_+, w \in Z^1,$$

which is the same as

$$z_{a_1}(b) = \max \sum_{j\in N} c_j x_j$$
$$\sum_{j\in N} a_j x_j = b$$
$$x_1 \in Z^1, x_j \in Z^1_+ \text{ for } j = 2, \dots, n.$$

When can we be sure that $\sum_{j\in N\setminus\{1\}} a_j x_j$ will be small enough to guarantee $x_1 \geqslant 0$? To answer this question, we use the fact that x_j for $j = 2, \dots, n$ is determined by solving a maximum-weight path problem on a graph with a_1 nodes and nonpositive weights on the arcs. Hence there is a maximum-weight path with no more than $a_1 - 1$ arcs; that is, the corresponding solution $\{x_j^0\}$ satisfies $\sum_{j\in N\setminus\{1\}} x_j^0 \leqslant a_1 - 1$. Thus

$$\sum_{j\in N\setminus\{1\}} a_j x_j^0 \leqslant (a_1 - 1)\bar{a},$$

where $\bar{a} = \max\{a_j: j \in N\setminus\{1\}\}$, so that $x_1^0 \geqslant (1/a_1)(b - (a_1 - 1)\bar{a})$. Consequently if $b \geqslant (a_1 - 1)\bar{a}$ and $k = a_1$, then GP yields an optimal solution to IP. Thus the relaxation GP is asymptotically exact in the sense that for suitably large b we obtain $z_{a_1}(b) = z(b)$.

Example 5.1 (continued). Suppose $u = \beta_1 = c_1/a_1$ and $k = a_1 = 78$. Here the digraph is too large to draw, but an optimal solution is easy to deduce. We have $\phi_{78}(a_2) = 10$, $\phi_{78}(a_3) = 37$, $\phi_{78}(a_4) = 1$, and $\phi_{78}(b) = 63$. With $(\bar{c}_2 \quad \bar{c}_3 \quad \bar{c}_4) = -\frac{1}{39}(20 \quad 35 \quad 2)$, an optimal solution is $x_2^0 = 0$, $x_3^0 = 1$, $x_4^0 = 26$, which yields

$$x_1^0 = \frac{1}{a_1}\left(b - \sum_{j\in N\setminus\{1\}} a_j x_j^0\right) = \frac{1}{78}(141 - 37 - 26) = 1.$$

Hence we obtain a feasible, and thus optimal, solution to IP given by $x^0 = (1 \quad 0 \quad 1 \quad 26)$. Note that $141 < 77 \cdot 37 = (a_1 - 1)\bar{a}$; that is, the condition $b \geqslant (a_1 - 1)\bar{a}$ is by no means necessary for an optimal solution of the group problem to solve IP.

The approach we have taken here generalizes straightforwardly to integer programs with more than one constraint. Consider an equality-constrained version of IP:

(IP) $$\max\{cx: Ax = b, x \in Z^n_+\},$$

where (A, b) is an integral $m \times (n + 1)$ matrix and c is an integral n-vector. We relax IP to the so-called *group problem*

(GP)
$$z_K(b) = \max cx - \beta w$$
$$Ax - Kw = b$$
$$x \in Z^n_+, w \in Z^p,$$

where $K = (k_1, \ldots, k_p)$ is an $m \times p$ integer matrix with $p \leqslant m$ and β is a p-vector. Our goal is to determine the canonical form or simplest possible path representation of GP.

Suppose that IP has bounded optimum value. To ensure that GP also has bounded optimum value, recall that a feasible integer program is bounded only if its linear programming relaxation is dual feasible. Thus we require that the dual of the linear programming relaxation of GP has a feasible solution, that is, there exists a $u \in R^m$ such that $uA \geqslant c$ and $uK = \beta$. Given such a u, we can rewrite GP as

$$z_K(b) = ub + \max(c - uA)x$$
$$x \in S_K(b),$$

where $S_K(b) = \{x \in Z_+^n: Ax - Kw = b \text{ for some } w \in Z^p\}$ and $c - uA \leqslant 0$.

To obtain a unique canonical representation of the maximum-weight path representation of GP, we use the Smith normal form of matrix K (see Theorem 4.11 of Section I.7.4), which is stated here in greater generality. We say that a square integral matrix R is *unimodular* if $|\det R| = 1$.

Theorem 5.1. *(Smith Normal Form). Given an $m \times p$ integer matrix K of rank $p \leqslant m$, there exist unimodular integer matrices R and C, where R is $m \times m$ and C is $p \times p$ such that $RKC = \Delta$. Matrix Δ is of the form $\delta_{ij} = 0$ for all $i \neq j$ and $i > p$; and the elements $\delta_{ii} = \delta_i$ for $i = 1, \ldots, p$, are positive integers such that δ_i is a divisor of δ_{i+1} for $i = 1, \ldots, p-1$. Moreover, matrix Δ is unique.*

For $d \in Z^m$, let $\phi_\Delta(d) = \tilde{d}$, where $\tilde{d}_i = d_i(\text{mod } \delta_i)$ for $i = 1, \ldots, p$, and $\tilde{d}_i = d_i$ for $i > p$. We can now give a canonical form of $S_K(b)$ and, hence, of GP.

Theorem 5.2. *A canonical representation of $S_K(b)$ is given by:*

$$S_K(b) = \left\{x \in Z_+^n: \sum_{j \in N} \phi_\Delta(Ra_j)x_j = \phi_\Delta(Rb) \quad (\text{mod } \Delta)\right\};$$

that is, for $i = 1, \ldots, p$, the ith equation must be satisfied mod δ_i, *and for $i > p$ the equation is an ordinary equality.*

Proof. We have

$$S_K(b) = \{x \in Z_+^n: Ax - Kw = b \text{ for some } w \in Z^p\}$$
$$= \{x \in Z_+^n: RAx - RKw = Rb \text{ for some } w \in Z^p\}$$

since R is a nonsingular matrix.

Now if $Cw' = w$, where C is a unimodular integer matrix, then $w \in Z^p$ if and only if $w' \in Z^p$. Hence

$$S_K(b) = \{x \in Z_+^n: RAx - RKCw' = Rb \text{ for some } w' \in Z^p\}$$
$$= \{x \in Z_+^n: RAx = Rb + \Delta w' \text{ for some } w' \in Z^p\}$$
$$= \{x \in Z_+^n: \phi_\Delta(RA)x = \phi_\Delta(Rb)\ (\text{mod } \Delta)\}.$$

■

Therefore GP can be stated as

$$z_K(b) = ub + \max \sum_{j\in N} (c_j - ua_j)x_j$$

(GP)
$$\sum_{j\in N} \phi_\Delta(Ra_j)x_j = \phi_\Delta(b) \pmod{\Delta}$$

$$x \in Z^n_+.$$

To make each of the m equations modular, we choose K to be $m \times m$ and nonsingular. Then Δ is an $m \times m$ diagonal matrix with $\delta_i \in Z^1_+$ for $i = 1, \ldots, m$ and $\prod_{i=1}^m \delta_i = |\det K|$.

Corollary 5.3. *If K is an $m \times m$ nonsingular integer matrix,* GP *is a maximum-weight path problem on a digraph with $\prod_{i=1}^m \delta_i = |\det K|$ nodes.*

Here $\mathcal{D}_K = (V_K, \mathcal{A}_K)$, where

$$V_K = \{\phi_\Delta(d): d \in R^m\} = \{d \in Z^m_+: d_i < \delta_i \text{ for } i = 1, \ldots, m\} \quad \text{and}$$

$$\mathcal{A}_K = \{(d, \phi_\Delta(d + Ra_j)): d \in V_K, j \in N\}.$$

Note that if $\delta_i = 1$, the ith modular equation holds trivially for all $x \in Z^n_+$ and can be omitted from the formulation. Correspondingly, in the digraph $\mathcal{D}_K$, if $\delta_i = 1$, then $d_i = 0$ for all $d \in V_K$. In particular, if $\delta_m = |\det K|$, we obtain a cyclic group as in the case of a single constraint problem.

By choosing $u = u^0$, an optimal solution to the dual of the linear programming relaxation of IP, we obtain the minimum value of $ub = u^0 b = z_{\text{LP}}(b)$. In this case, $z_K(b) = z_{\text{LP}}(b) + \psi_K(b)$, where

$$\psi_K(b) = -\min\left\{\sum_{j\in N} -(c_j - u^0 a_j)x_j: x \in S_K(b)\right\}.$$

So as before, $z_K(b) \leq z_{\text{LP}}(b)$ and GP yields a minimum-weight correction to the linear programming relaxation subject to $x \in S_K$. We also see that $\psi_K(d)$ is a cyclic function; that is, $\psi_K(d) = \psi_K(d + Kw)$ for all $d, w \in Z^m$.

By using the same argument as in the single constraint case, we obtain the following corollary.

Corollary 5.4. *$\psi_K(d)$ is superadditive for $d \in Z^m$.*

If, in addition, we choose K as an optimal basis matrix for the linear programming relaxation of IP, we may enhance the possibility of obtaining a feasible solution to IP. Suppose $A = (A_B, A_N)$ (where A_B is $m \times m$ and nonsingular), $x = (x_B, x_N)$, and $c = (c_B, c_N)$. Let $K = A_B$ and suppose that $A_B^{-1}b \geq 0$ and $c_N - c_B A_B^{-1} A_N \leq 0$. If Δ is the Smith Normal Form of A_B, then GP is

$$z_{A_B}(b) = z_{\text{LP}}(b) + \max(c_N - c_B A_B^{-1} A_N)x_N$$

(5.3)
$$\phi_\Delta(RA_N)x_N = \phi_\Delta(Rb) \pmod{\Delta}$$

$$x_N \in Z^{n-m}_+.$$

Problem (5.3) is the group problem originally considered by Gomory. Note that if x_N^0 is an optimal solution to (5.3), it yields an optimal solution to IP if $x_B^0 = A_B^{-1}(b - A_N x_N^0) \geq 0$.

We have seen that choosing $K = A_B$ is equivalent to dropping nonnegativity on x_B, which leads to the following corollary.

Corollary 5.5. *The group problem (5.3) is equivalent to the integer program*

$$\begin{aligned} z_{A_B}(b) = \max\ & c_B x_B + c_N x_N \\ & A_B x_B + A_N x_N = b \\ & x_B \in Z^m,\ x_N \in Z_+^{n-m}, \end{aligned} \tag{5.4}$$

where A_B is an optimal basis matrix for the linear programming relaxation of IP.

Note that a Gomory group problem can be generated from any dual feasible basis matrix A_B and that Corollary 5.5 holds for any such basis. Dual feasibility (i.e., $c_N - c_B A_B^{-1} A_N \leqslant 0$) is necessary to bound $z_{A_B}(b)$ from above.

Gomory began with problem (5.4) and derived the canonical form (5.3). The motivation for considering problem (5.4) is that at an optimal solution to the linear programming relaxation of IP, the constraints $x_B \geqslant 0$ are inactive—if there is degeneracy they may be tight. Thus, there is some hope that they will be inactive in an optimal solution to IP. In fact when b is "suitably large", this is true. This asymptotic behaviour of GP is given by the following proposition.

Theorem 5.6. *Let A_B be any dual feasible basis to the linear programming relaxation of* IP *and let $\omega = \max_{i,j} |(A_B^{-1} A_N)_{ij}|$. The group problem (5.3) defined by A_B solves* IP *for all $b \in Z^m$ such that $A_B^{-1} b \geqslant \omega\ |\det A_B|\ 1$, where 1 is the vector of all 1's. An optimal solution to* IP *is given by $x_B^0 = A_B^{-1}(b - A_N x_N^0)$ for some x_N^0 that is an optimal solution to (5.3).*

Proof. It suffices to show that there exists an optimal solution x_N^0 to (5.3) such that $x_B^0 = A_B^{-1}(b - A_N x_N^0) \geqslant 0$. Since GP is a maximum-weight path problem on a digraph with $|\det A_B|$ nodes, it follows that there is an optimal solution x_N^0 such that $\Sigma_j\, x_{N_j}^0 < |\det A_B|$. Hence $A_B^{-1} A_N x_N^0 < \omega\ |\det A_B|\ 1$. Since, by assumption, $A_B^{-1} b \geqslant \omega\ |\det A_B|\ 1$, it follows that $x_B^0 \geqslant 0$. ■

Example 5.2

$$\begin{aligned} \max\ & 7x_1 + 2x_2 \\ & -x_1 + 2x_2 + x_3 = 2 \\ & 5x_1 + x_2 + x_4 = 19 \\ & -2x_1 - 2x_2 + x_5 = -5 \\ & x \in Z_+^5. \end{aligned}$$

An optimal LP basis is given by $A_B = (a_1, a_2, a_5)$, and $u = c_B A_B^{-1} = (\frac{3}{11}\ \ \frac{16}{11}\ \ 0)$. It is readily checked that the Smith Normal Form of A_B is

$$\Delta = R A_B C = \begin{pmatrix} 1 & 0 & 0 \\ 0 & 1 & 0 \\ 0 & 0 & 11 \end{pmatrix}$$

with

$$R = \begin{pmatrix} -1 & 0 & 0 \\ -2 & 0 & 1 \\ 5 & 1 & 0 \end{pmatrix} \quad \text{and} \quad C = \begin{pmatrix} 1 & 0 & 2 \\ 0 & 0 & 1 \\ 0 & 1 & 6 \end{pmatrix}.$$

We have

$$RA = \begin{pmatrix} 1 & -2 & -1 & 0 & 0 \\ 0 & -6 & -2 & 0 & 1 \\ 0 & 11 & 5 & 1 & 0 \end{pmatrix} \quad \text{and} \quad Rb = \begin{pmatrix} -2 \\ -9 \\ 29 \end{pmatrix}.$$

Since $\delta_1 = \delta_2 = 1$, only the bottom row of $R(A, b)$ is needed to define the group problem. Because $c - uA = (0 \;\; 0 \;\; -\frac{3}{11} \;\; -\frac{16}{11} \;\; 0)$, the group problem is

$$z_B(b) = \frac{310}{11} + \max -\frac{3}{11}x_3 - \frac{16}{11}x_4$$

$$5x_3 + x_4 = 7 \qquad (\text{mod } 11)$$

$$(x_3, x_4) \in Z_+^2.$$

This problem can be solved on a digraph with 11 nodes, and the optimal solution is $x_3^0 = 8$, $x_4^0 = 0$. Thus

$$x_B^0 = \begin{pmatrix} x_1^0 \\ x_2^0 \\ x_5^0 \end{pmatrix} = A_B^{-1}b - 8A_B^{-1}a_3 = \frac{1}{11}\begin{pmatrix} 36 \\ 29 \\ 75 \end{pmatrix} - \frac{8}{11}\begin{pmatrix} -1 \\ 5 \\ 8 \end{pmatrix} = \begin{pmatrix} 4 \\ -1 \\ 1 \end{pmatrix}.$$

Since $x_2^0 < 0$, it follows that x^0 is not feasible to IP. Had we taken $b = \begin{pmatrix} 4 \\ 20 \\ -7 \end{pmatrix}$, we would have obtained the same group problem. But in this case we have

$$x_B^0 = A_B^{-1}(b - 8a_3) = \frac{1}{11}\begin{pmatrix} 36 \\ 40 \\ 75 \end{pmatrix} - \frac{8}{11}\begin{pmatrix} -1 \\ 5 \\ 8 \end{pmatrix} = \begin{pmatrix} 4 \\ 0 \\ 1 \end{pmatrix} \geqslant 0,$$

so x^0 is an optimal solution to the original problem.

Dropping nonnegativity on the basic variables yields the problem of finding an optimal integral solution in the cone generated by the active constraints. This is shown graphically in Figure 5.4, where $x_1 = 4$, $x_2 = -1$ is the optimal integral point in the "cone" defined by the constraints $-x_1 + 2x_2 \leqslant 2$ and $5x_1 + x_2 \leqslant 19$.

An interesting application of the asymptotic behavior of GP is that it can be used to show that by solving a finite number of group problems, $z_{IP}(b)$ can be obtained for all but a finite number of points in Z^m. Hence, as we remarked in Section 3, the complete value function of IP can be found by solving a finite number of integer programs.

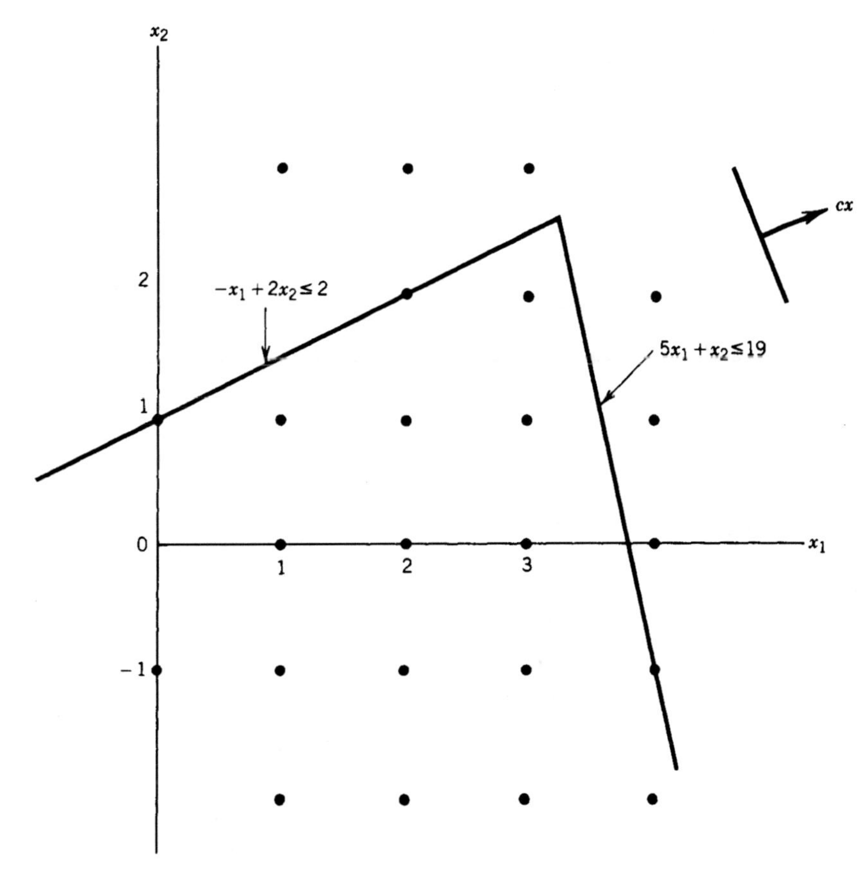

Figure 5.4. The optimal solution is $x_1 = 4$, $x_2 = -1$.

6. LAGRANGIAN RELAXATION AND DUALITY

Consider an integer program

$$z_{\mathrm{IP}} = \max\{cx: x \in S\}, \quad \text{where } S = \{x \in Z_+^n: Ax \leqslant b\},$$

which can be rewritten as

$$\begin{aligned} \text{(IP)} \qquad z_{\mathrm{IP}} = \max\ & cx \\ & A^1x \leqslant b^1 \qquad \text{(complicating constraints)} \\ & A^2x \leqslant b^2 \qquad \text{(nice constraints)} \\ & x \in Z_+^n, \end{aligned}$$

where $A = \binom{A^1}{A^2}$ and $b = \binom{b^1}{b^2}$. We suppose that $A^2x \leqslant b^2$ are $m - m_1$ "nice constraints", say those of a network problem. By dropping the m_1 complicating constraints $A^1x \leqslant b^1$ we obtain a relaxation that is easier to solve than the original problem. There are many problems for which the constraints can be partitioned in this way. We will give some examples later.

The idea of dropping constraints can be embedded into a more general framework called *Lagrangian relaxation*. It is convenient to consider a generalization of IP:

$$\text{IP}(Q) \qquad \begin{aligned} z_{\text{IP}} = \max\ & cx \\ & A^1x \leqslant b^1 \\ & x \in Q. \end{aligned}$$

However, when we are discussing results that are specific to IP, it is assumed that $Q = \{x \in Z_+^n: A^2x \leqslant b^2\} \neq \emptyset$. Again it has to be understood that the problem obtained from IP(Q) by dropping the complicating constraints is much easier to solve than IP(Q).

Now for any $\lambda \in R_+^{m_1}$, consider the problem

$$\text{LR}(\lambda) \qquad z_{\text{LR}}(\lambda) = \max\{z(\lambda, x): x \in Q\}, \quad \text{where } z(\lambda, x) = cx + \lambda(b^1 - A^1x).$$

The problem LR(λ) is called the *Lagrangian relaxation* of IP(Q) with respect to $A^1x \leqslant b^1$. This terminology is used because the vector λ plays a role in LR(λ) similar to the role of Lagrange multipliers in constrained continuous optimization problems.

LR(λ) does not contain the complicating constraints. Instead we have included these constraints in the objective function with the "penalty" term $\lambda(b^1 - A^1x)$. Since $\lambda \geqslant 0$, violations of $A^1x \leqslant b^1$ make the penalty term negative, and thus intuitively $A^1x \leqslant b^1$ will be satisfied if λ is suitably large.

Proposition 6.1. LR(λ) *is a relaxation of* IP(Q) *for all* $\lambda \geqslant 0$.

Proof. If x is feasible in IP(Q), then $x \in Q$ and hence x is feasible for LR(λ). Also, $z(\lambda, x) = cx + \lambda(b^1 - A^1x) \geqslant cx$ for all x feasible in IP(Q) since $A^1x \leqslant b^1$ and $\lambda \geqslant 0$. ■

As a consequence of Proposition 6.1, $z_{\text{LR}}(\lambda) \geqslant z_{\text{IP}}$ for all $\lambda \geqslant 0$. The least upper bound available from the infinite family of relaxations $\{\text{LR}(\lambda)\}_{\lambda\geq 0}$ is $z_{\text{LR}}(\lambda^*)$, where λ^* is an optimal solution to

$$\text{(LD)} \qquad z_{\text{LD}} = \min_{\lambda \geq 0} z_{\text{LR}}(\lambda).$$

Problem LD is called the *Lagrangian dual* of IP(Q) with respect to the constraints $A^1x \leqslant b^1$.

The following example from Section II.1.1, but with the constraints $x_1 \geqslant 2$, $x_2 \leqslant 4$ added, will be used throughout this section to illustrate the concepts and results presented.

Example 6.1

$$\begin{aligned} \max\ & 7x_1 + 2x_2 \\ & -x_1 + 2x_2 \leqslant 4 \\ & \left.\begin{aligned} 5x_1 + x_2 &\leqslant 20 \\ -2x_1 - 2x_2 &\leqslant -7 \\ -x_1 &\leqslant -2 \\ x_2 &\leqslant 4 \\ x \in Z_+^2. & \end{aligned}\right\} \quad Q = \{x \in Z_+^2: A^2x \leqslant b^2\}. \end{aligned}$$

The Lagrangian relaxation with respect to $-x_1 + 2x_2 \leqslant 4$ is

$$\begin{aligned} z_{\mathrm{LR}}(\lambda) &= \max_{x \in Q}[7x_1 + 2x_2 + \lambda(4 + x_1 - 2x_2)] \\ &= \max(7+\lambda)x_1 + (2-2\lambda)x_2 + 4\lambda \\ &\quad 5x_1 + x_2 \leqslant 20 \\ &\quad -2x_1 - 2x_2 \leqslant -7 \\ &\quad -x_1 \leqslant -2 \\ &\quad x_2 \leqslant 4 \\ &\quad x \in Z^2_+, \end{aligned}$$

where Q is the finite set of points

$$\{x^1, x^2, x^3, x^4, x^5, x^6, x^7, x^8\} = \{(2,2), (2,3), (2,4), (3,1), (3,2), (3,3), (3,4), (4,0)\}.$$

(see Figure 6.1).

The example suggests at least two different viewpoints. The first is to view $z(\lambda, x) = (c - \lambda A^1)x + \lambda b^1$ as an affine function of x for λ fixed. It then follows that $z_{\mathrm{LR}}(\lambda)$ can be determined by solving the linear program

$$z_{\mathrm{LR}}(\lambda) = \max\{z(\lambda, x) : x \in \mathrm{conv}(Q)\},$$

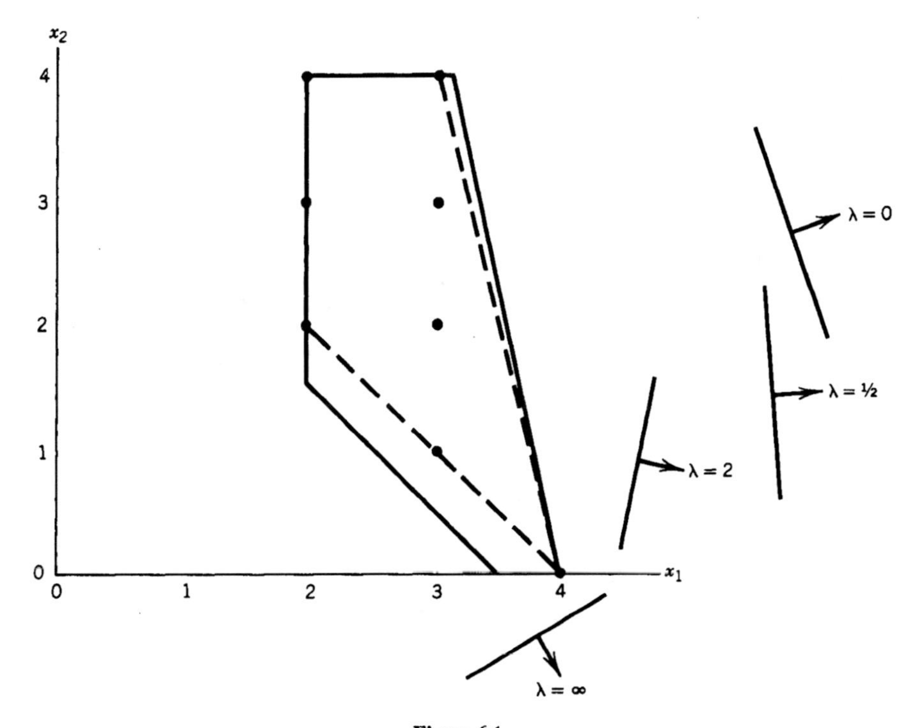

Figure 6.1

where as usual we assume that conv(Q) is a rational polyhedron.

In the example (see Figure 6.1),

$$\text{conv}(Q) = \{x \in R^2_+ : -x_1 \leqslant -2, x_2 \leqslant 4, -x_1 - x_2 \leqslant -4, 4x_1 + x_2 \leqslant 16\}.$$

Thus

$$z_{\text{LR}}(0) = \max\{7x_1 + 2x_2 : x \in \text{conv}(Q)\} = z(0, x^7) = 29$$

$$z_{\text{LR}}(1) = \max\{8x_1 + 0x_2 + 4 : x \in \text{conv}(Q)\} = z(1, x^8) = 36;$$

and as one increases λ from 0, $z_{\text{LR}}(\lambda)$ first decreases until $\lambda = \frac{1}{9}$ and then it increases. In general we obtain

$$z_{\text{LR}}(\lambda) = z(\lambda, x^7) = 29 - \lambda \quad \text{for } 0 \leqslant \lambda \leqslant \frac{1}{9}$$

$$z_{\text{LR}}(\lambda) = z(\lambda, x^8) = 28 + 8\lambda \quad \text{for } \lambda \geqslant \frac{1}{9}.$$

Hence $z_{\text{LD}} = z_{\text{LR}}(\frac{1}{9}) = z(\frac{1}{9}, x^7) = z(\frac{1}{9}, x^8) = 28\frac{8}{9}$ and $\lambda^* = \frac{1}{9}$.

All of these calculations can be seen in Figure 6.1, where we have shown the objective function $\max(c - \lambda A^1)x + \lambda b^1$ for different values of λ.

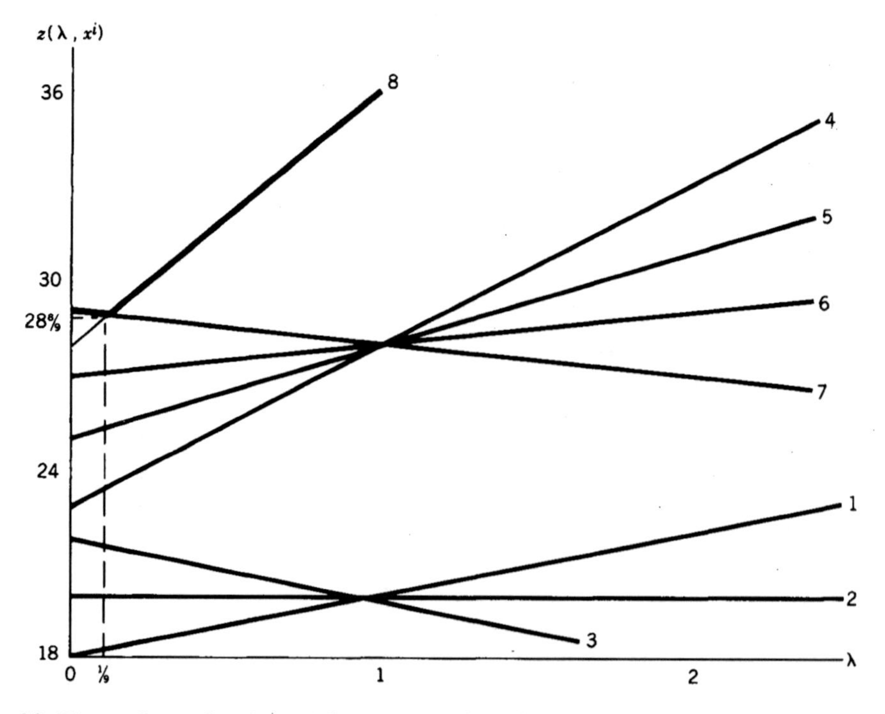

Figure 6.2. The numbers assigned to each line denote the following: (1) $18 + 2\lambda$; (2) 20; (3) $22 - 2\lambda$; (4) $23 + 5\lambda$; (5) $25 + 3\lambda$; (6) $27 + \lambda$; (7) $29 - \lambda$; (8) $28 + 8\lambda$.

The second viewpoint is to consider $z_{LR}(\lambda)$ to be determined by maximization over a set of discrete points, that is,

$$z_{LR}(\lambda) = \max_{x^i \in Q} z(\lambda, x^i),$$

and to observe that for fixed x^i, $z(\lambda, x^i) = cx^i + \lambda(b^1 - A^1 x^i)$ is an affine function of λ. See Figure 6.2, where we have drawn the affine functions $z(\lambda, x^i)$ for $x^i \in Q$.

In Figure 6.2 one can read off the value of $z_{LR}(\lambda)$ for any value of λ. We see that $z_{LR}(\lambda)$ is piecewise linear and convex (the heavy lines in Figure 6.2) and that $z_{LD} = 28\frac{8}{9}$.

Formally, one solves the linear program

$$z_{LR}(\lambda) = \min\{w\colon w \geq z(\lambda, x^i) \text{ for } i = 1, \ldots, 8\},$$

which shows that $z_{LR}(\lambda)$ is the maximum of a finite number of affine functions and is therefore piecewise linear and convex.

We now study how the solution of the Lagrangian dual relates to the solution of the original problem IP(Q). Returning to Figure 6.1, note that when $\lambda = \frac{1}{9}$ we obtain

$$\begin{aligned} 28\tfrac{8}{9} &= z(\tfrac{1}{9}, x^7) = z(\tfrac{1}{9}, x^8) \\ &= z(\tfrac{1}{9}, \tfrac{8}{9}x^7 + \tfrac{1}{9}x^8) \quad \text{since } z(\lambda, x) \text{ is affine in } x \\ &= z(\tfrac{1}{9}, \tfrac{8}{9}(3\ 4) + \tfrac{1}{9}(4\ \ 0)) \\ &= z(\tfrac{1}{9}, (\tfrac{28}{9}\ \tfrac{32}{9})) = z(\tfrac{1}{9}, x^*) \quad \text{with } x^* = (\tfrac{28}{9}\ \tfrac{32}{9}) \\ &= cx^* + \tfrac{1}{9}(4 + x_1^* - 2x_2^*) \\ &= cx^* \quad \text{since } x^* \text{ satisfies } -x_1 + 2x_2 = 4. \end{aligned}$$

In other words, by taking a convex combination of points in Q, in the example x^7 and x^8, we obtain a point x^* in conv(Q) satisfying the complicating constraint, for which $cx^* = z_{LD}$. This shows that for the example we obtain

$$z_{LD} = \max\{cx\colon A^1 x \leq b^1, x \in \text{conv}(Q)\}.$$

We now formalize the results suggested by the example. The major result is that the primal linear programming problem of finding a convex combination of points in Q that also satisfy the complicating constraint $A^1 x \leq b^1$ is dual to the Lagrangian dual.

Theorem 6.2. $z_{LD} = \max\{cx\colon A^1 x \leq b^1, x \in \text{conv}(Q)\}$.

Proof

$$\begin{aligned} z_{LR}(\lambda) &= \max_{x \in Q} (c - \lambda A^1)x + \lambda b^1 \\ &= \max_{x \in \text{conv}(Q)} (c - \lambda A^1)x + \lambda b^1 \\ &\quad \text{(since the objective function is linear)} \\ &= \max_{x \in \text{conv}(Q)} [cx + \lambda(b^1 - A^1 x)]. \end{aligned}$$

Hence

$$\begin{aligned} z_{\mathrm{LD}} &= \min_{\lambda \geq 0} z_{\mathrm{LR}}(\lambda) \\ &= \min_{\lambda \geq 0} \max_{x \in \operatorname{conv}(Q)} [cx + \lambda(b^1 - A^1 x)]. \end{aligned}$$

If $Q = \emptyset$ the inner max equals $-\infty$ for all λ. Hence $z_{\mathrm{LD}} = -\infty$ as desired. Otherwise, let $\{x^k \in R_+^n : k \in K\}$ and $\{r^j \in R_+^n : j \in J\}$, respectively, be the sets of extreme points and extreme rays of conv(Q). Thus

$$\max_{x \in \operatorname{conv}(Q)} [cx + \lambda(b^1 - A^1 x)] = \begin{cases} \infty \text{ if } (c - \lambda A^1) r^j > 0 & \text{for some } j \in J \\ cx^k + \lambda(b^1 - A^1 x^k) & \text{for some } k \in K \text{ otherwise.} \end{cases}$$

Hence

$$\begin{aligned} z_{\mathrm{LD}} = \min_{\lambda} & \left[\max_{k \in K} cx^k + \lambda(b^1 - A^1 x^k) \right] \\ & (c - \lambda A^1) r^j \leqslant 0 \quad \text{for } j \in J \\ & \lambda \geqslant 0, \end{aligned}$$

which can be restated as

$$\begin{aligned} z_{\mathrm{LD}} = \min_{\eta, \lambda} \ & \eta \\ & \eta + \lambda(A^1 x^k - b^1) \geqslant cx^k \quad \text{for } k \in K \\ & \lambda A^1 r^j \geqslant cr^j \quad \text{for } j \in J \\ & \lambda \geqslant 0. \end{aligned} \tag{6.1}$$

Thus by linear programming duality, we obtain

$$\begin{aligned} z_{\mathrm{LD}} = \max \ & c\left(\sum_{k \in K} \alpha^k x^k + \sum_{j \in J} \beta^j r^j \right) \\ & \sum_{k \in K} \alpha^k = 1 \\ & A^1\left(\sum_{k \in K} \alpha^k x^k + \sum_{j \in J} \beta^j r^j \right) \leqslant b^1 \left(\sum_{k \in K} \alpha^k \right) \\ & \alpha^k, \beta^j \geqslant 0 \quad \text{for } k \in K \text{ and } j \in J \\ = \max & \{cx : A^1 x \leqslant b^1, x \in \operatorname{conv}(Q)\}. \end{aligned} \tag{6.2}$$

■

Corollary 6.3. *z_{LD} can be calculated from the linear programs (6.1) or (6.2).*

The reader familiar with linear programming decomposition will recognize the linear program (6.2) as the reformulation obtained when Dantzig–Wolfe price decomposition is applied to

$$\max(cx : A^1 x \leqslant b^1, A^2 x \leqslant b^2, x \in R_+^n),$$

where $\text{conv}(Q) = \{x \in R_+^n: A^2x \leqslant b^2\}$ and λ are the "dual prices" associated with the constraints $A^1x \leqslant b^1$. It follows that (6.1) is the dual of the Dantzig–Wolfe reformulation. Alternatively, (6.1) is the reformulation obtained by applying resource or Benders' decomposition to the dual linear program

$$\min\{\lambda b^1 + \mu b^2: \lambda A^1 + \mu A^2 \geqslant c, \lambda \in R_+^{m_1}, \mu \in R_+^{m_2}\}$$

(see the next section).

Corollary 6.4. *$z_{\text{LR}}(\lambda)$ is piecewise linear and convex on the domain over which it is finite.*

Proof. $z_{\text{LR}}(\lambda)$ is finite if and only if λ lies in the polyhedron $\{\lambda \in R_+^{m_1}: \lambda A^1 r^j \geqslant cr^j$ for $j \in J\}$. On this polyhedron, $z_{\text{LR}}(\lambda) = \lambda b^1 + \max_{k \in K}(c - \lambda A^1)x^k$ and is the maximum of a finite number of affine functions. Convexity follows from Proposition 4.1 of Section I.2.4. ■

Since

$$\text{conv}(S) = \text{conv}\{Q \cap \{x \in R_+^n: A^1x \leqslant b^1\}\} \subseteq \text{conv}(Q) \cap \{x \in R_+^n: A^1x \leqslant b^1\},$$

we have that

$$z_{\text{IP}} = \max_{x \in S} cx \leqslant z_{\text{LD}} = \max\{cx: A^1x \leqslant b^1, x \in \text{conv}(Q)\}.$$

The duality gap $z_{\text{LD}} - z_{\text{IP}}$ depends on the relative sizes of conv(S), $\text{conv}(Q) \cap \{x: A^1x \leqslant b^1\}$, and the objective coefficients c.

Corollary 6.5. *$z_{\text{IP}} = z_{\text{LD}}$ for all c if and only if*

$$\text{conv}\{Q \cap \{x \in R_+^n: A^1x \leqslant b^1\}\} = \text{conv}(Q) \cap \{x \in R_+^n: A^1x \leqslant b^1\}.$$

When $Q = \{x \in Z_+^n: A^2x \leqslant b^2\}$, it is also of considerable interest to compare z_{LD} with $z_{\text{LP}} = \max\{cx: Ax \leqslant b, x \in R_+^n\}$. In Example 6.1, $z_{\text{IP}} = 28 < z_{\text{LD}} = 28\frac{8}{9} < z_{\text{LP}} = 30\frac{2}{11}$.

Corollary 6.6. *$z_{\text{LD}} = z_{\text{LP}}$ for all c if all the extreme points of $\{x \in R_+^n: A^2x \leqslant b^2\}$ are integral.*

Proof. Under the hypothesis of the corollary we obtain $\text{conv}(Q) = \{x \in R_+^n: A^2x \leqslant b^2\}$, and the result follows from Theorem 6.2. ■

In Example 6.1, a natural choice of the "complicating constraints" is

$$\{-x_1 + 2x_2 \leqslant 4, 5x_1 + x_2 \leqslant 20, -2x_1 - 2x_2 \leqslant -7\}.$$

Thus

$$Q = \{x \in Z_+^2: A^2x \leqslant b^2\} = \{x \in Z_+^2: -x_1 \leqslant -2, x_2 \leqslant 4\}.$$

Obviously $\{x \in R_+^2: -x_1 \leqslant -2, x_2 \leqslant 4\}$ only has integral extreme points, so that, by Corollary 6.6, this Lagrangian relaxation would terminate with $z_{\text{LD}} = z_{\text{LP}} = 30\frac{2}{11}$.

In summary,

$$\mathrm{conv}(S) \subseteq \mathrm{conv}(Q) \cap \{x \in R^n_+ : A^1x \leqslant b^1\} \subseteq \{x \in R^n_+ : Ax \leqslant b\}$$

and thus $z_{\mathrm{IP}} \leqslant z_{\mathrm{LD}} \leqslant z_{\mathrm{LP}}$. But because some faces of the respective polyhedra can coincide, we may obtain $z_{\mathrm{IP}} = z_{\mathrm{LD}}$ or $z_{\mathrm{LD}} = z_{\mathrm{LP}}$ for particular c even if the conditions of the two previous corollaries do not hold. Figure 6.3 illustrates this. The inner polytope is conv(S). The outer polytope is $\{x \in R^n_+ : Ax \leqslant b\}$. The inner polytope, together with the shaded region, is $\mathrm{conv}(Q) \cap \{x \in R^n_+ : A^1x \leqslant b^1\}$. Four different objective functions are indicated, and the results are summarized as follows:

Objective Functions	Objective Values
c^1	$z_{\mathrm{IP}} = z_{\mathrm{LD}} = z_{\mathrm{LP}}$
c^2	$z_{\mathrm{IP}} < z_{\mathrm{LD}} = z_{\mathrm{LP}}$
c^3	$z_{\mathrm{IP}} < z_{\mathrm{LD}} < z_{\mathrm{LP}}$
c^4	$z_{\mathrm{IP}} = z_{\mathrm{LD}} < z_{\mathrm{LP}}$

It is possible to characterize problems where $z_{\mathrm{IP}} = z_{\mathrm{LD}}$ in terms of a complementarity condition. We will obtain this result as a corollary to the following theorem.

Theorem 6.7. $z_{\mathrm{IP}} \geqslant z_{\mathrm{LD}} - \varepsilon$ *if and only if there exists* $\lambda^* \geqslant 0$ *and* $x^* \in S$ *such that* $\lambda^*(b^1 - A^1x^*) \leqslant \delta_1$, $z(\lambda^*, x^*) \geqslant z_{\mathrm{LR}}(\lambda^*) - \delta_2$, *and* $\delta_1 + \delta_2 \leqslant \varepsilon$.

Proof. To show sufficiency, we have

$$\begin{aligned} z_{\mathrm{LD}} &\leqslant z_{\mathrm{LR}}(\lambda^*) \leqslant z(\lambda^*, x^*) + \delta_2 \\ &\leqslant cx^* + \delta_1 + \delta_2 \leqslant cx^* + \varepsilon \\ &\leqslant z_{\mathrm{IP}} + \varepsilon \quad (\text{since } x^* \in S). \end{aligned}$$

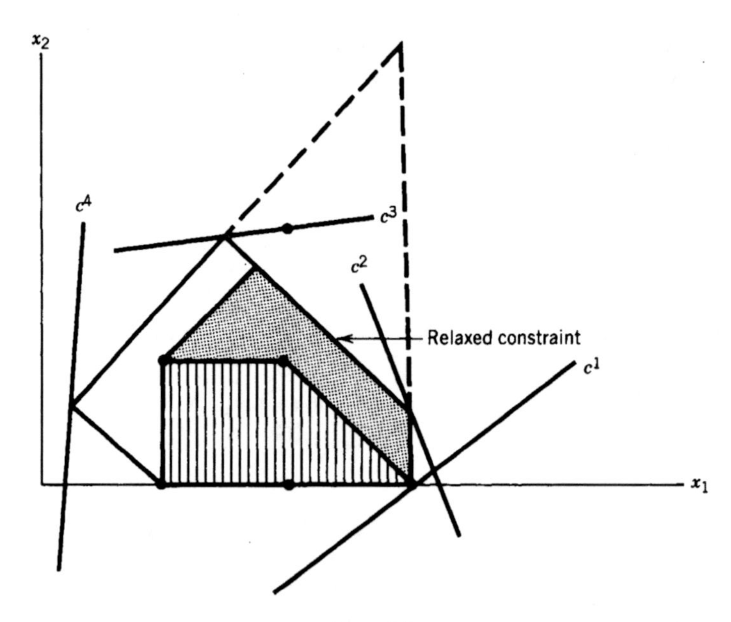

Figure 6.3

To show necessity, let x^* be an optimal solution of IP(Q) and let λ^* be an optimal solution of LD. We have

$$\begin{aligned} z_{\mathrm{LD}} = z_{\mathrm{LR}}(\lambda^*) &= z(\lambda^*, x^*) + z_{\mathrm{LR}}(\lambda^*) - z(\lambda^*, x^*) \\ &= cx^* + \lambda^*(b^1 - A^1x^*) + z_{\mathrm{LR}}(\lambda^*) - z(\lambda^*, x^*) \\ &= z_{\mathrm{IP}} + \lambda^*(b^1 - A^1x^*) + z_{\mathrm{LR}}(\lambda^*) - z(\lambda^*, x^*). \end{aligned}$$

Hence $z_{\mathrm{IP}} \geqslant z_{\mathrm{LD}} - \varepsilon$ implies

$$\lambda^*(b^1 - A^1x^*) + (z_{\mathrm{LR}}(\lambda^*) - z(\lambda^*, x^*)) \leqslant \varepsilon.$$

Now put $\delta_1 = \lambda^*(b^1 - A^1x^*)$ and $\delta_2 = z_{\mathrm{LR}}(\lambda^*) - z(\lambda^*, x^*)$. ■

By putting $\delta_1 = \delta_2 = \varepsilon = 0$ in Theorem 6.7, we obtain necessary and sufficient conditions for the duality gap to be 0.

Corollary 6.8. *$z_{\mathrm{IP}} = z_{\mathrm{LD}}$ if and only if there exists $\lambda^* \geqslant 0$ and $x^* \in S$ such that $\lambda^*(b^1 - A^1x^*) = 0$ and $z_{\mathrm{LR}}(\lambda^*) = z(\lambda^*, x^*)$.*

Theorem 6.7 can also be helpful in identifying (nearly) optimal solutions to IP(Q). For example, in the process of solving LR(λ) we may find an $x \in S$ that is nearly optimal in LR(λ) and nearly satisfies complementary slackness.

Corollary 6.9. *If $x^* \in S$ satisfies $\lambda(b^1 - A^1x^*) \leqslant \delta_1$ and $z(\lambda, x^*) \geqslant z_{\mathrm{LR}}(\lambda) - \delta_2$ for some $\lambda \geqslant 0$, then $cx^* \geqslant z_{\mathrm{IP}} - \delta_1 - \delta_2$.*

In Example 6.1, $x^6 = (3\ 3) \in S$. For $\lambda = \frac{1}{15}$, we obtain $z_{\mathrm{LR}}(\lambda) = 28\frac{14}{15}$, $\lambda(b^1 - A^1x^6) = \frac{1}{15} = \delta_1$, and $z_{\mathrm{LR}}(\lambda) - z(\lambda, x^6) = 1\frac{13}{15} = \delta_2$. Hence $cx^6 \geqslant z_{\mathrm{IP}} - 1\frac{14}{15}$.

The complementary slackness conditions are also useful in right-hand-side parametrics as shown in the following corollary to Theorem 6.7.

Corollary 6.10. *Let x^* be an optimal solution to LR(λ^*), where $\lambda^* \geqslant 0$, and define $d^* = A^1x^*$. Then x^* is an optimal solution to*

$$\max\{cx\colon A^1x \leqslant d^1, x \in Q\}$$

for all $d^1 \in D = \{d \in R^{m_1}\colon d_i = d_i^* \text{ if } \lambda_i^* > 0, d_i \geqslant d_i^* \text{ if } \lambda_i^* = 0 \text{ for } i = 1, \ldots, m_1\}$.

In Example 6.1, $x^7 = (3\ 4)$ is an optimal solution to LR($\frac{1}{10}$). Hence x^7 is an optimal solution when the first constraint is $-x_1 + 2x_2 \leqslant 5$.

Lagrangian relaxation and duality also apply to equality constraints. Suppose that $A^1x = b^1$ in Problem IP(Q). Then defining LR(λ) as before, we have the following proposition.

Proposition 6.11. *If $A^1x = b^1$ in IP, then LR(λ) is a relaxation of IP for all $\lambda \in R^{m_1}$.*

The only difference between the equality and inequality cases is that in the equality case the multipliers are unrestricted in sign.

We now give one problem to illustrate the formulation of Lagrangian relaxations. Others will be given later when we discuss computation.

Example 6.2 (A Flow Problem with Budget Constraint). Suppose there is a set of n jobs to be assigned to a set of n workers, with $N = \{1, \ldots, n\}$. Suppose that c_{ij} is the value of assigning worker i to job j, that t_{ij} is the cost of training worker i to do job j, and that we have a training budget of b units. We wish to maximize the total value of the assignment subject to the budget constraint, that is,

$$\max \sum_{i \in N} \sum_{j \in N} c_{ij} x_{ij}$$

$$\sum_{j \in N} x_{ij} = 1 \quad \text{for } i \in N \tag{1}$$

$$\sum_{i \in N} x_{ij} = 1 \quad \text{for } j \in N \tag{2}$$

$$\sum_{i \in N} \sum_{j \in N} t_{ij} x_{ij} \leqslant b \tag{3}$$

$$x \in B^{n^2}.$$

First we observe that the problem is $\mathcal{NP}$-hard. If we then wish to choose a Lagrangian relaxation, there are four options to consider. Note that in each option the relaxed problem LR(λ) is considerably easier to solve than the original problem.

1. *Lagrangian relaxation with respect to* (3). Then $\text{LR}_1(\lambda)$, $\lambda \in R^1_+$, is an assignment problem with objective function

$$\lambda b + \sum_{i \in N} \sum_{j \in N} (c_{ij} - \lambda t_{ij}) x_{ij}.$$

2. *Lagrangian relaxation with respect to* (1) *and* (2). Then $\text{LR}_2(u, v)$, $u \in R^n$, $v \in R^n$, is a knapsack problem with objective function

$$\sum_{i \in N} u_i + \sum_{j \in N} v_j + \sum_{i \in N} \sum_{j \in N} (c_{ij} - u_i - v_j) x_{ij}.$$

3. *Lagrangian relaxation with respect to* (1) *or* (2), *say* (1). Then $\text{LR}_3(u)$, $u \in R^n$, is a knapsack problem with generalized upper-bound constraints and objective function

$$\sum_{i \in N} u_i + \sum_{i \in N} \sum_{j \in N} (c_{ij} - u_i) x_{ij}.$$

4. *Lagrangian relaxation with respect to* (1) *or* (2) *and* (3), *say* (1) *and* (3). Only generalized upper-bound constraints remain. Thus the Lagrangian $\text{LR}_4(u, \lambda)$, $u \in R^n$, $\lambda \in R^1_+$, with objective function

$$\lambda b + \sum_{i \in N} u_i + \sum_{i \in N} \sum_{j \in N} (c_{ij} - u_i - \lambda t_{ij}) x_{ij}$$

is trivial to solve. For each j, an i is chosen to maximize $c_{ij} - u_i - \lambda t_{ij}$, and the corresponding x_{ij} is set to 1.

In choosing a relaxation there are two major questions to consider: How strong is the bound z_{LD}, and how difficult to solve is the Lagrangian dual (LD)? We defer discussion of the latter question until we discuss computation, and now we just consider the bounds.

When Q is a set of assignment constraints or a set of generalized upper-bound constraints, Corollary 6.6 applies and $z^1_{\rm LD} = z^4_{\rm LD} = z_{\rm LP}$. Since

$$Q^3 = \left\{x \in B^{n^2}: \sum_{i \in N} x_{ij} = 1 \text{ for } j \in N, \sum_{i \in N}\sum_{j \in N} t_{ij}x_{ij} \leq b\right\}$$

$$\subset Q^2 = \left\{x \in B^{n^2}: \sum_{i \in N}\sum_{j \in N} t_{ij}x_{ij} \leq b\right\}$$

and

$$\text{conv}(Q^2) \subset \left\{x \in R^{n^2}_+: \sum_{i \in N}\sum_{j \in N} t_{ij}x_{ij} \leq b, x_{ij} \leq 1 \text{ for } i, j \in N\right\},$$

we have

$$z_{\rm IP} \leq z^3_{\rm LD} \leq z^2_{\rm LD} \leq z^1_{\rm LD} = z^4_{\rm LD} = z_{\rm LP},$$

and each of the inequalities is strict for some objective function.

We now consider two ways of strengthening the Lagrangian dual of problem IP. The first approach yields a dual whose optimal value equals

$$\max\{cx: x \in \text{conv}(x \in Z^n_+: A^1x \leq b^1) \cap \text{conv}(x \in Z^n_+: A^2x \leq b^2)\}.$$

This dual is obtained by applying Lagrangian duality to the reformulation of IP given by

$$\text{(RIP)} \quad \begin{aligned} z_{\rm IP} = \max\ & cx^1 \\ & A^1x^1 \leq b^1 \\ & A^2x^2 \leq b^2 \\ & x^1 - x^2 = 0 \\ & x^1 \in Z^n_+, x^2 \in Z^n_+. \end{aligned}$$

Taking $x^1 - x^2 = 0$ as the complicating constraints, we obtain the Lagrangian dual of RIP:

$$\begin{aligned} z_{\rm CSD} = \min_u\ & \{\max\{(c-u)x^1 + ux^2\}\} \\ & A^1x^1 \leq b^1 \\ & A^2x^2 \leq b^2 \\ & x^1 \in Z^n_+, \quad x^2 \in Z^n_+ \\ = \min_{c^1+c^2=c}\ & \{\max c^1x^1 + \max c^2x^2\} \\ & A^1x^1 \leq b^1, \quad A^2x^2 \leq b^2 \\ & x^1 \in Z^n_+, \quad x^2 \in Z^n_+, \end{aligned}$$

where $u = c^2$.

From Theorem 6.2, we obtain a polyhedral interpretation of the dual.

Corollary 6.12

$$z_{\mathrm{CSD}} = \max\{cx: x \in \operatorname{conv}\{x \in Z_+^n: A^1x \leqslant b^1\} \cap \operatorname{conv}\{x \in Z_+^n: A^2x \leqslant b^2\}\}$$

and $z_{\mathrm{CSD}} \leqslant z_{\mathrm{LD}}$.

We have used the terminology CS since the technique has been called *cost splitting*. The technique is useful when:

1. $\operatorname{conv}\{x \in Z_+^n: A^1x \leqslant b^1\} \subset \{x \in R_+^n: A^1x \leqslant b^1\}$, so for some objective functions c we obtain $z_{\mathrm{CSD}} < z_{\mathrm{LD}}$.
2. The sets of constraints $A^ix \leqslant b^i$ are simple to deal with separately; that is, the difficulty is caused by their interaction.

In Example 6.2, we could take $A^1x \leqslant b^1$ to be constraint set (1) and (3) and take $A^2x \leqslant b^2$ to be constraint sets (2) and (3). This yields $z_{\mathrm{CSD}} \leqslant z_{\mathrm{LD}}^3$ with the inequality strict for some objective functions.

Another approach that dominates the Lagrangian dual is the "surrogate" dual. Starting from IP(Q), with weights $\lambda \in R_+^{m_1}$ for the complicating constraints, consider the problem

SD(λ) $$z_{\mathrm{SD}}(\lambda) = \max\{cx: \lambda A^1x \leqslant \lambda b^1, x \in Q\}.$$

The problem SD(λ) is called the *surrogate relaxation* of IP(Q) with respect to $A^1x \leqslant b^1$. SD(λ) contains a single "complicating" constraint. For instance, when $Q = Z_+^n$ the surrogate relaxation is a knapsack problem. The *surrogate dual* of IP(Q) is the problem

(SD) $$z_{\mathrm{SD}} = \min_{\lambda \geqslant 0} z_{\mathrm{SD}}(\lambda).$$

Proposition 6.13. LR(λ) *is a relaxation of* SD(λ) *for* $\lambda \geqslant 0$ *and* $z_{\mathrm{LD}} \geqslant z_{\mathrm{SD}}$.

Proof. The feasible region of SD(λ) is contained in that of LR(λ). In addition, when x is feasible in SD(λ) we obtain $\lambda(b^1 - A^1x) \geqslant 0$ and hence

$$z(\lambda, x) = cx + \lambda(b^1 - A^1x) \geqslant cx. \qquad \blacksquare$$

Although the surrogate dual can be used computationally, it does not have such nice theoretical properties as the Lagrangian dual.

We close this section by relating Lagrangian duality to the general duality theory of Section 2. Given the initial problem IP(Q), we define its value function z_Q by

$$z_Q(d^1) = \max\{cx: A^1x \leqslant d^1, x \in Q\}$$

for all $d^1 \in R^{m_1}$. Note that when $Q = \{x \in Z_+^n: A^2x \leqslant b^2\}$, it follows that z_Q is a projection of the IP value function z onto $d^2 = b^2$. Thus $z_Q(d^1) = z(d^1, b^2)$ for all $d^1 \in R^{m_1}$. Now using a similar approach to that of Section 2, with $S_Q(d^1) = \{x \in Q: A^1x \leqslant d^1\}$ in place of $S(d)$, we obtain as the equivalent of (2.4) the dual problem

$$
\begin{aligned}
&\min g(b^1) \\
(6.3)\qquad &g(A^1x) \geqslant cx \quad \text{for } x \in Q \\
&g \text{ nondecreasing, } g\colon R^{m_1} \to R^1.
\end{aligned}
$$

Example 6.1 (continued). The dual problem is

$$
\begin{aligned}
w = &\min g(4) \\
&g(-x_1 + 2x_2) \geqslant 7x_1 + 2x_2 \quad \text{for } x \in Q \\
&g \text{ nondecreasing, } g\colon R^1 \to R^1
\end{aligned}
$$

or

$$
\begin{aligned}
&\min g(4) \\
&g(2) \geqslant 18, \qquad x^1 = (2 \quad 2) \\
&g(4) \geqslant 20, \qquad x^2 = (2 \quad 3) \\
&g(6) \geqslant 22, \qquad x^3 = (2 \quad 4) \\
&g(-1) \geqslant 23, \qquad x^4 = (3 \quad 1) \\
&g(1) \geqslant 25, \qquad x^5 = (3 \quad 2) \\
&g(3) \geqslant 27, \qquad x^6 = (3 \quad 3) \\
&g(5) \geqslant 29, \qquad x^7 = (3 \quad 4) \\
&g(-4) \geqslant 28, \qquad x^8 = (4 \quad 0) \\
&g \text{ nondecreasing.}
\end{aligned}
$$

It is readily seen from Figure 6.4 that

$$
z_Q(d^1) = \begin{cases} -\infty & \text{if } d^1 < -4 \\ 28 & \text{if } -4 \leqslant d^1 < 5 \\ 29 & \text{if } d^1 \geqslant 5. \end{cases}
$$

Since $z_Q(0) = z(0, b^2)$, we cannot expect $z_Q(0) = 0$. Hence the simplest class of functions that are candidates for the dual (6.3) are affine functions $g(d^1) = \lambda_0 + \lambda d^1$, $\lambda \in R_+^{m_1}$. In particular, if we take g to be the affine function supporting z_Q and passing through the points $(-4 \quad 28)$ and $(5 \quad 29)$ (see Figure 6.4), then g is clearly dual feasible and $g(4) = 28\frac{8}{9} = z_{\mathrm{LD}}$.

This leads us to examine the restricted dual

$$
\begin{aligned}
w_{\mathrm{DR}} = &\min g(b^1) \\
(6.4)\qquad &g(A^1x) \geqslant cx \text{ for } x \in Q \\
&g \text{ affine and nondecreasing, } g\colon R^{m_1} \to R^1,
\end{aligned}
$$

which can be rewritten as

$$
w_{\mathrm{DR}} = \min_{\lambda \geqslant 0} w_{\mathrm{DR}}(\lambda),
$$

where $w_{\mathrm{DR}}(\lambda) = \min_{\lambda_0} \{\lambda_0 + \lambda b^1\colon \lambda_0 + \lambda A^1 x \geqslant cx \text{ for } x \in Q\}$.

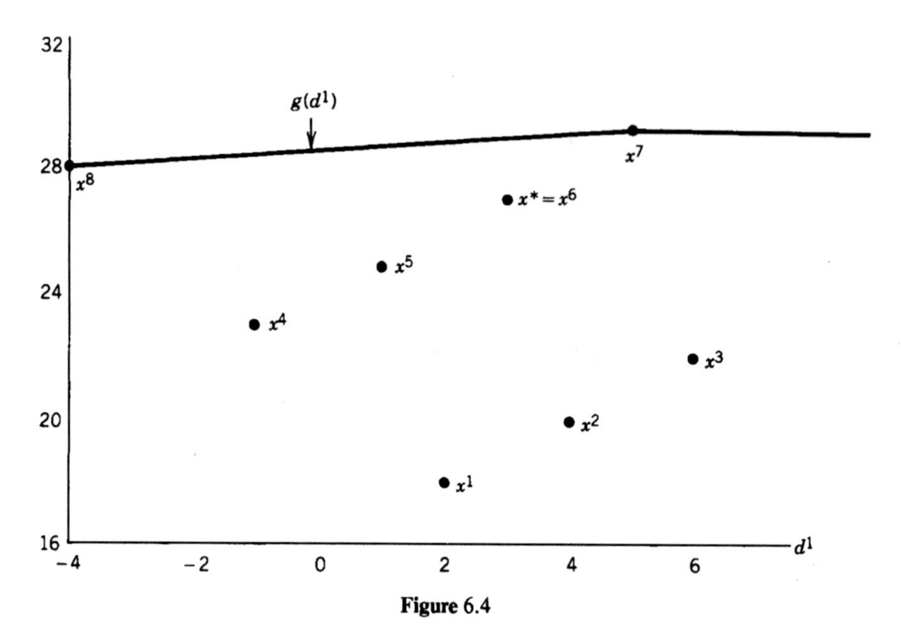

Figure 6.4

When $Q = \{x \in Z_+^n: A^2x \leqslant b^2\}$, it is also interesting to examine the dual (2.4) in R^m, where we restrict the dual functions $g: R^m \to R^1$ to be of the form $g(d^1, d^2) = \lambda d^1 + g_2(d^2)$ with $d^2 \in R^{m-m_1}$, and $\lambda \in R_+^{m_1}$. This gives the alternative dual

$$
\begin{aligned}
w_{\mathrm{D}} = \min\ & \lambda b^1 + g_2(b^2) \\
(6.5)\qquad & \lambda A^1x + g_2(A^2x) \geqslant cx \quad \text{for } x \in Z_+^n \\
& \lambda \geqslant 0,\ g_2 \text{ nondecreasing},\ g_2: R^{m_2} \to R^1,
\end{aligned}
$$

which can be rewritten as

$$w_{\mathrm{D}} = \min_{\lambda \geq 0} w_{\mathrm{D}}(\lambda),$$

where

$$w_{\mathrm{D}}(\lambda) = \lambda b^1 + \min\{g_2(b^2): g_2(A^2x) \geqslant cx - \lambda A^1x \text{ for } x \in Z_+^n,\ g_2 \text{ nondecreasing}\}.$$

We now compare these two restrictions [i.e., (6.4) and (6.5)] of the general dual with the Lagrangian dual.

Theorem 6.14. *The relationships among the Lagrangian dual and the restricted duals are given by:*

a. $z_{\mathrm{LR}}(\lambda) = w_{\mathrm{DR}}(\lambda)$ *for all* $\lambda \geqslant 0$. *Hence the Lagrangian dual and the restricted dual* (6.4) *are equivalent.*
b. *If* $Q = \{x \in Z_+^n: A^2x \leqslant b^2\}$, *then* $z_{\mathrm{LR}}(\lambda) = w_{\mathrm{DR}}(\lambda) = w_{\mathrm{D}}(\lambda)$. *Hence the Lagrangian dual and the two restricted duals* (6.4) *and* (6.5) *are equivalent.*

Proof

a. $w_{\text{DR}}(\lambda) = \min_{\lambda_0}\{\lambda_0 + \lambda b^1 \colon \lambda_0 + \lambda A^1 x \geqslant cx \quad \text{for } x \in Q\}$

$$= \lambda b^1 + \max_{x \in Q} (cx - \lambda A^1 x) = z_{\text{LR}}(\lambda).$$

b. $w_{\text{D}}(\lambda) = \lambda b^1 + \min g_2(b^2)$

$$g_2(A^2 x) \geqslant (c - \lambda A^1)x \quad \text{for } x \in Z^n_+$$

$$g_2 \text{ nondecreasing.}$$

Using (2.4), we obtain

$$w_{\text{D}}(\lambda) = \lambda b^1 + \max\{(c - \lambda A^1)x \colon A^2 x \leqslant b^2, x \in Z^n_+\}$$
$$= z_{\text{LR}}(\lambda). \qquad \blacksquare$$

The reader should now verify, for Example 6.1, that $g(d^1) = 28\frac{4}{9} + \frac{1}{9}d_1$ is an optimal solution to the restricted dual (6.4) and that

$$g(d) = \lambda d^1 + g_2(d^2) = \tfrac{1}{9}d_1 + \tfrac{16}{9}\lfloor \tfrac{4}{5}d_2 + \tfrac{1}{5}d_5 \rfloor$$

is an optimal solution to the restricted dual (6.5). Both evidently give the same objective value $z_{\text{LD}} = 28\frac{8}{9}$.

7. BENDERS' REFORMULATION

In the preceding section we gave a method for handling complicating constraints. We now consider the dual notion of complicating variables. In particular, in the mixed-integer program

$$\begin{aligned} z = \max\ & cx + hy \\ \text{(MIP)} \qquad & Ax + Gy \leqslant b \\ & x \in X \subseteq Z^n_+, y \in R^p_+, \end{aligned}$$

we can view the integer variables x as complicating variables to what would otherwise be a linear program, and we can view the continuous variables y as complicating variables to what would otherwise be a pure-integer program. For example, in a fixed-charge network flow problem where the integer variables represent decisions about which arcs to use in a network, the problem in the y-space is an ordinary network flow problem once x is specified.

The procedure described below shows how MIP can be reformulated as a problem in $X \times R^1$; that is, there is only one continuous variable. However, this formulation generally contains a huge number of linear constraints. Since one expects only a small subset of these constraints to be active in an optimal solution, a natural relaxation is obtained by dropping most of them.

As a first step, we suppose that the integer variables x have been fixed. The resulting linear program is

$$\text{LP}(x) \qquad z_{\text{LP}}(x) = \max\{hy\colon Gy \leqslant b - Ax,\ y \in R_+^p\}$$

and its dual is

$$\min\{u(b - Ax)\colon uG \geqslant h,\ u \in R_+^m\}.$$

We can characterize whether LP(x) is infeasible or has a bounded optimal value or has an unbounded optimal value by using the representation of the dual polyhedron in terms of its extreme points and extreme rays. Let $\{u^k \in R_+^m\colon k \in K\}$ be the set of extreme points of $Q = \{u \in R_+^m\colon uG \geqslant h\}$ and let $\{v^j \in R_+^m\colon j \in J\}$ be the set of extreme rays of $\{u \in R_+^m\colon uG \geqslant 0\}$. Note that if $Q \neq \emptyset$, then $\{v^j \in R_+^m\colon j \in J\}$ is also the set of extreme rays of Q. From Theorem 4.10 of Section I.4.4 we can characterize $z_{\text{LP}}(x)$.

Proposition 7.1. *The function $z_{\text{LP}}(x)$ is characterized as follows:*

i. *If $Q = \emptyset$, then $z_{\text{LP}}(x) = \infty$ if $v^j(b - Ax) \geqslant 0$ for all $j \in J$, and $z_{\text{LP}}(x) = -\infty$ otherwise.*
ii. *If $Q \neq \emptyset$, then $z_{\text{LP}}(x) = \min_{k\in K} u^k(b - Ax) < \infty$ if $v^j(b - Ax) \geqslant 0$ for all $j \in J$, and $z_{\text{LP}}(x) = -\infty$ otherwise.*

An immediate consequence of Proposition 7.1 is that when $Q \neq \emptyset$, MIP can be stated as

$$(7.1) \qquad \begin{aligned} z = \max_x \Big(cx + \min_{k\in K} u^k(b - Ax)\Big) & \\ v^j(b - Ax) \geqslant 0 \quad & \text{for } j \in J \\ x \in X. & \end{aligned}$$

This yields the Benders' representation of MIP given by the following theorem.

Theorem 7.2. MIP *can be reformulated as*

$$(\text{MIP}') \qquad \begin{aligned} z = \max \eta & \\ \eta \leqslant cx + u^k(b - Ax) \quad & \text{for } k \in K \\ v^j(b - Ax) \geqslant 0 \quad & \text{for } j \in J \\ x \in X, \eta \in R^1. & \end{aligned}$$

Proof. If there is no $x \in X$ such that $v^j(b - Ax) \geqslant 0$ for all $j \in J$, then $z_{\text{LP}}(x) = -\infty$ for all $x \in X$ and $z = -\infty$. If there is an $x \in X$ such that $v^j(b - Ax) \geqslant 0$ for all $j \in J$ and $Q = \emptyset$, then $K = \emptyset$ so that $z = \infty$; otherwise MIP′ is equivalent to (7.1). ■

MIP′ is Benders' reformulation. Since it typically has an enormous number of constraints, a natural approach is to consider relaxations obtained by generating only those constraints corresponding to a small number of extreme points and extreme rays. An algorithm based on such a relaxation will be discussed in the next chapter.

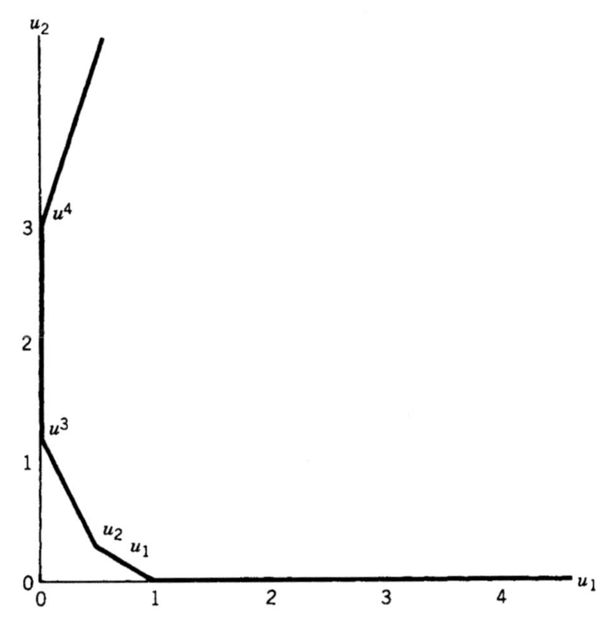

Figure 7.1

Example 7.1

$$
\begin{aligned}
\max\ & 5x_1 - 2x_2 + 9x_3 + 2y_1 - 3y_2 + 4y_3 \\
& 5x_1 - 3x_2 + 7x_3 + 2y_1 + 3y_2 + 6y_3 \leqslant -2 \\
& 4x_1 + 2x_2 + 4x_3 + 3y_1 - y_2 + 3y_3 \leqslant 10 \\
& x_j \leqslant 5 \quad \text{for } j = 1, 2, 3 \\
& x \in Z^3_+,\ y \in R^3_+.
\end{aligned}
$$

Here we suppose that $X = \{x \in Z^3_+ : x_j \leqslant 5 \text{ for } j = 1, 2, 3\}$.

In Figure 7.1 we show the polyhedron $\{u \in R^2_+ : uG \geqslant h\}$

$$
\begin{aligned}
2u_1 + 3u_2 &\geqslant 2 \\
3u_1 - u_2 &\geqslant -3 \\
6u_1 + 3u_2 &\geqslant 4 \\
u &\in R^2_+.
\end{aligned}
$$

The extreme points of this polyhedron are $u^1 = (1 \quad 0)$, $u^2 = (\frac{1}{2} \quad \frac{1}{3})$, $u^3 = (0 \quad \frac{4}{3})$, and $u^4 = (0 \quad 3)$, and its extreme rays are $v^1 = (1 \quad 0)$ and $v^2 = (1 \quad 3)$.

The resulting reformulation of the mixed-integer program is

$$
\begin{aligned}
z = \max\ & \eta \\
& \eta \leq 5x_1 - 2x_2 + 9x_3 + (-2 - 5x_1 + 3x_2 - 7x_3) \\
& \eta \leq 5x_1 - 2x_2 + 9x_3 + \tfrac{1}{2}(-2 - 5x_1 + 3x_2 - 7x_3) + \tfrac{1}{3}(10 - 4x_1 - 2x_2 - 4x_3) \\
& \eta \leq 5x_1 - 2x_2 + 9x_3 + \tfrac{4}{3}(10 - 4x_1 - 2x_2 - 4x_3) \\
& \eta \leq 5x_1 - 2x_2 + 9x_3 + 3(10 - 4x_1 - 2x_2 - 4x_3) \\
& (-2 - 5x_1 + 3x_2 - 7x_3) \geq 0 \\
& (-2 - 5x_1 + 3x_2 - 7x_3) + 3(10 - 4x_1 - 2x_2 - 4x_3) \geq 0 \\
& x_j \leq 5 \text{ for } j = 1, 2, 3 \\
& x \in Z^3_+,\ \eta \in R^1.
\end{aligned}
$$

The reader should check that an optimal solution is $x = (0 \quad 3 \quad 1)$ and $\eta = 3$.

Example 7.2 (Uncapacitated Facility Location). Here we use the alternative formulation of MIP given by

$$
\begin{aligned}
z = \max\ & cx + \eta' \\
& \eta' \leq u^k(b - Ax) \quad \text{for } k \in K \\
& v^j(b - Ax) \geq 0 \quad \text{for } j \in J \\
& x \in X,\ \eta' \in R^1.
\end{aligned}
$$

We consider the formulation given in Section I.1.3:

$$
\begin{aligned}
z = \max\ & -\sum_{j \in N} f_j x_j + \sum_{i \in I}\sum_{j \in N} c_{ij} y_{ij} \\
& \sum_{j \in N} y_{ij} = 1 \quad \text{for } i \in I \\
& -x_j + y_{ij} \leq 0 \quad \text{for } i \in I, j \in N \\
& x \in B^n,\ y \in R^{nm}_+,
\end{aligned}
$$

where $N = \{1, \ldots, n\}$ and $I = \{1, \ldots, m\}$.

In this case, LP(x) is

$$
\begin{aligned}
z_{\text{LP}}(x) = \max\ & \sum_{i \in I}\sum_{j \in N} c_{ij} y_{ij} \\
& \sum_{j \in N} y_{ij} = 1 \quad \text{for } i \in I \\
& y_{ij} \leq x_j \quad \text{for } i \in I, j \in N \\
& y \in R^{mn}_+.
\end{aligned}
$$

Now rather than applying the Benders' reformulation directly, we will take advantage of the fact that LP(x) can be decomposed into m subproblems. For $i \in I$, let

$$\text{LP}^i(x) \qquad \begin{aligned} z^i_{\text{LP}}(x) = \max & \sum_{j \in N} c_{ij} y_{ij} \\ & \sum_{j \in N} y_{ij} = 1 \\ & y_{ij} \leqslant x_j \quad \text{for } j \in N \\ & y \in R^n_+ \end{aligned}$$

and note that $z_{\text{LP}}(x) = \Sigma_{i \in I} z^i_{\text{LP}}(x)$.

Clearly, $\text{LP}^i(x)$ is feasible and bounded for $x \in B^n/\{0\}$. Hence to describe $z^i_{\text{LP}}(x)$, it suffices to find the extreme points of

$$U^i(x) = \{u_i \in R^1, w_i \in R^n_+ : u_i + w_{ij} \geqslant c_{ij} \quad \text{for } j \in N\},$$

where $w_i = (w_{i1}, \ldots, w_{in})$. It is easily seen that these extreme points are

$$u_i = c_{ik},\ w_i = ((c_{i1} - c_{ik})^+, \ldots, (c_{in} - c_{ik})^+) \quad \text{for } k \in N.$$

Hence

$$z^i_{\text{LP}}(x) = \min_{k \in N} \left[c_{ik} + \sum_{j \in N} (c_{ij} - c_{ik})^+ x_j \right].$$

As a result we can write the Benders' reformulation:

$$(7.2) \qquad \begin{aligned} z = \max - & \sum_{j \in N} f_j x_j + \sum_{i \in I} \eta_i \\ & \eta_i \leqslant c_{ik} + \sum_{j \in N} (c_{ij} - c_{ik})^+ x_j \quad \text{for } i \in I \text{ and } k \in N \\ & \sum_{j \in N} x_j \geqslant 1 \\ & x \in B^n,\ \eta \in R^m, \end{aligned}$$

which has no more than $mn + 1$ constraints. The standard Benders' reformulation, obtained directly from $\text{LP}(x)$ without decomposition, has an exponential number of constraints.

8. NOTES

Section II.3.1

The concepts of relaxation and weak duality might best be attributed to the folklore of the field. Geoffrion and Marsten (1972) were among the first to use the term *relaxation* explicitly in the context of discrete optimization. Nemhauser (1985) gave an annotated bibliography of the uses of duality in integer programming and combinatorial optimization.

Section II.3.2

The value function of a discrete optimization problem appeared in Everett's (1963) rather informal treatment of Lagrangian relaxation and duality. Its importance was brought out in much greater depth in Geoffrion's (1974) treatment of Lagrangian duality for integer programs.

The value function of a knapsack problem was studied and shown to be superadditive by Gilmore and Gomory (1966). Gomory (1965, 1967) extended these results to group problems.

The value functions of pure- and mixed-integer programs have been studied extensively by Blair and Jeroslow (1977, 1982, 1984, 1985).

The general dual problem (2.4) comes from Wolsey (1981a) and Tind and Wolsey (1981).

Section II.3.3

An explicit statement of a superadditive dual appears in Johnson (1973) in the context of a cyclic group problem. Obviously, superadditive duality is closely related to the superadditive characterization of all valid inequalities (see the notes for Sections II.2.4–II.2.7).

Blair and Jeroslow (1977) use the superadditivity of the value function to study the sensitivity of the optimal value as b varies. Jeroslow (1978), Wolsey (1981b), and Blair and Jeroslow (1982) studied the representation of the value function by a finite number of C–G functions. Cook, Gerards et al. (1986) generalized these results and also derived upper bounds on the Chvátal rank as a function of n, independent of the data.

Section II.3.4

This longest-path view of integer programs is based on Gilmore and Gomory's (1966) dynamic programming recursion for the knapsack problem. Also see Shapiro (1968a).

Section II.3.5

The group problem was introduced by Gomory (1965), and the results of this section are from that article. Also see Shapiro (1970) and Wolsey (1971a,b).

The literature on methods for solving the group problem and using it as relaxation for solving the general integer programming problem will be given in the notes for Section II.6.1.

Section II.3.6

Lorie and Savage (1955) proposed a simple heuristic for 0-1 integer programming that is equivalent to a Lagrangian relaxation with respect to all of the linear constraints. Nemhauser and Ullman (1968) showed that with respect to this relaxation, the problem of finding an optimal set of multipliers is equivalent to solving the dual of the linear programming relaxation; they also showed that this set of multipliers yields the same bound as that obtained from the linear programming relaxation.

Everett (1963) introduced the concept of Lagrangian relaxation for structured discrete optimization problems, and he proved Corollaries 6.5 and 6.6, Theorem 6.7, and Corollaries 6.8–6.10 without explicitly using Theorem 6.2.

Brooks and Geoffrion (1966) established the connection between Lagrangian relaxation and column generation methods for solving large-scale linear programs [see Dantzig and Wolfe 1960]. Geoffrion (1974) formalized the ideas of Lagrangian duality for general integer programs and, among other things, proved the main theorem (Theorem 6.2). Related articles are Shapiro (1971) and Fisher and Shapiro (1974).

Several approaches to closing the duality gap that arises in Lagrangian duality have been proposed [see, e.g., Bell and Shapiro (1977)].

The use of Lagrangian duality in solving structured combinatorial optimization problems was stimulated by Held and Karp's (1970, 1971) very successful application of it to the traveling salesman problem. Some of these applications will be elaborated on in Sections II.5.4 and II.6.1–II.6.3, and several others will be cited in the notes for Section II.5.4.

The idea of using cost splitting (Corollary 6.12) to obtain a Lagrangian dual problem equivalent to a linear program over the convex hull of the integer points in the intersection of two polyhedra appears in Ribeiro and Minoux (1985), Jornsten and Nasberg (1986), and Trick (1987).

In a somewhat different manner, this approach was used by Nemhauser and Weber (1979) to solve set partitioning problems using a matching relaxation and by Edmonds (1970) and Frank (1981) in matroid intersection problems (see Section III.3.5).

Surveys of the theory, computational aspects, and applications of Lagrangian duality are given by Shapiro (1979b) and Fisher (1981).

Surrogate duality is due to Glover (1968b, 1975) and Greenberg and Pierskalla (1970). Karwan and Rardin (1979) discussed the relationship between surrogate and Lagrangian duality. Fisher, Lageweg et al. (1983) applied surrogate duality to job shop scheduling problems.

Section II.3.7

Resource or Benders' decomposition for mixed-integer programming is described in Benders (1962). Lemke and Spielberg (1967) described a variation of Benders' algorithm that is designed for 0-1 MILPs. Geoffrion (1970, 1972) extended Benders' decomposition to handle a more general class of nonconvex optimization problems. Magnanti and Wong (1981) described techniques for obtaining stronger Benders-type reformulations. Wolsey (1981c) and Holm and Tind (1985) provided theoretical extensions to the decomposition of integer programs. Van Roy (1983, 1986) proposed a procedure called cross-decomposition, which simultaneously uses Lagrangian and Benders' decomposition.

9. EXERCISES

1. Formulate the packing and covering problems discussed in this chapter as integer programs and thereby show that they are dual problems. Do you know any cases where strong duality holds?

2. Find a maximum-weight node packing on the graph shown in Figure 9.1. The numbers on the nodes are the weights. Give a short proof that this packing is optimal.

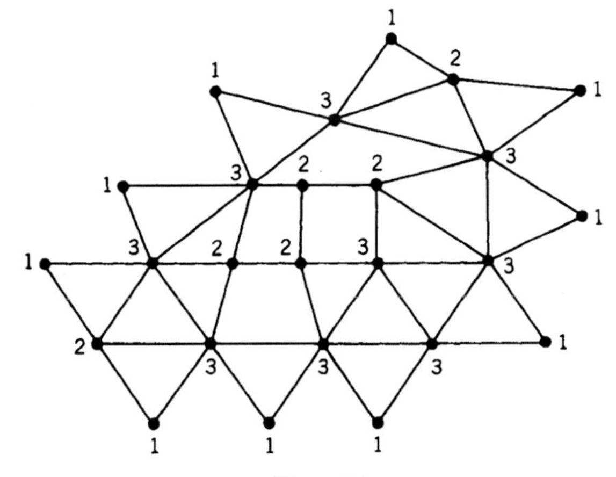

Figure 9.1

3. A *restriction* of IP is any maximization problem

$$z_T = \max\{z_T(x): x \in S_T\},$$

where (a) $S_T \subseteq S$, and (b) $z_T(x) \leqslant cx$ for $x \in S_T$.

i) What use is a restriction of IP?

ii) What can be said about its dual?

4. i) Calculate the value function of the knapsack problem

$$\begin{aligned} z(d) = \max\ & 7x_1 + 4x_2 + x_3 \\ & 5x_1 + 3x_2 + 2x_3 \leqslant d \\ & x \in Z_+^3. \end{aligned}$$

ii) Show that z is superadditive and nondecreasing for $d \in Z_+^1$.

iii) Express z in such a way that there is a short proof that it is superadditive and nondecreasing.

5. Let

(P) $$z = \max\{cx: Ax \leqslant b, x \in X\}$$

and

(P$_i$) $$z_i = \max\{cx: Ax \leqslant b, x \in X_i\} \quad \text{for } i = 1, \dots, n.$$

Show that if F_i is dual feasible for (P$_i$) for $i = 1, \dots, n$, and $X = \cup_{i=1}^n X_i$, then $F = \max_i \{F_i\}$ is dual feasible for (P).

6. Show that the problem $\min_{x \in B^n} f(x)$ has a dual problem given by

$$\begin{aligned} \max\ & y_0 + \sum_{j \in N} y_j \\ & y_0 + \sum_{j \in S} y_j \leqslant f(x^S) \quad \text{for all } S \subseteq N \\ & y_0 \in R^1, y \in R_+^n. \end{aligned}$$

Hint: Take $\mathscr{F} = \{F: F(d) = y_0 + yd\}$, the class of affine functions.

7. i) Show that the superadditive dual of

(IP) $$\max\{cx: Ax \leqslant b, x \in Z^n\}$$

is

(SD) $$\begin{aligned} & \min F(b) \\ & F(a_j) = c_j \quad \text{for } j \in N \\ & F \text{ superadditive and nondecreasing with } F(0) = 0. \end{aligned}$$

ii) Show that if F is feasible in (SD), then $F(Ax) = -F(-Ax)$ for all $x \in Z^n$.

iii) Show that $\Sigma_{j \in N} F(a_j)x_j \leqslant F(b)$ is a valid inequality for IP.

iv) Show that if IP is feasible for all d, the value function

$$z(d) = \max\{cx: Ax \leqslant d, x \in Z^n\}$$

is dual optimal for all d.

8. i) Give the superadditive dual of

$$\begin{aligned} z = \max\ & 7x_1 + 4x_2 + 1x_3 \\ & 5x_1 + 3x_2 + 2x_3 \leqslant b \\ & x \in Z^3_+ \end{aligned}$$

(see exercise 4).

ii) Find at least two dual feasible solutions when $b = 13$.

iii) Use these solutions to obtain bounds when $b = 15$.

9. i) Give the superadditive dual of

$$\begin{aligned} \max\ & 5x_1 + 11x_2 + 16x_3 + 20x_4 \\ & x_1 + 2x_2 + 3x_3 + 4x_4 \leqslant 14 \\ & x \in Z^4_+. \end{aligned}$$

ii) Use the superadditive description of conv(S) (see Example 3.1) to find an optimal dual solution.

10. i) Formulate the problem

$$\begin{aligned} \min\ & 7x_1 + 6x_2 + 2x_3 \\ & \begin{pmatrix} 3 \\ 3 \end{pmatrix} x_1 + \begin{pmatrix} 3 \\ 1 \end{pmatrix} x_2 + \begin{pmatrix} 1 \\ 2 \end{pmatrix} x_3 \geqslant \begin{pmatrix} 5 \\ 4 \end{pmatrix} \\ & x \in Z^3_+ \end{aligned}$$

as a shortest-path problem.

ii) Solve the problem by Dijkstra's shortest-path algorithm.

iii) Give a dual feasible solution.

11. Use the group problem to solve

$$\begin{aligned} \max\ & 7x_1 + 4x_2 + x_3 \\ & 5x_1 + 3x_2 + 2x_3 \leqslant b \\ & x \in Z^3_+ \end{aligned}$$

for $b = 217$, 495, and 621.

12. Use the group problem to solve

$$\begin{aligned}\max\ & 2x_1 + 5x_2 \\ & 4x_1 + \ \ x_2 \leqslant 28 \\ & x_1 + 4x_2 \leqslant 27 \\ & x_1 - \ \ x_2 \leqslant 1 \\ & x \in Z_+^2.\end{aligned}$$

(See exercises 1 and 13 of Section II.1.9).

13. Use Lagrangian duality to solve the problem of exercise 10 with $b = \binom{5}{4}$.

i) What bound is obtained by dualizing the first constraint?

ii) What bound is obtained by dualizing the second constraint?

iii) For what values of b is the optimal solution easily obtained? (See Corollary 6.10.)

14. Apply Lagrangian relaxation to the integer program in exercise 12.

i) Show that if any two constraints are dualized, the value of the Lagrangian dual equals the value of the linear programming relaxation.

ii) Find a different objective function for which i is false.

iii) Show that if any single constraint is dualized, the value of the Lagrangian dual is an improvement on the value of the linear programming relaxation.

iv) Apply cost splitting to get a better Lagrangian dual.

v) Demonstrate i–iv graphically.

15. Consider two different Lagrangian duals for the generalized assignment problem:

$$\begin{aligned}\max\ & \sum_i \sum_j c_{ij} x_{ij} \\ & \sum_j x_{ij} \leqslant 1 \quad \text{for } i \in M \\ & \sum_i l_i x_{ij} \leqslant b_j \quad \text{for } j \in N \\ & x \in B^{mn}.\end{aligned}$$

Discuss their relative merits according to the following three criteria:

i) ease of solution of the subproblem,

ii) ease of solution of the Lagrangian dual,

iii) strength of the upper bound obtained by solving the dual.

16. Discuss the merits of different Lagrangian duals for the capacitated facility location problem

$$\min \sum_{i\in M}\sum_{j\in N} h_{ij}y_{ij} + \sum_{j\in N} c_j x_j$$

$$\sum_{j\in N} y_{ij} \leq a_i \qquad \text{for } i \in M$$

$$\sum_{i\in M} y_{ij} \leq b_j x_j \qquad \text{for } j \in N$$

$$y_{ij} \leq \min(a_i, b_j)x_j \quad \text{for } i \in M, j \in N$$

$$y \in R_+^{mn}, x \in B^n.$$

17. Consider the problem of processing n jobs on one machine. Let p_j denote the processing time of job j, let r_j denote the earliest start time, and let w_j denote the weight associated with job j. The problem is to minimize $\Sigma_{j\in N} w_j t_j$, where t_j is the start time of job j. Without release dates ($r_j = 0$ for all j), the optimal job ordering is given by Smith's rule: Process the jobs in order $1, \ldots, n$, where $w_1/p_1 \geq \ldots \geq w_n/p_n$. How can Lagrangian relaxation be used to obtain a lower bound for the problem with release dates?

18. Consider the capacitated lot-sizing problem, that is, the uncapacitated problem formulated in Section I.1.4 (see also Section II.2.4) with additional capacity constraints on the production levels $y_t \leq u_t x_t$ for $t = 1, \ldots, T$. After dualizing these constraints, the Lagrangian subproblem is an uncapacitated problem that can be solved rapidly by dynamic programming (see Section II.5.5).
The Lagrangian dual is equivalent to a linear programming problem. Describe this linear programming problem in polyhedral terms.

19. Solving the Lagrangian duals in exercise 15 is equivalent to solving the dual problem

$$\min F\begin{pmatrix}1\\ b\end{pmatrix}$$

$$F\begin{pmatrix}e_i\\ l_i e_j\end{pmatrix} \geq c_{ij} \quad \text{for all } i \text{ and } j$$

$$F \in \mathcal{F}$$

for certain classes of functions $\mathcal{F}$, where e_i is the ith unit vector. For each of your proposed Lagrangians, what is $\mathcal{F}$?

20. Describe the class of dual functions that correspond to the cost splitting and surrogate duals, respectively. Show that neither dual dominates the other.

21. Suggest how to find optimal multipliers in the surrogate dual.

22. Apply Benders' reformulation to the fixed-charge network problem described in Section I.1.4. Discuss possible advantages of such a reformulation.

23. Apply Benders' reformulation to UFL (without separating the subproblems by client) and compare this formulation with (7.2).

24. Write out explicitly the Benders' reformulation of the mixed-integer program

$$\begin{aligned} \max\ & 2x_1 + x_2 + 3x_3 + 7y_1 + 5y_2 \\ & 9x_1 + 4x_2 + 14x_3 + 35y_1 + 24y_2 \leqslant 80 \\ & -x_1 - 2x_2 + 3x_3 - 2y_1 + 4y_2 \leqslant 10 \\ & x \in Z_+^3,\ y \in R_+^2. \end{aligned}$$

II.4

General Algorithms

1. INTRODUCTION

Here we discuss approaches for finding an optimal, or ε-approximate, solution of the linear integer programming problem

$$\text{(IP)} \qquad z_{\text{IP}} = \max\{cx: x \in S\}.$$

For simplicity, in this introductory discussion, we assume $S \neq \emptyset$ and $z_{\text{IP}} < \infty$. Therefore, to solve an instance of IP, an algorithm must produce a feasible solution $x^0 \in S$ and an upper bound w^0 on the value of all feasible solutions such that $cx^0 = w^0$. A general iterative scheme for finding x^0 and w^0 is shown in Figure 1.1.

Many integer programming algorithms focus on the dual step by systematically reducing the upper bound w^* but generally not producing an $x \in S$ until $w^* = z_{\text{IP}}$. Relaxation algorithms are of this type. At each iteration a relaxation of IP is solved and if an optimal solution of the relaxation does not yield an optimal solution of IP, the relaxation is refined. A general relaxation algorithm is the following.

General Relaxation Algorithm

Initialization: Set $t = 1$, $w^* = \infty$, and $z^* = -\infty$. Choose $S_R^1 \supseteq S$ and $z_R^1(x) \geqslant cx$ for $x \in S$.

Iteration t:

Step 1: Solve the relaxation of IP:

$$(\text{RP}^t) \qquad z_R^t = \max\{z_R^t(x): x \in S_R^t\}.$$

Step 2: Optimality test. Let the solution be x^t. If $x^t \in S$ and $z_R^t = cx^t$, then $w^* = cx^t = z^*$ and x^t is an optimal solution to IP.

Step 3: Refinement. Set $w^* = z_R^t$, $z^* = cx^t$ if $x^t \in S$, and $t \leftarrow t + 1$. Choose S_R^{t+1} to satisfy $S \subseteq S_R^{t+1} \subseteq S_R^t$ and $z_R^{t+1}(x)$ to satisfy $cx \leqslant z_R^{t+1}(x) \leqslant z_R^t(x)$ for $x \in S$ with either $S_R^{t+1} \neq S_R^t$ or $z_R^{t+1}(x) \neq z_R^t(x)$.

Note that in this algorithm, the sequence of upper bounds satisfies $z_R^{t+1} \leqslant z_R^t$ for all t. In many specific instances of the general relaxation algorithm, $z_R^t(x) = cx$ for all t so that optimality is achieved as soon as an $x^t \in S$ is produced. In this case, the refinement step satisfies $S_R^{t+1} \subset S_R^t$ for all t. It is then desirable to choose $S_R^{t+1} \subseteq S_R^t \setminus \{x^t\}$; otherwise $z_R^{t+1} = z_R^t$.

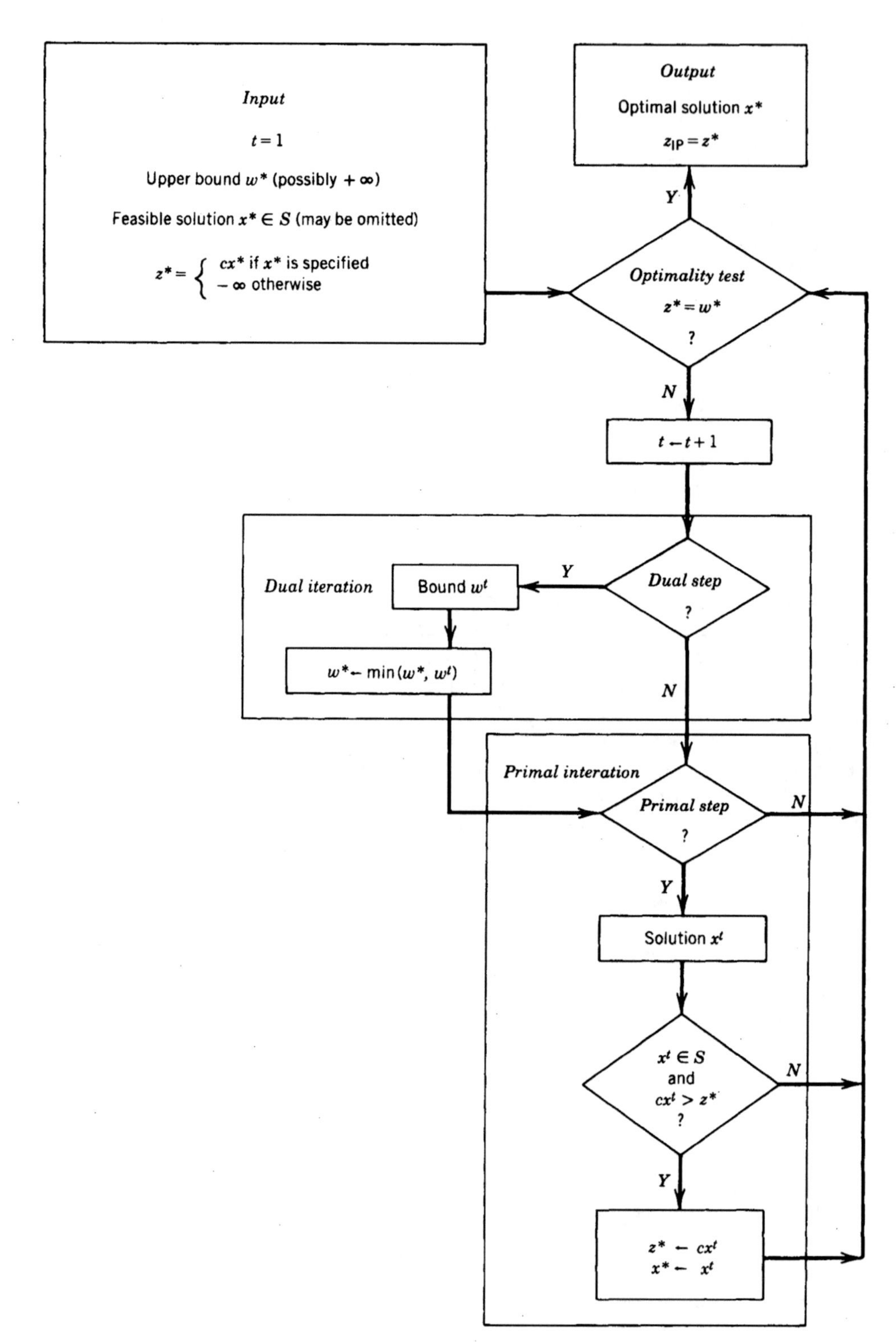

Input
t = 1
Upper bound w* (possibly + ∞)
Feasible solution x* ∈ S (may be omitted)
z* = { cx* if x* is specified
− ∞ otherwise
Output
Optimal solution x*
z_IP = z*
Y
Optimality test
z* = w*
?
N
t ← t + 1
Dual iteration
Bound w^t
Y
Dual step
?
w* ← min (w*, w^t)
N
Primal interation
Primal step
?
N
Y
Solution x^t
x^t ∈ S
and
cx^t > z*
?
N
Y
z* ← cx^t
x* ← x^t

Figure 1.1

An important example of this type of relaxation algorithm is a *fractional cutting-plane algorithm*. Here we assume that $S = \{x \in Z^n_+ : Ax \leqslant b\}$. Note that to specify the algorithm, it suffices to give the initial relaxation and the rule for constructing RP^{t+1} from RP^t.

Fractional Cutting-Plane Algorithm (FCPA)

Initialization: $z^1_R(x) = cx$ for all $x \in R^n_+$; $S^1_R = \{x \in R^n_+ : Ax \leqslant b\}$.

Refinement: $z^{t+1}_R(x) = z^t_R(x)$ for all $x \in R^n_+$; $S^{t+1}_R = S^t_R \cap \{x \in R^n_+ : \pi^t x \leqslant \pi^t_0\}$, where (π^t, π^t_0) is a valid inequality for S such that $\pi^t x^t > \pi^t_0$.

Observe that $\max\{cx: x \in S^t_R\}$ is a linear program whose optimal dual solution u^t is readily extended to the feasible solution $(u^t, 0)$ to the dual of $\max\{cx: x \in S^{t+1}_R\}$. The dual variable for the new constraint $\pi^t x \leqslant \pi^t_0$ equals zero. Thus it is desirable to solve the sequence of linear programs

$$(\text{LP}^t) \qquad \max\{cx: x \in S^t_R\}$$

by a dual algorithm. Hence we can interpret the fractional cutting-plane algorithm as a dual linear programming algorithm for solving IP. The dual of LP^t is weakly dual to IP. The generation of a valid inequality corresponds to the generation of a column in the dual space and, consequently, to a relaxation of the dual of LP^t.

Figure 1.2 illustrates the application of FCPA to the two-variable integer programming problem introduced in Chapter II.1.

We will not discuss here the very important question of how to choose the valid inequality (π^t, π^t_0) that separates x^t from S^{t+1}_R. In Section 3, we will give an FCPA for general integer programs that uses C–G inequalities. In Chapters II.5 and II.6 we will show how strong valid inequalities can be used in FCPAs for some structured integer programs and the general 0-1 integer program.

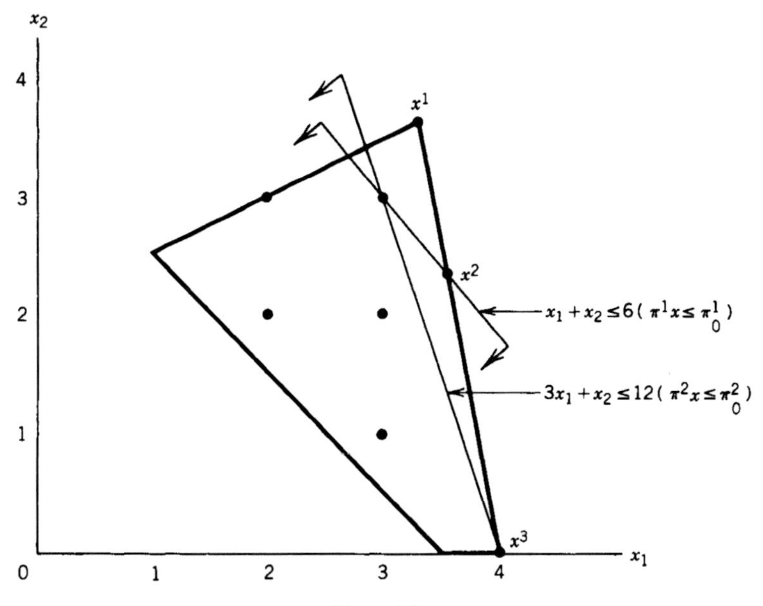

Figure 1.2

Another very important type of relaxation algorithm uses an enumerative approach. We say that $\{S^i: i = 1, \ldots, k\}$ is a *division* of S if $\bigcup_{i=1}^k S^i = S$. A division is called a *partition* if $S^i \cap S^j = \emptyset$ for $i, j = 1, \ldots, k$, $i \neq j$.

Proposition 1.1 *Let*

$$(\mathrm{IP}^i) \qquad z_{\mathrm{IP}}^i = \max\{cx: x \in S^i\},$$

where $\{S^i\}_{i=1}^k$ *is a division of* S. *Then* $z_{\mathrm{IP}} = \max_{i=1,\ldots,k} z_{\mathrm{IP}}^i$.

Proposition 1.1 expresses the familiar concept of *divide and conquer*. In other words, if it is too difficult to optimize over S, perhaps the problem can be solved by optimizing over smaller sets and then putting the results together.

The division is frequently done recursively as shown in the tree of Figure 1.3. Here the sons of a given node [e.g., (S^{11}, S^{12}, S^{13}) are the sons of S^1] represent a division of the feasible region of their father.

When $S \subseteq B^n$, a simple way of doing the recursive division is shown in Figure 1.4. Here $S^{\delta_1 \ldots \delta_k} = S \cap \{x \in B^n: x_j = \delta_j \in \{0, 1\}$ for $j = 1, \ldots, k\}$, and the division is a partition of S.

Carried to the extreme, division can be viewed as total enumeration of the elements of S. Total enumeration is not viable for problems with more than a very small number of variables. To have any hope of working, the enumerative approach needs to avoid dividing the initial set into too many subsets.

Suppose S has been divided into subsets $\{S^1, \ldots, S^k\}$. If we can establish that no further division of S^i is necessary, we say that the enumeration tree can be *pruned* at the node corresponding to S^i or, for short, that S^i can be pruned.

Proposition 1.2. *The enumeration tree can be pruned at the node corresponding to* S^i *if any one of the following three conditions holds.*

1. *Infeasibility:* $S^i = \emptyset$.
2. *Optimality: An optimal solution if* IP^i *is known.*
3. *Value dominance:* $z_{\mathrm{IP}}^i \leqslant z_{\mathrm{IP}}$.

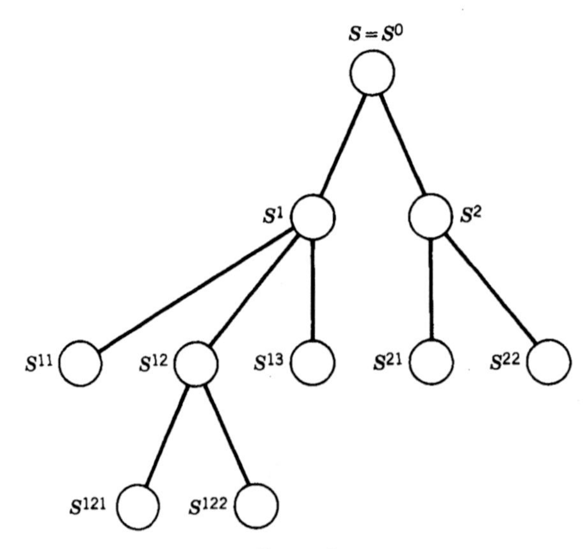

Figure 1.3

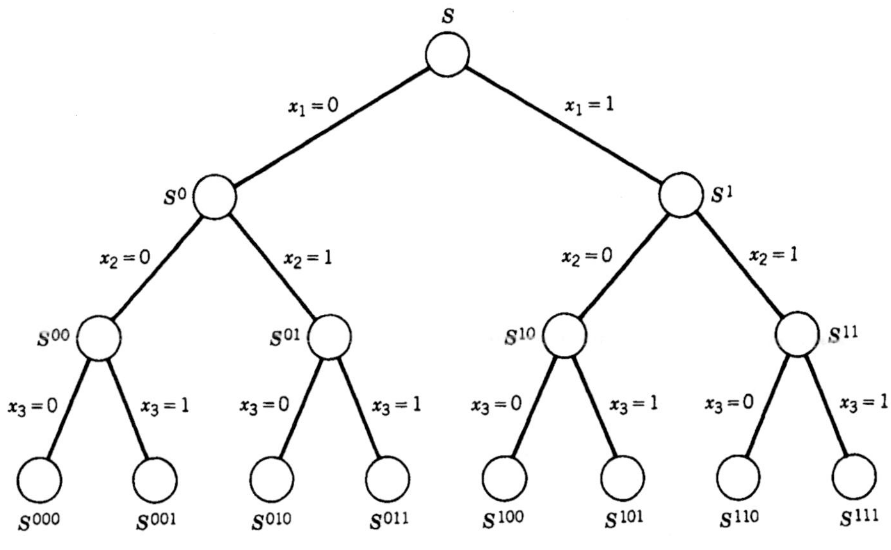

Figure 1.4

We would like to be able to apply Proposition 1.2 without necessarily having to solve IP^i. To accomplish this, we use relaxation or duality. Let RP^i be a relaxation of IP^i with $S^i \subseteq S_R^i$ and $z_R^i(x) \geqslant cx$ for $x \in S^i$.

Proposition 1.3. *The enumeration tree can be pruned at the node corresponding to S^i if any one of the following three conditions holds.*

1. RP^i *is infeasible.*
2. *An optimal solution x_R^i to* RP^i *satisfies $x_R^i \in S^i$ and $z_R^i = cx_R^i$.*
3. $z_R^i \leqslant \underline{z}_{\text{IP}}$, *where $\underline{z}_{\text{IP}}$ is the value of some feasible solution of* IP.

Proof. Condition 1 implies $S^i = \emptyset$. Condition 2 implies that x_R^i is an optimal solution to IP^i. Condition 3 implies $z_{\text{IP}}^i \leqslant \underline{z}_{\text{IP}}$. ■

Let DP^i be (weakly) dual to IP^i.

Proposition 1.4. *The enumeration tree can be pruned at the node corresponding to S^i if one of the following two conditions holds.*

1. *The objective value of* DP^i *is unbounded from below.*
2. DP^i *has a feasible solution of value equal to or less than $\underline{z}_{\text{IP}}$.*

Proof. Condition 1 implies $S^i = \emptyset$. Condition 2 implies $z_{\text{IP}}^i \leqslant \underline{z}_{\text{IP}}$. ■

Comparing Propositions 1.3 and 1.4, we see that RP^i must be solved to optimality before value dominance can be applied, but value dominance may be applicable with respect to dual feasible solutions that are not optimal. On the other hand, RP^i may yield a feasible solution to IP^i that establishes or improves the lower bound $\underline{z}_{\text{IP}}$.

Example 1.1

$$
\begin{aligned}
z_{\mathrm{IP}} = \max \; - 100x_1 + 72x_2 + 36x_3 & \\
-2x_1 + x_2 \quad & \leqslant 0 \\
-4x_1 \quad + x_3 & \leqslant 0 \\
x_1 + x_2 + x_3 & \geqslant 1 \\
x \in B^3. &
\end{aligned}
$$

A division is shown in the tree of Figure 1.5.

We use linear programming relaxation and Proposition 1.3 for pruning. The infeasibility condition holds for S^0 since

$$S_R^0 = \{x \in R_+^3: x_1 = 0, x_2 \leqslant 0, x_3 \leqslant 0, x_2 + x_3 \geqslant 1\} = \emptyset.$$

The optimality condition holds for S^{110} and S^{111} since these sets contain the unique solutions $(1 \quad 1 \quad 0)$ and $(1 \quad 1 \quad 1)$, respectively. Since $z_{\mathrm{IP}}^{110} < z_{\mathrm{IP}}^{111} = 8$, we have $\underline{z}_{\mathrm{IP}} = z_{\mathrm{IP}}^{111} = 8$. Now we can apply the value dominance criterion to S^{10} since $z_R^{10} = -64 < \underline{z}_{\mathrm{IP}}$. Hence $x^0 = (1 \quad 1 \quad 1)$ is an optimal solution to IP, and $z_{\mathrm{IP}} = 8$.

When relaxations are used for pruning, the enumerative approach fits into the context of the general relaxation algorithm. Suppose we have just solved a relaxation of IP^i. In the refinement step, we first divide S^i, say $S^i = \cup_{j=1}^k S^{ij}$. Then we form relaxations for the sets S^{ij} in such a way that $\cup_{j=1}^k S_R^{ij} \subset S_R^i$.

An enumerative relaxation algorithm is frequently called *branch-and-bound* or *implicit enumeration*. We now give a general branch-and-bound algorithm for solving IP. In the description of the algorithm, $\mathscr{L}$ is a collection of integer programs $\{\mathrm{IP}^i\}$, each of which is of the form $z_{\mathrm{IP}}^i = \max\{cx: x \in S^i\}$ where $S^i \subseteq S$. Associated with each problem in $\mathscr{L}$ is an upper bound $\overline{z}^i \geqslant z_{\mathrm{IP}}^i$.

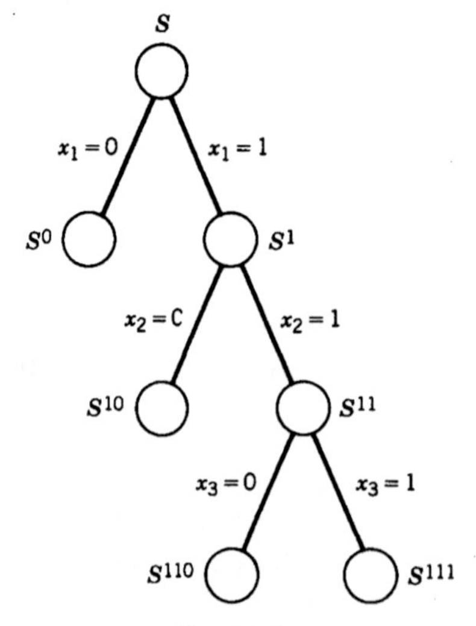

Figure 1.5

General Branch-and-Bound Algorithm

Step 1 (Initialization): $\mathscr{L} = \{\text{IP}\}$, $S^0 = S$, $\bar{z}^0 = \infty$, and $\underline{z}_{\text{IP}} = -\infty$.

Step 2 (Termination test): If $\mathscr{L} = \emptyset$, then the solution x^0 that yielded $\underline{z}_{\text{IP}} = cx^0$ is optimal.

Step 3 (Problem selection and relaxation): Select and delete a problem IP^i from $\mathscr{L}$. Solve its relaxation RP^i. Let z_R^i be the optimal value of the relaxation and let x_R^i be an optimal solution if one exists.

Step 4 (Pruning): a. If $z_R^i \leqslant \underline{z}_{\text{IP}}$, go to Step 2. (Note if the relaxation is solved by a dual algorithm, then the step is applicable as soon as the dual value reaches or falls below $\underline{z}_{\text{IP}}$.)

b. If $x_R^i \notin S^i$, go to Step 5.

c. If $x_R^i \in S^i$ and $cx_R^i > \underline{z}_{\text{IP}}$, let $\underline{z}_{\text{IP}} = cx_R^i$. Delete from $\mathscr{L}$ all problems with $\bar{z}^i \leqslant \underline{z}_{\text{IP}}$. If $cx_R^i = z_R^i$, go to Step 2; otherwise go to Step 5.

Step 5 (Division): Let $\{S^{ij}\}_{j=1}^k$ be a division of S^i. Add problems $\{\text{IP}^{ij}\}_{j=1}^k$ to $\mathscr{L}$, where $\bar{z}^{ij} = z_R^i$ for $j = 1, \ldots, k$. Go to Step 2.

Commercial codes for general mixed-integer programming problems use linear programming relaxations and division. We will study this class of algorithms in the next section. In Chapters II.5 and II.6 we will consider some special purpose branch-and-bound algorithms that use different relaxations or duals and other division tactics. We will also present cutting-plane algorithms that sometimes fail to find strong valid inequalities and then resort to branch-and-bound to complete the solution.

2. BRANCH-AND-BOUND USING LINEAR PROGRAMMING RELAXATIONS

Here we consider the general integer programming problem

$$\text{(IP)} \qquad z_{\text{IP}} = \max\{cx \colon x \in S\}, \quad \text{where } S = \{x \in Z_+^n \colon Ax \leqslant b\}.$$

We study its solution by a branch-and-bound algorithm that uses linear programming relaxations. This is the basic algorithm used by all commercial codes for solving mixed-integer programming problems. Merely for simplicity of notation, we confine the presentation to IP. Essentially, however, all of the ideas carry over unchanged to the mixed-integer program

$$\text{(MIP)} \qquad z_{\text{MIP}} = \max\{cx + hy \colon (x, y) \in T\}, \text{ where } T = \{x \in Z_+^n, y \in R_+^p \colon Ax + Gy \leqslant b\}.$$

This setting is simple but general enough to enable us to discuss various properties of branch-and-bound algorithms such as types of divisions, tree development strategies, finiteness of the resulting tree, the smallest possible tree, and so on.

In the initial relaxation, S is replaced by $S_{\text{LP}}^0 = \{x \in R_+^n \colon Ax \leqslant b\}$. We also take $z_R(x) = cx$ in each relaxation.

Pruning Criteria

When solving linear programming relaxations, the pruning criteria of infeasibility, optimality, and value dominance given in Propositions 1.3 and 1.4 are directly applicable. Suppose the linear programming relaxation at node i of the enumeration tree is

$$(\text{LP}^i) \qquad z_{\text{LP}}^i = \max\{cx \colon x \in S_{\text{LP}}^i\}, \quad \text{where } S_{\text{LP}}^i = \{x \in R_+^n \colon A^i x \leqslant b^i\}.$$

If LP^i has an optimal solution, we denote the one found by x^i.

The pruning conditions are:

1. $S^i_{\text{LP}} = \emptyset$ (infeasibility);
2. $x^i \in Z^n_+$ (optimality); and
3. $z^i_{\text{LP}} \leqslant \underline{z}_{\text{IP}}$ where $\underline{z}_{\text{IP}}$ is the value of a known feasible solution to IP (value dominance). Note that if LP^i is solved by a dual algorithm, we may be able to prune before an optimal solution to LP^i is found. Also, we may wish to use the weaker condition $z^i_{\text{LP}} \leqslant \underline{z}_{\text{IP}} + \varepsilon$ for some given tolerance $\varepsilon > 0$.

Division

Since we use a linear programming relaxation at each node, the division is done by adding linear constraints. An obvious way to do this is to take $S = S^1 \cup S^2$ with $S^1 = S \cap \{x \in R^n_+ : dx \leqslant d_0\}$ and $S^2 = S \cap \{x \in R^n_+ : dx \geqslant d_0 + 1\}$, where $(d, d_0) \in Z^{n+1}$. If x^0 is the solution to the relaxation

$$(\text{LP}^0) \qquad z^0_{\text{LP}} = \max\{cx : x \in R^n_+, Ax \leqslant b\},$$

we can choose (d, d_0) so that $d_0 < dx^0 < d_0 + 1$. This is highly desirable since it yields $x^0 \notin S^1_{\text{LP}} \cup S^2_{\text{LP}}$ and therefore gives the possibility that for $i = 1, 2$ we will obtain $z^i_{\text{LP}} = \max\{cx : x \in S^i_{\text{LP}}\} < z^0_{\text{LP}}$.

In practice, only very special choices of (d, d_0) are used.

i. *Variable dichotomy*. Here $d = e_j$ for some $j \in N$. Then x^0 will be infeasible in the resulting relaxations if $x^0_j \notin Z^1$ and $d_0 = \lfloor x^0_j \rfloor$ (see Figure 2.1). Note that if $x_j \in B^1$, then the left branch yields $x_j = 0$ and the right branch yields $x_j = 1$.

An important practical advantage of this division is that only simple lower- and upper-bound constraints are added to the linear programming relaxation. Thus it is only necessary to keep track of the bounds, and the size of the basis does not increase.

ii. *GUB dichotomy*. Suppose the problem contains the generalized upper-bound constraint $\Sigma_{j \in Q} x_j = 1$ for some $Q \subseteq N$. The division is shown in Figure 2.2. Note that x^0 will be infeasible in the resulting relaxations if $0 < \Sigma_{j \in Q_1} x^0_j < 1$, where Q_1 is a nonempty subset of Q.

iii. Assuming that x_j is bounded $(0 \leqslant x_j \leqslant k_j)$, we can consider each integral value of x_j separately (see Figure 2.3). This approach, however, is not used in commercial integer programming codes.

Note that each of the divisions i–iii is a partition.

We now consider the size of the enumeration tree. For most of the remainder of this section, we will assume that the division is done by variable dichotomy.

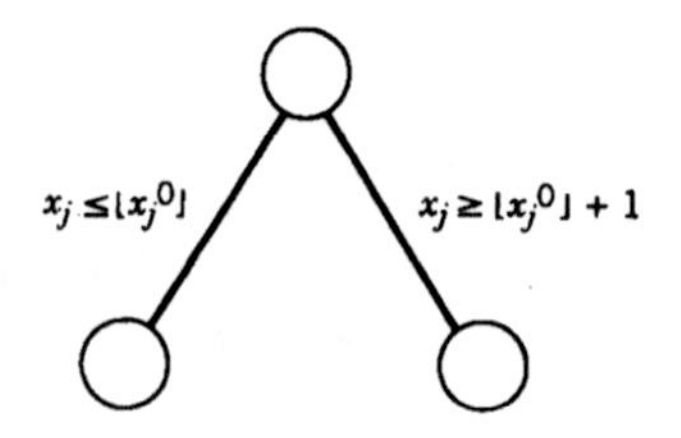

Figure 2.1

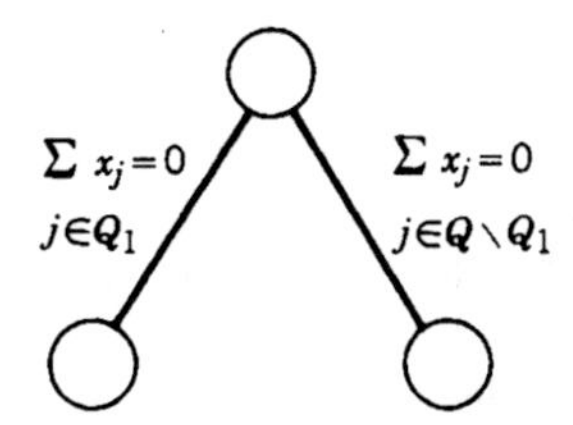

Figure 2.2

Proposition 2.1. *If $P=\{x\in R^n_+: Ax\leqslant b\}$ is bounded, an enumeration tree developed on variable dichotomies will be finite provided that at each node i that requires division, a dichotomy of the form $(x_j\leqslant\lfloor x^i_j\rfloor,\ x_j\geqslant\lfloor x^i_j\rfloor+1)$ is chosen where x^i_j is not integral. In particular, if $\omega_j=\lceil\max\{x_j: x\in P\}\rceil$, no path of the tree can contain more than $\Sigma_{j\in N}\,\omega_j$ edges.*

Proof. Once we have added the constraint $x_j\leqslant d$ for some $d\in\{0,\ldots,\omega_j-1\}$, the only other constraints that can subsequently appear on a path from the root to a leaf of the tree are $x_j\leqslant d'$ for $d'\in\{0,\ldots,d-1\}$ and $x_j\geqslant\bar{d}$ for $\bar{d}\in\{1,\ldots,d\}$. It follows that the largest number of constraints involving x_j will occur by adding $x_j\leqslant d$ for all $d\in\{0,\ldots,\omega_j-1\}$, or $x_j\geqslant d$ for all $d\in\{1,\ldots,\omega_j\}$, or $x_j\geqslant d$ for all $d\in\{1,\ldots,\alpha\}$ and $x_j\leqslant d$ for all $d\in\{\alpha,\ldots,\omega_j-1\}$. In each of these cases, we require ω_j constraints on x_j and hence $\Sigma_{j\in N}\,\omega_j$ in total on any path. ■

We can use Proposition 2.1 and the upper bounds given in Theorem 4.1 of Section I.5.4 to enforce the finiteness of the enumeration tree even when P is not bounded.

The size of the enumeration tree is very dependent on the quality of the bounds produced by the (linear programming) relaxation. In particular, we have the following proposition.

Proposition 2.2. *If node t of the enumeration tree with constraint set S^t is such that $\max\{cx: x\in S^t_R\}>z_{IP}$, then node t cannot be pruned.*

Proposition 2.2 indicates that, regardless of how we develop the tree, the bounds (quality of relaxations) are the primary factor in the efficiency of a branch-and-bound algorithm. Nevertheless, tree development strategies, such as which subproblem corresponding to an unpruned node should be considered next and which fractional variable should be selected for the dichotomous division, are also important. We now consider these problems.

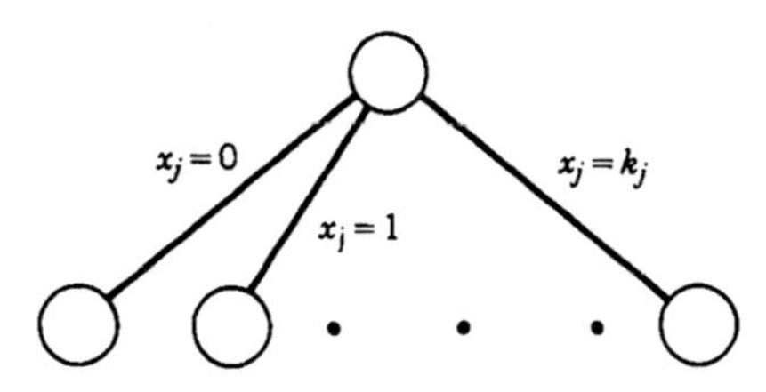

Figure 2.3

Node Selection

Given a list $\mathscr{L}$ of active subproblems or, equivalently, a partial tree of unpruned or *active* nodes, the question is to decide which node should be examined in detail next. Here there are two basic options: (1) *a priori rules* that determine, in advance, the order in which the tree will be developed; and (2) *adaptive rules* that choose a node using information (bounds, etc.) about the status of the active nodes.

A widely used (essentially) a priori rule is *depth-first search plus backtracking*, which is also known as *last in, first out* (LIFO). In depth-first search, if the current node is not pruned, the next node considered is one of its two sons. Backtracking means that when a node is pruned, we go back on the path from this node toward the root until we find the first node (if any) that has a son that has not yet been considered. Depth-first search plus backtracking is a completely a priori rule if we fix a rule for choosing branching variables and specify that the left son is considered before the right son. An example of depth-first search plus backtracking with left sons first is given in Figure 2.4. The nodes are numbered in the order in which they are considered. An underlined node is assumed to have been pruned.

Depth-first search has two principle advantages:

1. The linear programming relaxation for a son is obtained from the linear programming relaxation of its father by the addition of a simple lower- or upper-bound constraint. Hence given the optimal solution for the father node, we can directly reoptimize by the dual simplex algorithm without a basis reinversion or a transfer of data.
2. Experience seems to indicate that feasible solutions are more likely to be found deep in the tree than at nods near the root. The success of a branch-and-bound algorithm is very dependent on having a good lower-bound $\underline{z}_{\text{IP}}$ for value dominance pruning.

The default option in most commercial codes is depth first when the current node is not pruned. At least one son is considered immediately. Rules for choosing a son will be discussed later. However, when a node is pruned, the next node is not generally determined by the backtracking strategy. Before explaining how this selection is done, we mention one other essentially a priori rule, which is the opposite of depth-first search. The

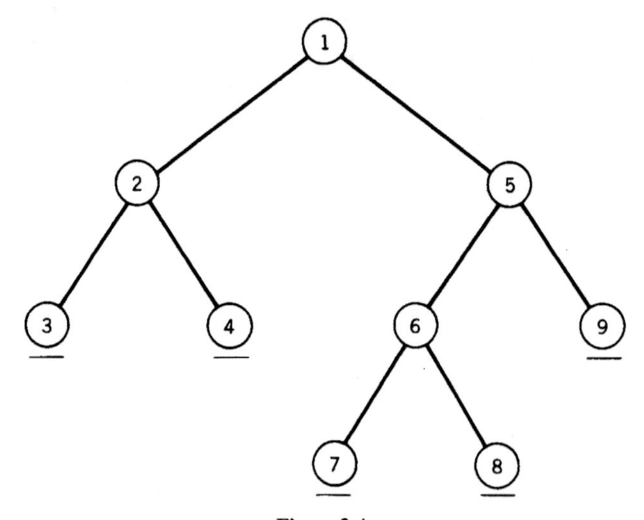

Figure 2.4

level of a node in an enumeration tree is the number of edges in the unique path between it and the root. In *breadth-first search*, all of the nodes at a given level are considered before any nodes at the next lower level. While this means of node selection is not practical for solving general integer programs using linear programming relaxations, it has some interesting properties, one of which is its use in heuristics.

Several reasonable criteria can be given for choosing an active node:

a. Choose a node that has to be considered in any case. By Proposition 2.2, if there is a unique node with the largest upper bound it must be considered. This argument mitigates for the rule *best upper bound*; that is, when a node has been pruned, next select from all active nodes one that has the largest upper bound. Thus if $\mathscr{L}$ is the set of active nodes, select an $i \in \mathscr{L}$ that maximizes $\overline{z}^i$.

b. Choose a node that is more likely to contain an optimal solution. The reason for this is that once we have found an optimal solution, even if we are unable to prove immediately that it is optimal, we will have obtained the largest possible value of $\underline{z}_{\text{IP}}$. This is very important for subsequent pruning. Later we will give a simple procedure for estimating z_{IP}^i. Suppose $\hat{z}^i \leqslant \overline{z}^i$ is an estimate of z_{IP}^i. The rule *best estimate* is to choose an $i \in \mathscr{L}$ that maximizes $\hat{z}^i$.

c. Although trying to find an optimal solution is highly desirable, it may be more practical to try to find quickly a feasible solution $\hat{x}$ such that $c\hat{x} > \underline{z}_{\text{IP}}$. The criterion

$$\max_{i \in \mathscr{L}} \frac{\overline{z}^i - \underline{z}_{\text{IP}}}{\overline{z}^i - \hat{z}^i}, \tag{2.1}$$

which we call *quick improvement*, attempts to achieve this objective. Note that node i with $\hat{z}^i > \underline{z}_{\text{IP}}$ will be preferred to node j with $\hat{z}^j \leqslant \underline{z}_{\text{IP}}$. Moreover, preference will be given to nodes for which $\overline{z}^i - \hat{z}^i$ is small. One expects that such nodes will yield a feasible solution quickly. Quick improvement is used in some commercial codes as the default option once a feasible solution is known.

Branching Variable Selection

Suppose we have chosen an active node i. Associated with it is the linear programming solution x^i. Next we must choose a variable to define the division. We restrict it to the index set $N^i = \{j \in N: x_j^i \notin Z^1\}$. Empirical evidence shows that the choice of a $j \in N^i$ can be very important to the running time of the algorithm. Frequently, there are a few variables that need to be fixed at integer values and then the rest turn out to be integer-valued in linear programming solutions. Because robust methods for identifying such variables have not been established, a common way of choosing a branching variable is by *user-specified priorities*. This means that an ordering of the variables is specified as part of the input and that branching variables are selected from N^i according to this order. For example, a 0-1 variable corresponding to whether a project should be done would be given higher priority than 0-1 variables corresponding to detailed decisions within the project.

Other possibilities involve *degradations* or *penalties*. Degradation attempts to estimate the decrease in $\overline{z}^i$ that is caused by requiring x_j to be integral. Suppose $x_j = x_j^i = \lfloor x_j^i \rfloor + f_j^i$ and $f_j^i > 0$. Then by branching on x_j, we estimate a decrease of $D_j^{-i} = p_j^{-i} f_j^i$ for the left son and $D_j^{+i} = p_j^{+i}(1 - f_j^i)$ for the right son. The coefficients $\{p_j^{-i}, p_j^{+i}\}$ can be specified as part of the input or estimated in several different ways (e.g., by using dual information at the node or by using information on previous branchings involving x_j).

Penalties involve more elaborate calculations to determine the coefficients $\{p_j^{-i}, p_j^{+i}\}$ and yield a lower bound on the decrease in $\bar{z}^i$. They were used in early commercial codes but are not in favor now because they are too costly to compute relative to the value of the information they give. An illustration of penalty calculations will be given in Example 2.1.

Given $\{D_j^{-i}, D_j^{+i}\}$ for $j \in N^i$, a common way to choose the branching variable is by the criterion

$$\max_{j \in N^i} \min\{D_j^{-i}, D_j^{+i}\}. \tag{2.2}$$

The idea is that a variable whose smallest degradation is largest is most important for achieving integrality. When $D_j^{-i} = f_j^i$ and $D_j^{+i} = 1 - f_j^i$, criterion (2.2) is called *maximum integer infeasibility*.

Other rules are also used, for example, $\max_{j \in N^i} \max\{D_j^{-i}, D_j^{+i}\}$. Here the idea is that one branch may easily be pruned by value dominance.

When the branching variable is chosen by (2.2), it is recommended that we next consider the subproblem corresponding to the son that yields the smaller degradation. Thus we select the subproblem corresponding to the left son if and only if $D_j^{-i} \leqslant D_j^{+i}$.

Now we can compute $\hat{z}^i$ by assuming that the degradations for each variable are independent. Thus if $D_j^{-i} \leqslant D_j^{+i}$, we estimate

$$\hat{z}^i = z_{\text{LP}}^i - D_j^{-i} - \sum_{k \in N^i \setminus \{j\}} \min\{D_k^{-i}, D_k^{+i}\}.$$

Note that if we are required to branch to the right son of node i, the estimate becomes

$$\hat{z}^i = z_{\text{LP}}^i - D_j^{+i} - \sum_{k \in N^i \setminus \{j\}} \min\{D_k^{-i}, D_k^{+i}\}.$$

Example 2.1

$$\begin{aligned} z_{\text{IP}} = \max\ & 7x_1 + 2x_2 \\ & -x_1 + 2x_2 \leqslant 4 \\ & 5x_1 + x_2 \leqslant 20 \\ & -2x_1 - 2x_2 \leqslant -7 \\ & x \in Z_+^2. \end{aligned}$$

We introduce slack variables $(x_3, x_4, x_5) \in R_+^3$. Although the slack variables will be integral when x_1, x_2 are integral, there is no need to require them to be integral.

Solving the linear programming relaxation gives the optimal basic solution

$$\begin{aligned} z_{\text{LP}} \qquad + \frac{3}{11}x_3 + \frac{16}{11}x_4 \qquad &= \frac{332}{11} \\ x_1 \qquad - \frac{1}{11}x_3 + \frac{2}{11}x_4 \qquad &= \frac{36}{11} \\ x_2 + \frac{5}{11}x_3 + \frac{1}{11}x_4 \qquad &= \frac{40}{11} \\ \frac{8}{11}x_3 + \frac{6}{11}x_4 + x_5 &= \frac{75}{11} \end{aligned}$$

$$(x_1, x_2) \in Z_+^2, \quad (x_3, x_4, x_5) \in R_+^3.$$

Thus $z_{LP}^0 = 30\frac{2}{11}$ and $x^0 = (\frac{36}{11} \ \frac{40}{11} \ 0 \ 0 \ \frac{75}{11})$.

Since $(x_1^0, x_2^0) \notin Z^2$, one must branch on either x_1 or x_2. We use (2.2) to choose between them.

Suppose we consider the criterion of maximum infeasibility; then $D_j^{-0} = f_j^0$ and $D_j^{+0} = 1 - f_j^0$ for $j = 1, 2$. Hence $D_1^{-0} = \frac{3}{11}$, $D_1^{+0} = \frac{8}{11}$, $D_2^{-0} = \frac{7}{11}$, and $D_2^{+0} = \frac{4}{11}$. By (2.2), we obtain

$$\max_{j\in\{1,2\}} \min(D_j^{-0}, D_j^{+0}) = \max\left(\frac{3}{11}, \frac{4}{11}\right) = \frac{4}{11} = D_2^{+0}.$$

Hence we would branch on x_2 ($x_2 \leqslant 3$, $x_2 \geqslant 4$) and examine the right son.

Now we illustrate the use of penalties to determine the branching variable. From the representation of the optimal solution, we see that if x_3 or x_4 increases, x_2 decreases. Hence we can set $p_2^{+0} = \infty$. Following this approach, z_{LP} decreases by $\frac{3}{5}$ ($\frac{16}{1}$) per unit decrease in x_2 if x_3 (x_4) is made basic; hence we can set $p_2^{-0} = \min(\frac{3}{5}, \frac{16}{1}) = \frac{3}{5}$. Similarly, $p_1^{+0} = 3$ and $p_1^{-0} = \frac{16}{2} = 8$. Hence

$$D_1^{-0} = \frac{3 \times 8}{11} = \frac{24}{11}, \quad D_1^{+0} = \frac{8 \times 3}{11} = \frac{24}{11}, \quad D_2^{-0} = \frac{3}{5} \times \frac{7}{11} = \frac{21}{55}, \quad D_2^{+0} = \infty.$$

Now

$$\max_{j\in\{1,2\}} \min(D_j^{-0}, D_j^{+0}) = \max\left(\frac{24}{11}, \frac{21}{55}\right) = \frac{24}{11} = D_1^- = D_1^+.$$

Thus we would branch on x_1. Empirical evidence indicates that these calculations are not worthwhile for large problems.

We choose to branch on x_2 (see Figure 2.5). Adding the constraint $x_2 \geqslant 4$ ($x_2 - t = 4, t \geqslant 0$) to the current optimal solution gives the node 1 relaxation. The full set of equations is given by

$$\begin{array}{lllllll}
z_{LP} & + \frac{3}{11}x_3 & + \frac{16}{11}x_4 & & & = & \frac{332}{11} \\
x_1 & - \frac{1}{11}x_3 & + \frac{2}{11}x_4 & & & = & \frac{36}{11} \\
x_2 & + \frac{5}{11}x_3 & + \frac{1}{11}x_4 & & & = & \frac{40}{11} \\
 & \frac{8}{11}x_3 & + \frac{6}{11}x_4 & + x_5 & & = & \frac{75}{11} \\
 & \frac{5}{11}x_3 & + \frac{1}{11}x_4 & & + t & = & -\frac{4}{11}
\end{array}$$

$$x, t \geqslant 0.$$

Note that in a computer system the bound constraints would not be added explicitly. The dual simplex algorithm shows immediately that this problem is primal infeasible (see the last constraint). Hence node 1 is pruned by infeasibility.

The only remaining node on the candidate list (node 2) corresponds to the original IP with $x_2 \leqslant 3$ ($x_2 + s = 3$, $s \geqslant 0$) added. The resulting linear programming relaxation is

$$\begin{array}{rrrrrrrr}
z_{\mathrm{LP}} & & + \frac{3}{11}x_3 & + \frac{16}{11}x_4 & & & = & \frac{332}{11} \\
& x_1 & - \frac{1}{11}x_3 & + \frac{2}{11}x_4 & & & = & \frac{36}{11} \\
& x_2 + & \frac{5}{11}x_3 & + \frac{1}{11}x_4 & & & = & \frac{40}{11} \\
& & \frac{8}{11}x_3 & + \frac{6}{11}x_4 & + x_5 & & = & \frac{75}{11} \\
& & - \frac{5}{11}x_3 & - \frac{1}{11}x_4 & & + s & = & -\frac{7}{11}
\end{array}$$

After one iteration of the dual simplex algorithm we obtain the optimal solution

$$\begin{array}{rrrrrrr}
z_{\mathrm{LP}} & & + \frac{7}{5}x_4 & & + \frac{3}{5}s & = & \frac{149}{5} \\
& x_1 & + \frac{1}{5}x_4 & & - \frac{1}{5}s & = & \frac{17}{5} \\
& x_2 & & & + s & = & 3 \\
& & \frac{2}{5}x_4 & + x_5 & + \frac{8}{5}s & = & \frac{29}{5} \\
& x_3 + & \frac{1}{5}x_4 & & - \frac{11}{5}s & = & \frac{7}{5}
\end{array}$$

Thus $z^2_{\mathrm{LP}} = 29\frac{4}{5}$ and $x^2 = (\frac{17}{5} \;\; 3 \;\; \frac{7}{5} \;\; 0 \;\; \frac{29}{5})$.

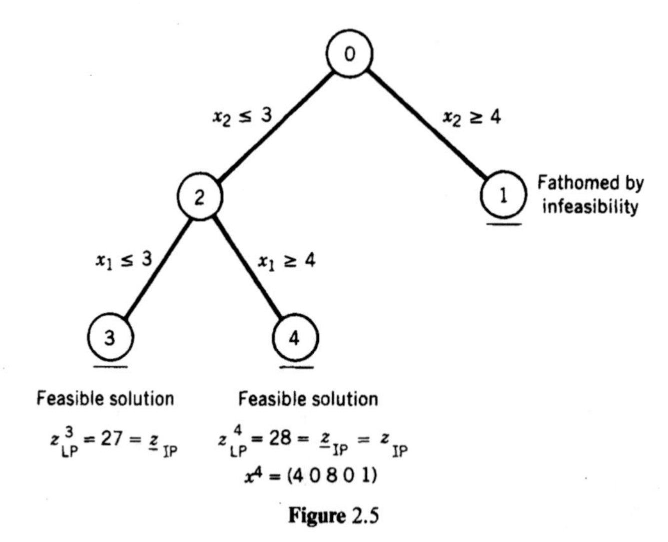

Figure 2.5

Since x_1^2 is not integral, we branch on x_1 and examine the left son first since $f_1^2 < \frac{1}{2}$. The tree is shown in Figure 2.5. Adding $x_1 \leqslant 3$ and reoptimizing by the dual simplex method gives an integral solution $x^3 = (3 \quad 3 \quad 1 \quad 2 \quad 5)$ with $z_{\text{LP}}^3 = 27$. Hence node 3 is pruned and we set $\underline{z}_{\text{IP}} = 27$.

The only remaining node is node 4. By solving the linear programming relaxation of the node 4 problem, we obtain $x^4 = (4 \quad 0 \quad 8 \quad 0 \quad 1)$ and $z_{\text{LP}}^4 = 28$. Hence node 4 is pruned and $\underline{z}_{\text{IP}} = 28$. The list of active nodes is now empty, so the algorithm terminates with the optimal solution $x = x^4$ and $z_{\text{IP}} = 28$.

Example 2.2

$$
\begin{aligned}
z_{\text{IP}} = \max \quad & 77x_1 + 6x_2 + 3x_3 + 6x_4 + 33x_5 + 13x_6 + 110x_7 + 21x_8 + 47x_9 \\
& 774x_1 + 76x_2 + 22x_3 + 42x_4 + 21x_5 + 760x_6 + 818x_7 + 62x_8 + 785x_9 \leqslant 1500 \\
& 67x_1 + 27x_2 + 794x_3 + 53x_4 + 234x_5 + 32x_6 + 797x_7 + 97x_8 + 435x_9 \leqslant 1500 \\
& x \in B^9
\end{aligned}
$$

We solved this problem by a branch-and-bound algorithm contained in a mathematical programming system. The tree is shown in Figure 2.6. Additional information about the nodes of the tree is given in Table 2.1. The linear programming relaxations are solved in the order given by the node numbers.

The algorithm begins by solving the initial linear programming relaxation. As indicated in Table 2.1, its value is $z_{\text{LP}}^0 = 225.7$ and there are two fractional variables. Associated with each fractional variable are two reduced costs. One is the reduced cost of the nonbasic variable that becomes basic if the fractional basic variable goes to its upper bound of 1; the other is the reduced cost of the nonbasic variable that becomes basic if the fractional variable goes to its lower bound of 0. By multiplying each of these costs by the distance that

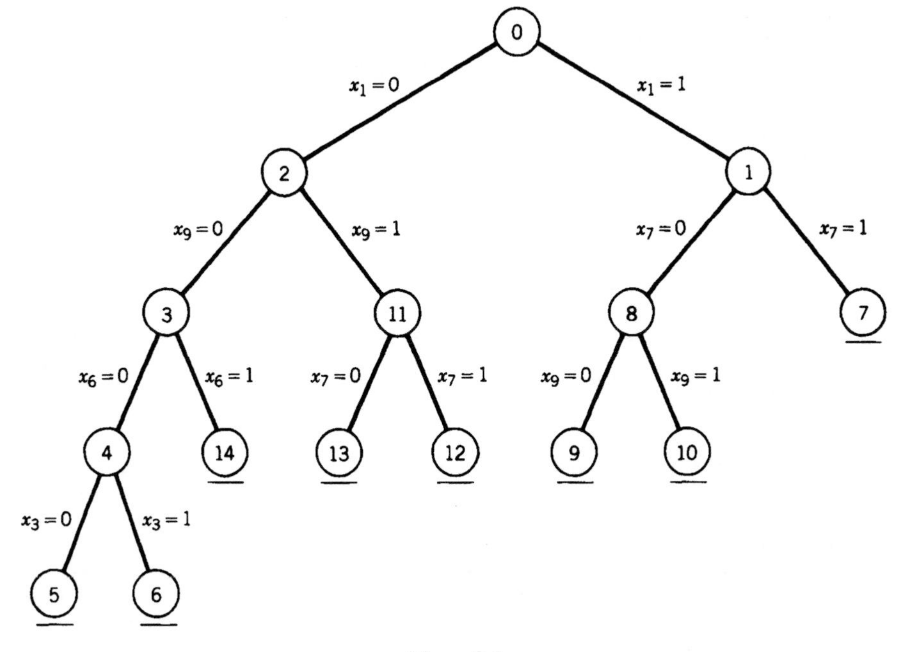

Figure 2.6

the fractional variable must move to achieve the corresponding bound, we estimate upward and downward costs. (If the reduced cost is less that 0.1, the algorithm uses 0.1 in place of the reduced cost.) The smaller of the upward and downward costs is used for the estimated cost of a fractional variable, and then the fractional variable with the largest estimated cost is chosen for branching. In the example, it is x_1.

Now we choose a direction for branching by comparing the upward and downward estimated costs for x_1, and we branch in the direction of the smaller of the two estimated costs. In the example, we set $x_1 = 1$ and solve the resulting linear program to obtain the solution at node 1. The algorithm next decides whether to consider the opposite branch $x_1 = 0$ or to branch from node 1. This is done by comparing the estimated solution value $\hat{z}^1$ with the estimated solution value at node 0 had the degradation been computed using the downward cost for x_1. The larger of these two values determines the next node. In the example, we solve the problem with $x_1 = 0$.

In general, after considering the first branch from a node, the algorithm either considers the opposite branch or branches down from the node just created. If the first branch is pruned, the algorithm next considers the opposite branch. If the first branch is not pruned, the algorithm chooses between the two possibilities as indicated above.

When both branches of a node have been considered in turn, there are three possibilities. If neither has been pruned, the algorithm selects the node corresponding to one of them. This selection is made by the criterion of higher estimated solution value until an integral solution has been found; thereafter it is made by the quick improvement rule given by (2.1). In the example, node 2 is chosen for division before node 1 because $\hat{z}^2 = 175.9 > 162.6 = \hat{z}^1$. If one of the two branch nodes has been pruned, the algorithm selects the node corresponding to the other. If both have been pruned, all active nodes are considered according to the criterion of highest estimated value until an integral solution has been found; thereafter they are considered according to the quick improvement rule.

Table 2.1.

Node i	LP Solution Value (z^i_{LP})	Number of Fractional Variables	Variable Chosen for Branching	First Direction	Estimated Solution $\hat{z}^i$
0	225.7	2	x_1	1	200.2
1	217.8	2	x_7	1	162.6
2	204.8	2	x_9	0	175.9
3	185.1	2	x_6	0	175.9
4	177.1	1	x_3	0	175.9
5	176	0			[b]
6	122.2[a]				
7	42.4[a]				
8	176.0	2	x_9	0	142.8
9	155.3[a]				
10	170.6[a]				
11	186.4	2	x_7	1	132.3
12	148[a]				
13	154.3[a]				
14	167.6[a]				

[a] The node was terminated without necessarily achieving primal feasibility because the LP value fell below the value of a feasible integer solution.

[b] The first feasible integer solution is found at node 5. It is $x^5 = (0 \quad 1 \quad 0 \quad 1 \quad 1 \quad 0 \quad 1 \quad 1 \quad 0)$.

In the example, we branch down from node 2 to node 5, where we find the integral solution $x = (0\ \ 1\ \ 0\ \ 1\ \ 1\ \ 0\ \ 1\ \ 1\ \ 0)$ of value 176. The opposite branch node 6 is pruned because of its linear programming bound. Now nodes 1, 2, and 3 are candidates and node 1 is selected by the quick improvement rule. The rest of the calculation is self-explanatory.

Generalized Upper-Bound Constraints

Many integer programs with binary variables have *generalized upper-bound constraints* of the form

$$\sum_{j \in Q_i} x_j = 1 \quad \text{for } i = 1, \ldots, p, \tag{2.3}$$

where the Q_i's are disjoint subsets of N. Here we explore the branching scheme given in Figure 2.2, which has proved to be a very efficient way of handling these constraints and is widely used in mathematical programming systems.

Suppose in a solution of a linear programming relaxation we have $0 < x_k < 1$ for some $k \in Q_i$. Conventional branching on x_k is equivalent to $x_k = 0$ or $\Sigma_{j \in Q_i \setminus \{k\}} x_j = 0$ since the latter equality is equivalent to $x_k = 1$. Now unless there is a good reason for singling out x_k as the variable that is likely to equal 1, the $x_k = 1$ branch probably contains relatively few solutions as compared to the $x_k = 0$ branch. If this is the case, almost no progress will have been made since the node with $x_k = 0$ corresponds to nearly the same feasible region as that of its father.

It appears to be more desirable to try to divide the feasible region of the father roughly equally between the sons. To accomplish this, we consider the branching rule

$$\sum_{j \in Q_i^1} x_j = 0 \quad \text{or} \quad \sum_{j \in Q_i \setminus Q_i^1} x_j = 0. \tag{2.4}$$

The conventional rule is the special case of (2.4) with $Q_i^1 = \{k\}$. We can use (2.4) for any Q_i^1 such that $k \in Q_i^1$ and $\Sigma_{j \in Q_i^1} x_j < 1$. It seems reasonable to take Q_i^1 and $Q \setminus Q_i^1$ of nearly equal cardinality.

A simple implementation of the branching rule (2.4) is obtained by indexing the variables in (2.3) as $x_{i_1}, x_{i_2}, \ldots, x_{i_t}$. The choice of Q_i^1 is then specified by an index j, $1 \leqslant j \leqslant t-1$, and $Q_i^1 = \{i_1, \ldots, i_j\}$.

Example 2.3

$$\begin{aligned} z_{\text{IP}} = \max\ & 50x_1 + 47x_2 + 44x_3 + 41x_4 + 38x_5 + 36x_6 + 31x_7 + 29x_8 + 27x_9 \\ & + 25x_{10} + 23x_{11} + 21x_{12} + 20x_{13} \\ & \sum_{j=1}^{13} (21 - j)x_j \leqslant 22 \\ & \sum_{j=1}^{12} x_j = 1 \\ & \quad x \in B^{13}. \end{aligned}$$

The solution is shown in the tree of Figure 2.7, where beside each node the solution of the linear programming relaxation is given.

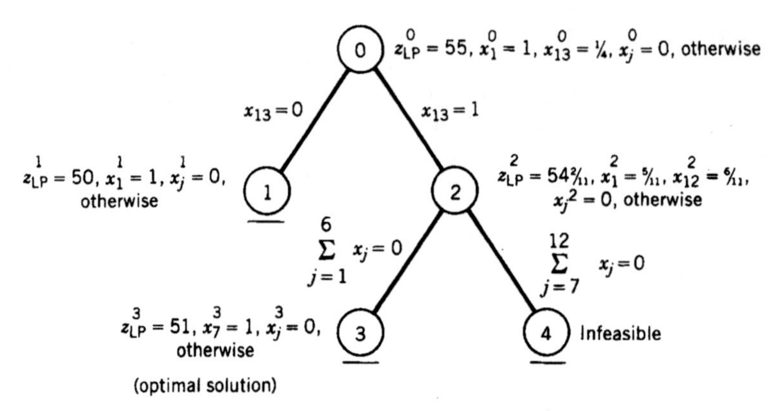

Figure 2.7

We leave it to the reader to show that if conventional branching had been used at node 2, a much larger enumeration tree would have resulted.

Piecewise Linear Functions

In Section I.1.4 we showed how a piecewise linear function $f(y)$ (see Figure 2.8) could be represented by a linear function with constraints on the variables.

For any $y = \Sigma_{j=0}^{t} a_j\lambda_j$, where

$$(2.5) \qquad \sum_{j=0}^{t} \lambda_j = 1 \quad \text{and} \quad \lambda_j \in R_+^1 \quad \text{for } j = 0, \ldots, t,$$

we have

$$f(y) = \sum_{j=0}^{t} f(a_j)\lambda_j,$$

provided that no more than two λ_j's are positive; and if $\lambda_j > 0$ and $\lambda_k > 0$, then $k = j - 1$ or $j + 1$.

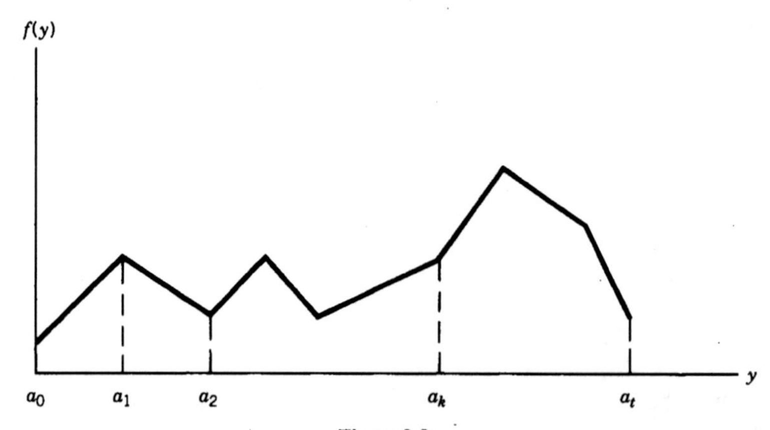

Figure 2.8

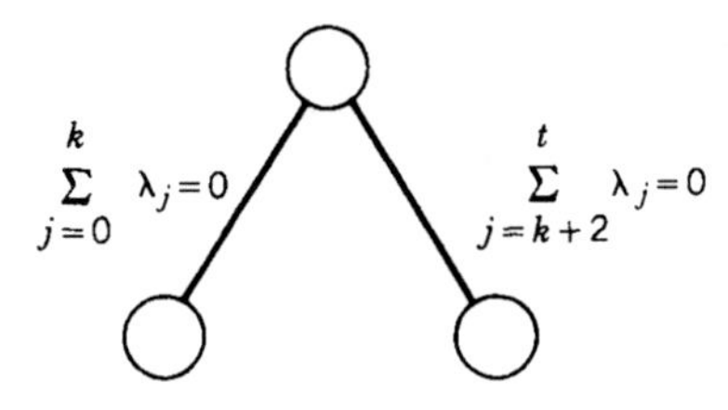

Figure 2.9

As noted in Section I.1.4, these conditions on the λ_j's can be represented using linear constraints and binary variables. But, within the scope of a branch-and-bound algorithm, it is more efficient to enforce the nonlinear constraints through branching. The approach is similar to the treatment of generalized upper-bound constraints.

If $\lambda_k > 0$, then either $\lambda_0 = \cdots = \lambda_{k-1} = 0$ or $\lambda_{k+1} = \cdots = \lambda_t = 0$. Hence

$$\sum_{j=0}^{k-1} \lambda_j = 0 \quad \text{or} \quad \sum_{j=k+1}^{t} \lambda_j = 0 \quad \text{for } k = 1, \ldots, t-1. \tag{2.6}$$

Moreover, if $\hat{\lambda} = (\hat{\lambda}_0, \ldots, \hat{\lambda}_t)$ satisfies (2.5) with $\hat{\lambda}_k > 0$ and $\hat{\lambda}_l > 0$ for some $l \geq k+2$, we can use (2.6) with index $k+1$ for branching (see Figure 2.9). It is important to note that the solution $\hat{\lambda}$ is infeasible along both branches.

Branching strategies for using the constraints (2.4) and (2.6) are left for the reader to develop.

3. GENERAL CUTTING-PLANE ALGORITHMS

We begin this section with a fractional cutting-plane algorithm (FCPA) for pure-integer programs that uses Gomory cuts. The main result is that the Gomory FCPA is finitely convergent. We then extend the algorithm to mixed-integer programs. The last topic of this section is a primal cutting-plane algorithm for pure-integer programs. It progresses by generating adjacent extreme points of the convex hull of feasible integral solutions.

Consider an equality-constrained integer program

$$\max \{cx: x \in S^e\}, \quad \text{where } S^e = \{x \in Z_+^n: Ax = b\},$$

which for the rest of this section will be written as

$$\text{(IP)} \quad \max\{x_0: (x_0, x) \in S^0\}, \quad \text{where } S^0 = \{x_0 \in Z^1, x \in Z_+^n: x_0 - cx = 0, Ax = b\}.$$

We suppose that an optimal basis for the linear programming relaxation has been obtained, so IP can be written as

$$\begin{aligned} &\max x_0 \\ &x_{B_i} + \sum_{j \in H} \bar{a}_{ij} x_j = \bar{a}_{i0} \quad \text{for } i = 0, 1, \ldots, m \\ &x_{B_0} \in Z, \quad x_{B_i} \in Z_+^1 \quad \text{for } i = 1, \ldots, m, \; x_j \in Z_+^1 \text{ for } j \in H, \end{aligned} \tag{3.1}$$

where $x_0 = x_{B_0}, x_{B_i}$ for $i = 1, \ldots, m$ are the basic variables and where x_j for

$j \in H \subset N = \{1, \ldots, n\}$ are the nonbasic variables. Since the basis is primal and dual feasible, we have $\bar{a}_{i0} \geqslant 0$ for $i = 1, \ldots, m$ and $\bar{a}_{0j} \geqslant 0$ for $j \in H$.

Suppose in (3.1) there exists an i such that $\bar{a}_{i0} \notin Z^1$. The results of Section II.1.3 yield the following proposition.

Proposition 3.1. *(Gomory fractional cut)* *If* $\bar{a}_{i0} \notin Z^1$, *then*

$$\sum_{j \in H} f_{ij} x_j = f_{i0} + x_{n+1}, \quad x_{n+1} \in Z^1_+,$$

where $f_{ij} = \bar{a}_{ij} - \lfloor \bar{a}_{ij} \rfloor$ for $j \in H$ *and* $f_{i0} = \bar{a}_{i0} - \lfloor \bar{a}_{i0} \rfloor$, *is a valid equality for* S^0.

Example 3.1. Our standard example written in equality form is

$$\begin{aligned}
&\max x_0 \\
&x_0 - 7x_1 - 2x_2 = 0 \\
&\quad - x_1 + 2x_2 + x_3 = 4 \\
&\quad 5x_1 + x_2 + x_4 = 20 \\
&\quad - 2x_1 - 2x_2 + x_5 = -7 \\
&x_0 \in Z^1, \quad x_j \in Z^1_+ \text{ for } j = 1, \ldots, 5.
\end{aligned}$$

An optimal solution to the linear programming relaxation is

$$\begin{aligned}
x_0 \qquad + \frac{3}{11}x_3 + \frac{16}{11}x_4 &= \frac{332}{11} \\
x_1 \qquad - \frac{1}{11}x_3 + \frac{2}{11}x_4 &= \frac{36}{11} \\
x_2 + \frac{5}{11}x_3 + \frac{1}{11}x_4 &= \frac{40}{11} \\
\frac{8}{11}x_3 + \frac{6}{11}x_4 + x_5 &= \frac{75}{11},
\end{aligned}$$

where $x_3 = x_4 = 0$.

Generating the fractional cut from row 0, we obtain

$$\frac{3}{11}x_3 + \frac{5}{11}x_4 = \frac{2}{11} + x_6, \qquad x_6 \in Z^1_+.$$

In terms of the original variables, the cut is $2x_1 + x_2 \leqslant 10$.

The Gomory FCPA is just the general FCPA given in Section 1, with all of the generated valid inequalities being Gomory cuts.

Initialization: Set $t = 1$, $z^1_R(x) = x_0$, $S^1_R = \{x_0 \in R^1, x \in R^n_+ : x_0 - cx = 0, Ax = b\}$.
Iteration t:

Step 1: Solution of the linear programming relaxation. Solve

$$(\mathrm{RP}^t) \qquad \max\{x_0: (x_0, x) \in S_R^t\}.$$

If RP^t is feasible and has an optimal solution, suppose the solution is (x_0^t, x^t). (See the remark below if RP^1 has unbounded optimal value.)

Step 2: Optimality test. If $x^t \in Z_+^n$, then x^t is an optimal solution.

Step 3: Infeasibility test. If RP^t is infeasible, then IP is infeasible.

Step 4: Addition of a cut. Choose a row $x_{B_i} + \Sigma_{j\in H^t} \overline{a}_{ij}^t x_j = \overline{a}_{i0}^t$ with $\overline{a}_{i0}^t \notin Z^1$. Let

$$\sum_{j\in H^t} f_{ij}x_j - x_{n+t} = f_{i0},\ x_{n+t} \in Z_+^1$$

be the fractional Gomory cut for the row. Set

$$S_R^{t+1} = S_R^t \cap \left\{x_0 \in R^1, x \in R_+^{n+t}: \sum_{j\in H^t} f_{ij}x_j - x_{n+t} = f_{i0}\right\}$$

Step 5: $t \leftarrow t + 1$.

When the cut is added, the new basis, which includes x_{n+t} as a basic variable, is dual feasible. Primal feasibility is violated only by $x_{n+t} < 0$. Hence it is natural to solve RP^{t+1} by the dual simplex method.

If RP^1 is unbounded, then by Corollary 6.8 of Section I.4.6 we have that IP is either unbounded or infeasible. Moreover, by Theorem 4.1 of Section I.5.4, if IP is feasible, then there is a feasible solution with $\Sigma_{j\in N} x_j \leqslant d$, where d is a suitably large integer. Hence we can add the constraint $\Sigma_{j\in N} x_j \leqslant d$ to RP^1. Then IP is unbounded if and only if the modified problem has a feasible solution.

Example 3.1 (continued). As noted above, the solution of the linear programming relaxation RP^1 is

$$(x_0^1, x^1) = (x_0^1, x_1^1, \ldots, x_5^1) = \left(\frac{332}{11}\ \ \frac{36}{11}\ \ \frac{40}{11}\ \ 0\ \ 0\ \ \frac{75}{11}\right).$$

Also, $x_2 + \frac{5}{11}x_3 + \frac{1}{11}x_4 = \frac{40}{11}$. Generating the fractional cut from this row yields

$$\frac{5}{11}x_3 + \frac{1}{11}x_4 = \frac{7}{11} + x_6, \qquad x_6 \in Z_+^1.$$

An optimal solution to RP^2 is

$$(x_0^2, x_1^2, \ldots, x_6^2) = \left(\frac{149}{5}\ \ \frac{17}{5}\ \ 3\ \ \frac{7}{5}\ \ 0\ \ \frac{29}{5}\ \ 0\right).$$

Also, $x_0 + \frac{7}{5}x_4 + \frac{3}{5}x_6 = 29\frac{4}{5}$. Generating the next fractional cut from this row yields

$$\frac{2}{5}x_4 + \frac{3}{5}x_6 = \frac{4}{5} + x_7, \qquad x_7 \in Z_+^1.$$

An optimal solution to RP^3 is $(29 \ \frac{11}{3} \ \frac{5}{3} \ \frac{13}{3} \ 0 \ \frac{11}{3} \ \frac{4}{3} \ 0)$. Also, $x_2 - \frac{2}{3}x_4 + \frac{5}{3}x_7 = \frac{5}{3}$. From this row, the fractional cut is

$$\frac{1}{3}x_4 + \frac{2}{3}x_7 = \frac{2}{3} + x_8, \qquad x_8 \in Z^1_+.$$

The optimal solution to RP^4 is $(28 \ 4 \ 0 \ 8 \ 0 \ 1 \ 3 \ 1 \ 0)$, which is also optimal to IP.

In terms of the variables x_1, and x_2, the three added cuts are $x_2 \leqslant 3$, $2x_1 + x_2 \leqslant 9$, and $3x_1 + x_2 \leqslant 12$ (see Figure 3.1).

Finite Convergence

We now give some additional specifications on the Gomory FCPA which guarantee that it converges finitely. We suppose that

$$\{x \in R^n_+ : Ax = b\} \subseteq \left\{x \in R^n_+ : \sum_{j \in N} x_j \leqslant d\right\}$$

for some suitably large $d \in Z^1_+$. As noted above, this is without loss of generality.

The convergence argument depends on a lexicographic decreasing sequence of solution vectors $\{x_0^t, x^t\}$, $t = 1, 2, \ldots$, which can be obtained, as will be explained soon, by solving a sequence of linear programs by a lexicographic dual simplex algorithm. Recall that $x \overset{L}{<} y$ if there exists a k such that $x_k < y_k$ and $x_i = y_i$ for $i < k$. Also, $x \overset{L}{\leqslant} y$ if $x \overset{L}{<} y$ or $x = y$.

The algorithm finds the lexicographically largest element in S^0 or shows that S^0 is empty. Since the objective value is the first component of (x_0, x), a lexicographically largest element is optimal. Let $c_j^+ = \max(0, c_j)$ and $c_j^- = \min(0, c_j)$. Since $0 \leqslant x_j \leqslant d$ for $j \in N$, it follows that if $(x_0, x) \in S^0$ we obtain

$$\left(d \sum_{j \in N} c_j^-, 0, \ldots, 0\right) \overset{L}{\leqslant} (x_0, x) \overset{L}{\leqslant} \left(d \sum_{j \in N} c_j^+, d, \ldots, d\right).$$

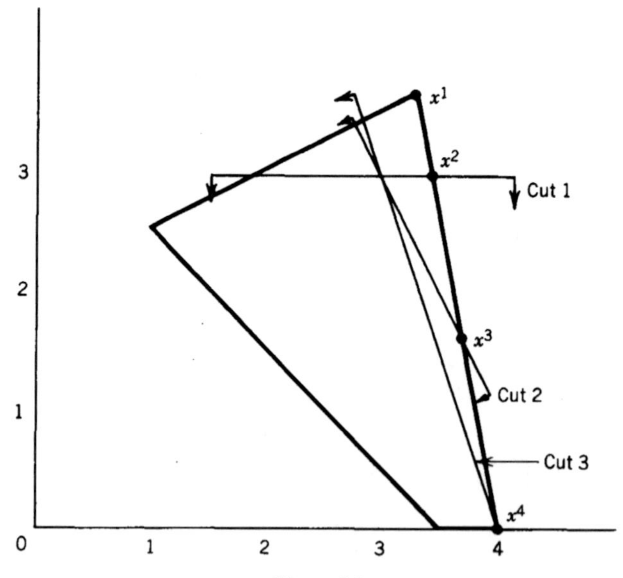

Figure 3.1

Let $\alpha^0 = (d \sum_{j \in N} c_j^+, d, \ldots, d)$. We will show that the cuts in the Gomory FCPA can be chosen so that after t cuts have been added it will follow that $(x_0, x) \in S^0$ implies $(x_0, x) \overset{L}{\leqslant} \alpha^t$, where $\alpha^t \in Z^{n+1}$ and $\alpha^t \overset{L}{<} \alpha^{t-1}$. It then follows that the total number of cuts is bounded.

The Lexicographic Dual Simplex Algorithm

Consider a basic solution to the linear programming relaxation of (3.1) written in the form

$$\max x_0$$

$$x_i + \sum_{j \in H} \bar{a}_{ij} x_j = \bar{a}_{i0} \quad \text{for } i = 0, 1, \ldots, n \tag{3.2}$$

$$x_i \geqslant 0 \quad \text{for } i = 1, \ldots, n,$$

where H is the index set of nonbasic variables. The representation (3.2) contains a row for each variable. Thus for $i \in H$ we have the trivial identity $x_i - x_i = 0$, that is, $\bar{a}_{ii} = -1$, $\bar{a}_{i0} = 0$, and $\bar{a}_{ij} = 0$ for $j \in H \setminus \{i\}$. The basic solution obtained from (3.2) is $x_i = \bar{a}_{i0}$ for $i = 0, \ldots, n$.

Since the constraint set is bounded, there is a dual feasible basis, that is, a basis with $\bar{a}_{0j} \geqslant 0$ for all $j \in H$. Thus if $\bar{a}_{0j} > 0$ or $(\bar{a}_{1j}, \ldots, \bar{a}_{nj}) \overset{L}{>} 0$, we have $\bar{a}_j = (\bar{a}_{0j}, \bar{a}_{1j}, \ldots, \bar{a}_{nj}) \overset{L}{>} 0$. However, if $\bar{a}_{0j} = 0$ and $(\bar{a}_{1j}, \ldots, \bar{a}_{nj}) \overset{L}{\ngtr} 0$ we add the redundant equation

$$y + \sum_{j \in H} x_j = d, \quad y \geqslant 0 \tag{3.3}$$

as the second row of (3.2). Now $(1, \bar{a}_{1j}, \ldots, \bar{a}_{nj}) \overset{L}{>} 0$ so that $(\bar{a}_{0j}, 1, \bar{a}_{1j}, \ldots, \bar{a}_{nj}) \overset{L}{>} 0$. Hence we assume that we have a basic solution to the linear program that satisfies $\bar{a}_j \overset{L}{>} 0$ for $j \in H$.

Proposition 3.2. *If* $\bar{a}_{i0} \geqslant 0$ *for* $i = 1, \ldots, n$ *and* $\bar{a}_j \overset{L}{>} 0$ *for* $j \in H$, *then* $(x_0, x) = \bar{a}_0 = (\bar{a}_{00}, \bar{a}_{10}, \ldots, \bar{a}_{n0})$ *is the lexicographically largest feasible solution to (3.2) and is optimal.*

Proof. By hypothesis, $(x_0, x) = \bar{a}_0$ is feasible. Moreover, $\bar{a}_0$ is the lexicographically largest feasible solution since any other feasible solution is of the form $\bar{a}_0 - \sum_{j \in H} \bar{a}_j x_j$ and $\bar{a}_j \overset{L}{>} 0$ with $x_j \geqslant 0$ for all $j \in H$. Finally, the lexicographically largest solution maximizes x_0. ∎

We now give a finite simplex algorithm for finding the lexicographically largest feasible solution to (3.2).

Proposition 3.3. *Suppose* $(x_0, x) = \bar{a}_0^p - \sum_{j \in H^p} \bar{a}_j^p x_j$ *is a basic solution with* $\bar{a}_j^p \overset{L}{>} 0$ *for* $j \in H^p$ *and* $\bar{a}_{i0}^p < 0$. *A dual simplex pivot that makes* x_i *nonbasic yields* $\bar{a}_0^{p+1} \overset{L}{<} \bar{a}_0^p$.

Proof. Suppose $k \in H^p$ and x_k is the variable to become basic. Then $\bar{a}_{ik}^p < 0$ and $\bar{a}_0^{p+1} = \bar{a}_0^p - (\bar{a}_{i0}^p / \bar{a}_{ik}^p)\bar{a}_k^p \overset{L}{<} \bar{a}_0^p$ since $\bar{a}_k^p \overset{L}{>} 0$ and $(\bar{a}_{i0}^p / \bar{a}_{ik}^p) > 0$. ∎

Thus we need to give a rule for choosing the variable to enter the basis so that $\bar{a}_j^{p+1} \overset{L}{>} 0$ for all $j \in H^{p+1}$.

Proposition 3.4. *Suppose $\overline{a}_{i0}{}^{p} < 0$, $\overline{a}_j^p \overset{L}{>} 0$ for $j \in H^p$, and x_i is chosen as the variable to leave the basis. Let $H_i^p = \{j \in H^p : \overline{a}_{ij}^p < 0\}$. If $H_i^p = \emptyset$, then there is no feasible solution. Otherwise choose $k \in H_i^p$ to satisfy*

$$(3.4) \qquad \frac{1}{\overline{a}_{ik}^p}\,\overline{a}_k^p \overset{L}{>} \frac{1}{\overline{a}_{ij}^p}\,\overline{a}_j^p \quad \text{for all } j \in H_i^p \setminus \{k\}$$

and also choose x_k as the variable to become basic. Then $\overline{a}_j^{p+1} \overset{L}{>} 0$ for all $j \in H^{p+1}$.

Proof. If $H_i^p = \emptyset$, then $x_i = \overline{a}_{i0}^p - \Sigma_{j \in H^p} \overline{a}_{ij}^p x_j < 0$ for all feasible solutions since $\overline{a}_{i0}^p < 0$ and $\overline{a}_{ij}^p \geq 0$ for all $j \in H^p$.

Now suppose $k \in H_i^p$ is chosen to satisfy (3.4). Note that because the system of equations contains the identities $x_j - x_j = 0$ for all $j \in H^p$, $\overline{a}_k^p$ cannot be a scalar multiple of $\overline{a}_j^p$ for any $j \in H_i^p \setminus \{k\}$. Hence (3,4) uniquely determines k. We have

a. $\overline{a}_i^{p+1} = -(1/\overline{a}_{ik}^p)\overline{a}_k^p \overset{L}{>} 0$ since $\overline{a}_{ik}^p < 0$ and $\overline{a}_k^p \overset{L}{>} 0$.
b. For $j \in H^p \setminus \{k\}$, we have $\overline{a}_j^{p+1} = \overline{a}_j^p - (\overline{a}_{ij}^p / \overline{a}_{ik}^p)\overline{a}_k^p$. There are two cases. If $j \in H^p \setminus H_i^p$, then $\overline{a}_j^{p+1} \overset{L}{>} 0$ since $\overline{a}_k^p \overset{L}{>} 0$, $\overline{a}_j^p \overset{L}{>} 0$ and $(\overline{a}_{ij}^p / \overline{a}_{ik}^p) \leq 0$. If $j \in H_i^p \setminus \{k\}$, then $\overline{a}_j^{p+1} \overset{L}{>} 0$ by (3.4). ∎

Theorem 3.5. *If we begin with a basic solution satisfying $\overline{a}_j^0 \overset{L}{>} 0$ for all $j \in H^0$ and apply the dual simplex pivoting rule given in Proposition 3.4, then in a finite number of pivots we either show that (3.2) has no feasible solution or find the lexicographically largest solution.*

Proof. Since the sequence $\{\overline{a}_0^p\}$ is lexicographically decreasing, no basis can be repeated. ∎

Now we return to the Gomory FCPA and suppose that we have found (x_0^t, x^t), the lexicographically largest solution to RP^t. If $(x_0^t, x^t) \in Z^{n+1}$, we have solved IP. So suppose this is not the case.

Proposition 3.6. *Let (x_0^t, x^t) be the lexicographically largest solution to RP^t and suppose $x_i^t \in Z^1$ for $i = 0, \ldots, s-1$ and $x_s^t \notin Z^1$. Let $\alpha^t = (x_0^t, \ldots, x_{s-1}^t, \lfloor x_s^t \rfloor, d, \ldots, d)$. If (x_0, x) is a feasible solution to* IP, *then $(x_0, x) \overset{L}{\leq} \alpha^t$.*

Proof. If (x_0, x) is feasible to IP, then $(x_0, x) \in Z^{n+1}$, $x_j \leq d$ for $j \in N$, and $(x_0, x) \overset{L}{\leq} (x_0^t, x^t)$. The vector α^t is the lexicographically largest vector that satisfies these properties. ∎

Now all we need to do is produce a Gomory cut so that (x_0^{t+1}, x^{t+1}), the lexicographically largest solution to RP^{t+1}, satisfies $(x_0^{t+1}, x^{t+1}) \overset{L}{<} \alpha^t$. Then, either $(x_0^{t+1}, x^{t+1}) \in Z^{n+1}$ and we are done or $\alpha^{t+1} \overset{L}{<} \alpha^t$.

Proposition 3.7. *Let (x_0^t, x^t) and α^t be defined as in Proposition 3.6. By adding the cut $\Sigma_{j \in H^t} f_{sj} x_j - x_{n+t} = f_{s0}$, $x_{n+t} \in Z_+^1$, and reoptimizing, we obtain $(x_0^{t+1}, x^{t+1}) \overset{L}{\leq} \alpha^t$.*

Proof. It suffices to consider the first pivot. In this pivot, the variable that becomes nonbasic is x_{n+t} since $x_{n+t} = -f_{s0} < 0$ and $x^t \geq 0$. Let x_k be the variable to become basic and let $(\hat{x}_0^t, \hat{x}^t)$ be the solution after one pivot. Then

$$\begin{pmatrix} \hat{x}_0^t \\ \hat{x}^t \end{pmatrix} = \begin{pmatrix} x_0^t \\ x^t \end{pmatrix} - \frac{f_{s0}}{f_{sk}} \overline{a}_k^t,$$

where $\overline{a}_k^t \overset{L}{>} 0$ and $f_{s0} / f_{sk} > 0$. There are two cases.

i. There exists an $i \leq s - 1$ such that $\overline{a}_{ik}^t \neq 0$. Since $\overline{a}_k^t \overset{L}{>} 0$, its first nonzero component $\overline{a}_{qk}^t$ is positive and $q \leq s - 1$. Hence, we obtain $\hat{x}_i^t = x_i^t$ for $i = 0, \ldots, q - 1$ and $\hat{x}_q^t < x_q^t$. Thus $(\hat{x}_0^t, \hat{x}^t) \overset{L}{<} \alpha^t$.
ii. Here $\overline{a}_{ik}^t = 0$ for $i = 1, \ldots, s - 1$. Since $f_{sk} \neq 0$ and $\overline{a}_k^t \overset{L}{>} 0$, we have $\overline{a}_{sk}^t \geq f_{sk} > 0$. Hence $\hat{x}_i^t = x_i^t$ for $i = 0, \ldots, s - 1$ and $\hat{x}_s^t \leq \lfloor x_s^t \rfloor$. Hence $(\hat{x}_0^t, \hat{x}^t) \overset{L}{\leq} \alpha^t$. ■

We preserve the order of the original equations by putting the equations for the cuts at the end. Moreover, since the slack variable x_{n+t} for the tth cut becomes nonbasic after the cut is added, we have the trivial equation $x_{n+t} - x_{n+t} = 0$. If x_{n+t} becomes basic in a subsequent pivot, its value is positive and the cut is no longer active. At this point, we drop the cut, and hence x_{n+t}, from the problem. This implies that, for computational purposes, we only need to keep the $n + 1$ equations $x_i + \Sigma_{j \in H^t} \overline{a}_{ij}^t x_j = \overline{a}_{i0}^t$ for $i = 0, \ldots, n$. Note that these equations will, in general, contain slack variables from cuts. The remaining equations are trivial identities. By Proposition 3.4, the vectors $\{\overline{a}_j^t\}_{j \in H}$ are lexicographically positive, so the addition and deletion of cut equations does not affect the properties of the lexicographic dual simplex method.

Theorem 3.8. *If the Gomory* FCPA *is executed by choosing the fractional cut from the row of lowest index whose corresponding variable is not an integer, and the resulting linear program is solved to obtain a lexicographically largest solution (i.e., by the lexicographic dual simplex method), then after at most* $(d + 1)^{n+1}(d\Sigma_{j \in N}(c_j^+ - c_j^- + 1))$ *cuts, the algorithm finds an optimal solution or shows that* IP *is infeasible.*

Proof. By Propositions 3.6 and 3.7, the number of cuts is bounded by the number of vectors $y \in Z^{n+1}$ that satisfy

$$\left(d \sum_{j \in N} c_j^-, 0, \ldots, 0\right) \overset{L}{\leq} y \overset{L}{\leq} \left(d \sum_{j \in N} c_j^+, d, \ldots, d\right).$$

In addition, we have added another factor of $(d + 1)$ to the bound to accommodate the upper-bound constraint $\Sigma_{j \in N} x_j \leq d$. ■

Example 3.1 (continued). Here we apply the finite Gomory FCPA. After pivoting, each cut row is discarded.

The solution to RP^1 is $(x_0^1, x^1) = (30\frac{2}{11} \quad \frac{36}{11} \quad \frac{40}{11} \quad 0 \quad 0 \quad \frac{75}{11})$. Since $x_0 + \frac{3}{11}x_3 + \frac{16}{11}x_4 = 30\frac{2}{11}$, we add the cut $\frac{3}{11}x_3 + \frac{5}{11}x_4 = \frac{2}{11} + x_6$, $x_6 \in Z_+^1$.

The solution to RP^2 is $(30 \quad \frac{10}{3} \quad \frac{10}{3} \quad \frac{2}{3} \quad 0 \quad \frac{19}{3})$. Since $x_1 + \frac{1}{3}x_4 - \frac{1}{3}x_6 = \frac{10}{3}$, we add the cut $\frac{1}{3}x_4 + \frac{2}{3}x_6 = \frac{1}{3} + x_7$, $x_7 \in Z_+^1$.

The solution to RP^3 is $(29\frac{1}{2} \quad \frac{7}{2} \quad \frac{5}{2} \quad \frac{5}{2} \quad 0 \quad 5)$. Since $x_0 + \frac{1}{2}x_4 + \frac{3}{2}x_7 = 29\frac{1}{2}$, we add the cut $\frac{1}{2}x_4 + \frac{1}{2}x_7 = \frac{1}{2} + x_8$, $x_8 \in Z_+^1$.

The solution to RP^4 is $(28\frac{8}{9} \quad \frac{28}{9} \quad \frac{32}{9} \quad 0 \quad \frac{8}{9} \quad \frac{57}{9})$. Since $x_0 + \frac{1}{9}x_3 + \frac{16}{9}x_8 = 28\frac{8}{9}$, we add the cut $\frac{1}{9}x_3 + \frac{7}{9}x_8 = \frac{8}{9} + x_9$, $x_9 \in Z_+^1$.

Reoptimizing yields the optimal solution $(28 \quad 4 \quad 0 \quad 8 \quad 0 \quad 1)$.

Note that the bounds $x_1 \leq 4$ and $x_2 \leq 4$ are easy to obtain from the original inequalities.

Hence $x_0 \leqslant 36, x_3 \leqslant 8, x_4 \leqslant 20$, and $x_5 \leqslant 9$. Thus any solution is lexicographically equal to or less than $\alpha^0 = (36 \quad 4 \quad 4 \quad 8 \quad 20 \quad 9)$. Now the solution to RP^1 yields $\alpha^1 = (30 \quad 4 \quad 4 \quad 8 \quad 20 \quad 9) \overset{L}{<} \alpha^0$. The successive linear programming solutions yield $\alpha^2 = (30 \quad 3 \quad 4 \quad 8 \quad 20 \quad 9) \overset{L}{<} \alpha^1$, $\alpha^3 = (29 \quad 4 \quad 4 \quad 8 \quad 20 \quad 9) \overset{L}{<} \alpha^2$, $\alpha^4 = (28 \quad 4 \quad 4 \quad 8 \quad 20 \quad 9) \overset{L}{<} \alpha^3$, and $\alpha^5 = (28 \quad 4 \quad 0 \quad 8 \quad 0 \quad 1) \overset{L}{<} \alpha^4$. Thus we see how the lexicographic upper bound is reduced at each iteration.

There is a nice interpretation of the sequence $\{\alpha^k\}_{k=0}^t$ on an enumeration tree (see Figure 3.2). Here we have enumerated all possible integral values for the variables where $x = (x_1, x_2, x_3) \in \text{B}^3$ and $0 \leqslant x_0 \leqslant 3$. Note that the leaves of the tree, read from left to right, give the possible values of (x_0, x) in increasing lexicographic order. Now suppose the solution of RP^1 gives $x_0^1 = 3$ and $0 < x_1^1 < 1$. Figure 3.2 shows the integral vectors that are eliminated by this solution and also shows those eliminated by RP^2 if $x_0^2 = 3$, $x_1^2 = 0$, and $0 < x_2^2 < 1$. Each cut eliminates at least the rightmost leaf that is still a candidate. Hence we can think of the cutting-plane procedure as a lexicographic search through the integral vectors until the lexicographically largest one that is feasible to IP is found. This suggests that it is important to choose a cut so that the subsequent pivot yields a large lexicographic decrease in $\bar{a}_0$. Insofar as we know, strategies of this type have not been systematically investigated. Perhaps some such strategy would improve the reputably poor performance of fractional cutting-plane algorithms.

Extension to Mixed-Integer Programming

The Gomory FCPA extends straightforwardly to mixed-integer programs. Suppose, in the solution of the linear programming relaxation of an MIP, $x_i \in Z_+^1$ is a basic variable given by

$$x_i + \sum_{h \in H_I} \bar{a}_{ij} x_j + \sum_{j \in H \setminus H_I} \bar{a}_{ij} y_j = \bar{a}_{i0},$$

where H_I is the index set of nonbasic integer variables and where $\bar{a}_{i0} \notin Z^1$. Here we use the Gomory mixed-integer cut

$$\sum_{\{j \in H_I : f_{ij} \leq f_{i0}\}} f_{ij} x_j + \frac{f_{i0}}{1 - f_{i0}} \sum_{\{j \in H_I : f_{ij} > f_{i0}\}} (1 - f_{ij}) x_j$$
$$+ \sum_{\{j \in H \setminus H_I : \bar{a}_{ij} > 0\}} \bar{a}_{ij} y_j - \frac{f_{i0}}{1 - f_{i0}} \sum_{\{j \in H \setminus H_I : \bar{a}_{ij} < 0\}} \bar{a}_{ij} y_j \geqslant f_{i0}$$

(see Proposition 7.4 of Section II.1.7). Everything else remains as above except for the finite convergence argument.

In mixed-integer programming, it is not reasonable to assume that the objective variable x_0 is integer-valued. Hence we cannot use the objective row for obtaining cuts. But our finite convergence argument depended on deriving a cut from the lowest-index fractional variable. In fact, by excluding x_0 as a candidate the convergence argument fails to hold (see the example given in exercise 12).

The only way we know to salvage finite convergence is to scale the problem so that x_0 is integral. But this is definitely unsatisfactory for computational purposes.

Primal Cutting-Plane Algorithm

A disadvantage of fractional cutting-plane algorithms is that no feasible solution is found until the algorithm terminates. Here we sketch a cutting-plane algorithm that circumvents

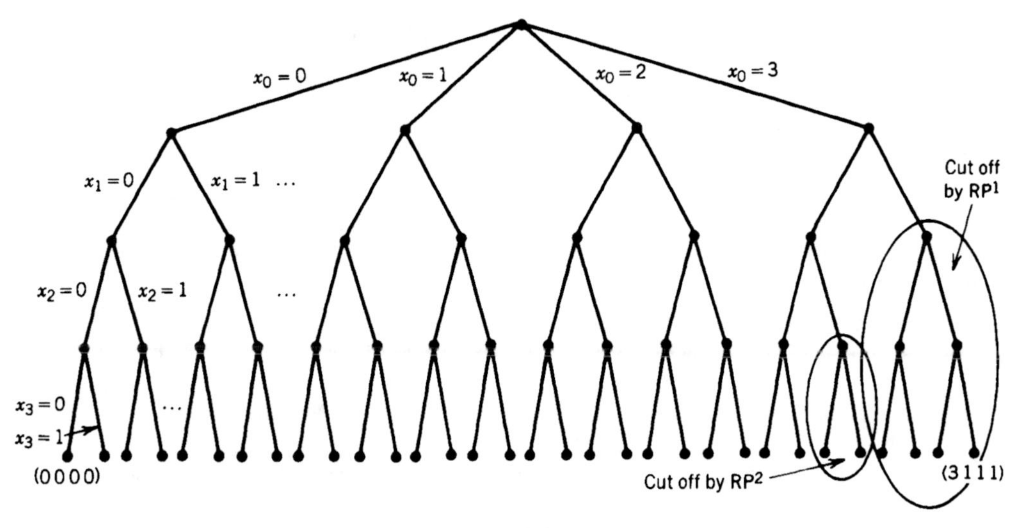

Figure 3.2

this problem. Unfortunately, it is not a practical algorithm because it tends to require an exorbitant number of cuts.

Suppose we have a nonoptimal extreme point of the convex hull of feasible integral solutions. The idea of a primal cutting-plane algorithm is to use cuts to enable pivoting to an adjacent extreme point of the convex hull whose objective value is greater.

The geometry is shown in Figure 3.3, where $S = \{x \in Z^2: Ax \leqslant b\}$ and the outer polytope is $P = \{x \in R^2: Ax \leqslant b\}$. If x^1 happens to be an integral extreme point of P, then it must also be an extreme point of conv(S). Given a basic representation of x^1 in which the active constraints are $x_k = 0$ and $x_{k'} = 0$, our objective is to pivot from x^1 to x^2 or to x^3. However, a standard simplex pivot will yield a fractional extreme point of P, either x^4 or x^5. To pivot from x^1 to x^2, the polytope that contains conv(S) must contain the facet-defining inequality $a^*x \leqslant b^*$ and any other valid inequality defining a face that supports conv(S) at x^2, say $a^0x \leqslant b^0$. By first adding the constraint $x_p = b^* - a^*x$, a degenerate pivot that makes x_p nonbasic and x_k basic can be performed, and we still have the extreme point x^1. We then add the constraint $x_{p'} = b^0 - a^0x$ and make $x_{p'}$ nonbasic and $x_{k'}$ basic. This yields the extreme point x^2. Thus, in two dimensions, we need the facet of conv(S) that defines the edge joining x^1 and x^2 to be able to pivot from x^1 to x^2.

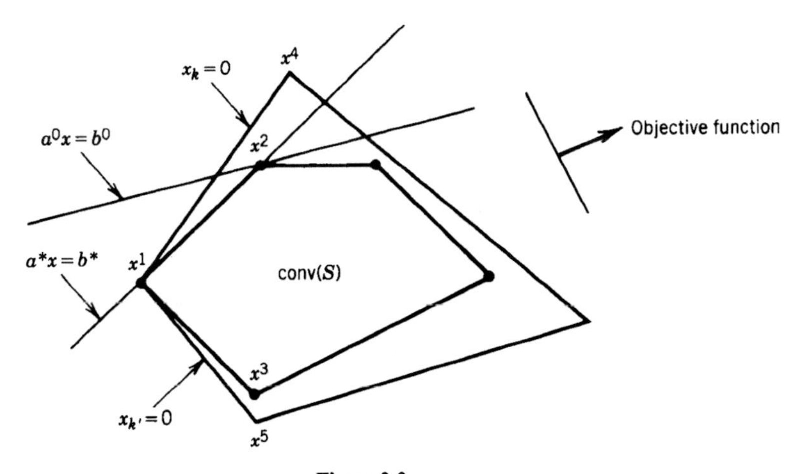

Figure 3.3

Analogously, in n dimensions we need $n - 1$ valid inequalities that contain the one-dimensional face joining x^1 to x^2 and another valid inequality that defines a face which supports conv(S) at x^2 to be able to pivot from x^1 to x^2. These very stringent requirements explain why a primal cutting-plane algorithm for general integer programming is likely to be very slow. Besides the problem of finding an initial integral point, it will be necessary to produce valid inequalities that contain the one-dimensional faces (edges) on a path from the initial point to an optimal point. In contrast, a fractional cutting-plane algorithm can succeed with a much weaker family of cuts, and a nondegenerate pivot occurs immediately after the addition of each cut.

We now study how these primal cuts can be derived algebraically. Consider a basis for the linear programming relaxation of IP given by (3.1) in which the coefficients $\overline{a}_{ij}$ are integral for $i = 0, 1, \ldots, m$ and all $j \in H$. A basis that satisfies these conditions is available if $A = (A', I)$ and $b \geqslant 0$. Otherwise, a Phase I procedure may be required.

If the basis is dual feasible, the integral solution $(x_0, x) = \overline{a}_0$ is optimal. So suppose $\overline{a}_{0k} < 0$ for $k \in H$. Consider a primal pivot in which x_k becomes basic. Suppose

$$\frac{\overline{a}_{r0}}{\overline{a}_{rk}} = \min_{i=1,\ldots,m:\overline{a}_{ik}>0} \frac{\overline{a}_{i0}}{\overline{a}_{ik}}.$$

If $\overline{a}_{rk} = 1$, we can pivot on the row $x_{B_r} + \Sigma_{j\in H}\, \overline{a}_{rj}x_j = \overline{a}_{r0}$ and maintain integrality. If $\overline{a}_{rk} \neq 1$, we add a C–G cut derived from the inequality $\Sigma_{j\in H}\, \overline{a}_{rj}x_j \leqslant \overline{a}_{r0}$. In particular, multiply this inequality by $1 / \overline{a}_{rk} > 0$ and then round to obtain

$$x_k + \sum_{j\in H\setminus\{k\}} \left\lfloor \frac{\overline{a}_{rj}}{\overline{a}_{rk}} \right\rfloor x_j \leqslant \left\lfloor \frac{\overline{a}_{r0}}{\overline{a}_{rk}} \right\rfloor.$$

Adding a slack variable yields the equation

$$(3.5) \qquad x_{n+1} + x_k + \sum_{j\in H\setminus\{k\}} \left\lfloor \frac{\overline{a}_{rj}}{\overline{a}_{rk}} \right\rfloor x_j = \left\lfloor \frac{\overline{a}_{r0}}{\overline{a}_{rk}} \right\rfloor, \qquad x_{n+1} \in Z^1_+.$$

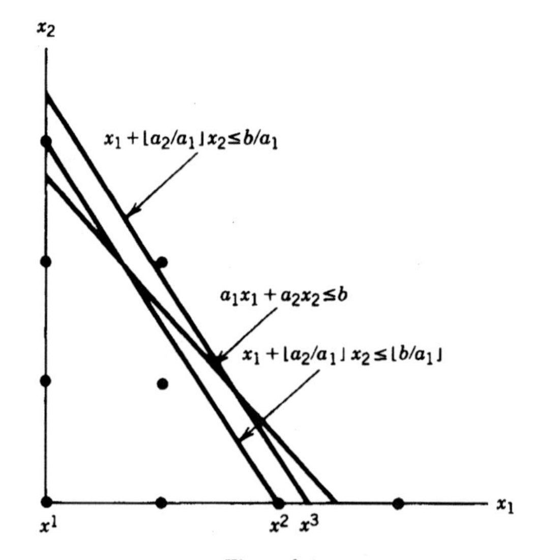

Figure 3.4

Since $\bar{a}_{rk} > 0$, $\lfloor \bar{a}_{r0}/\bar{a}_{rk} \rfloor \leqslant \bar{a}_{r0}/\bar{a}_{rk}$ and the coefficient of x_k in (3.5) is one, we can pivot on the row (3.5) and maintain integrality.

The geometry of the cut is shown in Figures 3.4 and 3.5. In Figure 3.4, assume we are at the point x^1 and we want to pivot along the line $x_2 = 0$. An ordinary simplex pivot would yield the fractional point x^3. Instead we introduce the cut $x_1 + \lfloor a_2 / a_1 \rfloor x_2 \leqslant \lfloor b/a_1 \rfloor$, which enables us to pivot to the integral point x^2. The cut is obtained by adding $(\lfloor a_2 / a_1 \rfloor - a_2 / a_1)x_2 \leqslant 0$ to $x_1 + (a_2 / a_1)x_2 \leqslant b/a_1$ and then rounding. Thus the cut gives the convex hull of the region $\{x \in R^2: x_1 + \lfloor a_2 / a_1 \rfloor x_2 \leqslant b/a_1\}$.

In Figure 3.5, there are no feasible integral points along the line $a_{21}x_1 + a_{22}x_2 = b_2$, so it is not possible to move from x^1 along this line. However from an appropriate nonnegative linear combination of $a_{i1}x_1 + a_{i2}x_2 \leqslant b_i$ for $i = 1, 2$, we obtain the inequality $\bar{a}_1 x_1 + \bar{a}_2 x_2 \leqslant \bar{b}$, with $0 < \bar{b} < \bar{a}_1$, $\bar{A}_2 < 0$. Now we proceed as above to obtain the inequality $x_1 + \lfloor \bar{a}_2 / \bar{a}_1 \rfloor x_2 \leqslant \bar{b}/\bar{a}_1$ and then the cut $x_1 + \lfloor \bar{a}_2 / \bar{a}_1 \rfloor x_2 \leqslant \lfloor \bar{b}/\bar{a}_1 \rfloor$.

Example 3.2

$$
\begin{aligned}
\max x_0 = \quad & x_1 + 2x_2 \\
-4x_1 + {} & x_2 + x_3 \qquad = 0 \\
7x_1 + {} & 4x_2 \qquad + x_4 = 14 \\
x \in {} & Z_+^4.
\end{aligned}
$$

An integral solution to the linear programming relaxation is

$$
\begin{aligned}
x_0 &= 0 + x_1 + 2x_2 \\
x_3 &= 0 + 4x_1 - x_2 \\
x_4 &= 14 - 7x_1 - 4x_2 \\
x_1 &= x_2 = 0.
\end{aligned}
$$

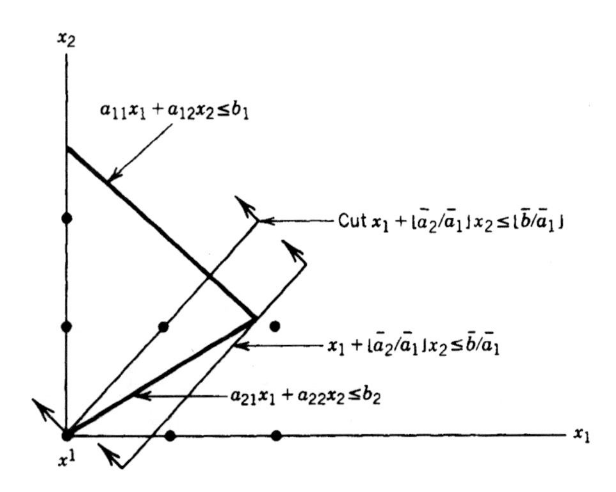

Figure 3.5

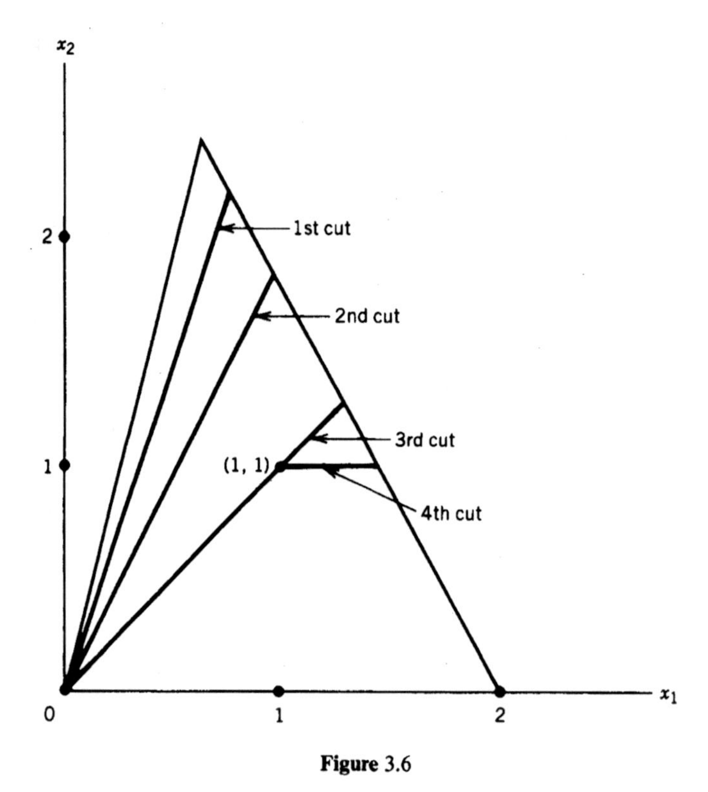

Figure 3.6

Suppose we choose to increase x_2. The variable that becomes nonbasic is x_3, and no cut is required since the coefficient of x_2 is one. This degenerate pivot yields the equations

$$x_0 = 0 + 9x_1 - 2x_3$$

$$x_2 = 0 + 4x_1 - x_3$$

$$x_4 = 14 - 23x_1 + 4x_3$$

$$x_1 = x_3 = 0.$$

Now we make x_1 basic and derive the cut $x_1 - x_3 \leq 0$ from $23x_1 - 4x_3 \leq 14$. We then do a degenerate pivot on the row $x_5 = 0 - x_1 + x_3$. Three more cuts and pivots are required (see Figure 3.6). The last pivot is the only nondegenerate one and it yields the optimal solution $(x_1, x_2) = (1 \quad 1)$.

It is a fact that with an appropriate choice of pivot columns and rows from which to generate the cuts, a finite algorithm can be obtained. But for the reasons stated above, primal cutting-plane algorithms are likely to be slower than fractional cutting-plane algorithms.

4. NOTES

Section II.4.1

Geoffrion and Marsten (1972) proposed a somewhat different framework for discrete optimization algorithms.

Enumerative methods go under the general names of *branch-and-bound*, *implicit enumeration*, and *divide-and conquer*. The latter term is frequently used in the computer science literature, and the first two terms are commonly used in the mathematical programming/operations research literature. Algorithms that focus on pruning by bounds that are obtained from relaxations or dual solutions are generally referred to as *branch-and-bound algorithms*, a term coined by Little et al. (1963). Those that focus on pruning based on logical tests and inequalities derived from Boolean implications (see Section I.1.6) are called *implicit enumeration algorithms*, a term apparently coined by Geoffrion (1967). Many algorithms use both of these ideas, so a classification of enumeration algorithms along these lines is not particularly relevant. Furthermore, although logical testing is important, pruning by bounds has emerged as the fundamental tool of enumerative algorithms.

Land and Doig (1960) gave the first branch-and-bound algorithm for general integer programs. However, the popularity of this approach increased substantially after the publication of the branch-and-bound algorithm for the traveling salesman problem by Little et al. (1963) because it demonstrated that large (at that time) problems could be solved by controlled enumeration. Balas (1965) gave the first implicit enumeration algorithm for general 0-1 IP's

General expositions and survey articles on enumerative methods are by Lawler and Wood (1966), Agin (1966), Mitten (1970), Tomlin (1970), Geoffrion and Marsten (1972), Beale (1979), Garfinkel (1979), and Spielberg (1979).

Sensitivity and parametric analysis for integer programs has been discussed by Geoffrion and Nauss (1977), Shapiro (1977), Nauss (1979), Holm and Klein (1984), Schrage and Wolsey (1985), and Cook et al. (1986).

Parallel processing presents new opportunities for computational advances in discrete optimization. Kindervater and Lenstra (1985, 1986) give an annotated bibliography and an introduction to parallelism in combinatorial optimization. In an empirical study, Pruul (1975) simulated parallel computation and showed that by exploring several nodes of an enumeration tree simultaneously it is possible to reduce substantially the total number of nodes that need to be considered. His results have been summarized in Pruul et al. (1988).

Another developing area is interactive optimization. Fisher (1985) surveyed results and opportunities for using interactive methods in discrete optimization.

Section II.4.2

Almost all general MILP codes use a branch-and-bound framework with linear programming relaxations. As noted above, the first algorithm of this type was described by Land and Doig (1960). They proposed the division scheme shown in Figure 2.3. The now commonly used variable dichotomy scheme (Figure 2.1) was proposed by Dakin (1965). Penalties were introduced by Driebeek (1966) and were sharpened by Tomlin (1971).

The treatment of generalized upper-bound constraints by the division scheme shown in Figure 2.2, together with the indexing scheme described below (2.4), was introduced by Beale and Tomlin (1970). They called such sets *specially ordered sets*. This terminology is now widely used, and the concept is very important in the global maximization of piecewise linear nonconcave functions. Beale and Forrest (1976) developed this approach, which enables the implementation of the division scheme (2.6) without the explicit use of auxiliary integer variables.

Various strategies for exploring the enumeration tree, together with experimental comparison, are given by Benichou et al. (1971, 1977), Breu and Burdet (1974), Forrest et al. (1974), Gauthier and Ribiere (1977), and Mitra (1973). The strategy described in the text is used in the commercial code SCICONIC as described by Beale (1979).

Eleven commercial mixed-integer programming systems, including all of the available options for branching, node, and variable selection as well as other characteristics of the codes, have been described by Land and Powell (1979). This article also includes a comparison of two of these codes on a small number of test problems, as well as brief descriptions of several "academic" codes. Powell (1985) is an annotated bibliography that updates the Land and Powell article.

The XMP system of Marsten (1981) has been updated to include branch-and-bound for general MIPs. It is a highly modular system, which makes it very useful for research, and is available in both microcomputer and main-frame versions. Several other linear programming systems for microcomputers also have branch-and-bound capabilities. One of the most widely used is the LINDO system of Schrage (1986).

The art involved in using commercial codes to solve large scale MIPs is discussed in the context of solving practical problems by Beale (1983) and Suhl (1985).

Jeroslow (1974), Ibaraki (1976, 1977), Rinnooy Kan (1976), and Fox et al. (1978) have presented some theoretical results on node selection and branching strategies. Jeroslow gives a family of problems for which the number of nodes that must be searched is exponential with respect to the size of the problem, regardless of which strategies are used.

Section II.4.3

Gomory (1958, 1960a, 1963a) shows that the FCPA is finitely convergent with appropriate use of the lexicographic dual simplex algorithm. The proof given here is based on that given by Nourie and Venta (1982), which is really just a reinterpretation of Gomory's proof that provides additional insight into the nature of the convergence.

Given a fractional LP solution, Gomory and Hoffman (1963) showed that the cuts $\Sigma_j x_j \geqslant 1$, where the sum is taken over all nonbasic variables cannot yield a finite FCPA. Bowman and Nemhauser (1970) proved that the stronger cuts given in exercise 11 yield a finite algorithm.

The mixed-integer cutting-plane algorithm and its finite convergence under the assumption that the objective function variable must be an integer is given in Gomory (1960b). Without this assumption, no finite cutting-plane algorithm for MILPs is known.

A primal cutting-plane algorithm for general integer programs was proposed by Ben-Israel and Charnes (1962). A finitely convergent primal cutting-plane algorithm was given by Young (1965), and simplified versions were obtained by Glover (1968a) and Young (1968). Because of poor computational experience, this line of research has been very inactive. An exception is a primal cutting-plane algorithm for the traveling salesman problem developed by Padberg and Hong (1980). Although this algorithm has been moderately successful, it seems to be inferior to an FCPA for the traveling salesman problem (see Section II.6.3).

Another strategy for cutting-plane algorithms is to maintain integrality and dual feasibility and then to use cuts to obtain primal feasibility. A finite algorithm of this type has been given by Gomory (1963b). Other such algorithms have been obtained by Glover (1965, 1967).

5. EXERCISES

1. Solve the integer program

$$\max\{2x_1 + 5x_2: 4x_1 + x_2 \leq 28, x_1 + 4x_2 \leq 27, x_1 - x_2 \leq 1, x \in Z_+^2\}$$

by branch-and-bound.
Examine the procedure graphically.
Investigate how the branch-and-bound tree changes with different branching strategies.

2. Solve the integer knapsack problem

$$\max\{16x_1 + 19x_2 + 23x_3 + 28x_4: 2x_1 + 3x_2 + 4x_3 + 5x_4 \leq 7, x \in Z_+^4\}$$

by branch-and-bound.

3. Solve the problem of exercise 2 with the additional constraint $x \in B^4$.

4. Propose various ways to estimate degradations. Test them on the above problems.

5. Solve Example 2.3 using a branch-and-bound algorithm, with the conventional branching rule. Draw the branch-and-bound tree.

6. Modify the general branch-and-bound algorithm if, instead of an optimum solution, we only want to find a feasible solution within a given percentage, say p% of the optimum value.

7. Consider the integer program

$$\begin{aligned} \max \quad & -x_{n+1} \\ & 2x_1 + 2x_2 + \cdots + 2x_n + x_{n+1} = n \\ & x \in B^{n+1}. \end{aligned}$$

Show that any branch-and-bound algorithm using the linear programming relaxation to compute upper bounds will require the enumeration of an exponential number of nodes when n is odd.

8. Consider a mixed-integer program with one integer variable. Show that the branch-and-bound tree for this problem will have no more than three nodes. Why?

9. Consider the integer program of exercise 1. The optimal linear programming basis gives the information

$$\text{(IP)}\quad \begin{aligned} &\max z \\ &z + \tfrac{1}{5}x_3 + \tfrac{6}{5}x_4 = 38 \\ &x_1 + \tfrac{4}{15}x_3 - \tfrac{1}{15}x_4 = \tfrac{17}{3} \\ &x_2 - \tfrac{1}{15}x_3 + \tfrac{4}{15}x_4 = \tfrac{16}{3} \\ &-\tfrac{1}{3}x_3 + \tfrac{1}{3}x_4 + x_5 = \tfrac{2}{3} \\ &x \in Z_+^5 \end{aligned}$$

i) Solve using Gomory's FCPA.

ii) Solve using Gomory's finitely convergent FCPA.

iii) Use the solutions of i and ii to give optimal dual solutions of IP.

iv) Use iii to find an upper bound on the optimal value of IP when

$$b = \begin{pmatrix} 28 \\ 27 \\ 1 \end{pmatrix} \quad \text{is replaced by} \quad b = \begin{pmatrix} 25 \\ 29 \\ 2 \end{pmatrix}$$

10. What is the maximum number of Gomory cuts needed to verify that a 0, 1 IP is feasible or infeasible. (*Hint*: Consider the enumeration tree.)

11. **i)** Let $S = \{x \in Z_+^n: \sum_{j \in N} a_j x_j = b\}$ with $a_j, b \in R^1$, and $b \notin Z_+^1$. Show that $\sum_{j \in N^*} x_j \geq 1$ is a valid inequality for S, where $N^* = \{j \in N: a_j \notin Z^1\}$.

ii) Show that a finitely convergent FCPA is obtained by using these cuts in place of Gomory cuts.

iii) Carry out several iterations on the IP of exercise 7, and compare the corresponding enumeration trees.

12. **i)** Use Gomory mixed-integer cuts to solve the integer program

$$\begin{aligned} \max \quad & y \\ & x_1 + x_2 + y \leq 2 \\ & -x_1 + y \leq 0 \\ & -x_2 + y \leq 0 \\ & x \in Z_+^2, \quad y \in Z_+^1. \end{aligned}$$

ii) Replace the constraint $y \in Z_+^1$ in i by $y \in R_+^1$ to give a mixed-integer program (see exercise 22 of Section II.1.9). What happens now using the Gomory mixed-integer cuts?

13. Describe a Phase 1 procedure for a primal cutting-plane algorithm.

II.5

Special-Purpose Algorithms

1. INTRODUCTION

The algorithms presented in the previous chapter have the great advantage of robustness. They can, in principle, be applied to all linear integer programs. However, there is often a heavy price to pay for this generality.

Three major reasons why a problem class may not be solved satisfactorily by a general algorithm are:

1. size of the formulation,
2. weakness of the bounds, and
3. speed of the algorithm.

On the other hand, when instances of a class of highly structured integer programs are to be solved, the structure can often be used to improve the performance substantially in one or all of the three areas cited above. In this chapter we will show how structure can be used either to devise special-purpose algorithms or to improve the performance of general algorithms for several classes of problems.

Integer programming formulations frequently have a very large number of variables or constraints. For example, in the strong formulation of the uncapacitated facility location problem described in Section I.1.3, an instance with $n = 50$ locations and $m = 200$ clients has more than $mn = 10{,}000$ variables and more than $mn = 10{,}000$ constraints. Similarly for the traveling salesman problem on m nodes, the formulation given in Chapter I.1 has $O(m^2)$ variables and $O(2^m)$ subtour elimination constraints.

The computation of bounds requires the *choice of a relaxation*. Typically this choice involves a tradeoff between the strength of the bound obtained from the relaxation and the speed with which it can be calculated. For the symmetric traveling salesman problem there is a large hierarchy of relaxations. For the uncapacitated facility location problem the tradeoff is between the linear programming relaxation of the weak formulation, which can be solved by a formula but gives weak bounds, and the linear programming relaxation of the strong formulation, which gives very good bounds but is much harder to solve.

Very often the relaxation is embedded in a branch-and-bound algorithm. Here structure may make it possible to find nearly optimal solutions to the dual of the relaxation very quickly, providing the upper bounds needed for the branch-and-bound algorithm. Structure may also help us to find good feasible solutions quickly, which are also important in pruning the branch-and-bound tree. Furthermore, it is often the case that nearly optimal primal and dual solutions are a satisfactory solution to the problem.

Structure frequently suggests decomposition. Both Lagrangian (row) decomposition and Benders (column) decomposition have been introduced in Chapter II.3. For a given problem, several Lagrangian relaxations may be available. In addition, algorithms for the Lagrangian dual provide alternative ways to solve large linear programs. However, both row and column decompositions typically lead to a large number of variables or constraints, so again the question of finding effective algorithms for large-sized problems is an issue. Finally for some structures, dynamic programming provides a decomposition which makes it possible to solve problems by a recursive algorithm that uses dominance to eliminate nonoptimal solutions.

Thus if we want to make efficient use of structure, we must deal effectively with the following three issues.

1. A choice of formulation and "strong" linear programming (or combinatorial) relaxation must be made.
2. The chosen relaxation typically has a very large number of constraints (and possibly columns). An algorithm that finds an optimal (or a good dual feasible) solution to the relaxation as quickly as possible has to be selected. It also may be desirable to have a heuristic for finding good primal feasible solutions rapidly.
3. Since the solution to the relaxation rarely solves the original problem, a procedure is needed, typically embedding the relaxation into branch-and-bound, to arrive at an optimal solution to the original problem.

In the following four sections we will present some methods that take advantage of structure. First we discuss strong cutting-plane (or constraint generation) algorithms. Then we present some ways of quickly finding nearly optimal dual and primal feasible solutions. Next we discuss the algorithms that can be used in combination with Lagrangian and Benders' decomposition, as well as some of the problems that arise in their implementation. Finally we describe dynamic programming and illustrate its application to certain discrete optimization problems.

The first four sections of the next chapter are each devoted to a particular structured problem, knapsack problems, 0-1 problems, the symmetric traveling salesman problem, and fixed-charge flow problems, respectively. For each problem we exhibit how some of the practical choices are made that lead to a relatively efficient special-purpose algorithm.

We will use the uncapacitated facility location problem as an example of a structured problem throughout this and the next three sections. As a starting point we know the two formulations presented in Chapter I.1, the so-called "strong formulation"

$$
\text{(UFL)}\qquad
\begin{aligned}
z = \max \sum_{i\in I}\sum_{j\in N} c_{ij}y_{ij} - \sum_{j\in N} f_j x_j & \\
\sum_{j\in N} y_{ij} = 1 \quad & \text{for } i \in I \\
y_{ij} - x_j \leq 0 \quad & \text{for } i \in I, j \in N \\
y \in R_+^{mn}, \quad x \in B^n, &
\end{aligned}
$$

where $I = \{1, \ldots, m\}$ and $N = \{1, \ldots, n\}$, and the "weak formulation"

$$\text{(WUFL)}\quad \begin{aligned} z = \max \sum_{i\in I}\sum_{j\in N} c_{ij}y_{ij} - \sum_{j\in N} f_j x_j \\ \sum_{j\in N} y_{ij} &= 1 \quad \text{for } i \in I \\ \sum_{i\in I} y_{ij} - m x_j &\leqslant 0 \quad \text{for } j \in N \\ y \in R_+^{mn},\quad x \in B^n. \end{aligned}$$

We have already remarked on the size of these formulations. Another important observation is that once we have decided which facilities $Q \subseteq N$ are to be opened, then the optimal allocation of the clients to the facilities is obvious. In particular, client i is served by a facility $k \in Q$ such that $c_{ik} = \max_{j\in Q} c_{ij}$. Hence there exists an optimal solution to UFL and WUFL with $y \in B^{mn}$.

Proposition 1.1. *The value of opening facilities at $Q \subseteq N$ is*

$$v(Q) = \sum_{i\in I} \max_{j\in Q} c_{ij} - \sum_{j\in Q} f_j.$$

Now suppose we wish to make a choice between the linear programming relaxations of UFL and WUFL. Note that there is a closed-form solution to the linear programming relaxation of WUFL.

Proposition 1.2. *When $f_j \geqslant 0$ for all $j \in N$, there exists an optimal solution (x^*, y^*) to the linear programming relaxation of* WUFL *with $y^*_{ij_i} = 1$, where $j_i = \arg\max_{j\in N} (c_{ij} - f_j/m)$ for all $i \in I$, $y^*_{ij} = 0$ otherwise, and where $x^*_j = (1/m)\, \Sigma_{i\in I}\, y^*_{ij}$ for all $j \in N$.*

From Proposition 1.2 it follows that $\Sigma_{j\in N}\, x^*_j = 1$, independent of the number of facilities opened in an optimal solution. Hence the fixed costs f_j for $j \in N$ are largely ignored, and the relaxation cannot provide a good upper bound on the optimal value z.

In contrast, the upper bound provided by the linear programming relaxation of UFL is usually very strong and no larger than the bound obtained from the linear programming relaxation of WUFL. The strength of this bound will be discussed further in Section 3.

Example 1.1. Consider the uncapacitated location problem with the following data:

$$m = 6, \qquad n = 5,$$

$$C = (c_{ij}) = \begin{pmatrix} 12 & 13 & 6 & 0 & 1 \\ 8 & 4 & 9 & 1 & 2 \\ 2 & 6 & 6 & 0 & 1 \\ 3 & 5 & 2 & 10 & 8 \\ 8 & 0 & 5 & 10 & 8 \\ 2 & 0 & 3 & 4 & 1 \end{pmatrix},$$

$$f = (f_j) = \quad (4 \quad 3 \quad 4 \quad 4 \quad 7).$$

Using Proposition 1.2, we obtain the optimal solution to the LP relaxation of WUFL given by $y^*_{12} = y^*_{23} = y^*_{32} = y^*_{44} = y^*_{54} = y^*_{64} = 1$, $x^*_2 = \frac{1}{3}$, $x^*_3 = \frac{1}{6}$, $x^*_4 = \frac{1}{2}$ with value $48\frac{1}{3}$.

The LP relaxation of UFL has an optimal solution (derived later) given by $y_{11}^* = y_{12}^* = y_{21}^* = y_{23}^* = y_{32}^* = y_{33}^* = \frac{1}{2}$, $y_{44}^* = y_{54}^* = y_{64}^* = 1$, $x_1^* = x_2^* = x_3^* = \frac{1}{2}$, $x_4^* = 1$ with value $z_{LP} = 41\frac{1}{2}$. The optimal value of UFL is $z = 41$.

From now on, we only consider the strong LP relaxation of UFL and concentrate on developing fast algorithms that solve it exactly or approximately.

2. A CUTTING-PLANE ALGORITHM USING STRONG VALID INEQUALITIES

Here we consider linear programs of the form

$$\text{(LP }(\mathscr{F}))\qquad \begin{aligned} z_{LP}(\mathscr{F}) = \max\ & cx \\ & Ax \leqslant b \\ & \pi x \leqslant \pi_0 \quad \text{for } (\pi, \pi_0) \in \mathscr{F} \\ & x \in R_+^n, \end{aligned}$$

where $\mathscr{F}$ generally contains a large number of constraints. Linear programs of this form arise as:

i. relaxations of an integer program $\max\{cx: x \in S\}$, where $S = \{x \in Z_+^n: Ax \leqslant b\}$ and $\mathscr{F}$ is a set of "strong" valid inequalities for S; and
ii. linear programs with a large number of constraints.

As an example of i, S is the set of solutions to a 0-1 knapsack problem, $\mathscr{F}$ represents the set of cover inequalities, and LP($\mathscr{F}$) represents the linear program consisting of the knapsack constraint and the cover inequalities.

As an example of ii, LP($\mathscr{F}$) is the LP relaxation of UFL, where $Ax \leqslant b$ are the constraints of the weak formulation WUFL without integrality and $\mathscr{F}$ represents the mn variable upper-bound constraints $y_{ij} - x_j \leqslant 0$ for all $i \in I, j \in N$.

The question we need to answer is how to solve the linear program LP($\mathscr{F}$). The "brute force" approach of adding all the constraints in $\mathscr{F}$ a priori is impractical when the number of inequalities in $\mathscr{F}$ is very large. Furthermore, most of the inequalities in $\mathscr{F}$ are unnecessary for the solution of LP($\mathscr{F}$). However, a priori addition of a subset of the inequalities from $\mathscr{F}$ may be very desirable.

A more general approach uses the inequalities of $\mathscr{F}$ as cutting planes. Only those inequalities in $\mathscr{F}$ that are likely to be active in the neighborhood of an optimal solution to LP($\mathscr{F}$) are generated.

Fractional Cutting-Plane Algorithm (FCPA) for LP($\mathscr{F}$)

Initialization: $S_R^1 = \{x \in R_+^n: Ax \leqslant b\}$. Set $t = 1$.
Iteration t:

Step 1: Solve the relaxation of LP($\mathscr{F}$)

$$\text{(LP}^t\text{)}\qquad z_R^t = \max\{cx: x \in S_R^t\}$$

and let x^t be an optimal solution. (Note that the dual solution complementary to x^{t-1} is a dual feasible solution to LP^t. Hence LP^t can be solved by a dual algorithm starting from the point x^{t-1}.)

Step 2: Optimality Test. If $\pi x^t \leqslant \pi_0$ for all $(\pi, \pi_0) \in \mathscr{F}$, then x^t is an optimal solution to LP($\mathscr{F}$). Stop.

Step 3: Refinement. Let $\mathscr{F}^t \subset \mathscr{F}$ be a set of one or more inequalities (π, π_0) with $\pi x^t > \pi_0$ and

$$S_R^{t+1} = S_R^t \cap \{x \in R_+^n: \pi x \leqslant \pi_0 \text{ for } (\pi, \pi_0) \in \mathscr{F}^t\}.$$

Step 4: $t \leftarrow t + 1$.

Given the solution x^t to the relaxation LP^t, we must show that x^t is a feasible solution to LP($\mathscr{F}$) or find a valid inequality $(\pi, \pi_0) \in \mathscr{F}$ for which $\pi x^t > \pi_0$. This is the separation problem that we introduced in Section I.6.3.

The Separation Problem for $\mathscr{F}$

Given a point $x^* \in R_+^n$, show that x^* satisfies all the valid inequalities in $\mathscr{F}$, or find one or more valid inequalities $(\pi, \pi_0) \in \mathscr{F}$ with $\pi x^* > \pi_0$.

Based on the polynomial equivalence of "separation" and "optimization" (see Theorem 3.3 of Section I.6.3), we make the following observations.

i. Under the assumption that we can check whether x^* satisfies the constraints $Ax \leqslant b$ in polynomial time, LP($\mathscr{F}$) can be solved in polynomial time if and only if the separation problem for LP($\mathscr{F}$) can be solved in polynomial time.
ii. If we are dealing with an integer program $\max\{cx: Ax \leqslant b, x \in Z_+^n\}$ that is $\mathcal{NP}$-hard, and $\mathscr{F}$ represents one or more families of facets, there may be some families of facets for which the separation problem is in $\mathscr{P}$, but there will be others for which the separation problem is $\mathcal{NP}$-hard. More precisely, based on Proposition 7.4 of Section I.5.7, we cannot expect to have a good characterization of all the facet classes for the problem, and hence there will certainly be problem instances for which FCPA will terminate with a solution x^t that is not integral.

To demonstrate the FCPA with separation, we return to the uncapacitated facility location problem. Suppose that the number of constraints in formulation UFL is too large to solve its linear programming relaxation directly. So we start with the LP relaxation of the formulation WUFL and let $\mathscr{F}$ consist of the mn constraints $y_{ij} \leqslant x_j$ for all $i \in I$ and $j \in N$.

Now given a point $(x^*, y^*) \in R_+^n \times R_+^{mn}$, the separation problem for $\mathscr{F}$ is to find whether one or more of the mn variable upper-bound constraints is violated. This is easily done by enumeration, and a violation occurs if and only if $\max_{i \in I} y_{ij}^* > x_j^*$ for some $j \in N$. Several implementations are possible, depending on the number of violated constraints added. In the implementation given below, for each j for which a violation occurs, we add one most violated constraint.

Example 1.1 (continued). We implement the FCPA with separation.

Iteration 1. It has been seen earlier that the optimal solution of LP^1 is $y_{12}^1 = y_{23}^1 = y_{32}^1 = y_{44}^1 = y_{54}^1 = y_{64}^1 = 1$, $x_2^1 = \frac{1}{3}$, $x_3^1 = \frac{1}{6}$, $x_4^1 = \frac{1}{2}$ with $z_{LP}^1 = 48\frac{1}{3}$. Since x^1 is not integral we apply the separation algorithm to (x^1, y^1) and find that three constraints from $\mathscr{F}$, namely $y_{12} \leqslant x_2$, $y_{23} \leqslant x_3$, and $y_{44} \leqslant x_4$, are violated. These constraints are now added to S_R^1.

Iteration 2. The linear program LP^2 is solved, giving the solution $y_{11}^2 = 0.8$, $y_{12}^2 = 0.2$, $y_{21}^2 = 1$, $y_{32}^2 = 1$, $y_{44}^2 = 0.4$, $y_{45}^2 = 0.6$, $y_{54}^2 = 1$, $y_{64}^2 = 1$, $x_1^2 = 0.3$, $x_2^2 = 0.2$, $x_4^2 = 0.4$, $x_5^2 = 0.1$ with $z_{LP}^2 = 44.9$. The separation algorithm now generates the four violated inequalities $y_{21} \leqslant x_1$, $y_{32} \leqslant x_2$, $y_{54} \leqslant x_4$, and $y_{45} \leqslant x_5$.

Iteration 3. The linear program LP^3 has an optimal solution $y_{11}^3 = 1$, $y_{21}^3 = 0.2$, $y_{23}^3 = 0.8$, $y_{33}^3 = 1$, $y_{44}^3 = 1$, $y_{54}^3 = 1$, $y_{64}^3 = 1$, $x_1^3 = 0.2$, $x_3^3 = 0.8$, $x_4^3 = 1$, and $z_{LP}^3 = 42.8$. The separation algorithm now generates the violated inequalities $y_{11} \leqslant x_1$ and $y_{33} \leqslant x_3$.

Iteration 4. The linear program LP^4 has an optimal solution $y_{11}^4 = y_{12}^4 = y_{21}^4 = y_{23}^4 = y_{32}^4 = y_{33}^4 = \frac{1}{2}$, $y_{44}^4 = y_{54}^4 = y_{64}^4 = 1$, $x_1^4 = x_2^4 = x_3^4 = \frac{1}{2}$, $x_4^4 = 1$, and $z_{LP}^4 = 41.5$. No violated inequalities are generated by the separation algorithm, so (x^4, y^4) is an optimal solution of the LP relaxation of UFL.

Note that only 9 out of the 30 possible variable upper-bound constraints $y_{ij} \leqslant x_j$ have been added in the course of the algorithm. Since the optimal solution (x^4, y^4) of LP($\mathscr{F}$) is not integral, we need to proceed further. One approach described below is to embed the FCPA into a branch-and-bound algorithm. Another is to enlarge the family $\mathscr{F}$ of strong valid inequalities (see exercise 5).

A Strong Cutting-Plane/Branch-and-Bound Algorithm for IP.
Given the problem

$$\text{(IP)} \qquad \max\{cx: Ax \leqslant b, x \in Z_+^n\}$$

and a class $\mathscr{F}$ of "strong" valid inequalities, we use the following 2-phase algorithm.

Phase 1. Solve LP($\mathscr{F}$) by the FCPA. On termination, let x^t be an optimal solution of LP($\mathscr{F}$). If $x^t \in Z_+^n$, stop. x^t solves IP. Otherwise, go to Phase 2.

Phase 2. Let $\mathscr{F}' \subset \mathscr{F}$ be the cuts generated in Phase 1, that is $\mathscr{F}' = \bigcup_{s=1}^{t} \mathscr{F}^s$. Solve the reformulation of IP,

$$\text{(IP}'\text{)} \qquad \max\{cx: Ax \leqslant b, \pi x \leqslant \pi_0 \text{ for } (\pi, \pi_0) \in \mathscr{F}', x \in Z_+^n\},$$

by branch-and-bound.

Example 1.1 (continued). The branch-and-bound tree for the problem IP′ with the nine inequalities added is shown in Figure 2.1.

On the first branch, node 2, x_1 is set to the value 1; the linear program has an optimal solution with x integer of value 40. The only remaining node is 3, where x_1 is set to zero. Here again the linear program has an optimal solution with x integer of value 41. The corresponding solution is $x_2 = x_3 = x_4 = 1$, $y_{12} = y_{23} = y_{32} = y_{44} = y_{54} = y_{64} = 1$. Since the tree has no active nodes, the solution found at node 3 is optimal.

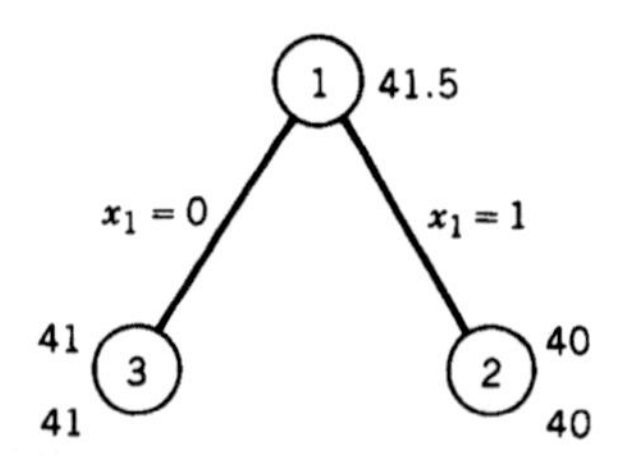

Figure 2.1

It may be possible to eliminate some variables from IP′. Given an optimal solution to LP($\mathcal{F}$), the reduced prices $\bar{c}_j$ are nonpositive for all nonbasic variables x_j at their lower bound, and nonnegative for all nonbasic variables at their upper bound. We suppose that $\underline{z}$ is the value of the best feasible solution known.

Proposition 2.1. *If x_j is nonbasic at its lower (upper) bound in the solution of* LP($\mathcal{F}$), *$x_j \in Z^1_+$, and $z_{\text{LP}(\mathcal{F})} + \bar{c}_j \leq \underline{z}$ ($z_{\text{LP}(\mathcal{F})} - \bar{c}_j \leq \underline{z}$), there exists an optimal solution to the integer program with x_j at its lower (upper) bound.*

This means that the set of variables needed in the branch-and-bound phase is $N^* = N \setminus \{j \in N: x_j \text{ is nonbasic}, z_{\text{LP}(\mathcal{F})} - |\bar{c}_j| \leq \underline{z}\}$.

Example 1.1 (continued). We exhibit the use of Proposition 2.1.

Suppose we have observed that the solution in which location 1, 2, and 4 are open has value $v(\{1, 2, 4\}) = 40$. Since y_{ij} equals 0 or 1 in an optimal solution, the reduced prices for the LP relaxation of UFL can be used to fix any variable with reduced price $|\bar{c}_{ij}| \geq 41.5 - 40 = 1.5$. In this case we can set $y_{13} = y_{14} = y_{15} = y_{22} = y_{24} = y_{25} = y_{31} = y_{34} = y_{35} = y_{41} = y_{42} = y_{43} = y_{52} = y_{53} = y_{55} = y_{62} = y_{65} = 0$ before entering the branch-and-bound phase.

Example 2.1. This is an instance of UFL with 33 facilities and clients. Each of the 33 cities is a client and a potential location for a facility. Here c_{ij} is the negative of the distance between cities i and j. The distances are given in Table 2.1, and the geographic locations are shown in Figure 2.2. The fixed costs are 2000 for each facility.

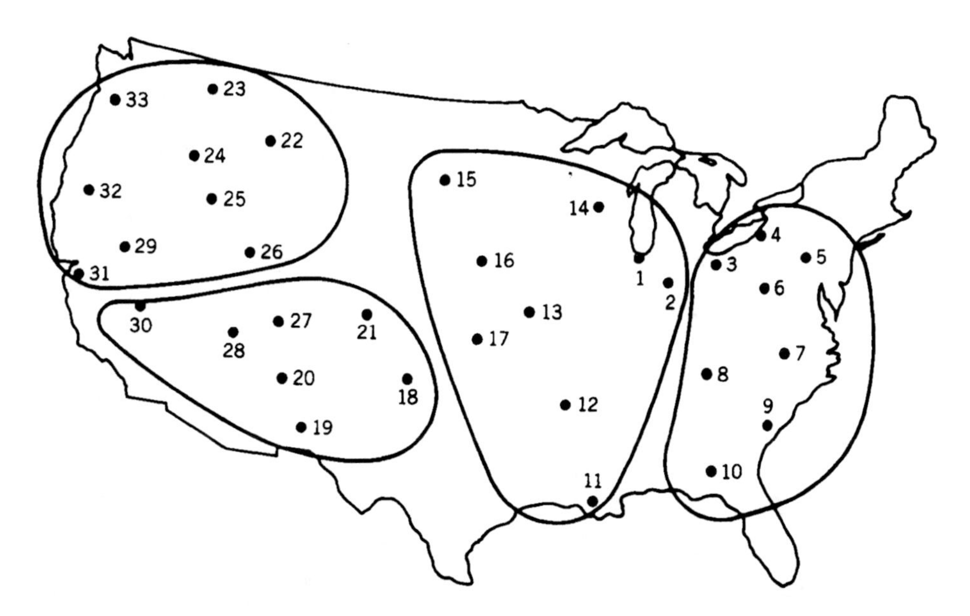

Figure 2.2. Thirty-three city problem: 1, Chicago, Ill.; 2, Indianapolis, Ind.; 3, Marion, Ohio; 4, Erie, Pa.; 5, Carlisle, Pa.; 6, Wana, W.V.; 7, Wilkesboro, N.C.; 8, Chattanooga, Tenn.; 9, Barnwell, S.C.; 10, Bainbridge, Ga.; 11, Baton Rouge, La.; 12, Little Rock, Ark.; 13, Kansas City, Mo.; 14, La Crosse, Wis.; 15, Blunt, S.D.; 16, Lincoln, Neb.; 17, Wichita, Kan.; 18, Amarillo, Tex.; 19, Truth or Consequences, N.M.; 20, Manuelito, N.M.; 21, Colorado Springs, Colo.; 22, Butte, Mont.; 23, Lewiston, Idaho; 24, Boise, Idaho; 25, Twin Falls, Idaho; 26, Salt Lake City, Utah; 27, Mexican Hat, Utah; 28, Marble Canyon, Ariz.; 29, Reno, Nev.; 30, Lone Pine, Calif.; 31, Gustine, Calif.; 32, Redding, Calif.; 33, Portland, Ore.

Table 2.1. Data for the 33-City Problem

	1	2	3	4	5	6	7	8	9	10	11	12	13	14	15	16	17	18	19	20	21	22	23	24	25	26	27	28	29	30	31	32	33
1	0																																
2	184	0																															
3	292	195	0																														
4	449	310	215	0																													
5	670	540	380	288	0																												
6	516	357	232	200	211	0																											
7	598	514	434	566	436	381	0																										
8	618	434	493	787	814	642	295	0																									
9	881	697	719	790	632	697	224	320	0																								
10	909	964	955	1020	974	952	541	341	318	0																							
11	978	892	1031	1246	1352	1180	843	538	747	441	0																						
12	654	597	803	1018	1154	1104	766	461	749	634	380	0																					
13	504	503	722	937	1043	806	986	722	1042	954	784	404	0																				
14	276	460	568	725	946	817	874	894	1214	1185	1218	660	452	0																			
15	780	964	1072	1229	1450	1321	1378	1326	1646	1672	1410	1030	626	476	0																		
16	529	644	789	1004	1184	1001	1214	950	1270	1213	1043	632	219	436	419	0																	
17	805	698	917	1132	1238	1055	1113	842	1162	1027	779	473	195	637	634	256	0																
18	1181	1007	1226	1441	1547	1364	1375	1080	1134	1138	783	611	563	1046	759	624	368	0															
19	1548	1444	1630	1845	1984	1801	1726	1431	1685	1477	1134	1033	906	1389	1094	967	711	404	0														
20	1547	1454	1668	1883	1994	1811	1879	1584	1776	1632	1267	1053	944	1427	1196	1005	749	442	251	0													
21	1239	1167	1353	1568	1707	1524	1584	1313	1633	1498	1151	979	614	988	600	525	471	368	512	507	0												
22	1538	1733	1830	2045	2208	2090	2136	2078	2398	2332	2110	1782	1378	1300	760	1229	1382	1319	1163	930	910	0											
23	1999	2158	2291	2448	2669	2515	2597	2500	2820	2675	2336	2164	1707	1860	1375	1582	1658	1545	1389	1156	1237	436	0										
24	1716	1875	2008	2165	2386	2232	2488	2217	2537	2392	2053	1881	1422	1577	1106	1244	1375	1262	1106	873	954	483	283	0									
25	1580	1738	1872	2029	2250	2095	2352	2081	2401	2256	1917	1745	1286	1473	988	1147	1239	1126	970	737	818	379	419	136	0								
26	1425	1569	1717	1874	2109	1926	2115	1844	2164	2019	1680	1508	1118	1335	862	913	1002	889	733	500	581	430	656	373	237	0							
27	1560	1549	1852	2009	2089	1906	2063	1792	2112	1967	1456	1274	1032	1540	1068	1007	944	665	521	282	491	768	994	711	575	358	0						
28	1918	1744	1963	2178	2284	2101	2174	1879	2071	1892	1562	1348	1239	1722	1258	1300	1044	737	526	295	802	816	1022	739	603	386	545	0					
29	2065	2102	2357	2514	2642	2459	2626	2355	2675	2530	2191	2019	1673	1905	1432	1570	1507	1320	1109	878	1124	842	715	432	465	533	849	739	0				
30	2284	2131	2326	2441	2671	2488	2418	2123	2437	2259	1929	1715	1606	1924	1451	1589	1411	1104	893	662	1143	1004	981	698	599	589	768	523	266	0			
31	2340	2348	2543	2658	2888	2705	2869	2598	2918	2773	2434	2262	1916	2148	1675	1813	1750	1468	1102	871	1367	1085	958	675	1033	778	1092	982	243	349	0		
32	2247	2327	2539	2696	2867	2684	2851	2580	2900	2755	2416	2244	1898	2130	1657	1795	1732	1545	1334	1103	1349	1014	814	531	1015	760	1047	964	225	497	266	0	
33	2163	2322	2455	2612	2833	2679	2761	2664	2984	2839	2500	2328	1871	2024	1539	1746	1822	1709	1553	1320	1391	693	346	447	583	820	1158	1355	581	847	710	444	0

Source: *Rand-McNally Road Atlas*, 38th edition, Rand-McNally Company (1962).

Table 2.2. The (i, j) Pairs for which Variable Upper-Bound Constraints Have Been Added

	$j=1$	2	3	4	5	6	7	8	9	10	11	12	13	14	15	16	17	18	19	20	21	22	23	24	25	26	27	28	29	30	31	32	33	t
$i=$	1	2	3	4	5	6	7	8	9	10	11	12	13	14	15	16	17	18	19	20	21	22	23	24	25	26	27	28	29	30	31	32	33	1
	14	3		6		5	8		10		12	11	17			15	13	21	20	27	18		33	23	22				30			29		2
	2	1		3	6	4	9	10	7	11			12	15		14	18			19			22	25	26	27	20		32		30	33	23	3
			2	5		3		7	8	9	10		16	1		13	21	19	18	28	15			33	23	22	26	20	31		29	30		4
	3		1	2		7	10	11		8		13	14			17	16	20		18	19	23		22	24	28	21	30	33	29	32	31		5
		14	4	1	3	2		9	11			10	15	13	16			17	28	21	27		24	29	33	30	19			31			32	6
	10	8	5		9		3	12				17	18	16	14		11		21			33		31	30	20		19		32			22	7
	16	9	6		4		5	2				8	1		22	21	19		27	30	20			32	29	24				28	33		31	8
	13	4	14			1						18	10				15	11			16			26		23	22	27						9
	4	5	7			9							11			1	14	15			17				27	25	28	26						10
	15	6			7		4	3	5					2			12	16																11
	5	13					6					16		17								15												12
		16													17			13							28	15								13

Table 2.3.

$\{j: x_j = 1\}$:	20, 24.
$\{j: x_j = \frac{1}{2}\}$:	3, 7, 8, 13, 16.
$\{(i, j): y_{ij} = 1\}$:	(18, 20), (19, 20), (20, 20), (21, 20), (22, 24), (23, 24), (24, 24), (25, 24), (26, 24), (27, 20), (28, 20), (29, 24), (30, 20), (31, 24), (32, 24), (33, 24).
$\{(i, j): y_{ij} = \frac{1}{2}\}$:	(1, 3), (1, 13), (2, 3), (2, 8), (3, 3), (3, 7), (4, 3), (4, 7), (5, 3), (5, 7), (6, 3), (6, 7), (7, 7), (7, 8), (8, 7), (8, 8), (9, 7), (9, 8), (10, 7), (10, 8), (11, 8), (11, 13), (12, 8), (12, 13), (13, 13), (13, 16), (14, 13), (14, 16), (15, 13), (15, 16), (16, 13), (16, 16), (17, 13), (17, 16).

When the FCPA/branch-and-bound algorithm is applied, the initial LP relaxation (with no variable upper-bound constraints) has value −2000. After adding 230 variable upper-bound constraints (see Table 2.2), the linear programming relaxation of UFL is solved with value −20,346. The solution is given in Table 2.3.

When branch-and-bound is applied, an integral solution of value −20,393 is found at node 2, and an integral solution of value −20,363 is found and proved optimal at node 3 (see Figure 2.3). The optimal solution is shown in Table 2.4

We have tacitly assumed that the number of variables in LP($\mathcal{F}$) does not cause computational difficulties. But this may not be the case. For example, if $m = 500$ and $n = 100$ in UFL, then FCPA could not be used directly to solve LP($\mathcal{F}$).

To handle a very large number of variables in addition to, perhaps, a very large number of constraints, we used a standard linear programming technique. We first solve LP($\mathcal{F}$) with a suitably chosen subset of variables eliminated; that is, we choose $N' \subset N$ and set $x_j = 0$ for $j \in N'$. After solving the restricted version of LP($\mathcal{F}$), we check for optimality with respect to LP($\mathcal{F}$) by calculating the reduced prices of the variables x_j with $j \in N'$. If all of these reduced prices are nonpositive, LP($\mathcal{F}$) is solved. Otherwise, we delete from N' all j such that the reduced price of x_j is positive. We then continue with the solution of LP($\mathcal{F}$). Note that after adding variables it is preferable to reoptimize with a primal algorithm since the current solution is primal feasible. Hence if this technique is used within an FCPA, it is desirable to have primal and dual linear programming algorithms available.

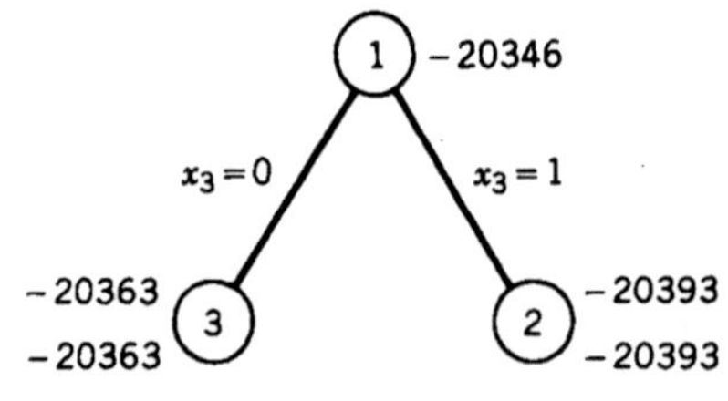

Figure 2.3

Table 2.4.

$\{j: x_j = 1\}$:	7, 13, 20, 24
$\{i: y_{i,7} = 1\}$:	3, 4, 5, 6, 7, 8, 9, 10
$\{i: y_{i,13} = 1\}$:	1, 2, 11, 12, 13, 14, 15, 16, 17
$\{i: y_{i,20} = 1\}$:	18, 19, 20, 21, 27, 28, 30
$\{i: y_{i,24} = 1\}$:	22, 23, 24, 25, 26, 29, 31, 32, 33

3. PRIMAL AND DUAL HEURISTIC ALGORITHMS

Heuristic or approximate algorithms are designed to find good, but not necessarily optimal, solutions quickly. For a variety of problems with structure, it is easy to devise heuristic algorithms to find primal and dual feasible solutions. It is particularly desirable to find both primal and dual feasible solutions since the dual solution provides an upper bound on the deviation from optimality of the primal solution. Depending on the quality of the solution required, an approximate solution can be the final answer to a problem or can be an input to an exact algorithm. The lower and upper bounds provided by approximate solutions can be of great help in decreasing the running time of branch-and-bound algorithms.

Though it is difficult to describe completely general heuristic algorithms, three ideas are applicable in a wide variety of cases. The first is that of a "greedy", alternatively called a "steepest ascent/descent" or "myopic", algorithm.

Greedy algorithms are frequently applied to the maximization of set functions. Let $v(Q)$ be a real-valued function defined on all subsets of $N = \{1, \ldots, n\}$ and consider the problem $\max\{v(Q): Q \subseteq N\}$.

A Greedy (Heuristic) Algorithm for Maximizing a Set Function

Initialization: $Q^0 = \emptyset$, $t = 1$.

Iteration t:

Step 1: Let $j_t = \arg\max_{j \in N \setminus Q^{t-1}} v(Q^{t-1} \cup \{j\})$ with ties broken arbitrarily.

Step 2: If $v(Q^{t-1} \cup \{j_t\}) \leqslant v(Q^{t-1})$, stop. Q^{t-1} is a greedy solution.

Step 3: If $v(Q^{t-1} \cup \{j_t\}) > v(Q^{t-1})$, set $Q^t = Q^{t-1} \cup \{j_t\}$.

Step 4: If $Q^t = N$, stop. N is a greedy solution. Otherwise let $t \leftarrow t + 1$.

The idea of this greedy algorithm is simple. Given a set Q^t, the next element chosen is one that gives the greatest immediate increase in value, provided that such an element exists. Moreover, once an element is chosen, it is kept throughout the algorithm. Recall that we used the greedy algorithm to find an optimal solution to the minimum-weight spanning tree problem (see Section I.3.3). In general, however, we cannot expect the greedy algorithm to yield an optimal solution.

In the uncapacitated facility location problem, we obtain

$$v(Q) = \begin{cases} \sum_{i \in I} \max_{j \in Q} c_{ij} - \sum_{j \in Q} f_j & \text{for } \emptyset \subset Q \subseteq N \\ -\infty & \text{for } Q = \emptyset \text{ (since } Q = \emptyset \text{ is infeasible).} \end{cases}$$

Example 1.1 (continued). We apply the greedy heuristic described above.

Iteration 1. $Q^0 = \emptyset$.

j:	1	2	3	4	5
$v(Q^0 \cup \{j\})$:	31	25	27	21	14

Iteration 2. $Q^1 = \{1\}$, $v(Q^1) = 31$.

j:	—	2	3	4	5
$v(Q^1 \cup \{j\})$:		35	33	38	29

Iteration 3. $Q^2 = \{1, 4\}$, $v(Q^2) = 38$.

j:	—	2	3	—	5
$v(Q^2 \cup \{j\})$:		40	39		31

Iteration 4. $Q^3 = \{1, 2, 4\}$, $v(Q^3) = 40$.

j:	—	—	3	—	5
$v(Q^3 \cup \{j\})$:			37		33

Since $v(Q^3 \cup \{j\}) \leqslant v(Q^3)$ for all $j \notin Q^3$, it follows that $Q^3 = \{1, 2, 4\}$ is a greedy solution with value 40.

There are generally several greedy heuristics for a given problem, and common sense must be used to convert the "greedy" idea into a reasonable greedy heuristic. An equally valid greedy approach for the uncapacitated facility location problem is to start with all facilities open and then, one-by-one, close a facility whose closing leads to the greatest increase in profit.

For the 0-1 packing problem $\max\{cx: Ax \leqslant 1, x \in B^n\}$, where A is a 0-1 matrix, one greedy approach is to recursively set that variable to one for which the resulting solution is still feasible and for which c_j is as large as possible. However, examination of a few examples quickly leads to the idea that c_j should be divided by the number of 1's in the column a_j; that is, the "improved" greedy criterion is to choose a column for which the average increase in profit per row covered, $c_j / \Sigma_{i \in M} a_{ij}$, is maximum.

The second important idea is that of "local search" or "interchange" heuristics. As the name implies, a heuristic of this type takes a given feasible solution and, by making only limited changes in it, tries to find a better feasible solution.

A k-Interchange Heuristic for $\max\{c(x): x \in S \subseteq B^n\}$.

Given a positive integer k, $k \leqslant n$, let

$$N_k(x) = \left\{ z \in B^n: \sum_{j \in N} |z_j - x_j| \leqslant k \right\} \quad \text{for } x \in B^n.$$

Initialization: Find a point $x^1 \in S$.

Iteration t: Given a point $x^t \in S$, if there is a point $x' \in N_k(x^t) \cap S$ with $c(x') > c(x^t)$, then let $x^{t+1} = x'$ and $t \leftarrow t + 1$. Otherwise stop; x^t is a k-interchange solution.

Clearly the amount of work per iteration in this algorithm depends crucially on k, and for the heuristic to be fast we typically limit k to values of 1, 2, or 3. Observe that when $k = n$, the algorithm asks for an examination of all the points in B^n. Again, depending on the problem structure, it is usual to make variations in the definition of $N_k(x)$. For the uncapacitated facility location problem, one reasonable choice (given a set Q of open facilities) is to look at the neighborhood in which either (a) one of the existing facilities is closed or (b) one new facility is opened, or where both a and b occur simultaneously, that is,

$$N_2'(Q) = \{F \subseteq N: |F \setminus Q| \leq 1 \text{ and } |Q \setminus F| \leq 1\}.$$

The third general principle is that often primal and dual heuristic solutions can be found in pairs. The complementary slackness conditions (see Section I.2.2) are one way of pairing heuristic solutions.

We use the 0-1 packing problem $\max\{cx: Ax \leq 1, x \in B^n\}$ to illustrate this idea. The dual of its linear programming relaxation is $\min\{\Sigma_{i=1}^m u_i: uA \geq c, u \in R_+^m\}$. Given a heuristic solution u^* to this dual, let $N^* = \{j \in N: \Sigma_{i=1}^m u_i^* a_{ij} = c_j\}$. Then the choice of an associated primal heuristic solution is restricted to the vectors x with $x_j = 1$ only if $j \in N^*$. Moreover, if such a primal feasible vector x^* can be found that also satisfies $\Sigma_{j\in N} a_{ij}x_j^* = 1$ for all i with $u_i^* > 0$, then by complementary slackness it follows that x^* and u^* are optimal solutions.

The pairing of primal and dual heuristics is now demonstrated for the uncapacitated facility location problem. First we consider the dual of the linear programming relaxation of UFL:

$$\begin{aligned} z_{\text{LP}} = \min \sum_{i\in I} u_i &+ \sum_{j\in N} t_j \\ u_i + w_{ij} &\geq c_{ij} \quad \text{for } i \in I, j \in N \\ -\sum_{i\in I} w_{ij} + t_j &\geq -f_j \quad \text{for } j \in N \\ w_{ij}, t_j &\geq 0 \quad \text{for } i \in I, j \in N. \end{aligned}$$

We can eliminate constraints and variables from this formation by observing that:

a. For given w_{ij}, the only constraints on t_j are nonnegativity and $t_j \geq \Sigma_{i\in I} w_{ij} - f_j$, and hence in any optimal solution we have $t_j = (\Sigma_{i\in I} w_{ij} - f_j)^+$, where x^+ denotes $\max(x, 0)$.
b. For given u_i, we have that $\Sigma_{j\in N} (\Sigma_{i\in I} w_{ij} - f_j)^+$ is minimized by setting w_{ij} as small as possible, that is $w_{ij} = (c_{ij} - u_i)^+$.

Hence the dual can be rewritten as

$$(3.1) \qquad z_{\text{LP}} = \min_{u\in R^m} w(u), \quad \text{where } w(u) = \sum_{i\in I} u_i + \sum_{j\in N} \left(\sum_{i\in I} (c_{ij} - u_i)^+ - f_j\right)^+.$$

Alternatively, if we assume that $f_j \geq 0$ for all $j \in N$, it is easy to see that the constraints $x_j \leq 1$ can be dropped from the linear programming relaxation of UFL. Thus t_j disappears from the dual and it becomes

$$(3.2) \qquad \begin{aligned} z_{\text{LP}} = \min &\sum_{i\in I} u_i \\ &\sum_{i\in I} (c_{ij} - u_i)^+ \leq f_j \quad \text{for } j \in N. \end{aligned}$$

The two condensed duals (3.1) and (3.2) are of interest because they only depend on $u \in R^m$. The dual (3.1) is particularly useful since it gives an upper bound for any $u \in R^m$.

Now we consider the association of primal and dual solutions. Given a primal solution with $Q \subseteq N$ being the set of open facilities, one way to associate a dual solution is to take u_i equal to the second largest c_{ij} over $j \in Q$. The motivation for this lies in the complementary slackness condition $(y_{ij} - x_j)w_{ij} = 0$. For $j \in Q$, we have $x_j = 1$ in the primal solution; and for each i, there is one $y_{ij} = 1$ with $j \in Q$. Hence if the complementary slackness condition is to hold, we must have no more than one $w_{ij} > 0$ for each $i \in I$ and $j \in Q$. Since $w_{ij} = (c_{ij} - u_i)^+$, this leads to the heuristic choice of u_i suggested above. Taking the greedy solution $Q = \{1, 2, 4\}$ obtained for Example 1.1, the associated dual solution is $u = (12 \quad 4 \quad 2 \quad 5 \quad 8 \quad 2)$. Using the formula in (3.1), we obtain $w(u) = 33 + (0 + 2 + 6 + 5 + 0) = 46$.

Now conversely, suppose we are given a dual solution u that is feasible to (3.2) and we wish to associate a primal feasible solution with it. The linear programming complementarity conditions suggest associating a primal solution in which $x_j = 0$ if $\Sigma_{i \in I} (c_{ij} - u_i)^+ < f_j$. Let $J(u) = \{j \in N: \Sigma_{i \in I} (c_{ij} - u_i)^+ = f_j\}$. The best solution that satisfies complementarity is obtained by solving

$$\max_{Q \subseteq J(u)} \left\{ \sum_{i \in I} \max_{j \in Q} c_{ij} - \sum_{j \in Q} f_j \right\}.$$

However, this problem may not be much easier to solve than the original problem UFL. Therefore we take as a primal heuristic solution $Q(u)$ any minimal set $Q(u) \subseteq J(u)$ satisfying

$$\max_{j \in Q(u)} c_{ij} = \max_{j \in J(u)} c_{ij} \quad \text{for all } i \in I. \tag{3.3}$$

The following proposition tells us when $(Q(u), u)$ are optimal to UFL and the dual of its linear programming relaxation, respectively.

Proposition 3.1. *Given a u that is feasible to (3.2) with $u_i \leqslant \max_{j \in J(u)} c_{ij}$ for $i \in I$, and a primal solution $Q(u)$ defined by (3.3), let $k_i = |\{j \in Q(u): c_{ij} > u_i\}|$. If $k_i \leqslant 1$ for all $i \in I$, then $Q(u)$ is an optimal set of open facilities.*

Proof

$$v(Q(u)) = \sum_{i \in I} \max_{j \in Q(u)} c_{ij} - \sum_{j \in Q(u)} f_j.$$

If $k_i = 0$, then

$$\max_{j \in Q(u)} c_{ij} = u_i = u_i + \sum_{j \in Q(u)} (c_{ij} - u_i)^+;$$

and if $k_i = 1$, then

$$\max_{j \in Q(u)} c_{ij} = u_i + \sum_{j \in Q(u)} (c_{ij} - u_i)^+.$$

Hence, if $k_i \leqslant 1$ for all $i \in I$, then

$$\begin{aligned} v(Q(u)) &= \sum_{i \in I} \sum_{j \in Q(u)} (c_{ij} - u_i)^+ - \sum_{j \in Q(u)} f_j + \sum_{i \in I} u_i \\ &= \sum_{j \in Q(u)} \left(\sum_{i \in I} (c_{ij} - u_i)^+ - f_j \right) + \sum_{i \in I} u_i \\ &= \sum_{i \in I} u_i \quad \text{(by definition of } J(u)) \\ &= z_{\text{LP}} \quad \text{(since } u \text{ is feasible in (3.2)).} \end{aligned}$$

Since $v(Q(u))$ achieves the upper bound of z_{LP}, it follows that $Q(u)$ is an optimal set of open facilities. ■

Now we present a heuristic algorithm for the dual problem (3.2) that uses the ideas of greedy and interchange. After finding a dual solution, the algorithm constructs a primal solution from (3.3) and then uses Proposition 3.1 to check optimality.

Dual Descent [A Greedy Algorithm for (3.2)]

Begin with $u_i^0 = \max_{j \in N} c_{ij}$ for $i \in I$. Cycle through the indices $i \in I$ one-by-one attempting to decrease u_i to the next smaller value of c_{ij}. If one of the constraints

$$(3.4) \qquad \sum_{i \in I} (c_{ij} - u_i)^+ \leqslant f_j \quad \text{for } j \in N$$

blocks the decrease of u_i to the next smaller c_{ij}, then decrease u_i to the minimum value allowed by the constraint. When all of the u_i's are blocked from further decreases, the procedure terminates.

A possible improvement of this greedy heuristic is obtained by modifying the order in which the u_i's are considered as candidates to decrease. The reasoning is the same as in the case of the 0-1 packing problem. Let $H_i(u) = \{j \in N: c_{ij} - u_i \geqslant 0\}$. Rather than just cycling through the u_i's, we choose u_s next if $|H_s(u)| \leqslant |H_i(u)|$ for all $i \in I$, since this implies the smallest increase in $\Sigma_{j \in N} \Sigma_{i \in I} (c_{ij} - u_i)^+$ per unit decrease in $\Sigma_{i \in I} u_i$. This discussion also justifies decreasing u_i only to the next smaller c_{ij} rather than to the smallest permissible value.

Now suppose that dual descent terminates with a solution u^* and that the associated primal solution $Q(u^*)$ given by (3.3) fails to verify the optimality conditions of Proposition 3.1. Then there exists an i such that $k_i > 1$. In an attempt to find an improved dual solution, we adopt the neighborhood search idea.

Interchange Step. Increase some u_i^* for which $k_i > 1$ to its previous value. Use the resulting u as the starting solution and reapply the dual greedy algorithm terminating with u' satisfying $\Sigma_{i \in I} u_i' \leqslant \Sigma_{i \in I} u_i^*$.

If $u^* = u'$, stop. u^* is the heuristic solution.

If u^* satisfies the optimality conditions, stop.

Otherwise, repeat the interchange step.

Example 1.1 (continued). Applying dual descent yields the results shown in Table 3.1. For the first five steps, u_i, $i = 1, \ldots, 5$, is decreased to the second maximum in the row.

Table 3.1.

	u						$f_j - \Sigma_{i\in I}(c_{ij} - u_i)^+$				
Step i:	1	2	3	4	5	6	j: 1	2	3	4	5
0	13	9	6	10	10	4	4	3	4	4	7
1	12	9	6	10	10	4	4	2	4	4	7
2	12	8	6	10	10	4	4	2	3	4	7
3	12	8	6	10	10	4	4	2	3	4	7
4	12	8	6	8	10	4	4	2	3	2	7
5	12	8	6	8	8	4	4	2	3	0	7
6	10	8	6	8	8	4	2	0	3	0	7
7	10	6	6	8	8	4	0	0	1	0	7

Now u_6 cannot be decreased because the constraint (3.4) for $j = 4$ would be violated. Next, u_1 is decreased toward the third maximum in the row but is only decreased by two units because constraint (3.4) becomes tight for $j = 2$. Finally, u_2 is decreased until (3.4) becomes tight for $j = 1$. This completes the dual descent with $u = (10 \quad 6 \quad 6 \quad 8 \quad 8 \quad 4)$, $w(u) = 42$, and $J(u) = \{1, 2, 4\}$. Now we associate a primal solution as described in (3.3) and obtain $Q(u) = J(u)$, with a primal solution value of 40.

The proposed modification given above to the order of decreasing the u_i's produces the same result. The solution $(u, Q(u))$ fails to satisfy the optimality conditions since $k_1 = |\{j \in Q(u): c_{1j} > u_1\}| = 2$. To apply the interchange step, we increase u_1 back to its previous value of 12 and then restart the dual descent. In Table 3.2, we see that u_3 can now be decreased by one unit and then no further move is possible except to decrease u_1 again. Hence $u = (11 \quad 6 \quad 5 \quad 8 \quad 8 \quad 4)$ and $w(u) = 42$ as before, but now $J(u) = \{2, 3, 4\}$. From (3.3) we obtain the associated primal solution $Q(u) = J(u)$ with the improved primal value of 41.

There is a branch-and-bound algorithm for the uncapacitated facility location problem, called DUALOC, that obtains primal and dual feasible solutions at each node of the branch-and-bound tree using dual descent, the primal heuristic given by (3.3), and interchange. If a node is not pruned by these heuristics, then branching is accomplished by taking a $j \in N$ and considering the two problems with $x_j = 0$ and $x_j = 1$.

Example 2.1 (continued). When DUALOC is applied, the algorithm iterates six times through the interchange step. The values of the corresponding dual lower bounds and primal upper bounds are shown in Table 3.3. Hence before entering the branch-and-bound phase we have $-20{,}503 \leqslant z \leqslant -20{,}340$. Branching on $x_3 = 0$, an optimal integer solution of value $-20{,}363$ is found. Branching on $x_3 = 1$, an optimal integer solution of value $-20{,}393$ is found. Hence an optimal solution has value $-20{,}363$ (see Figure 2.2 and Table 2.4).

Table 3.2.

	u						$f_j - \Sigma_{i\in I}(c_{ij} - u_i)^+$				
Step i:	1	2	3	4	5	6	j: 1	2	3	4	5
8	12	6	6	8	8	4	2	2	1	0	7
9	12	6	5	8	8	4	2	1	0	0	7
10	11	6	5	8	8	4	1	0	0	0	7

Table 3.3.

Iteration	Dual Bound	Primal Bound
1	−20,294	−20,503
2	−20,326	−21,553
3	−20,337	−20,503
4	−20,338	−21,021
5	−20,340	−20,503
6	−20,340	−20,853

Analysis of Heuristics

We have emphasized the importance of finding both primal and dual feasible solutions, particularly when a primal feasible solution is taken as an approximation to an optimal solution. The dual solution provides an upper bound on the deviation from optimality of the primal solution and thus gives an a posteriori evaluation of the quality of the primal solution.

In addition to this evaluation of an instance, it is frequently possible to give an a priori evaluation of a heuristic algorithm over all instances. One way to obtain results of this type is by *worst-case analysis*.

The essential idea of worst-case analysis is simple. Consider a maximization problem consisting of instances:

$$\text{P(I)} \qquad z(I) = \max\{c_I(x): x \in S_I \neq \emptyset\} \quad \text{for } I \in \mathcal{I}.$$

Suppose we have a heuristic algorithm (H) that finds a feasible solution of value $z_H(I)$. Worst-case analysis is based on calculating some maximum deviation between $z(I)$ and $z_H(I)$. The analysis depends crucially on the function that is used to measure deviation.

Perhaps the simplest function is just the absolute difference $z(I) - z_H(I)$. But for most problems the maximum value of the absolute difference is not a meaningful measure since it can be made arbitrarily large by scaling the objective function. (However, in Part III, we will consider a class of integer programs in which the objective function coefficients are all 1 and the difference between the optimal value of the linear programming relaxation and the optimal value of the integer program always is less than or equal to 1).

Relative values, which are independent of objective function scaling, are usually a more meaningful measure of deviation. To consider relative values, it is convenient to assume that $c_I(x) \geq 0$ for all $x \in S_I$ and all $I \in \mathcal{I}$. We say that heuristic algorithm H has a *worst-case relative performance* or *performance guarantee* of r_H if

$$\begin{aligned} r_H &= \inf_{I \in \mathcal{I}}\{r(I): z_H(I) = r(I)z(I)\} \\ &= \sup\{r: z_H(H) \geq rz(I) \text{ for all } I \in \mathcal{I}\}. \end{aligned}$$

If a heuristic algorithm is not completely specified (e.g., as a result of the absence of a tie-breaking rule), we assume the worst possible outcome. By definition, $0 \leq r_H \leq 1$ and H guarantees to find a feasible solution of value at least $r_H \times 100\%$ of the maximum value for all instances. Note that $r_H = 1 - \varepsilon_H$, where

$$\varepsilon_H = \inf\left\{\varepsilon: \varepsilon \geq \frac{z(I) - z_H(I)}{z(I)} \text{ for all } I \in \mathcal{I} \text{ with } z(I) > 0\right\},$$

is the largest possible *relative error*.

To keep the same scale for minimization problems of the form $z(I) = \min\{c_I(x): x \in S_I \neq \emptyset\}$ with $c_I(x) \geqslant 0$ for all $x \in S_I$ and $I \in \mathcal{I}$, we use the reciprocal ratio and define

$$r_H = \sup\{r: z(I) \geqslant r z_H(I) \text{ for all } I \in \mathcal{I}\}.$$

Thus $0 \leqslant r_H \leqslant 1$ and H guarantees to find a solution of value at most $r_H^{-1} \times 100\%$ of the minimum value for all instances. It is not at all unusual to obtain dramatically different results for maximization and minimization versions of what otherwise would be the same problem.

Worst-case analysis is a very conservative approach since it takes only one bad instance to give a poor result. The alternative approaches of a probabilistic or statistical analysis will be considered briefly at the end of this section.

There is one general principle that is used to obtain nearly all results on worst-case analysis. Consider a dual heuristic (DH) that produces an upper bound for $P(I)$; that is, $z_{\text{DH}}(I) \geqslant z(I)$ for all $I \in \mathcal{I}$. Let

$$r_{\text{DH}} = \sup\{r: z_H(I) \geqslant r z_{\text{DH}}(I) \text{ for all } I \in \mathcal{I}\}$$

so that $r_H \geqslant r_{\text{DH}}$. Now if there is a simple relationship between H and DH it is often possible to calculate r_{DH} directly and hence a lower bound on r_H. Moreover, in some cases, we can find an instance I_0 with $z_H(I^0) = r_{\text{DH}} z_{\text{DH}}(I^0)$ and $z_{\text{DH}}(I^0) = z(I^0)$ so that $r_H = r_{\text{DH}}$.

Example 3.1. We give a simple illustration of this approach by analyzing a greedy heuristic for the maximum-weight matching problem. This is the special case of maximum-weight packing in which matrix A is the node-edge incidence matrix of a graph $G = (V, E)$. Thus $\Sigma_{i=1}^{m} a_{ij} = 2$ for all j, and the greedy heuristic, introduced earlier in this section, chooses edges of maximum weight such that each chosen edge does not meet any of the edges chosen previously. The algorithm stops when no such edge exists. (We assume that all edge weights are positive.) Let E^H be the set of edges chosen by the greedy heuristic. Then E^H is a maximal matching; that is, $e \in E \setminus E^H$ meets an $e' \in E^H$.

The dual of the linear programming relaxation of the matching problem is

$$\min \sum_{i \in V} u_i$$

$$u_i + u_j \geqslant c_{ij} \quad \text{for } (i, j) \in E$$

$$u_i \geqslant 0 \quad \text{for } i \in V,$$

where c_{ij} is the weight of edge (i, j). Consider the dual solution u^H given by

$$u_i^H = \begin{cases} c_{ij} & \text{if } (i, j) \in E^H \\ 0 & \text{otherwise.} \end{cases}$$

Figure 3.1

We claim that u^H is dual feasible. Suppose $u_i^H + u_j^H < c_{ij}$ for some $(i, j) \in E$. Then $(i, j) \notin E^H$. Hence either i or j or both are met by an $e \in E^H$, and one of these edges, say (i, j'), has been chosen before (i, j) was considered. Hence $u_i^H = c_{ij'} \geq c_{ij}$, which contradicts $u_i^H + u_j^H < c_{ij}$.

Now

$$z_{\mathrm{DH}} = \sum_{i \in V} u_i^H = 2 \sum_{(i,j) \in E^H} c_{ij} \quad \text{and} \quad z_H = \sum_{(i,j) \in E^H} c_{ij} \leq z \leq z_{\mathrm{DH}},$$

so $r_H \geq r_{\mathrm{DH}} \geq \frac{1}{2}$. In the graph of Figure 3.1, in which each edge has weight equal to 1, by choosing $e = (2, 3)$ first, we obtain $E^H = \{(2, 3)\}$, $u^H = (0 \quad 1 \quad 1 \quad 0)$, $z_{\mathrm{DH}} = z = 2$, and $z_H = \frac{1}{2} z_{\mathrm{DH}}$. Hence $r_H = r_{\mathrm{DH}} = \frac{1}{2}$.

Although a solution whose value is only half of the optimal value is unlikely to be satisfactory, one must remember that a performance guarantee of 50% means that 50% is the worst possible outcome, and it is likely that for most instances the relative error will be much smaller. Moreover, it is not unusual for a heuristic to have performance guarantee of zero. Indeed, this is the case when the greedy heuristic is applied to the set-packing problem.

Consider the family of set-packing problems

$$z(k) = \max\left\{ \sum_{j=1}^{k+1} x_j \colon A_k x \leq 1, x \in B^{k+1} \right\} \quad \text{for } k = 1, 2, \ldots,$$

where

$$A_k = \begin{pmatrix} 1_k & I_k \\ 0_k & I_k \\ \vdots & \vdots \\ 0_k & I_k \end{pmatrix} \Bigg\} \; k^2 \text{ rows}$$

and I_k is the $k \times k$ identity matrix; 1_k and 0_k are $k \times 1$ matrices of all 1's and all 0's respectively. Since $\Sigma_{i=1}^m a_{i1} = k < \Sigma_{i=1}^m a_{ij} = k + 1$ for all $j > 1$, the greedy heuristic first sets $x_1 = 1$ and then stops. The optimal solution is $x_1 = 0$ and $x_j = 1$ for $j = 2, \ldots, k + 1$. Hence $z^H(k) / z(k) = 1 / k$ and $r_H = 0$. So, in the worst case, the greedy heuristic is arbitrarily bad for the set-packing problem.

This raises a question in computational complexity. Suppose $\mathcal{P} \neq \mathcal{NP}$. Given an $\mathcal{NP}$-hard optimization problem and $0 < r < 1$, does there exist a polynomial-time heuristic algorithm (H) with $r_H \geq r$? The answer to this question depends on the problem. In fact, both extremes are possible.

There are some problems for which the approximation problem is $\mathcal{NP}$-hard for $r_H \geq r$ for any $r > 0$. We will show below that this is the case for a minimization version of a p-facility location problem. For the problem of finding the minimum number of colors needed to color the nodes of a graph such that no pair of nodes joined by an edge have the same color, the approximation problem is $\mathcal{NP}$-hard for $r \geq \frac{1}{2}$. But no polynomial-time heuristic algorithm H is known that yields $r_H \geq r$ for any $r > 0$.

At the other extreme, there are $\mathcal{NP}$-hard optimization problems such that for any $0 < r < 1$, there is an algorithm with performance $r_H \geq r$, whose running time is polynomial in the length of the input and in $1 / (1 - r)$. We call such an algorithm a *fully polynomial approximation scheme*. In Section II.6.1, we will give a fully polynomial approximation scheme for the knapsack problem. A more modest result is a *polynomial approximation scheme*, which is an algorithm with performance $r_H \geqslant r$, whose running time is polynomial in the length of the input for any fixed r, $0 < r < 1$.

To illustrate some of these results, we now consider the analysis of some heuristics for a p-facility variation of the uncapacitated facility location problem. Here there are no fixed costs, but we can open no more than p facilities. Suppose $c_{ij} \geqslant 0$ is the cost of assigning client i to facility j so that the objective function is

$$v(Q) = \sum_{i \in I} \min_{j \in Q} c_{ij} \quad \text{for } \emptyset \subset Q \subseteq N.$$

We call the problem

$$z = \min\{v(Q): 1 \leqslant |Q| \leqslant p\}$$

the *p-facility minimization problem*.

Proposition 3.2. *The p-facility minimization problem with performance guarantee $r_H \geqslant r$ is $\mathcal{NP}$-hard for any $r > 0$.*

Proof. Given a graph $G = (V, E)$, we say that G has a node cover of its edges of size $|U|$ if there is a $U \subseteq V$ such that every edge of G is incident to a node of U. Given an integer $k < |V|$, the problem of determining whether a graph has a node cover of size k is $\mathcal{NP}$-complete (see exercise 14 of Section I.5.9). We now show that a polynomial-time algorithm for the p-facility minimization problem with performance guarantee $r_H \geqslant r$ for any $r > 0$ implies a polynomial-time algorithm for the node-cover problem.

Consider the family of p-facility minimization problems with $I = E$, $N = V$, $p = k$, and

$$c_{ij} = \begin{cases} 1 & \text{if } e_i \text{ is incident to node } j \\ \dfrac{|E| + 1}{r} & \text{otherwise.} \end{cases}$$

Now G contains a node cover of size k if and only if $z = |E|$. Hence an algorithm with a performance guarantee of $r > 0$ yields a solution with $v(Q) \leqslant |E|/r$. This implies that G contains a node cover of size k since any feasible solution that does not cover all of the edges has cost of at least

$$|E| - 1 + \frac{|E| + 1}{r} > \frac{|E|}{r}. \qquad \blacksquare$$

A dramatically different result is obtained for the *p-facility maximization problem*. Here $c_{ij} \geqslant 0$ is the profit obtained from assigning client i to facility j so that

$$v(Q) = \sum_{i \in I} \max_{j \in Q} c_{ij} \quad \text{for } \emptyset \subset Q \subseteq N,$$

and the problem is

$$z = \max\{v(Q): 1 \leqslant |Q| \leqslant p\}.$$

We analyze the greedy heuristic for this problem. To accommodate the constraint $|Q| \leqslant p$ in the greedy heuristic for maximizing a set function, we just modify the stopping rule: If $t = p$, stop; Q^p is a greedy solution. Otherwise, $t \leftarrow t + 1$. Moreover, since $c_{ij} \geqslant 0$ for all i and j, we can assume that the greedy heuristic produces a solution Q^t with $t = p$.

Let $\rho_1 = v(Q^1)$ and $\rho_t = v(Q^t) - v(Q^{t-1})$ for $t = 2, \ldots, p$. Thus the value of the greedy solution is $z_G = \sum_{t=1}^{p} \rho_t$.

Theorem 3.3

$$\frac{z_G}{z} \geqslant \left(1 - \left(\frac{p-1}{p}\right)^p\right) \geqslant \frac{e-1}{e} \cong 0.63 \quad \text{for } p = 1, 2, \ldots$$

(e is the base of the natural logarithm). Moreover, for each p there is an instance for which the bound is tight; that is, $r_G = 1 - ((p-1)/p)^p$ *for a p-facility maximization problem.*

We prove Theorem 3.3 using a series of propositions.

Proposition 3.4. *If* $S \subset T \subset N$ *and* $k \notin T$*, then* $v(T \cup \{k\}) - v(T) \leqslant v(S \cup \{k\}) - v(S)$.

Proof. We have $\max_{j \in S} c_{ij} \leqslant \max_{j \in T} c_{ij}$. Hence

$$\max\left(c_{ik} - \max_{j \in S} c_{ij}, 0\right) \geqslant \max\left(c_{ik} - \max_{j \in T} c_{ij}, 0\right)$$

and

$$v(S \cup \{k\}) - v(S) = \sum_{i \in I} \max\left(c_{ik} - \max_{j \in S} c_{ij}, 0\right) \geqslant \sum_{i \in I} \max\left(c_{ik} - \max_{j \in T} c_{ij}, 0\right)$$
$$= v(T \cup \{k\}) - v(T). \quad \blacksquare$$

This property of set functions is known as *submodularity* and is the essential property used to prove Theorem 3.3. In fact, Theorem 3.3 can be generalized to the maximization of submodular set functions (see Section III.3.9).

Proposition 3.5. $z \leqslant p\rho_1$ *and* $z \leqslant \sum_{i=1}^{t} \rho_i + p\rho_{t+1}$ *for* $t = 1, \ldots, p-1$.

Proof. Let Q^* be an optimal solution, that is, $z = v(Q^*)$. Since Q^t is the set obtained after t steps of the greedy algorithm, we have $v(Q^t) = \sum_{i=1}^{t} \rho_i$. Suppose $Q^* \setminus Q^t = \{j_1, \ldots, j_k\}$. We have $k \leqslant p$ since $|Q^*| \leqslant p$. Now

$$\begin{aligned} v(Q^*) &\leqslant v(Q^* \cup Q^t) \\ &= v(Q^t) + \sum_{i=1}^{k} (v(Q^t \cup \{j_1, \ldots, j_i\}) - v(Q^t \cup \{j_1, \ldots, j_{i-1}\})) \\ &\leqslant v(Q^t) + \sum_{i=1}^{k} (v(Q^t \cup \{j_i\}) - v(Q^t)) \\ &\leqslant v(Q^t) + p\rho_{t+1}, \end{aligned}$$

where the first inequality follows because v is nondecreasing, the second inequality follows by Proposition 3.4, and the last inequality follows since $p \geqslant k$ and $\rho_{t+1} \geqslant v(Q^t \cup \{j_i\}) - v(Q^t)$ by the definition of the greedy algorithm. $\blacksquare$

Proposition 3.6. *If* $z_G = 1 - ((p-1)/p)^p$, *then* $z \leqslant 1$.

Proof. By Proposition 3.5 we have

$$z \leqslant \max \eta$$

$$\eta \leqslant \sum_{i=1}^{t} \rho_i + p\rho_{t+1} \quad \text{for } t = 0, \ldots, p-1$$

$$\sum_{i=1}^{p} \rho_i = 1 - \left(\frac{p-1}{p}\right)^p$$

$$\rho_i \geqslant 0 \quad \text{for } i = 1, \ldots, p.$$

To show that $z \leqslant 1$, we consider the dual

$$\min u_p \left\{1 - \left(\frac{p-1}{p}\right)^p\right\}$$

$$\sum_{i=0}^{p-1} u_i = 1$$

$$-\sum_{i=0}^{t-1} u_i - pu_t + u_p \geqslant 0 \text{ for} \qquad t = 0, \ldots, p-1$$

$$u_i \geqslant 0 \qquad \text{for} \qquad i = 0, \ldots, p-1.$$

Now observe that a feasible solution to the dual is given by

$$u_p = \left(1 - \left(\frac{p-1}{p}\right)^p\right)^{-1} \quad \text{and} \quad u_i = \frac{(p - 1/p)^i\, u_p}{p} \text{ for } i = 0, \ldots, p-1.$$

This is true since

$$\sum_{i=0}^{p-1} u_i = \frac{u_p}{p}\left(1 - \left(\frac{p-1}{p}\right)^p\right)p$$

and

$$\sum_{i=0}^{t-1} u_i + pu_t = \frac{u_p}{p}\left(1 - \left(\frac{p-1}{p}\right)^t\right)p + \left(\frac{p-1}{p}\right)^t u_p = u_p.$$

∎

This proves the first part of Theorem 3.3. Now we show that the bound is tight for $p = 2, 3, \ldots$ ($p = 1$ is trivial).

Proposition 3.7. *For the family of p-facility location problems defined by* $C^p = (c_{ij}^p)$ *with* $|I| = p(p-1)$ *and* $|N| = 2p - 1$; *and with (a) for* $j = 1, \ldots, p-1$

$$c_{ij}^p = \begin{cases} (p-1)\, p^{p-2}\left(\dfrac{p-1}{p}\right)^{j-1} & \text{for} \quad i = (j-1)p + 1, \ldots, jp \\ 0 & \text{otherwise} \end{cases}$$

(b) for $j = p, \ldots, 2p-1$

$$c_{ij}^p = \begin{cases} p^{p-1} & \text{for } i = 1 + j + (l-2)p \text{ and } l = 1, \ldots, p-1 \\ 0 & \text{otherwise,} \end{cases}$$

we have

$$z_G = \left(1 - \left(\frac{p-1}{p}\right)^p\right) z.$$

(Table 3.4 gives C^2, C^3, and C^4.)

Proof. An optimal solution is given by the last p columns, so $z = p(p-1)p^{p-1} = (p-1)p^p$.

We now show that the greedy algorithm can select the first p columns. We have

$$\sum_{i \in I} c_{i1}^p = p(p-1)p^{p-2} \geq \sum_{i \in I} c_{ij}^p \quad \text{for } j = 2, \ldots, p-1$$

and

$$\sum_{i \in I} c_{ij}^p = (p-1)p^{p-1} \qquad \text{for } j = p, \ldots, 2p-1.$$

Hence the greedy algorithm can choose the first column first. Now suppose that the greedy algorithm has chosen the first $t-1$ columns and let $Q^{t-1} = \{1, \ldots, t-1\}$. Then

Table 3.4.

$$C^2 = \begin{pmatrix} 1 & 2 & 0 \\ 1 & 0 & 2 \end{pmatrix}, \quad C^3 = \begin{pmatrix} 6 & 0 & 9 & 0 & 0 \\ 6 & 0 & 0 & 9 & 0 \\ 6 & 0 & 0 & 0 & 9 \\ 0 & 4 & 9 & 0 & 0 \\ 0 & 4 & 0 & 9 & 0 \\ 0 & 4 & 0 & 0 & 9 \end{pmatrix}$$

$$C^4 = \begin{pmatrix} 48 & 0 & 0 & 64 & 0 & 0 & 0 \\ 48 & 0 & 0 & 0 & 64 & 0 & 0 \\ 48 & 0 & 0 & 0 & 0 & 64 & 0 \\ 48 & 0 & 0 & 0 & 0 & 0 & 64 \\ 0 & 36 & 0 & 64 & 0 & 0 & 0 \\ 0 & 36 & 0 & 0 & 64 & 0 & 0 \\ 0 & 36 & 0 & 0 & 0 & 64 & 0 \\ 0 & 36 & 0 & 0 & 0 & 0 & 64 \\ 0 & 0 & 27 & 64 & 0 & 0 & 0 \\ 0 & 0 & 27 & 0 & 64 & 0 & 0 \\ 0 & 0 & 27 & 0 & 0 & 64 & 0 \\ 0 & 0 & 27 & 0 & 0 & 0 & 64 \end{pmatrix}$$

$$v(Q^{t-1} \cup \{t\}) - v(Q^{t-1}) = p(p-1)p^{p-2}\left(\frac{p-1}{p}\right)^{t-1}$$
$$> v(Q^{t-1} \cup \{j\}) - v(Q^{t-1}), \quad \text{for } j = t+1, \ldots, p-1.$$

But for $j > p - 1$, we obtain

$$v(Q^{t-1} \cup \{j\}) - v(Q^{t-1}) = (p-1)p^{p-1} - (p-1)p^{p-2} \sum_{l=0}^{t-2} \left(\frac{p-1}{p}\right)^l$$
$$= (p-1)p^{p-1}\left(\frac{p-1}{p}\right)^{t-1}.$$

Hence the greedy heuristic can choose column t next. Thus

$$z_G = (p-1)p^{p-1} \sum_{t=0}^{p-2} \left(\frac{p-1}{p}\right)^t + (p-1)p^{p-1}\left(\frac{p-1}{p}\right)^{p-1} = (p-1)(p^p - (p-1)^p)$$

and

$$\frac{z_G}{z} = \frac{p^p - (p-1)^p}{p^p} = 1 - \left(\frac{p-1}{p}\right)^p. \qquad \blacksquare$$

Empirical evaluation of the greedy heuristic for the p-facility location problem shows that it performs reasonably well (above 80% of the optimal value) on most real and randomly generated instances. Moreover, the solution obtained by the greedy heuristic frequently can be improved by applying the interchange heuristic, which begins with a set of size p and recursively replaces an element in the set with one not in the set as long as the objective improves. However, there is a family of instances, like those given in Proposition 3.7, where the greedy heuristic obtains a solution that achieves its worst-case performance and the solution cannot be improved by applying the interchange heuristic. In addition, when the interchange heuristic begins with an arbitrary set of size p, its worst-case performance is inferior to that of the greedy heuristic.

Let z_I be the value of a solution produced by the interchange heuristic.

Proposition 3.8 $z_I \geq [p/(2p-1)]z$ *and for each* p *there is an instance for which the bound is tight, that is,* $r_I = p/(2p-1)$.

Proof. Let $Q^I \subset N$ be the set chosen by the interchange heuristic. Now apply the greedy heuristic to Q^I so that $z_I = v(Q^I) = \Sigma_{i=1}^p \rho_i$, where $\rho_i = v(Q^i) - v(Q^{i-1})$ for $i = 2, \ldots, p$, $\rho_1 = v(Q^1)$, and $Q^p = Q^I$. Let Q^* be an optimal solution and $Q^* \setminus Q^{p-1} = \{j_1, \ldots, j_k\}$. By the termination rule of the interchange heuristic, $v(Q^{p-1} \cup \{j_i\}) - v(Q^{p-1}) \leq \rho_p$. By Proposition 3.4 we have

$$v(Q^*) \leq v(Q^{p-1}) + \sum_{i=1}^{k} (v(Q^{p-1} \cup \{j_i\}) - v(Q^{p-1}))$$
$$\leq v(Q^{p-1}) + k\rho_p = z_I + (k-1)\rho_p \leq z_I + (p-1)\rho_p.$$

Now since $\rho_p \leq \rho_i$ for $i < p$, we have $\rho_p \leq (1/p)z_I$ and $z \leq [(2p-1)/p]z_I$.

See exercise 13 for a family of instances which establishes that the bound is tight. $\blacksquare$

Simulated Annealing

The interchange heuristic stops when it finds a "locally optimal" solution relative to the chosen neighborhood structure. As combinatorial optimization problems may have many local optima, it is typical to run the interchange heuristic many times with randomly chosen starting points.

A different approach for trying to obtain a global optimum using an interchange heuristic is called simulated annealing. Despite the fancy name, the idea is very simple. While the interchange heuristic produces a sequence of solutions with increasing objective value, here we allow the objective value to decrease occasionally to avoid getting stuck at a local optimum.

Consider the problem

$$(3.5) \qquad \max_{Q \subseteq M} \{c(Q) \colon Q \in \mathcal{F}\}.$$

Suppose Q^0 is the current solution and we find a point Q^1 in the neighborhood $N(Q^0)$ of Q^0. If $c(Q^1) > c(Q^0)$, we proceed as before by replacing Q^0 with Q^1. On the other hand if $c(Q^1) \leqslant c(Q^0)$, we replace Q^0 with Q^1 with probability p, where p is a decreasing function of $c(Q^0) - c(Q^1)$. The motivation for moving to a point with a smaller objective value is that if we are stuck in a shallow local optimum, there is a chance of escaping by moving to a neighbor having a lower objective value. The probability p can also be decreased as a function of the number of iterations. One reason for doing this is to obtain convergence; another reason is that as the global optimum is approached, making steps away from the optimum becomes less attractive.

Simulated Annealing Algorithm for (3.5)

Initialization: Let $\alpha_0 > 0$, $0 < \beta < 1$, $Q^0 \in \mathcal{F}$ and $i = 0$.

Step 1: Given Q^i, generate $Q' \subseteq N(Q^i)$.

Step 2: a) If $c(Q') > c(Q^i)$, then $Q^{i+1} = Q'$.

b) If $c(Q') \leqslant c(Q^i)$, then $Q^{i+1} = Q'$ with probability $p = \{\exp[c(Q') - c(Q^i)]/\alpha_i\}$ and $Q^{i+1} = Q^i$ with probability $1 - p$.

Step 3: $\alpha_{i+1} = \alpha_i(1 - \beta)$ and $i \leftarrow i + 1$.

Now provided that

i. it is possible to move from any set $Q \in \mathcal{F}$ to any other $Q' \in \mathcal{F}$ in a finite number of iterations,
ii. each set in a neighborhood is chosen with equal probability, and
iii. the neighborhoods are symmetric in the sense that $Q \in N(Q')$ if and only if $Q' \in N(Q)$,

it can be shown that the algorithm converges to the global optimum. However, the provable rate of convergence is exponential.

The empirical efficiency of simulated annealing depends on the neighborhood structure and the rate at which α is decreased. For some combinatorial optimization such as the traveling salesman problem and a variety of problems related to circuit design, simulated annealing has found much better solutions than those obtained by a random-start interchange algorithm.

Probabilistic Analysis

Experiments and statistical analysis can be done to draw conclusions about typical behavior of heuristics. In some cases, it is even possible to do a probabilistic analysis to obtain a priori results about average behavior. With this approach, one must be careful to use a probability distribution of the instances that is both realistic and mathematically tractable.

We mention three general types of stochastic models that are amenable to analysis. One such model deals with problems on graphs and uses random graphs as the underlying stochastic model. A *random graph* on n nodes is one in which the edges in the graph are selected at random. In the simplest of these models, the events of the graph containing any edge are identically and independently distributed random variables; that is, the probability that $(i, j) \in E$ is q for all $i, j \in V$. When $q = \frac{1}{2}$, all possible graphs on n nodes are equally likely. Then the probability of some property Q occuring on such a random graph with n nodes is simply the fraction of n-nodes graphs that possess property Q. We say that *almost all graphs possess property* Q if the probability approaches 1 as $n \to \infty$. For our purposes, property Q could be that a certain heuristic finds an optimal solution to a given problem whose instance is specified by an n-node graph.

To illustrate this idea, consider the p-facility maximization problem in which C is the edge-node incidence matrix of a graph. The problem is then to choose p nodes so that the number of edges incident to the chosen set is maximum. The greedy heuristic begins by choosing a node of maximum degree; then this node and all edges incident to it are deleted, and the process is repeated until p nodes have been chosen. It is a fact that for the random graph model given above, if p does not grow too fast as a function of n, then the greedy heuristic finds an optimal solution for almost all graphs. In addition, the greedy solution is optimal to the linear programming relaxation. We will not prove these results. But they are an easy consequence of an interesting theorem which says that in almost all graphs, if p does not grow too rapidly with n, then no two nodes in the set of p nodes of largest degree have the same degree.

Another stochastic model deals with problems in which the data are points in the plane. For example, the p-median problem in the plane is the special case of the p-facility minimization problem in which C is an $n \times n$ matrix and c_{ij} is the euclidean distance between points i and j. Here we assume that the points are placed randomly in a unit square using a two-dimensional uniform distribution.

For this problem, a very sharp estimate has been obtained on the asymptotic value of z_{IP}. By this we mean that as p and n approach infinity in a well-defined way, $z_{\text{IP}}(p, n)$ approaches $cf(p, n)$ with a probability that goes to 1 (almost surely), where c is a constant. Here $c = 0.377$ and $f(p, n) = n/p^{\frac{1}{2}}$. Results of this type are generally proved by comparing the asymptotic value of z_{IP} to the objective value of a continuous problem. These results can be used to analyze the asymptotic performance of heuristics since, as we have already shown, it is frequently not hard to analyze the behavior of the objective values produced by simple heuristics. For the p-median problem it has indeed been proved that there is a fast heuristic (H) that almost always finds a solution with $r_H \geqslant 1 - t$ for any $t > 0$. Similar analyses have been done for the linear programming relaxation of the p-median problem. Here it has been shown that $z_{\text{LP}}(p, n)$ converges to $0.376\, n/p^{\frac{1}{2}}$ almost surely. Consequently, for this stochastic model of the p-median problem, the asymptotic value of the absolute value of the duality gap is very small.

A third type of stochastic model deals with random objective or constraint coefficients. For example, in the p-median problem, we could assume that the c_{ij}'s are drawn randomly from a uniform distribution. Here it has been shown that $(z_{\text{IP}} - z_{\text{LP}}) / z_{\text{LP}}$ converges to $(p - 1) / 2p$ almost surely when p grows slowly with n, so a positive duality gap is to be expected. This has been confirmed by computational experiments as well.

A final comment on these models and results concerns what is deducible from $(z_{\text{IP}} - z_{\text{LP}}) / z_{\text{LP}}$ regarding the number of nodes L in a branch-and-bound algorithm that uses linear programming relaxation. For the euclidean problems it has been shown that $(z_{IP} - z_{LP}) / z_{IP}$ converges to 0.00284 almost surely. Nevertheless, it has also been shown that a branch-and-bound algorithm will almost surely explore $n^{p/200}$ nodes.

4. DECOMPOSITION ALGORITHMS

In this section, we will consider algorithms based on Lagrangian duality (Section II.3.6) and Benders' decomposition (Section II.3.7).

Solving the Lagrangian Dual by Subgradient Optimization

Recall that to obtain the Lagrangian dual of an integer programming problem, we partitioned the constraints into a set of linear constraints ($A^1x \leqslant b^1$) and a second set Q so that

$$\text{IP}(Q) \qquad z_{IP} = \max\{cx: A^1x \leqslant b^1, x \in Q\}.$$

Then we obtained the Lagrangian dual with respect to the constraints $A^1x \leqslant b^1$ given by

$$\text{(LD)} \qquad z_{LD} = \min_{\lambda \geq 0} z_{LR}(\lambda),$$

where the Lagrangian relaxation for a given λ is

$$(\text{LR}(\lambda)) \qquad z_{LR}(\lambda) = \max_{x \in Q} cx + \lambda(b^1 - A^1x).$$

It is essential to choose Q so that for fixed λ, LR(λ) is easy to solve. As we have already shown in Example 6.2 of Section II.3.6, there may be several ways of choosing Q and there generally is a tradeoff between the simplicity of solving LR(λ) and the quality of the bound z_{LD}. In this section, we will present a subgradient algorithm and a cut generation algorithm for solving the Lagrangian dual and use these algorithms to solve UFL. Applications to the traveling salesman problem will be presented in Section II.6.3.

It has been shown in Section II.3.6 (Corollary 6.4) that $z_{LR}(\lambda)$ is a piecewise linear convex function of λ. Furthermore, in Section I.2.4 we presented a subgradient algorithm for maximizing a piecewise linear concave function or, equivalently, minimizing a piecewise linear convex function. Here we use the subgradient algorithm to solve LD.

Proposition 4.1. *If x^0 is an optimal solution to* LR(λ^0), *then $s^0 = b^1 - A^1x^0$ is a subgradient of $z_{LR}(\lambda)$ at $\lambda = \lambda^0$.*

Proof. The result is a direct consequence of Proposition 4.2 of Section I.2.4. ∎

We now consider a Lagrangian dual for UFL. One option is to take the dual with respect to the constraints $\Sigma_{j \in N} y_{ij} = 1$ for $i \in I$. Then

$$Q = \{x \in B^n, y \in R^{mn}_+ : y_{ij} - x_j \leqslant 0 \quad \text{for} \quad i \in I, j \in N\},$$

$$z_{LR}(u) = \max_{(x,y) \in Q} \left(\sum_{i \in I} \sum_{j \in N} (c_{ij} - u_i) y_{ij} - \sum_{j \in N} f_j x_j + \sum_{i \in I} u_i \right)$$

and

$$z_{\mathrm{LD}} = \min_{u \in R^m} z_{\mathrm{LR}}(u).$$

Recall (Proposition 6.11 of Section II.3.6) that u is unconstrained since we have taken the dual with respect to equality constraints. Here we have used u rather than λ for the multipliers because of their connection with the dual variables introduced in the linear programming relaxation of UFL.

With this formulation it is very easy to solve the Lagrangian relaxation for fixed u.

Proposition 4.2. *For the Lagrangian relaxation of* UFL, *the following statements are true.*

a. $z_{\mathrm{LR}}(u) = \Sigma_{j \in N} (\Sigma_{i \in I} (c_{ij} - u_i)^+ - f_j)^+ + \Sigma_{i \in I} u_i$.

b. $z_{\mathrm{LD}} = z_{\mathrm{LP}}$.

c. *A subgradient of* $z_{\mathrm{LR}}(u)$ *at* $u = u^0$ *is given by* $s_i = 1 - \Sigma_{j \in N} y_{ij}(u^0)$ *for* $i \in I$, *where*

$$y_{ij}(u^0) = \begin{cases} 1 & \text{if } c_{ij} - u_i^0 > 0 \text{ and } \sum_{i \in I} (c_{ij} - u_i^0)^+ - f_j > 0 \\ 0 & \text{otherwise.} \end{cases}$$

Proof. Optimizing first over the y variables, we set $y_{ij} = 0$ if $c_{ij} - u_i \leq 0$ and $y_{ij} = x_j$ otherwise. Hence

$$z_{\mathrm{LR}}(u) = \max\left\{ \sum_{j \in N} \left(\sum_{i \in I} (c_{ij} - u_i)^+ - f_j \right) x_j + \sum_{i \in I} u_i \colon x \in B^n \right\}.$$

Now, maximizing over x_j, we set $x_j = 0$ if $\Sigma_{i \in I} (c_{ij} - u_i)^+ - f_j \leq 0$ and $x_j = 1$ otherwise. Hence statement a is true.

Statement b follows from the observation that $z_{\mathrm{LR}}(u) = w(u)$ as defined in (3.1).

Statement c follows from the optimal values of the variables given in the proof of a and Proposition 4.1. ■

Although the Lagrangian dual of UFL is not a stronger relaxation than the linear programming relaxation of UFL, the Lagrangian dual is still of interest since subgradient optimization is reportedly a very efficient algorithm for minimizing $z_{\mathrm{LR}}(u)$.

Example 1.1 (continued). We apply the subgradient algorithm to LD.

We start with $u^1 = (12 \quad 8 \quad 6 \quad 8 \quad 8 \quad 3)$, where u_i^1 is the second max in row i. We use the subgradient direction s given in Proposition 4.2 and a "geometric" sequence for determining step size. In particular, $u^{t+1} = u^t - \theta_t s^t$, where $\theta_1 = 2$ and θ_t is halved every three iterations thereafter.

The results of the first 11 iterations are shown in Table 4.1.

We observe that even though $w(u)$ has attained its minimum value of $41\frac{1}{2}$, the algorithm does not terminate since $s \neq 0$. To establish that $s = 0$ is a subgradient at u^{10}, we would need to generate multiple optimal solutions to LR(u^{10}) and then take a convex combination of the corresponding subgradients.

Table 4.1.

Iteration t	u^t						$w(u^t)$	s^t						θ_t
1	12	8	6	8	8	3	46	1	1	1	0	0	0	2
2	10	6	4	8	8	3	43	0	0	−1	0	0	0	2
3	10	6	6	8	8	3	42	1	1	1	0	0	0	2
4	8	4	4	8	8	3	47	−1	−1	−1	0	0	0	1
5	9	5	5	8	8	3	44	−1	−1	−1	0	0	0	1
6	10	6	6	8	8	3	42	1	1	1	0	0	0	1
7	9	5	5	8	8	3	44	−1	−1	−1	0	0	0	$\frac{1}{2}$
8	$9\frac{1}{2}$	$5\frac{1}{2}$	$5\frac{1}{2}$	8	8	3	$42\frac{1}{2}$	−1	0	0	0	0	0	$\frac{1}{2}$
9	10	$5\frac{1}{2}$	$5\frac{1}{2}$	8	8	3	42	−1	0	0	0	0	0	$\frac{1}{2}$
10	$10\frac{1}{2}$	$5\frac{1}{2}$	$5\frac{1}{2}$	8	8	3	$41\frac{1}{2}$	1	1	1	0	0	0	$\frac{1}{4}$
11	$10\frac{1}{4}$	$5\frac{1}{4}$	$5\frac{1}{4}$	8	8	3	$42\frac{1}{4}$	1	1	1	0	0	0	$\frac{1}{4}$

The example illustrates the difficulty of stopping the subgradient algorithm.

However, since the Lagrangian relaxation is to be embedded within a branch-and-bound algorithm, subgradient optimization can be used to obtain good bounds easily and quickly without having to wait for the algorithm to "converge". In particular, we use three criteria for stopping the subgradient algorithm at iteration t, namely:

a. $s^t = 0$;
b. if the data are integral, then $z_{LR}(u^t) - \underline{z} < 1$, where $\underline{z}$ is the value of the best available feasible solution; and
c. after a specific number of subgradient iterations has occurred, that is $t \geqslant t_{\max}$.

It is also important to use the multipliers to construct primal feasible solutions. For example, when solving UFL with a Lagrangian dual relaxation, for each u we can construct the solution $Q(u)$ given by (3.3).

Finally, the reduced prices and $\underline{z}$ can be used to fix variables at each node of the branch-and-bound tree (see Proposition 2.1).

Example 4.1. We consider a p-facility variant of Example 2.1 having the same matrix (c_{ij}) as before, with fixed-costs $f_j = 0$ for all $j \in N$ and where exactly four facilities must be opened. We solve this instance using a Lagrangian dual/subgradient optimization/branch-and-bound algorithm.

Using the greedy-interchange heuristic described in Section 3, a feasible solution of value −12,509 is found. The subgradient algorithm is then initialized with u_i^1 = second max c_{ij} for all i. The step size is halved every $n = 33$ iterations. After 102 iterations, an upper bound of −12,336 and a feasible solution of value −12,363 are found. By this stage, using reduced prices as in Proposition 2.1, two of the variables x_j can be fixed to 1, and 28 of the x_j variables can be set to 0. The remaining problem is to open facilities at two of the three remaining sites (a problem that is easily solved), and −12,363 is indeed shown to be the optimal value. The optimal solution is shown in Table 2.4.

Solving the Lagrangian Dual by Constraint Generation

It has been shown in Corollary 6.3 of Section II.3.6 that if $\{x^k \in R_+^n: k \in K\}$ and $\{r^j \in R_+^n: j \in J\}$ are the extreme points and extreme rays of conv(Q), then

$$
\text{(MLD)} \qquad
\begin{aligned}
z_{LD} = \min_{\eta,\lambda} \ & \eta \\
& \eta + \lambda(A^1x^k - b^1) \geq cx^k \quad \text{for } k \in K \\
& \lambda A^1 r^j \geq cr^j \quad \text{for } j \in J \\
& \lambda \geq 0.
\end{aligned}
$$

Since MLD has a very large number of rows, it is a suitable candidate for the FCPA of Section 2. Assuming that all but the nonnegativity constraints are in $\mathcal{F}$, we now describe the separation algorithm for $\mathcal{F}$.

Separation Algorithm for MLD. Given (η^*, λ^*), with $\lambda^* \geq 0$, calculate

$$z_{LR}(\lambda^*) = \max_{x \in \text{conv}(Q)} cx + \lambda^*(b^1 - A^1x).$$

If $\eta^* \geq z_{LR}(\lambda^*)$, stop. $\eta = z_{LR}(\lambda^*)$, $\lambda = \lambda^*$ is an optimal solution of MLD.
If $\eta^* < z_{LR}(\lambda^*)$, an inequality in $\mathcal{F}$ is violated.

a. If $z_{LR}(\lambda^*) \rightarrow \infty$, then there exists a ray r^j for $j \in J$ such that $(c - \lambda^* A^1)r^j > 0$. Hence the inequality $\lambda A^1 r^j \geq cr^j$ is violated.
b. If $z_{LR}(\lambda^*) < \infty$, then there exists an extreme point x^k for $k \in K$ such that $z_{LR}(\lambda^*) = cx^k + \lambda^*(b^1 - A^1x^k)$. Since $\eta^* < z_{LR}(\lambda^*)$, the inequality $\eta + \lambda(A^1x^k - b^1) \geq cx^k$ is violated.

For UFL, with $Q = \{x \in B^n, y \in R_+^{mn}: y_{ij} - x_j \leq 0 \quad \text{for} \quad i \in I, j \in N\}$,

the extreme points $\{x^k, y^k\}_{k \in K}$ of conv(Q) are

$$\{x \in B^n, y \in B^{mn}: y_{ij} - x_j \leq 0 \quad \text{for} \quad i \in I, j \in N\}.$$

Hence MLD is

$$
\begin{aligned}
z_{LD} = \min_{\eta,u} \ & \eta \\
& \eta + \sum_{i \in I} u_i\left(-1 + \sum_{j \in N} y_{ij}^k\right) \geq \sum_{i \in I}\sum_{j \in N} c_{ij}y_{ij}^k - \sum_{j \in N} f_j x_j^k \quad \text{for } k \in K,
\end{aligned}
$$

where $y_{ij}^k = 0$ if $x_j^k = 0$ and $y_{ij}^k \in \{0, 1\}$ if $x_j^k = 1$.

In the separation algorithm, the constraints are generated by solving for $z_{LR}(u)$ as indicated in Proposition 4.2.

Benders' Decomposition

We have seen in Section II.3.7 that the problem

$$
\text{(MIP)} \qquad
\begin{aligned}
z = \max \ & cx + hy \\
& Ax + Gy \leq b \\
& x \in X \subseteq Z_+^n, \\
& y \in R_+^p
\end{aligned}
$$

can be reformulated as

$$\text{(MIP}') \qquad \begin{aligned} z = \max \eta & \\ \eta \leqslant cx + u^k(b - Ax) & \quad \text{for } k \in K \\ v^j(b - Ax) \geqslant 0 & \quad \text{for } j \in J \\ x \in X, \ \eta \in R^1, & \end{aligned}$$

where $\{u^k \in R_+^m: k \in K\}$ are the extreme points of $Q = \{u \in R_+^m: uG \geqslant h\}$, and $\{v^j \in R_+^m: j \in J\}$ is the set of extreme rays of $\{u \in R_+^m: uG \geqslant 0\}$.

Though MIP′ is not a linear program, the large number of constraints suggests the use of a cut generation algorithm. It suffices to adapt the FCPA for LP($\mathscr{F}$) by replacing the linear programming relaxation LP^t with a mixed-integer programming relaxation MIP^t and to describe the separation algorithm.

Constraint Generation Algorithm for MIP′

Initialization: Find (possible empty) sets $K^1 \subseteq K$, $J^1 \subseteq J$. Let

$$S_R^1 = \{\eta \in R^1, x \in X: \eta \leqslant cx + u^k(b - Ax) \text{ for } k \in K^1, \; v^j(b - Ax) \geqslant 0 \text{ for } j \in J^1\}.$$

Set $t = 1$.

Iteration t:

Step 1: Solve the relaxation of MIP′:

$$(\text{MIP}^t) \qquad z^t = \max\{\eta: (\eta, x) \in S_R^t, x \in X\}$$

a. If MIP^t is infeasible, stop. MIP′ is infeasible.
b. If MIP^t is unbounded, find a feasible solution pair (η^t, x^t) with $\eta^t > \omega$ for some large value ω.
c. Otherwise let the optimal solution be (η^t, x^t).

Step 2: *Separation*. Solve the linear program (see Section II.3.7)

$$(\text{LP}(x^t)) \qquad \begin{aligned} z_{\text{LP}}(x^t) = \max hy & \\ Gy \leqslant b - Ax^t & \\ y \in R_+^p & \end{aligned}$$

or its dual.

a. If $z_{\text{LP}}(x^t) \to \infty$, stop. MIP′ is unbounded.
b. If $z_{\text{LP}}(x^t)$ is finite, let the primal solution be y^t, and the dual solution u^t.
c. If $\text{LP}(x^t)$ is infeasible, let v^t be a dual ray with $v^t(b - Ax^t) < 0$. (Note that at the indication of infeasibility, we also get a dual extreme point u^t.)
d. *Optimality test.* If $cx^t + hy^t \geqslant \eta^t$, stop. (x^t, y^t) is an optimal solution of MIP′.
e. *Violation.* If $cx^t + hy^t < \eta^t$ or $\text{LP}(x^t)$ is infeasible, at least one constraint of MIP′ is violated.

i. If $z_{LP}(x^t)$ is finite, $\eta \leq cx + u^t(b - Ax)$ is violated. Set $K^{t+1} = K^t \cup \{t\}$, that is,

$$S_R^{t+1} = S_R^t \cap \{(\eta, x): \eta \leq cx + u^t(b - Ax)\}.$$

ii. If $LP(x^t)$ is infeasible, $v^t(b - Ax) \geq 0$ is violated. Set $J^{t+1} = J^t \cup \{t\}$, that is,

$$S_R^{t+1} = S_R^t \cap \{(\eta, x): v^t(b - Ax) \geq 0\}.$$

(Although it is not necessary, we can also update K^t by setting $K^{t+1} = K^t \cup \{t\}$ so that

$$S_R^{t+1} = S_R^t \cap \{(\eta, x): v^t(b - Ax) \geq 0, \eta \leq cx + u^t(b - Ax)\}.)$$

f. $t \leftarrow t + 1$.

There are several difficulties in implementing Benders' decomposition that concern solving the relaxation

$$(\text{MIP}^t) \qquad \begin{aligned} z^t = \max\ & \eta \\ & \eta \leq cx + u^k(b - Ax) \quad \text{for } k \in K^t \\ & v^j(b - Ax) \geq 0 \quad \text{for } j \in J^t \\ & \eta \in R^1, x \in X \subseteq Z_+^n, \end{aligned}$$

where K^t and J^t are the index sets of inequalities available after the first t iterations.

One difficulty is that MIP^t is a mixed-integer program with one continuous variable. A way of alleviating this difficulty is to replace η by a threshold value η^* in MIP$'$. We then replace MIP$'$ by the pure-integer feasibility problem for the constraint set

$$\{x \in X: \eta^* \leq cx + u^k(b - Ax) \quad \text{for } k \in K, v^j(b - Ax) \geq 0 \quad \text{for } j \in J\}.$$

Then if the resulting problem is feasible (infeasible), η^* is increased (decreased) and the feasibility problem is solved again. If lower and upper bounds on z^t are known, then binary search can be used to specify the sequence of values for the parameter η^*.

A second difficulty is that very often there is primal degeneracy in the problem $LP(x^t)$, so there is not a unique dual solution u^t. The choice of the dual extreme point u^t leading to a "good" violated constraint can be very important. One approach is to generate cuts that are not dominated by any other constraint.

A third problem lies in the choice of the initial subset of constraints K^1, J^1. If care is not taken with this choice, very unstable behavior of the algorithm may be observed. One proposal is to solve the linear programming relaxation of MIP$'$ and to take K^1 and J^1 to be the index sets of the extreme points and extreme rays required to generate the optimal linear programming solution. A second alternative is to use a heuristic to generate a "good" solution (x^*, y^*) to MIP$'$ and then to derive initial cuts from the solution of $LP(x^*)$.

Example 7.1 of Section II.3.7 (continued)

$$\begin{aligned} \max\ & 5x_1 - 2x_2 + 9x_3 + 2y_1 - 3y_2 + 4y_3 \\ & 5x_1 - 3x_2 + 7x_3 + 2y_1 + 3y_2 + 6y_3 \leq -2 \\ & 4x_1 + 2x_2 + 4x_3 + 3y_1 - y_2 + 3y_3 \leq 10 \\ & x_j \leq 5 \quad \text{for } j = 1, 2, 3 \\ & x \in Z_+^3, y \in R_+^3. \end{aligned}$$

Initialization. $K^1 = J^1 = \emptyset$. $t = 1$.

Iteration 1

Step 1: (MIP^1) $\quad z^1 = \max\{\eta: \eta \in R^1, x \in Z^3_+, x_j \leqslant 5 \text{ for } j = 1, 2, 3\}$.

$\eta^1 \to \infty$. $x^1 = (0 \quad 0 \quad 0)$ is feasible.

Step 2: Separation. Solve the linear program

LP(x^1)

$$
\begin{aligned}
z_{LP}(x^1) = \max\ & 2y_1 - 3y_2 + 4y_3 \\
& 2y_1 + 3y_2 + 6y_3 \leqslant -2 \\
& 3y_1 - \ \ y_2 + 3y_3 \leqslant 10 \\
& y \in R^3_+.
\end{aligned}
$$

LP(x^1) is infeasible since its dual is unbounded, which is verified by the dual extreme point $u^1 = (1 \quad 0)$ and the extreme ray $v^1 = (1 \quad 0)$ (see Figure 7.1 of Section II.3.7). $K^2 = K^1 \cup \{1\}$, $J^2 = J^1 \cup \{1\}$.

Iteration 2

Step 1:

(MIP^2)

$$
\begin{aligned}
z^2 = \max\ & \eta \\
& \eta \leqslant -2 \qquad + \ x_2 + 2x_3 \\
& -2 - 5x_1 + 3x_2 - 7x_3 \geqslant 0 \\
& x_j \leqslant 5 \quad \text{for } j = 1, 2, 3 \\
& \eta \in R^1, \quad x \subseteq Z^3_+.
\end{aligned}
$$

An optimal solution is $z^2 = 5$, $x^2 = (0 \quad 5 \quad 1)$.

Step 2:

Dual of LP(x^2)

$$
\begin{aligned}
\min\ & 6u_1 - 4u_2 \\
& 2u_1 + 3u_2 \geqslant \ \ 2 \\
& 3u_1 - \ \ u_2 \geqslant -3 \\
& 6u_1 + 3u_2 \geqslant \ \ 4 \\
& u \in R^2_+.
\end{aligned}
$$

The dual is unbounded, which is verified by the extreme point $u^2 = (0 \quad 3)$ and the extreme ray $v^2 = (1 \quad 3)$. $K^3 = K^2 \cup \{2\}$, $J^3 = J^2 \cup \{2\}$.

Iteration 3

Step 1:

$$
(\text{MIP}^3)\quad
\begin{aligned}
z^3 = \max\ & \eta \\
\eta &\leqslant -2 \qquad\quad + 1x_2 + \ 2x_3 \\
\eta &\leqslant 30 - \ 7x_1 - 8x_2 - \ 3x_3 \\
& -2 - \ 5x_1 + 3x_2 - \ 7x_3 \geqslant 0 \\
& 28 - 17x_1 - 3x_2 - 19x_3 \geqslant 0 \\
& x_j \leqslant 5 \quad \text{for } j = 1, 2, 3 \\
& \eta \in R^1, x \in Z^3_+.
\end{aligned}
$$

An optimal solution is $z^3 = 3$, $x^3 = (0 \quad 3 \quad 1)$.

Step 2:

$$
\text{LP}(x^3)\quad
\begin{aligned}
z_{\text{LP}}(x^3) = \max\ & 2y_1 - 3y_2 + 4y_3 \\
& 2y_1 + 3y_2 + 6y_3 \leqslant 0 \\
& 3y_1 - \ y_2 + 3y_3 \leqslant 0 \\
& y \in R^3_+.
\end{aligned}
$$

An optimal solution is $z_{\text{LP}}(x^3) = 0$, $y^3 = (0 \quad 0 \quad 0)$ and an optimal dual solution is $u^3 = (0 \quad \frac{4}{3})$. $cx^3 + z_{\text{LP}}(x^3) = 3 = \eta^3$. Hence $(x^3, y^3) = (0 \quad 3 \quad 1 \quad 0 \quad 0 \quad 0)$ is an optimal solution.

As we observed in Section II.3.7, Benders' decomposition is useful algorithmically when LP(x) has structure. There we used the structure of UFL to obtain the reformulation (7.2) given by

$$
\begin{aligned}
z = \max - \sum_{j \in N} f_j x_j + \sum_{i \in I} \eta_i & \\
\eta_i \leqslant c_{ik} + \sum_{j \in N} (c_{ij} - c_{ik})^+ x_j & \quad \text{for } k \in N \text{ and } i \in I \\
\sum_{j \in N} x_j \geqslant 1 & \\
x \in B^n, \eta \in R^m. &
\end{aligned}
$$

We now illustrate the solution of this reformulation using the constraint generation algorithm.

Example 1.1 (continued)

Initialization

$$
S_R^1 = \left\{ (\eta, x)\colon \eta \in R^6, x \in B^5, \eta_i \leqslant \max_{j \in N} c_{ij}, \sum_{j \in N} x_j \geqslant 1 \right\}.
$$

Iteration 1

Step 1: $\eta_1^1 = 13$, $\eta_2^1 = 9$, $\eta_3^1 = 6$, $\eta_4^1 = 10$, $\eta_5^1 = 10$, $\eta_6^1 = 4$, $x^1 = (0 \quad 1 \quad 0 \quad 0 \quad 0)$. $z^1 = \Sigma_{i \in I} \eta_i - \Sigma_{j \in N} f_j x_j = 49$.
Step 2: *Separation* for each client i. $cx^1 + z_{LP}(x^1) = 25$:

$$
\begin{array}{llll}
i = 2: & \eta_2 \leqslant 1 + 7x_1 + 3x_2 + 8x_3 \qquad\qquad + x_5 & \text{is violated} \\
i = 4: & \eta_4 \leqslant 2 + x_1 + 3x_2 \qquad\qquad + 8x_4 + 6x_5 & \text{is violated} \\
i = 5: & \eta_5 \leqslant 0 + 8x_1 \qquad\qquad + 5x_3 + 10x_4 + 8x_5 & \text{is violated} \\
i = 6: & \eta_6 \leqslant 0 + 2x_1 \qquad\qquad + 3x_3 + 4x_4 + x_5 & \text{is violated}
\end{array}
$$

Iteration 2

Step 1: $\eta_1^2 = 13, \eta_2^2 = 9, \eta_3^2 = 6, \eta_4^2 = 10, \eta_5^2 = 10, \eta_6^2 = 4, x^2 = (0 \quad 0 \quad 1 \quad 1 \quad 0)$. $z^2 = 44$.
Step 2: *Separation*. $cx^2 + z_{\text{LP}}(x^2) = 37$.

$$\eta_1 \leqslant 0 + 12x_1 + 13x_2 + 6x_3 + x_5 \quad \text{is violated.}$$

Iteration 3

Step 1: $\eta_1^3 = 13, \eta_2^3 = 9, \eta_3^3 = 6, \eta_4^3 = 10, \eta_5^3 = 10, \eta_6^3 = 4, x^3 = (0 \quad 1 \quad 1 \quad 1 \quad 0)$. $z^3 = 41$.
Step 2: $cx^3 + z_{\text{LP}}(x^3) = 41$. Since the upper and lower bounds are equal, the solution x^3 is optimal.

5. DYNAMIC PROGRAMMING

Dynamic programming provides a framework for decomposing certain optimization problems into a nested family of subproblems. This nested structure suggests a recursive approach for solving the original problem from the solutions of the subproblems.

Dynamic programming was originally developed for the optimization of sequential decision processes. In a *discrete-time sequential decision process*, there are T periods, $t = 1, \ldots, T$. At the beginning of period t, the process is in *state* s_{t-1}, which depends on (a) the initial given state s_0, and (b) the decision variables $x_1, \ldots, x_{t-1}$ for periods $1, \ldots, t-1$. The significance of the state is that the contribution to the objective function in period t depends only on s_{t-1} and x_t, and the state in period $t+1$ depends only on s_{t-1} and x_t.

Formally we describe a sequential decision process by the model

$$
\begin{aligned}
z = \max_{x_1, \ldots, x_T} & \sum_{t=1}^{T} g_t(s_{t-1}, x_t) \\
& s_t = \phi_t(s_{t-1}, x_t) \quad \text{for } t = 1, \ldots, T-1 \\
& s_0 \text{ is given.}
\end{aligned}
\tag{5.1}
$$

The domains of the state and decision variables depend on the particular application being considered.

We can consider the 0-1 knapsack problem

$$z = \max\left\{\sum_{j=1}^{n} c_j x_j : \sum_{j=1}^{n} a_j x_j \leq b, x \in B^n\right\} \tag{5.2}$$

as a sequential decision process. An instance is specified by integers n and b and positive integral n-vectors $c = (c_1, \ldots, c_n)$ and $a = (a_1, \ldots, a_n)$.

In period k, $k = 1, \ldots, n$, the decision is whether to put the kth item into the knapsack. The state of the process in period k is the number of units of the knapsack that are available after we have made the decisions regarding items $1, \ldots, k-1$, that is,

$$s_{k-1} = b - \sum_{j=1}^{k-1} a_j x_j = s_{k-2} - a_{k-1} x_{k-1}.$$

Thus

$$\phi_k(s_{k-1}, x_k) = s_{k-1} - a_k x_k \text{ and } g_k(s_{k-1}, x_k) = c_k x_k \quad \text{for } k = 1, \ldots, n,$$

and $s_0 = b$. For $k = 1, \ldots, n$, the feasible domain is given by $0 \leq s_k \leq b$, and $x_k \in \{0, 1\}$.

Another problem that can be viewed as a sequential decision process is the uncapacitated lot-size problem (ULS), which has been formulated in Section I.1.5 as

$$\begin{aligned}
&\min \sum_{t=1}^{T} (p_t y_t + c_t x_t + h_t s_t) \\
&\qquad s_{t-1} + y_t = d_t + s_t \quad \text{for } t = 1, \ldots, T \\
&\qquad y_t \leq \omega x_t \qquad \text{for } t = 1, \ldots, T \\
&\qquad s_0 = 0, s_T = 0 \\
&\qquad s \in R_+^{T+1}, y \in R_+^T, x \in B^T.
\end{aligned} \tag{5.3}$$

The data are the unit production costs $\{p_t\}_{t=1}^T$, unit storage costs $\{h_t\}_{t=1}^T$, set-up costs $\{c_t\}_{t=1}^T$, and demands $\{d_t\}_{t=1}^T$. All of the data are assumed to be nonnegative and integral. The variable y_t is the production in period t. If $y_t > 0$, we must pay the set-up cost c_t. This is achieved by the constraint $y_t \leq \omega x_t$, where ω is a suitably large positive number. The variable s_t is the inventory available at the end of period t. Since demand cannot be backlogged, we have $s_t \geq 0$.

The formulation (5.3) is of the form (5.1) with

$$g_t(s_{t-1}, y_t, x_t) = p_t y_t + c_t x_t + h_t(s_{t-1} + y_t - d_t) \quad \text{for } t = 1, \ldots, T$$

and

$$\phi_t(s_{t-1}, y_t, x_t) = s_{t-1} + y_t - d_t \quad \text{for } t = 1, \ldots, T.$$

Here the decision variables are both x_t and y_t.

We now develop a recursive optimization scheme for the sequential decision process (5.1). For $k = T, T-1, \ldots, 1$, let

(5.4)
$$z_k(s_{k-1}) = \max_{x_k, \ldots, x_T} \sum_{t=k}^{T} g_t(s_{t-1}, x_t)$$
$$s_t = \phi_t(s_{t-1}, x_t) \quad \text{for } t = k, \ldots, T-1.$$

Thus $z = z_1(s_0)$ for the given value of s_0.

Proposition 5.1 $z_k(s_{k-1}) = \max_{x_k}\{g_k(s_{k-1}, x_k) + z_{k+1}(s_k)\}$

$$s_k = \phi_k(s_{k-1}, x_k).$$

Proof. By the definition of z_{k+1} given in (5.4), the term on the right equals

$$\max_{x_k}\left\{g_k(s_{k-1}, x_k) + \max_{x_{k+1}, \ldots, x_T} \sum_{t=k+1}^{T} g_t(s_{t-1}, x_t)\right\}$$
$$s_t = \phi_t(s_{t-1}, x_t) \quad \text{for } t = k, \ldots, T-1$$
$$= \max_{x_k} \max_{x_{k+1}, \ldots, x_T} \left\{g_k(s_{k-1}, x_k) + \sum_{t=k+1}^{T} g_t(s_{t-1}, x_t)\right\}$$
$$s_t = \phi_t(s_{t-1}, x_t) \quad \text{for } t = k, \ldots, T-1$$
$$= z_k(s_{k-1}).$$

The first equality holds since g_k is not a function of $x_{k+1}, \ldots, x_T$, and the second equality follows from (5.4). ■

The recursion given in Proposition 5.1 transforms the original optimization problem (5.1) with T decision variables, $T - 1$ state variables, and $T - 1$ state constraints into a sequence of T subproblems. The kth subproblem

(5.5)
$$z_k(s_{k-1}) = \max_{x_k}\{g_k(s_{k-1}, x_k) + z_{k+1}(s_k)\} \text{ for } k = T, \ldots, 1$$
$$s_k = \phi_k(s_{k-1}, x_k),\ z_{T+1}(s_T) = 0$$

has only one decision variable and one state constraint but must be solved for all possible values of s_{k-1}. Thus the efficiency of solving (5.5) depends on the number of values of s_{k-1}, unless it is possible to determine x_k analytically as a function of s_{k-1}.

The recursion expresses an intuitive *principle of optimality* for sequential decision processes; that is, once we have reached a particular state, a necessary condition for optimality is that the remaining decisions must be chosen optimally with respect to that state.

The shortest-path problem between specified nodes provides a nice illustration of the principle of optimality. Suppose that $p_{0,T}$ is a path from node 0 to node T and that node k, $k \neq 0, T$, is on $p_{0,T}$. Hence $p_{0,T}$ decomposes into the two paths $p_{0,k}$ and $p_{k,T}$. The principle of optimality says that a necessary condition for $p_{0,T}$ to be a shortest-path between nodes 0 and T is that $p_{k,T}$ be a shortest path between nodes k and T. We have used this fact in developing the shortest-path algorithms of Section I.3.2, which therefore may be considered as dynamic programming algorithms.

The principle of optimality is a means of excluding nonoptimal decisions by domination. In the general sequential decision process given by (5.1), if $x_1^0, \ldots, x_T^0$ is a feasible set of decisions that yields $s_k = s_k^0$, then a necessary condition for its optimality is that $x_{k+1}^0, \ldots, x_T^0$ be an optimal set of decisions with respect to the problem that begins in period $k+1$ with $s_k = s_k^0$.

A Dynamic Programming Algorithm for the 0-1 Knapsack Problem

We now demonstrate the solution of a recursive formulation by developing a classical dynamic programming algorithm for the 0-1 knapsack problem (5.2). Although the recursion (5.5) can be applied to the 0-1 knapsack problem, it is easier to develop the standard dynamic programming algorithm using a slightly different approach that reverses the order of the recursion.

For $k = 1, \ldots, n$, define $N_k = \{1, \ldots, k\}$ and

$$(5.6) \qquad z_k(d) = \max\left\{\sum_{j\in N_k} c_j x_j: \sum_{j\in N_k} a_j x_j \leqslant d, x \in B^k\right\} \quad \text{for } d = 0, 1, \ldots, b.$$

Thus $z_n(b) = z(b)$.

We will proceed recursively to calculate $z_n(b)$ from z_{n-1}, which in turn is calculated from z_{n-2}, and so on. The recursion is initialized with

$$z_1(d) = \begin{cases} c_1 & \text{if } a_1 \leqslant d \\ 0 & \text{if } a_1 > d. \end{cases}$$

Now observe that if $x_k = 1$ in an optimal solution to (5.6) then $d - a_k \geqslant 0$ and

$$\begin{aligned} z_k(d) &= c_k + \max\left\{\sum_{j\in N_{k-1}} c_j x_j: \sum_{j\in N_{k-1}} a_j x_j \leqslant d - a_k, x \in B^{k-1}\right\} \\ &= c_k + z_{k-1}(d - a_k). \end{aligned}$$

On the other hand, if $x_k = 0$ in an optimal solution to (5.6), then

$$z_k(d) = \max\left\{\sum_{j\in N_{k-1}} c_j x_j: \sum_{j\in N_{k-1}} a_j x_j \leqslant d, x \in B^{k-1}\right\} = z_{k-1}(d).$$

Hence for $k = 2, \ldots, n$ and $d = 0, \ldots, b$, we obtain

$$(5.7) \qquad z_k(d) = \begin{cases} z_{k-1}(d) & \text{if } a_k > d \\ \max\{z_{k-1}(d), c_k + z_{k-1}(d - a_k)\} & \text{if } a_k \leqslant d. \end{cases}$$

Relation (5.7) is the basic recursion for determining $z_n(b)$. It also applies for $k = 1$ by defining $z_0(d) = 0$ for $d \geqslant 0$. To put it in a slightly more compact form, we define $z_k(d) = -\infty$ for $d < 0$. Thus for $k = 1, \ldots, n$ and $d = 0, 1, \ldots, b$, we obtain

$$(5.8) \qquad z_k(d) = \max\{z_{k-1}(d), c_k + z_{k-1}(d - a_k)\},$$

where for all k, we have $z_k(d) = -\infty$ if $d < 0$.

For fixed k and d, a constant number of calculations is needed to solve (5.8); hence $O(nb)$ calculations are required to determine $z_n(b)$.

Given z_k for $k = 0, 1, \ldots, n$, a recursion in the opposite direction is used to determine an optimal solution $x^0 = (x_1^0, \ldots, x_n^0)$. We have

$$x_n^0 = \begin{cases} 0 & \text{if } z_n(b) = z_{n-1}(b) \\ 1 & \text{otherwise.} \end{cases}$$

Now let $d_k^0 = b - \sum_{j=k+1}^{n} a_j x_j^0$. Then, for $k = n - 1, \ldots, 1$, we obtain

$$x_k^0 = \begin{cases} 0 & \text{if } z_k(d_k^0) = z_{k-1}(d_k^0) \\ 1 & \text{otherwise.} \end{cases}$$

The amount of work required in the backward recursion to determine an optimal solution is dominated by the work in the forward recursion (5.8). Hence the overall running time of the algorithm is $O(nb)$. Thus we have obtained a pseudopolynomial-time algorithm (but not polynomial) for the 0-1 knapsack problem.

Example 5.1

$$\begin{aligned} \max\ & 16x_1 + 19x_2 + 23x_3 + 28x_4 \\ & 2x_1 + 3x_2 + 4x_3 + 5x_4 \leqslant 7 \\ & x \in B^4. \end{aligned}$$

$$z_1(d) = \begin{cases} 0 & \text{for } 0 \leqslant d \leqslant 1 \\ 16 & \text{for } 2 \leqslant d \leqslant 7 \end{cases}$$

$$z_2(d) = \begin{cases} 0 & \text{for } 0 \leqslant d \leqslant 1 \\ 16 & \text{for } d = 2 \\ 19 & \text{for } 3 \leqslant d \leqslant 4\ [\max(16, 19 + 0)] \\ 35 & \text{for } 5 \leqslant d \leqslant 7\ [\max(16, 19 + 16)] \end{cases}$$

$$z_3(d) = \begin{cases} z_2(d) & \text{for } 0 \leqslant d \leqslant 3 \\ 23 & \text{for } d = 4\ [\max(19, 23 + 0)] \\ 35 & \text{for } d = 5\ [\max(35, 23 + 0)] \\ 39 & \text{for } d = 6\ [\max(35, 23 + 16)] \\ 42 & \text{for } d = 7\ [\max(35, 23 + 19)]. \end{cases}$$

Finally,

$$z_4(7) = \max(42, 28 + z_3(2)) = 44.$$

Hence $x_4^0 = 1$, $x_3^0 = x_2^0 = 0$ since $z_3(2) = z_2(2) = z_1(2)$, and $x_1^0 = 1$ since $z_1(2) > 0$.

The recursion (5.7) can be interpreted as a method for solving a maximum-weight path formulation of the 0-1 knapsack problem. The directed graph has a node s and has nodes (k, d) for $k = 0, 1, \ldots, n$ and $d = 0, 1, \ldots, b$. For $1 \leqslant k \leqslant n - 1$ and all d, there is an arc of the form $((k - 1, d), (k, d))$ of weight 0 that corresponds to setting $x_k = 0$, given $\sum_{j \in N_{k-1}} a_j x_j = d$. For $1 \leqslant k \leqslant n - 1$ and all $d \geqslant a_k$, there is an arc of the form $((k - 1, d - a_k), (k, d))$ of weight c_k that corresponds to setting $x_k = 1$, given $\sum_{j \in N_{k-1}} a_j x_j = d - a_k$.

In addition, there are arcs $(s, (0, d))$ for all d of weight 0 (see Figure 5.1). The objective is to find a maximum-weight path that starts at node s and terminates at node (n, b). The recursion (5.7) chooses a maximum-weight path to node (k, d), given the weights of maximum-weight paths to nodes $(k-1, d)$ and $(k-1, d-a_k)$.

The algorithm given above generalizes straightforwardly to the *bounded variable knapsack problem*

$$z(b) = \max\left\{\sum_{j\in N} c_j x_j : \sum_{j\in N} a_j x_j \leq b,\ x_j \leq \beta_j \text{ for } j \in N,\ x \in Z_+^n\right\}, \tag{5.9}$$

where the a_j's, c_j's, β_j's, and b are positive integers. Note that (5.2) is the special case of (5.9) with $\beta_j = 1$ for all $j \in N$.

We simply replace (5.8) by the recursion

$$z_k(d) = \max\left\{c_k x_k + z_{k-1}(d - a_k x_k) : x_k \leq \min\left(\beta_k, \left\lfloor \frac{d}{a_k} \right\rfloor\right),\ x_k \in Z_+^1\right\} \tag{5.10}$$

for $k = 1, \ldots, n$ and $d = 0, 1, \ldots, b$, where $z_0(d) = 0$ for all $0 \leq d \leq b$.

The number of calculations needed to solve (5.10) for fixed k is $O(b(\beta_k + 1))$, which gives an overall running time of $O(nb^2)$. Finally, note that explicit bounds on the variables are not required since we can always take $x_j \leq \lfloor b/a_j \rfloor$ for all $j \in N$. However, in Section II.6.1 we will give an $O(nb)$ algorithm for the knapsack problem without explicit upper bounds on the variables.

A Dynamic Programming Algorithm for the Uncapacitated Lot-Size Problem (ULS)

For $k = T-1, \ldots, 1$, the recursion (5.4) for ULS is

$$z_k(s_{k-1}) = \min_{\substack{x_k \in \{0,1\} \\ 0 \leq y_k \leq \omega x_k}} \{p_k y_k + c_k x_k + h_k(s_{k-1} + y_k - d_k) + z_{k+1}(s_{k-1} + y_k - d_k)\} \tag{5.11}$$

and

$$z_T(s_{T-1}) = \min_{\substack{x_T \in \{0,1\} \\ 0 \leq y_T \leq \omega x_T}} \{p_T y_T + c_T x_T + h_T(s_{T-1} + y_T - d_T)\}. \tag{5.12}$$

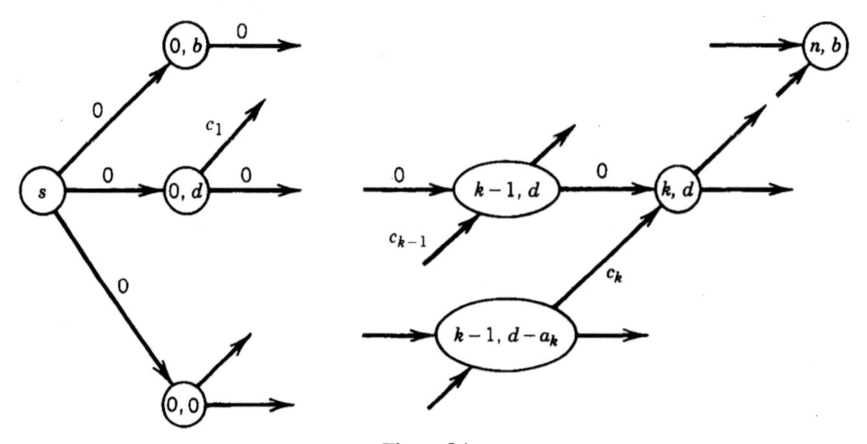

Figure 5.1

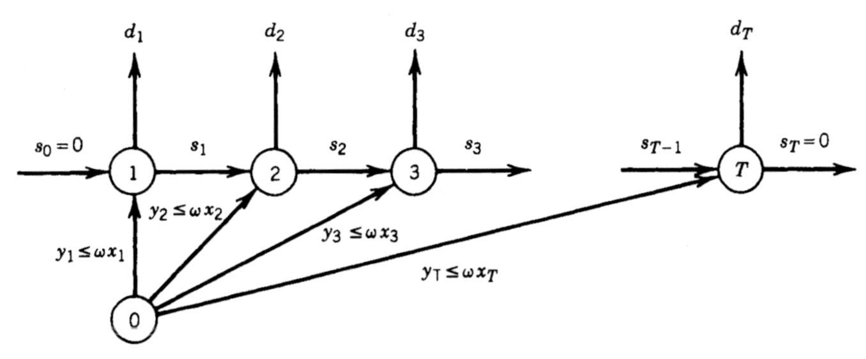

Figure 5.2

Since the demands are integral-valued, it can be shown that the production and storage variables will also be integers. The difficulty is that since demand in period k can be met by production in any period $t \leqslant k$, it follows that s_{k-1} can be as large as $\Sigma_{t=k}^{T} d_t$, and it appears that a very large number of combinations of (s_{k-1}, y_k) must be considered to solve (5.11). Fortunately, as the following theorem demonstrates, this is not the case.

Theorem 5.2. *There is an optimal solution to (5.3) is which*

a. $s_{t-1}y_t = 0$ *for* $t = 1, \ldots, T$.
b. *If* $y_t > 0$, *then* $y_t = \Sigma_{k=t}^{r} d_k$ *for some* $r, t \leqslant r \leqslant T$.
c. *If* $s_{t-1} > 0$, *then* $s_{t-1} = \Sigma_{k=t}^{q} d_k$ *for some* $q, t \leqslant q \leqslant T$.

Proof. We represent (5.3) as a fixed-charge flow problem on the network shown in Figure 5.2.

Let (x^*, y^*, s^*) be an optimal solution to (5.3). x^* specifies those arcs pointing out of node 0 that are available for flow. Thus, given x^*, we can delete arc $(0, j)$ if $x_j^* = 0$ and then determine (y^*, s^*) by solving a minimum-cost flow problem on the resulting network.

By Proposition 6.2 of Section I.3.6, the arcs with positive flow define a spanning forest rooted at node 0. Suppose $(j - 1, j)$ is in the forest. Then there is a path from 0 to $j - 1$ in the forest. If $(0, j)$ is also in the forest, we obtain a cycle. Hence it cannot be the case that both arcs $(j - 1, j)$ and $(0, j)$ are in the forest.

Parts b and c are simple consequences of a. ■

From Theorem 5.2, it follows that $2(T - t)$ combinations of (s_{t-1}, y_t) must be considered in solving (5.11) and (5.12). Thus the overall running time is $O(T^2)$, and recursive optimization yields a polynomial-time algorithm for ULS.

Example 5.2

t	1	2	3	4	5
p_t	3	3	4	5	5
h_t	1	1	1	2	2
c_t	30	30	30	30	30
d_t	32	41	48	36	20

$t = 5$: $s_4 \in \{0, 20\}$ and $s_4 + y_5 = 20$

$$z_5(0) = 20p_5 + c_5 = 130, \qquad y_5 = 20$$
$$z_5(20) = 0, \qquad y_5 = 0.$$

$t = 4$: $s_3 \in \{0, 36, 56\}$, $s_3 + y_4 \geq 36$, $s_3 y_4 = 0$

$$z_4(56) = h_4(s_3 - d_4) + z_5(20) = 2(20) = 40, \qquad y_4 = 0$$
$$z_4(36) = z_5(0) = 130, \qquad y_4 = 0$$
$$z_4(0) = \min\{36p_4 + c_4 + z_5(0), 56p_4 + 20h_4 + c_4 + z_5(20)\}$$
$$= \min\{340, 350\} = 340, \qquad y_4 = 36.$$

$t = 3$: $s_2 \in \{0, 48, 84, 104\}$, $s_2 + y_3 \geq 48$, $s_2 y_3 = 0$

$$z_3(104) = h_3(s_2 - d_3) + z_4(56) = 56 + 40 = 96, \qquad y_3 = 0$$
$$z_3(84) = 36 + z_4(36) = 36 + 130 = 166, \qquad y_3 = 0$$
$$z_3(48) = 0 + z_4(0) = 340, \qquad y_3 = 0$$
$$z_3(0) = \min\{48p_3 + c_3 + z_4(0), 84p_3 + c_3 + 36h_3 + z_4(36), 104p_3 + c_3 + 56h_3 + z_4(56)\}$$
$$= \min\{562, 532, 542\} = 532, \qquad y_3 = 84.$$

$t = 2$: $s_1 \in \{0, 41, 89, 125, 145\}$

$$\left.\begin{aligned} z_2(145) &= 104 + z_3(104) = 200 \\ z_2(125) &= 84 + z_3(84) = 250 \\ z_2(89) &= 48 + z_3(48) = 388 \\ z_2(41) &= 0 + z_3(0) = 532 \end{aligned}\right\} y_2 = 0$$

$$z_2(0) = \min\{41p_2 + c_2 + z_3(0), 89p_2 + c_2 + 48h_2 + z_3(48), 125p_2 + c_2 + 84h_2 + z_3(84), 145p_2 + c_2 + 104h_2 + z_3(104)\}$$
$$= \min\{685, 685, 655, 665\} = 655, \qquad y_2 = 125.$$

$t = 1$:
$$z_1(0) = \min\{32p_1 + c_1 + z_2(0), 73p_1 + c_1 + 41h_1 + z_2(41), 121p_1 + c_1 + 89h_1 + z_2(89), 157p_1 + c_1 + 125h_1 + z_2(125), 177p_1 + c_1 + 145h_1 + z_2(145)\}$$
$$= \min\{781, 822, 870, 876, 906\} = 781, \qquad y_1 = 32.$$

Hence the optimal solution is $y_1 = 32$, $x_1 = 1$, $y_2 = 125$, $x_2 = 1$, $y_3 = y_4 = x_3 = x_4 = 0$, $y_5 = 20$, $x_5 = 1$, and the optimal cost is $z = 781$.

6. NOTES

Section II.5.1

Efroymson and Ray (1966) have given a classical branch-and-bound algorithm for the uncapacitated facility location problem that uses bounds obtained from the weak formulation. Spielberg (1969a,b), among others, recognized the importance of the strong formulation. Bilde and Krarup (1977) gave a family of UFLs for which the linear programming relaxation of the strong formulation always has an integral optimal solution.

Survey articles on the uncapacitated facility location problem are by Krarup and Pruzan (1983) and Cornuejols, Nemhauser and Wolsey (1984). An annotated bibliography of articles on location problems that were published from 1980–1985 appears in Wong (1985). Francis and Mirchandani (1988) is a collection of survey articles on various aspects of discrete location models.

Section II.5.2

The strong cutting-plane algorithm described in this section was implemented by Morris (1978).

Schrage (1975) showed how variable upper-bound constraints could be treated implicitly within the simplex algorithm. Todd (1982) presented an alternative approach that circumvents degeneracy problems.

The test problem data (Table 2.1) is from Kuehn and Hamburger (1963).

Facets of the UFL polytope have been studied by Cornuejols, Fisher and Nemhauser (1977b), Guignard (1980), Cornuejols and Thizy (1982b), Cho et al. (1983), and Cho, Johnson et al. (1983).

Strong cutting planes and FCPAs have been developed for a variety of other hard combinatorial problems. These include the capacitated plant location problem, Leung and Magnanti. (1986), the matching problem, Grötschel and Holland (1985); the assignment problem with side constraints, Aboudi and Nemhauser (1987); the three-index assignment problem Balas and Saltzman (1986); the max cut problem, Barahona, Grötschel, and Mahjoub (1985) and Barahona and Mahjoub (1986); the linear ordering problem, Grötschel, Junger, and Reinelt (1984, 1985b), and the acyclic subgraph problem, Grötschel, Junger, and Reinelt (1985a). Several others will be cited in the notes for Chapter II.6.

Section II.5.3

Ball and Magazine (1981) and Rinnooy Kan (1986) gave general introductions on the design and analysis of heuristics for discrete optimization problems.

A greedy algorithm for maximizing a (constrained) set function was used by Kruskal (1956) to solve the maximum-weight spanning tree problem exactly. This is one of the first formal uses of the greedy algorithm in combinatorial optimization. However, it must have been used for centuries as a common sense tool for problem solving. Kuehn and Hamburger (1963), Spielberg (1969b), and others have used greedy heuristics to obtain solutions to UFL.

Much the same can be said for local search\interchange heuristics. Kuehn and Hamburger (1963), Manne (1964), and many other researchers have used interchange heuristics to obtain solutions to UFL. Reiter and Sherman (1965) described an interchange scheme for a rather general class of combinatorial optimization problems and carried out a statistical analysis of the results based on random starting solutions; also see Reiter and Rice (1966). Many other uses of the greedy and interchange heuristics will be cited later in these notes and in the notes for Chapter II.6.

Using primal and dual heuristics simultaneously is a more recent idea. The primal–dual heuristic described for UFL is essentially the DUALOC algorithm of Erlenkotter (1978). The projection algorithm of Conn and Cornuejols (1987) for UFL also can be given a primal–dual interpretation. A primal–dual vehicle routing algorithm has been given by Fisher and Jaikumar (1981).

The worst-case analysis of heuristics can be traced to a result of Graham (1966) on a scheduling problem. Another classical paper on worst-case analysis is the work of D.S. Johnson et al. (1974) on the bin-packing problem. Johnson (1974) also analyzed a variety of heuristics for several combinatorial optimization problems. Jenkins (1976), Korte and Hausmann (1976), and Hausmann et al. (1980) gave worst-case analyses of greedy-type algorithms for finding a maximum-weight subset in an independence system. Hausmann and Korte (1978) showed that in a certain well-defined way, no polynomial-time algorithm could improve on the performance of the greedy algorithm for this problem. Zemel (1981) discussed and evaluated various ways of measuring the quality of approximate solutions.

Surveys on the worst-case analysis of heuristics for combinatorial optimization problems are Sahni (1977), Chapter 6 of Garey and Johnson (1979), Korte (1979), and Fisher (1980). Wolsey (1980) explained how worst-case bounds are obtained from primal and dual feasible solutions and illustrated this idea with several examples.

Golden and Stewart (1985) presented general techniques for the statistical analysis of the performance of heuristics. Proposition 3.2 appeared in Fisher (1980). The results on the greedy and interchange heuristics for the p-facility maximization problem appeared in Cornuejols, Fisher, and Nemhauser (1977a). Babayev (1974) and Frieze (1974) showed the submodularity of the set function objective of UFL. Extensions of the Cornuejols et al. results to submodular set functions will be presented in Section III.3.9.

The term *simulated annealing* arises from an analogy with the physical process of cooling physical substances and how the state of the system depends on the rate at which the temperature is dropped. The method is described in Metropolis et al. (1953), Kirkpatrick et al. (1983), and Kirkpatrick (1984). Hajek (1985) gave a survey on theory and applications of simulated annealing. Lundy and Mees (1986) studied the convergence of the algorithm. Applications of simulated annealing were given by Bonomi and Lutton (1984) and Vecchi and Kirkpatrick (1983).

The probabilistic models for UFL come from Cornuejols, Nemhauser, and Wolsey (1980b), Fisher and Hochbaum (1980), Papadimitriou (1981b), and Ahn et al. (1988). Surveys of techniques and results in this field appeared in Karp (1976) and Karp and Steele (1985). An annotated bibliography was given by Karp, Lenstra et al. (1985).

Algorithms in which random or probabilistic choices are made are at an early stage of development in combinatorial optimization. Maffioli et al. (1985) gave an annotated bibliography on this subject, and Welsh (1983) and Maffioli (1986) gave surveys. Rabin (1976) published one of the first articles in this area.

Section II.5.4

Subgradient optimization and the Lagrangian dual for UFL have been used by Cornuejols, Fisher, and Nemhauser (1977a) to solve a float maximization problem, by Mulvey and Crowder (1979) to solve a problem in cluster analysis, and by Neebe and Rao (1983) to solve a problem of assigning users to sources. Geoffrion and McBride (1978) have used a similar approach to solve capacitated location problems.

The results of Example 4.1 were obtained by D. Peeters (private communication).

Applications of Lagrangian duality to the group problem, to set covering and partitioning, and to the traveling salesman problem will be presented or cited in Chapter II.6. Some other applications from the literature are: combinatorial scheduling [Fisher (1973, 1976), Fisher, Northup, and Shapiro (1975), and Potts (1985)]; multiperiod scheduling of power generators [Muckstadt and Koenig, 1977); generalized assignment problem [Ross and Soland (1975), Chalmet and Gelders (1977), and Fisher, Jaikumar, and Van Wassenhove (1986)]; and hierarchical production planning [Graves (1982)].

Grinold (1972) gave an alternative approach to solving the Lagrangian dual.

Other decomposition methods for solving UFL and the closely related p-median problem include: Dantzig–Wolfe decomposition [see Garfinkel, Neebe, and Rao (1974)]; Benders' reformulation [see Magnanti and Wong (1981) and Nemhauser and Wolsey (1981)]; primal subgradient optimization [see Cornuejols and Thizy (1982a)]; and a disaggregation scheme [see Cornuejols, Nemhauser, and Wolsey (1980a)].

Benders' decomposition has been applied by Geoffrion and Graves (1974) to the design of a multicommodity distribution system. A combined Lagrangian/Benders' scheme has been used by Van Roy (1986) to solve a capacitated location problem.

Section II.5.5

Richard Bellman coined the terms *dynamic programming* and *principle of optimality*, pioneered the development of the theory and applications, and wrote the first book on this subject [Bellman (1957)].

The recursion for the 0-1 knapsack problem appeared in Dantzig (1957) and Bellman (1957). Some computational improvements and generalizations to multiple constraints have been given by Weingartner and Ness (1967) and Nemhauser and Ullman (1969).

Dynamic programming algorithms for the lot-size problem have been given by Wagner and Whitin (1958) and Zangwill (1966).

Some general texts on dynamic programming are Bellman and Dreyfus (1962), Nemhauser (1966), White (1969), Dreyfus and Law (1977), and Denardo (1982).

7. EXERCISES

1. Solve the linear programming relaxation of WUFL for the problem instance with:

$$C = \begin{pmatrix} 3 & 3 & 5 & 0 & 6 & 0 \\ 1 & 9 & 7 & 7 & 7 & 7 \\ 6 & 2 & 4 & 6 & 4 & 5 \\ 6 & 6 & 6 & 6 & 0 & 8 \end{pmatrix}$$

and

$$f = (3 \quad 2 \quad 3 \quad 3 \quad 2 \quad 2).$$

2. Prove Proposition 1.2.

3. Show that: every noninteger extreme point (x, y) of the linear programming relaxation of UFL is of the form

i) $x_j = \max_i y_{ij}$ for all $j \in N_1$,

ii) there is, at most, one j with $0 < y_{ij} < x_j$ for each $i \in I$, and

iii) the rank of A equals $|N_1|$,

where $N_1 = \{j \in N: 0 < x_j < 1\}$, $I_1 = \{i \in I: y_{ij} = 0$ or x_j for all j, and $y_{ij} \notin Z^1$ for some $j\}$, and A is an $|I_1| \times |N_1|$ 0, 1 matrix with $a_{ij} = 1$ if $y_{ij} > 0$.

4. Consider the problem instance of UFL in exercise 1. Which variable upper-bound constraints are violated by the solution found in exercise 1? Use a linear programming system to solve the linear programming relaxation of UFL by adding such constraints.

5. **i)** Show that if $i \neq j \neq k$ and $r \neq s \neq t$, then

$$y_{ri} + y_{rj} + y_{sj} + y_{sk} + y_{tk} + y_{ti} \leqslant 1 + x_i + x_j + x_k$$

is a valid inequality for UFL.

ii) Find an inequality of this form that cuts off the fractional solution (y^4, x^4) of Example 1.1 (continued) in Section 2.

iii) Generalize to show that if, for $1 \leqslant t < l$, A^{lt} is a 0, 1 matrix with $\binom{l}{t}$ rows and l columns whose rows are all the different 0, 1 vectors with t 1's and $l - t$ 0's with $l \leqslant n$, $\binom{l}{t} \leqslant m$, then for any $I' \subseteq I$, $N' \subseteq N$ with $|I'| = \binom{l}{t}$, $|N'| = l$ it follows that

$$\sum_{i \in I'} \sum_{j \in N'} a_{ij}^{lt} y_{ij} - \sum_{j \in N'} x_j \leqslant \binom{l}{t} + t - l - 1$$

is a valid inequality for UFL.

iv) Find an inequality cutting off the point (x^*, y^*) given in Table 2.3.

6. Consider the problem that arises when we solve IP over $N' \subset N$ using valid inequalities $\Sigma_{j \in N'} \pi_j x_j \leqslant \pi_0$. To solve the problem over N we need to lift the inequalities so that $\Sigma_{j \in N'} \pi_j x_j + \Sigma_{j \in N \setminus N'} \pi_j x_j \leqslant \pi_0$ is valid for the IP over N. Suppose we have solved UFL with a subset x_j for $j \in N'$ and a subset of the y_{ij} for $i \in I'$, $j \in N'$. Is it easy to lift the inequalities of exercise 5?

7. Apply the following heuristics to the instance of UFL in exercise 1:

i) greedy;

ii) reverse-greedy (close one facility at a time);

iii) 1-interchange;

iv) 1-interchange plus greedy;

v) design your own heuristic.

8. The k-enumeration plus greedy heuristic for maximizing a set function can be described as follows:

1. Enumerate all subsets $S \subseteq N$ with $|S| = k$.
2. For each such S, apply the greedy heuristic to the problem $\max\{v^S(Q): Q \subseteq N \setminus S\}$, where $v^S(Q) = v(S \cup Q)$. Let Q^S be the greedy solution.
3. Let $S^* \cup Q^* = \arg\max\{v(S \cup Q^S): |S| = k, S \subseteq N\}$.

i) Apply the 1-enumeration plus greedy heuristic to the example of exercise 1.

ii) Show that the k-enumeration plus greedy heuristic for the p-facility maximization problem has worst-case behavior given by

$$\frac{z - z^H}{z} \leqslant \left(\frac{p-k}{p}\right)\left(\frac{p-k-1}{p-k}\right)^{p-k}.$$

9. i. Given $c_{ij} \geqslant 0$ for $i \in I$ and $j \in N$, show that the set function $v(Q) = \Sigma_{i \in I} \max_{j \in Q} c_{ij}$ for $Q \subseteq N$ can always be written in the canonical form

$$v(Q) = \sum_{T \cap Q \neq \emptyset} r_T \quad \text{with } r_T \geqslant 0 \text{ for } T \subseteq N.$$

ii) Write the set function arising from exercise 1 in canonical form.

iii) Write the resulting linear programming formulation and its dual.

iv) Which LP formulation would you choose and why?

v) Propose a branch-and-bound algorithm based on your choice.

10. **i)** Apply DUALOC to the instance of UFL in exercise 1.

ii) Apply DUALOC to the instance written in canonical form (see exercise 9).

11. Describe greedy and interchange heuristics for the capacitated facility location problem.

12. **i)** Formulate the problem of choosing k nodes to cover the maximum number of nodes in a graph $G = (V, E)$. (Note that $i \in V$ covers $j \in V$ if $(i, j) \in E$.)

ii) State and interpret a greedy heuristic.

iii) Study the performance of this heuristic when $|V|$ is large.

13. Show that for the family of instances of the p-facility maximization problem with $2p - 1$ clients and $2p$ facilities, and with weights C^p given by $c_j = e_j$ for $j = 1, \ldots, 2p - 1$, $c_{i,2p} = 1$ for $i = 1, \ldots, p$, and $c_{i,2p} = 0$ otherwise, for example,

$$C^4 = \begin{pmatrix} 1 & & & & & & & 1 \\ & 1 & & & & & & 1 \\ & & 1 & & & & & 1 \\ & & & 1 & & & & 1 \\ & & & & 1 & & & 0 \\ & & & & & 1 & & 0 \\ & & & & & & 1 & 0 \end{pmatrix},$$

the interchange heuristic satisfies $z_I = [p/(2p - 1)]z$ when it starts with $S = \{1, 2, 3, 4\}$.

14. Described a simulated annealing algorithm for UFL.

15. For the instance in exercise 1, solve the Lagrangian dual of UFL by using each of the following:

i) the subgradient algorithm;

ii) a constraint generation algorithm.

16. Solve the instance of exercise 1 by Benders' decomposition. Investigate the choice of violated constraints at each iteration.

17. Let

$$
\begin{array}{lll}
z = \min & \sum_{i \in I} \sum_{j \in N} c_{ij} y_{ij} + \sum_{j \in N} f_j x_j & \\
D: & \sum_{j \in N} y_{ij} = 1 & \text{for all } i \in I \\
C: & \sum_{i \in I} d_i y_{ij} \leqslant s_j x_j & \text{for all } j \in N \\
B: & 0 \leqslant y_{ij} \leqslant x_j & \text{for all } i \in I, j \in N \\
S: & \sum_{j \in N} s_j x_j \geqslant \sum_{i \in I} d_i & \\
I: & x_j \in \{0, 1\} & \text{for all } j \in N,
\end{array}
$$

where $c \geqslant 0, f \geqslant 0, d \geqslant 0$, and $s \geqslant 0$ are given. Denote by z^A the bound obtained from this formulation by deleting constraint A, and denote by z_A the bound given by the Lagrangian dual. For example,

$$z_D^S = \max_u z_D^S(u),$$

where

$$
\begin{array}{ll}
z_D^S(u) = \min \sum_{i \in I} \sum_{j \in N} c_{ij} y_{ij} + \sum_{j \in N} f_j x_j + \sum_{i \in I} u_i \left(1 - \sum_{j \in N} y_{ij}\right) & \\
\qquad \sum_{i \in I} d_i y_{ij} \leqslant s_j x_j & \text{for all } j \in N \\
\qquad 0 \leqslant y_{ij} \leqslant x_j & \text{for all } i \in I, j \in N \\
\qquad x_j \in \{0, 1\} & \text{for all } j \in N.
\end{array}
$$

i) Prove that $z^I = z_D^S$.

ii) Prove that $z_D^S \leqslant z_C^S$.

iii) Prove that $z_C^S = z_{SC}$.

iv) Show that $z_D^S < z_C^S$ for the following data:

$$
C = \begin{pmatrix} 1 & 0 & 0 \\ 0 & 1 & 0 \\ 0 & 0 & 1 \end{pmatrix}, \qquad d = \begin{pmatrix} 1 \\ 1 \\ 1 \end{pmatrix},
$$

$$
s = (3 \quad 3 \quad 3), \qquad f = (1 \quad 1 \quad 1).
$$

18. Let

$$
w(x) = \max\left\{ \sum_{j \in N} f_j x_j + \sum_{i \in I} \sum_{j \in N} c_{ij} y_{ij} \colon \sum_{j \in N} y_{ij} = 1 \text{ for } i \in I, \right.
$$

$$
\left. y_{ij} \leqslant x_j \text{ for } i \in I, j \in N, y \in R_+^{mn} \right\}.
$$

i) Show that $w(x)$ is concave.

ii) Show that $\max_{0 \leq x \leq 1} w(x) = z_{LP}$.

iii) Use the subgradient algorithm to solve $\max_{0 \leq x \leq 1} w(x)$ for the instance of exercise 1.

19. Consider the Dantzig–Wolfe formulation of UFL where $\Sigma_j y_{ij} = 1$ for $i \in I$ are taken as the global constraints and where the feasible region of subproblem j is

$$\mathrm{SP}(j)\colon \{(y_{1j}, \ldots, y_{mj}, x_j) \in R_+^{m+1}\colon y_{ij} \leq x_j \text{ for } i = 1, \ldots m, x_j \leq 1\}.$$

Interpret the columns and costs of the resulting master problem.

20. Derive a dynamic programming algorithm to solve

$$\max\left\{cx + hy\colon \sum_{j \in N} y_j \leq b, y_j \leq a_j x_j \text{ for } j \in N, x \in B^n, y \in R_+^n\right\}.$$

21. Derive an $O(T^2)$ dynamic programming algorithm for the uncapacitated lot-sizing problem with backlogging:

$$\min \sum_{t=1}^{T} (p_t y_t + c_t x_t + h_t s_t + g_t r_t)$$

$$s_{t-1} - r_{t-1} + y_t = d_t + s_t - r_t \quad \text{for } t = 1, \ldots, T$$

$$y_t \leq \omega x_t \quad \text{for } t = 1, \ldots, T$$

$$s_0 = r_0 = s_T = r_T = 0$$

$$s, r \in R_+^{T+1}, y \in R_+^T, x \in B^T.$$

See (5.3), where r_t denotes the amount backlogged at the end of period t.

22. Derive an $O(n^2 c_{\max})$ dynamic programming algorithm for the 0, 1 knapsack problem (5.2) where $c_{\max} = \max_j c_j$.

23. **i)** Derive a dynamic programming recursion for the traveling salesman problem using $f(S, j)$ = the length of the minimum-weight partial tour starting at node 1, traversing the nodes $S \subseteq N \setminus \{1, j\}$, and terminating at $j \in N \setminus S$.

ii) Use the recursion to solve the five-city problem with costs

$$C = \begin{pmatrix} - & 2 & 6 & 4 & 7 \\ 1 & - & 3 & 8 & 5 \\ 9 & 2 & - & 4 & 12 \\ 8 & 1 & 9 & - & 2 \\ 3 & 2 & 9 & 4 & - \end{pmatrix}.$$

24. (State Space Relaxation). Let

$$g(k, j) = \min_{S \subseteq N \setminus \{1, j\}} \{f(S, j)\colon |S| = k\},$$

where $f(S, j)$ is defined in exercise 23. What is $g(k, j)$? Write a recursion for $g(k, j)$, and show that $g(n, 1)$ is a lower bound on the weight of an optimal tour.

25. Apply the approach of exercise 24 to the multidimensional knapsack problem with

$$f_k(d) = \max\left\{\sum_{j=1}^{k} c_j x_j: \sum_{j=1}^{k} a_j x_j \leqslant d,\ x \in B^k\right\}$$
$$= \max\{f_{k-1}(d),\ f_{k-1}(d - a_k) + c_k\}.$$

26. Derive a dynamic programming algorithm for

$$\max\left\{\sum_{j\in N} h_j y_j: y_i + y_{i+1} \leqslant u_i \text{ for } i = 1, \ldots, n-1,\ y \in R_+^n\right\}.$$

27. The amount of work to multiply together a $p \times q$ and a $q \times r$ matrix is pqr. Given k matrices M_i of dimension $d_i \times d_{i+1}$ whose product $M_1 M_2 \ldots M_k$ must be formed, use dynamic programming to derive the optimal way in which to form the product.

II.6

Applications of Special-Purpose Algorithms

1. KNAPSACK AND GROUP PROBLEMS

The structure invoked in this section is that the problems have only one constraint other than bounds and integrality on the variables. We consider the integer knapsack problem, the group problem, and the 0-1 knapsack problem.

Many of the algorithms developed in Chapter II.5 can be specialized when there is only one constraint, and some other more specific approaches are also applicable.

The Integer Knapsack Problem

The *integer knapsack problem* is

$$z(b) = \max\left\{\sum_{j\in N} c_j x_j \colon \sum_{j\in N} a_j x_j \leqslant b,\, x \in Z_+^n\right\}, \tag{1.1}$$

where $c_j, a_j \in Z_+^1$ for $j \in N$, $b \in Z_+^1$, $a_j \leqslant b$ for $j \in N$, and there are no explicit upper bounds on the variables. In vector notation, (1.1) is stated as

$$z(b) = \max\{cx\colon ax \leqslant b,\, x \in Z_+^n\}.$$

We suppose throughout this section that $c_1/a_1 \geqslant c_2/a_2 \geqslant \cdots \geqslant c_n/a_n$, so the optimal solution of the linear programming relaxation is $x_1 = b/a_1$, $x_j = 0$ otherwise.

Dynamic Programming

Since $x_j \leqslant \lfloor b/a_j \rfloor \leqslant b$ in any feasible solution to (1.1), we can use the recursion (5.8) of Section II.5.5 to obtain an algorithm with worst-case running time $O(nb^2)$. However, it is possible to do better. The recursion we now describe directly calculates the value function

$$z(d) = \max\left\{\sum_{j\in N} c_j x_j \colon \sum_{j\in N} a_j x_j \leqslant d,\, x \in Z_+^n\right\} \quad \text{for } d \in D(b) = \{0, 1, \ldots, b\}.$$

We begin with $z(d) = 0$ for $d = 0, \ldots, \min_{j\in N} a_j - 1$, with corresponding optimal solution $x^0 = 0$. Given $z(d')$ for all $d' < d$, we claim that

$$z(d) = \max\{c_j + z(d - a_j)\colon j \in N,\, d \geqslant a_j\} \quad \text{for } d \in D(b),\, d \geqslant \min_{j\in N} a_j. \tag{1.2}$$

To prove the validity of (1.2), we first observe that if x^0 is an optimal solution to (1.1) with $b = d - a_j$, then $x^0 + e_j$ is a feasible solution to (1.1) with $b = d$. Hence

$$z(d) \geq c_j + z(d - a_j) \quad \text{for } j \in N, d \geq a_j.$$

On the other hand, if $\hat{x}$ is an optimal solution to (1.1) with $b = d \geq \min_{j \in N} a_j$, then $\hat{x}_k > 0$ for some k with $d \geq a_k$, and $\hat{x} - e_k$ is a feasible solution to (1.1) with $b = d - a_k$. Hence $z(d - a_k) \geq z(d) - c_k$, and (1.2) holds.

The recursion (1.2) requires $O(n)$ calculations for each d, $\min_{j \in N} a_j \leq d \leq b$. Hence the overall running time is $O(nb)$, which is better than the recursion (5.8) of Section II.5.5 by a factor of b.

Example 1.1

$$\begin{aligned} \max\ & 11x_1 + 7x_2 + 5x_3 + x_4 \\ & 6x_1 + 4x_2 + 3x_3 + x_4 \leq 25 \\ & x \in Z_+^4. \end{aligned}$$

$$\begin{aligned}
z(0) &= 0 \\
z(1) &= c_4 = 1 \\
z(2) &= c_4 + z(1) = 2 \\
z(3) &= \max(5 + z(0), 1 + z(2)) = 5 \\
z(4) &= \max(7 + z(0), 5 + z(1), 1 + z(3)) = 7 \\
z(5) &= \max(7 + z(1), 5 + z(2), 1 + z(4)) = 8 \\
z(6) &= \max(11 + z(0), 7 + z(2), 5 + z(3), 1 + z(5)) = 11 \\
z(7) &= \max(11 + 1, 7 + 5, 5 + 7, 1 + 11) = 12 \\
z(8) &= \max(11 + 2, 7 + 7, 5 + 8, 1 + 12) = 14 \\
z(9) &= 11 + z(3) = 16 \\
z(10) &= 11 + z(4) = 18 \\
z(d) &= 11 + z(d - 6) \text{ for } d \geq 11.
\end{aligned}$$

Hence $z(25) = 11 + z(19) = 22 + z(13) = 33 + z(7) = 44 + z(1) = 45$, and an optimal solution is $x^0 = (4 \quad 0 \quad 0 \quad 1)$.

As d increases, the recursion (1.2) has many redundant terms, since $z(d) = c_k + z(d - a_k)$ for any k for which $x_k > 0$ in some optimal solution to (1.1) with $b = d$. Let $p(d) = \min\{j \in N: x_j \text{ is positive in some optimal solution to (1.1) with } b = d\}$. Then, since $p(d - a_j) \geq p(d)$, it follows that

$$(1.3) \qquad z(d) = \max\{c_j + z(d - a_j): j \in N, j \leq p(d - a_j)\}.$$

For d sufficiently large, no comparisons at all are needed. Let $\bar{a} = \max_{j \in N \setminus \{1\}} a_j$.

Proposition 1.1. *If* $p(d - \bar{a}) = p(d - \bar{a} + 1) = \cdots = p(d - 1) = 1$, *then* $z(b) = c_1 + z(b - a_1)$ *for all* $b \geq d$.

Proof. For $j > 1$, we have $p(d - a_j) = 1 < j$. Hence by (1.3), we obtain $z(d) = c_1 + z(d - a_1)$. By induction we obtain the result. ■

In Example 1.1 it is readily checked that $p(d) = 1$ for $9 \leqslant d \leqslant 12$. It then follows from Proposition 1.1 that $p(d) = 1$ for all $d \geqslant 9$. It is important to observe that for any knapsack problem there exists a value of d for which the condition of Proposition 1.1 holds. The proof of the following proposition is a consequence of Proposition 5.6 of Section II.3.5.

Proposition 1.2. *In problem (1.1), we have $p(d) = 1$ for all $d \geqslant (a_1 - 1)\bar{a}$.*

Note that in Example 1.1, we have $(a_1 - 1)\bar{a} = 5 \times 4 = 20$ so that *a priori* we can conclude that $p(d) = 1$ for all $d \geqslant 20$. However, the computation establishes that $p(d) = 1$ for all $d \geqslant 9$.

It is interesting to observe that the recursions (1.2) and (1.3) are algorithms for the maximum-weight path formulation developed in Section II.3.4. For each $j \in N$ there is an arc of weight c_j from node $d - a_j$ to node d, and $z(d)$ is the weight of a maximum-weight path from node 0 to node d. Thus (1.2) states that the weight of a maximum-weight path from node 0 to node d is the maximum over $j \in N$ of the weight of a variable j arc plus the weight of a maximum-weight path from node 0 to node $d - a_j$.

A Superadditive Dual Algorithm

Here we give an algorithm that solves (1.1) and its superadditive dual

$$\min\{\pi(b): \pi(a_j) \geqslant c_j \text{ for } j \in N, \pi(0) = 0, \pi \geqslant 0, \pi \text{ superadditive}\}. \tag{1.4}$$

The idea of the algorithm is as follows. At each iteration, we have a dual feasible solution that also satisfies the complementarity slackness conditions

$$\pi(d) + \pi(b - d) = \pi(b) \quad \text{for all } d \in [0, b]$$

of Proposition 5.2 of Section II.1.5.

Let $S = \{x \in Z_+^n: \Sigma_{j \in N} a_j x_j \leqslant b\}$, and for any dual feasible π define $H_\pi = \{x \in S: \pi(ax) = cx\}$. Then if there exists $x^1, x^2 \in H_\pi$ such that $x = x^1 + x^2 \in S$ and $\pi(ax^1) + \pi(ax^2) = \pi(b)$, it follows that x is an optimal solution to (1.1) and that π is an optimal solution to the superadditive dual. This result is a consequence of

$$\pi(b) = \pi(ax^1) + \pi(ax^2) = cx^1 + cx^2 = cx.$$

At the ith iteration of the algorithm we have a feasible π^i that also satisfies complementary slackness and an $H^i \subseteq H_{\pi^i}$. The initial solution is given by

$$\pi^0(d) = \left(\frac{c_1}{a_1}\right)d \quad \text{where } \frac{c_1}{a_1} = \max_{j \in N} \frac{c_j}{a_j},$$

which is easily shown to be dual feasible, and $H^0 = \{0, e_1, \ldots, \lfloor b/a_1 \rfloor e_1\}$.

Suppose that (π^i, H^i) does not satisfy the optimality condition given above. Let $D^i = \{ax: x \in H^i\}$. The dual solution π^{i+1} is of the form

$$\pi^{i+1}(d) = \begin{cases} \pi^i(d) & \text{if } d \in D^i \\ \pi^i(d) - \theta & \text{if } b - d \in D^i \\ \alpha(d) & \text{otherwise, where } \alpha(d) \text{ is determined} \\ & \text{by linear interpolation between the points} \\ & \{d, b - d: d \in D^i\} \end{cases}$$

An example is shown in Figure 1.1, where $b = 10$ and $D^i = \{0, 4, 8\}$.
To specify the numerical value of θ, the algorithm works with a candidate set

$$C^i = \{x \in S: x \text{ is a minimal vector not in } H^i \text{ and } ax \neq d \text{ for any } d \in D^i\}.$$

Since the optimality condition is not satisfied, $C^i \neq \emptyset$. The algorithm sets $\theta = \theta^i$, where

$$\pi^{i+1}(ax) - cx - \geqslant 0 \quad \text{for all } x \in C^i$$

and

$$\pi^{i+1}(ax) - cx - = 0 \quad \text{for some } x \in C^i.$$

Let y^{i+1} be any point in C^i such that $\pi^{i+1}(ay^{i+1}) - cy^{i+1} = 0$. The algorithm sets $H^{i+1} = H^i \cup \{y^{i+1}\}$ and then checks the points $x + y^{i+1}$, $x \in H^i$, for optimality with respect to the function π^{i+1}. If optimality is not proved, then $C^{i+1} = C^i \cup (\{x + y^{i+1}: x \in H^{i+1}\} \setminus \{y^{i+1}\})$. Note that we can augment H^i by all points $y \in C^i$ such that $\pi^{i+1}(ay) - cy = 0$.

Although there is the possibility of a degenerate dual change (i.e., $\theta^i = 0$), by definition of C^i we have $ay^{i+1} \notin D^i$. Hence $D^{i+1} = D^i \cup \{ay^{i+1}\} \supset D^i$.

Now we claim that the algorithm stops after no more than $i^* = \lfloor (b + 1)/2 \rfloor + 1$ iterations. Otherwise, on iteration i^* we obtain $|D^{i^*}| > (b + 1)/2$, and hence there must exist values $d^1, d^2 \in D^{i^*}$ with $d^1 = b - d^2$. Now if x^1, x^2 are the associated points of H^{i^*}, the optimality criterion is satisfied for $x = x^1 + x^2$.

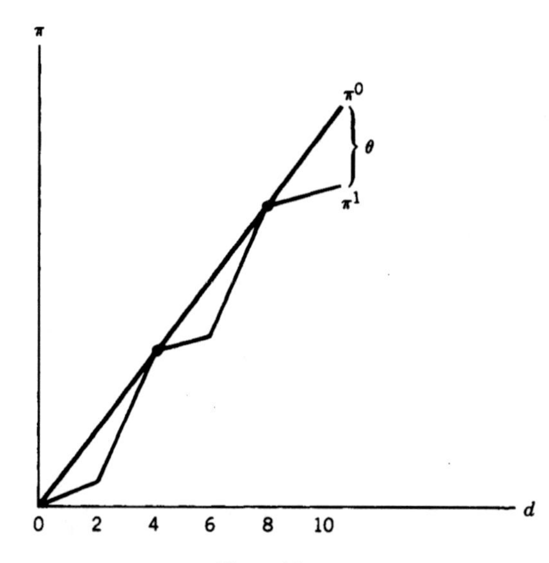

Figure 1.1

Theorem 1.3. *The algorithm terminates with an optimal solution.*

Proof. We only need to show that π^{i+1} is dual feasible and satisfies complementary slackness given that π^i has these properties. We have already said that π^0 has these properties. Also, for all i we have that $\pi^i(a_j) \geq c_j$ for $j \in N$ is satisfied since, (1) for all $j \in N$, either $a_j \in D^i$ or $e_j \in C^i$, and (2) $\pi(a_j) = c_j$ if $a_j \in D^i$, and $\pi(a_j) \geq c_j$ if $e_j \in C^i$.

Now consider superadditivity. We will show that

$$\pi^{i+1}(d_1) + \pi^{i+1}(d_2) \leq \pi^{i+1}(d_1 + d_2) \text{ for all } d_1, d_2, d_1 + d_2 \in [0, b].$$

However, it suffices to consider the subset of $[0, b]$ given by $\{d, b - d: d \in D^i\}$ since the result for all other points in $[0, b]$ can be shown to follow from linear interpolation. There are two cases.

Case 1 ($b - d_1$ or $b - d_2 \in D^i$). Then

$$\begin{aligned}\pi^{i+1}(d_1) + \pi^{i+1}(d_2) &\leq \pi^i(d_1) + \pi^i(d_2) - \theta^i \\ &\leq \pi^i(d_1 + d_2) - \theta^i \\ &\leq \pi^{i+1}(d_1 + d_2).\end{aligned}$$

The first and third inequalities follow from the construction of the dual solution, and the second inequality follows from the superadditivity of π^i.

Case 2 ($d_1, d_2 \in D^i$). Then $\pi^{i+1}(d_j) = \pi^i(d_j)$ for $j = 1, 2$, and there exists $x^1, x^2 \in H^i$ such that $ax^j = d_j$, $cx^j = \pi^i(d_j)$ for $j = 1, 2$. Let $x = x^1 + x^2$ and $d = d_1 + d_2$.

If $d \in D^i$, then

$$\pi^{i+1}(d_1) + \pi^{i+1}(d_2) = \pi^i(d_1) + \pi^i(d_2) \leq \pi^i(d_1 + d_2) = \pi^{i+1}(d_1 + d_2).$$

Also if $x \in C^i$, then

$$\pi^{i+1}(d) = \pi^{i+1}(ax) \geq cx \qquad \text{(by the choice of } \theta^i\text{)}$$

and

$$cx = cx^1 + cx^2 = \pi^i(d_1) + \pi^i(d_2) = \pi^{i+1}(d_1) + \pi^{i+1}(d_2).$$

The final possibility is $d \notin D^i$ and $x \notin C^i$. We have

$$\pi^{i+1}(d) \geq \pi^i(d) - \theta^i \qquad \text{(by the dual change)}$$

and, as above, $cx = \pi^{i+1}(d_1) + \pi^{i+1}(d_2)$. So it remains to show that $cx \leq \pi^i(d) - \theta^i$.

Since $x \in S \setminus (H^i \cup C^i)$, there exists $x' < x$ such that $x' \in C^i$ and $(x - x') \in S$. By the superadditivity of π^i, we have

$$\pi^i(d) = \pi^i(ax) \geq \pi^i(ax') + \pi^i(a(x - x')).$$

Also $cx = cx' + c(x - x')$. Hence

$$\pi^i(d) - cx \geq [\pi^i(ax') - cx'] + [\pi^i(a(x - x')) - c(x - x')].$$

By the choice θ^i, we have $\pi^i(ax') - cx' \geqslant \theta^i$; and by the feasibility of π^i, we obtain $\pi^i(a(x - x')) - c(x - x') \geqslant 0$. Hence $\pi^i(d) - cx \geqslant \theta^i$. ■

Example 1.2

$$\max\ 20x_1 + 9x_2 + 6x_3$$
$$10x_1 + 5x_2 + 4x_3 \leqslant 13$$
$$x \in Z^3_+.$$

Initialization. $\pi^0(d) = 2d$ for $0 \leqslant d \leqslant 13$, $H^0 = \{(0\ \ 0\ \ 0),\ (1\ \ 0\ \ 0)\}$, $C^0 = \{(0\ \ 1\ \ 0), (0\ \ 0\ \ 1)\}$, and $k = 0$. The dual functions $\pi^k(d)$ are shown in Figure 1.2.

Iteration 1. $\pi^1(13) = 26 - \theta$, $\pi^1(3) = 6 - \theta$, $\pi^1(10) = 20$, $\pi^1(5) \geqslant 9$, and $\pi^1(4) \geqslant 6$. Hence by linear interpolation we obtain $\frac{5}{7}(6 - \theta) + \frac{2}{7}(20) \geqslant 9$, so $\theta \leqslant \frac{7}{5}$, and $\frac{6}{7}(6 - \theta) + \frac{1}{7}(20) \geqslant 6$, so $\theta \leqslant \frac{12}{7}$. Hence $\theta = \frac{7}{5}$, $y^1 = (0\ \ 1\ \ 0)$, and

$$\pi^1(d) = \begin{cases} \frac{23}{15}d & \text{for } 0 \leqslant d \leqslant 3 \\ \frac{11}{5}d - 2 & \text{for } 3 \leqslant d \leqslant 10 \\ \frac{14}{3} + \frac{23}{15}d & \text{for } 10 \leqslant d \leqslant 13 \end{cases}$$

$$H^1 = H^0 \cup \{(0\ \ 1\ \ 0)\}, \quad C^1 = \{(0\ \ 0\ \ 1), (0\ \ 2\ \ 0)\}.$$

Iteration 2. $\pi^2(13) = 24\frac{3}{5} - \theta$, $\pi^2(3) = 4\frac{3}{5} - \theta$, $\pi^2(8) = 15\frac{3}{5} - \theta$, $\pi^2(4) \geqslant 6$, $\pi^2(10) \geqslant 18$, $\pi^2(10) = 20$, and $\pi^2(5) = 9$. Hence $\frac{1}{2}(4\frac{3}{5} - \theta) + \frac{1}{2}(9) = 6$, so $\theta = \frac{8}{5}$ and $y^2 = (0\ \ 0\ \ 1)$.
Now

$$\pi^2(d) = \begin{cases} d & \text{for } 0 \leqslant d \leqslant 3 \\ 3d - 6 & \text{for } 3 \leqslant d \leqslant 5 \\ \frac{5}{3}d + \frac{2}{3} & \text{for } 5 \leqslant d \leqslant 8 \\ 3d - 10 & \text{for } 8 \leqslant d \leqslant 10 \\ 10 + d & \text{for } 10 \leqslant d \leqslant 13 \end{cases}$$

$$H^2 = H^1 \cup \{0\ \ 0\ \ 1\}, \quad C^2 = \{(0\ \ 2\ \ 0), (0\ \ 1\ \ 1), (0\ \ 0\ \ 2)\}.$$

Iteration 3. $\pi^3(13) = 23 - \theta$, $\pi^3(3) = 3 - \theta$, $\pi^3(8) = 14 - \theta$, $\pi^3(9) = 17 - \theta$, $\pi^3(10) = 20$, $\pi^3(5) = 9$, $\pi^3(4) = 6$, $\pi^3(10) \geqslant 18$, $\pi^3(9) \geqslant 15$, and $\pi^3(8) \geqslant 12$. Hence $\theta = 14 - 12 = 17 - 15 = 2$, $y^3 = (0\ \ 1\ \ 1)$ or $(0\ \ 0\ \ 2)$, and $\pi(b) = 21$. Since $(0\ \ 1\ \ 0)$ and $(0\ \ 0\ \ 1)$ are in H, and $(0\ \ 1\ \ 0) + (0\ \ 0\ \ 2) = (0\ \ 0\ \ 1) + (0\ \ 1\ \ 1) = (0\ \ 1\ \ 2)$ is feasible with value 21, the algorithm stops with $x = (0\ \ 1\ \ 2)$. The optimal dual function is

$$\pi^3(d) = \begin{cases} \frac{1}{3}d & \text{for } 0 \leqslant d \leqslant 3 \\ -14 + 5d & \text{for } 3 \leqslant d \leqslant 4 \\ -6 + 3d & \text{for } 4 \leqslant d \leqslant 5 \\ 4 + d & \text{for } 5 \leqslant d \leqslant 8 \\ -12 + 3d & \text{for } 8 \leqslant d \leqslant 9 \\ -30 + 5d & \text{for } 9 \leqslant d \leqslant 10 \\ \frac{50}{3} + \frac{1}{3}d & \text{for } 10 \leqslant d \leqslant 13. \end{cases}$$

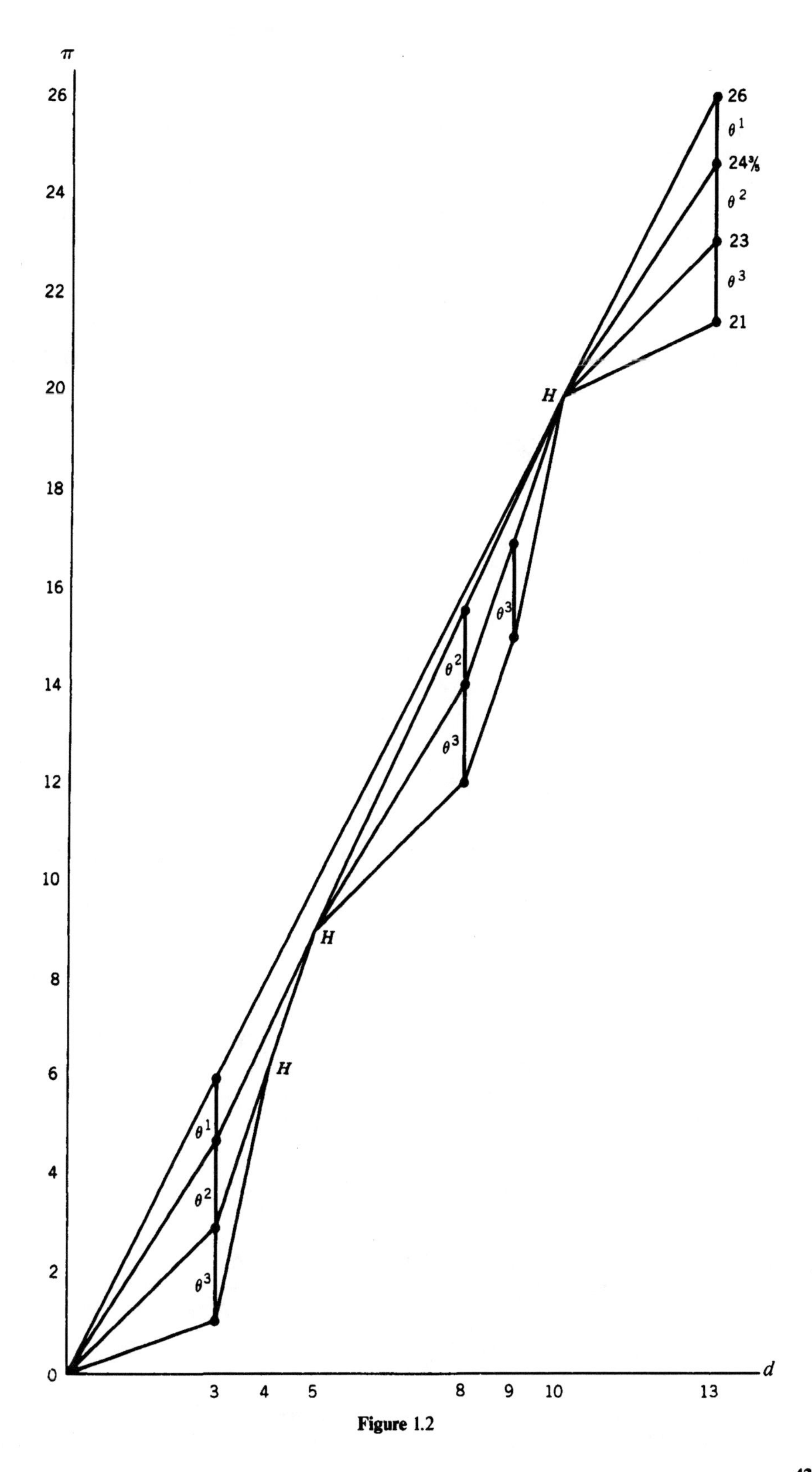

π
26
24
22
20
18
16
14
12
10
8
6
4
2
0
3
4
5
8
9
10
13
d
H
H
H
26
θ1
24⅗
θ2
23
θ3
21
θ2
θ3
θ3
θ1
θ2
θ3

Figure 1.2

Heuristic Algorithms

A very simple *greedy heuristic* for the knapsack problem (1.1) is obtained by considering the variables in order of decreasing c_j/a_j and then making each variable as large as possible. Since we have assumed $c_1/a_1 \geqslant \cdots \geqslant c_n/a_n$, a *greedy solution* is $x_j = \lfloor b^j/a_j \rfloor$ for $j \in N$, where $b^1 = b$ and $b^{j+1} = b^j - a_j \lfloor b^j/a_j \rfloor$ for $j = 1, \ldots, n-1$. Its value is $z_H = \Sigma_{j \in N} c_j \lfloor b^j/a_j \rfloor$. The running time is $O(n \log n)$, since the time required to sort the c_j/a_j's in decreasing order is the most time-consuming step.

Since an optimal solution to the linear programming relaxation of (1.1) is $x_1 = b/a_1$ and $x_j = 0$ otherwise, we have that $z_{\text{LP}} = z_{\text{LP}}(b) = c_1 b/a_1$. Now let $z_R = c_1 \lfloor b/a_1 \rfloor$ be the value of the *rounding heuristic* with $x_1 = \lfloor b/a_1 \rfloor$ and $x_j = 0$ otherwise. We have $z_H \geqslant z_R$ and $z_{\text{LP}} \geqslant z(b) = z_{\text{IP}}$.

We can use the optimal linear programming value to bound the worst-case relative errors of the greedy and rounding heuristics.

Proposition 1.4

a. $z_R > \frac{1}{2} z_{\text{LP}}$,

b. $z_R > z_{\text{LP}} - \max_{j \in N} c_j$.

Proof

a. Let $f = b/a_1 - \lfloor b/a_1 \rfloor < 1$. Now

$$\frac{z_R}{z_{\text{LP}}} = \frac{\lfloor b/a_1 \rfloor}{b/a_1} = 1 - \frac{f}{b/a_1} \geqslant 1 - \frac{f}{1+f} > \frac{1}{2}.$$

Note that the first inequality holds since $1 + f \leqslant b/a_1$ and that the second inequality is true since $f < 1$.

b. $z_{\text{LP}} - z_R = c_1 f \leqslant \max_{j \in N} c_j f < \max_{j \in N} c_j$. ■

Corollary 1.5. *If* $\max_{j \in N} c_j \leqslant \varepsilon z_{\text{LP}}$, *then* $z_R/z_{\text{LP}} > 1 - \varepsilon$.

To prepare for the presentation of a heuristic that always yields a relative error of no more than ε and that runs in time that is a polynomial function of n and ε^{-1}, we introduce a scaling heuristic that uses dynamic programming to solve a formulation of the knapsack problem with the roles of the objective function and constraint reversed. Let

$$(1.5) \qquad w(t) = \min\left\{ \sum_{j \in N} a_j x_j : \sum_{j \in N} c_j x_j \geqslant t, x \in Z_+^n \right\},$$

where t is a positive integer. Note that $w(t)$ is a nondecreasing function of t.

Analogous to (1.2), we have the recursion

$$(1.6) \qquad w(t) = \min_{j \in N} \{a_j + w(t - c_j)\}$$

for $t > 0$, and $w(t) = 0$ for $t \leqslant 0$. The work required to solve (1.6) is $O(nt)$.

Now we show that with a suitable choice of t, an optimal solution to (1.5) yields an optimal solution to (1.1).

Proposition 1.6. *Suppose x^0 is an optimal solution to (1.5) with $t = t^0$. Then x^0 is an optimal solution to the knapsack problem*

$$\max\left\{\sum_{j\in N} c_j x_j: \sum_{j\in N} a_j x_j \leqslant d, x \in Z_+^n\right\}$$

for all d satisfying $w(t^0) \leqslant d < w(t^0 + 1)$.

Proof. Suppose $w(t^0) \leqslant d < w(t^0 + 1)$. Then x^0 is feasible since $\Sigma_{j\in N} a_j x_j^0 = w(t^0) \leqslant d$. Now suppose that x^0 is not optimal; that is, there is a feasible $x^* \neq x^0$ and $\Sigma_{j\in N} c_j x_j^* \geqslant \Sigma_{j\in N} c_j x_j^0 + 1$. Hence $\Sigma_{j\in N} c_j x_j^* \geqslant t^0 + 1$ and $\Sigma_{j\in N} a_j x_j^* \leqslant d$, which contradicts the assumption that $w(t^0 + 1) > d$. ■

As a consequence of Proposition 1.6, we can take t in (1.5) equal to any known upper bound on z_{IP}; for example, $t = \lfloor z_{\text{LP}} \rfloor$. Then for some $t^0 \leqslant t$, we will obtain $w(t^0) \leqslant b < w(t^0 + 1)$. Hence to solve (1.1) using (1.6), the running time is $O(nz_{\text{LP}})$. This does not appear to be an improvement on the dynamic programming recursion (1.2) unless the c_j's are small relative to the a_j's.

The *scaling* heuristic works by replacing the objective function coefficients c_j by $\rho_j = \lfloor c_j/K \rfloor$ for some $K > 0$. The resulting knapsack problem

$$\max\left\{\sum_{j\in N} \rho_j x_j: \sum_{j\in N} a_j x_j \leqslant b, x \in Z_+^n\right\} \tag{1.7}$$

is solved using the recursion (1.6). We denote an optimal solution to (1.7) by $x(K)$ and say that $x(K)$ is a *scaling heuristic* solution. Its value is $z_S = \Sigma_{j\in N} c_j x_j(K)$.

Proposition 1.7. *If $K \leqslant \varepsilon \min_{j\in N} c_j$, then $z_S/z_{\text{IP}} > 1 - \varepsilon$.*

Proof. Since $\rho_j = \lfloor c_j/K \rfloor$, it follows that $\rho_j \leqslant c_j/K < \rho_j + 1$. Hence

$$z_S = \sum_{j\in N} c_j x_j(K) \geqslant K \sum_{j\in N} \rho_j x_j(K) \geqslant K \sum_{j\in N} \rho_j x_j^0,$$

where the last inequality holds because $x(K)$ is an optimal solution to (1.7). Also,

$$z_{\text{IP}} = \sum_{j\in N} c_j x_j^0 < K \sum_{j\in N} (\rho_j + 1) x_j^0.$$

Therefore

$$z_S - z_{\text{IP}} > -K \sum_{k\in N} x_k^0 \geqslant -\varepsilon \min_{k\in N} c_j \sum_{k\in N} x_k^0 \geqslant -\varepsilon \sum_{k\in N} c_k x_k^0 = -\varepsilon z_{\text{IP}}.$$ ■

The running time of the scaling heuristic is $O(nz_{\text{LP}}/K)$. Therefore it is of interest only if K is large—that is, if $\min_{j\in N} c_j$ is much larger than ε^{-1}. This, unfortunately, for any reasonable choice of ε requires large profit coefficients.

Observe that the greedy (or rounding) heuristic needs small profit coefficients to perform effectively and that the scaling heuristic requires large ones to run efficiently. By combining the two heuristics we are able to take advantage of the best features of each of them. The result is a heuristic that guarantees a relative error of no more than ε for any $\varepsilon > 0$ and whose running time is $O(n/\varepsilon^2)$.

We partition the set N into $(N^\theta, N \setminus N^\theta)$, where $N^\theta = \{j \in N: c_j > \theta\}$ and $\theta = (\varepsilon/4)\lfloor z_{LP} \rfloor$. The rounding heuristic is applied to a knapsack problem that contains only the items in $N \setminus N^\theta$, and the scaling heuristic with $K = (\varepsilon/2)\theta$ is applied to a knapsack problem that contains only the items in N^θ. The two solutions are then combined as explained below.

The Scaling/Rounding (SR) Heuristic

$$\theta = \frac{\varepsilon}{4}\lfloor z_{LP} \rfloor, \qquad K = \frac{\varepsilon}{2}\theta, \qquad \rho_j = \left\lfloor \frac{c_j}{K} \right\rfloor \quad \text{for all } j \in N.$$

Step 1: Solve the family of knapsack problems

$$w(t) = \min\left\{ \sum_{j \in N^\theta} a_j x_j : \sum_{j \in N^\theta} \rho_j x_j \geqslant t,\ x \in Z_+^{|N^\theta|} \right\}$$

by the recursion (1.6) for all nonnegative integers t with $w(t) \leqslant b$. Let $x^\theta(t)$ be the solution that yields $w(t)$.

Step 2: Let $c_r/a_r = \max_{j \in N \setminus N^\theta} (c_j/a_j)$. Define $x(t) \in Z_+^n$ by $x_j(t) = x_j^\theta(t)$ for $j \in N^\theta$, $x_r(t) = \lfloor (b - w(t))/a_r \rfloor$, and $x_j(t) = 0$ otherwise.

Step 3: Suppose $\max_t \{\Sigma_{j \in N^\theta} c_j x_j(t) + c_r x_r(t)\}$ is attained with $t = t^*$. Then $x(t^*)$ is the SR heuristic solution of value $z_{SR} = \Sigma_{j \in N} c_j x_j(t^*)$.

The SR heuristic produces a feasible solution to the knapsack problem since all of the variables are nonnegative integers and, by definition of $x_r(t)$, $\Sigma_{j \in N^\theta} a_j x_j(t) + a_r x_r(t) \leqslant b$ for all t.

Proposition 1.8. *The running time of the* SR *heuristic is* $O(n\varepsilon^{-2})$.

Proof. To solve the family of knapsack problems in Step 1 by the recursion (1.6), we need to consider no more than $\lfloor z_{LP}/K \rfloor = 1$ values of t. But

$$\frac{z_{LP}}{K} = \frac{2z_{LP}}{\varepsilon\theta} = 8\varepsilon^{-2}\frac{z_{LP}}{\lfloor z_{LP} \rfloor} \leqslant 16\varepsilon^{-2}.$$

Thus the running time of (1.6) is $O(n\varepsilon^{-2})$. Steps 2 and 3 take $O(\varepsilon^{-2})$ time, so the proof is complete. ■

Theorem 1.9. $z_{SR} \geqslant (1 - \varepsilon)z_{IP}$.

Proof. Suppose x^0 is an optimal solution to (1.1) and $\Sigma_{j \in N^\theta} c_j x_j^0 = t^0$. Since

$$\frac{K}{\min\{c_j: j \in N^\theta\}} \leqslant \frac{K}{\theta} = \frac{\varepsilon}{2},$$

Proposition 1.7, restricted to the variables in N^θ, yields

$$\sum_{j \in N^\theta} c_j x_j(t^0) \geqslant \left(1 - \frac{\varepsilon}{2}\right) \sum_{j \in N^\theta} c_j x_j^0.$$

Hence

$$\sum_{j\in N^{\theta}} c_j x_j(t^0) \geq \sum_{j\in N^{\theta}} c_j x_j^0 - \frac{\varepsilon}{2}\sum_{j\in N} c_j x_j^0 \tag{1.8}$$

since $N^{\theta} \subseteq N$.

Now by b of Proposition 1.4 and by $\theta \geq \max_{j\in N\setminus N^{\theta}} c_j$, the rounding heuristic yields

$$\sum_{j\in N\setminus N^{\theta}} c_j x_j(t^0) \geq \max\left\{\sum_{j\in N\setminus N^{\theta}} c_j x_j \colon \sum_{j\in N\setminus N^{\theta}} a_j x_j \leq b - w(t^0),\ x \in Z_+^{|N\setminus N^{\theta}|}\right\} - \theta$$

$$\geq \sum_{j\in N\setminus N^{\theta}} c_j x_j^0 - \frac{\varepsilon}{4}\lfloor z_{\mathrm{LP}}\rfloor$$

$$\left(\text{since } \sum_{j\in N\setminus N^{\theta}} a_j x_j^0 \leq b - w(t^0) \text{ and } \theta = \frac{\varepsilon}{4}\lfloor z_{\mathrm{LP}}\rfloor\right) \tag{1.9}$$

$$\geq \sum_{j\in N\setminus N^{\theta}} c_j x_j^0 - \frac{\varepsilon}{2}\sum_{j\in N} c_j x_j^0$$

$$\left(\text{since } \sum_{j\in N} c_j x_j^0 \geq \frac{1}{2}\lfloor z_{\mathrm{LP}}(b)\rfloor \text{ by a of Proposition 1.4}\right).$$

Adding the inequalities (1.8) and (1.9) yields

$$\sum_{j\in N} c_j x_j(t^0) \geq \sum_{j\in N} c_j x_j^0 - \varepsilon \sum_{j\in N} c_j x_j^0.$$

The proof is completed by observing that $z_{\mathrm{SR}} \geq \sum_{j\in N} c_j x_j(t^0)$. ■

There are some refinements of the SR heuristic that yield improvements on the relative error bound.

Example 1.3. We apply the SR heuristic to

$$\begin{aligned} z_{\mathrm{IP}} = \max\ & 592x_1 + 381x_2 + 273x_3 + 55x_4 + 48x_5 + 37x_6 + 23x_7 \\ & 3534x_1 + 2356x_2 + 1767x_3 + 589x_4 + 528x_5 + 451x_6 + 304x_7 \leq 119{,}567 \\ & x \in Z_+^7. \end{aligned}$$

Suppose we are given $\varepsilon = 0.2$. Hence $\theta = 0.05\lfloor z_{\mathrm{LP}}\rfloor = 1001.45$ and $K = 100.15$. Now observe that $c_j \leq \theta$ for all $j \in N$. Hence $N^{\theta} = \emptyset$, and the SR heuristic is trivial to execute. It yields $x_1 = \lfloor 119{,}567/3534\rfloor = 33$, $x_j = 0$ otherwise, and $z_{\mathrm{SR}} = 19{,}536$. It is not hard to show that an optimal solution is $x_1^0 = 33$, $x_2^0 = x_4^0 = 1$, and $x_j = 0$ otherwise and that $z_{\mathrm{IP}} = 19{,}972$. Hence the actual relative error is $436/19{,}972 = 0.0218$.

Note that if we replace the rounding heuristic by the greedy heuristic we obtain the optimal solution.

If $\varepsilon = 0.02$, then $\theta = 100.15$ and $K = 1.00$. Hence $N^{\theta} = \{1\quad 2\quad 3\}$. Now in Step 1, with $t = 19{,}917$ we obtain $x^{\theta}(t) = (33\quad 1\quad 0)$ and $w(t) = 118{,}978$. Since $b - w(t) = 589 = a_4$, this solution is completed to $x = (33\quad 1\quad 0\quad 1\quad 0\quad 0\quad 0)$, which we have already indicated is optimal.

The Group Problem

In Section II.3.5 we have shown how the problem

$$\text{(IP)} \qquad z_{\text{IP}} = \max\{cx\colon Ax = b,\ x \in Z_+^n\}$$

can be relaxed by choosing $u \in R^m$, an appropriate $m \times m$ unimodular matrix R, and a nonsingular diagonal matrix Δ with positive integer entries δ_i for $i = 1, \ldots, m$ to give the problem

$$z(u, \Delta) = ub + \max\{(c - uA)x\colon x \in S_\Delta(b)\},$$

where $S_\Delta(b) = \{x \in Z_+^n\colon RAx = Rb + \Delta w \text{ for some } w \in Z^m\}$.

In this section we consider what to do when the group problem for the given choices of u, R, and Δ does not yield a feasible solution to IP. We assume here that $z_{\text{IP}} < \infty$, and we have chosen a u such that $p = uA - c \geqslant 0$. In this case the group problem is a minimum-weight path problem on a digraph $\mathcal{D}_\Delta = (V_\Delta, \mathcal{A}_\Delta)$ having $|\det \Delta| = \Pi_{i=1}^m \delta_i$ nodes. We state the minimum-weight path problem as

$$\psi(d) = \min \sum_{j \in N} p_j x_j$$

$$\text{SP}(d) \qquad \sum_{j \in N} (Ra_j)x_j = d \qquad (\text{mod } \Delta)$$

$$x \in Z_+^n,$$

where $z(u, \Delta = ub - \psi(Rb)$. Here $V_\Delta = \{d \in Z_+^m\colon d_i < \delta_i \text{ for } i = 1, \ldots, m\}$, $\mathcal{A}_\Delta = \{(d, d + Ra_j \ (\text{mod } \Delta))\colon d \in V_\Delta \text{ for } j = 1, \ldots, n\}$, the weight of the arc $(d, d + Ra_j \ (\text{mod } \Delta))$ is p_j, and we seek a minimum-weight path from node 0 to node Rb (mod Δ).

The reader should recall that the relaxation simply replaces the ith equation $a^i x = b_i$ by the modular equation $a^i x = b_i$ (mod δ_i) and that u has been chosen so that the objective function of the relaxation is bounded from above. In particular, any feasible solution to IP corresponds to some path from node 0 to node Rb (mod Δ).

The connection between knapsack and group problems is motivated by taking $u = c_B A_B^{-1}$ and Δ to be the Smith normal form of A_B, where A_B is an optimal basis for the linear programming relaxation of IP [see (5.3) of Section II.3.5]. In this case, $p_j = 0$ and $Ra_j = 0$ (mod Δ) if x_j is a basic variable. Thus SP(d) only involves the nonbasic variables. Moreover, $\Delta = |\det A_B|$, and it is frequently the case that $\delta_1 = \cdots = \delta_{m-1} = 1$ and $\delta_m = |\det A_B|$. (This must be the case if $|\det A_B|$ is a prime number.) Since the ith equation of SP(d) is trivially satisfied when $\delta_i = 1$, this choice of u and Δ frequently leads to a single-constraint problem in nonnegative integer variables, which is an integer knapsack problem with ordinary arithmetic replaced by modular arithmetic.

When $|\det \Delta|$ is not too large, the minimum-weight path problem SP(d) is easily solved by Dijkstra's algorithm (see Section I.3.2). Here we consider how algorithms can be constructed for IP that make use of this shortest-path viewpoint.

The following proposition motivates the construction of an implicit path enumeration algorithm.

Proposition 1.10. *If* IP *is feasible and* $p_j > 0$ *for all* $j \in N$, *then there exists a positive integer* k *such that an optimal solution to* IP *corresponds to a* k*th best minimum-weight path in* $\mathcal{D}_\Delta$.

To enumerate we need to specify how to branch. One way is to subdivide the set of all solutions (paths) into sets in which each variable x_j (arc type) occurs at least y_j times. The following proposition tells us how to calculate an optimal solution at a node of the tree. Let $x(d)$ be an optimal solution to SP(d) such that the corresponding path in $\mathscr{D}_\Delta$ is acyclic.

Proposition 1.11. *Given* $y \in Z_+^n$, *an optimal solution to* SP(Rb) *satisfying* $x \geq y$ *is* $x^* = y + x\,(R(b - Ay))$, *with weight* $py + \psi(R(b - Ay))$.

Proof. Setting $x = y + x'$, $x' \in Z_+^n$, and substituting in SP(Rb) gives

$$\sum_{j \in N} p_j y_j + \min \sum_{j \in N} p_j x_j'$$

$$\sum_{j \in N} (Ra_j) x_j' = Rb - RAy \qquad (\text{mod } \Delta)$$

$$x' \in Z_+^n.$$

Hence to find an optimal solution x^* to SP(Rb) that satisfies $x \geq y$, we find a minimum-weight path from node 0 to node $R(b - Ay)$ mod Δ. ■

The next proposition tells us how to define the new nodes when we branch; and it uses the fact that if x^* is defined as in Proposition 1.11, then there is no vector $\hat{x}$ satisfying $y \leq \hat{x} \lneqq x^*$ that is feasible in SP(Rb).

Proposition 1.12. *If* $x^* = y + x(R(b - Ay))$, *then any* $x \neq x^*$ *that corresponds to a path from* 0 *to* Rb (*mod* Δ) *subject to the restriction* $x \geq y$ *must satisfy*

$$x_j \geq x_j^* + 1, \quad x_k \geq y_k \qquad \text{for } k \neq j$$

for some j.

Now we describe a straightforward path enumeration algorithm. We start from the group problem (5.3) of Section II.3.5 mentioned above. Hence we let A_B denote the basis of an optimal solution to the linear programming relaxation and let A_N denote the columns of the nonbasic variables. Now SP(Rb) only involves the nonbasic variables, which we suppose are numbered $1, 2, \ldots, n - m$, and the enumeration is carried out only over these variables. Also at this point, x, y, and p are dimensioned appropriately.

A Shortest-Path Enumeration Algorithm for IP

We begin by solving the shortest-path problem from node 0 to node d for all $d \in V_\Delta$ and let $x(d)$ be an acyclic optimal solution to SP(d) of cost $\psi(d)$. If there is no path from 0 to d, let $\psi(d) = \infty$. We then construct a branch-and-bound tree where the node labeled y corresponds to the feasible set $x \geq y$.

Initialization: $x(d)$ and $\psi(d)$ are given for all $d \in V_\Delta$. If $\psi(Rb) = \infty$, stop; IP is infeasible. Otherwise put $0 \in Z_+^{n-m}$ with lower bound $\underline{z}(0) = \psi(Rb)$ on the node list. $\bar{z} = +\infty$.

Iteration t

Step 1: If the node list is empty, stop. Then (a) if $\bar{z} = +\infty$, IP is infeasible and (b) if $\bar{z}$ is finite, $\hat{x}$ is an optimal solution. Otherwise choose a vector y on the node list and remove it from the list. Go to Step 2.

Step 2 (Feasibility check): If $\psi(R(b - Ay)) = \infty$, return to Step 1. Otherwise let $x^* = y + x(R(b - Ay))$. If $A_B^{-1}(b - A_N x^*) \geqslant 0$, go to Step 3. Otherwise go to Step 4.

Step 3 (Pruning by optimality): Set $\hat{x} \leftarrow x^*$, $\overline{z} \leftarrow py + \psi(R(b - Ay))$. Delete from the node list any node $\hat{y}$ with $\underline{z}(\hat{y}) \geqslant \overline{z}$. Return to Step 1.

Step 4 (Branching): For $i = 1, \ldots, n - m$, define y^i by

$$y_j^i = \begin{cases} y_j & \text{for } j \neq i \\ x_i^* + 1 & \text{for } j = i \end{cases}$$

and let $\underline{z}(y^i) = py^i + \psi(R(b - Ay^i))$. If $\underline{z}(y^i) < \overline{z}$, add $(y^i, \underline{z}(y^i))$ to the node list. Return to Step 1.

When the algorithm terminates, it solves IP. One way to guarantee finiteness is to impose upper bounds on the variables and to modify Step 4 accordingly.

Example 1.4 (Example 5.2 of Section II.3.5 continued)

$$\begin{aligned} \max\ 7x_1 &+ 2x_2 \\ -x_1 &+ 2x_2 + x_3 &&= 2 \\ 5x_1 &+ x_2 \quad + x_4 &&= 19 \\ -2x_1 &- 2x_2 \quad + x_5 &&= -5 \\ & x \in Z_+^5 \end{aligned}$$

leads to the relaxation SP(7):

$$\begin{aligned} \psi(7) = \min\ & \tfrac{3}{11}x_3 + \tfrac{16}{11}x_4 \\ & 5x_3 + x_4 = 7 \quad (\text{mod } 11) \\ & (x_3, x_4) \in Z_+^2 \end{aligned}$$

with $u = c_B A_B^{-1} = (\frac{3}{11}\ \ \frac{16}{11}\ \ 0)$, and $z(u, \Delta) = \frac{310}{11} - \psi(7)$.

The shortest paths and corresponding solution values found by Dijkstra's shortest-path algorithm are given in Table 1.1.

To test the feasibility of proposed paths involving the values of the nonbasic variables (x_3, x_4), we use $x_B = A_B^{-1}b - A_B^{-1}A_N x_N$ given by

$$\begin{pmatrix} x_1 \\ x_2 \\ x_5 \end{pmatrix} = \frac{1}{11}\left[\begin{pmatrix} 36 \\ 29 \\ 75 \end{pmatrix} - \begin{pmatrix} -1 \\ 5 \\ 8 \end{pmatrix} x_3 - \begin{pmatrix} 2 \\ 1 \\ 6 \end{pmatrix} x_4\right].$$

Initialization: $y = 0$, $\underline{z}(0) = \frac{24}{11}$, $\overline{z} = +\infty$.

Table 1.1.

d:	0	1	2	3	4	5	6	7	8	9	10
$x(d) = (x_3\ x_4)$:	(0 0)	(0 1)	(7 0)	(5 0)	(3 0)	(1 0)	(1 1)	(8 0)	(6 0)	(4 0)	(2 0)
$11\psi(d)$:	0	16	21	15	9	3	19	24	18	12	6

Iteration 1:

1. Pick $y = (0 \quad 0)$ from the node list.
2. $x^* = (x_3 \quad x_4) = (8 \quad 0)$.
   $x_B = (x_1 \quad x_2 \quad x_5) = (4 \quad -1 \quad 1) \ngeq 0$.
4. Add the nodes $(9 \quad 0)$ and $(0 \quad 1)$ to the list, with lower bounds $\underline{z}(y) = \frac{46}{11}, \frac{35}{11}$, respectively.

Iteration 2:

1. Pick $y = (9 \quad 0)$ from the node list. [Note that $(0 \quad 1)$ would be chosen if we followed the rule of smallest lower bound.]
2. $x^* = (9 \quad 0) + x(6) = (10 \quad 1)$. $x_B = (x_1 \quad x_2 \quad x_5) = (4 \quad -2 \quad -1) \ngeq 0$.
4. Add the nodes $(11 \quad 0)$, $(9 \quad 2)$ to the list with bounds $\underline{z}(y) = \frac{57}{11}, \frac{68}{11}$, respectively.

Iteration 3:

1. Pick $y = (0 \quad 1)$ from the list.
2. $x^* = (0 \quad 1) + (1 \quad 1) = (1 \quad 2)$. $x_B = (3 \quad 2 \quad 5)$.
3. $\bar{z} \leftarrow \frac{35}{11}$, $\hat{x} = (1 \quad 2)$. Delete the nodes $(11 \quad 0)$ and $(9 \quad 2)$ from the list since their lower bounds exceed $\bar{z}$.

Iteration 4:

1. The node list is empty. Stop. $(x_1 \quad x_2 \quad x_5) = x_B = (3 \quad 2 \quad 5)$, $x_N = (x_3 \quad x_4) = (1 \quad 2)$ is an optimal solution.

The enumeration tree is shown in Figure 1.3.

The relaxation SP(d) can also be used as a basis for several other algorithms. One obvious approach is to consider the dual problem $\min_{u,\Delta} z(u, \Delta)$. For fixed R and Δ, we can consider the standard Lagrangian dual $\min_{u \in R^m} z(u, \Delta)$. Based on Theorem 6.2 of Section II.3.6, we have the following result on the Lagrangian dual.

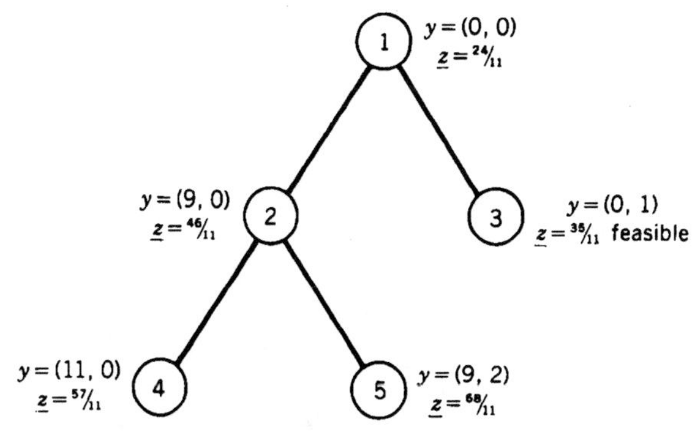

Figure 1.3

Proposition 1.13. $\min_{u \in R^m} z(u, \Delta) = \max\{cx: Ax = b, x \in \text{conv}(S_\Delta(b))\}$.

A different dual is obtained by fixing u and allowing Δ to vary over the set of $m \times m$ integer nonsingular diagonal matrices, giving the dual problem $\min_\Delta z(u, \Delta)$.

The algorithm described below for IP involves the solution of minimum-weight path problems over a series of digraph $\mathcal{D}_\Delta$ that increase in size from one iteration to the next but remain finite. The algorithm solves minimum-weight path problems of the form

$$\psi_\Delta = \min \sum_{j \in N} p_j x_j$$

$$\text{ID}(\Delta) \qquad \sum_{j \in N} (Ra_j)x_j = (Rb) \qquad (\text{mod } \Delta)$$

$$x \in Z_+^n,$$

where R and $p = uA - c > 0$ are fixed, Δ varies, and $z(u, \Delta) = ub - \psi_\Delta$. Also we no longer require $\delta_i \mid \delta_{i+1}$ for $i = 1, \ldots, m-1$.

The Increasing Group Algorithm

Initialization: Choose

$$\Delta^1 = \begin{pmatrix} \delta_1^1 & & \\ & \ddots & \\ & & \delta_m^1 \end{pmatrix}$$

with $\delta_i^1 \in Z_+^1 \setminus \{0\}$ for all i. Set $t = 1$. (A reasonable choice is Δ^1 equal to the Smith normal form of A_B, but this is not necessary.)

Iteration t:

Step 1: Solve the minimum-weight path problem $\text{ID}(\Delta^t)$. Let x^t be the resulting solution.

Step 2: If $RAx^t = Rb$, stop. x^t is an optimal solution of IP.

Step 3: If $RAx^t \neq Rb$, calculate

$$w^t = (\Delta^t)^{-1}(Rb - RAx^t) \in Z^m.$$

Step 4: Set $k_i = 1$ if $w_i^t = 0$, and otherwise let k_i be the smallest integer greater than 1 such that $\gcd\{|w_i^t|, k_i\} = 1$. Set

$$\Delta^{t+1} = \Delta^t \begin{pmatrix} k_1 & & \\ & \ddots & \\ & & k_m \end{pmatrix}.$$

and $t \leftarrow t + 1$.

Theorem 1.14. *If* IP *has a finite optimal value and* $p > 0$, *the increasing group algorithm terminates after a finite number of iterations with an optimal solution to* IP.

Proof. Let $S_{\Delta^t}(b) = \{x \in R_+^n: RAx = Rb \ (\text{mod } \Delta^t)\}$. As in the generic relaxation algorithm of Section II.4.1, we show that if x^t is not feasible for IP, then $S_{\Delta^{t+1}}(b) \subseteq S_{\Delta^t}(b) \setminus \{x^t\}$.

Since

$$\Delta^{t+1} = \Delta^t \begin{pmatrix} k_1 & & \\ & \ddots & \\ & & k_m \end{pmatrix}$$

with $k_i \in Z^1_+ \setminus \{0\}$ for all i, $RAx = Rb \pmod{\Delta^{t+1}}$ implies $RAx = Rb \pmod{\Delta^t}$, and hence $S_{\Delta^{t+1}}(b) \subseteq S_{\Delta^t}(b)$. Now $x^t \in S_{\Delta^{t+1}}(b)$ only if $(\Delta^{t+1})^{-1}(Rb - RAx^t) \in Z^m$. But

$$(\Delta^{t+1})^{-1}(Rb - Rax^t) = \begin{pmatrix} k_1 & & \\ & \ddots & \\ & & k_m \end{pmatrix}^{-1} (\Delta^t)^{-1}(Rb - RAx^t) = \left(\frac{w_1^t}{k_1}, \ldots, \frac{w_m^t}{k_m}\right).$$

Since $w_i^t \neq 0$ for some i, and k_i is chosen such that $w_i^t/k_i \notin Z^1$, $x^t \notin S_{\Delta^{t+1}}(b)$.

Finally as $p > 0$, we know from Proposition 1.10 that the optimal solution to IP is a qth best solution to $S_{\Delta^1}(b)$. Hence the algorithm must terminate after no more than q iterations. ■

Corollary 1.15. *Given $p > 0$, there exists a diagonal integer matrix Δ such that an optimal solution to* ID(Δ) *is an optimal solution to* IP.

Example 1.4 (continued)

$$\begin{aligned}
\max\ 7x_1 + \ 2x_2 & \\
-x_1 + 2x_2 + x_3 \qquad\qquad & = 2 \\
5x_1 + x_2 \qquad + x_4 \qquad & = 19 \\
-2x_1 - 2x_2 \qquad\qquad + x_5 & = -5 \\
x \in Z^5_+. &
\end{aligned}$$

Taking $u = (1 \quad 2 \quad \frac{1}{2})$, A_B, and $R = \begin{pmatrix} -1 & 0 & 0 \\ -2 & 0 & 1 \\ 5 & 1 & 0 \end{pmatrix}$ as previously, we obtain ID(Δ), namely

$$\begin{aligned}
\psi_\Delta = \min\ x_1 + x_2 + x_3 + 2x_4 + \tfrac{1}{2}x_5 & \\
x_1 - 2x_2 - x_3 \qquad\qquad & = -2 \pmod{\delta_1} \\
- 6x_2 - 2x_3 \qquad + x_5 & = -9 \pmod{\delta_2} \\
11x_2 + 5x_3 + x_4 \qquad & = 29 \pmod{\delta_3} \\
x \in Z^5_+, &
\end{aligned}$$

and $z(u, \Delta) = 37\frac{1}{2} - \psi_\Delta$.

Now we apply the increasing group algorithm.

Initialization: $\Delta^1 = \begin{pmatrix} 1 & & \\ & 1 & \\ & & 11 \end{pmatrix}$. $t = 1$.

Iteration 1:

Step 1: $x^1 = (0 \quad 0 \quad 1 \quad 2 \quad 0)$, $\psi_{\Delta^1} = px^1 = 5$.

Step 3: $Rb - RAx^1 = \begin{pmatrix} -1 \\ -7 \\ 22 \end{pmatrix} = \Delta^1 w^1$ with $w^1 = \begin{pmatrix} -1 \\ -7 \\ 2 \end{pmatrix}$.

Step 4: $k_1 = k_2 = 2, k_3 = 3, \Delta^2 = \begin{pmatrix} 2 & & \\ & 2 & \\ & & 33 \end{pmatrix}$.

Iteration 2:

Step 1: $x^2 = (1 \quad 2 \quad 1 \quad 2 \quad 1), \psi_{\Delta^2} = px^2 = 8\frac{1}{2}$.

Step 3: $Rb - RAx^2 = \begin{pmatrix} 2 \\ 4 \\ 0 \end{pmatrix} = \Delta^2 w^2$ with $w^2 = \begin{pmatrix} 1 \\ 2 \\ 0 \end{pmatrix}$.

Step 4: $k_1 = 2, k_2 = 3, k_3 = 1, \Delta^3 = \begin{pmatrix} 4 & & \\ & 6 & \\ & & 33 \end{pmatrix}$.

Iteration 3:

Step 1: $x^3 = (3 \quad 2 \quad 1 \quad 2 \quad 5), \psi_{\Delta^3} = px^3 = 12\frac{1}{2}$.

Step 2: $RAx^3 = Rb$. x^3 solves IP with $cx^3 = 37\frac{1}{2} - 12\frac{1}{2} = 25$.

The 0-1 Knapsack Problem

In many cases, 0-1 knapsack problems have to be solved repeatedly and quickly. For instance, in Example 6.2 of Section II.3.6, one of the Lagrangian relaxations resulted in a knapsack problem. In the next section, we will use the 0-1 knapsack problem as a subroutine in a fractional cutting-plane algorithm for 0-1 integer programs.

When the constraint coefficients are small integers, the dynamic programming recursion of Section II.5.5 is an efficient algorithm; and when the objective function coefficients are small integers, an efficient recursion is obtained by reversing the roles of the objective and constraint as in (1.5). In addition, there is a scaling/rounding heuristic similar to the one we have given for the integer knapsack problem with running time $O(n/\varepsilon^3)$ that guarantees a solution with a relative error of no more than ε for any $\varepsilon > 0$.

Nevertheless, a linear-programming-based branch-and-bound algorithm is still used to solve 0-1 knapsack problems. Here we examine the simple techniques that make such an algorithm effective.

Given the 0-1 knapsack problem

$$z_{\text{IP}} = \max\left\{ \sum_{j \in N} c_j x_j : \sum_{j \in N} a_j x_j \leq b, x \in B^n \right\}, \tag{1.10}$$

without loss of generality we suppose that $a_j, c_j > 0$ for all j and $\Sigma_{j \in N} a_j > b$. We note that if the variables are ordered so that $c_1/a_1 \geq \cdots \geq c_n/a_n$, an optimal solution of the linear programming relaxation is

$$x_j = 1 \quad \text{for } j = 1, \ldots, r-1, \qquad x_r = \frac{b - \Sigma_{j=1}^{r-1} a_j}{a_r}, \quad x_j = 0 \quad \text{for } j = r+1, \ldots, n,$$

where r is such that $\Sigma_{j=1}^{r-1} a_j \leq b$ and $\Sigma_{j=1}^{r} a_j > b$. Hence the solution is essentially characterized by r or, more definitely, by $\lambda^* = c_r/a_r$.

The optimal value function $z_{\text{LP}}(b)$ of the linear programming relaxation is shown in Figure 1.4. Note that λ^* is the slope of the function z_{LP} at the point b.

Since sorting the $\{c_j/a_j\}_{j \in N}$ into nondecreasing order can be done in $O(n \log n)$ time, there is an obvious $O(n \log n)$ algorithm for solving the linear programming relaxation.

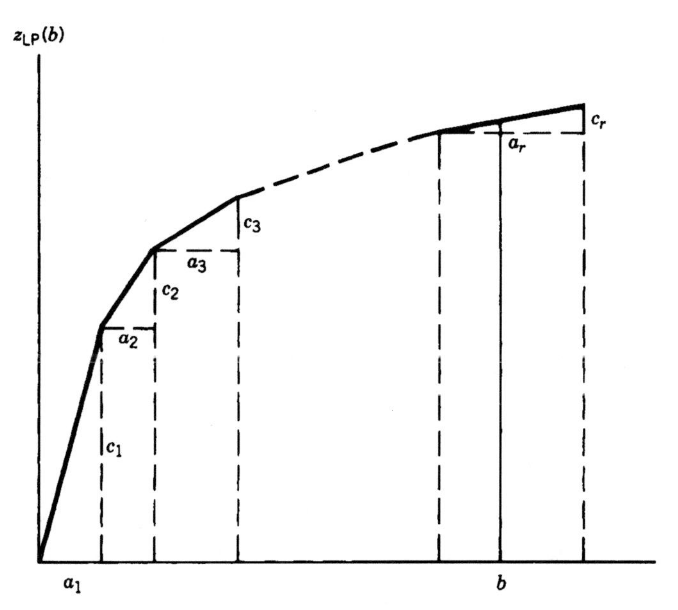

Figure 1.4

However, if λ^* is known, the linear programming relaxation can be solved in linear time since $x_j = 1$ if $c_j/a_j > \lambda^*$ and $x_j = 0$ if $c_j/a_j < \lambda^*$. We now give an algorithm that solves the linear programming relaxation in $O(n)$ time.

An Algorithm for the Linear Programming Relaxation

Let N^1 and N^0 denote the variables fixed to 1 and 0, respectively, and let N^f be the free variables. Given a candidate value λ, let

$$N^> = \left\{j \in N^f: \frac{c_j}{a_j} > \lambda\right\}, N^= = \left\{j \in N^f: \frac{c_j}{a_j} = \lambda\right\} \text{ and } N^< = \left\{j \in N^f: \frac{c_j}{a_j} < \lambda\right\}.$$

Also, let $S_1(\lambda) = \Sigma_{j \in N^>} a_j$ and let $S_2(\lambda) = \Sigma_{j \in N^> \cup N^=} a_j$.

Initialization: $N^1 = N^0 = \emptyset, N^f = N$.
Step 1: Let λ be the median of $\{c_j/a_j: j \in N^f\}$.
Step 2: Construct the sets $N^>$, $N^=$, $N^<$ and calculate $S_1(\lambda)$ and $S_2(\lambda)$.

i. $S_1(\lambda) > b$ implies that λ is too small. Let $N^0 \leftarrow N^0 \cup N^= \cup N^<$, $N^f = N^>$. Return to Step 1.
ii. $S_2(\lambda) < b$ implies that λ is too big. Let $N^1 \leftarrow N^1 \cup N^> \cup N^=$, $b \leftarrow b - \Sigma_{j \in N^> \cup N^=} a_j$, $N^f = N^<$. Return to Step 1.
iii. Otherwise, $S_1(\lambda) \leq b \leq S_2(\lambda)$. If $S_1(\lambda)$ or $S_2(\lambda) = b$, we immediately obtain an optimal integer solution. Otherwise, take the elements of $N^=$ in arbitrary order. If $N^= = \{j_1, \ldots, j_p\}$, find q such that $S_1(\lambda) = \Sigma_{i=1}^{q-1} a_{j_i} \leq b$ and $S_1(\lambda) = \Sigma_{i=1}^{q} a_{j_i} > b$. Set $N^1 \leftarrow N^1 \cup \{j_1, \ldots, j_{q-1}\}$, $r = j_q$, and $N^0 \leftarrow N^0 \cup \{j_{q+1}, \ldots, j_p\}$. Stop.

The algorithm terminates with an optimal solution to the linear program with $x_j = 1$ for $j \in N^1$, $x_j = 0$ for $j \in N^0$, and $x_r = (b - \Sigma_{j \in N^1} a_j)/a_r$. To verify that the algorithm has

$O(n)$ running time, we use the result that the median of k numbers can be found in $O(k)$ time. Because λ is chosen as the median, we have that $|N^f|$ is at least halved at each iteration since $|N^<| \leqslant \frac{1}{2}|N|$ and $|N^>| \leqslant \frac{1}{2}|N|$. Hence the total running time is $O(n) + O(n/2) + \cdots + O(n/2^t) \cdots = O(n)$.

Once N^1, N^0, and λ^* are determined, a natural greedy heuristic yields a solution to (1.10).

Primal Heuristic Algorithm

Step 1: Set $x_j^* = 1$ for all $j \in N^1$, and $x_r^* = 0$. Let $N^0 = \{r+1, \ldots, n\}$.
Step 2: Set $b \leftarrow b - \Sigma_{j \in N^1} a_j$.
Step 3: For $j \in N^0$, if $a_j > b$, set $x_j^* = 0$; otherwise, set $x_j^* = 1$ and $b \leftarrow b - a_j$. Return.

An obvious improvement of the heuristic is to order the elements of N^0 so that $c_{r+1}/a_{r+1} \geqslant c_{r+2}/a_{r+2} \geqslant \cdots \geqslant c_n/a_n$.

Given a lower bound $\underline{z}$ equal to the value of the best feasible solution found so far to (1.10) and z_{LP}, we now present two tests that may allow us to fix some variables. The first is just a restatement of Proposition 2.1 of Section II.5.2.

Variable Elimination Test 1. If $k \in N^1$ and $z_{\text{LP}} - (c_k - \lambda^* a_k) \leqslant \underline{z}$, then $x_k = 1$. Similarly if $k \in N^0$ and $z_{\text{LP}} + (c_k - \lambda^* a_k) \leqslant \underline{z}$, then $x_k = 0$.

Note that $c_k - \lambda^* a_k$ is just the reduced price of nonbasic variable x_k at either its upper or lower bound.

If $k \in N^1$ and we impose the condition $x_k = 0$, the new linear programming relaxation is

$$z_{\text{LP}}^k = \sum_{j \in N^1 \setminus \{k\}} c_j + \max \sum_{j \in N^0 \cup \{r\}} c_j x_j$$

$$\sum_{j \in N^0 \cup \{r\}} a_j x_j \leqslant b_k$$

$$0 \leqslant x_j \leqslant 1 \quad \text{for } j \in N^0 \cup \{r\},$$

where $b_k = b - \Sigma_{j \in N^1 \setminus \{k\}} a_j$. Thus we have the following test.

Variable Elimination Test 2. If $k \in N^1$ and $z_{\text{LP}}^k \leqslant \underline{z}$, x_k can be fixed at 1. A similar test exists for fixing $x_k = 0$ for $k \in N^0$.

A weakened version of Test 2 uses an upper bound on z_{LP}^k. Since

$$\max_{j \in N^0 \cup \{r\}} \frac{c_j}{a_j} = \frac{c_r}{a_r} = \lambda^* \quad \text{and} \quad z_{\text{LP}} = \sum_{j \in N^1} c_j + \lambda^*(b_k - a_k)$$

$$z_{\text{LP}}^k \leqslant \sum_{j \in N^1 \setminus \{k\}} c_j + \lambda^* b_k = z_{\text{LP}} - c_k + \lambda^* a_k,$$

and equality holds only if $\Sigma_j \{a_j : j \in N^0 \cup \{r\}, c_j/a_j = \lambda^*\} \geqslant b_k$. This validates Test 1 and shows that Test 2 dominates Test 1.

To obtain a better upper bound on z_{LP}^k, it suffices to find a set of variables in N^0 with the largest values of c_j/a_j—that is, a set $\{r+1, \ldots, q\} \subseteq N^0$ such that $\lambda^* = c_r/a_r \geqslant c_{r+1}/a_{r+1} \geqslant \cdots \geqslant c_q/a_q$ and $c_q/a_q \geqslant c_j/a_j$ for $j \in N^0 \setminus \{r+1, \ldots, q\}$. Then

$$z_{\text{LP}}^k = \begin{cases} \sum_{j \in N^1 \setminus \{k\}} c_j + \frac{c_r}{a_r} b_k & \text{if } a_r \geq b_k \\ \sum_{j \in N^1 \setminus \{k\}} c_j + c_r + \frac{c_{r+1}}{a_{r+1}} (b_k - a_r) & \text{if } a_{r+1} \geq b_k - a_r > 0 \\ \vdots & \\ \sum_{j \in N^1 \setminus \{k\}} c_j + c_r + c_{r+1} + \cdots + c_{q-1} + \frac{c_q}{a_q} (b_k - a_r - \cdots - a_{q-1}) & \\ \qquad \text{if } a_q \geq b_k - a_r - \cdots - a_{q-1} > 0 & \end{cases}$$

and

$$z_{\text{LP}}^k \leq \sum_{j \in N^1 \setminus \{k\}} c_j + c_r + \cdots + c_q + \frac{c_q}{a_q} (b_k - a_r - \cdots - a_q)$$
$$\text{if } a_q < b_k - a_r - \cdots - a_{q-1}.$$

These values can be used in Test 2.

The problem remaining after all the elimination tests have been carried out is called the *reduced problem*. Note that λ^* for the reduced problem is the same as for the original problem.

Example 1.5

$$\max\ 16x_1 + 12x_2 + 14x_3 + 17x_4 + 20x_5 + 27x_6 + 4x_7 + 6x_8 + 8x_9 + 20x_{10} + 11x_{11} + 10x_{12} + 7x_{13}$$

$$7x_1 + 6x_2 + 5x_3 + 6x_4 + 7x_5 + 10x_6 + 2x_7 + 3x_8 + 3x_9 + 9x_{10} + 3x_{11} + 5x_{12} + 5x_{13} \leq 48$$

$$x \in B^{13}.$$

First we solve the linear programming relaxation.

Initialization: $N^f = \{1, \ldots, 13\}$, $N^1 = N^0 = \emptyset$, $b = 48$.
Step 1: $\lambda = \frac{16}{7}$
Step 2: $N^> = \{3, 4, 5, 6, 9, 11\}$, $N^= = \{1\}$, $N^< = \{2, 7, 8, 10, 12, 13\}$.

i. $S_1(\lambda) = 34$.
ii. $S_2(\lambda) = 41 < b$. λ is too big.
$N^1 = \{1, 3, 4, 5, 6, 9, 11\}$, $b = 7$.

Step 1: $N^f = \{2, 7, 8, 10, 12, 13\}$, $\lambda = 2$.
Step 2: $N^> = \{10\}$, $N^= = \{2, 7, 8, 12\}$, $N^< = \{13\}$.

i. $S_1(\lambda) = 9 > b$. λ is too small.
$N^0 = \{2, 7, 8, 12, 13\}$.

Step 1: $N^f = \{10\}$, $\lambda = \frac{20}{9}$.
Step 2: $N^> = \emptyset$, $N^= = \{10\}$, $N^< = \emptyset$.

i. $S_1(\lambda) = 0 < b$.
ii. $S_2(\lambda) = 9 > b$.
iii. $r = 10$.

Hence the linear programming solution is $x_j = 1$ for $j = 1, 3, 4, 5, 6, 9, 11$, $x_{10} = \frac{7}{9}$ and $x_j = 0$ otherwise, with $z_{\text{LP}} = 128\frac{5}{9}$ and $\lambda^* = \frac{20}{9}$.

Applying the primal heuristic algorithm, we first set $x_j = 1$ for $j = 1, 3, 4, 5, 6, 9, 11$, $x_{10} = 0$, and then fill the remaining 7 units in greedy fashion. This gives $x_2 = 1$, with $x_j = 0$ otherwise. The solution has value 125. Hence $125 \leqslant z_{\text{IP}} \leqslant 128\frac{5}{9}$.

We calculate the reduced prices $c_j - \lambda^* a_j$ for $j \in N$:

j:	1	2	3	4	5	6	7	8	9	10	11	12	13
$c_j - \lambda^* a_j$:	$\frac{4}{9}$	$-\frac{12}{9}$	$\frac{26}{9}$	$\frac{33}{9}$	$\frac{40}{9}$	$\frac{43}{9}$	$-\frac{4}{9}$	$-\frac{6}{9}$	$\frac{12}{9}$	0	$\frac{39}{9}$	$-\frac{10}{9}$	$-\frac{37}{9}$

Applying Variable Elimination Test 1, we can fix $x_4 = x_5 = x_6 = x_{11} = 1$, and $x_{13} = 0$ as $|c_j - \lambda^* a_j| \geqslant z_{\text{LP}} - \underline{z} = \frac{32}{9}$.

To apply Variable Elimination Test 2, we observe that $c_j/a_j \leqslant 2$ for all $j \in N^0$. Thus we can take $q = r + 1$ with $c_{r+1}/a_{r+1} = 2$. For $k = 3$, we have $b_k = 12$; and we obtain $z_{\text{LP}}^3 \leqslant 99 + 20 + 2(12 - 9) = 125 \leqslant \underline{z}$, and hence we can fix $x_3 = 1$. None of the other variables can be fixed. Hence we are left with the reduced problem

$$\begin{aligned} z = 89 + \max\ & 16x_1 + 12x_2 + 4x_7 + 6x_8 + 8x_9 + 20x_{10} + 10x_{12} \\ & 7x_1 + 6x_2 + 2x_7 + 3x_8 + 3x_9 + 9x_{10} + 5x_{12} \leqslant 17 \\ & x_1, x_2, x_7, x_8, x_9, x_{10}, x_{12} \in B^1. \end{aligned}$$

Branch-and-Bound

We suppose that (1.10) is a reduced problem in which as many variables as possible have been fixed. The variables are now ordered so that $c_1/a_1 \geqslant \cdots \geqslant c_n/a_n$. The order of branching is fixed to be $x_1, x_2, \ldots, x_n$. Each variable is first set to 1 and then to 0.

A node t is completely specified by its level k, and a set $N^t \subseteq \{1, \ldots, k\}$. Node t represents the set of $x \in B^n$ for which $x_j = 1$ for $j \in N^t$ and $x_j = 0$ for $j \in \{1, \ldots, k\} \setminus N^t$. We let $z_t = \sum_{j \in N^t} c_j$ and let $b_t = b - \sum_{j \in N^t} a_j$. Note that node t corresponds to a nonempty set of feasible solutions if and only if $b_t \geqslant 0$, and when this holds z_t is a lower bound on the optimal value of the solutions in this set.

An upper bound is given by

$$\overline{z}_t = z_t + \max\left\{ \sum_{j=k+1}^{n} c_j x_j \colon \sum_{j=k+1}^{n} a_j x_j \leqslant b_t,\ 0 \leqslant x_j \leqslant 1 \text{ for } j = k+1, \ldots, n \right\}.$$

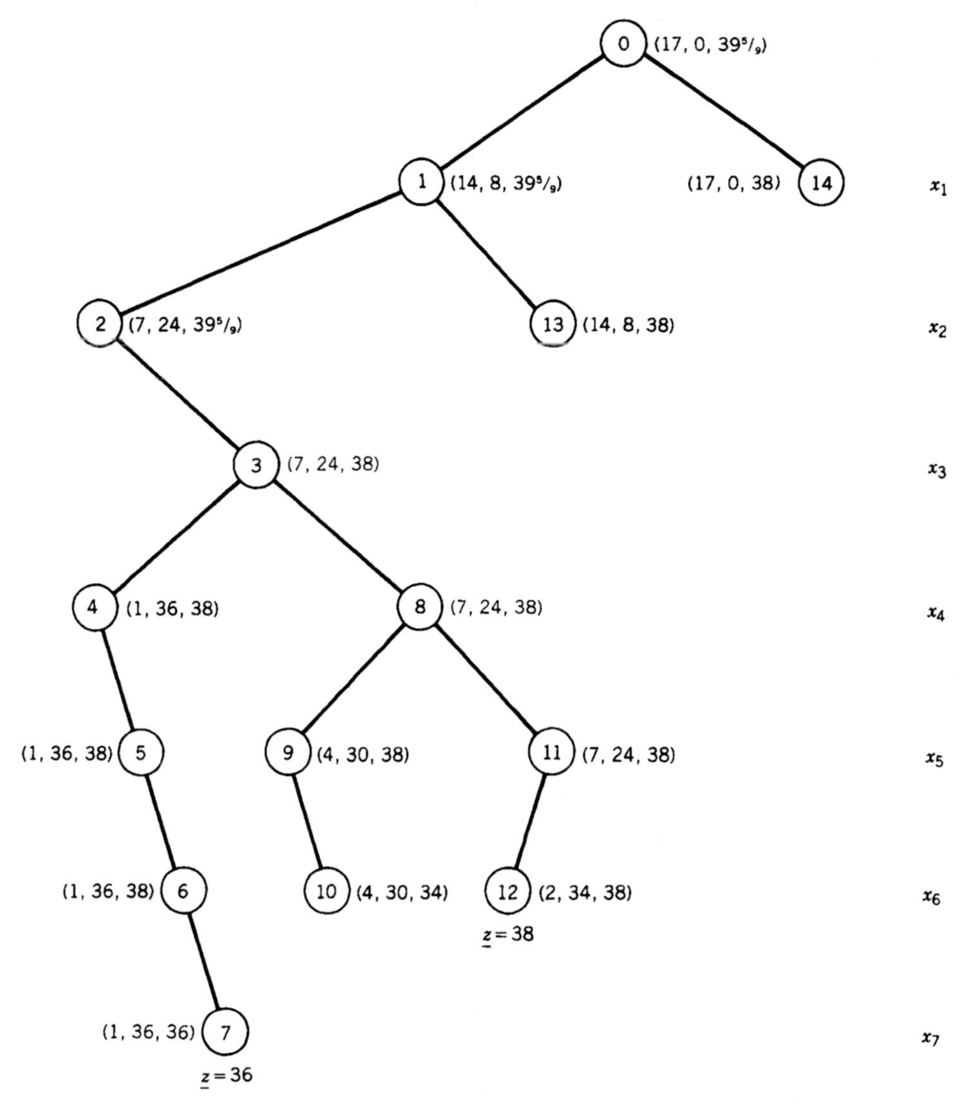

Figure 1.5

Since the variables are appropriately ordered, z_{LP} can be determined by a greedy algorithm. The node is pruned by bound if $\overline{z}_t \leqslant \underline{z}$ and is pruned by optimality if $\overline{z}_t = z_t$. If the node is not pruned by bound, there are three cases:

i. $a_{k+1} < b_t$. If $k + 1 < n$, we branch on $x_{k+1} = 1$. If $k + 1 = n$, an optimal solution for node t is $x_n = 1$. We set $\underline{z} \leftarrow \overline{z}_t$ and prune node t by optimality.

ii. $a_{k+1} = b_t$. An optimal solution for node t is obtained by setting $x_{k+1} = 1$ and $x_j = 0$ for $j > k + 1$. We set $\underline{z} \leftarrow \overline{z}_t$, and we prune node t by optimality.

iii. $a_{k+1} > b_t$. We prune the node with $x_{k+1} = 1$ by infeasibility, and we branch on $x_{k+1} = 0$.

Backtracking from t. $N^t = \{j_1, \ldots, j_r\} \subseteq \{1, \ldots, k\}$ with $j_1 < j_2 < \cdots < j_r$.

Case 1. $k \notin N^t$; that is, the last branch is $x_k = 0$. We move back up to level j_r and set $x_{j_r} = 0$. Hence node $t + 1$ is at level j_r with $N^{t+1} = \{j_1, \ldots, j_{r-1}\}$.

Case 2. $k \in N^t$; that is, the last branch is $x_k = 1$. Here we move back up to level j_{r-1} and set $x_{j_{r-1}} = 0$. Hence node $t + 1$ is at level j_{r-1} with $N^{t+1} = \{j_1, \ldots, j_{r-2}\}$.

Note that in Case 2 we do not branch on $x_k = 0$. To show that it is unnecessary to do so, first observe that because of the ordering of the variables the upper bound on the branch with $x_k = 1$ is at least as great as the upper bound on the branch with $x_k = 0$. Hence if node t is pruned by bound, the branch with $x_k = 0$ would have been as well. Alternatively, if node t is pruned by finding a feasible solution in i or ii, no better solution can be found on the branch with $x_k = 0$ because of the ordering of the variables.

The algorithm terminates when a node t with $N^t = \emptyset$ is pruned. We also repeat the variable elimination tests each time the value $\underline{z}$ of the best feasible solution found increases.

Example 1.5 (continued). After reordering and renaming the variables, the reduced problem is

$$\max 8x_1 + 16x_2 + 20x_3 + 12x_4 + 6x_5 + 10x_6 + 4x_7$$

$$3x_1 + 7x_2 + 9x_3 + 6x_4 + 3x_5 + 5x_6 + 2x_7 \leqslant 17$$

$$x \in B^7.$$

The optimal solution to the linear programming relaxation is $x_1 = x_2 = 1, x_3 = \frac{7}{9}, x_j = 0$ otherwise, $z_{\text{LP}} = 39\frac{5}{9}$, $N^1 = \{1, 2\}$, $r = 3$, $N^0 = \{4 \quad 5 \quad 6 \quad 7\}$, $\lambda^* = \frac{20}{9}$ as observed earlier.

The enumeration algorithm for the reduced problem leads to the tree shown in Figure 1.5. At each node t we give the values of b_t, z_t and $\overline{z}_t$.

The first feasible solution found at node 7 is precisely the primal heuristic solution. Node 10 is fathomed by bound. A feasible solution of value 38 is found at node 12 (with $x_7 = 1$). Nodes 13 and 14 are fathomed by bound. Hence $x_j = 1$ for $j = 1, 2, 6, 7$ and $x_j = 0$ otherwise is an optimal solution to the reduced problem of value 38.

2. 0-1 INTEGER PROGRAMMING PROBLEMS

The general 0-1 integer programming problem

$$\text{(BIP)} \qquad \max\{cx: Ax \leqslant b, x \in B^n\},$$

where A is an $m \times n$ integral matrix and $b \in Z^m$, typically is solved by a general branch-and-bound algorithm with linear programming relaxations (see Section II.4.2). However, BIP possesses a few properties that can be used to refine a general algorithm and make it more efficient.

As we have already noted in Section I.1.6, preprocessing can be quite useful for BIPs to reduce the number of variables and constraints. We assume here that preprocessing operations have already been done. But it is important to remember that they can be applied recursively and, perhaps, should be considered at each node of a branch-and-bound tree.

Linear programming relaxations can yield more information for BIPs than for general integer programs because of the following proposition.

Proposition 2.1. *Every feasible solution to* BIP *is an extreme point of* $P = \{x \in R^n_+: Ax \leqslant b, x \leqslant 1\}$.

Proof. If x is not extreme, then $x = \frac{1}{2}x^1 + \frac{1}{2}x^2$, $x^1, x^2 \in P$ with $x^1 \neq x^2$, which implies $0 < x_j < 1$ for some $j \in N$; that is, $x \notin B^n$. ■

This result motivates a heuristic that systematically searches the integral extreme points of P in the neighborhood of an optimal solution to the linear programming relaxation for good feasible solutions to BIP.

Another useful fact is that by complementing variables, the individual constraints of BIP can be written as the constraint sets of 0-1 knapsack problems. Specifically, the ith constraint can be restated as

$$\sum_{j \in N} |a_{ij}| \, \tilde{x}_j \leqslant b_i - \sum_{\{j \in N:\, a_{ij} < 0\}} a_{ij}, \qquad \tilde{x} \in B^n, \tag{2.1}$$

where $\tilde{x}_j = x_j$ if $a_{ij} > 0$ and $\tilde{x}_j = 1 - x_j$ if $a_{ij} < 0$. This transformation enables us to use strong valid inequalities for the 0-1 knapsack constraint set (see Section II.2.2) as valid inequalities for BIP in an FCP \ branch-and-bound algorithm.

After developing these ideas, we will invoke a bit more structure and consider set-covering and -packing problems in which A is a 0-1 matrix and $b_i = 1$ for $i \in M = \{1, \ldots, m\}$.

A Simplex-Based Heuristic for BIP

Suppose we solve the linear programming relaxation of BIP by a simplex algorithm that treats the upper bounds $x_j \leqslant 1$ for $j \in N$ as implicit constraints. If, in an optimal solution, x_j is nonbasic for all $j \in N$ or, equivalently, the slack variables x_{n+i} are basic for all $i \in M$, then the solution is integral.

This suggests the idea of finding good integral solutions by pivoting out of the basis the *regular* (non-slack) variables and replacing them by slack variables. These pivots, other related ones, and the rounding of the values of the fractional basic variables are attempted, with the objective of finding a feasible integral solution. If a feasible integral solution is found, then we try to improve it by local search. This is done by complementing nonbasic regular variables (switching their values from 0 to 1 and vice versa).

Algorithm

Phase 0. Solve the linear programming relaxation. If the solution is integral or there is no feasible solution, stop. Otherwise go to Phase I. Let $x_i = \overline{a}_{i0}$ be the value of the ith basic regular variable; then let $q = \Sigma_{i \in N'} \min(\overline{a}_{i0}, 1 - \overline{a}_{i0})$ be the value of integer infeasibility, where $N' = \{i: x_i \text{ is basic and a regular variable}\}$.

Phase I (Feasibility Search)

Step 1: If there is at least one pivot that maintains primal feasibility and reduces the number of basic regular variables, then do that pivot which yields the largest value of the objective function. If the resulting solution is integral, go to Phase II; otherwise return. If no such pivot exists, go to Step 2.

Step 2: If there is at least one pivot that maintains primal feasibility, leaves unchanged the number of basic regular variables, and reduces q, then do the first one found. If the resulting solution is integral, then go to Phase II and otherwise return to Step 1. If no such pivot exists, go to Step 3.

Step 3: Round each basic regular variable to the nearest integer. If the solution is feasible, go to Phase II; otherwise reduce each fractional regular variable to zero. If the solution is integral, go to Phase II; otherwise go to Step 4.

Step 4: Among those pivots that make a slack variable basic and positive and that make a regular variable nonbasic, do the one that minimizes the resulting primal infeasibility given by $h = \Sigma_{i \in N'} \max(0, -\bar{a}_{i0}, \bar{a}_{i0} - 1)$. Go to Step 5.

Step 5: If there is a nonbasic regular variable that can be complemented to reduce h, complement the one that yields the largest reduction in infeasibility. Then if $h = 0$, go to Step 3; otherwise return. If no such variable exists, go to Step 6.

Step 6: If there is a pair of nonbasic regular variables that can be simultaneously complemented to reduce h, then do the first such complementation that is found. Then if $h = 0$, go to Step 3; otherwise go to Step 5. If no pair exists, the feasibility search has failed.

Phase I either produces a feasible solution and we go to Phase II, or it ends in failure and the heuristic terminates.

Phase II (Local Search for Improvement)

Step 1: Fix variables using the reduced-profit criterion of Proposition 2.1 of Section II.5.2. Go to Step 2.

Step 2: If a better feasible solution can be found by complementing one nonbasic regular variable, do the complementation that yields the largest improvement and go to Step 1. Otherwise go to Step 3.

Step 3: For $i = 2, 3$, if a better feasible solution can be found by complementing i nonbasic regular variables, do the first such complementation found and go to Step 1. Otherwise terminate.

This heuristic has performed well in practice on a variety of types and sizes of binary integer programs. It is typical for such heuristics to work reasonably well for the larger, more complicated instances where other alternatives are not available; however, for small instances and restricted problem classes, such heuristics usually fail or do what much simpler heuristics are capable of doing.

For example, in a 0-1 knapsack problem, Step 1 of Phase I immediately pivots out the fractional variable and pivots in the slack variable, yielding the solution that would have been obtained by the greedy algorithm if it were stopped upon first encountering an item that did not fit into the knapsack. Such a solution would be completed to a greedy solution in Step 1 of Phase II.

Example 2.1

$$\begin{aligned} \max z = 9x_1 + 10x_2 & \\ 3x_1 + 3x_2 + x_3 \qquad & = 4 \\ 4x_1 + 5x_2 \qquad + x_4 & = 6 \\ (x_1, x_2) \in B^2, (x_3, x_4) \in R_+^2, & \end{aligned}$$

where x_3 and x_4 are slack variables.

The optimal solution to the linear programming relaxation is

$$x_1 = \tfrac{2}{3} - \tfrac{5}{3}x_3 + x_4$$
$$x_2 = \tfrac{2}{3} + \tfrac{4}{3}x_3 - x_4$$
$$x_3 = x_4 = 0.$$

In Step 1 if x_3 becomes basic, the pivot yields $x_2 = 1$, $x_1 = \frac{1}{4}$, and $z = \frac{49}{4}$. If x_4 becomes basic, the pivot yields $x_1 = 1$, $x_2 = \frac{1}{3}$, and $z = \frac{37}{3}$. Hence we choose to make x_4 basic. The resulting basic solution is

$$x_4 = \tfrac{1}{3} + \tfrac{5}{3}x_3 - (1 - x_1)$$
$$x_2 = \tfrac{1}{3} - \tfrac{1}{3}x_3 + (1 - x_1)$$
$$x_1 = 1,\ x_3 = 0.$$

Step 1 is repeated, and the next pivot yields the integral solution $x_1 = 1$, $x_2 = 0$. Hence Phase I terminates.

An FCP/Branch-and-Bound Algorithm

We have observed that the individual constraints of BIP can be stated in the form (2.1) and in Section II.2.2 we studied strong valid inequalities for $S = \{x \in B^n\colon \Sigma_{j\in N} a_j x_j \leqslant b\}$, where $a_j \in Z^1_+$ for $j \in N$ and $b \in Z^1_+$. In particular, we gave the class $\mathscr{F}$ of cover inequalities

$$\sum_{j\in C} x_j \leqslant |C| - 1,$$

where $C \subseteq N$ is a cover if $\Sigma_{j\in C}\, a_j > b$.

Now to be able to apply the FCPA of Section II.5.2, we formalize the separation problem for the class $\mathscr{F}$. Here C is an unknown subset of N, and given a point $x^* \in R^n_+ \setminus B^n$ we want to find a C (assuming that one exists) with $\Sigma_{j\in C}\, a_j > b$ and $\Sigma_{j\in C}\, x^*_j > |C| - 1$. Introducing a vector $z \in B^n$ to represent the unknown set C, we attempt to choose z such that $\Sigma_{j\in N}\, a_j z_j > b$ and $\Sigma_{j\in N}\, x^*_j z_j > \Sigma_{j\in N}\, z_j - 1$. The second inequality is equivalent to $\Sigma_{j\in N}\,(1 - x^*_j)z_j < 1$.

Thus we obtain the *Separation Problem for Cover Inequalities:*

$$\zeta = \min\left\{\sum_{j\in N} (1 - x^*_j)z_j\colon \sum_{j\in N} a_j z_j > b,\ z \in B^n\right\}. \tag{2.2}$$

Note that, since the constraint coefficients are integral, $\Sigma_{j\in N}\, a_j z_j > b$ is equivalent to $\Sigma_{j\in N}\, a_j z_j \geqslant b + 1$. Let z^C be the characteristic vector of $C \subseteq N$.

Proposition 2.2. *Let (ζ, z^C) be an optimal solution to (2.2). Then:*

a. *if $\zeta \geqslant 1$, then x^* satisfies all the cover inequalities for S; and*

b. *if* $\zeta < 1$, *then* $\Sigma_{j\in C} x_j \leqslant |C| - 1$ *is a most violated cover inequality for S, and it is violated by the amount* $1 - \zeta$.

Proof. If $\zeta \geqslant 1$, then all $z \in B^n$ satisfying $\Sigma_{j\in N} a_j z_j > b$ also satisfy $\Sigma_{j\in N} x_j^* z_j \leqslant \Sigma_{j\in N} z_j - 1$; that is, for all covers C, the corresponding cover inequality is satisfied by x^*.
If $\zeta < 1$, then $\Sigma_{j\in N} (1 - x_j^*) z_j^C = \zeta < 1$; hence

$$\sum_{j\in C} x_j^* = |C| - \zeta > |C| - 1.$$

Since z^C is optimal in (2.2), the maximum violation is by the amount $1 - \zeta$. ■

Example 2.2 $S = \{x \in B^5\colon 47x_1 + 45x_2 + 79x_3 + 53x_4 + 53x_5 \leqslant 178\}$ and $x^* = (0 \;\; 0 \;\; 1 \;\; 1 \;\; \frac{46}{53})$. To check whether there is a cover inequality for S violated by x^*, we solve:

$$\begin{aligned} \zeta = \min\ & 1z_1 + 1z_2 + 0z_3 + 0z_4 + \tfrac{7}{53}z_5 \\ & 47z_1 + 45z_2 + 79z_3 + 53z_4 + 53z_5 \geqslant 179 \\ & z \in B^5, \end{aligned}$$

having optimal solution $\zeta = \frac{7}{53}$, $z^C = (0 \;\; 0 \;\; 1 \;\; 1 \;\; 1)$. As $\zeta < 1$, the cover inequality $x_3 + x_4 + x_5 \leqslant 2$ is violated by x^*.

It is now straightforward to implement the FCPA with separation for BIP. As the initial relaxation we take $S_R^1 = \{x \in R_+^n\colon Ax \leqslant b, x \leqslant 1\}$. The separation algorithm for BIP involves the solution of the knapsack separation problem (2.2) for each constraint $\Sigma_{j\in N} a_{ij} x_j \leqslant b_i$, restated as the knapsack set (2.1). Note that if we find a violated cover inequality specified by C, we can easily strengthen it to $\Sigma_{j\in E(C)} x_j \leqslant |C| - 1$, where $E(C) = \{j \notin C\colon a_j \geqslant a_k$ for all $R \in C\}$, (see Section II.2.2). Thus when A and b are nonnegative, the algorithm will terminate with a solution satisfying $\Sigma_{j\in C} x_j \leqslant |C| - 1$ for all C with $\Sigma_{j\in C} a_j \nleqslant b$, $x \in R_+^n$, where a_j is the jth column of A, and the original constraints.

Example 2.3. We apply the FCPA of Section II.5.2 to the BIP

$$\begin{aligned} \max\ & 77x_1 + 6x_2 + 3x_3 + 6x_4 + 33x_5 + 13x_6 + 110x_7 + 21x_8 + 47x_9 \\ & 774x_1 + 76x_2 + 22x_3 + 42x_4 + 21x_5 + 760x_6 + 818x_7 + 62x_8 + 785x_9 \leqslant 1500 \\ & 67x_1 + 27x_2 + 794x_3 + 53x_4 + 234x_5 + 32x_6 + 797x_7 + 97x_8 + 435x_9 \leqslant 1500 \\ & x \in B^9. \end{aligned}$$

Iteration 1. Solution of the linear programming relaxation LP^1 of BIP yields $x_4^1 = x_5^1 = x_7^1 = x_8^1 = 1$, $x_1^1 = 0.71$, $x_3^1 = 0.35$, $x_j^1 = 0$ otherwise, and $z_{LP}^1 = 225.7$.

Solution of the separation problem (2.2) for row 1 yields $\zeta = 0.29$, $z_1 = z_7 = 1$, and $z_j = 0$ otherwise, giving the violated cover inequality $x_1 + x_7 \leq 1$. Here $E(C) = C$.

Solution of the separation problem (2.2) for row 2 yields $\zeta = 0.65$, $z_3 = z_7 = 1$, and $z_j = 0$ otherwise, giving the violated cover inequality $x_3 + x_7 \leq 1$. Again $E(C) = C$.

Iteration 2. Solution of the linear programming relaxation LP^2 of BIP with the two additional constraints $x_1 + x_7 \leq 1$ and $x_3 + x_7 \leq 1$ yields $x_2^2 = x_4^2 = x_5^2 = x_7^2 = x_8^2 = 1$, $x_9^2 = 0.61$, $x_j^2 = 0$ otherwise, and $z_{\text{LP}}^2 = 204.8$.

Solution of the separation problem (2.2) gives the violated cover inequalities $x_7 + x_9 \leq 1$ for row 1 and $x_4 + x_5 + x_7 + x_9 \leq 3$ for row 2.

Iteration 3. $x^3 = (0.63, 1, 0.60, 1, 1, 0, 0.37, 1, 0.63)$. The separation routines give $C = \{1, 9\}$ and the extended cover inequality $x_1 + x_7 + x_9 \leq 1$ for row 1, and $C = \{3, 5, 8, 9\}$ and the extended cover inequality $x_3 + x_5 + x_7 + x_8 + x_9 \leq 3$ for row 2.

Iteration 4. $x^4 = (0, 1, 0, 1, 1, 0.63, 1, 1, 0)$. The cover inequality $x_6 + x_7 \leq 1$ is added.

Iteration 5. $x_5 = (0, 1, 0, 1, 1, 0, 1, 1, 0)$ is integer and thus solves BIP.

Example 2.3 raises two issues. Given that the separation problem (2.2) is a knapsack problem, which is an $\mathcal{NP}$-hard problem, should we solve (2.2) exactly or use a fast heuristic algorithm? In practice, heuristics have been used very effectively. But this, of course, means that some cover inequalities may be missed by the separation procedure.

The second issue stems from the observation that the first two cuts generated from row 1 in the course of the algorithm, namely $x_1 + x_7 \leq 1$, $x_7 + x_9 \leq 1$, are dominated by the third cut $x_1 + x_7 + x_9 \leq 1$. Hence, we could speed up the algorithm if we could obtain this stronger cut from row 1 on the first iteration.

To obtain the stronger cuts, remember from Proposition 2.5 of Section II.2.2 that every cover inequality generated from a minimal cover C gives rise to a *lifted cover inequality* of the form

$$\sum_{j \in N \setminus C} \alpha_j x_j + \sum_{j \in C_2} \gamma_j x_j + \sum_{j \in C_1} x_j \leq |C_1| - 1 + \sum_{j \in C_2} \gamma_j, \tag{2.3}$$

where $C_1 \cap C_2 = \emptyset$ and $C_1 \cup C_2 = C$. Moreover, $\{\alpha_j\}$ and $\{\gamma_j\}$ can be chosen so that (2.3) defines a facet of the knapsack convex hull.

The coefficients in (2.3) are obtained by sequential lifting. Unfortunately we know of no efficient way to consider all possible ordering of the elements of $N \setminus C$ that can be used in sequential lifting. From a practical point of view, we avoid this difficulty by choosing an ordering of the elements of $N \setminus C$ in a greedy fashion.

A Lifting Heuristic to obtain a lifted cover inequality of the form (2.3) with $C_2 = \emptyset$

Initialization: Given x^*, solve the knapsack problem (2.2) to obtain a cover C. (Note that the cover inequality may not be violated.) Let $L^1 = N \setminus C$ and let $k = 1$. Set $\alpha_j = 1$ for all $j \in C$.

Iteration k: For all $j \in L^k$ find β_j, which is the maximum value of π_j such that $\pi_j x_j + \Sigma_{i \in N \setminus L^k} \alpha_i x_i \leq |C| - 1$ is valid. Let $j^* = \arg\max_{j \in L^k} \beta_j x_j^*$. Set $L^{k+1} = L^k \setminus \{j^*\}$ and $\alpha_{j^*} = \beta_{j^*}$. If $L^{k+1} = \emptyset$, test whether $\Sigma_{j \in N} \alpha_j x_j^* > |C| - 1$. If so, add the cut $\Sigma_{j \in N} \alpha_j x_j \leq |C| - 1$. If $L^{k+1} \neq \emptyset$, $k \leftarrow k + 1$. Return.

As shown in Section II.2.2, we have $\beta_j = |C| - 1 - \zeta_j$, where

$$\zeta_j = \max\left\{ \sum_{i \in N \setminus L^k} \alpha_i x_i : \sum_{i \in N \setminus L^k} a_i x_i \leq b - a_j,\ x \in B^{|N \setminus L^k|} \right\}. \tag{2.4}$$

Note that because of the small size of the coefficients $\alpha_j \leq |C| - 1$, the knapsack problem (2.4) can be solved efficiently by dynamic programming (see Proposition 1.6).

A simple extension of the lifting heuristic suggests how we can also search for extended cover inequalities of the form (2.3) with $C_2 \neq \emptyset$.

Separation Algorithm to obtain lifted cover inequalities (2.3)

Step 1: Apply the lifting heuristic described above. If a violated inequality is found, stop.

Step 2: If not, choose $k = \arg(\max_{j \in C} a_j x_j^*)$. Set $C_2 = \{k\}$, and use the lifting heuristic to generate a facet-defining inequality for $\text{conv}(S^k)$ from the cover $C \setminus \{k\}$, where $S^k = \{x \in B^{n-1}: \Sigma_{j \in N \setminus \{k\}} a_j x_j \leq b - a_k\}$.

Step 3: Convert this inequality into a facet-defining inequality of the form (2.3) for S by lifting back in the variable x_k. (See Example 2.2 of Section II.2.2).

Step 4: Check the resulting inequality for violation. Stop.

Example 2.2 (continued)

$$S = \{x \in B^5: 47x_1 + 45x_2 + 79x_3 + 53x_4 + 53x_5 \leq 178\}$$

and $x^* = (\frac{1}{2}\ \frac{1}{2}\ 1\ \frac{1}{2}\ \frac{1}{2})$. The knapsack problem (2.1) gives the cover inequality $x_3 + x_4 + x_5 \leq 2$, which is not violated by x^*. The separation algorithm starts with $C = \{3, 4, 5\}$.

Step 1: The lifting heuristic leads to the same inequality.

Step 2: $C_2 = \{3\}$ is chosen, and the lifting heuristic is called, starting with the cover inequality $x_4 + x_5 \leq 1$ for $S^3 = \{x \in B^4: 47x_1 + 45x_2 + 53x_4 + 53x_5 \leq 99\}$.

Iteration 1: $L^1 = \{1, 2\}$ and $\beta_1 = 1$, $\beta_2 = 0$. Hence $x_{j^*} = x_1$ is lifted with coefficient $\alpha_1 = 1$.

Iteration 2: $L^2 = \{2\}$, $\beta_2 = 0$. The resulting inequality for S^3 is $x_1 + x_4 + x_5 \leq 1$.

Step 3: Variable x_3 is lifted in giving the inequality $x_1 + 2x_3 + x_4 + x_5 \leq 3$, which defines a facet of $\text{conv}(S)$ that is violated by x^*.

Given the heuristic nature of the above separation algorithm, we can no longer determine a priori what problem will be solved at the termination of the FCPA with separation. We can only assert that the cuts generated at least include all the cover inequalities. Remember that even this assertion may be false if we use a heuristic algorithm for the knapsack problem (2.2). However, as the example below suggests, and as computational experience shows, the use of the lifted cover inequalities (2.3) in place of the cover inequalities leads to significant improvements in performance.

Example 2.3 (continued). We apply the FCP/branch-and-bound algorithm, where the separation algorithm for extended cover inequalities is applied to each row of BIP.

Phase 1 (FCPA)

Iteration 1: Solution of the relaxation LP^1 of BIP yields $x_4^1 = x_5^1 = x_7^1 = x_8^1 = 1$, $x_1^1 = 0.71$, $x_3^1 = 0.35$, $x_j^1 = 0$ otherwise, and $z_{LP}^1 = 225.7$.
Row 1. Cut $x_1 + x_6 + x_7 + x_9 \leq 1$ is generated.
Row 2. Cut $x_3 + x_7 \leq 1$ is generated.

Iteration 2: Solution of the relaxation LP^2 yields $x_2^2 = x_4^2 = x_5^2 = x_7^2 = x_8^2 = 1$, $x_j^2 = 0$ otherwise, and $z_{LP}^2 = 176$. Because x^2 is integer, it is an optimal solution of BIP.

An alternative or complement to the use of the heuristic lifting algorithms is to use Proposition 2.6 of Section II.2.2, which provides upper and lower bounds on the values taken by α_j for $j \in N \setminus C$ in the lifting heuristic. In particular, we obtain conditions for the existence of a violated inequality (2.3) when C is a minimal cover and $C_2 = \emptyset$.

Proposition 2.3. *Let $C = \{j_1, \ldots, j_r\}$ be a minimal cover with $a_{j_1} \geq a_{j_2} \geq \cdots \geq a_{j_r}$, and for $h = 0, \ldots, r$ let*

$$Q_h = \{j \in N \setminus C: \mu_h \leq a_j \leq \mu_{h+1} - \lambda\} \text{ and } R_h = \{j \in N \setminus C: \mu_{h+1} - \lambda + 1 \leq a_j \leq \mu_{h+1} - 1\},$$

where $\mu_h = \sum_{k=1}^h a_{j_k}$, $\mu_0 = 0$, and $\lambda = \mu_r - b > 0$.

1. *If $\sum_{j \in C} x_j^* + \sum_h \sum_{j \in Q_h} hx_j^* + \sum_h \sum_{j \in R_h} (h+1)x_j^* \leq |C| - 1$, there is no violated lifted inequality for C with $C_2 = \emptyset$.*
2. *If $\sum_{j \in C} x_j^* + \sum_h \sum_{j \in Q_h} hx_j^* + \sum_h \sum_{j \in R_h} hx_j^* + \max_{j \in \cup_h R_h} x_j^* > |C| - 1$, then*

$$\sum_{j \in C} x_j + \sum_h \sum_{j \in Q_h} hx_j + \sum_h \sum_{j \in R_h} hx_j + x_t \leq |C| - 1,$$

where $t = \arg(\max_{j \in \cup_h R_h} x_j^*)$, is a valid inequality violated by x^*.

The proof is an immediate application of Proposition 2.6 of Section II.2.2. This proposition can be used to speed up the lifting heuristic by stopping the algorithm if condition 1 is satisfied, or otherwise fixing the values of α_j for $j \in \cup_h Q_h$. Alternatively, we can simply use the valid inequality given in condition 2.

Example 2.3 (continued). At iteration 1 we have $x_4^1 = x_5^1 = x_7^1 = x_8^1 = 1$, $x_1^1 = 0.71$, $x_3^1 = 0.35$, $x_j = 0$ otherwise. For row 1 the knapsack problem (1.1) gives the cover $C = \{1,7\}$. From Proposition 2.3 we have $Q_0 = \{2, 3, 4, 5, 8\}$ and $r_0 = \{6, 9\}$. It follows without further calculations that both $x_1 + x_6 + x_7 \leq 1$ and $x_1 + x_7 + x_9 \leq 1$ are valid inequalities. To establish the validity of $x_1 + x_6 + x_7 + x_9 \leq 1$, we must lift one of the above inequalities.

Example 2.4. This is a 0-1 minimization problem with 15 constraints and 33 variables. The data, as well as the 20 cuts added in seven iterations (six sets of cuts) of the FCPA, are given in Table 2.1. Note that the value of the initial LP relaxation is $z_{\text{LP}}^1 = 2520.7$, and after adding the cuts the lower bound given by the LP relaxation of the reformulation $\max\{cx\colon x \in S_{\text{LP}}^7\}$ is $z_{\text{LP}}^7 = 2962.2$.

The corresponding solution x^7 for which no cuts are found is

1	2	3	4	5	6	7	8	9	10	11	12	13	14	15	16	17
(0	1	0	0	0	.83	.17	.83	.83	0	.83	0	.17	.66	0	1	0

18	19	20	21	22	23	24	25	26	27	28	29	30	31	32	33
1	0	0	0	0	1	0	1	.67	1	.50	1	1	0	0	.83).

Applying branch-and-bound to the reformulated problem, a solution of value 3095 is found at node 17, and an optimal solution of value 3089 is found at node 65. Optimality is proved (i.e., the search is completed) at node 77. The optimal solution is $x_1 = x_7 = x_8 = x_{10} = x_{14} = x_{18} = x_{21} = x_{23} = x_{25} = x_{26} = x_{27} = x_{28} = x_{29} = x_{30} = 1$, and $x_j = 0$ otherwise.

If branch-and-bound is applied without adding cuts, the best solution found after 1000 nodes has value 3095, and the tree still contains 163 active nodes.

Set Covering and Packing

When (A, b) is a 0-1 matrix, each individual constraint is already in the form of a covering or packing inequality, and no mileage can be gained from the cutting-plane approach developed above. Some simple combinatorial ideas can yield cuts. For example, in a packing problem the constraints

$$\begin{aligned} x_1 + x_2 \quad\quad &\leq 1 \\ x_1 \quad\quad + x_3 &\leq 1 \\ x_2 + x_3 &\leq 1 \\ x \in B^3 & \end{aligned}$$

imply the inequality $x_1 + x_2 + x_3 \leq 1$. And in a covering problem, the constraints

$$\begin{aligned} x_1 + x_2 \quad\quad &\geq 1 \\ x_1 \quad\quad + x_3 &\geq 1 \\ x_2 + x_3 &\geq 1 \\ x \in B^3 & \end{aligned}$$

imply the inequality $x_1 + x_2 + x_3 \geq 2$. More generally, if we have constraints for all sets of size k from $k + 1$ variables, then we can derive a nontrivial valid inequality involving all $k + 1$ variables.

The disjunctive approach can also be used to derive valid inequalities for covering and packing problems. Here we leave the cutting-plane approach and consider some other features of covering and packing problems.

Table 2.1.

	1	2	3	4	5	6	7	8	9	10	11	12	13	14	15	16	17	18	19	20	21	22	23	24	25	26	27	28	29	30	31	32	33	j	
	171	171	171	171	163	162	163	69	69	183	183	183	183	49	183	258	517	250	500	250	500	159	318	159	318	159	318	159	318	114	228	159	318	c_j	
1	1	1	1	1																														≤ 1	
2					1	1	1																											≤ 1	
3								1	1																									≤ 1	
4										1	1	1	1		1																			≤ 1	
5										230						200	400																	≥ 5	
6			300	300	285	285		265	265			230	230	190								200	400	200	400	200	400	200	400	200	400			≤ 2700	
7			300	300	285	285		265	265			230	230	190								200	400	200	400	200	400	200	400	200	400			≥ 2600	
8				300																										200	400			≥ 100	
9	300					285			265					190												200	400							≥ 900	
10	300		300			285			265				230	190												200	400	200	400					≥ 1656	
11					285			265			230											200	400											≥ 335	
12					285			265			230	230										200	400	200	400									≥ 1026	
13		300																200	400															≥ 5	
14		300																200	400	200	400													≥ 500	
15							285																									200	400	≥ 270	
										1						1	1																	≥ 1	Row 5
						1			1					1									1	1	1		1	1	1		1			≤ 8	6
				1																										1	1			≥ 1	8
	1		1			1							1													1	1		2					≥ 4	10
					1			1			1	1										1	1		1									≥ 3	12
		1																1	1	1	1													≥ 2	14
							1																										1	≥ 1	15
			1						1					1								1	1		1	1	1	1	1	1	1			≤ 9	Row 6
	1		1			1							1														1		1					≥ 3	10
					1			1			1	1											1	1	1									≥ 3	12
								1	1				1	1									1		1	1	1	1	1	1	1	1		≤ 9	Row 6
			1	1	1	1			1			1										1	1	1	1		1		1		1			≥ 4	7
	1					1			1																	1	1							≥ 3	9
	1		1			1			1																		1		1					≥ 3	10
					1							1										1	1	1	1									≥ 2	12
		1																	1		1													≥ 1	14
					1			1			1	1										1	2	1	2									≥ 5	Row 12
						1			1					1									1		1	1	1		1	1	1			≤ 8	Row 6
			1	1	1	1		1	1			1	1									1	1	1	1	1	1	1	1		2			≥ 7	7
	1		1			1			1				1													1	1	1	1					≥ 5	Row 10

The greedy heuristic has a natural realization for the set-covering problem

$$\text{(SC)} \qquad z_{SC} = \min\left\{\sum_{j\in N} c_j x_j \colon \sum_{j\in N} a_{ij}x_j \geq 1 \text{ for } i \in M, x \in B^n\right\},$$

where $a_{ij} \in \{0, 1\}$ for all i and j. We assume that $\Sigma_{j\in N}\, a_{ij} \geq 1$ for $i \in M$, which is necessary and sufficient for a feasible solution. Let $M_j = \{i \colon a_{ij} = 1\}$.

Greedy Heuristic for Set Covering

Initialization: $M^1 = M, N^1 = N, t = 1$.

Iteration t: Select $j^t \in N^t$ to $\min\{c_j/|M_j \cap M^t|\}$. Let $N^{t+1} = N^t\backslash\{j^t\}$ and $M^{t+1} = M^t\backslash M_{j^t}$. If $M^{t+1} = \emptyset$, the greedy solution is given by $x_j = 1$ for $j \notin N^{t+1}$ and by $x_j = 0$ otherwise. Its cost is $z_G = \Sigma_{j\notin N^{t+1}}\, c_j$. If $M^{t+1} \neq \emptyset$, then let $t \leftarrow t + 1$ and return.

We see that at each step the greedy heuristic selects the column that meets the largest number of uncovered rows per unit cost and then stops when a feasible solution has been found.

Although we cannot give a positive, data-independent performance guarantee for the greedy heuristic, we will show that it has a performance guarantee that is independent of n and the objective coefficients, and that decreases only logarithmically with $|M|$.

For any positive integer k, let $H(k) = 1 + \frac{1}{2} + \cdots + \frac{1}{k}$ and let $d = \max_{j\in N} \Sigma_{i\in M}\, a_{ij}$. We will use the following elementary result.

Proposition 2.4. *Let* $u = (u_1, \ldots, u_n) \in R^n_+$ *and* $v = (v_1, \ldots, v_n) \in Z^n_+$. *If* $0 < u_1 \leq u_2 \leq \cdots \leq u_n$ *and* $v_1 \geq v_2 \geq \cdots \geq v_n$ *then*

$$\sum_{i=1}^{n-1} u_i(v_i - v_{i+1}) + u_n v_n \leq \max_i(u_i v_i)H(v_1).$$

Theorem 2.5. $z_{SC}/z_G \geq 1/H(d)$.

Proof. We use the approach presented in Section II.5.3 for worst-case analysis of heuristics. In particular, we construct a feasible solution u^* to the dual of the linear programming relaxation of SC. Then, by duality, we obtain $z_{SC} \geq \Sigma_{i=1}^m\, u_i^*$. The result then follows by showing that $\Sigma_{i=1}^m\, u_i^* = z_G/H(d)$.

Suppose that the greedy heuristic terminates on iteration T, and let $\theta^t = \min_j \{c_j/|M_j \cap M^t|\}$. The dual vector $u^*$is defined by $u_i^* = \theta^t/H(d)$ for $i \in M^t\backslash M^{t+1}$.

We will show dual feasibility for each $j \in N$ by using Proposition 2.4. Since $M^t \supset M^{t+1}$, it follows that $|M_j \cap M^t| \geq |M_j \cap M^{t+1}|$ for all t. Hence if $v_t = |M_j \cap M^t|$, then $v_1 \geq v_2 \geq \cdots \geq v_T \geq 0$. We also have from its definition that $0 < \theta^1 \leq \cdots \leq \theta^T$.

Now

$$\sum_{i\in M} u_i^* a_{ij} = \sum_{t=1}^{T} \frac{\theta^t}{H(d)}\left(\sum_{i\in M^t\backslash M^{t+1}} a_{ij}\right)$$

$$= \frac{1}{H(d)} \sum_{t=1}^{T} \theta^t\,(|M^t \cap M_j| - |M^{t+1} \cap M_j|)$$

$$= \frac{1}{H(d)} \sum_{t=1}^{T} \theta^t\,(v_t - v_{t+1}),$$

and applying Proposition 2.4 to the last term gives

$$\sum_{i\in M} u_i^* a_{ij} \leq \frac{1}{H(d)} H(|M^1 \cap M_j|) \max_t\{\theta^t(|M^t \cap M_j|)\}.$$

Since $|M^1 \cap M_j| = \Sigma_{i=1}^m a_{ij} \leq d$, and $\theta^t(|M^t \cap M_j|) \leq c_j$ for all t by definition of θ^t, we obtain $\Sigma_{i\in M} u_i^* a_{ij} \leq c_j$, and $u^*$is dual feasible.

Finally, the dual objective value is

$$\sum_{i\in M} u_i^* = \frac{1}{H(d)} \sum_{t=1}^{T} \theta^t (|M^t| - |M^{t+1}|) = \frac{1}{H(d)} \sum_{t=1}^{T} c_{j^t} = \frac{z^G}{H(d)}.$$ ■

We leave it as an exercise to show that the bound of Theorem 2.5 can be asymptotically achieved.

We now turn to set-packing problems and, in particular, to the node-packing problem. An instance of the node-packing problem is given by a graph $G = (V, E)$ and a weight function $c: V \to R^1$. A feasible solution is any subset of nodes such that no pair in the subset is joined by an edge. The weight of a solution $U \subseteq V$ is $c(U) = \Sigma_{i\in U} c_i$, and the objective is to find a solution of maximum weight. Node packing is $\mathcal{NP}$-hard (see Section I.5.6). Moreover, any set-packing problem is easily transformed to a node-packing problem on the intersection graph of the family of sets.

Here we are going to present a rather unusual property of node packing that does not appear to be shared by any other $\mathcal{NP}$-hard problem and that may yield a substantial reduction in the size of an instance once the linear programming relaxation has been solved. If the linear programming relaxation of an integer program has an optimal integral solution, that solution is, of course, also optimal to the integer program. But if just one variable is fractional in the optimal linear programming solution, we can no longer deduce anything about the variables in the integer program. On the other hand, in node packing, all of the variables that are integral in the solution of the linear programming relaxation (if any) keep these same integer values in some optimal solution to the integer program. Hence, having solved the linear programming relaxation, we can fix the values of the integral variables and then eliminate them from the problem.

A binary integer programming formulation of the node-packing problem on $G = (V, E)$ is

$$\text{(NP)} \qquad \begin{aligned} &\max \sum_{j\in V} c_j x_j \\ &x_i + x_j \leq 1 \quad \text{for } (i, j) \in E \\ &x \in B^n, \end{aligned}$$

where $n = |V|$. Its linear programming relaxation (LNP) is obtained by replacing $x \in B^n$ by $x \in R_+^n$.

We need the following proposition that relates local and global optimality. For $U \subset V$, the neighbors of U are the set $N(U) = \{i \in V: i \notin U, (i, j) \in E \text{ for some } j \in U\}$. Let $S(U) = U \cup N(U)$, $\overline{S}(U) = V \backslash (U \cup N(U))$, and let $G(S(U))$ be the subgraph induced by $S(U)$. A property that we use several times is the following:

(P.1) There are no edges joining a node of U and $\overline{S}(U)$.

Proposition 2.6. *If U is an optimum packing on $G(S(U))$, then there is an optimum packing V^0 on G with $V^0 \supseteq U$.*

Proof. Let $V^* = V_1^* \cup V_2^*$ be an optimum packing on G, where $V_1^* = V^* \cap S(U)$. By (P.1), $U \cup V_2^*$ is a packing on G. By hypothesis, $c(U) \geqslant c(V_1^*)$; hence $c(U) + c(V_2^*) \geqslant c(V^*)$.

Theorem 2.7. *If x^0 is an optimal solution to* LNP, *then there is an optimal solution x^* to* NP *with $x_j^* = x_j^0$ for all j such that x_j^0 is integral.*

Proof. The result is trivial if x^0 is integral, so we suppose that it is not. We first show that if $U = \{j: x_j^0 = 1\}$, there exists an optimal solution x^* to NP with $x_j^* = 1$ for all $j \in U$. By Proposition 2.6, we need to show that U is an optimal packing on $G(S(U))$. Note that $x_j^0 = 0$ for $j \in N(U)$. Also, for some $k \in \overline{S}(U)$, $x_k^0 > 0$; otherwise x^0 is integral.

Suppose that $\hat{U} \neq U$ is an optimal packing on $G(S(U))$ and that $c(\hat{U}) > c(U)$. We will show that this contradicts the optimality of x^0. Let

$$\hat{x}_j = \begin{cases} 1 & \text{for } j \in \hat{U} \\ 0 & \text{for } j \in S(U)\backslash\hat{U} \\ x_j^0 & \text{for } j \in \overline{S}(U) \end{cases}$$

and $\overline{x} = \lambda x^0 + (1-\lambda)\hat{x}$, where $\lambda = \max\{x_j^0: j \in \overline{S}(U)\}$. Since $0 \leqslant x_j^0 < 1$ for all $j \in \overline{S}(U)$, we obtain $0 < \lambda < 1$. We claim that $\overline{x}$ is a feasible solution to LNP; that is, $\overline{x}_i + \overline{x}_j \leqslant 1$ for $(i, j) \in E$. This is clear if $i, j \in S(U)$ (since U and $\hat{U}$ are packings) and if $i, j \in \overline{S}(U)$. By (P.1), the remaining case is $i \in N(U)$ and $j \in \overline{S}(U)$. Then $\overline{x}_i = \lambda x_i^0 + (1-\lambda)\hat{x}_i \leqslant 1 - \lambda$ since $x_i^0 = 0$, and $\overline{x}_j = x_j^0$ since $x_j^0 = \hat{x}_j$. Hence $\overline{x}_i + \overline{x}_j \leqslant (1-\lambda) + x_j^0 \leqslant 1$ by the definition of λ. Now

$$\sum_{j \in V} c_j \overline{x}_j = \lambda c(U) + (1-\lambda)c(\hat{U}) + \sum_{j \in \overline{S}(U)} c_j x_j^0$$

$$> c(U) + \sum_{j \in \overline{S}(U)} c_j x_j^0 = \sum_{j \in V} c_j x_j^0,$$

which contradicts the hypothesis that x^0 is an optimal solution to LNP.

Finally, $x_j^* = 1$ for $j \in U$ implies $x_j^* = x_j^0 = 0$ for $j \in N(U)$, and if $x_j^0 = 0$ for $j \in \overline{S}(U)$, then $c_j \leqslant 0$ so that $x_j^* = 0$ as well. ■

The use of Theorem 2.7 is enhanced by the fact that LNP can be solved in polynomial time, essentially as an assignment problem on a graph with $2n$ nodes. On the other hand, the theorem will be useful only if an optimal solution to LNP contains a significant number of integer-valued variables. It is also important to observe that the bound obtained from the linear programming relaxation of a set-packing problem is stronger than the bound obtained from LNP when the set-packing problem is transformed to a node-packing problem.

These advantages and disadvantages must be balanced, but if we decide to use LNP as a relaxation to NP in a branch-and-bound algorithm, then Theorem 2.7 should be applied at every node of the branch-and-bound tree.

Example 2.5. Consider the node-packing problem on the graph of Figure 2.1 with $c = (3 \;\; 1 \;\; 1 \;\; 2 \;\; 2 \;\; 3)$. An optimal solution to LNP is $x = (1 \;\; 0 \;\; 0 \;\; \frac{1}{2} \;\; \frac{1}{2} \;\; \frac{1}{2})$. Hence there is an optimal solution to NP with $x_1 = 1$ and $x_2 = x_3 = 0$. Now it is trivial to solve NP

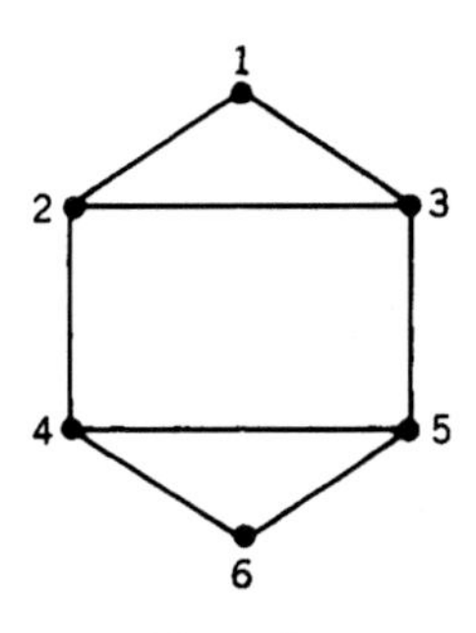

Figure 2.1

on the subgraph induced by nodes {4, 5, 6}; that is, the solution is $x_4 = x_5 = 0$ and $x_6 = 1$. Hence an optimal solution to NP on the whole graph is $x = (1 \quad 0 \quad 0 \quad 0 \quad 0 \quad 1)$.

3. THE SYMMETRIC TRAVELING SALESMAN PROBLEM

An instance of the symmetric traveling salesman problem is given by a graph $G = (V, E)$ and a weight vector $c \in R^{|E|}$. A *tour* T of G is a subgraph of G that is a cycle on V; that is, if $T = (V, E_T)$, then each node of T is of degree 2, T is connected and $|E_T| = |V|$. The feasible solutions are all of the tours of G (if any), and assuming that G contains at least one tour, the objective is to find a tour of minimum weight. The weight of a tour T with edge set $E_T \subset E$ is $\Sigma_{e \in E_T} c_e$. Let

$$z_{\text{TS}} = \min\left\{ \sum_{e \in E_T} c_e \colon T = (V, E_T) \text{ is a tour of } G \right\}.$$

Several special-purpose algorithms originally were developed to solve the traveling salesman problem, which has become a prototype problem for illustrating, testing, and comparing algorithms. We begin this section by describing and comparing various relaxations. We then present and analyze some heuristics for obtaining good feasible solutions. Finally, we give some algorithms that use a heuristic for finding feasible solutions and upper bounds, a relaxation or dual problem for finding lower bounds, and, if necessary, a branch-and-bound phase for finding an optimal solution and proving optimality.

Relaxations

We first consider two relaxations that can be obtained by considering families of subgraphs that contain all of the tours of G. First we drop connectedness and consider the family of subgraphs of G that contain $|V|$ edges and in which each node is of degree 2. These subgraphs are called *2-matchings* (see Figure 3.2 of Section II.2.3).

Let

$$z_{\text{M}} = \min\left\{ \sum_{e \in E_{\text{M}}} c_e \colon M = (V, E_M) \text{ is a 2-matching of } G \right\}.$$

Since every tour is a 2-matching, we have

$$z_{\text{TS}} \geqslant z_{\text{M}}. \tag{3.1}$$

Next we drop the degree-2 requirement on all nodes except node 1, but we keep connectedness and the requirement that the subgraph contains $|V|$ edges. This means that the subgraph on nodes $V \setminus \{1\}$ is connected and contains $|V| - 2$ edges. By Proposition 1.2 of Section I.3.1, it is a tree. Hence the subgraph on G is a spanning tree on $V \setminus \{1\}$, together with 2 edges incident to node 1. These subgraphs are called *1-trees* (see Figure 3.1). Let

$$z_{1T} = \min\left\{ \sum_{e \in E_T} c_e : T = (V, E_T) \text{ is a 1-tree of } G \right\}.$$

Note that a 1-tree is a tour if and only if each node of the 1-tree is of degree 2. Since every tour is a 1-tree, we have

$$z_{TS} \geqslant z_{1T} \tag{3.2}$$

The above discussion implies the following proposition.

Proposition 3.1. *$T = (V, E_T)$ is a tour of G if and only if T is both a 2-matching and a 1-tree.*

We now consider integer programming formulations of these relaxations. For $F \subset E$, let $x^F \in B^{|E|}$ be the characteristic vector of F; that is, $x_e^F = 1$ if $e \in F$, and $x_e^F = 0$ if $e \notin F$.

The characteristic vectors of 2-matchings are simply described by

$$x \in B^{|E|} \tag{3.3}$$

$$\sum_{e \in \delta(\{v\})} x_e = 2 \quad \text{for } v \in V \qquad \text{(degree constraints)}, \tag{3.4}$$

where for any $U \subset V$, $\delta(U)$ is the set of edges with one end in U.

Let $E(U)$ be the set of edges with both ends in U. The characteristic vectors of 1-trees are described by (3.3), (3.4) for $v = 1$,

$$\sum_{e \in E(U)} x_e \leqslant |U| - 1 \quad \text{for all } U \subseteq V \setminus \{1\} \text{ with } 3 \leqslant |U| \quad \text{(subtour elimination constraints)} \tag{3.5}$$

and

$$\sum_{e \in E} x_e = |V|. \tag{3.6}$$

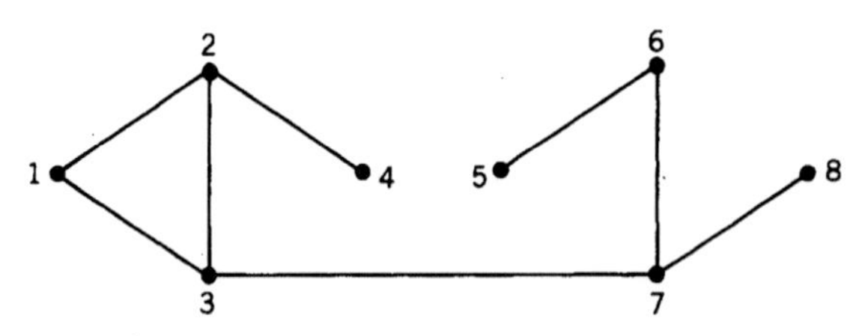

Figure 3.1

Thus by applying Proposition 3.1, we get that tours are described by (3.3)–(3.6). However, there are redundancies that can be eliminated. First observe that (3.4) implies (3.6). Then, as we observed in Section II.2.3, the subtour elimination constraint for $V \setminus U$ is implied by the subtour elimination constraint for U. Hence the characteristic vectors of tours are given by (3.3), (3.4), and (3.5) for $U \subset V$ with $3 \leq |U| \leq \lfloor \frac{1}{2}|V| \rfloor$.

Two relaxations that are themselves relaxations of 2-matchings are fractional 2-matchings and integer 2-matchings. In *fractional 2-matchings*, the variables are not required to be integral, so (3.3) is replaced by

$$x \in R_+^{|E|} \tag{3.7}$$

and

$$x_e \leq 1 \quad \text{for } e \in E. \tag{3.8}$$

In *integer 2-matchings*, the variables are not required to be binary, so (3.3) is replaced by

$$x \in Z_+^{|E|}. \tag{3.9}$$

We have

$$z_{\text{FM}} = \min\left\{\sum_{e \in E} c_e x_e \colon x \text{ satisfies (3.4), (3.7) and (3.8)}\right\} \leq z_M$$

and

$$z_{\text{IM}} = \min\left\{\sum_{e \in E} c_e x_e \colon x \text{ satisfies (3.4) and (3.9)}\right\} \leq z_M.$$

Furthermore, we will prove in Chapter III.1 that all of the extreme points of the polytope $\{x \in R_+^{|E|} \colon x \text{ satisfies (3.4)}\}$ are integral. Hence

$$z_{\text{IM}} = \min\left\{\sum_{e \in E} c_e x_e \colon x \text{ satisfies (3.4) and (3.7)}\right\} \leq z_{\text{FM}}.$$

Example 3.1 Consider the graph shown in Figure 3.2. The numbers on the edges are their weights. Figure 3.3 shows an optimal tour, an optimal 2-matching, an optimal fractional 2-matching, an optimal integer 2-matching, and an optimal 1-tree.

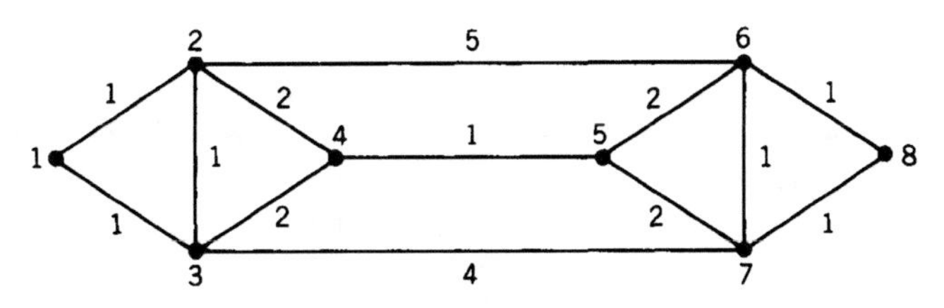

Figure 3.2

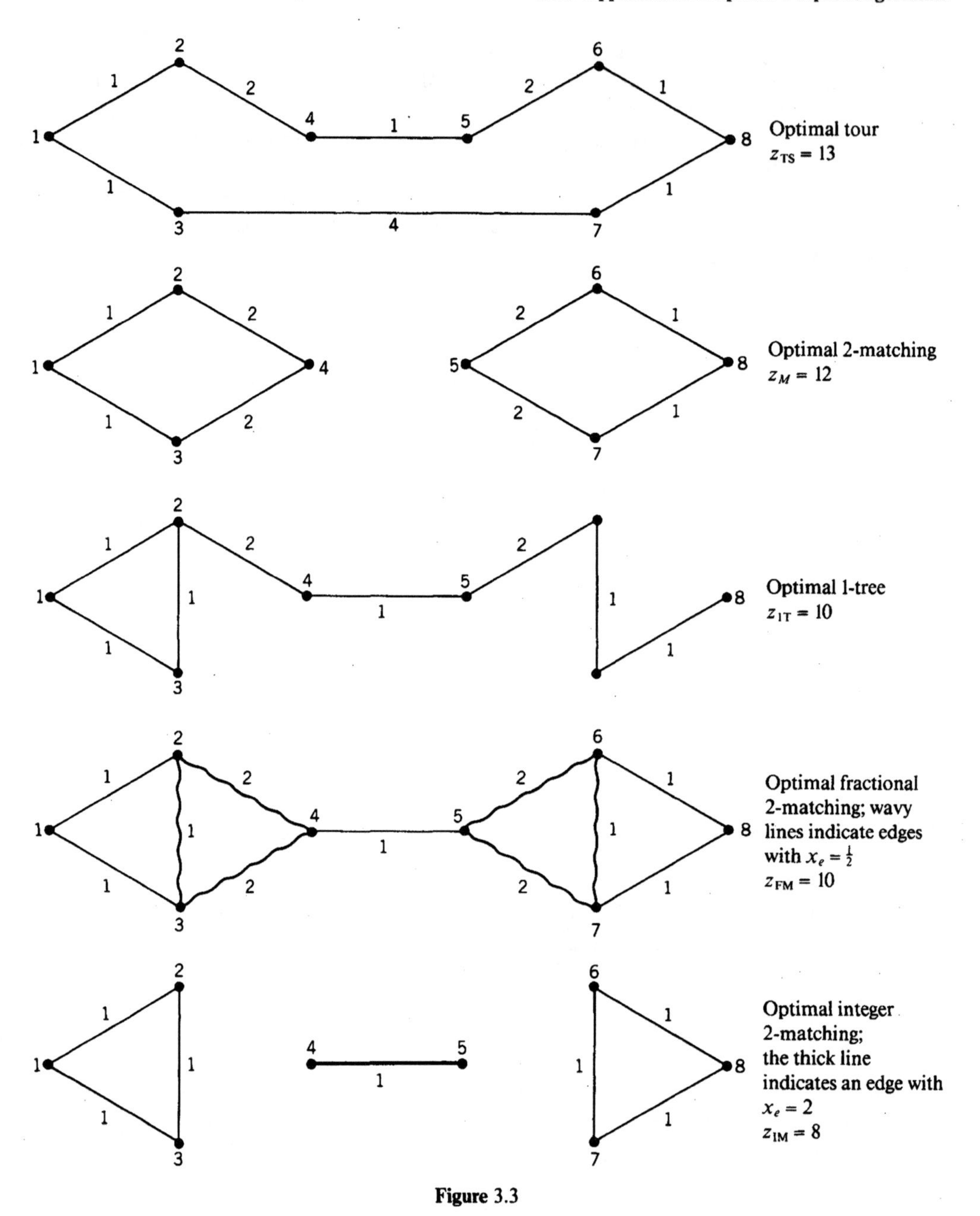

Figure 3.3

We now consider two more powerful relaxations that combine 2-matchings and 1-trees.

In the first of these, we seek a minimum-weight convex combination of 1-trees that satisfies the degree constraints. To formulate this problem, let $x^i \in B^{|E|}$ be the characteristic vector of the ith 1-tree for $i = 1, \ldots, p$, where p is the number of 1-trees of G, and let $c^i = \Sigma_{e \in E} c_e x_e^i$ be the weight of the ith 1-tree. The problem is

$$z_{\mathrm{MIT}} = \min \sum_{i=1}^{p} \lambda_i c^i$$

$$\sum_{i=1}^{p} \lambda_i \left(\sum_{e \in \delta(\{v\})} x_e^i \right) = 2 \quad \text{for } v \in V \tag{3.10}$$

$$\sum_{i=1}^{p} \lambda_i = 1$$

$$\lambda \in R_+^p.$$

The linear program (3.10) is a relaxation of the traveling salesman problem because if x^i is the characteristic vector of a tour, then $\lambda_i = 1$ and $\lambda_k = 0$ for $k \neq i$ is a feasible solution. The problem contains an enormous number of variables, since p is generally exponential in the size of the graph.

Figure 3.4 shows a feasible solution to (3.10) that is not a tour. Note, however, that it is a fractional 2-matching.

To see the relationship between (3.10) and the previous relaxations, we substitute $x = \Sigma_{i=1}^{p} \lambda_i x^i$ and use the fact that for all i, x^i satisfies (3.5). Thus x satisfies (3.4), (3.5), (3.7), and (3.8), so

$$z_{\mathrm{MIT}} \geqslant \min \left\{ \sum_{e \in E} c_e x_e \colon x \text{ satisfies (3.4), (3.5), (3.7), and (3.8)} \right\}.$$

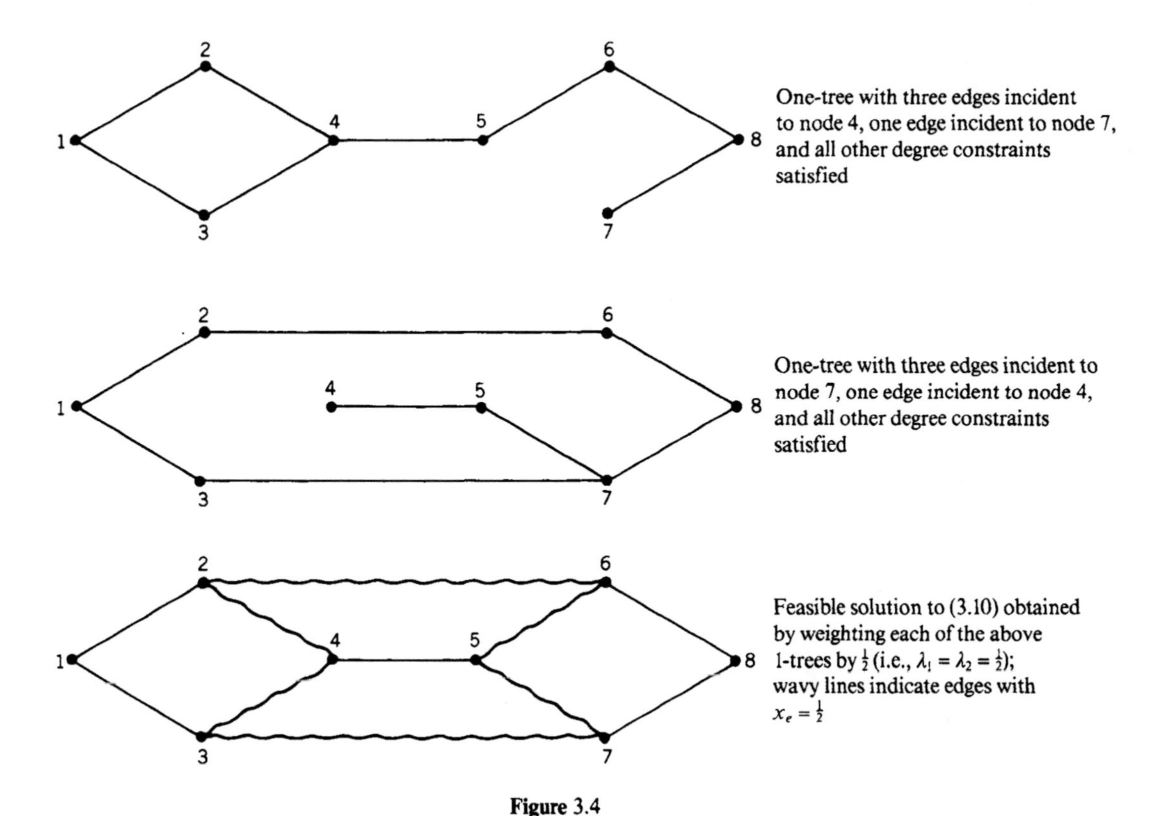
2
6
1
4
5
8
3
7
One-tree with three edges incident to node 4, one edge incident to node 7, and all other degree constraints satisfied
One-tree with three edges incident to node 7, one edge incident to node 4, and all other degree constraints satisfied
Feasible solution to (3.10) obtained by weighting each of the above 1-trees by ½ (i.e., λ₁ = λ₂ = ½); wavy lines indicate edges with xₑ = ½

Figure 3.4

Moreover, we will prove in Section III.3.3 that the convex hull of 1-trees is given by the polytope $\{x \in R_+^{|E|}: x$ satisfies (3.4) for $v = 1$, (3.5) and (3.8)$\}$. Hence if x satisfies (3.4), (3.5), (3.7), and (3.8), there is a λ that satisfies the constraints of (3.10) such that $x = \Sigma_{i=1}^p \lambda_i x^i$. Hence

$$z_{\text{MIT}} = \min\left\{\sum_{e \in E} c_e x_e: x \text{ satisfies (3.4), (3.5), (3.7), and (3.8)}\right\}, \tag{3.11}$$

and we obtain the result that the linear program (3.10) is equivalent to the linear programming relaxation of the integer programming formulation with the degree constraints and subtour elimination inequalities.

Also note that

$$z_{\text{MIT}} \geqslant \min_{i=1,\dots,p} c^i = z_{\text{IT}} \quad \text{and} \quad z_{\text{MIT}} \geqslant z_{\text{FM}}$$

since fractional 2-matching is a relaxation of (3.11) with the constraints (3.5) omitted.

Example 3.2. In Example 3.1, it can be shown that $z_{\text{MIT}} = z_{\text{TS}}$. The graph of Figure 3.5 provides an example with $z_{\text{MIT}} < z_{\text{TS}}$.

Analogous to the previous relaxation, we can consider the problem of finding a minimum-weight convex combination of 2-matchings that satisfies the constraints (3.5). This relaxation, as we will see, yields a bound that dominates all of the ones given above. Let $y^i \in B^{|E|}$ be the characteristic vector of the ith 2-matching for $i = 1, \dots, s$, where s is the number of 2-matchings of G, and let $d^i = \Sigma_{e \in E} c_e y_e^i$ be the weight of the ith 2-matching.

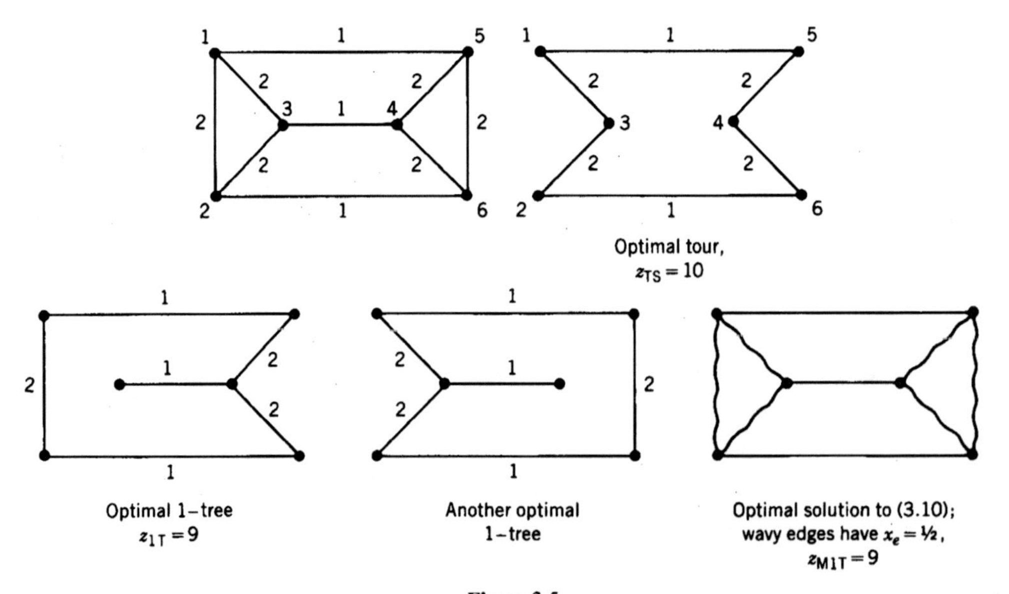

Figure 3.5

The problem is

$$z_{TM} = \min \sum_{i=1}^{s} \alpha_i d^i$$

(3.12)
$$\sum_{i=1}^{s} \alpha_i \left(\sum_{e \in E(U)} y_e^i \right) \leqslant |U| - 1 \quad \text{for } 3 \leqslant |U| \leqslant \left\lfloor \frac{|V|}{2} \right\rfloor \text{ and } U \subseteq V \backslash \{1\},$$

$$\sum_{i=1}^{s} \alpha_i = 1$$

$$\alpha \in R_+^s.$$

The linear program (3.12) is a relaxation of the traveling salesman problem because if y^i is the characteristic vector of a tour, then $\alpha_i = 1$ and $\alpha_k = 0$ for $k \neq i$ is a feasible solution.

To see the relationship between (3.12) and the previous relaxations, we first substitute $y = \Sigma_{i=1}^{s} \alpha_i y^i$ and use the fact that for all i, y^i satisfies the degree constraints. This yields $z_{TM} \geqslant z_{MIT}$ [see (3.11)].

Moreover, additional valid inequalities for the convex hull of 2-matchings are the 2-matching inequalities (3.6) of Section II.2.3. It can be shown that the 2-matching inequalities, together with (3.4), (3.7), and (3.8), define the convex hull of 2-matchings. Hence

(3.13)
$$z_{TM} = \min \left\{ \sum_{e \in E} c_e x_e \colon x \text{ satisfies (3.4), (3.5), (3.7), (3.8), and the 2-matching inequalities} \right\}.$$

Example 3.3. In Example 3.2, we have $z_{TM} = z_{TS} > z_{MIT}$, which shows that (3.13) may give a strictly better bound than (3.11). The graph of Figure 3.6 shows that it is possible to have $z_{TS} > z_{TM}$. An optimal solution to (3.13) is obtained by taking $\frac{1}{2}$ of each of the 2-matchings in Figure 3.6. Wavy edges have value of $\frac{1}{2}$, and $z_{TM} = 21$.

Figure 3.7 summarizes the bound information from the various relaxations.

The two relaxations that are most interesting are (3.11) and (3.13) since they alone use both the degree constraints and the subtour elimination constraints. In fact, we will see later in this section that both of these relaxations can be solved in polynomial time. Unfortunately, the only polynomial-time algorithms known for solving them require combining a cutting-plane or separation algorithm with an ellipsoid linear programming algorithm. Although this is not practical, a good (but not polynomial) approach is to use an FCPA for the subtour elimination constraints and to solve the resulting linear programs by a simplex algorithm. The other four relaxations can be solved efficiently by combinatorial polynomial-time algorithms.

Primal Heuristics

The general heuristic approaches proposed in Section II.5.3 are applicable to the symmetric traveling salesman problem. Several greedy-type algorithms can be constructed.

1. Nearest Neighbor. Start at an arbitrary node i_1 and construct a path $i_1, i_2, \ldots, i_j, i_{j+1}, \ldots, i_n$, where $i_{j+1} = \arg(\min\{c_{i_j k} \colon k \in V \setminus \{i_1, i_2, \ldots, i_j\})$, with ties broken arbitrarily.

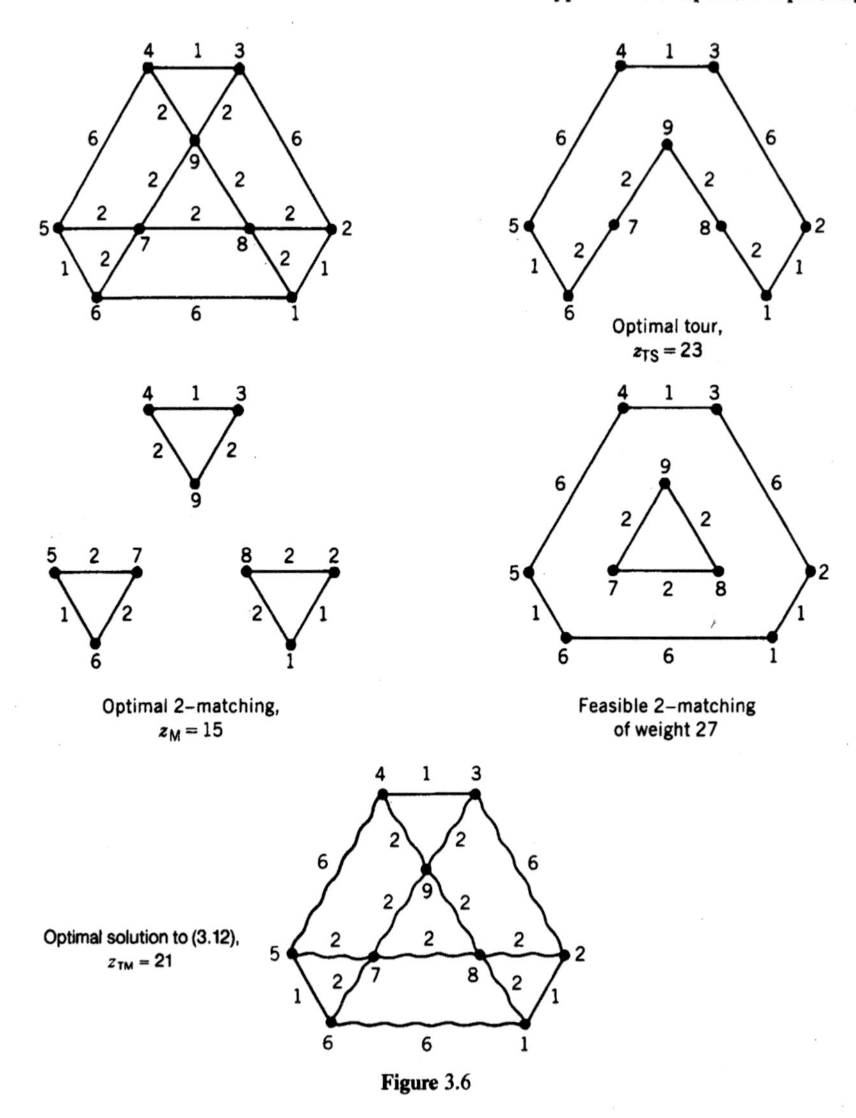

Figure 3.6

Complete the path to a tour by adding the edge (i_1, i_n). Note that unless the graph is complete, the procedure may fail to find a tour even if one exists. Moreover, even on complete graphs it can perform very badly by being forced to choose edges of very large weight in the last steps. In Example 3.1, nearest neighbor, starting at node 4, can choose the optimal tour (4 5 6 8 7 3 1 2 4), but it can also get stuck at (4 5 6 7 8).

2. Greedy Feasible. Start with $E^0 = \emptyset$. Given a set E^t at step $t < n - 1$ such that (i) (V, E^t) is acyclic and (ii) each node is of degree equal to or less than 2, add a minimum-weight edge $e \in E \setminus E^t$ (if one exists) so that $(V, E^t \cup \{e\})$ has properties i and ii. Complete (V, E^{n-1}) to a tour (if possible) by joining the two nodes of degree 1. The remarks we made about nearest neighbor also apply to greedy feasible. In particular, in Example 1.1, greedy feasible can find the optimal tour by taking edges in the order (1 2), (1 3), (6 8), (7 8), (4 5), (2 4), (5 6), (3 7), but it can also fail to find a tour by beginning with the edges (6 7), (6 8).

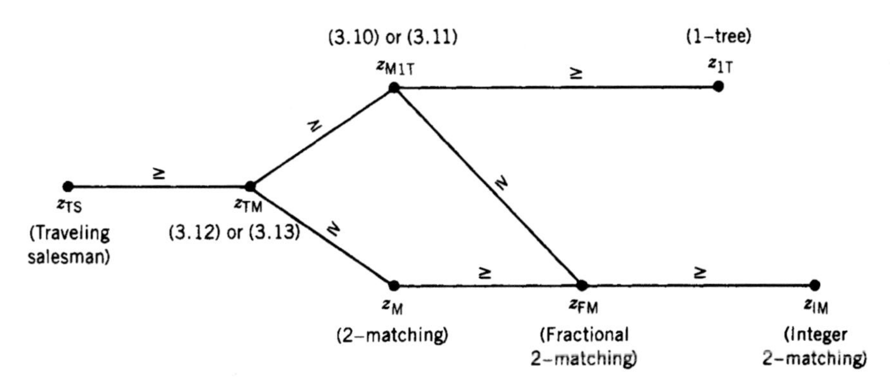

Figure 3.7

3. Nearest Insertion. (Here we suppose that G is a complete graph.) Given a subtour T and a node $i \in V \setminus T$, let $d(i, T) = \min_{j \in T} c_{ij}$, and let $i^* = \arg(\min\{d(i, T): i \in V \setminus T\})$. Suppose $j^* = \arg(\min\{c_{i^*j}: j \in T\})$. Thus i^* is the "closest" node to T, and j^* is the node in T that is closest to i^*. Now construct a subtour on $T \cup \{i^*\}$ by inserting i^* between j^* and one of its neighbors in T; that is, if (j_1, j^*) and (j_2, j^*) are edges of T and $c_{i^*j_1} \leq c_{i^*j_2}$, insert i^* between j_1 and j^*. This process terminates with a tour, but again we cannot guarantee that it will be a good tour.

4. k-Interchange. Local search heuristics are also useful for the traveling salesman problem. Given a tour, the *k-interchange* heuristic replaces k edges in the tour by k edges that are not in the tour if such a change yields a tour of lower weight. When $k = 2$, the two edges to be replaced cannot be adjacent, and there is a unique pair of replacement edges (if they exist) (see Figure 3.8) where the edges (i, j) and $(i + 1, j + 1)$ replace $(i, i + 1)$ and $(j, j + 1)$. Unfortunately, it is possible for a locally optimal tour to be poor for any k that is small relative to $|V|$.

The negative remarks we have made about each of the heuristics is to be expected. In fact, for complete graphs and arbitrary edge weights, we cannot expect any fast heuristic to provide a good performance guarantee. The proof of the following proposition, which is similar to the proof of Proposition 3.2 of Section II.5.3, is left as an exercise.

Proposition 3.2. *The traveling salesman problem with performance guarantee $r_H \geq r$ for any $r > 0$ is $\mathcal{NP}$-hard.*

Example 3.4. We apply the four heuristics given above to the traveling salesman problem on the 10-city distance matrix given in Table 2.1 of Section I.3.2.

1. *Nearest neighbor starting at city 1.* This yields the tour (1 8 9 4 7 10 6 2 3 5 1) of weight or distance 349.

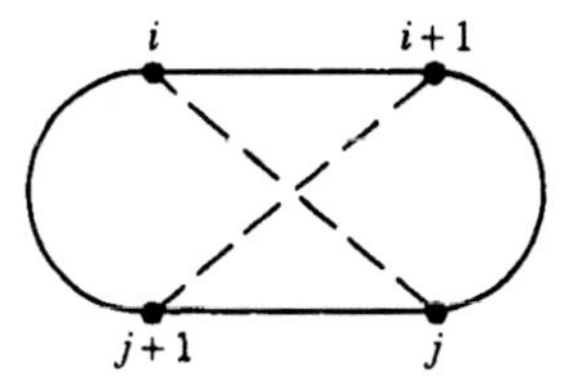

Figure 3.8

2. *Greedy feasible*. This yields the edge set (6 10), (4 9), (4 10), (2 6), (8 9), (1 8), (5 7), (1 5), (3 7), (2 3) and the tour (1 8 9 4 10 6 2 3 7 5 1) of distance 323.
3. *Nearest insertion beginning with the triangle*. (4 9 10 4). The successive subtours are (4 9 6 10 4), (4 9 6 10 7 4), (4 9 6 2 10 7 4), (4 8 9 6 2 10 7 4), (4 1 8 9 6 2 10 7 4). The resulting tour (1 8 9 6 2 10 3 7 5 4 1) has weight 372.
4. *2-Interchange beginning with the tour produced by nearest insertion.* We find the following sequence of improving tours (1 8 9 2 6 10 3 7 5 4 1) of weight 353, (1 8 9 2 6 10 3 7 4 5 1) of weight 328, and (1 8 9 2 6 10 3 4 7 5 1) of weight 325.

To obtain performance guarantees on the performance of the heuristics, the weight matrix must have structure. A natural structure to impose is nonnegativity and the triangle inequality, that is,

$$c_{ij} + c_{jk} \geqslant c_{ik} \quad \text{for all } i, j, k \in V.$$

The triangle inequality is, for example, satisfied by euclidean and rectilinear distances. We use the following property implied by the triangle inequality which is easily proved by induction.

Proposition 3.3. *If the triangle inequality is satisfied, then* $\Sigma_{t=0}^{k-1} c_{i_t i_{t+1}} \geqslant c_{i_0 i_k}$.

When the triangle inequality is satisfied, performance guarantees can be established for several heuristics. Some of these results are given as exercises. Here we present the polynomial-time heuristic, called *spanning tree-matching*, that has a performance guarantee of two-thirds. No other polynomial-time heuristic is known that has a performance guarantee that is as good. Moreover, it is not known if a polynomial-time heuristic with a better performance guarantee exists.

Before describing and analyzing this heuristic and a related one, we need to present a few additional definitions and results from graph theory. A graph $G = (V, E)$ in which there may be more than one edge joining a pair of nodes is called a *multigraph*. A *eulerian cycle* of a multigraph is a walk with the same beginning and end points that contains each edge of the graph exactly once. The graph of Figure 3.9 contains the eulerian cycle with node sequence (1 2 3 4 2 3 1) and edge sequence (e_1 e_3 e_6 e_5 e_4 e_2).

The following classic result from graph theory will be used to establish the performance bounds.

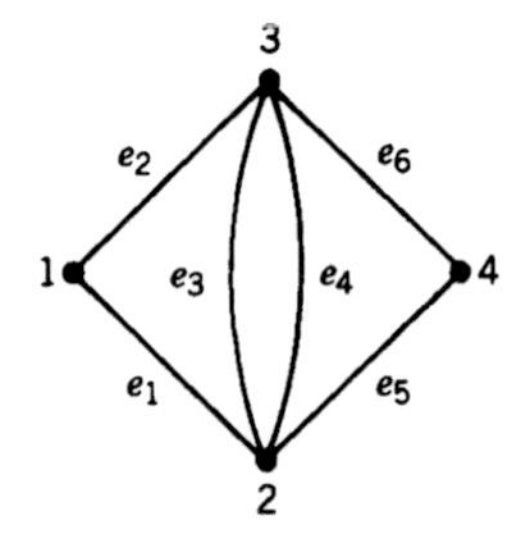

Figure 3.9

Proposition 3.4. *A multigraph contains a eulerian cycle if and only if each node is of even degree.*

Moreover, there is a simple and fast procedure (linear in the number of edges) for finding a eulerian cycle when one exists.

Now suppose we are given a complete graph $G = (V, E)$ and a spanning tree $G' = (V, E')$ of G. Here is a procedure for constructing a tour on G. Construct the multigraph $\hat{G}$ from G' by duplicating each $e \in E'$. Since each node of $\hat{G}$ is of even degree, $\hat{G}$ contains a eulerian cycle Q. Delete all node repetitions from Q except for the final return to the first node. The resulting node sequence T is a tour on G.

The procedure is illustrated in Figure 3.10. The node sequence of a eulerian cycle on $\hat{G}$ is $Q = (1\ \ 2\ \ 3\ \ 4\ \ 5\ \ 4\ \ 6\ \ 7\ \ 6\ \ 8\ \ 6\ \ 4\ \ 3\ \ 2\ \ 9\ \ 2\ \ 1)$. Hence $T = (1\ \ 2\ \ 3\ \ 4\ \ 5\ \ 6\ \ 7\ \ 8\ \ 9\ \ 1)$.

Double Spanning-Tree Heuristic. Find a minimum-weight spanning tree $G' = (V, E')$ of G. Duplicate each $e \in E$ and find a eulerian cycle Q on the resulting graph. Extract a tour T on G from Q by deleting node repetitions.

Theorem 3.5. *If the edge weights are nonnegative and satisfy the triangle inequality, then any tour produced by the double spanning-tree heuristic is of weight not greater than twice the weight of an optimal tour.*

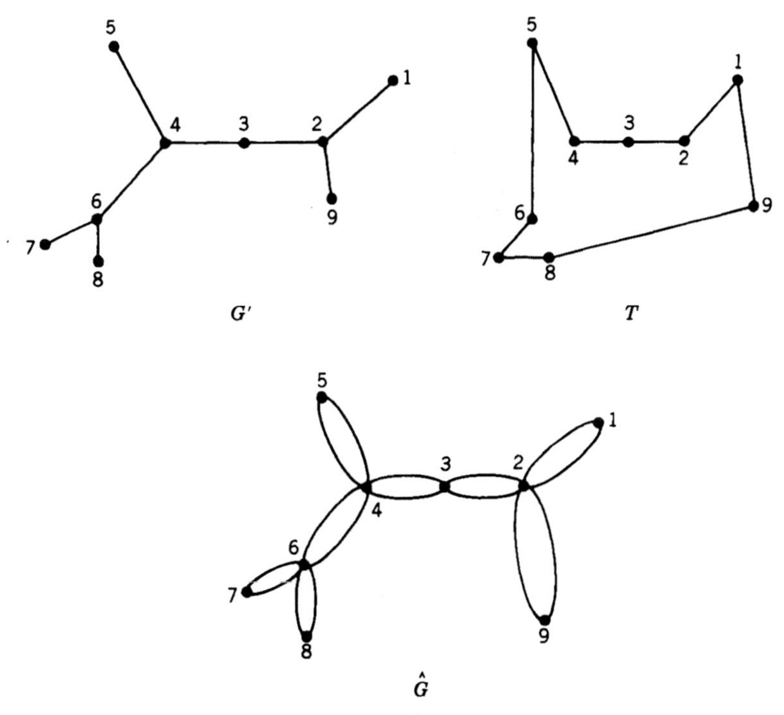

Figure 3.10

Proof. Let T^0 be an optimal tour with edge set E_{T^0}. Let E_T be the edge set formed by the heuristic, let E_Q be the edge set of the eulerian cycle, and let E' be the edge set of a minimum-weight spanning tree. Then

$$\sum_{e \in E_T} c_e \leqslant \sum_{e \in E_Q} c_e = 2 \sum_{e \in E'} c_e \leqslant 2 \sum_{e \in E_{T^0}} c_e,$$

where the first inequality follows from the triangle inequality, and the second one follows from nonnegativity because if an edge is deleted from a tour, the resulting subgraph is a spanning tree. ■

To produce a heuristic of this type that has a better performance guarantee, we need to find a smaller-weight set of edges to add to the minimum-weight spanning tree while maintaining the property that the resulting subgraph is eulerian.

Consider the nodes $U \subseteq V$ of $G' = (V, E')$ that are of odd degree. Since the sum of the node degrees for any graph is even, $|U|$ is even. Hence if we add $|U|/2$ edges to G', each of which is incident to two nodes of U, the resulting graph is eulerian. To find a minimum-weight set of such edges, we find a minimum-weight perfect matching M on the induced subgraph $G(U) = (U, E(U))$ of G. (In a perfect matching, each node is of degree 1.) This can be done in polynomial time (see Section III.2.3).

Now observe that a tour T is a sequence of paths $P_1 \cup P_2 \cup \cdots \cup P_{|U|}$, where P_i joins the ith and $(i+1)$st nodes j_i and j_{i+1} of U on the tour T (see Figure 3.11). By the triangle inequality, the length of path P_k is greater than or equal to $c_{j_k j_{k+1}}$. Moreover, edge sets $M_1 = \{(j_1, j_2), (j_3, j_4), \ldots, (j_{|U|-1}, j_{|U|}\}$ and $M_2 = \{(j_2, j_3), \ldots, (j_{|U|}, j_1)\}$ are both perfect matchings on $G(U)$. Hence

$$2 \sum_{e \in M} c_e \leqslant \sum_{e \in M_1} c_e + \sum_{e \in M_2} c_e \leqslant \sum_{e \in E_T} c_e.$$

Spanning-Tree/Perfect-Matching Heuristic. Find a minimum-weight spanning tree $G' = (V, E')$ of G. Find a minimum-weight perfect matching on the induced subgraph $G(U)$ of G, where $U \subseteq V$ is the set of nodes of V that are of odd degree in G'. Let M be the edge set of the perfect matching. Find a eulerian cycle Q on the multigraph $\tilde{G} = (V, E' \cup M)$. Extract a tour T on G from Q by deleting node repetitions.

The heuristic is illustrated in Figure 3.12.

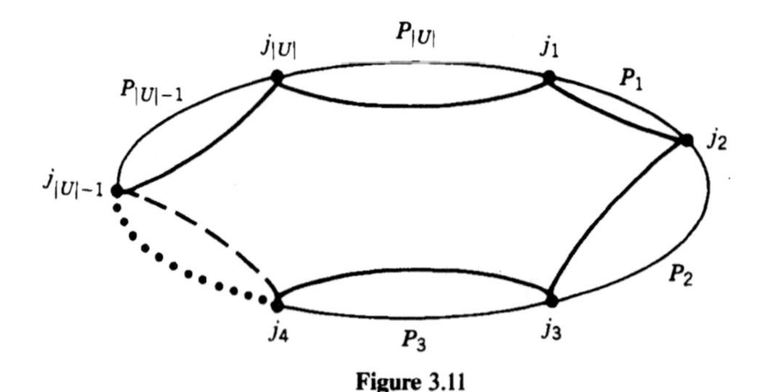

Figure 3.11

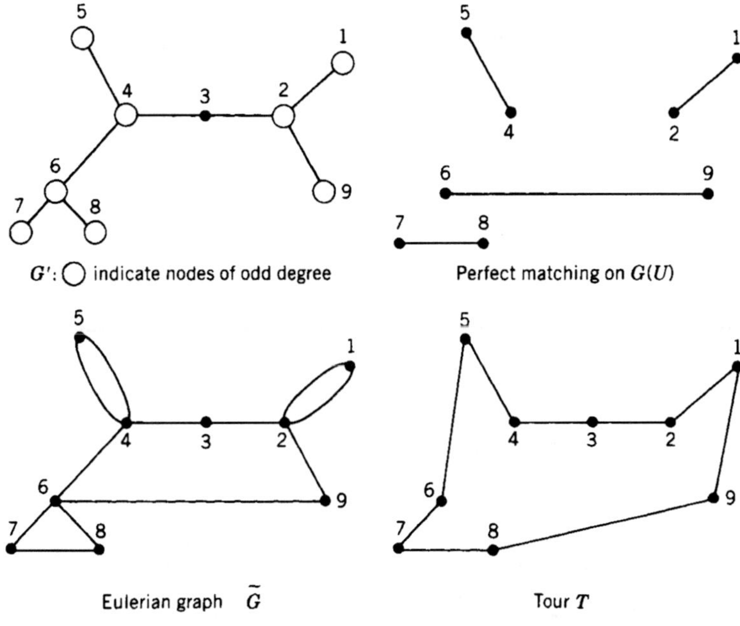

Figure 3.12

We have sketched a proof of the following theorem:

Theorem 3.6. *If the edge weights are nonnegative and satisfy the triangle inequality, then any tour produced by the spanning-tree/perfect-matching heuristic is of weight not greater than three-halves the weight of an optimal tour.*

In fact there are families of graphs for which the bound is asymptotically achieved.

Example 3.4. (continued). A minimum-weight spanning tree and a minimum-weight perfect matching on the nodes of odd degree in the tree are shown in Figure 3.13.

A eulerian cycle obtained from the double spanning-tree heuristic is (1 8 9 4 10 6 2 6 10 3 10 4 7 5 7 4 9 8 1), yielding the tour (1 8 9 4 10 6 2 3 7 5 1) of distance 323.

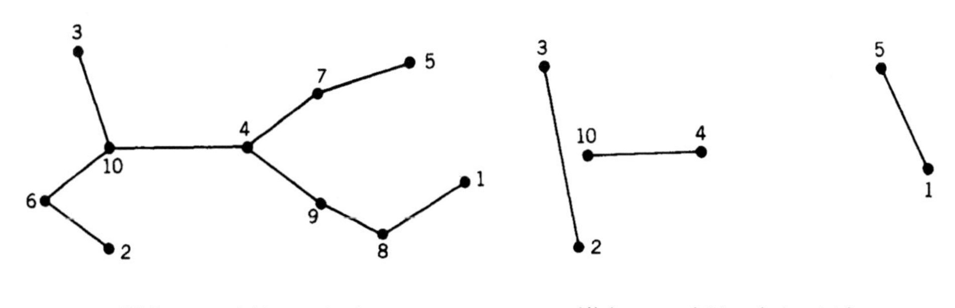

Figure 3.13

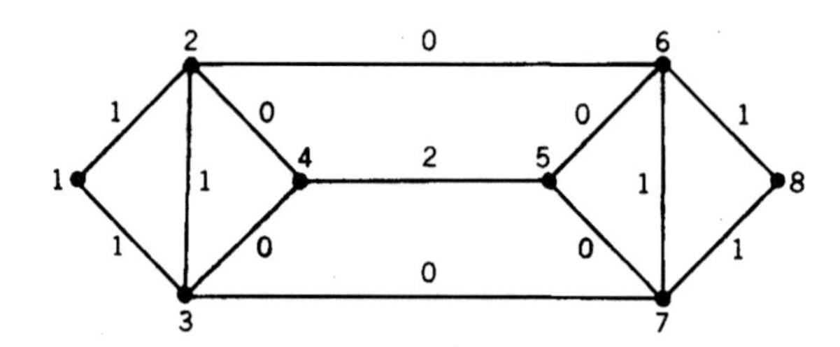

G and an integer 2–matching

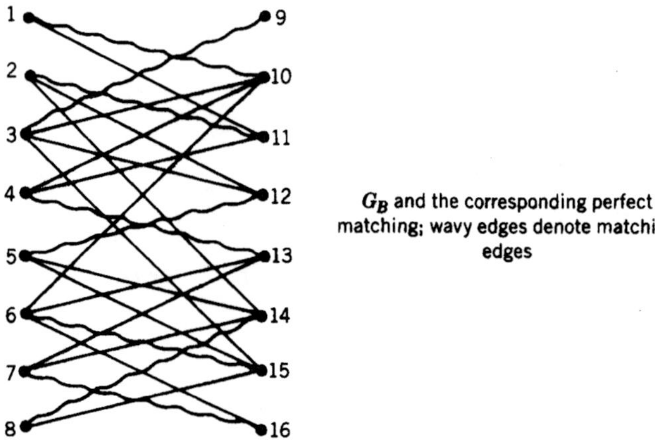

G_B and the corresponding perfect matching; wavy edges denote matching edges

Figure 3.14

A eulerian cycle obtained from the spanning-tree/perfect-matching heuristic is (1 8 9 4 10 6 2 3 10 4 7 5 1), which yields the same tour. Note that each of these heuristics could have produced several other tours, depending on the eulerian cycle chosen.

Relaxation/Branch-and-Bound Algorithms

Here we use the relaxations developed earlier in the section, together with primal heuristics and branch-and-bound, to develop algorithms for the traveling salesman problem that are capable of finding an optimal solution and proving optimality.

An Assignment Problem/Branch-and-Bound Algorithm. One of the earliest approaches for solving the traveling salesman problem used the integer 2-matching relaxation. In fact, the integer 2-matching relaxation can be solved as a $|V| \times |V|$ assignment problem or, equivalently, as a perfect-matching problem on a bipartite graph.

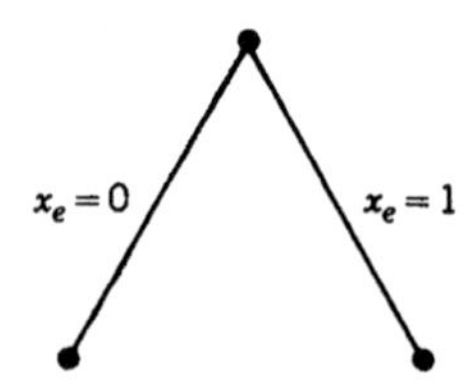

Figure 3.15

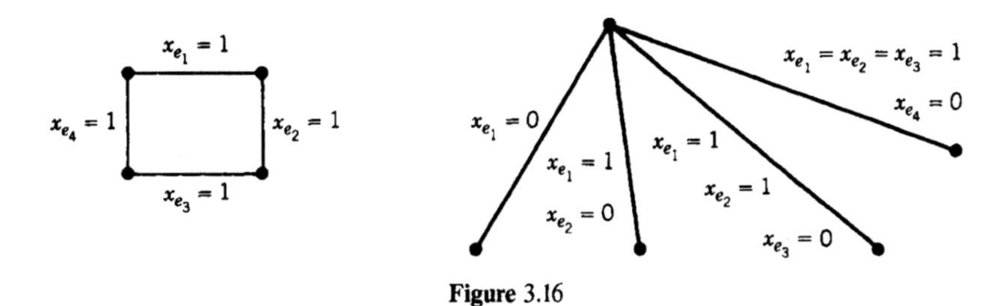

Figure 3.16

The bipartite graph $G_B = (V^L \cup V^R, E^*)$ is constructed from G as follows. Given $V = \{1, 2, \dots, m\}$, then $V^L = V$ and $V^R = \{m + 1, m + 2, \dots, 2m\}$. Corresponding to each edge $e = (i, j) \in E$, G_B contains two edges, $e^L = (i, m + j)$ and $e^R = (j, m + i)$. Also $c_{e^L} = c_{e^R} = c_e$ for all $e \in E$. The construction is illustrated in Figure 3.14. It is easy to see that if $y^0 \in B^{2|E|}$ is the characteristic vector of an optimal perfect matching in G_B, then x^0 with $x_e^0 = y_{e^L}^0 + y_{e^R}^0$ is an optimal integer 2-matching on G. Figure 3.14 also shows an integer 2-matching on G and a corresponding matching on G_B.

If we want to consider an integer 2-matching on G with $x_e = 0$, where $e = (i, j)$, then in G_B we delete e^L and e^R. Similarly, to obtain an integer 2-matching on G with $x_e = 1$, we delete nodes i and $j + m$ and all of the edges adjacent to them.

Now suppose we have solved the integer 2-matching problem and it is not a tour. To eliminate a solution with $x_e = 2$, we branch as shown in Figure 3.15.

To eliminate a subtour, we branch as shown in Figure 3.16. Multibranching is necessary in the case of a subtour in order to produce a tree in which the current infeasible solution is violated along every branch. Note that in the kth branch with $x_{e_k} = 0$, $k \geqslant 2$, we set $x_{e_1} = x_{e_2} = \cdots = x_{e_{k-1}} = 1$ since any tour that does not contain all of the edges $e_1, \dots, e_{k-1}$ is contained in one of the branches $1, \dots, k - 1$. To avoid creating many branches, it is desirable to choose a subtour containing the fewest number of edges.

Example 3.1 (continued). We solve this problem using the integer 2-matching relaxation. The initial solution of weight 8 is shown in Figure 3.3. We choose to branch on the edge (4, 5) since $x_{45} = 2$ (see Figure 3.17).

The node 1 and 2 solutions are shown in Figure 3.18. Branching from node 1, as shown in Figure 3.19, we find that none of the remaining nodes are feasible. Hence the solution at node 2 is optimal.

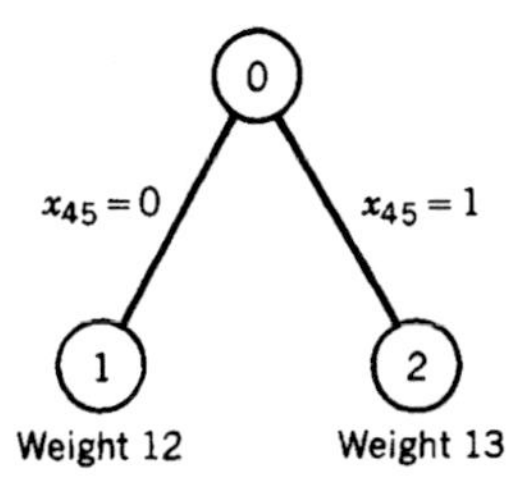

Figure 3.17

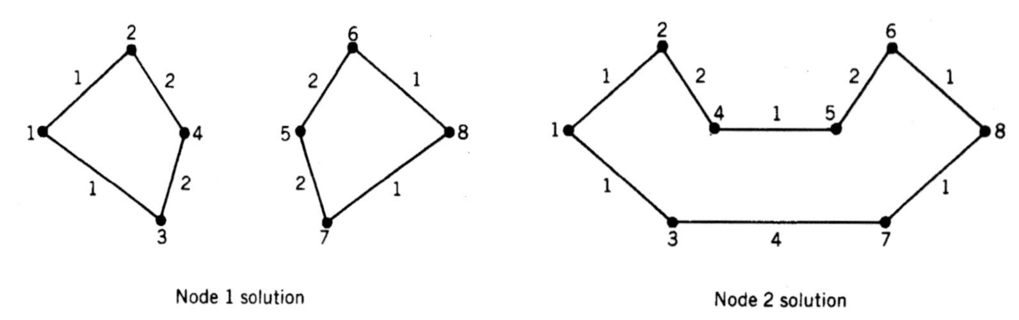

Figure 3.18

A 1-Tree, Subgradient Optimization, Branch-and-Bound Algorithm. Now we consider a branch-and-bound algorithm that uses a Lagrangian dual relaxation. For $\lambda = (\lambda_1 = 0, \lambda_2, \ldots, \lambda_{|V|}) \in R^{|V|}$, let

$$z_{1\mathrm{T}}(\lambda) = 2 \sum_{i \in V} \lambda_i + \min_{x \in B^{|E|}} \left\{ \sum_{(i,j) \in E} (c_{ij} - \lambda_i - \lambda_j) x_{ij} \right\}, \tag{3.14}$$

where x satisfies (3.5) and, for node 1, also satisfies (3.4); that is, x is the characteristic vector of a 1-tree. Let

$$z_{\mathrm{LD}} = \max_{\substack{\lambda \in R^{|V|} \\ \lambda_1 = 0}} z_{1\mathrm{T}}(\lambda).$$

As noted above, the vertices of the polytope $\{x \in R_+^{|E|}$: x satisfies (3.5), (3.4) for node 1, and (3.8)$\}$ are precisely the 1-trees. Hence from Corollary 6.6 of Section II.3.6, we have

$$z_{\mathrm{LD}} = z_{\mathrm{M1T}}. \tag{3.15}$$

For a given λ, problem (3.14) is to find a minimum-weight 1-tree with respect to the weights $c_{ij} - \lambda_i - \lambda_j$. In Section I.3.3, we gave an efficient "greedy" algorithm for finding a minimum-weight spanning tree of a graph. To find a minimum-weight 1-tree, we first find a minimum-weight spanning tree for the subgraph induced by nodes $V \setminus \{1\}$ and then we add the two smallest-weight edges incident to node 1.

If the resulting 1-tree is a tour, then by Corollary 6.8 of Section II.3.6, $z_{1\mathrm{T}}(\lambda) = z_{\mathrm{LD}} = z_{\mathrm{TS}}$. If the resulting 1-tree is not a tour, we can iterate on the λ's. An intuitive scheme, suggested by the objective function in (3.14), is to increase λ_i when the degree of node i in the 1-tree is equal to 1 and to decrease λ_i when the degree of node i in the 1-tree is greater than 2. In fact,

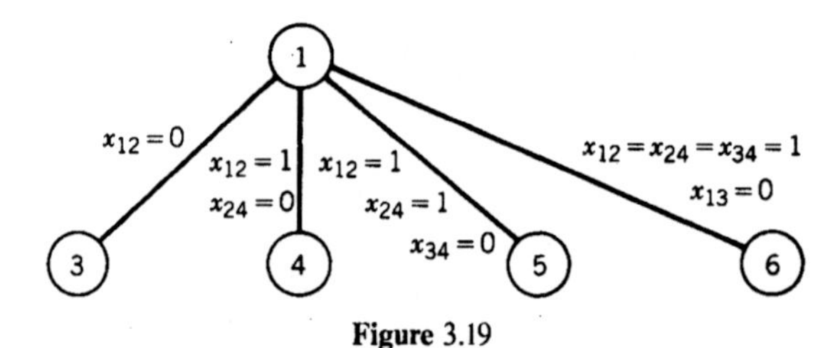

Figure 3.19

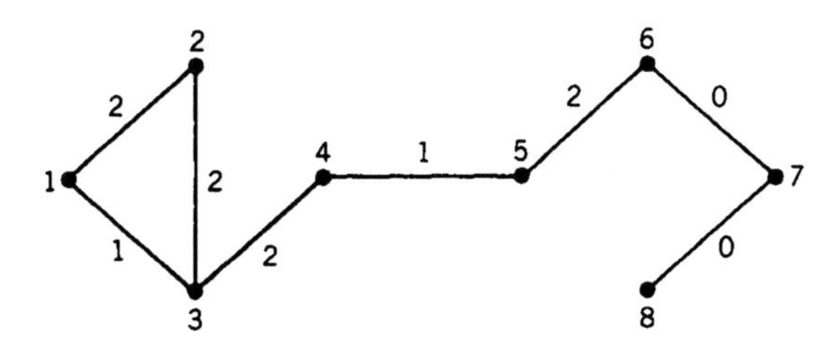

Figure 3.20

for a given λ^*, the vector $\delta(\lambda^*)$ with $\delta_i(\lambda^*) = (2 -$ degree of node i in an optimal 1-tree) is a subgradient to the objective function $z_{1T}(\lambda)$ at $\lambda = \lambda^*$. Hence we only need to specify a step size to solve the Lagrangian dual by subgradient optimization (see Section I.2.4). An intuitive explanation of the Lagrangian relaxation is that by transforming the edge weights to $c'_{ij} = c_{ij} - \lambda_i - \lambda_j$, the weight of all tours decreases by $2 \sum \lambda_i$. Thus we get an equivalent problem with weight vector c'. However, minimum-weight 1-trees are a function of λ, so the objective is to find a λ such that the minimum-weight 1-tree is a tour.

It may be difficult to solve the Lagrangian dual to optimality, particularly when $z_{LD} < z_{TS}$. Corollary 6.9 of Section II.3.6 can be used to find a nearly optimal λ; alternatively, we can stop with $z_{1T}(\lambda^*)$ if $|z_{1T}(\lambda^*) - \bar{z}_{TS}| < \varepsilon$, where $\varepsilon > 0$ is a prescribed tolerance and $\bar{z}_{TS}$ is the weight of some feasible tour.

When we terminate without having found an optimal tour, the calculations can be continued using branch-and-bound. Suppose $z_{1T}(\lambda^*)$ is the largest known value of $z_{1T}(\lambda)$, and let x^* be the characteristic vector of the 1-tree obtained from solving (3.14) with $\lambda = \lambda^*$. This 1-tree contains a subtour. Thus we can proceed as we did with the integer matching relaxation algorithm to develop a branch-and-bound tree.

Example 3.1 (continued). With $\lambda^0 = 0$, an optimal 1-tree is shown in Figure 3.3 and we obtain $z_{1T}(\lambda^0) = 10$. Since node 2 is of degree 3 and node 8 is of degree 1, we decrease λ_2 and increase λ_8. Let $\lambda^1 = (0 \ \ -1 \ \ 0 \ \ 0 \ \ 0 \ \ 0 \ \ 0 \ \ 1)$. An optimal 1-tree is shown in Figure 3.20 and we obtain $z_{1T}(\lambda^1) = 10$.

Continuing in this manner, after several iterations, we find $\lambda^* = (0 \ \ -2 \ \ -2 \ \ -1 \ \ 0 \ \ 2 \ \ 1 \ \ 2)$. The weights $c_{ij} - \lambda_i^* - \lambda_j^*$ and an optimal 1-tree are shown in Figure 3.21. Thus we have found an optimal tour.

An FCP/Branch-and-Bound Algorithm. Here we consider an FCP/branch-and-bound algorithm of the type described in Section II.5.2. As shown above, the characteristic vectors of tours are given by (3.3), (3.4), and (3.5) for $U \subset V$ with $3 \leqslant |U| \leqslant \lfloor \frac{1}{2}|V| \rfloor$. Hence the formulation we work with is

$$\min \sum_{e \in E} c_e x_e$$

(STSP)

$$\sum_{e \in \delta(v)} x_e = 2 \qquad \text{for } v \in V \tag{3.4}$$

$$\sum_{e \in E(U)} x_e \leqslant |U| - 1 \qquad \text{for } U \subset V,\ |3| \leqslant |U| \leqslant \left\lfloor \frac{|V|}{2} \right\rfloor \tag{3.5}$$

$$x \in B^{|E|}. \tag{3.3}$$

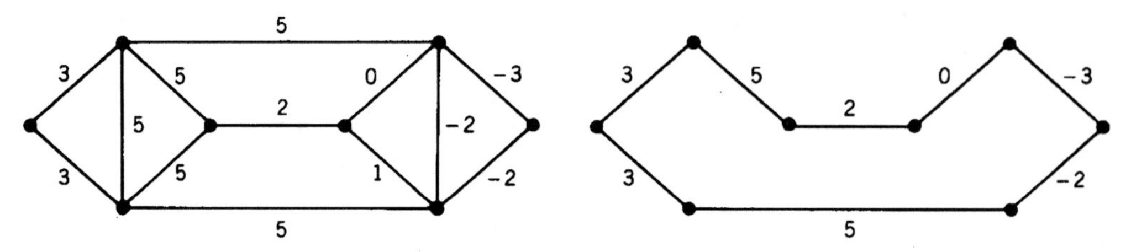

Figure 3.21

In Section II.2.3 we derived some classes of facets for the convex hull of solutions to STSP, so we now investigate the separation problems for these classes. First we examine the separation problem for the subtour elimination inequalities (3.5). Though these appear in our formulation of STSP, the exponential number of these inequalities makes it impossible to consider all of them as part of the initial LP relaxation. Therefore we typically start with the relaxation LP^1 involving just the degree constraints (3.4), nonnegativity, and the upper bounds (3.8), namely,

$$S_R^1 = \{x \in R_+^{|E|}: x \text{ satisfies (3.4) and (3.8)}\}.$$

Proposition 3.7. *If $x^* \in S_R^1$, then $\Sigma_{e \in E(W)} x_e^* = |W| - 1 + \varepsilon$ if and only if $\Sigma_{e \in \delta(W)} x_e^* = 2 - 2\varepsilon$.*

Proof. From (3.4), we obtain

$$2|W| = 2\left(\sum_{e \in E(W)} x_e^*\right) + \sum_{e \in \delta(W)} x_e^*;$$

or in other words,

$$2 - \sum_{e \in \delta(W)} x_e^* = 2\left(\sum_{e \in E(W)} x_e^* - (|W| - 1)\right). \qquad \blacksquare$$

It follows that a subtour inequality (3.5) is violated by x^* if and only if some *cut-set inequality*

$$\sum_{e \in \delta(W)} x_e \geq 2$$

is violated by x^*. Hence to determine whether there exists $W \subset V$ with $|W| \geq 3$ for which $\Sigma_{e \in \delta(W)} x_e^* < 2$, it suffices to solve

$$\text{(3.16)} \qquad \zeta = \min\left\{\sum_{e \in \delta(U)} x_e^*: U \subset V, 3 \leq |U| \leq \left\lfloor \frac{|V|}{2} \right\rfloor\right\},$$

and check whether $\zeta < 2$ or not.

Now if we impose $s \in U$, and $t \in \overline{U}$, then

$$\min\left\{\sum_{e \in \delta(U)} x_e^*: U \subset V, s \in U, t \in \overline{U}\right\}$$

is a minimum $s-t$ cut problem and can be solved by the maximum $s-t$ flow algorithm (see Section I.3.4). It follows that (3.16) can be solved efficiently by solving a set of maximum $s-t$ flow problems.

Based on the symmetry $\Sigma_{e\in\delta(U)} x_e^* = \Sigma_{e\in\delta(\overline{U})} x_e^*$, an alternative to (3.16) is

$$\zeta = \min\left\{\sum_{e\in\delta(U)} x_e^*: 3 \leqslant |U| \leqslant |V| - 3,\ 1 \in U\right\}. \tag{3.17}$$

Note that the choice of the node fixed in U is arbitrary. To solve (3.17), let

$$\zeta_j = \min\left\{\sum_{e\in\delta(U)} x_e^*: \{1, 2, \ldots, j-1\} \subset U, j \in \overline{U}, 3 \leqslant |U| \leqslant |V| - 3\right\} \tag{3.18}$$

$$\text{for } j = 2, \ldots, |V| - 2.$$

Then $\zeta = \min_{j=2,\ldots,|V|-2} \zeta_j$, since the minimum cut is a $1-j$ cut for some j. Imposing the condition $\{2, \ldots, j-1\} \subset U$ in the $1-j$ cut problem is easily carried out by replacing the capacities x_{1k}^* by ∞ for $k = 2, \ldots, j-1$. Thus the separation algorithm is to solve the maximum $1-j$ flow problem for $j = 2, \ldots, |V| - 2$.

Proposition 3.8. *Let (ζ, U) be an optimal solution resulting from the separation algorithm:*

a. *If $\zeta \geqslant 2$, no subtour elimination constraint is violated.*
b. *If $\zeta < 2$, the subtour elimination inequality (3.5) with $W = U$ is a most violated inequality.*

It is very often possible to reduce the size of the separation problem for subtour elimination constraints. Let x^* be a feasible solution of S_R^1 and let $G(x^*) = (V, E(x^*))$, where $e \in E(x^*)$ only if $x_e^* > 0$. The simplest case is when $G(x^*)$ is not connected (see, e.g., Figure 3.22). For each component with node set U, we obtain $\Sigma_{e\in E(U)} x_e = |U|$ because of the degree constraints, and hence the violated inequalities are found by testing $G(x^*)$ for connectedness.

The second case is where $G(x^*)$ is connected, but $x_e^* = 1$ for some $e \in E$. All the edges with $x_e^* = 1$ can be shrunk by the following procedure.

Shrinking an Edge $e = (i, j)$ of $G(x^*)$ with $x_e^* = 1$

Step 1: Replace nodes i and j by a single node l.

Step 2: Every pair of edges $e_1 = \{i, k\}$, $e_2 = \{j, k\}$ is replaced by a single edge $e^* = (l, k)$ with edge weight $x'_{e^*} = x_{e_1}^* + x_{e_2}^*$.

Step 3: All other edges (i, p) and (j, q) are replaced by the edges (l, p) and (l, q), respectively, with the same weight as before.

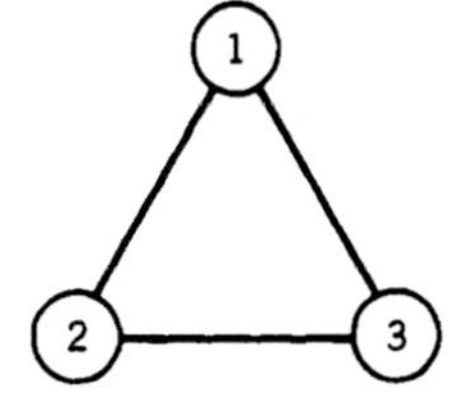

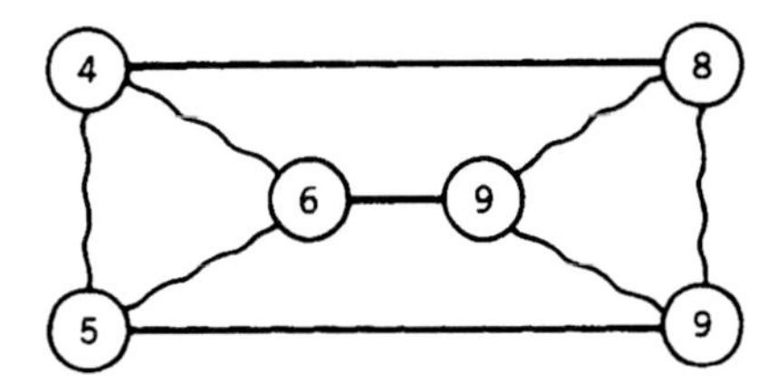

Figure 3.22. Wavy lines indicate $x_e = \frac{1}{2}$.

Let $G(x') = (V', E(x'))$ be the new graph obtained after shrinking.

Proposition 3.9. *There exists a $W \subset V$, $W \neq \{i, j\}$, such that $\Sigma_{e \in E(W)} x_e^* > |W| - 1$ if and only if there exists a $W' \subset V'$ such that $\Sigma_{e \in E(W')} x_e' > |W'| - 1$ in the reduced graph.*

Proof. If $\{i, j\} \subset W$, then it suffices to take $W' = (W \setminus \{i, j\}) \cup l$.

If $i \in W, j \notin W$ and the subtour inequality constraint for W is violated, then the one for $\tilde{W} = W \cup \{j\}$ is violated by at least as much. Now $\{i, j\} \subset W$, and the argument is as above.

If $i \notin W, j \notin W$, then take $W' = W$. ■

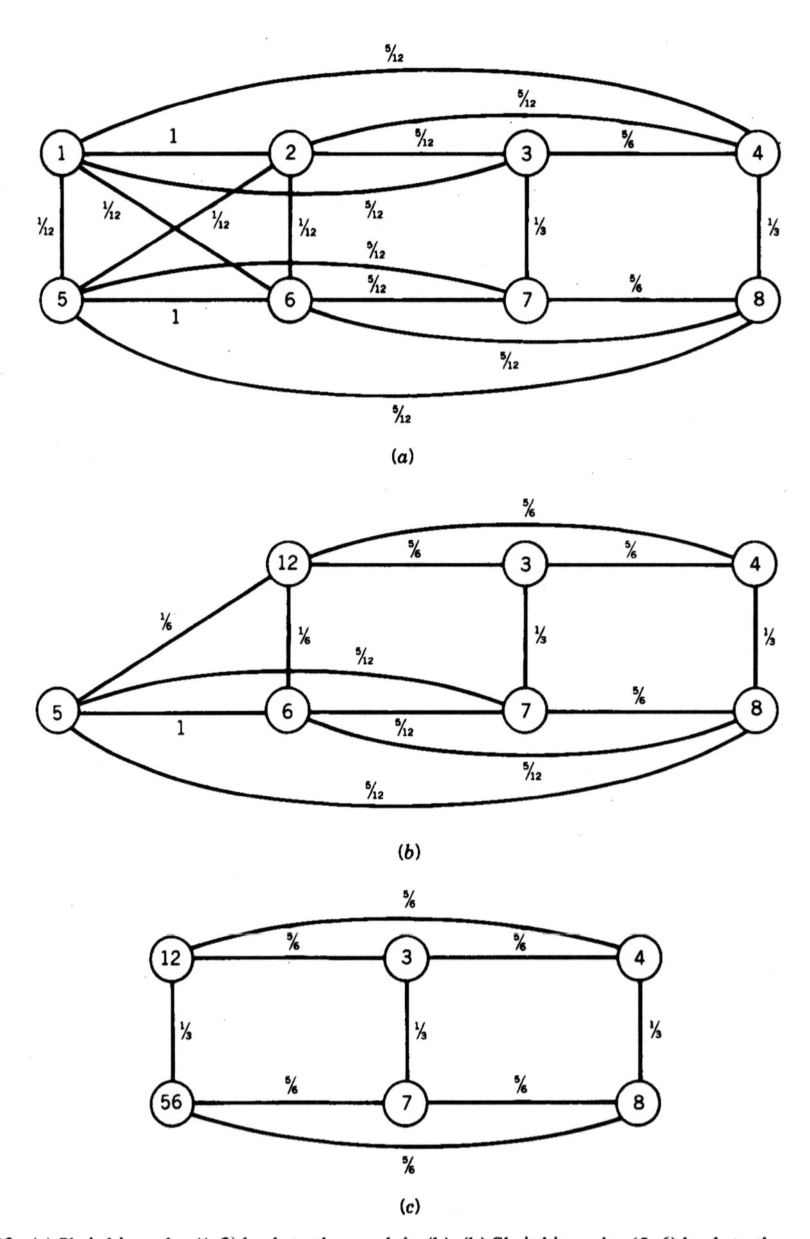

Figure 3.23. (a) Shrinking edge (1, 2) leads to the graph in (b). (b) Shrinking edge (5, 6) leads to the graph in (c).

Obviously this procedure can be applied iteratively, so the separation algorithm need only be applied to the reduced graph in which all the initial edges with $x_e^* = 1$ (and possibly others created during the procedure) have been shrunk (see Figure 3.23).

There is also no doubt that the human eye is very good at detecting anomalies in tours, routes, and so on, and several researchers have very successfully found violated inequalities in this way. The reader should therefore have no difficulty in finding a violated subtour inequality in the last graph of Figure 3.23, which can then be converted into a violated inequality for the initial graph.

Now we describe a modification of the FCP/branch-and-bound algorithm of Section II.5.2 with $\mathcal{F}$ equal to the set of subtour elimination inequalities. A modification is required because all the subtour elimination constraints are needed to correctly describe the integer programming formulation of STSP, and the branch-and-bound algorithm is applied to a formulation involving only a subset of these constraints. Thus the linear programming relaxation at any node other than the initial node may yield an integer solution that is a 2-matching but not a tour. (At the initial node 2-matchings are always cut off by subtour elimination inequalities.)

We describe three options that differ only in their treatment of the problem at nodes of the tree other than the initial node. In option 1, the remaining nodes are pruned when an integer solution is found. Hence the branch-and-bound phase terminates with an integer solution that may be a tour or a 2-matching. If it is a tour, it is an optimal solution of STSP. Otherwise we add the subtour elimination inequalities that are violated by the 2-matchings that have been found and not pruned by bounding, and we restart the branch-and-bound algorithm from the beginning.

Option 2 is to apply the separation algorithm at each node of the tree. Then the linear programming relaxation of STSP is solved exactly, and the lower bound obtained at each node is identical with that obtained by Lagrangian duality.

A third option, which is a compromise between the first two, is to apply the separation algorithm only at those nodes of the tree that yield an integer solution that is not a tour. A justification for this option is that the separation routine for integral solutions only involves a test for connectedness and allows us to exclude infeasible integral solutions.

Example 3.3 (continued). We apply the modified FCP/branch-and-bound algorithm.

Phase 1

Iteration 1: The solution is the optimal 2-matching given in Figure 3.6 and we obtain $z_{LP}^1 = 15$. Because $G(x^1)$ is not connected, the connected components immediately give the cuts

$$x_{12} + x_{18} + x_{28} \leqslant 2,$$

$$x_{56} + x_{57} + x_{67} \leqslant 2,$$

$$x_{34} + x_{39} + x_{49} \leqslant 2.$$

Iteration 2: $z_{LP}^2 = 21$. Applying the separation algorithm, no violated subtour elimination inequalities are found. The solution x^2 is the fractional solution shown in Figure 3.6.

Phase 2. The branch-and-bound tree has three nodes (see Figure 3.24). With $x_{16} = 0$, the relaxation has an integer optimal solution that is a tour of weight 23, given in Figure 3.6. With $x_{16} = 1$, the branch is pruned by bounding.

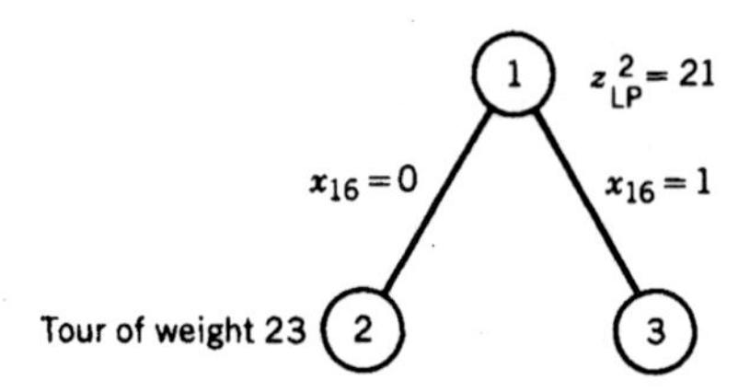

Figure 3.24

The second class of facets of interest for STSP are the 2-matching inequalities [see (3.6) of Section II.2.3]. There is a polynomial algorithm, again involving the solution of maximum-flow problems, to detect whether a point x^* feasible in S_R^1 violates a 2-matching inequality. This separation algorithm is based on the fact that the 2-matching inequalities are valid inequalities for the set

$$S_{2M} = \left\{x \in B^{|E|}: \sum_{e\in\delta(v)} x_e = 2 \text{ for all } v \in V\right\}.$$

Note that if FCPA is applied with both subtour elimination and 2-matching constraints, it terminates with an optimal solution to (3.13) with value z_{TM}. The same modifications as before must be made in the branch-and-bound phase.

For more general comb inequalities and clique tree inequalities (see Section II.2.3), no polynomial-time separation algorithm is known. However, using heuristics to reduce the size of the problem and inspection is sometimes a viable way of finding violated comb inequalities.

Example 3.3 (continued). We apply the FCPA with separation where subtour elimination, 2-matching, and comb inequalities are added.

Iteration 1: As before, $z^1_{LP} = z_{2M} = 15$.

Iteration 2: After adding subtour elimination inequalities, we obtain $z^2_{LP} = z_{MIT} = 21$. Applying a separation algorithm for 2-matching inequalities to the solution x^2 in Figure 3.6, no violated inequalities are found. Hence $z^2_{LP} = z_{TM}$. However, the comb inequality (3.7) of Section II.2.3 with $H = \{7 \quad 8 \quad 9\}$, $W_1 = \{1 \quad 2 \quad 8\}$, $W_2 = \{3 \quad 4 \quad 9\}$, $W_3 = \{5 \quad 6 \quad 7\}$,

$$(x_{78} + x_{79} + x_{89}) + (x_{12} + x_{19} + x_{29}) + (x_{34} + x_{38} + x_{48}) + (x_{56} + x_{57} + x_{67}) \leqslant 7$$

is violated by x^2.

Iteration 3: After adding this constraint, we obtain $z^3_{LP} = 23$; the resulting solution is the optimal tour shown in Figure 3.6.

Solution of a Large Problem

Example 3.5. The problem is to find the shortest tour through 67 cities in Belgium. The intercity distances, in kilometers, are shown in Table 3.1.

The nearest-neighbor heuristic, starting at city 1, leads to a tour of length 2045 km. The greedy heuristic gives a tour of length 1805; and when the 2-interchange heuristic is applied to the greedy tour, a solution of length 1691 is found (see Figure 3.25).

Table 3.1.

	AALST	AARSCHOT	ANTWERPEN	ARLON	ATH	BASTOGNE	BEAUMONT	BLANKENBERGE	BOUILLON	BRUGGE	BRUSSEL	CHARLEROI	CHIMAY	DEINZE	DENDERMONDE	DIEST	DIKSMUIDE	DINANT	DURBUY	EEKLO	ENGHIEN	EUPEN	GERAARDSBERGEN	GENT	HASSELT	HUY	HOUFFALIZE	IEPER	KORTRIJK	LA ROCHE	LEUVEN	LEUZE	LIEGE	LIER	LUXEMBURG	MAASEIK	MALMEDY	MARCHE	MECHELEN	MONS	NAMUR	NEUFCHATEAU	NIVELLES	OOSTENDE	OUDENAARDE	PHILIPPEVILLE	QUEVY	ROESELARE	RONSE	SOIGNIES	ST. HUBERT	ST. VITH	ST. NIKLAAS	ST. TRUIDEN	SPA	STAVELOT	THUIN	TIELT	TIENEN	TONGEREN	TORHOUT	TOURNAI	TURNHOUT	VERVIERS	VEURNE	VIRTON
AALST
AARSCHOT	72
ANTWERPEN	51	43
ARLON	222	217	229
ATH	52	92	97	219
BASTOGNE	182	163	199	40	179
BEAUMONT	99	116	121	156	57	130
BLANKENBERGE	83	143	115	285	123	265	168
BOUILLON	186	161	203	63	159	65	91	269
BRUGGE	72	141	101	299	100	251	154	14	255
BRUSSEL	30	42	47	192	50	152	74	113	156	99
CHARLEROI	76	89	94	174	61	135	27	165	104	151	47
CHIMAY	118	140	145	150	81	139	24	206	87	192	98	51
DEINZE	46	118	71	283	60	243	114	54	232	40	76	108	138
DENDERMONDE	13	61	38	221	65	181	107	92	185	78	29	76	131	48
DIEST	85	16	50	187	105	147	117	168	162	154	55	80	131	131	77
DIKSMUIDE	96	166	121	296	100	256	145	47	247	41	124	143	169	50	98	179
DINANT	120	95	137	110	117	92	53	203	66	189	80	55	62	166	119	96	198
DURBUY	155	110	172	99	161	59	103	243	81	229	120	107	102	196	168	97	231	50
EEKLO	47	112	74	269	98	229	141	39	233	25	77	124	165	37	51	132	66	167	202
ENGHIEN	32	72	77	226	20	178	62	120	182	106	30	44	86	60	45	85	110	116	153	79
EUPEN	173	122	165	144	200	104	166	261	169	247	103	139	176	219	172	106	269	114	74	220	173
GERAARDSBERGEN	27	83	78	235	25	195	82	97	184	83	41	61	106	43	40	99	93	116	168	73	17	184
GENT	27	92	54	249	70	202	121	51	213	52	55	104	145	17	31	110	67	147	182	20	59	200	41
HASSELT	107	38	78	165	130	195	120	193	169	179	77	93	146	153	99	22	206	95	92	152	107	84	118	132
HUY	124	77	108	128	121	88	103	207	114	193	67	76	108	163	123	67	198	46	33	171	120	68	137	144	57
HOUFFALIZE	184	146	201	57	181	17	132	267	89	253	154	136	141	245	183	130	258	79	51	231	180	87	197	211	108	84
IEPER	101	164	126	309	92	270	132	70	239	56	126	135	156	55	103	180	23	190	242	71	102	269	85	68	202	211	280
KORTRIJK	68	140	93	272	63	242	103	57	228	43	87	107	127	22	70	156	42	162	214	59	73	246	56	39	173	183	243	29
LA ROCHE	159	136	176	69	156	29	107	242	72	218	129	111	116	220	158	126	233	54	26	206	155	112	172	184	113	59	25	246	218
LEUVEN	56	16	47	201	76	149	100	139	145	125	26	79	124	102	55	29	162	79	94	103	56	117	67	81	54	61	145	152	122	120
LEUZE	64	104	109	231	12	191	66	102	168	88	62	70	90	48	77	117	86	129	175	87	32	212	37	58	142	133	193	80	51	168	88
LIEGE	128	77	129	126	155	86	136	211	129	197	98	109	144	174	127	61	222	79	48	175	128	45	139	153	39	33	69	224	195	74	72	167
LIER	57	26	17	229	89	189	113	132	195	118	39	86	137	94	46	42	138	121	132	94	69	148	84	71	64	109	172	149	116	162	42	101	103
LUXEMBURG	252	227	269	26	246	70	191	335	89	321	222	204	176	313	251	217	322	136	125	299	248	141	265	277	195	158	87	339	298	99	186	271	156	259
MAASEIK	180	88	118	184	177	144	182	235	187	220	127	152	188	203	149	72	256	136	120	193	157	103	168	172	41	87	127	253	223	132	101	189	59	116	214
MALMEDY	182	130	173	115	192	75	152	258	128	244	145	156	159	221	174	114	299	97	60	222	186	29	203	200	86	91	58	291	274	61	119	224	47	156	112	105
MARCHE	139	114	156	83	136	43	87	222	65	208	109	91	96	200	138	115	213	34	16	186	135	92	152	159	103	49	45	224	198	20	98	148	64	148	113	122	63
MECHELEN	42	30	24	224	74	193	98	113	180	109	24	71	122	88	31	46	129	102	117	82	54	140	65	66	68	84	176	134	101	143	23	86	107	15	246	118	149	121
MONS	62	96	101	210	25	183	32	136	134	122	54	36	56	82	75	109	113	91	143	109	30	175	50	95	129	112	172	114	71	145	80	37	145	93	227	181	203	125	78
NAMUR	92	67	109	130	89	90	71	175	94	161	62	44	79	138	91	68	166	28	63	139	88	100	105	119	67	32	92	170	151	67	51	101	65	89	160	108	123	47	74	80
NEUFCHATEAU	194	169	217	36	192	28	120	249	37	263	165	129	124	256	193	175	262	74	72	242	190	132	184	220	159	105	45	258	238	50	153	204	120	217	62	178	103	56	176	165	102
NIVELLES	54	74	79	180	47	133	42	145	119	136	12	15	66	93	61	87	143	70	106	109	33	153	50	89	109	75	135	127	98	110	54	59	108	71	203	159	158	90	56	38	43	144
OOSTENDE	98	153	126	295	113	275	160	21	279	24	123	158	184	65	92	169	26	213	243	49	125	266	108	69	203	217	277	49	57	252	149	101	221	143	345	256	268	249	123	128	185	259	159
OUDENAARDE	40	112	82	260	42	221	96	72	190	58	66	86	120	18	53	128	69	140	190	48	42	204	28	28	150	125	222	61	32	178	92	30	164	97	290	197	211	158	62	64	118	214	75	83
PHILIPPEVILLE	103	115	120	139	81	121	24	183	78	172	73	26	33	138	102	116	169	29	79	156	70	143	106	128	115	75	108	156	133	83	99	90	111	112	165	156	126	63	97	56	46	103	41	193	111
QUEVY	75	99	114	208	38	194	27	149	147	135	67	49	69	95	88	122	126	98	154	112	43	182	57	117	147	125	177	113	84	158	93	47	158	108	236	194	200	132	93	13	77	172	51	141	77	51
ROESELARE	78	141	103	295	76	256	123	47	225	33	113	121	147	32	80	157	22	176	228	48	89	256	72	49	185	197	249	23	20	232	139	64	211	113	325	240	258	212	111	91	165	250	122	37	47	147	104
RONSE	44	106	94	238	30	220	84	90	186	76	84	73	108	30	63	119	70	128	178	60	40	209	23	40	141	147	207	62	33	182	90	18	164	103	264	191	225	162	92	52	117	202	65	85	12	108	65	48
SOIGNIES	46	80	85	195	25	177	48	127	151	127	38	30	72	74	59	93	113	85	130	93	14	169	31	73	115	97	176	105	76	139	64	37	130	77	234	165	168	119	82	16	65	158	22	139	56	56	29	91	43
ST. HUBERT	170	153	186	61	167	34	121	253	47	239	140	122	120	231	169	147	265	58	47	215	157	120	174	195	134	80	42	258	229	25	137	179	95	175	87	153	100	31	160	149	86	25	138	263	208	87	156	234	186	143
ST. VITH	196	151	212	94	232	54	149	296	119	282	166	171	178	242	220	128	314	96	75	241	203	50	220	221	107	108	37	306	278	62	165	231	68	177	91	133	21	82	170	207	129	82	172	289	247	125	220	294	242	194	79
ST. NIKLAAS	30	63	21	241	82	201	124	95	205	81	49	93	147	50	17	79	100	139	150	56	62	173	57	33	99	117	203	105	72	178	56	94	157	38	267	151	175	158	33	92	111	213	78	105	61	119	105	82	90	86	189	203
ST. TRUIDEN	93	41	84	170	113	130	121	176	154	162	83	76	127	139	92	25	187	88	75	140	93	80	104	120	17	42	108	189	160	101	37	125	35	67	195	58	82	91	60	112	50	147	83	186	133	96	125	171	129	101	122	126	93
SPA	165	111	154	134	201	79	144	253	134	239	126	137	153	211	164	95	281	91	41	212	189	33	206	192	67	61	62	291	262	62	107	216	35	137	131	86	19	57	130	182	93	122	143	271	231	120	195	288	229	176	87	40	163	63
STAVELOT	193	128	179	126	213	66	135	281	151	267	150	152	144	239	192	112	289	82	43	240	184	44	201	220	91	76	49	278	249	51	137	222	52	162	123	110	15	48	160	188	108	94	138	299	226	111	201	273	210	173	76	32	193	87	24
THUIN	96	108	103	184	59	163	18	175	120	161	66	19	42	121	95	99	147	74	124	158	64	158	81	126	112	95	153	141	105	128	92	88	128	105	223	153	171	108	90	34	63	148	34	162	103	42	77	139	86	40	132	170	102	95	165	171
TIELT	61	124	86	283	75	243	129	44	247	30	31	133	153	15	63	140	35	181	216	31	75	234	58	32	162	185	245	40	21	220	117	63	189	96	313	212	236	200	94	97	153	250	110	50	33	159	110	17	45	88	231	257	65	154	217	24'	131
TIENEN	75	35	66	170	95	130	95	148	140	144	45	58	109	121	74	22	180	74	75	131	75	98	86	105	35	42	126	171	142	101	19	107	53	61	206	94	107	93	42	94	46	147	65	168	111	94	107	158	109	83	122	144	75	18	81	118	77	136
TONGEREN	113	58	104	145	133	105	140	196	148	182	83	96	145	159	112	42	209	83	67	160	113	64	124	140	20	37	88	209	180	93	57	145	19	84	175	50	66	83	88	149	69	139	112	206	153	112	146	196	147	121	114	94	113	20	47	71	132	174	38
TORHOUT	80	145	107	315	96	281	150	34	252	20	110	154	174	36	84	161	21	209	241	45	106	253	89	53	183	197	264	31	33	263	136	84	208	124	341	233	266	245	115	118	178	283	135	24	60	174	131	13	66	109	276	276	86	175	247	275	152	21	157	195
TOURNAI	81	121	126	245	29	227	76	84	201	70	79	80	100	61	94	134	69	135	192	98	49	219	54	71	156	156	214	56	27	189	105	17	189	118	271	206	236	169	103	44	124	211	76	84	43	106	57	47	31	54	202	253	104	158	226	232	78	48	124	178	60
TURNHOUT	93	44	42	227	129	187	159	157	204	143	79	122	174	113	80	42	163	138	142	119	109	148	124	96	65	109	190	168	135	168	60	141	103	40	259	76	158	157	55	140	110	214	111	168	137	158	153	145	147	117	189	172	63	67	137	163	141	128	84	84	149	158
VERVIERS	158	106	159	152	185	94	166	244	140	230	128	139	174	204	157	90	254	109	59	203	171	15	169	183	69	63	77	254	225	80	102	197	30	147	149	82	37	75	125	175	95	130	138	251	194	141	172	241	194	160	105	58	158	65	18	[illegible]	158	219	83	49	238	214	135
VEURNE	112	168	148	312	116	272	163	47	263	50	142	167	187	66	114	197	16	214	247	71	126	285	109	83	219	214	274	31	60	249	168	104	240	154	338	269	305	229	154	131	182	288	159	26	85	187	144	38	86	129	272	338	116	205	277	305	163	51	187	225	37	87	179	270
VIRTON	233	208	245	28	211	68	154	217	52	303	203	156	139	284	232	211	209	113	111	280	229	172	236	258	198	144	84	291	275	89	192	220	159	242	51	217	145	95	231	186	141	39	183	262	242	130	199	286	238	186	64	121	252	186	162	153	172	294	192	178	304	248	251	315	327
VISE	136	80	127	144	158	104	154	219	137	205	108	127	182	182	135	64	230	97	66	183	136	47	147	163	42	51	87	232	203	92	80	168	18	106	170	47	55	82	110	163	83	132	137	229	172	126	176	212	140	144	112	88	136	43	46	73	146	195	61	23	218	207	108	29	172	151

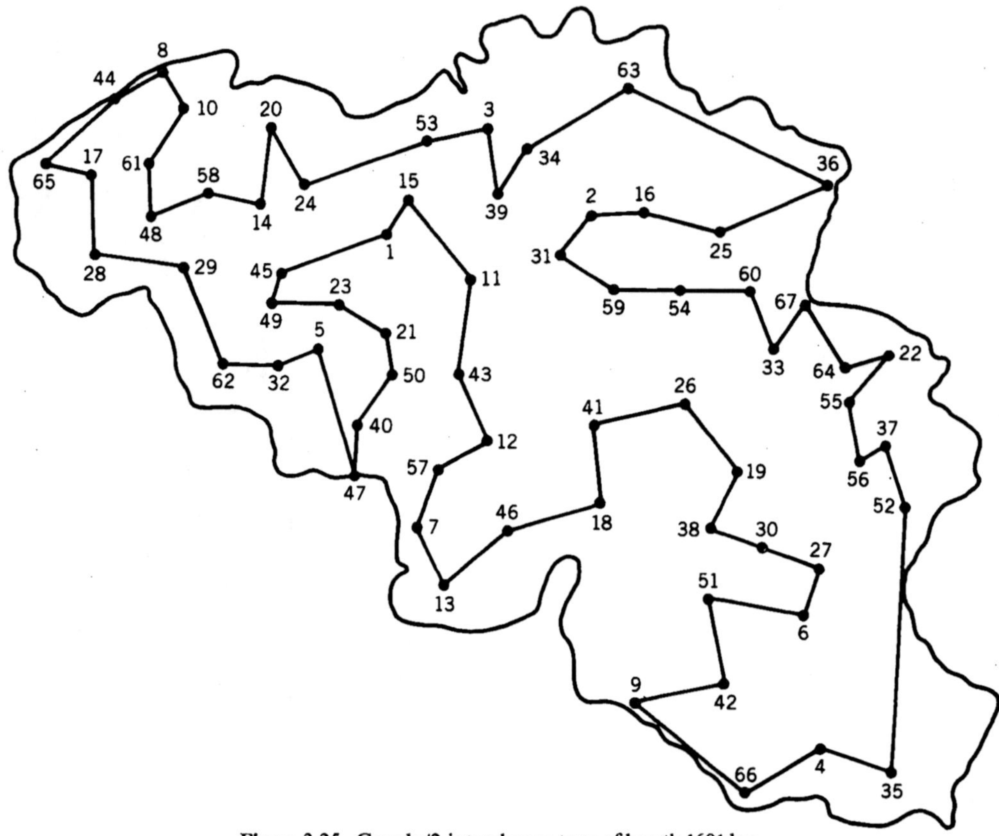

Figure 3.25. Greedy/2-interchange tour of length 1691 km.

Table 3.2.

Subtour	W
1.	W = {1 2 3 11 15 16 25 31 33 34 36 37 39 52 53 54 56 59 60 63 67}
2.	{4 22 35 55 64 66}
3.	{4 35 66}
4.	{22 55 64}
5.	{1 2 3 11 15 16 25 31 33 34 36 39 53 54 59 60 63 67}
6.	{4 47 50}
7.	{2 16 25 31 33 36 54 59 60 67}
8.	{2 16 63}
9.	{5 14 21 23 29 32 40 45 47 48 49 50 58 61 62}
10.	{5 14 21 23 29 32 40 45 47 49 50 58 62}
11.	{2 16 25 31 33 36 54 59 60 63 67}
12.	{12 43 57}
13.	{17 28 65}
14.	{25 33 36 54 59 60 67}
15.	{5 8 10 14 17 20 21 23 24 28 29 32 44 45 48 49 58 61 62 65}
16.	{5 8 10 14 17 20 21 23 24 28 29 32 40 44 45 47 48 49 50 58 61 62 65}
17.	{25 33 36 54 60 67}
18.	{22 37 55 56 64}
19.	{22 37 52 55 56 64}

To find the optimal tour, we apply the FCP/branch-and-bound algorithm using subtour elimination inequalities and the first option described above so that subtour elimination constraints are only added at the initial node.

Phase 0. The initial LP problem with the degree constraints (3.4) and the upper-bound constraints (3.8) has value $z_{\text{FM}} = 1571.5$.

Phase 1. LP($\mathcal{F}$) is solved after adding 19 subtour elimination inequalities (3.5), with the sets W given in Table 3.2. $z_{\text{LP}} = z_{\text{MIT}} = 1606.75$.

Phase 2. $t = 1$. Branch-and-bound applied to LP($\mathcal{F}$) finds a tour of length 1615 at node 22, a 2-matching of length 1614 at node 37, and a 2-matching of length 1613 at node 55, and then it terminates at node 78.

$t = 2$. Two subtour elimination constraints are added, with $W = \{1 \quad 15 \quad 53\}$ and $W = \{8 \quad 10 \quad 14 \quad 17 \quad 20 \quad 24 \quad 28 \quad 29 \quad 32 \quad 44 \quad 45 \quad 48 \quad 49 \quad 58 \quad 61 \quad 62 \quad 65\}$ eliminating the 2-matching solutions of length 1613 and 1614, respectively. Branch-and-bound is now applied with a cutoff of 1615; and the search terminates after 73 nodes, with no solution of value less than 1615. Hence an optimal tour is of length 1615 (see Figure 3.26).

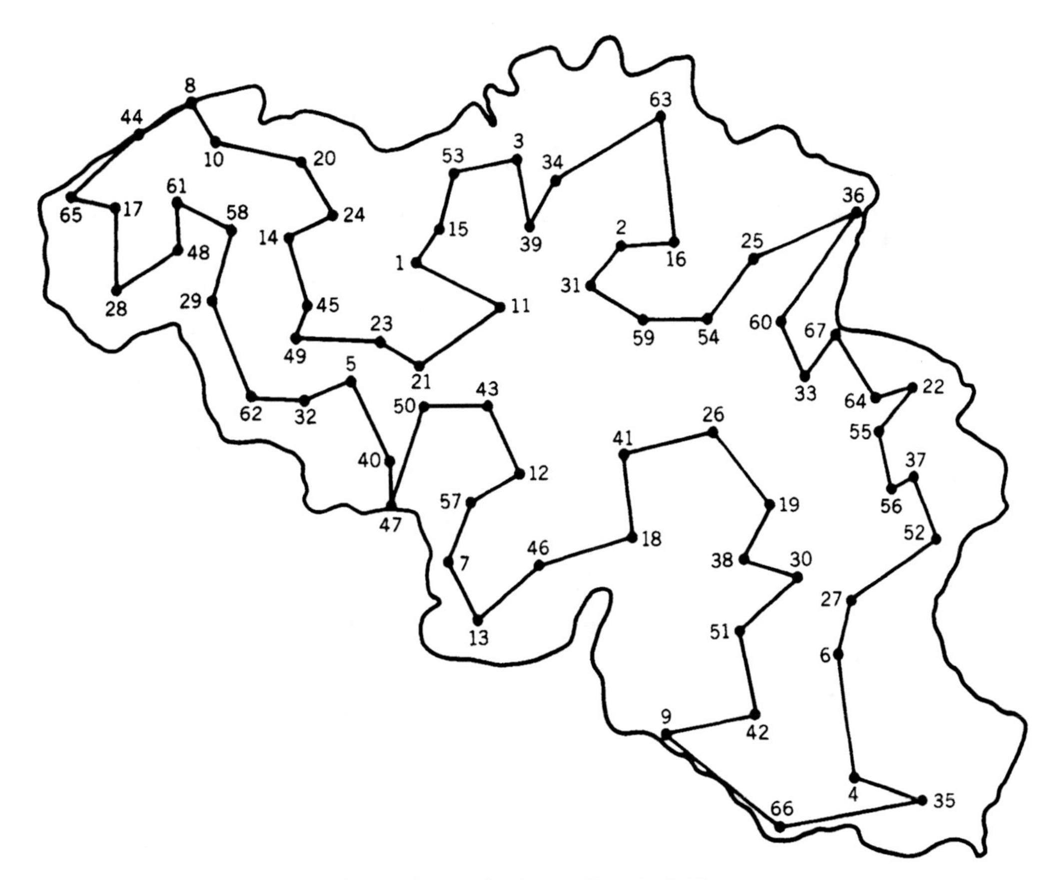

Figure 3.26. Optimal tour of length 1615 km.

Table 3.3.

z_{1T}	1-Tree relaxation	1401
z_{FM}	Fractional matching	1571.5
z_{M1T}	Linear programming (Lagrangian 1-tree)	1606.75
z_{TS}	Optimal tour	1615
	Greedy + 2-interchange heuristic	1691
	Greedy heuristic	1805
	Nearest neighbor	2045

A smaller branch-and-bound tree would be obtained if the 1-tree Lagrangian relaxation was used to obtain the bounds at each node. The subgradient algorithm was used to solve the Lagrangian dual having value $z_{\mathrm{M1T}} = 1606.75$ at node 1. The length of the initial 1-tree with the multipliers at zero is $z_{\mathrm{1T}} = 1401$. Using an initial step-size of 5 and decreasing by a factor of 2 every $N = 67$ iterations, a bound exceeding 1600 was first obtained on iteration 197, a bound exceeding 1605 was obtained on iteration 276, and a bound of 1606.23 was obtained on iteration 399. A summary of the bounds obtained is given in Table 3.3.

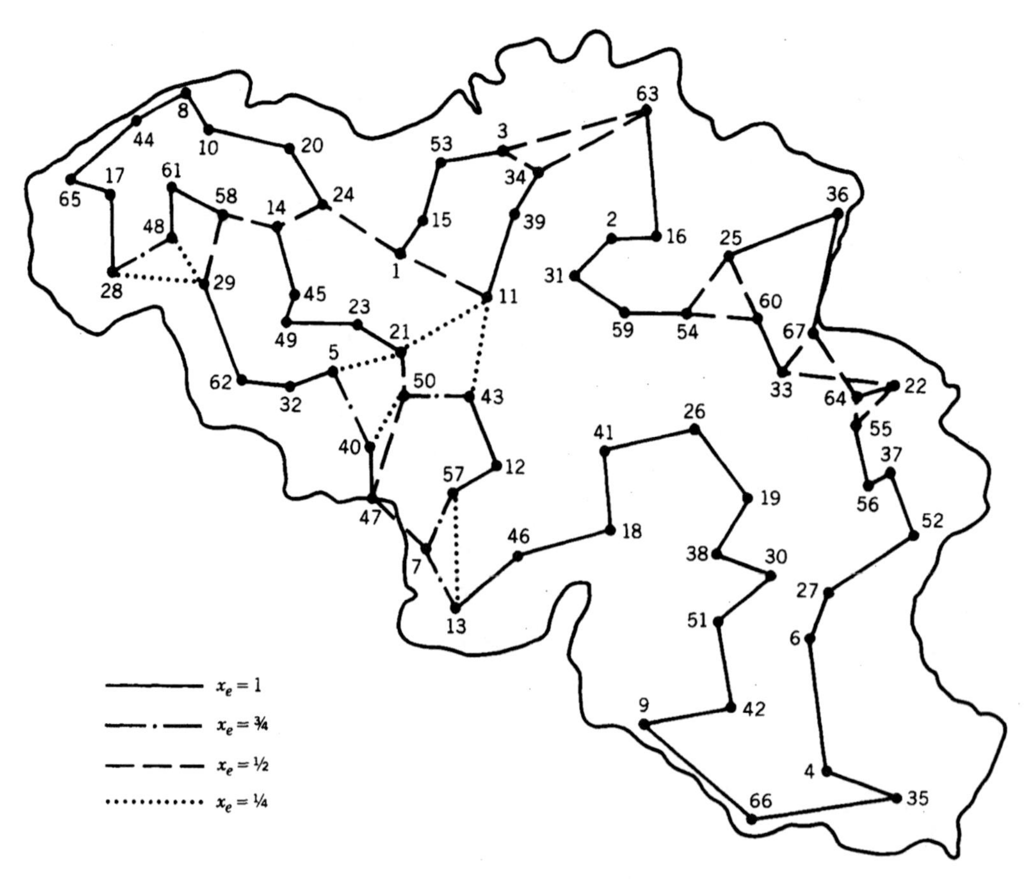

Figure 3.27. $z^{?}_{\mathrm{LP}} = z_{\mathrm{M1T}} = 1606.75$.

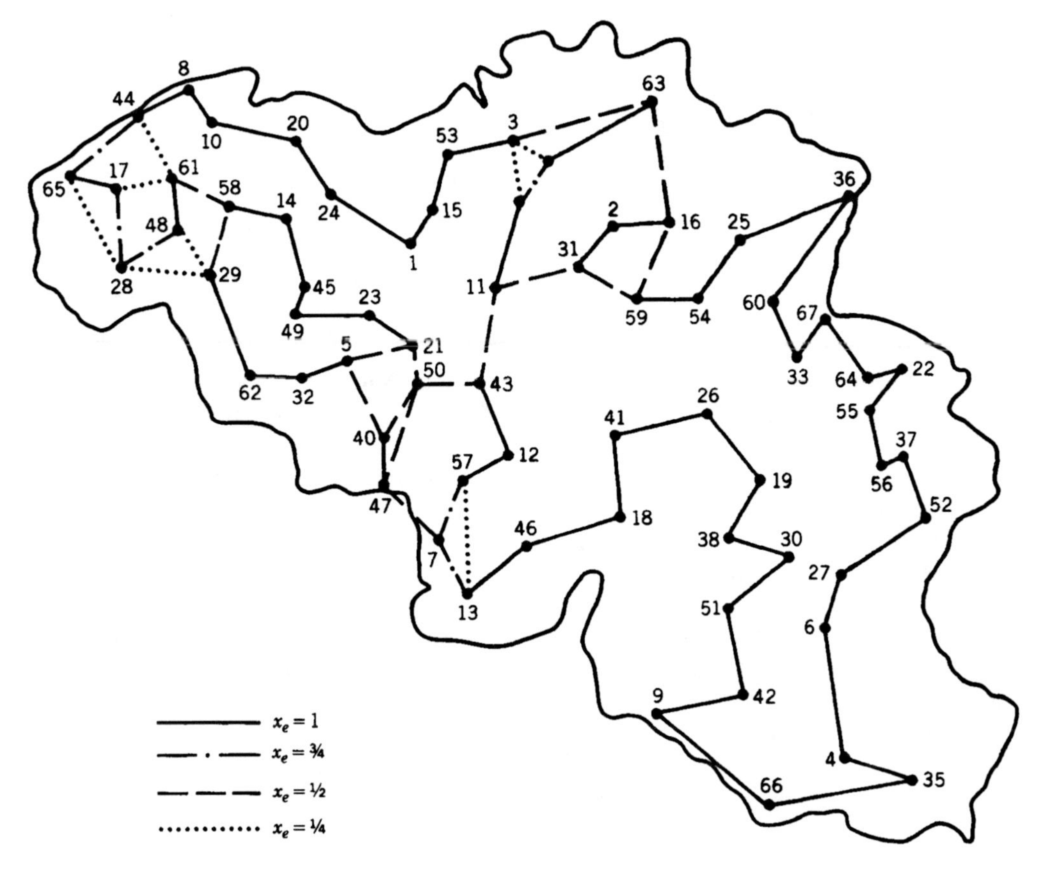

Figure 3.28. Length = 1609.75.

Finally we consider briefly the addition of 2-matching and comb inequalities which are found by inspection. The fractional solution to LP($\mathscr{F}$), obtained by the addition of the subtour elimination constraints, is shown in Figure 3.27. Inspection of the figure readily reveals that at least four 2-matching inequalities are violated. Adding the inequalities with

1. $H = \{3 \quad 34 \quad 63\}$ $\qquad \hat{E} = \{(3 \quad 53), (34 \quad 39), (16 \quad 63)\}$
2. $H = \{28 \quad 29 \quad 48\}$ $\qquad \hat{E} = \{(17 \quad 28), (29 \quad 62), (48 \quad 61)\}$
3. $H = \{25 \quad 54 \quad 60\}$ $\qquad \hat{E} = \{(25 \quad 36), (54 \quad 59), (33 \quad 60)\}$
4. $H = \{22 \quad 33 \quad 55 \quad 64 \quad 67\}$ $\qquad \hat{E} = \{(33 \quad 60), (55 \quad 56), (36 \quad 67)\}$

leads to the solution of value 1609.75 shown in Figure 3.28. It is left to the reader to find further violated inequalities.

4. FIXED-CHARGE NETWORK FLOW PROBLEMS

So far in this chapter, we have considered classes of pure-integer programming problems. Here we consider an important class of mixed-integer problems. The *fixed-charge network flow problem* (FN) was formulated in Section I.1.3 as

$$
(4.1)\quad \begin{aligned}
&\min \sum_{(i,j)\in\mathcal{A}} c_{ij}x_{ij} + \sum_{(i,j)\in\mathcal{A}} h_{ij}y_{ij} \\
&(4.1\text{a}) \qquad \sum_{j\in\delta^+(i)} y_{ij} - \sum_{j\in\delta^-(i)} y_{ji} = b_i \quad \text{for } i \in V \\
&(4.1\text{b}) \qquad y_{ij} \leqslant u_{ij}x_{ij} \quad \text{for } (i,j)\in\mathcal{A} \\
&\qquad\qquad y \in R_+^{|\mathcal{A}|}, \quad x \in B^{|\mathcal{A}|},
\end{aligned}
$$

where $\mathcal{D} = (V, \mathcal{A})$ is a digraph, $\delta^+(i) = \{j \in V: (i,j) \in \mathcal{A}\}$, $\delta^-(i) = \{j \in V: (j,i) \in \mathcal{A}\}$, b_i is the supply at node i, c_{ij} is the fixed cost of having flow on arc (i,j), h_{ij} is the variable cost per unit of flow on arc (i,j), and u_{ij} is the capacity of arc (i,j). Recall that the difference between FN and the linear minimum-cost flow problem is that in FN if $y_{ij} > 0$, then the cost of the flow is $c_{ij} + h_{ij}y_{ij}$. This is achieved by the capacity constraints (4.1b), which force $x_{ij} = 1$ when $y_{ij} > 0$.

A necessary condition for feasibility, assumed throughout this section, is $\Sigma_{i\in V} b_i = 0$. We also assume that $c_{ij} \geqslant 0$ for all $(i,j) \in \mathcal{A}$ since if $c_{ij} < 0$, we can set $x_{ij} = 1$ and eliminate x_{ij} from the problem. Similarly, we assume that with respect to the h_{ij} there are no negative-cost directed cycles. This assures that the objective function is bounded from below.

Besides being an important model in its own right for a variety of network design problems, several special cases of FN are of substantial interest. A simple way to obtain special cases is to restrict the network structure (e.g., as in the transportation problem).

Another simplification concerns the capacity constraints. When u_{ij} is sufficiently large—for instance, $u_{ij} \geqslant \frac{1}{2}\Sigma_{i\in V} |b_i|$—there is no feasible flow with $y_{ij} > u_{ij}$, and the capacity constraint only serves to force the fixed cost to be included in the objective function when the flow is positive. We call such problems *uncapacitated* and denote them by (UFN).

Yet another important subclass of FNs are those in which $|i \in V: b_i > 0| = 1$. We call these problems *single source* (SFN) and use the notation (SUFN) for *single-source uncapacitated problems*.

Several interesting problems can be modeled as SUFNs—for instance, the uncapacitated facility location problem (UFL) considered in Chapter II.5. Figure 4.1a gives the digraph for a UFL with $m = 2$ and $n = 3$. The arcs joining the dummy node to the facility nodes are uncapacitated and have only the fixed cost of opening the jth facility. The arcs that join facility i to customer j are also uncapacitated and have the variable cost h_{ij}. Note that since UFL is $\mathcal{NP}$-hard, SUFN is $\mathcal{NP}$-hard.

Another interesting SUFN is the Steiner r-branching problem. Given a subset $D \subseteq V$, a root $r \in D$, and weights on the arcs, a Steiner *r-branching* is a minimum-weight branching that spans D. Here for the root node $r \in D$ we let $b_r = |D| - 1$, $b_i = -1$ for $i \in D \setminus \{r\}$, and $b_i = 0$ for $i \in V \setminus D$. The objective function is accommodated by letting c_{ij} be the weight of arc (i,j) and letting $h_{ij} = 0$ for all (i,j). It then follows that feasible solutions without directed cycles are branchings that span D (see Figure 4.1b). Note that when $D = V$ we obtain the minimum-weight directed r-branching problem (see Section III.3.5), and when $|D| = 2$ we obtain the shortest-path problem (see Section I.3.2). Although both of these problems can be solved in polynomial time, the general Steiner branching problem is $\mathcal{NP}$-hard.

In several practical models, FN or special cases arise as subproblems. For example, production planning problems frequently contain the uncapacitated lot-size problem, which is an SUFN (see Figure 5.2 of Section II.5.5). Thus algorithms based on Lagrangian relaxations and techniques for solving FNs can be used to solve practical problems that are FNs with additional constraints. This hierarchy of FNs is displayed in Figure 4.2.

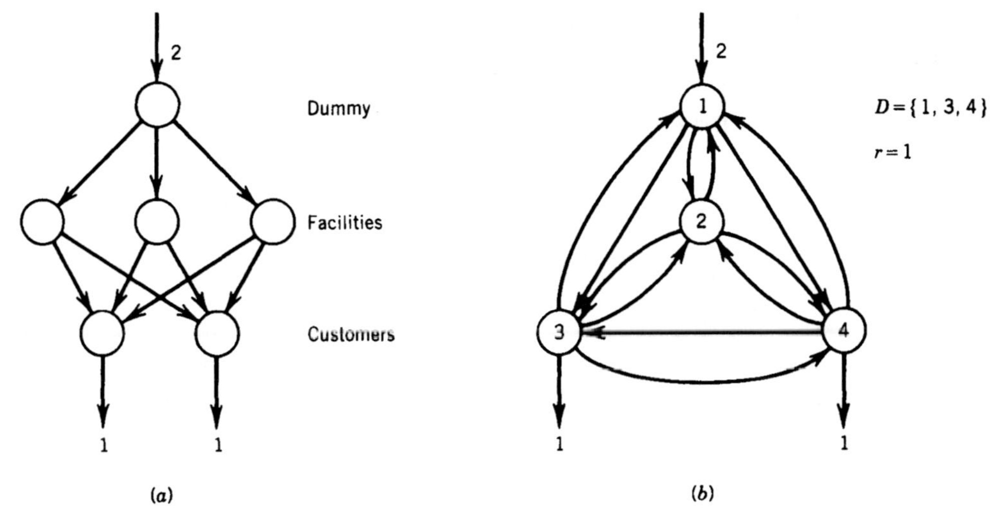

Figure 4.1

In this section, we begin by mentioning briefly a standard branch-and-bound algorithm for FN primarily to point out its advantages and limitations. We then propose an FCPA for FN and apply it to the fixed-cost uncapacitated transportation problem.

Next we given another IP formulation for SFN and show that its linear programming relaxation is stronger than the linear programming relaxation of (4.1). This formulation simplifies for SUFNs, and we illustrate it with the Steiner branching problem and the uncapacitated lot-size problem. In the case of the uncapacitated lot-size problem, this serves as the basis for other reformulations, one of which is a shortest-path problem. The shortest-path reformulation has been used in a linear programming relaxation of multi-item lot-size problems and appears to have the capability of solving quite large instances.

A Branch-and-Bound Algorithm for FN

An obvious way to solve (4.1) is by a branch-and-bound algorithm that uses linear programming relaxations. An advantage of this approach is that the linear programming

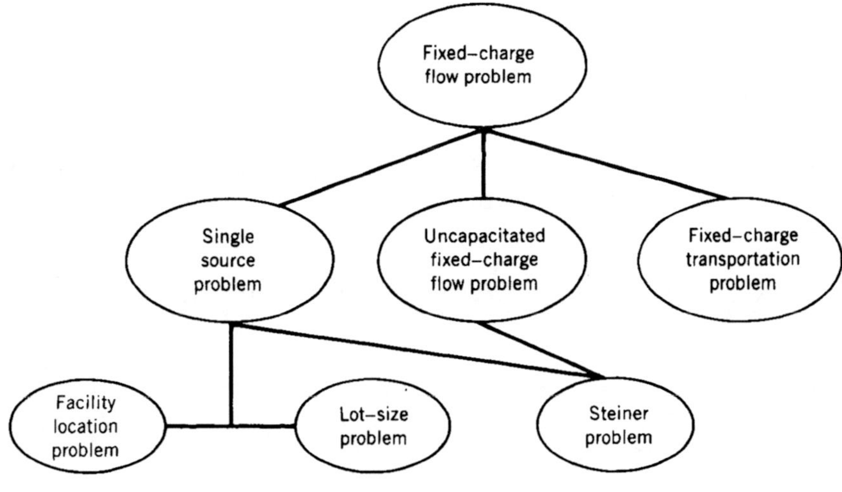

Figure 4.2

relaxation of (4.1) is a network flow problem. Network flow problems can be solved very efficiently by, for instance, the network simplex algorithm (see Section I.3.6). In addition, several other parameters used in a branch-and-bound algorithm (e.g., penalties) are easy to obtain from optimal basic solutions to network flow problems.

Proposition 4.1. *The linear programming relaxation of* FN *is the network flow problem*

$$\min\left\{\sum_{(i,j)\in\mathcal{A}}\left(h_{ij}+\frac{c_{ij}}{u_{ij}}\right)y_{ij}\colon (4.1\text{a}),\ y_{ij}\leqslant u_{ij} \text{ for } (i,j)\in\mathcal{A},\ y\in R_{+}^{|\mathcal{A}|}\right\}. \tag{4.2}$$

Proof. Replacing $x \in B^{|\mathcal{A}|}$ by $x_{ij} \leqslant 1$, the only constraints on x_{ij} are $y_{ij}/u_{ij} \leqslant x_{ij} \leqslant 1$. Because $c_{ij} \geqslant 0$, there exists an optimal solution with $x_{ij} = y_{ij}/u_{ij}$. This substitution gives the network flow problem (4.2). ■

Unfortunately, the bounds obtained from these relaxations are frequently very poor primarily because they do not accurately represent the fixed costs. This is true, as we have noted earlier, because if the optimal solution y has $0 < y_{ij} < u_{ij}$, then only the fraction y_{ij}/u_{ij} of the fixed cost is included in the objective function. Another disadvantage of this approach is its inflexibility in accommodating additional constraints. If the problem to be solved has additional constraints, the network structure of the linear programming relaxation will be destroyed unless another technique such as Lagrangian relaxation is used.

An FCPA for FN

To improve the bounds obtained from the network flow relaxation and to accommodate additional constraints within the scope of a linear programming relaxation, we now consider an FCPA for FN that uses strong cutting planes.

We will use three classes of valid inequalities for FN. Observe first that for fixed $i \in V$, any solution of (4.1) satisfies

(i) $$\sum_{j\in\delta^{+}(i)} u_{ij}x_{ij} \geqslant \sum_{j\in\delta^{+}(i)} y_{ij} = b_i + \sum_{j\in\delta^{-}(i)} y_{ji} \geqslant b_i$$

and

(ii) $$\sum_{j\in\delta^{-}(i)} u_{ji}x_{ji} \geqslant \sum_{j\in\delta^{-}(i)} y_{ji} = -b_i + \sum_{j\in\delta^{+}(i)} y_{ij} \geqslant -b_i.$$

Using the separation procedure described in Section II.6.2, violated extended cover inequalities can be generated from knapsack set (i) if $b_i > 0$ and from (ii) if $b_i < 0$.

Also observe that any solution of (4.1) satisfies

$$\sum_{j\in\delta^{+}(i)} y_{ij} - \sum_{j\in\delta^{-}(i)} y_{ji} = b_i,\ y_{ij} \leqslant u_{ij}x_{ij},\ y_{ij}\in R_{+}^{1},\ x_{ij}\in B^{1} \quad \text{for } (i,j)\in\delta^{+}(i),\ (j,i)\in\delta^{-}(i).$$

Replacing this equality by two inequalities gives sets having the form of the single-node flow model introduced in Section II.2.4. Thus to obtain the second and third types of inequalities we consider the region

$$(4.3) \qquad T = \left\{ y \in R^n_+, x \in B^n: \sum_{j\in N^+} y_j - \sum_{j\in N^-} y_j \leq b, y_j \leq a_j x_j \text{ for } j \in N^+ \cup N^- \right\}$$

where $n = |N^+ \cup N^-|$.

For the set of T, we have derived the class of valid inequalities [see (4.4) of Section II.2.4]

$$(4.4) \qquad \sum_{j\in C} [y_j + (a_j - \lambda)^+ (1 - x_j)] \leq b + \sum_{j\in L} \lambda x_j + \sum_{j\in N^- \setminus L} y_j,$$

where $C \subseteq N^+$ is a dependent set (i.e., $\lambda = \sum_{j\in C} a_j - b > 0$ and $L \subseteq N^-$).

These can be generalized to the following larger class of valid inequalities, called *generalized flow cover inequalities* (GFC)

$$(4.5) \qquad \sum_{j\in C^+} [y_j + (a_j - \lambda)^+(1 - x_j)] \leq b + \sum_{j\in C^-} a_j + \sum_{j\in L} \lambda x_j + \sum_{j\in N^- \setminus (L\cup C^-)} y_j$$

where $C^+ \subseteq N^+, C^- \subseteq N^-, L \subseteq N^- \setminus C^-$ and $\lambda = \Sigma_{j\in C^+} a_j - \Sigma_{j\in C^-} a_j - b > 0$. We leave it as an exercise to show that these are valid for T.

The separation problem for the family (4.5) is: Given a point (x^*, y^*), check whether for any sets C^+, C^-, and L, inequality (4.5) is violated.

We let $\alpha \in B^{|N^+|}$ be the characteristic vector of $C^+ \subseteq N^+$ and let $\beta \in B^{|N^-|}$ be the characteristic vector of $C^- \subseteq N^-$. The definition of λ yields the equality knapsack constraint

$$(4.6) \qquad \sum_{j\in N^+} a_j\alpha_j - \sum_{j\in N^-} a_j\beta_j = b + \lambda, \quad \text{subject to } \lambda > 0.$$

The violation to be maximized is

$$(4.7) \qquad \sum_{j\in N^+} [y_j^* + (a_j - \lambda)^+ (1 - x_j^*)]\alpha_j - b - \sum_{j\in N^-} a_j\beta_j - \sum_{j\in N^-} \min(\lambda x_j^*, y_j^*)(1 - \beta_j),$$

where the last term is derived from the observation that for $j \in N^- \setminus C^-$, any violation is maximized by taking $j \in L$ if $\lambda x_j^* < y_j^*$ and $j \in N^- \setminus (C^- \cup L)$ if $\lambda x_j^* > y_j^*$.

The resulting separation problem is the nonlinear integer program $\max\{(4.7): (4.6), \alpha \in B^{|N^+|}, \beta \in B^{|N^-|}\}$. This problem is equivalent to solving the family of equality knapsack problems

$$(4.8) \qquad \zeta_\lambda = \max\{(4.7): (4.6), \alpha \in B^{|N^+|}, \beta \in B^{|N^-|}\}$$

for all positive integral values of λ. Hence we have shown the following:

Proposition 4.2. *An inequality of the form (4.5) with $\lambda = \lambda^* > 0$ is violated by the point (x^*, y^*) if and only if $\zeta_{\lambda^*} > 0$.*

Unfortunately, there are two difficulties with this separation problem. Equality knapsack problems are hard to solve, and the function ζ_λ is not well behaved as a function of λ. Therefore we look for a heuristic solution to the problem of choosing the sets C^+ and C^- for which inequality (4.5) is most violated by (x^*, y^*). As a first step we consider a subclass of the inequalities in which $L = N^- \setminus C^-$, and then we relax these inequalities by reducing the term $(a_j - \lambda)^+ (1 - x_j)$ to $(a_j - \lambda)(1 - x_j)$. The resulting valid inequalities are

$$\sum_{j\in C^+}[y_j + (a_j - \lambda)(1 - x_j)] \leqslant b + \sum_{j\in C^-} a_j + \sum_{j\in N^-\setminus C^-} \lambda x_j. \tag{4.9}$$

Finding the sets C^+ and C^- for which (4.9) is most violated is still not computationally easy, so we take a second heuristic step which is to work with an upper bound on the violation for any set C^+, C^- in (4.9).

Because $y_j \leqslant a_j x_j$ for all j, an upper bound on the violation of (4.9) is obtained by replacing y_j^* by the possibly larger value $a_j x_j^*$. This gives the upper bound

$$\sum_{j\in C^+}[a_j x_j^* + (a_j - \lambda)(1 - x_j^*)] - b - \sum_{j\in C^-} a_j - \sum_{j\in N^-\setminus C^-} \lambda x_j^*.$$

Substituting $b = \Sigma_{j\in C^+} a_j - \Sigma_{j\in C^-} a_j - \lambda$ and canceling terms, the upper bound on the violation of (4.9) is equal to

$$\lambda\left[-\sum_{j\in C^+}(1 - x_j^*) + \sum_{j\in C^-} x_j^* - \left(\sum_{j\in N^-} x_j^* - 1\right)\right].$$

To find the maximum value of this upper bound, we solve the knapsack problem

$$\begin{aligned} \xi = \max&\left\{\sum_{j\in N^+}(x_j^* - 1)\alpha_j + \sum_{j\in N^-} x_j^*\beta_j\right\} \\ &\sum_{j\in N^+} a_j\alpha_j - \sum_{j\in N^-} a_j\beta_j > b \\ &\alpha \in B^{|N^+|}, \quad \beta \in B^{|N^-|}. \end{aligned} \tag{4.10}$$

We now know that if some inequality (4.9) is violated, then $\lambda[\xi - (\Sigma_{j\in N^-} x_j^* - 1)]$, which is an upper bound on the value of the violation, must be positive.

Proposition 4.3. *A necessary condition for the violation of an inequality of the form (4.9) is $\xi > \Sigma_{j\in N^-} x_j^* - 1$. This condition is also sufficient if $y_j^* = a_j x_j^*$ for all $j \in C^+$, where C^+ is determined by an optimal solution to (4.10).*

The above discussion leads to the following heuristic separation algorithm for generalized flow cover inequalities.

Separation Algorithm for Generalized Flow Cover Inequalities

Step 1: Solve the knapsack problem (4.10) exactly or approximately to obtain an optimal or "near-optimal" pair C^+, C^-.

Step 2: Given C^+ and C^-, test whether (x^*, y^*) violates the inequality (4.5), where for $j \in N^- \setminus C^-$ we put $j \in L$ if $\lambda x_j^* < y_j^*$ and $j \in N^- \setminus (L \cup C^-)$ otherwise.

Note that even if $\xi < \Sigma_{j\in N^-} x_j^* - 1$, the inequality (4.5) may be violated because our arguments have been based on approximations to the violation of (4.5).

Example 4.1. Consider the mixed 0-1 constraint

$$y_1 \leqslant 2250x_2 + 4500x_3 + 6750x_4$$

with $y_1 \leqslant 6500$, $y_1 \in R^1_+$, $x \in B^3$, and the point $(y_1^*, x_2^*, x_3^*, x_4^*) = (6500\ 0\ 1\ 0.296)$. We can rewrite the above constraints in the form of a single-node variable upper-bound set T:

$$y_1 \leqslant y_2 + y_3 + y_4,\ y_1 \leqslant 6500x_1,\ y_2 \leqslant 2250x_2,\ y_3 \leqslant 4500x_3,\ y_4 \leqslant 6750x_4,\ y \in R^4_+,\ x \in B^4,$$

with the additional constraints $x_1 = 1$, $y_2 = 2250x_2$, $y_3 = 4500x_3$, $y_4 = 6750x_4$.

Applying the heuristic separation algorithm for generalized flow cover inequalities (4.5), we obtain the knapsack problem (4.10):

$$\begin{aligned} \xi = \max \quad & 0\alpha_1 + 0\beta_2 + 1\beta_3 + 0.296\beta_4 \\ & 6500\alpha_1 - 2250\beta_2 - 4500\beta_3 - 6750\beta_4 > 0 \\ & \alpha \in B^1, \quad \beta \in B^3 \end{aligned}$$

with optimal solution $\alpha_1 = 1$, $\beta_3 = 1$, and $\xi = 1$.

With $C^+ = \{1\}$ and $C^- = \{3\}$, we have $\lambda = 2000$ and $L = \{2, 4\}$, and the resulting inequality (4.5) is

$$y_1 + (6500 - 2000)(1 - x_1) \leqslant 0 + 4500 + 2000x_2 + 2000x_4;$$

or, using the additional constraint $x_1 = 1$, we obtain

$$y_1 \leqslant 4500 + 2000x_2 + 2000x_4,$$

which is violated by (x^*, y^*).

Note that because $\xi > \Sigma_{j=2}^4 x_j^* - 1 = 0.296$, the necessary condition of Proposition 4.3 is, in fact, satisfied. Since y_1^* is at its upper bound, the sufficient condition also happens to hold.

Valid inequalities of the third class are called *extended* GFCs and are of the form

$$\begin{aligned} & \sum_{j\in C^+\cup L^+} y_j + \sum_{j\in C^+} (a_j - \lambda)^+ (1 - x_j) - \sum_{j\in L^+} (\bar{a}_j - \lambda)^+ x_j \\ & \leqslant b + \sum_{j\in C^-} a_j - \sum_{j\in C^-} \min\{\lambda, [a_j - (\bar{a} - \lambda)]^+\} (1 - x_j) \qquad (4.11) \\ & \quad + \sum_{j\in L^-} \max\{\lambda, a_j - (\bar{a} - \lambda)\} x_j + \sum_{j\in N^-\setminus(C^-\cup L^-)} y_j, \end{aligned}$$

where $\bar{a} = \max_{j\in C^+} a_j$, $\bar{a}_j = \max(\bar{a}, a_j)$, $L^+ \subseteq N^+ \setminus C^+$, and $L^- \subseteq N^- \setminus C^-$, and we require $\bar{a} \geqslant \lambda > 0$.

We do not develop a separation routine for the extended GFCs. Instead we use the sets C^+ and C^- derived in the separation routine for GFCs, together with sets L^+ and L^- constructed by

$$L^+ = \{j \in N^+ \setminus C^+ : y_j^* > (\bar{a}_j - \lambda)^+ x_j^*\}$$

and

$$L^- = \{j \in N^- \setminus C^- : \max\{\lambda, a_j - (\bar{a} - \lambda)\} x_j^* < y_j^*\},$$

to find a violated inequality of the form (4.11).

In summary, for each constraint (4.1a) of (4.1), we try to find violated extended cover inequalities, GFCs, and extended GFCs as indicated above. Note that since sets of the form T given by (4.3) arise in relaxations of general mixed 0-1 programs (see Section II.2.4), the FCPA can be used to generate violated inequalities for general mixed 0-1 models.

We now illustrate the FCPA by applying it to the fixed-charge uncapacitated transportation problem.

Solving a Fixed-Charge Uncapacitated Transportation Problem by an FCPA and Branch-and-Bound

For a transportation problem we obtain $V = (V_1 \cup V_2)$, and all arcs are directed from V_1 to V_2. Hence (4.1) simplifies to

$$\min \sum_{i\in V_1}\sum_{j\in V_2} h_{ij}y_{ij} + \sum_{i\in V_1}\sum_{j\in V_2} c_{ij}x_{ij}$$

$$\sum_{j\in V_2} y_{ij} = b_i \quad \text{for } i \in V_1$$

$$\sum_{i\in V_1} y_{ij} = d_j \quad \text{for } j \in V_2$$

$$y_{ij} \leqslant u_{ij}\,x_{ij}, \quad x_{ij} \in \{0, 1\} \quad \text{for } i \in V_1, j \in V_2,$$

where $u_{ij} = \min(b_i, d_j)$ for $i \in V_1$ and $j \in V_2$; $V_1 = \{1, \ldots, m\}$; $V_2 = \{1, \ldots, n\}$; and $\Sigma_{i\in V_1} b_i = \Sigma_{j\in V_2} d_j$.

The initial linear programming relaxation LP^1 is obtained by replacing the integrality constraints by $0 \leqslant x_{ij} \leqslant 1$ for $i \in V_1, j \in V_2$. Note that LP^1 is a transportation problem with $h'_{ij} = h_{ij} + c_{ij}/u_{ij}$ because we can set $x_{ij} = y_{ij}/u_{ij}$ for $i \in V_1$ and $j \in V_2$. However, once cutting planes are added we no longer have a transportation problem.

In the cutting-plane part of the algorithm we add three types of cuts:

Step a: Extended cover inequalities are obtained from the knapsack sets

$$\sum_{j\in V_2} u_{ij}x'_{ij} \leqslant \sum_{j\in V_2} u_{ij} - b_i, \quad (x'_{i1}, \ldots, x'_{in}) \in B^n \quad \text{for } i \in V_1$$

and

$$\sum_{i\in V_1} u_{ij}x'_{ij} \leqslant \sum_{i\in V_1} u_{ij} - d_j, \quad (x'_{1j}, \ldots, x'_{mj}) \in B^m \quad \text{for } j \in V_2$$

where $x'_{ij} = 1 - x_{ij}$. These constraints are obtained from

$$\sum_{j\in V_2} u_{ij}x_{ij} \geqslant b_i \quad \text{and} \quad \sum_{i\in V_1} u_{ij}x_{ij} \geqslant d_j.$$

Steps b and c: GFC and extended GFC flow inequalities are obtained from the following sets of inequalities:

(i) $$\sum_{j\in V_2} y_{ij} \leqslant b_i,\ y_{ij} \leqslant u_{ij}x_{ij} \text{ for } j \in V_2, \qquad \text{for } i \in V_1$$

(ii) $$-\sum_{j\in V_2} y_{ij} \leqslant -b_i,\ y_{ij} \leqslant u_{ij}x_{ij} \text{ for } j \in V_2, \qquad \text{for } i \in V_1$$

(iii) $$\sum_{i\in V_1} y_{ij} \leqslant d_j,\ y_{ij} \leqslant u_{ij}x_{ij} \text{ for } i \in V_1, \qquad \text{for } j \in V_2$$

(iv) $$-\sum_{i\in V_1} y_{ij} \leqslant -d_j,\ y_{ij} \leqslant u_{ij}x_{ij} \text{ for } i \in V_1, \qquad \text{for } j \in V_2.$$

Example 4.2. We solve the instance of the fixed-charge uncapacitated transportation problem given by the following data:

$$m = 4, \quad n = 6$$

$$(h_{ij}) = \begin{pmatrix} 0.69 & 0.64 & 0.71 & 0.79 & 1.70 & 2.83 \\ 1.01 & 0.75 & 0.88 & 0.59 & 1.50 & 2.63 \\ 1.05 & 1.06 & 1.08 & 0.64 & 1.22 & 2.37 \\ 1.94 & 1.50 & 1.56 & 1.22 & 1.98 & 1.98 \end{pmatrix}$$

$$(c_{ij}) = \begin{pmatrix} 11 & 16 & 18 & 17 & 10 & 20 \\ 14 & 17 & 17 & 13 & 15 & 13 \\ 12 & 13 & 20 & 17 & 13 & 15 \\ 16 & 19 & 16 & 11 & 15 & 12 \end{pmatrix}$$

$$b = (45 \quad 35 \quad 20 \quad 15), \quad d = (35 \quad 30 \quad 25 \quad 15 \quad 5 \quad 5).$$

Phase 1

Iteration 1: $z^1_{\mathrm{LP}} = 185.6$. The corresponding solution is shown in Figure 4.3, where (y^1_{ij}, x^1_{ij}) is indicated for each edge.

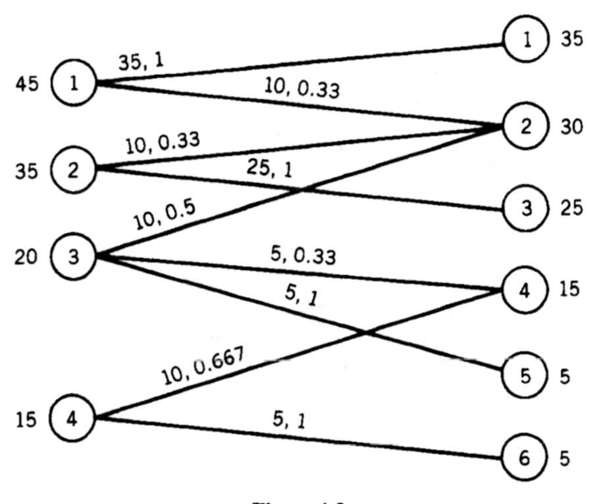

Figure 4.3

Applying the separation routine at source row 1, we combine the constraint

$$y_{11} + y_{12} + y_{13} + y_{14} + y_{15} + y_{16} = 45$$

with the variable upper-bound constraints $y_{11} \leqslant 35x_{11}$, $y_{12} \leqslant 30x_{12}$, $y_{13} \leqslant 25x_{13}$, $y_{14} \leqslant 15x_{14}$, $y_{15} \leqslant 5x_{15}$, and $y_{16} \leqslant 5x_{16}$ to obtain

$$35x_{11} + 30x_{12} + 25x_{13} + 15x_{14} + 5x_{15} + 5x_{16} \geqslant 45,$$

which is the knapsack inequality

$$35x'_{11} + 30x'_{12} + 25x'_{13} + 15x'_{14} + 5x'_{15} + 5x'_{16} \leqslant 70.$$

The separation routine for extended cover inequalities with $x' = (0 \;\; 0.67 \;\; 1 \;\; 1 \;\; 1 \;\; 1)$ then gives a violated constraint

$$x'_{11} + x'_{12} + x'_{13} + x'_{14} + x'_{16} \leqslant 3$$

or

$$x_{11} + x_{12} + x_{13} + x_{14} + x_{16} \geqslant 2.$$

Similarly, from source rows 2, 3, and 4, we obtain the violated inequalities

$$\begin{aligned} x_{21} + x_{22} \phantom{+ x_{33}} + x_{24} + x_{26} &\geqslant 1 \\ x_{31} + x_{32} + x_{33} + x_{34} &\geqslant 1 \\ x_{41} + x_{42} + x_{43} + x_{44} &\geqslant 1 \end{aligned}$$

and from demand row 2, we obtain the violated inequality

$$x_{12} + x_{22} + x_{42} \geqslant 1.$$

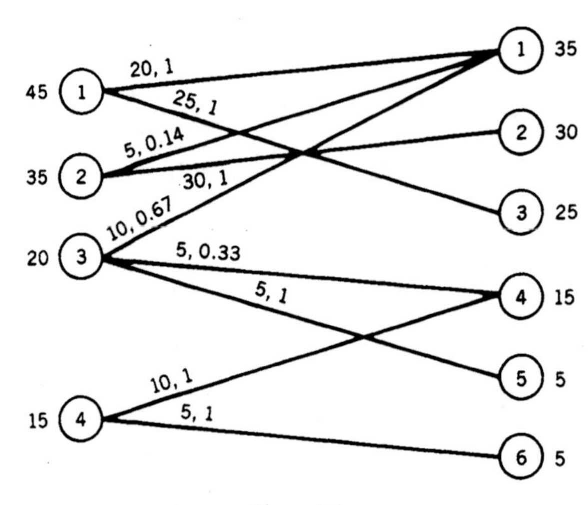

Figure 4.4

Iteration 2: After addition of the above constraints and reoptimizing the linear program, we obtain $z_{LP}^2 = 198.67$ and the solution shown in Figure 4.4.

Now the knapsack inequality

$$35x_{21} + 30x_{22} + 25x_{23} + 15x_{24} + 5x_{25} + 5x_{26} \geqslant 35$$
$$(x_{21}, \ldots, x_{26}) \in B^6$$

yields the violated cover inequality

$$x_{21} + x_{23} + x_{24} + x_{25} + x_{26} \geqslant 1.$$

Also, in Step b the set

$$y_{21} + y_{22} + \cdots + y_{26} \leqslant 35,$$
$$y_{21} \leqslant 35x_{21}, \quad y_{22} \leqslant 30x_{22}, \quad y_{23} \leqslant 25x_{23},$$
$$y_{24} \leqslant 15x_{24}, \quad y_{25} \leqslant 5x_{25}, \quad y_{26} \leqslant 5x_{26}$$

yields the violated GFC inequality (4.5)

$$y_{21} + y_{22} \leqslant 30 + 5x_{21},$$

which is obtained with $C^+ = \{1,2\}$ and $C^- = \emptyset$.

Iterations 3 and 4: We obtain $z_{LP}^3 = 200.4$. The cuts (4.11)

$$y_{22} + y_{23} + y_{24} \leqslant 20 + 10x_{22} + 5x_{23} + 10x_{24}$$

and

$$y_{21} + y_{22} + y_{23} + y_{24} \leqslant 20 + 15x_{21} + 10x_{22} + 5x_{23} + 10x_{24}$$

are both derived from source row 2, the first with $C^+ = \{2, 3\}$, $C^- = \emptyset$, $L^+ = \{4\}$, and $L^- = \emptyset$ and the second with $C^+ = \{2, 3\}$, $C^- = \emptyset$, $L^+ = \{1, 4\}$, and $L^- = \emptyset$.

Iteration 5: The lower bound increases to $z_{LP}^5 = 200.61$. On Iteration 5 no more cuts are generated, so the cut generation phase terminates.

Phase 2. Branch-and-bound is now applied. The solution shown in Figure 4.5 is found at node 3, and it is proved to be optimal at node 5. Its cost is 202.35.

If the problem is solved directly by branch-and-bound, a tree containing 129 nodes is needed to prove optimality.

For larger fixed-charge transportation problems, this FCPA is often successful in substantially increasing the lower bounds obtained from the linear programming relaxation. However, it remains an open question to find and develop separation algorithms for other classes of valid inequalities that will make it possible to obtain lower bounds that are reliably close to the optimal cost.

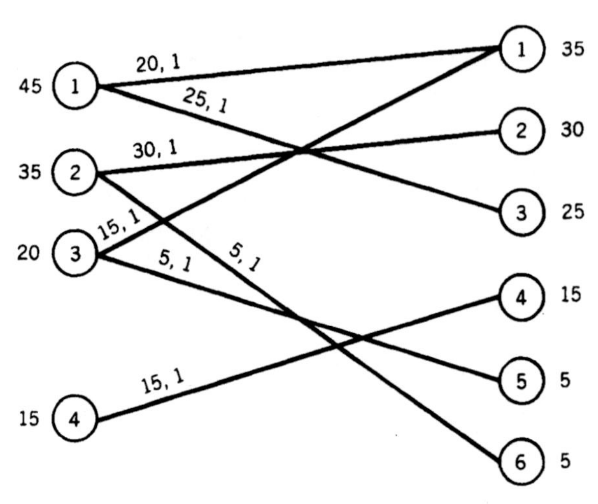

Figure 4.5

A Reformulation of the Single Source Problem (SFN)

The idea of the reformulation is to decompose the flows by destination. We suppose that node 1 is the source and let $U = \{k \in V: b_k < 0\}$. Thus $b_1 = \Sigma_{k \in U} |b_k|$. Now let z_{ijk} be the flow in arc (i, j) destined for node $k \in U$. The reformulation of (4.1) is

$$\min \sum_{(i,j)\in \mathscr{A}} h_{ij}y_{ij} + \sum_{(i,j)\in \mathscr{A}} c_{ij}x_{ij}$$

$$\begin{array}{llrll}
 & \text{(4.12a)} & \displaystyle\sum_{j\in\delta^+(i)} z_{ijk} - \sum_{j\in\delta^-(i)} z_{jik} & = 0 & \text{for } i \in V\setminus\{1,k\} \text{ and } k \in U \\
 & \text{(4.12b)} & \displaystyle - \sum_{j\in\delta^-(k)} z_{jkk} & = b_k & \text{for } k \in U \\
\text{(4.12)} & \text{(4.12c)} & z_{ijk} - \min(|b_k|, u_{ij})x_{ij} & \leqslant 0 & \text{for } (i,j) \in \mathscr{A} \text{ and } k \in U \\
 & \text{(4.12d)} & \displaystyle\sum_{k\in U} z_{ijk} - y_{ij} & = 0 & \text{for } (i,j) \in \mathscr{A} \\
 & & y_{ij} - u_{ij}x_{ij} & \leqslant 0 & \text{for } (i,j) \in \mathscr{A}
\end{array}$$

$$z \in R_+^{|\mathscr{A}||U|}, \quad y \in R_+^{|\mathscr{A}|}, \quad x \in B^{|\mathscr{A}|}.$$

The important difference between (4.12) and (4.1) is the upper-bound constraints $z_{ijk} \leqslant |b_k| x_{ij}$, which, for fractional x_{ij}, can restrict the flows more than $y_{ij} \leqslant u_{ij}x_{ij}$ can.

Proposition 4.4. *For* SFN, *the optimal cost of the linear programming relaxation of formulation (4.12) is not less than the optimal cost of the linear programming relaxation of formulation (4.1), and it may be strictly greater.*

Proof. By summing the constraints (4.12a), (4.12b) over k for fixed i, we see that if z_{ijk} is feasible in (4.12), then $y_{ij} = \Sigma_{k\in U} z_{ijk}$ is feasible in (4.1). It follows that every solution $(z^*_{ijk}, y^*_{ij}, x^*_{ij})$ to the linear programming relaxation of (4.12) gives rise to a feasible solution of the linear programming relaxation of (4.1).

Example 4.3 shows that the linear programming relaxation of (4.12) can yield a larger lower bound, and thus it completes the proof. ■

Example 4.3. The graph and data are shown in Figure 4.6.

An optimal solution to the linear programming relaxation of (4.1) is $x_{12} = x_{13} = \frac{1}{2}$, $x_{23} = 0$, $y_{12} = y_{13} = 1$, $y_{23} = 0$ with cost $\frac{5}{2}$. An optimal solution to the linear programming relaxation of (4.12) is $x_{12} = x_{23} = 1$, $x_{13} = 0$, $y_{12} = 2$, $y_{23} = 1$, $y_{13} = 0$ with cost 4. Note that the constraint $z_{133} \leq x_{13}$ makes it infeasible to have $x_{12} = x_{13} = \frac{1}{2}$ and $y_{13} = 1$.

Thus from the point of view of bounds, (4.12) is preferable to (4.1). However, (4.12) has one major disadvantage—its size—which makes it impractical for all but very small problems. Benders' decomposition sometimes provides a way around this problem. We will illustrate this approach with the Steiner branching problem.

For uncapacitated single-source problems, the reformulation (4.12) can be simplified because the variables y_{ij} can be eliminated. Thus for SUFNs we obtain the reformulation

$$\text{(4.13)} \quad \min\Big\{\sum_{k\in U}\sum_{(i,j)\in\mathscr{A}} h_{ij}z_{ijk} + \sum_{(i,j)\in\mathscr{A}} c_{ij}x_{ij}\text{: (4.12a), (4.12b),}$$
$$z_{ijk} - b_k x_{ij} \leq 0 \text{ for } (i,j)\in\mathscr{A} \text{ and } k\in U,\ z\in R_+^{|\mathscr{A}||U|},\ x\in B^{|\mathscr{A}|}\Big\}.$$

Proposition 4.5. *Formulation (4.13) is stronger than the formulation (4.1) for* SUFNs.

Steiner Branchings

We now consider the formulation (4.13) for the Steiner r-branching problem that we defined at the beginning of the section. Recall that we want to find a minimum-weight r-branching on a subset of nodes $D \subseteq V$. Let $r = 1$ and $U = D \setminus \{1\}$. The formulation (4.13) yields the constraints for each $k \in U$:

$$\text{(4.14)} \quad \begin{aligned} \sum_{j\in\delta^+(i)} z_{ijk} - \sum_{j\in\delta^-(i)} z_{jik} &= 0 \quad \text{for } i\in U\setminus\{k\} \text{ and } k\in U \\ \sum_{j\in\delta^-(k)} z_{jkk} &= 1 \quad \text{for } k\in U \\ z_{ijk} &\leq x_{ij} \quad \text{for } (i,j)\in\mathscr{A} \text{ and } k\in U \\ z\in R_+^{|\mathscr{A}||U|},\quad & x\in B^{|\mathscr{A}|}. \end{aligned}$$

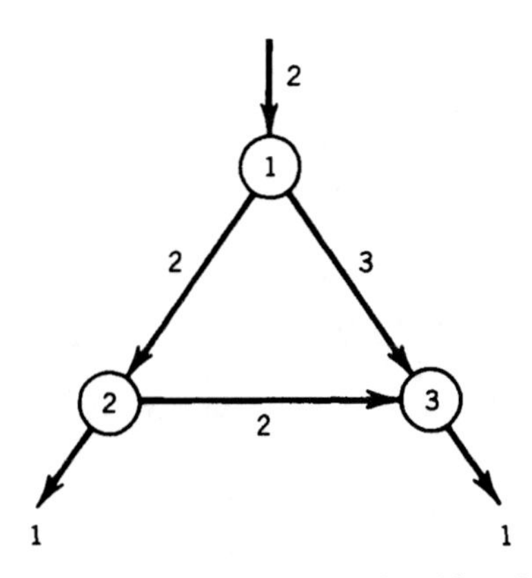

Figure 4.6. The c_{ij} appear on each arc; $u_{ij} = 2$ and $h_{ij} = 0$ for $(i,j)\in\mathscr{A}$.

Observe that if we fix x, say $x = \bar{x}$, (4.14) decomposes into $|U|$ separate feasibility problems. The kth problem is to determine whether there exists a feasible flow of one unit from node 1 to node k with arc capacities $\bar{x}_{ij}$. By the max-flow–min-cut theorem of Section I.3.4, such a flow exists if and only if $\Sigma_{i\in S}\Sigma_{j\in V\setminus S}\bar{x}_{ij} \geqslant 1$ for all S with $1 \in S$, and $k \in V\setminus S$. Hence if Benders' decomposition is applied to the linear programming relaxation of (4.14), the resulting master problem is

$$\min \sum_{(i,j)\in\mathscr{A}} c_{ij}x_{ij}$$

$$\sum_{(i,j)\in\delta^+(S)} x_{ij} \geqslant 1 \quad \text{for all } S \subseteq V \text{ with } 1 \in S, (V\setminus S)\cap U \neq \emptyset$$

$$x_{ij} \in R_+^{|\mathscr{A}|},$$

which states that every cutset having $1 \in S$ and $(V\setminus S)\cap U \neq \emptyset$ has weight at least 1. For $x \in B^{|\mathscr{A}|}$ this is precisely the requirement that the subgraph induced by the arcs with $x_{ij} = 1$ contains a 1-branching that spans the nodes of D.

Reformulations of the Uncapacitated Lot-Size Problem (ULS)

We first introduced the uncapacitated lot-size problem (ULS) in Section I.1.5 using the formulation

$$\min \sum_{t=1}^{T} (p_t y_t + h_t s_t + c_t x_t)$$

$$(4.15) \qquad \begin{aligned} y_1 &= d_1 + s_1 \\ s_{t-1} + y_t &= d_t + s_t \quad \text{for } t = 2, \ldots, T \\ y_t &\leqslant \omega x_t \qquad \text{for } t = 1, \ldots, T \\ s_T &= 0 \\ s, y \in R_+^T,& \quad x \in B^T, \end{aligned}$$

and then we reformulated it as an uncapacitated facility location problem.

We let $S \subseteq R_+^{3T}$ denote the set of feasible solutions to (4.15). In Section II.2.4 we described the convex hull of S, and in Section II.5.5 we gave an $O(T^2)$ dynamic programming algorithm for solving ULS. Here we take a different point of view. We assume that ULS is part of a more complicated problem. For example, we can add capacity constraints on the productions or inventories or assume that the actual problem to be solved involves several items. In this case, the model contains a copy of ULS for each product, and they are linked together by capacity constraints.

Our objective is to solve the complicated model by a linear-programming/branch-and-bound algorithm. Thus it is important that a tight formulation of ULS, with respect to the bounds obtained from linear programming relaxation, be used in the overall model. One possibility is to use the description of conv(S) given in Section II.2.4. However, since this description contains an exponential number of constraints, an FCPA will be required.

Here we consider some other options that are derived from the SUFN formulation (4.13) on the network of Figure 5.2 of Section II.5.5 (see Figure 4.7).

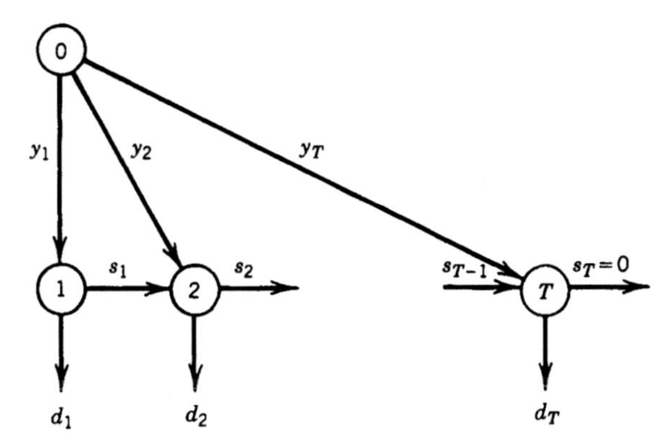

Figure 4.7

To obtain the formulation (4.13), we introduce variables y_{jk} equal to the amount produced in period j to satisfy demand in period k and s_{jk} equal to the stock at the end of period j destined for period k. This yields

$$\min \sum_{j=1}^{T} \sum_{k=j}^{T} p_j y_{jk} + \sum_{j=1}^{T-1} \sum_{k=j+1}^{T-1} s_{jk} + \sum_{j=1}^{T} c_j x_j$$

(4.16)

$$\text{(4.16a)} \quad s_{j-1,k} + y_{jk} = s_{jk} \quad \text{for all } j \text{ and } k > j$$

$$\text{(4.16b)} \quad s_{k-1,k} + y_{kk} = d_k \quad \text{for all } k$$

$$y_{jk} \leqslant d_k x_j \quad \text{for all } j \text{ and } k \geqslant j$$

$$y \in R_+^{T(T+1)/2}, \quad s \in R_+^{T(T-1)/2}, \quad x \in B^T.$$

To simplify the presentation we will not bother to explicitly state the constraints $y_{11} = d_1$ and $y_{1k} = s_{1k}$ for $k > 1$, and we assume $d_k > 0$ for $k = 1, \ldots, T$.

Now we use $s_{jk} = \sum_{i=1}^{j} y_{ik}$ to eliminate the inventory variables from the objective function and constraints. This yields

$$\min \sum_{j=1}^{T} \sum_{k=j}^{T} (p_j + h_j + \cdots + h_{k-1}) y_{jk} + \sum_{j=1}^{T} c_j x_j$$

(4.17)

$$\sum_{j=1}^{k} y_{jk} = d_k \quad \text{for } k = 1, \ldots, T$$

$$y_{jk} \leqslant d_k x_j \quad \text{for all } j \text{ and } k \geqslant j$$

$$y \in R_+^{T(T+1)/2}, \quad x \in B^T.$$

Since $s_{jk} \geqslant 0$ are implied by $y_{ij} \geqslant 0$, no other conditions are needed. Note that (4.17) is precisely the formulation (5.5) given in Section I.1.5. As noted there, we obtain a formulation of ULS as an uncapacitated facility location problem by letting $w_{jk} = y_{jk}/d_k$ for all j and k for which y_{jk} is defined. This yields

$$\min \sum_{j=1}^{T}\sum_{k=j}^{T}(p_j + h_j + \cdots + h_{k-1})d_k w_{jk} + \sum_{j=1}^{T} c_j x_j$$

(4.18) (4.18a) $$\sum_{j=1}^{k} w_{jk} = 1 \quad \text{for } k = 1, \ldots, T$$

(4.18b) $$w_{jk} \leqslant x_j \quad \text{for all } j \text{ and } k \geqslant j$$

$$w \in R_+^{T(T+1)/2}, \quad x \in B^T.$$

Now if we were to solve the formulation (4.18), the original variables y_t, s_t for $t = 1, \ldots, T$ would be obtained from

(4.19)
$$y_t = \sum_{k=t}^{T} y_{tk} = \sum_{k=t}^{T} d_k w_{jk}$$

$$s_t = \sum_{k=t+1}^{T} s_{tk} = \sum_{k=t+1}^{T}\sum_{i=1}^{t} y_{ik} = \sum_{k=t+1}^{T}\sum_{i=1}^{t} d_k w_{ik}.$$

The observation we need to make is that corresponding to any feasible solution in the original variables (s, y, x) there can be an infinity of feasible solutions in the variables (w, x) with the same cost. For example, corresponding to the solution shown in Figure 4.8, we note that $x_1 = x_2 = 1$, $x_3 = 0$, $w_{11} = 1$, $w_{12} = w_{23} = \alpha$, $w_{13} = w_{22} = 1 - \alpha$ is a feasible solution to (4.17) of cost $c_1 + c_2 + 2p_1 + p_2 + h_1 + h_2$ for any $\alpha \geqslant 0$.

We claim that we will still have a valid formulation of ULS by adding the constraints

(4.20) $$w_{jT} \leqslant w_{j,T-1} \leqslant \cdots \leqslant w_{jj} \quad \text{for } j = 1, \ldots, T-1$$

to (4.18). In other words, we claim that the formulation

$$\min \sum_{j=1}^{T}\sum_{k=j}^{T}(p_j + h_j + \cdots + h_{k-1})d_k w_{jk} + \sum_{j=1}^{T} c_j x_j$$

(4.21) (4.21a) $$\sum_{j=1}^{k} w_{jk} = 1 \qquad \text{for } k = 1, \ldots, T$$

(4.21b) $$w_{jT} \leqslant w_{j,T-1} \leqslant \cdots \leqslant w_{jj} \leqslant x_j \quad \text{for } j = 1, \ldots, T$$

$$w \in R_+^{T(T+1)/2}, \quad x \in B^T.$$

is valid for ULS.

To establish this claim, note that we have already shown in Section II.5.5 that every extreme point of conv(S) is of the following form: For some subset $\{i_1, \ldots, i_r\} \subseteq \{1, \ldots, n\}$ with $1 = i_1 < i_2 < \cdots < i_r$, we obtain

$$x_{i_l} = 1 \quad \text{for } l = 1, \ldots, r, \quad x_j \in \{0, 1\} \quad \text{otherwise}$$

$$y_{i_l} = \sum_{t=i_1}^{i_{l+1}-1} d_t, \; y_j = 0 \quad \text{otherwise.}$$

Using (4.19), this corresponds to a feasible solution (x,w) of (4.21) with the same values of x_j for $j = 1, \ldots, T$, and with $w_{i_l t} = 1$ for $t = i_l, \ldots, i_{l+1} - 1$ and $l = 1, \ldots, r$ and $w_{ij} = 0$ otherwise.

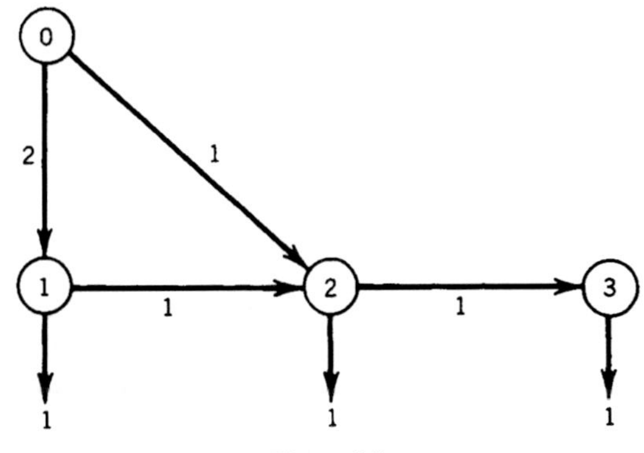

Figure 4.8

Now let $Q(Q_{\text{LP}})$ be the image in (y, s, x)-space under transformation (4.19) of the points (w, x) feasible in (4.21) [the linear programming relaxation of (4.21)]. The above discussion shows that:

Proposition 4.6. $S = Q$ (*and* $\text{conv}(S) \subseteq Q_{\text{LP}}$).

Our interest in (4.21) arises from a final formulation of ULS as a minimum-weight path problem. For $k = 2, \ldots, T$, we subtract the $(k-1)$st constraint from the kth constraint in (4.21a). This leads to the $T - 1$ constraints

$$\text{(4.22)} \qquad w_{kk} - \sum_{j=1}^{k-1} (w_{j,k-1} - w_{jk}) = 0 \quad \text{for } k = 2, \ldots, T.$$

Now define $z_{jk} = w_{jk} - w_{j,k+1} \geqslant 0$ for $1 \leqslant j \leqslant k < T$, and define $z_{jT} = w_{jT} \geqslant 0$ for $j = 1, \ldots, T$. Then $w_{kk} = \Sigma_{l=k}^{T} z_{kl}$, and we obtain the reformulation

$$\min \sum_{j=1}^{T} \sum_{k=j}^{T} \left(\sum_{t=j}^{k} (p_j + h_j + \cdots + h_{t-1}) d_t \right) z_{jk} + \sum_{j=1}^{T} c_j x_j$$

$$\text{(4.23a)} \qquad \sum_{l=1}^{T} z_{1l} = 1$$

$$\text{(4.23)} \quad \text{(4.23b)} \qquad -\sum_{j=1}^{k-1} z_{j,k-1} + \sum_{l=k}^{T} z_{kl} = 0 \quad \text{for } k = 2, \ldots, T$$

$$\text{(4.23c)} \qquad \sum_{l=k}^{T} z_{kl} - x_k \leqslant 0 \quad \text{for } k = 1, \ldots, T$$

$$z \in R_+^{T(T+1)/2}, \quad x \in B^T$$

where (4.23c) comes from $w_{kk} \leqslant x_k$ for all k.

If $c_k \geqslant 0$, then in the linear programming relaxation of (4.23) we can take (4.23c) as an equality. Substituting this equality into (4.23a) and (4.23b) yields

$$(4.24)\quad \begin{array}{lll} (4.24\text{a}) & x_1 = 1 & \\ (4.24\text{b}) & -\sum_{j=1}^{k-1} z_{j,k-1} + x_k = 0 & \text{for } k = 2, \ldots, T \\ (4.24\text{c}) & \sum_{l=k}^{T} z_{kl} - x_k = 0 & \text{for } k = 1, \ldots, T. \end{array}$$

Now observe that by constructing a digraph with node set $\{1, \ldots, T+1, 1', \ldots, T'\}$ and by letting x_k be the flow from k to k' and letting z_{jk} be the flow from j' to $k+1$, then (4.24b) and (4.24c) are the flow conservation equations at nodes $k = 2, \ldots, T$ and at nodes $k = 1', \ldots, T'$, respectively. See Figure 4.9 for $T = 3$.

Moreover, if $c_k < 0$ we can set $x_k = 1$, and (4.23c) is superfluous. In terms of the graph, the arc corresponding to x_k is deleted, and the nodes k and k' are coalesced.

Proposition 4.7. *The linear programming relaxations of* (4.21) *and* (4.23) *have optimal solutions with* $x \in B^T$ *for any objective function* (p, h, c).

An important consequence of Propositions 4.6 and 4.7 concerns the polyhedron Q_{LP} representing the set of feasible solutions to the linear programming relaxation of (4.21) in terms of the original variables.

Theorem 4.8. $Q_{\text{LP}} = \text{conv}(S)$.

Proof. We show that $Q_{\text{LP}} \subseteq \text{conv}(S)$. If not, let (y^*, s^*, x^*) obtained from (y^*, s^*, x^*, w^*) be an extreme point of Q_{LP} that is not in conv(S). There exists an objective function (p, h, c) for which (y^*, s^*, x^*) is the unique optimal solution to $\min\{py + hs + cx: (y, s, x) \in Q_{\text{LP}}\}$. But this implies that (x^*, w^*) is a feasible solution to (4.21) whose objective value is less than that of any point (x, w) corresponding to an extreme point of conv(S). Thus $x^* \notin B^T$, which contradicts Proposition 4.7. Hence $Q_{\text{LP}} \subseteq \text{conv}(S)$. $Q_{\text{LP}} \supseteq \text{conv}(S)$ was shown in Proposition 4.6. ■

It can be shown that Proposition 4.7 also holds for formulation (4.18). Thus we can conclude that the corresponding version of Theorem 4.8 holds for formulation (4.18).

We now consider the problem stated at the beginning of this section of choosing a formulation to embed in a more complicated model. To formalize the problem, we wish to solve

$$(4.25)\qquad z' = \min\{py + hs + cx: (y, s, x) \in S \cap P'\},$$

where $P' \subseteq R_+^{3T}$ represents the set of complicating constraints. For each of the three models (4.18), (4.21), (4.23), it is easy to represent the constraints of P' in terms of the new

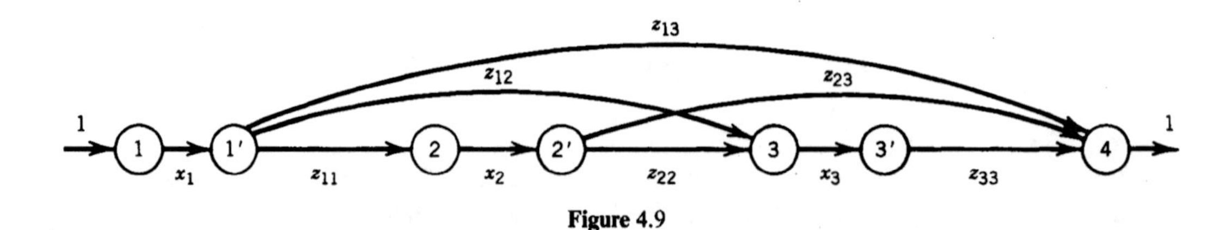

Figure 4.9

Table 4.1.

Formulation	Number of variables	Number of constraints
(4.18) and (4.21)	$O(T^2)$	$O(T^2)$
(4.23)	$O(T^2)$	$O(T)$
(4.15) & conv(S)	$O(T)$	$O(2^T)$

variables, so we then obtain three reformulations of problem (4.25). The values of their respective linear programming relaxations are denoted by z^i_{LP} for $i = 1, 2, 3$.

For each of these formulations we are interested in two things, namely, the tightness of the linear programming relaxation and the size of the formulation. Considering the bounds first, we have, by Theorem 4.8 and the identical result for formulation (4.18), the following proposition:

Proposition 4.9. $z^i_{\text{LP}} = \min\{py + hs + cx\colon (y, s, x) \in \text{conv}(S) \cap P'\}$ *for* i = 1, 2, 3.

Hence each of the formulations is as tight as it can be made without studying the structure of the complicating constraints P'. Therefore to choose among the formulations we turn to the question of size, and we consider the number of variables and constraints in each formulation (see Table 4.1). The last formulation shown in Table 4.1 is to add the facet-defining inequalities described in Section II.2.4 to the formulation (4.15) of ULS.

The figures in the table suggest that a model based on (4.23) significantly dominates formulations (4.18) and (4.21) with respect to the number of constraints. Recently, problems with 200 items and $T = 10$ periods have been solved by a standard linear-programming/branch-and-bound algorithm using reformulation (4.23).

The last formulation, which involves only $O(T)$ variables but an exponential number of constraints, might be competitive with (4.23) using an FCP algorithm because the number of facet-defining inequalities needed at an optimal extreme point is bounded by the number of variables.

5. APPLICATIONS OF BASIS REDUCTION

The use of basis reduction in lattices is new to integer programming. To indicate its potential, we outline two applications. The first is a simple heuristic algorithm to find a feasible solution to a 0-1 equality knapsack constraint. The second is an algorithm for integer programming that is polynomial for fixed n. Although this is an important theoretical result, the algorithm is not practical. The result has, however, motivated the application of basis reduction techniques to a variety of problems.

The Subset Sum Problem

Here we consider the $\mathcal{NP}$-hard problem of finding a feasible solution to a 0-1 knapsack equality constraint:

$$(5.1) \qquad \sum_{j=1}^{n} a_j x_j = M, \qquad x \in B^n.$$

This problem is of particular interest in cryptography where problems of the form (5.1) are constructed to have a unique solution that corresponds to a message to be transmitted. In such a system the coefficients a_j for $j \in N$ are public information, the message transmitted is M, and the problem (5.1) must have "very large" coefficients and be "impossible" to solve, except by the receiver who knows a trick for any M.

Here we describe a fast heuristic algorithm for (5.1), which uses the reduced basis algorithm of Section I.7.5. Let

$$C^1 = \begin{pmatrix} I & 0 \\ -a & M \end{pmatrix},$$

where $a = (a_1, \ldots, a_n)$ and I is the $n \times n$ identity matrix. Consider the lattice $L(C^1) \subseteq R^{n+1}$ given by $\{v \in Z^{n+1}: v = C^1 y, y \in Z^{n+1}\}$. Now observe that if $x \in B^n$ is a feasible solution to (5.1), then

$$v^1 = C^1 \begin{pmatrix} x \\ 1 \end{pmatrix} = \begin{pmatrix} x \\ 0 \end{pmatrix}$$

is an element of the lattice. Moreover, v^1 is a short vector in $L(C^1)$ because $\|v^1\| \leq n^{1/2}$, which is much smaller than the bound given in Theorem 5.5 of Section I.7.5.

In addition, by setting $\overline{x}_j = 1 - x_j$ for $j = 1, \ldots, n$ and by treating

$$\sum_{j=1}^{n} a_j \overline{x}_j = M' = \sum_{j=1}^{n} a_j - M, \qquad \overline{x} \in B^n$$

similarly, we see that $v^2 = \binom{\overline{x}}{0}$ is a short vector in the associated lattice $L(C^2)$, where

$$C^2 = \begin{pmatrix} I & 0 \\ -a & M' \end{pmatrix}.$$

Now $\min(\|v^1\|, \|v^2\|) \leq (n/2)^{1/2}$.

The idea of the algorithm is that if v^i is a very short, and possibly the shortest, vector in $L(C^i)$ for $i = 1$ or 2, there is a good chance that it will appear in a reduced basis for $L(C^i)$. Thus it suffices to check whether the reduced basis contains a vector of the form $\binom{\pm x}{0}$ with $x \in B^n$.

The Reduced Basis Algorithm to Find a Solution of (5.1)

Step 1: Consider the lattice $L(C^1) \subseteq R^{n+1}$, where C^1 is the matrix given above.

Step 2: Find a reduced basis $\tilde{B}^*$ of $L(C)$.

Step 3: Check if $\tilde{B}^*$ contains a column of the form $\binom{\pm x^1}{0}$ with $x^1 \in B^n$. If so, stop. x^1 solves (5.1).

Step 4: Repeat Steps 1 to 3 with C^1 replaced by C^2. If a vector $x^2 \in B^n$ is found, $1 - x^2$ solves (5.1). Otherwise, stop. No solution has been found.

The reduced basis algorithm has a very high probability of finding a feasible solution for certain classes of knapsack problems. We define the density $d(a)$ of a set of weights $(a_1, \ldots, a_n)$ by

$$d(a) = \frac{n}{\log(\max_j a_j)}.$$

It can be shown, under appropriate distribution assumptions, that there exist constants α and β such that:

a. For "nearly all" feasible instances (5.1) with $d(a) < \alpha$, (5.1) has a unique solution $x \in B^n$; this solution is the shortest nonzero vector in $L(C)$.
b. For "nearly all" feasible instances with $d(a) < \beta/n$, the reduced basis algorithm finds a solution.

The proof of statement b is demonstrated by showing that all other vectors in the lattice $L(C)$ are much longer than $v^1 = \binom{x}{0}$. In particular, if $\|w\| \geqslant 2^{n-1}\|v^1\|$ for all $w \in L(C) \setminus \{0, v^1\}$, then we know by Theorem 5.5(iii) of Section I.7.5 that $\pm v^1$ is in the reduced basis.

The Linear Inequality Integer Feasibility Problem

Here we outline an algorithm for the linear inequality integer feasibility problem

$$\text{Find } x \in P \cap Z^n \text{ or show } P \cap Z^n = \emptyset, \tag{5.2}$$

where $P = \{x \in R^n: Ax \leq b\}$ and n is fixed. From Section I.5.4, it can be assumed that if $P \neq \emptyset$ there exists $\omega_{A,b}$ such that $|x_j| \leq \omega_{A,b}$ for some $x \in P \cap Z^n$.

The algorithm is essentially enumerative. If we could show, for all $x \in P$, that $|x_j| \leq \gamma$ where γ is any function polynomial in $\log \theta_{A,b}$, where $\theta_{A,b}$ is the largest coefficient in (A, b), then we would immediately obtain a polynomial algorithm by enumerating the $(2\gamma + 1)^n$ points with $|x_j| \leq \gamma$ and $x \in Z^n$. Since the bound $\omega_{A,b}$ is not polynomial in $\log \theta_{A,b}$, this simple approach does not work. However, by using a reduced basis it is possible to obtain a polynomial-time enumeration algorithm.

The first important concept in the algorithm is the idea of a family of polytopes being "round". Let $S(p, r)$ be an n-dimensional sphere with center p and radius r.

Definition 5.1. A family of full-dimensional polytopes in R^n is *round* if there exist a function c^1 such that for each P in the family, there exist rationals $p \in R^n$, $r, q \in R^1_+$ such that

i. $S(p, r) \subseteq P \subseteq S(p, q)$ and
ii. $q/r \leq c^1$.

To motivate the algorithm, let us first consider the solution of problem (5.2) for a full-dimensional and round family of polytopes. Here we will see that straightforward enumeration is polynomial. There are two cases to be considered, as demonstrated in Figure 5.1.

Case 1. $r \geq \frac{1}{2}n^{1/2}$. In this case, the unit hypercube with center p is contained in $S(p, r)$ and hence in P. Now let p^* be a closest integer point to p; that is, $p_j = \lfloor p_j \rfloor + f_j$ for $j \in N$, $p_j^* = \lfloor p_j \rfloor$ if $f_j \leq \frac{1}{2}$, and $p_j^* = \lfloor p_j \rfloor + 1$ otherwise. Then $p^* \in P \cap Z^n$, and hence (5.2) is solved.

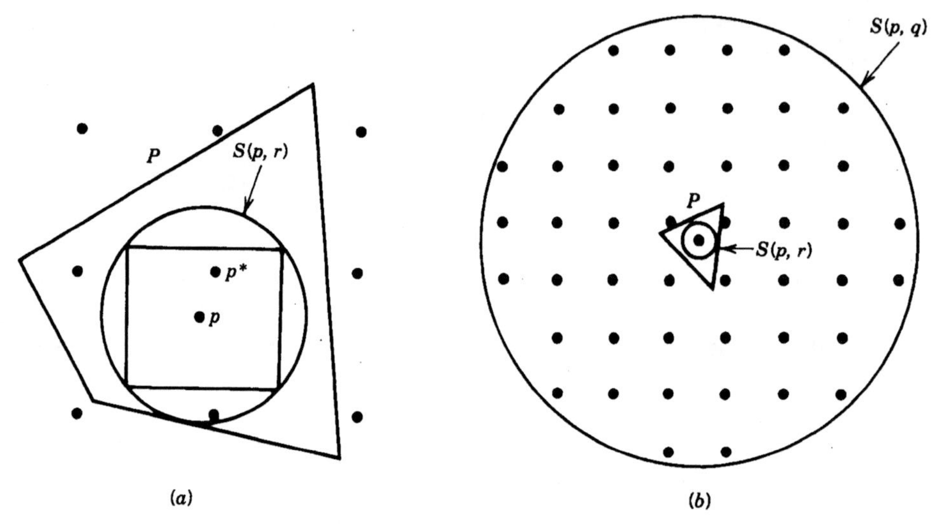

Figure 5.1. (a) *Case 1*: The closest integer point p^* to p is feasible. (b) *Case 2*. Enumerate the integer points in $S(p, q)$.

Case 2. $r < \frac{1}{2} n^{1/2}$. In this case, $q < \frac{1}{2}n^{1/2}c^1$. But now because $P \subseteq S(p, q)$, we have that if $x \in P$, then

$$p_j - \tfrac{1}{2}n^{1/2}c^1 \leq x_j \leq p_j + \tfrac{1}{2}n^{1/2}c^1$$

for all j. Thus total enumeration gives a polynomial algorithm for fixed n.

Now we indicate how for any polytope P, we can find a linear transformation $K: R^n \to R^n$ depending on P such that the transformed family of polytopes $\{K(P)\}$ is round. For simplicity, we consider only the case of full-dimensional polytopes.

Using linear programming and Gaussian elimination, we start by finding $n + 1$ affinely independent extreme points $\{v^i\}_{i=0}^n$ of P. The convex hull of $n + 1$ affinely independent points in R^n is called an *n-simplex*. Thus, $\{v^0, v^1, \ldots, v^n\}$ is an n-simplex $Q \subseteq P$.

Next we find a "large" n-simplex $Q' \subseteq P$. In particular, for each $i = 0, 1, \ldots, n$, we attempt to find a new extreme point $\tilde{v}^i$ of P so that the simplex $\{v^0, v^1, \ldots, v^{i-1}, \tilde{v}^i, v^{i+1}, \ldots, v^n\}$ has a volume more than 50% larger than that of Q. To do this, for each i we find the facet $\pi^i x = \alpha_i$ of the simplex opposite the vertex v^i. We find $\pi^i \in R^n$ by using Gaussian elimination to solve the linear system $\pi^i v^j = \alpha_i$ for all $j \neq i$. We then solve the linear program $\max\{\pi^i x: x \in P\}$ whose optimal solution is $\tilde{v}^i$.

If $|\pi^i \tilde{v}^i - \alpha_i| > \frac{3}{2}|\pi^i v^i - \alpha_i|$, we replace v^i by $\tilde{v}^i$ and start again with a larger simplex. If not, we replace π^i by $-\pi^i$ and resolve the linear program. Every time the simplex changes, its volume increases by at least 50% (see Figure 5.2a). Thus the number of linear programs that need to be solved cannot be too large. We stop when no change occurs for any $i = 0, 1, \ldots, n$. The final simplex $Q' = \{v^0, v^1, \ldots, v^n\}$ is a "large" simplex within P. Furthermore, we know that P lies inside a polytope with no more than $2n + 2$ facets, namely, the polytope

$$P^* = \{x \in R^n: |\pi^i x - \alpha_i| \leq \tfrac{3}{2}|\pi^i v^i - \alpha_i| \text{ for } i = 0, \ldots, n\}$$

(see Figure 5.2.b).

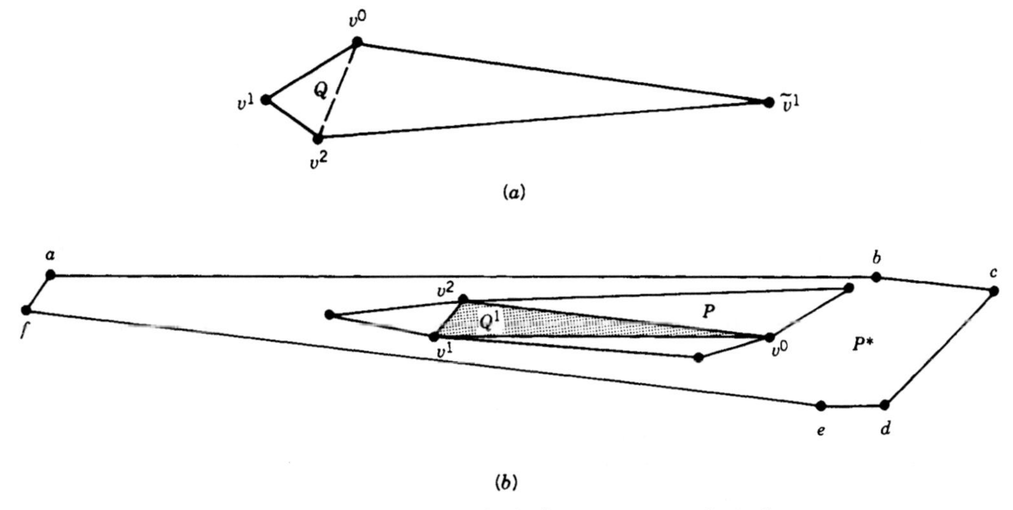

Figure 5.2. In part a, $Q = \{v^0, v^1, v^2\}$ is replaced by $\{v^0, \tilde{v}^1, v^2\}$.

Now a linear transformation K can be found so that $K(Q')$ becomes a regular n-simplex. Clearly $K(Q') \subseteq K(P) \subseteq K(P^*)$.

What has been achieved? Taking $p = [1/(n + 1)] \sum_{i=0}^{n} K(v^i)$, we can construct a hypersphere $S(p, r)$ inside $K(Q')$. $K(P^*)$ has no more than $2n + 2$ facets, so its vertices can be computed, and we can construct a hypersphere $S(p, q)$ containing $K(P^*)$. Hence $S(p, r) \subseteq K(P) \subseteq S(p, q)$. Simple calculations give that $q/r < kn^{3/2}$ for some constant k. Figure 5.3 shows Figure 5.2 after the transformation K; it also shows the spheres $S(p, r)$ and $S(p, q)$.

Although many technical details have been omitted, we have motivated the result that the family of all n-dimensional polytopes can be made round by a suitable linear transformation.

Proposition 5.1. *There exists a constant c^1 such that for any n-dimensional polytope $P = \{x \in R^n: Ax \leq b\}$, there exists a rational nonsingular matrix K and rationals $p \in R^n$, $r \in R^1_+$ and $q \in R^1_+$ such that $K(P) = \{y \in R^n: AK^{-1}y \leq b\}$ satisfies*

i. $S(p, r) \subseteq K(P) \subseteq S(p, q)$ *and*
ii. $q/r \leq c^1$.

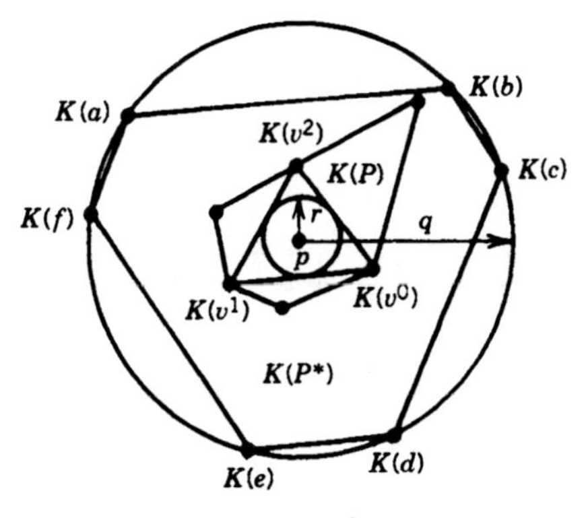

Figure 5.3

The initial problem (5.2) has now been transformed to the problem

(5.3) Find a vector $y \in K(P) \cap L(K)$,

where $L(K)$ is the lattice with basis K. In addition, we have rationals p, q, and r such that i and ii of Proposition 5.1 hold. This resembles the earlier situation in that we have a family of polytopes that is round, but now $L(K)$ has replaced Z^n.

The second transformation we introduce involves finding a reduced basis B for the lattice $L(K)$. Problem (5.2) is now equivalent to the problem

(5.4) Find a vector $y \in K(P) \cap L(B)$.

The geometry for $n = 2$ is shown in Figure 5.4, where $S(p, r) \subseteq K(P) \subseteq S(p, q)$ and the points of the lattice $L(B)$ are given.

We will now show how (5.4) can be solved by enumeration in polynomial time. As before, the algorithm breaks up into the case where r is large and the case where r is small. Previously when r was large, we used rounding to find a lattice point "close" to p. Now we will use a simple construction underlying the proof of the following proposition.

Proposition 5.2. *Given a lattice* $L(B)$ *and* $p \in R^n$, *there exists* $z \in L(B)$ *such that* $\|z - p\|^2 \leq \frac{1}{4}\sum_{j=1}^{n} \|b_j\|^2$.

We now consider the two cases of r large and small. Without loss of generality we assume that the columns of B are ordered so that $\max_{j=1,\ldots,n} \|b_j\| = \|b_n\|$.

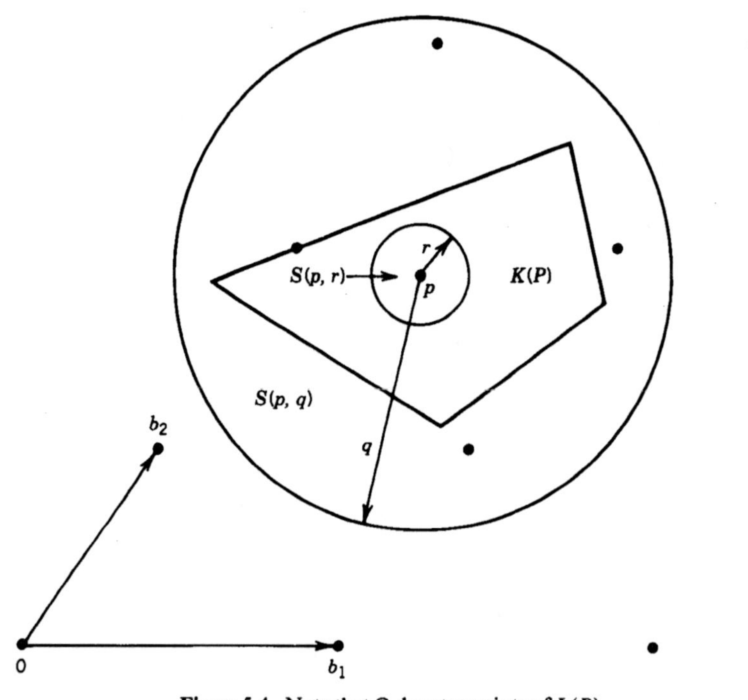

Figure 5.4. Note that ⊙ denotes points of $L(B)$.

Case 1. If $r \geq \frac{1}{2}n^{1/2}\|b_n\|$, apply Proposition 5.2 to find a point $y \in L(B)$ such that

$$\|y - p\|^2 \leq \tfrac{1}{4}\sum_{j=1}^{n} \|b_j\|^2 \leq \tfrac{1}{4}n\|b_n\|^2.$$

Since

$$\|y - p\| \leq \tfrac{1}{2}n^{1/2}\|b_n\| \leq r,$$

it follows that $y \in S(p, r) \subseteq K(P)$. Therefore $y \in K(P) \cap L(K)$, and $x = K^{-1}y$ is feasible in (5.2).

Case 2. If $r < \frac{1}{2}n^{1/2}\|b_n\|$, we will show that it is possible to enumerate in the direction of b_n and only test feasibility for a polynomial number of points. Let $L^{n-1} = L(b_1,\ldots, b_{n-1})$, and let H^{n-1} be the associated subspace. We let h denote the distance from b_n to H^{n-1}. By Definition 5.2 of Section I.7.5, we have $\det B = h \det(b_1, \ldots, b_{n-1})$. Since B is a reduced basis, we know from Theorem 5.5iii of that section that

$$\prod_{j=1}^{n} \|b_j\| \leq 2^{n(n-1)/4}d(L) = 2^{n(n-1)/4}h \det(b_1, \ldots, b_{n-1}),$$

and we know from Hadamard's inequality that $\det(b_1, \ldots, b_{n-1}) \leq \prod_{j=1}^{n-1} \|b_j\|$. Therefore, by canceling terms we obtain

$$h \geq \|b_n\|2^{-n(n-1)/4}.$$

Now observe that if $y \in L(B)$ with $y = Bz$, $z \in Z^n$, then $y = y^{n-1} + b_n z_n$, $z_n \in Z^1$, where $y^{n-1} \in L^{n-1}$, and hence $y \in H^{n-1} + b_n z_n$ for some $z_n \in Z^1$. $H^{n-1} + b_n z_n$, $z_n \in Z^1$, is a family of hyperplanes separated by a distance h. The number γ of such hyperplanes that can intersect $S(p, q)$ is no more than $2q/h + 1$ (see Figure 5.5).

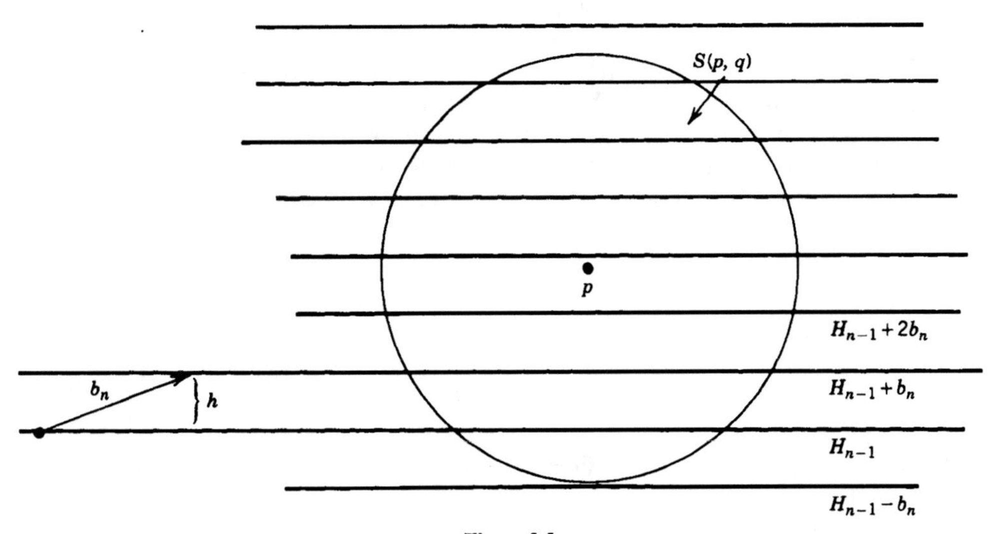

Figure 5.5

We now have that $\gamma \leq 2q/h + 1$ and $q/r \leq c^1$ because $K(P)$ is round, $h \geq \|b_n\| 2^{-n(n-1)/4}$ because the basis is reduced, and $r < \frac{1}{2}n^{1/2}\|b_n\|$ by assumption. Thus we have

$$\gamma - 1 \leq \frac{2q}{h} \leq \frac{2r}{h} c^1 \leq \frac{n^{1/2}\|b_n\|}{h} c^1 \leq n^{1/2} c^1 2^{n(n-1)/4}.$$

Therefore it is possible to enumerate over these γ possible values of $z \in Z^1$, and each of the resulting problems reduces to finding an integer point in a polytope whose dimension is no greater than $n - 1$.

In integer programming terms we have shown that

$$\max\{z_n: y = Bz, y \in K(P)\} - \min\{z_n: y = Bz, y \in K(P)\} \leq n^{1/2} c^1 2^{n(n-1)/4} + 1.$$

So we have outlined a basic inductive step. Either P is not full-dimensional, or we find a point in $P \cap Z^n$, or we reduce to a polynomial number of similar problems in $n - 1$ variables. It can be verified that for fixed n, all the steps indicated above can be carried out in time that is polynomial in the input length. Hence, we obtain Lenstra's theorem:

Theorem 5.3. *For fixed n, there is a polynomial algorithm for the linear inequality integer feasibility problem (5.2).*

Using bisection on the objective function value and the bounds given in Theorem 4.1 of Section I.5.4, Theorem 5.3 leads immediately to a result for integer programs.

Theorem 5.4. *For fixed n, there is a polynomial algorithm for the integer programming problem.*

Another immediate consequence of Theorem 5.3 is:

Theorem 5.5. *For fixed m, there is a polynomial algorithm for the linear inequality integer feasibility problem (5.2) and for the integer programming problem.*

Proof. If $m \geq n$, the claim is immediate. If $m < n$, we find the Hermite normal form of A given by $(H, 0) = AC$. This can be done in polynomial time (see Section I.7.4). Now if $y = C^{-1}x$, then $\{x: Ax \leq b, x \in Z^n\} \neq \emptyset$ if and only if $\{y: ACy \leq b, y \in Z^n\} \neq \emptyset$. Hence the problem is reduced to the feasibility problem for $\{u \in Z^m: Hu \leq b\}$, where $u = (y_1, \ldots, y_m)$. Thus by Theorems 5.3 and 5.4 the claim follows. ■

6. NOTES

Section II.6.1

The dynamic programming recursion (1.2) and the asymptotic properties given in Propositions 1.1 and 1.2 appeared in Gilmore and Gomory (1966); also see Shapiro and Wagner (1967).

The superadditive dual algorithm is due to Johnson (1973, 1980b).

The heuristic analysis presented here also applies with some small variations to the 0-1 knapsack problem, and the references cited analyze either the integer or 0-1 knapsack problem or both. Sahni (1975) combined the greedy heuristic with enumeration to obtain a polynomial approximation scheme for the knapsack problem. Ibarra and Kim (1975) used scaling to obtain a fully polynomial approximation scheme. The scaling/rounding heuristic is due to Lawler (1979). By adding the rounding feature, Lawler improved the running time of Ibarra and Kim's heuristic by a multiplicative factor of epsilon. A different fully polynomial approximation scheme for the 0-1 knapsack problem was given by Magazine and Oguz (1981).

Gomory (1965) gave a dynamic programming algorithm for solving the group problem. Shortest-path algorithms for the group problem were given by Shapiro (1968a), Glover (1969), and Hu (1970). A comparison of algorithms for solving the group problem was presented by Chen and Zionts (1976).

Shapiro (1968b) used the group problem and branch-and-bound to obtain an algorithm for general pure-integer programs [also see Gorry and Shapiro (1971), Gorry, Shapiro, and Wolsey (1972), and Crowder and Johnson (1973)]. An extensive computational study with this type of algorithm was carried out by Gorry, Northup, and Shapiro (1973). A shortest-path enumeration scheme was described in general terms by Lawler (1972) and was developed in the context of the group/branch-and-bound algorithm by Wolsey (1973).

The increasing group algorithm is a variant of an algorithm of Bell and Shapiro (1977).

Shapiro (1971), Fisher and Shapiro (1974), Bell and Fisher (1975), and Fisher, Northup, and Shapiro (1975) investigated how the group theoretic and Lagrangian dual approaches can be combined to solve general integer programs.

Kolesar (1967) gave one of the first branch-and-bound algorithms for the 0-1 knapsack problem. The computational efficiency of the basic algorithm has been improved by many researchers who have refined the node selection, branching and pruning rules, the variable fixing tests, and the method of solving the linear programming relaxation [see, among others, Ingargiola and Korsh (1973), Fayard and Plateau (1975, 1982), Lauriere (1978), Suhl (1978), and Balas and Zemel (1980)]. The presentation given here is largely based on the article by Lauriere (1978). Martello and Toth (1979) have given a comprehensive survey of methodology and an empirical comparison of algorithms. Another survey was given by Salkin and de Kluyver (1975).

Despite the excellent empirical results that have been obtained in solving knapsack problems by branch-and-bound, there are difficult families of knapsack problems for which any branch-and-bound algorithm with linear programming relaxations will enumerate an exponential number of nodes of the search tree [see Chvátal (1980)].

Some of the approaches and results for knapsack problems have been generalized to deal with problems having more than one knapsack-type constraint (i.e., the multidimensional knapsack problem). Polynomial approximation schemes have been obtained by Chandra et al. (1976) and Frieze and Clarke (1984). However, the problem of finding a fully polynomial approximation scheme for the multidimensional knapsack problem is $\mathcal{NP}$-hard. Korte and Schrader (1980) showed this for the 0-1 problem, and Magazine and Chern (1984) obtained the result for bounded and unbounded integer variables. Various practical heuristics have been proposed and evaluated [see, e.g., Loulou and Michaelides (1979), and Martello and Toth (1981b)]. Martello and Toth (1981a) also have given a branch-and-bound algorithm.

There have also been many studies of the knapsack problem with general upper-bound constraints including heuristics, branch-and-bound algorithms, and efficient methods for solving the linear programming relaxation [see Frieze (1976), Sinha and Zoltners (1979), Zemel (1980, 1984), Johnson and Padberg (1981), and Dyer (1984)].

Section II.6.2

The heuristic for BIP, called *pivot and complement*, is due to Balas and Martin (1980).

The FCP/branch-and-bound algorithm is from Crowder, Johnson and Padberg (1983) [also see Johnson and Padberg (1983), Johnson, Kostreva, and Suhl (1985), and Hoffman and Padberg (1985)]. The algorithm has been implemented in the mathematical programming systems PIPEX of IBM, MPSARX of Scicon, and XMP of Marsten (1981). Example 2.4 is a test problem from Crowder et al. (1983), and the results were obtained using the MPSARX system [see Van Roy and Wolsey (1987)].

Earlier approaches for solving BIPs emphasized implicit enumeration [see Balas (1965), Geoffrion (1967), and Petersen (1967)]. This type of algorithm was improved by the addition of surrogate constraints [see Balas (1967), Glover (1968c), and Geoffrion (1969)]. Spielberg (1979) gave a survey of these algorithms.

Specialized versions of the implicit enumeration approach have been used to solve set-partitioning and -covering problems [see Pierce (1968), Garfinkel and Nemhauser (1969), Pierce and Lasky (1973), and Marsten (1974)].

Several implicit enumeration and branch-and-bound algorithms for set-covering and -partitioning problems have incorporated special techniques for solving the linear programming relaxation and tightening it. Etcheberry (1977) gave an implicit enumeration algorithm that uses Lagrangian relaxation and subgradient optimization. Nemhauser, Trotter, and Nauss (1974) used a combinatorial relaxation based on finding a minimum-weight chain decomposition in a partially ordered set. This relaxation can be solved as a network flow problem. Combined with Lagrangian duality to accommodate side constraints, it yields the same bound as the linear programming relaxation. Nemhauser and Weber (1979) used a weighted matching problem relaxation that, when combined with Lagrangian duality to accommodate side constraints, yields a tighter bound than the linear programming relaxation. Ali and Thiagarajan (1986) reformulated the set-covering problem as a network flow problem with side constraints. Again, Lagrangian duality is used to accommodate the side constraints. This relaxation yields a bound equal to the bound obtained from the linear programming relaxation. Marsten and Shepardson (1981) gave a linear programming based branch-and-bound algorithm.

Fulkerson, Nemhauser, and Trotter (1974) gave a family of set-covering problems arising in the statistical design of experiments that are difficult to solve. Avis (1980) showed that these problems cannot be solved in polynomial time by branch-and-bound algorithms that use linear programming relaxations.

Some work has been done on using Gomory cuts to solve covering and partitioning problems [see, e.g., Balinski and Quandt (1964) and Salkin and Koncal (1973)]. Other cutting-plane approaches have been investigated by Bellmore and Ratliff (1971) and Balas (1980). Balas' cutting-plane approach, which is based on conditional bounds, has been implemented into an algorithm that also uses heuristics and subgradient optimization [see Balas and Ho (1980)].

Another approach to solving the set-partitioning problem uses Proposition 2.1. Balas and Padberg (1972, 1975) have given an algorithm that starts with an integer feasible solution and then uses pivoting to obtain a sequence of integer solutions of increasing weight which terminates with an optimal solution. The sequence is short, and its length is bounded by m, but exponential time may be required to find the appropriate pivots. Ikura and Nemhauser (1985) extended these ideas on pivoting from an integer solution to an adjacent one. For the set-packing problem, they showed that starting with any integer feasible solution and an associated unimodular basis matrix (see Section III.1.2), there exists a short sequence of primal simplex pivots, where each pivot element equals 1, to an optimal solution. This result also applies to set partitioning since, as shown by Lemke et al. (1971), by a linear transformation of the weight vector, a set-partitioning problem can be reformulated as either a set-packing or a set-covering problem.

Heuristics have been used to obtain good solutions to very large set-covering problems [see Baker (1981) and Baker and Fisher (1981)]. Worst-case analyses of the bounds between heuristic, optimal, and dual solutions have been given by Lovasz (1975), Chvatal (1979), Dobson (1982), Fisher and Wolsey (1982), Hochbaum (1982), and Wolsey (1982a). The analysis of the greedy heuristic given here comes from Fisher and Wolsey.

Crew-scheduling problems have been a fertile application area for set-covering and -partitioning models [see Arabeyre et al. (1969) and Marsten and Shepardson (1981)].

Theorem 2.7 is due to Nemhauser and Trotter (1975). They used this result to develop a branch-and-bound algorithm for the node-packing problem. This property has been studied further by Picard and Queyranne (1977). Grimmett and Pulleyblank (1985) showed that in large random graphs, LNP with a cardinality objective function is very unlikely to have an optimal solution with any of the variables equal to an integer.

Nemhauser and Sigismondi (1988) gave an FCP/branch-and-bound algorithm for node and set packing. The algorithm uses classes of facets for the convex hull of node packings [see Padberg (1973, 1975a, 1977), Nemhauser and Trotter (1974), and Trotter (1975)].

Facets of the convex hull of set covers have been studied by Sassano (1985), Balas and Ng (1985), and Cornuejols and Sassano (1986).

The literature on packing and covering problems for which the polyhedron of the linear programming relaxation has only integer extreme points will be presented in the notes for Chapter III.1.

Surveys on covering and partitioning problems have been given by Garfinkel and Nemhauser (1972b), Christofides and Korman (1975), Balas and Padberg (1976), and Padberg (1979). An annotated bibliography on combinatorial aspects of packing and covering was given by Trotter (1985).

Section II.6.3

All of the material presented in this section and much more can be found in the collection of survey articles on the traveling salesman problem, edited by Lawler, Lenstra, Rinnooy Kan, and Shmoys [LLRS (1985)].

Although the section treats the symmetric traveling salesman problem (STSP), many of the articles cited here deal with the slightly more general asymmetric problem (ATSP)—that is, the problem on a directed graph. We generally do not distinguish between these two versions in the citations and just use the acronym TSP.

As we have observed previously, the linear programming relaxation of (3.3)–(3.5) was introduced by Dantzig, Fulkerson, and Johnson (1954). The integer 2-matching or assignment problem relaxation was used by Eastman (1958) and Little et al. (1963). Although this is the weakest of our bounds, Balas and Toth (1985) reported a statistical experiment with 400 randomly generated problems in which the ratio of the cost of an optimal assignment solution to the cost of an optimal TSP solution is 99.2%. A modified assignment problem relaxation that tends to avoid the difficulty of creating numerous small subtours was given by Jonker, Deleve et al. (1980). Bellmore and Malone (1971) introduced the 2-matching relaxation. A tighter relaxation is the 2-matching problem where triangles are excluded. Cornuejols and Pulleyblank (1982, 1983) gave a polynomial-time algorithm for the integer 2-matching problem where triangles are excluded. They also showed that the problem of finding 2-matchings with no circuits of size 5 or smaller is $\mathcal{NP}$-hard [also see Cornuejols, Naddef, and Pulleyblank (1983)].

Held and Karp (1970, 1971) introduced the 1-tree relaxation and, by combining it with a Lagrangian relaxation with respect to the degree constraints, arrived at the relaxation (3.11). Related work on this approach was done by Christofides (1970) and Helbig-Hansen and Krarup (1974). Balas and Christofides (1981) used the 2-matching relaxation in conjunction with a Lagrangian relaxation with respect to the subtour elimination constraints to obtain the relaxation (3.13).

The tightest relaxations have been used by Padberg and Grötschel (1985) and Padberg and Rinaldi (1987a,b). Their cutting-plane algorithms use the degree constraints, all of the active subtour elimination constraints, and some 2-matching and comb inequalities.

The LLRS collection of articles contains three surveys on the analysis of heuristics for the TSP: empirical analysis by Golden and Stewart (1985), worst-case analysis by Johnson and Papadimitriou (1985b), and probabilistic analysis by Karp and Steele (1985) [also see Golden et al. (1980)].

Interchange heuristics for the TSP were developed by Croes (1958), Lin (1965), and Lin and Kernighan (1973). The k-interchange heuristic of Lin and Kernighan, where k varies by iteration, has proved to be very powerful. It is, however, much more complicated than using $k = 2$ or 3 and repeating the procedure from several initial tours. An alternative way of using an interchange heuristic is to combine it with simulated annealing [see Bonomi and Lutton (1984)].

Insertion procedures were introduced by Clarke and Wright (1964). Some rules for choosing the next node to insert and where to insert it are described by Rosenkrantz et al. (1977) and Norback and Love (1979).

Several composite heuristics that begin with a tour construction procedure followed by an interchange procedure were investigated by Golden and Stewart (1985). They also discussed the statistical comparison of heuristics.

Karp (1972) proved that determining whether an arbitrary graph contains a Hamiltonian circuit is $\mathcal{NP}$-complete. Subsequently, many special cases have been shown to be $\mathcal{NP}$-complete [see Johnson and Papadimitriou (1985a) for a survey of these results]. Sahni and Gonzales (1976) proved Proposition 3.2.

Papadimitriou and Steiglitz (1977, 1978) have analyzed the worst-case behavior of interchange algorithms. They have shown that, if $\mathcal{P} \neq \mathcal{NP}$, interchange algorithms whose neighborhood search time is polynomially bounded cannot be guaranteed to find an optimal solution, even with an exponential number of iterations.

Rosenkrantz et al. (1977) have analyzed the worst-case behavior of several tour construction heuristics for TSPs that satisfy the triangle inequality. The spanning-tree/perfect-matching heuristic and Theorem 3.6 are due to Christofides (1975b). Cornuejols and Nemhauser (1978) showed that this bound is tight.

Fisher, Nemhauser, and Wolsey (1979) gave worst-case bounds for several heuristics for the maximum-weight Hamiltonian circuit problem; and Jonker, Kaas, and Volgenant (1980) gave data-dependent bounds for the general TSP. Frieze, Galbiati, and Maffioli (1982) analyzed the worst-case performance of some algorithms for the ATSP. Much more information on the worst-case analysis of heuristics for the TSP is contained in Johnson and Papadimitriou (1985b).

The probabilistic analysis of TSP algorithms was surveyed by Karp and Steele. Karp gave two polynomial-time algorithms that asymptotically have a very high probability of finding an optimal solution. The first algorithm, by Karp (1977) [also see Halton and Terada (1982)], is for random euclidean problems on a d-dimensional cube. (Originally, Karp considered random points on a unit square.) The idea of the algorithm is to divide the square into a large number of very small subsquares. On each subsquare the problem can solved for an optimal solution in polynomial time. Finally, the small cycles are assembled into a tour. The second algorithm, by Karp (1979) and Karp and Steele (1985), deals with the ATSP with costs taken from the uniform distribution. Here the idea is to solve the assignment problem relaxation and then to patch the subtours together.

Surveys of branch-and-bound algorithms for the traveling salesman problem have been presented by Carpento and Toth (1980) and Balas and Toth (1985). The branching rule shown in Figure 3.16 is due to Garfinkel (1973). Cutting-plane/branch-and-bound algorithms for the TSP were initiated by Dantzig, Fulkerson, and Johnson (1954, 1959). Systematic algorithms of this type were developed by Miliotis (1976, 1978), Padberg and Hong (1980), Crowder and Padberg (1980), Grötschel (1980a), Padberg and Grötschel (1985), and Padberg and Rinaldi (1987a,b). The Padberg–Grötschel article surveyed these results and reported computational experience. The Padberg–Hong algorithm uses primal cutting planes; the other algorithms are FCPAs of the type described in this section. The Padberg–Rinaldi FCPA has solved a 2,392-city problem to optimality.

The separation algorithm for subtour elimination constraints is due to Gomory and Hu (1961). The shrinking procedure illustrated in Figure 3.23 is taken from Padberg and Grötschel (1985). Padberg and Rao (1982) have given a polynomial-time separation algorithm for 2-matching inequalities. In particular, they have shown that the separation problem is a minimum odd-cut problem (see Section III.3.7).

An interactive computer package with various TSP heuristics and exact algorithms has been developed by Boyd et al. (1987).

Some very restricted families of TSPs can be solved in polynomial time. An application of this type of result appeared in Ratliff and Rosenthal (1983), and a survey of these results was given by Gilmore et al. (1985).

A generalization of the traveling salesman problem is the vehicle routing problem in which there are k salesmen located at a given city, and each must choose a subtour so that all cities are covered. Bodin et al. (1983), Christofides (1985a,b), and Golden and Assad (1986) surveyed results on this problem. Cullen et al. (1981) presented an approach that formulates routing problems as set-partitioning problems. Also see Laporte et al. (1985), Fisher, Greenfield et al. (1982), and Kolen et al. (1987).

Another generalization is the quadratic assignment problem [see Burkhard (1984) for a survey of this topic].

Section II.6.4

The fixed-charge network flow problem belongs to a family of problems known as network design problems. Magnanti and Wong (1984) gave a survey of models and algorithms in this area, and Wong (1985) gave an annotated bibliography. A Benders' decomposition approach to network design has been given by Magnanti et al. (1986), and a heuristic approach has been given by Lin (1975).

A branch-and-bound algorithm of the type described can be found in Barr et al. (1981) [also see Cabot and Erenguc (1984), Guignard (1982), MacKeown (1981), Neebe and Rao (1983), and Suhl (1985)].

The generalized flow cover inequalities (see the notes for Section II.2.4) and their separation heuristics come from Padberg, Van Roy, and Wolsey (1985), Van Roy and Wolsey (1986, 1987), and Wolsey (1987).

Example 4.2 is problem 2 from Gray (1971). The results were obtained using MPSARX.

The multicommodity reformulation is part of the folklore. It can be found explicitly in Rardin and Choe (1979).

Algorithms for finding optimal Steiner trees and branchings appeared in Shore et al. (1982), Beasley (1984), Wong (1984), and Prodon et al. (1985).

Reformulations of the lot-size problem were discussed in Sections I.1.5 and II.5.5. The shortest-path reformulation given here is due to Eppen and Martin (1988), and our development is based on Pochet and Wolsey (1988) [also see Martin (1987)]. Versions of Theorem 4.8 appeared in Rosling (1983) and Barany et al. (1984).

The idea of introducing auxiliary variables to tighten a formulation has recently attracted considerable attention. Balas and Pulleyblank (1983) gave an example, like the lot-size problem, where the convex hull of solutions has an exponential number of facets, but the enlarged system contains a polynomial number of constraints and variables. Martin (1984) discussed how a dynamic programming algorithm can be used to derive a tight formulation with auxiliary variables, and in Martin (1987) it is observed how a linear programming separation algorithm also leads to a reformulation with auxiliary variables.

Section II.6.5

The feasibility algorithm for the subset sum problem is due to Lagarios and Odlyzko (1985). Frieze (1986) gave simpler proofs of these results. Related results have been obtained by Furst and Kannan (1987).

See Lenstra (1984) for a general discussion of integer programming and cryptography.

The polynomial-time algorithm for the integer feasibility problem for fixed n is due to Lenstra (1983) [also see Kannan (1983)]. Earlier results for $n = 2$ were obtained by Kannan (1980) and Scarf (1981a,b). Rubin (1985) gave a polynomial-time algorithm for $m \times (m+1)$ integer programs.

7. EXERCISES

1. Solve the knapsack problem

$$\begin{aligned} \max\ & 18x_1 + 7x_2 + 5x_3 + x_4 \\ & 9x_1 + 4x_2 + 3x_3 + 2x_4 \leq b \\ & x \in Z_+^4 \end{aligned}$$

by dynamic programming for all values of b from 1 to 100.

2. Prove Proposition 1.2.

3. Apply the superadditive dual algorithm to the instance in exercise 1 with $b = 16$.

4. i) Suggest other superadditive functions to be used in the superadditive dual algorithm.

ii) Interpret the dynamic programming algorithm as a superadditive dual algorithm.

5. Use the SR heuristic to find a solution to the knapsack problem:

$$\begin{aligned} \max\ & 537x_1 + 636x_2 + 849x_3 + 712x_4 + 834x_5 + 219x_6 + 832x_7 \\ & 924x_1 + 1123x_2 + 1501x_3 + 1402x_4 + 1579x_5 + 498x_6 + 1649x_7 \leq 23{,}762 \\ & x \in Z_+^7, \end{aligned}$$

which is within 1% of optimal.

6. Solve the integer program (see exercise 13 of section II.1.9)

$$\begin{aligned} \max\ & 2x_1 + 5x_2 \\ & 4x_1 + x_2 \le 28 \\ & x_1 + 4x_2 \le 27 \\ & x_1 - x_2 \le 1 \\ & x \in Z_+^2 \end{aligned}$$

i) by the shortest-path enumeration algorithm and

ii) by the increasing group algorithm.

7. Describe a fully polynomial approximation scheme based on a scaling/rounding heuristic for the 0-1 knapsack problem.

8. Solve the 0-1 knapsack problem

$$\begin{aligned} \max\ & 43x_1 + 41x_2 + 27x_3 + 32x_4 + 15x_5 + 50x_6 + 19x_7 + 21x_8 \\ & 20x_1 + 19x_2 + 14x_3 + 16x_4 + 7x_5 + 28x_6 + 12x_7 + 14x_8 \le 61 \\ & x \in B^8 \end{aligned}$$

by the branch-and-bound algorithm of Section 1.

9. For the 0-1 knapsack problem,

i) propose a neighborhood search algorithm,

ii) propose a simulated annealing algorithm, and

iii) suggest alternative neighborhoods for use in i and ii.

10. Consider the 0-1 knapsack problems with c_j/a_j = constant for all $j \in N$. Why might these be difficult? Suggest a way to solve such problems.

11. Propose heuristic algorithms for the 0-1 multidimensional knapsack problem

$$\max\left\{\sum_{j\in N} c_j x_j : \sum_{j\in N} a_{ij} x_j \le b_i \text{ for } i \in M, x \in B^n\right\},$$

where $a_{ij} \in Z_+^1$ for all $i \in M, j \in N$.

12. Describe an efficient algorithm for the linear programming relaxation of the multiple-choice knapsack problem

$$\begin{aligned} \max\ & \sum_{j\in N} c_j x_j \\ & \sum_{i\in I^+}\sum_{j\in Q_i} a_j x_j - \sum_{i\in I^-}\sum_{j\in Q_i} a_j x_j \le b \\ & \sum_{j\in Q_i} x_j \le 1 \quad \text{for } i \in I^+ \cup I^- \\ & x \in R_+^n, \end{aligned}$$

where $I^+ \cap I^- = \emptyset$, $Q_i \cap Q_k = \emptyset$ for $i \neq k$, and $N = \cup_{i\in I^+\cup I^-} Q_i$.

13. Apply the simplex-based heuristic BIP to the following problems.

i)

$$\max 9x_1 + 2x_2 - 3x_3$$
$$4x_1 + x_2 - 5x_3 \leq 1$$
$$4x_1 - 2x_2 + 6x_3 \leq 7$$
$$x \in B^3;$$

ii) the covering problem $\min\{1x: Ax \geq 1, x \in B^9\}$, where

$$A = \begin{pmatrix} Z & I & 0 \\ 0 & Z & I \\ I & 0 & Z \\ I & I & I \end{pmatrix} \quad \text{and} \quad Z = \begin{pmatrix} 0 & 1 & 1 \\ 1 & 0 & 1 \\ 1 & 1 & 0 \end{pmatrix}.$$

14. Let $S = \{x \in B^6: 40x_1 + 40x_2 + 35x_3 + 35x_4 + 15x_5 + 15x_6 \leq 100\}$. Find violated inequalities for S that cut off

i) $x^a = (1 \;\; \tfrac{1}{2} \;\; 0 \;\; \tfrac{1}{2} \;\; \tfrac{1}{2} \;\; 1)$,

ii) $x^b = (\tfrac{1}{3} \;\; 1 \;\; \tfrac{1}{3} \;\; \tfrac{1}{3} \;\; \tfrac{1}{3} \;\; 1)$.

15. Apply the FCP/branch-and-bound algorithm to the problem

$$\max \; 43x_1 + 10x_2 + 18x_3 + 12x_4 + 36x_5 + 22x_6$$
$$12x_1 + 2x_2 + 3x_3 + 2x_4 + 4x_5 + 3x_6 \leq 20$$
$$3x_1 + 8x_2 + 12x_3 + 13x_4 + 20x_5 + 14x_6 \leq 36$$
$$x \in B^6.$$

16. In Section II.2.2 the extended cover inequalities

$$\sum_{j \in E(C)} x_j \leq |C| - 1$$

were defined for 0-1 knapsack problems. Formulate the separation problem for extended cover inequalities and propose a heuristic algorithm to solve it.

17. Let

$$S = \{x \in B^6: 7x_1 + 3x_2 + 4x_3 + 6x_4 + 7x_5 + 7x_6 \leq 10$$
$$x_1 + x_2 \leq 1, x_3 + x_4 \leq 1, x_5 + x_6 \leq 1\}.$$

i) Show that $x_1 + x_2 + x_4 + x_5 + x_6 \leq 2$ is a valid inequality.

ii) Find a valid inequality that cuts off $x^* = (0 \;\; 1 \;\; 1 \;\; 0 \;\; \tfrac{3}{7} \;\; 0)$.

iii) Formulate the separation problem for the families of valid inequalities in exercise 9 of Section II.2.6.

18. Consider the set $\bar{S} = S \cap \{x \in B^n\colon cx \leq c_0\}$ with $S = \{x \in B^n\colon Ax \geq b\}$. Given t inequalities $\Sigma_{j \in Q_k} x_j \geq 1$ for $k = 1, \ldots, t$, suppose there exists $v \in R^t_+$ such that

$$\sum_{\{k: Q_k \ni j\}} v_k \leq c_j \quad \text{for } j \in N \quad \text{and} \quad \sum_{k=1}^{t} v_k > c_0.$$

Show the following:

i) If $x \in \bar{S}$, then for some $k \in \{1, \ldots, t\}$, we have $x_j = 0$ for all $j \in Q_k$ (i.e., at least one of the t inequalities is violated by every point $x \in \bar{S}$).

ii) If $S = \{x \in B^n\colon \Sigma_{j \in N_i} x_j \geq 1 \text{ for } i \in M\}$, then for any subset $\{i(1), \ldots, i(t)\}$ of M,

$$\sum_{j \in \bigcup_{k=1}^{t} (N_{i(k)} \setminus Q_k)} x_j \geq 1$$

is a valid inequality for $\bar{S}$.

iii) Apply these observations to the covering problem with an additional constraint

$$\begin{aligned} 3x_1 + x_2 + x_3 + x_4 &\leq 1 \\ x_1 \qquad\quad x_3 + x_4 &\geq 1 \\ x_1 + x_2 \qquad\quad + x_4 &\geq 1 \\ x \in B^4. \end{aligned}$$

Derive the valid inequalities $x_1 \leq 0$ and $x_4 \geq 1$.

19. Apply the greedy heuristic to the set-covering problem

$$\begin{aligned} \min\ 9x_1 + 4x_2 + 7x_3 + 2x_4 + 8x_5 & \\ x_1 + \qquad + x_3 \qquad\quad + x_5 &\geq 1 \\ x_1 \qquad\qquad\qquad\qquad + x_5 &\geq 1 \\ x_2 + x_3 \qquad\qquad\quad &\geq 1 \\ x_1 \qquad\qquad\quad x_4 \qquad\quad &\geq 1 \\ x \in B^5. & \end{aligned}$$

What lower bounds on the optimal value are given by the heuristic? Can you derive stronger lower bounds?

20. Prove Proposition 2.4.

21. Show that the bound of Theorem 2.5 can be achieved asymptotically.

22. Solve the weighted node-packing problem on the graph shown in Figure 7.1

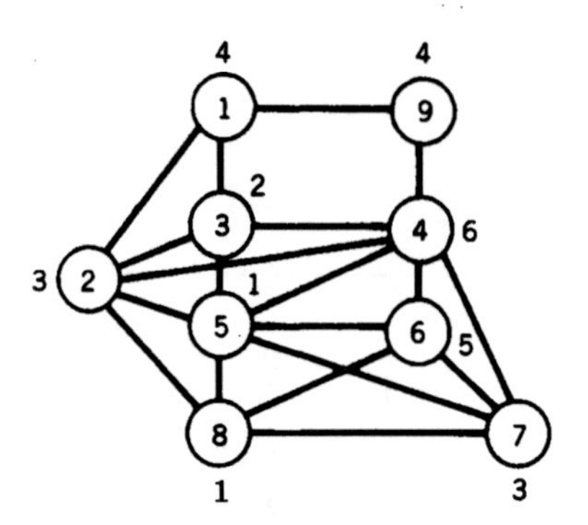

Figure 7.1

i) by solving LNP and fixing variables,

ii) by adding cuts of the type discussed Section II.2.1.

23. Show that LNP can be solved as an assignment problem.

24. Find a minimum road distance tour of the midwest visiting each city exactly once and returning to the city from which you started. Distances are in tens of miles (revised since Chapter I.3):

		2	3	4	5	6	7	8	9	10
1.	Chicago	92	99	50	41	79	46	29	50	70
2.	Dallas		78	49	94	21	64	63	42	37
3.	Denver			60	84	61	54	86	76	51
4.	Kansas City				45	35	20	26	17	20
5.	Minneapolis					80	36	55	59	64
6.	Oklahoma City						46	50	29	16
7.	Omaha							45	37	30
8.	St. Louis								21	45
9.	Springfield (Mo.)									25
10.	Wichita									

i) Use any method that you like, but you must prove optimality of your solution. Your grade will be decreasing function of the length of your proof.

ii) Calculate as many of the bounds given in Figure 3.7 as possible.

iii) Test the primal heuristics given in Section 3.

iv) Solve by an FCP algorithm.

25. Some other heuristic algorithms for the symmetric traveling salesman problem include:

i) Furthest insertion.

ii) Sweep. Locate an "origin" in the center of the map, and then denote each city by its rectangular coordinates (r, θ). Order the cities by increasing θ.

Apply them to the examples in Section 3.

26. Prove Proposition 3.2.

27. Devise a simulated annealing algorithm for the asymmetric traveling salesman problem.

28. Find a family of graphs for which the worst-case bound of the spanning-tree/perfect-matching heuristic is asymptotically achieved.

29. Find one or more violated inequalities for the fractional solution shown in Figure 3.28.

30. i) Solve the uncapacitated fixed-charge network problem exhibited in Figure 7.2 by branch-and-bound.

ii) Find generalized flow cover inequalities that cut off the initial linear programming solution.

iii) Solve the multicommodity reformulation of the problem.

31. Consider the following fixed-charge transportation problem with 3 suppliers and 7 customers. The supplies are 15, 25, and 33; the demands are 5, 7, 8, 10, 12, 15, and 16; and the variable and fixed costs are

$$h_{ij} = \begin{pmatrix} 1 & 1 & 2 & 2 & 1 & 1 & 1 \\ 2 & 1 & 3 & 2 & 1 & 3 & 1 \\ 4 & 2 & 3 & 3 & 2 & 2 & 1 \end{pmatrix}, \qquad c_{ij} = \begin{pmatrix} 51 & 31 & 10 & 6 & 19 & 12 & 14 \\ 10 & 12 & 32 & 46 & 29 & 11 & 14 \\ 9 & 32 & 17 & 16 & 15 & 24 & 12 \end{pmatrix}.$$

Solve this problem by an FCP/branch-and-bound algorithm.

32. Show that the inequalities (4.5) are valid.

33. Another way to reformulate a fixed-charge network flow problem is to define variables y_{ij}^p, where y_{ij}^p is the flow along the path p passing through arc (i, j). Write out and solve the arc-path reformulation for Example 4.3.

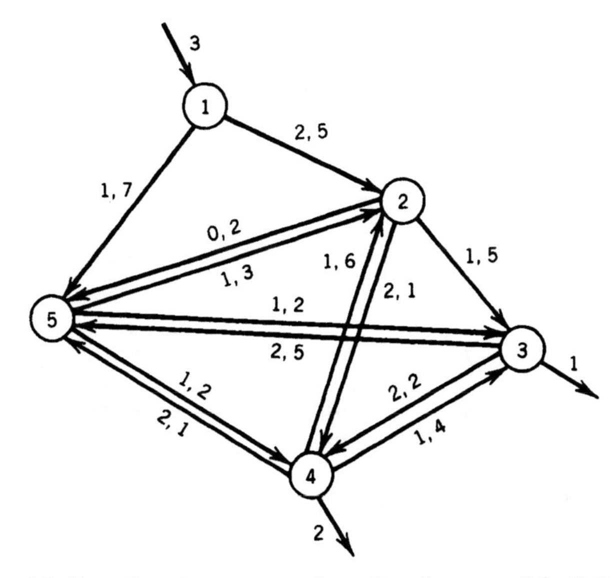

Figure 7.2. Costs (h_{ij}, c_{ij}) appear on each arc. Bounds are $u_{ij} = 3$ for $(i, j) \in \mathcal{A}$.

34. Show that it is always possible to convert a capacitated fixed-charge network problem into an uncapacitated problem by increasing the number of nodes and arcs.

35. Consider the problem of finding a minimum-weight 1-branching. It can be shown that the linear program

$$\min \sum_{(i,j)\in\mathcal{A}} w_{ij}x_{ij}$$

$$\sum_{(i,j)\in\delta^+(S)} x_{ij} \geq 1 \quad \text{for } S \subset V \text{ with } 1 \in S$$

$$x \in R_+^{|\mathcal{A}|}$$

always has an optimal solution with $x \in B^n$ when $w \in R_+^{|\mathcal{A}|}$, and it is unbounded otherwise.

i) Show that this linear program solves the minimum-weight branching problem.

ii) Give a linear program having a polynomial in $|V|$ number of constraints and variables that solves the minimum-weight branching problem.

iii) Give a linear program with similar characteristics that solves the minimum-weight spanning-tree problem.

36. Consider the problem of finding a minimum-weight Steiner 1-branching when there are two demand nodes ($|D| = 2$):

i) What structure do the branchings have?

ii) Give a polynomial algorithm to solve this problem.

37. Prove that the linear programming relaxation of (4.18) always has an optimal solution with $x \in B^T$.

Part III
COMBINATORIAL OPTIMIZATION

III.1
Integral Polyhedra

1. INTRODUCTION

In Part III we will continue to study feasible regions of the form $S = \{x \in Z_+^n: Ax \leq b\}$, where (A, b) is an $m \times (n + 1)$ integral matrix, and integer programs of the form $\max\{cx: x \in S\}$. However, for most of the problems considered here a nice description of $\text{conv}(S)$ is known. This is the essential distinction between the Part II and Part III problems.

We will encounter some problems with the property that $\text{conv}(S) = \{x \in R_+^n: Ax \leq b\}$, and we will encounter others for which we can specify an explicit set of constraints $A'x \leq b'$ such that

$$\text{conv}(S) = \{x \in R_+^n: Ax \leq b, A'x \leq b'\}$$

Frequently, in these cases, we also obtain an efficient combinatorial algorithm for solving the linear optimization problem. Conversely, such an algorithm for solving the optimization problem may provide a proof that the inequalities define the convex hull.

The minimum-weight $s-t$ path problem on a digraph $\mathcal{D} = (V, \mathcal{A})$ (see Section I.3.2) is an example of a Part III problem. It is a network flow problem where we require one unit of flow out of node s, one unit of flow into node t, and conservation of flow at all other nodes. For this formulation, we gave an algorithm in Section I.3.6 which, for an arbitrary weight function, either (a) yields an integral optimal solution and thus provides a minimum-weight $s-t$ path or (b) shows that the objective value is unbounded. Thus, the algorithm provides a proof that the polyhedron of feasible solutions only has integral extreme points. In Section 2, we will establish an important property of node-arc incidence matrices, which gives a different proof of this result.

A second formulation arises from considering the relationship between $s-t$ dipaths and $s-t$ dicuts. Let $(U, \overline{U})$ be any partition of V with $s \in U$ and $t \in \overline{U}$. Then the set of arcs whose tail is in U and whose head is in $\overline{U}$ is an $s-t$ dicut. Let $|\mathcal{A}| = n, M = \{1, \ldots, m\}$ be the index set of all $s-t$ dicuts, and let $a^i \in B^n$ for $i \in M$ be the incidence vectors of the $s-t$ dicuts. Now a dipath must intersect each dicut. Thus, if $x \in B^n$ is the incidence vector of an $s-t$ dipath, then $a^i x \geq 1$ for all $i \in M$. It is not difficult to show that all of the incidence vectors of $s-t$ dipaths are extreme points of the polyhedron

$$\{x \in R_+^n: a^i x \geq 1 \text{ for all } i \in M\}.$$

Much more significantly, they are the only extreme points.

Now let K be the index set of all $s-t$ dipaths, and let $x^i \in B^n$ for $i \in K$ be the corresponding incidence vectors of the $s-t$ dipaths. Then, by polarity, we obtain another integral polyhedron

$$\{a \in R^n_+: x^i a \geqslant 1 \text{ for all } i \in K\}$$

whose extreme points are the $s-t$ dicuts. This approach will be pursued further in Section 6.

In both formulations of the minimum-weight path problem, we obtain a polyhedron having only integral extreme points. We now state this property precisely.

Definition 1.1. A nonempty polyhedron $P \subseteq R^n$ is said to be *integral* if each of its nonempty faces contains an integral point.

It is sufficient to consider the minimal faces. Now, since each minimal nonempty face is an extreme point if and only if $\text{rank}(A) = n$ (see Proposition 4.2 of Section I.4.4), we have

Proposition 1.1. *A nonempty polyhedron* $P = \{x \in R^n: Ax \leqslant b\}$ *with* $\text{rank}(A) = n$ *is integral if and only if all of its extreme points are integral.*

Also, if $P = \{x \in R^n: Ax \leqslant b\} \subseteq R^n_+$ and is not empty, then $\text{rank}(A) = n$. Thus, we have the following corollary:

Corollary 1.2. *A nonempty polyhedron* $P \subseteq R^n_+$ *is integral if and only if all of its extreme points are integral.*

We assume hereafter, unless otherwise stated, that nonempty polyhedra have extreme points.

Consider the linear programming problem over the polyhedron P given by

$$\text{(LP)} \qquad z_{\text{LP}} = \max\{cx: x \in P\}.$$

Integral polyhedra can be characterized by optimal solutions to LP.

Proposition 1.3. *The following statements are equivalent.*

1. *P is integral.*
2. LP *has an integral optimal solution for all* $c \in R^n$ *for which it has an optimal solution.*
3. LP *has an integral optimal solution for all* $c \in Z^n$ *for which it has an optimal solution.*
4. z_{LP} *is integral for all* $c \in Z^n$ *for which* LP *has an optimal solution.*

Proof. $1 \to 2$. If LP has an optimal solution, it has an optimal solution at an extreme point of P (see Theorem 4.5 of Section I.4.4).

$2 \to 3$ and $3 \to 4$ are obvious.

$4 \to 1$. We prove the contrapositive using the fact that if $x \in P$ is an extreme point, there is a $c \in Z^n$ such that x is the unique optimal solution to LP (see Theorem 4.6 of Section I.4.4).

Thus if statement 1 is false, there exists $\tilde{c} \in Z^n$ such that x' is the unique optimal solution to LP, and some component of x', say x'_j, is fractional. Now it follows that there exists a (suitably large) integer q such that x' is also optimal for the objective vector $c' = \tilde{c} + (1/q)e_j$ and for the objective vector $qc' = q\tilde{c} + e_j$. But $qc'x' - q\tilde{c}x' = x'_j$, which means that z_{LP} is fractional for at least one of the objectives qc' or $q\tilde{c}$. Hence statement 4 is false. ■

Statement 4 of Proposition 1.3 provides a technique for establishing the integrality of P by studying the dual polyhedron.

Definition 1.2. A system of linear inequalities $Ax \leq b$ is called *totally dual integral* (TDI) if, for all integral c such that $z_{\text{LP}} = \max\{cx: Ax \leq b\}$ is finite, the dual $\min\{yb: yA = c, y \in R^m_+\}$ has an integral optimal solution.

Note that the definition is not given in terms of a polyhedron P but, instead, is given more specifically in terms of a linear inequality description of it.

Corollary 1.4. *If $Ax \leq b$ is* TDI *and b is integral, then $P = \{x \in R^n: Ax \leq b\}$ is integral.*

Proof. Since the dual has an integral optimal solution and b is integral, the optimal objective value of the dual is integral. Hence for all c for which z_{LP} is finite, z_{LP} is integral. Now the result follows from Proposition 1.3. ■

Example 1.1. We are given a complete bipartite graph with node partition $V_1 = \{1, \ldots, m\}$, $V_2 = \{m+1, \ldots n\}$, node weights c_j for $j \in V_1 \cup V_2 = V$, and edge weights b_{ij} for $i \in V_1$ and $j \in V_2$. The problem is to assign node numbers x_j for $j \in V$ to solve the linear program

$$\max \sum_{j \in V} c_j x_j$$

$$x_i + x_j \leq b_{ij} \quad \text{for } i \in V_1 \text{ and } j \in V_2.$$

Its dual is

$$\min \sum_{i \in V_1} \sum_{j \in V_2} b_{ij} y_{ij}$$

$$\sum_{j \in V_2} y_{ij} = c_i \quad \text{for } i \in V_1$$

$$\sum_{i \in V_1} y_{ij} = c_j \quad \text{for } j \in V_2$$

$$y_{ij} \geq 0 \quad \text{for } i \in V_1 \text{ and } j \in V_2.$$

The dual problem is the transportation problem (see Section I.3.5), which has an integral optimal solution if it is feasible and if c_j is integral for all $j \in V$. Hence, the linear system $x_i + x_j \leq b_{ij}$ for $i \in V_1, j \in V_2$ is TDI. Hence, if the b_{ij}'s are integers, the polyhedron $\{x \in R^n: x_i + x_j \leq b_{ij} \text{ for } i \in V_1, j \in V_2\}$ is integral.

The fact that $Ax \leq b$ with $b \in R^m$ is a TDI system says nothing about integrality unless $b \in Z^m$. In fact, the TDI property is sensitive to scaling the rows of A.

Proposition 1.5. *If $Ax \leqslant b$ is any linear system with rational coefficients, there exists a positive integer q such that $(1/q)Ax \leqslant (1/q)b$ is* TDI.

Proof. Consider the dual constraints $yA = c$, $y \in R^m_+$ with $c \in Z^n$. By Proposition 3.1 of Section I.5.3, there is a positive integer q such that every extreme point can be written as $y = (1/q)(p_1, \ldots, p_m)$, where p_i is an integer for $i = 1, \ldots, m$. Now let $y' = qy$. Hence every extreme point of $(1/q)y'A = c$, $y' \in R^m_+$ is integral, and the dual system $(1/q)Ax \leqslant (1/q)b$ is TDI. ■

Corollary 1.6. *Any polyhedron $P = \{x \in R^n: Ax \leqslant b\}$ can be represented by a* TDI *linear inequality system.*

Integral polyhedra are distinguished by the existence of a TDI representation with an integral right-hand side.

Proposition 1.7. *If $P = \{x \in R^n: Ax \leqslant b\}$ is an integral polyhedron, then P can be represented as $P = \{x \in R^n: A'x \leqslant b'\}$, where $A'x \leqslant b'$ is* TDI *and b' is integral.*

Proof. Consider a $c \in Z^n$ for which $\max\{cx: x \in P\}$ is bounded. Let F be the face of P of optimal solutions with equality set $M^=_F$. Now consider the polyhedral cone

$$C(F) = \{d \in R^n: d = \sum_{i \in M^=_F} u_i a^i,\ u \in R^{|M^=_F|}_+\}.$$

By Theorem 6.1(ii) of Section I.4.6, $C(F) \cap Z^n$ is finitely generated with generators $\pi^k \in Z^n$ for $k \in K(F)$; that is,

$$C(F) \cap Z^n = \{d \in R^n: d = \sum_{k \in K(F)} y_k \pi^k,\ y \in Z^{|K(F)|}_+\}.$$

Also, since P is integral and $\pi^k \in Z^n$, we obtain $\max\{\pi^k x: x \in P\} = \pi^k_0 \in Z^1$. In addition since $\pi^k \in C(F)$, we have $\pi^k x = \pi^k_0$ for all $x \in F$.

We now add the finite set of inequalities $\pi^k x \leqslant \pi^k_0$ for $k \in K(F)$ to the description of P for each of the finite number of faces F of P. This gives the dual problem

$$\min \sum_i u_i b_i + \sum_F \sum_{k \in K(F)} y_k \pi^k_0$$

$$\sum_i u_i a^i + \sum_F \sum_{k \in K(F)} y_k \pi^k = c$$

$$u \in R^m_+, \qquad y \in R_+^{\sum_F |K(F)|}$$

Now since $c \in C(F) \cap Z^n$, there exists $y^*_k \in Z^1_+$, $k \in K(F)$ such that $c = \Sigma_{k \in K(F)} y^*_k \pi^k$. Finally, y^* is an optimal dual solution because for any optimal primal solution $x^* \in F$, we have $cx^* = \Sigma_{k \in K(F)} y^*_k \pi^k x^* = \Sigma_{k \in K(F)} y^*_k \pi^k_0$ since $\pi^k x^* = \pi^k_0$. ■

A linear inequality description of an integral polyhedron may not be TDI. This is illustrated in the following example.

Example 1.2. The problem is to find a minimum cardinality covering of the nodes of a graph by its edges. In particular, given a complete graph on four nodes we consider the linear program:

$$
\begin{array}{lllllll}
\min x_{12} & + x_{13} & + x_{14} & + x_{23} & + x_{24} & + x_{34} & \\
x_{12} & + x_{13} & + x_{14} & & & & \geq 1 \\
x_{12} & & & + x_{23} & + x_{24} & & \geq 1 \\
& x_{13} & & + x_{23} & & + x_{34} & \geq 1 \\
& & x_{14} & & + x_{24} & + x_{34} & \geq 1 \\
& x \in R^6_+. & & & & &
\end{array} \tag{1.1}
$$

There are three optimal solutions: $x_{12} = x_{34} = 1$, $x_{ij} = 0$ otherwise; $x_{13} = x_{24} = 1$, $x_{ij} = 0$ otherwise; $x_{14} = x_{23} = 1$, $x_{ij} = 0$ otherwise.

It can be shown that these solutions, together with the four solutions obtained by setting the edge variables for the edges incident to node i equal to 1 and the others equal to zero, are the only extreme points of the linear system (1.1). We leave the details to the reader.

Now consider the dual, which is the fractional node-packing problem

$$\max\left\{\sum_{i=1}^{4} y_i \colon y_i + y_j \leq 1 \text{ for all } i \text{ and } j \text{ with } j > i,\ y \in R^4_+\right\}.$$

Its unique optimal solution is $y_i = \frac{1}{2}$ for $i = 1, \ldots, 4$. Hence the linear system $Ax \geq 1$, $x \geq 0$ is not TDI, but the polyhedron $P = \{x \in R^n \colon Ax \geq 1, x \geq 0\}$ is integral.

All of the results of this section hold regardless of whether P has extreme points or not. However, for full-dimensional integral polyhedra, Proposition 1.7 can be strengthened.

Proposition 1.8. *For a full-dimensional integral polyhedron, there exists a unique minimal (with respect to removing constraints)* TDI *representation with an integral right-hand side.*

We now outline the rest of this chapter and briefly describe the topics of the following two chapters. In Section 2, we describe a class of matrices for which the integrality of $P(b) = \{x \in R^n_+ \colon Ax \leq b\}$ holds for all integral b. A subset of these matrices, including node-arc incidence matrices of digraphs, are studied in Section 3. We provide a recognition algorithm for these matrices and observe that the associated linear programming problem can be solved by network flow algorithms.

Thereafter, we consider packing polytopes of the form $P = \{x \in R^n_+ \colon Ax \leq 1\}$ and covering polyhedra of the form $Q = \{x \in R^n_+ \colon Ax \geq 1\}$, where A is a (0, 1) matrix. In Section 4, we describe matrices, in terms of forbidden submatrices, such that both the packing and covering polyhedra are integral. Here the linear systems are TDI, and for a subclass of the matrices we give a recognition algorithm and an efficient combinatorial algorithm for solving the associated linear programming problems.

In Section 5 we give a complete description of integral packing polytopes $P = \{x \in R_+^n : Ax \leqslant 1\}$. Here the matrices A are defined by incidence vectors of cliques of a class of graphs, and the extreme points of the polytopes are incidence vectors of node packings. Then by invoking antiblocking polarity we obtain a proof of the famous perfect graph theorem. In Section 6, we study blocking polarity and obtain results of the type exemplified by the polarity between incidence vectors of paths and cuts in a graph.

Chapters III.2 and III.3 deal with combinatorial objects known as matchings and matroids, respectively. Matchings generalize network flows and matroids generalize forests of a graph. Both of these combinatorial settings yield interesting polyhedral results and efficient optimization algorithms.

2. TOTALLY UNIMODULAR MATRICES

Definition 2.1. An $m \times n$ integral matrix A is *totally unimodular* (TU) if the determinant of each square submatrix of A is equal to 0, 1, or −1.

It is evident that $a_{ij} = 0, 1$, or -1 if A is TU; that is, A is a $(0, 1, -1)$ matrix.

Example 2.1. The matrix

$$A = \begin{pmatrix} 1 & 1 & 0 & 0 \\ 1 & 0 & 1 & 1 \\ 0 & 1 & 1 & 0 \\ 1 & 1 & 0 & 1 \end{pmatrix}$$

is not TU since $|\det A'| = 2$, where A' is the submatrix of A consisting of the first three rows and columns.

Note that the example illustrates that recognizing TU matrices is in $\mathscr{C}o\mathscr{N}\mathscr{P}$. That is, to give a short proof that a matrix is not TU, we only need to give an appropriate submatrix because determinants can be calculated in polynomial time (see Section I.5.3). On the other hand, the definition does not give a clue about how to give a short proof that a matrix is TU, since the number of square submatrices is exponential in the description of the matrix. We will discuss the recognition question in the next section.

The following proposition, which follows directly from the definition of total unimodularity, provides ways of constructing other TU matrices from a given TU matrix.

Proposition 2.1. *The following statements are equivalent.*

1. *A is* TU.
2. *The transpose of A is* TU.
3. *(A, I) is* TU.
4. *A matrix obtained by deleting a unit row (column) of A is* TU.
5. *A matrix obtained by multiplying a row (column) of A by −1 is* TU.
6. *A matrix obtained by interchanging two rows (columns) of A is* TU.
7. *A matrix obtained by duplicating columns (rows) of A is* TU.
8. *A matrix obtained by a pivot operation on A is TU.*

Proof. We will only prove statement 8. Suppose $|a_{ij}| = 1$. Recall from Section I.2.3 that a simplex pivot on a $(0, \pm 1)$ matrix A with pivot element a_{ij} involves the following steps.

1. If $a_{ij} = -1$, multiply the ith row of A by -1. Call the new row $\overline{a}^i$.
2. For $k \neq i$, we obtain

$$\overline{a}^k = \begin{cases} a^k & \text{if } a_{kj} = 0 \\ a^k - \overline{a}^i & \text{if } a_{kj} = 1 \\ a^k + \overline{a}^i & \text{if } a_{kj} = -1. \end{cases}$$

Now consider a square submatrix B of $\underline{A}$. Let $\overline{B}$ be the matrix obtained after the pivot has been executed. We will prove that $\det \overline{B} \in \{-1, 0, 1\}$.

Case 1. The ith row of A appears in B. Then $|\det \overline{B}| = |\det B|$.

Case 2. The jth column, but not the ith row, appears in B. Then $|\det \overline{B}| = 0$.

Case 3. Neither the ith row nor the jth column appear in B.

Let

$$C = \begin{pmatrix} a_{ij} & \ldots & a_{ip} \\ \vdots & B & \\ a_{lj} & & \end{pmatrix}.$$

Then after pivoting we have

$$\overline{C} = \begin{pmatrix} 1 & \ldots & \overline{a}_{ip} \\ 0 & & \\ \vdots & \overline{B} & \\ 0 & & \end{pmatrix}.$$

Hence $|\det \overline{B}| = |\det \overline{C}| = |\det C|$. ∎

Proposition 2.2. *If A is TU, then $P(b) = \{x \in R^n_+ : Ax \leqslant b\}$ is integral for all $b \in Z^m$ for which it is not empty.*

Proof. Consider the linear program with constraint set $Ax + Iy = b, x \in R^n_+, y \in R^m_+$, where A is TU and b is integral. Let $(A, I) = (A_B, A_N)$, where A_B is a basis matrix for the linear program. By statement 8 of Proposition 2.1, it follows that A_B^{-1} is an integral matrix. Thus $A_B^{-1}b$ is integral, so the correspondence between basic feasible solutions and extreme points yields the result. ∎

A similar argument yields the following generalization of Proposition 2.2.

Proposition 2.3. *If A is TU, if b, b', d, and d' are integral, and if $P(b, b', d, d') = \{x \in R^n : b' \leqslant Ax \leqslant b, d' \leqslant x \leqslant d\}$ is not empty, then $P(b, b', d, d')$ is an integral polyhedron.*

Because the transpose of a TU matrix is TU, the dual polyhedron is also integral.

Corollary 2.4. *If A is TU, c is integral, and $Q(c) = \{u \in R^m_+: uA \geqslant c\}$ is not empty, then $Q(c)$ is an integral polyhedron.*

The sufficiency of total unimodularity for $P(b)$ to be integral is not the least bit surprising. But the converse is not so obvious.

Theorem 2.5. *If $P(b) = \{x \in R^n_+: Ax \leqslant b\}$ is integral for all $b \in Z^m$ for which it is not empty, then A is TU.*

Proof. Let A_1 be an arbitrary $k \times k$ nonsingular submatrix of A, and let

$$\tilde{A} = \begin{pmatrix} A_1 & 0 \\ A_2 & I_{m-k} \end{pmatrix}$$

be the $m \times m$ nonsingular submatrix of (A, I) generated from A_1 by taking the appropriate $m-k$ unit vectors from I. Let $b = \tilde{A}z + e_i$, where $z \in Z^m$ and e_i is the ith unit vector. Then $\tilde{A}^{-1}b = z + \tilde{a}_i^{-1}$, where $\tilde{a}_i^{-1}$ is the ith column of $\tilde{A}^{-1}$. Choose z so that $z + \tilde{a}_i^{-1} \geqslant 0$. Thus $z + \tilde{a}_i^{-1}$ is the vector of basic variables of an extreme point of $P(b)$. By hypothesis, $z + \tilde{a}_i^{-1} \in Z^m$ and $z \in Z^m$; hence $\tilde{a}_i^{-1} \in Z^m$ and $\tilde{A}^{-1}$ is an integral matrix. Thus A_1^{-1} is an integral matrix.

Finally, $\det A_1$ and $\det A_1^{-1}$ are integers and

$$|\det A_1| \cdot |\det A_1^{-1}| = |\det (A_1 A_1^{-1})| = 1.$$

Thus, $|\det A_1| = 1$. ■

Theorem 2.5 is false if $P(b) = \{x \in R^n_+; Ax = b\}$. A counterexample is given in exercise 5.

Now we consider sufficient conditions for a matrix to be totally unimodular.

Proposition 2.6. *If the $(0, 1, -1)$ matrix A has no more than two nonzero entries in each column, and if $\sum_i a_{ij} = 0$ if column j contains two nonzero coefficients, then A is TU.*

This result is very easy to prove; but rather than giving a direct proof, we will establish it as a corollary to a much more general result. Its significance is that it implies that a node-arc incidence matrix of any digraph is TU, thus establishing that the sets of feasible solutions to a network flow problem and its dual are integral polyhedra. Consequently, linear programming duality yields integral min–max results such as the max-flow–min-cut theorem (see Theorem 4.1 of Section I.3.4).

We now present a characterization of total unimodularity that yields Proposition 2.6 and some other sufficient conditions as corollaries.

Theorem 2.7. *The following statements are equivalent.*

i. *A is TU.*
ii. *For every $J \subseteq N = \{1, \ldots, n\}$, there exists a partition J_1, J_2 of J such that*

$$\left| \sum_{j \in J_1} a_{ij} - \sum_{j \in J_2} a_{ij} \right| \leqslant 1 \quad \textit{for } i = 1, \ldots, m.$$

Proof. i → ii. Let J be an arbitrary subset of N. Define z by $z_j = 1$ if $j \in J$, $z_j = 0$ otherwise. Also let $d' = 0$, $d = z$, $g = Az$, $b_i' = b_i = \frac{1}{2}g_i$ if g_i is even, and $b_i' = \frac{1}{2}(g_i - 1)$, $b_i = b_i' + 1$ if g_i is odd. Now consider

$$P(b, b', d, d') = \{x \in R^n_+ : b' \leqslant Ax \leqslant b, d' \leqslant x \leqslant d\}.$$

Note that $x = z/2 \in P(b, b', d, d')$. Since A is TU, we have $b', b \in Z^m$, $d', d \in Z^n$ and $P \neq \emptyset$. Proposition 2.3 states that P is integral. Thus there exists $x^0 \in P \cap B^n$ with $x_j^0 = 0$ for $j \in N \setminus J$ and $x_j^0 \in \{0, 1\}$ for $j \in J$. Note that $z_j - 2x_j^0 = \pm 1$ for $j \in J$.

Let $J_1 = \{j \in J: z_j - 2x_j^0 = 1\}$ and $J_2 = \{j \in J: z_j - 2x_j^0 = -1\}$. We have

$$\sum_{j \in J_1} a_{ij} - \sum_{j \in J_2} a_{ij} = \sum_{j \in J} a_{ij}(z_j - 2x_j^0) = \begin{cases} g_i - g_i = 0 & \text{if } g_i \text{ is even} \\ g_i - (g_i \pm 1) = \pm 1 & \text{if } g_i \text{ is odd.} \end{cases}$$

Thus

$$\left| \sum_{j \in J_1} a_{ij} - \sum_{j \in J_2} a_{ij} \right| \leqslant 1 \quad \text{for } i = 1, \ldots, m.$$

ii → i. $|J| = 1$ in statement ii yields $a_{ij} \in \{0, \pm 1\}$ for all i and j. The proof is by induction on the size of the nonsingular submatrices of A using the hypothesis that the determinant of every $(k-1) \times (k-1)$ submatrix of A equals $0, \pm 1$.

Let B be a $k \times k$ nonsingular submatrix of A, and let $r = |\det B|$. Our objective is to prove that $r = 1$.

By the induction hypothesis and Cramer's rule, we have $B^{-1} = B^*/r$, where $b_{ij}^* = \{0, \pm 1\}$. By the definition of B^*, we have $Bb_1^* = re_1$, where b_1^* is the first column of B^*.

Let $J = \{i: b_{i1}^* \neq 0\}$ and $J_1' = \{i \in J: b_{i1}^* = 1\}$. Hence for $i = 2, \ldots, k$, we have

$$(Bb_1^*)_i = \sum_{j \in J_1'} b_{ij} - \sum_{j \in J \setminus J_1'} b_{ij} = 0.$$

Thus $|\{i \in J: b_{ij} \neq 0\}|$ is even; so for any partition $(\tilde{J}_1, \tilde{J}_2)$ of J, it follows that $\sum_{j \in \tilde{J}_1} b_{ij} - \sum_{j \in \tilde{J}_2} b_{ij}$ is even for $i = 2, \ldots, k$. Now by hypothesis, there is a partition (J_1, J_2) of J such that $|\sum_{j \in J_1} b_{ij} - \sum_{j \in J_2} b_{ij}| \leqslant 1$. Hence

$$\sum_{j \in J_1} b_{ij} - \sum_{j \in J_2} b_{ij} = 0 \quad \text{for } i = 2, \ldots, k.$$

Now consider the value of $\alpha_1 = |\sum_{j \in J_1} b_{1j} - \sum_{j \in J_2} b_{1j}|$. If $\alpha_1 = 0$, define $y \in R^k$ by $y_i = 1$ for $i \in J_1$, $y_i = -1$ for $i \in J_2$, and $y_i = 0$ otherwise. Since $By = 0$ and B is nonsingular, we have $y = 0$, which contradicts $J \neq \emptyset$. Hence by hypothesis we have $\alpha_1 = 1$ and $By = \pm e_1$. However, $Bb_1^* = re_1$. Since y and b_1^* are $(0, \pm 1)$ vectors, it follows that $b_1^* = \pm y$ and $|r| = 1$.■

Note that because A is TU if and only if its transpose is TU, statement ii can equivalently be phrased in terms of partitions of subsets of rows of A; that is, for every $Q \subseteq M = \{1, \ldots, m\}$, there exists a partition Q_1, Q_2 of Q such that

$$\left| \sum_{i \in Q_1} a_{ij} - \sum_{i \in Q_2} a_{ij} \right| \leqslant 1 \text{ for } j = 1, \ldots, n.$$

Corollary 2.8. *Let A be a $(0, 1, -1)$ matrix with no more than two nonzero elements in each column. Then A is* TU *if and only if the rows of A can be partitioned into two subsets Q_1 and Q_2 such that if a column contains two nonzero elements, the following statements are true:*

a. *If both nonzero elements have the same sign, then one is in a row contained in Q_1 and the other is in a row contained in Q_2.*
b. *If the two nonzero elements have opposite sign, then both are in rows contained in the same subset.*

Proof. The partitioning of statement ii of Theorem 2.7 is applied to the rows of A. Conditions a and b provide the partition for any $Q \subseteq M$. ■

Corollary 2.8 immediately yields Proposition 2.6 as well as the following corollary:

Corollary 2.9. *The node-edge incidence matrix of a bipartite graph is* TU.

Another consequence of Corollary 2.8 is a linear-time algorithm for recognizing whether a (0, 1, –1) matrix A with, at most, two nonzero entries per column is TU. Without loss of generality, assume that every column of A contains two nonzero elements and every row of A contains at least one nonzero element. Let $B(j) = \{i: a_{ij} \neq 0\}$. Arbitrarily put row 1 in Q_1. Then Corollary 2.8 fixes the assignment of all rows i such that there exists j with $B(j) = \{1, i\}$. Once i is assigned, Corollary 2.8 fixes the assignment of all rows k such that there exists j with $B(j) = \{i, k\}$. The process is repeated in this way until either the partition is completed or an incompatibility with the conditions of the corollary is discovered. The latter occurs when a row already assigned is required to be placed in the complementary set. Note that for a (0, 1) matrix, the procedure simply tests whether the graph of the given node-edge matrix is bipartite.

Definition 2.2. An $m \times n$ (0, 1) matrix A is called an *interval matrix* if in each column the 1's appear consecutively; that is, if $a_{ij} = a_{kj} = 1$ and $k > i + 1$, then $a_{lj} = 1$ for all l with $i < l < k$.

Corollary 2.10. *Interval matrices are* TU.

Proof. This follows from statement ii of Theorem 2.7 by observing that the interval property of a matrix is closed under row deletions and, for $Q = \{1, \ldots, m\}$, taking $Q_1 = \{i: i \text{ is odd}\}$ and $Q_2 = Q \setminus Q_1$. ■

An integer programming problem that involves assigning workers to shifts can be modeled using an interval matrix. Suppose the work day consists of m hours. A shift is a set of consecutive hours. Suppose that n different shifts are possible. The jth shift is represented by a 0-1 m-vector a_j, where $a_{ij} = 1$ if hour i is in the jth shift. Thus A is an $m \times n$ interval matrix of specified shifts. Let $b \in Z_+^m$, where b_i is the minimum number of workers required in the ith hour. The set of feasible solutions is given by $S = \{x \in Z_+^n: Ax \geqslant b\}$, where x_j is the number of workers assigned to the jth shift.

In the next section we will show that if A is an interval matrix and $b \in Z_+^m$, then integer programs of the form

$$\min\{cx: Ax \geqslant b, x \in Z_+^n\} \tag{2.1}$$

are network flow problems. Moreover, for any (0, 1) matrix, a problem of the form (2.1) can be relaxed to another nontrivial problem of the form (2.1) in which the constraint matrix is an interval matrix.

To see this, suppose that the (0, 1) matrix A' is not an interval matrix. Then some column of A' is not an $m \times 1$ interval matrix (column). However, any noninterval column can be uniquely written as the sum of p interval columns, where $p < m/2$ (see Figure 2.1). Now replace each noninterval column a_j' by the p interval columns defined in its decomposition, and give the new columns an objective function coefficient c_{jk} for $k = 1, \ldots, p$ with $c_j = \sum_{k=1}^{p} c_{jk}$. We then obtain a problem of the form (2.1) in which the matrix A is an $m \times s$ interval matrix with $s \leq mn/2$. This is a relaxation of the original problem since we have omitted the constraints that each of the variables associated with the p interval columns that have replaced a_j' must be equal.

We leave as an exercise the comparison of the bounds obtained from this relaxation with those obtained from the linear programming relaxation. The advantage of this relaxation lies in the efficiency of solving flow problems with side constraints.

We close this section by presenting a composition procedure for TU matrices that is used in Section 3 to describe a characterization of TU matrices.

Proposition 2.11. *Let*

$$\begin{pmatrix} A & a & a \\ c & 0 & 1 \end{pmatrix} \quad and \quad \begin{pmatrix} 1 & 0 & b \\ d & d & B \end{pmatrix}$$

be $m \times n$ *and* $n \times m$ TU *matrices respectively, where* A *is* $(m-1) \times (n-2)$, a *is* $(m-1) \times 1$, c *is* $1 \times (n-2)$, B *is* $(n-1) \times (m-2)$, b *is* $1 \times (m-2)$, d *is* $(n-1) \times 1$, *and* 0 *and* 1 *are scalars. Then the* $(m+n-2) \times (m+n-4)$ *matrix*

$$\left(\begin{array}{c|c} A & ab \\ \hline dc & B \end{array}\right)$$

is TU.

This proposition can be proved by applying Theorem 2.7.

$$\begin{pmatrix} 1 \\ 1 \\ 0 \\ 1 \\ 1 \\ 1 \\ 0 \\ 1 \end{pmatrix} = \begin{pmatrix} 1 \\ 1 \\ 0 \\ 0 \\ 0 \\ 0 \\ 0 \\ 0 \end{pmatrix} + \begin{pmatrix} 0 \\ 0 \\ 0 \\ 1 \\ 1 \\ 1 \\ 0 \\ 0 \end{pmatrix} + \begin{pmatrix} 0 \\ 0 \\ 0 \\ 0 \\ 0 \\ 0 \\ 0 \\ 1 \end{pmatrix}$$

Figure 2.1

Example 2.2. Given that

$$A_1 = \begin{pmatrix} 1 & 0 & 1 & -1 & -1 \\ -1 & 1 & 0 & 0 & 0 \\ 0 & -1 & -1 & 0 & 1 \end{pmatrix} \quad \text{and} \quad A_2 = \begin{pmatrix} 1 & 0 & 0 \\ 1 & 1 & 0 \\ 1 & 1 & 1 \\ 0 & 0 & 1 \\ 0 & 0 & 1 \end{pmatrix}$$

are TU, Proposition 2.11 yields the TU matrix

$$A_3 = \begin{pmatrix} 1 & 0 & 1 & 0 \\ -1 & 1 & 0 & 0 \\ 0 & -1 & -1 & 0 \\ 0 & -1 & -1 & 1 \\ 0 & 0 & 0 & 1 \\ 0 & 0 & 0 & 1 \end{pmatrix}.$$

3. NETWORK MATRICES

This section relies on, and is motivated by, the graphical representation of a system of equations $A'x = b$, where A' is the node-arc incidence matrix of a digraph (see Section I.3.6). We begin with a brief review of the results of Section I.3.6 that are needed here.

Let $\mathscr{D} = (V, \mathscr{A})$ be a digraph with $m + 1$ nodes and n arcs, and let A' be the node-arc incidence matrix of $\mathscr{D}$. Suppose the graph underlying $\mathscr{D}$ is connected.

1. $\text{rank}(A') = m$. Since it convenient to work with a matrix of full row rank, we let A be the $m \times n$ matrix obtained by deleting any row of A'.
2. Let $A = (A_1, A_2)$ where A_1 is an $m \times m$ nonsingular submatrix of A. The arcs $(e_1, \ldots, e_m)$ that correspond to the columns of A_1 induce a spanning tree in $\mathscr{D}$, denoted by $\mathscr{T} = (V, \mathscr{A}_1)$.
3. The representation of a column of A_2, corresponding to the arc $e_j = (u, v)$ as a linear combination of the columns of A_1, is given by the incidence vector $\overline{a}_j$ of the unique dipath p_j in $\mathscr{T}$ from u to v, where

$$\overline{a}_{ij} = \begin{cases} 1 & \text{if } p_j \text{ passes } e_i \text{ in a forward direction} \\ -1 & \text{if } p_j \text{ passes } e_i \text{ in a backward direction} \\ 0 & \text{otherwise.} \end{cases}$$

Using the terminology of linear programming, A_1 is a basis matrix and $\overline{A}_2 = A_1^{-1}A_2$ is the incidence matrix of the dipaths corresponding to the columns $\overline{a}_j$. Since we are not concerned with primal or dual feasibility here, b and an objective vector are both irrelevant.

Definition 3.1. Given a directed tree $\mathscr{T} = (V, \mathscr{A}_1)$ and a digraph $\mathscr{D} = (V, \mathscr{A}_2)$, where $|V| = m + 1$, $|\mathscr{A}_1| = m$, and $|\mathscr{A}_2| = n$, the $m \times n$ arc-dipath incidence matrix $M(\mathscr{T}, \mathscr{D})$ corresponding to the dipaths in $\mathscr{T}$ whose endpoints are defined by the arcs of $\mathscr{A}_2$ is called a *network matrix*. (For convenience, it is desirable to allow $\mathscr{A}_2$ to contain edge repetitions.)

Note that in this definition the arcs of $\mathcal{T}$ may or may not be arcs of $\mathcal{D}$. This gives us the freedom to avoid, if we wish, having an identity matrix as a submatrix of every network matrix.

4. Let $e_r \in \mathcal{A}_1$ and $e_s \in \mathcal{A}_2$, and suppose that $\mathcal{T}' = (V, (\mathcal{A}_1 \setminus \{e_r\}) \cup \{e_s\})$ is acyclic. A pivot in $M(\mathcal{T}, \mathcal{D})$ corresponds to forming the tree $\mathcal{T}'$ and the digraph $\mathcal{D}' = (V, (\mathcal{A}_2 \setminus \{e_s\}) \cup \{e_r\})$ and then computing the updated incidence matrix $M(\mathcal{T}', \mathcal{D}')$.

Hence a network matrix is precisely a matrix whose columns represent arcs of a node-arc incidence matrix of a digraph after one row has been deleted and any number of simplex pivots have been executed.

Example 3.1. Consider the digraph $\mathcal{D}$ and tree $\mathcal{T}$ shown in Figure 3.1.
The incidence matrix of $\mathcal{D}$ is

$$A' = \begin{pmatrix} -1 & 0 & 0 & -1 & 1 & 0 \\ 1 & 1 & 0 & 0 & 0 & 1 \\ 0 & -1 & -1 & 0 & -1 & 0 \\ 0 & 0 & 1 & 1 & 0 & -1 \end{pmatrix} \begin{matrix} 1 \\ 2 \\ 3 \\ 4 \end{matrix}$$
$$\begin{matrix} e_1 & e_2 & e_3 & e_4 & e_5 & e_6 \end{matrix}$$

Let A be the submatrix consisting of the first three rows of A', and let

$$A_1 = \begin{pmatrix} -1 & 0 & 0 \\ 1 & 1 & 0 \\ 0 & -1 & -1 \end{pmatrix}$$

be the submatrix consisting of the first three rows and columns of A. Then we obtain the network matrix $M(\mathcal{T}, \mathcal{D})$ given by

$$A_1^{-1}A = \begin{pmatrix} 1 & 0 & 0 & 1 & -1 & 0 \\ 0 & 1 & 0 & -1 & 1 & 1 \\ 0 & 0 & 1 & 1 & 0 & -1 \end{pmatrix} \begin{matrix} e_1 \\ e_2 \\ e_3 \end{matrix}$$
$$\begin{matrix} p_1 & p_2 & p_3 & p_4 & p_5 & p_6 \end{matrix}$$

Network matrix $M(\mathcal{T}, \mathcal{D})$

Forming a new tree $\mathcal{T}'$ by adding e_5 to $\mathcal{T}$ and excluding e_2 as shown in Figure 3.1, we obtain the network matrix $M(\mathcal{T}', \mathcal{D}')$ given by

$$\begin{pmatrix} 1 & 1 & 0 & 0 & 0 & 1 \\ 0 & 1 & 0 & -1 & 1 & 1 \\ 0 & 0 & 1 & 1 & 0 & -1 \end{pmatrix} \begin{matrix} e_1 \\ e_5 \\ e_3 \end{matrix}$$
$$\begin{matrix} p_1 & p_2 & p_3 & p_4 & p_5 & p_6 \end{matrix}$$

Note that $M(\mathcal{T}', \mathcal{D}')$ is obtained by pivoting on the second row and fifth column of $M(\mathcal{T}, \mathcal{D})$.

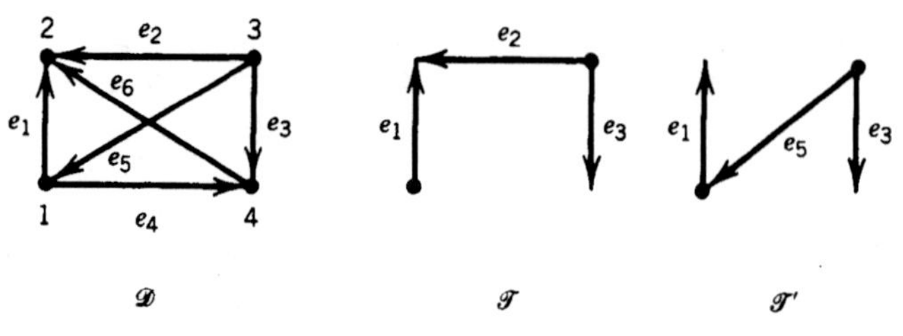

Figure 3.1

Proposition 3.1. *Network matrices have the following properties:*

1. *They are closed under row and column deletions and duplications.*
2. *They are closed under multiplication of a column by −1.*
3. *If A is a network matrix, then (A, I) is a network matrix.*
4. *They are closed under pivoting.*
5. *They are TU.*

Proof.

1. Deleting a column means just to ignore the corresponding dipath. Duplicating a column means simply to repeat the representation of the corresponding dipath. Removing a row is equivalent to removing the corresponding arc [say, $e = (u, v)$] from $\mathscr{T}$ and then constructing the tree $\mathscr{T}'$ by "identifying" nodes u and v as shown in Figure 3.2a. This operation is called a *contraction* of e. Duplicating a row is equivalent to splitting the corresponding arc as shown in Figure 3.2b.
2. Multiplying a column by −1 means to reverse the direction of the corresponding path.

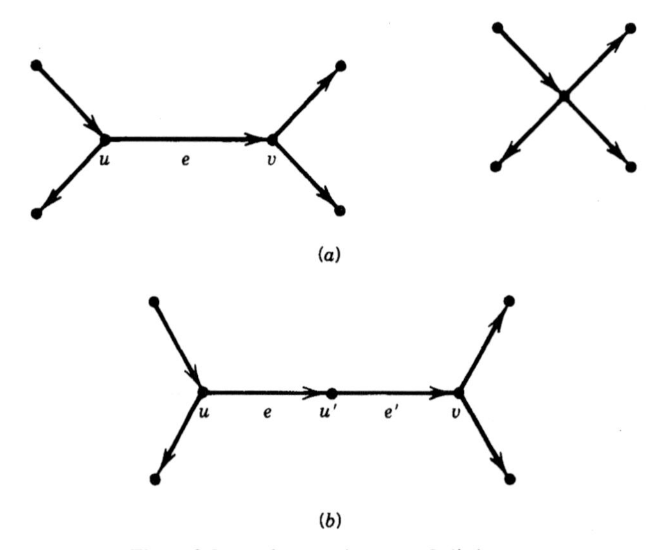

Figure 3.2. (a) Contracting e. (b) Splitting e.

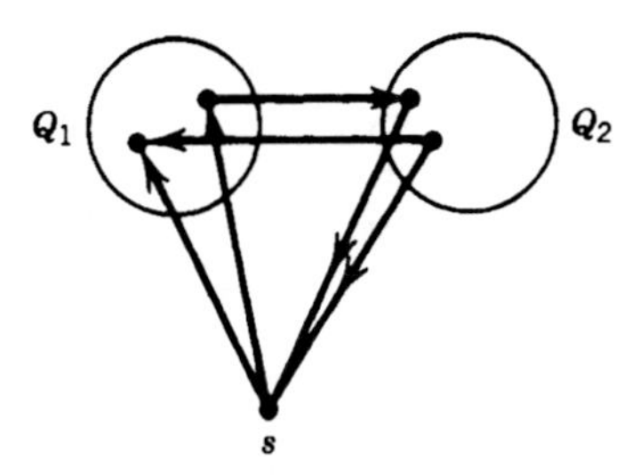

Figure 3.3

3. Here we add a path for each tree arc.

4 and 5. These have been shown above. ■

Two classes of TU matrices presented in Section 2 are network matrices. In each case, we obtain the result simply by giving the appropriate class of trees.

Proposition 3.2. *If A is* TU *and contains no more than two nonzero elements in each column, then A is a network matrix.*

Proof. Let Q_1, Q_2 be the partition of the rows of A defined in Corollary 2.8, and let $\mathcal{T} = (\{s\} \cup Q_1 \cup Q_2, \mathcal{A}_1)$, where

$$\mathcal{A}_1 = \{(u, s)\text{: for all } u \in Q_1\} \cup \{(s, v)\text{: for all } v \in Q_2\}$$

(see Figure 3.3). Let $\mathcal{D} = (\{s\} \cup Q_1 \cup Q_2, \mathcal{A}_2 \cup \mathcal{A}_3 \cup \mathcal{A}_4)$. All of the arcs in $\mathcal{A}_2$ are from one node in Q_1 to another in Q_1 and correspond to the columns of A with two elements of opposite sign. The arcs in $\mathcal{A}_3$ are from a node in Q_1 to a node in Q_2 (or vice versa) and correspond to those columns of A with two elements of the same sign. The arcs in $\mathcal{A}_4$ are arcs of $\mathcal{T}$ and correspond to columns with only one nonzero entry. ■

Proposition 3.3. *Interval matrices are network matrices.*

Proof. Let $V = \{1, \ldots, m+1\}$. $\mathcal{T}$ is a path from node 1 to node $m+1$; that is, $\mathcal{A}_1 = \{(i, i+1)\text{: } i = 1, \ldots, m\}$. A column of A whose first 1 is in row p and whose last 1 is in row q is represented by the arc $(p, q+1) \in \mathcal{A}_2$. ■

Example 3.2. Consider a linear program with constraint set $\{(x, y) \in R_+^{m+n}\text{: } Ax + Iy = b\}$, where

$$A = \begin{pmatrix} 0 & 1 & 0 \\ 1 & 1 & 0 \\ 1 & 1 & 1 \\ 1 & 0 & 1 \\ 1 & 0 & 1 \end{pmatrix}.$$

This is a network flow problem over the network shown in Figure 3.4. For the basic solution with $y = b, x = 0$, the tree arcs corresponding to basic variables are $\{e_1, \ldots, e_5\}$; the digraph arcs corresponding to nonbasic variables are $\{e_6, e_7, e_8\}$. Note that with the supplies shown in Figure 3.4, it follows that $y_i = b_i$ for $i = 1, \ldots, 5$ is a feasible flow.

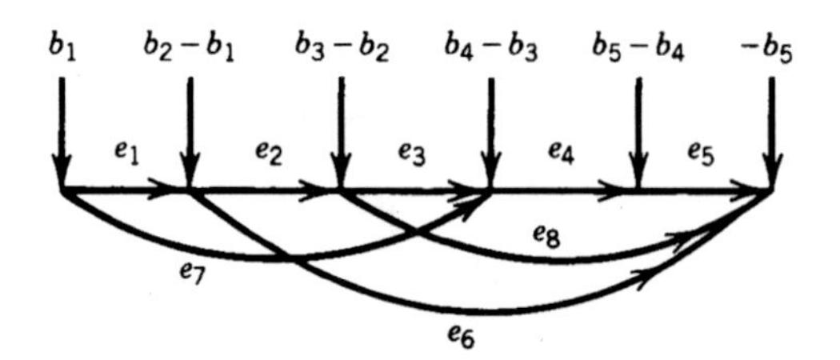

Figure 3.4

Example 3.2 illustrates that if A is a network matrix associated with a known tree $\mathcal{T} = (V, \mathcal{A}_1)$ and digraph $\mathcal{D} = (V, \mathcal{A}_2)$, we can model and solve the linear program $\max\{cx: Ax \leqslant b, x \in R_+^n\}$ as a network flow problem. Furthermore, there is no need to transform A into a node-arc incidence matrix. We immediately obtain a basic solution $y = b, x = 0$ by setting $y_i = b_i$, where y_i is the flow on the tree arc e_i for $i = 1, \ldots, m$. The digraph arcs $\mathcal{A}_2$ represent the nonbasic variables x. Then if $b \geqslant 0$, we have an initial primal feasible basic solution for the network simplex algorithm of Section I.3.6.

We now turn to the question of recognizing network matrices. This problem is in $\mathcal{NP}$ since, given the appropriate digraph and spanning tree, it is easy to verify that the matrix is the desired arc-dipath matrix. On the other hand, it is not so obvious how to give a short proof that a matrix is not a network matrix. There are, however, polynomial-time algorithms for recognizing network matrices. Before describing one, we note that it is extremely unlikely for a random $\{0, 1, -1\}$ matrix to be a network matrix. Thus it would not be wise to use a network recognition algorithm unless there was some reason to believe that the matrix being checked had appropriate structure.

By Proposition 3.2 and the algorithm based on Corollary 2.8 for recognizing TU matrices with no more than two nonzero elements in each column, we have a polynomial-time algorithm for recognizing whether a matrix with no more than two nonzero elements per column is a network matrix. The following recursive algorithm uses this result, by reducing the general question to a suitably small set of recognition problems in which each matrix has no more than two nonzero elements in each column. In the following presentation, we assume for simplicity that A has no zero rows or columns and no row or column duplications.

The algorithm has two parts. In the first part, we ignore the signs of the coefficients and determine whether the matrix is an edge-path incidence matrix of a tree (i.e., a connected forest). In the second part, we consider the orientations of the edges.

Let $M = \{1, \ldots, m\}$ and $N = \{1, \ldots, n\}$.

Definition 3.2. The $m \times n$ (0, 1) matrix A is the *edge-path incidence matrix of a tree* (an EPT matrix) if there is a tree T on $m + 1$ nodes such that each column of A is the characteristic vector of the edges of a path in T.

Definition 3.3. The *row intersection graph* $G(A)$ of an $m \times n$ (0, 1) matrix A is the graph with node set M that has an edge between nodes i and k if there is a column j of A with $a_{ij} \neq 0$ and $a_{kj} \neq 0$.

If $G(A)$ contains $k > 1$ components, then A has the structure

$$\begin{pmatrix} A_1 & 0 & \cdots & 0 \\ 0 & A_2 & \cdots & 0 \\ \vdots & \vdots & & \vdots \\ 0 & 0 & \cdots & A_k \end{pmatrix}.$$

Thus the A_i for $i = 1, \ldots, k$ can be considered separately. So we assume $k = 1$.

By ignoring edge orientations, we see that any (0, 1) network matrix is an EPT matrix. However, as shown below, the converse is false.

Note that any (0, 1) matrix with no more than two 1's in each column is an EPT matrix of a *star* (see Figure 3.5). The reader can easily check that the EPT matrix

$$\begin{pmatrix} 1 & 1 & 0 \\ 1 & 0 & 1 \\ 0 & 1 & 1 \end{pmatrix}$$

is not a network matrix because the required orientations cannot be achieved.

We need to establish some properties of EPT matrices. The following proposition is analogous to statement 1 of Proposition 3.1. Its proof is left as an exercise.

Proposition 3.4. *If A is an* EPT *matrix, then every submatrix of A is an* EPT *matrix.*

Every edge of a tree T is a *cut edge* in the sense that if $e = (u, v)$ is deleted (not contracted) from T, then the resulting subgraph is a forest with two components T_u and T_v. An edge is called a *proper cut edge* if each component of the resulting forest contains at least one edge. Otherwise, the edge is called an *end edge*.

Let B^i be the submatrix of A with row i deleted and all columns j with $a_{ij} = 1$ deleted, and let $G(B^i)$ be the row intersection graph of B^i. If $a_{ij} = 1$ for all j, take B^i to be an identity matrix of size $m - 1$.

Proposition 3.5. *If A is an* EPT *matrix and $\Sigma_i a_{ij} \geq 3$ for some j, then:*

1. *There exists a row k such that $G(B^k)$ contains at least two components.*
2. *For any k such that $G(B^k)$ contains at least two components, A is an* EPT *matrix of some tree T for which e_k is a proper cut edge.*

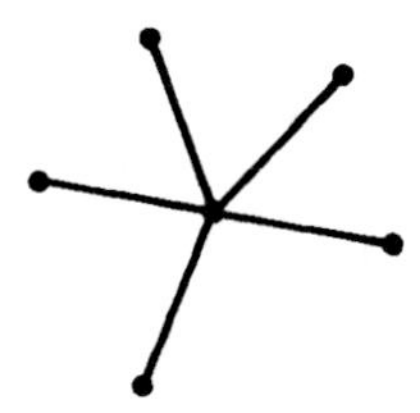

Figure 3.5. Star graph.

Proof.

1. By hypothesis, any tree T for which A is an EPT matrix contains at least one path of length at least 3. Hence T contains a proper cut edge, say e_k. Now, since B^k is an EPT matrix of the tree obtained from T by contracting e_k and ignoring all of the paths that contain e_k, it follows immediately that $G(B^k)$ contains at least two components.
2. Suppose A is an EPT matrix of T'. If $e_k = (u, v)$ is a proper cut edge of T', there is nothing to prove. So suppose that e_k is an end edge of T'. Since $G(B^k)$ contains at least two components, there exists a partition (M_1, M_2) of the rows of $M \setminus \{k\}$ such that no path contains edges from both M_1 and M_2. Furthermore, without loss of generality, it can be assumed that the subforests obtained from the edges of M_1 and M_2 meet at u and that v is of degree 1 (see Figure 3.6a). Hence A is also an EPT matrix for the tree T shown in Figure 3.6b, and e_k is a proper cut edge of T. ■

Example 3.3

$$A = \begin{pmatrix} 1 & 1 & 1 & 0 \\ 1 & 1 & 0 & 1 \\ 1 & 0 & 1 & 1 \end{pmatrix}$$

$B^i = \binom{1}{1}$ for $i = 1, 2, 3$. Hence $G(B^i)$ is connected for all i, and A is not an EPT matrix.

In the following presentation the index k is fixed since we are assuming that e_k is a proper cut edge. For simplicity of notation, the dependence on k is suppressed.

Let $U = \{1, \ldots, t\}$ index the components of $G(B^k)$ where $t \geq 2$. The components induce a partition of $M \setminus \{k\}$. Let $Q^q = \{i \in M: i \text{ is in the } q\text{th component}\}$. Let $R_i = \{j: a_{ij} = 1\}$, let $R_i' = R_i \cap R_k$ for $i \in M \setminus \{k\}$, and let $R^q = \{\cup R_i': i \in Q^q\}$ for all $q \in U$. If A is an EPT matrix of a tree T, then for each q the set of edges indexed by Q^q is the edge set of a subtree T^q of T, R_i is the set of paths containing e_i, R_i' is the set of paths containing e_i and e_k, and R^q is the set of paths that contain e_k and some edge from T^q. Note that $R^q \neq \emptyset$ for any q since $G(A)$ is connected.

Now if A is an EPT matrix of T and $e_k = (u, v)$ is a proper cut edge of T, there exists at least one bipartition of U—say, (U_u, U_v) with $U_w \neq \emptyset$ for $w \in \{u, v\}$—such that if $q \in U_w$, then T^q is on the w side of e_k.

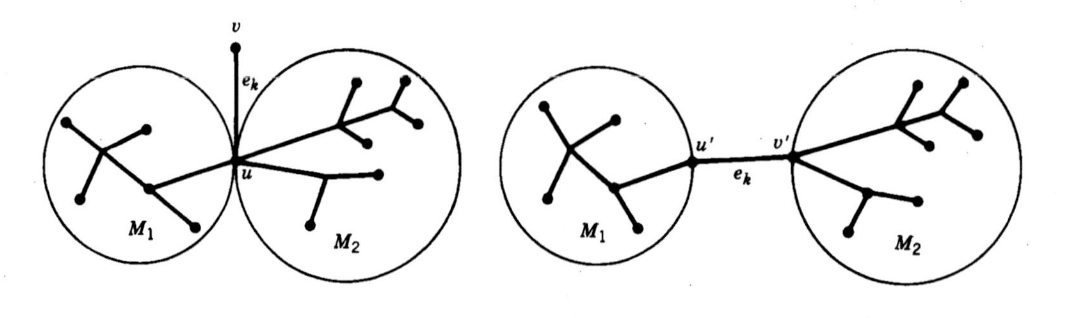

Figure 3.6

Now we try to decide whether two subtrees T^q and $T^{q'}$ can lie on the same side of e_k. Suppose

(3.1) $\qquad R'_i \cap R^{q'} \neq \emptyset$ and $R^{q'} \setminus R'_i \neq \emptyset$ $\qquad$ for some $i \in Q^q$ and $q \neq q'$.

Then there is a path containing e_k, e_i, and an edge of $T^{q'}$; and there is another path containing e_k and an edge of $T^{q'}$, but not e_i. Note that (3.1) does not preclude T^q and $T^{q'}$ from being on the same side of e_k (see Figure 3.7). But if (3.1) is true and T^q and $T^{q'}$ are on the same side of e_k, every path that contains e_k and an edge of T^q must contain precisely the same set of edges from $T^{q'}$. This establishes an ordering between T^q and $T^{q'}$, since e_k must be closer to $T^{q'}$ than to T^q. We say that q' *precedes* q when (3.1) holds.

Similarly, when

(3.2) $\qquad R'_l \cap R^q \neq \emptyset$ and $R^q \setminus R'_l \neq \emptyset$ $\qquad$ for some $l \in Q^{q'}$ and $q \neq q'$

holds, we say that q *precedes* q'. Now it follows that if (3.1) and (3.2) hold, then T^q and $T^{q'}$ must be on opposite sides of e_k.

This discussion motivates the use of the graph $H^k = (U, E^k)$ to determine which pairs of subtrees must lie on opposite sides of e_k, where $(q, q') \in E^k$ if and only if (3.1) and (3.2) are true for the pair (q, q').

We now give necessary and sufficient conditions for A to be an EPT matrix. Moreover, the conditions yield an efficient and constructive algorithm for determining whether A is an EPT matrix.

Theorem 3.6. *A is an* EPT *matrix if and only if, for any k such that $G(B^k)$ contains at least two components, the following statements are true:*

a. *H^k is bipartite.*
b. *The submatrices A^q with column index set N and row index set $Q^q \cup \{k\}$ are* EPT *matrices for all $q \in U$.*

Proof. Suppose A is an EPT matrix. By Proposition 3.4, condition b must hold.

We have already shown that $e_k = (u, v)$ is a proper cut edge and that if $(q, q') \in E^k$, then T^q and $T^{q'}$ must be on opposite sides of e_k. But this cannot hold for all such pairs of subtrees if H^k contains an odd cycle. Hence if A is an EPT matrix, condition a must be true.

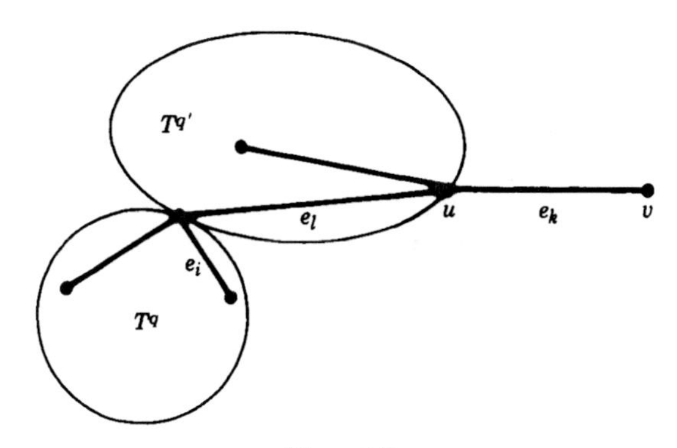

Figure 3.7

Now we show that if conditions a and b are true, then A is an EPT matrix. From condition a, there exists a bipartition of U with the property that if $(q, q') \in E^k$, then q and q' are in different subsets; let (U_u, U_v) be any such partition. For $w \in \{u, v\}$, let A_w be the submatrix of A consisting of the column index set N and row index set $Q_w = \cup_{q \in U_w} Q^q \cup \{k\}$.

The substance of the proof is to show that A_w is an EPT matrix of a tree T_w with e_k as an *end edge*. If this is true, it then follows immediately that A is an EPT matrix of the tree T obtained by joining T_u and T_v together on e_k as shown in Figure 3.8.

We now show how to construct T_w from A_w. We begin by constructing a partial order on the set U_w. Consider $q, q' \in U_w$. Since $(q, q') \notin E^k$, either q and q' are unrelated, or q precedes q' or q' precedes q, but not both.

We claim that if q precedes q' and q' precedes q'', then q precedes q''. Since q precedes q' and q' does not precede q, (3.2) and the complement of (3.1) yield:

i. $R_l' \cap R^q \neq \emptyset$ for some $l \in Q^{q'}$
ii. Either $R_i' \cap R^{q'} = \emptyset$ or $R^{q'} \setminus R_i' = \emptyset$ for all $i \in Q^q$.

By statement i, we obtain $R_l' \cap R^{q'} \neq \emptyset$. Hence by statement ii, we have $R_l' \supseteq R^{q'}$. Since $R^q \supseteq R_l'$, we have $R^q \supseteq R^{q'}$ if q precedes q'. Now, since q' precedes q'' and $R^q \supseteq R^{q'}$, it follows that $R_r' \cap R^q \neq \emptyset$ and $R^q \setminus R_r' \neq \emptyset$ for some $r \in Q^{q''}$. Hence q precedes q''.

So we have a partial order of the elements of U_w. We represent the partial order by any sequence $q_1, q_2, \ldots, q_{t_w}$ with the property that for $2 \leq r \leq t_w$ and $r' < r$, q_r does not precede $q_{r'}$ (see Figure 3.9).

By hypothesis b of Theorem 3.6, we have that for all r, the matrix A^{q_r} with column index set N and row index set $Q^{q_r} \cup \{k\}$ is an EPT matrix of some tree. Furthermore, by the choice of k and Q^{q_r}, there exists some such tree, say T^r, with the property that e_k is an end edge of T^r. This is true because the row intersection graph of the matrix A^{q_r} with row k deleted is connected; that is, it defines a component of $G(B^k)$.

Let $\tilde{Q}^r = \cup_{l=1}^{r} Q^{q_l}$. Now we proceed by induction with the hypothesis that the matrix with row index set $\tilde{Q}^{t_w-1} \cup \{k\}$ and column index set N is an EPT matrix of some tree $\tilde{T}^{t_w-1}$ in which e_k is an end edge.

We must show how to construct $\tilde{T}^{t_w}$ from $\tilde{T}^{t_w-1}$ and $T^{q_{t_w}}$. Since q_{t_w} does not precede q_r for any $r < t_w$, we have either $R_i' \cap R^{t_w} = \emptyset$ or $R^{t_w} \setminus R_i' = \emptyset$ for all $i \in \tilde{Q}^{t_w-1}$. For those i satisfying $R_i' \cap R^{t_w} = \emptyset$, we have $a_{ij} = 0$ for all $j \in R^{t_w}$; for those i satisfying $R^{t_w} \setminus R_i' = \emptyset$, we have $a_{ij} = 1$ for all $j \in R^{t_w}$. In other words, if $j, j' \in R^{t_w}$, then $a_{ij} = a_{ij'}$ for all $i \in \tilde{Q}^{t_w-1}$.

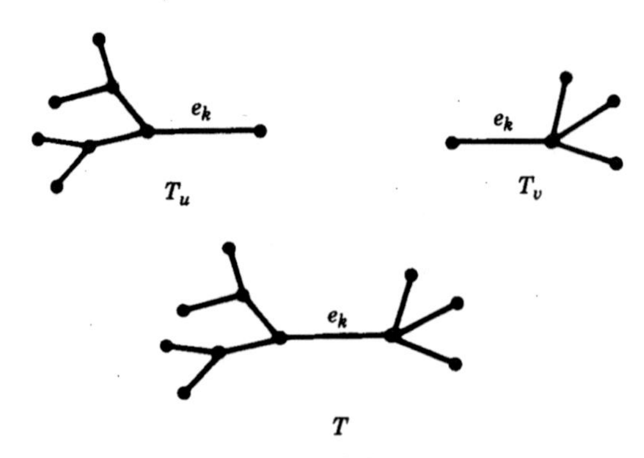

Figure 3.8

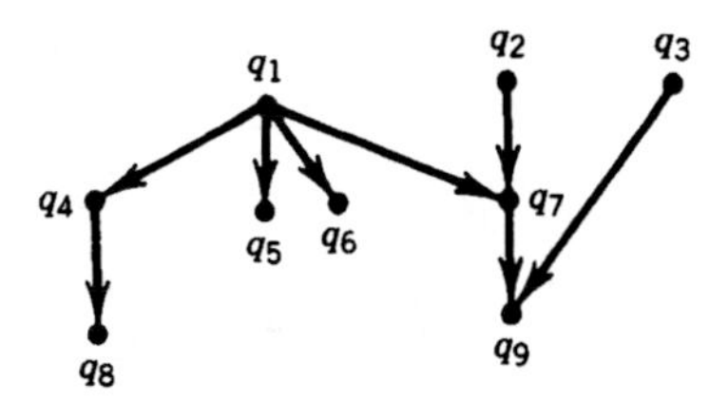

Figure 3.9

Let $S = \{i \in \tilde{Q}^{t_{w-1}}: R^{t_w} \setminus R_i' = \emptyset\}$. ($S$ may be empty.) By the induction hypothesis, there is a path p^* in $\tilde{T}^{t_{w-1}}$ containing precisely the edges e_i for $i \in S$. One end node of p^* is u, call the other end node u^*. (If S is empty, then $u^* = u$.)

The construction of $\tilde{T}^{t_w}$ is shown in Figure 3.10.

Finally, since $T_u = \tilde{T}^{t_u}$ and $T_v = \tilde{T}^{t_v}$, the proof is complete. ∎

Proposition 3.5 and Theorem 3.6 yield a recursive polynomial-time algorithm for recognizing EPT matrices. The algorithm has two fundamental subroutines. The first one finds the components of a row intersection graph. The second one checks whether a graph is bipartite.

Algorithm for Recognizing an EPT Matrix

Step 1: Given a (0, 1) matrix A: (a) If $\Sigma_i a_{ij} \leq 2$ for all j, then A is an EPT matrix of a star; (b) otherwise, partition A according to the components of its row intersection graph, and treat each component separately.

Step 2 (Component Finding): Let $k = 1$, and let B^k be the matrix obtained from A by deleting row k and all columns j with $a_{kj} = 1$ unless $a_{kj} = 1$ for all j. In the latter case, let B^k be an $(m - 1) \times (m - 1)$ identity matrix. Let $G(B^k)$ be the row intersection graph of

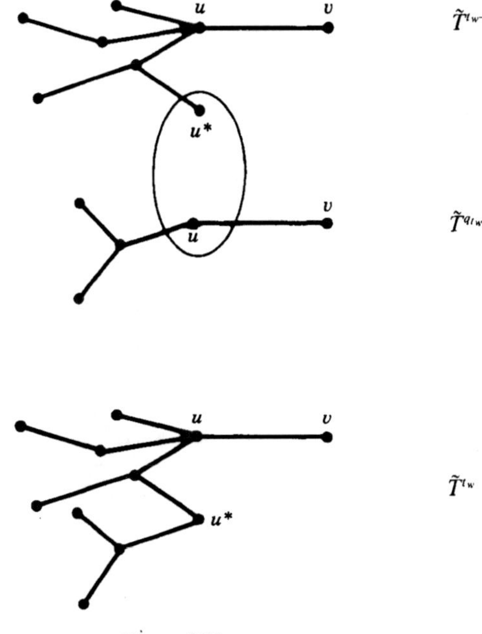

Figure 3.10

B^k. Determine t, the number of components of $G(B^k)$. If $t > 1$, go to Step 3. Otherwise, if $k < m$, then $k \leftarrow k + 1$, and go to Step 2; and if $k = m$, then A is not an EPT matrix.

Step 3 (Bipartite Test): Construct the graph $H^k = (U, E^k)$, where $U = \{1, \ldots, t\}$ and E^k is determined by (3.1) and (3.2). If H^k is not bipartite, A is not an EPT matrix. Otherwise, let (U_u, U_v) be any bipartition of U with the property that if $(q, q') \in E$, then either $q \in U_u$ or $q' \in U_u$, but not both.

Step 4 (Recursion): Construct the matrices A_w for $w \in \{u, v\}$, where A_w consists of row k and the rows i of A, with i in the qth component of $G(B^k)$ and $q \in U_w$. Mark row k of A_w, and call the algorithm for the matrices A_u and A_v, with the exception that marked rows may not be selected in Step 2.

Step 5 (Constructing the Tree): If the recursion ends in Step 1, each pair of terminal submatrices is joined on the edge specified by the marked row. This procedure is applied recursively to determine some tree represented by A.

To show that the algorithm runs in polynomial time, we first calculate $f(m)$, the maximum number of passes through Steps 1–3 for a matrix with m rows. In Step 4, a matrix with m rows is split into two matrices, one with i rows and the other with $m - i + 1$ rows where $2 \leqslant i \leqslant m - 1$. Hence for $m \geqslant 3$, we obtain

$$f(m) = \max_{2 \leqslant i \leqslant m-1} [f(i) + f(m - i + 1)] + 1 \tag{3.3}$$

and $f(2) = 1$. It is a simple exercise to show that the unique solution to (3.3) is $f(m) = 2m - 3$.

Both Steps 2 and 3 can be executed in polynomial time by well-known algorithms. Step 2 dominates. It may require up to m executions of forming a row intersection graph and finding its components. The dominant step in each execution is the pairwise comparison of the rows of A. Hence Step 2 is $O(m^3 n)$ for an $m \times n$ matrix, and the overall time complexity is $O(m^4 n)$.

As suggested by Theorem 3.6, the algorithm can be modified to yield a finer decomposition of A at each step. In particular, instead of decomposing A into A_u and A_v, we decompose A into A^q for $q \in U$. Then if A^q for all $q \in U$ are EPT matrices, we use the partial orders of the nodes in U_u and U_v to construct T. This will be illustrated in Example 3.5.

Example 3.4

$$A = \begin{pmatrix} 0 & 0 & 0 & 1 & 1 & 1 & 1 \\ 1 & 0 & 0 & 1 & 1 & 0 & 0 \\ 0 & 1 & 0 & 1 & 0 & 1 & 0 \\ 0 & 0 & 1 & 1 & 0 & 0 & 1 \end{pmatrix}.$$

The components of $G(B^1)$ are $Q^1 = \{2\}$, $Q^2 = \{3\}$, $Q^3 = \{4\}$. $R_i' = R^{i-1} = \{4, 3 + i\}$ for $i = 2, 3, 4$; for each pair of indices, (3.1) and (3.2) are true. Hence H^1 is a triangle, and A is not an EPT matrix.

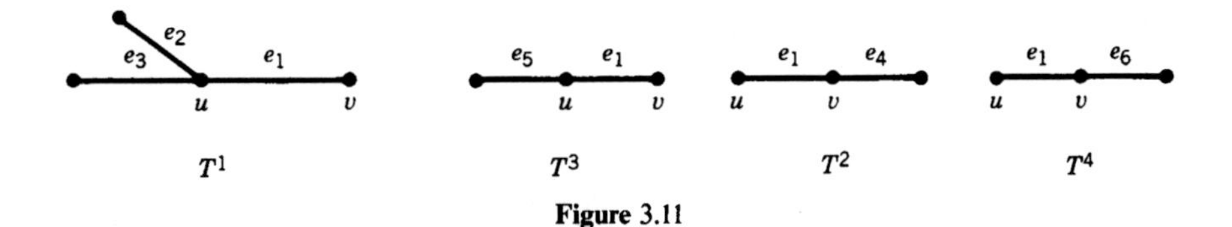

Figure 3.11

Example 3.5

$$A = \begin{pmatrix} 0 & 0 & 0 & 0 & 0 & 1 & 1 & 1 & 1 & 1 \\ 1 & 1 & 0 & 0 & 0 & 1 & 1 & 0 & 0 & 0 \\ 1 & 0 & 0 & 0 & 0 & 0 & 0 & 1 & 0 & 0 \\ 0 & 0 & 1 & 0 & 0 & 0 & 1 & 1 & 0 & 1 \\ 0 & 0 & 0 & 1 & 0 & 0 & 0 & 0 & 1 & 1 \\ 0 & 0 & 0 & 0 & 1 & 0 & 1 & 1 & 0 & 0 \end{pmatrix} \begin{matrix} 1 = k \\ 2 \\ 3 \\ 4 \\ 5 \\ 6 \end{matrix}$$

(columns numbered 1 2 3 4 5 6 7 8 9 10)

$Q^1 = \{2, 3\}$, $Q^2 = \{4\}$, $Q^3 = \{5\}$, $Q^4 = \{6\}$. $R_1' = \{6, 7, 8\}$, $R_2' = \{7, 8, 10\}$, $R_3' = \{9, 10\}$, $R_4' = \{7, 8\}$.

$H^1 = (U, E^1)$, where $U = \{1, 2, 3, 4\}$ and $E^1 = \{(1, 2), (1, 4), (2, 3)\}$. H^1 is bipartite with bipartition $U_u = \{1, 3\}$ and $U_v = \{2, 4\}$. In the set U_u, nodes 1 and 3 are unrelated; and in the set U_v, node 2 precedes node 4. The matrices A^q for $q = 1, \ldots, 4$ yield the stars shown in Figure 3.11.

Since nodes 1 and 3 are unrelated and node 2 precedes node 4, the trees are put together as shown in Figure 3.12.

Only a small modification of the EPT matrix recognition algorithm is required to obtain a recognition algorithm for network matrices.

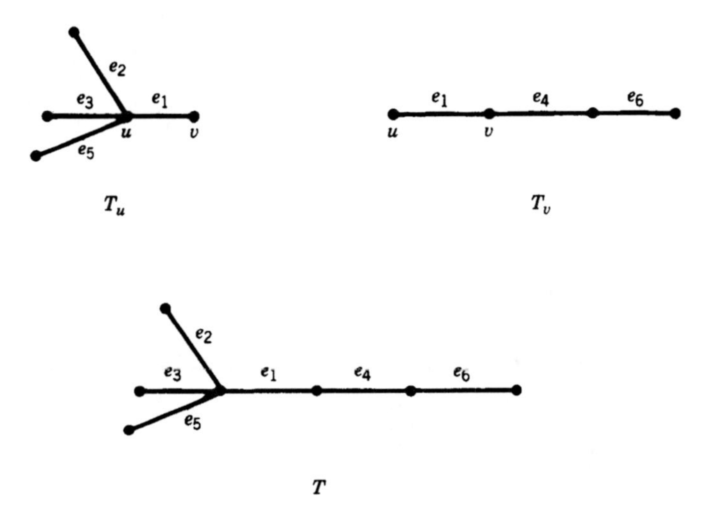

Figure 3.12

Theorem 3.7. *The* $(0, 1, -1)$ *matrix* C *is a network matrix if and only if:*

a. *the matrix* A *obtained from* C *by replacing each element of* C *by its absolute value is an* EPT *matrix; and*
b. *the submatrices of* C *with no more than two nonzero elements in each column, corresponding to the submatrices of* A *with no more than two 1's in each column that are produced in the* EPT *recognition algorithm, are network matrices.*

Proof. The necessity of condition b follows, since all submatrices of network matrices are network matrices. The necessity of condition a follows, since a dipath in a directed tree must be a path in the underlying tree.

To prove sufficiency we only need to show that two directed subtrees $\mathcal{T}_u$ and $\mathcal{T}_v$ can be merged as in Figure 3.8. This is clear if both directed trees contain the arc (u, v) or both contain (v, u). So suppose that $\mathcal{T}_u$, the arc-dipath matrix of A_u, contains (u, v) and that $\mathcal{T}_v$, the arc-dipath matrix of A_v, contains (v, u). Now observe that by reversing the direction of every arc in $\mathcal{T}_v$, we obtain another directed tree $\mathcal{T}'_v$ that also is an arc-dipath matrix of A_v, since a dipath in $\mathcal{T}'_v$ represented by the arc (r, s) corresponds to a dipath in $\mathcal{T}'_v$ represented by the arc (s, r). ■

Example 3.6

$$
\begin{array}{cc}
 & \begin{array}{ccccccccc} 1 & 2 & 3 & 4 & 5 & 6 & 7 & 8 & 9 \end{array} \\
C = & \begin{pmatrix}
0 & 0 & 0 & 0 & 1 & 1 & -1 & 1 & -1 \\
-1 & 1 & 0 & 0 & 0 & 1 & 0 & 0 & 0 \\
1 & 0 & 0 & 0 & 1 & 0 & 0 & 0 & 0 \\
0 & 1 & 0 & 0 & 0 & 0 & 1 & 0 & 0 \\
0 & 0 & 1 & 0 & 1 & 0 & 0 & 1 & 0 \\
0 & 0 & -1 & 0 & 0 & 1 & 0 & 0 & 0 \\
0 & 0 & 0 & 1 & 0 & 0 & 1 & 0 & 1 \\
0 & 0 & 0 & 1 & 0 & 0 & 1 & 0 & 0 \\
0 & 0 & 0 & 0 & 0 & 0 & 0 & -1 & 1
\end{pmatrix}
\begin{array}{c} 1 \\ 2 \\ 3 \\ 4 \\ 5 \\ 6 \\ 7 \\ 8 \\ 9 \end{array}
\end{array}
$$

Let A be the matrix obtained from C by ignoring the signs of the coefficients.

$$
\begin{array}{cc}
 & \begin{array}{ccccccccc} 1 & 2 & 3 & 4 & 5 & 6 & 7 & 8 & 9 \end{array} \\
A = & \begin{pmatrix}
0 & 0 & 0 & 0 & 1 & 1 & 1 & 1 & 1 \\
1 & 1 & 0 & 0 & 0 & 1 & 0 & 0 & 0 \\
1 & 0 & 0 & 0 & 1 & 0 & 0 & 0 & 0 \\
0 & 1 & 0 & 0 & 0 & 0 & 1 & 0 & 0 \\
0 & 0 & 1 & 0 & 1 & 0 & 0 & 1 & 0 \\
0 & 0 & 1 & 0 & 0 & 1 & 0 & 0 & 0 \\
0 & 0 & 0 & 1 & 0 & 0 & 1 & 0 & 1 \\
0 & 0 & 0 & 1 & 0 & 0 & 1 & 0 & 0 \\
0 & 0 & 0 & 0 & 0 & 0 & 0 & 1 & 1
\end{pmatrix}
\begin{array}{l} 1 \\ \left.\begin{array}{c} 2 \\ 3 \\ 4 \end{array}\right\} Q^1 \\ \left.\begin{array}{c} 5 \\ 6 \end{array}\right\} Q^2 \\ \left.\begin{array}{c} 7 \\ 8 \end{array}\right\} Q^3 \\ 9 \,\} \; Q^4 \end{array}
\end{array}
$$

Now we determine if A is an EPT matrix. Since $G(B^1)$ has four components, the edge e_1 corresponding to row 1 must be a proper cut edge. To see which subtrees lie on opposite sides of e_1, we let $H^1 = (U, E^1)$, where $U = \{1, 2, 3, 4\}$ and $E^1 = \{(1, 2), (1, 3), (2, 4), (3, 4)\}$. H^1 is bipartite, and $U_{u_1} = \{1, 4\}$, $U_{v_1} = \{2, 3\}$ is a bipartition of U. The components Q^1 and Q^4 are unrelated, as are the components Q^2 and Q^3. Hence we carry on, under the assumption that subtrees T^1 and T^4 lie on one side of e_1 and that subtrees T^2 and T^3 lie on the other side of e_1.

We now need to show that the matrices A^i corresponding to potential subtrees $T^i \cup \{e_1\}$ are tree matrices with e_1 an end edge.

$$
\begin{array}{c} \\ A^1 = \end{array}
\begin{array}{c} \begin{array}{ccccccc} 1 & 2 & 5 & 6 & 7 & 8 & 9 \end{array} \\ \begin{pmatrix} 0 & 0 & 1 & 1 & 1 & 1 & 1 \\ 1 & 1 & 0 & 1 & 0 & 0 & 0 \\ 1 & 0 & 1 & 0 & 0 & 0 & 0 \\ 0 & 1 & 0 & 0 & 1 & 0 & 0 \end{pmatrix} \end{array}
\begin{array}{l} \\ 1 \quad * \\ 2 \\ 3 \\ 4 \end{array}
$$

$$
C^1 = \begin{pmatrix} 0 & 0 & 1 & 1 & 1 & 1 & -1 \\ -1 & 1 & 0 & 1 & 0 & 0 & 0 \\ 1 & 0 & 1 & 0 & 0 & 0 & 0 \\ 0 & 1 & 0 & 0 & 1 & 0 & 0 \end{pmatrix}
\begin{array}{l} 1 \quad * \\ 2 \\ 3 \\ 4 \end{array}
$$

$$
\begin{array}{c} \\ A^4 = \end{array}
\begin{array}{c} \begin{array}{ccccc} 5 & 6 & 7 & 8 & 9 \end{array} \\ \begin{pmatrix} 1 & 1 & 1 & 1 & 1 \\ 0 & 0 & 0 & 1 & 1 \end{pmatrix} \end{array}
\begin{array}{l} \\ 1 \quad * \\ 9 \end{array}
$$

$$
C^4 = \begin{pmatrix} 1 & 1 & -1 & 1 & -1 \\ 0 & 0 & 0 & -1 & 1 \end{pmatrix}
\begin{array}{l} 1 \quad * \\ 9 \end{array}
$$

$$
\begin{array}{c} \\ A^2 = \end{array}
\begin{array}{c} \begin{array}{cccccc} 3 & 5 & 6 & 7 & 8 & 9 \end{array} \\ \begin{pmatrix} 0 & 1 & 1 & 1 & 1 & 1 \\ 1 & 1 & 0 & 0 & 1 & 0 \\ 1 & 0 & 1 & 0 & 0 & 0 \end{pmatrix} \end{array}
\begin{array}{l} \\ 1 \quad * \\ 5 \\ 6 \end{array}
$$

$$
C^2 = \begin{pmatrix} 0 & 1 & 1 & -1 & 1 & -1 \\ 1 & 1 & 0 & 0 & 1 & 0 \\ -1 & 0 & 1 & 0 & 0 & 0 \end{pmatrix}
\begin{array}{l} 1 \quad * \\ 5 \\ 6 \end{array}
$$

Since A^1, A^2, and A^4 contain no more than two 1's in each column, we can immediately check that the corresponding submatrices of C are all dipath incidence matrices. The corresponding trees are shown in Figure 3.13.

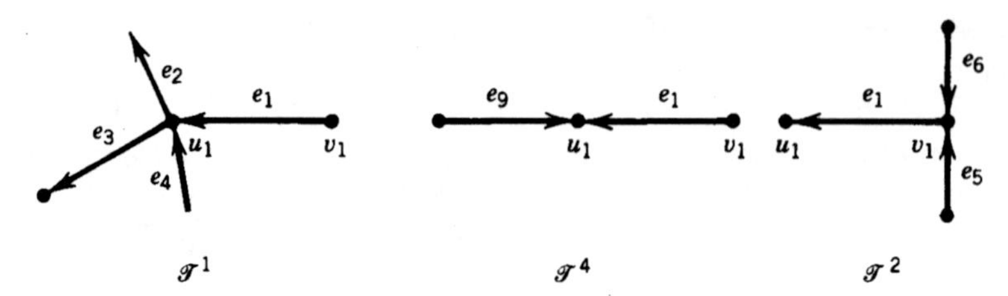

Figure 3.13

$$\begin{array}{c c l}
 & \begin{array}{cccccc} 4 & 5 & 6 & 7 & 8 & 9 \end{array} & \\
A^3 = & \begin{pmatrix} 0 & 1 & 1 & 1 & 1 & 1 \\ 1 & 0 & 0 & 1 & 0 & 1 \\ 1 & 0 & 0 & 1 & 0 & 0 \end{pmatrix} & \begin{array}{ll} 1 & * \\ 7 & \\ 8 & \end{array}
\end{array}$$

needs to be decomposed further. $G(B^7)$ has two components; $Q^5 = \{1\}$ and $Q^6 = \{8\}$.

$H^7 = (U, E^7)$, where $U = (5, 6)$ and $E^7 = \{(5, 6)\}$. H^7 is bipartite; $U_{u_7} = \{5\}$ and $U_{v_7} = \{6\}$.

$$\begin{array}{c c l}
 & \begin{array}{cccccc} 4 & 5 & 6 & 7 & 8 & 9 \end{array} & \\
A^5 = & \begin{pmatrix} 0 & 1 & 1 & 1 & 1 & 1 \\ 1 & 0 & 0 & 1 & 0 & 1 \end{pmatrix} & \begin{array}{ll} 1 & * \\ 7 & * \end{array} \\
C^5 = & \begin{pmatrix} 0 & 1 & 1 & -1 & 1 & -1 \\ 1 & 0 & 0 & 1 & 0 & 1 \end{pmatrix} & \begin{array}{ll} 1 & * \\ 7 & * \end{array}
\end{array}$$

$$\begin{array}{c c l}
 & \begin{array}{ccc} 4 & 7 & 9 \end{array} & \\
A^6 = & \begin{pmatrix} 1 & 1 & 1 \\ 1 & 1 & 0 \end{pmatrix} & \begin{array}{ll} 7 & * \\ 8 & \end{array} \\
C^6 = & \begin{pmatrix} 1 & 1 & 1 \\ 1 & 1 & 0 \end{pmatrix} & \begin{array}{ll} 7 & * \\ 8 & \end{array}
\end{array}$$

The corresponding trees and their merger are shown in Figure 3.14.

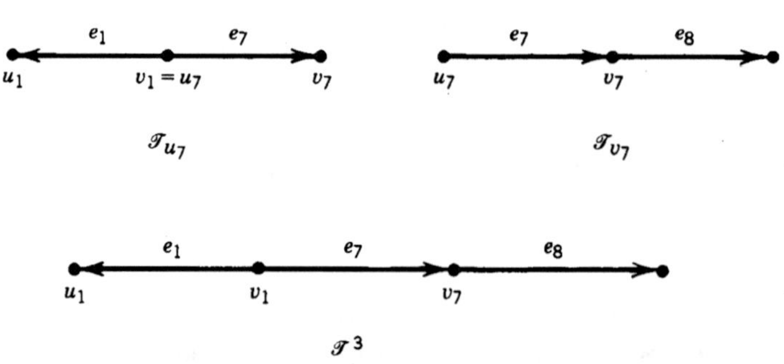

Figure 3.14

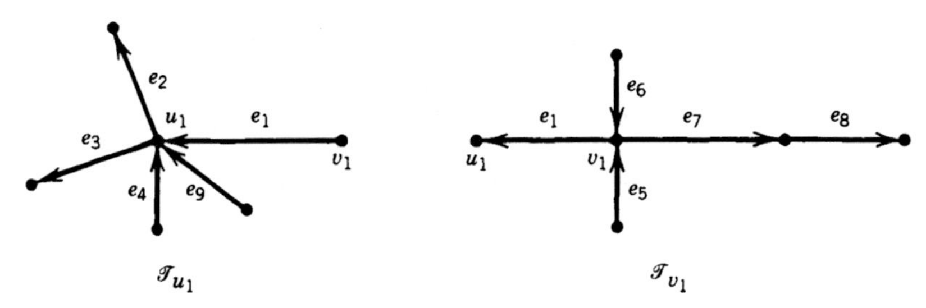

Figure 3.15

The merging of $\mathcal{T}^1$ and $\mathcal{T}^4$ and of $\mathcal{T}^2$ and $\mathcal{T}^3$ are shown in Figure 3.15.

The final merging of $\mathcal{T}_{u_1}$ and $\mathcal{T}_{v_1}$ is shown in Figure 3.16, and C is a network matrix of this tree.

The final topic of this section is a brief discussion of the recognition problem for totally unimodular matrices. The following two matrices are not network matrices but are TU:

$$\begin{pmatrix} 1 & -1 & 0 & 0 & -1 \\ -1 & 1 & -1 & 0 & 0 \\ 0 & -1 & 1 & -1 & 0 \\ 0 & 0 & -1 & 1 & -1 \\ -1 & 0 & 0 & -1 & 1 \end{pmatrix}, \quad \begin{pmatrix} 1 & 1 & 1 & 1 & 1 \\ 1 & 1 & 1 & 0 & 0 \\ 1 & 0 & 1 & 1 & 0 \\ 1 & 0 & 0 & 1 & 1 \\ 1 & 1 & 0 & 0 & 1 \end{pmatrix}.$$

These two matrices and network matrices are the fundamental building blocks for constructing all TU matrices. This result is a deep theorem whose proof is beyond the scope of this presentation.

Theorem 3.8. *Every* TU *matrix that is not a network matrix or one of the two matrices given above can be constructed from these matrices using the rules of Proposition 2.1 and Proposition 2.11.*

A consequence of this theorem is that the TU recognition problem is in $\mathcal{NP}$, since a short proof of total unimodularity for matrix A is to give easily recognizable TU matrices and the rules to construct A from them. Theorem 3.8 also yields a polynomial-time algorithm for the recognition problem and a polynomial-time algorithm for solving linear programs with TU constraint matrices. But the conclusion of practical importance to be drawn from Theorem 3.8 is that "nearly all" TU matrices are network matrices.

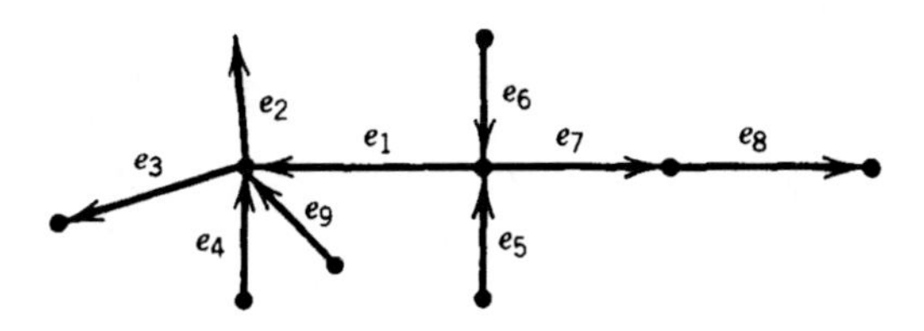

Figure 3.16

4. BALANCED AND TOTALLY BALANCED MATRICES

In the remainder of this chapter, we will study packing and covering problems. Let A be an $m \times n$ (0, 1) matrix.

The *fractional packing* problem we consider is the linear program

$$\text{(FP)} \qquad \max\{cx: x \in P\},$$

where $P = \{x \in R^n_+: Ax \leq 1\}$. Its dual is

$$\text{(DFP)} \qquad \min\{y1: yA \geq c, y \in R^m_+\}.$$

Here it makes sense to eliminate primal unboundedness and dual infeasibility by assuming that A contains no zero columns; that is, $a_j \neq 0$ for $j \in N = \{1, \ldots, n\}$. We also assume that $c_j > 0$ for $j \in N$ since if $c_j \leq 0$, there is an optimal solution to FP with $x_j = 0$.

The *fractional covering* problem we consider is the linear program

$$\text{(FC)} \qquad \min\{cx: x \in Q\},$$

where $Q = \{x \in R^n_+: Ax \geq 1\}$. Its dual is

$$\text{(DFC)} \qquad \max\{y1: yA \leq c, y \in R^m_+\}.$$

Here it is sensible to eliminate primal infeasibility and dual unboundedness by assuming that A contains no zero rows; that is, $a^i \neq 0$ for $i \in M = \{1, \ldots, m\}$. We also assume that $c_j > 0$ for $j \in N$ because if $c_j < 0$, FC is unbounded, and if $c_j = 0$, we can set $x_j = 1$.

We can view the rows of A as incidence vectors of a family of subsets $N_i \subseteq N$ for $i \in M$. To describe P, the maximal rows are necessary and sufficient; and to describe Q, the minimal rows are necessary and sufficient. Hence, in both cases, we can assume that the rows of A are incomparable (0, 1) vectors; that is, they are the incidence vectors of a set of subsets called a *clutter*.

Our goal is to determine classes of matrices and classes of combinatorial optimization problems for which these linear programs have integral optimal solutions. Thus the fundamental questions are:

1. When is P an integral polytope?
2. When is the system $Ax \leq 1, x \geq 0$ TDI?
3. When is Q an integral polyhedron?
4. When is the system $Ax \geq 1, x \geq 0$ TDI?

As we have already seen, total unimodularity of A is a correct answer to all four questions. But, as we shall see, there are larger classes of (0, 1) matrices for which P and Q are integral, and the packing and covering systems are TDI. Let $P(b) = \{x \in R^n_+: Ax \leq b\}$ and $Q(b) = \{x \in R^n_+: Ax \geq b\}$. Note that if P is integral, then $P(b)$ is integral for all $b \in B^m$. This is true since if $b \in B^m$ and $M(b) = \{i \in M: b_i = 0\}$, then $P(b)$ is the face of P determined by setting $x_j = 0$ for all j with $a_{ij} = 1$ for some $i \in M(b)$.

The integrality of P and Q are, in general, unrelated. To relate them it is necessary to consider families of polyhedra that are obtained by eliminating constraints. For the constraint $a^i x \geq b_i$, setting $b_i = 0$ is just a way of saying that the ith constraint is superfluous or has been eliminated. Similarly, for the constraint $a^i x \leqslant b_i$, setting $b_i \geqslant \Sigma_{j=1}^n a_{ij}$ has the same effect. Here we use the notation $b_i = \infty$.

Proposition 4.1. *Let A be a $(0, 1)$ matrix with no zero rows or columns. The following two statements are equivalent.*

1. *$P(b)$ is integral for all b with $b_i \in \{1, \infty\}$ for $i \in M$.*
2. *$Q(b)$ is integral for all $b \in B^m$.*

Proof. Each member of each of the families is nonempty since $0 \in P(b)$ and $1 \in Q(b)$. Consider $Q(b)$ with $b_i = 1$ for $i \in K \subseteq M$ and with $b_i = 0$ otherwise. Suppose x is a fractional extreme point of $Q(b)$. Then there exist $N^= \subseteq N$ and $K^= \subseteq K$ such that $|N^=| + |K^=| = n$ and x is a solution to

$$a^i x = \begin{cases} 1 & \text{for } i \in K^= \\ \alpha_i \geqslant 1 & \text{otherwise,} \end{cases} \qquad x_j = \begin{cases} 0 & \text{for } j \in N^= \\ \beta_j \geqslant 0 & \text{otherwise.} \end{cases}$$

But then x is an extreme point of $P(\overline{b})$, where $\overline{b}_i = b_i$ for $i \in K^=$ and $\overline{b}_i = \infty$ otherwise. Thus $1 \Rightarrow 2$. The proof of $2 \Rightarrow 1$ is similar. ∎

The matrices we study in this section are precisely those for which statements 1 and 2 of Proposition 4.1 hold. Let $\mathcal{M}_k$, $k \geqslant 3$, be the family of $k \times k$ $(0, 1)$ matrices, all of whose row and column sums equal 2, that do not contain the submatrix

$$\begin{pmatrix} 1 & 1 \\ 1 & 1 \end{pmatrix}.$$

Definition 4.1. A $(0, 1)$ matrix is *totally balanced* (TB) if it does not contain a submatrix in $\mathcal{M}_k$ for any $k \geqslant 3$.

Definition 4.2. A $(0, 1)$ matrix is *balanced* if it does not contain a submatrix in $\mathcal{M}_k$ for any $k \geqslant 3$ and odd.

Note that a matrix A in $\mathcal{M}_k$, $k \geqslant 3$, that does not contain a submatrix in $\mathcal{M}_l$, $l < k$, is a node-edge incidence matrix of a cycle (see Figure 4.1). By permuting rows and columns, we can write such a matrix in a canonical form with $a_{jj} = a_{j+1,j} = 1$ for $j = 1, \ldots, k-1$, $a_{kk} = a_{1k} = 1$, and $a_{ij} = 0$ otherwise. Then $|\det A| = 2$ when k is odd, and $|\det A| = 0$ when k is even.

$$\begin{pmatrix} 1 & 0 & 0 & 0 & 1 \\ 0 & 0 & 1 & 1 & 0 \\ 1 & 1 & 0 & 0 & 0 \\ 0 & 0 & 0 & 1 & 1 \\ 0 & 1 & 1 & 0 & 0 \end{pmatrix}$$

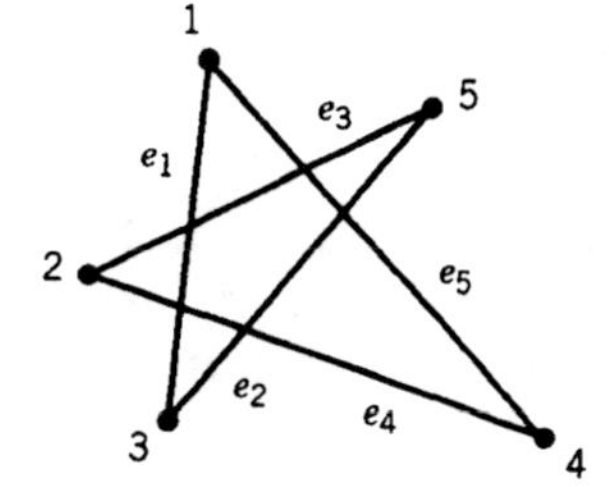

Figure 4.1

We now give simple consequences of the above definitions.

Proposition 4.2. *If A is a (totally) balanced matrix, then the following matrices are (totally) balanced.*

1. (A, I).
2. *The transpose of A.*
3. *Any matrix obtained by permuting rows or columns of A.*
4. *Any submatrix of A.*

Proposition 4.3. *If A is a* (0, 1) TU *matrix, then A is balanced.*

Proof. If A is not balanced, then it contains a submatrix $A' \in \mathcal{M}_k$ where $k \geqslant 3$ and odd. Hence $|\det A'| = 2$. ■

On the other hand, a (0, 1) TU matrix may not be a TB matrix. For example, any matrix in $\mathcal{M}_4$ is a TU matrix. The following example illustrates some of the properties of TB matrices.

Example 4.1. Let

$$A = \begin{pmatrix} 1 & 1 & 1 & 1 \\ 1 & 1 & 0 & 0 \\ 1 & 0 & 1 & 0 \\ 1 & 0 & 0 & 1 \end{pmatrix}.$$

The reader should check the following statements.

1. A is not a TU matrix.
2. A is a TB matrix.
3. P is integral. (Its extreme points are the null vector and the four unit vectors.)
4. $P(b)$ with $b = (2 \quad 1 \quad 1 \quad 1)$ contains the extreme point $(\frac{1}{2} \quad \frac{1}{2} \quad \frac{1}{2} \quad \frac{1}{2})$.
5. The matrix

$$\begin{pmatrix} O & A \\ A' & O \end{pmatrix}$$

where $A' \in \mathcal{M}_4$ is balanced, but not a TU or TB matrix.

The relationship among these classes is given in Figure 4.2.

Although there are nice polyhedral results for balanced matrices, no polynomial-time combinatorial methods are known for solving the corresponding linear programming problems. Moreover, the recognition problem also is unsolved except for the obvious result that it is in $\mathcal{C}o\mathcal{NP}$.

In contrast, the results for TB matrices are much richer because both optimization and recognition problems can be solved by efficient combinatorial methods. Hence, for most of the remainder of this section, we consider TB matrices. Also, since the theory and algorithms for FP (the fractional packing problem) and FC (the fractional covering

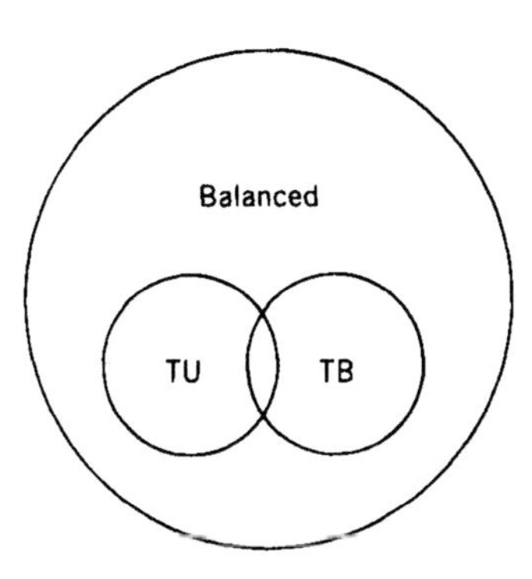

Figure 4.2

problem) are essentially the same for TB matrices, we only need to consider one of them in detail. Since more general results for FP will be given in the next section, we consider FC and its dual DFC here. We will give a polynomial-time algorithm that obtains integral optimal solutions to FC and DFC. We first need some preliminary results.

Definition 4.3. A (0, 1) matrix A is called a *row inclusion matrix* if it does not contain the submatrix

$$F = \begin{pmatrix} 1 & 1 \\ 1 & 0 \end{pmatrix}.$$

In other words, all of the rows i with $a_{ij} = 1$ are ordered by inclusion with respect to the columns $j, \ldots, n$. The reader should keep in mind that the row inclusion property of a matrix is sensitive to the ordering of its rows and columns. Later in this section we will address the issue of whether the rows and columns of a given (0, 1) matrix can be permuted to obtain a row inclusion matrix.

Two obvious properties of row inclusion matrices are given in the following propositions.

Proposition 4.4. *The recognition problem for row inclusion matrices is solvable in polynomial time.*

Proposition 4.5. *Row inclusion matrices are totally balanced.*

Proof. Suppose A contains a submatrix B in $\mathcal{M}_k$, $k \geqslant 3$. Then there exists i, j, k, l with $i < k$ and $j < l$ such that $b_{ij} = b_{il} = b_{kj} = 1$. Then $b_{kl} = 0$, since otherwise B contains the submatrix $\begin{pmatrix} 1 & 1 \\ 1 & 1 \end{pmatrix}$. Hence A contains the submatrix F. ■

The converse of Proposition 4.5 obviously is false, but later we will show that by row-and-column permutations of a TB matrix we can obtain a row inclusion matrix.

Our next objective is to show that when A is a row inclusion matrix, the fractional covering problem (FC) and its dual (DFC) are easily solved. DFC is solved by greedily packing the rows of A into the vector c. That is, we first take as much as possible of row a^1, then row a^2, and so on. When a positive multiple of a row is taken, we note the largest column index j for which the remaining amount of c_j is reduced to zero. These columns are then used to find a primal optimal solution. The primal solution also is constructed greedily by processing these columns in the reverse order from which they were selected.

Algorithm for DFC and FC for Row Inclusion Matrices

DFC:

Initialization: Let $N_i = \{j \in N: a_{ij} = 1\}$ for $i = 1, \ldots, m$, $i = 1$, $c^1 = c > 0$, $J_0 = \emptyset$.

Iteration i: Let $y_i = \min\{c_j^i: j \in N_i\}$. If $y_i > 0$, then let $c^{i+1} = c^i - y_i a^i$ and $J_i = J_{i-1} \cup \{k\}$, where $k = \max\{j \in N_i: y_i = c_j^i\}$. Let $\alpha(k) = i$. Otherwise $J_i = J_{i-1}$ and $c^{i+1} = c^i$. If $i = m$, stop; $(y_1, \ldots, y_m)$ is an optimal solution. Otherwise $i \leftarrow i + 1$.

FC:

Initialization: Let $b = 1$, $J_m = \{k_1, \ldots, k_p\}$ (from DFC), where $\alpha(k_i) < \alpha(k_{i+1})$ for $i = 1, \ldots, p - 1$. Set $x_j = 0$ for $j \notin J_m$, and set $l = p$.

Iteration l: Set $x_{k_l} = \max(0, b_{\alpha(k_l)})$ and $b \leftarrow b - a_{k_l} x_{k_l}$. If $l = 1$, stop; $x = (x_1, \ldots, x_n)$ is an optimal solution. Otherwise $l \leftarrow l - 1$.

Theorem 4.6. *The algorithm gives integral optimal solutions to* DFC *and* FC *when A is a row inclusion matrix.*

Proof. It is clear that the solutions are integral and that the dual solution is feasible. Throughout the proof, we use the following facts: (a) If $J_m = \{k_1, \ldots, k_p\}$, then $\alpha(k_i) < \alpha(k_{i+1})$ for $i = 1, \ldots, p - 1$, and (b) if $y_i > 0$, then there is a $k \in J_m$ with $\alpha(k) = i$, and either $c_j^i > 0$ or $a_{ij} = 0$ for all j.

We now consider primal feasibility. By construction, we have $x \geqslant 0$. The proof of $a^i x \geqslant 1$ for $i = 1, \ldots, m$ is divided into two cases.

Case 1 ($y_i > 0$). Then for some $k_l \in J_m$, we have $\alpha(k_l) = i$ and $a_{ik_l} = 1$. At Step l, we set $x_{k_l} = \max(0, \tilde{b}_i)$, where $\tilde{b}_i$ is the current value of b_i. If $\tilde{b}_i \leqslant 0$, then $a^i x \geqslant 1$. If $\tilde{b}_i = 1$, then $x_{k_l} = 1$ and $a^i x \geqslant 1$.

Case 2 ($y_i = 0$). Let $N_i^0 = N_i \cap \{j \in N: c_j^i = 0\}$. Since $y_i = 0$, we obtain $N_i^0 \neq \emptyset$. Let $j_1 = \max\{j: j \in N_i^0\}$. Suppose $j_1 \notin J_{i-1}$. Now, since $c_{j_1}^i = 0$, there is an $i_1 < i$ such that $y_{i_1} = c_{j_1}^{i_1} > 0$, and $a_{i_1 j_1} = 1$. Since $j_1 \notin J_{i-1}$, there is a $j_2 \in J_{i-1}$ with $j_2 > j_1$ such that $i_1 = \alpha(j_2)$, $y_{i_1} = c_{j_2}^{i_1}$, and $a_{i_1 j_2} = 1$. Also, by the definition of j_1, we have $a_{ij_1} = 1$ and $a_{ij_2} = 0$. Hence we have the following submatrix of A:

$$\begin{array}{cc} j_1 & j_2 \end{array}$$
$$\begin{pmatrix} 1 & 1 \\ 1 & 0 \end{pmatrix} \begin{array}{c} i_1 \\ i \end{array}$$

Since A is a row inclusion matrix, this is not possible, so $j_1 \in J_{i-1}$.

Now we can define i_1 by $\alpha(j_1) = i_1 < i$. Hence if $x_{j_1} = 1$, then $a^i x \geqslant 1$.

On the other hand, if $x_{j_1} = 0$, let $j_3 \in J$ be such that $x_{j_3} = 1$ and row i_1 was first covered by column j_3. This means that $a_{i_1 j_3} = a_{i_3 j_3} = 1$, where $i_3 = \alpha(j_3) > i_1$. In addition, $a_{i_3 j_1} = 0$ since $\alpha(j_1) = i_1$ and $y_{i_3} > 0$. Now if $j_3 < j_1$, we have

$$\begin{array}{cc} j_3 & j_1 \end{array}$$
$$\begin{pmatrix} 1 & 1 \\ 1 & 0 \end{pmatrix} \begin{array}{c} i_1 \\ i_3 \end{array}$$

Hence $j_3 > j_1$. Then, if $a_{ij_3} = 0$, we have

$$\begin{array}{cc} j_1 & j_3 \end{array}$$
$$\begin{pmatrix} 1 & 1 \\ 1 & 0 \end{pmatrix} \begin{array}{c} i_1 \\ i \end{array}$$

Hence $a_{ij_3} = 1$ and $a^i x \geqslant a_{ij_3} x_3 = 1$.

Next we establish the complementarity conditions

$$x_j\left(c_j - \sum_{i=1}^{m} a_{ij} y_i\right) = 0 \quad \text{for } j = 1, \ldots, n, \tag{4.1}$$

$$y_i\left(\sum_{j=1}^{n} a_{ij} x_j - 1\right) = 0 \quad \text{for } i = 1, \ldots, m. \tag{4.2}$$

The conditions (4.1) are satisfied by the construction of J_m and $x_j = 0$ if $j \notin J_m$. Now consider (4.2) and suppose $y_i > 0$ and $\alpha(k) = i$. We need to show that

$$\sum_{j \in J_m} a_{ij} x_j = \sum_{j \in J_m:\, \alpha(j) < i} a_{ij} x_j + a_{ik} x_k + \sum_{j \in J_m:\, \alpha(j) > i} a_{ij} x_j = 1.$$

1. $\Sigma_{j \in J_m:\, \alpha(j) < i}\, a_{ij} x_j = 0$ since $\alpha(j) = i_1 < i$ implies $c_j^i = 0$, and thus $y_i > 0$ implies $a_{ij} = 0$.
2. Since $\alpha(k) = i$, it follows that $a_{ik} = 1$. By the construction of the primal solution, $x_k = 1$ if and only if $\Sigma_{j \in J_m:\alpha(j) > i}\, a_{ij} x_j = 0$.
3. Now it suffices to show that $\Sigma_{j \in J_m:\, \alpha(j) > i}\, a_{ij} x_j \leqslant 1$. If not, there exists $j_1, j_2 \in J_m$ such that $a_{ij_1} x_{j_1} + a_{ij_2} x_{j_2} = 2$ and $\alpha(j_2) > \alpha(j_1) > i$. Then $a_{\alpha(j_2), j_1} = 0$ since $c_{j_1}^{\alpha(j_2)} = 0$, and $a_{\alpha(j_1), j_2} = 0$ since $x_{j_1} = x_{j_2} = 1$. Hence we have

$$\begin{array}{cc} j_1 & j_2 \end{array} \qquad\qquad \begin{array}{cc} j_2 & j_1 \end{array}$$
$$\begin{pmatrix} 1 & 1 \\ 1 & 0 \end{pmatrix} \begin{array}{l} i \\ \alpha(j_1) \end{array} \quad \text{or} \quad \begin{pmatrix} 1 & 1 \\ 1 & 0 \end{pmatrix} \begin{array}{l} i \\ \alpha(j_2). \end{array}$$

Since neither of these is possible, we have $\Sigma_{j \in J_m:\, \alpha(j) > i}\, a_{ij} x_j \leqslant 1$. ■

Example 4.2. $\min\{cx: Ax \geqslant 1, x \in R_+^n\}$ with $c = (2 \quad 3 \quad 3 \quad 1)$ and

$$A = \begin{pmatrix} 1 & 1 & 0 & 0 \\ 0 & 1 & 1 & 0 \\ 0 & 0 & 1 & 1 \\ 1 & 1 & 1 & 1 \end{pmatrix}.$$

It is easy to check that A is a row inclusion matrix.

DFC:

1. $y_1 = 2, c^2 = (2 \quad 3 \quad 3 \quad 1) - (2 \quad 2 \quad 0 \quad 0) = (0 \quad 1 \quad 3 \quad 1), k = 1, J_1 = \{1\}, \alpha(1) = 1.$
2. $y_2 = 1, c^3 = (0 \quad 0 \quad 2 \quad 1), k = 2, J_2 = \{1, 2\}, \alpha(2) = 2.$
3. $y_3 = 1, c^4 = (0 \quad 0 \quad 1 \quad 0), k = 4, J_3 = \{1, 2, 4\}, \alpha(4) = 3.$
4. $y_4 = 0, J_4 = \{1, 2, 4\}; y = (2 \quad 1 \quad 1 \quad 0)$ is an optimal solution.

FC:

1. $J_4 = \{1, 2, 4\}, x_3 = 0, b = (1 \quad 1 \quad 1 \quad 1)$.
2. $l = 3, k_3 = 4, x_4 = b_3 = 1, b = (1 \quad 1 \quad 0 \quad 0)$.
3. $l = 2, k_2 = 2, x_2 = b_2 = 1, b = (0 \quad 0 \quad 0 \quad -1)$.
4. $l = 1, k_1 = 1, x_1 = b_1 = 0$. Stop; $x = (0 \quad 1 \quad 0 \quad 1)$ is an optimal solution of value 4.

Next we show that the rows and columns of a TB matrix can be permuted so that the resulting TB matrix is a row inclusion matrix.

Given a (0, 1) matrix A, let $\tilde{a}^i = (a_{in}, \ldots, a_{i1})$ for $i = 1, \ldots, m$ be the elements of row i in reverse order, and let $\tilde{a}_j = (a_{mj}, \ldots, a_{1j})$ for $j = 1, \ldots, n$ be the elements of column j in reverse order.

Definition 4.4. The (0, 1) matrix A is called *totally reverse lexicographic* (TRL) if $\tilde{a}^{i+1} \overset{L}{\geqslant} \tilde{a}^i$ for $i = 1, \ldots, m-1$ and if $\tilde{a}_{j+1} \overset{L}{\geqslant} \tilde{a}_j$ for $j = 1, \ldots, n-1$.

We now give an algorithm which shows that:

Proposition 4.7. *By permuting rows and columns, any* (0, 1) *matrix can be transformed to a* TRL *matrix in polynomial time.*

Proof. For any partition $M_1, \ldots, M_t$ of the rows of A, let $d_j \in Z_+^t$ be given by

$$d_j = \left(\sum_{i \in M_t} a_{ij}, \ldots, \sum_{i \in M_1} a_{ij}\right) \quad \text{for } j \in N.$$

Initially, let $t = 1$ and $M_1 = M = \{1, \ldots, m\}$. Hence $d_j = \Sigma_{i \in M} a_{ij}$. Suppose $d_{j_n} = \max_{j \in N} d_j$.

We now begin to construct the TRL permutation of A by making the following row and column permutations:

1. j_n is the last column.
2. $M_1 = \{i \in M: a_{ij_n} = 0\}$ and $M_2 = M \setminus M_1$. Hereafter, all rows in M_1 precede those in M_2.

Hence we have the $m \times 1$ TRL matrix

$$\begin{array}{c} j_n \\ \begin{pmatrix} 0 \\ \vdots \\ 0 \\ 1 \\ \vdots \\ 1 \end{pmatrix} \end{array} \begin{array}{l} \left.\begin{matrix} \\ \\ \\ \end{matrix}\right\} M_1 \\ \left.\begin{matrix} \\ \\ \\ \end{matrix}\right\} M_2 \end{array}$$

and regardless of how we permute the rows within M_1 and M_2, we have $\tilde{a}_j \overset{L}{\leqslant} \tilde{a}_{j_n}$ for $j \neq j_n$.

Now for $j \neq j_n$, let $d_j = (\Sigma_{i \in M_2} a_{ij}, \Sigma_{i \in M_1} a_{ij})$ and suppose $d_{j_{n-1}} \overset{L}{\geqslant} d_j$ for $j \in N \setminus \{j_n\}$. Partition M_k for $k = 1, 2$ into $M_k' = \{i \in M_k: a_{ij_{n-1}} = 0\}$ and $M_k \setminus M_k'$. Now put the rows in

M'_k before those in $M_k \setminus M'_k$ and call the new partition $M_1, \ldots, M_t$, where $t \leqslant 4$. Also move column J_{n-1} to the $(n-1)$st position. Now we have an $m \times 2$ TRL matrix

$$\begin{array}{c} \begin{matrix} j_{n-1} & j_n \end{matrix} \\ \left(\begin{matrix} 0 & 0 \\ \vdots & \vdots \\ 0 & 0 \\ 1 & 0 \\ \vdots & \vdots \\ 1 & 0 \\ 0 & 1 \\ \vdots & \vdots \\ 0 & 1 \\ 1 & 1 \\ \vdots & \vdots \\ 1 & 1 \end{matrix}\right) \begin{matrix} \left.\vphantom{\begin{matrix}0\\ \vdots\\ 0\end{matrix}}\right\} M_1 \\ \left.\vphantom{\begin{matrix}0\\ \vdots\\ 0\end{matrix}}\right\} M_2 \\ \left.\vphantom{\begin{matrix}0\\ \vdots\\ 0\end{matrix}}\right\} M_3 \\ \left.\vphantom{\begin{matrix}0\\ \vdots\\ 0\end{matrix}}\right\} M_4 \end{matrix} \end{array}$$

with $\tilde{a}_j \overset{L}{\lneqq} \tilde{a}_{j_{n-1}}$ for $j \in N \setminus \{j_{n-1}, j_n\}$.

The process can be continued by choosing a $j \in N \setminus \{j_{n-1}, j_n\}$ such that d_j is lexicographically largest and then partitioning each of the M_i to maintain the lexicographic ordering of the rows. ■

Example 4.3.

$$A = \begin{array}{c} \begin{matrix} 1 & 2 & 3 & 4 & 5 & 6 & 7 \end{matrix} \\ \left(\begin{matrix} 0 & 1 & 0 & 0 & 0 & 0 & 1 \\ 0 & 0 & 1 & 1 & 1 & 0 & 0 \\ 1 & 0 & 1 & 0 & 1 & 1 & 1 \\ 1 & 0 & 1 & 0 & 0 & 1 & 1 \\ 0 & 1 & 1 & 0 & 1 & 0 & 0 \\ 1 & 1 & 0 & 1 & 0 & 0 & 0 \end{matrix}\right) \end{array} \begin{matrix} 1 \\ 2 \\ 3 \\ 4 \\ 5 \\ 6 \end{matrix}$$

Step 1: $d_3 = \max_{j=1,\ldots,7} d_j = 4$. Hence $j_7 = 3$, $M_1 = \{1, 6\}$, and $M_2 = \{2, 3, 4, 5\}$.
Step 2: $d_5 = (3\ 0) \overset{L}{\gneqq} d_j, j \neq 3$. Hence $j_6 = 5$, $M_1 = \{1, 6\}$, $M_2 = \{4\}$, and $M_3 = \{2, 3, 5\}$.
Step 3: $d_1 = (1\quad 1\quad 1) \overset{L}{\gneqq} d_2, d_4, d_6, d_7$. Hence $j_5 = 1$, $M_1 = \{1\}$, $M_2 = \{6\}$, $M_3 = \{4\}$, $M_4 = \{2, 5\}$, and $M_5 = \{3\}$.

Continuing in this manner, we obtain the TRL matrix

$$\begin{array}{c} \begin{matrix} 4 & 2 & 6 & 7 & 1 & 5 & 3 \end{matrix} \\ \left(\begin{matrix} 0 & 1 & 0 & 1 & 0 & 0 & 0 \\ 1 & 1 & 0 & 0 & 1 & 0 & 0 \\ 0 & 0 & 1 & 1 & 1 & 0 & 1 \\ 1 & 0 & 0 & 0 & 0 & 1 & 1 \\ 0 & 1 & 0 & 0 & 0 & 1 & 1 \\ 0 & 0 & 1 & 1 & 1 & 1 & 1 \end{matrix}\right) \end{array} \begin{matrix} 1 \\ 6 \\ 4 \\ 2 \\ 5 \\ 3 \end{matrix}$$

Now we can establish the equivalence of totally balanced TRL matrices and row inclusion matrices.

Proposition 4.8. *Let A be a* (0, 1) TRL *matrix. A is totally balanced if and only if it does not contain the submatrix F.*

Proof. If A is not TB, then it contains F (see Proposition 4.5). Now suppose A contains

$$F = \begin{pmatrix} 1 & 1 \\ 1 & 0 \end{pmatrix} \begin{matrix} i_1 \\ i_2 \end{matrix} \qquad (\text{columns } j_1, j_2)$$

Consider rows i_1 and i_2, and let j_3 be the last column of A with $a_{i_1j_3} \neq a_{i_2j_3}$. Since A is TRL, we obtain $j_3 > j_2$, $a_{i_1j_3} = 0$, and $a_{i_2j_3} = 1$. Similarly, considering columns j_1 and j_2, with i_3 being the last row with $a_{i_3j_1} \neq a_{i_3j_2}$, we obtain $i_3 > i_2$, $a_{i_3j_1} = 0$, and $a_{i_3j_2} = 1$. Let

$$A_3 = \begin{pmatrix} 1 & 1 & 0 \\ 1 & 0 & 1 \\ 0 & 1 & a_{i_3j_3} \end{pmatrix} \begin{matrix} i_1 \\ i_2 \\ i_3 \end{matrix} \qquad (\text{columns } j_1, j_2, j_3)$$

If $a_{i_3j_3} = 1$, then $A_3 \in \mathcal{M}_3$ and A is not TB. If $a_{i_3j_3} = 0$, we repeat the argument using rows i_2 and i_3. So we obtain $j_4 > j_3$ with $a_{i_2j_4} = 0$, $a_{i_3j_4} = 1$, and $a_{i_2j} = a_{i_3j}$ for all $j > j_4$. Now we also observe that by the definition of j_3, it follows that $a_{i_1j_4} = a_{i_2j_4} = 0$. Similarly, from columns j_2 and j_3, we obtain $i_4 > i_3$ with $a_{i_4j_1} = a_{i_4j_2} = 0$ and $a_{i_4j_3} = 1$. Let

$$A_4 = \begin{pmatrix} 1 & 1 & 0 & 0 \\ 1 & 0 & 1 & 0 \\ 0 & 1 & 0 & 1 \\ 0 & 0 & 1 & a_{i_4j_4} \end{pmatrix} = \begin{pmatrix} & & & 0 \\ & A_3 & & 0 \\ & & & 1 \\ 0 & 0 & 1 & a_{i_4j_4} \end{pmatrix}.$$

Again, if $a_{i_4j_4} = 1$, then A is not TB; and if $a_{i_4j_4} = 0$, then i_4 and j_4 cannot be the last row and column of A.

After k steps, we get

$$A_k = \begin{pmatrix} & & & 0 \\ & A_{k-1} & & \vdots \\ & & & 1 \\ 0 & \dots & 1 & a_{i_kj_k} \end{pmatrix}.$$

But this process is finite, so for some k we have $A_k \in \mathcal{M}_k$. ■

Now we have a polynomial-time algorithm for recognizing TB matrices.

Algorithm for Recognizing TB Matrices

Step 1: Given an arbitrary (0, 1) matrix A, permute its rows and columns to obtain a TRL matrix A'. (By Proposition 4.1, A is TB if and only if A' is TB. It can be shown that the algorithm in the proof of Proposition 4.7 runs in $O(n^2m)$ time).

Step 2: Check all 2×2 submatrices of A' for the matrix F and then apply Proposition 4.8. (This takes $O(m^2n^2)$ time.)

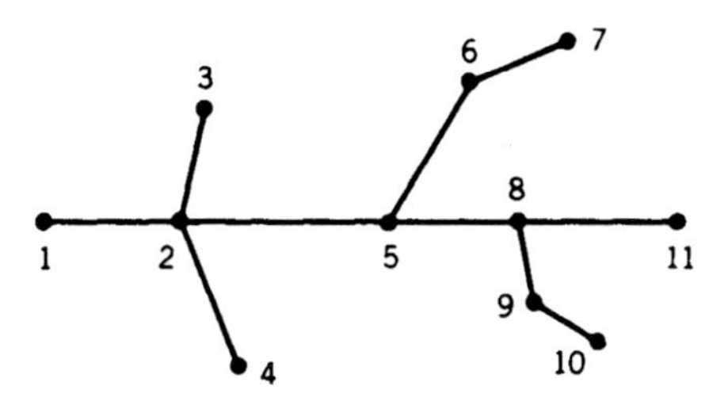

Figure 4.3

Suppose A is TB and TRL. Then, by Proposition 4.8, A is a row inclusion matrix, and FC and DFC for A can be solved for integral optimal solutions by the algorithm given on p. 566. Furthermore, since submatrices of TB matrices are also TB, it follows that FC and DFC have integral optimal solutions for all $b \in B^m$.

Theorem 4.9. *If A is a* TB *matrix, then $Q(b) = \{x \in R^n_+ : Ax \geqslant b\}$ is integral, and* DFC *has an integral optimal solution for all $b \in B^m$.*

By proceeding in exactly the same way, we obtain analogous results for FP and DFP.

Theorem 4.10. *If A is a* TB *matrix, then $P(b) = \{x \in R^n_+ : Ax \leqslant b\}$ is integral, and* DFP *has an integral optimal solution for all b, with $b_i \in \{1, \infty\}$ for all i.*

Totally balanced matrices arise in the formulation of some location problems as set-covering problems. Let $T = (V, E)$ be a tree with nonnegative weights on its edges. The weight of the unique path joining nodes i and j, denoted by d_{ij}, is the sum of the edge weights over all edges in the path. In addition, for each $j \in V$, there is an $r_j \geqslant 0$ called the *radius* of node j.

A *neighborhood subtree* of T rooted at node j is an induced subgraph $T_j = (V_j, E_j)$, where $V_j = \{i \in V: d_{ij} \leqslant r_j\}$. V_j is the set of nodes that can be served by a facility placed at node j. Let c_j be the cost of T_j.

The problem of finding a minimum-cost set of neighborhood subtrees that covers V is the set-covering problem $\min\{cx: Ax \geqslant 1, x \in B^n\}$, where $a_{ij} = 1$ if $i \in V_j$ and $a_{ij} = 0$ otherwise.

Example 4.4. We are given the tree in Figure 4.3. Let $r_j = 1$ for all $j \in V$, and let d_{ij} be the number of edges on the unique path joining nodes i and j. Then each neighborhood subtree contains a node and all of the nodes adjacent to it, and A is the node-star incidence matrix of T.

$$A = \begin{pmatrix} 1 & 1 & 0 & 0 & 0 & 0 & 0 & 0 & 0 & 0 & 0 \\ 1 & 1 & 1 & 1 & 1 & 0 & 0 & 0 & 0 & 0 & 0 \\ 0 & 1 & 1 & 0 & 0 & 0 & 0 & 0 & 0 & 0 & 0 \\ 0 & 1 & 0 & 1 & 0 & 0 & 0 & 0 & 0 & 0 & 0 \\ 0 & 1 & 0 & 0 & 1 & 1 & 0 & 1 & 0 & 0 & 0 \\ 0 & 0 & 0 & 0 & 1 & 1 & 1 & 0 & 0 & 0 & 0 \\ 0 & 0 & 0 & 0 & 0 & 1 & 1 & 0 & 0 & 0 & 0 \\ 0 & 0 & 0 & 0 & 1 & 0 & 0 & 1 & 1 & 0 & 1 \\ 0 & 0 & 0 & 0 & 0 & 0 & 0 & 1 & 1 & 1 & 0 \\ 0 & 0 & 0 & 0 & 0 & 0 & 0 & 0 & 1 & 1 & 0 \\ 0 & 0 & 0 & 0 & 0 & 0 & 0 & 1 & 0 & 0 & 1 \end{pmatrix}.$$

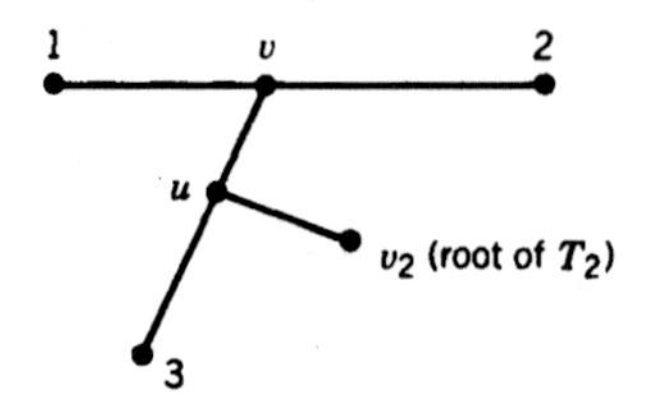

Figure 4.4

Proposition 4.11. *If A is a node by neighborhood subtree incidence matrix, then A is totally balanced.*

Proof. Suppose A is not TB. Then we can assume that for some $k \geq 3$, A contains the $k \times k$ matrix of a cycle; that is, $a_{jj} = a_{j+1,j} = 1$ for $j = 1, \ldots, k-1$, $a_{kk} = a_{1k} = 1$, and $a_{ij} = 0$ otherwise. So for $j = 1, \ldots, k-1$, T_j contains nodes j and $j+1$ but no other nodes from $\{1, \ldots, k\}$, and T_k contains nodes 1 and k but no nodes in $\{2, \ldots, k-1\}$. Let p_j be the unique path in T joining nodes j and $j+1$, and let p_k be the unique path joining nodes 1 and k.

Suppose $k = 3$. Then since T_1, T_2, and T_3 are neighborhood subtrees, nodes 1, 2, and 3 cannot lie on a common path. Hence the paths p_1, p_2, and p_3 intersect at some node v other than nodes 1, 2, or 3 (see Figure 4.4).

Now let $d_{lv} = \min_{i=1,2,3} d_{iv}$. Suppose $l = 1$. Let u be the node closest to v_2 on p_2. Hence $d_{1u} \leq \max(d_{2u}, d_{3u})$, and T_2 contains node 1. By symmetry the same argument applies if $l = 2$ or 3.

Now suppose $k > 3$. Define v to be the node closest to node 3 on the path p_1. (Now we can have $v = 1$ or 2 since T_3 contains nodes 3 and 4.) Define

$$j \doteq \begin{cases} \min\{i: v \text{ is on the path joining nodes 3 and } i+1, 3 \leq i \leq k-1\} \\ k \quad \text{otherwise.} \end{cases}$$

If $j < k$, the path from 3 to j does not contain v, and the path from 3 to $j+1$ does. Thus v is on the path p_j (see Figure 4.5a). If $j = k$, the path from 3 to k does not contain v. But since v is on the path joining nodes 1 and 3, v is on p_k (see Figure 4.5b).

In Figure 4.5a let $d_{qv} = \min_{i=1,2,j,j+1} d_{iv}$. Then T_1 and T_j contain q, which is a contradiction. Similarly, in Figure 4.5b, let $d_{qv} = \min_{i=1,2,3,k} d_{iv}$. Then T_2 and T_k contain q, which again is a contradiction. So the $k \times k$ cycle matrix cannot occur and A is totally balanced. ■

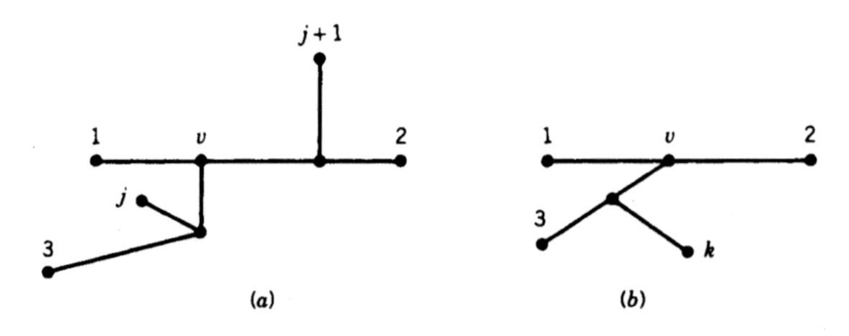

Figure 4.5

The nodes of a graph are trivial neighborhood subtrees $T_i = ((v_i), \emptyset)$ for $i = 1, \ldots, m$. Hence a node by neighborhood-subtree incidence matrix is a special case of a neighborhood-subtree by neighborhood-subtree incidence matrix. Given two families of neighborhood subtrees of a tree T, namely, $T_i = (V_i, E_i)$ for $i = 1, \ldots, m$ and $T'_j = (V'_j, E'_j)$ for $j = 1, \ldots, n$, let

$$a_{ij} = \begin{cases} 1 & \text{if } V_i \cap V'_j \neq \emptyset \\ 0 & \text{otherwise.} \end{cases}$$

By using an argument similar to the one given in the proof of Proposition 4.11, we obtain the following generalization:

Proposition 4.12 *If A is a neighborhood subtree by neighborhood subtree incidence matrix, then A is totally balanced.*

We conclude this section by mentioning some results about balanced matrices. Note that if A is not balanced, and therefore contains a submatrix $A' \in \mathcal{M}_{2k+1}$ for some $k \geq 1$, then Theorem 4.9 is false. This is an immediate consequence of the fact that the unique solution to $A'x = 1$ is $x = \frac{1}{2}(1 \ldots 1)$. Hence with $b_i = 1$ for the rows of A' and $b_i = 0$ otherwise, $Q(b)$ is not integral. The main result, which we will not prove here, is that Theorems 4.9 and 4.10 are still true when A is balanced.

Theorem 4.13. *Let A be a* (0, 1) *matrix with no zero rows or columns. The following statements are equivalent.*

1. *A is balanced.*
2. *$P(b) = \{x \in R^n_+: Ax \leq b\}$ is integral for all b with $b_i \in \{1, \infty\}$.*
3. *$Q(b) = \{x \in R^n_+: Ax \geq b\}$ is integral for all $b \in B^m$.*

In Section 5 we will study matrices A for which $P = \{x \in R^n_+: Ax \leq 1\}$ is integral but where A is not balanced. In Section 6 we will consider some matrices A for which $Q = \{x \in R^n_+: Ax \geq 1\}$ is integral but where A is not balanced.

5. NODE PACKING AND PERFECT GRAPHS

Integrality results for the fractional packing polytope $P = \{x \in R^n_+: Ax \leq 1\}$ can be generalized to a larger class of (0, 1) matrices than that considered in Section 4. These matrices are clique matrices of a family of graphs known as *perfect graphs*. Recall that in Section II.2.1 we used clique matrices in the formulation of the node-packing problem.

For completeness, some definitions are repeated here.

Definition 5.1. A *node packing* on a graph $G = (V, E)$ is a $U \subseteq V$ with the property that no pair of nodes in U is joined by an edge.

Definition 5.2. A *clique* in a graph $G = (V, E)$ is a $C \subseteq V$ with the property that every pair of nodes in C is joined by an edge.

Unless otherwise specified, when we use the term *clique* we mean a *maximal* clique.

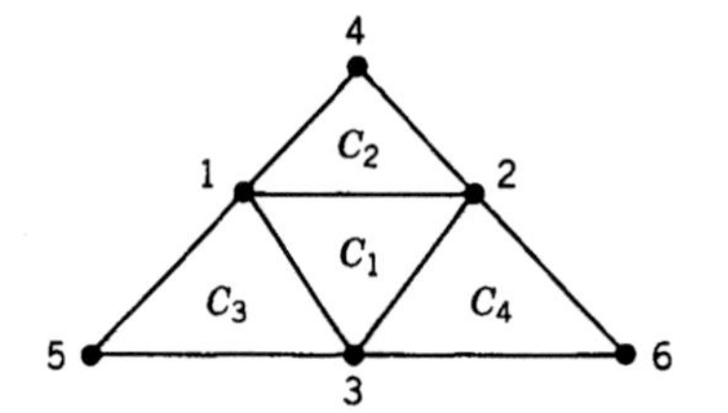

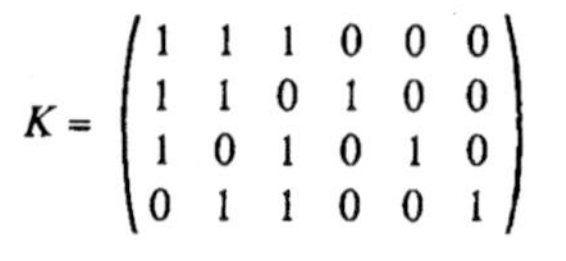

$K = \begin{pmatrix} 1 & 1 & 1 & 0 & 0 & 0 \\ 1 & 1 & 0 & 1 & 0 & 0 \\ 1 & 0 & 1 & 0 & 1 & 0 \\ 0 & 1 & 1 & 0 & 0 & 1 \end{pmatrix}$

Figure 5.1

Definition 5.3. The *clique matrix* K of a graph G is the (0, 1) incidence matrix whose rows correspond to all of the cliques of G and whose columns correspond to the nodes of G.

Definition 5.4. The *fractional node-packing polytope* of a graph G is $P = \{x \in R^n_+: Kx \leqslant 1\}$, where $n = |V|$.

Definition 5.5. A graph G is *perfect* if its fractional node-packing polytope is integral.

This polyhedral definition of a perfect graph is not the standard definition. Later in this section, we will show that it is equivalent to the standard definition, which is given purely in graphical terms. Originally, graphs that satisfied Definition 5.5 were called *pluperfect*.

Example 5.1. A graph and its clique matrix are shown in Figure 5.1.

Matrix K is not balanced. Nevertheless, G is perfect since P is integral. The reader can check that its only extreme points are $x = 0$, $x = e_j$ for $j = 1, \dots, 6$, $x = e_j + e_k$ with $j + k = 7$, and $x = (0 \ \ 0 \ \ 0 \ \ 1 \ \ 1 \ \ 1)$. This is not inconsistent with Theorems 4.10 or 4.13 since with $b_1 = \infty$ and $b_i = 1$ otherwise, $P(b)$ contains the extreme point $\frac{1}{2}(1 \ \ 1 \ \ 1 \ \ 0 \ \ 0 \ \ 0)$, which is also an extreme point of $Q = \{x \in R^6_+: Kx \geqslant 1\}$.

Before studying some classes of perfect graphs, we explain why it suffices to consider clique matrices. In particular, we will show that if A is the incidence matrix of clutter that is not a clique matrix, then $P = \{x \in R^n_+: Ax \leqslant 1\}$ is not integral.

Proposition 5.1. *Let A be the $m \times n$ incidence matrix of a clutter. The following statements are equivalent.*

1. *A is a clique matrix.*
2. *If A contains a $p \times p$ submatrix A' where $p \geqslant 3$ and all of the row and column sums of A' equal $p - 1$, then A' is contained in a $(p + 1) \times p$ submatrix that contains a row of all 1's.*

Proof. $1 \to 2$. Let $G(A)$ be the intersection graph of A and, without loss of generality, let A' be the submatrix of A consisting of the first p rows and columns of A. If statement 2 is false, then the sets $\{1, \dots, p\} \setminus \{i\}$ are contained in cliques for $i = 1, \dots, p$, but no clique contains $\{1, \dots, p\}$. This is impossible for a clique matrix.

$2 \to 1$. Let $N_i = \{j: a_{ij} = 1\}$ for $i = 1, \dots, m$. If statement 1 is false, there exists a minimal $C \subseteq V$ with $|C| \geqslant 3$ such that the subgraph of $G(A)$ induced by C is complete, and there is no i such that $N_i \supseteq C$. Since C is minimal, for each $j \in C$, there is a distinct $i(j)$ such that $N_{i(j)} \cap C = C \setminus \{j\}$. Hence A contains a $k \times k$ submatrix A', all of whose row and column sums equal $k - 1$, but there is no $(k + 1) \times k$ submatrix that contains A' and has a k-vector of ones. Thus statement 2 is false. ■

Proposition 5.2. *Let A be an $m \times n$ incidence matrix of a clutter. If A is not a clique matrix, then $P = \{x \in R^n_+ : Ax \leqslant 1\}$ is not integral.*

Proof. A contains the submatrix A' of Proposition 5.1. Suppose the columns of A' are indexed $1, \ldots, p$. Then it is easy to see that $x_j = 1/(p - 1)$ for $j = 1, \ldots, p$, and $x_j = 0$ otherwise is an extreme point of P. ■

We now consider two well-known classes of perfect graphs and a necessary condition for a graph to be perfect.

Proposition 5.3. *Bipartite graphs are perfect.*

Proof. The cliques of a bipartite graph are its edges. Hence K is the edge-node incidence matrix of the graph. Thus, by Corollary 2.9, K is totally unimodular and, by Proposition 2.2, P is integral. ■

Definition 5.6 A *chord* of a cycle is an edge joining two nodes of the cycle that are not adjacent on the cycle. A graph with k nodes, $k \geqslant 4$, corresponding to a cycle without chords is called a *k-hole*. A graph that is the complement of a k-hole is called a *k-antihole*. A hole or antihole is *odd* (*even*) if k is odd (even).

A 5-hole and a 7-antihole are shown in Figure 5.2.

Proposition 5.4. *If a graph G contains a node-induced subgraph that is an odd hole or an odd antihole, then G is not perfect.*

Proof. Suppose G contains an odd hole on nodes $\{1, \ldots, 2k + 1\}$. Then we can write K as

$$K = \begin{pmatrix} K_1 & K_2 \\ K_3 & K_4 \end{pmatrix},$$

where K_1 is the edge-node incidence matrix of the odd hole, and K_3 contains at most two positive elements in each row. Hence $x_j = \frac{1}{2}$ for $j = 1, \ldots, 2k + 1$, and $x_j = 0$ otherwise is an extreme point of P.

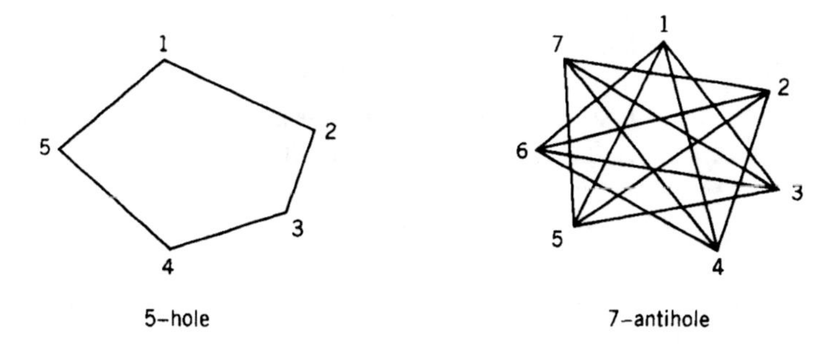

Figure 5.2

On the other hand, if G contains an odd antihole on nodes $\{1, \ldots, 2k+1\}$, then each maximum clique of the subgraph is of size k. Hence $x_j = 1/k$ for $j = 1, \ldots, 2k+1$, and $x_j = 0$ otherwise is an extreme point of P. ■

Many classes of graphs without induced odd holes or antiholes are known to be perfect. However, the converse of Proposition 5.4 is unresolved. It is known as the *perfect graph conjecture* and is considered to be one of the most challenging and difficult problems in graph theory and polyhedral combinatorics.

We now consider a fundamental class of perfect graphs.

Definition 5.7. A graph is called *chordal* if it does not contain any k-holes for $k \geqslant 4$.

Proposition 5.5. *If matrix A is a totally balanced incidence matrix of a clutter, then it is the clique matrix of a chordal graph.*

Proof. If A is not a clique matrix, then by Proposition 5.2 it follows that $P = \{x \in R_+^n: Ax \leqslant 1\}$ is not integral. Hence by Theorem 4.10, A is not TB. If A is a clique matrix of a graph that is not chordal, then by Definition 5.7 it follows that A contains a member of $\mathcal{M}_k$ for some $k \geqslant 4$. Hence A is not TB. ■

The chordal graph of Figure 5.1 shows that the converse is false.

There is a nice characterization of chordal graphs that yields an efficient recognition algorithm as well as an algorithm that gives an integer solution (node packing) to the linear programming problem over the fractional node-packing polytope. Let $N(v) = \{u \in V: (u, v) \in E\}$ be the nodes adjacent to v—that is, the set of *neighbors* of v.

Definition 5.8. A node v of $G = (V, E)$ is called *simplicial* if it and its neighbors form a clique.

In the graph of Figure 5.1, nodes 4, 5, and 6 are simplicial, but nodes 1, 2, and 3 are not.

Definition 5.9. An ordering of V, $\sigma = [v_1, v_2, \ldots, v_n]$ is a *perfect elimination scheme* (PES) if, for $i = 1, \ldots, n-1$, v_i is a simplicial node of the subgraph induced by $\{v_i, \ldots, v_n\}$.

In the graph of Figure 5.1, $[6 \quad 5 \quad 4 \quad 3 \quad 2 \quad 1]$ is a PES.

It is easy to see that the existence of a PES implies that a graph is chordal. For graph G let v be the first node of a PES that is contained in a k-hole with $k \geqslant 4$. If v does not exist, G is chordal. Otherwise, v and the two nodes u and w adjacent to v on the cycle are contained in a clique. Hence $(u, w) \in E$ so that the cycle contains a chord, and G is chordal.

Moreover, it can be shown that every chordal graph has a PES.

Theorem 5.6. *A graph is chordal if and only if it contains a* PES.

By considering the nodes sequentially, a PES can be constructed in polynomial time or can be shown not to exist.

Now suppose that $[1 \quad 2 \quad \ldots \quad n]$ is a PES for G. Let $\tilde{k}^i$ be the characteristic vector of the clique containing node i in the subgraph induced by the nodes $\{i, \ldots, n\}$. Then the $n \times n$ matrix $\tilde{K}$ whose rows are $\tilde{k}^1, (0, \tilde{k}^2), \ldots, e_n$ contains the incidence vectors of all of the (maximal) cliques of G. $\tilde{K}$ can, of course, also contain dominated rows corresponding to nonmaximal cliques. However, it is convenient to work with $\tilde{K}$ since $\tilde{k}_{ii} = 1$ for $i = 1, \ldots, n$ and $\tilde{k}_{ij} = 0$ for $j < i$. Let $S_1, \ldots, S_m$ be any partition of $\{1, \ldots, n\}$ with the property that if $l \in S_i$, then $\tilde{k}_l \leqslant k_i$.

For any $c \in Z^n_+$, we can write the fractional node-packing problem as

$$\text{(FNP)} \qquad \max\{cx: Kx \leqslant 1, x \in R^n_+\} = \max\{cx: \tilde{K}x \leqslant 1, x \in R^n_+\}$$

and its dual as

$$\text{(DFNP)} \qquad \min\{1u: u\tilde{K} \geqslant c, u \in R^n_+\} = \min\{1y: yK \geqslant c, y \in R^m_+\}.$$

We obtain a feasible y from a feasible u with $1u = 1y$ by $y_i = \Sigma_{l \in S_i} u_l$ for $i = 1, \ldots, m$.

We now give a greedy algorithm that finds integral optimal solutions to DFNP and FNP for chordal graphs. The algorithm is very similar to the greedy algorithm given in the previous section for the fractional covering problem with totally balanced matrices.

Algorithm for DFNP and FNP for Chordal Graphs

DFNP:

Initialization: $i = 1, c^1 = c, J_0 = \varnothing$ $[1\,2 \ldots n]$ is a PES.

Iteration i: $u_i = \max(0, c^i_i)$. If $u_i > 0$, let $J_i = J_{i-1} \cup \{i\}$ and $c^{i+1} = c^i - u_i\tilde{k}^i$. If $u_i = 0$, then $J_i = J_{i-1}$ and $c^{i+1} = c^i$. If $i = n$, stop; $(u_1, \ldots, u_n)$ is an optimal solution. Otherwise, $i \leftarrow i + 1$.

FNP:

Initialization: $J = J_n$ (from DFNP), $x_j = 0$ for $j \notin J$.

Iteration: Let l be the last element of J. Set $x_l = 1$ and $J \leftarrow J \setminus \{i: \tilde{k}_{il} = 1\}$. If $J = \varnothing$, stop; $x = (x_1, \ldots, x_n)$ is an optimal solution. Otherwise repeat.

Proposition 5.7. *The algorithm gives integral optimal solutions to* DFNP *and* FNP.

Proof. By construction, the solutions are integral and the dual solution is feasible. Suppose that the ith constraint of FNP (with respect to the matrix $\tilde{K}$) is violated. Then there exists $j_1 \neq j_2$ with $\tilde{k}_{ij_1}x_{j_1} = \tilde{k}_{ij_2}x_{j_2} = 1$. Suppose j_2 follows j_1 in the PES. Then $i = j_1$ or i precedes j_1 in the PES. Since j_1 was not deleted from J when we set $x_{j_2} = 1$, it follows that $\tilde{k}_{j_1j_2} = 0$. But this contradicts the assumption that i is a simplicial node of the subgraph induced by $\{i, \ldots, j_1, \ldots, j_2, \ldots n\}$.

To complete the proof, we show that complementary slackness is satisfied. By construction, we have $\Sigma_{i=1}^{j-1} \tilde{k}_{ij}u_i > c_j$ only if $c^j_j < 0$. But then $J_j = J_{j-1}$. Hence $j \notin J$ and $x_j = 0$. Now suppose $u_i > 0$, $i \in J$. By the construction of the primal solution, either $x_i = 1$ or there exists an l that comes after i in the PES with $\tilde{k}_{il}x_l = 1$. Hence, $\Sigma_{j=1}^n \tilde{k}_{ij}x_j = 1$. ■

Corollary 5.8. *Chordal graphs are perfect.*

Example 5.1. (continued). We use the PES [4 5 6 1 2 3]. Then

$$\tilde{K} = \begin{array}{c} \begin{matrix} 4 & 5 & 6 & 1 & 2 & 3 \end{matrix} \\ \begin{pmatrix} 1 & 0 & 0 & 1 & 1 & 0 \\ 0 & 1 & 0 & 1 & 0 & 1 \\ 0 & 0 & 1 & 0 & 1 & 1 \\ 0 & 0 & 0 & 1 & 1 & 1 \\ 0 & 0 & 0 & 0 & 1 & 1 \\ 0 & 0 & 0 & 0 & 0 & 1 \end{pmatrix} \end{array}.$$

Let $c = (3 \quad 1 \quad 2 \quad 4 \quad 6 \quad 3) = c^1$.

DFNP:

1. $u_1 = 3, c^2 = (0 \quad 1 \quad 2 \quad 1 \quad 3 \quad 3), J_1 = \{4\}$.
2. $u_2 = 1, c^3 = (0 \quad 0 \quad 2 \quad 0 \quad 3 \quad 2), J_2 = \{4, 5\}$.
3. $u_3 = 2, c^4 = (0 \quad 0 \quad 0 \quad 0 \quad 1 \quad 0), J_3 = \{4, 5, 6\}$.
4. $u_4 = 0, c^5 = c^4, J_4 = J_3$.
5. $u_5 = 1, c^6 = (0 \quad 0 \quad 0 \quad 0 \quad 0 \quad -1), J_5 = \{4, 5, 6, 2\}$.
6. $u_6 = 0, J_6 = J_5$.

FNP:

1. $J = \{4, 5, 6, 2\}, x_2 = 1$.
2. $J = \{5\}, x_5 = 1$.

An optimal solution to FNP is $x = (0 \quad 1 \quad 0 \quad 0 \quad 1 \quad 0)$. Let $y_i = u_i$ for $i = 1, 2, 3$ and let $y_4 = u_4 + u_5 + u_6$. An optimal solution to DFNP is $y = (3 \quad 1 \quad 2 \quad 1)$.

We now give some general properties of perfect graphs and the corresponding polytopes.

Let $\overline{E} = \{e: e \notin E\}$; then $\overline{G} = (V, \overline{E})$ is the complement of G. Let $\overline{K}$ be the clique matrix of $\overline{G}$. Perfect graphs always come in pairs because:

Proposition 5.9. *G is perfect if and only if $\overline{G}$ is perfect.*

Proof. If G is perfect, then by definition it follows that $P = \{x \in R^n_+: Kx \leq 1\}$ is an integral polyhedron whose extreme points are the incidence vectors of node packings of G. Since there is a one-to-one correspondence between cliques of $\overline{G}$ and maximal node packings of G, the maximal extreme points of P are the rows of $\overline{K}$.

Now by antiblocking polarity—in particular, Proposition 5.8 of Section I.4.5—$\overline{P} = \{x \in R^n_+: \overline{K}x \leq 1\}$ is an integral polyhedron. Hence $\overline{G}$ is perfect.

The converse follows trivially since G is the complement of $\overline{G}$. ∎

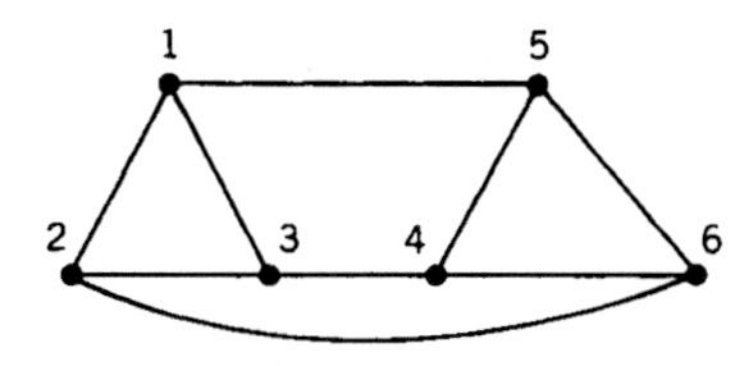

Figure 5.3

Example 5.2. In the graph of Figure 5.3, $V_1 = \{1, 2, 3\}$ and $V_2 = \{4, 5, 6\}$ are cliques. Hence this graph is the complement of a bipartite graph. By Proposition 5.3 and Proposition 5.9, it is perfect.

The fractional node-packing polytope on the subgraph induced by $V \setminus U$ is $P \cap \{x \in R^n_+: x_j = 0 \text{ for } j \in U\}$, which is a face of P and therefore is integral if P is integral. Hence we have the following proposition:

Proposition 5.10. *Node-induced subgraphs of perfect graphs are perfect.*

Now consider $P(b) = \{x \in R^n_+: Kx \leqslant b\}$ with $b \in B^m$. $P(b)$ is the face of P with $x_j = 0$ if there is an i with $b_i = 0$ and $k_{ij} = 1$. Hence if $P = P(1)$ is integral, then $P(b)$ is integral for all $b \in B^m$.

Let $\alpha(G)$ be the *size of a maximum cardinality node packing* in G, and let $\theta(G)$ be the *minimum number of cliques required to cover all the nodes* of G. For any graph G with n nodes and m cliques, by relaxation and duality we have

$$\begin{aligned}\alpha(G) &\leqslant \max\{1x: Kx \leqslant 1, x \in R^n_+\}\\ &= \min\{1y: yK \geqslant 1, y \in R^m_+\} \leqslant \theta(G).\end{aligned}$$

For perfect graphs, the first inequality is an equality; for an odd hole on $2k + 1$ nodes, however, $\alpha(G) = k$, $\theta(G) = k + 1$, and the linear programming relaxations have value $k + \frac{1}{2}$.

To generalize α and θ for graphs with node weights, let

$$\begin{aligned}\alpha(G, c) &= \max\{cx: Kx \leqslant 1, x \in B^n\}\\ z(G, c) &= \max\{cx: Kx \leqslant 1, x \in R^n_+\}\\ &= \min\{1y: yK \geqslant c, y \in R^m_+\}\\ \theta(G, c) &= \min\{1y: yK \geqslant c, y \in Z^m_+\}.\end{aligned}$$

Hence $\alpha(G, 1) = \alpha(G)$, $\theta(G, 1) = \theta(G)$, and for any $c \in B^n$ with $U = \{j \in V: c_j = 1\}$ we have $\alpha(G, c) = \alpha(H)$ and $\theta(G, c) = \theta(H)$, where H is the subgraph of G induced by U. By duality and relaxation, we obtain

$$\alpha(G, c) \leqslant z(G, c) \leqslant \theta(G, c).$$

For perfect graphs, we know that the first inequality is an equality for all c and, in particular, for $c \in B^n$. The following result, which is the fundamental theorem of this section, establishes the second equality for $c \in B^n$ and also yields some interesting corollaries.

Theorem 5.11. *The following statements are equivalent.*

1. $P = \{x \in R^n_+: Kx \leqslant 1\}$ *is integral. (G is perfect.)*
2. $\alpha(H) = \theta(H)$ *for all node-induced subgraphs H of G.*

Proof. $1 \Rightarrow 2$. Given that G is perfect, by Proposition 5.10 every subgraph H of G is perfect. Hence if H is the subgraph induced by U and if $c_j = 1$ for $j \in U$ and $c_j = 0$ otherwise, we obtain

$$\text{(FNP)} \qquad \alpha(H) = \max\{cx: Kx \leqslant 1, x \in R^n_+\}.$$

Thus, by linear programming duality, we need to prove that

$$\text{(DFNP)} \qquad \min\{1y: yK \geqslant c, y \in R^m_+\}$$

has an integral optimal solution for all $c \in B^n$.

The proof is by induction on the number of positive components of c, which equals $|U|$. Note that if $c = 0$, then $y = 0$ is an optimal solution to DFNP. Now it suffices to assume the hypothesis for all proper subgraphs of G and to prove that $\min\{1y: yK \geqslant 1, y \in R^m_+\}$ has an integral optimal solution.

Let the rows of K be k^i for $i = 1, \ldots, m$. Since $y = 0$ is not feasible to DFNP, there is an i, say $i = r$, such that $y_r > 0$ in an optimal dual solution. Hence by complementary slackness, $k^r x = 1$ for every optimal solution to FNP.

If $k^r = 1$, then $r = 1$ and an optimal solution to DFNP is $y_1 = 1$. If $k^r < 1$, by the induction hypothesis, $\min\{1y: yK \geqslant 1 - k^r, y \in R^m_+\}$ has an integral optimal solution, say y^0.

Claim 1: $y^0 + e_r$ is an optimal solution to DFNP. Note it is feasible since $(y^0 + e_r)K \geqslant 1 - k^r + k^r = 1$. Because G is perfect, any maximum cardinality node packing on G is an optimal solution to FNP, and hence, by the definition of r, every maximum cardinality node packing on G contains a node in the clique $C = \{j \in V: k_{rj} = 1\}$. So for the subgraph H induced by $V \setminus C$, a maximum cardinality packing is obtained by deleting a node from C from a maximum cardinality packing on G. Hence $\alpha(H) = \alpha(G) - 1$, and

$$1y^0 + 1 = \theta(H) + 1 = \alpha(H) + 1 = \alpha(G) \leqslant \theta(G) \leqslant 1(y^0 + e_r) = 1y^0 + 1.$$

So $y^0 + e_r$ is an integral optimal solution to DFNP.

$2 \Rightarrow 1$. We will prove that $\alpha(G, c) = z(G, c)$ for all $c \in Z^n_+$. Statement 2 says that $\alpha(G, c) = z(G, c)$ for all $c \in B^n$. Now for any $c \in Z^n \setminus Z^n_+$ let $\bar{c}_j = c_j$ if $c_j \geqslant 1$ and let $\bar{c}_j = 0$ otherwise. Since $c_j < 0$ implies $x_j = 0$ in both the fractional and integer node-packing problems, $\alpha(G, \bar{c}) = \alpha(G, c)$ and $z(G, \bar{c}) = z(G, c)$. Hence from statement 2 we have $\alpha(G, c) = z(G, c)$ for all $c \in Z^n$ with $c \leqslant 1$.

The proof for $c \in Z^n_+$ is by induction with the hypothesis $\alpha(G, c') = z(G, c')$ for all $c' < c$. Consider $c \in Z^n_+$ with $c_j \geqslant 2$ for some j. Let $c' = c - e_j$. Since $c'_j \geqslant 1$, it follows from complementary slackness that there is an r such that $k_{rj} = 1$ and $k^r x = 1$ in every optimal solution to $\max\{c'x: x \in P\}$. Let $\tilde{c} = c - k^r$. Hence by the induction hypothesis, we have $\alpha(G, \tilde{c}) = z(G, \tilde{c})$.

Claim 2: $\alpha(G, c) = \alpha(G, \tilde{c}) + 1$. Since $c > \tilde{c}$, we have $\alpha(G, c) \geqslant \alpha(G, \tilde{c})$.

Let $\tilde{y}$ be an optimal solution to

$$z(G, \tilde{c}) = \min\{1y: yK \geqslant \tilde{c}, y \in R^m_+\}.$$

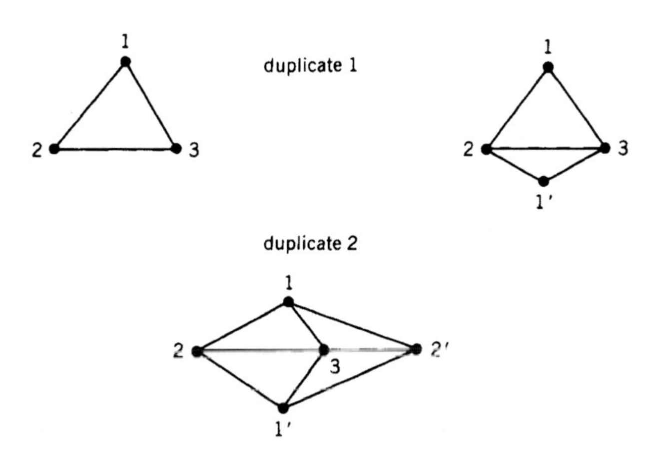

Figure 5.4

Since $(\tilde{y} + e_r)K \geqslant \tilde{c} + k^r = c$, we have $z(G, c) \leqslant z(G, \tilde{c}) + 1$. Since $\alpha(G, c) \leqslant z(G, c)$ and $\alpha(G, \tilde{c}) = z(G, \tilde{c})$, we have

$$\alpha(G, \tilde{c}) \leqslant \alpha(G, c) \leqslant \alpha(G, \tilde{c}) + 1.$$

Finally, since the α's are integers, $\alpha(G, c) = \alpha(G, \tilde{c})$ or $\alpha(G, c) = \alpha(G, \tilde{c}) + 1$.

Suppose $\alpha(G, c) = \alpha(G, \tilde{c})$ and let $\tilde{x}$ be the characteristic vector of any node packing with $\tilde{c}\tilde{x} = \alpha(G, \tilde{c})$. Then $k^r\tilde{x} = 0$ and $c'\tilde{x} = \alpha(G, c')$ since $\alpha(G, \tilde{c}) \leqslant \alpha(G, c') \leqslant \alpha(G, c)$. This is a contradiction because we have already shown that $k^r x = 1$ for any node packing x with $c'x = \alpha(G, c')$. Hence $\alpha(G, c) = \alpha(G, \tilde{c}) + 1$.

Now we have

$$\alpha(G, c) \leqslant z(G, c) \leqslant z(G, \tilde{c}) + 1 = \alpha(G, \tilde{c}) + 1 = \alpha(G, c).$$

Hence $z(G, c) = \alpha(G, c)$, and the theorem is proved. ■

The standard definition of a perfect graph is statement 2 of Theorem 5.11. In the proof of 2 ⇒ 1 we first used the trivial implication 2 ⇒ 3, where

3. $\max\{cx: x \in P\}$ has an integral optimal solution for all $c \in B^n$,

and thus we proved 3 ⇒ 1. Hence the real content of 2 ⇒ 1 is the following result.

Corollary 5.12. *$P = \{x \in R^n_+: Kx \leqslant 1\}$ is integral if $\max\{cx: x \in P\}$ has an integral optimal solution for all $c \in B^n$.*

Corollary 5.12 is rather surprising since, in general, we need integral optimal solutions for all $c \in Z^n$ to conclude that a polytope is integral.

There is another interesting interpretation of this result, which involves duplicating the nodes of a graph. By duplicating a node v of a graph G, we mean that a new node v' is added to G and that v' is joined to all of the neighbors of v but not to v (see Figure 5.4).

It is easy to see that if G' is the graph obtained by duplicating node j $c_j - 1$ times in the graph G, then $\alpha(G')$ is the weight of a maximum-weight node packing in G with weight c_j on node j. Hence if statement 2 holds for G and all of the graphs obtained from G by duplicating nodes, then statement 1 is true. Thus Corollary 5.12 can also be interpreted in graphical terms.

Corollary 5.13. *If G is a perfect graph and G' is obtained from G by duplicating nodes, then G' is perfect.*

In polyhedral terms, since G' is perfect if and only if $\alpha(H') = \theta(H')$ for all subgraphs H' of G', we have that if G is perfect, then $\alpha(G, c) = \theta(G, c)$ for all $c \in Z^n$. In other words:

Corollary 5.14. *If G is a perfect graph, the linear system $Kx \leqslant 1, x \geqslant 0$ is* TDI.

Yet another corollary to Theorem 5.11 is obtained from Proposition 5.9. Let $\omega(G)$ be the size of a maximum cardinality clique of G. Since cliques in G correspond to maximal node packings in $\overline{G}$ and conversely, we have

$$\omega(G) = \alpha(\overline{G}) \quad \text{and} \quad \omega(\overline{G}) = \alpha(G).$$

Also define the *chromatic number* of G, denoted by $\gamma(G)$, to be the minimum number of colors required to color the nodes of G so that no adjacent nodes have the same color. The celebrated four-color theorem says that every planar graph (a graph that can be drawn in the plane without crossing edges) has $\gamma(G) \leqslant 4$. The complete graph on 4 nodes is a planar graph with $\gamma(G) = 4$. A minimum cardinality node coloring for the graph of Figure 5.1 is shown in Figure 5.5.

Note that in any feasible coloring, all of the nodes of the same color form a node packing. Thus $\gamma(G)$ is the minimum number of node packings needed to cover all of the nodes. Hence we have

$$\gamma(G) = \theta(\overline{G}) \quad \text{and} \quad \gamma(\overline{G}) = \theta(G).$$

Now from Proposition 5.9 and Theorem 5.11, we immediately obtain the following theorem:

Theorem 5.15. *The following statements are equivalent.*

1. *$\alpha(H) = \theta(H)$ for all node-induced subgraphs H of G.*
2. *$\omega(H) = \gamma(H)$ for all node-induced subgraphs H of G.*

Theorem 5.15 is known as the *perfect graph theorem*.

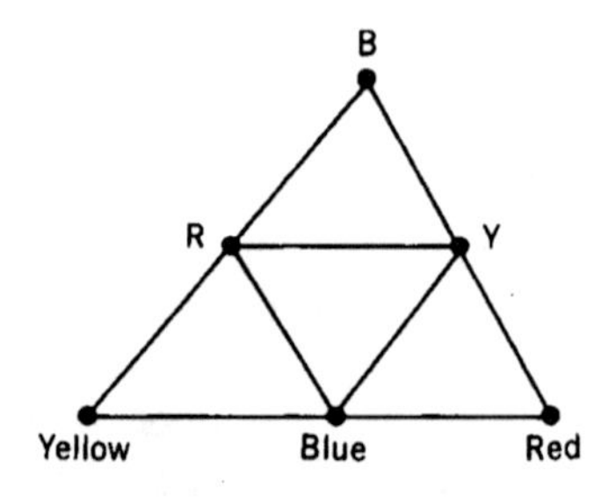

Figure 5.5. $\omega(G) = \gamma(G) = 3$.

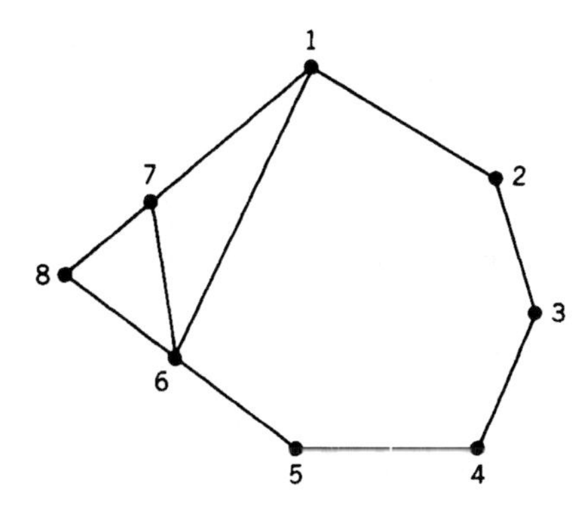

Figure 5.6 G is perfect by Theorem 5.17.

It would take us too far from the subject matter of this book to study additional classes of perfect graphs. The following two theorems, given without proofs, illustrate some of the progress that has been made in identifying classes of perfect graphs. They do not, however, give the most general results.

Theorem 5.16. *G is perfect if each odd cycle of length at least 5 contains at least two chords.*

Theorem 5.17. *G is perfect if for each odd cycle there is an edge* (i, j) *of the cycle with the property that every clique that contains i and j also contains another node of the cycle.*

Theorem 5.17 is illustrated in Figure 5.6. Edge (1, 7) satisfies the hypothesis of the theorem, since the only clique that contains (1, 7) is $C = \{1, 6, 7\}$. However, the edge (6, 7) does not satisfy the hypothesis since it is contained in the clique {6, 7, 8}.

Theorem 5.17 is an immediate corollary to Theorem 4.13, since the class of graphs defined in Theorems 5.17 is balanced. This follows since the hypothesis of the theorem forbids a submatrix in $\mathcal{M}_k$ for $k \geqslant 3$ and odd by requiring that any $(2k + 1) \times (2k + 1)$ submatrix with all row and column sums at least 2 has at least one row with row sum at least 3.

Although no characterization of perfect graphs is known in graphical terms, there is an important result which characterizes (0, 1) matrices that are clique matrices of perfect graphs in terms of forbidden submatrices. We will not prove this theorem.

Theorem 5.18. *Let A be the* $m \times n$ *incidence matrix of a clutter. The following statements are equivalent.*

1. *A is the clique matrix of a perfect graph.*
2. *If A contains a* $p \times p$ *nonsingular submatrix* A' *whose row and column sums are all equal to* β, $2 \leqslant \beta \leqslant \lfloor n/2 \rfloor$, then there is a $(p + 1) \times p$ *submatrix that contains* A' *and also contains a row with row sum greater than* β *or a row with row sum* β *that is not equal to any row of* A'.

The implication $1 \Rightarrow 2$ is easy to prove since if statement 2 is false we obtain a fractional extreme point with $x_j = \beta^{-1}$ for each column of A', and $x_j = 0$ otherwise.

If statement 2 is false, then $\beta = p - 1$ implies that A is not a clique matrix (see Proposition 5.1), and $\beta = 2$ or $\lfloor p/2 \rfloor$ for p odd implies that the graph contains an odd hole or an odd antihole (see Proposition 5.4).

Thus, one approach to the perfect graph conjecture is to consider *minimal imperfect graphs*—that is, graphs that are imperfect but all of whose node-induced subgraphs are perfect. If the perfect graph conjecture were true, Theorem 5.18 says that for a minimally imperfect graph, statement 2 must have p odd and $\beta = 2$ or $\lfloor p/2 \rfloor$.

For some classes of graphs, the perfect graph conjecture is known to be true. For example, the planar graphs without odd holes and odd antiholes are perfect.

Theorem 5.18 also establishes that the recognition problem for imperfect graphs is in $\mathcal{NP}$. This follows since (a) the clique matrix of an imperfect graph must have a $(p + 1) \times p$ submatrix for which statement 2 is false and (b) such a matrix can be validated in polynomial time. However, it is not known whether the recognition problem for perfect graphs is in $\mathcal{NP}$.

In addition, it is not known whether recognizing graphs that contain no odd hole or antihole is in $\mathcal{NP}$. Obviously these two recognition problems are equivalent if the perfect graph conjecture holds.

We close this section with a brief discussion of algorithms for solving node-packing problems. For general graphs, the maximum cardinality node-packing is $\mathcal{NP}$-hard (see Section I.5.6), and even the maximum-weight fractional node-packing problem is $\mathcal{NP}$-hard (see Section I.6.3). However, strong fractional cutting-plane algorithms (which use heuristics to find violated clique and other inequalities, and good feasible solutions) are quite successful in solving a variety of instances.

For general perfect graphs, there is an ellipsoid algorithm that solves the maximum-weight node-packing problem in polynomial time. However, the fractional node-packing polytope is not the basis of the algorithm since the separation problem for clique inequalities is another weighted node-packing problem on a perfect graph. Instead, the algorithm uses a convex constraint set which, for a general graph, is contained in the fractional node-packing polytope and contains the convex hull of node packings. Hence for perfect graphs it coincides with the convex hull of node packings. The separation problem for this convex constraint set is solvable in polynomial time. But it is necessary to use a generalization of the ellipsoid algorithm to accommodate the nonlinear constraints.

For some classes of perfect graphs, efficient combinatorial algorithms are known for the recognition problem and for solving the maximum-weight node-packing problem. We have already solved these problems for bipartite and chordal graphs. More generally, for the perfect graphs given in Theorem 5.16, the recognition problem and the maximum-weight node-packing problem can be solved in polynomial time.

Efficient node-packing algorithms are not restricted to perfect graphs.

Definition 5.10. A *line graph* $L(G)$ of a graph G is obtained by replacing each edge of G by a node and joining two nodes by an edge if the two edges in G are incident to a common node (see Figure 5.7).

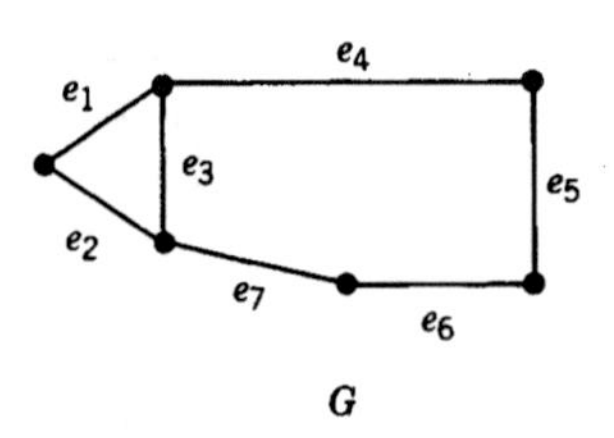

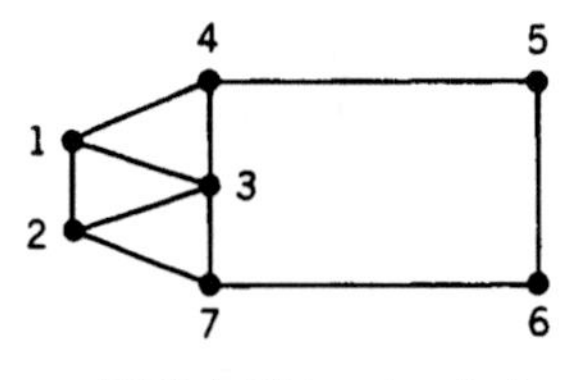

Figure 5.7

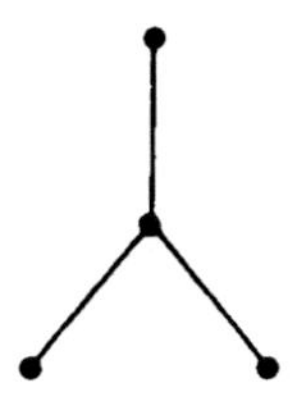

Figure 5.8

It is easy to see that a subset of nodes in $L(G)$ is a packing if and only if the corresponding set of edges in G is a matching. Hence for line graphs, the maximum-weight node-packing problem in $L(G)$ is equivalent to a maximum-weight matching problem in G (see Chapter III.2).

The graph in Figure 5.8 is called a *claw*. A graph is called *claw-free* if it does not contain a claw as a node-induced subgraph. By drawing a few pictures, the reader can establish that line graphs are claw-free, but the converse is false.

It is easy to see that claw-free graphs need not be perfect since a 5-hole is claw-free. An interesting property of claw-free graphs is illustrated in Figure 5.9. The black nodes of the graph are a node packing. Nodes $\{1, 2, 3, 4, 5\}$ induce a path whose nodes alternate between white and black and whose end nodes are white. By interchanging the colors of the nodes on this path, we increase the cardinality of the packing.

This means of increasing the size of a packing works for claw-free graphs because if there were any edges between the nodes $\{1, 3, 5\}$ or between one of these nodes and a black node not on the path, the graph would contain a claw. This approach leads to an efficient algorithm for solving the maximum-weight node-packing problem on claw-free graphs. The algorithm is closely related to the matching algorithm discussed in Section III.2.3.

Claw-free graphs without odd holes and odd antiholes are perfect; that is, the perfect graph conjecture is true for these graphs. However, no description of the convex hull of node packings is known for claw-free graphs.

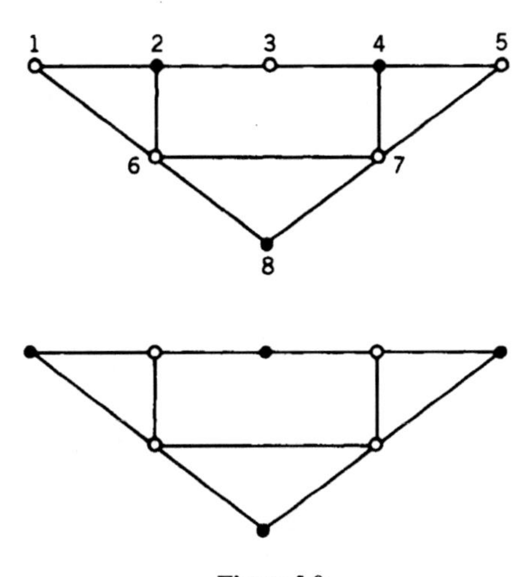

Figure 5.9

6. BLOCKING AND INTEGRAL POLYHEDRA

In the previous sections of this chapter, we have considered the following types of questions: (1) Given a family of polyhedra of the form $P = \{x \in R^n_+ : Ax \leqslant b\}$, under what conditions on (A, b) will P be integral? (2) When does the dual linear program $\min\{yb: yA \geqslant c, y \in R^m_+\}$ have an integral optimal solution? In particular, in the last section we completely characterized when $P = \{x \in R^n_+ : Ax \leqslant 1\}$ is integral when A is a 0-1 matrix. Here we consider the question of when $Q = \{x \in R^n_+ : Ax \geqslant 1\}$ is integral, and when the corresponding inequality system is TDI. However, there is no nice characterization known, so we start from a different point of view. Given a finite set $N = \{1, \ldots, n\}$ and a set $\mathcal{F}$ of subsets of N, we consider the problem

$$\min\{w(F): F \in \mathcal{F}\}, \tag{6.1}$$

where $w \in R^n_+$ is a weight function on the elements of N and $w(F) = \Sigma_{j \in F} w_j$.

We consider two questions:

a. How can we formulate (6.1) as an integer program?
b. How can we formulate (6.1) as a linear program?

We will formulate (6.1) as an integer program of the form

$$\min(wx: Ax \geqslant 1, x \in B^n\}$$

where A is a 0-1 matrix, and then we will ask when the polyhedron $Q = \{x \in R^n_+ : Ax \geqslant 1\}$ is integral.

If Q is not integral, then to formulate (6.1) as a linear program

$$\min\{wx: x \in Q^*\} \tag{6.2}$$

we will describe the polyhedron Q^* whose extreme points are the characteristic vectors x^F for $F \in \mathcal{F}$, and such that $\min\{wx: x \in Q^*\}$ is unbounded if and only if $w \in R^n \setminus R^n_+$.

Many familiar examples of (6.1) are associated with graphs. Let $G = (V, E)$ be a complete graph, let $N = E$ and w_j be the weight of $e_j \in E$. Some problems are given below.

1. *The minimum-weight s–t path problem.* $F \in \mathcal{F}$ if F is the edge set of an s–t path.
2. *The minimum-weight s–t cut problem.* $F \in \mathcal{F}$ if F is the set of edges of a minimal s–t cut.
3. *The minimum-weight covering of nodes by edges.* $F \in \mathcal{F}$ if F is a minimal set of edges with the property that every node is met by some edge in F.
4. *The minimum-weight star problem.* $F \in \mathcal{F}$ if F is the set of edges incident to a node. F is called a *star*.
5. *The traveling salesman problem.* $F \in \mathcal{F}$ if F is the edge set of a Hamiltonian cycle.

A significant difference between problem 4 and the others is that in problem 4, $|\mathcal{F}| = |V|$, while in the others $|\mathcal{F}|$ grows exponentially with $|V|$. Hence problem 4 is easily solved by enumeration. Problems 1 and 2 are network flow problems (see Sections I.3.2 and I.3.4). Problem 3 is closely related to the matching problems considered in Chapter III.2 and will be considered in Section III.2.4. It can be solved in polynomial time. Problem 5 is $\mathcal{NP}$-hard.

To develop integer and linear programming formulations of (6.1), we consider another clutter.

Definition 6.1. The *blocking clutter* of $\mathcal{F}$ is the clutter $B(\mathcal{F})$ whose members H satisfy the following two conditions.

1. *Intersection*: $H \cap F \neq \emptyset$ for all $F \in \mathcal{F}$.
2. *Minimality*: If $H' \subset H$, then $H' \cap F = \emptyset$ for some $F \in \mathcal{F}$.

Example 6.1. Suppose $\mathcal{F}$ is represented by the rows of the matrix

$$\begin{pmatrix} 1 & 1 & 1 & 0 \\ 0 & 1 & 0 & 1 \\ 0 & 0 & 1 & 1 \end{pmatrix}.$$

The reader can check that its blocking clutter is specified by the rows of the matrix.

$$\begin{pmatrix} 1 & 0 & 0 & 1 \\ 0 & 1 & 1 & 0 \\ 0 & 1 & 0 & 1 \\ 0 & 0 & 1 & 1 \end{pmatrix}.$$

Proposition 6.1. $B(B(\mathcal{F})) = \mathcal{F}$.

Proof. For any clutter $\mathcal{F}$, let $\mathcal{F}^+ = \{R: R \supseteq F \text{ for some } F \in \mathcal{F}\}$. Suppose $F \in \mathcal{F}$. By the definition of $B(\mathcal{F})$ we have that if $H \in B(\mathcal{F})$, then $F \cap H \neq \emptyset$. Hence $F \in (B(B(\mathcal{F})))^+$. Now we need to prove that the members of $\mathcal{F}$ are the minimal elements of $(B(B(\mathcal{F})))^+$.

Suppose $T \notin \mathcal{F}^+$. Then for any $G \in \mathcal{F}$, we obtain $G \nsubseteq T$. Hence $G \cap (N \setminus T) \neq \emptyset$ for all $G \in \mathcal{F}$. So $N \setminus T \in (B(\mathcal{F}))^+$ and thus $T \notin (B(B(\mathcal{F})))^+$. Hence the minimal elements of $(B(B(\mathcal{F})))^+$ are precisely the members of $\mathcal{F}$; that is, $\mathcal{F} = B(B(\mathcal{F}))$. ■

Thus we can interchange the roles of $\mathcal{F}$ and $B(\mathcal{F})$ and simply refer to a pair of clutters $\mathcal{F}$ and $\mathcal{H}$ as blocking clutters when $\mathcal{H} = B(\mathcal{F})$ or $\mathcal{F} = B(\mathcal{H})$.

The proof of Proposition 6.1 establishes the following theorem of the alternative, which characterizes blocking pairs of clutters.

Corollary 6.2. *The clutters $\mathcal{F}$ and $\mathcal{H}$ are a pair of blocking clutters if and only if for all $T \subseteq N$, there is either an $F \in \mathcal{F}$ with $F \subseteq T$ or an $H \in \mathcal{H}$ with $H \subseteq N \setminus T$ but not both.*

Proof. We have already shown that if $T \notin \mathcal{F}^+$, then $N \setminus T \in (B(\mathcal{F}))^+$. Both statements cannot be true because of the intersection condition. The converse is proved similarly. ■

Example 6.2. Suppose $\mathcal{F}$ is the clutter of s–t paths in a connected graph. We have proved in Section I.3.4 that G contains an s–t path if and only if every s–t cut is nonempty. Hence every s–t path contains an edge belonging to every s–t cut and conversely. Thus $B(\mathcal{F})$ is the clutter of minimal s–t cuts. Figure 6.1 shows a graph and the matrices of incidence vectors of s–t paths and minimal s–t cuts.

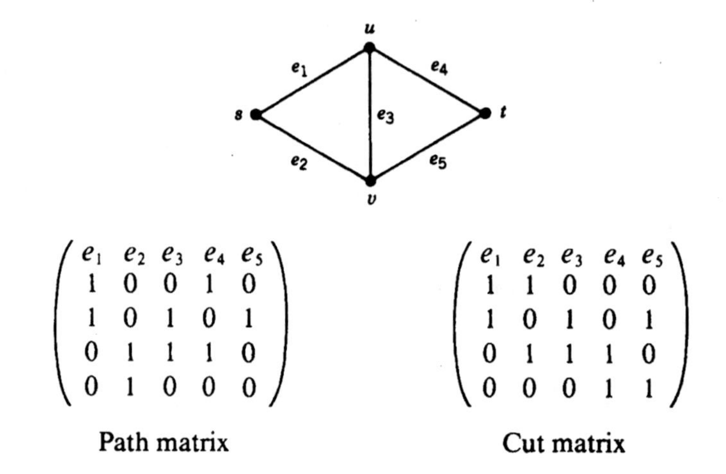

Path matrix

e_1	e_2	e_3	e_4	e_5
1	0	0	1	0
1	0	1	0	1
0	1	1	1	0
0	1	0	0	0

Cut matrix

e_1	e_2	e_3	e_4	e_5
1	1	0	0	0
1	0	1	0	1
0	1	1	1	0
0	0	0	1	1

Figure 6.1

Example 6.3. Let $\mathscr{F}$ be the clutter of edge covers (covers of nodes by edges) in a graph $G = (V, E)$ without isolated nodes. $E' \subseteq E$ is an edge cover if and only if every node in the subgraph $G = (V, E')$ has degree at least 1. Hence $B(\mathscr{F})$ are the stars of G. Note that the star matrix is the node-edge incidence matrix of G. For the graph of Figure 6.1, the incidence matrices of minimal edge covers and stars are given below.

Minimal edge cover matrix

e_1	e_2	e_3	e_4	e_5
1	0	1	1	0
1	0	0	0	1
0	1	1	0	1
0	1	0	1	0

Star matrix

e_1	e_2	e_3	e_4	e_5
1	1	0	0	0
1	0	1	1	0
0	1	1	0	1
0	0	0	1	1

Example 6.4. There are some obvious members of the blocking clutter of tours. For example, every tour contains at least two edges incident to every node. Thus stars with an edge deleted are members of the blocking clutter. But a complete description of the minimal edge sets whose deletion would make the graph non-Hamiltonian is not known.

From the perspective of integer programming, the importance of knowing $B(\mathscr{F})$ is that it gives a formulation of (6.1) as a set-covering problem. We use binary vectors x^F for $F \in \mathscr{F}$ to represent the elements of $\mathscr{F}$ and use binary vectors a^H for $H \in B(\mathscr{F})$ to represent elements of $B(\mathscr{F})$. Let

$$Q = \{x \in R_+^n: a^H x \geq 1 \text{ for all } H \in B(\mathscr{F})\}.$$

$$Q^* = \text{conv}\{x \in Z_+^n: x \geq x^F \text{ for some } F \in \mathscr{F}\}.$$

$$Q_B = \{a \in R_+^n: x^F a \geq 1 \text{ for all } F \in \mathscr{F}\}.$$

$$Q_B^* = \text{conv}\{a \in Z_+^n: a \geq a^H \text{ for some } H \in B(\mathscr{F})\}.$$

Example 6.1 (continued). Q is the polyhedron given by

$$\begin{aligned} x_1 \quad\quad\quad + x_4 &\geq 1 \\ x_2 + x_3 \quad\quad &\geq 1 \\ x_2 \quad\quad + x_4 &\geq 1 \\ x_3 + x_4 &\geq 1 \\ x \in R^4_+. \end{aligned}$$

The reader can check that (a) its extreme points are the incidence vectors of the members of $\mathcal{F}$ and the point $(\frac{1}{2}\ \frac{1}{2}\ \frac{1}{2}\ \frac{1}{2})$ and (b) its extreme rays are the 4 unit vectors. Hence $Q \cap Z^4$ is the set of integer vectors equal to or greater than some incidence vector of a member of $\mathcal{F}$. Thus Q_B is the polyhedron given by

$$\begin{aligned} a_1 + a_2 + a_3 \quad\quad &\geq 1 \\ a_2 \quad\quad + a_4 &\geq 1 \\ a_3 + a_4 &\geq 1 \\ a \in R^4_+. \end{aligned}$$

Its extreme points are the incidence vectors of the members of $\mathcal{H}$ and the point $(0\ \frac{1}{2}\ \frac{1}{2}\ \frac{1}{2})$. Again $Q_B \cap Z^4$ is the set of integer vectors equal to or greater than some incidence vector of a member of $\mathcal{H}$, and $Q_B^* = \text{conv}(Q_B \cap Z^4)$.

Proposition 6.3. *The following statements are true.*

1. $Q \cap Z^n = \{x \in Z^n: x \geq x^F$ *for some* $F \in \mathcal{F}\}$.
2. $Q^* = \text{conv}(Q \cap Z^n)$.
3. $Q_B \cap Z^n = \{x \in Z^n: a \geq a^H$ *for some* $H \in B(\mathcal{F})\}$.
4. $Q_B^* = \text{conv}(Q_B \cap Z^n)$.

Proof. We will establish statement 1. Statement 2 follows immediately from statement 1. Statements 3 and 4 are proved similarly.

If $x \in Z^n$ and $x \geq x^F$ for some $F \in \mathcal{F}$, then $a^H x \geq 1$ for all $H \in B(\mathcal{F})$. Hence $x \in Q \cap Z^n$. Conversely, if $x \in Q \cap Z^n$ but $x \geq x^F$ fails to hold for all $F \in \mathcal{F}$, let $T = \{j: x_j > 0\}$. Then it follows from Corollary 6.2 that there exists $H \in B(\mathcal{F})$ with $H \subseteq N \setminus T$. Hence $a^H x = 0$ and $x \notin Q \cap Z^n$. ■

Since $w \in R^n_+$ and any $x \in Q \cap Z^n$ satisfies $x \geq x^F$ for some $F \in \mathcal{F}$, (6.1) can be reformulated as the set-covering problem $\min\{wx: x \in Q \cap B^n\}$. Moreover, since the extreme points of Q^* are precisely x^F for $F \in \mathcal{F}$, we obtain

$$\min\{wx: x \in Q^*\} = \min\{wx: x \in Q \cap B^n\}.$$

We also obtain analogous results for the problem $\min\{w(H)\colon\ H \in B(\mathcal{F})\}$. In particular,

$$\begin{aligned}\min\{w(H)\colon H \in B(\mathcal{F})\} &= \min\{wa\colon a \in Q_B \cap B^n\} \\ &= \min\{wa\colon a \in Q_B^*\}.\end{aligned}$$

Example 6.1 (continued). Note that $\min\{w(F)\colon F \in \mathcal{F}\}$ can be formulated as the set-covering problem

$$\begin{aligned}\min\ & w_1x_1 + w_2x_2 + w_3x_3 + w_4x_4 \\ & x_1 + x_4 \geq 1 \\ & x_2 + x_3 \geq 1 \\ & x_2 + x_4 \geq 1 \\ & x_3 + x_4 \geq 1 \\ & x \in B^4.\end{aligned}$$

Also $\min\{w(H)\colon H \in \mathcal{H}\}$ can be formulated as the set-covering problem

$$\begin{aligned}\min\ & w_1a_1 + w_2a_2 + w_3a_3 + w_4a_4 \\ & a_1 + a_2 + a_3 \geq 1 \\ & a_2 + a_4 \geq 1 \\ & a_3 + a_4 \geq 1 \\ & a \in B^4.\end{aligned}$$

We now investigate the relationships among Q, Q^*, Q_B, and Q_B^*.

Proposition 6.4. *The following statements are true.*

a. *Q^* and Q_B are a blocking pair of polyhedra.*
b. *Q and Q_B^* are a blocking pair of polyhedra.*

Proof. The extreme points of Q^* are x^F for $F \in \mathcal{F}$. Hence by Proposition 5.7 of Section I.4.5, its blocker is

$$(Q^*)^B = \{a \in R_+^n\colon x^Fa \geq 1 \quad \text{for all} \quad F \in \mathcal{F}\} = Q_B.$$

An identical argument yields statement b. ■

The relationships are summarized in Figure 6.2.

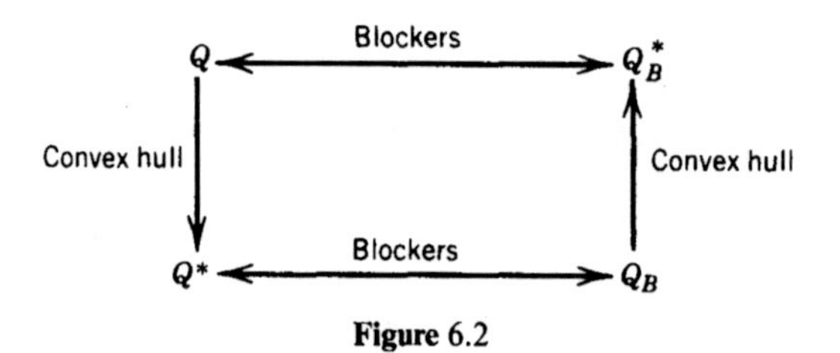

Figure 6.2

Example 6.1. (continued). We have shown that the extreme points of Q_B are the incidence vectors of the members of $\mathscr{H}$ and the point $(0\ \frac{1}{2}\ \frac{1}{2}\ \frac{1}{2})$. Hence $\min\{w(F)\colon F \in \mathscr{F}\}$ can be reformulated as the linear program

$$\begin{aligned}
\min\ & w_1x_1 + w_2x_2 + w_3x_3 + w_4x_4 \\
& x_1 + x_4 \geqslant 1 \\
& x_2 + x_3 \geqslant 1 \\
& x_2 + x_4 \geqslant 1 \\
& x_3 + x_4 \geqslant 1 \\
& x_2 + x_3 + x_4 \geqslant 2 \\
& x \in R^4_+.
\end{aligned}$$

since these constraints define the polyhedron Q^*.

Similarly, $\min\{w(H)\colon H \in \mathscr{H}\}$ can be reformulated as

$$\begin{aligned}
\min\ & w_1a_1 + w_2a_2 + w_3a_3 + w_4a_4 \\
& a_1 + a_2 + a_3 \geqslant 1 \\
& a_2 + a_4 \geqslant 1 \\
& a_3 + a_4 \geqslant 1 \\
& a_1 + a_2 + a_3 + a_4 \geqslant 2 \\
& a \in R^4_+.
\end{aligned}$$

since these constraints define the polyhedron Q_B^*.

Now when Q is integral, $Q = Q^*$. Hence their respective blockers Q_B^* and Q_B are equal (see Figure 6.2). Thus we obtain (see Theorem 5.10 of Section I.4.5) a pair of max–min relationships.

Theorem 6.5. *The following statements are equivalent.*

1. *Q is integral.*
2. *Q_B is integral.*
3. *For all $w \in R^n_+$, we have*

$$\max\left\{\sum_{F\in\mathscr{F}} y_F\colon \sum_{F\in\mathscr{F}} y_F x^F \leqslant w,\ y \in R_+^{|\mathscr{F}|}\right\} = \min_{H\in B(\mathscr{F})} a^H w.$$

4. *For all $w \in R^n_+$, we have*

$$\max\left\{\sum_{H\in B(\mathscr{F})} y_H\colon \sum_{H\in B(\mathscr{F})} y_H a^H \leqslant w,\ y \in R_+^{|B(\mathscr{F})|}\right\} = \min_{F\in\mathscr{F}} x^F w.$$

The max–min equality of statement 4 says that for all $w \in R^n_+$, the weight of a minimum-weight element of $\mathscr{F}$ equals the maximum number of elements of the blocking clutter that can be packed fractionally into the weight vector w. Can more be said for $w \in Z^n_+$? In general, the answer is no; we will consider some examples later. However, when the packing problem has an integral optimal solution for all $w \in Z^n_+$, we say that the *max–min equality holds strongly*. This is equivalent to the system $a^H x \geqslant 1$ for $H \in B(\mathcal{F})$ and $x \geqslant 0$ being TDI, since if $w \in Z^n \setminus Z^n_+$, the packing problem is infeasible. By Proposition 6.1, all of the remarks made in this paragraph about statement 4 also apply to statement 3.

The results of Proposition 6.4 and Theorem 6.5, together with the polynomial equivalence of optimization and separation (see Theorem 3.3 of Section I.6.3), relate the computational complexity of the linear programs over Q, Q^*, Q_B, and Q^*_B.

Theorem 6.6. *Each member of the following pairs of problems is solvable in polynomial time if and only if the other member of the pair is solvable in polynomial time.*

1. *The linear programs over Q^* and Q_B for all $w \in R^n_+$.*
2. *The linear programs over Q^*_B and Q for all $w \in R^n_+$.*
3. *The linear programs over Q and Q_B when Q is integral.*

Example 6.2 (continued). The problem

$$\max\left\{\sum_{F\in\mathcal{F}} y_F \colon \sum_{F\in\mathcal{F}} y_F x^F \leqslant w,\ y \in R^{|\mathcal{F}|}_+\right\} \tag{6.3}$$

can be interpreted as the maximum number of s–t paths that can be packed fractionally into the weight or capacity vector w. Since we can think of each path as a flow of one unit from s to t, (6.3) is a formulation of the max-flow problem. Hence by the max-flow–min-cut theorem (see Theorem 4.1 of Section I.3.4), statement 3 of Theorem 6.5 holds for all $w \in R^n_+$. Thus Q and Q_B are integral polyhedra. The extreme points of Q_B are the incidence vectors of all minimal s–t cuts, and the extreme points of Q are the incidence vectors of all s–t paths. Note from Figure 6.1 that neither the incidence matrix of s–t paths nor the matrix of minimal s–t cuts is balanced, which means that if certain rows were dropped from those matrices the corresponding polyhedra would no longer be integral.

The weighted min-cut problem formulation given by the dual of (6.3),

$$\min\{wa\colon ax^F \geqslant 1 \text{ for } F \in \mathcal{F},\ a \in R^n_+\}, \tag{6.4}$$

can be solved by a constraint generation algorithm since for any $a^* \in R^n_+$, it follows that $a^* x^F \geqslant 1$ for all $F \in \mathcal{F}$ if and only if the weight of a minimum-weight s–t path with weight vector a^* is at least 1. Although this algorithm is not practical, it illustrates the connection between optimization and separation and how the ellipsoid algorithm is used to prove that a combinatorial linear program with a large number of constraints can be solved in polynomial time.

The max–min equality in statement 4 of Theorem 6.5 also holds strongly; that is, the maximum number of s–t cuts that can be packed into $w \in R^n_+$ equals the weight of a minimum-weight s–t path, and the packing problem has an integral optimal solution. Moreover, Dijkstra's algorithm can be used to construct an integral optimal solution to the cut packing problem.

To show this, we refer to the algorithm in Section I.3.2, and we replace each edge by a pair of directed arcs.

Let $g(j)$ be the weight of a minimum-weight path from node s to node j, and let $g(s) = 0$. Let $U^0 = \{s\}$. At iteration i, we have a set U^i with

$$\max_{j \in U^i} g(j) \leqslant \min_{j \in \overline{U}^i} g(j) \quad \text{with } \overline{U}^i = V \setminus U^i.$$

Let $s = j_0$, and define j_i to be any $j \in U^i$ that satisfies $g(j_i) = \max_{j \in U^i} g(j)$. Hence

$$g(j_{i+1}) = \max_{j \in U^{i+1}} g(j) \leqslant \min_{j \in \overline{U}^i} g(j).$$

The cut $(U^i, \overline{U}^i)$ is assigned the weight $y_{U^i} = g(j_{i+1}) - g(j_i)$ for $i = 0, 1, \ldots$. Thus

$$\sum_{l=1}^{i} y_{U^{l-1}} = \sum_{l=1}^{i} (g(j_l) - g(j_{l-1})) = g(j_i). \tag{6.5}$$

Now if $t = j_k$, we claim that an optimal integral solution to the cut packing problem is given by $y_{U^i} = g(j_{i+1}) - g(j_i)$ for $i = 0, \ldots, k-1$ and by $y_U = 0$ otherwise.

Given $w \in Z_+^n$, we have $y_{U^i} \in Z_+^1$, and by (6.5), we obtain $\Sigma_{i=1}^k y_{U^{i-1}} = g(t)$. Thus it remains to be shown that

$$\sum_{\{i : U^i \ni e, i \leqslant k\}} y_{U^{i-1}} \leqslant w_e \quad \text{for all } e \in E.$$

Let $e = (j_p, j_q)$, where $q > p$. By definition of $g(j)$, it follows that $g(j_q) \geqslant g(j_p)$ and $w_e \geqslant g(j_q) - g(j_p)$. By (6.5), we have $g(j_q) - g(j_p) = \Sigma_{i=p+1}^q y_{U^{i-1}}$ and

$$\sum_{\{i : U^i \ni e, i \leqslant k\}} y_{U^{i-1}} \leqslant \sum_{i=p+1}^{\min(k,q)} y_{U^{i-1}}$$

$$= \begin{cases} 0 & \text{if } p \geqslant k \\ g(j_k) - g(j_p) \leqslant g(j_q) - g(j_p) & \text{if } q > k \\ g(j_q) - g(j_p) & \text{otherwise.} \end{cases}$$

Let $w = (3 \quad 1 \quad 1 \quad 2 \quad 4)$ be a weight vector for the graph of Figure 6.1. Dijkstra's algorithm yields $U^0 = \{s\}$, $g(s) = 0$; $U^1 = \{s, v\}$, $g(v) = 1$; $U^2 = \{s, v, u\}$, $g(u) = 2$; $U^3 = \{s, v, u, t\}$, $g(t) = 4$. Hence an optimal integral solution to the cut packing problem is obtained by assigning weight $g(v) - g(s) = 1$ to the cut $(U^0, \overline{U}^0) = \{e_1, e_2\}$, weight $g(u) - g(v) = 1$ to the cut $(U^1, \overline{U}^1) = \{e_1, e_3, e_5\}$, and weight $g(t) - g(u) = 2$ to the cut $(U^2, \overline{U}^2) = \{e_4, e_5\}$.

Example 6.2 shows the nicest possible behavior. Q and Q_B are integral, and both polyhedra are represented by TDI systems. Example 6.3 reveals other possibilities.

Example 6.3 (continued)

A. *Bipartite Graphs.* Since the matrix whose rows are the incidence vectors of stars in G is the node-edge incidence matrix, it is totally unimodular (Corollary 2.9). Hence, the polyhedron Q is integral and the linear system of inequalities is TDI. Since the packing of stars is the same as node packing, we obtain from statement 4 of Theorem 6.5 that the weight of a minimum-weight edge cover equals the maximum number of stars or nodes that can be packed into $w \in Z_+^n$. In particular, for $w = 1$, this is the classical result that the minimum number of edges needed to cover all of the nodes equals the size of a maximum-cardinality node packing.

Since Q is an integral polyhedron, so is Q_B. It can be shown that the packing problem in statement 4 of Theorem 6.5 has an integral optimal solution for $w = 1$. This says that the maximum number of edge disjoint edge covers equals the degree of the minimum degree node.

B. *General Graphs.* Q is not integral for all graphs. For example, if G is a triangle, Q contains the extreme point $(\frac{1}{2}\ \frac{1}{2}\ \frac{1}{2})$.

The edge-covering problem on the complete graph on 4 nodes, which we considered in Example 1.2, is interesting in that it reveals that the packing problems in statements 3 and 4 of Theorem 6.5 can have different behavior. Q is integral, but with $w = 1$ the star packing problem has a unique optimal fractional solution. On the other hand, it can be shown that the problem of fractionally packing the edge covers has an integral optimal solution for all $w \in Z_+^n$. Thus, we have an example of a blocking pair of integral polyhedra for which the max–min equality holds strongly for one but not for the other.

There is an analogous theory, which we consider only briefly, for finding a maximum-weight element of a clutter $\mathcal{F}$.

Definition 6.2. The *antiblocking* clutter of $\mathcal{F}$ is the clutter $A(\mathcal{F})$ whose members H satisfy the following two conditions.

1. Minimum intersection: $|H \cap F| \leq 1$ for $F \in \mathcal{F}$.
2. Maximality: If $H' \supset H$, then $|H' \cap F| > 1$ for some $F \in \mathcal{F}$.

A familiar example of the antiblocking relation arises in the maximum-weight node-packing problem. Here $\mathcal{F}$ is the set of maximal node packings in a graph G, and $A(\mathcal{F}) = \mathcal{C}$ is the set of maximal cliques. Given the weight vector $w \in R_+^n$ on the nodes, the maximum-weight node packing problem is

$$\max\{w(F): F \in \mathcal{F}\} = \max\{wx: x \in P \cap B^n\},$$

where $P = \{x \in R_+^n: k^C x \leq 1 \text{ for all } C \in \mathcal{C}\}$, and k^C is the incidence vector of the clique C. P is the fractional node-packing polytope for G.

In Section 5, we showed that if P is integral (G is perfect), then the system $k^C x \leq 1$ for $C \in \mathcal{C}, x \geq 0$ is TDI. We used the antiblocking theorem for packing polytopes corresponding to Theorem 6.5 (see Proposition 5.8 and Theorem 5.10 of Section I.4.5) to show that G is perfect if and only if the complement of G is perfect. We also established that these results for perfect graphs characterize antiblocking pairs of integral polyhedra. In contrast, no simple characterization of blocking pairs of integral polyhedra is known.

Integer Rounding

We close this section by considering a related integrality issue regarding the packing problems

(6.6) $$z(w) = \max\{1y: yA \leqslant w, y \in R_+^m\}$$

(6.7) $$z_{IP}(w) = \max\{1y: yA \leqslant w, y \in Z_+^m\},$$

where the rows of A, namely, $a^i \in Z_+^n \setminus 0$ for $i = 1, \ldots, m$, are incomparable vectors and $w \in Z_+^n$. The problem is to determine when $z_{IP}(w) = \lfloor z(w) \rfloor$ for all $w \in Z_+^n$.

Definition 6.3. The system $\{y \in R_+^m: yA \leqslant w\}$ is IRD (*integer round down*) if $z_{IP}(w) = \lfloor z(w) \rfloor$ for all $w \in Z_+^n$.

Let $Q = \{w \in Z_+^n: w \geqslant \Sigma_{i=1}^m \lambda_i a^i, \Sigma_{i=1}^m \lambda_i = 1$ for some $\lambda \in R_+^m\}$, and let $kQ = \{kw: w \in Q\}$, where k is a positive integer. Note that $kQ \supseteq (k+1)Q$ for $k = 1, 2, \ldots$.

Proposition 6.7. *For any positive integer r, $z(w) \geqslant r$ if and only if $w \in rQ$.*

Proof. $z(w) \geqslant r \Leftrightarrow$ for some $y \in R_+^m$,

$$\sum_{i=1}^m y_i = r \quad \text{and} \quad \sum_{i=1}^m y_i a^i \leqslant w \quad [\text{by (6.6)}]$$

$\Leftrightarrow$ for some $\lambda \in R_+^m$,

$$\sum_{i=1}^m \lambda_i = 1 \quad \text{and} \quad \sum_{i=1}^m \lambda_i (ra^i) \leqslant w \quad \left(\lambda_i = \frac{y_i}{r} \quad \text{for all } i\right)$$

$\Leftrightarrow w \in rQ$. ■

Corollary 6.8. *$r \leqslant z(w) < r + 1$ if and only if $w \in rQ \setminus (r+1)Q$.*

Hence IRD holds if and only if for all $w \in Z_+^n$, $w \in (rQ \setminus (r+1)Q) \cap Z_+^n$, implies $z_{IP}(w) = r$.

Let $S_k = kQ \cap Z^n$ for $k = 1, 2, \ldots$.

Definition 6.4. Q is *integrally decomposable* if for each integer $k \geqslant 1$ and each $w \in S_k$, there exist $\tilde{a}^i, \ldots, \tilde{a}^k \in S_1$ (not necessarily distinct) such that $w = \Sigma_{i=1}^k \tilde{a}^i$.

To show that Q is integrally decomposable, it suffices to show that the minimal integral points of kQ can be expressed as a sum of k integral points of Q. This follows since if w^1, $w^2 \in S_k$, $w^2 > w^1$, and $w^1 = \Sigma_{i=1}^k \tilde{a}^i$, where $\tilde{a}^i \in S_1$ for $i = 1, \ldots, k$, then

$$w^2 = \sum_{i=1}^{k-1} \tilde{a}^i + (\tilde{a}^k + w^2 - w^1),$$

where $(\tilde{a}^k + w^2 - w^1) \in S_1$.

Theorem 6.9. *The system $\{y \in R_+^m: yA \leqslant w\}$ is* IRD *if and only if Q is integrally decomposable.*

Proof. We show that if $r \leqslant z(w) < r + 1$ and Q is integrally decomposable, then $z_{IP}(w) = r$. For $r = 0$, we have $0 = z_{IP}(w) \leqslant z(w) < 1$. Now suppose that r is a positive integer. By Proposition 6.7, $w \in rQ$. Hence there are $\tilde{a}^i \in Q \cap Z_+^n$ for $i = 1, \ldots, r$ such that $\Sigma_{i=1}^r \tilde{a}^i = w$, and there are minimal points $a^{l(i)} \in Q \cap Z_+^n$, not necessarily distinct, such that $a^{l(i)} \leqslant \tilde{a}^i$ for $i = 1, \ldots, r$ and $\Sigma_{i=1}^r a^{l(i)} \leqslant w$. Now let y_i^* be the number of times that $a^{l(i)}$ appears in $\Sigma_{i=1}^r a^{l(i)}$. Hence y^* is a feasible solution to (6.7), and $\Sigma_{i=1}^m y_i^* = r = \lfloor z(w) \rfloor$.

To prove the converse, we observe that a feasible solution of value r to (6.7), together with the remark that preceded the statement of Theorem 6.9, yields a suitable decomposition. ■

Example 6.5. Suppose

$$A = \begin{pmatrix} 1 & 1 & 0 \\ 1 & 0 & 1 \\ 0 & 1 & 1 \end{pmatrix}.$$

Note that with $w = (1 \quad 1 \quad 1)$, the unique solution to (6.6) is $y = (\frac{1}{2} \quad \frac{1}{2} \quad \frac{1}{2})$ and $z(w) = \frac{3}{2}$.

Now we show that Q is integrally decomposable. It is easy to check that all minimal points of kQ are of the form

$$\begin{aligned} &(\lambda_1 + \lambda_2, \lambda_1 + \lambda_3, \lambda_2 + \lambda_3), \ \lambda_1 + \lambda_2 + \lambda_3 = k, \ \lambda \geqslant 0 \\ &= (\alpha_1, \alpha_2, 2k - \alpha_1 - \alpha_2), \ 0 \leqslant \alpha_1, \alpha_2 \leqslant k, \ \alpha_1 + a_2 \geqslant k. \end{aligned}$$

So we need to show that for $\alpha_1, \alpha_2 \in Z_+^1, \alpha_1, \alpha_2 \leqslant k, \alpha_1 + \alpha_2 \geqslant k$, there is a $y \in Z_+^3$ such that $\Sigma_{i=1}^3 y_i = k$ and

$$\begin{aligned} y_1 + y_2 &= \alpha_1 \\ y_1 + y_3 &= \alpha_2 \\ y_2 + y_3 &= 2k - \alpha_1 - \alpha_2. \end{aligned}$$

A solution is $y_1 = \alpha_1 + \alpha_2 - k$, $y_2 = k - \alpha_2$, and $y_3 = k - \alpha_1$.

Different behavior is observed for the matrix

$$A = \begin{pmatrix} 1 & 1 & 1 & 0 & 0 & 0 \\ 1 & 0 & 0 & 1 & 1 & 0 \\ 0 & 1 & 0 & 0 & 1 & 1 \\ 0 & 0 & 1 & 1 & 0 & 1 \end{pmatrix}.$$

Note that

$$\tfrac{1}{2}(a^1 + a^2 + a^3 + a^4) = w = (1 \quad 1 \quad 1 \quad 1 \quad 1 \quad 1) \in 2Q.$$

But there are not two integral vectors in Q whose sum is w. Hence Q is not integrally decomposable. In particular, $z_{\text{IP}}(w) = 1$ and $z(w) = 2$.

We now consider a network flow model whose integral solutions define a matrix A such that $\{y \in R_+^m: yA \leqslant w\}$ is IRD. Let $\mathcal{D} = (V, \mathcal{A})$ be a directed graph with $|\mathcal{A}| = n$. A vector $b \in Z^{|V|}$ with $\Sigma_{v \in V} b_v = 0$ is called a *supply–demand vector*. The nodes $L = \{v \in V: b_v > 0\}$ are called *supply nodes*, and the nodes $T = \{v \in V: b_v < 0\}$ are called *demand nodes*. A *feasible flow* is a vector $a \in R_+^n$ that satisfies the conservation equations

$$b_v + \sum_{u \in \delta^-(v)} a_{uv} - \sum_{u \in \delta^+(v)} a_{vu} = 0 \quad \text{for } v \in V. \tag{6.8}$$

Let A be the matrix whose rows are the vectors of minimal, integral feasible flows in $\mathscr{D}$. The problem we consider is packing the rows of A into $w \in Z_+^n$.

Example 6.6. Consider the data given in Figure 6.3. It can be shown that the matrix of minimal integral feasible flows is

$$A = \begin{pmatrix} 2 & 0 & 1 & 2 & 0 \\ 2 & 0 & 2 & 1 & 1 \\ 2 & 0 & 3 & 0 & 2 \\ 0 & 2 & 0 & 1 & 1 \\ 0 & 2 & 1 & 0 & 2 \\ 1 & 1 & 0 & 2 & 0 \\ 1 & 1 & 1 & 1 & 1 \\ 1 & 1 & 2 & 0 & 2 \end{pmatrix}.$$

It is easy to see that the packing problem does not have an integral optimal solution for all $w \in Z_+^n$ for which it is feasible; for example, take $w = (1 \quad 0 \quad 1 \quad 1 \quad 0)$.

To show that the system $\{y \in R_+^m: yA \leqslant w\}$ is IRD, we need to introduce a capacity vector $d \in Z_+^n$ on the arcs of $\mathscr{D}$.

Proposition 6.10. *Given any $d \in Z_+^n$, the following two statements are equivalent.*

i. *There exists an $a \in Z_+^n$ that satisfies (6.8) and $a \leqslant d$.*
ii. *For all $U \subseteq V$,*

$$\sum_{v \in U} b_v \leqslant \sum_{e \in \delta^+(U)} d_e. \tag{6.9}$$

Proof. i $\Rightarrow$ ii is obvious since for any $U \subseteq V$, the flow out of U must be at least $\Sigma_{v \in U} b_v$.
The proof of ii $\Rightarrow$ i uses the max-flow–min-cut theorem on the graph $\mathscr{D}' = (V \cup \{s, t\}, \mathscr{A}')$, where

$$\mathscr{A}' = \mathscr{A} \cup \{(s, v): v \in L\} \cup \{(v, t): v \in T\},$$

The capacity of $e \in \mathscr{A}$ is d_e, the capacity of e_{sv} for $v \in L$ is b_v, and the capacity of e_{vt} for $v \in T$ is $-b_v$. We only sketch the proof.

If (6.9) holds for all $U \subseteq V$, then it can be shown that a minimum-weight cut in $\mathscr{D}'$ is given by the set of arcs $\{(s, v): v \in L\}$—that is, the cut generated by the node partition $(\{s\} \cup (V \setminus L), \{t\} \cup L)$. Then by the max-flow–min-cut theorem of Section I.3.4, there is

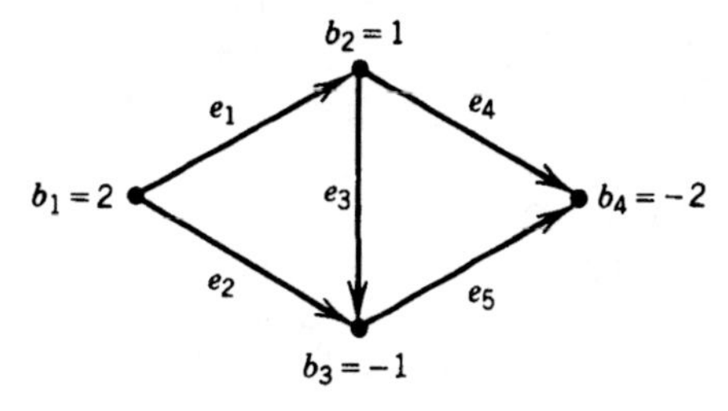

Figure 6.3

an integral $s-t$ flow in $\mathscr{D}'$ of size $\Sigma_{v\in L} b_v$. Thus in every maximum flow, the flow on e_{sv} is b_v for all $v \in L$. It then follows that statement i is true. ■

Theorem 6.11. *If A is an $m \times n$ matrix whose rows are the minimal integral flows in a digraph $\mathscr{D}$ with supply-demand vector $b \in Z^{|V|}$, then $\{y \in R^m_+ : yA \leqslant w\}$ is* IRD.

Proof. Let $S_k = kQ \cap Z^n$ for $k = 1, 2, \ldots$. By Theorem 6.9, it suffices to prove that for any k and $w \in S_k$, w can be written as the sum of k integral points in S_1. This is a triviality for $k = 1$. Now suppose it is true for S_{k-1} where $k \geqslant 2$.

We must show that for any $w \in S_k$, there exist an $a \in S_1$ such that $w - a \in S_{k-1}$. Note that $w \in S_k$ means that for the supply-demand vector kb, there is a flow $\tilde{a}^k \leqslant w$. But since the supply-demand system is totally unimodular, we can choose $\tilde{a}^k \in Z^n_+$. Hence

$$kb_v + \sum_{u\in\delta^-(v)} \tilde{a}^k_{uv} - \sum_{u\in\delta^+(v)} \tilde{a}^k_{vu} = 0 \quad \text{for } v \in V,$$

and for any $U \subseteq V$ with $\Sigma_{v\in U} b_v \geqslant 0$, we have

$$\sum_{e\in\delta^+(U)} \tilde{a}^k_e = k \sum_{v\in U} b_v + \sum_{e\in\delta^-(U)} \tilde{a}^k_e \geqslant k \sum_{v\in U} b_v \geqslant \sum_{v\in U} b_v.$$

Now taking $\tilde{a}^k$ to be the capacity vector in Proposition 6.10, there exists an $a \in Z^n_+$ that satisfies (6.8) and $a \leqslant \tilde{a}^k$. Thus $a \in S_1$, $\tilde{a}^k - a \in Z^n_+$, and

$$(k-1)b_v + \sum_{e\in\delta^-(v)} (\tilde{a}^k_{uv} - a_{uv}) - \sum_{u\in\delta^+(v)} (\tilde{a}^k_{vu} - a_{vu}) = 0 \quad \text{for } v \in V.$$

Hence $(\tilde{a}^k - a) \in S_{k-1}$ and, since $w \geqslant \tilde{a}^k$, we have $(w - a) \in S_{k-1}$. ■

Theorem 6.11 generalizes to capacitated supply-demand systems where, in addition to (6.8), the flow must satisfy $a \leqslant c$ where $c \in Z^n_+$. This can be shown by transforming a capacitated supply-demand system to an uncapacitated one. It also generalizes to circulations; that is, $b_v = 0$ for all $v \in V$, and $l \leqslant a \leqslant c$ where $l, c \in Z^n_+$. Thus a circulation is a solution to $Ga = 0, l \leqslant a \leqslant c$, where G is a node-arc incidence matrix. Finally, packing the minimal solutions of $Ga = 0, l \leqslant a \leqslant c$ is IRD for any totally unimodular matrix G.

7. NOTES

Section III.1.1

The study of integral polyhedra has its roots in the theory of network flows [see Ford and Fulkerson (1962)] and, in particular, in the max-flow-min-cut theorem. Two early proofs of this theorem illustrate fundamental techniques in the theory of integral polyhedra. Dantzig and Fulkerson (1956) proved it using linear programming duality, and Ford and Fulkerson (1956) proved it by giving an algorithm that produces a feasible flow and an $s-t$ cut of weight equal to the value of the flow (see Section I.3.4).

Proposition 1.3 is due to Hoffman (1974). Edmonds and Giles (1977) independently proved Proposition 1.3 and Corollary 1.4, and they coined the term *total dual integrality* and expounded upon its significance. They also developed the notion of box TDI systems: A system $Ax \leqslant b, c \leqslant x \leqslant d$ is *box TDI* if it is TDI for all vectors c, d.

Giles and Pulleyblank (1979) proved Proposition 1.7. Schrijver (1981) proved Proposition 1.8.

Cook (1983a) studied operations that preserve total dual integrality [also see Cook (1986) for box TDI systems]. Computational issues regarding TDI systems have been studied by Chandrasekaran (1981) and Cook, Lovasz, and Schrijver (1984).

Edmonds and Giles (1984) gave a survey of theoretical results on total dual integrality and classes of TDI systems.

Schrijver (1986b) gave a survey of proof techniques for establishing integrality and related properties of polyhedra.

Section III.1.2

Hoffman and Kruskal (1956) proved Theorem 2.5 and thus established the fundamental part of the connection between total unimodularity and integer programming [also see Hoffman (1979)]. A substantially simpler proof, the one presented in the text, was discovered by Veinott and Dantzig (1968).

Theorem 2.7 was proved by Ghouila-Houri (1962). The results on characterizations of totally unimodular matrices with no more than two nonzero elements in each column are due to Heller and Tompkins (1956), Hoffman and Kruskal (1956), and Dantzig and Fulkerson (1956).

Interval matrices were studied by Fulkerson and Gross (1965). The relaxation of a set-covering problem to a problem with an interval constraint matrix was given by Nemhauser, Trotter, and Nauss (1974).

Other conditions for total unimodularity were given by Camion (1965), Chandrasekaran (1969), Heller (1957, 1963), Heller and Hoffman (1962), Padberg (1976a, 1988), Tamir (1976), and Truemper (1977, 1978). See Padberg (1975b) for a survey.

In a study of the integrality of the matching polytope, Hoffman and Oppenheim (1978) proposed the idea of local unimodularity and thus gave another technique for establishing the integrality of a polyhedron.

Section III.1.3

The significance of recognizing network structure has been stimulated, in part, by a number of practical linear programming models that can be reformulated as network flow problems [see Zangwill (1966), Cunningham (1983), and Bland (1988)] and was also motivated by the efficiency of network codes (see the notes for Chapter I.3).

The definition of network matrices was proposed by Tutte in his study of graphic matroids [see Tutte (1965)]. Further references to matroids will be given in the notes for Chapter III.3.

Iri (1966) gave a polynomial-time algorithm for recognizing network matrices. A much more efficient algorithm was obtained by Bixby and Cunningham (1980). Their presentation is in terms of matroids. The algorithm given here is adapted from Schrijver (1986a).

Recently, attention has been given to finding large network submatrices [see Bixby and Cunningham (1980) and Bixby (1984)]. Several researchers have developed heuristics for this problem [see Brown and Wright (1984) and Gunawardane et al. (1981)]. The problem of finding a largest network submatrix is $\mathcal{NP}$-complete [see Bartholdi (1981)].

Theorem 3.8 and the algorithm for recognizing totally unimodular matrices are due to Seymour (1980). For a restricted class of totally unimodular matrices, Yannakakis (1985) gave efficient recognition and optimization algorithms.

Section III.1.4

Balanced matrices were introduced by Berge (1972). He proved the fundamental result given by Theorem 4.13. Several other results on the integrality of polyhedra associated with balanced matrices were obtained by Fulkerson, Hoffman, and Oppenheim (1974). In particular, they showed that if A is balanced and the system $Ax = 1, x \geqslant 0$ is feasible, then the polytope defined by this system is integral. This result on set-partitioning polytopes can be used to prove Theorem 4.13.

The restriction to totally balanced matrices was apparently proposed by Lovasz (1979b). The main results on totally balanced matrices given here (Proposition 4.4 through Theorem 4.10) come from Hoffman, Kolen, and Sakarovitch (1985). Proposition 4.11 is due to Giles (1978) and was used by Kolen (1983) to obtain integrality results for a class of uncapacitated facility location problems. Tamir (1983) gave the generalization stated in Proposition 4.12. Further generalizations were given by Tamir (1987).

Farber (1983) and Anstee and Farber (1984) independently obtained nearly the same results as Hoffman et al. (1985). Their characterization of totally balanced matrices is in terms of node–node incidence matrices of graphs. Extensions have been obtained by Lubiw (1982) and Chang and Nemhauser (1984, 1985). Also see Sakarovitch (1975, 1976) and Farber (1984).

Section III.1.5

The concept of perfect graphs is due to Berge (1960). It has led to a vast literature, mainly on graph theory, which we barely cite here. Instead, we refer the reader to the book by Golumbic (1980), the collection of articles edited by Berge and Chvátal (1984), and the chapter entitled "Stable Sets in Graphs" in the book by Grötschel, Lovasz, and Schrivjer (1987).

Duchet (1984) presented a survey of classic results on perfect graphs. Fulkerson (1970b, 1971, 1972, 1973) made the connection between perfect graphs and polyhedral combinatorics, and he introduced the concept of pluperfect graphs.

Dirac (1961) established the connection between simplicial nodes and chordal graphs. Gavril (1972) solved the cardinality node-packing problem and the corresponding clique-covering problem for chordal graphs. Frank (1975) solved the weighted versions of these problems essentially by the linear-time algorithm given in the text.

Theorem 5.11 and the perfect graph theorem, Theorem 5.15, were proved by Lovasz (1972). However, he acknowledges that much credit should be given to Fulkerson who had already shown that these theorems were true if and only if Corollary 5.12 was true. The proof of Theorem 5.11 given here comes from Chvátal (1975).

Theorem 5.15 was proved by Meyniel (1976, 1984). A polynomial-time agorithm for recognizing these graphs has been obtained by Burlet and Fonlupt (1984). Theorem 5.16 was proved by Berge (1972).

Theorem 5.17 was proved by Padberg (1974). Some other articles related to Padberg's work on minimally imperfect graphs are by Padberg (1975b, 1976b, 1984), Bland, Huang, and Trotter (1984), and Whitesides (1984).

A polynomial-time ellipsoid algorithm for maximum-weight node packing in perfect graphs was given by Grötschel, Lovasz, and Schrijver (1984a). Recently, they have obtained a polynomial-time ellipsoid algorithm for maximum-weight node packing in graphs for which the node-packing polytope is described by the clique and odd hole constraints [Grötschel, Lovasz, and Schrijver (1988)]. These graphs are called t-perfect.

Hsu (1984) gave a survey of graphs for which the strong perfect graph conjecture is true. It was proved for claw-free graphs by Parthasarathy and Ravindra (1976). Polynomial-time algorithms for solving the weighted node-packing problem on claw-free graphs have been given independently by Minty (1980) and Sbihi (1980). The convex hull of node packings for these graphs has been studied by Giles and Trotter (1981). Polynomial-time algorithms for maximum-weight cliques, minimum-weight clique covers, and minimum colorings for claw-free perfect graphs have been obtained by Hsu (1981) and Hsu and Nemhauser (1981, 1982, 1984). These problems are $\mathcal{NP}$-hard for general claw-free graphs.

Section III.1.6

The theory of blocking and antiblocking polyhedra was developed in a series of articles by Fulkerson (1968, 1970a, 1971, 1972). Fulkerson's work was motivated by a 1965 paper of Lehman which was not published until 1979. A survey of results obtained in the 1970s has been presented by Tind (1979). [Also see Tind (1974, 1977), Johnson (1978), and Huang and Trotter (1980).]

Proposition 6.1 was proved by Edmonds and Fulkerson (1970). Propositions 6.3 and 6.4 and Theorem 6.5 were proved by Fulkerson (1970a).

Fulkerson (1968) showed that the max–min inequality holds strongly for the $s-t$ path and $s-t$ cut clutters. There are several interesting pairs of clutters for which the max–min inequality holds, but not strongly, and for which one or both of the dual problems has an optimal solution that is half-integer for all nonnegative integers w. (A vector is said to be half-integer if each of its components is either an integer or an integer divided by 2.) An example where both of the clutters have this property is 2-commodity cuts and flows in graphs [see Hu (1969) and Seymour (1978)]. The max–min inequality holds for the T-join, T-cut clutters to be studied in Section III.2.4. However, here one of the packing problems has the half-integer property and the other does not [see Edmonds and Johnson (1973) and Seymour (1979)].

In general, the problem of characterizing pairs of clutters for which the max–min inequality holds (or holds strongly) or for which the half-integer property is obtained for one or both of the packing problems is unresolved. However, Seymour (1977) characterized the strong max–min inequality for an interesting class of clutters known as *binary clutters*. Some other blocking relations will be studied in Section III.2.4 and Chapter III.3.

The connection between the integer round-down property and integral decomposability was established by Baum and Trotter (1977, 1981). Further results along these lines were obtained by McDiarmid (1983).

The IRD property for network flows given in Theorem 6.11 is due to Fulkerson and Weinberger (1975). Additional integer-rounding results for network flow problems have been obtained by Weinberger (1976) and Trotter and Weinberger (1978).

Marcotte (1985, 1986a) has established some families of knapsack problems for which the cutting stock problem has the integer-rounding property and has also given an instance of the cutting stock problem where the gap is equal to 1.

Some literature on integer-rounding results for matroid problems will be cited in the notes for Section III.3.8.

Computational complexity issues associated with problems with the IRD property have been studied by Baum and Trotter (1982) and Orlin (1982).

8. EXERCISES

1. Consider the polytope P described by the linear inequality system

$$x_1 + 2x_2 + 4x_3 \leqslant 4, \qquad x \geqslant 0.$$

i) Show that P is an integral polytope.

ii) Show that the linear inequality system is not TDI.

iii) Find the unique minimal TDI representation with an integral right-hand side.

2. Find a TDI representation for the polytope

$$P = \{x \in R^2_+ : 4x_1 + x_2 \leqslant 28, x_1 + 4x_2 \leqslant 27, x_1 - x_2 \leqslant 1\}.$$

3. A linear inequality system $Ax \leqslant b$ is *box* TDI if $Ax \leqslant b$, $l \leqslant x \leqslant u$ is a TDI system for all l and $u \in R^n$.

i) Show that the system of exercise 1(iii) is not box TDI.

ii) Show that the system $x_1 + x_2 + x_3 \leqslant 4$, $x \geqslant 0$ is box TDI.

iii) Show that the system of Example 1.1 is box TDI.

4. Verify that the top two matrices are TU but the bottom two are not.

$$\begin{pmatrix} 1 & -1 & 0 & 0 & -1 \\ -1 & 1 & -1 & 0 & 0 \\ 0 & -1 & 1 & -1 & 0 \\ 0 & 0 & -1 & 1 & -1 \\ -1 & 0 & 0 & -1 & 1 \end{pmatrix} \qquad \begin{pmatrix} 1 & 1 & 1 & 1 & 1 \\ 1 & 1 & 1 & 0 & 0 \\ 1 & 0 & 1 & 1 & 0 \\ 1 & 0 & 0 & 1 & 1 \\ 1 & 1 & 0 & 0 & 1 \end{pmatrix}$$

$$\begin{pmatrix} 1 & 1 & 0 & 1 & 0 \\ 0 & 0 & 1 & 0 & 1 \\ 0 & 1 & 1 & 1 & 0 \\ 1 & 1 & 0 & 0 & 0 \\ 0 & 1 & 0 & 0 & 1 \end{pmatrix} \qquad \begin{pmatrix} 1 & 0 & -1 & 0 \\ 0 & 1 & 1 & 0 \\ 1 & 0 & 0 & -1 \\ 0 & 1 & -1 & 1 \end{pmatrix}.$$

5. Show that

$$A = \begin{pmatrix} 1 & 1 & 1 \\ -1 & 1 & 0 \\ 1 & 0 & 0 \end{pmatrix}$$

is not TU. Then show that $P(b) = \{x \in R^n_+ : Ax = b\}$ is integral for all $b \in Z^n$ for which it is nonempty.

6. Show that if A is a 0, 1, -1 matrix in which the sum of the entries of every square submatrix with even row and column sums is divisible by 4, then A is TU.

7. Suppose that the 0, 1 matrix A is not an interval matrix and that the integer program (2.1) is relaxed by splitting columns as described. If each column is split into, at most, p columns, compare the bound from this relaxation with that from the standard linear programming relaxation.

8. Prove Proposition 2.11.

9. Verify whether the following are network matrices or not.

i)

$$\begin{pmatrix} 0 & -1 & 0 & 1 & -1 & 1 & 0 & -1 & 0 \\ 0 & 1 & 0 & 0 & -1 & 0 & -1 & 1 & 0 \\ 1 & 0 & 1 & 0 & 0 & 0 & 0 & 0 & 0 \\ 0 & 1 & -1 & 0 & 0 & 0 & 0 & 1 & 0 \\ 1 & 1 & 0 & 0 & 0 & 0 & 0 & 0 & 0 \\ 0 & 0 & 0 & 0 & 1 & 0 & 1 & 0 & 1 \\ 0 & 0 & 0 & 0 & -1 & 1 & -1 & 0 & 0 \\ 0 & -1 & 0 & 0 & 0 & 0 & 0 & -1 & -1 \\ 0 & 0 & -1 & 1 & 0 & 1 & 0 & 0 & 0 \end{pmatrix}.$$

ii)

$$\begin{pmatrix} 1 & 0 & 0 & 1 & -1 & 0 \\ 0 & 1 & -1 & 0 & 0 & 0 \\ 0 & 0 & 0 & 1 & -1 & -1 \\ 1 & 0 & 0 & 0 & 0 & -1 \\ 1 & 1 & 0 & 0 & 1 & 0 \\ 0 & -1 & 0 & -1 & 0 & 0 \\ 0 & 0 & 1 & 1 & 0 & 0 \end{pmatrix}.$$

iii)

$$\begin{pmatrix} 1 & -1 & 0 & 0 & -1 \\ -1 & 1 & -1 & 0 & 0 \\ 0 & -1 & 1 & -1 & 0 \\ 0 & 0 & -1 & 1 & -1 \\ -1 & 0 & 0 & -1 & 1 \end{pmatrix}.$$

10. Modify the network recognition algorithm so as to find a maximal network submatrix.

11. Let A be a 0, 1 matrix with no zero rows or columns. Show that $\{x \in R^n_+ : Ax = 1\}$ is integral if and only if statement 1 or statement 2 of Proposition 4.1 holds.

12. Are interval matrices (i) balanced, (ii) TB?

13. **i)** Show that

$$A = \begin{pmatrix} 1 & 1 & 1 & 0 & 0 & 0 \\ 0 & 0 & 1 & 0 & 1 & 0 \\ 0 & 1 & 1 & 1 & 1 & 0 \\ 1 & 1 & 1 & 1 & 1 & 1 \\ 0 & 1 & 1 & 1 & 1 & 1 \\ 0 & 0 & 0 & 1 & 1 & 1 \end{pmatrix}$$

is a row inclusion matrix.

ii) Solve $\min\{cx: Ax \geqslant 1, x \in Z^6_+\}$ with $c = (4 \quad 2 \quad 7 \quad 1 \quad 3 \quad 5)$.

14. Convert the following matrix to a TRL matrix.

$$\begin{pmatrix} 0 & 1 & 0 & 1 & 0 & 0 & 1 & 1 & 0 & 0 & 0 & 0 \\ 1 & 1 & 0 & 0 & 0 & 0 & 0 & 0 & 0 & 0 & 0 & 0 \\ 0 & 0 & 1 & 1 & 0 & 0 & 0 & 0 & 0 & 0 & 0 & 0 \\ 1 & 1 & 1 & 1 & 0 & 0 & 0 & 0 & 0 & 0 & 0 & 0 \\ 0 & 1 & 0 & 0 & 0 & 1 & 1 & 1 & 0 & 0 & 0 & 0 \\ 0 & 0 & 0 & 1 & 1 & 0 & 1 & 1 & 0 & 0 & 0 & 0 \\ 1 & 0 & 0 & 0 & 0 & 0 & 0 & 0 & 0 & 1 & 1 & 1 \\ 0 & 0 & 1 & 0 & 0 & 0 & 0 & 0 & 1 & 0 & 1 & 1 \\ 1 & 0 & 1 & 0 & 0 & 0 & 0 & 0 & 0 & 0 & 1 & 1 \\ 0 & 0 & 0 & 0 & 0 & 1 & 0 & 1 & 0 & 1 & 0 & 1 \\ 0 & 0 & 0 & 0 & 1 & 0 & 0 & 1 & 1 & 0 & 0 & 1 \\ 0 & 0 & 0 & 0 & 0 & 0 & 0 & 1 & 0 & 0 & 0 & 1 \end{pmatrix}.$$

Then give a short proof that the matrix is not totally balanced.

15. Solve the problem of finding a minimum-weight set of nodes that can serve every node on the graph shown in Figure 4.3, where $r_j = 1$ when j is even, $r_j = 2$ when j is odd, and d_{ij} is the number of edges on the path joining i and j. The weights are given by $c = (4 \quad 2 \quad 7 \quad 3 \quad 12 \quad 8 \quad 10 \quad 5 \quad 7 \quad 3 \quad 12)$.

16. Prove Proposition 4.12.

17. Prove Theorem 4.13.

18. Let

$$A^* = \begin{pmatrix} I & A \\ A^T & E \end{pmatrix},$$

where E is the matrix of all 1's.

i) Prove that A^* is a neighborhood matrix.

ii) Prove that A is TB if and only if A^* is TB.

19. A *strong elimination ordering* of a graph $G = (V, E)$ is a perfect elimination ordering $v_1, \ldots, v_n$ of V satisfying the following additional conditions for each i, j, k,

l: If $i < j < k < l$ and (v_i, v_k), (v_i, v_l), $(v_j, v_k) \in E$, then $(v_j, v_l) \in E$. A graph is *strongly chordal* if it has a strong elimination ordering.

i) Show that the graph of Figure 5.1 is not strongly chordal

ii) Show that the graph of Figure 8.1 is strongly chordal.

iii) Show that a graph is strongly chordal if and only if its neighborhood matrix is balanced.

20. Is $P = \{x \in R^n_+: Ax \leqslant 1\}$ integral for the following matrices? Why?

i)

$$A = \begin{pmatrix} 1 & 1 & 0 & 0 & 0 & 1 \\ 0 & 1 & 1 & 0 & 0 & 1 \\ 1 & 0 & 0 & 1 & 0 & 0 \\ 0 & 1 & 0 & 0 & 1 & 1 \\ 0 & 0 & 0 & 1 & 1 & 0 \end{pmatrix}.$$

ii)

$$A = \begin{pmatrix} 0 & 1 & 0 & 1 & 0 & 1 & 1 \\ 0 & 0 & 0 & 1 & 1 & 1 & 1 \\ 0 & 0 & 0 & 1 & 1 & 0 & 1 \\ 0 & 0 & 0 & 0 & 1 & 0 & 0 \\ 1 & 0 & 1 & 0 & 0 & 0 & 0 \\ 1 & 1 & 1 & 0 & 0 & 1 & 0 \\ 1 & 1 & 0 & 1 & 0 & 1 & 0 \end{pmatrix}.$$

21. A graph $G = (V, E)$ is called an *interval graph* if there is an assignment of an interval of the real line to each $v \in V$ such that $(u, v) \in E$ if and only if the intervals corresponding to u and v intersect. Show that interval graphs are chordal and that there is a greedy algorithm for solving $\max\{cx: Ax \leqslant 1, x \in B^n\}$ when A is an interval matrix.

22. Describe a polynomial algorithm to check whether A is the clique matrix of some graph.

23. i) Give a polynomial algorithm to find an odd cycle in a graph.

ii) Use this to devise a heuristic algorithm to detect odd holes in a graph.

24. Give an $O(|E|)$ algorithm to construct a PES or to show that a graph is not chordal.

25. Let N be a set of subtrees of a tree. Let A be the resulting node-tree incidence matrix. Show that A is the clique matrix of a chordal graph and conversely.

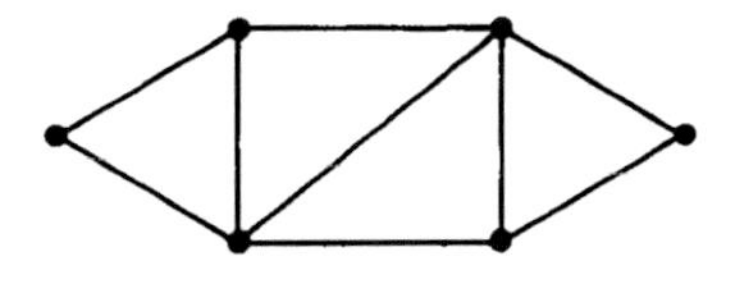

Figure 8.1

26. Let A be the clique matrix of a chordal graph with PES $= \{1, \ldots, n\}$. Give an algorithm to solve $\max\{cx: Ax \leqslant b, x \in Z_+^n\}$, where $b_1 \leqslant b_2 \leqslant \cdots \leqslant b_n$.

27. Give a polynomial algorithm for node coloring of chordal graphs.

28. $G = (V, E)$ is a *comparability* graph if there is an orientation of each edge $e \in E$ giving a digraph $\mathcal{D} = (V, \mathcal{A})$ having the properties that if $(i, j), (j, k) \in \mathcal{A}$, then $(i, k) \in \mathcal{A}$. Show that comparability graphs are perfect.

29. i) What is the rank 1 hull of the node-packing problem, where $S = P \cap Z^n$ and $P = \{x \in R_+^n: x_i + x_j \leqslant 1 \text{ for } e = (i, j) \in E\}$?

ii) Show that the rank 1 hull is not integral if G contains a node-induced subgraph of the form shown in Figure 8.2.

30. Describe the convex hull of incidence vectors of node packings for line graphs.

31. A clutter $\mathcal{F}$ is represented by the rows of the matrix

$$A = \begin{pmatrix} 1 & 1 & 0 & 0 & 0 \\ 1 & 0 & 1 & 0 & 0 \\ 0 & 1 & 0 & 1 & 0 \\ 0 & 0 & 1 & 1 & 1 \\ 0 & 1 & 0 & 0 & 1 \end{pmatrix}.$$

i) Find its blocking clutter $B(\mathcal{F})$.

ii) Find the polyhedron Q_B^* of the form $\{x \in R_+^n: Bx \geqslant 1\}$ with $B \geqslant 0$ having the incidence vectors of the members of $B(\mathcal{F})$ as extreme points.

iii) Find a polyhedron Q^* of the above form having the incidence vectors of the members of $\mathcal{F}$ as extreme points.

32. Given a connected graph $G = (V, E)$, let A be the incidence matrix of spanning trees by edges; that is, each row of A is the incidence vector of a spanning tree.

i) Give a polynomial-time ellipsoid algorithm for solving the linear program $\min\{cx: Ax \geqslant 1, 0 \leqslant x \leqslant 1\}$.

ii) Specify r, q, and any other information needed by the ellipsoid algorithm.

iii) Give a combinatorial interpretation of the problem. Do you know an efficient combinatorial algorithm for solving it?

33. Let $\mathcal{F}$ be the clutter of spanning trees of a graph G.

i) Find its blocking clutter $B(\mathcal{F})$.

ii) Give an example to show that $Q = \{x \in R_+^{|E|}: a^H x \geqslant 1 \text{ for } H \in B(\mathcal{F})\}$ is not integral.

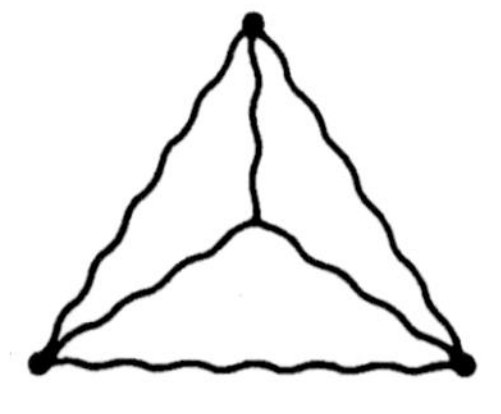

Figure 8.2. "Odd K^4"; each wavy line denotes a path with an odd number of edges.

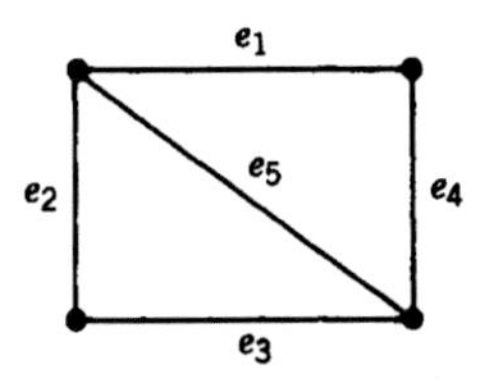

Figure 8.3

34. Let $\mathscr{F}$ be the clutter of branchings rooted at node 1 in $\mathscr{D}$.

i) Find its blocking clutter $B(\mathscr{F})$.

ii) Show that $Q = \{x \in R_+^{|\mathscr{A}|}: a^H x \geq 1 \text{ for } H \in B(\mathscr{F})\}$ is integral.

iii) Let A be the branching by arc incidence matrix.

a) Does the max–min inequality hold strongly for rooted branchings? That is, does

$$\max\{1y: yA \leq w,\ y \in Z_+^m\} = \min_{H \in B(\mathscr{F})} \{a^H w : w \in Z_+^{|\mathscr{A}|}\}?$$

b) Does the IRD property hold for $\{y \in R_+^m: yA \leq w\}$?

35. Let $\mathscr{F}$ be the clutter of cycles in a graph G.

i) Find its blocking clutter $B(\mathscr{F})$.

ii) Use the graph in Figure 8.3 with $w = (4 \quad 3 \quad 2 \quad 1 \quad 8)$ to show that Q is not integral.

iii) Give a polynomial combinatorial algorithm to find a minimum-weight cycle when $w \in R_+^{|E|}$.

iv) Give a polynomial combinatorial algorithm to find a minimum-weight element of $B(\mathscr{F})$.

36. For the graph of Figure 8.4, find the maximum number of s–t cuts that can be packed into w, where w is indicated in the figure.

37. Let $\mathscr{F}$ be the set of minimal feasible solutions to $S = \{x \in B^n: \Sigma_{j \in N} a_j x_j \geq b\}$.

i) Find the blocking clutter $B(\mathscr{F})$.

ii) Use this to give a reformulation of $\min\{cx: x \in S\}$ with $c \in R_+^n$.

iii) Compare the inequalities of this form with the valid inequalities generated in Section II.2.2.

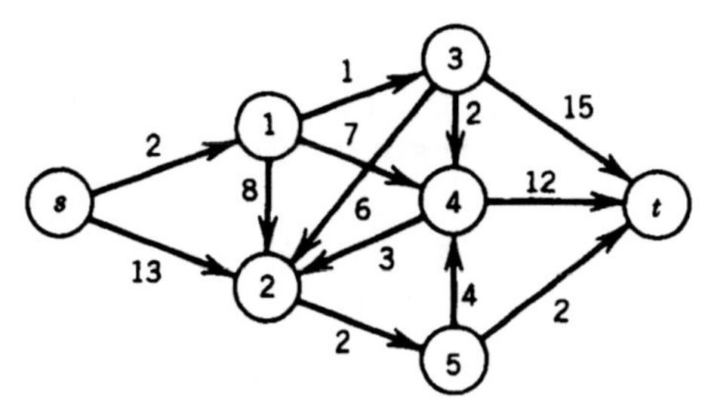

Figure 8.4

III.2
Matching

1. INTRODUCTION

In a graph $G = (V, E)$, the number of edges that meet node i is called the *degree* of node i. Matching problems involve choosing a subset of the edges subject to degree constraints on the nodes. The simplest case is *1-matching* (or just matching). A *matching* $M \subseteq E$ is a subset of edges with the property that each node in the subgraph $G(M) = (V, M)$ is met by no more than one edge. Every graph G contains a matching, namely $M = \emptyset$. An obvious generalization of 1-matching is *b-matching* in which node i is met by no more than b_i edges, where b_i is a positive integer. In a b-matching problem, we may impose the restriction that each edge is chosen no more than once (*0-1 b-matching*) or allow an edge to be chosen a nonnegative integer number of times (*integer b-matching*). A b-matching is called *perfect* if each of the degree constraints holds with equality. In particular, in a *perfect 1-matching* each node is met by exactly one edge. Another variation on matchings is to require that each node i be met by at least b_i edges. These problems are called *node covering by edges*.

Let c_e be the weight of $e \in E$ and let $c(E') = \Sigma_{e \in E'} c_e$ be the weight of $E' \subseteq E$. The *weighted b-matching problem* is to find a b-matching of maximum weight. In the case of perfect matchings, it also makes sense to consider minimum-weight matchings. When $c_e = 1$ for all $e \in E$, the optimization problem is called an *unweighted or cardinality problem*.

An integer programming formulation of the weighted 0-1 b-matching problem is

$$\max cx$$
$$Ax \leqslant b$$
$$x \in B^n,$$

where A is the node-edge incidence matrix of the graph, $|E| = n$, and $x_e = 1$ means that e is in the matching.

The important property of A for matching problems is that *each of its columns contains exactly two 1's*; in other words, $\Sigma_i a_{ij} = 2$ for all $j \in E$. Note that if the graph is bipartite, then A is totally unimodular so that the extreme points of $\{x \in R^n_+ : Ax \leqslant b\}$ are precisely the b-matchings. However, when G contains an odd cycle, the constraint set of the linear programming relaxation can contain fractional extreme points. For example, in the graph of Figure 1.1, $x = (\frac{1}{2} \;\; \frac{1}{2} \;\; \frac{1}{2})$ is the unique optimal solution to the linear programming relaxation with $c = (1 \;\; 1 \;\; 1)$ and $b = (1 \;\; 1 \;\; 1)$.

The classic application of matching deals with the pairing of objects from two disjoint sets (e.g., workers with jobs, men with women, etc.). The perfect matching problem

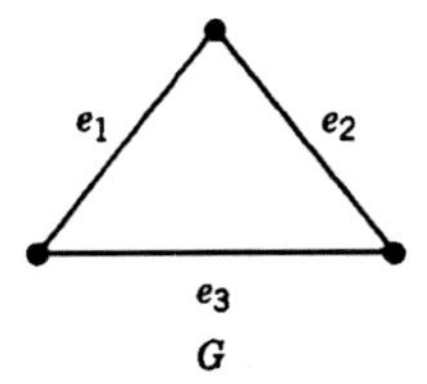

Figure 1.1

associated with such pairings is on a bipartite graph, and the optimization problem is the assignment problem (see Section I.3.5). Pairings, however, do not necessarily involve disjoint sets (e.g., the selection of roommates in a college dormitory). So we see that the weighted perfect 1-matching problem is a meaningful generalization of the assignment problem.

We have already mentioned some other applications of matching in connection with relaxations and heuristics for the traveling salesman problem (see Section II.6.3). For example, a perfect 0-1 2-matching is a relaxation of the traveling salesman problem. We have also used weighted 1-matching in the spanning-tree matching heuristic for the euclidean traveling salesman problem.

Another application of weighted 1-matching is to the postman problem. Given a graph G with weights on the edges, the *postman problem* is to find a minimum-weight set of edges to add to G so that the resulting multigraph MG contains a eulerian cycle (i.e., a closed walk containing each edge of MG exactly once; see Section II.6.3). The eulerian cycle on MG translates to a minimum-weight closed walk on G in which each edge is visited at least once and therefore generates a minimum-weight delivery route for the postman.

Recall that a multigraph is eulerian if and only if each node is of even degree. Let V' be the nodes of odd degree in G, and let c_{ij} be the weight of a minimum-weight path between nodes i and j in V' (see Figure 1.2). Now consider the complete graph $G' = (V', E')$, and

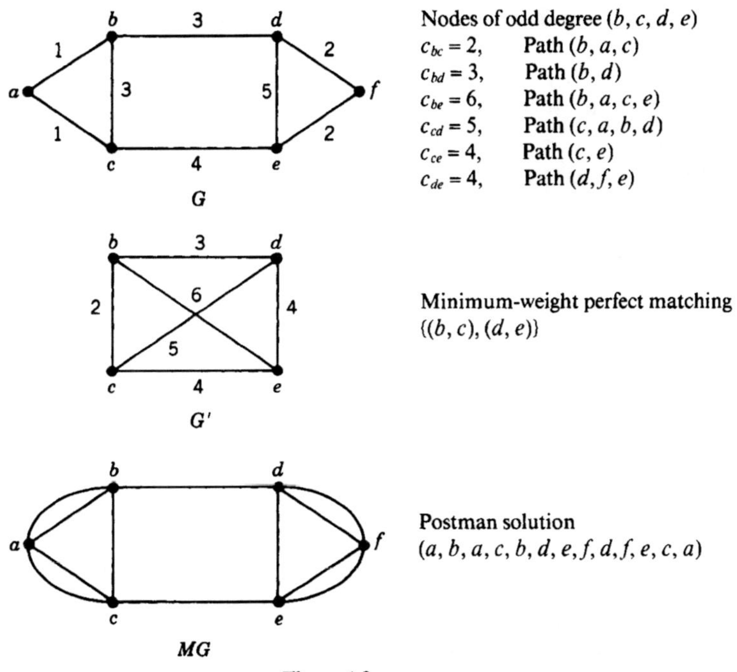

Figure 1.2

let c_{ij} be the weight of $e = (i, j)$ in E'. Let M' be a minimum-weight perfect matching on G'. For each edge $(i, j) \in M'$, add to G a minimum-weight path joining i and j. We leave it as an exercise to show that the resulting multigraph generates an optimal solution to the postman problem.

Matching problems are celebrated in the history of combinatorial optimization as the first true integer programs (i.e., integer programs that cannot be solved merely from the linear programming relaxation) for which polynomial-time algorithms were obtained. Moreover, these algorithms use a class of valid inequalities for the convex hull of matchings and, in fact, prove that these inequalities, together with the degree and nonnegativity constraints, give a linear inequality description of the convex hull of matchings.

We will mainly study the weighted 1-matching problem stated as the integer program

$$\text{(WM)} \qquad \begin{aligned} &\max \sum_{e \in E} c_e x_e \\ &\sum_{e \in \delta(v)} x_e \leqslant 1 \quad \text{for } v \in V \\ &x \in B^n. \end{aligned}$$

The more general problem of weighted b-matching, as well as other generalizations that allow any constraint matrix A with $\Sigma_i \; |a_{ij}| \leqslant 2$ for all j, are examined in Section 4.

We say that $U \subseteq V$ is an odd set if $|U| \geqslant 3$ and is odd. We have already given the valid inequalities, called *odd-set constraints*:

$$\text{(1.1)} \qquad \sum_{e \in E(U)} x_e \leqslant \left\lfloor \frac{|U|}{2} \right\rfloor \quad \text{for all odd sets } U \subseteq V$$

for the convex hull of matchings. Recall that they are valid since each matching edge "uses up" two nodes. They can be obtained by one iteration of the integer-rounding procedure (see Sections II.1.1 and II.1.2). They are, of course, also valid when $|U|$ is even, but they are not interesting because they can be obtained as a nonnegative linear combination of the degree and nonnegativity constraints.

The main results of this chapter are a polynomial-time algorithm for WM and a linear inequality description of the convex hull of matchings. The algorithm solves the linear program

$$\text{(1.2)} \qquad \begin{aligned} &\max \sum_{e \in E} c_e x_e \\ &\sum_{e \in \delta(v)} x_e \leqslant 1 \quad \text{for } v \in V \\ &\sum_{e \in E(U)} x_e \leqslant \left\lfloor \frac{|U|}{2} \right\rfloor \quad \text{for all odd sets } U \subseteq V \\ &x \in R^n_+ \end{aligned}$$

and obtains an integral optimal solution, and therefore a matching, for any objective function vector c. Hence it provides a proof that the convex hull of matchings is given by the degree, nonnegativity, and odd-set constraints.

An algorithm for maximum-weight matching can also be used to find a maximum-weight perfect matching, when one exists, by a simple transformation of the objective function. Let $k = |V|/2$, $a = \max_{e \in E} \max(c_e, 0)$, $b = \min_{e \in E} c_e$, $\theta = k(a - b) + 1$, and $c'_e = c_e + \theta$ for $e \in E$. The ranking of perfect matchings by weight is the same for c and c'. Moreover, with respect to the objective function c', a lower bound on the weight of a perfect matching is $k(b + \theta)$, and an upper bound on the weight of an imperfect matching is $(k - 1)(a + \theta)$. Now

$$k(b + \theta) - (k - 1)(a + \theta) = \theta + kb - (k - 1)a = a + 1 > 0$$

so that any perfect matching has greater weight than any imperfect matching.

Our approach to solving (1.2) is by a primal–dual algorithm similar to the algorithm we gave for the transportation problem in Section I.3.5. The main difficulty to overcome is the exponential number of odd-set constraints.

The dual of (1.2) is

$$w = \min \sum_{v \in V} \pi_v + \sum_{\text{odd sets } U} \left\lfloor \frac{|U|}{2} \right\rfloor y_U$$

$$(1.3) \qquad \begin{aligned} \sum_{v: e \in \delta(v)} \pi_v + \sum_{U: e \in E(U)} y_U &\geq c_e \quad \text{for } e \in E \\ \pi_v &\geq 0 \quad \text{for } v \in V \\ y_U &\geq 0 \quad \text{for all odd sets } U. \end{aligned}$$

The algorithm to be presented maintains primal and dual feasibility and achieves optimality when the complementary slackness conditions are satisfied. At each major iteration the cardinality of the matching is increased. This is done by solving a cardinality matching problem. So we begin the presentation of the general algorithm by studying maximum cardinality matching.

2. MAXIMUM-CARDINALITY MATCHING

In our study of the maximum-flow problem (see Section I.3.4), we gave necessary and sufficient conditions for a flow to be maximum in terms of augmenting paths. That is, the flow could be increased if and only if an augmenting path existed with respect to the current flow. We then gave an efficient procedure for finding an augmenting path or showing that none existed. We use the same idea to find a maximum-cardinality matching. Thus we begin by defining an augmenting path with respect to a matching.

Given a graph G and a matching M, a path in G is said to be *alternating* relative to M if its edges alternate between M and $E \setminus M$. (See Figure 2.1, where edges in M are represented by wavy lines.) A node v is said to be *exposed* relative to M if no edge of M meets v. A path in G is *augmenting* relative to M if it is alternating and both of its end nodes are exposed. This definition is natural since, if there is an augmenting path relative to M, a new matching M' with one more edge is obtained by deleting from M the matching edges in the path and adding to M the nonmatching edges in the path (see Figure 2.2).

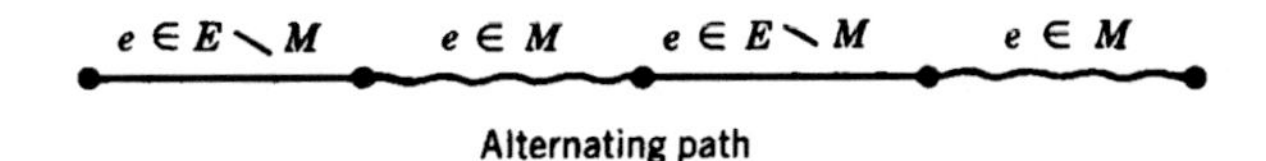

Figure 2.1

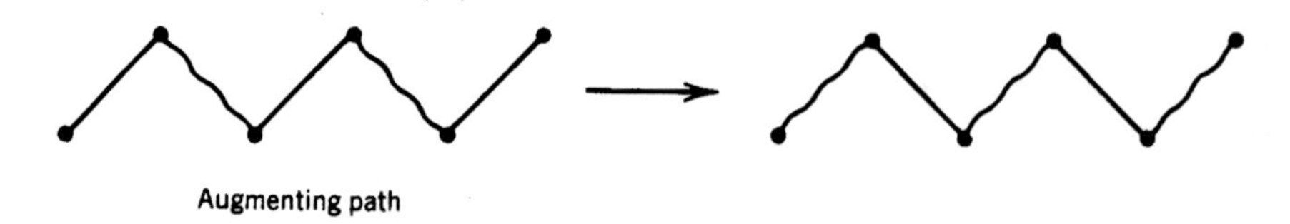

Figure 2.2

The interesting result is that if there is no augmenting path, the matching is maximum.

Theorem 2.1. *A matching M is not maximum if and only if there exists an augmenting path relative to M.*

Proof. Let E' be the edge set of the augmenting path, and let $M' = (M \cup E') \setminus (M \cap E')$. Then M' is a matching, and $|M'| = |M| + 1$. This formally establishes our claim that the existence of an augmenting path implies that the matching is not maximum.

We now show that if M is not maximum, then there exists an augmenting path relative to M. If M is not maximum, there exists a matching M' with $|M'| = |M| + 1$. Let D be the symmetric difference of M and M'; that is, $D = (M \cup M') \setminus (M \cap M')$. Thus

$$|D| = |M| + |M'| - 2|M \cap M'| = 2|M| + 1 - 2|M \cap M'|.$$

Hence $|D|$ is odd.

Consider the subgraph $G(D) = (V, D)$. Since M and M' are matchings, the degree of each node is no more than 2; and if the degree is 2, then one edge is from M and the other is from M'. Hence each component of $G(D)$ is either an isolated node, a cycle containing an even number of edges, or an alternating path relative to both M and M'. Since $|D|$ is odd, there must be at least one alternating path of odd length. Moreover, since $|M'| = |M| + 1$, one of these alternating paths of odd length must be augmenting with respect to M. ■

The basic idea of the augmenting-path algorithm is to grow a tree of alternating paths rooted at an exposed node. Then if a leaf of the tree is also exposed, an augmenting path has been found. We begin by describing an augmenting-path algorithm for bipartite graphs. Finding an augmenting path in a bipartite graph is much simpler than finding one in a general graph. In fact, in the primal–dual algorithm for the transportation problem, we have shown that an augmenting path in a bipartite graph can be found by finding a flow augmentation in a maximum-flow problem. The algorithm given below is essentially a flow-augmentation algorithm described with augmenting-path terminology. This terminology will be useful in the description of the general algorithm.

Cardinality Matching Algorithm for Bipartite Graphs

Initialization: M is an arbitrary matching. All nodes are unlabeled and unscanned.

Step 1 (Optimality Test): If no nodes are exposed and unlabeled, the current matching is maximum. Otherwise choose an exposed and unlabeled node r. Label it $(E, -)$. (Here E stands for even and should not be confused with the usual use of E for an edge set. The first component of a node label is either E or O. A labeled node is said to be *even* if the first component of its label is E; otherwise it is *odd*.)

Step 2 (Grow an Alternating Tree): Choose a labeled and unscanned node i. If it is even, let $J = (j \in V: j$ is an unlabeled neighbor of $i)$. Label all $j \in J$ with (O, i). Node i is scanned; go to Step 3. If i is odd and exposed, go to Step 4. If i is odd and not exposed, label the node joined to i by a matching edge (E, i). Node i is scanned; go to Step 3.

Step 3: If there is a labeled and unscanned node, go to Step 2; otherwise go to Step 1.

Step 4 (Augmenting-Path Identification): Use the second components of the labels to identify the augmenting path from node r to node i. Remove all labels, update the matching, and return to Step 1.

Theorem 2.2. *The algorithm produces a maximum-cardinality matching on a bipartite graph.*

Proof. In Steps 2 and 3, we grow a forest of alternating paths. An odd node i yields an alternating path between r and i with an odd number of edges. Hence if i is exposed, the path is augmenting.

Now we show that if there are no exposed and unlabeled nodes, the final matching M^0 is maximum. We do this by giving a feasible solution to the dual problem (1.3) with $w = |M^0|$. One way to obtain a feasible solution to (1.3) is to find a subset of nodes $W \subseteq V$ such that each $e \in E$ is incident to a node in W. Then we set $\pi_v = 1$ for all $v \in W$, $\pi_v = 0$ otherwise, and $y_U = 0$ for all odd sets. Here our objective is to produce a dual feasible solution of this form with $|W| = |M^0|$.

When the algorithm terminates, we have a set of labeled trees $T_i = (V_i, E_i)$ for $i = 1, \ldots, s-1$, and we also have a set of unlabeled nodes V_s (see Figure 2.3).

Since no nodes in V_s are exposed, the subgraph induced by V_s, (V_s, E_s) contains a perfect matching. Let (V_s^1, V_s^2) be a bipartition of V_s, and let $V_i^0 \subset V_i$ be the odd nodes of T_i for $i = 1, \ldots, s-1$. Set $W = \bigcup_{i=1}^{s-1} V_i^0 \cup V_s^1$.

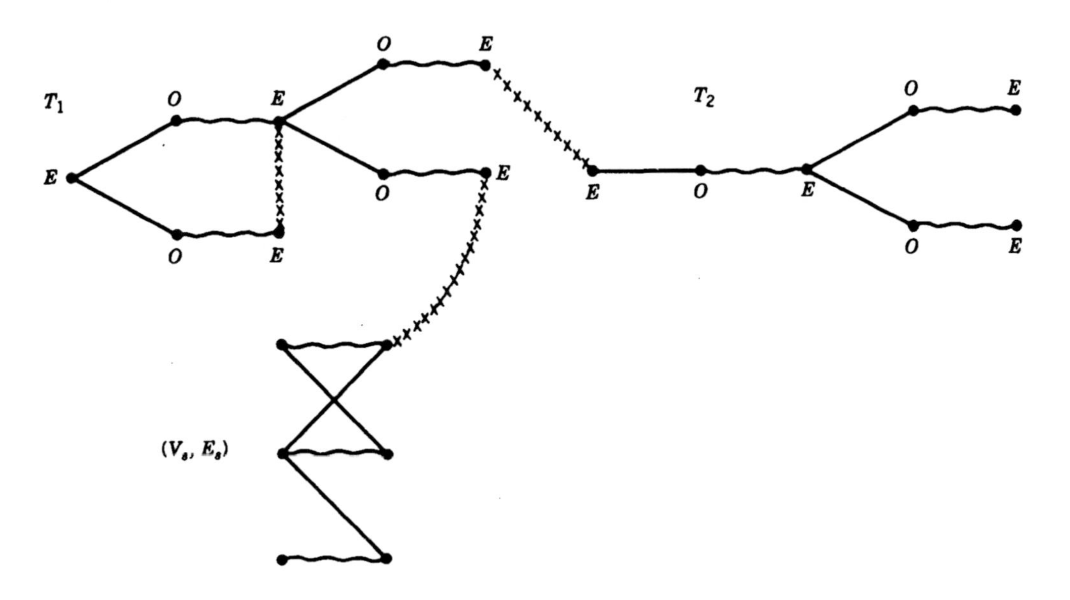

Figure 2.3. Crosses mean that the edge cannot be in E.

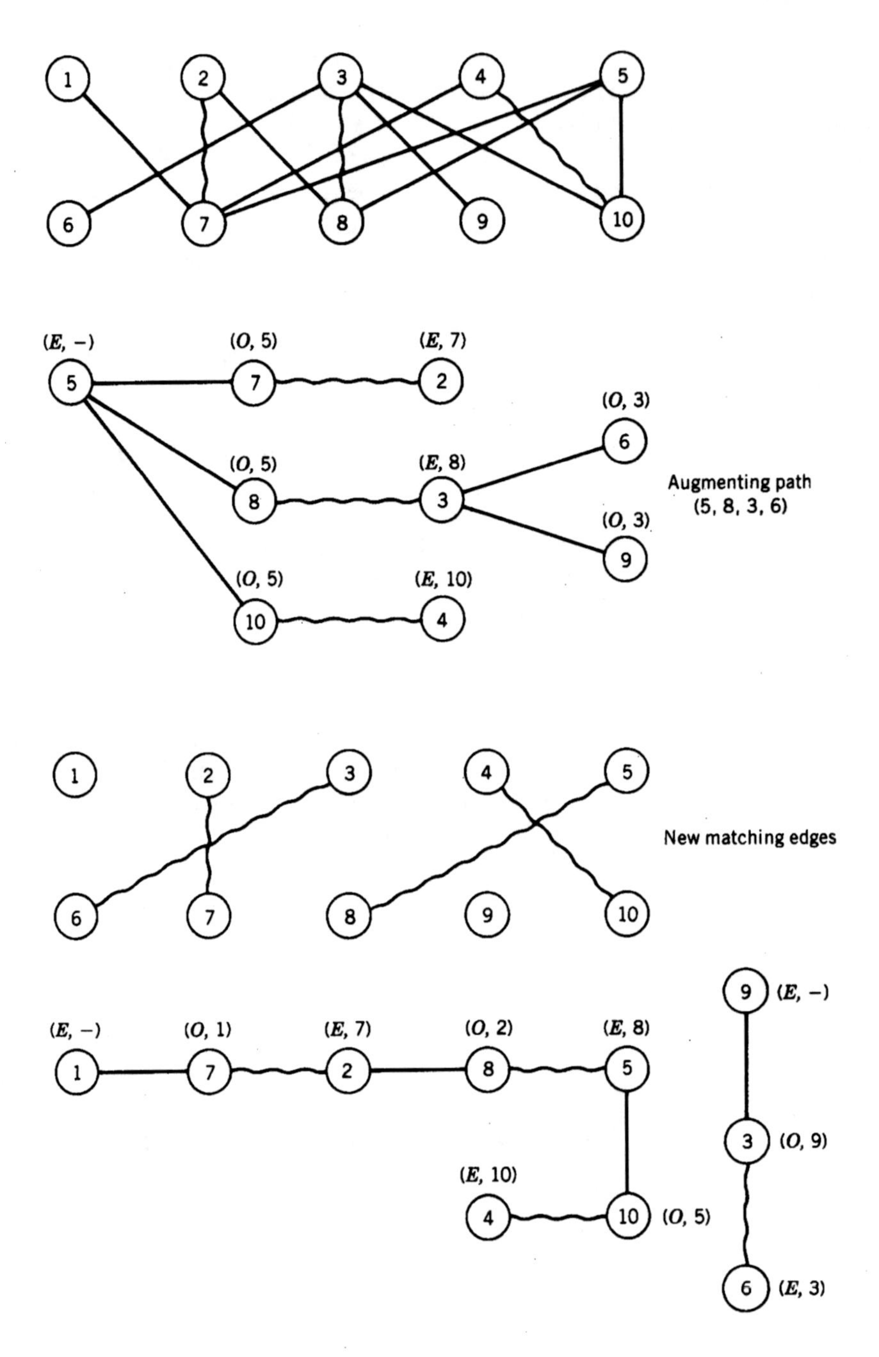

Maximum matching: $M^0 = \{(2, 7), (3, 6), (4, 10), (5, 8)\}$

Optimal dual solution: $\pi_v = \begin{cases} 1 & \text{for } v = 3, 7, 8, 10 \\ 0 & \text{otherwise.} \end{cases}$

Figure 2.4

W generates a dual feasible solution since:

a. Each $e \in \cup_{i=1}^{s} E_i$ is incident to a node in W.
b. In the subgraph induced by V_i, $i = 1, \ldots, s-1$, there cannot be an edge joining two even nodes; otherwise there would be an odd cycle.
c. There cannot be an edge joining even nodes in different trees; otherwise one of these nodes would have been labeled from the other.
d. There cannot be an edge joining an even node and an unlabeled node; otherwise the unlabeled node would have been labeled from the even node.

To show strong duality, note that

$$|V_i^0| = |E_i \cap M^0| \quad \text{for } i = 1, \ldots, s-1 \text{ and } |V_s^1| = |E_s \cap M^0|.$$

Hence $|W| = \Sigma_{i=1}^{s} |E_i \cap M^0| = |M^0|$. ■

Example 2.1. An example of the algorithm is given in Figure 2.4.

The algorithm may fail to find an augmenting path if the graph is not bipartite. An example is shown in Figure 2.5. Here there are two paths between nodes 1 and 4. The odd-length path is augmenting, but we find it only by labeling in a particular way.

We now develop a procedure that circumvents this problem. Let M be a matching. Suppose in the process of growing an alternating tree using the algorithm given above, we find that there are two alternating paths to node i, one of even length and the other of odd length. This can happen in two ways (see Figure 2.6):

a. Node i is even and is adjacent to another even node in the tree;
b. Node i is odd, adjacent to another odd node in the tree, and the edge that joins them is in M.

By tracing the two paths back toward the root of the tree until the node where they intersect is reached, we identify a set of labeled nodes $U \subseteq V$ with $|U|$ odd and $|M \cap E(U)| = \lfloor |U/2| \rfloor$. Note that in both cases the intersection node, denoted by $b(U)$, is even.

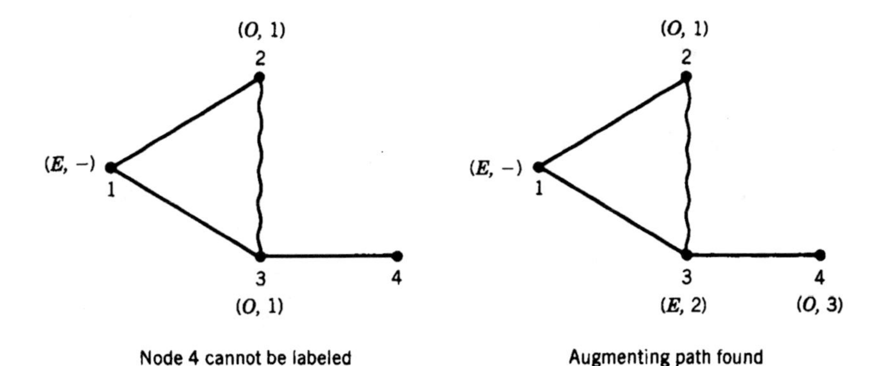

Figure 2.5

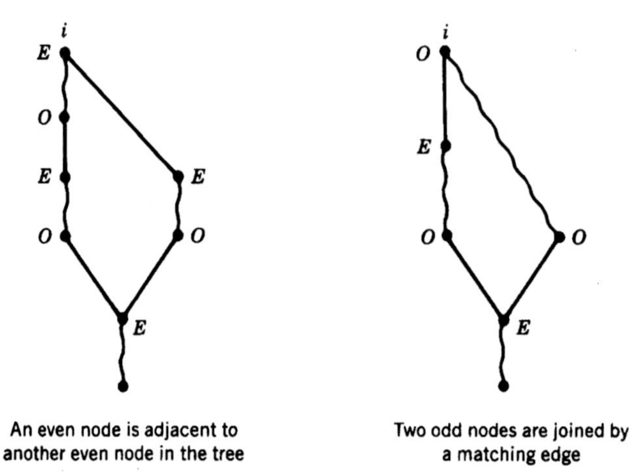

Figure 2.6

Thus relative to M, the odd-set constraint for U is satisfied at equality. The subgraph $(U, E(U))$ is called a *blossom* relative to M. Each $u \in U \setminus b(U)$ is met by an edge in $M \cap E(U)$. Node $b(U)$, which is called the *base of the blossom*, is either the root of the tree or is adjacent to a matching edge in the tree.

Now we shrink the blossom as described below and illustrated in Figure 2.7.

Procedure for Shrinking a Blossom

Construct a *reduced graph* $\tilde{G}$ by replacing $(U, E(U))$ by a node $B(U)$ called a *pseudonode*. In $\tilde{G}$, each node that is adjacent to a node in U in the original graph is joined to the pseudonode. All of these edges are nonmatching edges unless there is a matching edge adjacent to $b(U)$, in which case that edge remains a matching edge in the shrunken graph. The remainder of the graph remains the same. The resulting reduced matching on $\tilde{G}$ is denoted $\tilde{M}$. $B(U)$ receives the label previously assigned to $b(U)$, and any node not in U that has been labeled from a node in U has the second component of its label changed to $B(U)$. Also record the triple $(B(U),\ b(U),\ U)$. After a blossom is shrunk, the labeling process continues on the reduced graph $\tilde{G}$. A reduced graph may be shrunk again, and it may happen that a blossom to be shrunk contains a pseudonode. In this case the pseudonode is treated like an ordinary node. Both the terminology *reduced graph* and the notation $\tilde{G}$ are used for any graph that contains a pseudonode; and correspondingly, $\tilde{M}$ is used to indicate a matching on the reduced graph.

When an augmenting path is found in a reduced graph $\tilde{G}$ with matching $\tilde{M}$, we also find an augmenting path in G with respect to M. A procedure for finding an augmenting path in G is given below and illustrated in Figure 2.8.

Procedure for Obtaining an Augmenting Path in G

Let $B(U)$ be a pseudonode on the augmenting path in $\tilde{G}$. Let $a(U)$ be the node adjacent to $B(U)$ on the augmenting path that is joined to $B(U)$ by a nonmatching edge, let G' be the graph obtained by replacing $B(U)$ by the blossom $(U, E(U))$, and let $b'(U)$ be a node in U that is adjacent to $a(U)$ in G'. By construction, $(U, E(U))$ contains an even-length alternating path p joining $b(u)$ and $b'(u)$. Replace $B(U)$ in the augmenting path on $\tilde{G}$ by the path p. This yields an augmenting path in G'. The procedure is repeated for each pseudonode on the augmenting path. Note that old pseudonodes may reappear when a pseudonode is replaced by an alternating path.

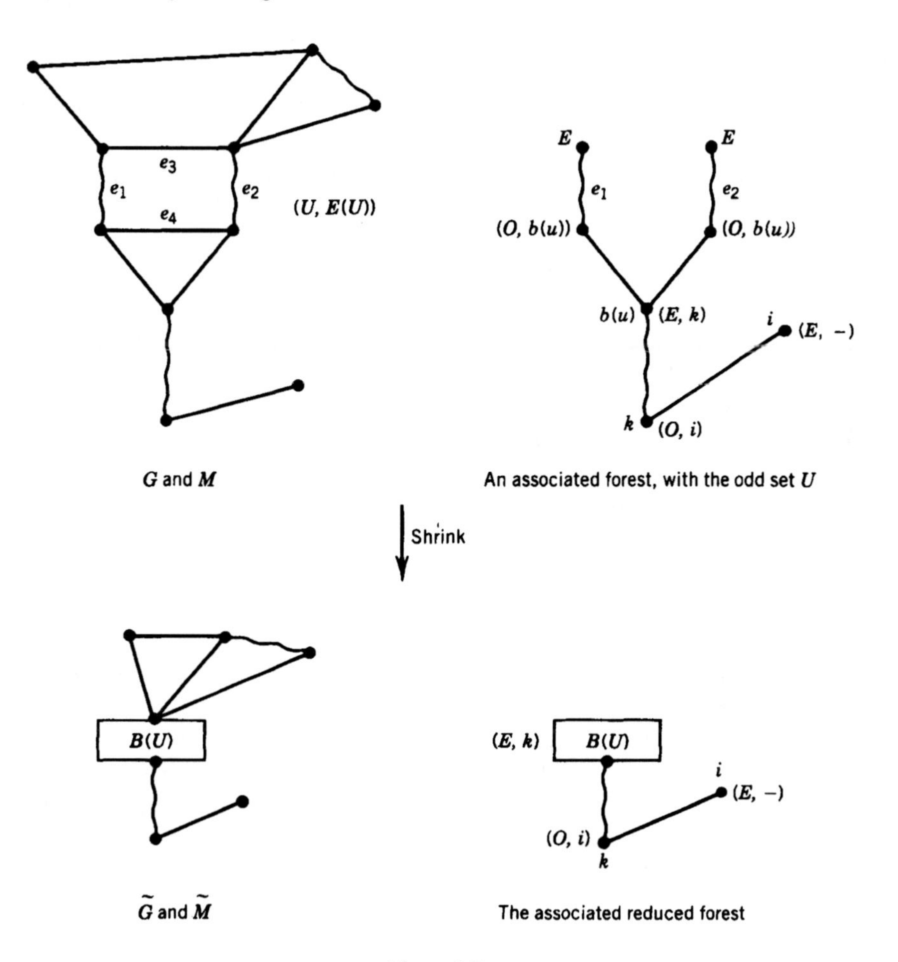

Figure 2.7

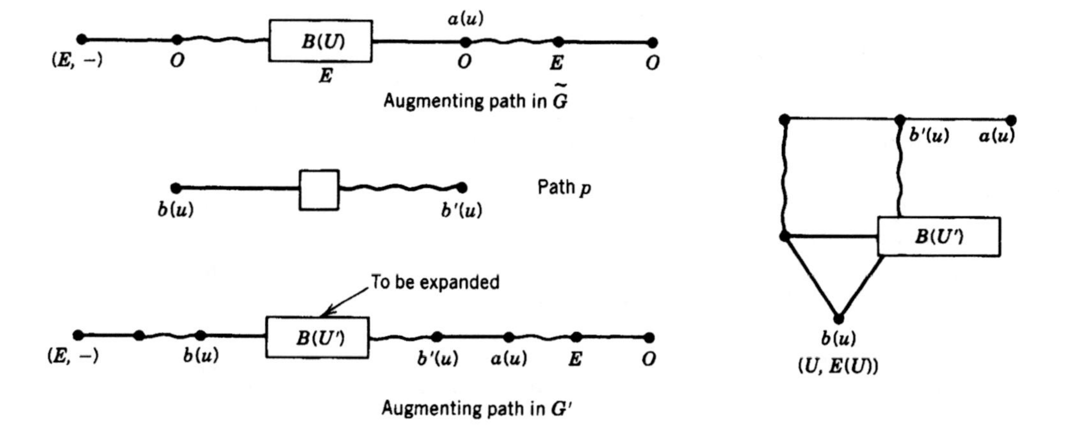

Figure 2.8

General Cardinality Matching Algorithm I

Initialization: M is an arbitrary matching. $\tilde{G} = G$. $\tilde{M} = M$. All nodes are unlabeled and unscanned.

Step 1 (Optimality Test): If no nodes in $\tilde{G}$ are exposed and unlabeled, $\tilde{M}$ is maximum in $\tilde{G}$ and M is maximum in G. Otherwise choose an exposed and unlabeled node r in $\tilde{V}$. Label it $(E, -)$.

Step 2 (Grow an Alternating Forest): Choose a labeled and unscanned node or pseudonode $i \in \tilde{V}$. If there is none, go to Step 1.

a. If i is even and has an even neighbor, go to Step 4.
b. If i is even and does not have an even neighbor, label all unlabeled neighbors of i with (O, i). Node i is scanned; go to Step 3.
c. If i is odd and is exposed, go to Step 5.
d. If i is odd and is not exposed, label the endpoint of the matching edge adjacent to node i with (E, i). Node i is scanned; go to Step 3.

Step 3: If there is a labeled and unscanned node or pseudonode in $\tilde{G}$, go to Step 2. Otherwise go to Step 1.

Step 4 (Shrink a Blossom): Use the second components of the labels on node i and its neighbors to identify a blossom $(U, E(U))$ and its base $b(U)$. Use the shrinking procedure described above to replace $(U, E(U))$ by a pseudonode $B(U)$. Complete the scanning of $B(U)$ as in Step 2 [$B(U)$ has an even label] and then go to Step 3.

Step 5 (Augmentation in G): Use the second components of the labels to identify an augmenting path in $\tilde{G}$, and use the procedure given above for identifying an augmenting path in G. Find a new matching M' in G, and $M \leftarrow M'$, $\tilde{G} \leftarrow G$, and $\tilde{M} \leftarrow M$. Return to Step 1.

Theorem 2.3. *The algorithm produces a maximum-cardinality matching.*

Proof. When the algorithm terminates in Step 1, we have a reduced graph $\tilde{G}$, the associated matching $\tilde{M}$, and the matching M^0 in the original graph. We also have a set of labeled trees $\tilde{T}_i = (\tilde{V}_i, \tilde{E}_i)$ for $i = 1, \ldots, s - 1$ and a set of unlabeled nodes $\tilde{V}_s$. The subgraph $\tilde{G}_s$ induced by $\tilde{V}_s$ contains a perfect matching since no nodes in $\tilde{V}_s$ are exposed. Moreover, all of the matching edges $e \in \tilde{M}$ are either tree edges, edges internal to a shrunken blossom, or edges internal to $\tilde{G}_s$.

We show that M^0 is maximum by giving a feasible solution to the dual linear program (1.3) with $w = |M^0|$. The dual solution and the proof of its feasibility and optimality are similar to those given in the proof of Theorem 2.2. So only the details that are different are given here.

When the algorithm terminates, all labeled pseudonodes in the shrunken graph $\tilde{G}$ are even. Let $B(U)$ be a labeled pseudonode in $\tilde{G}$, and let

$$R(U) = \{v \in V: v \in U \text{ or } v \text{ is in a blossom nested in } B(U)\}.$$

Since U is an odd set and $R(U)$ is obtained from U by replacing pseudonodes by odd sets, $R(U)$ is an odd set. The dual constraints for the edges in the graph induced by $R(U)$ are satisfied by setting $y_{R(U)} = 1$.

Now consider $\tilde{G}_s$, and let

$$Q(\tilde{V}_s) = \{v \in V: v \in \tilde{V}_s \text{ or } v \text{ is in a blossom nested in a pseudonode in } \tilde{G}_s\}.$$

Note that since $|\tilde{V}_s|$ is even and $Q(\tilde{V}_s)$ is obtained from $\tilde{V}_s$ by replacing pseudonodes by blossoms, $|Q(\tilde{V}_s)|$ is even and there is a perfect matching on the subgraph induced by $Q(\tilde{V}_s)$. If $|Q(\tilde{V}_s)| = 2$, the dual constraints for edges in the subgraph are satisfied by setting $\pi_q = 1$ for any one $q \in Q(\tilde{V}_s)$. Otherwise let q be an arbitrary node in $Q(\tilde{V}_s)$. Then $Q(\tilde{V}_s) \setminus \{q\}$ is an odd set containing $(|Q(\tilde{V}_s)| - 2)/2$ matching edges. Hence the dual constraints for the edges in the subgraph induced by $Q(\tilde{V}_s) \setminus \{q\}$ are satisfied by setting $y_{Q(\tilde{V}_s)\setminus\{q\}} = 1$.

Let $\tilde{V}_i^0 \subseteq \tilde{V}_i$ be the odd nodes of T_i for $i = 1, \ldots, s-1$. A feasible solution to (1.3) is given by

$\pi_v = 1$	if $v \in \tilde{V}_i^0$ for $i = 1, \ldots, s-1$
$\pi_q = 1$	for any one $q \in Q(\tilde{V}_s)$
$\pi_v = 0$	otherwise
$y_{R(U)} = 1$	if $B(U)$ is a labeled pseudonode in the final graph
$y_{Q(\tilde{V}_s)\setminus\{q\}} = 1$	if $\lvert Q(\tilde{V}_s)\rvert > 2$
$y_U = 0$	otherwise.

Now following the argument in the proof of Theorem 2.2, we obtain

$$\sum_{v \in V} \pi_v + \sum_{\text{odd sets } U} \left\lfloor \frac{|U|}{2} \right\rfloor y_U \leq |M^0|.$$

So by weak duality, $w = |M^0|$. ■

Let $\lfloor |U/2| \rfloor$ be the weight of the odd set U.

Corollary 2.4. *The maximum number of edges in a matching equals the minimum number of nodes plus weighted odd sets needed to cover all the edges.*

Now we consider the complexity of the algorithm, where $m = |V|$ and $n = |E|$. The number of augmentations is no more than $m/2$. Between augmentations we need to create an alternating forest, contract pseudonodes, and then reexpand to find the new matching. Using the labels and storing blossoms appropriately, these steps can be carried out in such a way that each edge is considered only a constant number of times.

Proposition 2.5. *The complexity of the cardinality matching algorithm is $O(mn)$.*

Example 2.2

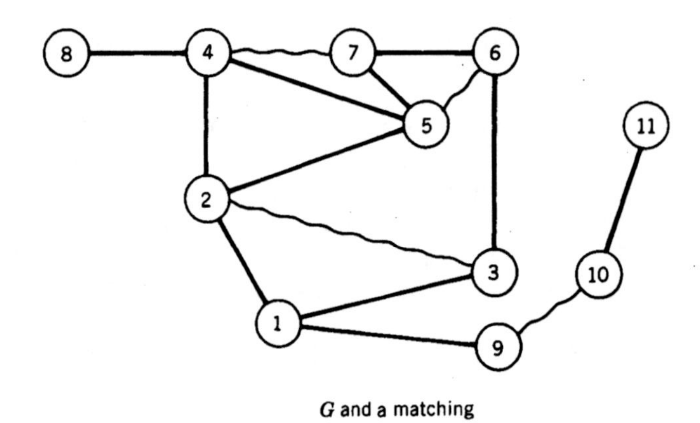

G and a matching

1. We grow an alternating tree rooted at node 1. Node 2 is chosen next, and $(2, 3) \in M$ indicates the blossom with $U_1 = \{1, 2, 3\}$ and $b(U_1) = 1$.

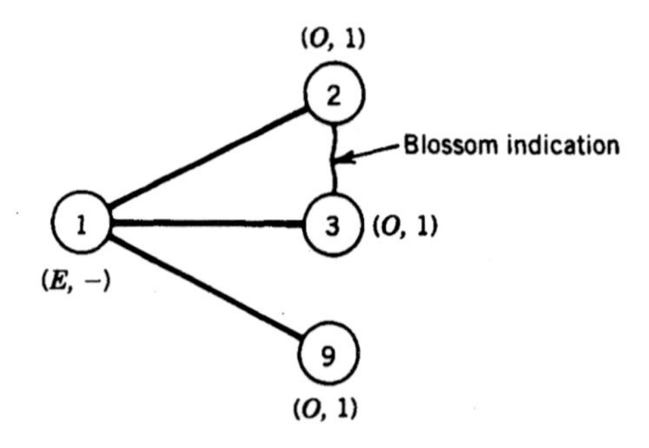

2. The pseudonode $B(U_1) = B_1$ is created and becomes the root of the tree. B_1 is scanned. Node 4 is scanned. In scanning node 5, a blossom is identified with node set $U_2 = \{B_1, 5, 6\}$ and $b(U_2) = B_1$.

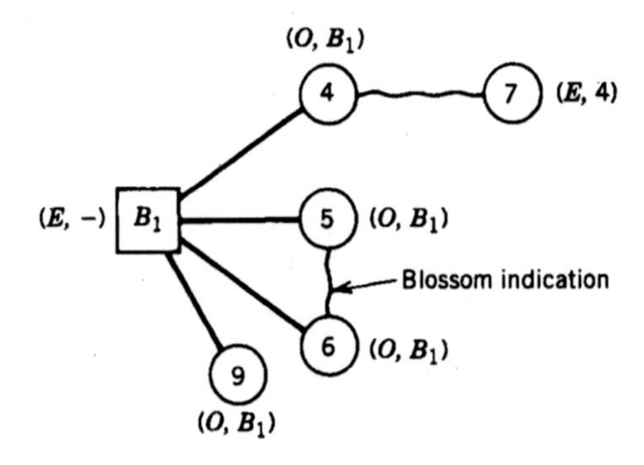

3. The pseudonode $B_2 = B(U_2)$ is created and becomes the root of the tree. In scanning B_2, we find the blossom with $U_3 = \{B_2, 4, 7\}$ and $b(U_3) = B_2$.

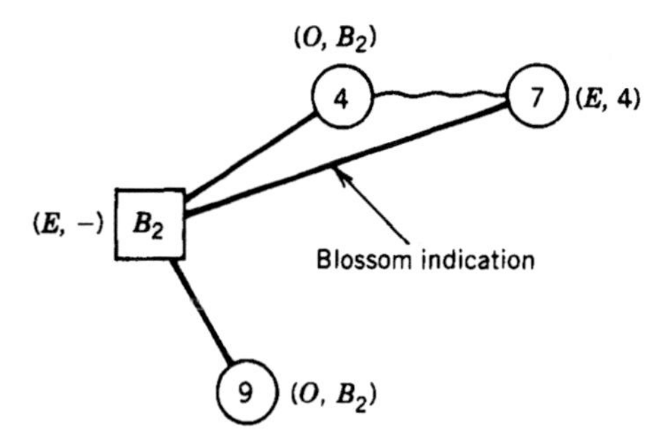

4. The pseudonode $B_3 = B(U_3)$ is created and scanned.

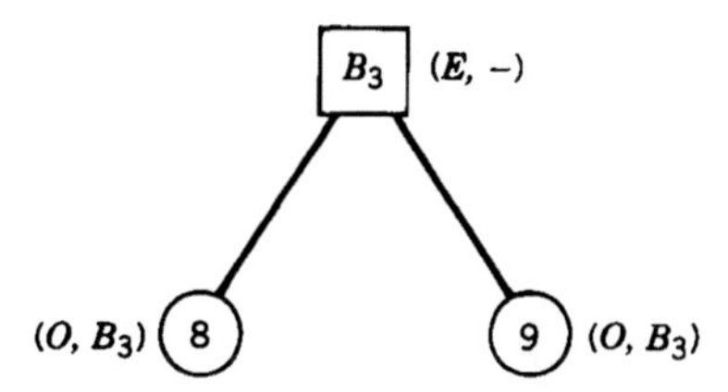

Node 8 is odd and exposed. Hence we find the augmenting path $(B_3, 8)$.

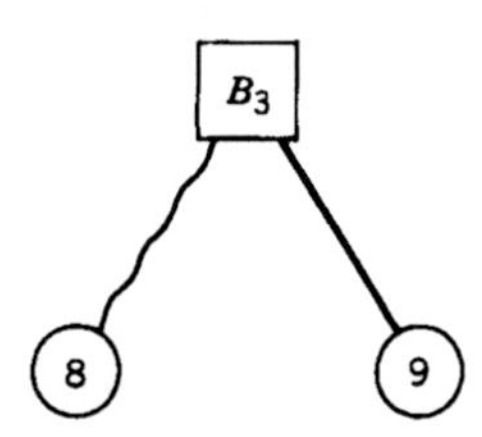

5. The graph $\tilde{G}$ and the new matching $\tilde{M}$ are shown below.

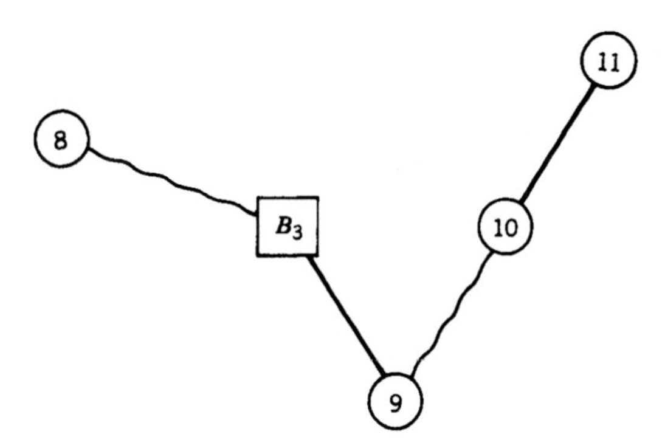

To find the corresponding matching M, we start with $\tilde{M} = \{(8, B_3), (9, 10)\}$ and then expand B_3. Node 8 is joined to node 4 of B_3. So next we find an even-length path from node 4 to B_2, which is the base of B_3. From the path $(4, 7, B_2)$ we identify the matching edges $(8, 4)$ and $(7, B_2)$. Hence $\tilde{M} = \{(8, 4), (7, B_2), (9, 10)\}$.

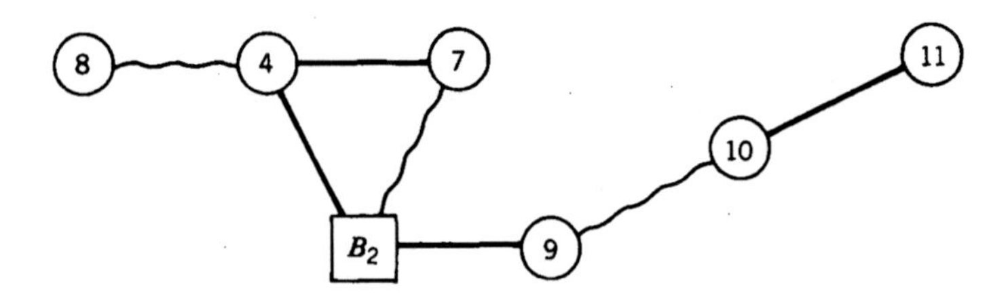

Next we expand B_2. Node 7 is joined to node 6 of B_2, and the base of B_2 is B_1. From the even-length path $(6, 5, B_1)$, we identify the matching edges $(6, 7)$ and $(5, B_1)$, and we find the matching $\tilde{M} = \{(8, 4), (6, 7), (5, B_1), (9, 10)\}$.

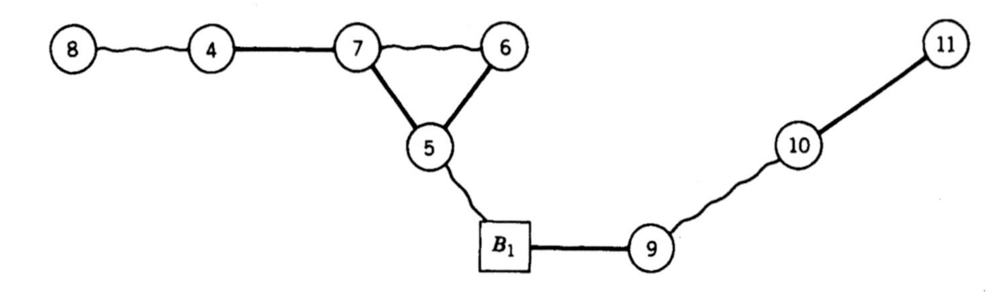

Finally by expanding B_1, we obtain $M = \tilde{M} = \{(8, 4), (7, 6), (5, 2), (3, 1), (9, 10)\}$. The new matching, along with the blossoms B_1, B_2, B_3, is shown below.

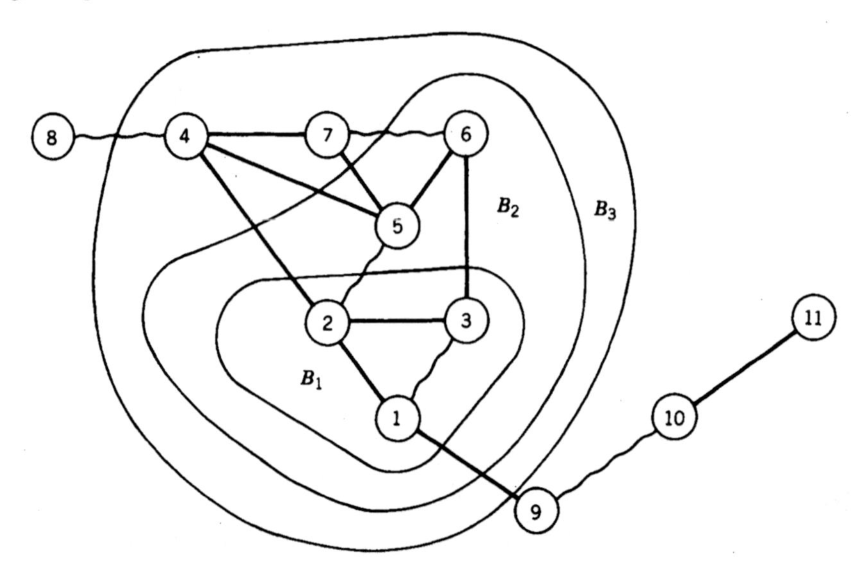

6. We grow an alternating tree rooted at node 11. Nodes 11, 10, 9, 1, 3, 2, and 5 are scanned. In the process of scanning node 6, a blossom with $U_4 = \{2, 3, 5, 6, 7\}$ and $b(U_4) = 3$ is identified.

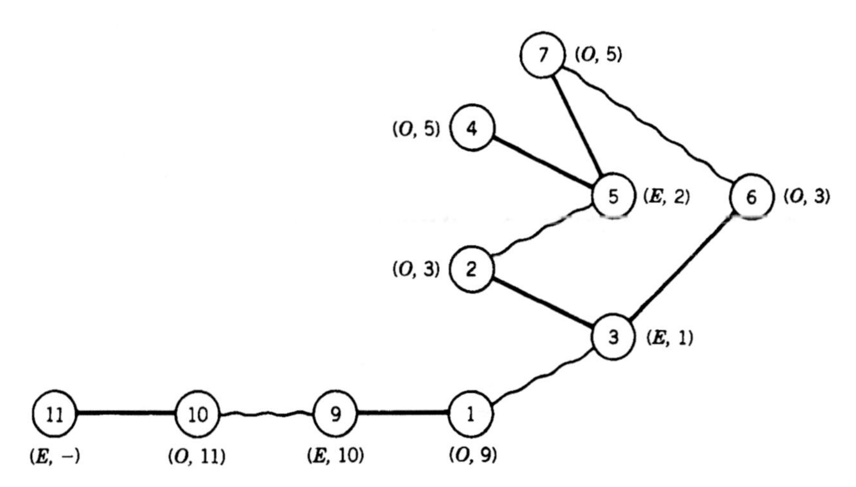

7. The pseudonode $B_4 = B(U_4)$ is created, and labeling continues.

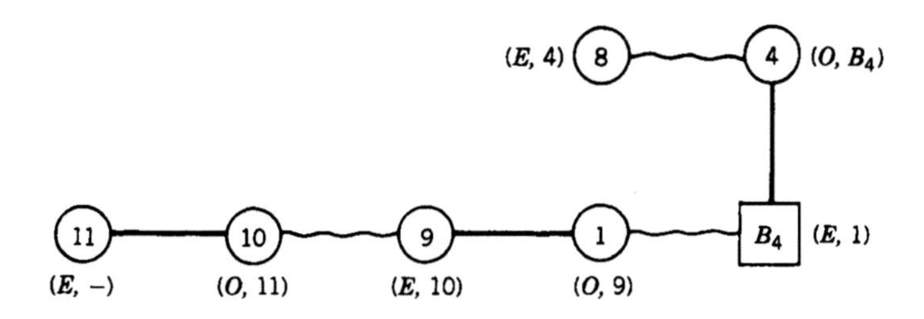

8. All nodes are labeled so that the current matching M is maximum.

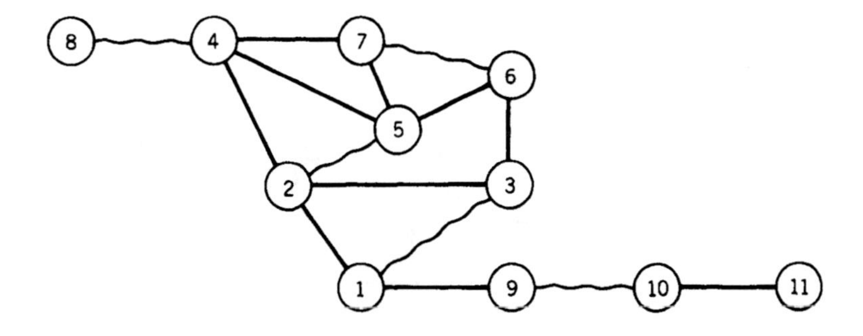

An optimal solution to the dual is given by

$$\pi_1 = \pi_4 = \pi_{10} = 1, \qquad \pi_i = 0 \quad \text{otherwise}$$

$$y_{U_4} = 1, \qquad y_U = 0 \quad \text{otherwise.}$$

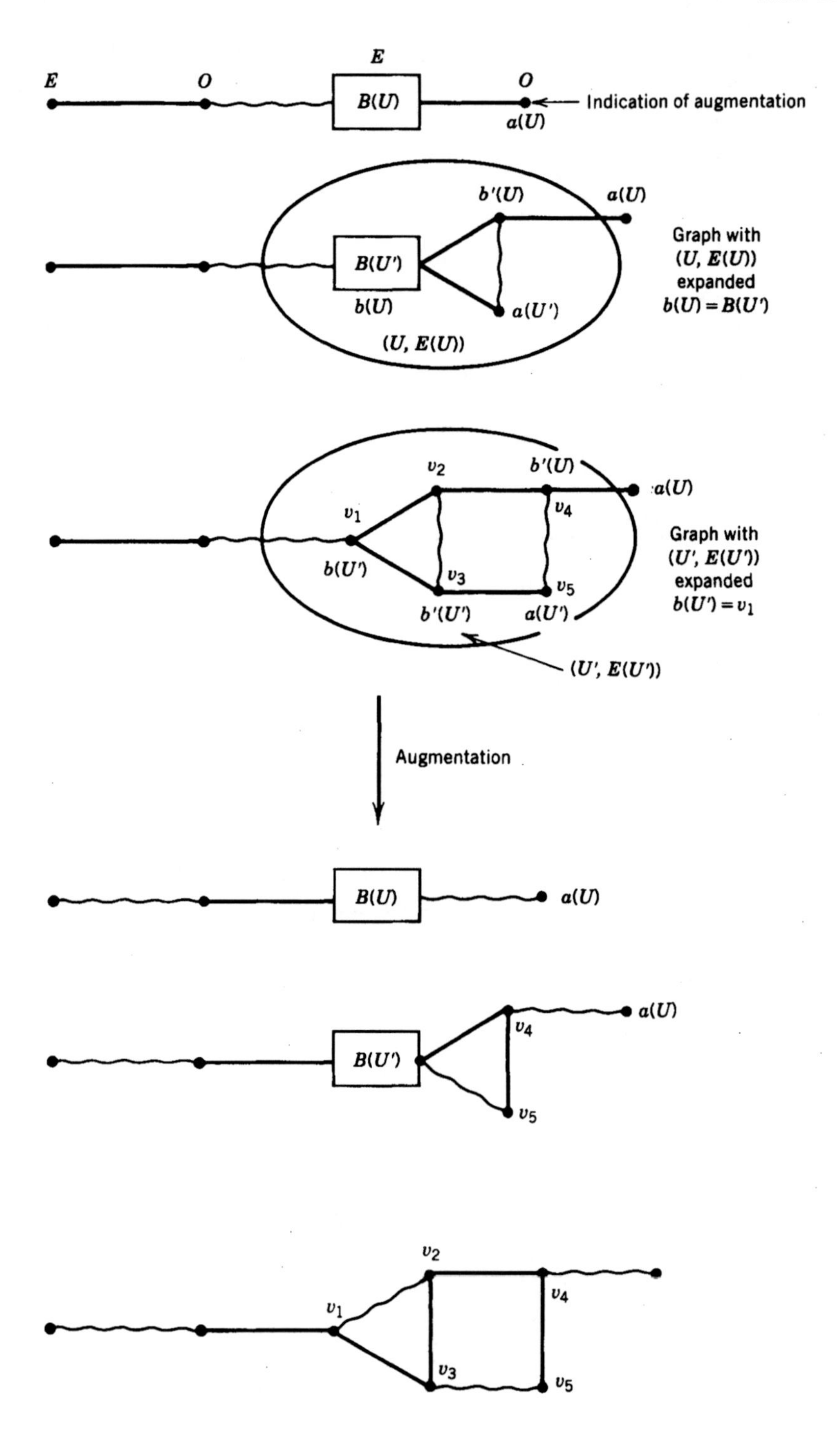

E
E
O
O
B(U)
Indication of augmentation
a(U)
b'(U)
a(U)
B(U')
b(U)
a(U')
(U, E(U))
Graph with
(U, E(U))
expanded
b(U) = B(U')
v2
b'(U)
a(U)
v1
v4
b(U')
v3
v5
b'(U')
a(U')
Graph with
(U', E(U'))
expanded
b(U') = v1
(U', E(U'))
Augmentation
B(U)
a(U)
a(U)
v4
B(U')
v5
v2
v1
v4
v3
v5

Figure 2.9

We have two reasons for modifying Algorithm I. First of all, restarting from scratch with a new matching after an augmentation is found can be inefficient, since many of the pseudonodes we had before may be recreated. Secondly, in the weighted algorithm given in the next section, we will need to keep some pseudonodes after an augmentation is found.

In Algorithm II given below, when an augmentation is found in $\tilde{G}$ we update the matching in G, but we grow a new alternating forest in $\tilde{G}$ with respect to the new matching $\tilde{M}$ in $\tilde{G}$. Algorithm II has the same complexity as Algorithm I; however, for the reason mentioned above, it is likely to be more efficient. Some additional steps are needed as explained below.

After an augmentation and a new matching $\tilde{M}'$ are found in $\tilde{G}$, the bases of the pseudonodes in $\tilde{G}$ on the augmenting path, as well as all of the pseudonodes nested within these pseudonodes, must be updated. This is done during the recursive process of finding the new matching M' in G and is illustrated in Figure 2.9.

In addition, when we grow a new alternating forest for $\tilde{G}$, a previously created pseudonode may receive an odd label. In this case there may be no augmenting path in the reduced graph, whereas there is an augmenting path in the graph in which this odd pseudonode is expanded. Hence when a pseudonode $B(U)$ receives an odd label, we expand it as shown in Figure 2.10; and in the alternating forest, we replace $B(U)$ by a node $\alpha(U) \in U$ that is joined by a nonmatching edge to the node from which $B(U)$ was labeled. Note that in this case, the new matching contains fewer than $\lfloor |U|/2 \rfloor$ edges from the blossom $B(U)$. This observation is important for the weighted matching algorithm that will be described in the next section.

General Cardinality Matching Algorithm II

Steps 1, 3, and 4: These are the same as in Algorithm I.

Step 2′: Modify Step 2c,d by: If i is an odd pseudonode go to Step 6.

Step 5′ (Augmentation in G and $\tilde{G}$): Modify Step 5 with the following additions:

a. Use the augmenting path in $\tilde{G}$ to find a new matching $\tilde{M}'$ in $\tilde{G}$. Update the bases of all the pseudonodes on the augmenting path and all of the pseudonodes nested within these pseudonodes.
b. $M \leftarrow M'$, $\tilde{M} \leftarrow \tilde{M}'$, and return to Step 1.

Step 6′ (Expand a Pseudonode): Pseudonode $i = B(U)$ has the label (O, j). Change $\tilde{G}$ by expanding the blossom $B(U) = (U, E(U))$. Find $\alpha(U) \in U$ with $(j, \alpha(U)) \in \tilde{E} \setminus \tilde{M}$. Replace $B(U)$ by $\alpha(U)$ in the alternating forest and give $\alpha(U)$ the label (O, j). If $\alpha(U)$ is a pseudonode, the process is repeated. Go to Step *2′*.

Example 2.2 (continued). Here we apply Algorithm II. The first five steps are the same as before. Referring to the graph at the end of Step 5, we see that $b(U_3) = 4$, $b(U_2) = 6$, and $b(U_1) = 2$.

6′. We grow an alternating tree rooted at node 11.

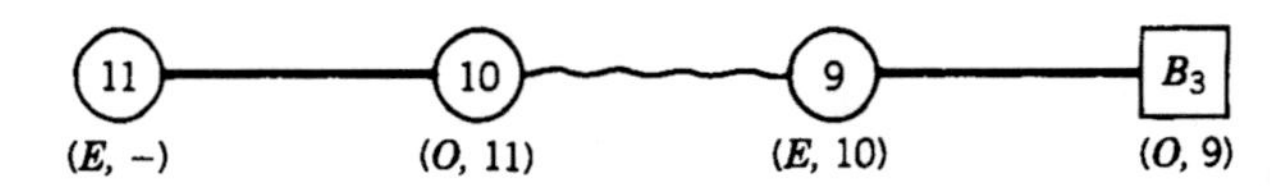

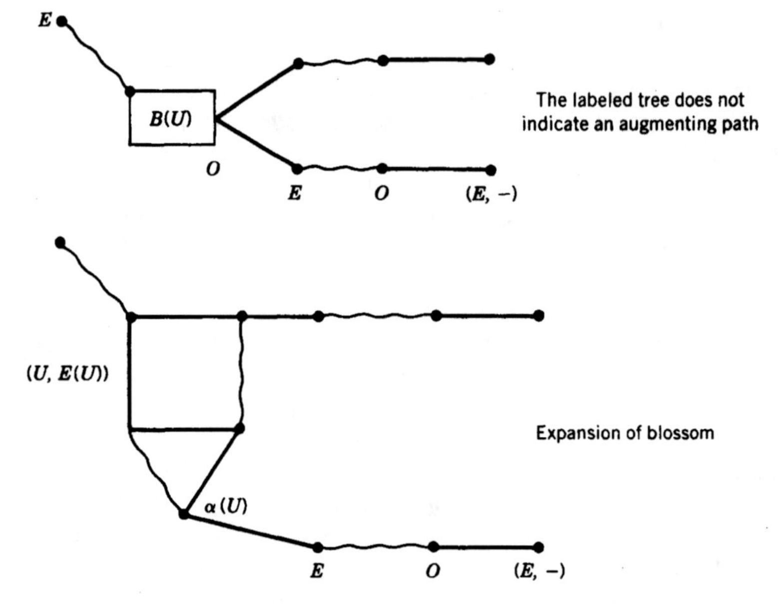

Labeling continues and augmenting path is found

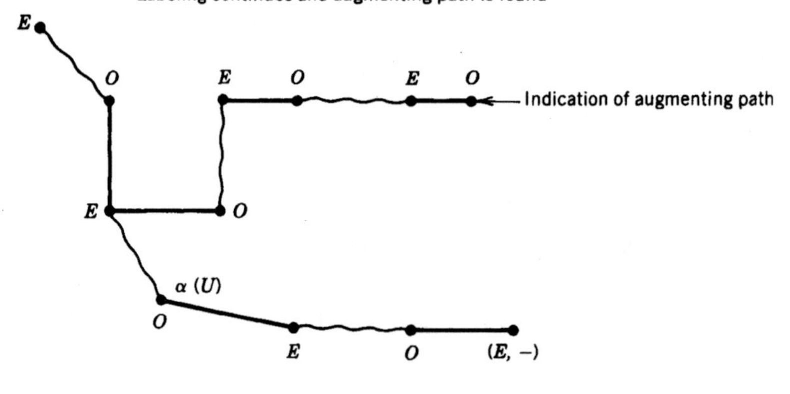

Figure 2.10

7′. B_3 is odd, so it is expanded. $\alpha(B_3) = B_2$; hence B_2 replaces B_3 in the tree. B_2 is odd, so it is expanded. $\alpha(B_2) = B_1$; hence B_1 replaces B_2 in the tree. B_1 is odd, so it is expanded. $\alpha(B_1) = 1$; hence 1 replaces B_1 in the tree. Now the tree is

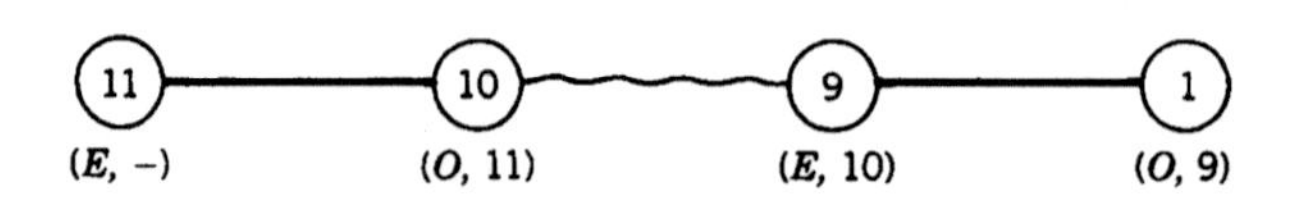

8′. Nodes 1, 3, 2, and 5 are scanned. In the process of scanning node 6,——(carry on with point 6 of Example 2.2).

3. MAXIMUM-WEIGHT MATCHING

Here we give a primal–dual algorithm for the linear program (1.2) and prove that the solution is integral for any objective function vector c. Such a solution is therefore a solution to the weighted matching problem. We can assume $c_e > 0$ for all $e \in E$ since $c_e \leqslant 0$ implies that there is an optimal solution with $x_e = 0$.

Given a matching M, let $x_e = 1$ for $e \in M$, let $x_e = 0$ otherwise, and let

$$c'_e = \sum_{v: e \in \delta(v)} \pi_v + \sum_{\text{odd sets } U: e \in E(U)} y_U - c_e.$$

The complementary slackness conditions for the linear programs (1.2) and (1.3) are

(3.1) $\quad c'_e x_e = 0 \quad$ for $e \in E$ (either $c'_e = 0$ or $e \notin M$)

(3.2) $\quad \left(\lfloor |U|/2 \rfloor - \sum_{e \in E(U)} x_e\right) y_U = 0 \quad$ for odd sets U

(either $y_U = 0$ or $M \cap E(U) = \lfloor |U|/2 \rfloor$)

(3.3) $\quad \left(1 - \sum_{e \in \delta(v)} x_e\right) \pi_v = 0 \quad$ for $v \in V$ (either $\pi_v = 0$ or v is met by an $e \in M$).

The primal–dual algorithm maintains primal and dual feasibility and also maintains the conditions (3.1) and (3.2). Therefore, optimality is achieved when (3.3) is satisfied.

An initial integral primal feasible solution and a dual feasible solution that satisfy (3.1) and (3.2) are given by

$$\begin{aligned} x_e &= 0 && \text{for } e \in E \\ y_U &= 0 && \text{for odd sets } U \\ \pi_v &= \frac{1}{2} \max_{e \in E} c_e && \text{for } v \in V. \end{aligned} \tag{3.4}$$

Note that $c'_{e'} = 0$ for all $e' \in E$ such that $c_{e'} = \max_{e \in E} c_e$.

Let $E' = \{e \in E: c'_e = 0\}$. The graph $G' = (V, E')$ is called the *equality-constrained subgraph*. Throughout the course of the algorithm, (3.1) is maintained by setting $x_e = 0$ for $e \in E \setminus E'$. We maintain (3.2) by requiring $y_U = 0$ unless $(U, E(U))$ is a blossom in the equality-constrained subgraph that has been shrunk into a pseudonode.

To see if (3.3) can be satisfied, we find a maximum-cardinality matching in the equality-constrained subgraph G'. Again, we will be dealing with reduced subgraphs $\tilde{G}'$ of G' that contain pseudonodes. There are two possibilities:

i. A matching $\tilde{M}$ is found in $\tilde{G}'$ with $\pi_v = 0$ for all exposed nodes. The corresponding matching M in G' is an optimal solution to the weighted matching problem.
ii. For the reduced graph $\tilde{G}'$ and matching $\tilde{M}$, no augmenting path is found.

In the latter case a dual change is made that maintains dual feasibility and also maintains (3.1) and (3.2). Then the equality-constrained subgraph G' and its reduced subgraph $\tilde{G}'$ are updated. In addition the edges of the alternating forest $\tilde{F}'$ in $\tilde{G}'$ still have $c'_e = 0$, so that $\tilde{F}'$ is kept. After a small number of dual changes, either an augmentation is obtained or $\pi_v = 0$ for all exposed nodes.

After either a dual change or augmentation, all pseudonodes $B(U)$ with dual variables $y_U = 0$ are expanded. However, pseudonodes with $y_U > 0$ are not expanded. The implication of this is that an augmenting path of the type shown in Figure 2.10 may not be found immediately. It will be necessary to reduce y_U to zero before such an augmenting path can be found. The reason for this change is to maintain the complementary slackness condition (3.2).

When new edges are added to $\tilde{G}'$ after a dual change, we continue with the development of the alternating forest $\tilde{F}'$ by adding edges, labeling nodes, and creating pseudonodes as described previously unless an edge (u, v) is added where u and v are both even and contained in different trees of $\tilde{F}'$. In this case, $\tilde{F}'$ contains an augmenting path joining the roots of the two trees as shown in Figure 3.1.

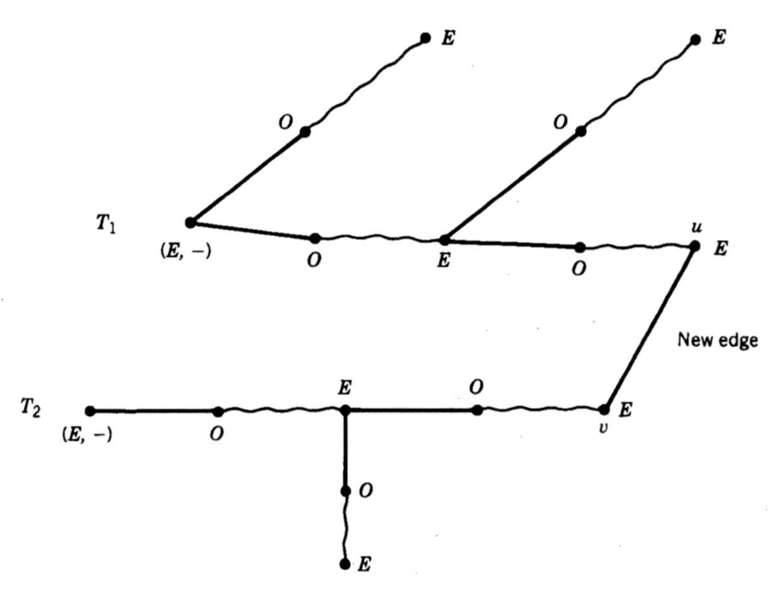

Figure 3.1

Weighted Matching Algorithm

Initialization: Start with the primal and dual solutions given by (3.4). Let $E' = \{e \in E: c'_e = 0\}$, $G' = (V, E')$, $\tilde{G}' = G'$, $\tilde{M} = M = \emptyset$, and $\tilde{F}' = \emptyset$.

Step 1: Continue with the construction of the alternating forest $\tilde{F}'$. If an augmenting path is found, go to Step 2. Otherwise, go to Step 3.

Step 2 (Augmentation): Update the primal solution M and expand all pseudonodes $B(U)$ with $y_U = 0$. Update the bases of the remaining blossoms. Let $\tilde{G}'$ be the reduced equality-constrained subgraph with matching $\tilde{M}'$. If $\pi_v = 0$ for all exposed nodes, the current primal and dual solutions are optimal. Otherwise set $\tilde{F}' = \emptyset$ and go to Step 1.

Step 3 (Dual Change): Apply the dual change given by (3.5) and (3.6) below. If $\pi_v = 0$ for all exposed nodes, the current primal and dual solutions are optimal. Otherwise update $\tilde{G}'$ and expand all pseudonodes $B(U)$ with $y_U = 0$. If an $e = (u, v)$ has been added to $\tilde{G}'$ where u and v are both even and contained in different trees of $\tilde{F}'$, then identify an augmenting path and go to Step 2. Otherwise keep $\tilde{F}'$ intact and go to Step 1.

We need to ensure that (3.1) and (3.2) remain satisfied if an augmentation or a dual change occurs. In addition, we must ensure that dual feasibility and that part of the equality subgraph corresponding to the alternating forest are preserved when a dual change occurs.

Proposition 3.1. *If conditions (3.1) and (3.2) are satisfied prior to an augmentation, then they are satisfied by the matching obtained from the augmentation.*

Proof. Since $c'_e = 0$ for all edges in the equality-constrained subgraph, it is clear that (3.1) remains satisfied. The only way that (3.2) can be violated is by an augmentation that reduces the number of matching edges in $E(U)$, where U is an odd set with $y_U > 0$. However, if $y_U > 0$, $(U, E(U))$ is represented by a pseudonode in $\tilde{G}'$. Any augmentation in $\tilde{G}'$ translates into a new matching M with $|M \cap E(U)| = |U|/2$. ∎

We now describe the dual change used in Step 3. Define the following sets:

$\mathcal{U}^+$ = {odd set U: U is the node set of a shrunken blossom represented by an even pseudonode}

$\mathcal{U}^-$ = {odd set U: U is the node set of a shrunken blossom represented by an odd pseudonode}

$\mathcal{U}^\sim$ = {odd set U: U is the node set of a shrunken blossom represented by an unlabeled pseudonode}

$V^+ = \{v \in V: v \text{ is even or } v \in U \text{ for some } U \in \mathcal{U}^+\}$

$V^- = \{v \in V: v \text{ is odd or } v \in U \text{ for some } U \in \mathcal{U}^-\}$

$V^\sim = \{v \in V: v \text{ is unlabeled or } v \in U \text{ for some } U \in \mathcal{U}^\sim\}$

$E^+ = \{e = (i,j) \in E: i \in V^+, j \in V^\sim\}$

$E^{++} = \{e = (i,j) \in E: i \in V^+, j \in V^+, \text{ no } U \in \mathcal{U}^+ \text{ contains both } i \text{ and } j\}$.

Let

$$\delta_1 = \min_{v \in V^+} \pi_v$$

$$\delta_2 = \begin{cases} \frac{1}{2} \min_{U \in \mathcal{U}^-} y_U & \text{if } \mathcal{U}^- \neq \emptyset \\ \infty & \text{if } \mathcal{U}^- = \emptyset \end{cases}$$

$$\delta_3 = \begin{cases} \frac{1}{2} \min_{e \in E^{++}} c'_e & \text{if } E^{++} \neq \emptyset \\ \infty & \text{if } E^{++} = \emptyset \end{cases}$$

$$\delta_4 = \begin{cases} \min_{e \in E^+} c'_e & \text{if } E^+ \neq \emptyset \\ \infty & \text{if } E^+ = \emptyset \end{cases}$$

and $\delta = \min(\delta_1, \delta_2, \delta_3, \delta_4)$.

The dual change is given by

$$\hat{\pi}_v = \begin{cases} \pi_v - \delta & \text{for } v \in V^+ \\ \pi_v + \delta & \text{for } v \in V^- \\ \pi_v & \text{otherwise} \end{cases} \tag{3.5}$$

and

$$\hat{y}_U = \begin{cases} y_U + 2\delta & \text{for } U \in \mathcal{U}^+ \\ y_U - 2\delta & \text{for } U \in \mathcal{U}^- \\ y_U & \text{otherwise.} \end{cases} \tag{3.6}$$

The effect of the dual change on node weights π_v and edge weights c'_e is shown in Figure 3.2.

Proposition 3.2. *The dual change is bounded and not degenerate; that is,* $0 < \delta \leqslant \frac{1}{2} \max_{e \in E} c_e$.

Proof. 1. ($\delta_1 > 0$). Before the dual change, there is an exposed node $u \in V^+$ with $\pi_u > 0$; otherwise the algorithm would have terminated prior to the dual change. Moreover, u has been exposed before all previous dual changes since once a node is met by a matching edge, it remains covered thereafter. Thus $\pi_u = \min_{v \in V^+} \pi_v$ since π_u has been decreased at every dual change and we started with π_v equal to a constant for all $v \in V$. Hence $\{v \in V^+ : \pi_v = 0\} = \emptyset$.

2. ($\delta_2 > 0$). All pseudonodes with $y_U = 0$ are expanded after dual changes and augmentations. Thus the only possibility for a labeled pseudonode with $y_U = 0$ is that U has been shrunk since the last augmentation or dual change. But then $U \in \mathcal{U}^+$. Hence $\{U \in \mathcal{U}^- : y_U = 0\} = \emptyset$.

3. ($\delta_3 > 0$). If $c'_e = 0$, then e is in the equality-constrained subgraph. If both ends of E are even and $c'_e = 0$, then either an augmenting path is identified or both endnodes of e are contained in a shrunken blossom. Thus $\{e \in E^{++} : c'_e = 0\} = \emptyset$.

4. ($\delta_4 > 0$). If $e \in E^+$ and $c'_e = 0$, then the other end of e receives an odd label. Hence $\{e \in E^+ : c'_e = 0\} = \emptyset$. Finally $\delta = \min(\delta_1, \delta_2, \delta_3, \delta_4) > 0$ and $\delta \leqslant \delta_1 \leqslant \frac{1}{2} \max_{e \in E} c_e$. ∎

Proposition 3.3. *If the primal and dual solutions satisfy (3.1) and (3.2), and $c'_e = 0$ for all edges of the alternating forest, then these conditions are satisfied after a dual change.*

Proof. a. ($\hat{\pi}_v \geqslant 0$ for $v \in V$). We have

$$\begin{aligned} \hat{\pi}_v &\geqslant \pi_v - \delta && \text{(by (3.5))} \\ &\geqslant \pi_v - \delta_1 && \text{(by the definition of } \delta) \\ &\geqslant 0 && \text{(by the definition of } \delta_1). \end{aligned}$$

b. ($\hat{y}_U \geqslant 0$ for odd sets U). We have

$$\begin{aligned} \hat{y}_U &\geqslant y_U - 2\delta && \text{(by (3.6))} \\ &\geqslant y_U - 2\delta_2 && \text{(by the definition of } \delta) \\ &\geqslant 0 && \text{(by the definition of } \delta_2). \end{aligned}$$

c. ($\hat{c}'_e \geqslant 0$ for $e \in E$). By (3.5) and (3.6) we only need to consider $e \in E^+ \cup E^{++}$ or e is in a blossom whose pseudonode is labeled; otherwise $\hat{c}'_e \geqslant c'_e \geqslant 0$.

i. ($e \in E^+$). We have

$$\begin{aligned} \hat{c}'_e &= c'_e - \delta && \text{(by (3.5) and the definition of } c'_e) \\ &\geqslant c'_e - \delta_4 && \text{(by the definition of } \delta) \\ &\geqslant 0 && \text{(by the definition of } \delta_4). \end{aligned}$$

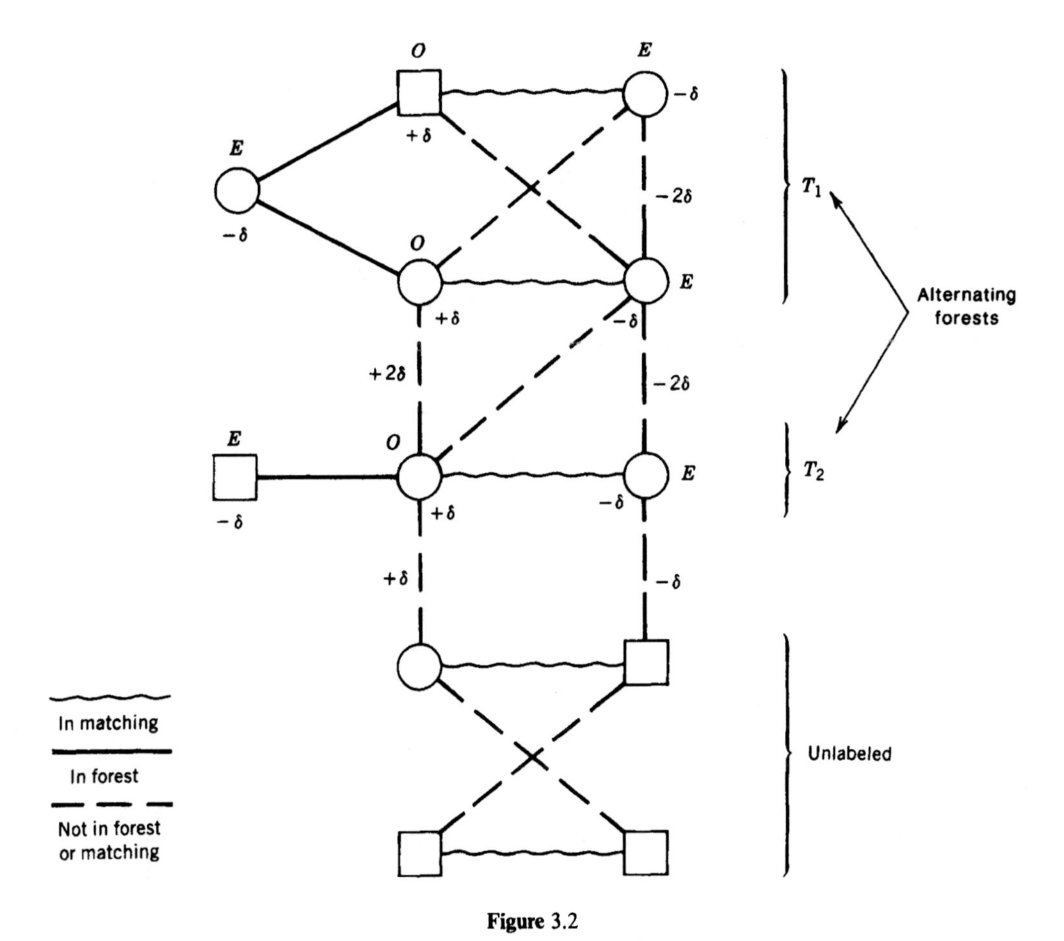

Figure 3.2

ii. ($e \in E^{++}$). We have

$$\begin{aligned} \hat{c}'_e &= c'_e - 2\delta && \text{(by (3.5) and the definition of } c'_e) \\ &\geqslant c'_e - 2\delta_3 && \text{(by the definition of } \delta) \\ &\geqslant 0 && \text{(by the definition of } \delta_3). \end{aligned}$$

iii. (e is in a blossom whose pseudonode is even). Then both endnodes of e are in V^+. Thus

$$\hat{c}'_e = c'_e - \delta - \delta + 2\delta = c'_e \quad \text{(by (3.5), (3.6), and the definition of } c'_e).$$

iv. (e is in a blossom whose pseudonode is odd). Then both endnodes of e are in V^-. Thus

$$\hat{c}'_e = c'_e + \delta + \delta - 2\delta = c'_e \quad \text{(for the reason given in iii)}.$$

Thus the new solution is dual feasible.

Next, we establish that (3.1) is satisfied; that is, $e \in M$ implies $\hat{c}'_e = 0$. Since $c'_e = 0$ for $e \in M$, it suffices to show that $\hat{c}'_e = c'_e$ for $e \in M$. Consider an $e \in M$ and first suppose that e has not been shrunk; then either one endnode of e is odd and the other is even, or both endnodes are unlabeled. Then by (3.5), we have $\hat{c}'_e = c'_e$. On the other hand, if e has been shrunk, then $\hat{c}'_e = c'_e$ by iii and iv above if the pseudonode is labeled, whereas $\hat{c}'_e = c'_e$ by (3.5) and (3.6) if the pseudonode is unlabeled.

Next we show that (3.2) is satisfied; that is, if U is an odd set and $|E(U) \cap M| < \lfloor |U|/2 \rfloor$, then $\hat{y}_U = y_U = 0$. This follows directly from (3.6) since if $|E(U) \cap M| < \lfloor |U|/2 \rfloor$, then $(U, E(U))$ cannot be shrunk into a pseudonode.

Finally, consider the edges of the alternating forest. If e is in a labeled blossom, we have shown above that $\hat{c}'_e = c'_e = 0$. Otherwise, one end of e is in V^+ and the other end is in V^-. Thus

$$\hat{c}'_e = c'_e - \delta + \delta = c'_e = 0. \qquad \blacksquare$$

Proposition 3.4. *There are no more than $5m/3$ dual changes between augmentations.*

Proof. We will use the fact that the alternating forest $\tilde{F}'$ is kept between augmentations. The effect of the dual change depends on $\arg(\min\{\delta_1, \delta_2, \delta_3, \delta_4\})$.

a. $(\delta = \delta_1)$. Since all exposed nodes are even in $\tilde{G}'$, and $\pi_v = \delta_1$ for all exposed nodes, we have $\hat{\pi}'_v = 0$ for all expoded nodes. In this case the current primal and dual solutions are optimal.
b. $(\delta = \delta_2)$. Then $\hat{y}_U = 0 < y_U$ for some odd pseudonode. After an augmentation there can be no more than $m/3$ odd pseudonodes, and all new pseudonodes created between augmentations are even. Hence $\delta = \delta_2$ can happen no more than $m/3$ times between augmentations.
c. $(\delta = \delta_3)$. Then $\hat{c}'_e = 0 < c'_e$ for some e whose two endnodes are even. This edge is added to the equality-constrained subgraph. The result is an augmentation or the creation of a pseudonode. The latter can happen no more than $m/3$ times between augmentations because a newly created pseudonode $B(U)$ is even so that its dual variable y_U can only increase between augmentations.
d. $(\delta = \delta_4)$. Then $\hat{c}'_e = 0 < c'_e$ for some e with one endnode even and the other unlabeled. Then the unlabeled node becomes odd. This can happen no more than $m - 1$ times because there are no more than $m - 1$ unlabeled nodes. ■

Theorem 3.5. *The weighted matching algorithm finds an integral optimal solution to (1.2) and also finds an optimal solution to its dual (1.3). Its complexity is $O(m^2 n)$.*

Proof. Integrality of the primal solution is maintained throughout the algorithm because each solution is a matching. When the algorithm terminates, both the primal and dual solutions are feasible and satisfy complementary slackness.

The work between successive dual changes is $O(n)$. By Proposition 3.4 the maximum number of dual changes between an augmentation is $O(m)$, and the number of augmentations is $O(m)$.

Finally, observe that after p dual changes, it follows that π, y and c' are rationals with denominator 2^k for some integer k, $0 \leqslant k \leqslant p$. Hence the numbers involved in the calculations remain polynomially bounded. ■

Theorem 3.6. *The polytope defined by the constraint set of (1.2) is the convex hull of matchings.*

Proof. By Theorem 3.5, (1.2) has an integral optimal solution that is a matching for any objective function vector c. Thus, by Proposition 1.1 of Section III.1.1, each extreme point of the polytope defined by the constraints of (1.2) is integral. ∎

Example 3.1

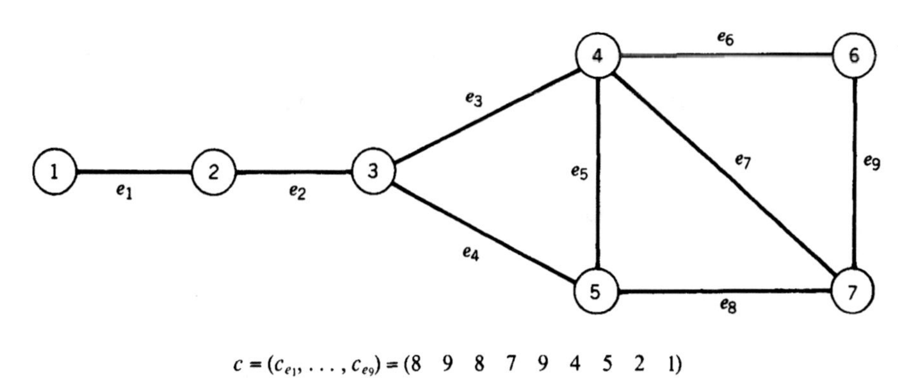

$$c = (c_{e_1}, \ldots, c_{e_9}) = (8 \quad 9 \quad 8 \quad 7 \quad 9 \quad 4 \quad 5 \quad 2 \quad 1)$$

1. *Initialization*

$$\pi_v = 4.5 \quad \text{for all } v \in V$$
$$y_U = 0 \quad \text{for all } U$$
$$c' = (1 \quad 0 \quad 1 \quad 2 \quad 0 \quad 5 \quad 4 \quad 7 \quad 8)$$

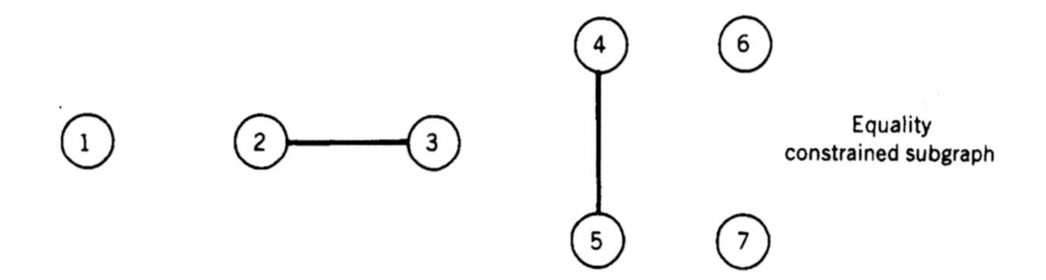

2. *Equality-constrained subgraph and labels with* $M = \{e_2, e_5\}$

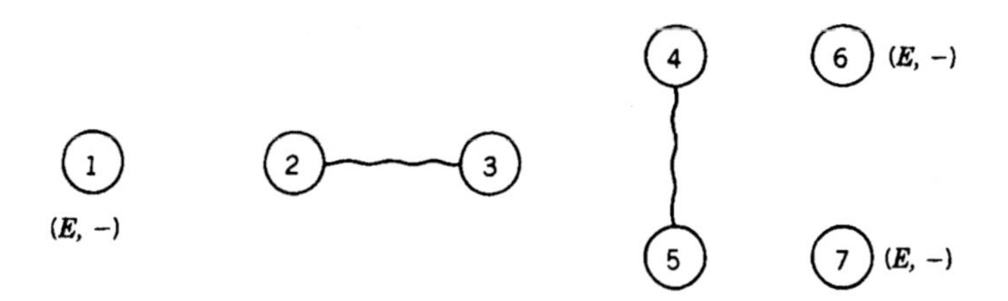

3. *Dual change*

$$\delta_1 = \min\{\pi_1, \pi_6, \pi_7\} = 4.5, \qquad \delta_2 = \infty, \qquad \delta_3 = \tfrac{1}{2}c'_{e_9} = 4$$

$$\delta_4 = \min\{c'_{e_1}, c'_{e_6}, c'_{e_7}, c'_{e_8}\} = 1, \qquad \delta = \delta_4 = 1$$

$\pi = (3.5 \quad 4.5 \quad 4.5 \quad 4.5 \quad 4.5 \quad 3.5 \quad 3.5)$

$y_U = 0$ for all U

$c' = (0 \quad 0 \quad 1 \quad 2 \quad 0 \quad 4 \quad 3 \quad 6 \quad 6)$

e_1 is added to the equality constrained subgraph.

4. *Equality-constrained subgraph and labels*

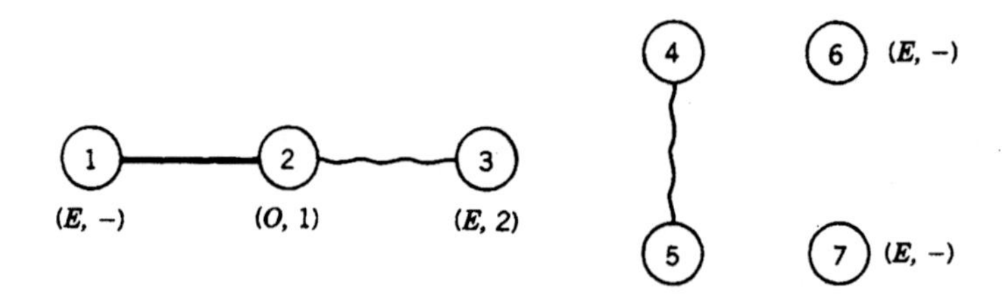

5. *Dual change*

$\delta_1 = 3.5, \quad \delta_2 = \infty, \quad \delta_3 = 3, \quad \delta_4 = \min\{1 \quad 2 \quad 4 \quad 3 \quad 6\} = c'_{e_3} = 1, \delta = \delta_4 = 1$

$\pi = (2.5 \quad 5.5 \quad 3.5 \quad 4.5 \quad 4.5 \quad 2.5 \quad 2.5)$

$y_U = 0$ for all U

$c' = (0 \quad 0 \quad 0 \quad 1 \quad 0 \quad 3 \quad 2 \quad 5 \quad 4)$

e_3 is added to the equality constrained subgraph.

6. *Equality-constrained subgraph and labels*

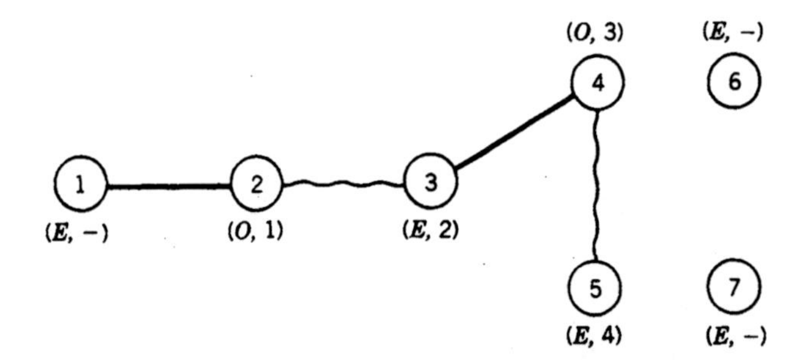

7. *Dual change*

$$\delta_1 = 2.5, \qquad \delta_2 = \infty, \qquad \delta_3 = \tfrac{1}{2}\min\{1 \quad 5 \quad 4\} = \tfrac{1}{2}, \qquad \delta_4 = \infty, \qquad \delta = \delta_3 = \tfrac{1}{2}$$

$\pi = (2 \quad 6 \quad 3 \quad 5 \quad 4 \quad 2 \quad 2)$

$y_U = 0$ for all U

$c' = (0 \quad 0 \quad 0 \quad 0 \quad 0 \quad 3 \quad 2 \quad 4 \quad 3)$

e_4 is added to the equality-constrained subgraph.

8. *Reduced equality-constrained subgraph and labels*

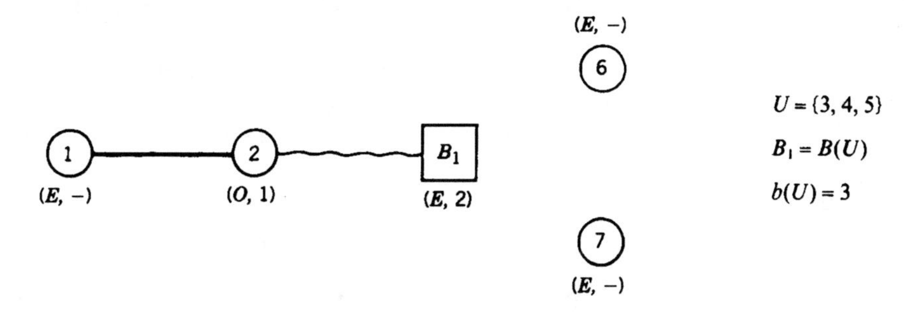

9. *Dual change*

$$\delta_1 = 2, \qquad \delta_2 = \infty, \qquad \delta_3 = \tfrac{1}{2}\min_{i=6,7,8,9}\{c'_{e_i}\} = 1, \qquad \delta_4 = \infty, \qquad \delta = \delta_3 = 1$$

$\pi = (1 \quad 7 \quad 2 \quad 4 \quad 3 \quad 1 \quad 1)$

$y_U = 2$ for $U = \{3, 4, 5\}$, $\quad y_U = 0$ otherwise

$c' = (0 \quad 0 \quad 0 \quad 0 \quad 0 \quad 1 \quad 0 \quad 2 \quad 1)$

e_7 is added to the equality-constrained subgraph.

10. *Augmentation in the reduced graph and new labeling.* $M = \{e_1, e_4, e_7\}$

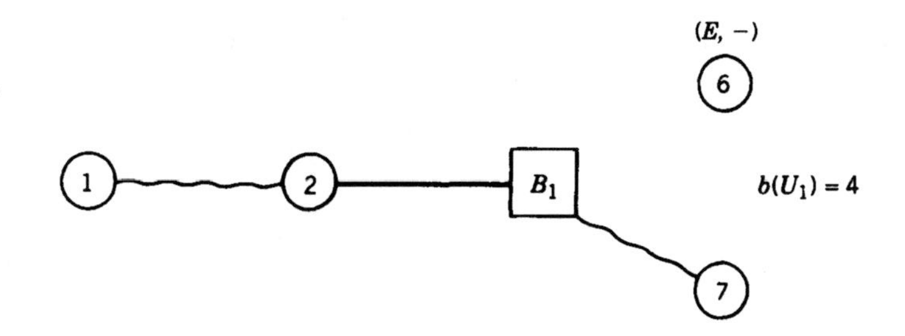

11. *Dual change*

$$\delta_1 = \pi_6 = 1, \quad \delta_2 = \infty, \quad \delta_3 = \infty, \quad \delta_4 = \min(c_{e_6}, c_{e_9}) = 1$$

$$\pi = (1 \quad 7 \quad 2 \quad 4 \quad 3 \quad 0 \quad 1)$$

$$y_U = 2 \quad \text{for } U_1 = \{3, 4, 5\}, \qquad y_U = 0 \quad \text{otherwise.}$$

12. *Optimal solution*

$$\text{Primal: } x_{e_i} = 1 \quad \text{for } i = 1, 4, 7, \qquad x_{e_i} = 0 \quad \text{otherwise}$$

$$\text{Dual: } \pi = (1 \quad 7 \quad 2 \quad 4 \quad 3 \quad 0 \quad 1)$$

$$y_{U_1} = 2 \quad \text{for } U_1 = \{3, 4, 5\}, \qquad y_U = 0 \quad \text{otherwise.}$$

4. ADDITIONAL RESULTS ON MATCHING AND RELATED PROBLEMS

This section contains a potpourri of topics related to matchings. We begin by presenting further results on the convex hull of matchings. Then we describe the polytope of the convex hull of perfect matchings and relate matchings to the problem of covering nodes by edges.

The next topic is the reduction of integer and (0, 1) b-matching problems to matching problems. These reductions may also be viewed as a technique for obtaining linear inequality descriptions of b-matching polytopes.

We then introduce a pair of combinatorial objects known as T-joins and T-cuts. T-joins include perfect matchings, s–t paths, and eulerian subgraphs.

The final topic of this section is the problem of coloring the edges of a graph so that no pair of edges that are incident to the same node have the same color. This edge coloring problem is equivalent to partitioning the edges of a graph into matchings.

The Matching Polytope

Here we demonstrate an interesting nonalgorithmic proof technique for showing that a set of inequalities describes the convex hull of a set S by proving Theorem 3.6; that is, the convex hull of matchings in a graph $G = (V, E)$ is given by

$$\sum_{e \in \delta(v)} x_e \leq 1 \quad \text{for } v \in V$$

$$\text{(4.1)} \qquad \sum_{e \in E(U)} x_e \leq \left\lfloor \frac{|U|}{2} \right\rfloor \text{ for all odd sets } U \text{ with } |U| \geq 3$$

$$x \in R^n_+.$$

Let $\mathcal{M}$ be the set of matchings on G, let w be a weight vector on the edges of G, let $w(M) = \Sigma_{e \in M} w_e$, let

$$z(w) = \max\{w(M): M \in \mathcal{M}\}.$$

and let $\mathcal{M}(w)$ be the set of maximum-weight matchings. We use the following property of $\mathcal{M}(w)$.

Proposition 4.1. *If $w > 0$ and G is connected, then either*

1. *there exists a $v \in V$ such that $\delta(v) \cap M \neq \emptyset$ for all $M \in \mathcal{M}(w)$, or*
2. *$|M| = \lfloor |V|/2 \rfloor$ for all $M \in \mathcal{M}(w)$, and $|V|$ is odd.*

Proof. Suppose that statement 1 is false and there exists an $M \in \mathcal{M}(w)$ with $|M| < \lfloor |V|/2 \rfloor$. Since $|M| < \lfloor |V|/2 \rfloor$, there are at least two exposed nodes relative to M. Now choose an $M \in \mathcal{M}(w)$ so that there are exposed nodes u and v as close together as possible. Then $(u, v) \notin E$; otherwise $M \cup \{(u, v)\} \in \mathcal{M}$ and $w(M \cup \{(u, v)\}) > z(w)$.

Let t be any internal node on a minimum-length path joining u and v. By the choice of u and v, t is not exposed relative to M. Also, since statement 1 is false, there is another matching $M' \in \mathcal{M}(w)$ such that t is exposed relative to M'.

Now the graph $\tilde{G} = (V, M \cup M')$ consists of a node disjoint union of paths and cycles in which the degrees of nodes t, u, and v are equal to 1. (If u or v was exposed relative to M', we would have a contradiction to the choice of M, u, and v.) The component $\tilde{G}_t = (V_t, E_t)$ of $\tilde{G}$ containing t is therefore a path with t as one endpoint. Hence this component cannot contain both u and v. Since the edges of E_t alternate between M and M', it follows that

$$\tilde{M} = (M \cup (M' \cap E_t)) \setminus (M \cap E_t) \quad \text{and} \quad \tilde{M}' = (M' \cup (M \cap E_t)) \setminus (M' \cap E_t)$$

are matchings, and

$$w(\tilde{M}) + w(\tilde{M}') = w(M) + w(M') = 2z(w).$$

Since $w(\tilde{M}), w(\tilde{M}') \leqslant z(w)$, we have $w(\tilde{M}) = z(w)$ and $\tilde{M} \in \mathcal{M}(w)$. This is a contradiction because (a) u and t are exposed relative to $\tilde{M}$ and (b) the path between u and t is shorter than the path between u and v. So either statement 1 is true or $|M| = \lfloor |V|/2 \rfloor$ for all $M \in \mathcal{M}(w)$.

Finally, if $|M| = \lfloor |V|/2 \rfloor$ and $|V|$ is even, M is a perfect matching, so statement 1 is true. Thus if statement 1 is false, statement 2 is true. ∎

Proof of Theorem 3.6. Let S be the set of incidence vectors of matchings in $G = (V, E)$. Suppose

$$\sum_{e \in E} w_e x_e \leqslant w_0 \tag{4.2}$$

defines a facet of conv(S). We consider two cases:

1. $w_0 \leqslant 0$ or $w_e < 0$ for some $e \in E$. Since the set of matchings form an independence system, the only inequalities that define facets with $w_0 \leqslant 0$ or $w_e < 0$ are $-x_e \leqslant 0$ for $e \in E$ (see Section II.1.5).
2. $w_0 > 0$ and $w_e \geqslant 0$ for $e \in E$. (i) Since (4.2) defines a facet, $w_0 = z(w)$ and $x^M \in S$ satisfies $wx^M = w_0$ if and only if $M \in \mathcal{M}(w)$; and (ii) since conv(S) is full-dimensional, by Theorem 3.6 of Section I.4.3, the set of equations

$$\sum_{e \in E} w_e x_e^M = w_0 \quad \text{for } M \in \mathcal{M}(w) \tag{4.3}$$

has a unique solution up to scalar multiplication.

Suppose there is a $v \in V$ such that $|\delta(v) \cap M| = 1$ for all $M \in \mathcal{M}(w)$. Then a solution to (4.3) is $w_e = 1$ for $e \in \delta(v)$, $w_e = 0$ otherwise, and $w_0 = 1$. Hence (4.2) is of the form $\Sigma_{e\in\delta(v)} x_e \leqslant 1$.

If no such v exists, let $E' = \{e \in E: w_e > 0\}$ and $G' = (V', E')$ be the subgraph of G induced by E'. G' is connected; otherwise (4.2) is the sum of valid inequalities for G and thus cannot define a facet. Define w' on G' by $w'_e = w_e$ for $e \in E'$, and let $\mathcal{M}'(w')$ be the set of maximum-weight matchings on G'. Hence $M' \in \mathcal{M}'(w')$ if and only if $M' = M \cap E'$ for some $M \in \mathcal{M}(w)$. Hence, by hypothesis, statement 1 of Proposition 4.1 is false for the pair (G', w'). Thus $|M'| = \lfloor |V'|/2 \rfloor$ for all $M' \in \mathcal{M}'(w')$, and $|V'|$ is odd. Now a solution to (4.3) is $w_e = 1$ for $e \in E(V')$, $w_e = 0$ otherwise, and $w_0 = \lfloor |V'|/2 \rfloor$. Thus (4.2) is of the form $\Sigma_{e\in E(V')} x_e \leqslant \lfloor |V'|/2 \rfloor$, where $|V'| \geqslant 3$ and is odd. ■

For any pair (G, w), we have $z(w - 1) \geqslant z(w) - \lfloor |V|/2 \rfloor$. However, when statement 1 of Proposition 4.1 is false and w is integral, it can be shown that $z(w - 1) = z(w) - \lfloor |V|/2 \rfloor$, which implies $|M| = \lfloor |V|/2 \rfloor$ for all $M \in \mathcal{M}(w)$. Thus we can state a stronger version of Proposition 4.1 for integral w.

Proposition 4.2. *If $w \geqslant 1$ and is integral, and G is connected, either*

a. *there exists a $v \in V$ such that $\delta(v) \cap M \neq \emptyset$ for all $M \in \mathcal{M}(w)$, or*
b. *$|V|$ is odd and $z(w - 1) = z(w) - \lfloor |V|/2 \rfloor$.*

By using Proposition 4.2, we obtain a simple proof that the dual of $\max\{\Sigma_{e\in E} w_e x_e$: x satisfies (4.1)$\}$ has an integral optimal solution for all $w \in Z^n$.

Theorem 4.3. *The system of inequalities (4.1) is* TDI.

Proof. The proof is by induction on $|V| + |E| + \Sigma_{e\in E} w_e$. Clearly the result is true for a graph with two nodes and one edge. We can assume that G is connected; otherwise the induction hypothesis can be applied separately to each component. We can also assume $w \geqslant 1$; otherwise an edge can be deleted. Hence the hypotheses of Proposition 4.2 hold. Let $\pi \in R^m_+$ and $y \in R^p_+$ be the dual variables for the degree constraints and odd-set constraints, respectively. There are two cases according to Proposition 4.2.

1. Statement a of Proposition 4.2 is true for v. Let w' be defined by $w'_e = w_e - 1$ for $e \in \delta(v)$ and by $w'_e = w_e$ otherwise. Clearly, $z(w') \geqslant z(w) - 1$; but if $z(w') = z(w)$, then statement a of Proposition 4.2 is false. Hence $z(w') = z(w) - 1$. Now by the induction hypothesis, there is an optimal dual solution $(\pi', y') \in Z^{m+p}_+$ of cost $z(w) - 1$. Now define $(\pi, y) \in Z^{m+p}_+$ by $\pi_v = \pi'_v + 1$, $\pi_u = \pi'_u$ otherwise, and $y = y'$. Then it is a simple calculation to show that (π, y) is an optimal dual solution for the weight vector w.
2. Statement b of Proposition 4.2 is true. Let $w' = w - 1$. Hence $z(w') = z(w) - \lfloor |V|/2 \rfloor$, and $|V|$ is odd. By the induction hypothesis, there is an optimal dual solution $(\pi', y') \in Z^{m+p}_+$. Now define $(\pi, y) \in Z^{m+p}_+$ by $\pi = \pi'$, $y_V = y'_V + 1$, and $y_U = y'_U$ otherwise. Again, it is easy to check that $(\pi, y) \in Z^{m+p}_+$ is an optimal dual solution for the weight vector w. ■

Perfect Matchings

We now consider perfect matchings. Clearly, if $|V|$ is odd, there are no perfect matchings.

Theorem 4.4. *The convex hull of perfect matchings on a graph $G = (V, E)$ with $|V|$ even is given by*

$$(4.4)\qquad \begin{aligned} &\text{(a) } x \in R^n_+ \\ &\text{(b) } \sum_{e\in\delta(v)} x_e = 1 \quad \textit{for } v \in V \\ &\text{(c) } \sum_{e\in E(U)} x_e \leq \left\lfloor \frac{|U|}{2} \right\rfloor \textit{for all odd sets } U \subseteq V \textit{ with } |U| \geq 3 \end{aligned}$$

or by (a), (b), *and*

$$\text{(d) } \sum_{e\in\delta(U)} x_e \geq 1 \textit{ for all odd sets } U \subseteq V \textit{ with } |U| \geq 3.$$

Proof. Since the convex hull of perfect matchings is the face of the matching polytope with $\Sigma_{e\in\delta(v)} x_e = 1$ for all $v \in V$, the claim for (a), (b), and (c) follows from Theorem 3.6. We now show that an x satisfies (a), (b), and (c) if and only if it satisfies (a), (b), and (d).

By summing the constraints of (b), we obtain

$$\sum_{e\in E} x_e = \tfrac{1}{2}|V|.$$

Now since $|V|$ is even and U is an odd set, $V \setminus U$ is an odd set. Hence (c) yields the inequalities

$$-\sum_{e\in E(U)} x_e \geq -\left\lfloor \frac{|U|}{2} \right\rfloor$$

and

$$-\sum_{e\in E(V\setminus U)} x_e \geq -\left\lfloor \frac{|V\setminus U|}{2} \right\rfloor = -\tfrac{1}{2}|V| + \left\lfloor \frac{|U|}{2} \right\rfloor + 1.$$

Summing the last three constraints yields (d). ■

The system (4.4 (a), (b), (c)) is TDI since it is obtained from the TDI system (4.1) by changing some inequalities to equalities. The system (4.4 (a), (b), (d)) is *not* TDI for all graphs (see exercise 9). However, it can be shown that the dual problem always has an optimal solution in which each variable is an integer or an integer divided by 2.

Edge Coverings

The theory and algorithmic aspects of edge coverings completely parallel those for matching. We illustrate this with two results.

Proposition 4.5. *Let M be a maximum-cardinality matching, and let C be a minimum-cardinality covering of the nodes by edges in a graph $G = (V, E)$. Then $|M| + |C| = |V|$.*

Proof. Given M, let U be the set of nodes of degree zero relative to M. Thus $|U| = |V| - 2|M|$. Since we obtain a cover by adding $|U|$ edges to M, we have

$$|C| \leqslant |M| + |U| = |M| + |V| - 2|M| = |V| - |M|.$$

Given C, let $\hat{M}$ be a maximum-cardinality matching in (V, C), and let $\hat{U}$ be the set of nodes of degree zero relative to $\hat{M}$ in (V, C). Then

$$|C| = |\hat{M}| + |\hat{U}| = |\hat{M}| + |V| - 2|\hat{M}| = |V| - |\hat{M}|$$

and

$$|M| \geqslant |\hat{M}| = |V| - |C|.$$

Hence $|C| + |M| = |V|$. ■

To cover an odd set of nodes U, we need at least $\lfloor |U|/2 \rfloor + 1$ edges. Thus we obtain the valid inequalities

$$\sum_{e \in E(U)} x_e + \sum_{e \in \delta(U)} x_e \geqslant \left\lfloor \frac{|U|}{2} \right\rfloor + 1 \quad \text{for all odd sets } U.$$

These inequalities, together with the degree and nonnegativity constraints, yield the convex hull of edge covers.

Theorem 4.6. *The convex hull of edge covers in a graph $G = (V, E)$ is given by*

$$\sum_{e \in \delta(v)} x_e \geqslant 1 \quad \textit{for } v \in V$$

$$\sum_{e \in E(U) \cup \delta(U)} x_e \geqslant \left\lfloor \frac{|U|}{2} \right\rfloor + 1 \quad \textit{for all odd sets } U$$

$$x \in R^n_+.$$

b-Matching

The next topic deals with the reduction of b-matching problems to 1-matching problems. These reductions may be viewed as modeling devices for transforming harder problems to easier ones, and they can be used in contexts other than matching. Although they are not necessarily polynomial reductions, they serve three useful purposes.

1. The transformed problem may yield theoretical results—for example, polyhedral descriptions of the convex hull of solutions in the original space.
2. The transformed problem can be solved by a standard matching algorithm. This may be preferred, even when the transformation is not polynomial, to constructing a new algorithm.
3. An efficient algorithm can often be developed for the original problem by studying the (possibly nonpolynomial) algorithm on the transformed problem.

We first consider the integer b-matching problem on $G = (V, E)$. Its constraints are $\sum_{e\in\delta(v)} x_e \leq b_v$ for $v \in V$ and $x \in Z_+^n$.

Suppose that $b_v = 2$ for all $v \in V$. In this case the edges $M = \{e \in E: x_e > 0\}$ of a feasible solution produce a graph (V, M) whose components are paths in which $x_e = 1$ for all edges of the path, cycles in which $x_e = 1$ for all edges of the cycle, isolated nodes, and single edges with $x_e = 2$. In perfect integer 2-matchings, only the cycles and single edges with $x_e = 2$ can occur.

In Section II.6.3, we used perfect 2-matchings as a relaxation for the traveling salesman problem, and we reduced the perfect integer 2-matching problem on $G = (V, E)$ to a perfect matching problem on a bipartite graph $H = (V^1 \cup V^2, E')$, where V^1 and V^2 are copies of V, and $e^1 = (u^1, v^2)$, $e^2 = (v^1, u^2)$ are in E' if and only if $e = (u, v) \in E$. The reduction does not depend on the matching being perfect.

Let $y \in B^{2n}$ be the incidence vector of a matching on H, and let $x_e = y_{e^1} + y_{e^2}$. Then $x \in Z_+^n$ and

$$\sum_{e\in\delta(v^1)} y_e + \sum_{e\in\delta(v^2)} y_e = \sum_{e\in\delta(v)} x_e.$$

Thus to find a maximum-weight integer 2-matching on G with weight vector $w \in R^n$, we can find a matching on H with weight vector $w' \in R^{2n}$, where $w'_{e^1} = w'_{e^2} = w_e$ for all $e \in E$.

The reduction also yields a linear inequality description of the convex hull of integer 2-matchings.

Proposition 4.7. *The convex hull of integer 2-matchings on $G = (V, E)$ is given by*

$$(4.5)\qquad \sum_{e\in\delta(v)} x_e \leq 2 \quad \textit{for } v \in V, \qquad x \in R_+^n.$$

Proof. The convex hull of matchings on the bipartite graph H is given by

$$(4.6)\qquad \sum_{e\in\delta(v^i)} y_e \leq 1 \quad \text{for } v^i \in V^i \text{ and } i = 1, 2, \qquad y \in R_+^{2n}.$$

We need to show that the projection onto R^n of the points that satisfy (4.6) and $x_e = y_{e^1} + y_{e^2}$ for $e \in E$ is precisely those points in R^n that satisfy (4.5).

Every point x of the projection lies in R_+^n because $y_{e^1}, y_{e^2} \in R_+^n$ and $x_e = y_{e^1} + y_{e^2}$. Also, every such point satisfies (4.5) because

$$\sum_{e\in\delta(v)} x_e = \sum_{e\in\delta(v^1)} y_e + \sum_{e\in\delta(v^2)} y_e \leq 2$$

by (4.6). It remains to show that every point $x \in R_+^n$ satisfying (4.5) is a point of the projection. For this it suffices to take $y_{e^1} = y_{e^2} = \frac{1}{2}x_e$ for $e \in E$. ∎

This approach readily extends to integer b-matchings with b_v even for all $v \in V$ and yields the result that when b_v is even for all $v \in V$, integer b-matching is a network flow problem on a graph with $2|V|$ nodes and $2|E|$ edges.

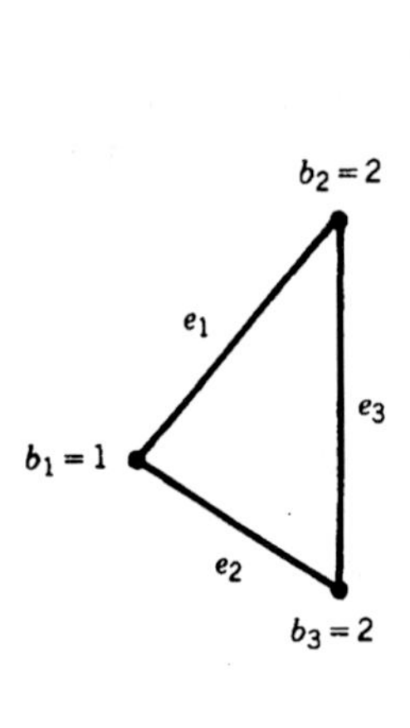

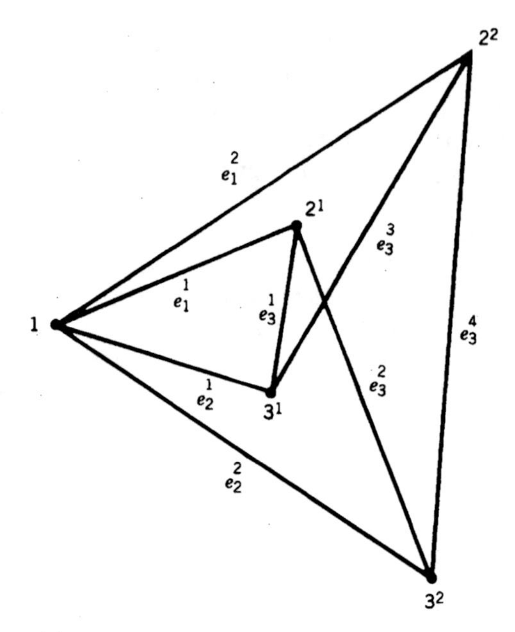

Figure 4.1

For general $b \in Z_+^n$, the transformation of integer b-matching on G to 1-matching is more complicated. Here each node is replaced by b_v copies of itself, and for each pair of adjacent nodes in the original graph all the resulting copies are joined to form a complete bipartite graph. Formally, we construct a new graph $H = (\bigcup_{v \in V} V^v, \bigcup_{(u,v) \in E} E^{u,v})$, where $V^v = (v^1, \ldots, v^{b_v})$ for $v \in V$, $E^{u,v} = (e^1, \ldots, e^{b_u b_v})$ for $(u, v) \in E$, and $(V^u \cup V^v, E^{u,v})$ is a complete bipartite graph with $b_u + b_v$ nodes and $b_u b_v$ edges. Hence H contains $\Sigma_{v \in V} b_v$ nodes and $n^* = \Sigma_{(u,v) \in E} b_u b_v$ edges. An example is given in Figure 4.1. This is not a polynomial reduction, since the new description of the problem is not a polynomial function of $\Sigma_{v \in V} \log b_v$.

Let $y \in B^{n^*}$ be the incidence vector of a matching on H, and let $x_e = \Sigma_{i=1}^{b_{uv}} y_{e^i}$ for $e \in E$. Then $x \in Z_+^n$ and $\Sigma_{e \in \delta(v)} x_e = \Sigma_{e \in \delta(V^v)} y_e \leqslant b_v$ for $v \in V$. Hence x is an integer b-matching on G.

Conversely, if $x \in Z_+^n$ is the vector of an integer b-matching on G, then we get a matching on H by setting $y_{e^i} = 1$ for x_e node disjoint edges for all $e \in E$.

For the graph of Figure 4.1, Table 4.1 gives the maximal b-matchings on G and the corresponding matchings on H.

Table 4.1.

x_{e_1}	x_{e_2}	x_{e_3}	$y_{e_1}^1$	$y_{e_1}^2$	$y_{e_2}^1$	$y_{e_2}^2$	$y_{e_3}^1$	$y_{e_3}^2$	$y_{e_3}^3$	$y_{e_3}^4$
1	0	1	1	0	0	0	0	0	1	0
			1	0	0	0	0	0	0	1
			0	1	0	0	1	0	0	0
			0	1	0	0	0	1	0	0
0	1	1	0	0	1	0	0	1	0	0
			0	0	1	0	0	0	0	1
			0	0	0	1	1	0	0	0
			0	0	0	1	0	0	1	0
0	0	2	0	0	0	0	1	0	0	1
			0	0	0	0	0	1	1	0

We also obtain a linear inequality description of the convex hull of integer b-matchings.

Theorem 4.8. *The convex hull of integer b-matchings is given by*

$$\sum_{e \in \delta(v)} x_e \leqslant b_v \quad \text{for } v \in V$$

$$(4.7) \qquad \sum_{e \in E(U)} x_e \leqslant \left\lfloor \frac{1}{2} \sum_{v \in U} b_v \right\rfloor \quad \text{for } U \subseteq V \text{ with } \sum_{v \in U} b_v \text{ odd}$$

$$x \in R_+^n.$$

Proof. Let $y \in R_+^{n^*}$ be a point in the convex hull of matchings on H; and for $e = (u, v) \in E$, let $x_e = \Sigma_{i=1}^{b_u b_v} y_{e^i}$. Then $x \in R_+^n$ and, as above,

$$\sum_{e \in \delta(v)} x_e = \sum_{e \in \delta(V^v)} y_e \leqslant b_v.$$

Now suppose $\Sigma_{v \in U} b_v$ is odd. Let $S = \cup_{v \in U} V^v$ so that $|S| = \Sigma_{v \in U} b_v$. Hence from the odd-set constraints we obtain

$$\sum_{e \in E(U)} x_e = \sum_{e \in E(S)} y_e \leqslant \left\lfloor \frac{1}{2} |S| \right\rfloor = \left\lfloor \frac{1}{2} \sum_{v \in U} b_v \right\rfloor.$$

So x satisfies (4.7).

Conversely, suppose x satisfies (4.7). We need to show that for each such x there is a $y \in R^{n^*}$ lying in the convex hull of 1-matchings on H. For $e = (u, v) \in E$, let $y_{u^i, v^j} = x_{u,v}/b_u b_v$ for $i = 1, \ldots, b_u$ and $j = 1, \ldots, b_v$. Then $y \in R_+^{n^*}$ and

$$\sum_{u \in V \setminus \{v\}} \sum_{u^i \in V^u} y_{u^i, v^j} = \sum_{u \in V \setminus \{v\}} \frac{x_{u,v}}{b_v} \leqslant 1 \quad \text{for } v^j \in V^v \text{ and } v \in V.$$

Now consider an odd set S of nodes in H.

Case 1. $S = \cup_{v \in U} V^v$ and $\Sigma_{v \in U} b_v$ is odd. Hence $|S| = \Sigma_{v \in V} b_v$ and

$$\sum_{e \in E(S)} y_e = \sum_{e \in E(U)} x_e \leqslant \left\lfloor \frac{1}{2} \sum_{v \in V} b_v \right\rfloor = \left\lfloor \frac{1}{2} |S| \right\rfloor$$

Case 2. S contains $k_v > 0$ nodes from V^v for $v \in U$ and for some $w \in U$, $k_w < b_w$. Hence $|S| = \Sigma_{v \in U} k_v$ and

$$\sum_{e \in E(S)} y_e = \sum_{u,v \in U} \frac{k_u k_v}{b_u b_v} x_{u,v}.$$

We will show that

$$\sum_{u,v \in U} \frac{k_u k_v}{b_u b_v} x_{u,v} \leqslant \frac{1}{2} \left(\sum_{v \in U \setminus \{w\}} k_v + (k_w - 1) \right) = \left\lfloor \frac{1}{2} |S| \right\rfloor.$$

For $v \in U \setminus \{w\}$, multiply the constraint $\Sigma_{u\in\delta(v)} x_{u,v} \leqslant b_v$ by $k_v/2b_v$ and multiply the degree constraint for w by $(k_w - 1)/2b_w$. By summing these inequalities and using $x \in R^n_+$ to eliminate the coefficients of edges not in $E(U)$, we obtain

$$\sum_{u,v\in U\setminus\{w\}} \left(\frac{k_u}{2b_u} + \frac{k_v}{2b_v}\right) x_{u,v} + \sum_{u\in U\setminus\{w\}} \left(\frac{k_u}{2b_u} + \frac{k_w - 1}{2b_w}\right) x_{u,w} \leqslant \left(\sum_{v\in U\setminus\{w\}} k_v + (k_w - 1)\right)$$

Now for $u, v \in U \setminus \{w\}$, we obtain

$$\frac{k_u}{2b_u} + \frac{k_v}{2b_v} = \frac{k_u b_v + k_v b_u}{2b_u b_v} \geqslant \frac{2k_u k_v}{2b_u b_v} \quad \text{(since } k_u \leqslant b_u \text{ and } k_v \leqslant b_v\text{)}.$$

For $u \in U \setminus \{w\}$, we obtain

$$\frac{k_u}{2b_u} + \frac{k_w - 1}{2b_w} - \frac{2k_u k_w}{2b_u b_w} = \frac{b_w k_u + b_u(k_w - 1) - 2k_u k_w}{2b_u b_w}$$

and

$$b_w k_u + b_u(k_w - 1) - 2k_u k_w = k_u(b_w - k_w - 1) + (k_w - 1)(b_u - k_u) \geqslant 0$$

since $1 \leqslant k_w \leqslant b_w - 1$ and $1 \leqslant k_u \leqslant b_u$. Hence,

$$\sum_{u,v\in U} \frac{k_u k_v}{b_u b_v} x_{u,v} \leqslant \sum_{u,v\in U\setminus\{w\}} \left(\frac{k_u}{2b_u} + \frac{k_v}{2b_v}\right) x_{u,v} + \sum_{u\in U\setminus\{w\}} \left(\frac{k_u}{2b_u} + \frac{k_w - 1}{2b_w}\right) x_{u,w}$$

$$\leqslant \frac{1}{2}\left(\sum_{v\in U\setminus\{w\}} k_v + (k_w - 1)\right) = \left\lfloor \frac{1}{2}|S| \right\rfloor. \qquad \blacksquare$$

A triangle with $b_v = 2$ for $v = 1, 2, 3$ and $c = (1 \quad 1 \quad 1)$ shows that the system (4.7) is not TDI. An interesting result that we will not prove is that by adding the superfluous constraints $\Sigma_{e\in E(U)} x_e \leqslant \frac{1}{2} \Sigma_{v\in U} b_v$ for $U \subseteq V$ with $\Sigma_{v\in U} b_v$ even, we obtain a TDI system.

Analogous to Theorem 4.4 and by an identical proof, which uses Theorem 4.8, we obtain the convex hull of perfect b-matchings.

Corollary 4.9. *The convex hull of perfect b-matchings is given by*

$$\sum_{e\in\delta(v)} x_e = b_v \quad \textit{for } v \in V$$

$$\text{(4.8)} \qquad \sum_{e\in\delta(U)} x_e \geqslant 1 \quad \textit{for } U \subseteq V \textit{ with } \sum_{v\in U} b_v \textit{ odd}$$

$$x \in R^n_+.$$

This result will be used later to establish the convex hull of perfect binary 2-matchings.

Binary b-matching problems can be reduced to integer b-matching problems. We will only study binary perfect 2-matching, denoted by BP2M. Here the feasible solutions are cycles that cover all of the nodes. We showed in Section II.6.3 that BP2M gives a tighter relaxation for the traveling salesman problem than does integer perfect 2-matching.

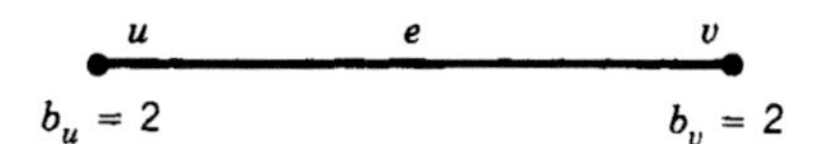

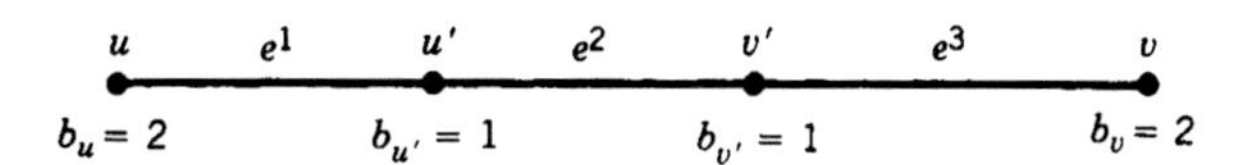

Figure 4.2

Given a BP2M problem on $G = (V, E)$, we construct a new graph $G' = (V \cup V', E \cup E')$, where each $e \in E$ is replaced by a path with three edges as shown in Figure 4.2. Hence $|V'| = |E'| = 2|E|$. We let $b_v = 2$ for $v \in V$ and $b_v = 1$ for $v \in V'$.

Now every perfect b-matching on G' has either (i) $x_{e^1} = x_{e^3} = 1$ and $x_{e^2} = 0$ or (ii) $x_{e^2} = 1$ and $x_{e^1} = x_{e^3} = 0$. So there is a one-to-one correspondence between BP2M's on G and perfect b-matchings on G' given by $x_e = x_{e^1} = x_{e^3} = 1 - x_{e^2}$ for $e \in E$.

This reduction, together with the reduction of perfect integer b-matching with $b_v \in (1, 2)$ to perfect matching, gives a polynomial-time algorithm for BP2M (i.e., the algorithm of Section 3). Figure 4.3 shows the transformation of BP2M for a triangle to perfect 1-matching, and it also shows a perfect 1-matching on the resulting graph. This reduction also yields a linear inequality description of the convex hull of BP2M.

In Section II.2.3, we derived the rank 1 inequalities

$$\sum_{e \in E(H)} x_e + \sum_{e \in \hat{E}} x_e \leqslant |H| + \left\lfloor \frac{|\hat{E}|}{2} \right\rfloor \quad \text{for } H \subset V, \tag{4.9}$$

where $\hat{E} \subseteq \delta(H)$ is an odd set of node disjoint edges. Here we will show that these inequalities, together with the degree constraints and $0 \leqslant x \leqslant 1$, define the convex hull of 0-1 perfect 2-matchings.

First we restate (4.9) by subtracting $\frac{1}{2} \sum_{e \in \delta(v)} x_e = 1$ for $v \in H$. This yields

$$\frac{1}{2} \sum_{e \in \hat{E}} x_e - \frac{1}{2} \sum_{e \in \delta(H) \setminus \hat{E}} x_e \leqslant \left\lfloor \frac{|\hat{E}|}{2} \right\rfloor$$

or

$$\sum_{e \in \hat{E}} x_e - \sum_{e \in \delta(H) \setminus \hat{E}} x_e \leqslant |\hat{E}| - 1$$

or

$$\sum_{e \in \hat{E}} (1 - x_e) + \sum_{e \in \delta(H) \setminus \hat{E}} x_e \geqslant 1 \quad \text{for } H \subset V, \hat{E} \subseteq \delta(H), |\hat{E}| \text{ odd}.$$

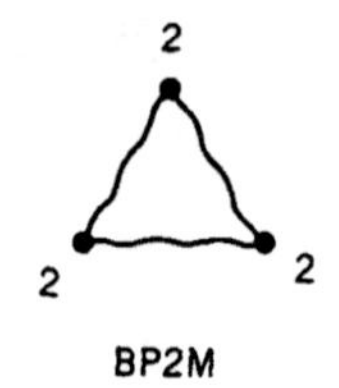

BP2M

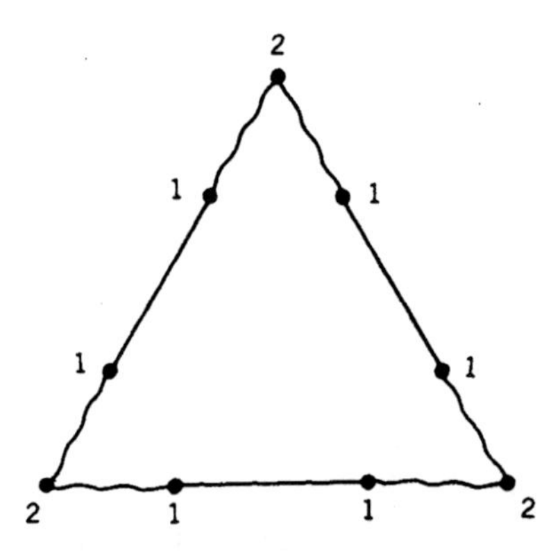

Perfect integer b-matching with $b_v \in \{1, 2\}$

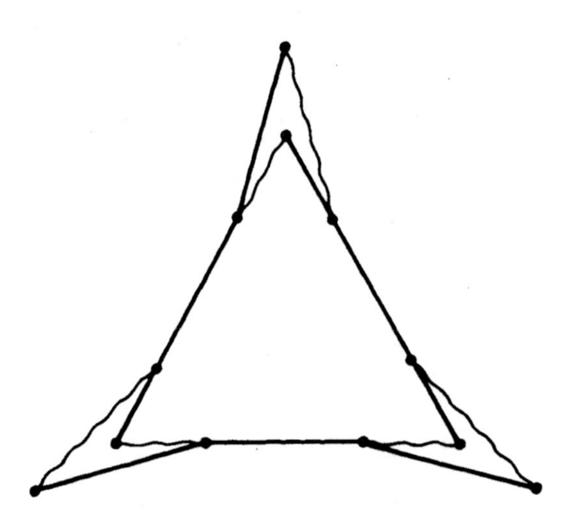

Perfect 1-matching

Figure 4.3

Theorem 4.10. *The convex hull of binary perfect 2-matchings is given by*

$$\sum_{e \in \delta(v)} x_e = 2 \quad \textit{for } v \in V$$

$$(4.10) \qquad \sum_{e \in \hat{E}} (1 - x_e) + \sum_{e \in \delta(H) \setminus \hat{E}} x_e \geq 1 \quad \textit{for } H \subset V, \textit{ where } \hat{E} \subseteq \delta(H) \textit{ is an odd set of node disjoint edges}$$

$$x \leq 1, \qquad x \in R^n_+.$$

Proof. We transform BP2M to an integer perfect b-matching problem with $b_i \in \{1, 2\}$ as shown in Figure 4.3, and then we apply Corollary 4.9 to the graph $G' = (V \cup V', E \cup E')$.

Suppose $y \in R^{3n}_+$ satisfies (4.8) for the graph G'. For $e \in E$, let $x_e = 1 - y_{e^2} = y_{e^1} = y_{e^3}$ (see Figure 4.3). Since $0 \leq y_{e^2} \leq 1$, we have $0 \leq x_e \leq 1$ for all $e \in E$. Also $\Sigma_{e \in \delta(v)} x_e = \Sigma_{e \in \delta(v)} y_{e^1} = 2$.

Now for $H \subset V$ and $\hat{E} \subset E$ with $|\hat{E}|$ odd, define $W \subset V \cup V'$ by $W = H \cup \{u' : (u, v) \in \hat{E}\}$. Then $\Sigma_{v \in W} b_v$ is odd. Also

$$\delta(W) = \{(u', v') : (u, v) \in \hat{E}\} \cup \{(u, u') : (u, v) \in \delta(H) \setminus \hat{E}\}.$$

Thus $\Sigma_{e \in \delta(W)} y_e \geq 1$ yields

$$\sum_{e^2 = (u', v') : (u,v) \in \hat{E}} y_{e^2} + \sum_{e^1 = (u,v) \in \delta(H) \setminus \hat{E}} y_{e^1} \geq 1.$$

Transforming to the variables $x \in R^n_+$ yields

$$\sum_{e \in \hat{E}} (1 - x_e) + \sum_{e \in \delta(H) \setminus \hat{E}} x_e \geq 1.$$

Hence x satisfies (4.10).

Conversely, if x satisfies the constraints (4.10), then with $y_{e^1} = y_{e^3} = 1 - y_{e^2} = x_e$, we have $y \in R^{3n}_+$; also, y satisfies the degree constraints for G'. Now if $\Sigma_{v \in W} b_v$ is odd, $|W \cap V'|$ is odd and, in particular, $|\{u' \in W : v' \notin W\}$ is odd. Define $H = W \cap V$ and $\hat{E} = \{(u, v) \in E : u, u' \in W, \; v, v' \notin W\}$. Hence $|\hat{E}|$ is odd and $\delta(H) = \{(u, v) \in E : u \in W, \; v \notin W\}$. Now from

$$\sum_{e \in \hat{E}} (1 - x_e) + \sum_{e \in \delta(H) \setminus \hat{E}} x_e \geq 1,$$

we obtain

$$\sum_{e^2 = (u', v') : (u,v) \in \hat{E}} y_{e^2} + \sum_{e^1 = (u,v) \in \delta(H) \setminus \hat{E}} y_{e^1} = \sum_{e \in \delta(W)} y_e \geq 1.$$

Hence y satisfies (4.8). ∎

Theorem 4.10 generalizes to perfect 0-1 matchings. The convex hull of these matchings is given by

$$x \leqslant 1, \qquad x \in R^n_+$$

$$\sum_{e \in \delta(v)} x_e = b_v \quad \text{for } v \in V$$

$$\sum_{e \in \hat{E}} (1 - x_e) + \sum_{e \in \delta(H) \setminus \hat{E}} x_e \geq 1$$

where $H \subseteq V$ and $\hat{E} \subseteq \delta(H)$ are such that $\Sigma_{v \in H} b_v + |\hat{E}|$ is odd.

Moreover, these results can be generalized further to include the constraints $x \leqslant d$ for any $d \in Z^n_+$.

T-Joins and T-Cuts

Our next topic introduces parity conditions into matching problems and includes the postman problem and the minimum-weight $s-t$ path problem.

Definition 4.1. Given $G = (V, E)$ and $T \subseteq V$ with $|T|$ even, a subset of edges $E' \subseteq E$ is a *T-join* if, in the subgraph $G' = (V, E')$, the degree of v is odd if and only if $v \in T$.

Proposition 4.11. *Minimal T-joins are forests.*

Proof. By deleting all of the edges from all cycles of the T-join, we obtain a smaller T-join. ■

Example 4.1. In the graph of Figure 4.4, if $T = V$, the T-joins are $\{e_1, e_4\}$ and $\{e_2, e_3, e_4\}$. If $T = \{1, 4\}$, the T-joins are $\{e_1, e_2, e_4\}$ and $\{e_3, e_4\}$. If $T = \{3, 4\}$ the T-joins are $\{e_4\}$ and $\{e_1, e_2, e_3, e_4\}$.

By choosing different types of sets T, the minimal T-joins yield forests with interesting properties.

1. If $T = \{s, t\}$ and $E' \subseteq E$ is a minimal T-join, then the forest is an $s-t$ path.
2. If $T = V$, $E' \subseteq E$ is a minimal T-join, and the edges of E' form a matching, then the forest is a perfect matching.
3. If $T = \{u: u \text{ is of odd degree}\}$ and $E' \subseteq E$ is a minimal T-join, then E' is a minimal set of edges with the property that the multigraph obtained by duplicating E' is eulerian.

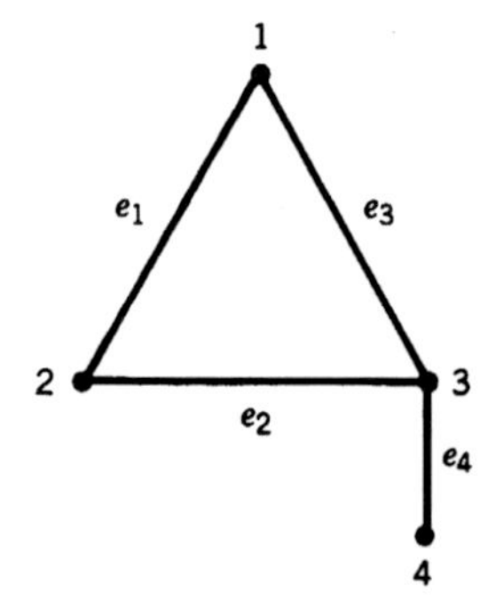

Figure 4.4

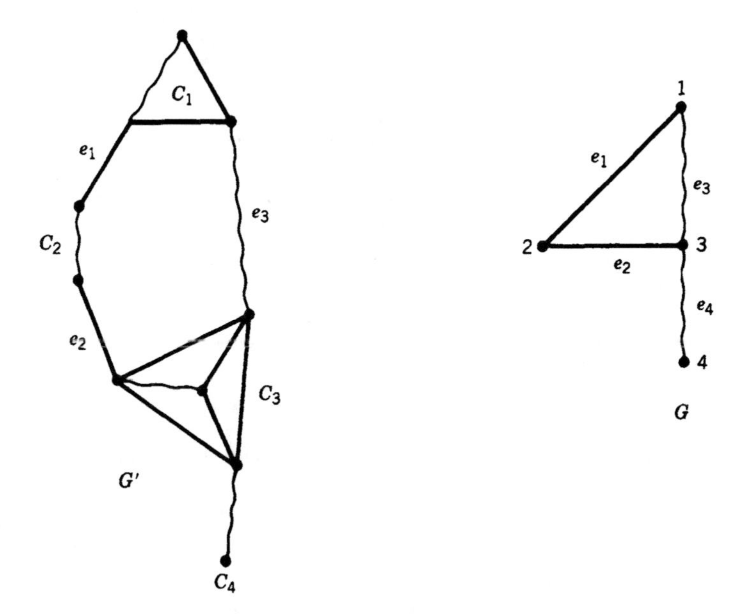

Figure 4.5

The minimum-weight T-join problem is solvable in polynomial time. We show this by reducing it to a perfect matching problem. Given $G = (V, E)$ and T, replace each $v \in V$ by a clique C_v containing $|\delta(v)| + \alpha_v$ nodes, where

$$\alpha_v = \begin{cases} 0 & \text{if } v \in T \text{ and } |\delta(v)| \text{ is odd, or } v \notin T \text{ and } |\delta(v)| \text{ is even} \\ 1 & \text{otherwise.} \end{cases}$$

Then for each $(u, v) \in E$, join a node in C_u to a node in C_v in such a way that no two of these edges are incident to the same node. Call the new graph $G' = (V', E \cup E')$, where E' are the clique edges.

For the graph of Example 4.1 and $T = \{1, 4\}$, we have $\alpha_1 = \alpha_3 = 1$ and $\alpha_2 = \alpha_4 = 0$. A perfect matching on G' and the corresponding T-join are shown in Figure 4.5.

Proposition 4.12. *If $\hat{E} \subseteq E \cup E'$ is a perfect matching in G', then $\hat{E} \cap E$ is a T-join in G. Conversely, if E^* is a T-join in G, there exists an $\tilde{E} \subseteq E'$ so that $\tilde{E} \cup E^*$ is a perfect matching in G'.*

Proof. Suppose $v \in T$. By the definition of α_v, $|C_v|$ is odd. Hence a perfect matching in G' contains an odd number of edges in $\delta(C_v)$. Similarly, if $v \notin T$, then $|C_v|$ is even and a perfect matching in G' contains an even number of edges in $\delta(C_v)$.

The argument for the converse is similar. ■

Next we consider a class of valid inequalities for the convex hull of T-joins. The following definition generalizes the definition of $s-t$ cuts in a graph.

Definition 4.2. Given $G = (V, E)$ and $T \subseteq V$ with $|T|$ even, $\delta(U)$ for $U \subseteq V$ is a *T-cut* if $|U \cap T|$ is odd.

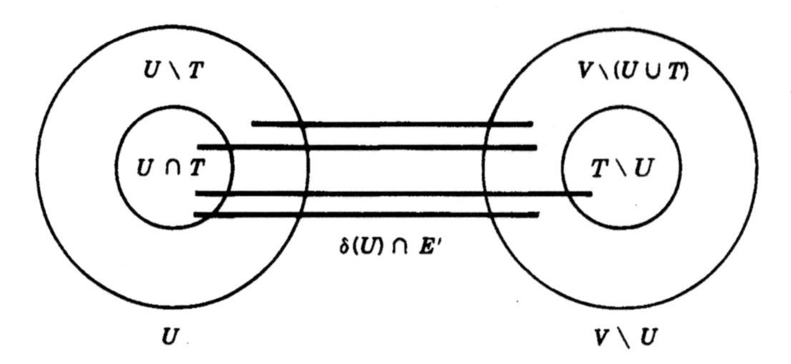

Figure 4.6

Example 4.1 (continued). The minimal T-cuts with $T = V$ are $\{e_1, e_2\}$, $\{e_1, e_3\}$, and $\{e_4\}$.

Suppose $\delta(U)$ is a T-cut and $E' \subseteq E$ is a T-join. Consider the graph $\tilde{G} = (V, E')$ shown in Figure 4.6. Since $\delta(U)$ is a T-cut, $|U \cap T|$ is odd. Since E' is a T-join, the degree of each node in $U \cap T$ is odd, and the degree of each node in $U \setminus T$ is even. Now if $\delta(U) \cap E' = \emptyset$, the graph $(U, E(U) \cap E')$ would have an odd number of nodes of odd degree, which is impossible for any graph. Hence $\delta(U) \cap E' \neq \emptyset$, and

$$\sum_{e \in \delta(U)} x_e \geqslant 1 \quad \text{for } \delta(U) \text{ a } T\text{-cut}, \ U \subseteq V$$

is a valid inequality for the convex hull of T-joins. Moreover, these inequalities yield a polyhedron where extreme points are the minimal T-joins.

Theorem 4.13. *A linear inequality description of the polyhedron whose extreme points are the incidence vectors of minimal T-joins in $G = (V, E)$ and whose extreme rays are the n unit vectors is given by*

$$\begin{aligned} &\sum_{e \in \delta(U)} x_e \geqslant 1 \quad \textit{for } U \subseteq V \textit{ with } |U \cap T| \textit{ odd} \\ &x \in R^n_+. \end{aligned} \tag{4.11}$$

The proof involves showing that y satisfies $y \geqslant 0$ and the odd-set constraints of (4.2) for G' if and only if x satisfies (4.11) for G. The details are left as an exercise.

We now show that blocking polarity can be used to determine a polyhedron whose extreme points are the minimal T-cuts.

Proposition 4.14. *For any graph G, the set of minimal T-joins and T-cuts are a pair of blocking clutters.*

Proof. The proof is by Corollary 6.2 of Section III.1.6. In particular, we show that if $E' \subset E$ does not contain a T-join, then $E \setminus E'$ contains a T-cut. Note that it suffices to take a maximal set E' that does not contain a T-join; that is, for any $e = (u, v) \in E \setminus E'$, $E' \cup \{e\}$ contains a minimal T-join $\overline{E}$.

Each component of $(V, \overline{E})$, and hence each component of $(V, E' \cup \{e\})$, contains an even number of nodes of T. Now let $\tilde{G} = (U, E(U) \cap (E' \cup \{e\}))$ be the component of $(V, E' \cup \{e\})$ containing e.

We claim that $(U, \underline{E}(U) \cap E')$ is disconnected. If not, there exists a cycle C in $\tilde{G}$ containing e. But then $\overline{E} \setminus C \subseteq E$ is a T-join, contradicting the definition of E'.

Now let V_1, V_2 be a bipartition of V according to the components of $(U, E(U) \cap E')$. Since $|U \cap T|$ is even and $e \in \overline{E}$, it follows that $|V_1 \cap T|$ and $|V_2 \cap T|$ must be odd. Finally, since $\delta(V_1) \cap E' = \emptyset$, $\delta(V_1)$ is a T-cut with $\delta(V_1) \subseteq E \setminus E'$. ■

From Theorem 4.13, Proposition 4.14, and Theorem 6.5 of Section III.1.6, we obtain a description of a T-cut polyhedron.

Theorem 4.15. *A linear inequality description of the polyhedron whose extreme points are the incidence vectors of minimal T-cuts in G = (V, E) and whose extreme rays are the n unit vectors is given by*

$$\begin{aligned} &\sum_{e \in E'} x_e \geqslant 1 \quad \text{for all minimal } T\text{-joins } E' \subseteq E \\ &x \in R_+^n. \end{aligned} \tag{4.12}$$

Since we have already given a polynomial-time algorithm for finding minimum-weight T-joins, it follows from the polynomial-time equivalence between optimization and separation that:

Corollary 4.16. *The minimum-weight T-cut problem is solvable in polynomial time.*

In fact, there is an efficient combinatorial algorithm for solving the minimum-weight T-cut problem. It uses a max-flow algorithm as a subroutine and is closely related to the algorithm given in Section II.6.3 for finding violated subtour inequalities.

Edge Coloring

The last topic of this section is the *edge-coloring problem*: Given $G = (V, E)$, color the edges of G, with a minimum number of colors subject to the restriction that no pair of edges incident to a common node has the same color.

Edge coloring is related to matching since an edge coloring is feasible if and only if all of the edges with the same color are a matching. Hence we can formulate the edge-coloring problem as one of covering the edges of a graph with a minimum number of maximal matchings. This yields a minimum-cardinality set-covering problem with a huge number of variables of the form

$$\begin{aligned} \chi(G) = \min\ &1y \\ &yA \geqslant 1 \\ &y \in B^m, \end{aligned} \tag{4.13}$$

where the rows of A correspond to the maximal matchings in G, and $\chi(G)$ is the minimum number of colors needed to obtain a feasible edge coloring. $\chi(G)$ is called the *chromatic index* of G.

Let $\Delta(G) = \max_{v \in V} |\delta(v)|$; that is $\Delta(G)$ is the degree of a node v^* of maximum degree. Since all of the edges incident to v^* require a different color, we have $\chi(G) \geqslant \Delta(G)$ for all graphs G.

Proposition 4.17. $\chi(G) = \Delta(G)$ *for bipartite graphs.*

Proof. If $\Delta(G) \leqslant 2$, the result is trivial. That is, if $\Delta(G) = 2$, then G contains disjoint paths and even cycles, so two colors suffice.

Now suppose that $\Delta(G) \geqslant 3$. Attempt to construct a feasible coloring with $\Delta(G)$ colors by coloring the edges in any order and not using a new color unless it is necessary to do so. Suppose that we have already used $\Delta(G)$ colors and that $e = (u, v)$ requires a new color. This means that all of the $\Delta(G)$ colors except i have been used to color edges incident to u, i has been used to color an edge adjacent to v, and some color j has not been used to color edges adjacent to v (see Figure 4.7). Now consider the subgraph generated by e and all of the edges already colored either i or j. In this subgraph, each node is of degree no larger than 2 and there are no odd cycles; hence it is possible to color these edges with i and j alone. So now we have a coloring with no more than $\Delta(G)$ colors that includes e. ■

Note that the proof gives a polynomial-time algorithm for the edge-coloring problem on bipartite graphs.

Next we consider the edge-coloring problem in general graphs. The graph of Figure 4.8 has $\Delta(G) = 3 < \chi(G) = 4$. A 4-coloring is shown in Figure 4.8; $\chi(G) \geqslant 4$ since $|E| = 7$, and each maximal matching has two edges.

Surprisingly, this example gives the largest possible value of $\chi(G) - \Delta(G)$. The following theorem, which we will not prove, is known as *Vizing's theorem*.

Theorem 4.18. *For any graph G, $\chi(G)$ equals $\Delta(G)$ or $\Delta(G) + 1$.*

We now comment on its implications on solving the edge-coloring problem. Since $\Delta(G)$ can be found for any graph in $O(|E|)$ time and Vizing's proof provides a fast algorithm to color the edges with $\Delta(G) + 1$ colors, we might hope that Theorem 4.18 could be used to find $\chi(G)$ efficiently. Unfortunately, this is not the case since the decision problem "Does $\chi(G) = \Delta(G)$?" is $\mathcal{NP}$-complete. Moreover, determining the chromatic index is difficult even if $\chi(G)$ is small.

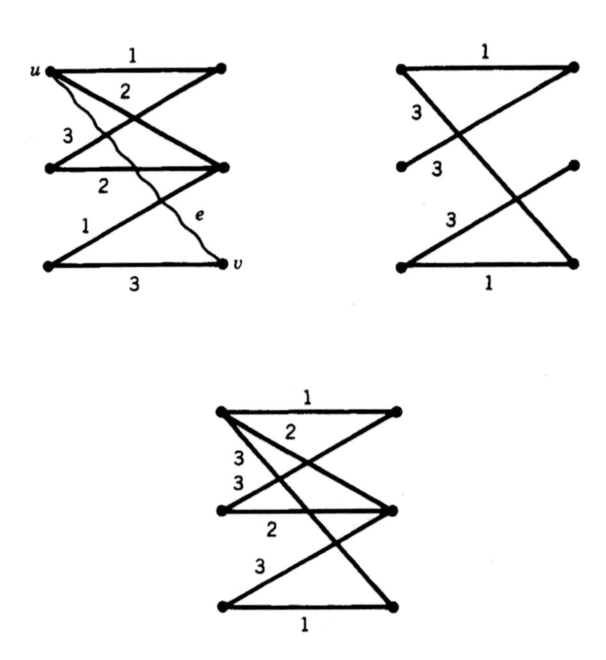

Figure 4.7. $i = 3, j = 1$.

Theorem 4.19. *The problem of deciding whether $\chi(G) \leqslant 3$ is $\mathcal{NP}$-complete.*

In other words, there is an infinite family of graphs with $\Delta(G) = 3$, for which the problem of deciding whether $\chi(G) = 3$ or 4 is $\mathcal{NP}$-complete. An immediate consequence of this result is:

Corollary 4.20. *Unless $\mathcal{P} = \mathcal{NP}$, no polynomial-time algorithm can yield a feasible edge coloring that requires fewer than $\lfloor \frac{4}{3}\chi(G) \rfloor$ colors for all graphs.*

Despite these negative results regarding the polynomial solvability of the edge-coloring problem, we now show how Theorem 4.18 and the ellipsoid algorithm can, in certain cases, yield a polynomial-time algorithm for finding $\chi(G)$. The linear programming relaxation of (4.13) is

$$\chi_{\mathrm{LP}}(G) = \min 1y$$

$$(4.14) \qquad yA \geqslant 1$$

$$y \in R_+^m$$

and its dual is

$$\Delta_{\mathrm{LP}}(G) = \max 1x$$

$$(4.15) \qquad Ax \leqslant 1$$

$$x \in R_+^n.$$

Although problem (4.15) has a constraint for each maximal matching, it can be solved in polynomial time since the separation problem is a maximum-weight matching problem. That is, x^*, $0 \leqslant x^* \leqslant 1$, is a feasible solution to (4.15) if and only if a maximum-weight matching in G with edge weights x^* has value no greater than 1.

Proposition 4.21. *If $\Delta_{\mathrm{LP}}(G) > \Delta(G)$, then $\chi(G) = \Delta(G) + 1$.*

Proof. Note that $\Delta(G) \leqslant \Delta_{\mathrm{LP}}(G)$ since a feasible solution to (4.15) is $x_e = 1$ for all $e \in \delta(v^*)$, where v^* is a node of maximum degree. Now by linear programming duality, $\Delta_{\mathrm{LP}}(G) = \chi_{\mathrm{LP}}(G) \leqslant \chi(G)$. Hence if $\Delta_{\mathrm{LP}}(G) > \Delta(G)$, then $\chi(G) > \Delta(G)$. Then by Theorem 4.18, we have $\chi(G) = \Delta(G) + 1$. ∎

In the graph of Figure 4.8 we have $\Delta_{\mathrm{LP}}(G) > 3$, which implies $\chi(G) = 4$.

Heuristics provide a practical approach for finding good colorings of large graphs. In fact, there are heuristics that achieve the performance bound of $\lfloor \frac{4}{3}\chi(G) \rfloor$, and it is also possible to realize asymptotic bounds of the form $\alpha\chi(G) + p$ with $\alpha < \frac{4}{3}$. We will not give

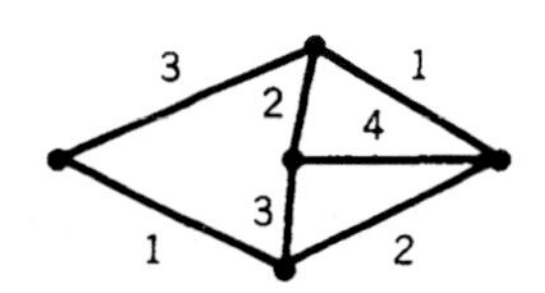

Figure 4.8

any details, but the basic idea of many heuristic coloring schemes has been used in the proof of Proposition 4.17. Namely, in a sequential coloring scheme, whenever we encounter an edge $e = (u, v)$ that requires a "new" color, we try to adjust the present coloring so that a new color is not required for e. A simple rule of this type is to consider two colors, say red and blue. Now we try to recolor all of the red and blue edges by red and blue so that feasibility is maintained and every edge adjacent to u and v is colored red. Then e can be colored blue.

5. NOTES

Section III.2.1

Matching theory predates mathematical programming. Remarks on the early literature, which was primarily concerned with bipartite graphs, appear in Pulleyblank (1983) and Schrijver (1983a). Lovasz and Plummer (1986) is a recent book on matching theory that emphasizes the graph-theoretic aspects of matching.

The application to the postman problem was given by Edmonds and Johnson (1973). Fujii et al. (1969) and Coffman and Graham (1972) gave an application to a scheduling problem. Network flow problems in which an arc can have two heads or two tails can be modeled as matching problems [see Edmonds and Johnson (1970)]. Nemhauser and Weber (1979) used weighted matching in the solution of set-partitioning problems. Ball, Bodin and Dial (1983) gave a matching-based algorithm for the scheduling of mass transit crews and vehicles.

Section III.2.2

The augmenting-path proposition is due to Berge (1957) and Norman and Rabin (1959).

A fast cardinality matching algorithm for bipartite graphs was given by Hopcroft and Karp (1973).

The algorithmic aspects of the 1-matching problem on general graphs were initiated by Edmonds (1965a). In this article, he gave a polynomial-time algorithm for the cardinality problem. The Hopcroft–Karp algorithm for bipartite graphs was extended to general graphs by Even and Kariv (1975).

Section III.2.3

The maximum-weight matching algorithm was developed by Edmonds (1965c). The algorithm given here is a slight variation of the one by Edmonds. Another variation is given in Lawler (1976).

Other weighted matching algorithms have been given by Cunningham and Marsh (1978), Derigs (1986), and Grötschel and Holland (1985). The latter is a fractional cutting-plane approach that uses the simplex method and an efficient separation routine for finding violated blossom inequalities. The separation routine is based on a polynomial-time algorithm by Padberg and Rao (1982) for finding minimum-weighted T-cuts (see Section III.2.4).

Ball and Derigs (1983) presented alternative strategies for implementing matching algorithms. Burkhard and Derigs (1980) gave FORTRAN listings of matching and assignment algorithms.

Pulleyblank and Edmonds (1975) characterized the blossom inequalities that are facets of the matching polytope.

Sensitivity analysis in weighted matching has been considered by Weber (1981), Derigs (1985), and Ball and Taverna (1985).

Avis (1983) presented a survey of heuristics for solving weighted matching problems.

Edmonds and Johnson (1970) described an algorithm for weighted b-matching problems. The first polynomial-time algorithm for this class of problems is attributed to Cunningham and Marsh (1978).

Section III.2.4

The nonalgorithmic proof technique given here for the convex hull of 1-matchings is due to Lovasz (1979a). Other nonalgorithmic proofs have been given by Balinski (1972), Hoffman and Oppenheim (1978), and Schrijver (1983b). The proof of total dual integrality comes from Schrijver (1983a). A different proof is given by Cunningham and Marsh (1978).

Relationships between matching and edge covering have been studied by Norman and Rabin (1959) and Balinski (1970b).

The transformations used to obtain the b-matching results come from Schrijver (1983a), who attributed them to Tutte (1954).

Theorem 4.8 on the b-matching polytope is due to Edmonds and Pulleyblank and appears in Pulleyblank (1973). Pulleyblank (1980, 1981) established that this system is TDI. Further results regarding a minimal TDI system have been obtained by Cook (1983b). Cook and Pulleyblank (1987) provided a minimal linear inequality representation of the convex hull of capacitated b-matchings.

The reduction of the minimum T-join problem to a perfect matching problem comes from Edmonds and Johnson (1973). They also used the connection with matchings to prove Theorem 4.15. Also see Gastou and Johnson (1986) and Johnson and Mosterts (1987).

Generalizations of matching problems have been studied by Gerards and Schrijver (1986), Cornuejols and Hartvigsen (1986), and Cornuejols (1986).

The edge-coloring result for bipartite graphs is a classic theorem of Konig. The proof given here can be found in many graph theory texts [e.g., Bondy and Murty (1976)].

Theorem 4.18 is due to Vizing (1964). Marcotte (1986b) showed that Vizing's theorem is true in the weighted case for a restricted class of graphs.

Proposition 4.19 and Theorem 4.20 are due to Holyer (1981).

The result cited on the worst-case bounds of edge-coloring heuristics is due to Hochbaum et al. (1986).

6. EXERCISES

1. Find a maximum-cardinality matching in the bipartite graph of Figure 6.1, and give a short proof of optimality.

2. Find a maximum-cardinality matching in the graph of Figure 6.2, and give a short proof of optimality.

3. A graph $G = (V, E)$ is said to have a *perfect matching* if there exists $M \subseteq E$ such that no node is exposed relative to M. Let $P(U)$, $U \subseteq V$, denote the number of components with an odd number of nodes in the graph G_U induced by $V \setminus U$. Prove that G has a perfect matching if and only if $P(U) \leqslant |U|$ for all $U \subseteq V$.

4. Find a maximum-weight matching in

   i) the graph of Figure 6.2 with weights as shown,

   ii) the graph of Figure 6.3 with weights as shown.

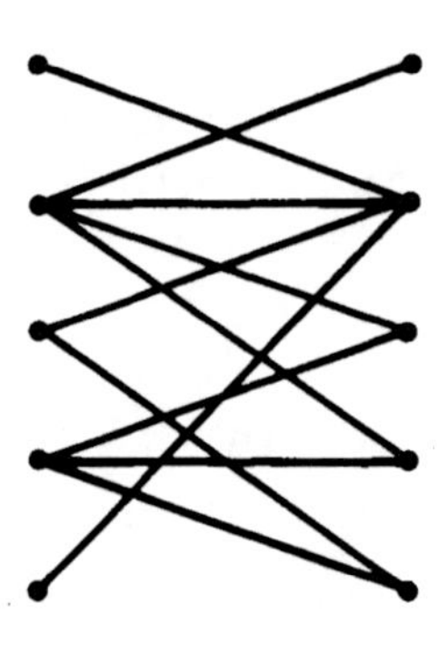

Figure 6.1

5. **i)** Devise a fast heuristic algorithm to find violated blossom inequalities.

ii) Use this in an FCPA to find a maximum-weight matching in the graph of Figure 6.2.

6. Prove that the dual solution is always half-integer in the maximum-weight matching algorithm.

7. For the maximum-weight matching problem, define an *augmenting path p*, relative to M, to be an alternating path or alternating cycle having no edge of $M \setminus P$ incident to P and having the property

$$\sum_{e_j \in P \setminus M} c_j - \sum_{e_j \in M \setminus P} c_j > 0,$$

where P is the set of edges contained in the path p. Prove that M is optimal if and only if M admits no augmenting path.

8. Find an optimal postman route for the graph of Figure 6.2 with the distances as shown.

9. Show that the system (4.4) (a), (b), (d) is not TDI for the complete graph on four nodes.

10. Show that the weighted b-matching problem reduces to a network flow problem when b_v is even for all $v \in V$.

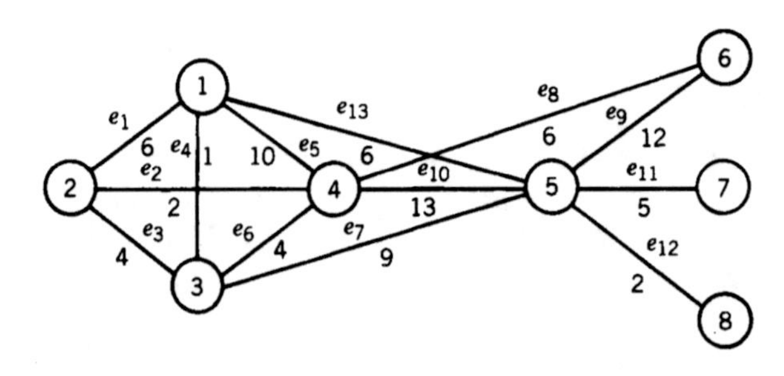

Figure 6.2

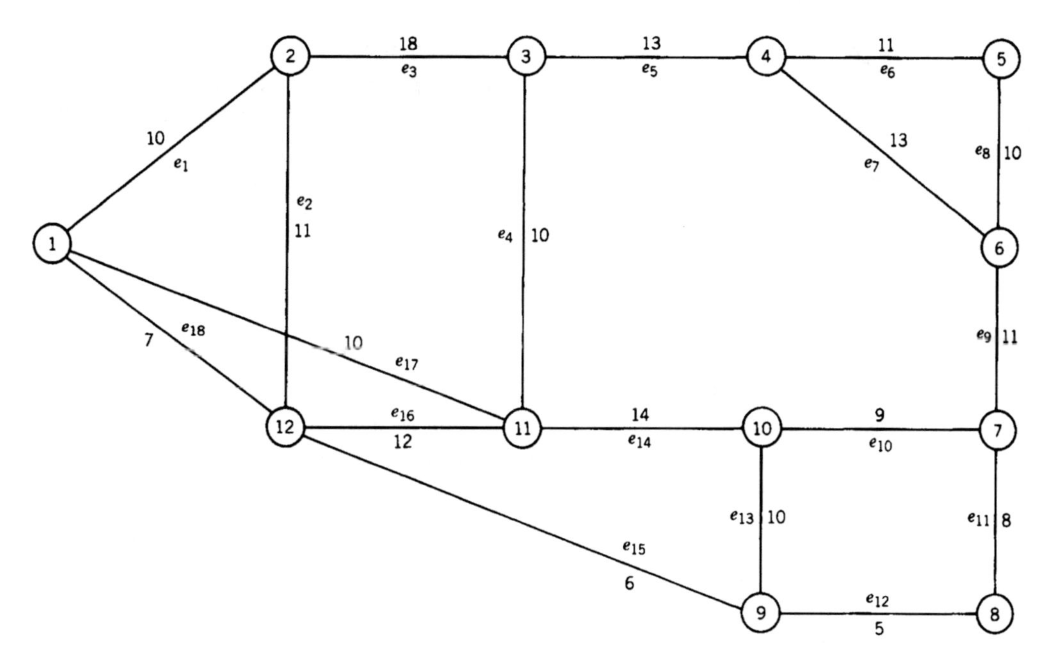

Figure 6.3

11. Solve the weighted b-matching problem on the graph of Figure 6.4 by reducing it to a 1-matching problem.

12. For the graph of Figure 6.2, solve the minimum-weight T-join problem by reducing it to a perfect matching problem for:

i) $T = \{2, 7\}$;

ii) $T = V$.

13. Prove Theorem 4.13.

14. Describe an efficient combinatorial algorithm for the minimum-weight T-cut problem.

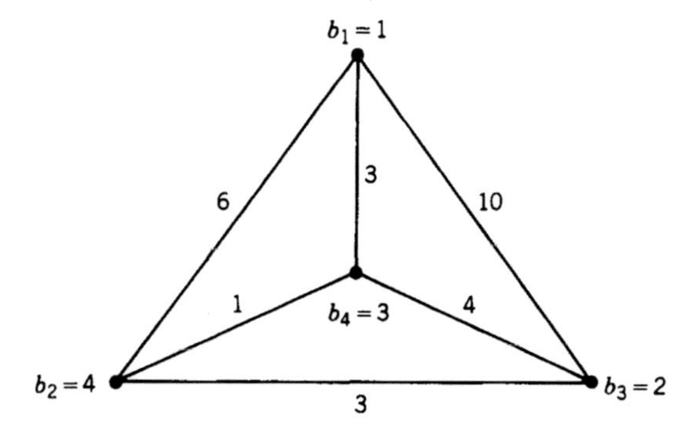

Figure 6.4

15. Given a graph G, suppose that an efficient combinatorial algorithm is known for the separation problem for the 1-matching polytope $P(1)$ (see Section III.3.7). Let $\chi_{LP}(w) = \min\{1y: yA \geq w, y \in R^m_+\}$, where A is the matching-edge incidence matrix of G.

i) Verify that $\chi_{LP}(w) \leq 1$ if and only if w lies in the 1-matching polytope $P(1)$ and that

$$\chi_{LP}(w) = \max_S \left\{\frac{w(S)}{(|S|-1)/2}: |S| \text{ odd}\right\}, \quad \text{where } w(S) = \sum_{e \in E(S)} w_e.$$

ii) Verify that $\zeta(w) = \max\{\lambda: \lambda w \in P(1)\} = 1/\chi_{LP}(w)$.

iii) Consider the following algorithm to calculate $\zeta(w)$, using as a subroutine an efficient separation algorithm for the polytope $P(1)$; that is,

$$\max\{w(S) - (|S|-1)/2: |S| \text{ odd}\}$$

Algorithm: Choose λ^0 with $\lambda^0 w \notin P(1)$. Set $t = 0$.

Iteration t:

a) Solve the separation problem for $\lambda^t w$.

b) Stop if $\lambda^t w \in P(1)$.

c) Set λ^{t+1} such that $\lambda^{t+1} w(S^t) = (|S^t| - 1)/2$.

d) Augment t.

Verify that

a) $\lambda^{t+1} < \lambda^t$,

b) $[1/\lambda^{t+1} - 1/\lambda^t]((|S^t| - 1)/2 - (|S^{t+1}| - 1)/2) > 0$.

c) $|S^t|$ is strictly decreasing.

d) The algorithm terminates after, at most, $|V|/2$ iterations.

iv) Use this algorithm to calculate $\chi_{LP}(G)$ and $\chi(G)$ for the graph of Figure 4.8.

16. Show that the maximum-weight assignment problem with the following conditions can be formulated as a matching problem: $c_{ii} = -\infty$ for all i, $c_{ij} = c_{ji}$ for all i and j, and $x_{ij} = x_{ji}$ for all i and j.

III.3.

Matroid and Submodular Function Optimization

1. INTRODUCTION

Matroids and submodular functions are the foundations for some combinatorial optimization problems that generalize both network flow problems and the spanning tree problem treated in Chapter I.3. Matroids can be viewed as prototypes of independence systems and 0-1 integer programs with "nice" properties that can be used to obtain efficient algorithms for the corresponding optimization problems.

Definition 1.1. Let $N = \{1, \ldots, n\}$ be a finite set, and let $\mathcal{F}$ be a set of subsets of N. $\mathcal{I} = (N, \mathcal{F})$ is an *independence system* if $F_1 \in \mathcal{F}$, and $F_2 \subseteq F_1$ implies $F_2 \in \mathcal{F}$. Elements of $\mathcal{F}$ are called *independent sets*, and the remaining subsets of N are called *dependent sets*.

Let $\mathcal{F}_T = \{F \in \mathcal{F}: F \subseteq T\}$. Then if $\mathcal{I} = (N, \mathcal{F})$ is an independence system, $\mathcal{I}_T = (T, \mathcal{F}_T)$ is an independence system for all $T \subseteq N$.

Definition 1.2. Given an independence system $\mathcal{I} = (N, \mathcal{F})$, we say that $F \in \mathcal{F}$ is a *maximal* independent set if $F \cup \{j\} \notin \mathcal{F}$ for all $j \in N \setminus F$. A maximal independent set T is *maximum* if $|S| \leq |T|$ for all $S \in \mathcal{F}$.

In describing independence systems, we use the notation

$$m(T) = \max_{S \subseteq T}\{|S|: S \in \mathcal{F}\} \quad \text{for } T \subseteq N$$

to denote the size of a maximum-cardinality independent set in T. Note that $m(T) \leq |T|$ and $\mathcal{F} = \{T \subseteq N: m(T) = |T|\}$. Hence $\mathcal{I}$ can also be specified as $\mathcal{I} = (N, m)$.

Matroids are those independence systems for which all maximal independent sets in T are maximum for any subset $T \subseteq N$.

Definition 1.3. $M = (N, \mathcal{F})$ is a *matroid* if M is an independence system in which for any subset $T \subseteq N$, every independent set in T that is maximal in T has cardinality $m(T)$.

The following proposition is an immediate consequence of the fact that maximal sets must be maximum not just in N but also for all subsets $T \subseteq N$.

Proposition 1.1. *If $M = (N, \mathcal{F})$ is a matroid, then the independence system $\mathcal{I}_T = (T, \mathcal{F}_T)$ is a matroid for $T \subseteq N$.*

Matroids were originally developed from matrices to generalize the properties of linear independence and bases in a vector space. This generalization has yielded several classes of matroids.

(a) *Matric Matroids*. Let A be an $m \times n$ matrix, and let N be the index set of the columns of A. Define the independence system $(N, \mathcal{F})$ by $F \in \mathcal{F}$ if the set of columns defined by F is linearly independent. For any submatrix A_T with columns a_j for $j \in T$, it is well known that every maximal set of linearly independent columns contains $m(T) = \text{rank}(A_T)$ columns. Hence $(N, \mathcal{F})$ is a matroid. If M is a matroid and there exists a matrix A such that the independent sets of M correspond to the linearly independent columns of A, then M is called a *matric matroid*.

(b) *Graphic Matroids*. Let $G = (V, E)$ be a graph, and let $F \subseteq E$ be a subset of the edges. Let $F \in \mathcal{F}$ if $G_F = (V, F)$ contains no cycles. For any subset $T \subseteq E$, the cardinality of a maximal set of edges that is acyclic in G_T is $m(T) = |V|$ − number of connected components of G_T. Hence $(E, \mathcal{F})$ is a matroid. If M is a matroid and there exists a graph G such that the independent sets of M correspond to the acyclic edge sets of G, then M is a *graphic matroid*. We leave it as an exercise to show that graphic matroids are matric.

(c) *Partition Matroids*. Given m disjoint finite sets E_i for $i \in I = \{1, \ldots, m\}$, let $E = \bigcup_{i=1}^{m} E_i$. $F \subseteq E$ is independent if $|F \cap E_i| \leq 1$ for all $i \in I$. For any $T \subseteq E$, the cardinality of a maximal independent set contained in T is $\Sigma_{i \in I} \delta_i$, where $\delta_i = 1$ if $T \cap E_i \neq \emptyset$ and $\delta_i = 0$ otherwise. Hence $(E, \mathcal{F})$ is a matroid.

The set of matchings in a graph do not form a matroid. For a path e_1, e_2, e_3, both the sets $\{e_2\}$ and $\{e_1, e_3\}$ are maximal matchings in $\{e_1, e_2, e_3\}$, but they differ in cardinality.

In the context of combinatorial optimization, the most striking property of matroids—and indeed, another way to define them—is that, given a weight vector $c \in R^{|E|}$, a greedy algorithm (see Section I.3.3 for trees and Section II.5.3 for general independence systems) always gives an optimal-weight independent set. This will be demonstrated in Section 3.

Submodular functions are closely related to matroids. We will see that for a matroid, the cardinality function m is submodular. Such functions have already appeared in Section II.5.3, where the uncapacitated location problem was shown to be a problem of maximizing a submodular function.

Definition 1.4. Let N be a finite set, and let f be a real-valued function on the subsets of N.

a. f is *nondecreasing* if $f(S) \leq f(T)$ for $S \subseteq T \subseteq N$.
b. f is *submodular* if $f(S) + f(T) \geq f(S \cup T) + f(S \cap T)$ for $S, T \subseteq N$.
c. f is *supermodular* if $-f$ is submodular.
d. r is a *submodular rank function* if $r(\emptyset) = 0$, r is integer-valued, nondecreasing, and submodular, and $r(\{j\}) \leq 1$ for all $j \in N$.

Example 1.1. Given a digraph $\mathcal{D} = (V, \mathcal{A})$ and weights $c \in R_+^{|\mathcal{A}|}$, for $S \subseteq V$ let

$$c(S) = \sum_{\{(i,j) \in \delta^+(S)\}} c_{ij} = \sum_{\substack{i \in S \\ j \in V \setminus S}} c_{ij}.$$

The cut function $c(S)$ is submodular because

$$c(S) + c(T) - c(S \cup T) - c(S \cap T) = \sum_{\substack{i \in S \setminus T \\ j \in T \setminus S}} c_{ij} + \sum_{\substack{i \in T \setminus S \\ j \in S \setminus T}} c_{ij} \geq 0.$$

Having introduced both matroids and submodular functions, we now briefly indicate some of the other problems to be studied in this chapter. In the next section we will establish the equivalence between a matroid $M = (N, \mathcal{F})$ and a submodular rank function r on N, and we will introduce and develop some elementary matroid properties for later use.

In Section 3 we will consider the *matroid optimization problem*: An instance is given by a matroid $M = (N, \mathcal{F})$ and a weight vector $c \in R^n$. The problem is

$$\max_S \left\{ \sum_{j \in S} c_j \colon S \in \mathcal{F} \right\}.$$

Formulating this problem as an integer program leads us to study polytopes of the form

$$P(f) = \left\{ x \in R^n_+ \colon \sum_{j \in S} x_j \leqslant f(S) \text{ for } S \subseteq N \right\},$$

where f is a submodular function.

An important generalization of the matroid optimization problem is the *k-matroid intersection problem*: Given k matroids $M_i = (N, \mathcal{F}_i)$ for $i = 1, \ldots, k$ and a weight vector $c \in R^n$, the problem is

$$\max_S \left\{ \sum_{j \in S} c_j \colon S \in \bigcap_{i=1}^{k} \mathcal{F}_i \right\}.$$

Thus, feasible solutions correspond to sets that are independent in each of the matroids.

Remember that a branching in a digraph $\mathcal{D} = (V, \mathcal{A})$ is a set of arcs $\mathcal{A}' \subseteq \mathcal{A}$ such that $\mathcal{D}' = (V, \mathcal{A}')$ is a spanning tree and no more than one arc enters each node. Hence a set of arcs forms part of a branching if and only if it is independent in both a partition and a graphic matroid. In Sections 4 and 5 we will study efficient algorithms for the 2-matroid intersection problem.

Now consider the arc sets that form part of a branching in a digraph and intersect these sets with a second partition matroid specifying that no more than one arc leaves each node. The resulting objects of maximum cardinality are Hamiltonian paths. Because it is known that the question of whether a graph contains a Hamiltonian path is $\mathcal{NP}$-complete, it follows that the k-matroid intersection problem is $\mathcal{NP}$-hard for all $k \geqslant 3$.

In Sections 6 and 7 we will consider, in more detail, polytopes $P(f)$ where f is submodular and nondecreasing. It will be shown that the separation problem for $P(f)$ is equivalent to the problem of minimizing another submodular function; that is,

$$\min_S \{f'(S) \colon S \subseteq N\}, \qquad f' \text{ submodular.}$$

Thus we study algorithms for this minimization problem and some special cases where f' has more structure.

In Section 8 we will study a covering problem of the form: Given a matroid $M = (N, \mathcal{F})$, what is the minimum number of independent sets whose union is N? This problem has the integer-rounding property and can be solved efficiently.

Finally, we consider the problem of maximizing a submodular function:

$$\max_S \{f(S) \colon S \subseteq N\}, \qquad f \text{ submodular.}$$

In contrast to the earlier problems of the chapter, this model includes $\mathcal{NP}$-hard problems, such as the uncapacitated location problem. Hence we examine different integer programming formulations and heuristics.

2. ELEMENTARY PROPERTIES

There are many ways of defining and viewing both matroids and submodular functions. Here we introduce the definitions and the fundamental results that we will use later. First we study submodular functions (see Definition 1.4).

Proposition 2.1

i. *f is submodular if and only if*

$$\text{(a)} \qquad f(S \cup \{j\}) - f(S) \geq f(S \cup \{j, k\}) - f(S \cup \{k\}) \quad \text{for } j, k \in N, j \neq k, \text{ and } S \subseteq N \setminus \{j, k\}.$$

ii. *f is submodular and nondecreasing if and only if*

$$\text{(b)} \qquad f(T) \leq f(S) + \sum_{j \in T \setminus S} [f(S \cup \{j\}) - f(S)] \quad \text{for } S, T \subseteq N.$$

Proof. i. If f is submodular we obtain (a) by setting $S \leftarrow S \cup \{j\}$ and $T \leftarrow S \cup \{k\}$ in Definition 1.4.

If (a) holds, let $S = A \cap B$, $A \setminus B = \{j_1, \ldots, j_r\}$, and $B \setminus A = \{k_1, \ldots, k_s\}$. Then

$$\begin{aligned}
&f(B) - f(A \cap B) \\
&= \sum_{t=1}^{s} [f(S \cup \{k_1, \ldots, k_t\}) - f(S \cup \{k_1, \ldots, k_{t-1}\})] \\
&\geq \sum_{t=1}^{s} [f(S \cup \{k_1, \ldots, k_t\} \cup \{j_1\}) - f(S \cup \{k_1, \ldots, k_{t-1}\} \cup \{j_1\})] \\
&\qquad \vdots \\
&\geq \sum_{t=1}^{s} [f(S \cup \{k_1, \ldots, k_t\} \cup \{j_1, \ldots, j_r\}) - f(S \cup \{k_1, \ldots, k_{t-1}\} \cup \{j_1, \ldots, j_r\})] \\
&= \sum_{t=1}^{s} [f(A \cup \{k_1, \ldots, k_t\}) - f(A \cup \{k_1, \ldots, k_{t-1}\})] \\
&= f(A \cup B) - f(A).
\end{aligned}$$

ii. Let $T \setminus S = \{j_1, \ldots, j_r\}$. Then

$$\begin{aligned}
f(T) \leq f(S \cup T) &= f(S) + \{f(S \cup T) - f(S)\} \\
&= f(S) + \sum_{t=1}^{r} \{f(S \cup \{j_1, \ldots, j_t\}) - f(S \cup \{j_1, \ldots, j_{t-1}\})\} \\
&\leq f(S) + \sum_{t=1}^{r} \{f(S \cup \{j_t\}) - f(S)\},
\end{aligned}$$

where the first inequality holds if f is nondecreasing, and the second one holds if f is submodular. Taking $T = S \cup \{j,k\}$ and $T = S \setminus \{k\}$ in (b) gives the converse. ∎

Complex submodular functions are often constructed from simple submodular functions.

Proposition 2.2. *The following conditions yield submodular functions.*

a. *If $a_j \in R^1$ for $j \in N$ and $a_0 \in R^1$, then $f(S) = a_0 + \Sigma_{j\in S} a_j$ for $S \subseteq N$ is submodular on N.*
b. *If f is submodular on N, then $\bar{f}(S) = f(N \setminus S)$ for $S \subseteq N$ is submodular on N.*
c. *If f is submodular on N and $k \in R^1$, then $f'(S) = \min(f(S), k)$ is submodular on N.*
d. *If f_1 and f_2 are submodular on N, then $f(S) = f_1(S) + f_2(S)$ is submodular on N.*

Proposition 2.3. *If f is integer valued, submodular, and nondecreasing with $f(\emptyset) = 0$, and $r(S) = \min_{Q\subseteq S} \{f(Q) + |S \setminus Q|\}$, then r is a submodular rank function.*

Proof. Suppose $r(S) = f(A) + |S \setminus A|$ and $r(T) = f(B) + |T \setminus B|$. Then

$$\begin{aligned}
r(S) + r(T) &= f(A) + f(B) + |S \setminus A| + |T \setminus B| \\
&\geqslant f(A \cup B) + f(A \cap B) + |S \setminus A| + |T \setminus B| \\
&\geqslant f(A \cup B) + f(A \cap B) + |(S \cup T) \setminus (A \cup B)| + |(S \cap T) \setminus (A \cap B)| \\
&= f(A \cup B) + |(S \cup T) \setminus (A \cup B)| + f(A \cap B) + |(S \cap T) \setminus (A \cap B)| \\
&\geqslant r(S \cup T) + r(S \cap T).
\end{aligned}$$

Hence r is submodular.

Now suppose

$$r(S \cup \{j\}) = f(Q^*) + |(S \cup \{j\}) \setminus Q^*|, \quad \text{where } Q^* \subseteq S \cup \{j\}.$$

If $j \in Q^*$, we obtain

$$\begin{aligned}
r(S \cup \{j\}) &\geqslant f(Q^* \setminus \{j\}) + |S \setminus (Q^* \setminus \{j\})| \quad (\text{since } f(Q^*) \geqslant f(Q^* \setminus \{j\})) \\
&\geqslant r(S) \quad (\text{since } Q^* \setminus \{j\} \subseteq S).
\end{aligned}$$

If $j \notin Q^*$, we obtain

$$r(S \cup \{j\}) = f(Q^*) + |S \setminus Q^*| + 1 \geqslant r(S) + 1.$$

Hence r is nondecreasing.

Finally, $r(\emptyset) = f(\emptyset) = 0$ and $r(\{j\}) \leqslant f(\emptyset) + |\{j\}| = 1$ for all j. ■

Theorem 2.4. *If $M = (N, \mathcal{F})$ is a matroid, its cardinality function $m(T) = \max_{S\subseteq T} \{|S| : S \in \mathcal{F}\}$ is submodular. If $(N, \mathcal{F})$ is an independence system whose cardinality function $m(T)$ is submodular, then $M = (N, \mathcal{F})$ is a matroid.*

Proof. Clearly $m(\emptyset) = 0$, m is nondecreasing, and $m(S \cup \{j\}) - m(S) \leqslant 1$, since $(N, \mathcal{F})$ is an independence system. To prove that m is submodular we will show that

$$m(S \cup \{j\}) - m(S) \geqslant m(S \cup \{j, k\}) - m(S \cup \{k\}).$$

The inequality is obvious when $m(S \cup \{j\}) - m(S) = 1$, so suppose that $m(S \cup \{j\}) = m(S) = t$ and $m(S \cup \{j, k\}) - m(S \cup \{k\}) = 1$. There are now two cases to consider.

Case 1. $m(S \cup \{j, k\}) = t + 2$. Then $m(S \cup \{j, k\}) - m(S \cup \{j\}) = 2$, which is impossible.

Case 2. $m(S \cup \{j, k\}) = t + 1$. Let Q be a maximal independent set in S. It follows that $Q \cup \{l\} \notin \mathcal{F}$ for all $l \in S \setminus Q$. Also, since $m(S \cup \{j\}) = m(S \cup \{k\}) = t$, we have $Q \cup \{j\} \notin \mathcal{F}$ and $Q \cup \{k\} \notin \mathcal{F}$. Hence Q is maximal in $S \cup \{j, k\}$ so that $m(S \cup \{j, k\}) = t$, which is a contradiction.

Let $T \subseteq N$, and suppose that S_1 and S_2 are maximal independent sets in T with $|S_1| < |S_2|$. Thus $m(S_1) = |S_1| < m(S_2) = |S_2|$. Using (b) of Proposition 2.1, we have

$$m(S_2) \leqslant m(S_1) + \sum_{j \in S_2 \setminus S_1} [m(S_1 \cup \{j\}) - m(S_1)],$$

which implies that $m(S_1 \cup \{j\}) > m(S_1)$ for some $j \in S_2 \setminus S_1$. Hence $m(S_1 \cup \{j\}) = |S_1 \cup \{j\}|$, contradicting the maximality of S_1. Therefore $(N, \mathcal{F})$ is a matroid. ■

From now on we will represent a matroid M as either $(N, \mathcal{F})$ or (N, r), where r is its submodular rank function, depending on which is more convenient.

The last part of the proof of Theorem 2.4 establishes an important exchange property of matroids that is well known for matrices.

Proposition 2.5. *If $M = (N, \mathcal{F})$ is a matroid and $S_1, S_2 \in \mathcal{F}$ satisfy $|S_1| < |S_2|$, then there exists $j \in S_2 \setminus S_1$ such that $S_1 \cup \{j\} \in \mathcal{F}$.*

There are various other important properties of matroids, most of which are familiar from matrices.

Definition 2.1. Let $M = (N, \mathcal{F})$ be a matroid with rank function r.

a. A is a *basis* of the matroid if $A \in \mathcal{F}$ and $r(A) = r(N)$.
b. A is a *circuit* of the matroid if A is a minimal dependent set (i.e., $A \notin \mathcal{F}$, but $A \setminus \{j\} \in \mathcal{F}$ for all $j \in A$).
c. For $A \subseteq N$, the *span or closure* of A is the set $\text{sp}(A) = \{j \in N: r(A \cup \{j\}) = r(A)\}$.

Bases are evidently the maximal independent sets in the matroid, all of which are of cardinality $r(N)$. Circuits are minimal dependent sets. Hence if A is a circuit, then $r(A) = |A| - 1$. From the submodularity of the rank function, we observe that $\text{sp}(A)$ is the maximal set $B \supseteq A$ for which $r(A) = r(B)$.

For a graphic matroid, bases are the edge sets of spanning trees, circuits are the edge sets of cycles, and the span of an edge set E' contains E' plus any edge that, when added to E', yields a new cycle.

One of the most useful properties of a matroid, which we have already seen to be true for cycles in a graph, is:

Proposition 2.6. *If $F \in \mathcal{F}$ and $F \cup \{j\} \notin \mathcal{F}$, there exists a unique circuit $C \subseteq F \cup \{j\}$. This implies $F \cup \{j\} \setminus \{k\} \in \mathcal{F}$ for all $k \in C \setminus \{j\}$.*

Proof. Suppose there exist distinct circuits C_1, C_2 in $F \cup \{j\}$. Now $C_1 \cap C_2 \in \mathcal{F}$ by the minimality of circuits. Also, $(C_1 \cup C_2) \setminus \{j\} \in \mathcal{F}$ because $(C_1 \cup C_2) \setminus \{j\} \subseteq F$. But $r(C_i) = |C_i| - 1$ for $i = 1, 2$, $r(C_1 \cup C_2) = |C_1 \cup C_2| - 1$, and $r(C_1 \cap C_2) = |C_1 \cap C_2|$, contradicting the submodularity of r.

If $(F \cup \{j\}) \setminus \{k\} \notin \mathcal{F}$, then $(F \cup \{j\}) \setminus \{k\}$ contains a circuit C' different from C, contradicting the uniqueness of C. ∎

Example 2.1. Consider the matric matroid $M = (N, \mathcal{F})$ defined by the matrix

$$\begin{pmatrix} 2 & 1 & 4 & -1 & 0 & -2 \\ 1 & 1 & 3 & 2 & 3 & 4 \\ 3 & 2 & 7 & 4 & 6 & 8 \end{pmatrix},$$

where $N = \{1, \ldots, 6\}$ is the index set for the columns, and $S \in \mathcal{F}$ if and only if the set of columns indexed by S is linearly independent. The rank function r takes the following values:

1. $r(\emptyset) = 0$, $r(\{j\}) = 1$ for all j.
2. $r(\{j, k\}) = 2$ for all $j \neq k$ except that $r(\{4, 6\}) = 1$.
3. $r(S) = 3$ for all S with $|S| \geqslant 3$, except that $r(\{1, 2, 3\}) = r(\{2, 4, 5\}) = r(\{2, 4, 5, 6\}) = r(\{4, 6, k\}) = 2$ for all $k \in \{1, 2, 3, 5\}$.

Since $r(N) = 3$, the bases are the independent sets S with $|S| = 3$. The circuits are $\{4, 6\}$, $\{1, 2, 3\}$, $\{2, 4, 5\}$, and all 4-tuples that do not contain any of these circuits. Also, $\text{sp}(2, 4) = \{2, 4, 5, 6\}$, and so on.

The last important concept that we introduce is matroid duality.

Proposition 2.7. *If r is the rank function of a matroid $M = (N, r)$ and $r^D(S) = |S| + r(N \setminus S) - r(N)$, then r^D is the rank function of a matroid.*

Proof. It follows immediately from Proposition 2.2 that r^D is submodular. Also $r^D(\emptyset) = 0$, and since

$$r^D(S \cup \{j\}) - r^D(S) = 1 - (r(N \setminus S) - r(N \setminus (S \cup \{j\}))),$$

we have $0 \leqslant r^D(S \cup \{j\}) - r^D(S) \leqslant 1$. Thus the result follows from Theorem 2.4. ∎

Definition 2.2. $M^D = (N, r^D)$ is the *dual matroid* associated with $M = (N, r)$.

It is readily seen that A is a basis of M^D if and only if $N \setminus A$ is a basis of M. Moreover, the dual of M^D is again M.

The dual of a graphic matroid is called a *cographic matroid*. Its bases are the complements of spanning trees—that is, the maximal sets that do not disconnect the graph. It follows that the circuits of this dual matroid are the minimal disconnecting sets, or minimal cuts.

Example 2.2. Consider the graphic matroid M associated with the graph in Figure 2.1 and its dual M^D.

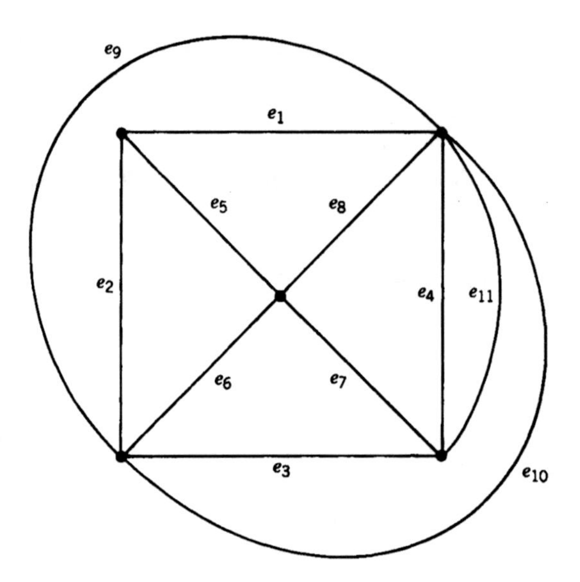

Figure 2.1

The circuits of M are the cycles such as $\{e_i: i = 1, 5, 8\}$, $\{e_4, e_{11}\}$, $\{e_i: i = 1, 2, 3, 4\}$. The bases of M are the spanning trees containing $|V| - 1 = 4$ edges.

The circuits of M^D are the minimal cut-sets such as $\{e_i: i = 3, 4, 7, 11\}$, $\{e_i: i = 3, 4, 5, 6, 8, 11\}$. The bases of M^D are the complements of the spanning trees (i.e., $\{e_i: i = 4, 6, 7, 8, 9, 10, 11\}$, etc.) containing $|E| - |V| + 1 = 7$ edges.

3. MAXIMUM-WEIGHT INDEPENDENT SETS

Matroid Representation

We have already seen two ways to represent matroids: One is by listing the set $\mathcal{F}$ of independent sets, and the other is by the rank function r. However, using either $\mathcal{F}$ or r, $O(2^n)$ sets or values typically must be specified, where n is the number of elements of the matroid. This contrasts strongly with the representation of matroids that interest us. For example, a graphic matroid on $G = (V, E)$ is completely described by its graph, so the length of the input description is $O(n)$.

The reader will see that the algorithms we describe contain *independence tests* of the form: "Is $S \subseteq N$ an independent set in M, or not?"

We avoid the representation issue by simply reporting the number of independence tests in an algorithm as a function of n. Note that answering standard questions such as "Is $S \subseteq N$ a basis of M?" or "Given that $S \in \mathcal{F}$, $S \cup \{j\} \notin \mathcal{F}$, find the circuit $C \subseteq S \cup \{j\}$." can be answered with $O(n)$ independence tests.

The Greedy Algorithm

Given a matroid $M = (N, \mathcal{F})$ and $c \in R^n$, the problem of finding a maximum-weight independent set is

$$\max_S \left\{ \sum_{j \in S} c_j : S \in \mathcal{F} \right\}. \tag{3.1}$$

The algorithm to find a maximum-weight forest in Section I.3.3 is an instance of the greedy algorithm we now describe for solving (3.1).

The Greedy Algorithm for $M = (N, \mathcal{F})$

Initialization: Order the elements of N so that $c_1 \geqslant c_2 \geqslant \cdots \geqslant c_n$. $S^0 = \emptyset$, $t = 1$.

Iteration t: If $c_t \leqslant 0$, stop. S^{t-1} is optimal. If $c_t > 0$ and $S^{t-1} \cup \{t\} \in \mathcal{F}$, set $S^t = S^{t-1} \cup \{t\}$. If $c_t > 0$ and $S^{t-1} \cup \{t\} \notin \mathcal{F}$, set $S^t = S^{t-1}$. If $t = n$, stop. S^n is optimal. If $t < n$, set $t \leftarrow t + 1$.

Although for general independence systems the greedy algorithm does not necessarily yield an optimal solution, for matroids it does.

Theorem 3.1. *The greedy algorithm for matroids terminates with a maximum-weight independent set.*

Proof. Let the greedy solution be $S^G = \{j_1, \ldots, j_p\}$ with $j_1 < j_2 < \cdots < j_p$. Suppose the greedy solution is not optimal, and let $S^L = \{k_1, \ldots, k_q\}$, $k_1 < k_2 < \cdots < k_q$ be the lexicographically smallest optimal solution. Suppose $j_1 = k_1, j_2 = k_2, \ldots, j_{s-1} = k_{s-1}$, but $j_s \neq k_s$. From the greedy algorithm, we have $j_s < k_s$ and hence $c_{j_s} \geqslant c_{k_s} > 0$. Now $S^L \cup \{j_s\} \notin \mathcal{F}$ since otherwise S^L is not optimal. Hence $S^L \cup \{j_s\}$ contains a unique circuit C with $j_s \in C$. Also, since $\{j_1, \ldots, j_{s-1}\} = \{k_1, \ldots, k_{s-1}\} \in \mathcal{F}$, we have $k_t \in C$ for some $t \geqslant s$. But by Proposition 2.6, we have that $(S^L \cup \{j_s\}) \setminus \{k_t\} \in \mathcal{F}$, its value is at least that of S^L, and it is lexicographically smaller than S^L, which is a contradiction. ■

Note that there no more than n independence calls by the algorithm and that the sorting of the initialization step requires $O(n \log n)$ comparisons. For a specific class of matroids, we can use this to calculate the running time of the algorithm. For graphic matroids the independence test involves testing for a cycle in a graph, which requires $O(n)$ calculations. Hence the running time of the simplistic algorithm given above is $O(n^2)$.

Example 3.1. Given the graph $G = (V, E)$ shown in Figure 3.1, the problem is to find a maximum-weight independent set in the cographic matroid (i.e., a set of edges whose removal does not disconnect the graph). We have ordered the edges so that $c_1 \geqslant c_2 \geqslant \cdots \geqslant c_8$. Applying the greedy algorithm, we obtain $S^G = (e_1, e_3, e_5)$, provided that $c_5 > 0$.

The converse of Theorem 3.1 also holds:

Theorem 3.2. *If $(N, \mathcal{F})$ is an independence system but not a matroid, there exists a weight function $c \in R^n$ for which the greedy algorithm does not yield an optimal solution to (3.1).*

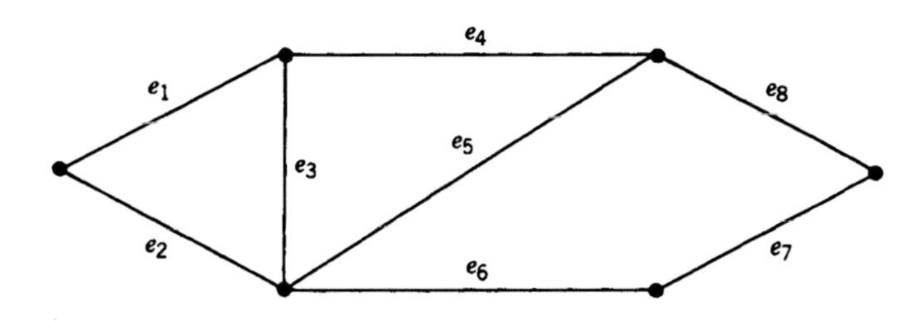

Figure 3.1

Proof. Since $(N, \mathscr{F})$ is not a matroid, there exists an $S \subseteq N$ such that all maximal independent sets in S are not of the same cardinality. Let $A \subseteq S$ be a maximal independent set in S of minimum cardinality, and suppose $|A| = k$. Let

$$c_j = \begin{cases} k+2 & \text{for } j \in A \\ k+1 & \text{for } j \in S \setminus A \\ 0 & \text{otherwise.} \end{cases}$$

The greedy algorithm yields the set A of value $k(k+2)$. But an optimal solution has value at least $(k+1)^2 > k(k+2)$ for $k \geqslant 1$. ■

Several variants of problem (3.1) can be solved by simple modifications of the greedy algorithm. These include the problems of finding a maximum-weight basis and a maximum-weight independent set of cardinality not greater than k. Another useful observation is that a maximum-weight basis is the complement of a minimum-weight basis in the dual matroid.

The Matroid Polytope

Here we consider an integer programming formulation of (3.1) and its linear programming relaxation. Let x^T be the characteristic vector of $T \subseteq N$. By the definition of an independence system $\mathscr{I} = (N, \mathscr{F})$ with cardinality function m, it follows that $T \in \mathscr{F}$ if and only if $|T| \leqslant m(T)$ if and only if $\Sigma_{j \in S} x_j^T = |S \cap T| \leqslant m(S)$ for all $S \subseteq N$.

Let

$$P(m) = \left\{ x \in R_+^n \colon \sum_{j \in S} x_j \leqslant m(S) \text{ for } S \subseteq N \right\}.$$

Then an integer programming formulation of the problem of finding a maximum-weight independent set in $\mathscr{I}$ is

$$\max\{cx \colon x \in P(m), x \in B^n\}.$$

We now show that if the independence system is a matroid $M = (N, r)$, then $P(r)$ is the convex hull of the characteristic vectors of its independent sets.

Consider the linear program

$$\max \sum_{j \in N} c_j x_j$$

$$(3.2) \qquad \sum_{j \in S} x_j \leqslant r(S) \quad \text{for } S \subseteq N$$

$$x \in R_+^n$$

and its dual

$$\min \sum_{S \subseteq N} r(S) y_S$$

$$(3.3) \qquad \sum_{S: S \ni j} y_S \geqslant c_j \quad \text{for } j \in N$$

$$y_S \geqslant 0 \quad \text{for } S \subseteq N.$$

Proposition 3.3. *Let $S^G = \{j_1, \ldots, j_p\}$ be the greedy solution to (3.1) with the ordering $c_1 \geqslant c_2 \geqslant \cdots \geqslant c_n$. Let $J_t = \{j_1, \ldots, j_t\}$ for $t \leqslant p$, and let $K_t = \mathrm{sp}(J_t)$. Then an optimal solution to (3.3) is*

$$y_{K_t} = c_{j_t} - c_{j_{t+1}} \quad \text{for } t = 1, \ldots, p-1,$$
$$y_{K_p} = c_{j_p},$$
$$y_K = 0 \quad \text{otherwise}.$$

Proof. Clearly the dual solution is nonnegative. If $c_j > 0$, then $j \in K_t \setminus K_{t-1}$ for some $t \leqslant p$. Also, if $j \in K_t \setminus K_{t-1}$, then $c_j \leqslant c_{j_t}$. Hence if $j \in K_t \setminus K_{t-1}$, we obtain

$$\sum_{S:S\ni j} y_S = \sum_{l \geqslant t} y_{K_l} = c_{j_t} \geqslant c_j.$$

Now note that since $r(K_t) = t$, we have

$$\sum_{S \subseteq N} r(S) y_S = \sum_{t=1}^{p-1} t(c_{j_t} - c_{j_{t+1}}) + p c_{j_p} = \sum_{t=1}^{p} c_{j_t} = \sum_{j \in S^G} c_j,$$

so the primal and dual objective functions are equal. ■

We have shown that the linear system

$$\sum_{j \in S} x_j \leqslant r(S) \quad \text{for } S \subseteq N, x \in R^n_+$$

is totally dual integral.

Theorem 3.4. *$P(r)$ is an integral polytope.*

Example 3.2. For a graphic matroid, the associated tree polytope $P(r)$ is of the form

$$\sum_{e \in E'} x_e \leqslant r(E') \quad \text{for } E' \subseteq E$$
$$x \in R^{|E|}_+,$$

where, as was shown in Section 1, $r(E') = |V| -$ number of components of $G_{E'} = (V, E')$.

When U is the set of nodes attained by E', and $G' = (U, E')$ is connected, the corresponding inequality is dominated by the inequality with $E' = E(U)$. When G' itself has several components, the corresponding inequality is dominated by the inequalities from the components. Hence we obtain a polyhedral description of a graphic matroid given by

$$\sum_{e \in E(U)} x_e \leqslant |U| - 1 \quad \text{for } U \subseteq V \text{ with } |U| \geqslant 2$$
$$x \in R^{|E|}_+.$$

This example raises the question of which inequalities describing the tree polytope are facets and the more general question of describing the facets of any matroid polytope $P(r)$.

Note that if an inequality does not define a facet, we can delete the corresponding dual variable from (3.3).

Definition 3.1. A set $S \subseteq N$ with $S = \text{sp}(S)$ is *separable* (or *disconnected*) if there exists a partition (A, B) of S, that is, $A \neq \emptyset$, $B \neq \emptyset$, $A \cup B = S$, and $A \cap B = \emptyset$, with $r(A) + r(B) = r(S)$.

Now it is easy to see that the inequality $\Sigma_{j \in S} x_j \leqslant r(S)$ is dominated by $\Sigma_{j \in \text{sp}(S)} x_j \leqslant r(S)$ when $S \neq \text{sp}(S)$ and that it is the sum of two inequalities when S is separable. These turn out to be the only redundant inequalities.

Proposition 3.5. *Suppose* $\{j\} \in \mathcal{F}$ *for all* $j \in N$. *A minimal description of* $P(r)$ *is given by* $P(r) = \{x \in R^n_+ : \Sigma_{j \in S} x_j \leqslant r(S)$ *for* $S \subseteq N$ *with* $S = \text{sp}(S)$ *and* S *nonseparable*$\}$.

Example 3.3. Consider the maximum-weight spanning tree problem on the weighted graph shown in Figure 3.2.

The greedy algorithm with (3, 6) preceding (5, 6) in the ordering gives the solution $S^G = \{(3, 4), (1, 5), (1, 3), (3, 6), (6, 7), (2, 6)\}$ of weight 57.

The optimal dual solution specified by Proposition 3.3 is

$$\begin{aligned}
y_{S_1} &= 13 - 11 = 2 && \text{with} \quad S_1 = \{3, 4\} \\
y_{S_2} &= 11 - 10 = 1 && \text{with} \quad S_2 = \{(3, 4), (1, 5)\} \\
y_{S_3} &= 10 - 9 = 1 && \text{with} \quad S_3 = E(\{1, 3, 4, 5\}) \\
y_{S_4} &= 9 - 8 = 1 && \text{with} \quad S_4 = E(\{1, 3, 4, 5, 6\}) \\
y_{S_5} &= 8 - 6 = 2 && \text{with} \quad S_5 = E(\{1, 3, 4, 5, 6, 7\}) \\
y_{S_6} &= 6 && \text{with} \quad S_6 = E
\end{aligned}$$

of value $(2 \times 1) + (1 \times 2) + (1 \times 3) + (1 \times 4) + (2 \times 5) + (6 \times 6) = 57$.

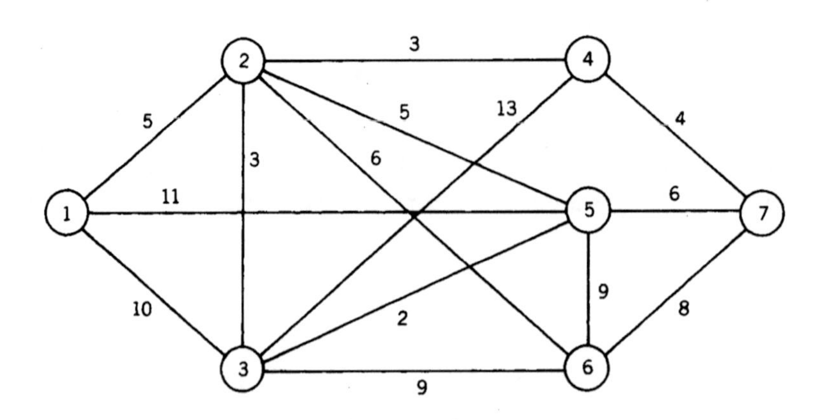

Figure 3.2

Using Proposition 3.5, we see that the edge sets $\{(3, 4), (1, 5)\}$, $E(\{1, 3, 4, 5\})$, and $E(\{1, 3, 4, 5, 6\})$ do not define facets. Decomposing each of these edge sets, we have

$$\begin{array}{ll} \{(3, 4), (1, 5)\} & \text{yields } \{3, 4\} \text{ and } \{1, 5\} \\ E(\{1, 3, 4, 5\}) & \text{yields } \{(3, 4)\} \text{ and } E(\{1, 3, 5\}) \\ E(\{1, 3, 4, 5, 6\}) & \text{yields } \{(3, 4)\} \text{ and } E(\{1, 3, 5, 6\}), \end{array}$$

and we obtain the alternative dual solution

$$\begin{array}{ll} y_S = 3 + 1 + 1 + 1 & \text{for } S = E(\{3, 4\}) \\ y_S = 1 & \text{for } S = E(\{1, 5\}) \\ y_S = 1 & \text{for } S = E(\{1, 3, 5\}) \\ y_S = 1 & \text{for } S = E(\{1, 3, 5, 6\}) \\ y_S = 2 & \text{for } S = E(\{1, 3, 4, 5, 6, 7\}) \\ y_S = 6 & \text{for } S = E \end{array}$$

in which only the dual variables associated with facets are positive.

4. MATROID INTERSECTION

We have already seen that the branchings on a digraph can be viewed as the edge sets that are independent in two matroids simultaneously. Feasible solutions to matching problems on a bipartite graph $G = (V_1, V_2, E)$ can also be viewed in this way. Let $M_i = (E, \mathcal{F})$ for $i = 1, 2$, be partition matroids where $F \subseteq E$ is independent in M_i if there is no more than one edge of F adjacent to each node of V_i. F is a matching in G if and only if $F \in \mathcal{F}_1 \cap \mathcal{F}_2$. In polyhedral terms, we have

$$P(r_1) = \left\{x \in R_+^{|E|}: \sum_{j \in V_2} x_{ij} \leqslant 1 \text{ for } i \in V_1\right\}$$

and

$$P(r_2) = \left\{x \in R_+^{|E|}: \sum_{j \in V_1} x_{ij} \leqslant 1 \text{ for } j \in V_2\right\},$$

and $P(r_1) \cap P(r_2)$ describes the convex hull of matchings in a bipartite graph.

Here we consider the general *maximum-cardinality matroid intersection problem* for the matroids $M_i = (N, \mathcal{F}_i)$ for $i = 1, 2$, formulated as

$$z = \max_S\{|S|: S \in \mathcal{F}_1 \cap \mathcal{F}_2\}. \tag{4.1}$$

Throughout the text, we have stressed the importance of duality. Consider the problem

$$w = \min_T\{r_1(T) + r_2(N \setminus T)\}. \tag{4.2}$$

Proposition 4.1. *Problem (4.2) is a weak dual of problem (4.1).*

Proof. Suppose $S \in \mathcal{F}_1 \cap \mathcal{F}_2$. Then for any set $T \subseteq N$, we obtain

$$|S| = |S \cap T| + |S \setminus T| = r_1(S \cap T) + r_2(S \setminus T) \leqslant r_1(T) + r_2(N \setminus T).$$ ■

Example 4.1. Consider the two graphic matroids defined on the graphs shown in Figure 4.1. Taking $S = \{a, c\} \in \mathcal{F}_1 \cap \mathcal{F}_2$, we see that $z \geqslant 2$. Taking $T = \{c, d\}$, we see that

$$w \leqslant r_1(\{c, d\}) + r_2(\{a, b, e\}) = 1 + 1 = 2.$$

Hence, using weak duality, we obtain $z = w = 2$, and $\{a, c\}$ is a maximum-cardinality set independent in both matroids.

The major aim of this section is to develop an algorithm which shows constructively that (4.2) is, in fact, a *strong dual* of problem (4.1). As is the case for the maximum-flow problem and the matching problem, the algorithm is based on finding augmenting paths.

To motivate and explain this idea, consider the following example involving the two graphic matroids of Figure 4.2.

Example 4.2. We are given an independent set $S = \{a, b, k_1, k_2\}$. The additional information we have is that

$$S \cup \{j_1\} \notin \mathcal{F}, \quad (S \cup \{j_1\}) \setminus \{k_1\} \in \mathcal{F}$$

and

$$S \cup \{j_2\} \notin \mathcal{F}, \quad (S \cup \{j_2\}) \setminus \{k_2\} \in \mathcal{F}.$$

A question that we need to answer in searching for common independent sets of greater cardinality is: "Is $(S \cup \{j_1, j_2\}) \setminus \{k_1, k_2\} \in \mathcal{F}$?" Note that the answer is "yes" in matroid 1 but "no" in matroid 2, and observe that $(S \cup \{j_1\}) \setminus \{k_2\} \notin \mathcal{F}$ in matroid 1.

The following proposition explains why the answer is "yes" in matroid 1, and it is fundamental to what follows. If $S \in \mathcal{F}$ and $S \cup \{j\} \notin \mathcal{F}$, let $C(S, j)$ denote the unique circuit contained in $S \cup \{j\}$.

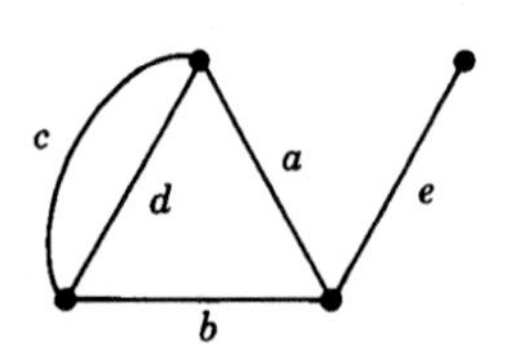

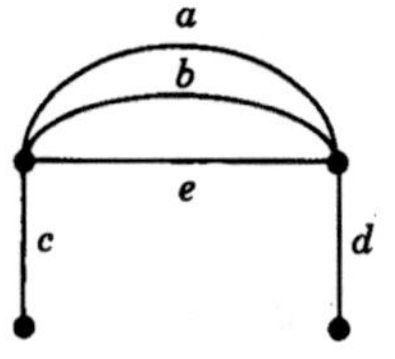

Figure 4.1

Proposition 4.2. *Let $M = (N, \mathscr{F})$ be a matroid, let $S \in \mathscr{F}$, and let $j_1, k_1, \ldots, j_p, k_p$ be a sequence of distinct elements of N with $j_1, \ldots, j_p \in N \setminus S$ and $k_1, \ldots, k_p \in S$ satisfying*

(a) $S \cup \{j_i\} \notin \mathscr{F}, \quad (S \cup \{j_i\}) \setminus \{k_i\} \in \mathscr{F}$ *for* $i = 1, \ldots, p$

(b) $(S \cup \{j_i\}) \setminus \{k_l\} \notin \mathscr{F}$ *for* $1 \leq i < l \leq p$;

then

$$(S \cup \{j_i, \ldots, j_l\}) \setminus \{k_i, \ldots, k_l\} \in \mathscr{F} \quad \textit{for } 1 \leq i < l \leq p.$$

Proof. Note first that (a) is equivalent to $k_i \in C(S, j_i)$, and (b) is equivalent to $k_l \notin C(S, j_i)$. First we will establish that

$$C(S, j_i) = C((S \cup \{j_{i+1}, \ldots, j_l\}) \setminus \{k_{i+1}, \ldots, k_l\}, j_i) \quad \text{for } 1 \leq i \leq l \leq p. \tag{4.3}$$

The proof is by induction on the number of pairs in the sequence $j_1, k_1, \ldots, j_p, k_p$. When $p = 1$, condition (b) is void and (4.3) reduces to $C(S, j_1) = C(S, j_1)$. Now suppose that the result holds for all sequences involving $p \leq t - 1$ pairs (j_i, k_i). It now suffices to show that (4.3) holds with $i = 1$ and $l = t$.

Therefore we must show that

$$\begin{aligned} C(S, j_1) &= C((S \cup \{j_2, \ldots, j_p\}) \setminus \{k_2, \ldots, k_p\}, j_1) \\ &= C(I \cup \{j_2\}) \setminus \{k_2\}, j_1), \end{aligned}$$

where $I = S \cup \{j_3, \ldots, j_p\} \setminus \{k_3, \ldots, k_p\}$.

By the induction hypothesis applied to the sequence $j_1, k_1, j_3, k_3, \ldots, j_p, k_p$, we have $C(S, j_1) = C(I, j_1)$. Since $k_2 \notin C(S, j_1)$ by (b), we obtain $k_2 \notin C(I, j_1)$ and hence $C(I, j_1) \subseteq (I \cup \{j_1\}) \setminus \{k_2\}$.

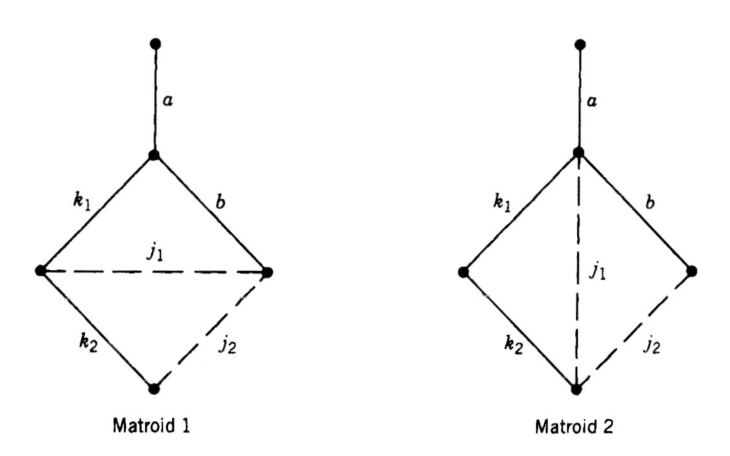

Figure 4.2

Now applying the induction hypothesis to the sequence $j_2, k_2, \ldots, j_p, k_p$, we get that $C(S, j_2) = C(I, j_2)$ and hence $(I \cup \{j_2\}) \setminus \{k_2\} \in \mathcal{F}$ since $k_2 \in C(S, j_2)$. But $(I \cup \{j_1, j_2\}) \setminus \{k_2\}$ contains no more than one circuit. Since it contains the circuits $C((I \cup \{j_2\}) \setminus \{k_2\}, j_1)$ and $C(I, j_1)$, they must be identical, so

$$C(S, j_1) = C(I, j_1) = C((I \cup \{j_2\}) \setminus \{k_2\}, j_1)$$
$$= C((S \cup \{j_2, \ldots, j_p\}) \setminus \{k_2, \ldots, k_p\}, j_1)$$

and (4.3) is established.

Finally we observe that (4.3) and $k_i \in C(S, j_i)$ imply $S \cup \{j_i, \ldots, j_l\} \setminus \{k_i, \ldots, k_l\} \in \mathcal{F}$. ■

We will be particularly interested in sequences of odd length.

Corollary 4.3. *Let $M = (N, \mathcal{F})$ be a matroid, let $S \in \mathcal{F}$, and let $j_1, k_1, \ldots, j_{p-1}, k_{p-1}, j_p$ be a sequence of distinct elements of N satisfying*

(a) $S \cup \{j_i\} \notin \mathcal{F}, \quad (S \cup \{j_i\}) \setminus \{k_i\} \in \mathcal{F}$ *for* $i = 1, \ldots, p-1$

(b) $(S \cup \{j_i\}) \setminus \{k_l\} \notin \mathcal{F}$ *for* $1 \leq i < l \leq p-1$

(c) $S \cup \{j_p\} \in \mathcal{F}$,

then $S' = (S \cup \{j_1, \ldots, j_p\}) \setminus \{k_1, \ldots, k_{p-1}\} \in \mathcal{F}$.

Proof. Consider $\tilde{S} = S \cup \{j_p\} \in \mathcal{F}$. Now $S \cup \{j_i\} \notin \mathcal{F}$ implies $\tilde{S} \cup \{j_i\} \notin \mathcal{F}$. On the other hand, since $\tilde{S} \in \mathcal{F}$, it follows that $\tilde{S} \cup \{j_i\}$ contains a unique circuit that must be the circuit $C(S, j_i)$ containing k_i. Hence $(\tilde{S} \cup \{j_i\}) \setminus \{k_i\} \in \mathcal{F}$, and (a) holds for $\tilde{S}$. Clearly (b) also holds for $\tilde{S}$. Therefore, we can apply Proposition 4.2 to $\tilde{S}$ and the sequence $j_1, k_1, \ldots, j_{p-1}, k_{p-1}$ to conclude that $S' = (\tilde{S} \cup \{j_1, \ldots, j_{p-1}\}) \setminus \{k_1, \ldots, k_{p-1}\} \in \mathcal{F}$. ■

Since $|S'| = |S| + 1$, Corollary 4.3 provides a scheme for finding a larger cardinality independent set in a matroid, but it is obviously unnecessary because $\tilde{S} = S \cup \{j_p\}$ suffices. However, for the problem of increasing the cardinality of a set S that is a common independent set in two matroids, Corollary 4.3 gives us a sufficient condition.

Proposition 4.4. *Given two matroids $M_i = (N, \mathcal{F}_i)$ for $i = 1, 2$, a set $S \subseteq N$ with $S \in \mathcal{F}_1 \cap \mathcal{F}_2$, and a sequence $j_1, k_1, \ldots, j_{p-1}, k_{p-1}, j_p$ of distinct elements with $j_1, \ldots, j_p \in N \setminus S$, $k_1, \ldots, k_{p-1} \in S$ satisfying*

(a1) $S \cup \{j_i\} \notin \mathcal{F}_1, \quad S \cup \{j_i\} \setminus \{k_{i-1}\} \in \mathcal{F}_1$ *for* $i = 2, \ldots, p$

(b1) $S \cup \{j_i\} \setminus \{k_l\} \notin \mathcal{F}_1$ *for* $1 \leq l < i - 1 \leq p - 1$

(c1) $S \cup \{j_1\} \in \mathcal{F}_1$

(a2) $S \cup \{j_i\} \notin \mathcal{F}_2, \quad S \cup \{j_i\} \setminus \{k_i\} \in \mathcal{F}_2$ *for* $i = 1, \ldots, p-1$

(b2) $S \cup \{j_i\} \setminus \{k_l\} \notin \mathcal{F}_2$ *for* $1 \leq i < l < p$

(c2) $S \cup \{j_p\} \in \mathcal{F}_2$,

then $S' = (S \cup \{j_1, \ldots, j_p\}) \setminus \{k_1, \ldots, k_{p-1}\} \in \mathcal{F}_1 \cap \mathcal{F}_2$.

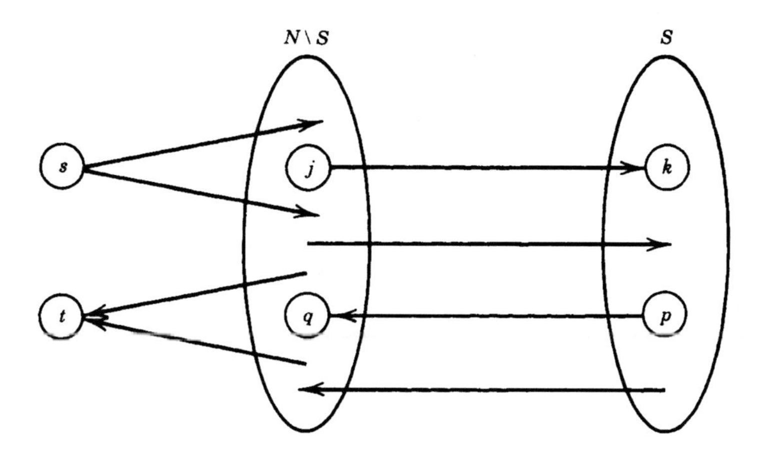

Figure 4.3

Proof. First we apply Corollary 4.3 to the sequence $j_1, k_1, \ldots, j_{p-1}, k_{p-1}, j_p$, observing that conditions (a2), (b2), and (c2) are precisely the conditions of the corollary. Therefore $S' \in \mathcal{F}_2$.

To show that $S' \in \mathcal{F}_1$, we consider the reverse sequence $j_p, k_{p-1}, \ldots, j_2, k_1, j_1$. Conditions (a1), (b1), and (c1) are precisely conditions (a), (b), and (c) of Corollary 4.3 with respect to this sequence. Hence $(S \cup \{j_p, \ldots, j_1\}) \setminus \{k_{p-1}, \ldots, k_1\} = S' \in \mathcal{F}_1$. ∎

Now we construct a digraph $\mathcal{D}_S = (N \cup \{s, t\}, \mathcal{A})$ (see Figure 4.3) that will allow us to find a sequence of the type described in Proposition 4.4. The arcs are defined as follows:

$$(s, j) \in \mathcal{A} \quad \text{if } S \cup \{j\} \in \mathcal{F}_1$$

$$(j, t) \in \mathcal{A} \quad \text{if } S \cup \{j\} \in \mathcal{F}_2$$

$$(j, k) \in \mathcal{A} \quad \text{if } S \cup \{j\} \notin \mathcal{F}_2, (S \cup \{j\}) \setminus \{k\} \in \mathcal{F}_2$$

$$(k, j) \in \mathcal{A} \quad \text{if } S \cup \{j\} \notin \mathcal{F}_1, (S \cup \{j\}) \setminus \{k\} \in \mathcal{F}_1$$

Note that an arc $(j, k), j \in N \setminus S, k \in S$, refers to a replacement of j by k to achieve $(S \cup \{k\}) \setminus \{j\} \in \mathcal{F}_2$ and that an arc (k, j), $k \in S, j \in N \setminus S$ refers to a replacement of j by k to achieve $(S \cup \{j\}) \setminus \{k\} \in \mathcal{F}_1$. This can be interpreted graphically.

Proposition 4.5. *If $(s, j_1, k_1, \ldots, j_p, t)$ is an s–t dipath in $\mathcal{D}_S$ and $\mathcal{D}_S$ contains no arcs of the form (j_i, k_l) and (k_i, j_{l+1}) for $l > i$, then $(j_1, k_1, \ldots, j_p)$ is a sequence satisfying the conditions of Proposition 4.4.*

In this and the next section we will be interested in the existence of certain dipaths in $\mathcal{D}_S$. Such dipaths may not exist even though an s–t path may exist in the underlying undirected graph. We therefore keep the term *dipath* (see Section I.3.1). An s–t dipath $(s, l_1, \ldots, l_p, t)$ is *node minimal* if there is no subsequence $\{l_{j_1}, \ldots, l_{j_q}\} \subseteq \{l_1, \ldots, l_p\}$ with $1 \leq j_1 < j_2 < \cdots < j_q \leq p$ such that $(s, l_{j_1}, \ldots, l_{j_q}, t)$ is an s–t dipath.

An s–t dipath satisfying the condition of Proposition 4.5 is node minimal (see Figure 4.4). Now we observe that a minimum-length dipath from s to t (i.e, a dipath with a minimum number of arcs) is necessarily node minimal. Hence such a node minimal dipath can be found by breadth-first search or a standard shortest-path algorithm.

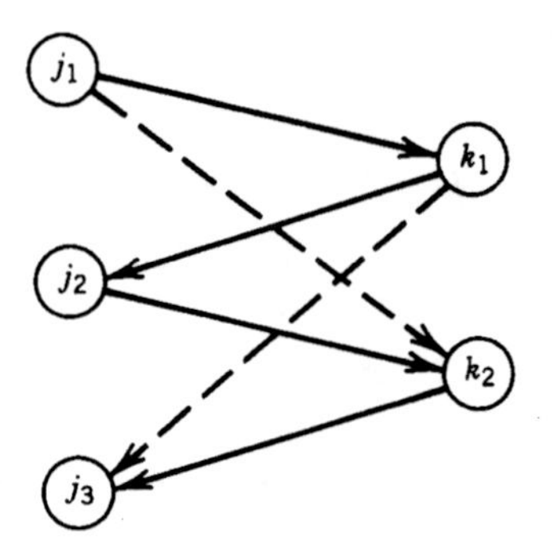

Figure 4.4

Example 4.3. Two graphic matroids are exhibited in Figure 4.5a, and the digraph $\mathcal{D}_S$ for $S = \{b, d\}$ is shown in Figure 4.5b. Note that (s, a, d, e, t) is a node minimal s–t dipath in $\mathcal{D}_S$ giving the set $S' = \{a, b, e\} \in \mathcal{F}_1 \cap \mathcal{F}_2$. Note also that the s–t dipath (s, a, b, c, d, e, t) is not node minimal because $(a, d) \in \mathcal{A}$ and does not lead to a larger common independent set since $S'' = \{a, c, e\} \notin \mathcal{F}_2$.

The final step in developing an algorithm for problem (4.1) is to show that if $\mathcal{D}_S$ contains no s–t dipath, then S is a maximum-cardinality set independent in both matroids. Let $N_L = \{i \in N$: there exists a dipath from s to i in $\mathcal{D}_S\}$, $S_L = N_L \cap S$, $N_R = N \setminus N_L$, and $S_R = N_R \cap S$.

Proposition 4.6. *If $\mathcal{D}_S$ contains no s–t dipath, then S is a maximum-cardinality common independent set.*

Proof. Suppose that $\mathcal{D}_S$ contains no s–t dipath. We show first that $N_L \subseteq \mathrm{sp}_2(S_L)$. Since $S_L \subseteq \mathrm{sp}_2(S_L)$, we consider $j \in N_L \setminus S_L$. Now $S \cup \{j\} \notin \mathcal{F}_2$, because otherwise there would be an arc $(j, t) \in \mathcal{A}$ and an (s, t) dipath would exist. If $k \in C_2(S, j) \setminus \{j\}$, then $\mathcal{D}_S$ contains the arc (j, k) and so $C_2(S, j) \setminus \{j\} \subseteq S_L$. In other words, $j \in \mathrm{sp}_2(S_L)$, and hence $N_L \subseteq \mathrm{sp}_2(S_L)$.

Next we show that $N_R \subseteq \mathrm{sp}_1(S_R)$. We consider $j \in N_R \setminus S_R$. First we observe that $S \cup \{j\} \notin \mathcal{F}_2$, because otherwise there would be an arc (s, j), which is impossible as

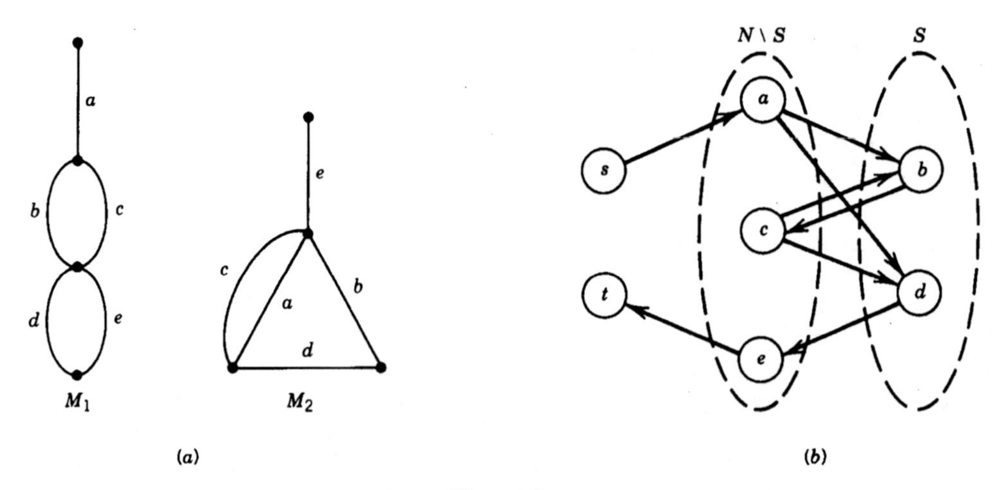

Figure 4.5

$j \notin N_L$. If $k \in C_1(S, j) \setminus \{j\}$, then $\mathscr{D}_S$ contains the arc (k, j). Since $j \notin N_L$, we have $k \notin S_L$, and hence $C_1(S, j) \setminus \{j\} \subseteq S_R$. Thus $N_R \subseteq \text{sp}_1(S_R)$.

Finally, we use weak duality in problems (4.1) and (4.2):

$$z \geqslant |S|, \quad \text{and} \quad w \leqslant r_1(N_R) + r_2(N_L) \quad \text{(by Proposition 4.1)}$$
$$\leqslant r_1(\text{sp}_1(S_R)) + r_2(\text{sp}_2(S_L))$$
$$= |S_R| + |S_L| = |S|. \quad \blacksquare$$

Theorem 4.7. *Problem (4.2) is a strong dual of problem (4.1).*

Now we can describe the algorithm for problem (4.1).

Maximum-Cardinality Matroid Intersection Algorithm

Initialization: Start with $S^1 \in \mathscr{F}_1 \cap \mathscr{F}_2$. $q = 1$.

Iteration q: Construct the digraph $\mathscr{D}_{S^q}$. If there is no s–t dipath in $\mathscr{D}_{S^q}$, stop; S^q is an optimal solution. Otherwise, find a shortest s–t dipath $(s, j_1, k_1, \ldots, j_p, t)$. Set $S^{q+1} = (S^q \cup \{j_1, \ldots, j_p\}) \setminus \{k_1, \ldots, k_{p-1}\}$ and $q \leftarrow q + 1$.

Example 4.4. Consider the two graphic matroids shown in Figure 4.6. $S = \{e_1, e_2, e_5\}$. The digraph $\mathscr{D}_S$ is shown in Figure 4.7.

Since $\mathscr{D}_S$ contains no s–t dipath, we obtain

$$N_L = \{e_2, e_5, e_7, e_8\}, \quad N_R = \{e_1, e_3, e_4, e_6\}$$
$$S_L = \{e_2, e_5\}, \quad S_R = \{e_1\}$$
$$N_L \subseteq \text{sp}_2(S_L) = \{e_2, e_4, e_5, e_7, e_8\}$$
$$N_R \subseteq \text{sp}_1(S_R) = \{e_1, e_3, e_4, e_6\}$$
$$r_1(N_R) + r_2(N_L) = 3 = |S|.$$

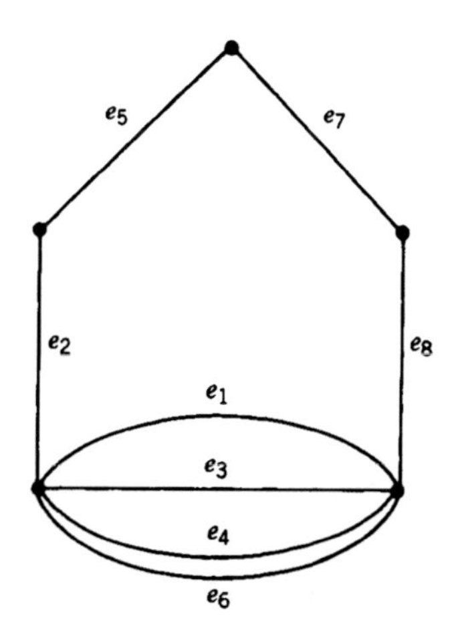

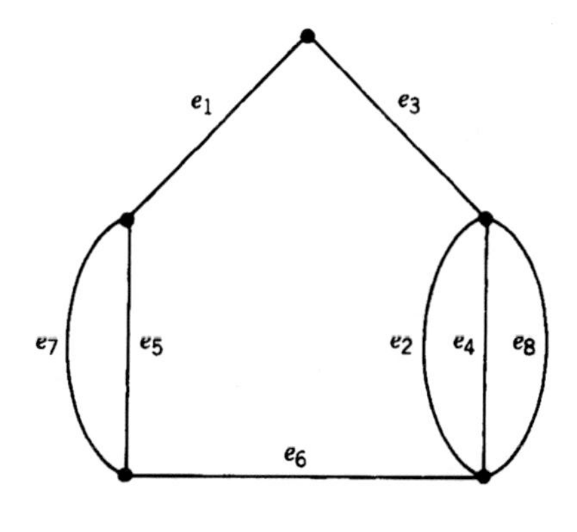

Figure 4.6

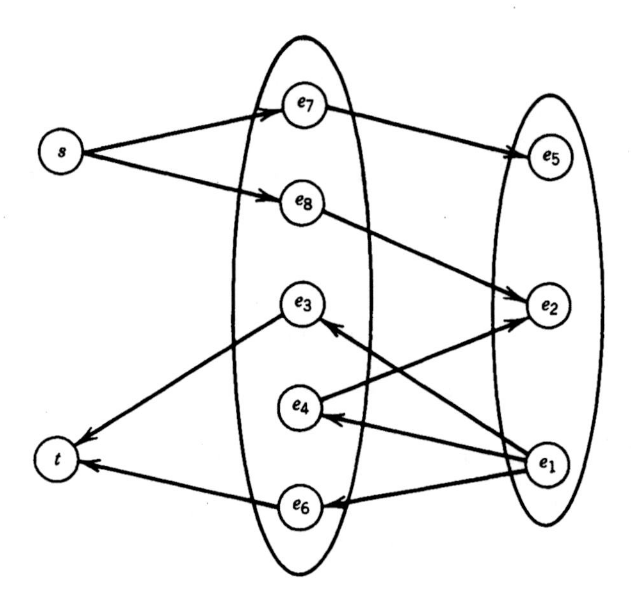

Figure 4.7

Finally, we consider the complexity of the cardinality matroid intersection algorithm

Proposition 4.8. *The cardinality matroid intersection algorithm terminates after no more than $O(n^3)$ independence tests.*

Proof. There are no more than $z < n$ iterations. At each iteration the digraph $\mathcal{D}_S$ has to be constructed. Deciding if (i, j) is an arc of the digraph requires no more than two independence tests, so there are $O(n^2)$ tests at each iteration. ■

5. WEIGHTED MATROID INTERSECTION

Given two matroids $M_i = (N, \mathcal{F}_i)$ for $i = 1, 2$ and a weight vector $c \in R^n$, we consider the weighted matroid intersection problem

$$z = \max_S \left\{ \sum_{j \in S} c_j \colon S \in \mathcal{F}_1 \cap \mathcal{F}_2 \right\}. \tag{5.1}$$

It is convenient to introduce the following notation:

$$\mathcal{F}^q = \{S \subseteq N \colon S \in \mathcal{F},\ |S| \leqslant q\}, \quad \mathcal{F}_i^q = \{S \subseteq N \colon S \in \mathcal{F}_i,\ |S| \leqslant q\}$$
$$\mathcal{F}_{12} = \mathcal{F}_1 \cap \mathcal{F}_2, \quad \mathcal{F}_{12}^q = \{S \in \mathcal{F}_{12} \colon |S| \leqslant q\}$$
$$r_{12}(S) = \max_{T \subseteq S} \{|T| \colon T \in \mathcal{F}_{12}\}.$$

The algorithm we describe actually solves the family of problems

$$z^q = \max_S \left\{ \sum_{j \in S} c_j \colon S \in \mathcal{F}_{12}^q \right\} \text{ for } q = 0, 1, \ldots, r_{12}(N). \tag{5.2}$$

We will solve (5.2) for increasing values of q and base our proof of optimality on the following dual pair of linear programs:

$$\max \sum_{j\in N} c_j x_j$$

$$\sum_{j\in A} x_j \leq r_1(A) \quad \text{for } A \subseteq N$$

$$(5.3) \qquad \sum_{j\in A} x_j \leq r_2(A) \quad \text{for } A \subseteq N$$

$$\sum_{j\in N} x_j \leq q$$

$$x_j \geq 0 \quad \text{for } j \in N$$

and

$$\min \sum_{A\subseteq N} r_1(A)y_1(A) + \sum_{A\subseteq N} r_2(A)y_2(A) + qt$$

$$(5.4) \qquad \sum_{A\ni j} y_1(A) + \sum_{A\ni j} y_2(A) + t \geq c_j \quad \text{for } j \in N$$

$$y_1(A),\ y_2(A) \geq 0 \quad \text{for } A \subseteq N,\ t \geq 0.$$

For all values of q we will show that (5.3) has an integral optimal solution by giving an integral primal feasible solution and a dual feasible solution of the same value. The primal solution is constructed using cost splitting (see Section II.3.6).

Proposition 5.1. *Given c, c^1, $c^2 \in R^n$ with $c^1 + c^2 = c$, if $S^q \subseteq N$ is an optimal solution to the problems*

$$(5.5) \qquad \max_S \left\{ \sum_{j\in S} c_j^i \colon S \in \mathscr{F}_i^q \right\} \text{ for } i = 1, 2,$$

then S^q is an optimal solution to (5.2).

Proof. For any $S \in \mathscr{F}_{12}^q$, we obtain

$$\sum_{j\in S} c_j = \sum_{j\in S} c_j^1 + \sum_{j\in S} c_j^2$$

$$\leq \sum_{j\in S^q} c_j^1 + \sum_{j\in S^q} c_j^2 = \sum_{j\in S^q} c_j. \qquad \blacksquare$$

The greedy algorithm for matroids gives a characterization of an optimal-weight solution in $\mathscr{F}^q$.

Proposition 5.2. *Given a matroid $M = (N, \mathscr{F})$ with weight vector c, a set S with $|S| = q$ is optimal in $\mathscr{F}^q$ if and only if:*

i. *$c_j \geq 0$ for $j \in S$;*
ii. *if $j \notin S$ and $S \cup \{j\} \in \mathscr{F}$, then $c_k \geq c_j$ for $k \in S$; and*
iii. *if $j \notin S$ and $S \cup \{j\} \notin \mathscr{F}$, then $c_k \geq c_j$ for $k \in C(S, j) \setminus \{j\}$.*

Given c^1, c^2, and S^q as in Proposition 5.1, it is simple to give an optimal solution to (5.4). Consider the problem

$$\max_S \left\{ \sum_{j \in S} (c_j^i - m_i) \colon S \in \mathscr{F}_i^q \right\} \quad \text{for } i = 1, 2, \tag{5.6}$$

where $m_i = \max\{c_j^i \colon j \notin S^q, S^q \cup \{j\} \in \mathscr{F}_i\}$ and $m_i = -\infty$ if S^q is a basis of M_i. Its dual is

$$\begin{aligned} &\min \sum_{A \subseteq N} r_i(A)\, y_i(A) \\ &\sum_{A \ni j} y_i(A) \geqslant c_j^i - m_i \quad \text{for } j \in N \\ &y_i(A) \geqslant 0 \qquad\qquad \text{for } A \subseteq N. \end{aligned} \tag{5.7}$$

Proposition 5.3. *If (a) c^1, c^2, S^q satisfy the conditions of Proposition 5.1 with $|S^q| = q$, (b) $m_1 + m_2 \geqslant 0$, and (c) y_i^* is an optimal solution to (5.7) for $i = 1, 2$, then an optimal solution to (5.4) is $y_i = y_i^*$ for $i = 1, 2$ and $t = m_1 + m_2$.*

Proof. The proposed solution is feasible to (5.4). By hypothesis, S^q is an optimal solution to (5.5). Hence, by Proposition 5.2, we have $c_j^i \geqslant m_i$ for $j \in S^q$. It follows that S^q is also an optimal solution to (5.6). Hence, equating the optimal values of (5.6) and (5.7), we obtain $\sum_{j \in S^q} (c_j^i - m_i) = \sum_{A \subseteq N} r_i(A) y_i^*(A)$. Now the value of the proposed solution to (5.4) is

$$\begin{aligned} &\sum_{A \subseteq N} r_1(A) y_1^*(A) + \sum_{A \subseteq N} r_2(A) y_2^*(A) + q(m_1 + m_2) \\ &= \sum_{j \in S^q} (c_j^1 - m_1) + \sum_{j \in S^q} (c_j^2 - m_2) + q(m_1 + m_2) \\ &= \sum_{j \in S^q} c_j. \end{aligned}$$

The characteristic vector of S^q is feasible to (5.3), so the claim follows. ■

Example 5.1. A digraph is shown in Figure 5.1, along with associated arc weights. We wish to find a branching of maximum weight. Thus the underlying edge set must be independent in the graphic matroid M_1 and the partition matroid M_2, where the number of arcs entering each node is restricted to be no greater than 1.

Suppose we have the following split of the arc weights c_j^1, c_j^2:

	1	2	3	4	5	6	7	8	9	10	11
c_j	4	3	1	4	1	7	2	6	−5	−1	1
c_j^1	4	3	1	4	1	5	2	4	−5	−1	1
c_j^2	0	0	0	0	0	2	0	2	0	0	0

Observing that $S^2 = \{e_1, e_6\}$ is optimal in $\mathscr{F}_1^2$ with weight c^1 and that S^2 is optimal in $\mathscr{F}_2^2$ with weight c^2, we have by Proposition 5.1 that S^2 is optimal in $\mathscr{F}_{12}^2$ with weight c.

Now we consider the question of how to pass from S^q to S^{q+1}, a maximum-weight independent set in $\mathscr{F}_{12}^{q+1}$. We know that if we construct the digraph $\mathscr{D}_{S^q}$ used in the

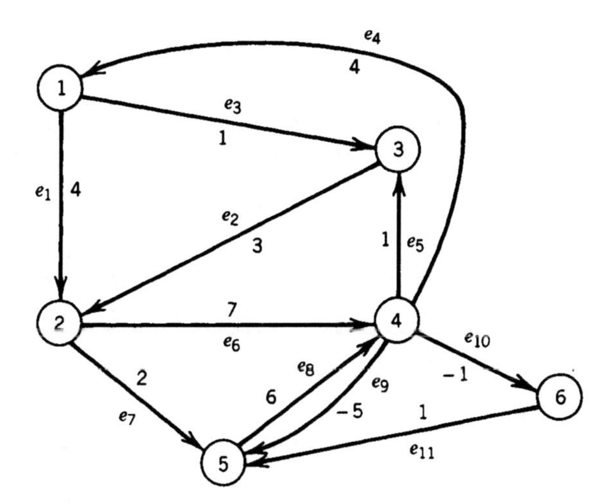

Figure 5.1

cardinality algorithm, then s–t dipaths with a minimum number of edges give us sets $S' \in \mathcal{F}_{12}^{q+1}$. However, because we want to obtain a maximum-weight independent set in $\mathcal{F}_{12}^{q+1}$, we can only use a subset of the arcs of $\mathcal{D}_{S^q}$. The following proposition suggests, on the basis of the weight vector c, those arcs that should be kept.

Proposition 5.4. *Let $M^q = (N, \mathcal{F}^q)$, let S be an optimal-weight solution in M^q, and let $j_1, k_1, \ldots, j_p, k_p$ be distinct elements of N with $j_1, \ldots, j_p \in N \setminus S$, $k_1, \ldots, k_p \in S$, and*

a. $S \cup \{j_i\} \notin \mathcal{F}$, $(S \cup \{j_i\}) \setminus \{k_i\} \in \mathcal{F}$ *for* $i = 1, \ldots, p$
b. $c_{j_i} = c_{k_i}$ *for* $i = 1, \ldots, p$
c. $c_{j_i} = c_{j_l}$ *and* $i < l$ *implies* $(S \cup \{j_i\}) \setminus \{k_l\} \notin \mathcal{F}$.

Then $S' = (S \cup \{j_1, \ldots, j_p\}) \setminus \{k_1, \ldots, k_p\}$ *is also an optimal-weight solution in* M^q.

Proof. We can reorder the elements $j_1, \ldots, j_p$ so that $c_{j_1} \geq \cdots \geq c_{j_p}$ and conditions a–c still hold. This is possible because conditions a and b are unaffected by the ordering, and condition c only affects pairs such as (j_i, j_l) with $i < l$ if $c_{j_i} = c_{j_l}$. But if the ordering of such pairs is preserved in the new ordering, condition c holds.

We now claim that $(S \cup \{j_i\}) \setminus \{k_l\} \notin \mathcal{F}^q$ for all $i < l$. If not, $S^* = (S \cup \{j_i\}) \setminus \{k_l\} \in \mathcal{F}^q$ for some $i < l$ with $c_{j_i} > c_{k_l}$. But then $\Sigma_{j \in S^*} c_j > \Sigma_{j \in S} c_j$, contradicting the optimality of S. Now the conditions of Proposition 4.2 are satisfied and $S' \in \mathcal{F}^q$. By condition b the weights of S and S' are identical, and hence S' is optimal. ∎

Now given S optimal in $\mathcal{F}_{12}^q$, c^1, c^2 as in Proposition 5.1 and m_1, m_2 defined below (5.6), we construct a digraph $\mathcal{D}_S(c^1, c^2) = (N \cup \{s, t\}, \mathcal{A})$ with arcs of four types (note that these are a subset of the arcs of $\mathcal{D}_S$ given in the previous section):

$$(s, j) \in \mathcal{A} \text{ if } j \notin S,\ S \cup \{j\} \in \mathcal{F}_1, \text{ and } m_1 = c_j^1$$

$$(j, t) \in \mathcal{A} \text{ if } j \notin S,\ S \cup \{j\} \in \mathcal{F}_2, \text{ and } m_2 = c_j^2$$

$$(j, k) \in \mathcal{A} \text{ if } j \notin S,\ S \cup \{j\} \notin \mathcal{F}_2,\ (S \cup \{j\}) \setminus \{k\} \in \mathcal{F}_2, \text{ and } c_j^2 = c_k^2$$

$$(k, j) \in \mathcal{A} \text{ if } j \notin S,\ S \cup \{j\} \notin \mathcal{F}_1,\ (S \cup \{j\}) \setminus \{k\} \in \mathcal{F}_1, \text{ and } c_j^1 = c_k^1.$$

First we consider what happens when $\mathscr{D}_S(c^1, c^2)$ contains an s–t dipath.

Proposition 5.5. *Let S be optimal in $\mathscr{F}_{12}^q$, and let $s, j_1, k_1, \ldots, k_{p-1}, j_p, t$ be an s–t dipath in $\mathscr{D}_S(c^1, c^2)$ with a minimum number of arcs. Then $(S \cup \{j_1, \ldots, j_p\}) \setminus \{k_1, \ldots, k_{p-1}\}$ is optimal in $\mathscr{F}_{12}^{q+1}$.*

Proof. We will apply Proposition 5.4 twice, first to M_2 and then to M_1. Note first that by definition of m_2, we have that $S^* = S \cup \{j_p\}$ is optimal in $\mathscr{F}_2^{q+1}$ with weight c^2.

First we apply Proposition 5.4 to $S^* \in \mathscr{F}_2^{q+1}$ with the sequence $j_1, k_1, \ldots, j_{p-1}, k_{p-1}$ and weights c^2. From the construction of $\mathscr{D}_S(c^1, c^b)$, condition b holds for any sequence derived from an s–t dipath. Also, $S^* \cup \{j_i\} \notin \mathscr{F}_2$ and $(S^* \cup \{j_i\}) \setminus \{k_i\} \in \mathscr{F}_2$, so condition a holds for any such sequence.

Now we use the fact that the dipath has a minimum number of edges and hence is node minimal. This means that if $c^2_{j_i} = c^2_{k_l}$ for some $i < l$, there is no arc (j_i, k_l) and hence $(S \cup \{j_1\}) \setminus \{k_l\} \notin \mathscr{F}_2$. It follows that $(S^* \cup \{j_i\}) \setminus \{k_l\} \notin \mathscr{F}_2$, and hence condition c holds. Therefore $S' = (S \cup \{j_1, \ldots, j_p\}) \setminus \{k_1, \ldots, k_p\}$ is optimal in $\mathscr{F}_2^{q+1}$ with weight c^2.

Taking the path in the reverse order, a similar argument shows that S' is optimal in $\mathscr{F}_1^{q+1}$ with weight c^1. Now by Proposition 5.1, S' is optimal in $\mathscr{F}_{12}^{q+1}$. ■

The other possibility is that there is no s–t dipath in $\mathscr{D}_S(c^1, c^2)$. In this case we make a *dual change* by changing (c^1, c^2). We let $N_L = \{j \in N$: there exists an s–j dipath in $\mathscr{D}_S(c^1, c^2)\}$ and $N_R = N \setminus N_L$.

The dual change is given by

$$\begin{aligned} \tilde{c}^1_j &= c^1_j - \varepsilon_1 && \text{if } j \in N_L \\ \tilde{c}^2_j &= c^2_j - \varepsilon_2 && \text{if } j \in N_R \\ \tilde{c}^1_j &= c_j - \tilde{c}^2_j && \text{if } j \in N_R \\ \tilde{c}^2_j &= c_j - \tilde{c}^1_j && \text{if } j \in N_L \end{aligned}$$

where $\varepsilon_1, \varepsilon_2$ are calculated from δ_i, $i = 1, \ldots, 4$, which are the minimum-cost changes needed to add an arc of each of the four possible types to $\mathscr{D}_S(c^1, c^2)$. Their values are

$$\begin{aligned} \delta_1 &= \min\{m_1 - c^1_j : j \notin S,\ S \cup \{j\} \in \mathscr{F}_1, j \in N_R\} \\ \delta_2 &= \min\{m_2 - c^2_j : j \notin S,\ S \cup \{j\} \in \mathscr{F}_2, j \in N_L\} \\ \delta_3 &= \min\{c^2_k - c^2_j : j \notin S,\ S \cup \{j\} \notin \mathscr{F}_2,\ (S \cup \{j\}) \setminus \{k\} \in \mathscr{F}_2, j \in N_L,\ k \in N_R\} \\ \delta_4 &= \min\{c^1_k - c^1_j : j \notin S,\ S \cup \{j\} \notin \mathscr{F}_1,\ (S \cup \{j\}) \setminus \{k\} \in \mathscr{F}_2, j \in N_R,\ k \in N_L\} \end{aligned}$$

with $\delta_i = \infty$ if the corresponding set is empty. Then

$$\varepsilon = \min\{m_1 + m_2, \min(\delta_1, \delta_2, \delta_3, \delta_4)\}$$

$$\varepsilon_1 = \min\{\varepsilon, m_1\} \quad \text{and} \quad \varepsilon_2 = \varepsilon - \varepsilon_1$$

First we check that a real change occurs.

Proposition 5.6. *If $m_1 + m_2 > 0$, then $\varepsilon > 0$ in the dual change.*

Proof. By the definition of m_1, we have that $m_1 > c_j^1$ if $S \cup \{j\} \in \mathcal{F}_1$. Hence $\delta_1 \geqslant 0$. If $\delta_1 = 0$, then $\mathcal{D}_S(c^1, c^2)$ contains an arc (s, j), contradicting $j \in N_R$. Hence $\delta_1 > 0$. A similar argument holds for δ_2.

If $S \cup \{j\} \notin \mathcal{F}_2$, $(S \cup \{j\}) \setminus \{k\} \in \mathcal{F}_2$, and $c_k^2 < c_j^2$, then S is not optimal in $\mathcal{F}_2^q$ with weights c^2. Hence $\delta_3 \geqslant 0$. If $\delta_3 = 0$, then $\mathcal{D}_S(c^1, c^2)$ contains an arc (j, k) joining $j \in N_L$ to $k \in N_R$, which is impossible. Hence $\delta_3 > 0$. A similar argument holds for δ_4. ■

Now we show that the conditions of Proposition 5.1 still apply with the new weights $(\tilde{c}^1, \tilde{c}^2)$.

Proposition 5.7. *After a dual change based on $\mathcal{D}_S(c^1, c^2)$, S is still optimal in $\mathcal{F}_i^q$ with weights $\tilde{c}^i$ for $i = 1, 2$.*

Proof. We verify that S satisfies the optimality conditions of Proposition 5.2 with the new weights $(\tilde{c}^1, \tilde{c}^2)$. We consider matroid $M_2^q = (N, \mathcal{F}_2^q)$.

i. The condition $\tilde{c}_j^2 \geqslant 0$ for $j \in S$ holds because $\tilde{m}_2 \geqslant m_2 - \varepsilon_2 \geqslant 0$, and S is optimal before the dual change.
ii. Suppose $j \notin S$, $S \cup \{j\} \in \mathcal{F}$, $k \in S$, and $\tilde{c}_k^2 < \tilde{c}_j^2$. Because $c_k^2 \geqslant c_j^2$, this can only happen if $\tilde{c}_j^2 = c_j^2 + \varepsilon_1$ and $\tilde{c}_k^2 = c_k^2 - \varepsilon_2$, so that $j \in N_L$ and $k \in N_R$. But because $k \in S$, we obtain $c_k^2 \geqslant m_2$. Moreover, since $j \in N_L$, we have $\varepsilon \leqslant \delta_2 \leqslant m_2 - c_j^2$. Hence
$$\tilde{c}_j^2 = c_j^2 + \varepsilon_1 \leqslant m_2 - \varepsilon + \varepsilon_1 \leqslant c_k^2 - \varepsilon_2 = \tilde{c}_k^2,$$
which is a contradiction.
iii. Suppose $j \notin S$, $S \cup \{j\} \notin \mathcal{F}_2$, $k \in C_2(S, j) \setminus \{j\}$, and $c_k^2 < c_j^2$. This implies that $\tilde{c}_j^2 = c_j^2 + \varepsilon_1$ and $\tilde{c}_k^2 = c_k^2 - \varepsilon_2$ with $j \in N_L$ and $k \in N_R$. But $\varepsilon \leqslant \delta_3$, and since $j \in N_L$, $k \in N_R$, and $(S \cup \{j\}) \setminus \{k\} \in \mathcal{F}_2$, we obtain $\delta_3 \leqslant c_k^2 - c_j^2$. Hence
$$\tilde{c}_j^2 = c_j^2 + \varepsilon_1 \leqslant c_k^2 - \varepsilon + \varepsilon_1 = \tilde{c}_k^2,$$
which is a contradiction.

A similar argument shows that S is optimal in $\mathcal{F}_1^q$ with weights $\tilde{c}^1$. ■

It remains to establish that after a finite number of dual changes, either a larger common independent set is found or the algorithm terminates.

Proposition 5.8. *After no more than n dual changes, either an s–t dipath is found, or S is optimal for all $q' \geqslant q$.*

Proof. We consider the different possibilities for a dual change, together with the successive digraphs $\mathcal{D}_S(c^1, c^2)$ and $\mathcal{D}_S(\tilde{c}^1, \tilde{c}^2)$. We will establish two claims. First we claim that every arc in $\mathcal{D}_S(c^1, c^2)$ with both its head and tail in N_L is also an arc in $\mathcal{D}_S(\tilde{c}^1, \tilde{c}^2)$. Then we also show that a new arc appears in $\mathcal{D}_S(\tilde{c}^1, \tilde{c}^2)$ whose tail is in N_L and whose head is in N_R. Together these imply that $\tilde{N}_L \supset N_L$, and hence no more than n dual changes can occur.

To establish the first claim, we examine each type of arc in turn. Let $T = \{j : j \notin S, S \cup \{j\} \in \mathcal{F}_1\}$. Consider (s, j) arcs for which $c_j^1 = m_1$. Note that if $j \in T$ and $c_j^1 < m_1$, then $j \in N_R$ because the only arcs of $\mathcal{A}$ entering a node $j \in T$ are those of the form (s, j). Now we know from the definition of δ_1 that $c_j^1 \leqslant m_1 - \delta_1 \leqslant m_1 - \varepsilon_1$. After the dual change, we obtain

$$\begin{aligned}\tilde{m}_1 &= \max\{\tilde{c}_j^1 : j \in T\} \\ &= \max\{\max\{c_j^1 - \varepsilon_1 : j \in N_L \cap T\}, \max\{c_j^1 : j \in N_R \cap T\}\} \\ &= \max\{m_1 - \varepsilon_1, \max\{c_j^1 : j \in N_R \cap T\}\} \\ &= m_1 - \varepsilon_1\end{aligned}$$

It follows that if (s, j) is an arc of $\mathcal{D}_S(c_1^1, c_2^2)$, then (s, j) is an arc of $\mathcal{D}_S(\tilde{c}_1^1, \tilde{c}_2^2)$.

Now consider an arc of the form (j, k) in $\mathcal{D}_S(c_1^1, c_2^2)$, where $j \notin S$, $S \cup \{j\} \notin \mathcal{F}_2$, $(S \cup \{j\}) \setminus \{k\} \in \mathcal{F}_2$, and $c_j^2 = c_k^2$ with both $j, k \in N_L$. After the dual change, we obtain $\tilde{c}_j^2 = c_j^2 - \varepsilon$ and $\tilde{c}_k = c_k^2 - \varepsilon$, so $\tilde{c}_j^2 = \tilde{c}_k^2$ and the arc is in $\mathcal{D}_S(\tilde{c}_1^1, \tilde{c}_2^2)$. An identical argument holds for the (k, j) arcs based on matroid M_1, and hence the first claim is established.

To establish the second claim, observe that the algorithm terminates if $\varepsilon = m_1 + m_2$ based on $\mathcal{D}_S(c^1, c^2)$, and an s–t dipath is created if $\varepsilon = \delta_2$. Hence a dual change can only occur if $\varepsilon = \delta_1$, δ_3, or δ_4. In each of these cases a new arc appears in $\mathcal{D}_S(\tilde{c}^1, \tilde{c}^2)$ whose tail is in N_L and whose head is in N_R. ■

The Weighted Matroid Intersection Algorithm

Initialization: Start with c^1, c^2, S^q as described in Propositon 5.1, if such a solution is known for any $q \geqslant 1$. Alternatively let $q = 0$, let $S^q = \varnothing$, and choose any c^1, c^2 satisfying $c^1 + c^2 = c$. (The simple choice in this case is $c^1 = c$, $c^2 = 0$.)

Step 1: Calculate m_1, m_2. If $m_1 + m_2 \leqslant 0$, stop. S^q is optimal for all $q' \geqslant q$. Otherwise if $m_i < 0$ for some i (say $i = 1$), then $c_j^1 \leftarrow c_j^1 - m_1$, $c_j^2 \leftarrow c_j^2 + m_1$ for all $j \in N$.

Step 2: Construct $\mathcal{D}_{S^q}(c^1, c^2)$.

Step 3: If there is no s–t dipath in $\mathcal{D}_{S^q}(c^1, c^2)$, go to Step 5. Otherwise find a shortest s–t dipath in $\mathcal{D}_{S^q}(c^1, c^2)$ and go to Step 4.

Step 4 (Augmentation): Use the s–t dipath in $\mathcal{D}_{S^q}(c^1, c^2)$ to find S^{q+1} optimal in $\mathcal{F}_2^{q+1}$. Set $q \leftarrow q + 1$. Go to Step 1.

Step 5 (Dual Change): Change (c^1, c^2) as described in (5.6). If $\varepsilon = m_1 + m_2$, stop. S^q is optimal for $q' \geqslant q$. Otherwise, calculate $(\tilde{c}^1, \tilde{c}^2)$, $\tilde{m}_1$ and $\tilde{m}_2$, and construct $\mathcal{D}_{S^q}(\tilde{c}^1, \tilde{c}^2)$. Go to Step 3.

Example 5.1. (continued). We apply the weighted matroid intersection algorithm starting with $S^2 = \{e_1, e_6\}$ and c^1, c^2 given below.

	1	2	3	4	5	6	7	8	9	10	11
c_j	4	3	1	4	1	7	2	6	−5	−1	1
c_j^1	4	3	1	4	1	5	2	4	−5	−1	1
c_j^2	0	0	0	0	0	2	0	2	0	0	0

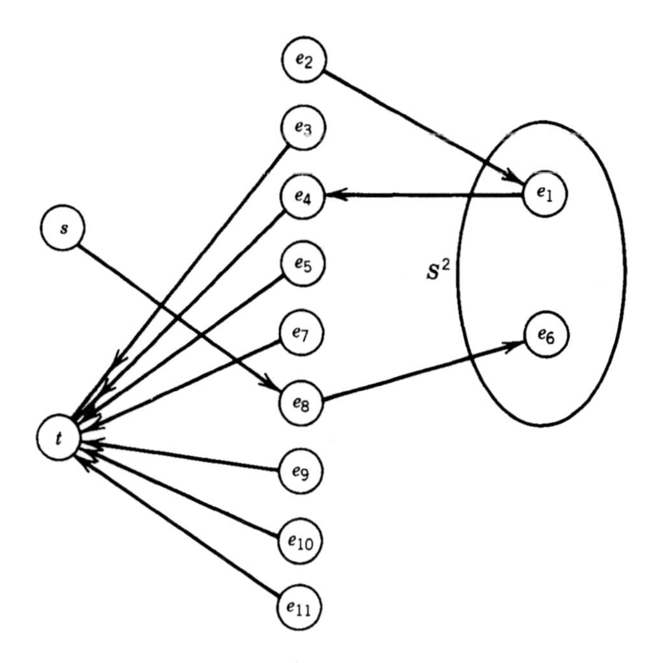

Figure 5.2.

$q = 2$:

Step 1: $m_1 = c_8^1 = 4$, $m_2 = c_3^2 = c_4^2 = c_5^2 = c_7^2 = c_9^2 = c_{10}^2 = c_{11}^2 = 0$. $N_L = \{e_6, e_8\}$
Step 2: $\mathcal{D}_{S^2}(c^1, c^2)$ is shown in Figure 5.2.

Step 5: $\delta_1 = m_1 - c_2^1 = 4 - 3 = 1$, $\delta_2 = \infty$, $\delta_3 = \infty$, $\delta_4 = c_6^1 - c_4^1 = 1$. $\varepsilon = \delta_1 = \delta_4 = 1$. With $\varepsilon_1 = 1$, $\varepsilon_2 = 0$, and $c_j^i \leftarrow \tilde{c}_j^i$, we obtain

	1	2	3	4	5	6	7	8	9	10	11
c_j	4	3	1	4	1	7	2	6	−5	−1	1
c_j^1	4	3	1	4	1	4	2	3	−5	−1	1
c_j^2	0	0	0	0	0	3	0	3	0	0	0

$m_1 \leftarrow 3$, $m_2 \leftarrow 0$. The new $\mathcal{D}_{S^2}(c^1, c^2)$ is shown in Figure 5.3.

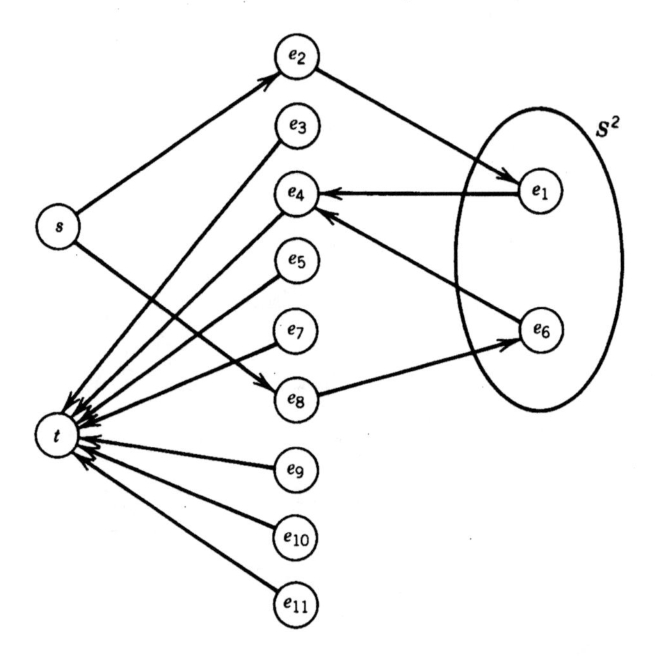

Figure 5.3

Two $s-t$ dipaths with a minimum number of edges are found. We use the dipath (s, e_8, e_6, e_4, t).

$q = 3$:

Step 1: $S^3 = \{e_1, e_4, e_8\}$. $m_1 = c_2^1 = 3$, $m_2 = c_3^2 = c_5^2 = c_7^2 = c_9^2 = c_{10}^2 = c_{11}^2 = 0$. $\mathscr{D}_{S^3}(c^1, c^2)$ is shown in Figure 5.4. $N_L = \{e_1, e_2, e_6, e_8\}$

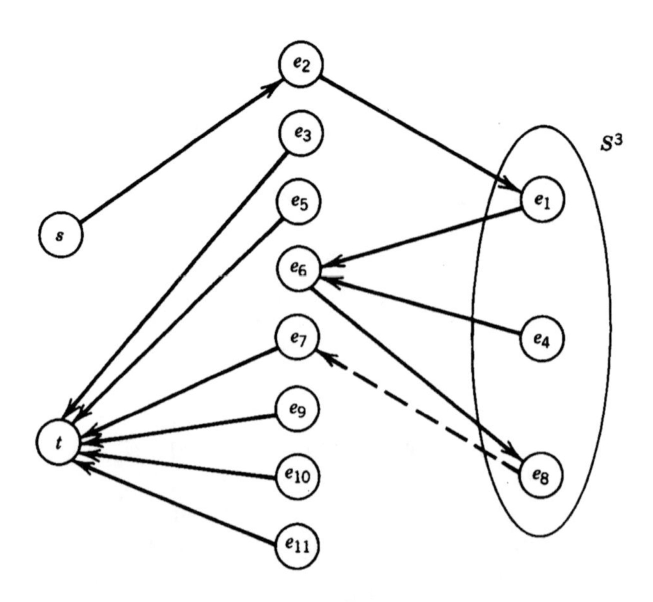

Figure 5.4

Step 5: $\delta_1 = m_1 - c_3^1 = 3 - 1 = 2, \delta_2 = \infty, \delta_3 = \infty, \delta_4 = c_8^1 - c_7^1 = 1. \varepsilon = \delta_4 = 1, \varepsilon_1 = 1, \varepsilon_2 = 0.$

	1	2	3	4	5	6	7	8	9	10	11
c_j	4	3	1	4	1	7	2	6	−5	−1	1
c_j^1	3	2	1	4	1	3	2	2	−5	−1	1
c_j^2	1	1	0	0	0	4	0	4	0	0	0

Adding the arc (e_8, e_7), an s–t dipath is found.

$q = 4$:

Step 1: $S^4 = \{e_2, e_4, e_6, e_7\}$. $m_1 = c_{11}^1 = 1$, $m_2 = c_3^2 = c_5^2 = c_{10}^2 = 0$. $\mathscr{D}_{S^4}(c^1, c^2)$ is shown in Figure 5.5. $N_L = \{e_1, e_2, e_6, e_7, e_8, e_{11}\}$

Step 5: $\delta_1 = m_1 - c_{10}^1 = 2, \delta_2 = \infty, \delta_3 = \infty, \delta_4 = c_6^1 - c_5^1 = c_6^1 - c_3^1 = 1, \varepsilon = \delta_4 = m_1 + m_2 = 1.$ $\varepsilon_1 = 1, \varepsilon_2 = 0.$

	1	2	3	4	5	6	7	8	9	10	11
c_j	4	3	1	4	1	7	2	6	−5	−1	1
c_j^1	2	1	1	4	1	2	1	1	−5	−1	0
c_j^2	2	2	0	0	0	5	1	5	0	0	1

$m_1 = c_{11}^1 = 0, m_2 = c_3^2 = c_5^2 = c_{10}^2 = 0.$

$S = \{e_2, e_4, e_6, e_7\}$ is a maximum-weight branching in $\mathscr{D}$.

Now we consider the number of independence tests required in the algorithm. Because the set S^q does not change through a sequence of dual changes, only $O(n^2)$ independence tests are required between augmentations to construct the digraphs $\mathscr{D}_{S^q}$ from which a subset of the arcs are selected to give $\mathscr{D}_{S^q}(c^1, c^2)$. Since there are, at most, n augmentations, no more than $O(n^3)$ independence tests are required.

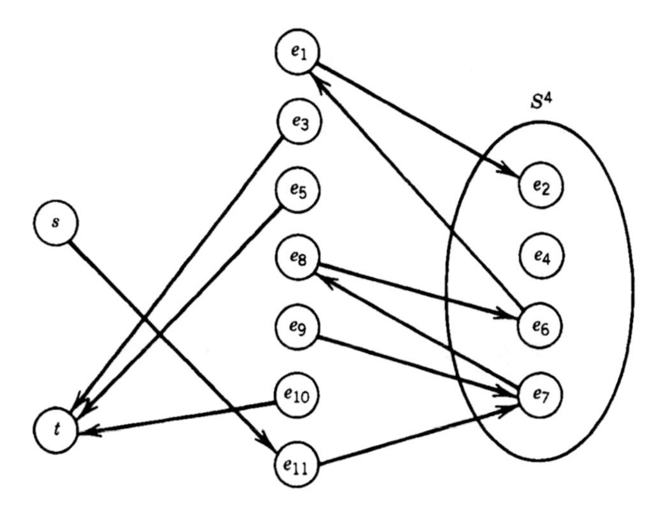

Figure 5.5

We conclude this section with a polyhedral result. As a consequence of the weighted matroid intersection algorithm and Proposition 5.3, we have shown:

Theorem 5.9. *Given matroids $M_1 = (N, r_1)$ and $M_2 = (N, r_2)$, the polytope*

$$\sum_{j \in A} x_j \leqslant r_1(A) \quad \text{for } A \subseteq N$$

$$\sum_{j \in A} x_j \leqslant r_2(A) \quad \text{for } A \subseteq N$$

$$\sum_{j \in N} x_j \leqslant q$$

$$x \in R_+^n$$

is the convex hull of the common independent sets of cardinality not greater than q. The inequality set is totally dual integral.

6. POLYMATROIDS, SEPARATION, AND SUBMODULAR FUNCTION MINIMIZATION

Two of the problems we study in this section are (1) the problem of minimizing a submodular function and (2) the separation problem for a matroid polytope $P(r)$. To understand better the relationship between these problems, we introduce a generalization of a matroid.

Definition 6.1. Given a finite set N and a nondecreasing submodular function f on N with $f(\emptyset) = 0$, the polytope

$$P(f) = \left\{ x \in R_+^n \colon \sum_{j \in S} x_j \leqslant f(S) \text{ for } S \subseteq N \right\}$$

is the *polymatroid* associated with (N, f).

The property of independence systems generalizes to:

If $x, y \in Z_+^n$, $x \in P(f)$, and $y \leqslant x$, then $y \in P(f)$.

Definition 6.2. Let $r_{P(f)}\colon R_+^n \to R_+^1$ be defined by

$$r_{P(f)}(a) = \max\left\{ \sum_{j \in N} x_j \colon x \in P(f), x \leqslant a \right\},$$

$r_{P(f)}$ is called the *polymatroid rank function* associated with $P(f)$.

The next proposition shows how the "maximal = maximum" property of matroids generalizes.

Proposition 6.1. *Given a polymatroid $P(f)$ and $a \in R_+^n$, any point x that is maximal in $P(f) \cap \{x \in R_+^n \colon x \leqslant a\}$ satisfies $\Sigma_{j \in N}\, x_j = r_{P(f)}(a)$.*

Proof. Suppose the claim is false, so there exists $a \in R^n_+$ and u, v maximal in $P(f) \cap \{x \in R^n_+: x \leqslant a\}$ with $\Sigma_{j\in N} u_j < \Sigma_{j\in N} v_j$. Let $V = \{j \in N: v_j > u_j\}$. Since u is maximal, for each $k \in V$ there exists U_k with $k \in U_k$ such that $\Sigma_{j\in U_k} u_j = f(U_k)$.

Let U be a maximal subset of N such that $\Sigma_{j\in U} u_j = f(U)$. By submodularity, we have for each $k \in V$:

$$\begin{aligned}\sum_{j\in U\cup U_k} u_j &= \sum_{j\in U} u_j + \sum_{j\in U_k} u_j - \sum_{j\in U\cap U_k} u_j \\ &\geqslant f(U) + f(U_k) - f(U \cap U_k) \\ &\geqslant f(U \cup U_k).\end{aligned}$$

But since $u \in P(f)$, we obtain $\Sigma_{j\in U\cup U_k} u_j \leqslant f(U \cup U_k)$. Hence $\Sigma_{j\in U\cup U_k} u_j = f(U \cup U_k)$. Since U is maximal, we have $U_k \subseteq U$ for all $k \in V$. Hence $V \subseteq U$.

But now

$$f(U) = \sum_{j\in N} u_j - \sum_{j\in N\backslash U} u_j < \sum_{j\in N} v_j - \sum_{j\in N\backslash U} v_j = \sum_{j\in U} v_j,$$

since $\Sigma_{j\in N} u_j < \Sigma_{j\in N} v_j$ and $u_j \geqslant v_j$ for $j \in N \setminus V$, and $N \setminus V \supseteq N \setminus U$. This contradicts the assumption $v \in P(f)$. ■

Example 6.1. Suppose a set N of jobs is to be processed on one machine and all processing of job j must be terminated by the deadline d_j. If x_j is the machine time allocated to job j, then $x \in R^n_+$ is a feasible set of allocation times if and only if $\Sigma_{j\in S} x_j \leqslant \max_{j\in S} \{d_j\}$ for all $S \subseteq N$. It is easily checked that $f(S) = \max_{j\in S} \{d_j\}$ defines a polymatroid function if $d \in R^n_+$ and $f(\varnothing) = 0$.

There is a converse to Proposition 6.1 giving an alternative definition of a polymatroid.

Proposition 6.2. *Suppose $P \subseteq R^n_+$ satisfies*

i. *if $x \in P$, $y \in R^n_+$ and $y \leqslant x$, then $y \in P$, and*
ii. *for all $a \in R^n$ if x^1 and x^2 are maximal points in $P \cap \{x \in R^n_+: x \leqslant a)$, then $\Sigma_{j\in N} x^1_j = \Sigma_{j\in N} x^2_j$.*

Then P is a polymatroid.

As is the case for matroid polytopes, the greedy algorithm solves the linear optimization problem over a polymatroid. Consider the dual pair of linear programs

$$\max\left\{\sum_{j\in N} c_j x_j: \sum_{j\in S} x_j \leqslant f(S) \text{ for } S \subseteq N, x \in R^n_+\right\} \tag{6.1}$$

and

$$\min\left\{\sum_{S\subseteq N} f(S) y_S: \sum_{S\ni j} y_S \geqslant c_j \text{ for } j \in N, y \in R^{2^n}_+\right\}. \tag{6.2}$$

Suppose $c_1 \geqslant c_2 \geqslant \cdots \geqslant c_k > 0 \geqslant c_{k+1} \geqslant \cdots \geqslant c_n$, and $S^j = \{1, \ldots, j\}$ for $j \in N$ with $S^0 = \varnothing$.

Proposition 6.3. *An optimal solution to (6.1) is*

$$x_j = \begin{cases} f(S^j) - f(S^{j-1}) & \textit{for } 1 \leqslant j \leqslant k \\ 0 & \textit{for } j > k. \end{cases}$$

An optimal solution to (6.2) is

$$y_{S^j} = \begin{cases} c_j - c_{j+1} & \textit{for } 1 \leqslant j < k \\ c_k & \textit{for } j = k \\ 0 & \text{otherwise.} \end{cases}$$

Proof. The proposed x is primal feasible, because f nondecreasing implies $x_j \geqslant 0$, and because we have for all $T \subseteq N$

$$\begin{aligned} \sum_{j \in T} x_j &= \sum_{\{j: j \in T, j \leqslant k\}} [f(S^j) - f(S^{j-1})] \\ &\leqslant \sum_{\{j: j \in T, j \leqslant k\}} [f(S^j \cap T) - f(S^{j-1} \cap T)] \text{ (by the submodularity of } f) \\ &= f(S^k \cap T) - f(\varnothing) \leqslant f(T) - f(\varnothing) = f(T). \end{aligned}$$

The proposed y is dual feasible because $y_S \geqslant 0$ and because $\Sigma_{S \ni j}\, y_S = y_{S^j} + \cdots + y_{S^k} = c_j$ if $j \leqslant k$, and $\Sigma_{S \ni j}\, y_S \geqslant 0 \geqslant c_j$ if $j > k$.

The primal objective value is $\Sigma_{j=1}^{k} c_j\, (f(S^j) - f(S^{j-1}))$, and the dual objective value is

$$\sum_{j=1}^{k-1} (c_j - c_{j+1}) f(S^j) + c_k f(S^k) = \sum_{j=1}^{k} c_j\, (f(S^j) - f(S^{j-1})). \qquad \blacksquare$$

Note that in the special case where $f = r$ is the rank function of a matroid, Proposition 6.3 gives a different solution than the one given in Proposition 3.3. Also, as for matroids, we obtain:

Corollary 6.4. *The inequality system*

$$\left\{ x \in R^n_+ : \sum_{j \in S} x_j \leqslant f(S) \text{ for } S \leqslant N \right\}$$

is totally dual integral.

Example 6.1 (continued). Suppose four jobs are to be completed with respective deadlines given by $d = (10 \quad 7 \quad 8 \quad 4)$, and the profit from each job is proportional to the time devoted to processing it, with weights $c = (2 \quad 3 \quad 5 \quad 6)$. Since $c_4 > c_3 > c_2 > c_1 > 0$, the greedy algorithm yields

$$\begin{aligned} x_4 &= \max\{d_4\} && = 4 \\ x_3 &= \max\{d_3, d_4\} - \max\{d_4\} && = 4 \\ x_2 &= \max\{d_2, d_3, d_4\} - \max\{d_3, d_4\} && = 0 \\ x_1 &= \max\{d_1, d_2, d_3, d_4\} - \max\{d_2, d_3, d_4\} && = 2 \end{aligned}$$

with objective value $cx = 48$.

The integrality of the matroid intersection polyhedron also carries over to polymatroids.

Theorem 6.5. *If f_1 and f_2 are two polymatroid functions on N, the linear system* $\{\sum_{j\in A} x_j \leq f_i(A)$ for $A \subseteq N$ and $i = 1, 2, x \in R^n_+\}$ *is totally dual integral.*

Proof. We consider the dual problem

$$\min \sum_{A\subseteq N} f_1(A)y_1(A) + \sum_{A\subseteq N} f_2(A)y_2(A)$$

$$(6.3) \qquad \sum_{A\ni j} y_1(A) + \sum_{A\ni j} y_2(A) \geq c_j \quad \text{for } j \in N,$$

$$y_1(A), y_2(A) \geq 0 \quad \text{for } A \subseteq N.$$

Let y_1^*, y_2^* be an optimal solution, and let $c_j^i = \sum_{A\ni j} y_i^*(A)$ for $i = 1, 2$ and $j \in N$. Thus there is an optimal solution to (6.3) with $\sum_{A\ni j} y_i(A) \geq c_j^i$ for $i = 1, 2$ and $j \in N$. Hence we can decompose (6.3) into the problems

$$\min \sum_{A\subseteq N} f_i(A)y_i(A)$$

$$\sum_{A\ni j} y_i(A) \geq c_j^i \quad \text{for } j \in N$$

$$y_i(A) \geq 0 \quad \text{for } A \subseteq N$$

for $i = 1, 2$. These are duals of polymatroid optimization problems. Hence, by Proposition 6.2, there exist optimal solutions $\hat{y}_i$ of the form

$$\{S: \hat{y}_i(S) > 0\} = \{S_i^1, \ldots, S_i^{l_i}\},$$

with $S_i^1 \subseteq \cdots \subseteq S_i^{l_i} \subseteq N$, and $(\hat{y}_1, \hat{y}_2)$ is an alternate optimal solution to (6.3).

Now setting $y_i(A) = 0$ for $A \neq S_i^t$ for some t, $1 \leq t \leq l_i$, $i = 1, 2$, (6.3) reduces to a problem of the form

$$(6.4) \qquad \min\{fy: By \geq c, y \geq 0\},$$

where the columns of B are the characteristic vectors of $\{S_i^1, \ldots, S_i^{l_i}\}$ for $i = 1, 2$. By arranging these columns in the order $S_1^1, S_1^2, \ldots, S_1^{l_1}, S_2^{l_2}, S_2^{l_2-1}, \ldots, S_2^2, S_2^1$, we see that if $j \in S_i^{k_i} \setminus S_i^{k_i-1}$ for $i = 1, 2$, then $j \in S_1^{k_1}, S_1^{k_1+1}, \ldots, S_1^{l_1}, S_2^{l_2}, \ldots, S_2^{k_2}$ but j is in no other sets. So row j of B has the consecutive 1's property. Hence B is an interval matrix and is totally unimodular (see Definition 2.2 and Corollary 2.10 of Section III.1.2). So whenever c is integer, (6.3) has an optimal solution in integers, and hence the given inequality description of $P(f_1) \cap P(f_2)$ is totally dual integral. ■

Theorem 6.5 allows us to establish some important properties of polymatroids very easily. Generalizing the duality result for maximum-cardinality matroid intersection yields:

Corollary 6.6 *If $P(f_1)$ and $P(f_2)$ are polymatroids, then*

$$\max\left\{\sum_{j\in N} x_j : x \in P(f_1) \cap P(f_2)\right\} = \min_{S\subseteq N} \{f_1(S) + f_2(N \setminus S)\}.$$

Proof. Take $c_j = 1$ for all $j \in N$ in (6.3), and let (y_1, y_2) be an optimal solution. By Theorem 6.5, we can assume the solutions are of the form $y_1(S_1^l) = 1$ for $l = 1, \ldots, l_1$, $y_1(A) = 0$ otherwise, $y_2(T_2^l) = 1$ for $l = 1, \ldots, l_2$, $y_2(A) = 0$ otherwise. Let $S = \bigcup_{l=1}^{l_1} S_1^l$ and $T = \bigcup_{l=1}^{l_2} T_2^l$. Since f_1 and f_2 are nondecreasing and submodular, we obtain $f_1(S) \leqslant \Sigma_{l=1}^{l_1} f_1(S_1^l)$ and $f_2(T) \leqslant \Sigma_{l=1}^{l_2} f_2(T_2^l)$. Hence an alternate optimal solution is $y_1(S) = y_2(T) = 1$, $y_1(A), y_2(A) = 0$ otherwise. Feasibility implies $S \cup T = N$. Finally, since f_1 and f_2 are nondecreasing, we can take $N \setminus S = T$. ■

Corollary 6.7. *If $P(f)$ is a polymatriod with rank function r_P, then*

$$r_P(a) = \min_{T\subseteq N}\left\{f(T) + \sum_{j\in N\setminus T} a_j\right\}.$$

Proof. By definition, we have

$$r_P(a) = \max\left\{\sum_{j\in N} x_j : x \in P(f) \cap \{x \in R_+^n : x \leqslant a\}\right\}.$$

But $\{x \in R_+^n : x \leqslant a\}$ is a polymatroid with underlying submodular function $f_2(S) = \Sigma_{j\in S}\, a_j$ for all $S \subseteq N$. Hence the result follows from Corollary 6.6. ■

Example 6.1 (continued). Suppose that $a = (1 \quad 4 \quad 6 \quad 2)$ gives upper bounds on the processing times of the four jobs. Writing out a polyhedral description of $P(f) \cap \{x : x \leqslant a\}$ and removing the inequalities that are redundant, we obtain

$$\begin{aligned}
x_2 \qquad\quad + x_4 &\leqslant 7\\
x_2 + x_3 + x_4 &\leqslant 8\\
x_1 + x_2 + x_3 + x_4 &\leqslant 10\\
x_1 \qquad\qquad\qquad &\leqslant 1\\
x_2 \qquad\qquad &\leqslant 4\\
x_3 \qquad &\leqslant 6\\
x_4 &\leqslant 2\\
x \in R_+^4&.
\end{aligned}$$

Since $x = (1 \quad 3 \quad 3 \quad 2)$ satisfies all the equalities, we obtain $9 = \Sigma_{j=1}^{4} x_j \leqslant r_P(a)$. But by Corollary 6.7, we have

$$r_P(a) = \min_{T \subseteq N} \left\{ f(T) + \sum_{j \in N \setminus T} a_j \right\} = \min_{T \subseteq N} \left\{ \max_{j \in T} d_j + \sum_{j \in N \setminus T} a_j \right\},$$

and taking $T = \{2, 3, 4\}$, we obtain $r_P(a) \leqslant 9$. Hence $r_P(a) = 9$.

Now we are ready to tackle the problems mentioned at the beginning of this section—in particular, *submodular function minimization:*

$$\text{(6.5)} \qquad \min_{S \subseteq N} \{f(S)\}, \quad \text{with } f \text{ submodular},$$

polymatroid separation:

(6.6) Given $x^* \in R^n_+$, is $x^* \in P(f)$? If not, find $S \subseteq N$ so that the violation of $\sum_{j \in S} x_j \leqslant f(S)$ is maximized,

and *polymatroid rank function calculation:*

(6.7) Given $P(f)$ and $a \in R^n_+$, calculate $r_{P(f)}(a)$.

First we consider problem (6.5), where f is an arbitrary submodular function. By adding a constant, we can assume without loss of generality that $f(\varnothing) = 0$. Furthermore, since $f(S \cup \{j\}) - f(S)$ is nonincreasing in S, it follows that if $f(N) - f(N \setminus \{j\}) > 0$, then j is not contained in any optimal solution of (6.5), so the problem reduces to $\min\{f(S): S \subseteq N \setminus \{j\}\}$. Now define $k \in R^n$ by $k_j = f(N \setminus \{j\}) - f(N)$ for $j \in N$, and define a modified function f^* by

$$f^*(S) = f(S) + \sum_{j \in S} k_j \quad \text{for } S \subseteq N.$$

Based on the above remark, we assume without loss of generality that $k \in R^n_+$.

It is easily verified that:

Proposition 6.8. *If f is submodular, then f^* is nondecreasing and submodular.*

Theorem 6.9. *The following statements are equivalent:*

1. *S^* is an optimal solution of (6.5).*
2. *For the separation problem (6.6) with respect to $P(f^*)$ and the point $k \in R^n_+$, $\Sigma_{j \in S^*} x_j \leqslant f^*(S^*)$ is a most violated inequality.*

3. $r_{P(f^*)}(k) = f^*(S^*) + \Sigma_{j\in N\setminus S^*} k_j$ *for* $k \in R^n_+$.

Proof $f(S^*) \leqslant f(T)$ for $T \subseteq N$ if and only if

$$f^*(S^*) - \sum_{j\in S^*} k_j \leqslant f^*(T) - \sum_{j\in T} k_j$$

if and only if

$$f^*(S^*) + \sum_{j\in N\setminus S^*} k_j \leqslant f^*(T) + \sum_{j\in N\setminus T} k_j.$$

But the first inequality is equivalent to statement 1, the second inequality is equivalent to statement 2, and the third inequality is equivalent to statement 3. ■

Corollary 6.10. *The following statements are equivalent:*

1. $\min_{S\subseteq N} f(S) = 0$.
2. $k \in P(f^*)$.
3. $r_{P(f^*)}(k) = \Sigma_{j\in N} k_j$.

Hence we have shown that problems (6.5), (6.6), and (6.7) are equivalent. In the next section we will consider algorithms for these problems.

Example 6.1 (continued). We consider the separation problem with $f(S) = \max_{j\in S}\{d_j\}$ and $d = (10 \quad 7 \quad 8 \quad 4)$. Is $x^* = (\frac{1}{2} \quad 3\frac{1}{2} \quad 3 \quad 2) \in P(f)$? If not, find a most violated inequality.

Using Theorem 6.9, we have a choice of solving the maximum violation problem $\max_{S\subseteq N}\{\Sigma_{j\in S} x^*_j - f(S)\}$, or, equivalently, of solving the problem of minimizing a submodular function, namely, $\min_{S\subseteq N}\{f(S) - \Sigma_{j\in S} x^*_j\}$, or of calculating

$$r_{P(f)}(x^*) = \min_{S\subseteq N}\left\{f(S) + \sum_{j\in N\setminus S} x^*_j\right\}.$$

Now it is easy to check that

$$r_{P(f)}(x^*) = 8\tfrac{1}{2} = \max_{j=2,3,4}\{d_j\} + x^*_1 < \sum_{j=1}^{4} x^*_j = 9.$$

Hence $x^* \notin P(f)$, and $x_2 + x_3 + x_4 \leqslant 8$ is a most violated inequality.

7. ALGORITHMS TO MINIMIZE A SUBMODULAR FUNCTION

Here we discuss polynomial-time algorithms for the problem (6.5) of minimizing a submodular function, and we also discuss the related problem (6.6) of separation for polymatroids.

First we consider an important class of submodular functions that includes many of the submodular functions encountered in practical models.

Proposition 7.1. *If* $c_T, r_T \geqslant 0$ *for* $T \subseteq N$, *then*

$$f(S) = -\sum_{T \subseteq S} c_T + \sum_{T \cap S \neq \emptyset} r_T \tag{7.1}$$

is a submodular function.

Proof. We have

$$f(S \cup \{j\}) - f(S) = -\sum_{T \subseteq S} c_{T \cup \{j\}} + \sum_{\{T: T \cap S = \emptyset, j \in T\}} r_T.$$

If $S' \supset S$ and $j \notin S'$, then $\{T: T \subseteq S\} \subseteq \{T: T \subseteq S'\}$ and $\{T: T \cap S' = \emptyset, j \in T\} \subseteq \{T: T \cap S = \emptyset, j \in T\}$. Hence $f(S \cup \{j\}) - f(S)$ is nonincreasing in S. ∎

Functions of the form (7.1) can be used to represent Boolean functions. In particular, consider a quadratic Boolean function

$$g(x) = dx - x^T Q x, \quad x \in B^n \quad \text{with } d \geqslant 0,\ Q \geqslant 0, \text{ and symmetric, and } q_{ii} = 0 \quad \text{for all } i.$$

Let x^S be the characteristic vector of S. Then

$$f(S) = \sum_{j \in S} d_j - \sum_{i,j \in S} q_{ij} = g(x^S)$$
$$= -\sum_{T \subseteq S} c_T + \sum_{T \cap S \neq \emptyset} r_T,$$

where

$$c_T = \begin{cases} q_{ij} + q_{ji} & \text{for } T = \{i, j\} \\ 0 & \text{otherwise,} \end{cases}$$

$$r_T = \begin{cases} d_j & \text{for } T = \{j\} \\ 0 & \text{otherwise.} \end{cases}$$

On a graph $G = (V, E)$ with weights $w \in R_+^{|E|}$ on the edges, the function $f(S) = |S| - \Sigma_{e \in E(S)} w_e$ for $S \subseteq V$ can be modeled this way with $d_j = 1$ for all $j \in V$, $q_{ij} = q_{ji} = \frac{1}{2} w_e$ for $e = (i, j)$, and $q_{ij} = 0$ otherwise. Note that $f(S) - 1$ for $S \neq \emptyset$ is the function needed to solve the separation problem for the tree polytope.

We now show that when $f(S)$ is of the form (7.1), problem (6.5) can be solved as a maximum-flow problem. Consider the digraph $\mathscr{D} = (V_1 \cup V_2 \cup (s, t), \mathscr{A})$, where $V_1 = \{S \subseteq N: c_S > 0\}$, $V_2 = \{T \subseteq N: r_T > 0\}$, and

$$\mathscr{A} = \{(S, T): S \in V_1, T \in V_2, S \cap T \neq \emptyset\} \cup \{(s, S): S \in V_1\} \cup \{(T, t): T \in V_2\}$$

with capacities d_e for $e \in \mathscr{A}$, where

$$e = (S, T) \text{ has capacity } d_e = \infty$$
$$e = (s, S) \text{ has capacity } d_e = c_S$$
$$e = (T, t) \text{ has capacity } d_e = r_T$$

(see Figure 7.1).

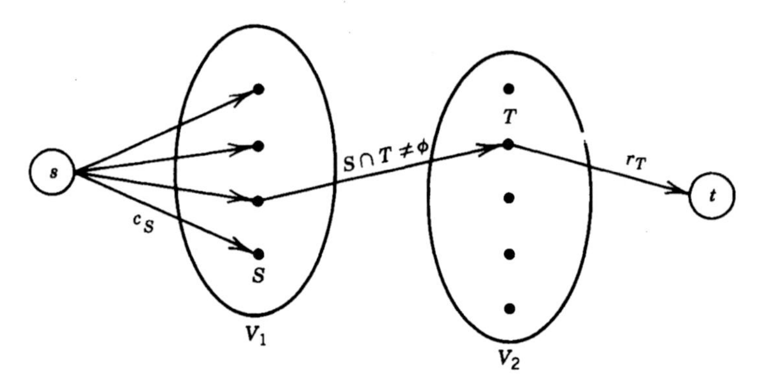

Figure 7.1

Now we consider s–t cuts. Let $W_1 \subseteq V_1$, $W_2 \subseteq V_2$, and (W_1, W_2) be the cut

$$\{(i, j) \in \mathcal{A}: i \in \{s\} \cup W_1 \cup W_2, j \in \{t\} \cup (V_1 \setminus W_1) \cup (V_2 \setminus W_2)\}.$$

The capacity of (W_1, W_2) is

$$d(W_1, W_2) = \sum_{S \in V_1 \setminus W_1} c_S + \sum_{T \in W_2} r_T$$

if all pairs of sets (S, T) with $S \in W_1$, $T \in V_2 \setminus W_2$ are disjoint; otherwise it is $d(W_1, W_2) = \infty$ (see Figure 7.2).

Now for any cut (W_1, W_2) we have:

i. Let $R = \cup_{S \in W_1} S$. $d(W_1, W_2)$ is finite if and only if for all $T \in V_2$ with $T \cap R \neq \emptyset$ we have $T \in W_2$.
ii. If $S \in V_1 \setminus W_1$ and $S \subseteq R$, we can reduce $d(W_1, W_2)$ by c_S by including S in W_1.
iii. If $T \in W_2$ and $T \subseteq N \setminus R$, we can reduce $d(W_1, W_2)$ by r_T by removing T from W_2.

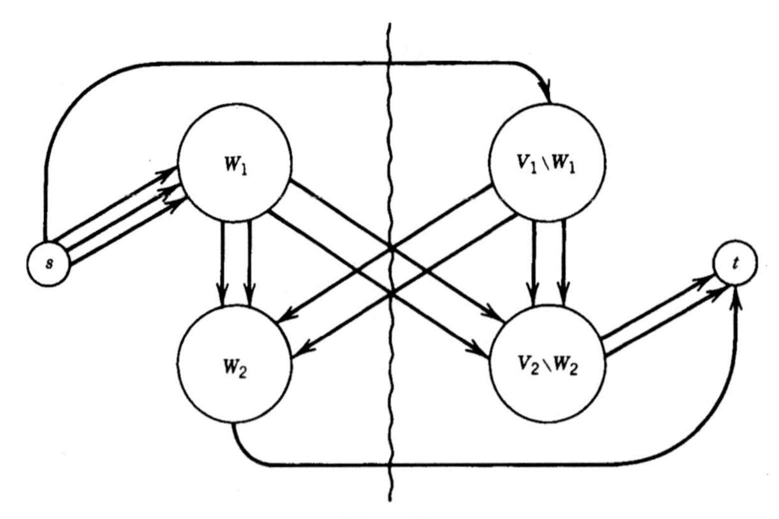

Figure 7.2

Thus we have established:

Proposition 7.2. *Every minimal capacity s–t cut* (W_1, W_2) *in* $\mathcal{D}$ *can be characterized by a set* $R \subseteq N$ *where* $W_1 = \{S \in V_1 : S \subseteq R\}$ *and* $W_2 = \{T \in V_2 : T \cap R \neq \emptyset\}$. *The cut has capacity*

$$d(W_1, W_2) = d(R) = \sum_{S \cap (N \setminus R) \neq \emptyset} c_S + \sum_{T \cap R \neq \emptyset} r_T.$$

Hence the problem $\min_{R \subseteq N} d(R)$ can be solved by finding a maximum flow in $\mathcal{D}$. Since

$$\begin{aligned} d(R) &= \sum_{S \subseteq N} c_S - \sum_{S \subseteq R} c_S + \sum_{T \cap R \neq \emptyset} r_T \\ &= f(R) + \sum_{S \subseteq N} c_S, \end{aligned}$$

Proposition 7.2 is applicable to the minimization of submodular fractions of the form (7.1).

Theorem 7.3. *If f is a submodular function of the form (7.1),* $\min_{S \subseteq N} f(S)$ *can be solved by finding a maximum s–t flow in a digraph* $\mathcal{D}$ *with* $n' + 2$ *nodes, where*

$$n' = |\{S \subseteq N: c_S > 0\}| + |\{T \subseteq N: r_T > 0\}|.$$

Corollary 7.4. $\min_{x \in B^n} (cx - x^T Q x)$ *with* $Q > 0$ *can be solved as a maximum-flow problem in a digraph with* $O(n^2)$ *nodes.*

Example 7.1. We solve the quadratic Boolean problem

$$\min_{x \in B^4} \{9x_1 + 4x_2 + 2x_3 + 6x_4 - 4x_1x_2 - 4x_1x_3 - 7x_2x_3 - 2x_1x_4 - 4x_2x_4\}.$$

Alternatively, we can solve $\min_{S \subseteq N} f(S)$, where $f(S)$ is of the form (7.1) with $r_{\{1\}} = 9$, $r_{\{2\}} = 4$, $r_{\{3\}} = 2$, $r_{\{4\}} = 6$, $r_T = 0$ otherwise, and $c_{\{1,2\}} = c_{\{1,3\}} = c_{\{2,4\}} = 4$, $c_{\{2,3\}} = 7$, $c_{\{1,4\}} = 2$, $c_S = 0$ otherwise.

We construct the digraph $\mathcal{D}$ shown in Figure 7.3 and solve the maximum-flow problem giving the s–t cut indicated with $R = \{2, 3\}$ and $d(R) = 20$. By Corollary 7.4, it follows that

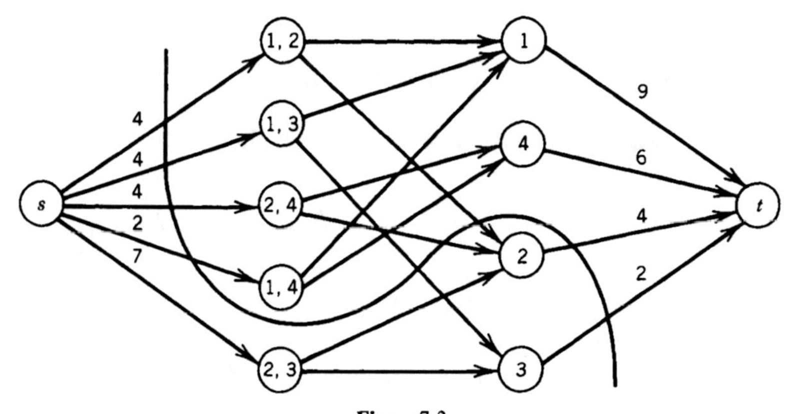

Figure 7.3

$x^R = (0\ \ 1\ \ 1\ \ 0)$ solves the problem with value $g(x^R) = d(R) - \Sigma_{S \subseteq N} c_S = 20 - 21 = -1$.

When f is a general submodular function, the *ellipsoid algorithm* provides a very different approach to the minimization of a submodular function.

Theorem 7.5. *There exists an ellipsoid algorithm for the problem (6.5) of minimizing a submodular function, requiring a polynomial number of evaluations of the function f.*

Proof

i. By Theorem 6.9, it suffices to give a polynomial-time algorithm for the separation problem for polymatroid polytopes.
ii. By the polynomial equivalence of linear programming optimization and separation (Theorem 3.3 of Section I.6.3), it suffices to give a polynomial-time algorithm for the linear programming problem over polymatroid polytopes.
iii. Proposition 6.2 gives a polynomial-time (greedy) algorithm to solve the linear programming problem over the family of polymatroid polytopes. ■

Theorem 7.5 motivated the search for a purely combinatorial algorithm for problem (6.5). An augmenting-path algorithm has been developed for the polymatroid separation problem, but the bound on the number of function evaluations is polynomial in n and $f(N)$. This gives a purely combinatorial separation algorithm with a polynomial number of function evaluations for any matroid polytope $P(r)$ because $r(N) \leqslant n$.

The final topic of this section is the minimization of a submodular function subject to some simple constraints. First suppose that $S = \emptyset$ is not feasible. This yields the problem

$$(7.2) \qquad \min_{\emptyset \subset S \subseteq N} \{f(S)\}, \quad \text{where } f \text{ is submodular.}$$

Proposition 7.6. *Problem (7.2) can be solved by solving problem (6.5) no more than $|N|$ times.*

Proof. Since $S \neq \emptyset$, it follows that $j \in S$ for some $j \in N$. Therefore it suffices to solve the problem

$$\min_{T \subseteq N_j} \{f_j(T)\} \quad \text{for } j \in N,$$

where $N_j = N \setminus \{j\}$, and $f_j(T) = f(T \cup \{j\})$. Because f_j is submodular, the claim follows. ■

Proposition 7.6 is applicable to the tree polytope on the graph $G = (V, E)$, namely,

$$\left\{x \in R_+^{|E|}: \sum_{e \in E(S)} x_e \leqslant |S| - 1 \text{ for } S \subseteq V,\ |S| \geqslant 2\right\}.$$

Corollary 7.7. *The separation problem for the tree polytope can be solved by solving no more than $|V|$ maximum-flow problems on a digraph with $|V| + |E| + 2$ nodes.*

Proof. Let $f(S) = |S| - \Sigma_{e \in E(S)} x_e^*$. Then $x^* \in R_+^{|E|}$ lies in the tree polytope if and only if $\min_{\emptyset \subset S \subseteq V} f(S) \geq 1$. Since $f(S)$ is of the form (7.1), the claim follows from Theorem 7.3 and Proposition 7.6. ■

Note that the algorithm given in Section II.6.3 for finding violated subtree elimination constraints is a special case of the separation problem for the tree polytope.

Now consider the problem

$$(7.3) \qquad \min_{\emptyset \subset S \subset N} \{f(S): |S \cap T| \text{ odd}\},$$

where f is submodular, and $T \subseteq N$ with $|T|$ even. Problem (7.3) is important because odd sets arise in the constraints and therefore occur in the separation problems for some combinatorial optimization problems. These include the minimum-weight T-join problem and the matching problem. In matching we take $T = N$, and we can always assume that N is even by adding a dummy node to the graph.

Let $n = |N|$. We will show how (7.3) can be reduced to solving (6.5) n^3 times. First consider the relaxation of (7.3):

$$(7.4) \qquad \min_{\emptyset \subset S \subset N} \{f(S): 1 \leq |S \cap T| \leq |T| - 1\}.$$

Note that when $|T| = 2$, the problems (7.3) and (7.4) are equivalent.

Proposition 7.8. *Problem (7.4) can be reduced to solving (6.5) n^2 times.*

Proof. For $j \in T$, let $f_j(S) = f(S \cup j)$. An optimal solution to (7.4) is obtained by solving

$$\min_{\emptyset \subseteq S \subseteq N \setminus \{j,k\}} f_j(S)$$

for each j $k \in T$ with $j \neq k$ and then taking the best of these solutions. ■

Next we show how to reduce (7.3) to solving (7.4) n times. Let S^e be any optimal solution to (7.4). If $|S^e \cap T|$ is odd, then S^e is an optimal solution to (7.3). The next result imposes restrictions on an optimal solution to (7.3) when $|S^e \cap T|$ is even.

Proposition 7.9. *If S^e is an optimal solution to (7.4) and $|S^e \cap T|$ is even, then there exists an optimal solution S^0 to (7.3) satisfying one of the four following conditions:*

1. $S^0 \cap T \subset T \setminus S^e$,
2. $S^0 \cap T \supset T \setminus S^e$,
3. $S^0 \cap T \supset S^e \cap T$,
4. $S^0 \cap T \subset S^e \cap T$.

Proof. If $(S^0 \cap S^e) \cap T = \emptyset$, then condition 1 holds. Also, if $(S^0 \cup S^e) \cap T = T$, then condition 2 holds.

Now suppose that $(S^0 \cap S^e) \cap T \neq \emptyset$ and $(S^0 \cup S^e) \cap T \neq T$. Since

$$|S^0 \cap T| + |S^e \cap T| = |(S^0 \cup S^e) \cap T| + |(S^0 \cap S^e) \cap T|,$$

either $|(S^0 \cup S^e) \cap T|$ or $|(S^0 \cap S^e) \cap T|$ is odd, but not both. Suppose $|(S^0 \cup S^e) \cap T|$ is odd.

By submodularity, we have

$$f(S^0) + f(S^e) \geqslant f(S^0 \cup S^e) + f(S^0 \cap S^e).$$

But since S^e is optimal in (7.4) and $S^0 \cap S^e$ is feasible in (7.4), we have $f(S^e) \leqslant f(S^0 \cap S^e)$. Hence $f(S^0) \geqslant f(S^0 \cup S^e)$. But because S^0 is optimal in (7.3) and $S^0 \cup S^e$ is feasible in (7.3), we obtain $f(S^0) \leqslant f(S^0 \cup S^e)$. Hence $f(S^0) = f(S^0 \cup S^e)$, and $S^0 \cup S^e$ is an alternative optimal solution to (7.3). Thus there is an optimal solution to (7.3) that strictly contains S^e, and condition 3 holds.

Finally, when $|(S^0 \cap S^e) \cap T|$ is odd, a similar argument yields condition 4. ■

As a consequence of Proposition 7.9, we have reduced (7.3) to four subproblems. In each of these problems, T has been replaced by a smaller even set, either $T' = T \setminus S^e$ or $T'' = T \cap S^e$.

Next we show how to recombine the four subproblems into two problems of the form (7.3), one with $T \leftarrow T'$ and the other with $T \leftarrow T''$.

If condition 4 holds, the subproblem is

$$\min_{\emptyset \subset S \subset N \setminus T'} \{f(S) \colon |S \cap T''| \text{ odd}\};$$

and if condition 2 holds, the subproblem is

$$\min_{\emptyset \subset S \subset N \setminus T'} \{f(S \cup T') \colon |S \cap T''| \text{ odd}\}.$$

Now let n' represent T', let $N' = (N \setminus T') \cup \{n'\}$, and for $\emptyset \subseteq S \subseteq N'$ let

$$f'(S) = \begin{cases} f(S) & \text{if } n' \notin S \\ f(S \cup T') & \text{if } n' \in S. \end{cases}$$

It is easily verified that f' is submodular, and

$$\min_{\emptyset \subset S \subset N \setminus T'} \{\min(f(S), f(S \cup T')) \colon |S \cap T''| \text{ odd}\}$$

is equal to

$$\min_{\emptyset \subset S \subset N'} \{f'(S) \colon |S \cap T''| \text{ odd}\}. \tag{7.5}$$

Similarly, if either conditions 1 or 3 of Proposition 7.9 holds, we obtain the subproblem

$$\min_{\emptyset \subset S \subset N''} \{f''(S) \colon |S \cap T'| \text{ odd}\}, \tag{7.6}$$

where n'' represents T''; $N'' = (N \setminus T'') \cup \{n''\}$; and for $\emptyset \subseteq S \subseteq N''$, we have that $f''(S)$ is the submodular function given by

$$f''(S) = \begin{cases} f(S) & \text{if } n'' \notin S \\ f(S \cup T'') & \text{if } n'' \in S. \end{cases}$$

Hence we have reduced (7.3) to the smaller problems (7.5) and (7.6) of the same form where: $|N'|$, $|N''| \leq n - 2$; $|T'| + |T''| = |T|$; and $|T'|$, $|T''| \geq 2$ and even. Now we proceed recursively by relaxing (7.5) and (7.6). In each case, either an optimal solution is found or the subproblem is decomposed again. Since (7.3) and (7.4) are equivalent when $|T| = 2$, in the worst case the original problem will finally decompose into $\frac{1}{2}|T|$ problems of the form (7.4); and in each of these problems, a feasible solution must contain exactly one element from a subset of size 2.

Theorem 7.10. *Problem (7.3) can be reduced to solving problem (7.4) n^3 times.*

Proof. Let $g(2k)$ be the maximum number of calls of problem (7.4) when $|T| = 2k$. Then $g(2) = 1$ and

$$g(2k) = \max_{1 \leq l \leq k-1} \{g(2l) + g(2(k - l))\}.$$

It can be shown that the unique solution is $g(2k) = 2k - 1$. Since $2k \leq n$, the result follows from Proposition 7.8. ■

When $f(S)$ represents a quadratic Boolean function, the functions f' and f'' in (7.5) and (7.6) also are of this form. In particular, if

$$g(x) = \sum_{j=1}^{n} r_j x_j - \sum_{i=1}^{j-1} \sum_{j=2}^{n} c_{ij} x_i x_j, \quad x \in B^n$$

and $T' = \{k + 1, \ldots, n\}$, then setting $x_{n'} = x_{k+1} = \cdots = x_n$, we obtain

$$g'(x) = \sum_{j=1}^{k} r_j x_j + \left(\sum_{j=k+1}^{n} r_j - \sum_{i=k+1}^{j-1} \sum_{j=k+2}^{n} c_{ij} \right) x_{n'}$$
$$- \sum_{i=1}^{j-1} \sum_{j=2}^{k} c_{ij} x_i x_j - \sum_{i=1}^{k} \left(\sum_{j=k+1}^{n} c_{ij} \right) x_i x_{n'}$$

with $(x_1, \ldots, x_k, x_{n'}) \in B^{k+1}$.

Example 7.1 (continued)

$$\min\{9x_1 + 4x_2 + 2x_3 + 6x_4 - 4x_1x_2 - 4x_14x_3 - 7x_2x_3 - 2x_1x_4 - 4x_2x_4 : x \in B^4\}$$

subject to the constraint $\Sigma_{j=1}^4 x_j$ odd.

Let $T = N = \{1, 2, 3, 4\}$. The first step is to solve the relaxation (7.4). As shown previously, the optimal solution is given by $S^e = \{2, 3\}$. Hence $T' = \{1, 4\}$ and $T'' = \{2, 3\}$.

Now the problem is reduced to solving (7.5) and (7.6) where (7.5) is

$$\min\{4x_2 + 2x_3 + 13x_5 - 7x_2x_3 - 8x_2x_5 - 4x_3x_5: x_2 + x_3 = 1, x_2, x_3, x_5 \in B^1\}$$

and (7.6) is

$$\min\{9x_1 + 6x_4 - x_6 - 2x_1x_4 - 8x_1x_6 - 4x_4x_6: x_1 + x_4 = 1, x_1, x_4, x_6 \in B^1\}.$$

The former has optimal solution $x_3 = 1, x_2 = x_5 = 0$ with $f(3) = 2$, and the latter has optimal solution $x_1 = x_6 = 1, x_4 = 0$ with $f(123) = 0$. Hence $S = \{1, 2, 3\}$ is an optimal solution to the original problem.

Both the separation problems for minimum-weight 0-1 b-matchings and minimum-weight T-joins correspond to the minimization of a quadratic Boolean function subject to an odd set constraint where the corresponding set function is submodular.

Proposition 7.11. *The separation problems for minimum-weight 0-1 b-matchings and minimum-weight T-joins on a graph $G(V, E)$ can be reduced to solving no more than $|V|^3$ max-flow problems.*

8. COVERING WITH INDEPENDENT SETS AND MATROID PARTITION

Here we consider the problem of finding the minimum number of independent sets needed to cover each element of a matroid a given number of times. Let A be the $m \times n$ matrix whose rows are the characteristic vectors of the independent sets of the matroid, and let $w \in R^n_+$ specify the number of times each element must be covered. Then the fractional version of this problem can be formulated as

$$\zeta_w = \{\min 1y: yA \geqslant w, y \in R^m_+\}, \tag{8.1}$$

and the integer version can be formulated as

$$z_w = \{\min 1y: yA \geqslant w, y \in Z^m_+\}. \tag{8.2}$$

First we consider the fractional covering problem (8.1).

Proposition 8.1. *Given a matroid polytope $P(r)$, the following statements are true for the fractional covering problem (8.1):*

1. $\zeta_w \leqslant 1$ *if and only if* $w \in P(r)$,
2. $\zeta_w = \max_{S \subseteq N} \{\Sigma_{j \in S} w_j / r(S)\}$,
3. $w/\zeta_w \in P(r)$, *and if y^* is an optimal solution to (8.1), then y^*/ζ_w expresses w/ζ_w as a convex combination of points in $P(r)$.*

Proof

1. Since the rows of A correspond to the independent sets in the matroid (N, r), we know that the convex hull of these rows is $P(r)$. So by Proposition 5.8 of Section I.4.5, the antiblocker of $P(r)$ is $\Pi^A = \{\pi \in R^m_+: A\pi \leqslant 1\}$. Therefore, for $w \in R^n_+$,

we have that $w \in P(r)$ if and only if $w\pi \leq 1$ for all $\pi \in \Pi^A$ or $\max\{w\pi: \pi \in \Pi^A\} \leq 1$. However, by linear programming duality, $\zeta_w = \max\{w\pi: A\pi \leq 1, \pi \geq 0\}$, and therefore the claim follows.

2. Since $P(r) = \{x \in R_+^n: (1/r(S)) \sum_{j\in S} x_j \leq 1 \text{ for } S \subseteq N\}$, the maximal extreme points of Π^A are of the form $\pi_j = 1/r(S)$ for $j \in S$, $\pi_j = 0$ for $j \in N \setminus S$. Hence $\zeta_w = \max_{S\subseteq N} \{\sum_{j\in S} w_j/r(S)\}$.
3. $\min\{1y: yA \geq w/\zeta_w, y \in R_+^m\} = 1$, and hence by statement 1 we have $w/\zeta_w \in P(r)$. Letting $\lambda_i = y_i^*/\zeta_w$, we have that $\sum_{i=1}^m \lambda_i = 1, \lambda \geq 0$, and $\lambda A \geq w/\zeta_w$. Since the rows of A form an independence system, we can obtain $\lambda A = w/\zeta_w$ by modifying the solution by replacing $a^i \in Z^m$ by $\bar{a}^i < a^i$ if $\lambda_i > 0$. Thus the claim follows. ∎

Statements 1 and 2 suggest that there is a link between (a) the separation problem (6.6) for $P(r)$ and (b) problem (8.1). In fact, by solving problem (6.6) a polynomial number of times, we can obtain algorithms to compute the value of ζ_w, to find the set S maximizing $\sum_{j\in S} w_j/r(S)$, and to find an optimal solution y^* in (8.1).

For the special case of $w = 1$, we obtain the following result:

Corollary 8.2. *For a matroid $M = (N, r)$, the minimum number of independent sets needed to cover N fractionally is* $\max_{S\subseteq N} \{|S|/r(S)\}$.

Example 8.1. Consider the graph of Figure 8.1 with the values of w as shown. The problem is to find a fractional covering of the weighted edges of G with subgraphs that are forests.

Suppose the feasible solution y^* to (8.1) shown in Figure 8.2 has been found. Since $\sum y_j^* = \frac{5}{2}$, it follows that $\zeta_w \leq \frac{5}{2}$. Now taking $S = \{e_4, e_5, e_6\}$, we see that $\zeta_w \geq \sum_{j\in S} w_j/r(S) = \frac{5}{2}$, and hence y^* is an optimal solution to the fractional covering problem (8.1).

Now we consider the integer covering problem (8.2). To solve this problem, we introduce the concepts of matroid union and matroid partition.

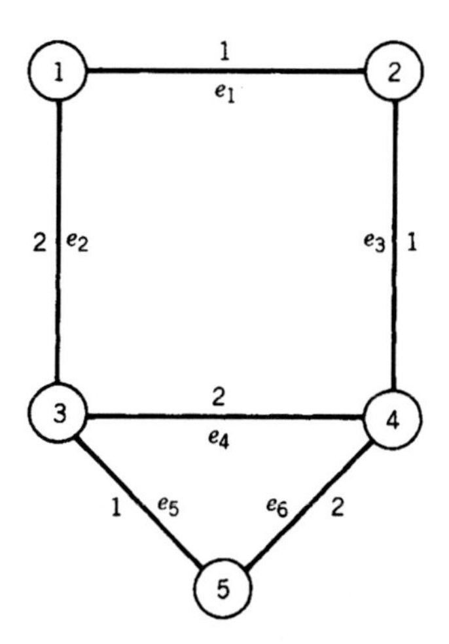

Figure 8.1

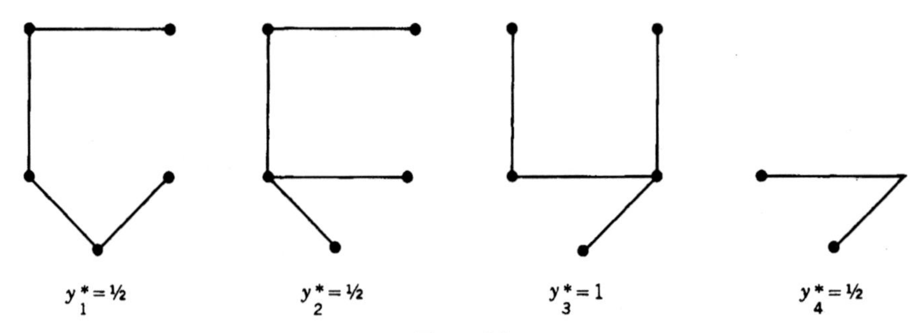

Figure 8.2

Definition 8.1. Given matroids $M_i = (N, \mathscr{F}_i)$ for $i = 1, \ldots, k$, the independence system $M(k) = (N, \mathscr{F}(k))$ is the *matroid union* where $S \in \mathscr{F}(k)$ if and only if there exist $S_i \in \mathscr{F}_i$ for $i = 1, \ldots, k$ such that $\cup_{i=1}^{k} S_i = S$.

This definition motivates the *matroid partition problem* for a matroid union $M(k)$:

(8.3) Given $S \subseteq N$, determine whether $S \in \mathscr{F}(k)$.

Problem (8.3) is related to (8.2) with $w = 1$ because when the matroids M_i are identical, we obtain $N \subseteq \mathscr{F}(k)$ if and only if $z_1 \leq k$. Below we will show how the matroid partition problem can be solved as a matroid intersection problem, and we will also show how to reduce problem (8.2) with $w \in Z_+^n$ to a problem of the same type with $w = 1$. This will enable us to establish the integer-rounding property for the clutter consisting of the bases of a matroid.

Example 8.2. We are given a set N of jobs, each requiring unit processing time. Job j has a deadline d_j. $S \in \mathscr{F}$ if there is some ordering of the jobs of S such that each job is finished by its deadline.

Taking $f(S) = \max_{j \in S} d_j$ as in Example 6.1, we see that $T \in \mathscr{F}$ if and only if $x^T \in P(f)$. But by Corollary 6.6 this holds if and only if

$$r_P(x^T) = \min_{S \subseteq T}\{f(S) + |T \setminus S|\} = |T|.$$

Now it follows from Proposition 2.3 that $(N, \mathscr{F})$ is a matroid with rank function $r(T) = \min_{S \subseteq T}\{f(S) + |T \setminus S|\}$.

Thus, two people working together can accomplish the set S of jobs if $S = S_1 \cup S_2$ with $S_i \in \mathscr{F}$ for $i = 1, 2$, or, in other words, if and only if S is independent in the matroid union $M(2) = (N, \mathscr{F}(2))$.

Suppose there are 10 jobs, and the deadlines are as follows:

Job j:	1	2	3	4	5	6	7	8	9	10
d_j:	1	1	1	2	2	3	3	3	3	4

We have $\{1, 4, 6, 10\}$, $\{2, 5, 7\} \in \mathscr{F}$, so that $\{1, 2, 4, 5, 6, 7, 10\} \in \mathscr{F}(2)$. On the other hand, $\{1, 2, 3, 4, 5\} \notin \mathscr{F}(2)$, since $\{i, j, k\} \notin \mathscr{F}$ for any choice of $1 \leq i < j < k \leq 5$.

Finally, note that $\{1, 4, 6, 10\}$, $\{2, 5, 7\}$, $\{3, 8, 9\} \in \mathscr{F}$. Hence $N \in \mathscr{F}(3)$. Since $N \notin \mathscr{F}(2)$, it follows that the optimal value of problem (8.2) with $w = 1$ is $z_1 = 3$.

Given k matroids $M_i = (N_i, \mathcal{F}_i)$ on distinct sets N_i, their *sum* is $M^* = (\bigcup_{i=1}^k N_i, \bigcup_{i=1}^k \mathcal{F}_i)$. It is easily checked that the sum of matroids is a matroid. The partition matroid (see Section 1) is a simple example of such a sum.

We now show how the k-matroid partition problem for $M = (N, \mathcal{F})$ can be viewed as a matroid intersection problem. Consider the set $N^* = \{(i, j): i \in K, j \in N\}$ where $K = \{1, \dots, k\}$. Any subset $F^* \subseteq N^*$ can be written as $F^* = \bigcup_{i\in K} \bigcup_{j\in F_i} (i, j)$, denoted $(F_1, \dots, F_k)$, where $F_i = \{j: (i, j) \in F^*\} \subseteq N$. We now consider two matroids over the set N^*. The first one, $M_1^* = (N^*, \mathcal{F}_1^*)$, is just the sum of k copies of the original matroid $M = (N, \mathcal{F})$, so $F^* \in \mathcal{F}_1^*$ if and only if $F_i \in \mathcal{F}$ for $i = 1, \dots, k$. The second one, $M_2^* = (N^*, \mathcal{F}_2)$, is the partition matroid where $F^* \in \mathcal{F}_2^*$ if and only if $F_i \cap F_j = \emptyset$ for all $1 \leq i < j \leq k$.

Example 8.1 (continued). We are given the graphic matroid $M = (N, \mathcal{F})$ for the graph $G = (V, E)$ of Figure 8.1. Taking the sum of three copies of $G = (V, E)$ gives the graphic matroid M_1^* shown in Figure 8.3.

A set $E^* \subseteq N^*$ of edges is independent in the partition matroid M_2^* if no edge of the same type appears more than once; that is, no more than one copy of edge (1, 2)—$e_{1,1}$ or $e_{2,1}$ or $e_{3,1}$—is allowed.

Now we consider the independence system $(N^*, \mathcal{F}_1^* \cap \mathcal{F}_2^*)$ of common independent sets in the matroids M_1^* and M_2^*, and we investigate how it relates to the matroid union $(N, \mathcal{F}(k))$ and the matroid partition problem. Let r_i^* be the rank function of M_i^* for $i = 1, 2$, let m^* be the rank function of $(N^*, \mathcal{F}_1^* \cap \mathcal{F}_2^*)$, and let m^k be the rank function of the matroid union $(N, \mathcal{F}(k))$. By definition of M_i^* and $M(k)$, we have:

Proposition 8.3. *The following statements are true.*

1. $r_1^*(F^*) = \sum_{i=1}^k r_i(F_i)$.
2. $r_2^*(F^*) = |\bigcup_{i=1}^k F_i|$.
3. *If* $F^* = (F_1, \dots, F_k) \in \mathcal{F}_1^* \cap \mathcal{F}_2^*$, *then* $S = \bigcup_{i=1}^k F_i \in \mathcal{F}(k)$.
4. *If* $S \in \mathcal{F}(k)$, *then there exists* $F^* = (F_1, \dots, F_k) \in \mathcal{F}_1^* \cap \mathcal{F}_2^*$ *such that* $\bigcup_{i=1}^k F_i = S$.
5. $m^k(S) = m^*(S, \dots, S)$.

Now we can show that $M(k)$ is, in fact, a matroid.

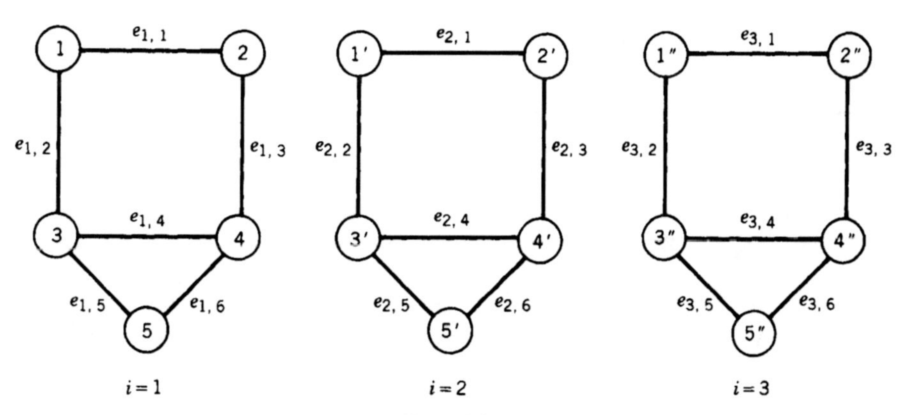

Figure 8.3

Proposition 8.4. *The matroid union $M(k) = (N, \mathcal{F}(k))$ is a matroid with rank function $m^k(S) = \min_{T \subseteq S} \{\sum_{i=1}^{k} r_i(T) + |S \setminus T|\}$.*

Proof. We just consider the case $S = N$. By Proposition 4.9, we have

$$m^*(N^*) = m^*(N, \ldots, N) = \min_{F^* \subseteq N^*} \{r_1^*(F^*) + r_2^*(N^* \setminus F^*)\}$$

$$= \min_{F_i \subseteq N} \left\{ \sum_{i=1}^{k} r_i(F_i) + \left| \bigcup_{i=1}^{k} (N \setminus F_i) \right| \right\}$$

$$= \min_{F_i \subseteq N} \left\{ \sum_{i=1}^{k} r_i(F_i) + \left| N \setminus \bigcap_{i=1}^{k} F_i \right| \right\}$$

$$= \min_{F_i \subseteq N} \left\{ \sum_{i=1}^{k} r_i\left(\bigcap_{j=1}^{k} F_j \right) + \left| N \setminus \bigcap_{i=1}^{k} F_i \right| \right\}$$

because the r_i are nondecreasing.

It follows that the minimum is attained by a set F^* of the form $F^* = (T, \ldots, T)$, where $T = \bigcap_{j=1}^{k} F_j$. Hence

$$m^*(N, \ldots, N) = \min_{T \subseteq N} \left\{ \sum_{i=1}^{k} r_i(T) + |N \setminus T| \right\}.$$

Now by statement 5 of Proposition 8.3, we have $m^k(N) = m^*(N, \ldots, N)$, and hence m^k is of the required form. But by Proposition 2.3, m^k is a submodular rank function; and hence by statement (ii) of Theorem 2.4, the matroid union is a matroid. ■

By Proposition 8.4, we can solve the matroid partition problem by applying the cardinality matroid intersection algorithm to $(N^*, \mathcal{F}_1^* \cap \mathcal{F}_2^*)$, demonstrating either that $S \in \mathcal{F}(k)$ or that there exists a set $T \subseteq S$ with $\sum_{i=1}^{k} r_i(T) < |T|$. A more efficient and direct algorithm is the matroid partition algorithm.

Conversely, we can use the matroid partition algorithm to solve the cardinality intersection problem for matroids $M_1 = (N, r_1)$ and $M_2 = (N, r_2)$. It suffices to consider the matroid union $(M(2), m^2)$ of M_1 and M_2^D. Then because the partition algorithm can be used to find $m^2(N)$ and we have

$$m^2(N) = \min_{T \subseteq N} \{r_1(T) + r_2(N \setminus T)\} + |N| - r_2(N),$$

the claim follows.

Now we return to the covering problem (8.2). When $w = 1$, the problem can be restated as:

> Given a matroid $M = (N, \mathcal{F})$, determine a minimum number z_1 of independent sets whose union is the whole set.

As we have already observed, matroid unions give a method to solve this problem. Taking k identical copies of $M = (N, r)$, the matroid union $M(k)$ has rank function

$$m^k(N) = \min_{T} \{kr(T) + |N \setminus T|\}.$$

Proposition 8.5. $z_1 = \max_S \lceil |S|/r(S) \rceil$.

Proof. $z_1 \leqslant k$ if and only if N is independent in the matroid union $M(k)$ or $m^k(N) = |N|$, or

$$kr(T) + |N \setminus T| \geqslant |N| \quad \text{for all } T \subseteq N.$$

Hence $z_1 \leqslant k$ if and only if $k \geqslant |T|/r(T)$ for all $T \subseteq N$. ■

In addition, we have seen that either the matroid intersection algorithm applied to $M(k)$ or the matroid partition algorithm gives the z_1 bases required.

One approach to problem (8.2) for general $w \in Z^n_+$ is to construct a matroid M^w from M by duplicating each element j w_j times for $j \in N$. Let $N^w = \{(j, i): j \in N, i = 1, \ldots, w_j\}$. Given $T^w \subseteq N^w$, we let $T^w_j = \{i \in \{1, \ldots, w_j\}: (j, i) \in T^w\}$ so that T^w_j is the set of different copies of j in T^w, and we write $T^w = (T^w_1, \ldots, T^w_n)$. Now $\overline{T}^w = \{j \in N: |T^w_j| \geqslant 1\}$ is the set of elements $j \in N$ of which at least one copy appears in T^w.

We now define an independence system $M^w = (N^w, \mathscr{F}^w)$ such that $T^w \in \mathscr{F}^w$ if $|T^w_j| \leqslant 1$ for $j \in N$ and $\overline{T}^w \in \mathscr{F}$. $M^w = (N, r^w)$ is easily seen to be a matroid with $r^w(T^w) = r(\overline{T}^w)$. By construction of N^w, we have:

Proposition 8.6. *z_w is the optimal value of problem (8.2) for the matroid $M = (N, \mathscr{F})$ if and only if z_w is the optimal value of problem (8.2) for the matroid $M^w = (N^w, \mathscr{F}^w)$ with a right-hand-side vector of 1's.*

Corollary 8.7. $z_w = \max_T \lceil (\sum_{j \in T} w_j)/r(T) \rceil$.

Proof. By Proposition 8.5, $z_w = \max_{T^w \subseteq N^w} \lceil |T^w|/r^w(T^w) \rceil$. Since $r^w(T^w) = r(\overline{T}^w)$, we obtain

$$z_w = \max_{\overline{T}^w \subseteq N} \left\lceil \frac{\sum_{j \in \overline{T}^w} w_j}{r(\overline{T}^w)} \right\rceil = \max_T \left\lceil \frac{\sum_{j \in T} w_j}{r(T)} \right\rceil.$$ ■

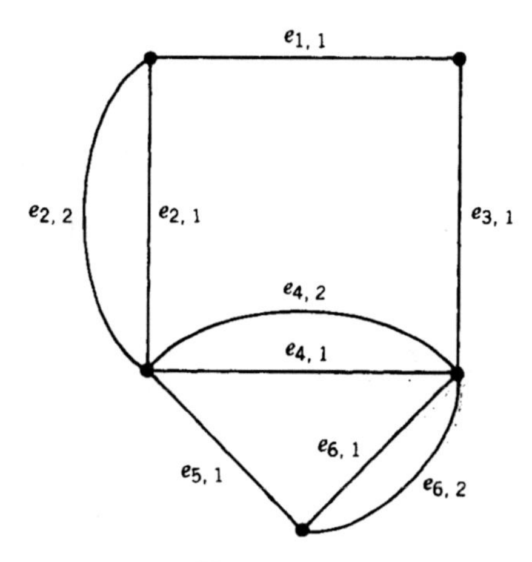

Figure 8.4

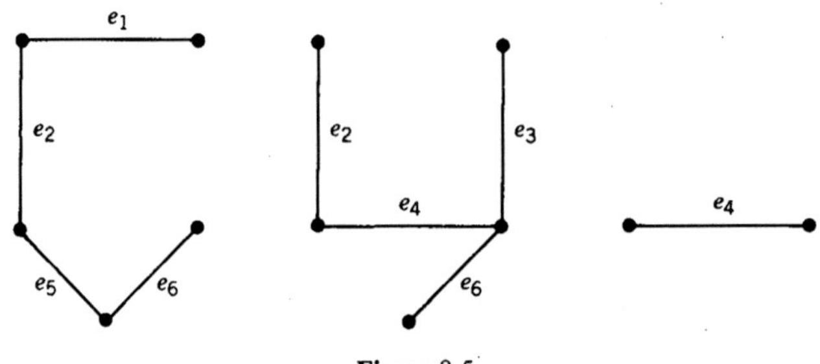

Figure 8.5

From Proposition 8.1, it follows that problem (8.1), the linear programming relaxation of problem (8.2), has value $\zeta_w = \max_T (\Sigma_{j \in T} w_j)/r(T)$. Hence, since $z_w = \lceil \zeta_w \rceil$, the matroid covering problem provides an example of the integer-rounding property discussed in Section III.1.6.

Similar results can be obtained for packing bases. However, for general $w \in Z^n_+$ the above construction does not lead to a polynomial algorithm.

Example 8.1 (continued). The problem is to find an integer covering of the weighted edges of G by forests (see Figure 8.1).

Since $\zeta_w = \frac{5}{2}$, we know from the integer-rounding property that $z_w = \lceil \zeta_w \rceil = 3$. Now we construct an optimal solution. In Figure 8.4, we show the graph $G^w = (V, E^w)$ underlying the graphic matroid $M^w = (E^w, \mathcal{F}^w)$ for which we need to solve problem (8.2) with weights $w_{e_{j,i}} = 1$.

Now we construct an independence system $M^* = (N^*, \mathcal{F}^*)$ consisting of three copies of M^w, such that $E^* = (E_1, E_2, E_3) \subseteq (E^w, E^w, E^w)$ and $E^* \in \mathcal{F}^*$ if the edge sets E_1, E_2, E_3 are disjoint and each edge set E_i is a forest in G^w.

Applying the cardinality matroid intersection algorithm to $M^* = (N^*, \mathcal{F}^*)$, we obtain the solution shown in Figure 8.5, which provides an optimal solution $\{e_1, e_2, e_5, e_6\}$, $\{e_2, e_3, e_4, e_6\}$, $\{e_4\}$ for the covering problem.

9. SUBMODULAR FUNCTION MAXIMIZATION

Whereas there is a polynomial algorithm for the minimization of a submodular function, the problem of maximizing a submodular function is $\mathcal{NP}$-hard. Here we investigate three problems that are natural generalizations of problems treated either earlier in this chapter or earlier in the book.

The three problems are:

(1) *maximizing an arbitrary submodular function*:

$$z_1 = \max_S \{f(S)\} \quad \text{with } f \text{ submodular,} \tag{9.1}$$

(2) *maximizing a nondecreasing submodular function subject to a cardinality constraint*:

$$z_2 = \max_S \{f(S): |S| \leqslant p\} \quad \text{with } f \text{ submodular and nondecreasing,} \tag{9.2}$$

(3) *minimizing a linear function subject to a submodular constraint*:

$$(9.3) \qquad z_3 = \min_S \left\{ \sum_{j \in S} c_j : f(S) = f(N) \right\} \quad \text{with } f \text{ submodular and nondecreasing.}$$

A typical example of problem (9.1) is the uncapacitated location problem discussed in Chapter II.5 with

$$f(S) = \sum_{i \in I} \max_{j \in S} c_{ij} - \sum_{j \in S} f_j.$$

If capacity constraints limiting the amount of demand that can be met by an open facility are added, the problem can still be formulated as the maximization of a submodular function.

The p-facility location problem studied in Section II.5.3, namely,

$$\max_S \left\{ \sum_{i \in I} \max_{j \in S} c_{ij} : |S| \leq p \right\},$$

is an example of problem (9.2).

An example of problem (9.3) is the integer covering problem

$$\min \left\{ \sum_{j \in N} c_j x_j : Ax \geqslant b, x \in B^n \right\},$$

where (A, b) is an $m \times (n + 1)$ nonnegative integer matrix. Here we set

$$f(S) = \sum_{i=1}^{m} \min \left\{ \sum_{j \in S} a_{ij}, b_i \right\}.$$

Observe that $f(N) = \Sigma_{i=1}^m b_i$ and $f(S) = f(N)$ if and only if $\Sigma_{j \in S} a_j \geqslant b$. Note that we can assume that the $\{x \in B^n : Ax \geqslant b\}$ is nonempty by adding, if necessary, an artificial variable x_{n+1} with $a_{n+1} = b$ and $c_{n+1} > \Sigma_{j=1}^n c_j$.

Another important application is the k-matroid intersection problem

$$z = \max_S \left\{ \sum_{j \in S} c_j : |S| \leq r_i(S) \text{ for } i = 1, \ldots, k \right\}.$$

To see that it fits this model, remember from Section 2 that $S \in \mathcal{F}$ if and only if $N \setminus S$ contains a basis in the dual matroid. Hence

$$z = \sum_{j \in N} c_j - \min_S \left\{ \sum_{j \in N \setminus S} c_j : r_i^D(N \setminus S) = r_i^D(N) \text{ for } i = 1, \ldots, k \right\}$$

$$= \sum_{j \in N} c_j - \min_T \left\{ \sum_{j \in T} c_j : f(T) = f(N) \right\},$$

where $f = \Sigma_{i=1}^k r_i^D$.

Since Problems (9.1)–(9.3) are $\mathcal{NP}$-hard, we consider two approaches. The first is to formulate and solve them as integer programs, and the second is to apply heuristic algorithms.

We first consider an integer programming formulation of (9.1). We assume $f(\emptyset) = 0$; then we set $f(S) = f^*(S) - \Sigma_{j\in S} k_j$, where $k_j = f(N \setminus \{j\}) - f(N)$ for all $j \in N$. Hence f is written as the difference of a nondecreasing submodular function f^* and a linear function. Now consider the polyhedron

$$Q(f) = \Big\{(\eta, x) \in R^1 \times R^n_+ \colon\ \eta \leq f^*(S) + \sum_{j\in N\setminus S} [f^*(S \cup \{j\}) - f^*(S)]x_j - \sum_{j\in N} k_j x_j \text{ for all } S \subseteq N\Big\}.$$

Proposition 9.1. *Given* $(\eta, x^T) \in R^1 \times B^n$, *we obtain* $(\eta, x^T) \in Q(f)$ *if and only if* $\eta \leq f(T)$.

Proof. If $(\eta, x^T) \in Q(f)$, then

$$\eta \leq f^*(T) + \sum_{j\in N\setminus T} [f^*(T \cup \{j\}) - f^*(T)]x_j^T - \sum_{j\in N} k_j x_j^T$$
$$= f^*(T) - \sum_{j\in T} k_j = f(T).$$

Now suppose $\eta \leq f(T)$. By Proposition 2.1(b), we have

$$f(T) = f^*(T) - \sum_{j\in T} k_j$$
$$\leq f^*(S) + \sum_{j\in T\setminus S} [f^*(S \cup \{j\}) - f^*(S)] - \sum_{j\in T} k_j$$

for all $S \subseteq N$, and hence

$$\eta \leq f^*(S) + \sum_{j\in N\setminus S} [f^*(S \cup \{j\}) - f^*(S)]x_j^T - \sum_{j\in N} k_j x_j^T$$

for all $S \subseteq N$, so $(\eta, x^T) \in Q(f)$. ■

As a consequence of Proposition 9.1, an alternative formulation for (9.1) is

$$z_1 = \max\{\eta \colon (\eta, x) \in Q(f),\ x \in B^n,\ \eta \in R^1\}. \tag{9.4}$$

Since $Q(f)$ has an exponential number of constraints, we suggest a cutting-plane algorithm, similar to that of Benders, for solving (9.4) (see Section II.5.4).

Example 9.1. The problem is to maximize the quadratic function from Example 7.1, namely,

$$\max_{x \in B^n}\{9x_1 + 4x_2 + 2x_3 + 6x_4 - 4x_1x_2 - 4x_1x_3 - 7x_2x_3 - 2x_1x_4 - 4x_2x_4\},$$

or $\max_S \{f^*(S) - \Sigma_{j \in S} k_j\}$, where f^* is defined above and $k = (1 \quad 11 \quad 9 \quad 0)$.

Generating the constraints of $Q(f)$ for $S = \emptyset$ and $S = \{1\}$, we obtain the relaxation

$$\begin{aligned} &\max \eta \\ &\eta \leq 0 + 9x_1 + 4x_2 + 2x_3 + 6x_4, \qquad S = \emptyset \\ &\eta \leq 10 - x_1 - 2x_3 + 4x_4, \qquad S = \{1\} \\ &x \in B^4 \end{aligned}$$

with optimal solution $\eta = 13$, $x^* = (1 \quad 0 \quad 0 \quad 1)$. Hence $z_1 \leq 13$. However, $f(\{1,4\}) = 13$, and hence x^* is optimal.

A formulation for problem (9.3) is derived similarly with

$$R(f) = \left\{x \in R^n_+ : \sum_{j \in N \setminus S} [f(S \cup \{j\}) - f(S)]x_j \geq f(N) - f(S) \text{ for } S \subseteq N\right\}.$$

Proposition 9.2

$$z_3 = \min\left\{\sum_{j \in N} c_j x_j : x \in R(f), x \in B^n\right\}. \tag{9.5}$$

Proof. If $x^T \in B^n$ is the characteristic vector of T and if $x^T \in R(f)$, then

$$0 = \sum_{j \in N \setminus T} [f(T \cup \{j\}) - f(T)]x_j^T \geq f(N) - f(T).$$

Hence $f(T) = f(N)$, and T is feasible in (9.3).

Conversely, if T is feasible in (9.3), then by submodularity (Proposition 2.1, statement ii) we have

$$f(N) = f(T) \leq f(S) + \sum_{j \in T \setminus S} [f(S \cup \{j\}) - f(S)] \quad \text{for } S \subseteq N$$

or, in other words,

$$\sum_{j \in N \setminus S} [f(S \cup \{j\}) - f(S)]x_j^T \geq f(N) - f(S) \quad \text{for all } S \subseteq N,$$

so $x^T \in R(f)$. ■

Now we turn to heuristic algorithms. The greedy heuristic algorithm for problem (9.2) has already been analyzed in Theorem 3.3 of Section II.5.3. Repeating the theorem, we have:

Theorem 9.3. *If $f(\emptyset) = 0$, and z_G is the value of a greedy heuristic solution to problem (9.2), and z is the value of an optimal solution, then $z_G/z \geqslant 1 - [(p-1)/p]^p$.*

The greedy heuristic for the 0-1 covering problem was analyzed in Theorem 2.5 of Section II.6.2. As indicated above, this is a special case of problem (9.3). The same result holds for the general problem. First we describe the heuristic.

The Greedy Algorithm for Problem (9.3)

Initialization: $S^0 = \emptyset, N^1 = N, t = 1.$

Iteration t: Let

$$\theta^t = \min_{j \in N^t} \frac{c_j}{f(S^{t-1} \cup \{j\}) - f(S^{t-1})},$$

with the minimum attained at $j_t \in N^t$. Let $N^{t+1} = N^t \setminus \{j_t\}$, and let $S^t = S^{t-1} \cup \{j_t\}$. If $f(S^t) = f(N)$, then S^t is the greedy solution with $z_G = \Sigma_{j \in S^t} c_j$. If $f(S^t) < f(N)$, let $t \leftarrow t + 1$ and return.

Using a relaxation of the formulation (9.5), and using a dual heuristic as in the proof of Theorem 2.5 of Section II.6.2, we obtain the following theorem:

Theorem 9.4. *Let f be integer-valued, let $f(\emptyset) = 0$, $d = \max_{j \in N} (f\{j\})$, and let z_G be the value of a greedy heuristic solution to (9.3). Then $z_G/z \leqslant H(d)$, where $H(d) = \Sigma_{i=1}^{d} (1/i)$.*

10. NOTES

Sections III.3.1 and III.3.2

Matroids were introduced by Whitney (1935). Further early developments are due to Tutte (1965). Detailed developments of matroid, polymatroid, and submodular function theory are contained in the books by Tutte (1971), Welsh (1976), and Recski (1988). Recski's book also gives many applications of matroids in the physical sciences and engineering, as does the survey article by Iri (1983).

The importance of matroids in combinatorial optimization was established by Edmonds (1965b, 1970, 1971). Chapters 8 and 9 of the book by Lawler (1976) present the work done in matroid optimization through the mid-1970s. A survey of matroid results tailored to the operations research community was presented by Bixby (1982).

Lovasz (1983) surveyed the relationships between submodularity and convexity. Topkis (1978) studies properties of submodular functions that are of interest in optimization.

Section III.3.3

The optimality of the greedy algorithm was first discovered by Rado (1957) and independently by Gale (1968), Welsh (1968), and Edmonds (1971). A more general combinatorial structure for which the greedy algorithm works, known as a *greedoid*, has been studied by Korte and Lovasz (1984).

The matroid polytope was studied by Edmonds (1970, 1971).

Sections III.3.4 and III.3.5

The matroid and polymatroid intersection theorems are due to Edmonds (1970). Algorithms for maximum-cardinality matroid intersections have been developed by Lawler (1975) and Edmonds (1979). The algorithm for maximum-weighted matroid intersections is based on an algorithm of Frank (1981). [Also see Lawler (1975), Cunningham (1986), and Brezovec et al. (1986).]

Edmonds (1967b) gave an algorithm for minimum-weight branchings. [Also see Chu and Liu (1965), Bock (1971), Karp (1971), Murchland (1973), and Tarjan (1977).]

Fulkerson (1974) gave an algorithm for the problem of packing rooted directed cuts in a weighted digraph and established a blocking relationship between these cuts and branchings. His results yielded a TDI system for the convex hull of edge sets that contain a branching.

Edmonds and Giles (1977) studied a model, now known as the *submodular flow problem*, that generalizes both network flows and polymatroid intersection. Min–max results and algorithms for this class of problems have been obtained by Frank (1982, 1984), Hassin (1982), Lawler and Martel (1982a,b), Schrijver (1984a,b), and Cunningham and Frank (1985). An application to a scheduling problem has been given by Martel (1982).

An even more robust model, known as the *matroid matching problem* or *matroid parity problem*, that generalizes polymatroid intersections and matchings has been studied by Lovasz (1980, 1981). He gave a polynomial-time algorithm for the case of matric matroids and showed that the general problem is $\mathcal{NP}$-hard. Related results were given by Tong et al. (1984).

An annotated bibliography on these problems was given by Lawler (1985).

Section III.3.6

Polymatroid and submodular rank functions have been studied by Edmonds (1970), and the role of these functions in combinatorial optimization has been examined by Lovasz (1983). [Also see McDiarmid (1975).]

Section III.3.7

The max-flow reduction algorithm for submodular function minimization given in Section 7 is due to Rhys (1970) [also see Picard and Ratliff (1975)]. Its applications to graphic matroids was given by Picard and Queyranne (1982) and Padberg and Wolsey (1983, 1984).

Crama (1986) gave an efficient recognition algorithm for certain classes of submodular functions representable in the form (7.1). A general treatment of Boolean functions, primarily of historical interest now, is the book by Hammer and Rudeanu (1966).

The polynomiality of submodular function minimization has been established by Grötschel, Lovasz, and Schrijver (1981, 1984b). They have also developed the procedure for minimizing over odd sets [Grötschel, Lovasz, and Schrijver (1984c)]. These developments and many related results are presented in Grötschel, Lovasz, and Schrijver (1988).

Purely combinatorial algorithms for the separation problem for the matroid polytope have been given in Cunningham (1984), and for submodular function minimization in Cunningham (1985).

Section III.3.8

The problem of covering and packing with independent sets was studied by Edmonds and Fulkerson (1965) and Edmonds (1965b). The matroid partition algorithm is due to Edmonds (1965b) [also see Cunningham (1986)].

Edmonds (1973) considered the packing of branchings, and Cunningham (1977) described the blocking polyhedron of the convex hull of the common independent sets in two matroids. These results were generalized by Baum and Trotter (1981).

A polynomial-time algorithm for problem (8.1) with general w is obtained using the separation algorithm of Section 7 [see Cunningham (1984, 1985)].

Section III.3.9

Submodular function maximization has been studied by Nemhauser, Wolsey, and Fisher (1978), Fisher, Nemhauser and Wolsey (1978), Nemhauser and Wolsey (1979, 1981), Wolsey (1982a,b), and Conforti and Cornuejols (1984). In Nemhauser and Wolsey (1979), it was shown that within a large class of algorithms the greedy algorithm is the best possible one for problem (9.2).

11. EXERCISES

1. Show that graphic matroids are matric.
2. Given a family of subsets $\{S_i\}_{i=1}^m$ of a finite set N, we define a *transversal* of the family to be a set $T = \{i_1, \ldots, i_m\}$ with the following properties:

   a) $|T| = m$;

   b) $i_j \in S_j$ for $j = 1, \ldots, m$.

   If $R \subseteq T$ is a transversal, R is a *partial transversal*.

   i) Show that a family has a transversal if and only if the maximum $s-t$ flow in the digraph of Figure 11.1, with $(j, S_i) \in \mathcal{A}$ if $j \in S_i$, has value m, where: $(s, j) \in \mathcal{A}$ has capacity 1 for $j \in N$, $(S_i, t) \in \mathcal{A}$ has capacity 1 for $i \in \{1, \ldots, m\}$, and (j, S_i) has infinite capacity.

   ii) Show that the set of partial transversals forms a matroid on N.
3. Given an $m \times n$ 0, 1 matrix A, a set $J \subseteq N$ of columns is called dependent if there exists a $J' \subseteq J$ such that $\Sigma_{j \in J'} a_{ij} = 0 (\text{mod } 2)$ for all i.

   i) Show that the sets of independent columns form a matroid. Such matroids are called *binary* matroids.

   ii) Show that a graphic matroid is a binary matroid.

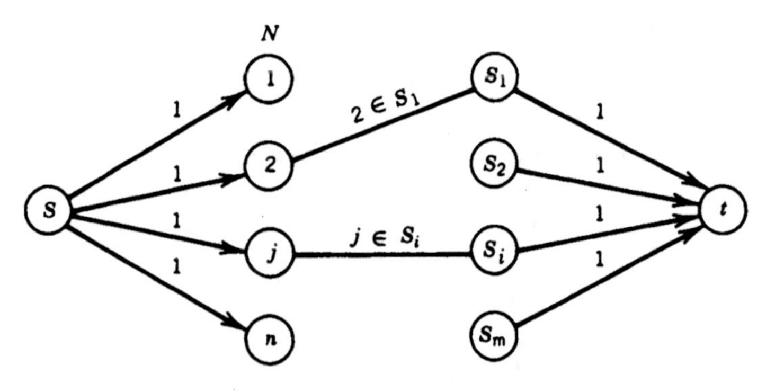

Figure 11.1

4. Show that the following set functions are submodular:

i) $v(S) = \max\{\Sigma_{j \in T} c_j\colon |T| \leqslant k, T \subseteq S\}$ for $S \subseteq N$;

ii) $v(S) = |\{j \in V\colon j \in S \text{ or } (i, j) \in E \text{ for some } i \in S\}|$ for $S \subseteq V$, where $G = (V, E)$ is a graph (this is the cardinality of the neighborhood of S);

iii) $v(S) = \Sigma_{j \in S} c_j - \Sigma_{i \in S, j \in S} q_{ij}$ for $S \subseteq N$, where $q_{ij} \geqslant 0$ for $1 \leqslant i < j \leqslant n$.

5. What is the complexity of the greedy algorithm for problem (3.1) when M is a partition matroid?

6. Prove Proposition 3.5 directly by showing that if $\{j\} \in \mathcal{F}$ for all $j \in N$, every facet of the convex hull of independent sets of a matroid is either of the form

(a) $$x_j \geqslant 0 \quad \text{for } j \in N$$

or

(b) $$\sum_{j \in A} x_j \leqslant r(A) \quad \text{for } A \subseteq N.$$

Hint: (i) Show that if $\pi x \leqslant \pi_0$ is facet-inducing, either it is the inequality $-x_j \leqslant 0$, or $\pi_j \geqslant 0$ for all $j \in N$. (ii) Show that if $\pi_j > 0$ for $j \in A$, and the characteristic vector of $S \in F$ lies on the facet, then $|S \cap A| = r(A)$.

7. Characterize the facets of the tree polytope—that is, when is $\Sigma_{e \in E(U)} x_e \leqslant |U| - 1$ a facet-defining inequality?

8. Show that the following set functions are submodular:

i) Given a matroid $M = (N, \mathcal{F})$ with $c \in R^n$, let

$$v(S) = \max\left\{\sum_{e \in T} c_e\colon T \subseteq S, T \in \mathcal{F}\right\} \quad \text{for } S \subseteq N.$$

ii) Given M as in part i, let $\{Q_i\}_{i \in I}$ be subsets of N and let

$$v(J) = \max\left\{\sum_{e \in T} c_e\colon T \subseteq \bigcup_{i \in J} Q_i, T \in \mathcal{F}\right\} \quad \text{for } J \subseteq I.$$

(Q_i is the set of elements with color i.)

iii)

$$v(S) = \max\left\{\sum_{i \in I}\sum_{j \in S} c_{ij} y_{ij}\colon \sum_{j \in S} y_{ij} \leqslant a_i \text{ for } i \in I,\right.$$

$$\left.\sum_{i \in I} y_{ij} \leqslant b_j \text{ for } j \in S,\ y_{ij} \geqslant 0 \text{ for } i \in I, j \in S\right\}.$$

(This function arises in the capacitated facility location problem where $S \subseteq N$ is the set of open facilities.)

9. $L: R^m \to R^1$ is *submodular* if

$$L(u) + L(v) \geqslant L(u \vee v) + L(u \wedge v) \quad \text{for } u, v \in R^m,$$

where $(u \vee v)_i = \max(u_i, v_i)$ and $(u \wedge v)_i = \min(u_i, v_i)$. Show that if $L(u, y)$ is submodular on $R^m \times R^p$, then $W(y) = \min_u L(u, y)$ is submodular on R^p.

10. Consider the clutter of bases of a matroid $M = (N, \mathcal{F})$.

i) Prove that

$$Q^* = \text{conv}\{x \in R^n_+ : x \geqslant x^F \text{ for } F \text{ a basis of } M\}$$
$$= \left\{x \in R^n_+ : \sum_{j \in S} x_j \geqslant r(N) - r(S)\right\}.$$

ii) Show that for the clutter of spanning trees, this gives

$$Q^* = \left\{x \in R^n_+ : \sum_{\{(i,j): i<j\}} \sum_{e \in \delta(V_i) \cap \delta(V_j)} x_e \geqslant f - 1 \text{ for all } V_1, \ldots, V_f \text{ that are disjoint subsets of } V \text{ and all } f > 1\right\}.$$

11. Apply the maximum-cardinality matroid intersection algorithm to the pair of graphic matroids in Figure 11.2, starting with $S = \{e_1, e_2, e_4\}$

12. Show that the lengths of the shortest s–t dipaths at successive iterations of the cardinality matroid intersection algorithm are nondecreasing.

13. Apply the weighted matroid intersection algorithm to find a maximum-weight set of arcs forming part of a branching (with no specified root) in the digraph of Figure 11.3.

14. **i)** Show that the polytope of the arc sets in exercise 13 is given by

$$\left\{x \in R^n_+ : \sum_{i \in \delta^-(j)} x_{ij} \leqslant 1 \text{ for } j \in V, \sum_{i \in S, j \in S} x_{ij} \leqslant |S| - 1 \text{ for } \emptyset \subset S \subseteq V\right\}$$

ii) Devise a more efficient algorithm for the maximum-weight branching problem. What is the complexity of your algorithm?

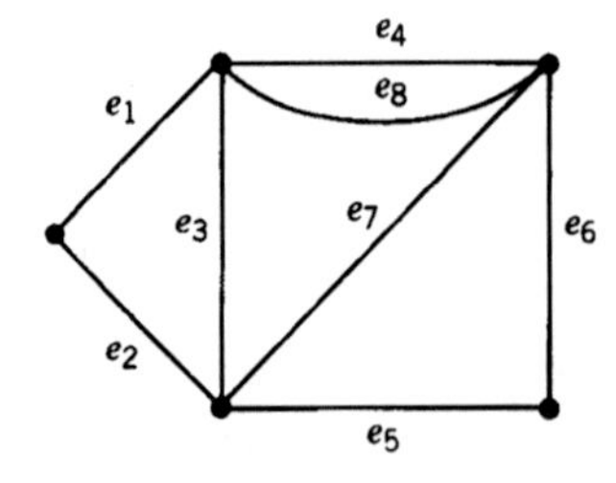

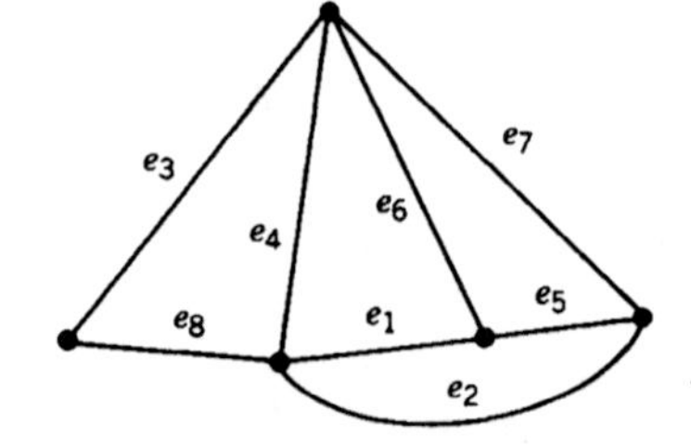

Figure 11.2

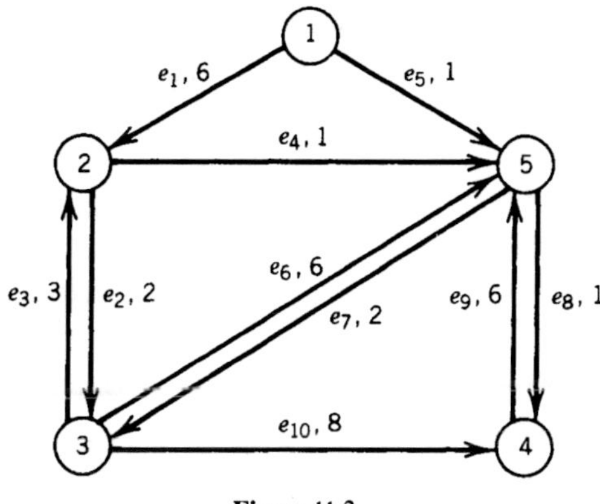

Figure 11.3

15. For the clutter of rooted branchings, prove that

$$Q^* = \left\{x \in R_+^n: \sum_{(i,j)\in\delta^+(S)} x_{ij} \geqslant 1 \text{ for all } S \subseteq V \text{ with } \{1\} \in S\right\}$$

is integral and that the max–min inequality holds strongly where $\delta^+(S)$ are the rooted dicuts (see exercise 34 of Section III.1.8).

16. Show that the matroid intersection polyhedron is box TDI (see exercise 3 of Section III.1.8).

17. Show that if $P = \{x \in R^n: Ax \leqslant b\}$ is box TDI, there exists a 0, 1, −1 matrix A' and a vector b' such that $P = \{x \in R^n: A'x \leqslant b'\}$. *Hint*: Observe that $w \in P$ if and only if

$$\max\{1x' - 1x'': x' \leqslant 0, x'' \geqslant 0, x = w, Ax + Ax' + Ax'' = b\} \geqslant 0.$$

18. Prove Proposition 6.2.

19. Transform the problem

$$\min_{x\in B^4}(6x_1 + x_2 + 4x_3 + 3x_4 - 8x_1x_2 - 2x_1x_3 - x_2x_3 - 2x_3x_4)$$

into a polymatroid separation problem. Write out the polymatroid explicitly.

20. **i)** Show that a polymatroid P_f is integral (has integral extreme points) if f is integer-valued.

ii) Show that if $P \subseteq R_+^n$ is an integral polymatroid, $r_p(a) \in Z_+^1$ for all $a \in Z_+^n$.

21. Let $\tilde{P}(f) = \{x \in R^n: \Sigma_{j\in S}\, x_j \leqslant f(S) \text{ for } S \subseteq N\}$ with $f(\emptyset) = 0$. Prove the following:

i) For any $T \subseteq N$, there exists $x \in \tilde{P}(f)$ such that $\Sigma_{j\in T}\, x_j = f(T)$.

ii) All maximal points y such that $y \leqslant x$, $y \in \tilde{P}(f)$ have the same value of $\Sigma_{j\in N}\, y_j$, denoted by $\tilde{r}_f(x)$.

iii) If f_1 and f_2 are submodular with $f_1(\emptyset) = f_2(\emptyset) = 0$, then

$$\max\left\{\sum_{j\in N} x_j: x \in \tilde{P}(f_1) \cap \tilde{P}(f_2)\right\} = \min_{S\subseteq N}\{f_1(S) + f_2(N \setminus S)\}.$$

22. Solve the problem of exercise 19 by a maximum-flow algorithm.

23. Consider the general problem of minimizing a quadratic boolean function $z = \min_{x \in B^n} f(x)$, where

$$f(x) = \sum_{j \in N} c_j x_j + \sum_{(i,j) \in P \cup Q} q_{ij} x_i x_j,$$

$$P = \{(i,j): 1 \leqslant i < j \leqslant n \text{ and } q_{ij} > 0\},$$

and

$$Q = \{(i,j): 1 \leqslant i < j \leqslant n \text{ and } q_{ij} < 0\}.$$

i) Show that the problem can be reformulated as the mixed-integer program

$$z = \min \sum_{j \in N} c_j x_j + \sum_{(i,j) \in P} q_{ij} y_{ij} \sum_{(i,j) \in Q} q_{ij} y_{ij}$$

$$y_{ij} \leqslant x_i,\ y_{ij} \leqslant x_j \quad \text{for } (i,j) \in Q$$

$$x_i + x_j - 1 \leqslant y_{ij} \quad \text{for } (i,j) \in P$$

$$y \in R_+^{|P|+|Q|}, \qquad x \in B^n.$$

ii) Show that f is submodular if and only if $P = \emptyset$.

iii) Show that the problem matrix is TU when $P = \emptyset$.

24. Let $f(S) = \sum_{T \subseteq S} c_T$. Show that f is submodular if and only if

$$\sum_{T \subseteq S} c_{T \cup \{i,j\}} \leqslant 0 \text{ for all } S \subseteq N \setminus \{i,j\} \text{ and all } i,j \in N.$$

25. Show that if f is cubic [i.e., $f(S) = \sum_{T \subseteq S} c_T$, and $c_T = 0$ for $|T| > 3$]:

i) There is a polynomial algorithm to test if f is submodular.

ii) f is submodular if and only if it can be put in the form (7.1).

26. Show that the recognition problem: "Is the quartic function ($f(S) = \sum_{T \subseteq S} c_T$ with $c_T = 0$ for $|T| > 4$) not submodular?" is $\mathcal{NP}$-complete.

27. Let P be the convex hull of 1-matchings for the complete graph on 5 nodes. Is the point shown in Figure 11.4 in P? If not, find a most violated facet-defining inequality (not by inspection).

28. Consider the min-cut problem $\min_{\emptyset \subset S \subset V} \sum_{e \in \delta(S)} c_e$ on a graph $G = (V, E)$ with $c \in R_+^{|E|}$.

i) Show that if S_1, S_2 are minimum cuts with $S_1 \cap S_2 \neq \emptyset$, $S_1 \cup S_2 \neq V$, then $S_1 \cap S_2$ and $S_1 \cup S_2$ are minimum cuts.

ii) Suppose $|V|$ is even, and there exists a minimum cut S^* with $|S^*|$ even. Show that there exists a solution S^0 to the problem

$$\min_{\emptyset \subset S \subset V} \left\{ \sum_{e \in \delta(S)} c_e: |S| \text{ odd} \right\}$$

such that either $S^0 \subset S^*$ or $S^0 \subset N \setminus S^*$.

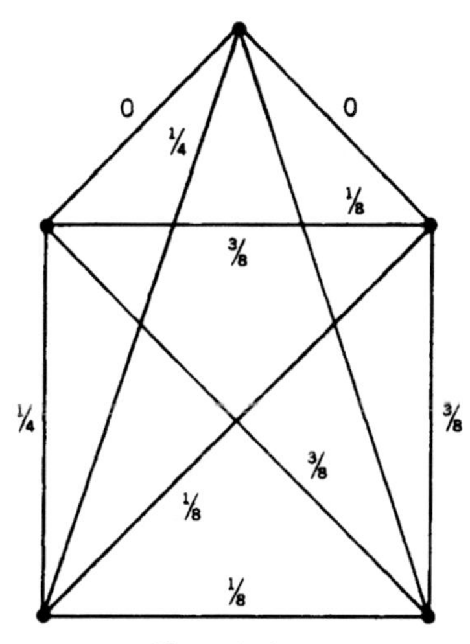

Figure 11.4

29. Describe a polynomial algorithm to (a) compute ζ_w in problem (8.1) and (b) find an optimal solution S and y^* as given in Proposition 8.1.

30. Consider the fractional packing problem given by

$$\eta_w = \{\max\ 1y\colon yA \leqslant w,\ y \in R_+^m\},$$

where A is the basis-element incidence matrix of a matroid M, and also consider the integer packing problem given by

$$\zeta_w = \{\max\ 1y\colon yA \leqslant w,\ y \in Z_+^m\}.$$

i) Express η_w as the minimum of a set of objects.

ii) Give polynomial algorithms to calculate η_1 and ξ_1.

iii) Does the integer round-up property hold, that is $\lceil \eta_w \rceil = \xi_w$?

iv) Apply these results to the graphic matroid of Figure 8.1 with $w = 1$.

31. Solve the max-cut problem $\max_{\emptyset \subset S \subset V} \Sigma_{e \in \delta(S)}\, c_e$ in the graph of Figure 11.5 using formulation $Q(f)$ and a cutting-plane algorithm.

32. **i)** Prove Theorem 9.4.

ii) Deduce that when f is a matroid rank function, $R(f)$ is integral. What are its extreme points (see exercise 10)?

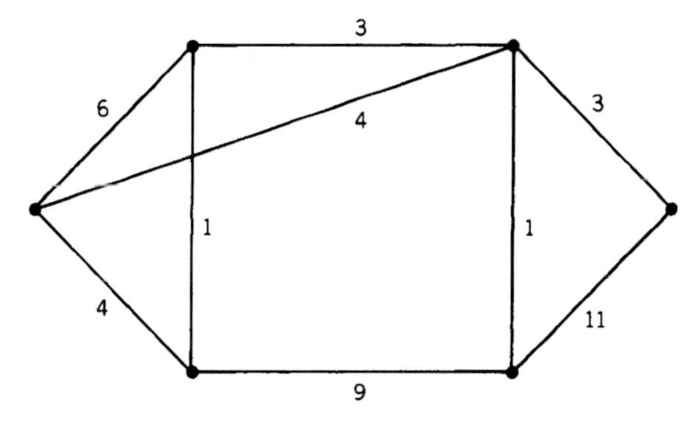

Figure 11.5

References

R. Aboudi and G. L. Nemhauser (1987). A Strong Cutting Plane Algorithm for an Assignment Problem with Side Constraints, Report J-87-3, Industrial and Systems Engineering, Georgia Institute of Technology.

N. Agin (1966). Optimum Seeking with Branch-and-Bound, *Management Science* **13**, B176–B185.

S. Ahn, C. Cooper, G. Cornuejols, and A. M. Frieze (1988). Probabilistic Analysis of a Relaxation for the k-Median Problem, *Mathematics of Operations Research* **13**, 1–31.

A. V. Aho, J. E. Hopcroft, and J. D. Ullman (1974). *The Design and Analysis of Computer Algorithms*, Addison-Wesley.

A. I. Ali and H. Thiagarajan (1986). A Network Based Enumeration Algorithm for Set Partitioning, Department of General Business, University of Texas at Austin.

R. P. Anstee and M. Farber (1984). Characterization of Totally Balanced Matrices, *Journal of Algorithms* **5**, 215–230.

J. P. Arabeyre, J. Fearnley, F. C. Steiger, and W. Teather (1969). The Airline Crew Scheduling Problem: A Survey, *Transportation Science* **3**, 140–163.

J. Araoz (1973). Polyhedral Neopolarities, Doctoral Thesis, University of Waterloo, Waterloo, Ontario.

J. Araoz and E. L. Johnson (1981). Some Results on Polyhedra of Semigroup Problems, *SIAM Journal on Algebraic and Discrete Methods* **3**, 244–258.

D. Avis (1980). A Note on Some Computationally Difficult Set Covering Problems, *Mathematical Programming* **8**, 138–145.

D. Avis (1983). A Survey of Heuristics for the Weighted Matching Problem, *Networks* **13**, 475–494.

D. A. Babayev (1974). Comments on a Note of Frieze, *Mathematical Programming* **7**, 249–252.

A. Bachem and M. Grötschel (1982). New Aspects of Polyhedral Theory, in B. Korte, ed., *Modern Applied Mathematics, Optimization and Operations Research*, North-Holland, pp. 51–106.

A. Bachem, M. Grötschel, and B. Korte, eds. (1983). *Mathematical Programming: The State of the Art*, Springer.

A. Bachem, E. L. Johnson, and R. Schrader (1982). A Characterization of Minimal Valid-Inequalities for Mixed Integer Programs, *Operations Research Letters* **1**, 63–66.

A. Bachem and R. Kannan (1984). Lattices and the Basis Reduction Algorithm. Report CMU-CS-84-112, Department of Computer Science, Carnegie-Mellon University.

A. Bachem and R. Schrader (1980). Minimal Inequalities and Subadditive Duality, *SIAM Journal on Control and Optimization* **18**, 437–443.

E. K. Baker (1981). Efficient Heuristic Algorithms for the Weighted Set Covering Problem, *Computers and Operations Research* **8**, 303–310.

E. K. Baker and M. L. Fisher (1981). Computational Results for Very Large Air Crew Scheduling Problems, *Omega* **9**, 613–618.

E. Balas (1965). An Additive Algorithm for Solving Linear Programs with Zero-One Variables, *Operations Research* **13**, 517–546.

E. Balas (1967). Discrete Programming by the Filter Method, *Operations Research* **15**, 915–957.

E. Balas (1975a). Facets of the Knapsack Polytope, *Mathematical Programming* **8**, 146–164.

E. Balas (1975b). Disjunctive Programming: Cutting Planes from Logical Conditions, in *Nonlinear Programming 2*, O. L. Mangasarian et al., eds., Academic Press, pp. 279–312.

E. Balas (1979). Disjunctive Programming, *Annals of Discrete Mathematics* **5**, 3–51.

E. Balas (1980). Cutting Planes from Conditional Bounds: A New Approach to Set Covering, *Mathematical Programming Study*, **12**, 19–36.

E. Balas and N. Christofides (1981). A Restricted Lagrangean Approach to the Traveling Salesman Problem, *Mathematical Programming* **21**, 19–46.

E. Balas and A. Ho (1980). Set Covering Algorithms Using Cutting Planes, Heuristics, and Subgradient Optimization: A Computational Study, *Mathematical Programming Study* **12**, 37–60.

E. Balas and R. Martin (1980). Pivot and Complement: A Heuristic for 0-1 Programming, *Management Science* **26**, 86–96.

E. Balas and S. M. Ng (1985). On the Set Covering Polytope I: All Facets with Coefficients in {0, 1, 2}, MSSR-522, Graduate School of Industrial Administration, Carnegie–Mellon University.

E. Balas and M. Padberg (1972). On the Set Covering Problem, *Operations Research* **20**, 1152–1161.

E. Balas and M. Padberg (1975). On the Set Covering Problem: II. An Algorithm for Set Partitioning, *Operations Research* **23**, 74–90.

E. Balas and M. Padberg (1976). Set Partitioning: A Survey, *SIAM Review* **18**, 710–760.

E. Balas and W. R. Pulleyblank (1983). The Perfectly Matchable Subgraph Polytope of a Bipartite Graph, *Networks* **13**, 486–516.

E. Balas and M. J. Saltzman (1986). Facets of the Three-Index Assignment Polytope, MSRR-529, Graduate School of Industrial Administration, Carnegie–Mellon University.

E. Balas and P. Toth (1985). Branch and Bound Methods, in Lawler, Lenstra et al., pp. 361–403.

E. Balas and E. Zemel (1978). Facets of the Knapsack Polytope from Minimal Covers, *SIAM Journal on Applied Mathematics* **34**, 119–148.

E. Balas and E. Zemel (1980). An Algorithm for Large Zero–One Knapsack Problems, *Operations Research* **28**, 1130–1145.

E. Balas and E. Zemel (1984). Lifting and Complementing Yields all the Facets of Positive Zero-One Programming Polytopes, in *Mathematical Programming, Proceedings of the International Conference on Mathematical Programming*, R. W. Cottle et al., eds., pp. 13–24.

M. L. Balinski (1965). Integer Programming: Methods, Uses, Computation, *Management Science* **14**, 253–313.

M. L. Balinski (1967). Some General Methods in Integer Programming, in *Nonlinear Programming*, J. Abadie, ed., Wiley, pp. 221–247.

M. L. Balinski (1970a). On Recent Developments in Integer Programming, in *Proceedings of the Princeton Symposium on Mathematical Programming*, H. Kuhn, ed., pp. 267–302.

M. L. Balinski (1970b). On Maximum Matching, Minimum Covering and Their Connections, in *Proceedings of the Princeton Symposium on Mathematical Programming*, H. Kuhn, ed., pp. 303–312.

M. L. Balinski (1972). Establishing the Matching Polytope, *Journal of Combinatorial Theory* **B13**, 1–13.

M. L. Balinski, ed., (1974). *Approaches to Integer Programming*, Mathematical Programming Study 2.

M. L. Balinski and A. J. Hoffman, eds. (1978). *Polyhedral Combinatorics*, Mathematical Programming Study 8.

M. L. Balinski and R. E. Quandt (1964). On an Integer Program for a Delivery Problem, *Operations Research* **12**, 300–304.

M. L. Balinski and K. Spielberg (1969). Methods for Integer Programming: Algebraic, Combinatorial, and Enumerative, in *Progress in Operations Research, Relationship Between Operations and the Computer, Volume III*, J. S. Aranofsky, ed., Wiley, pp. 195–292.

M. O. Ball, L. D. Bodin and R. Dial (1983). A Matching Based Heuristic for Scheduling Mass Transit Crews and Vehicles, *Transportation Science* **17**, 4–31.

M. O. Ball and U. Derigs (1983). An Analysis of Alternative Strategies for Implementing Matching Algorithms, *Networks* **13**, 517–550.

M. Ball and M. Magazine (1981). The Design and Analysis of Heuristics, *Networks* **11**, 215–219.

M. O. Ball and R. Taverna (1985). Sensitivity Analysis for the Matching Problem and Its Use in Solving Matching Problems with a Single-Side Constraint, *Annals of Operations Research* **4**, 25–56.

F. Barahona, M. Grötschel, and A. R. Mahjoub (1985), Facets of the Bipartite Subgraph Polytope, *Mathematics of Operations Research* **10**, 340–358.

F. Barahona and A. R. Mahjoub (1986). On the Cut Polytope, *Mathematical Programming* **36**, 157–173.

I. Barany, J. Edmonds, and L. A. Wolsey (1986). Packing and Covering a Tree by Subtrees, *Combinatorica* **6**, 245–257.

I. Barany, T. J. Van Roy, and L. A. Wolsey (1984). Uncapacitated Lot-Sizing: The Convex Hull of Solutions, *Mathematical Programming Study* 22, 32–43.

R. S. Barr, F. Glover, and D. Klingman (1981). A New Optimization Method for Large Scale Fixed Charge Transportation Problems, *Operations Research* **29**, 448–463.

J. J. Bartholdi III (1981). A Good Submatrix is Hard to Find, *Operations Research Letters* 1, 190–193.

S. Baum and L. E. Trotter, Jr. (1977). Integer Rounding and Polyhedral Decomposition of Totally Unimodular Systems, in *Optimization and Operations Research*, R. Henn, B. Korte, and W. Oettli, eds., Lecture Notes in Economics and Mathematical Systems 157, Springer, pp. 15–23.

S. Baum and L. E. Trotter, Jr. (1981). Integer Rounding for Polymatroid and Branching Optimization Problems, *SIAM Journal on Algebraic and Discrete Methods* **2**, 416–425.

S. Baum and L. E. Trotter, Jr. (1982). Finite Checkability for Integer Rounding Properties in Combinatorial Programming Problems, *Mathematical Programming* **22**, 141–147.

M. S. Bazarra and J. J. Jarvis (1977). *Linear Programming and Network Flows*, Wiley.

E. M. L. Beale (1965). Survey of Integer Programming, *Operational Research Quarterly* **16**, 219–228.

E. M. L. Beale (1968). *Mathematical Programming in Practice*, Wiley.

E. M. L. Beale (1979). Branch and Bound Methods for Mathematical Programming Systems, *Annals of Discrete Mathematics* **5**, 201–219.

E. M. L. Beale (1983). A Mathematical Programming Model for the Long-Term Development of an Off-Shore Gas Field. *Discrete Applied Mathematics* **5**, 1–10.

E. M. L. Beale and J. J. H. Forrest (1976). Global Optimization Using Special Ordered Sets, *Mathematical Programming* **10**, 52–69.

E. M. L. Beale and J. A. Tomlin (1970). Special Facilities in a General Mathematical Programming System for Nonconvex Problems using Ordered Sets of Variables, in *Proceedings of the Fifth International Conference on Operational Research*, J. Lawrence, ed., Tavistock Publications, pp. 447–454.

J. E. Beasley (1984). An Algorithm for the Steiner Problem in Graphs, *Networks* **14**, 147–160.

D. E. Bell (1977). A Theorem Concerning the Integer Lattice. *Studies in Applied Mathematics* **56**, 187–188.

D. E. Bell and M. L. Fisher (1975). Improved Integer Programming Bounds Using Intersections of Corner Polyhedra, *Mathematical Programming* **8**, 345–368.

D. E. Bell and J. F. Shapiro (1977). A Convergent Duality Theory for Integer Programming, *Operations Research* **25**, 419–434.

R. Bellman (1957). *Dynamic Programming*, Princeton University Press.

R. E. Bellman (1958). On a Routing Problem, *Quarterly of Applied Mathematics* **16**, 87–90.

R. E. Bellman and S. E. Dreyfus (1962). *Applied Dynamic Programming*, Princeton University Press.

M. Bellmore and J. F. Malone (1971). Pathology of Traveling Salesman Subtour Elimination Algorithms. *Operations Research* **19**, 278–307.

M. Bellmore and H. D. Ratliff (1971). Set Covering and Involutory Bases, *Management Science* **18**, 194–206.

J. F. Benders (1962). Partitioning Procedures for Solving Mixed Variables Programming Problems, *Numerische Mathematik* **4**, 238–252.

M. Benichou, J. M. Gauthier, P. Girodet, G. Hentges, G. Ribiere and O. Vincent (1971). Experiments in Mixed Integer Linear Programming, *Mathematical Programming* **1**, 76–94.

M. Benichou, J. M. Gauthier, G. Hentges, and G. Ribiere (1977). The Efficient Solution of Large Scale Linear Programming Problems—Some Algorithmic Techniques and Computation Results, *Mathematical Programming* **13**, 280–322.

A. Ben-Israel and A. Charnes (1962). On Some Problems of Diophantine Programming, *Cahiers du Centre d'Etudes de Recherche Operationelle* **4**, 215–280.

C. Berge (1957). Two Theorems in Graph Theory, *Proceedings of the National Academy of Science* **43**, 842–844.

C. Berge (1960). Les Problemes de Colorations en Theorie des Graphes, *Publication of the Institute of Statistics, University of Paris* **9**, 123–160.

C. Berge (1972). Balanced Matrices, *Mathematical Programming* **2**, 19–31.

C. Berge (1973). *Graphs and Hypergraphs*, North-Holland.

C. Berge and V. Chvátal, eds. (1984). *Topics on Perfect Graphs* (*Annals of Discrete Mathematics* **21**).

D. P. Bertsekas (1985). A United Framework for Primal-Dual Methods in Minimum Cost Network Flow Problems, *Mathematical Programming* **32**, 125–145.

O. Bilde and J. Krarup (1977). Sharp Lower Bounds and Efficient Algorithms for the Simple Plant Location Problem, *Annals of Discrete Mathematics* **1**, 79–97.

R. E. Bixby (1982). Matroids and Operations Research, in *Advanced Techniques in the Practice of Operations Research*, H. J. Greenberg, F. H. Murphy, and S. H. Shaw, eds., North–Holland, pp. 333–458.

R. E. Bixby (1984). Recent Algorithms for Two Versions of Graph Realization and Remarks on Applications to Linear Programming, in Pulleyblank, pp. 39–67.

R. E. Bixby and W. H. Cunningham (1980). Converting Linear Programs to Network Problems, *Mathematics of Operations Research* **5**, 321–357.

C. E. Blair (1976). Two Rules for Deducing Valid Inequalities for 0-1 Problems, *SIAM Journal of Applied Mathematics* **31**, 614–617.

C. E. Blair (1978). Minimal Inequalities for Mixed Integer Programs, *Discrete Mathematics* **24**, 147–151.

C. E. Blair and R. G. Jeroslow (1977). The Value Function of a Mixed Integer Program 1, *Discrete Mathematics* **19**, 121–138.

C. E. Blair and R. G. Jeroslow, (1982). The Value Function of a Mixed Integer Program, *Mathematical Programming* **23**, 237–273.

C. E. Blair and R. G. Jeroslow (1984). Constructive Characterizations of the Value Function of a Mixed-Integer Program I, *Discrete Applied Mathematics* **9**, 217–233.

C. E. Blair and R. G. Jeroslow (1985). Constructive Characterizations of the Value Function of a Mixed-Integer Program II, *Discrete Applied Mathematics* **10**, 227–240.

C. E. Blair, R. G. Jeroslow, and J. K. Lowe (1986). Some Results and Experiments in Programming Techniques for Propositional Logic, School of Management, Georgia Institute of Technology.

R. G. Bland (1988). A Class of Production Planning Problems Solvable by Network Flows, to appear in *Operations Research*.

R. G. Bland, D. Goldfarb, and M. J. Todd (1981). The Ellipsoid Method: A Survey, *Operations Research* **29**, 1039–1091.

R. G. Bland, H. C. Huang, and L. E. Trotter (1984). Graphical Properties Related to Minimal Imperfection, *Annals of Discrete Mathematics* **21**, 181–192.

R. G. Bland and D. L. Jensen (1987). On the Computational Behavior of a Polynomial-Time Network Flow Algorithm, School of Operations Research and Industrial Engineering, Cornell University.

F. Bock (1971). An Algorithm to Construct a Minimum Directed Spanning Tree in a Directed Network, in *Developments in Operations Research, Vol. 1*, B. Avi-Itzhak, ed., Gordon and Breach, pp. 29–44.

L. Bodin, B. Golden, A. Assad, and M. Ball (1983). Routing and Scheduling of Vehicles and Crews: The State of the Art, *Computers and Operations Research* **10**, 69–211.

J. A. Bondy and U. S. R. Murty (1976). *Graph Theory with Applications*, Macmillan.

E. Bonomi and J. L. Lutton (1984). The *N*-City Travelling Salesman Problem: Statistical Mechanics and the Metropolis Algorithm, *SIAM Review* **26**, 551–568.

K. H. Borgwardt (1982a). Some Distribution-Independent Results about the Asymptotic Order of the Average Number of Pivot Steps of the Simplex Method, *Mathematics of Operations Research* **7**, 441–462.

K. H. Borgwardt (1982b). The Average Number of Pivot Steps Required by the Simplex-Method Is Polynomial, *Zeitschrift für Operations Research* **26**, 157–177.

I. Borosh and L. L. Treybig (1976). Bounds on Positive Integral Solutions of Linear Diophantine Equations, *Proceedings of the American Mathematical Society* **55**, 299–304.

V. J. Bowman, Jr., and G. L. Nemhauser (1970). A Finiteness Proof for Modified Dantzig Cuts in Integer Programming, *Naval Research Logistics Quarterly* **17**, 309–313.

S. Boyd, W. R. Pulleyblank, and G. Cornuejols (1987). TRAVEL—An Interactive Traveling Salesman Package for the IBM Personal Computer. *Operations Research Letters* **6**, 141–144.

G. H. Bradley, G. G. Brown, and G. W. Graves (1977). Design and Implementation of Large Scale Primal Transshipment Algorithms, *Management Science* **24**, 1–34.

G. H. Bradley, P. L. Hammer, and L. A. Wolsey (1974). Coefficient Reduction for Inequalities in 0-1 Variables, *Mathematical Programming* **7**, 263–282.

A. L. Brearley, G. Mitra, and H. P. Williams (1975). An Analysis of Mathematical Programming Problems Prior to Applying the Simplex Method, *Mathematical Programming* **8**, 54–83.

R. Breu and C. A. Burdet (1974). Branch and Bound Experiments in Zero–One Programming, *Mathematical Programming Study* **2**, 1–50.

C. Brezovec, G. Cornuejols, and F. Glover (1986). Two Algorithms for Weighted Matroid Intersection, *Mathematical Progrmaming* **36**, 39–53.

R. Brooks and A. Geoffrion (1966). Finding Everett's Lagrange Multipliers by Linear Programming, *Operations Research* **14**, 1149–1153.

G. G. Brown and W. Wright (1984). Automatic Identification of Embedded Network Rows in Large Scale Optimization Models, *Mathematical Programming* **29**, 41–46.

C. A. Burdet and E. L. Johnson (1974). A Subadditive Approach to the Group Problem of Integer Programming, *Mathematical Programming* **2**, 51–71.

C. A. Burdet and E. L. Johnson (1977). A Subadditive Approach to Solve Linear Integer Programs, *Annals of Discrete Mathematics* **1**, 117–144.

R. E. Burkhard (1984). Quadratic Assignment Problems, *European Journal of Operations Research* **15**, 283–289.

R. E. Burkhard and V. Derigs (1980). *Assignment and Matching Problems: Solutions Methods with Fortran Programs*, Springer.

M. Burlet and J. Fonlupt (1984). Polynomial Algorithm to Recognize a Meyniel Graph, *Annals of Discrete Mathematics* **21**, 225–252.

A. V. Cabot and S. S. Erenguc (1984). Some Branch-and-Bound Procedures for Fixed-cost Transportation Problems, *Naval Research Logistics Quarterly* **31**, 145–154.

P. M. Camerini, L. Fratta, and F. Maffioli (1975). On Improving Relaxation Methods by Modified Gradient Techniques, *Mathematical Programming* **3**, 26–34.

P. Camion (1965). Characterization of Totally Unimodular Matrices, *Proceedings of the American Mathematical Society* **16**, 1068–1073.

G. Carpaneto and P. Toth (1980). Some New Branching and Bounding Criteria for the Asymmetric Travelling Salesman Problem. *Management Science* **26**, 736–743.

J. W. S. Cassels (1971). *An Introduction to the Theory of Numbers*, Springer.

L. Chalmet and L. F. Gelders (1977). Lagrangean Relaxations for a Generalized Assignment-Type Problem, in *Advances in Operations Research*, M. Roubens, ed., North-Holland, pp. 103–110.

A. K. Chandra, D. S. Hirshberg, and C. K. Wong (1976). Approximate Algorithms for some Generalized Knapsack Problems, *Theoretical Computer Science* **3**, 293–304.

R. Chandrasekaran (1969). Total Unimodularity of Matrices, *SIAM Journal* **11**, 1032–1034.

R. Chandrasekaran (1981). Polynomial Algorithms for Totally Dual Integral Systems and Extensions, *Annals of Discrete Mathematics* **11**, 39–51.

G. J. Chang and G. L. Nemhauser (1984). The k-Domination and k-Stability Problems on Sun-Free Chordal Graphs, *SIAM Journal on Algebraic and Discrete Methods* **5**, 332–345.

G. J. Chang and G. L. Nemhauser (1985). Covering, Packing and Generalized Perfection, *SIAM Journal on Algebraic and Discrete Methods* **6**, 109–132.

A. Charnes and W. W. Cooper (1961). *Management Models and Industrial Application of Linear Programming*, Vols. I and II, Wiley.

D. S. Chen and S. Zionts (1976). Comparison of Some Algorithms for Solving the Group Theoretic Programming Problem, *Operations Research* **24**, 1120–1128.

D. C. Cho, E. L. Johnson, M. W. Padberg, and M. R. Rao (1983), On the Uncapacitated Plant Location Problem I: Valid Inequalities and Facets, *Mathematics of Operations Research* **8**, 579–589.

D. C. Cho, M. W. Padberg, and M. R. Rao (1983), On the Uncapacitated Plant Location Problem II: Facets and Lifting Theorems, *Mathematics of Operations Research* **8**, 590–612.

N. Christofides (1970). The Shortest Hamiltonian Chain of a Graph, *SIAM Journal of Applied Mathematics* **19**, 689–696.

N. Christofides (1975a). *Graph Theory: An Algorithmic Approach*, Academic Press.

N. Christofides (1975b). Worst-Case Analysis of a New Heuristic for the Travelling Salesman Problem, Report 388, Graduate School of Industrial Administration, Carnegie–Mellon University.

N. Christofides (1985a). Vehicle Routing, in Lawler, Lenstra et al., pp. 431–448.

N. Christofides (1985b). Vehicle Routing, in O'hEigeartaigh et al., pp. 148–163.

N. Christofides and S. Korman (1975). A Computational Survey of Methods for the Set Covering Problem, *Management Science* **21**, 591–599.

N. Christofides, A. Mingozzi, P. Toth, and M. Sandi, eds. (1979). *Combinatorial Optimization*, Wiley.

Y. J. Chu and T. H. Liu (1965). On the Shortest Arborescence of Directed Graphs, *Scientia Sinica* **14**, 1390–140.

V. Chvátal (1973a). Edmonds Polytopes and a Hierarchy of Combinatorial Problems, *Discrete Mathematics* **4**, 305–337.

V. Chvátal (1973b). Edmonds Polytopes and Weakly Hamiltonian Graphs, *Mathematical Programming* **5**, 29–40.

V. Chvátal (1975). On Certain Polytopes Associated with Graphs, *Journals of Combinatorial Theory* **B13**, 138–154.

V. Chvátal (1979). A Greedy Heuristic for the Set Covering Problem, *Mathematics of Operations Research* **4**, 233–235.

V. Chvátal (1980). Hard Knapsack Problems, *Operations Research* **28**, 1402–1411.

V. Chvátal (1983). *Linear Programming*, Freeman.

G. Clarke and J. W. Wright (1964). Scheduling of Vehicles from a Central Depot to a Number of Delivery Points, *Operations Research* **12**, 568–581.

E. G. Coffman and R. L. Graham (1972). Optimal Scheduling for Two-Processor Systems, *Acta Informatica* **1**, 200–213.

M. Conforti and G. Cornuejols (1984). Submodular Set Functions, Matroids and the Greedy Algorithm: Tight Worst-Case Bounds and Some Generalizations of the Rado-Edmonds Theorem, *Discrete Applied Mathematics* 7, 251–274.

A. R. Conn and G. Cornuejols (1987). A Projection Method for the Uncapacitated Facility Location Problem, WP No. 26-86-87, Graduate School of Industrial Administration, Carnegie-Mellon University.

S. A. Cook (1971). The Complexity of Theorem-Proving Procedures, *Proceedings of the 3rd Annual ACM Symposium on Theory of Computing Machinery*, pp. 151–158, ACM.

W. Cook (1983a). Operations that Preserve Total Dual Integrality, *Operations Research Letters* **2**, 31–35.

W. Cook (1983b). A Minimal Totally Dual Integral Defining System for the b-Matching Polyhedron, *SIAM Journal on Algebraic and Discrete Methods* **4**, 212–220.

W. Cook (1986). On Box Totally Dual Integral Polyhedra. *Mathematical Programming* **34**, 48–61.

W. Cook, A. M. H. Gerards, A. Schrijver, and E. Tardos (1986). Sensitivity Results in Integer Programming, *Mathematical Programming* **34**, 251–264.

W. Cook, L. Lovasz, and A. Schrijver (1984). A Polynomial-Test for Total Dual Integrality in Fixed Dimension, *Mathematical Programming Study* **22**, 64–69.

W. Cook and W. R. Pulleyblank (1987). Linear Systems for Constrained Matching Problems, *Mathematics of Operations Research* **12**, 97–120.

G. Cornuejols (1986). General Factors of a Graph, Graduate School of Industrial Administration, Carnegie-Mellon University.

G. Cornuejols, M. L. Fisher, and G. L. Nemhauser (1977a). Location of Bank Accounts to Optimize Float: An Analytic Study of Exact and Approximate Algorithms, *Management Science* **23**, 789–810.

G. Cornuejols, M. L. Fisher, and G. L. Nemhauser (1977b). On the Uncapacitated Location Problem, *Annals of Discrete Mathematics* **1**, 163–177.

G. Cornuejols, J. Fonlupt, and D. Naddef (1985). The Traveling Salesman Problem on A Graph and Some Related Integer Polyhedra, *Mathematical Programming* **33**, 1–27.

G. Cornuejols and D. Hartvigsen (1986). An Extension of Matching Theory, *Journal of Combinatorial Theory* **B40**, 285–296.

G. Cornuejols, D. Naddef, and W. R. Pulleyblank (1983). Halin Graphs and the Traveling Salesman Problem, *Mathematical Programming* **26**, 287–294.

G. Cornuejols and G. L. Nemhauser (1978). Tight Bounds for Christofides' Traveling Salesman Heuristic, *Mathematical Programming* **14**, 116–121.

G. Cornuejols, G. L. Nemhauser, and L. A. Wolsey (1980a). A Canonical Representation of Simple Plant Location Problems and its Applications, *SIAM Journal on Algebraic and Discrete Methods* **1**, 261–272.

G. Cornuejols, G. L. Nemhauser and L. A. Wolsey (1980b). Worst Case and Probabilistic Analysis of Algorithms for a Location Problem, *Operations Research* **28**, 847–858.

G. Cornuejols, G. L. Nemhauser, and L. A. Wolsey (1984). The Uncapacitated Facility Location Problem, Report No. 605, Operations Research and Industrial Engineering, Cornell University (to appear in Francis and Mirchandini).

G. Cornuejols and W. R. Pulleyblank (1982). The Travelling Salesman Polytope and {0, 2}-Matchings, *Annals of Discrete Mathematics* **16**, 27–55.

G. Cornuejols and W. R. Pulleyblank (1983). Critical Graphs, Matchings and Tours, or a Hierarchy of Relaxations for the Traveling Salesman Problem, *Combinatorica* **3**, 35–52.

G. Cornuejols and A. Sassano (1986). On the 0,1 Facets of the Set Covering Polytope, Report 153, Instituto di Analisi dei Sistemi ed Informatica del C.N.R., Rome.

G. Cornuejols and J. M. Thizy (1982a). A Primal Approach to the Simple Plant Location Problem, *SIAM Journal on Algebraic and Discrete Methods* **3**, 504–510.

G. Cornuejols and J. M. Thizy (1982b). Some Facets of the Simple Plant Location Polytope, *Mathematical Programming* **23**, 50–74.

Y. Crama (1986). Recognition Problems for Special Classes of Pseudo-Boolean Functions, RUTCOR, Rutgers University.

G. A. Croes (1958). A Method for Solving Traveling Salesman Problems, *Operations Research* **6**, 791–812.

H. P. Crowder and E. L. Johnson (1973). Use of Cyclic Group Methods in Branch and Bound, in *Mathematical Programming*, T. C. Hu and S. M. Robinson, eds., Academic Press, pp. 213–216.

H. P. Crowder, E. L. Johnson, and M. W. Padberg (1983). Solving Large-Scale Zero-One Linear Programming Problems, *Operations Research* **31**, 803–834.

H. P. Crowder and M. W. Padberg (1980). Solving Large-Scale Symmetric Traveling Salesman Problems to Optimality, *Management Science*, **26**, 495–509.

F. H. Cullen, J. J. Jarvis, and H. D. Ratliff (1981). Set Partitioning Heuristics for Interactive Routing, *Networks* **11**, 125–143.

W. H. Cunningham (1977). An Unbounded Matroid Intersection Polyhedron, *Linear Algebra and Its Applications* **16**, 209–215.

W. H. Cunningham (1983). A Class of Linear Programs Convertible to Network Problems, *Operations Research* **32**, 387–391.

W. H. Cunningham (1984). Testing Membership in Matroid Polyhedra, *Journal of Combinatorial Theory* **36B**, 161–188.

W. H. Cunningham (1985). On Submodular Function Minimization, *Combinatorica* **5**, 185–192.

W. H. Cunningham (1986). Improved Bounds for Matroid Partition and Intersection Algorithms, *SIAM Journal on Computing* **15**, 948–957.

W. Cunningham and A. Frank (1985). A Primal–Dual Algorithm for Submodular Flows, *Mathematics of Operations Research* **10**, 251–262.

W. H. Cunningham and A. B. Marsh III (1978). A Primal Algorithm for Optimum Matching, *Mathematical Programming Study* **8**, 50–72.

R. J. Dakin (1965). A Tree Search Algorithm for Mixed Integer Programming Problems, *Computer Journal* **8**, 250–255.

G. B. Dantzig (1957). Discrete-Variable Extremum Problems, *Operations Research* **5**, 266–277.

G. B. Dantzig (1959). Note on Solving Linear Programs in Integers, *Naval Research Logistics Quarterly* **6**, 75–76.

G. B. Dantzig (1960). On the Significance of Solving Linear Programming Problems with Some Integer Variables, *Econometrica* **28**, 30–44.

G. B. Dantzig (1963). *Linear Programming and Extensions*, Princeton University Press.

G. B. Dantzig and D. R. Fulkerson (1956). On the Max-Flow Min-Cut Theorem of Networks, in *Linear Inequalities and Related Systems*, H. W. Kuhn and A. W. Tucker, eds., Princeton University Press, pp. 215–221.

G. B. Dantzig, D. R. Fulkerson, and S. M. Johnson (1954). Solution of a Large-Scale Traveling Salesman Problem, *Operations Research* **2**, 393–410.

G. B. Dantzig, D. R. Fulkerson, and S. M. Johnson (1959). On a Linear-Programming, Combinatorial Approach to the Traveling Salesman Problem, *Operations Research* **7**, 58–66.

G. B. Dantzig, and P. Wolfe (1960). Decomposition Principle for Linear Programs, *Operations Research* **8**, 101–111.

M. Davis and H. Putnam (1960). A Computing Procedure for Quantification Theory, *Journal of the Association for Computing Machinery* **7**, 201–215.

G. de Ghellinck and J. P. Vial (1986). A Polynomial Newton Method for Linear Programming, *Algorithmica* **1**, 425–453.

G. de Ghellinck and J. P. Vial (1987). An Extension of Karmarkar's Algorithm for Solving a System of Linear Homogeneous Equations on the Simplex, *Mathematical Programming* **39**, 79–92.

E. V. Denardo (1982). *Dynamic Programming Models and Applications*, Prentice–Hall.

U. Derigs (1985). Postoptimal Analysis for Matching Problems, *Methods of Operations Research* **49**, 215–221.

U. Derigs (1986). Solving Matching Problems Efficiently: A New Primal Approach, *Networks* **16**, 1–16.

E. W. Dijkstra (1959). A Note on Two Problems in Connection with Graphs, *Numerische Mathematik* **1**, 269–271.

G. A. Dirac (1961). On Rigid Circuit Graphs, *Abh. Math. Sem., Univ. Hamburg* **25**, pp. 71–76.

G. Dobson (1982). Worst-Case Analysis of Greedy Heuristics for Integer Programming with Nonnegative Data, *Mathematics of Operations Research* **7**, 515–531.

P. D. Domich, R. Kannan, and L. E. Trotter (1987). Hermite Normal Form Computation Using Modulo Determinant Arithmetic, *Mathematics of Operations Research* **12**, 50–59.

S. E. Dreyfus and A. M. Law (1977). *The Art and Theory of Dynamic Programming*, Academic Press.

N. J. Driebeek (1966). An Algorithm for the Solution of Mixed Integer Programming Problems, *Management Science* **12**, 576–587.

P. Duchet (1984). Classical Perfect Graph Conjecture on Special Graphs—A Survey, *Annals of Discrete Mathematics* **21**, 67–96.

M. E. Dyer (1984), An $O(n)$ Algorithm for the Multiple Choice Knapsack Linear Program, *Mathematical Programming* **29**, 57–63.

W. L. Eastman (1958). Linear Programming with Pattern Constraints, Ph.D. Thesis, Harvard University.

J. Edmonds (1965a). Paths, Trees and Flowers, *Canadian Journal of Mathematics* **17**, 449–467.

J. Edmonds (1965b). Minimum Partition of a Matroid into Independent Subsets, *Journal of Research of the National Bureau of Standards* **69B**, 67–72.

J. Edmonds (1965c), Maximum Matching and a Polyhedron with 0-1 Vertices, *Journal of Research of the National Bureau of Standards* **69B**, 125–130.

J. Edmonds (1967a). Systems of Distinct Representatives and Linear Algebra, *Journal of Research of the National Bureau of Standards* **71B**, 241–245.

J. Edmonds (1967b). Optimum Branchings, *Journal of Research of the National Bureau of Standards* **71B**, 233–240.

J. Edmonds (1970). Submodular Functions, Matroids and Certain Polyhedra, in *Combinatorial Structures and Their Applications*, R. Guy et al., eds., Gordon and Breach, pp. 69–87.

J. Edmonds (1971). Matroids and the Greedy Algorithm, *Mathematical Programming* **1**, 127–136.

J. Edmonds (1973). Edge-Disjoint Branchings, in *Combinatorial Algorithms*, R. Rustin, ed., Academic Press, pp. 91–96.

J. Edmonds (1979). Matroid Intersection, *Annals of Discrete Mathematics* **4**, 39–49.

J. Edmonds and D. R. Fulkerson (1965). Transversals and Matroid Partition, *Journal of Research of the National Bureau of Standards* **69B**, 147–153.

J. Edmonds and D. R. Fulkerson (1970). Bottleneck Extrema, *Journal of Combinatorial Theory* **8**, 299–306.

J. Edmonds and R. Giles (1977). A Min–Max Relation for Submodular Functions on Graphs, *Annals of Discrete Mathematics* **1**, 185–204.

J. Edmonds and R. Giles (1984). Total Dual Integrality of Linear Inequality Systems, in Pulleyblank, pp. 117–129.

J. Edmonds and E. L. Johnson (1970). Matching: A Well-Solved Class of Integer Linear Programs, in *Proceedings of the Calgary International Conference on Combinatorial Structures and Their Applications*, R. K. Guy et al. eds., Gordon and Breach, pp. 89–92.

J. Edmonds and E. L. Johnson (1973). Matching, Euler Tours and the Chinese Postman, *Mathematical Programming* **5**, 88–124.

J. Edmonds and R. M. Karp (1972). Theoretical Improvements in Algorithmic Efficiency for Network Flow Problems, *Journal of the Association for Computing Machinery* **19**, 248–264.

M. A. Efroymson and T. L. Ray (1966). A Branch-and-Bound Algorithm for Plant Location, *Operations Research* **14**, 361–368.

G. D. Eppen and R. K. Martin (1988). Solving Multi-Item Capacitated Lot-Sizing Problems Using Variable Redefinition, to appear in *Operations Research*.

D. Erlenkotter (1978). A Dual-Based Procedure for Uncapacitated Facility Location, *Operations Research* **26**, 992–1009.

J. Etcheberry (1977). The Set Covering Problem: A New Implicit Enumeration Algorithm, *Operations Research* **25**, 760–772.

S. Even and O. Kariv (1975). An $O(n^{2.5})$ Algorithm for Maximum Matching in General Graphs, *Proceedings of the 16th Annual Symposium on Foundations of Computer Science*, 100–112.

H. Everett III (1963). Generalized Lagrange Multiplier Method for Solving Problems of Optimum Allocation of Resources, *Operations Research* **11**, 399–417.

M. Farber (1983). Characterization of Strongly Chordal Graphs, *Discrete Mathematics* **43**, 173–189.

M. Farber (1984). Domination, Independent Domination and Duality in Strongly Chordal Graphs, *Discrete Applied Mathematics* **7**, 115–130.

D. Fayard and G. Plateau (1975). Resolution of the 0-1 Knapsack Problem: Comparison of Methods, *Mathematical Programming* **3**, 272–307.

D. Fayard and G. Plateau (1982). An Algorithm for the Solution of the 0-1 Knapsack Problem, *Computing* **28**, 269–287.

M. L. Fisher (1973). Optimal Solution of Scheduling Problems Using Lagrange Multipliers: Part I, *Operations Research* **21**, 1114–1127.

M. L. Fisher (1976). A Dual Algorithm for the One-Machine Scheduling Problem, *Mathematical Programming* **11**, 229–251.

M. L. Fisher (1980). Worst-Case Analysis of Heuristic Algorithms, *Management Science* **26**, 1–18.

M. L. Fisher (1981). The Lagrangian Relaxation Method for Solving Integer Programming Problems, *Management Science* **27**, 1–18.

M. L. Fisher (1985). Interactive Optimization, *Annals of Operations Research* **4**, 541–556.

M. L. Fisher, A. Greenfield, R. Jaikumar, and J. T. Lester (1982). A Computerized Vehicle Routing Application, *Interfaces* **12**, 42–52.

M. L. Fisher and D. S. Hochbaum (1980). Probabilistic Analysis of the Planar *K*-Median Problem, *Mathematics of Operations Research* **5**, 27–34.

M. L. Fisher and R. Jaikumar (1981). A Generalized Assignment Heuristic for Vehicle Routing, *Networks* **11**, 109–124.

M. L. Fisher, R. Jaikumar, and L. N. Van Wassenhove (1986). A Multiplier Adjustment Method for the Generalized Assignment Problem, *Management Science* **32**, 1095–1103.

M. L. Fisher, B. J. Lageweg, J. K. Lenstra, and A. H. G. Rinnooy Kan (1983). Surrogate Duality Relaxation for Job Shop Scheduling, *Discrete Applied Mathematics* **5**, 65–76.

M. L. Fisher, G. L. Nemhauser and L. A. Wolsey (1978). Analysis of Approximation Algorithms for Maximizing a Submodular Set Function II, *Mathematical Programming Study* **8**, 73–87.

M. L. Fisher, G. L. Nemhauser, and L. A. Wolsey (1979). An Analysis of Approximations for Finding a Maximum Weight Hamiltonian Circuit, *Operations Research* **27**, 799–809.

M. L. Fisher, W. D. Northup, and J. F. Shapiro (1975). Using Duality to Solve Discrete Optimization Problems: Theory and Computational Experience, *Mathematical Programming Study* **3**, 56–94.

M. L. Fisher and J. F. Shapiro (1974). Constructive Duality in Integer Programming, *SIAM Journal on Applied Mathematics* **27**, 31–52.

M. L. Fisher and L. A. Wolsey (1982). On the Greedy Heuristic for Covering and Packing Problems, *SIAM Journal on Algebraic and Discrete Methods* **3**, 584–591.

L. R. Ford and D. R. Fulkerson (1956). Maximal Flow Through a Network, *Canadian Journal of Mathematics* **8**, 399–404.

L. R. Ford, Jr. and D. R. Fulkerson (1962). *Flows in Networks*, Princeton University Press.

J. J. H. Forrest, J. P. H. Hirst, and J. A. Tomlin (1974). Practical Solution of Large Mixed Integer Programming Problems with UMPIRE, *Management Science* **20**, 736–773.

B. L. Fox, J. K. Lenstra, A. H. G. Rinnooy Kan, and L. E. Schrage (1978). Branching from the Largest Upper Bound: Folklore and Facts, *European Journal of Operations Research* **2**, 191–194.

R. L. Francis and P. Mirchandani, eds. (1988). *Discrete Location Theory*, Wiley.

A. Frank (1975). Some Polynomial Algorithms for Certain Graphs and Hypergraphs, *Proceedings of the 5th British Combinatorial Conference*, pp. 211–226.

A. Frank (1981). A Weighted Matroid Intersection Theorem, *Journal of Algorithms* **2**, 328–336.

A. Frank (1982). An Algorithm for Submodular Functions on Graphs, *Annals of Discrete Mathematics* **16**, 97–120.

A. Frank (1984). Submodular Flows, in Pulleyblank, pp. 147–166.

A. Frank and E. Tardos (1987). An Application of Simultaneous Approximation in Combinatorial Optimization, *Combinatorica* **7**, 49–66.

A. M. Frieze (1974). A Cost Function Property for Plant Location Problems, *Mathematical Programming* **7**, 245–248.

A. M. Frieze (1976). Shortest Path Algorithms for Knapsack Type Problems, *Mathematical Programming* **11**, 150–157.

A. M. Frieze (1986). On the Lagarias–Odlyzko Algorithm for the Subset Sum Problem, *SIAM Journal on Computing* **15**, 536–540.

A. M. Frieze and M. R. B. Clarke (1984). Approximation Algorithm for the m-Dimensional 0-1 Knapsack Problem: Worst Case and Probabilistic Analyses, *European Journal of Operations Research* **15**, 100–109.

A. M. Frieze, G. Galbiati, and F. Maffioli (1982). On the Worst-Case Performance of Some Algorithms for the Asymmetric Traveling Salesman Problem, *Networks* **12**, 23–39.

K. R. Frisch (1955). The Logarithmic Potential Method of Convex Programming, Institute of Economics, University of Oslo.

M. Fujii, T. Kasami, and K. Ninomiya (1969). Optimal Sequencing of Two Equivalent Processors, *SIAM Journal of Applied Mathematics* **17**, 784–789.

S. Fujishige (1986). A Capacity-Rounding Algorithm for the Minimum Cost Circulation Problem: A Dual Framework of the Tardos Algorithm, *Mathematical Programming* **35**, 298–308.

D. R. Fulkerson (1968). Networks, Frames, Blocking Systems, in *Mathematics of the Decision Sciences: Part 1*, G. F. Dantzig and A. F. Veinott, Jr., eds., American Mathematical Society, pp. 303–334.

D. R. Fulkerson (1970a). Blocking Polyhedra, in *Graph Theory and Its Applications*, B. Harris, ed., Academic Press, pp. 93–112.

D. R. Fulkerson (1970b). The Perfect Graph Conjecture and Pluperfect Graph Theorem, in *Proceedings of the Second Chapel Hill Conference on Combinatorial Mathematics and Its Applications*, R. C. Bose et al., eds., University of North Carolina Press, pp. 171–175.

D. R. Fulkerson (1971). Blocking and Antiblocking Pairs of Polyhedra, *Mathematical Programming* **1**, 168–194.

D. R. Fulkerson (1972). Antiblocking Polyhedra, *Journal of Combinatorial Theory* **B12**, 56–71.

D. R. Fulkerson (1973). On the Perfect Graph Theorem, *Mathematical Programming*, T. C. Hu and S. M. Robinson, eds., Academic Press, pp. 69–76.

D. R. Fulkerson (1974). Packing Rooted Directed Cuts in a Weighted Directed Graph, *Mathematical Programming* **6**, 1–13.

D. R. Fulkerson and D. A. Gross (1965). Incidence Matrices and Interval Graphs, *Pacific Journal of Mathematics* **15**, 833–835.

D. R. Fulkerson, A. J. Hoffman, and R. Oppenheim (1974). On Balanced Matrices, *Mathematical Programming Study* **1**, 120–132.

D. R. Fulkerson, G. L. Nemhauser, and L. E. Trotter, Jr. (1974). Two Computationally Difficult Set Covering Problems that Arise in Computing the 1-Width of Incidence Matrices of Steiner Triple Systems, *Mathematical Programming Study* **2**, 72–81.

D. R. Fulkerson and D. B. Weinberger (1975). Blocking Pairs of Polyhedra Arising from Network Flows, *Journal of Combinatorial Theory* **B18**, 265–283.

M. L. Furst and R. Kannan (1987). Succinct Certificates for Almost All Subset Sum Problems, Computer Science Department, Carnegie–Mellon University.

H. P. Gabow, Z. Galil, T. Spencer, and R. E. Tarjan (1986). Efficient Algorithms For Finding Minimum Spanning Trees in Undirected and Directed Graphs, *Combinatorica* **6**, 109–122.

H. P. Gacs and L. Lovasz (1981). Khachiyan's Algorithm for Linear Programming, *Mathematical Programming Study* **14**, 61–68

D. Gale (1968). Optimal Assignments in an Ordered Set: An Application of Matroid Theory, *Journal of Combinatorial Theory* **4**, 176–180.

G. Gallo and S. Pallottino (1986). Shortest Path Methods: A Unifying Approach, *Mathematical Programming Study* **26**, 38–64.

M. R. Garey and D. S. Johnson (1979). *Computers and Intractibility: A Guide to the Theory of $\mathcal{NP}$-Completeness*, Freeman.

R. S. Garfinkel (1973). On Partitioning the Feasible Set in a Branch-and-Bound Algorithm for the Asymmetric Traveling-Salesman Problem, *Operations Research* **21**, 340–343.

R. S. Garfinkel (1979). Branch and Bound Methods for Integer Programming, in Christofides, Mingozzi et al., pp. 1–20.

R. S. Garfinkel, A. W. Neebe, and M. R. Rao (1974). An Algorithm for the M-median Plant Location Problem, *Transportation Science* **8**, 217–236.

R. S. Garfinkel and G. L. Nemhauser (1969). The Set Partitioning Problem: Set Covering with Equality Constraints, *Operations Research* **17**, 848–856.

R. S. Garfinkel and G. L. Nemhauser (1972a). *Integer Programming*, Wiley.

R. S. Garfinkel and G. L. Nemhauser (1972b). Optimal Set Covering: A Survey, in *Perspectives On Optimization: A Collection of Expository Articles*, A. M. Geoffrion, ed., pp. 164–183.

R. S. Garfinkel and G. L. Nemhauser (1973). A Survey of Integer Programming Emphasizing Computation and Relations Among Models, in *Mathematical Programming*, T. C. Hu and S. M. Robinson, eds., Academic Press, pp. 77–155.

S. Gass (1975). *Linear Programming*, 4th ed. McGraw–Hill.

G. Gastou and E. L. Johnson (1986). Binary Group and Chinese Postman Polyhedra, *Mathematical Programming* **34**, 1–33.

J. M. Gauthier and G. Ribiere (1977). Experiments in Mixed-Integer Programming Using Pseudo-Costs, *Mathematical Programming* **12**, 26–47.

F. Gavril (1972). Algorithms for Minimum Coloring, Maximum Clique, Minimum Covering by Cliques, and Maximum Independent Set of a Chordal Graph, *SIAM Journal on Computing* **1**, 183–191.

A. M. Geoffrion (1967). Integer Programming by Implicit Enumeration and Balas Method, *SIAM Review* **9**, 178–190.

A. M. Geoffrion (1969). An Improved Implicit Enumeration Approach for Integer Programming, *Operations Research* **17**, 437–454.

A. M. Geoffrion (1970). Primal Resource-Directive Approaches for Optimizing Nonlinear Decomposable Systems, *Operations Research* **18**, 375–403.

A. M. Geoffrion (1972). Generalized Benders Decomposition, *Journal of Optimization Theory and Applications* **10**, 237–260.

A. M. Geoffrion (1974). Lagrangean Relaxation for Integer Programming, *Mathematical Programming Study* **2**, 82–114.

A. M. Geoffrion (1976). A Guided Tour of Recent Practical Advances in Integer Linear Programming, *Omega* **4**, 49–57.

A. M. Geoffrion and G. Graves (1974). Multicommodity Distribution System Design by Benders Decomposition, *Management Science* **20**, 822–844.

A. M. Geoffrion and R. E. Marsten (1972). Integer Programming Algorithms: A Framework and State-of-the-Art Survey, *Management Science* **18**, 465–491.

A. M. Geoffrion and R. McBride (1978). Lagrangian Relaxation Applied to Capacitated Facility Location Problems, *AIIE Transactions* **10**, 40–47.

A. M. Geoffrion and R. Nauss (1977). Parametric and Postoptimality Analysis in Integer Linear Programming, *Management Science* **18**, 453–466.

A. M. Gerards and A. Schrijver (1986). Matrices with the Edmonds–Johnson Property, *Combinatorica* **6**, 365–379.

A. Ghouila-Houri (1962). Caracterisation des Matrices Totalement Unimodulaires, *C.R. Academy of Sciences of Paris* **254**, 1192–1194.

R. Giles (1978). A Balanced Hypergraph Defined by Certain Subtrees of a Tree, *ARS Combinatoria* **6**, 179–183.

R. Giles and W. R. Pulleyblank (1979). Total Dual Integrality and Integral Polyhedra, *Linear Algebra and Its Applications* **25**, 191–196.

R. Giles and L. E. Trotter, Jr. (1981). On Stable Set Polyhedra for $K_{1,3}$-Free Graphs, *Journal of Combinatorial Theory* **B31**, 313–316.

P. C. Gilmore and R. E. Gomory (1966). The Theory and Computation of Knapsack Functions, *Operations Research* **14**, 1045–1074.

P. C. Gilmore, E. L. Lawler, and D. B. Shmoys (1985). Well-Solved Special Cases, in Lawler, Lenstra et al., pp. 87–144.

F. Glover (1965). A Multiphase-Dual Algorithm for the Zero-One Integer Programming Problem, *Operations Research* **13**, 879–919.

F. Glover (1967). A Pseudo Primal-Dual Integer Programming Algorithm, *Journal of Research of the National Bureau of Standards* **71B**, 187–195.

F. Glover (1968a). A New Foundation for a Simplified Primal Integer Programming Algorithm, *Operations Research* **16**, 727–740.

F. Glover (1968b). Surrogate Constraints, *Operations Research* **16**, 741–749.

F. Glover (1968c). A Note on Integer Programming and Integer Feasibility, *Operations Research* **16**, 1212–1216.

F. Glover (1969). Integer Programming over a Finite Additive Group, *SIAM Journal of Control* **7**, 213–231.

F. Glover (1975). Surrogate Constraint Duality in Mathematical Programming, *Operations Research* **23**, 434–451.

F. Glover (1985). Future Paths for Integer Programming and Links to Artificial Intelligence, Report 85-8. Center for Applied Artificial Intelligence, University of Colorado.

F. Glover, D. Karney, and D. Klingman (1974). Implementation and Computational Comparisons of Primal, Dual and Primal-Dual Computer Codes for Minimum Cost Network Flow Problems, *Networks* **4**, 191–212.

J. L. Goffin (1977). On the Convergence Rates of Subgradient Optimization Methods, *Mathematical Programming* **13**, 329–347.

B. L. Golden and A. A. Assad (1986). Perspectives on Vehicle Routing, Exciting New Developments, *Operations Research* **34**, 803–810.

B. L. Golden, L. D. Boden, T. Doyle, and W. R. Stewart (1980). Approximate Traveling Salesman Algorithms, *Operations Research* **28**, 694–711.

B. L. Golden and W. R. Stewart (1985). Empirical Analysis of Heuristics, in Lawler, Lenstra et al., pp. 207–250.

M. C. Golumbic (1980). *Algorithmic Graph Theory and Perfect Graphs*, Academic Press.

R. E. Gomory (1958). Outline of an Algorithm for Integer Solutions to Linear Programs, *Bulletin of the American Mathematical Society* **64**, 275–278.

R. E. Gomory (1960a). Solving Linear Programming Problems in Integers, in *Combinatorial Analysis*, R. E. Bellman and M. Hall, Jr., eds., American Mathematical Society, pp. 211–216.

R. E. Gomory (1960b). An Algorithm for the Mixed Integer Problem, RM-2597, The Rand Corporation.

R. E. Gomory (1963a). An Algorithm for Integer Solutions to Linear Programs, in *Recent Advances in Mathematical Programming*, R. Graves and P. Wolfe, eds., McGraw-Hill, pp. 269–302.

R. E. Gomory (1963b). An All-Integer Programming Algorithm, in *Industrial Scheduling*, J. F. Muth and G. I. Thompson, eds., Prentice-Hall, pp. 193–206.

R. E. Gomory (1965). On the Relation between Integer and Non-Integer Solutions to Linear Programs, *Proceedings of the National Academy of Science* **53**. 260–265.

R. E. Gomory (1967). Faces of an Integer Polyhedron, *Proceedings of the National Academy of Science* **57**, 16–18.

R. E. Gomory (1969). Some Polyhedra Related to Combinatorial Problems, *Linear Algebra and Its Applications* **2**, 451–558.

R. E. Gomory (1970). Properties of a Class of Integer Polyhedra, in *Integer and Nonlinear Programming*, J. Abadie, ed., North-Holland, pp. 353–365.

R. E. Gomory and A. J. Hoffman (1963). On the Convergence of an Integer-Programming Process, *Naval Research Logistics Quarterly* **10**, 121–123.

R. E. Gomory and T. C. Hu (1961). Multi-Terminal Network Flows, *SIAM Journal* **9**, 551–570.

R. E. Gomory and E. L. Johnson (1972). Some Continuous Functions Related to Corner Polyhedra, *Mathematical Programming* **3**, 23–85.

R. E. Gomory and E. L. Johnson (1973). The Group Problem and Subadditive Functions, in *Mathematical Programming*, T. C. Hu and S. M. Robinson, eds., Academic Press, pp. 157–184.

M. Gondran and M. Minoux (1984). *Graphs and Algorithms*, Wiley–Interscience.

G. A. Gorry, W. D. Northup, and J. F. Shapiro (1973). Computational Experience with a Group Theoretic Integer Programming Algorithm, *Mathematical Programming* **4**, 171–192.

G. A. Gorry and J. F. Shapiro (1971). An Adaptive Group Theoretic Algorithm for Integer Programming Problems, *Management Science* **7**, 285–306.

G. A. Gorry, J. F. Shapiro, and L. A. Wolsey (1972). Relaxation Methods for Pure and Mixed Integer Programming Problems, *Management Science* **18**, 229–239.

R. L. Graham (1966). Bounds for Certain Multiprocessing Anomalies, *Bell System Technical Journal* **45**, 1563–1581.

S. C. Graves (1982). Using Lagrangean Techniques to Solve Hierarchial Production Planning Problems, *Management Science* **28**, 260–274.

P. Gray (1971). Exact Solution of the Fixed-Charge Transportation Problem, *Operations Research* **19**, 1529–1538.

H. Greenberg (1971). *Integer Programming*, Academic Press.

H. J. Greenberg and W. P. Pierskalla (1970). Surrogate Mathematical Programming, *Operations Research* **18**, 924–939.

G. R. Grimmett and W. R. Pulleyblank (1985). Random Near-Regular Graphs and the Node Packing Problem, *Operations Research Letters* **4**, 169–174.

R. C. Grinold (1970). Lagrangean Subgradients, *Management Science* **17**, 185–188.

R. C. Grinold (1972). Steepest Ascent for Large Scale Linear Programs, *SIAM Review* **14**, 447–464.

M. Grötschel (1980a). On the Symmetric Travelling Salesman Problem: Solution of a 120-City Problem, *Mathematical Programming Study* **12**, 61–77.

M. Grötschel (1980b). On the Monotone Symmetric Travelling Salesman Problem: Hypohamiltonian/Hypotraceable Graphs and Facets, *Mathematics of Operations Research* **5**, 285–292.

M. Grötschel (1984). Developments in Combinatorial Optimization, in *Perspectives in Mathematics: Anniversary of Oberwolfach 1984*, W. Jager, J. Moser, and R. Remmert, eds., Birkhauser, pp. 249–294.

M. Grötschel (1985). Polyhedral Combinatorics, in O'hEigeartaigh et al., pp. 1–10.

M. Grötschel and O. Holland (1985). Solving Matching Problems with Linear Programming, *Mathematical Programming* **33**, 243–259.

M. Grötschel, M. Junger, and G. Reinelt (1984). A Cutting Plane Algorithm for the Linear Ordering Problem, *Operations Research* **32**, 1195–1220.

M. Grötschel, M. Junger and G. Reinelt (1985a). On the Acyclic Subgraph Polytope, *Mathematical Programming* **33**, 1–27.

M. Grötschel, M. Junger, and G. Reinelt (1985b). Facets of the Linear Ordering Polytope, *Mathematical Programming* **33**, 43–60.

M. Grötschel, L. Lovasz, and A. Schrijver (1981). The Ellipsoid Method and Its Consequences in Combinatorial Optimization, *Combinatorica* **1**, 169–197.

M. Grötschel, L. Lovasz, and A. Schrijver (1984a). Polynomial Algorithms for Perfect Graphs, *Annals of Discrete Mathematics* **21**, 325–356.

M. Grötschel, L. Lovasz, and A. Schrijver (1984b). Geometric Methods in Combinatorial Optimization, in Pulleyblank (1984), pp. 167–184.

M. Grötschel, L. Lovasz, and A. Schrijver (1984c). Corregendum to Our Paper "The Ellipsoid Method and Its Consequences in Combinatorial Optimization", *Combinatorica* **4**, 291–295.

M. Grötschel, L. Lovasz and A. Schrijver (1988). *Geometric Algorithms and Combinatorial Optimization*, Springer.

M. Grötschel and M. W. Padberg (1975). Partial Linear Characterizations of the Asymmetric Traveling Salesman Polytope, *Mathematical Programming* **8**, 378–381.

M. Grötschel and M. W. Padberg (1979a). On the Symmetric Travelling Salesman Problem I: Inequalities, *Mathematical Programming* **16**, 265–280.

M. Grötschel and M. W. Padberg (1979b). On the Symmetric Travelling Salesman Problem II: Lifting Theorems and Facets, *Mathematical Programming* **16**, 281–302.

M. Grötschel and M. W. Padberg (1985). Polyhedral Theory, in Lawler, Lenstra et al., pp. 251–302.

M. Grötschel and W. R. Pulleyblank (1986). Clique Tree Inequalities and the Symmetric Travelling Salesman Problem, *Mathematics of Operations Research* **11**, 537–569.

M. Grötschel and Y. Wakabayashi (1981a). On the Structure of the Monotone Asymmetric Travelling Salesman Polytope I: Hypohamiltonian Facets, *Discrete Mathematics* **34**, 43–59.

M. Grötschel and Y. Wakabayashi (1981b). On the Structure of the Asymmetric Travelling Salesman Polytope II: Hypotraceable Facets, *Mathematical Programming Study* **14**, 77–97.

B. Grunbaum (1967). *Convex Polytopes*, Wiley.

M. Guignard (1980). Fractional Vertices, Cuts and Facets of the Simple Plant Location Problem, *Mathematical Programming* **12**, 150–162.

M. Guignard (1982). Preprocessing and Optimization in Network Flow Problems with Fixed Charges, *Methods of Operations Research* **45**, 235–256.

M. Guignard and K. Spielberg (1977). Reduction Methods for State Enumeration Integer Programming, *Annals of Discrete Mathematics* **1**, 273–286.

M. Guignard and K. Spielberg (1981), Logical Reduction Methods in Zero–One Programming (Minimal Preferred Variables), *Operations Research* **29**, 49–74.

G. Gunawardane, S. Hoff, and L. Schrage (1981). Identification of Special Structure Constraints in Linear Programs, *Mathematical Programming* **21**, 90–97.

G. Hadley (1962), *Linear Programming*, Addison–Wesley.

B. Hajek (1985), A Tutorial Survey of Theory and Applications of Simulated Annealing, *Proceedings of the 24th IEEE Conference on Decision and Control*, 755–760.

P. R. Halmos (1959). *Finite Dimensional Vector Spaces*, Van Nostrand.

J. H. Halton and R. Terada (1982). A Fast Algorithm for the Euclidean Traveling Salesman Problem, Optimal with Probability One, *SIAM Journal on Computing* **11**, 28–46.

P. L. Hammer, E. L. Johnson, and B. Korte, eds. (1979a). *Discrete Optimization I* (*Annals of Discrete Mathematics* **4**).

P. L. Hammer, E. L. Johnson, and B. Korte, eds. (1979b). *Discrete Optimization II* (*Annals of Discrete Mathematics* **5**).

P. L. Hammer, E. L. Johnson, B. Korte, and G. L. Nemhauser, eds. (1977). *Studies in Integer Programming* (*Annals of Discrete Mathematics* **1**).

P. L. Hammer, E. L. Johnson, and U. N. Peled (1975). Facets of Regular 0-1 Polytopes, *Mathematical Programming* **8**, 179–206.

P. L. Hammer and S. Rudeanu (1966). *Boolean Methods in Operations Research and Related Areas*, Springer.

P. Hansen, ed. (1981), *Studies on Graphs and Discrete Programming* (*Annals of Discrete Mathematics* **11**).

R. Hassin (1982). Minimum Cost Flow with Set Constraints, *Networks* **12**, 1–21.

D. Hausmann, ed., (1978). *Integer Programming and Related Areas: A Classified Bibliography 1976–1978*, Springer.

D. Hausmann, T. A. Jenkins, and B. Korte (1980). Worst Case Analysis of Greedy Algorithms for Independence Systems, *Mathematical Programming Study* **12**, 120–131.

D. Hausmann and B. Korte (1978). Lower Bounds on the Worst Case Complexity of Some Oracle Algorithms, *Discrete Mathematics* **24**, 261–276.

K. Helbig-Hansen and J. Krarup (1974). Improvements of the Held–Karp Algorithm for the Symmetric Traveling Salesman Problem, *Mathematical Programming* **7**, 87–96.

M. Held and R. M. Karp (1970). The Traveling Salesman Problem and Minimum Spanning Trees, *Operations Research* **18**, 1138–1162.

M. Held and R. M. Karp (1971). The Traveling Salesman Problem and Minimal Spanning Trees: Part II, *Mathematical Programming* **1**, 6–25.

M. Held, P. Wolfe, and H. P. Crowder (1974). Validation of Subgradient Optimization, *Mathematical Programming* **6**, 62–88.

I. Heller (1957). On Linear Systems with Integral Valued Solutions, *Pacific Journal of Mathematics* **7**, 1351–1364.

I. Heller (1963). On Unimodular Sets of Vectors, in *Recent Advances in Mathematical Programming*, R. L. Graves and P. Wolfe, eds., McGraw–Hill, pp. 39–53.

I. Heller and A. J. Hoffman. (1962). On Unimodular Matrices, *Pacific Journal of Mathematics* **12**, 1321–1327.

I. Heller and C. B. Tompkins (1956). An Extension of a Theorem of Dantzig, in *Linear Inequalities and Related Systems*, H. W. Kuhn and A. W. Tucker, eds., Princeton University Press, pp. 247–254.

D. S. Hochbaum (1982). Approximation Algorithms for the Set Covering and Vertex Cover Problems, *SIAM Journal on Computing* **11**, 555–556.

D. S. Hochbaum, T. Nishizeki, and D. B. Shmoys (1986). A Better than "Best Possible" Algorithm to Edge Color Multigraphs, *Journal of Algorithms* **7**, 79–104.

A. J. Hoffman (1974). A Generalization of Max–Flow Min–Cut Theorem, *Mathematical Programming* **6**, 352–359.

A. J. Hoffman (1979). The Role of Unimodularity in Applying Linear Inequalities to Combinatorial Theorems, *Annals of Discrete Mathematics* **4**, 73–84.

A. J. Hoffman, A. Kolen, and M. Sakarovitch (1985). Totally Balanced and Greedy Matrices, *SIAM Journal on Algebraic and Discrete Methods* **6**, 721–730.

A. J. Hoffman and J. B. Kruskal (1956). Integral Boundary Points of Convex Polyhedra, in *Linear Inequalities and Related Systems*, H. W. Kuhn and A. W. Tucker, eds., Princeton University Press, pp. 223–246.

A. J. Hoffman and R. Oppenheim (1978). Local Unimodularity in the Matching Polytope, *Annals of Discrete Mathematics* **2**, 201–209.

K. Hoffman and M. Padberg (1985). LP-based Combinatorial Problem Solving, *Annals of Operations Research* **4**, 145–194.

S. Holm and D. Klein (1984). Three Methods for Postoptimal Analysis in Integer Linear Programming, *Mathematical Programming Study* **21**, 97–109.

S. Holm and J. Tind (1985). Decomposition in Integer Programming by Superadditive Functions, WP 32-85-86, Graduate School of Industrial Administration, Carnegie-Mellon University.

I. Holyer (1981). The $\mathcal{NP}$-Completeness of Edge Coloring, *SIAM Journal on Computing* **10**, 718–720.

J. Hopcroft and R. M. Karp (1973). An $n^{2.5}$ Algorithm for Maximum Matching in Bipartite Graphs, *SIAM Journal on Computing* **2**, 223–231.

W. L. Hsu (1981). How To Color Claw-Free Perfect Graphs, *Annals of Discrete Mathematics* **11**, 189–197.

W. L. Hsu (1984). The Perfect Graph Conjecture on Special Graphs—A Survey, *Annals of Discrete Mathematics* **21**, 103–114.

W. L. Hsu and G. L. Nemhauser (1981). Algorithms for Minimum Coverings by Cliques and Maximum Cliques in Claw-Free Perfect Graphs, *Discrete Mathematics* **37**, 181–191.

W. L. Hsu and G. L. Nemhauser (1982). A Polynomial Algorithm for the Minimum Weighted Clique Cover Problem on Claw-Free Perfect Graphs, *Discrete Mathematics* **38**, 65–71.

W. L. Hsu and G. L. Nemhauser (1984). Algorithms for Maximum Weight Cliques, Minimum Weighted Clique Covers and Minimum Colorings of Claw-Free Perfect Graphs, *Annals of Discrete Mathematics* **21**, 357–369.

T. C. Hu (1969). *Integer Programming and Network Flows*, Addison-Wesley.

T. C. Hu (1970). On the Asymptotic Integer Algorithm, *Linear Algebra and Its Applications* **3**, 279–294.

H. C. Huang and L. E. Trotter, Jr. (1980). A Technique for Determining Blocking and Antiblocking Polyhedral Descriptions, *Mathematical Programming Study* **12**, 197–205.

P. Huard (1967). Resolution of Mathematical Programming with Nonlinear Constraints by the Method of Centers, in *Nonlinear Programming*, J. Abadie, ed., North-Holland, pp. 209–219.

T. Ibaraki (1976). Theoretical Comparisons of Search Strategies in Branch-and-Bound Algorithms, *Journal of Computer and Information Science* **5**, 315–344.

T. Ibaraki (1977). Power of Dominance Relations in Branch-and-Bound Algorithms, *Journal of the Association for Computing Machinery* **24**, 264–279.

O. H. Ibarra and C. E. Kim (1975). Fast Approximation Algorithms for the Knapsack and Sum of Subset Problems, *Journal of the Association for Computing Machinery* **22**, 463–468.

Y. Ikura and G. L. Nemhauser (1985). Simplex Pivots on the Set Packing Polytope, *Mathematical Programming* **33**, 123–138.

Y. Ikura and G. L. Nemhauser (1986). Computational Experience with a Polynomial-Time Dual Simplex Algorithm for the Transportation Problem, *Discrete Applied Mathematics* **13**, 232–248.

J. P. Ingargiola and J. F. Korsch (1973). A Reduction Algorithm for Zero-One Single Knapsack Problems, *Management Science* **20**, 460–463.

M. Iri (1966). A Criterion for the Reducibility of a Linear Programming Problem to a Network Flow Problem, *RAAG Research Notes*, Third Series, No. 98.

M. Iri (1983). Applications of Matroid Theory, in Bachem, Grötschel and Korte, pp. 158–201.

T. A. Jenkins (1976). The Efficacy of the Greedy Algorithm, *Proceedings of the 7th Southeastern Conference on Combinatories, Graph Theory and Computing*, F. Hoffman et al., eds., Utilitas Mathematica, pp. 341–350.

P. A. Jensen and J. W. Barnes (1980). *Network Flow Programming*, Wiley.

R. G. Jeroslow (1971). Comments on Integer Hulls of Two Linear Constraints, *Operations Research* **19**, 1061–1069.

R. G. Jeroslow (1972). There Cannot be Any Algorithm for Integer Programming with Quadratic Constraints, *Operations Research* **21**, 221–224.

R. G. Jeroslow (1974). Trivial Integer Programs Unsolvable by Branch-and-Bound, *Mathematical Programming* **6**, 105–109.

R. G. Jeroslow (1977). Cutting Plane Theory: Disjunctive Methods, *Annals of Discrete Mathematics* **1**, 293–330.

R. G. Jeroslow (1978). Cutting Plane Theory: Algebraic Methods, *Discrete Mathematics* **23**, 121–150.

R. G. Jeroslow (1979a). An Introduction to the Theory of Cutting Planes, *Annals of Discrete Mathematics* **5**, 71–95.

R. G. Jeroslow (1979b). Minimal Inequalities, *Mathematical Programming* **17**, 1–15.

R. G. Jeroslow (1979c). The Theory of Cutting-Planes, in Christofides, Mingozzi et al., pp. 21–72.

R. G. Jeroslow (1985). Representability in Mixed Integer Programming II: A Lattice of Relaxations, College of Management, Georgia Institute of Technology.

R. G. Jeroslow and K. O. Kortanek (1971). On an Algorithm of Gomory, *SIAM Journal* **21**, 55–60.

R. G. Jeroslow and J. K. Lowe (1984). Modelling with Integer Variables, *Mathematical Programming Study* **22**, 167–184.

D. S. Johnson (1974). Approximation Algorithms for Combinatorial Problems, *Journal of Computer and System Science* **9**, 256–278.

D. S. Johnson, A. Demers, J. D. Ullman, M. R. Garey, and R. L. Graham (1974). Worst-Case Performance Bounds for Simple One-Dimensional Packing Problems, *SIAM Journal on Computing*, **3**, 299–325.

D. S. Johnson and C. H. Papadimitriou (1985a). Computational Complexity, in Lawler, Lenstra et al., pp. 37–86.

D. S. Johnson and C. H. Papadimitriou (1985b). Performance Guarantees for Heuristics, in Lawler, Lenstra et al., pp. 87–144.

E. L. Johnson (1973). Cyclic Groups, Cutting Planes and Shortest Paths, in *Mathematical Programming*, T. C. Hu and S. Robinson, eds., Academic Press, pp. 185–211.

E. L. Johnson (1974). On the Group Problem for Mixed Integer Programming, *Mathematical Programming Study* **2**, 137–179.

E. L. Johnson (1978). Support Functions, Blocking Pairs, and Anti-Blocking Pairs, *Mathematical Programming* **8**, 167–196.

E. L. Johnson (1979), On the Group Problem and a Subadditive Approach to Integer Programming, *Annals of Discrete Mathematics* **5**, 97–112.

E. L. Johnson, (1980a). *Integer Programming—Facets, Subadditivity, and Duality for Group and Semi-Group Problems*, SIAM Publications.

E. L. Johnson (1980b). Subadditive Lifting Methods for Partitioning and Knapsack Problems, *Journal of Algorithms* **1**, 75–96.

E. L. Johnson (1981a). On the Generality of the Subadditive Characterization of Facets, *Mathematics of Operations Research* **6**, 101–112.

E. L. Johnson (1981b). Characterization of Facets for Multiple Right-Hand Choice Linear Programs, *Mathematical Programming Study* **14**, 112–142.

E. L. Johnson, M. M. Kostreva, and U. H. Suhl (1985), Solving 0-1 Integer Programming Problems Arising from Large Scale Planning Models, *Operations Research* **33**, 803–819.

E. L. Johnson and S. Mosterts (1987). On Four Problems in Graph Theory, *SIAM Journal on Algebraic and Discrete Methods* **8**, 163–185.

E. L. Johnson and M. W. Padberg (1981). A Note on the Knapsack Problem with Special Ordered Sets, *Operations Research Letters* **1**, 18–22.

E. L. Johnson and M. W. Padberg (1983). Degree-two Inequalities, Clique Facets, and Bipartite Graphs, *Annals of Discrete Mathematics* **16**, 169–188.

E. L. Johnson and U. H. Suhl (1980). Experiments in Integer Programming, *Discrete Applied Mathematics* **2**, 39–55.

R. Jonker, G. Deleve, J. Vandervelde, and T. Volgenant (1980). Rounding Symmetric Traveling Salesman Problems with an Asymmetric Assignment Problem, *Operations Research* **28**, 623–627.

R. Jonker, R. Kaas, and T. Volgenant (1980). Data-dependent Bounds for Heuristics to find a Minimum Weight Hamiltonian Circuit, *Operations Research* **28**, 1219–1221.

K. O. Jornsten and M. Nasberg (1986). A New Lagrangian Relaxation Approach to the Generalized Assignment Problem, *European Journal of Operations Research* **27**, 313–323.

R. Kannan (1980). A Polynomial Algorithm for the Two Variable Integer Programming Problem, *Journal of the Association for Computing Machinery* **27**, 118–122.

R. Kannan (1983). Improved Algorithms for Integer Programming and Related Lattice Problems, *Proceedings of the 1983 Symposium on the Theory of Computing*, 193–206.

R. Kannan (1987a). Minkowski's Convex Body Theorem and Integer Programming, *Mathematics of Operations Research* **12**, 415–440.

R. Kannan (1987b). Algebraic Geometry of Numbers, Report 87453-OR, Institute for Econometrics and Operations Research, University of Bonn.

R. Kannan and A. Bachem (1979). Polynomial Algorithms for Computing the Smith and Hermite Normal Forms of an Integer Matrix, *SIAM Journal on Computing* **8**, 499–507.

R. Kannan and C. L. Monma (1978). On the Computational Complexity of Integer Programming Problems, in *Optimization and Operations Research*, Lecture Notes in Economics and Mathematical Systems 157, Springer, pp. 161–172.

N. Karmarkar (1984). A New Polynomial Time Algorithm for Linear Programming, *Combinatorica* **4**, 375–395.

R. M. Karp (1971). A Simple Derivation of Edmonds Algorithm for Optimum Branchings, *Networks* **1**, 265–272.

R. M. Karp (1972). Reducibility among Combinatorial Problems, in *Complexity of Computer Computations*, R. E. Miller and J. W. Thatcher, eds., Plenum Press, pp. 85–103.

R. M. Karp (1975). On the Complexity of Combinatorial Problems, *Networks* **5**, 45–68.

R. M. Karp (1976). The Probabilistic Analysis of Some Combinatorial Search Algorithms, in *Algorithms and Complexity: New Directions and Recent Results*, J. F. Traub, ed., Academic Press, pp. 1–19.

R. M. Karp (1977). Probabilistic Analysis of Partitioning Algorithms for the Traveling Salesman Problem in the Plane, *Mathematics of Operations Research* **2**, 209–224.

R. M. Karp (1979). A Patching Algorithm for the Nonsymmetric Traveling Salesman Problem, *SIAM Journal on Computing* **8**, 561–573.

R. M. Karp, J. K. Lenstra, C. J. H. McDiarmid, and A. H. G. Rinnooy Kan (1985). Probabilistic Analysis, in O'hEigeartaigh et al., pp. 52–88.

R. M. Karp and C. H. Papadimitriou (1982). On Linear Characterizations of Combinatorial Optimization Problems, *SIAM Journal on Computing* **11**, 620–632.

R. M. Karp and J. M. Steele (1985). Probabilistic Analysis of Heuristics, in Lawler, Lenstra et al., pp. 207–250.

M. H. Karwan and R. L. Rardin (1979). Some Relationships Between Lagrangian and Surrogate Duality in Integer Programming, *Mathematical Programming* **17**, 320–324.

C. Kastning, ed. (1976). *Integer Programming and Related Areas, A Classified Bibliography*, Lecture Notes in Economics and Mathematical Systems 128, Springer.

J. Kennington and R. V. Helgason (1980). *Algorithms for Network Programming*, Wiley.

L. G. Khachian (1979). A Polynomial Algorithm in Linear Programming, *Soviet Mathematics Doklady* **20**, 191–194.

A. Khintchine (1930). *Continued Fractions*, Noordhoff (1963), English translation.

G. A. P. Kindervater and J. K. Lenstra (1985). Parallel Algorithms, in O'hEigeartaigh et al., pp. 106–128.

G. A. P. Kindervater and J. K. Lenstra (1986). An Introduction to Parallelism in Combinatorial Optimization, *Discrete Applied Mathematics* **14**, 135–156.

S. Kirkpatrick (1984). Optimization by Simulated Annealing: Quantitative Studies, *Journal of Statistical Physics* **34**, 975–986.

S. Kirkpatrick, C. D. Gelatt, Jr., and M. P. Vecchi (1983). Optimization by Simulated Annealing, *Science* **220**, 671–680.

V. Klee (1980). Combinatorial Optimization: What Is the State of the Art? *Mathematics of Operations Research* **5**, 1–26.

V. Klee and G. J. Minty (1972). How Good is the Simplex Algorithm?, in *Inequalities III*, O. Shisha, ed., Academic Press, pp. 159–175.

D. E. Knuth (1979). *The Art of Computer Programming, Vol 1: Fundamental Algorithms*, 2nd ed., Addison-Wesley.

D. E. Knuth (1981). *The Art of Computer Programming, Vol. 2: Seminumerical Algorithms*, 2nd ed., Addison-Wesley.

A. Kolen (1983). Solving Covering Problems and the Uncapacitated Plant Location Problem on Trees, *European Journal of Operations Research* **12**, 266–278.

A. Kolen, A. H. G. Rinnooy Kan, and H. Trienekeus (1987). Vehicle Routing with Time Windows, *Operations Research* **35**, 266–273.

P. J. Kolesar (1967). A Branch and Bound Algorithm for the Knapsack Problem, *Management Science* **13**, 723–735.

B. Korte (1979). Approximative Algorithms for Discrete Optimization Problems, *Annals of Discrete Mathematics* **4**, 85–120.

B. Korte and D. Hausmann (1976). An Analysis of the Greedy Heuristic for Independence Systems, *Annals of Discrete Mathematics* **2**, 65–74.

B. Korte and L. Lovasz (1984). Greedoids—A Structural Framework for the Greedy Algorithm, in Pulleyblank, pp. 221–244.

B. Korte and R. Schrader (1980). On the Existence of Fast Approximation Schemes, in *Nonlinear Programming 4*, O. L. Mangasarian, R. R. Meyer, and S. M. Robinson, eds., Academic Press, pp. 415–437.

J. Krarup and O. Bilde (1977). Plant Location, Set Covering, and Economic Lot-Size: An $O(mn)$ Algorithm for Structured Problems, in *Numerische Methoden bei Optimierungsaufgaben, Band 3: Optimierung bei Graphentheoretischen und Ganzzahlligen Problemen*, Birkhauser, pp. 155–186.

J. Krarup and P. M. Pruzan (1983). The Simple Plant Location Problem: Survey and Synthesis, *European Journal of Operations Research* **12**, 36–81.

J. B. Kruskal (1956). On the Shortest Spanning Subtree of a Graph and the Traveling Salesman Problem, *Proceedings of the American Mathematical Society* **7**, 48–50.

A. A. Kuehn and M. J. Hamburger (1963). A Heuristic Program for Locating Warehouses, *Management Science* **9**, 643–666.

J. C. Lagarias (1985). The Computational Complexity of Simultaneous Diophantine Approximation Problems, *SIAM Journal on Computing* **14**, 196–209.

J. C. Lagarias and A. M. Odlyzko (1985). Solving Low-density Subset Sum Problems, *Journal of the Association for Computing Machinery* **32**, 229–246.

A. H. Land and A. G. Doig (1960). An Automatic Method for Solving Discrete Programming Problems, *Econometrica* **28**, 497–520.

A. H. Land and S. Powell (1979). Computer Codes for Problems of Integer Programming, *Annals of Discrete Mathematics* **5**, 221–269.

G. Laporte, Y. Norbet, and M. Desrochers (1985). Optimal Routing under Capacity and Distance Restrictions, *Operations Research* **33**, 1050–1073.

M. Lauriere (1978). An Algorithm for the 0/1 Knapsack Problem, *Mathematical Programming* **14**, 1–10.

E. L. Lawler (1972). A Procedure for Computing the K Best Solutions to Discrete Optimization Problems and Its Application to the Shortest Path Problem, *Management Science* **18**, 401–405.

E. L. Lawler (1975). Matroid Intersection Algorithms, *Mathematical Programming* **9**, 31–56.

E. L. Lawler (1976). *Combinatorial Optimization: Networks and Matroids*, Holt, Rinehart and Winston.

E. L. Lawler (1979). Fast Approximation Algorithms for Knapsack Problems, *Mathematics of Operations Research* **4**, 339–356.

E. L. Lawler (1980). The Great Mathematical Sputnik of 1979, *The Sciences*, September Issue.

E. L. Lawler (1985). Submodular Functions and Polymatroid Optimization, in O'hEigeartaigh et al., pp. 32–38.

E. L. Lawler, J. K. Lenstra, A. H. G. Rinnooy Kan, and D. B. Shmoys (1985) eds., *The Traveling Salesman Problem: A Guided Tour of Combinatorial Optimization*, Wiley.

E. L. Lawler and C. U. Martel (1982a). Computing Maximal Polymatroidal Network Flows, *Mathematics of Operations Research* **7**, 334–347.

E. L. Lawler and C. U. Martel (1982b). Flow Network Formulations of Polymatroid Optimization Problems, *Annals of Discrete Mathematics* **16**, 189–200.

E. L. Lawler and D. E. Wood (1966). Branch-and-Bound Methods: A Survey, *Operations Research* **14**, 699–719.

A. Lehman (1979). On the Width-Length Inequality, *Mathematical Programming* **16**, 245–259; **17**, 403–417 (with proof corrections).

C. E. Lemke, H. M. Salkin, and K. Spielberg (1971). Set Covering by Single-Branch Enumeration with Linear Programming Subproblems, *Operations Research* **19**, 998–1022.

C. E. Lemke and K. Spielberg (1967). Directed Search Algorithms for Zero-One and Mixed-Integer Programming, *Operations Research* **15**, 892–914.

A. K. Lenstra, H. W. Lenstra, Jr., and L. Lovasz (1982). Factoring Polynomials with Rational Coefficients, *Mathematics Annals* **261**, 515–534.

H. W. Lenstra, Jr. (1983). Integer Programming with a Fixed Number of Variables, *Mathematics of Operations Research* **8**, 538–547.

H. W. Lenstra Jr. (1984). Integer Programming and Cryptography, *Mathematical Intelligencer* **6**, 14–21.

J. K. Lenstra and A. H. G. Rinnooy Kan (1979). Computational Complexity of Discrete Optimization Problems, *Annals of Discrete Mathematics* **4**, 121–140.

J. Leung and T. L. Magnanti (1986). Valid Inequalities and Facets of the Capacitated Plant Location Problem, Working Paper OR149-86, Operations Research Center, Massachusetts Institute of Technology.

H. R. Lewis and C. H. Papadimitriou (1981). *Elements of the Theory of Computation*, Prentice-Hall.

S. Lin (1965). Computer Solutions of the Traveling Salesman Problem, *Bell System Technical Journal* **44**, 2245–2269.

S. Lin (1975). Heuristic Programming as an Aid to Network Design, *Networks* **5**, 33–43.

S. Lin and B. W. Kernighan (1973). An Effective Heuristic Algorithm for the Traveling Salesman Problem, *Operations Research* **21**, 498–516.

J. D. C. Little, K. G. Murty, D. W. Sweeney, and C. Karel (1963). An Algorithm for the Traveling Salesman Problem, *Operations Research* **11**, 972–989.

J. Lorie and L. J. Savage (1955). Three Problems in Capital Rationing, *Journal of Business* **28**, 229–239.

R. Loulou and E. Michaelides (1979). New Greedy-like Heuristics for the Multidimensional 0-1 Knapsack Problem, *Operations Research* **27**, 1101–1114.

L. Lovasz (1972). Normal Hypergraphs and the Perfect Graph Conjecture, *Discrete Mathematics* **2**, 253–267.

L. Lovasz (1975). On the Ratio of Optimal Integral and Fractional Covers, *Discrete Mathematics* **13**, 383–390.

L. Lovasz (1979a). Graph Theory and Integer Programming, *Annals of Discrete Mathematics* **4**, 141–158.

L. Lovasz (1979b). *Combinatorial Problems and Exercises*, Akademiai Kiado, p 528.

L. Lovasz (1980). Matroid Matching and Some Applications, *Journal of Combinatorial Theory* **B28**, 208–236.

L. Lovasz (1981). The Matroid Matching Problem, in *Algebraic Methods in Graph Theory*, L. Lovasz and V. T. Sos, eds., North-Holland, pp. 495–517.

L. Lovasz (1983). Submodular Functions and Convexity, in Bachem, Grötschel and Korte, pp. 235–257.

L. Lovasz and M. D. Plummer (1986). *Matching Theory*, Akademiai Kiado.

A. Lubiw (1982). Gamma-Free Matrices, M. S. Thesis, University of Waterloo.

M. Lundy and A. Mees (1986). Convergence of an Annealing Algorithm, *Mathematical Programming* **34**, 111–124.

P. G. MacKeown (1981). A Branch-and-Bound Algorithm for Solving Fixed Charge Problems, *Naval Research Logistics Quarterly* **28**, 607–618.

F. Maffioli (1986). Randomized Algorithms in Combinatorial Optimization: A Survey, *Discrete Applied Mathematics* **14**, 157–170.

F. Maffioli, M. G. Speranza, and C. Vercellis (1985). Randomized Algorithms, in O'hEigeartaigh et al., pp. 89–105.

M. J. Magazine and M. S. Chern (1984). A Note on Approximation Schemes for Multidimensional Knapsack Problems, *Mathematics of Operations Research* **9**, 244–247.

M. J. Magazine and O. Oguz (1981). A Fully Polynomial Approximation Algorithm for the 0-1 Knapsack Algorithm, *European Journal of Operations Research* **8**, 270–273.

T. L. Magnanti, P. Mireault, and R. T. Wong (1986). Tailoring Benders' Decomposition for Uncapacitated Network Design, *Mathematical Programming Study* **26**, 112–154.

T. L. Magnanti and R. T. Wong (1981). Accelerated Benders Decomposition: Algorithmic Enhancement and Model Section Criteria, *Operations Research* **29**, 464–484.

T. L. Magnanti and R. T. Wong (1984). Network Design and Transportation Planning: Models and Algorithms, *Transportation Science* **18**, 1–55.

A. S. Manne (1964). Plant Location under Economies of Scale-Decentralization and Computation, *Management Science* **11**, 213–235.

O. Marcotte (1985). The Cutting Stock Problem and Integer Rounding, *Mathematical Programming* **33**, 89–92.

O. Marcotte (1986a). An Instance of the Cutting Stock Problem for Which the Rounding Property Does Not Hold, *Operations Research Letters* **4**, 239–243.

O. Marcotte (1986b). On the Chromatic Index of Multigraphs and a Conjecture of Seymour, *Journal of Combinatorial Theory* **41B**, 306–331.

H. M. Markowtiz and A. S. Manne (1957). On the Solution of Discrete Programming Problems, *Econometrica* **25**, 84–110.

R. E. Marsten (1974). An Algorithm for Large Set Partitioning Problems, *Management Science* **20**, 774–787.

R. E. Marsten (1981). XMP: A Structured Library of Subroutines for Experimental Mathematical Programming, *ACM Transactions on Mathematical Software* **7**, 481–497.

R. E. Marsten and F. Shepardson (1981). Exact Solution of Crew Scheduling Problems Using the Set Partitioning Model: Recent Successful Applications, *Networks* **11**, 165–178.

C. U. Martel (1982). Preemptive Scheduling with Release Times, Deadlines and Due Times, *Journal of the Association for Computing Machinery* **29**, 812–829.

S. Martello and P. Toth (1979). The 0-1 Knapsack Problem, in Christofides, Mingozzi et al., pp. 237–279.

S. Martello and P. Toth (1981a). A Branch and Bound Algorithm for the Zero-One Multiple Knapsack Problem, *Discrete Applied Mathematics* **3**, 275–288.

S. Martello and P. Toth (1981b). Heuristic Algorithms for the Multiple Knapsack Problem, *Computing* **27**, 93–112.

R. K. Martin (1984). Generating Alternative Mixed-Integer Linear Programming Models, Graduate School of Business, University of Chicago.

R. K. Martin (1987). Using Separation Algorithms to Generate Mixed Integer Model Reformulations, Graduate School of Business, University of Chicago.

R. K. Martin and L. Schrage (1985). Subset Coefficient Reduction Cuts for 0-1 Mixed Integer Programming, *Operations Research* **33**, 505–526.

J. F. Maurras (1975). Some Results on the Convex Hull of Hamiltonian Cycles of Symmetric Complete Graphs, in *Combinatorial Programming: Methods and Applications*, B. Roy, ed., Reidel, pp. 179–190.

C. J. H. McDiarmid (1975). Rado's Theorem for Polymatroids, *Proceedings of the Cambridge Philosophical Society* **78**, 263–281.

C. J. H. McDiarmid (1983). Integral Decomposition in Polyhedra, *Mathematical Programming* **25**, 183–198.

N. Metropolis, A. Rosenbluth, M. Rosenbluth, A. Teller, and E. Teller (1953). Equations of State Calculations by Fast Computing Machines, *Journal of Chemical Physics* **21**, 1087–1091.

R. R. Meyer (1974). On the Existence of Optimal Solutions to Integer and Mixed-Integer Programming Problems, *Mathematical Programming* **7**, 223–235.

R. R. Meyer (1975). Integer and Mixed Integer Programming Models: General Properties, *Journal of Optimization Theory and Applications* **16**, 191–206.

R. R. Meyer and M. L. Wage (1978). On the Polyhedrality of the Convex Hull of the Feasible Set of an Integer Program, *SIAM Journal on Control and Optimization* **16**, 682–687.

H. Meyniel (1976). On the Perfect Graph Conjecture, *Discrete Mathematics* **16**, 339–342.

H. Meyniel (1984). The Graphs Whose Odd Cycles Have at Least Two Chords, *Annals of Discrete Mathematics* **21**, 103–114.

P. Miliotis (1976). Integer Programming Approaches to the Traveling Salesman Problem, *Mathematical Programming* **10**, 367–378.

P. Miliotis (1978). Using Cutting Planes to Solve the Symmetric Travelling Salesman Problem, *Mathematical Programming* **15**, 177–188.

C. E. Miller, A. W. Tucker, and R. A. Zemlin (1960). Integer Programming Formulations and Traveling Salesman Problems, *Journal of the Association for Computing Machinery* **7**, 326–329.

G. J. Minty (1980). On Maximal Independent Sets of Vertices in a Claw-Free Graph, *Journal of Combinatorial Theory* **B28**, 284–304.

G. Mitra (1973). Investigations of some Branch and Bound Strategies for the Solution of Mixed Integer Linear Programs, *Mathematical Programming* **4**, 155–170.

L. G. Mitten (1970). Branch-and-Bound Methods: General Formulation and Properties, *Operations Research* **18**, 24–34.

C. L. Monma, ed. (1986). *Algorithms and Software for Optimization–Part I* (*Annals of Operations Research* **4**).

J. G. Morris (1978). On the Extent to which Certain Fixed Depot Location Problems Can Be Solved by LP, *Journal of the Operational Research Society* **29**, 71–76.

J. A. Muckstadt and S. A. Koenig (1977). An Application of Lagrangian Relaxation to Scheduling in Power Generation Systems, *Operations Research* **25**, 387–403.

J. M. Mulvey and H. M. Crowder (1979). Cluster Analysis: An Application of Lagrangian Relaxation, *Management Science* **25**, 329–340.

J. D. Murchland (1973). Historical Note on Optimal Spanning Arborescences, *Networks* **3**, 287–288.

K. G. Murty (1976). *Linear and Combinatorial Programming*, Wiley.

R. M. Nauss (1979). *Parametric Integer Programming*, University of Missouri Press, Columbia.

A. W. Neebe and M. R. Rao (1983). An Algorithm for the Fixed Charge Assignment of Users to Sources Problem, *Journal of the Operational Research Society* **34**, 1107–1115.

G. L. Nemhauser (1966). *Introduction to Dynamic Programming*, Wiley.

G. L. Nemhauser (1985). Duality for Integer Optimization, in O'hEigeartaigh et al., pp. 11–20.

G. L. Nemhauser and G. Sigismondi (1988). A Constraint Generation Algorithm for Node Packing, School of Industrial and Systems Engineering, Georgia Institute of Technology.

G. L. Nemhauser and L. E. Trotter (1974). Properties of Vertex Packing and Independence System Polyhedra, *Mathematical Programming* **6**, 48–61.

G. L. Nemhauser and L. E. Trotter (1975). Vertex Packings: Structural Properties and Algorithms, *Mathematical Programming* **8**, 232–248.

G. L. Nemhauser, L. E. Trotter, and R. M. Nauss (1974). Set Partitioning and Chain Decomposition, *Management Science* **20**, 1413–1423.

G. L. Nemhauser and Z. Ullman (1968). A Note on the Generalized Multiplier Solution to an Integer Programming Problem, *Operations Research* **16**, 450–452.

G. L. Nemhauser and Z. Ullman (1969). Discrete Dynamic Programming and Capital Allocation, *Management Science* **15**, 494–505.

G. L. Nemhauser and G. M. Weber (1979). Optimal Set Partitioning, Matchings and Lagrangian Duality, *Naval Research Logistics Quarterly* **26**, 553–563.

G. L. Nemhauser and L. A. Wolsey (1979). Best Algorithms for Approximating the Maximum of a Submodular Set Function, *Mathematics of Operations Research* **3**, 177–188.

G. L. Nemhauser and L. A. Wolsey (1981). Maximizing Submodular Set Functions: Formulations and Analysis of Algorithms, *Annals of Discrete Mathematics* **11**, 279–301.

G. L. Nemhauser and L. A. Wolsey (1984). A Recursive Procedure for Generating all Cuts for 0-1 Mixed Integer Programs, Core DP 8439, Université Catholique du Louvain.

G. L. Nemhauser, L. A. Wolsey, and M. L. Fisher (1978). An Analysis of Approximations for Maximizing Submodular Set Functions—I, *Mathematical Programming* **14**, 265–294.

J. P. Norback and R. F. Love (1979). Heuristic for the Hamiltonian Path Problem in Euclidean Two Space, *Operations Research* **30**, 363–368.

R. Z. Norman and M. D. Rabin (1959). An Algorithm for the Minimum Cover of a Graph, *Proceedings of the American Mathematical Society* **10**, 315–319.

F. J. Nourie and E. R. Venta (1982). An Upper Bound on the Number of Cuts Needed in Gomory's Method of Integer Forms, *Operations Research Letters* **1**, 129–133.

M. O'hEigertaigh, J. K. Lenstra and A. H. G. Rinnooy Kan, eds. (1985). *Combinatorial Optimization: Annotated Bibliographies*, Wiley.

J. Orlin (1982). A Polynomial Algorithm for Integer Programming Covering Problems Satisfying the Integer Round-Up Property, *Mathematical Programming* **22**, 231–235.

J. B. Orlin (1984). Genuinely Polynomial Simplex and Non-Simplex Algorithms for the Minimum Cost Flow Problem, WP 1615-84, Sloan School of Management, Massachusetts Institute of Technology.

J. B. Orlin (1986). A Dual Version of Tardos's Algorithm for Linear Programming, *Operations Research Letters* **5**, 221–226.

M. W. Padberg (1973). On the Facial Structure of Set Packing Polyhedra, *Mathematical Programming* **5**, 199–215.

M. W. Padberg (1974), Perfect Zero–One Matrices, *Mathematical Programming* **6**, 180–196.

M. W. Padberg (1975a). A Note on Zero–One Programming, *Operations Research* **23**, 833–837.

M. W. Padberg (1975b). Characterizations of Totally Unimodular, Balanced and Perfect Matrices, in *Combinatorial Programming: Methods and Applications*, B. Roy, ed., Reidel, pp. 275–284.

M. W. Padberg (1976a). A Note on the Total Unimodularity of Matrices, *Discrete Mathematics* **14**, 273–278.

M. W. Padberg (1976b). Almost Integral Polyhedra Related to Certain Combinatorial Optimization Problems, *Linear Algebra and Its Applications* **15**, 69–88.

M. W. Padberg (1977). On the Complexity of Set Packing Polyhedra, *Annals of Discrete Mathematics* **1**, 421–434.

M. W. Padberg (1979). Covering, Packing and Knapsack Problems, *Annals of Discrete Mathematics* **4**, 265–287.

M. W. Padberg, ed. (1980a). *Combinatorial Optimization*, Mathematical Programming Study 12.

M. W. Padberg (1980b). (1, k)-Configurations and Facets for Packing Problems, *Mathematical Programming* **18**, 94–99.

M. W. Padberg (1984). A Characterization of Perfect Matrices, *Annals of Discrete Mathematics* **21**, 169–178.

M. W. Padberg (1988). Total Unimodularity and the Euler-Subgraph Problem. To appear in *Operations Research Letters*.

M. W. Padberg and M. Grötschel (1985). Polyhedral Computations, in Lawler, Lenstra et al., pp. 307–360.

M. W. Padberg and S. Hong (1980). On the Symmetric Traveling Salesman Problem: A Computational Study, *Mathematical Programming Study* **12**, 78–107.

M. W. Padberg and M. R. Rao (1982). Odd Minimum Cut-Sets and b-Matchings, *Mathematics of Operations Research* **7**, 67–80.

M. W. Padberg and G. Rinaldi (1987a). Optimization of a 532-City Traveling Salesman Problem by Branch and Cut, *Operations Research Letters* **6**, 1–8.

M. W. Padberg and G. Rinaldi (1987b). Facet Indentification for the Symmetric Traveling Salesman Polytope, New York University.

M. W. Padberg, T. J. Van Roy and L. A. Wolsey (1985). Valid Linear Inequalities for Fixed Charge Problems, *Operations Research* **33**, 842–861.

M. W. Padberg and L. A. Wolsey (1983). Trees and Cuts, *Annals of Discrete Mathematics* **17**, 511–517.

M. W. Padberg and L. A. Wolsey (1984). Fractional Covers for Forests and Matchings, *Mathematical Programming* **29**, 1–14.

C. H. Papadimitriou (1981a). On the Complexity of Integer Programming, *Journal of the Association for Computing Machinery* **28**, 765–768.

C. H. Papadimitriou (1981b). Worst-Case and Probabilistic Analysis of a Geometric Location Problem, *SIAM Journal on Computing* **10**, 542–557.

C. H. Papadimitriou (1984). Polytopes and Complexity, in Pulleyblank, pp. 295–306.

C. H. Papadimitriou (1985). Computational Complexity, in O'hEigeartaigh et al., pp. 39–51.

C. H. Papadimitriou and K. Stieglitz (1977). On the Complexity of Local Search for the Traveling Salesman Problem, *SIAM Journal on Computing* **6**, 76–83.

C. H. Papadimitriou and K. Stieglitz (1978). Some Examples of Difficult Traveling Salesman Problems, *Operations Research* **26**, 434–443.

C. H. Papadimitriou and K. Stieglitz (1982). *Combinatorial Optimization: Algorithms and Complexity*, Prentice–Hall.

C. H. Papadimitriou and M. Yannakakis (1984). The Complexity of Facets (and Some Facets of Complexity), *Journal of Computing and System Science* **28**, 244–259.

K. R. Parthasarathy and G. Ravindra (1976). The Strong Perfect Graph Conjecture Is True for $K_{1,3}$-Free Graphs, *Journal of Combinatorial Theory* **B21**, 212–223.

C. C. Petersen (1967). Computational Experience with Variants of the Balas Algorithm Applied to the Selection of R and D Projects, *Management Science* **13**, 736–750.

J. C. Picard and M. Queyranne (1977). On the Integer-Valued Variables in the Linear Vertex Packing Problem, *Mathematical Programming* **12**, 97–101.

J. C. Picard and M. Queyranne (1982). Selected Applications of Minimum Cuts in Networks, *INFOR* **20**, 394–422.

J. C. Picard and H. D. Ratliff (1975). Minimum Cuts and Related Problems, *Networks* **5**, 357–370.

J. F. Pierce (1968). Application of Combinatorial Programming to a Class of All Zero–One Integer Programming Problems, *Management Science* **13**, 736–750.

J. F. Pierce and J. Lasky (1973). Improved Combinatorial Programming Algorithms for a Class of All Zero–One Integer Programming Problems, *Management Science* **19**, 528–543.

Y. Pochet (1988). Valid Inequalities and Separation for Capacitated Economic Lot Sizing, to appear in *Operations Research Letters.*

Y. Pochet and L. A. Wolsey (1988). Lot-Size Models with Backlogging: Strong Formulations and Cutting Planes, to appear in *Mathematical Programming*.

C. N. Potts (1985). A Lagrangian Based Branch and Bound Algorithm for Single Machine Scheduling with Precedence Constraints to Minimize Total Weighted Completion Time, *Management Science* **31**, 1300–1311.

S. Powell (1985). Software, in O'hEigeartaigh et al., pp. 190–194.

R. C. Prim (1957). Shortest Connection Networks and Some Generalizations, *Bell System Technological Journal* **36**, 1389–1401.

A. Prodon, T. M. Liebling, and H. Groflin (1985). Steiner's Problem on Two-Trees, RO 850315, Departement de Mathematiques, Ecole Polytechnique Federale de Lausanne.

E. Pruul (1975). Parallel Processing and a Branch-and-Bound Algorithm, M. S. Thesis, Cornell University.

E. Pruul, G. L. Nemhauser, and R. Rushmeier (1988). Parallel Processing and Branch-and-Bound: A Historical Note, to appear in *Operations Research Letters*.

W. R. Pulleyblank (1973). Faces of Matching Polyhedra, Ph.D. Thesis, University of Waterloo.

W. R. Pulleyblank (1980). Dual Integrality in b-Matching Problems, *Mathematical Programming Study* **12**, 176–196.

W. R. Pulleyblank (1981). Total Dual Integrality and b-Matchings, *Operations Research Letters* **1**, 28–30.

W. R. Pulleyblank (1983). Polyhedral Combinatorics, in Bachem, Grötschel and Korte, pp. 312–345.

W. R. Pulleyblank, ed. (1984). *Progress in Combinatorial Optimization*, Academic Press.

W. R. Pulleyblank and J. Edmonds (1975). Facets of 1-Matching Polyhedra, in *Hypergraph Seminar*, C. Berge and D. Ray-Chaudhuri, eds., Springer, pp. 214–242.

M. O. Rabin (1976). Probabilistic Algorithms, in *Algorithms and Complexity: New Directions and Recent Results*, J. F. Traub, ed., Academic Press, pp. 21–40.

R. Rado (1957). Note on Independence Functions, *Proceedings of the London Mathematical Society* 7, 300–320.

R. L. Rardin and U. Choe (1979). Tighter Relaxations of Fixed Charge Network Flow Problems, Industrial and Systems Engineering Report J-79-18, Georgia Institute of Technology.

H. D. Ratliff and A. S. Rosenthal (1983). Order Picking in a Rectangular Warehouse: A Solvable Case of the T. S. P., *Operations Research* **31**, 507–521.

A. Recski (1988). *Matroid Theory and Its Applications*, Springer.

S. Reiter, and D. B. Rice (1966). Discrete Optimizing Solution Procedures for Linear and Nonlinear Integer Programming Problems, *Management Science* **12**, 829–850.

S. Reiter and G. Sherman (1965). Discrete Optimizing, *SIAM Journal* **13**, 864–899.

J. M. W. Rhys (1970). A Selection Problem of Shared Fixed Costs and Network Flows, *Management Science* **17**, 200–207.

C. Ribeiro and M. Minoux (1985). Solving Hard Constrained Shortest Path Problems by Lagrangian Relaxation and Branch and Bound Algorithms, in *Proceedings of X Symposium on Operations Research*, M. Beckmann et al., eds., Methods of Operations Research 53, Anton Hain, pp. 303–316.

A. H. G. Rinnooy Kan (1976). On Mitten's Axioms for Branch and Bound, *Operations Research* **24**, 1176–1178.

A. H. G. Rinnooy Kan (1986). An Introduction to the Analysis of Approximation Algorithms, *Discrete Applied Mathematics* **14**, 111–134.

T. Rockafellar (1970). *Convex Analysis*, Princeton University Press.

D. J. Rosenkrantz, R. E. Stearns, and P. M. Lewis (1977). An Analysis of Several Heuristics for the Traveling Salesman Problem, *SIAM Journal on Computing* **6**, 563–581.

K. Rosling (1983). The Dynamic Inventory Model and the Uncapacitated Facility Location Problem, S-581 83, Department of Production Economics, Linkoping Institute of Technology.

G. T. Ross and R. M. Soland (1975). A Branch and Bound Algorithm for the Generalized Assignment Problem, *Mathematical Programming* **8**, 91–103.

D. S. Rubin (1985). Polynomial Algorithms for $m \times (m + 1)$ Integer Programs and $m \times (m + k)$ Diophantine Systems, *Operations Research Letters* **3**, 289–291.

S. Sahni (1975). Approximate Algorithms for the 0-1 Knapsack Problem, *Journal of the Association for Computing Machinery* **22**, 115–124.

S. Sahni (1977). General Techniques for Combinatorial Approximation, *Operations Research* **25**, 920–936.

S. Sahni and T. Gonzalez (1976). *P*-Complete Approximation Problems, *Journal of the Association for Computing Machinery* **23**, 555–565.

M. Sakarovitch (1975). Quasi-Balanced Matrices, *Mathematical Programming* **8**, 382–386.

M. Sakarovitch (1976). Quasi-Balanced Matrices—an Addendum, *Mathematical Programming* **10**, 405–407.

H. M. Salkin (1975). *Integer Programming*, Addison–Wesley.

H. M. Salkin and C. A. de Kluyver (1975). The Knapsack Problem: A Survey, *Naval Research Logistics Quarterly* **22**, 127–144.

H. M. Salkin and R. D. Koncal (1973). Set Covering by an All-Integer Algorithm: Computational Experience, *Journal of the Association for Computing Machinery* **31**, 336–345.

C. Sandi (1979). Subgradient Optimization, in Christofides, Mingozzi et al., pp. 73–91.

A. Sassano (1985). On the Facial Structure of the Set Covering Polytope, R139, Instituto di Analisi dei Sistemi ed Informatica del CNR. Rome.

N. Sbihi (1980). Algorithme de Recherche d'un Stable de Cardinalite Maximum dans un Graphe sans Etoile, *Discrete Mathematics* **29**, 53–76.

H. E. Scarf (1981a). Production Sets with Indivisibilities Part I, *Econometrica* **49**, 1–32.

H. E. Scarf (1981b). Production Sets with Indivisibilities Part II, *Econometrica* **49**, 395–423.

L. Schrage (1975). Implicit Representation of Variable Upper Bounds in Linear Programming, *Mathematical Programming Study*, **4**, 118–132.

L. Schrage (1986). *Linear, Integer and Quadratic Programming with LINDO*, Scientific Press.

L. Schrage and L. A. Wolsey (1985). Sensitivity Analysis for Branch and Bound Integer Programming, *Operations Research* **33**, 1008–1023.

A. Schrijver (1980). On Cutting Planes, *Annals of Discrete Mathematics* **9**, 291–296.

A. Schrijver (1981). On Total Dual Integrality, *Linear Algebra and Its Applications* **38**, 27–32.

A. Schrijver (1983a). Min–Max Results in Combinatorial Optimization, in Bachem, Grötschel and Korte, pp. 439–500.

A. Schrijver (1983b). Short Proofs on the Matching Polytope, *Journal of Combinatorial Theory* **B34**, 104–108.

A. Schrijver (1984a). Proving Total Dual Integrality with Cross-Free Families—A General Framework, *Mathematical Programming* **29**, 15–27.

A. Schrijver (1984b). Total Dual Integrality from Directed Graphs, Crossing Families and Sub- and Supermodular Functions, in Pulleyblank, pp. 315–362.

A. Schrijver (1986a). *Linear and Integer Programming*, Wiley.

A. Schrijver (1986b). Polyhedral Proof Methods in Combinatorial Optimization. *Discrete Applied Mathematics* **14**, 111–134.

P. D. Seymour (1977). The Matroids with the Max-Flow Min-Cut Property, *Journal of Combinatorial Theory* **B26**, 189–222.

P. D. Seymour (1978). A Two-Commodity Cut Theorem, *Discrete Mathematics* **23**, 177–181.

P. D. Seymour (1979). On Multi-Colourings of Cubic Graphs, and Conjectures of Fulkerson and Tutte, *Proceedings of the London Mathematical Society* **38**, 423–460.

P. D. Seymour (1980). Decomposition of Regular Matroids, *Journal of Combinatorial Theory* **B28**, 305–359.

R. Shamir (1987). The Efficiency of the Simplex Method: A Survey, *Management Science* **33**, 301–334.

J. F. Shapiro (1968a). Dynamic Programming Algorithms for the Integer Programming Problem 1: The Integer Programming Problem Viewed as a Knapsack Type Problem, *Operations Research* **16**, 103–121.

J. F. Shapiro (1968b). Group Theoretic Algorithms for the Integer Programming Problem-II: Extensions to a General Algorithm, *Operations Research* **18**, 103–121.

J. F. Shapiro (1970). Turnpike Theorems for Integer Programs, *Operations Research* **18**, 432–440.

J. F. Shapiro (1971). Generalized Lagrange Multipliers in Integer Programming, *Operations Research* **19**, 68–76.

J. F. Shapiro (1977). Sensitivity Analysis in Integer Programming, *Annals of Discrete Mathematics* **1**, 467–477.

J. F. Shapiro (1979a). *Mathematical Programming, Structures and Algorithms*, Wiley.

J. F. Shapiro (1979b). A Survey of Lagrangian Techniques for Discrete Optimization, *Annals of Discrete Mathematics* **5**, 113–138.

J. F. Shapiro and H. M. Wagner (1967). A Finite Renewal Algorithm for the Knapsack and Turnpike Models, *Operations Research* **15**, 319–341.

M. L. Shore, L. R. Foulds and P. B. Gibbons (1982). An Algorithm for the Steiner Problem in Graphs, *Networks* **12**, 323–333.

P. Sinha and A. A. Zoltners (1979). The Multiple-Choice Knapsack Problem, *Operations Research* **27**, 503–515.

S. Smale (1983a). The Problem of the Average Speed of the Simplex Method, in Bachem, Grötschel and Korte, pp. 530–539.

S. Smale (1983b). On the Average Number of Steps of the Simplex Method of Linear Programming, *Mathematical Programming* **27**, 241–262.

K. Spielberg (1969a). Plant Location with Generalized Search Origin, *Management Science* **16**, 165–178.

K. Spielberg (1969b). Algorithms for the Simple Plant Location Problem with Some Side Conditions, *Operations Research* **17**, 85–111.

K. Spielberg (1979). Enumerative Methods in Integer Programming, *Annals of Discrete Mathematics* **5**, 139–183.

J. Stoer and C. Witzgall (1970). *Convexity and Optimization in Finite Dimensions*, Springer.

G. Strang (1976). *Linear Algebra and Its Applications*, Academic Press.

U. Suhl (1978). Algorithm and Efficient Data Structures for the Binary Knapsack Problem, *European Journal of Operations Research* **2**, 420–428.

U. Suhl (1985), Solving Large Scale Mixed Integer Programs with Fixed Charge Variables, *Mathematical Programming* **32**, 165–182.

H. A. Taha (1975). *Integer Programming, Theory, Applications, and Computations*, Academic Press.

A. Tamir (1976). On Totally Unimodular Matrices, *Networks* **6**, 373–382.

A. Tamir (1983). A Class of Balanced Matrices Arising From Location Problems, *SIAM Journal of Algebraic and Discrete Methods* **4**, 363–370.

A. Tamir (1987). Totally Balanced and Totally Unimodular Matrices Defined by Center Location Problems, *Discrete Applied Mathematics* **16**, 245–264.

E. Tardos (1985). A Strongly Polynomial Minimum Cost Circulation Algorithm, *Combinatorica* **5**, 247–255.

E. Tardos (1986). A Strongly Polynomial Algorithm to Solve Combinatorial Linear Programs, *Operations Research* **34**, 250–256.

E. Tardos, C. A. Tovey, and M. A. Trick (1986), Layered Augmenting Path Algorithms, *Mathematics of Operations Research* **11**, 362–370.

R. E. Tarjan (1977). Finding Optimum Branchings, *Networks* **7**, 25–35.

R. E. Tarjan (1983). *Data Structures and Network Algorithms*, SIAM Publications.

R. E. Tarjan (1986). Algorithms for Maximizing Network Flow, *Mathematical Programming* **26**, 1–11.

J. Tind (1974). Blocking and Antiblocking Sets, *Mathematical Programming* **6**, 157–166.

J. Tind (1977). On Antiblocking Sets and Polyhedra, *Annals of Discrete Mathematics* **1**, 507–515.

J. Tind (1979). Blocking and Antiblocking Polyhedra, *Annals of Discrete Mathematics* **4**, 159–174.

J. Tind and L. A. Wolsey (1981). An Elementary Survey of General Duality Theory in Mathematical Programming, *Mathematical Programming* **21**, 241–261.

M. J. Todd (1982). An Implementation of the Simplex Method for Linear Programming with Variable Upper Bounds, *Mathematical Programming* **23**, 34–49.

M. J. Todd (1987). Polynomial Algorithms for Linear Programming, in *Proceedings of the Optimization Days 1986*, H. A. Eiselt, ed.

M. J. Todd and B. P. Burrell (1986). An Extension of Karmarkar's Algorithm for Linear Programming Using Dual Variables, *Algorithmica* **1**, 409–424.

J. A. Tomlin (1970). Branch and Bound Methods for Integer and Non-Convex Programming, in *Integer and Nonlinear Programming*, J. Abadie, ed., American Elsevier, 437–450.

J. A. Tomlin (1971). An Improved Branch and Bound Method for Integer Programming, *Operations Research* **19**, 1070–1075.

P. Tong, E. L. Lawler, and V. V. Vazirani (1984). Solving the Weighted Parity Problem for Gammoids by Reduction to Graphic Matching, in Pulleyblank, pp. 363–374.

D. Topkis (1978). Minimizing a Subadditive Function on a Lattice, *Operations Research* **26**, 305–321.

M. Trick (1987). Networks with Additional Structured Constraints, Ph.D. Thesis, School of Industrial and Systems Engineering, Georgia Institute of Technology.

L. E. Trotter, Jr. (1975). A Class of Facet Producing Graphs for Vertex Packing Polyhedra, *Discrete Mathematics* **12**, 373–388.

L. E. Trotter, Jr. (1985). Discrete Packing and Covering, in O'hEigeartaigh et al., pp. 21–31.

L. E. Trotter and D. B. Weinberger (1978). Symmetric Blocking and Antiblocking Relations for Generalized Circulations, *Mathematical Programming* **8**, 141–158.

K. Truemper (1977). Unimodular Matrices of Flow Problems with Additional Constraints, *Networks* **7**, 343–358.

K. Truemper (1978). Algebraic Characterizations of Unimodular Matrices, *SIAM Journal on Applied Mathematics* **35**, 328–332.

W. T. Tutte (1954). A Short Proof of the Factor Theorem for Finite Graphs, *Canadian Journal of Mathematics* **6**, 347–352.

W. T. Tutte (1965). Lectures on Matroids, *Journal of Research of the National Bureau of Standards* **69B**, 1–48.

W. T. Tutte (1971). *Introduction to the Theory of Matroids*, American Elsevier.

P. Van Emde Boas (1981). Another $\mathcal{NP}$-Complete Partition Problem and the Complexity of Computing Short Vectors in a Lattice, Report 81-04, Mathematical Institute, University of Amsterdam.

T. J. Van Roy (1983). Cross Decomposition for Mixed Integer Programming, *Mathematical Programming* **25**, 46–63.

T. J. Van Roy (1986). A Cross Decomposition Algorithm for Capacitated Facility Location, *Operations Research* **34**, 145–163.

T. J. Van Roy and L. A. Wolsey (1985). Valid Inequalities and Separation for Uncapacitated Fixed Charge Networks, *Operations Research Letters* **4**, 105–112.

T. J. Van Roy and L. A. Wolsey (1986). Valid Inequalities for Mixed 0-1 Programs, *Discrete Applied Mathematics* **14**, 199–213.

T. J. Van Roy and L. A. Wolsey (1987). Solving Mixed 0-1 Programs by Automatic Reformulation, *Operations Research* **35**, 45–57.

M. P. Vecchi and S. Kirkpatrick (1983). Global Wiring by Simulated Annealing, IEEE Transactions on Computer-Aided Design **2**, 215–222.

A. F. Veinott, Jr. and G. B. Dantzig (1968). Integral Extreme Points, *SIAM Review* **10**, 371–372.

V. G. Vizing (1964), On an Estimate of the Chromatic Class of a *P*-graph *Diskretnyi Analiz* **3**, 25–30 (in Russian).

R. von Randow, ed. (1982). *Integer Programming and Related Areas, A Classified Bibliography 1978–1981*, Lecture Notes in Economics and Mathematical Systems 197, Springer.

R. von Randow, ed. (1985). *Integer Programming and Related Areas, A Classified Bibliography 1981–1984*, Lecture Notes in Economics and Mathematical Systems 243, Springer.

J. Von zur Gathen and M. Sieveking (1978). A Bound on Solutions of Linear Integer Equalities and Inequalities, *Proceedings of the American Mathematical Society* **72**, 155–158.

H. M. Wagner (1959). On a Class of Capacitated Transportation Problems, *Management Science*, **5**, 304–318.

H. M. Wagner and T. M. Whitin (1958). Dynamic Version of the Economic Lot Size Model, *Management Science* **5**, 89–96.

G. M. Weber (1981). Sensitivity Analysis of Optimal Matchings, *Networks* **11**, 41–56.

D. B. Weinberger (1976). Network Flows, Minimum Coverings, and the Four-Color Conjecture, *Operations Research* **24**, 272–290.

H. M. Weingartner and D. N. Ness (1967). Methods for the Solution of Multidimensional 0/1 Knapsack Problems, *Operations Research* **15**, 83–103.

D. J. A. Welsh (1968). Kruskal's Theorem for Matroids, *Proceedings of the Cambridge Philosophical Society* **64**, 3–4.

D. J. A. Welsh (1976). *Matroid Theory*, Academic Press.

D. J. A. Welsh (1983). Randomised Algorithms, *Discrete Applied Mathematics* **5**, 133–146.

D. J. White (1969). *Dynamic Programming*, Holden–Day.

W. W. White (1961). On Gomory's Mixed Integer Algorithm, Senior Thesis, Department of Mathematics, Princeton University.

S. H. Whitesides (1984). A Classification of Certain Graphs with Minimum Imperfection Properties. *Annals of Discrete Mathematics* **21**, 207–218.

H. Whitney (1935). On the Abstract Properties of Linear Dependence, *American Journal of Mathematics* **57**, 509–533.

H. P. Williams (1974). Experiments in the Formulation of Integer Programming Problems, *Mathematical Programming Study* **2**, 180–197.

H. P. Williams (1978a). *Model Building in Mathematical Programming*, Wiley.

H. P. Williams (1978b). The Reformulation of Two Mixed Integer Programming Problems, *Mathematical Programming* **14**, 325–331.

L. A. Wolsey (1971a), Group-Theoretic Results in Mixed Integer Programming, *Operations Research* **19**, 1691–1697.

L. A. Wolsey (1971b). Extensions of the Group Theoretic Approach in Integer Programming, *Management Science* **18**, 74–83.

L. A. Wolsey (1973). Generalized Dynamic Programming Methods in Integer Programming, *Mathematical Programming* **4**, 222–232.

L. A. Wolsey (1975). Faces for a Linear Inequality in 0-1 Variables, *Mathematical Programming* **8**, 165–178.

L. A. Wolsey (1976). Facets and Strong Valid Inequalities for Integer Programs, *Operations Research* **24**, 367–372.

L. A. Wolsey (1977). Valid Inequalities and Superadditivity for 0-1 Integer Programs, *Mathematics of Operations Research* **2**, 66–77.

L. A. Wolsey (1980). Heuristic Analysis, Linear Programming and Branch and Bound, *Mathematical Programming Study* **13**, 121–134.

L. A. Wolsey (1981a). Integer Programming Duality: Price Functions and Sensitivity Analysis, *Mathematical Programming* **20**, 173–195.

L. A. Wolsey (1981b). The b-Hull of an Integer Program, *Discrete Applied Mathematics* **3**, 193–201.

L. A. Wolsey (1981c). A Resource Decomposition Algorithm for General Mathematical Programs, *Mathematical Programming Study* **14**, 244–257.

L. A. Wolsey (1982a). An Analysis of the Greedy Algorithm for the Submodular Set Covering Problem, *Combinatorica* **2**, 417–425.

L. A. Wolsey (1982b). Maximizing Real Valued Submodular Functions: Primal and Dual Heuristics for Location Problems, *Mathematics of Operations Research* **7**, 410–425.

L. A. Wolsey (1987). Strong Formulations for Mixed Integer Programming: A Survey, Ecole Polytechnique Federale de Lausanne.

R. T. Wong (1984). A Dual Ascent Approach for Steiner Tree Problems on Directed Graphs, *Mathematical Programming* **28**, 271–287.

R. T. Wong (1985). Location and Network Design, in O'hEigeartaigh et al., pp. 129–147.

M. Yannakakis (1985). On a Class of Totally Unimodular Matrices, *Mathematics of Operations Research* **10**, 280–304.

R. D. Young (1965). A Primal (All Integer), Integer Programming Algorithm, *Journal of Research of the National Bureau of Standards* **69B**, 213–250.

R. D. Young (1968). A Simplified Primal (All-Integer) Integer Programming Algorithm, *Operations Research* **16**, 750–782.

W. I. Zangwill (1966). A Deterministic Multi-Period Production Scheduling Model with Backlogging, *Management Science* **13**, 105–119.

E. Zemel (1978). Lifting the Facets of 0-1 Polytopes, *Mathematical Programming* **15**, 268–277.

E. Zemel (1980). The Linear Multiple Choice Knapsack Problem, *Operations Research* **28**, 1412–1423.

E. Zemel (1981). Measuring the Quality of Approximate Solutions to Zero–One Programming Problems, *Mathematics of Operations Research* **6**, 319–332.

E. Zemel (1984). An $O(n)$ Algorithm for the Linear Multiple Choice Knapsack Problem and Related Problems, *Information Processing Letters* **18**, 123–128.

E. Zemel (1986). On the Computational Complexity of Facets of the Knapsack Problem, Working Paper No. 713, Graduate School of Management, Northwestern University.

AUTHOR INDEX

SUBJECT INDEX

WILEY-INTERSCIENCE
SERIES IN DISCRETE MATHEMATICS AND OPTIMIZATION

ROOS AND VIAL • Ph. Theory and Algorithms for Linear Optimization: An Interior Point Approach
SCHEINERMAN AND ULLMAN • Fractional Graph Theory: A Rational Approach to the Theory of Graphs
SCHRIJVER • Theory of Linear and Integer Programming
TOMESCU • Problems in Combinatorics and Graph Theory *(Translated by R. A. Melter)*
TUCKER • Applied Combinatorics, Second Edition
WOLSEY • Integer Programming
YE • Interior Point Algorithms: Theory and Analysis